D0114667

Progress in Inorganic Chemistry

Volume 47

Advisory Board

JACQUELINE K. BARTON
CALIFORNIA INSTITUTE OF TECHNOLOGY, PASADENA, CALIFORNIA
THEODORE J. BROWN
UNIVERSITY OF ILLINOIS, URBANA, ILLINOIS
JAMES P. COLLMAN
STANFORD UNIVERSITY, STANFORD, CALIFORNIA
F. ALBERT COTTON
TEXAS A & M UNIVERSITY, COLLEGE STATION, TEXAS
ALAN H. COWLEY
UNIVERSITY OF TEXAS, AUSTIN, TEXAS
RICHARD H. HOLM
HARVARD UNIVERSITY, CAMBRIDGE, MASSACHUSETTS
EIICHI KIMURA
HIROSHIMA UNIVERSITY, HIROSHIMA, JAPAN
NATHAN S. LEWIS
CALIFORNIA INSTITUTE OF TECHNOLOGY, PASDENA, CALIFORNIA
STEPHEN J. LIPPARD
MASSACHUSETTS INSTITUTE OF TECHNOLOGY, CAMBRIDGE, MASSACHUSETTS
TOBIN J. MARKS
NORTHWESTERN UNIVERSITY, EVANSTON, ILLINOIS
EDWARD I. STIEFEL
EXXON RESEARCH & ENGINEERING CO., ANNANDALE, NEW JERSEY
KARL WIEGHARDT
MAX-PLANCK-INSTITUT, MÜLHEIM, GERMANY

QD
151
.P76
v.47

PROGRESS IN INORGANIC CHEMISTRY

Edited by

KENNETH D. KARLIN

DEPARTMENT OF CHEMISTRY
JOHNS HOPKINS UNIVERSITY
BALTIMORE, MARYLAND

VOLUME 47

AN INTERSCIENCE® PUBLICATION
JOHN WILEY & SONS, INC.
New York • Chichester • Weinheim • Brisbane • Singapore • Toronto

JC ABF8364

Cover Illustration of "a molecular ferric wheel" was adapted from Taft, K. L. and Lippard, S. J., *J. Am. Chem. Soc.*, **1990,** 112, 9629.

Indiana University

JAN 07 1998

Library Northwest

This book is printed on acid-free paper. ♾

Copyright © 1998 by John Wiley & Sons, Inc. All rights reserved.

Published simultaneously in Canada.

No part of this publication may be reproduced, stored in a retrieval system or transmitted in any form or by any means, electronic, mechanical, photocopying, recording, scanning, or otherwise, except as permitted under Sections 107 or 108 of the 1976 United States Copyright Act, without either the prior written permission of the Publisher, or authorization through payment of the appropriate per-copy fee to the Copyright Clearance Center, 222 Rosewood Drive, Danvers, MA 01923, (508) 750-8400, fax (508) 750-4744. Requests to the Publisher for permission should be addressed to the Permissions Department, John Wiley & Sons, Inc., 605 Third Avenue, New York, NY 10158-0012, (212) 850-6011, fax (212) 850-6008, E-Mail: PERMREQ @ WILEY.COM.

Library of Congress Catalog Card Number 59-13035
ISBN 0-471-24039-7

Printed in the United States of America

10 9 8 7 6 5 4 3 2 1

Contents

Progress in Inorganic Chemistry

Volume 47

Terminal Chalcogenido Complexes of the Transition Metals

GERARD PARKIN

Department of Chemistry
Columbia University
New York

CONTENTS

Progress in Inorganic Chemistry, Vol. 47, Edited by Kenneth D. Karlin.
ISBN 0-471-24039-7 © 1998 John Wiley & Sons, Inc.

I. INTRODUCTION

Transition metal complexes with multiply bonded ligands are an important class of molecules that have received considerable attention over recent years. Such focus is a consequence of the fundamental interest associated with the electronic structure and intrinsic reactivity of metal–ligand multiple bonds, and also the roles that such moieties may play in a variety of chemical systems. The principal activity in this area has centered on transition metal complexes that exhibit multiple bonds to second-row elements (viz carbon, nitrogen, and oxygen). Indeed, the already classic 1988 text *Metal–Ligand Multiple Bonds* by Nugent and Mayer (1) was devoted entirely to transition metal complexes with metal–carbon, metal–nitrogen, and metal–oxygen multiple bonds (2). By comparison, very few studies have been reported for complexes with multiple bonds between the transition metals and the heavier nonmetals, despite the inherent interest in such complexes. The purpose of this chapter is to provide an account of the structures and chemistry of complexes with multiple bonds between the transition metals and chalcogens, that is, the Group 16 (VI A) elements: oxygen, sulfur, selenium, tellurium, and polonium (E = O, S, Se, Te, or Po) (3). It is, however, worth noting that there is some controversy as to whether or not oxygen should be considered as a chalcogen, with some chemists preferring to exclude it. The name chalcogen has developed directly from the words "chalkosphäre'' and ''chalophile,'' as introduced by a geochemist named Goldschmidt during the earlier part of this century (4a, b). Specifically, ''chalkosphäre'' was used to describe a region of the earth that is rich in metal oxides and sulfides (i.e., ores), while ''chalkophile'' was used to describe elements found in this region (e.g., Fe, Cu, Zn). Since these names are derived from the Greek word ''chalkos'' ($\chi\alpha\lambda\kappa o\sigma$), meaning copper, a literal translation for chalcogen is ''copper-former.'' However, Jensen (4c) has rationalized that since ''chalkos'' may also be interpreted as copper alloy or ore, and since copper is not the only element considered to be a ''chalkophile,'' a more pertinent interpretation of the term chalcogen is ''ore-former.'' In this sense, oxygen is appropriately considered to be a chalcogen, a term that has also been approved by IUPAC as the collective name for the Group 16 (VI A) elements (5), and will also be adopted as such in this chapter.

Interest in complexes with multiple bonds between the transition metals and chalcogens derives from their possible participation in (a) metal-based oxidations (6), (b) hydrodesulfurization (7–9), (c) biological systems (10–15), and (d) the syntheses of materials with applications in the electronics industry (16–18). At a more fundamental level, however, terminal chalcogenido complexes are of particular interest since the chalcogens are the first group of nonmetals for which all of the members (with the exception of radioactive polonium) are known to form multiple bonds to the transition metals. In contrast, even though terminal imido complexes are very common (19), only the first three members of the pnictogens are known to form isolable complexes with multiple bonds to transition metals (20), as in $(Bu^t{}_3SiO)_3Ta{=}EPh$ (E = N, P, or As), the first series of such complexes (21). Moreover, whereas terminal nitrido complexes are common (1), the first terminal phosphido complexes $[(3,5\text{-}Me_2\text{-}C_6H_3)NBu^t]_3Mo{\equiv}P$ (22) and $[\eta^4\text{-}N(CH_2CH_2NSiMe_3)_3]M{\equiv}P$ (M = Mo and W) (23) have only recently been isolated (24).

Since metal–oxo and, to a certain extent, metal–sulfido chemistry has been extensively reviewed (1, 25, 26), the scope of this chapter will mainly be limited to a discussion of the heavier chalcogens, selenium and tellurium. Oxo and sulfido complexes will be discussed only where appropriate, such as to provide comparisons with their heavier analogues. This chapter will also be specifically concerned with well-defined molecular complexes with terminal chalcogenido ligands, as opposed to (a) polynuclear species with bridging chalcogenido ligands, or (b) complexes with polychalcogenido ligands, each of which are areas that have been well reviewed in recent years (27–30). It is hoped that this chapter will serve to highlight interesting similarities and differences in the chemistry of systems as a function of the chalcogen and metal. For this reason, particular emphasis will be given to complexes for which a series of structurally related derivatives are known.

II. DISTRIBUTION OF TERMINAL CHALCOGENIDO COMPLEXES

The distribution of terminal chalcogenido complexes of the transition metals as a function of both the chalcogen and transition metal has recently been analyzed via a search of the Cambridge Structural Database (CSD), the results of which are summarized in Table I and Figs. 1–7 (31). Despite the fact that such a survey excludes complexes that have not been structurally characterized, the

TABLE I
Structurally Characterized Terminal Chalcogenido Complexes of the Transition Metals[a]

	O	S	Se	Te	Total
Sc	0	0	0	0	0
Ti	13	0	0	0	13
	1.648[43]				
V	399	9	1	0	409
	1.608[71]	2.098[26]	2.196(3)		
Cr	50	0	0	0	50
	1.595[42]				
Mn	6	0	0	0	6
	1.511[71]				
Fe	0	0	0	0	0
	1.60–1.67[b]				
Co	0	0	0	0	0
Ni	0	0	0	0	0
Y	0	0	0	0	0
Zr	1	2	1	2	6
	1.804(4)	2.326[12]	2.480(1)	2.690[56]	
Nb	36	17	0	0	53
	1.738[109]	2.205[66]			
Mo	783	71	7	2	863
	1.706[79]	2.154[87]	2.301[58]	2.566[31]	
Tc	101	0	0	0	101
	1.676[43]				
Ru	17	0	0	0	17
	1.721[38]				
Rh	0	0	0	0	0
Pd	0	0	0	0	0
La	0	0	0	0	0
Hf	1	1	1	2	5
	1.826(9)	2.311(3)	2.467(1)	2.677[40]	
Ta	1	3	3	2	9
	1.720[30]	2.217[24]	2.341[26]	2.578[14]	
W	212	80	14	3	309
	1.718[68]	2.150[72]	2.306[51]	2.546[34]	
Re	286	8	0	0	294
	1.700[64]	2.099[16]			

TABLE I (*Continued*)

	O	S	Se	Te	Total
Os	48 1.719[32]	0	0	0	48
Ir	1 1.725(9)	0	0	0	1
Pt	0	0	0	0	0
Total	1955	191	27	11	2184

[a]The data are taken from (31) and are the result of a search of the CSD (Version 5.11, April 1996) restricting the coordination number of E to 1 and using the ''insist-on-coordinates'' and ''insist-perfect-match'' parameters in order to help eliminate inaccurate data. As a result of these constraints, not all complexes that have been structurally characterized may be retrieved in a particular search. For each transition metal, the number of terminal chalcogenido complexes is listed in the first line, with the mean M = E bond length (in Å) listed below. The number following the mean bond length in brackets indicates the sample standard deviation, while a number in parentheses indicates the estimated standard deviation for a single measurement.

[b]Although no molecular terminal iron oxo complexes are listed in the CSD, the structure of the ferrate derivative K_2FeO_4, with a bond length of 1.67[1] Å, has been determined and extended X-ray absorption fine structure (EXAFS) studies on horseradish peroxidase and model compounds indicate Fe—O bond lengths in the range 1.60–1.65 Å.

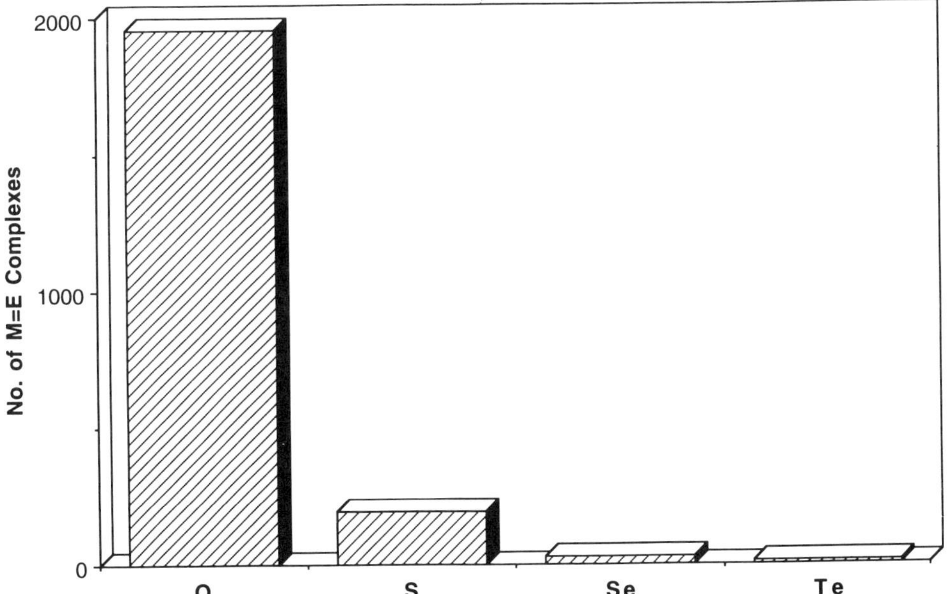

Figure 1. Structurally characterized terminal chalcogenido complexes as a function of the chalcogen. [Reprinted from *Polyhedron*, vol. 16, T. M. Trnka and G. Parkin, ''Survey of Terminal Chalcogenido Complexes of the Transition Metals,'' p. 1031, Copyright © 1997, with kind permission from Elsevier Science Ltd., The Boulevard, Langford Lane, Kidlington OX5 1GB, UK.]

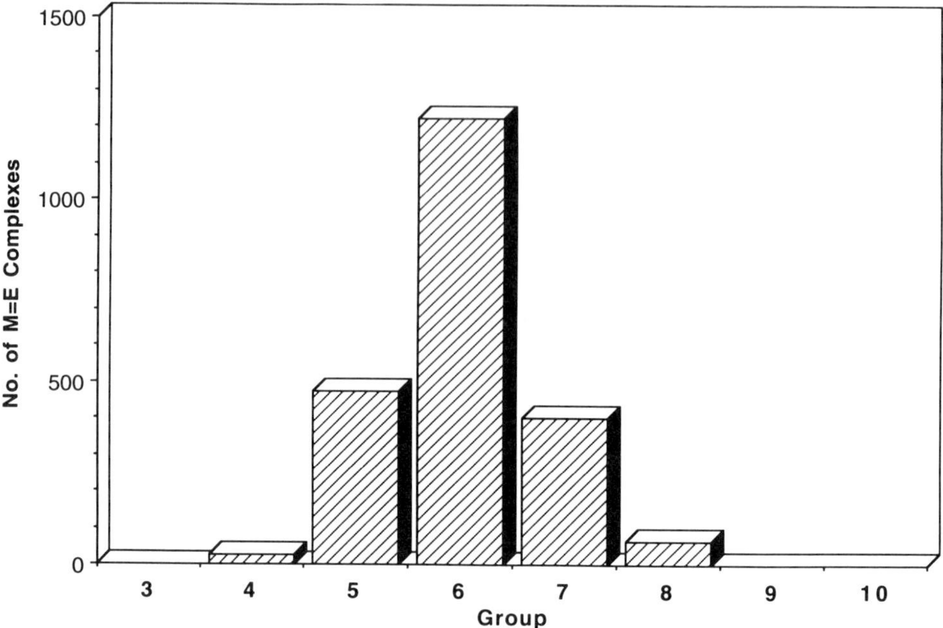

Figure 2. Structurally characterized terminal chalcogenido complexes as a function of periodic group. [Reprinted from *Polyhedron*, vol. 16, T. M. Trnka and G. Parkin, "Survey of Terminal Chalcogenido Complexes of the Transition Metals," p. 1031, Copyright © 1997, with kind permission from Elsevier Science Ltd., The Boulevard, Langford Lane, Kidlington OX5 1GB, UK.]

distributions illustrated in Figs. 1–7 are presumably representative of all terminal chalcogenido complexes, with the caution that the number of terminal oxo complexes is likely to be significantly underestimated because definitive characterization of terminal [M=O] moieties does not normally require X-ray diffraction techniques. For example, the [M=O] group, with characteristic absorptions in the range of about 750–1100 cm^{-1} (1, 32), is readily amenable to infrared (IR) spectroscopic analysis. In contrast, structural authentication by X-ray diffraction is usually essential for complexes of the heavier chalcogens (S, Se, and Te), since the terminal chalcogenido [M=E] moieties in these complexes are not as readily identified by IR and other spectroscopies. Additionally, X-ray diffraction is often required to eliminate alternative structures, such as those involving bridging chalcogenide and polychalogenide ligands, both of which are common for the heavier chalcogens (27–29).

Several trends are evident from Figs. 1–7, of which the most obvious is that the occurrence of terminal chalcogenido complexes decreases rapidly in the

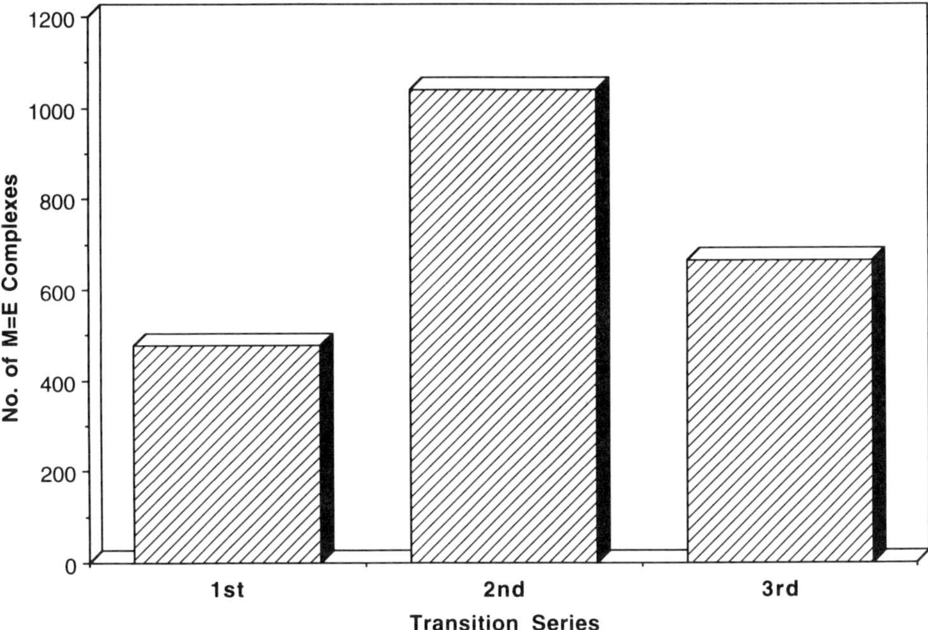

Figure 3. Structurally characterized terminal chalcogenido complexes as a function of the transition series. [Reprinted from *Polyhedron*, vol. 16, T. M. Trnka and G. Parkin, "Survey of Terminal Chalcogenido Complexes of the Transition Metals," p. 1031, Copyright © 1997, with kind permission from Elsevier Science Ltd., The Boulevard, Langford Lane, Kidlington OX5 1GB, UK.]

order O >> S > Se > Te (Fig. 1). Thus, by comparison with metal–oxo complexes, very few (<1%) terminal tellurido complexes exist. Indeed, until very recently, terminal tellurido complexes of the transition metals were unknown, with the first example, namely, *trans*-$W(PMe_3)_4(Te)_2$, being reported in 1991. However, as the result of the efforts of a number of research groups, several terminal tellurido complexes have now been synthesized for a variety of the transition metals, as summarized in Table II.

It is important to emphasize that the distribution of structurally characterized terminal chalcogenido complexes should not be used in an absolute sense to indicate the relative stabilities of M=E double and M—E single bonds, since the distribution may be skewed for a number of reasons. For example, to a certain extent, the preponderance of terminal oxo complexes is a consequence of the oxygen and water in the environment providing a plentiful source of oxygen atoms; in fact, a significant number of terminal oxo complexes has been

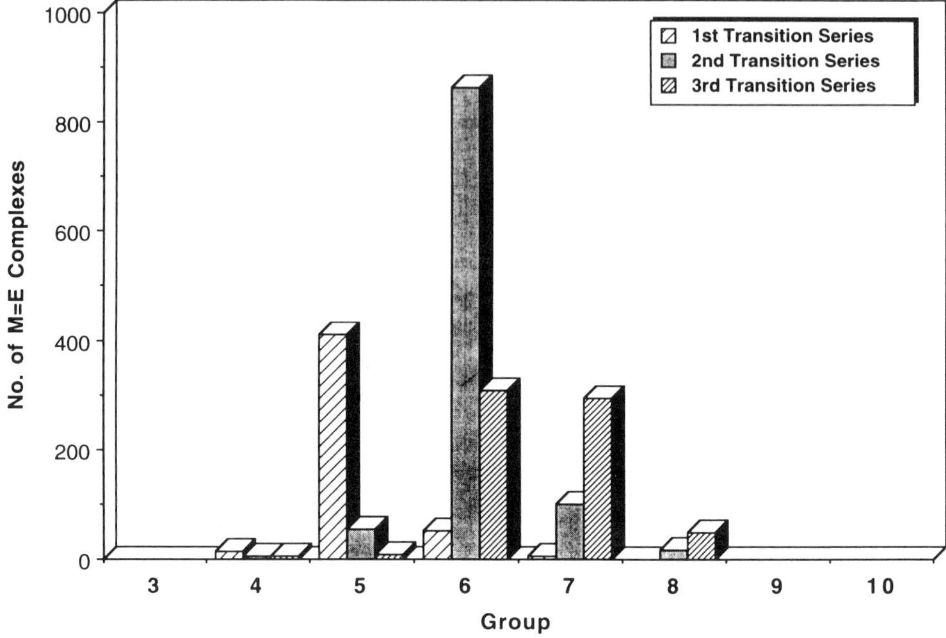

Figure 4. Structurally characterized terminal chalcogenido complexes as a function of periodic group and transition series. [Reprinted from *Polyhedron*, vol. 16, T. M. Trnka and G. Parkin, "Survey of Terminal Chalcogenido Complexes of the Transition Metals," p. 1031, Copyright © 1997, with kind permission from Elsevier Science Ltd., The Boulevard, Langford Lane, Kidlington OX5 1GB, UK.]

obtained adventitiously! Alternatively, the relative dearth of terminal tellurido complexes of the transition metals may reflect less research effort into the chemistry of tellurium compared to its light congeners; nevertheless, it should be noted that there are, in fact, a significant number of complexes with transition metal–tellurium bonds (33, 34). Despite the fact that the distribution of terminal chalcogenido complexes should not be used as a thermodynamic statement concerning the strengths of multiple bonds, it is apparent that the scarcity of terminal tellurido complexes does nonetheless reflect the preference of the heavier main group elements to form single rather than multiple bonds (35, 36). Such tendencies have been associated with the difference in σ- and π-bond energies decreasing down a group, in part due to reduced π overlap for the heavier elements (35d). Thus, it is much more common for the heavier chalcogens to form single bonds to transition metals, via either single atom bridges or polychalcogenide ligands (28, 29), rather than to form terminal multiple bonds.

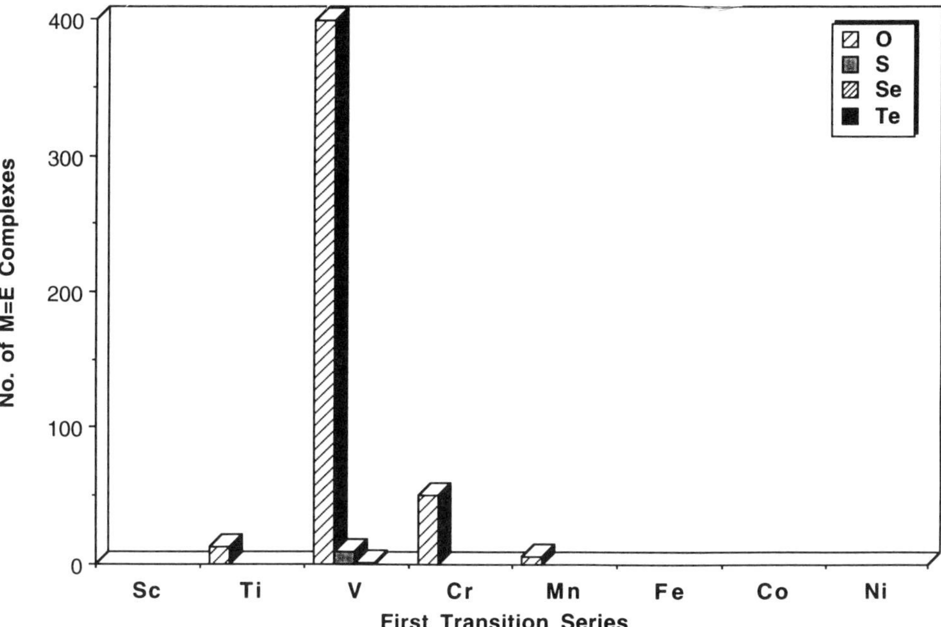

Figure 5. Structurally characterized terminal chalcogenido complexes of the first transition series. [Reprinted from *Polyhedron*, vol. 16, T. M. Trnka and G. Parkin, "Survey of Terminal Chalcogenido Complexes of the Transition Metals," p. 1031, Copyright © 1997, with kind permission from Elsevier Science Ltd., The Boulevard, Langford Lane, Kidlington OX5 1GB, UK.]

Whereas the distribution of terminal chalcogenido complexes is a monotonic function of the chalcogen, the same is not true for the distribution with respect to the transition metals, as illustrated in Figs. 2–7. As a function of group, their occurrence reaches a maximum with the Group 6 (VI B) metals (Cr, Mo, and W), and decreases rapidly for the earlier and later groups (Fig. 2). Consequently, there are no isolated terminal chalcogenido complexes of either the Group 3 (III B) (Sc, Y, and La) or Group 10 (VIII) (Ni, Pd, and Pt) transition metals, with only one example for the Group 9 (IX) metals, namely (2,4,6-$Me_3C_6H_2)_3IrO$ (37). In this regard, Mayer (38) has suggested that the lack of terminal oxo complexes for the late transition metals is a consequence of the high *d* electron count, which results in antibonding interactions with the π orbitals on the oxo ligand, thereby destabilizing a terminal [M=O] moiety with respect to a bridged structure (39).

Although the Group 6 (VI B) metals form the greatest number of terminal chalcogenido complexes, it is apparent from Figs. 4–7 that the peak in the

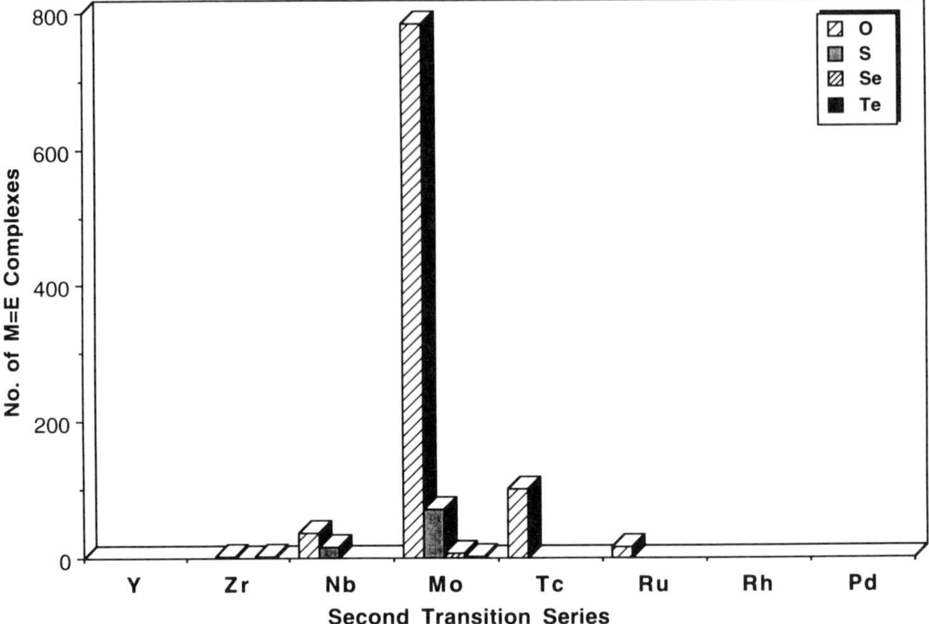

Figure 6. Structurally characterized terminal chalcogenido complexes of the second transition series. [Reprinted from *Polyhedron*, vol. 16, T. M. Trnka and G. Parkin, "Survey of Terminal Chalcogenido Complexes of the Transition Metals," p. 1031, Copyright © 1997, with kind permission from Elsevier Science Ltd., The Boulevard, Langford Lane, Kidlington OX5 1GB, UK.]

distribution shifts from the earlier to later transition metals upon progressing from the first to third transition series. The majority of terminal chalcogenido complexes are therefore concentrated on the vanadium–molybdenum–rhenium diagonal, as noted previously for metal–ligand multiple bonds in general (1).

III. STRUCTURAL STUDIES

The mean d(M=E) bond lengths for all structurally characterized terminal chalcogenido complexes of the transition metals are summarized in Table I (31), and are intended to provide a basis for structural comparisons with other complexes. However, no effort is made here to subclassify the bond lengths according to either coordination number or oxidation state of the metal, since Mayer has previously noted that M=O bond lengths are generally insensitive to such factors (40, 41).

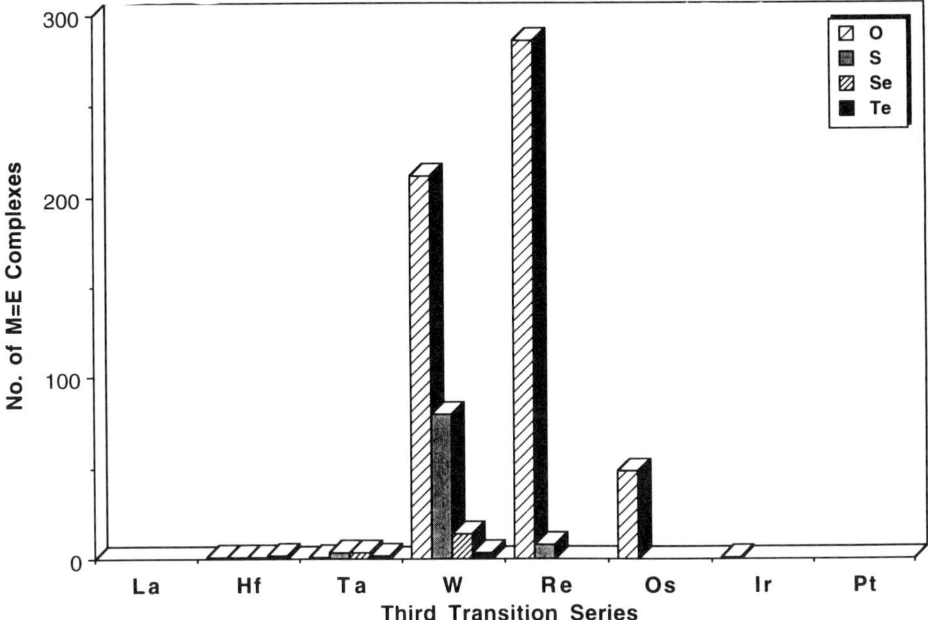

Figure 7. Structurally characterized terminal chalcogenido complexes of the third transition series. [Reprinted from *Polyhedron*, vol. 16, T. M. Trnka and G. Parkin, "Survey of Terminal Chalcogenido Complexes of the Transition Metals," p. 1031, Copyright © 1997, with kind permission from Elsevier Science Ltd., The Boulevard, Langford Lane, Kidlington OX5 1GB, UK.]

A. Electronic Nature of the Metal Center

Metal–chalcogenido interactions may be considered to be a composite of three resonance structures, with formal single $[\overset{+}{M}-\overset{-}{E}]$, double [M=E], and triple $[\overset{-}{M}\equiv\overset{+}{E}]$ bonds.

$$\overset{+}{M}-\overset{-}{E} \longleftrightarrow M=E \longleftrightarrow \overset{-}{M}\equiv\overset{+}{E}$$

For simplicity, however, unless a specific aspect of bonding is being discussed, the representation [M=E] will be used in this chapter to refer generally to any terminal chalcogenido moiety, regardless of bond order. In this respect, the majority of terminal chalcogenido complexes appear to be best described as containing a $[\overset{-}{M}\equiv\overset{+}{E}]$ triple bond (1, 42); however, there are several examples where particular metal–chalcogenido functionalities are better described with reduced bond orders as a result of the electronic nature of the metal centers (see

TABLE II
Terminal Tellurido Complexes of the Transition Metals

Complex	Reference
$Cp^*_2Zr(Te)(NC_5H_5)$	*a*
$Cp_2^{Et*}Zr(Te)(NC_5H_5)$	*a*
$(dmpe)_2Zr[TeSi(SiMe_3)_3]_2(Te)$	*b*
$Cp^*_2Hf(Te)(NC_5H_5)$	*c*
$Cp_2^{Et*}Hf(Te)(NC_5H_5)$	*c*
$(dmpe)_2Hf[TeSi(SiMe_3)_3]_2(Te)$	*b*
$[\eta^4\text{-}N(CH_2CH_2NSiMe_3)_3]V(Te)$	*d*
$Cp^*Nb(PMe_3)(NAr)(Te)$ (Ar = 2,6-$Pr^i_2C_6H_3$)	*e*
$[\eta^4\text{-}N(CH_2CH_2NSiMe_3)_3]TaTe$	*f*
$Cp^*_2Ta(Te)H$	*g*
$Cp^*_2Ta(Te)Me$	*g*
$Mo(PMe_3)_4(Te)_2$	*h*
$Mo(dppee)_2(Te)_2$	*i*
$[(3,5\text{-}Me_2C_6H_3)NBu^t]_3MoTe$	*j*
$W(PMe_3)_4(Te)_2$	*k*
$W(PMe_3)_2(Te)_2(\eta^2\text{-}OCHR)$ (R = H and Ph)	*k*
$(Ph_4P)_2[W(O)(Te)_3]$	*l*

[a] W. A. Howard and G. Parkin, *J. Am. Chem. Soc.*, *116*, 606 (1994).
[b] V. Christou and J. Arnold, *J. Am. Chem. Soc.*, *114*, 6240 (1992). C. P. Gerlach, V. Christou, and J. Arnold, *Inorg. Chem.*, *35*, 2758 (1996).
[c] W. A. Howard and G. Parkin, *J. Organomet. Chem.*, *472*, C1 (1994).
[d] C. C. Cummins, R. R. Schrock, and W. M. Davis, *Inorg. Chem.*, *33*, 1448 (1994).
[e] U. Siemeling and V. C. Gibson, *J. Chem. Soc. Chem. Commun.*, 1670 (1992).
[f] V. Christou and J. Arnold, *Angew. Chem. Int. Ed. Engl.*, *32*, 1450 (1993).
[g] J. H. Shin and G. Parkin, *Organometallics*, *13*, 2147 (1994).
[h] V. J. Murphy and G. Parkin, *J. Am. Chem. Soc.*, *117*, 3522 (1995).
[i] F. A. Cotton and X. Feng, *Inorg. Chem.*, *35*, 4921 (1996). F. A. Cotton and G. Schmid, *Inorg. Chem.*, *36*, 2278 (1997).
[j] C. C. Cummins, personal communication.
[k] D. Rabinovich and G. Parkin, *Inorg. Chem.*, *34*, 6341 (1996).
[l] D. R. Gardner, J. C. Fettinger, and B. W. Eichorn, *Angew. Chem. Int. Ed. Engl.*, *33*, 1859 (1994).

below). For a closely related series of such complexes, the M=E bond lengths would be expected to correlate with bond order. An illustration of such a correlation is provided by the series of tungsten sulfido and selenido complexes listed in Table III (43), for which the W=E bond lengths progressively decrease as the formal bond order increases from 2 in the 18-electron complexes $W(PMe_3)_2(L)_2(E)_2$ (L = PMe_3, Bu^tNC and $W(PMe_3)_4(Se)H_2$ to 3 in the "12"-electron complexes $W(E)X_4$ (E = S, Se; X = F, Cl, Br). Another example of this effect has been noted by Mayer (40), who explained the presence of long

TABLE III
Variation of W = E Bond Length with Bond Order[a]

Complex	Bond Order	S	Se
$W(PMe_3)_4(E)H_2$	2		2.445(2)
$W(PMe_3)_4(E)_2$	2	2.253[3]	2.380[1]
$W(PMe_3)_2(CNBu^t)_2(E)_2$	2	2.248(2)	2.375(2)
$W(PMe_3)_2(E)_2(\eta^2\text{-}OCHPh)$	2.5	2.186[2]	2.317[4]
$W(E)F_4$	3	2.104(7)	2.226(7)
$W(E)Cl_4$	3	2.086(6)	2.203(4)
$W(E)Br_4$	3	2.109(11)	2.220(22)

[a]Taken from (31).

ruthenium–oxo bond lengths in 18-electron d^4 octahedral ruthenium oxo complexes as a consequence of the bond order being restricted to 2 (40).

The bent metallocene system is also one in which a metal–chalcogenido interaction would formally be expected to be less than 3 on the basis of the commonly used molecular orbital (MO) description for such derivatives, as illustrated for Cp_2MoO in Fig. 8 (44). Specifically, since the frontier orbitals of the $[Cp_2M]$ moiety all lie in the equatorial plane, a chalcogenido ligand may form only a single π bond in addition to the σ bond, thereby restricting the bond order to 2 (i.e., [M=E]) in a first-order approximation. Nevertheless, triple-bond contributions of the type $[\overset{-}{M}\equiv\overset{+}{E}]$ could occur at the expense of the metal–cyclopentadienyl interactions. In this regard, MO calculations on molybdenocene oxo derivatives indicate that there is an interaction between a b_1 (d_{xz}) orbital on molybdenum and the oxygen p_x orbital (45a, b). However, since the b_1 orbital on molybdenum is also involved in bonding to the cyclopentadienyl ligands, the result is the generation of three molecular orbitals: one strongly bonding, one almost nonbonding, and one antibonding, with the first two being occupied. In fact, the "nonbonding" b_1 orbital is slightly Mo—O antibonding, so that its occupation results in the overall Mo—O bond order being reduced from 3. Consequently, the triple-bond formalism $\overset{-}{Mo}\equiv\overset{+}{O}$ for such complexes may not be the most appropriate description of the bonding. Furthermore, computational studies on hypothetical $[Cp_2TiE]$ (E = O, S, Se, Te) and $[Cp_2ZrE]$ (E = O, Te) species have demonstrated that the out-of-plane b_1 interaction is less important than the in-plane interaction for these complexes (45c, 46).

Evidence supporting the notion that the triple-bond formalism $[\overset{-}{Mo}\equiv\overset{+}{O}]$ may not be the best description of the bonding in molybdenocene oxo derivatives is provided by X-ray diffraction and infrared (IR) spectroscopic studies. Thus, the Mo=O bond lengths in $Cp_2^{Me}MoO$ [1.721(2) Å] (47a) and $Cp_2^{Bu^t}MoO$ [1.706(4) Å] (48) are slightly longer than the mean value of 1.678[40] Å for other Mo^{IV}

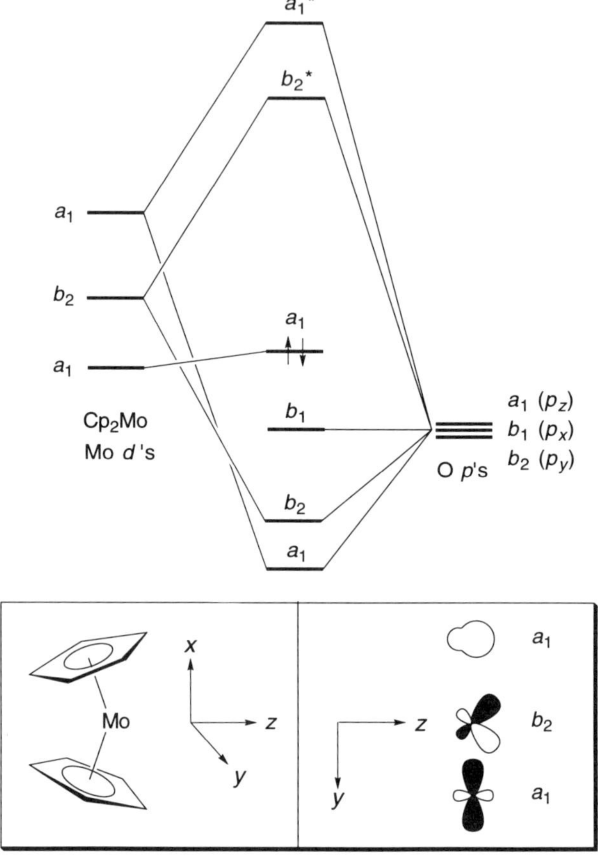

Figure 8. Qualitative MO diagram for Cp_2MoO.

oxo complexes, which exhibit a considerable degree of triple bond [$\bar{Mo}\equiv\overset{+}{O}$] character (49). Likewise, the ν(Mo=O) stretching frequencies in these molybdenocene complexes (see Section VI.D.2) are significantly lower in energy than those for other molybdenum oxo complexes, with the result that they have been described as "deviant" (50). In this regard, Bercaw and co-worker (54) have categorized metal oxo complexes with stretching frequencies in the range of approximately 930–1000 cm^{-1} as "class *a*" (i.e., those in which the metal–oxo bond order is close to 3), and those with stretching frequencies below approximately 930 cm^{-1} as "class *b*" (i.e., those in which the metal–oxo bond order is close to 2).

In contrast to the molybdenocene oxo complexes described above, it should be noted that the *tert*-butylimido ligand in the molybdenocene derivatives

Cp_2MoNBu^t and $[Cp_2^{Me}Mo(NBu^t)Me]^+$ has been suggested to compete effectively with the coordination of the cyclopentadienyl ligands, such that the molybdenum–imido interactions are best described as triply bonded, $[\overset{-}{Mo}{\equiv}\overset{+}{N}R]$ (46g). In contrast, however, the triply bonded resonance structure is considered to be less important for the tantalocene imido derivatives $Cp_2^*Ta(NPh)H$ and $Cp_2^*Ta(NPh)Cl$ (46a, b). As such, it would seem prudent not to generalize, but to consider the nature of the interaction between a metallocene fragment and a multiply bonded ligand on a case-by-case basis.

It is evident from the above discussion that the electronic nature of the metal center may have a pronounced effect on the character of the metal–chalcogenido interaction by influencing its bond order. In this regard, Nugent was the first to comment upon the electrophilic versus nucleophilic character of complexes with multiply bonded alkylidene, imido, and oxo ligands, and rationalized their differences in terms of the relative energies of the ligand $p\pi$ and metal $d\pi$ orbitals (52). Thus, a metal with diffuse high-energy d orbitals (as for an early transition metal) interacts poorly in a π fashion with the p orbitals of a multiply bonded ligand resulting in a highest occupied molecular orbital (HOMO) that is largely localized on the ligand, which therefore manifests nucleophilic reactivity [Fig. 9, (*a*)]. As the d orbitals become less diffuse and lower in energy (as with the later transition metals), transfer of electron density to the metal center is pro-

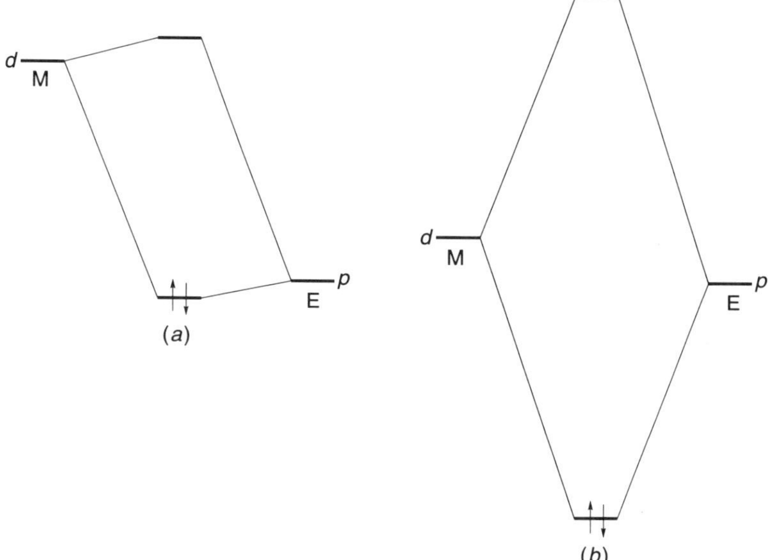

Figure 9. Nucleophilic versus electrophilic character of terminal chalcogenido complexes as a function of the relative energies of the chalcogen $p\pi$ and metal $d\pi$ orbitals.

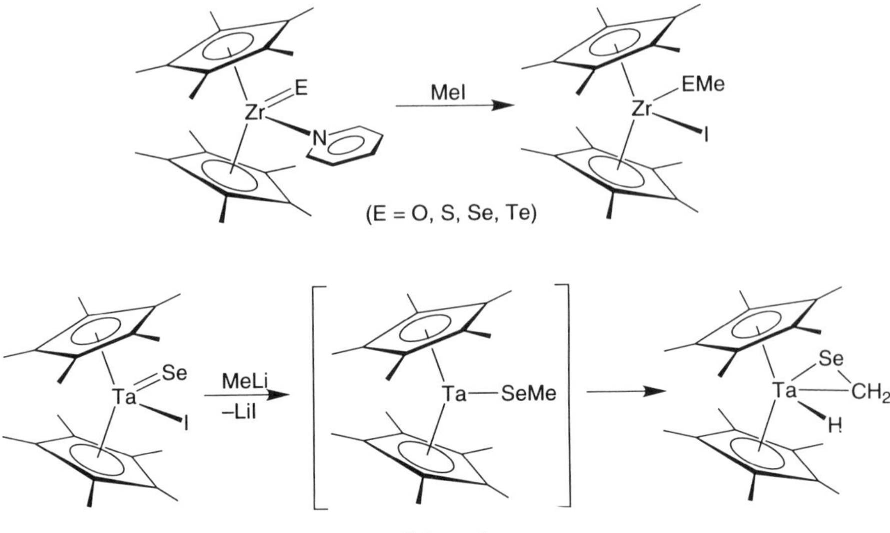

Scheme 1

moted, with a corresponding decrease in nucleophilicity [Fig. 9, (*b*)]. An illustration of nucleophilic reactivity of terminal chalcogenido ligands is provided by the chemistry of $Cp^*_2Zr(E)(NC_5H_5)$ in which chalcogenolate derivatives $Cp^*_2Zr(EMe)I$ are obtained upon reaction with MeI (Scheme 1) (32). Likewise, electrophilic character of a terminal chalcogenido ligand is illustrated by the functionalization of the selenido ligand of $Cp^*_2Ta(Se)I$ upon treatment with MeLi to give the selenoformaldehyde complex $Cp^*_2Ta(\eta^2\text{-}SeCH_2)H$ (53).

B. Variation in M=E Bond Length as a Function of the Chalcogen

Only five of the transition elements (Zr, Hf, Ta, Mo, W) are known to form terminal chalcogenido complexes with all of the chalcogens (excluding Po) and their variation in M=E bond length is illustrated in Fig. 10. Overall, the change in M=E bond length as a function of the chalcogen correlates reasonably well with the change in double-bond covalent radii. However, it is apparent from the data in Table IV and Fig. 10 that, by comparison with their heavier counterparts, the metal–oxo bond lengths are anomalously short (i.e., the increment in bond length between oxo and sulfido derivatives is greater than the difference in covalent radii). In the case of zirconium, for which the known chalcogenido complexes are almost exclusively the cyclopentadienyl derivatives $Cp^R_2Zr(E)(NC_5H_5)$ (Cp^R = Cp* or Cp^{Et*}) (54), the observation has been rationalized in terms of the resonance structure $[\overset{+}{Zr}\text{—}\overset{-}{O}]$, with bond shortening being attributed to an ionic component (55). Since the heavier chalcogens (S, Se, Te)

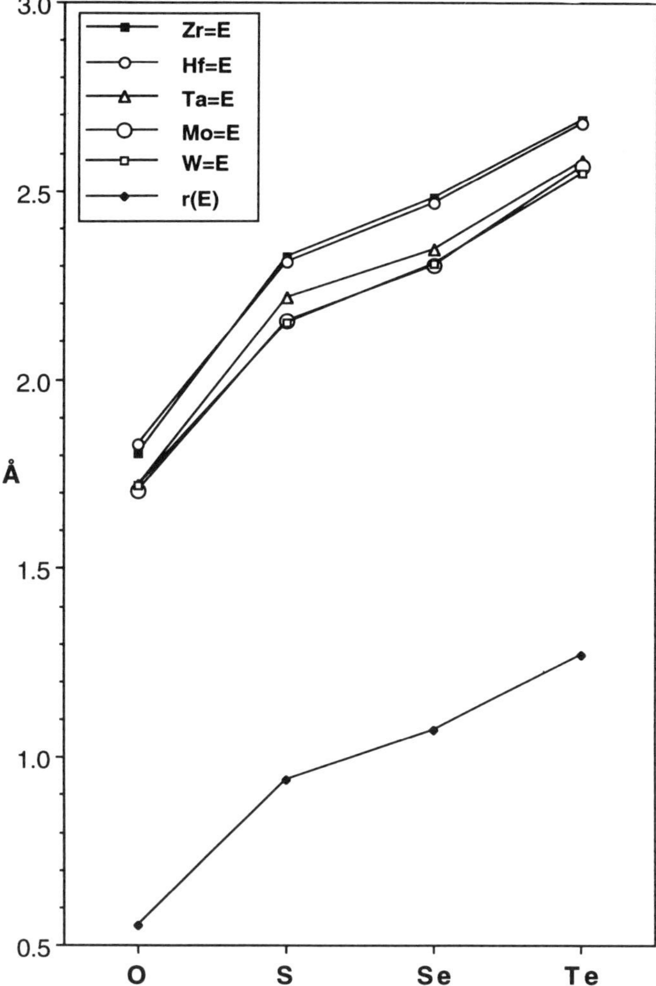

Figure 10. Variation of mean M=E bond lengths as a function of the chalcogen. [Reprinted from *Polyhedron*, vol. 16, T. M. Trnka and G. Parkin, "Survey of Terminal Chalcogenido Complexes of the Transition Metals," p. 1031, Copyright © 1997, with kind permission from Elsevier Science Ltd., The Boulevard, Langford Lane, Kidlington OX5 1GB, UK.]

are both less electronegative and larger than oxygen, the corresponding zirconium–chalcogenido interactions would be subject to less ionic shortening and the resonance from $\overset{+}{\text{Zr}}-\overset{-}{\text{E}}$ is considered to play a less significant role in describing the bonding within the zirconium–chalcogenido moiety. The zirconium–sulfido, zirconium–selenido, and zirconium–tellurido bond lengths are thus more com-

TABLE IV

Comparison of M=E Bond Length Increments with Respect to the M=O Derivative[a]

	r_{db}(E) (Å)[b]	[r_{db}(E) − r_{db}(O)] (Å)	d(M=E) − d(M=O) (Å)				
			Zr	Hf	Ta	Mo	W
O	0.55	0	0	0	0	0	0
S	0.94	0.39	0.52	0.49	0.50	0.45	0.43
Se	1.07	0.52	0.67	0.64	0.62	0.60	0.59
Te	1.27	0.72	0.89	0.85	0.86	0.86	0.83

[a]Taken from (31).

[b]r_{db} is the double bond covalent radius.

parable to the values predicted by the sum of covalent radii than is the zirconium–oxo bond length. Furthermore, computational studies by Cundari and coworkers (56) on [$Cp_2Zr{=}E$] (E = O, S, Se, Te) demonstrate that the oxo species is unique and shows a significant contribution from a singly bonded $\overset{+}{Zr}{-}\overset{-}{O}$ structure, with the π bonds localized largely on the oxo ligand, that is, [$\overset{+}{Zr}{-}\overset{-}{O}$]. Finally, as described in Section III.A, the more commonly encountered triply bonded structure [$\overset{-}{M}{\equiv}\overset{+}{E}$] is not appropriate for $Cp_2^RZr(E)(NC_5H_5)$ in view of both the symmetries of the frontier orbitals and the 18-electron nature of the zirconium centers.

C. Variation in M=E Bond Length as a Function of the Metal

The variation in M=E bond length as a function of the metal is best illustrated for the oxo derivatives because not many of the transition metals are known to form terminal chalcogenido complexes with the heavier chalcogens. Thus, for the oxo derivatives, there is a general decrease in M=O bond length upon crossing a period, with the largest incremental change for the earlier metals (Fig. 11). Such changes correlate reasonably well with the variation in co-

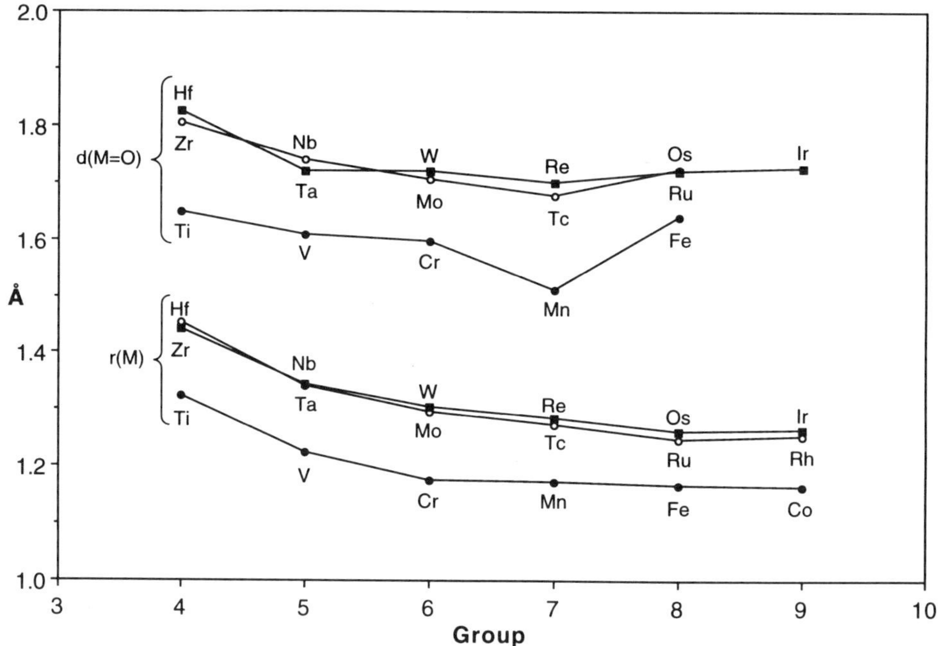

Figure 11. Variation of mean M=O bond lengths as a function of the metal. [Reprinted from *Polyhedron*, vol. 16, T. M. Trnka and G. Parkin, "Survey of Terminal Chalcogenido Complexes of the Transition Metals," p. 1031, Copyright © 1997, with kind permission from Elsevier Science Ltd., The Boulevard, Langford Lane, Kidlington OX5 1GB, UK.]

valent radii of the transition metals. Cundari and co-workers (56) and Ziegler and co-workers (57) have also recently calculated the M=O bond lengths for a series of terminal oxo complexes. It is anticipated that the variation in bond lengths for the heavier chalcogenido complexes will be analogous to that for the oxo derivatives.

D. Cis versus Trans Preferences for Octahedral Bis(chalcogenido) Complexes

Octahedral transition metal complexes with two terminal chalcogenido ligands may adopt either a cis or trans arrangement, depending on the electronic configuration of the metal center. In general, d^0 metal centers favor a cis arrangement of chalcogenido ligands, whereas d^2 metal centers favor a trans arrangement (58).

As first noted by Griffith and Wickins (59), these geometrical preferences are normally rationalized in terms of maximizing π bonding while minimizing the occupation of antibonding orbitals (60–62). Thus, a trans arrangement is favored for a d^2 metal center since it allows the two d electrons to occupy a nonbonding orbital, rather than an antibonding orbital that would result from a cis configuration. The situation is illustrated by the qualitative MO diagram of Fig. 12, which focuses attention only on π interactions; the important bonding

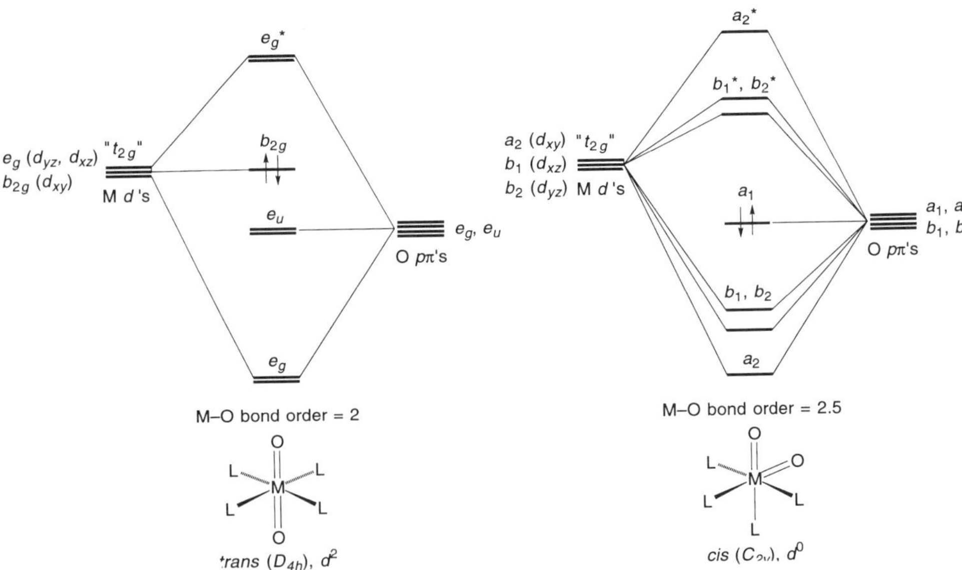

Figure 12. Qualitative MO diagrams illustrating only the π interactions for octahedral trans- and cis-dioxo complexes.

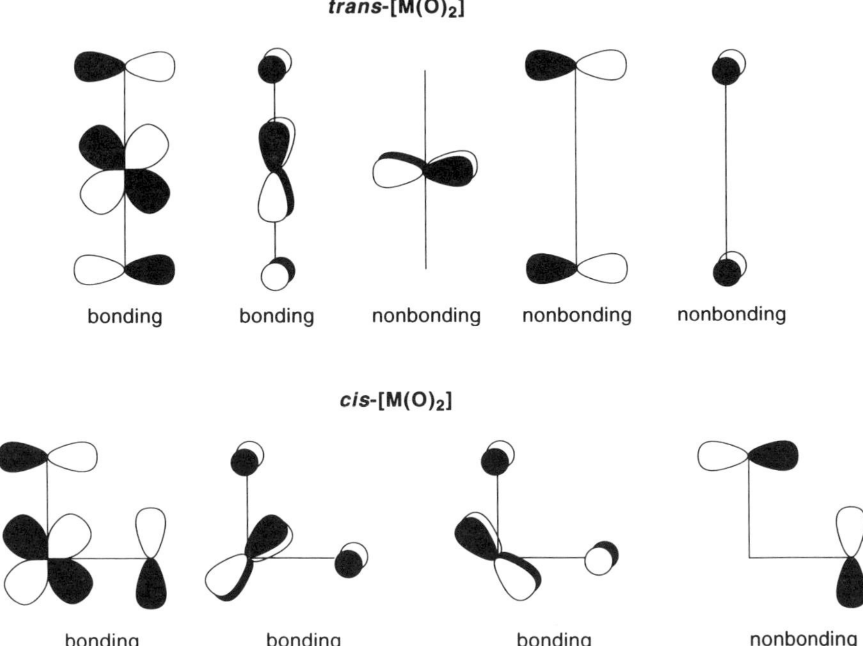

Figure 13. Bonding and nonbonding MO interactions for octahedral trans- and cis-dioxo complexes. For the trans derivative, the z axis is coincident with the two M—O bonds, while for the cis derivative the z axis bisects the two M—O bonds.

and nonbonding molecular orbitals are illustrated in Fig. 13. With only two π bonding molecular orbitals occupied for the trans-[M(O)$_2$] moiety, the M—O bond order is appropriately described as 2. In contrast to the octahedral d^2 case, a cis geometry becomes favored for a d^0 metal center because such an arrangement maximizes π interactions without suffering the consequence of forcing metal-based electrons (which are necessarily absent on a d^0 metal center) to occupy antibonding orbitals. With three π-bonding molecular orbitals occupied for the d^0 cis-[M(O)$_2$] moiety, the M—O bond order is increased from 2 to 2.5.

Illustrative examples of d^0 and d^2 bis(chalcogenido) complexes, which exhibit the aforementioned cis versus trans preferences, are provided in Fig. 14. For instance, the formally 18-electron d^2 complexes W(PMe$_3$)$_4$(E)$_2$ and W(PMe$_3$)$_2$(CNBu^t)$_2$(E)$_2$ exhibit trans geometries, whereas the formally 16-electron (in the absence of chalcogen lone-pair donation) d^0 complexes W(PMe$_3$)$_2$(E)$_2$(η^2-OCHR) exhibit cis geometries. Furthermore, with an average W=E bond order of 2.5 for the d^0 cis complexes, the W=E bond lengths are shorter than the corresponding values for 18-electron complexes with formal W=E double bonds (Table III).

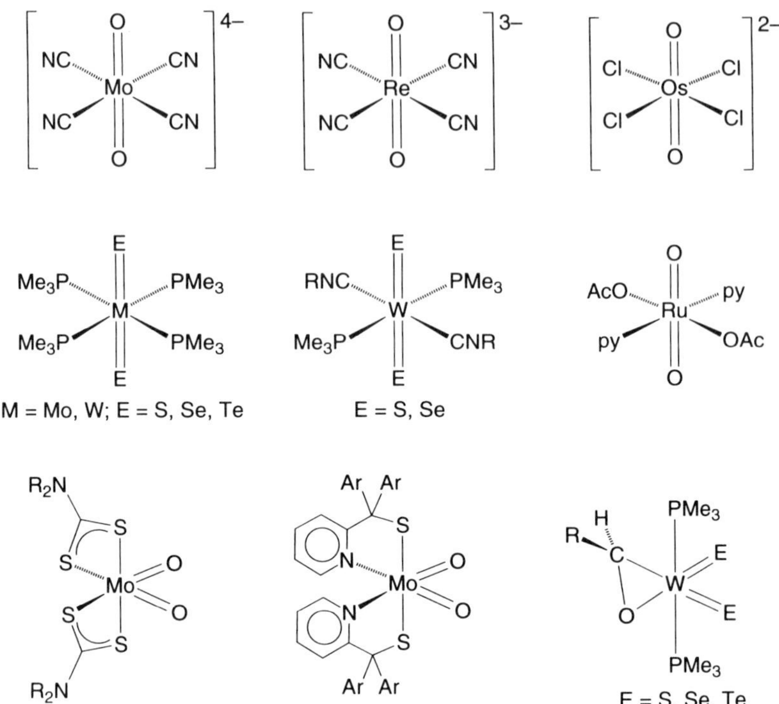

Figure 14. Cis versus trans preferences for d^0 and d^2 bis(chalcogenido) complexes.

There are, however, a growing number of exceptions to these structural preferences. For example, a number of d^2 dioxo complexes, for example, $[Ru(\eta^2\text{-}OAc)Cl_2(O)_2]^-$ (63), $[Re(bpy)(py)_2(O)_2]^-$ (64), $[Os(\eta^2\text{-}OAc)(\eta^1\text{-}OAc)_2(O)_2]^-$ (65), and $[Os(bpy)_2(O)_2]^{2+}$ (66), are known to exhibit a cis arrangement of oxo ligands; the latter complex is, however, unstable with respect to the trans derivative. The existence of several of these d^2 cis-dioxo complexes has been attributed to the presence of bidentate ligands (60e–g). Specifically, extended Hückel MO (EHMO) calculations have demonstrated that reducing the bite angle of the co-ligands results in the stabilization of the d_{xz} orbital of the cis geometry. Consequently, the trans versus cis electronic preferences for a nonoctahedral d^2 complex may be reversed from that of a d^2 complex with approximate octahedral geometry.

In addition to the above complexes, $\{(\eta^3\text{-}Me_3tacn)Ru[\eta^1\text{-}OC(O)CF_3](O)_2\}^+$ (67) and $[\{\eta^4\text{-}Me_2NCH_2CH_2N(Me)CH_2CH_2N(Me)CH_2CH_2NMe_2\}Ru(O)_2]^{2+}$ (68) also exhibit a cis arrangement of oxo ligands; however, for these examples, the cis geometries are presumably a consequence of steric constraints imposed by the supporting ligands, since closely related derivatives are known to exhibit trans geometries (69). Indeed, Drago and co-workers (70) here employed steric interactions between the methyl substituents of 2,9-dimethyl-1,10-phenanthroline (dmp) to enforce a cis geometry in $[Ru(dmp)_2(O)_2]^{2+}$.

It is noteworthy that related cis and trans dioxo complexes exhibit some interesting differences. For example, the Ru=O bond length of 1.759(9) Å for cis-$[\{\eta^4\text{-}Me_2NCH_2CH_2N(Me)CH_2CH_2N(Me)CH_2CH_2NMe_2\}Ru(O)_2]^{2+}$ (68b) is longer than the value of 1.705(7) Å in the trans counterpart, trans-[{*cyclo*-$(MeNCH_2CH_2CH_2)_4\}Ru(O)_2]^{2+}$ (69a). Furthermore, cis and trans ruthenium dioxo complexes exhibit different reactivities, with the cis derivatives being the better oxidants (68a, 70b). For example, cis ruthenium dioxo complexes react with olefins to effect cleavage of the C=C double bond to give carbonyl complexes, whereas the trans counterparts predominantly achieve epoxidation; the origin of these differences has been addressed theoretically (71).

E. Structural Anomalies

The bond length data described in Sections III B and III C and presented in Table I correspond to the mean values for complexes listed in the CSD. However, their reliability and usefulness is highly dependent on both the accuracy of the individual observations and the number of such observations. As an example, the distribution of bond lengths for the [W=S] moiety is illustrated in Fig. 15, from which it is evident that the principal distribution is fairly narrow. It must, however, be recognized that there are complexes such as $[W_4S_8(H_2NCH_2CH_2NH_2)_4]S$ (72), which exhibit bond lengths [in this case 2.54(1) Å] far removed from the mean value (2.150 Å). In such instances, the significant deviation from the mean raises the issue as to whether or not the complex actually contains a multiply bonded terminal sulfido ligand. More generally, where observed bond lengths deviate significantly from the mean, it seems appropriate to consider other explanations, including incorrect atom or ligand assignment and crystal purity. As an example of the latter effect, the anomalously long Mo=O bond lengths reported for the so-called "distortional" or "bond-stretch" isomers *cis-mer*-$MoOCl_2(PR_3)_3$ (73, 74) are now known to be an artifact due to compositional disorder (75).

IV. SYNTHESES AND REACTIVITY OF TERMINAL CHALCOGENIDO COMPLEXES OF THE TRANSITION METALS

A. Terminal Chalcogenido Complexes of Group 3 (Sc, Y, La)

As is evident from Figs. 4–7, structurally authenticated terminal chalcogenido complexes of Sc, Y, and La are unknown. The development of this area is clearly of interest and represents a challenge for the synthetic inorganic chemist.

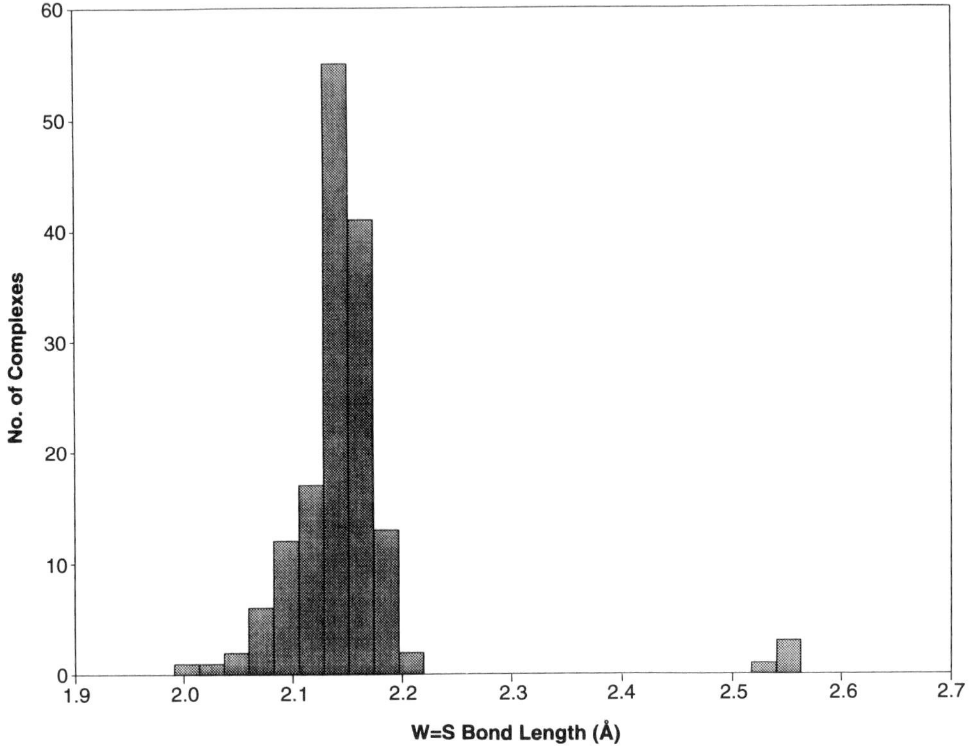

Figure 15. Distribution of W=S bond lengths in tungsten sulfido complexes. [Reprinted from *Polyhedron*, vol. 16, T. M. Trnka and G. Parkin, "Survey of Terminal Chalcogenido Complexes of the Transition Metals," p. 1031, Copyright © 1997, with kind permission from Elsevier Science Ltd., The Boulevard, Langford Lane, Kidlington OX5 1GB, UK.]

B. Terminal Chalcogenido Complexes of Group 4 (Ti, Zr, Hf)

The distribution of structurally characterized terminal chalcogenido complexes for the Group 4 transition metals is illustrated in Figs. 4–7 and 16. Although precedented, such complexes are very rare compared to those of the Groups 5–8 transition metals.

1. Titanium Chalcogenido Complexes

By comparison with its heavier congeners (Zr and Hf), terminal chalcogenido complexes are relatively common for titanium, but are presently limited to oxo, sulfido, and selenido derivatives (Fig. 16).

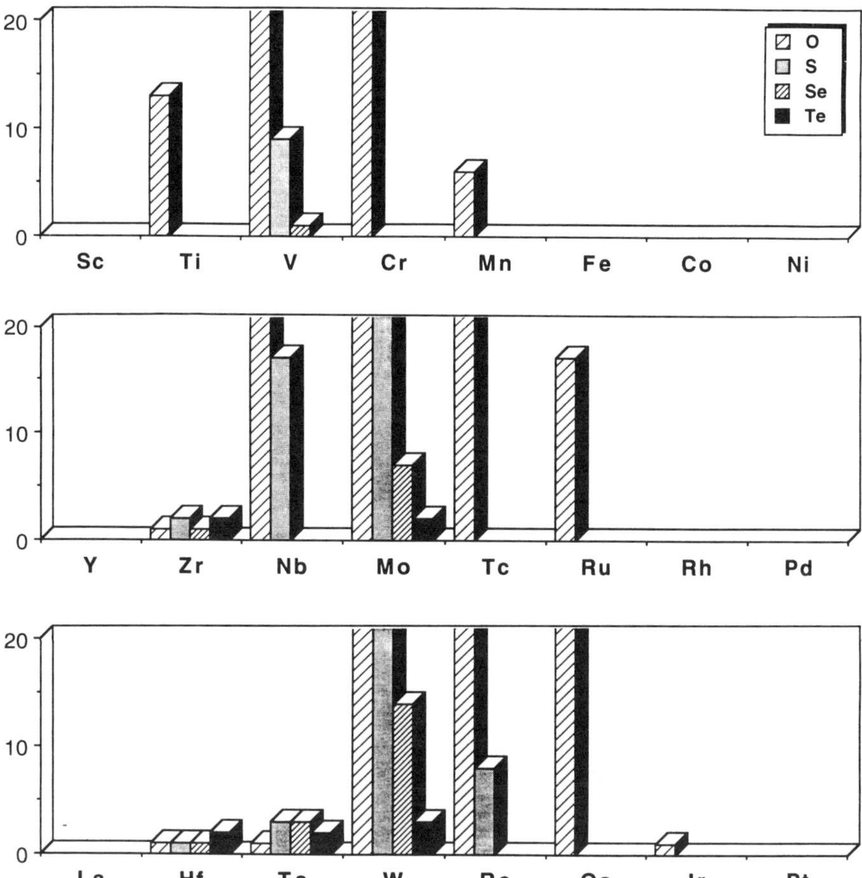

Figure 16. Structurally characterized terminal chalcogenido complexes of the first, second, and third transition series. [Reprinted from *Polyhedron*, vol. 16, T. M. Trnka and G. Parkin, "Survey of Terminal Chalcogenido Complexes of the Transition Metals," p. 1031, Copyright © 1997, with kind permission from Elsevier Science Ltd., The Boulevard, Langford Lane, Kidlington OX5 1GB, UK.]

a. Titanium Oxo Complexes. A variety of terminal titanium oxo complexes supported by a range of different ligand types have been synthesized, as illustrated by the examples listed in Table V.

i. Macrocyclic Derivatives. The largest class of terminal titanium oxo complexes is one in which the titanyl moiety is stabilized by a tetradentate macrocyclic ligand, such as porphyrin, phthalocyanine, and tetraazaannulene

TABLE V

Bond Length and IR Data for Some Terminal Titanium Oxo Complexes

Complex	d(Ti = O) (Å)	ν(It = O) (cm^{-1})	References
(OEP)TiO	1.619(4)		*a*
	1.613(5)		*b*
(Me_2OEP)TiO	1.619(4)		*c*
(TPP)TiO	1.64(5)	980	*d*
(PC)TiO	1.650(4)	972	*e, f*
	1.626(7)		
[η^4-Me_4taa]TiO	1.653(3)	930	*g*
[η^4-$Et_8C_4(C_4H_2N)_2(C_5H_3N)(p\text{-}MeC_5H_2N)$]TiO	1.628(2)	968	*h*
($edtaH_2$)Ti(O)·H_2O		950	*i*
[η^3-Me_3tacn]Ti(O)Cl_2	1.637(3)		*j*
[η^3-Me_3tacn]Ti(O)$(NCS)_2$	1.638(3)	943	*k*
[η^3-Pr^i_3tacn]Ti(O)$(NCS)_2$		950	*j*
[η^3-Pr^i_3tacn]Ti(O)$(NCO)_2$		949	*j*
Cp^*_2Ti(O)(NC_5H_4Ph)	1.665(3)		*l*
Cp^*_2Ti(O)(NC_5H_5)		852	*l*
$[C(NH_2)_3]_4[(\eta^2\text{-}OCO_2)_3Ti(O)]$	1.680(2)	875	*m*
$[Et_4N]_2[Cl_4TiO]$	1.79	975	*n*

[a] C. Lecomte, J. Protas, and R. Guilard, *C. R. Hebd. Seances Acad. Sci. Paris, Ser. C*, *281*, 921 (1975).
[b] R. Guilard, J.-M. Latour, C. Lecomte, J.-C. Marchon, J. Protas, and D. Ripoll, *Inorg. Chem.*, *17*, 1228 (1978).
[c] P. N. Dwyer, L. Puppe, J. W. Buchler, and W. R. Scheidt, *Inorg. Chem.*, *14*, 1782 (1975).
[d] R. Guilard and C. Lecomte, *Coord. Chem. Rev.*, *65*, 87 (1985).
[e] W. Hiller, J. Strähle, W. Kobel, and M. Hanack, *Z. Kristollogr.*, *159*, 173 (1982).
[f] B. P. Block and E. G. Meloni, *Inorg. Chem.*, *4*, 111 (1965).
[g] C.-H. Yang, J. A. Ladd, and V. L. Goedken, *J. Coord. Chem.*, *19*, 235 (1988).
[h] R. Crescenzi, E. Solari, C. Floriani, A. Chiesi-Villa, and C. Rizzoli, *Organometallics 15*, 5456 (1996).
[i] F. J. Kristine, R. E. Shepherd, and S. Siddiqui, *Inorg. Chem.*, *20*, 2571 (1981).
[j] P. Jeske, G. Haselhorst, W. Weyhermüller, and B. Nuber, *Inorg. Chem.*, *33*, 2462 (1994).
[k] A. Bodner, P. Jeske, W. K. Weyhermüller, E. Dubler, H. Schmalle, and B. Nuber, *Inorg. Chem.*, *31*, 3737 (1992).
[l] M. R. Smith, III, P. T. Matsunaga, and R. A. Andersen, *J. Am. Chem. Soc.*, *115*, 7049 (1993).
[m] L. Peng-Ju, H. Sheng-Hua, H. Kun-Yao, W. Ru-Ji, and T. C. W. Mak, *Inorg. Chim. Acta*, *175*, 105 (1990).
[n] W. Haase, and H. Hoppe, *Acta Crystallogr. Sect. 13*, *24*, 282 (1968).

derivatives, as illustrated in Fig. 17 (76). The relationship between Ti=O bond length and the O=Ti—N bond angle in this type of complex has been suggested to be "direct," that is, the Ti=O bond length increases as the O=Ti—N bond angle increases (77). Such an effect has been rationalized in terms of a decrease in Ti—O overlap population as the O=Ti—N bond angle increases.

Titanium porphyrin oxo complexes have been known since 1968 (78) and are commonly prepared by methods that include (a) reaction of $(acac)_2Ti(O)$ with $(POR)H_2$ (79), (b) reaction of $(POR)H_2$ with $TiCl_4$ followed by addition

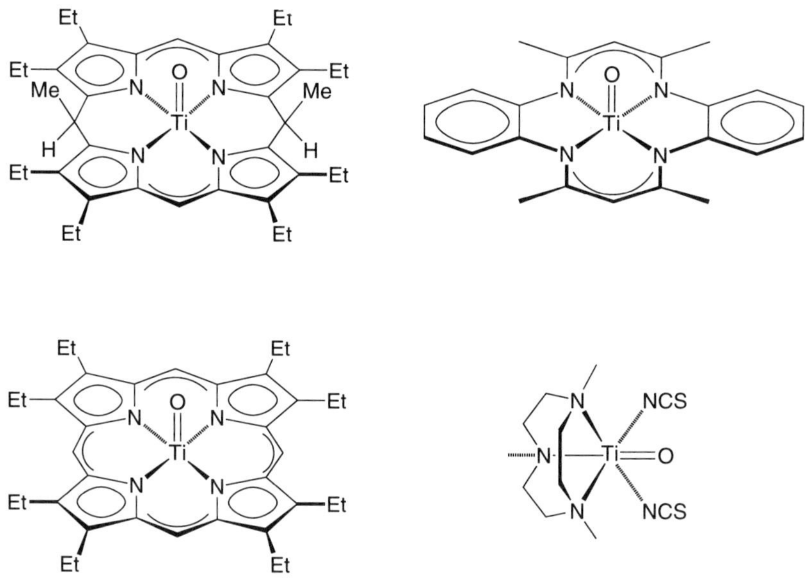

Figure 17. Representative examples of macrocyclic titanium oxo complexes.

of water (80), and (c) hydrolysis of [Ti=NR] species (81, 82). However, relatively few studies have been reported for (POR)TiO complexes, an observation that has been attributed to the low reactivity of the titanyl moiety in these complexes (82). Nevertheless, some reactivity has been described. For example, the peroxo complex (OEP)Ti(η^2-O_2) has been obtained upon reaction of (OEP)TiO with either H_2O_2 or benzoyl peroxide (83). Likewise, (TPP)Ti(η^2-O_2) and (MPOEP)Ti(η^2-O_2) may be obtained by a similar method (83). As expected, the Ti—O bond lengths in (OEP)Ti(η^2-O_2) [1.822(4) and 1.827(4) Å] are considerably longer than that in the oxo derivative [1.613(5) Å] (83). The oxo complex (OEP)TiO may be regenerated from the peroxo complex (OEP)Ti(η^2-O_2) by both electrochemical reduction (83, 84) and oxygen-atom abstraction utilizing the diphenylacetylene complex (TPP)Ti(η^2-Ph_2C_2) (81, 85). A further example of the reversible nature of the reactions involving porphyrin titanium oxo complexes is illustrated by the reversible formation of (POR)TiX_2 (X = F, Cl, Br) via reaction of (POR)TiO with HX.

The most studied macrocyclic titanium oxo complex is that supported by the tetramethyldibenzotetraaza[14]annulene ligand, [η^4-Me_4taa]TiO, and was first synthesized by hydrolysis of [η^4-Me_4taa]TiCl_2 with NH_3(aq) (Scheme 2) (86). The titanyl moiety of [η^4-Me_4taa]TiO is characterized by a ν(Ti=O) absorption at 930 cm^{-1}, a d(Ti=O) bond length of 1.653(3) Å (87), and a resonance at δ 965 ppm in the ^{17}O nuclear magnetic resonance (NMR) spectrum (88).

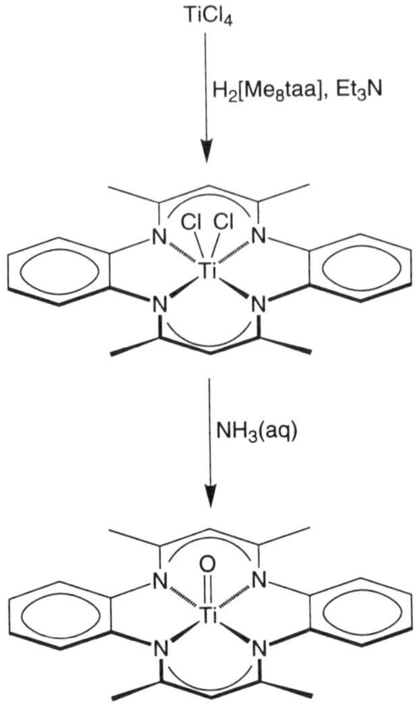

$TiCl_4$

$H_2[Me_8taa]$, Et_3N

$NH_3(aq)$

Scheme 2

The complex $[\eta^4\text{-}Me_4taa]TiO$ has been studied most extensively by Geoffroy and co-workers (88). For example, $[\eta^4\text{-}Me_4taa]TiO$ reacts with 2 equiv of CF_3SO_3H to give the bis triflate derivative $[\eta^4\text{-}Me_4taa]Ti(OSO_2CF_3)_2$ (Scheme 3) (88). In the presence of only 1 equiv of CF_3SO_3H, a bridging oxo complex $[\{[\eta^4\text{-}Me_4taa]Ti]_2(\mu\text{-}O)]^{2+}$ is obtained. The triflate ligands in $[\eta^4\text{-}Me_4taa]Ti(OSO_2CF_3)_2$ are labile and are readily displaced by water to give the aqua complex $\{[\eta^4\text{-}Me_4taa]Ti(OH_2)_2\}[CF_3SO_3]_2$, which has been characterized by X-ray diffraction (88). As would be expected, the Ti—O bond lengths in $\{[\eta^4\text{-}Me_4taa]Ti(OH_2)_2\}^{2+}$ [2.093(7) and 2.112(7) Å] and $[\{[\eta^4\text{-}Me_4taa]Ti]_2(\mu\text{-}O)]^{2+}$ [1.806(11) and 1.830(11) Å] are significantly longer than the Ti=O double bond [1.653(3)] in the oxo complex $[\eta^4\text{-}Me_4taa]TiO$. The aqua complex is easily deprotonated by Et_3N to regenerate the oxo complex, $[\eta^4\text{-}Me_4taa]TiO$, possibly via the hydroxy intermediate $\{[\eta^4\text{-}Me_4taa]Ti(OH)_2\}$. Likewise, the dinuclear complex $[\{[\eta^4\text{-}Me_4taa]Ti\}_2(\mu\text{-}O)]^{2+}$ is cleaved by $NH_3(aq)$ to give $[\eta^4\text{-}Me_4taa]TiO$.

In addition to the above reactions that involve protonation of the oxo group, $[\eta^4\text{-}Me_4taa]TiO$ undergoes a related reaction with $(Me_3Si)_2NH$ to give $\{[\eta^4\text{-}Me_4taa]Ti(OSiMe_3)\}^+$, which has been isolated as the tetraphenylborate deriv-

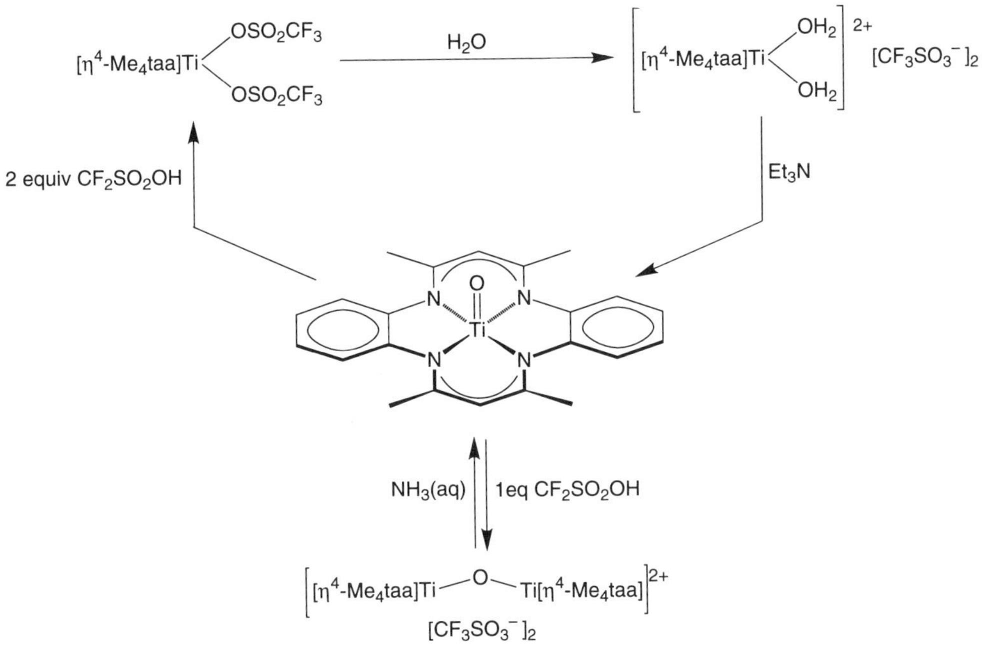

$[\eta^4\text{-}Me_4taa]Ti(OSO_2CF_3)_2$

H_2O

$[[\eta^4\text{-}Me_4taa]Ti(OH_2)_2]^{2+}$ $[CF_3SO_3^-]_2$

Et_3N

2 equiv CF_2SO_2OH

$NH_3(aq)$

1eq CF_2SO_2OH

$[[\eta^4\text{-}Me_4taa]Ti\text{—}O\text{—}Ti[\eta^4\text{-}Me_4taa]]^{2+}$ $[CF_3SO_3^-]_2$

Scheme 3

ative following addition of $NaBPh_4$ (87). In the absence of $NaBPh_4$, the bis(trimethylsiloxide) complex $[\eta^4\text{-}Me_4taa]Ti(OSiMe_3)_2$ is obtained, presumably as a result of ligand redistribution within $[\eta^4\text{-}Me_4taa]Ti(OSiMe_3)\text{-}[N(H)SiMe_3]$.

The titanyl moiety of $[\eta^4\text{-}Me_4taa]TiO$ undergoes a number of [2 + 2] cycloaddition-type reactions with a variety of X=Y double bonds, including the S=O bonds of SO_2 and SO_3 and the C=O bonds of certain activated ketones, as illustrated in Scheme 4 (89). Specifically, the ketones that were observed to react with $[\eta^4\text{-}Me_4taa]Ti{=}O$, for example, $(CF_3)_2CO$, $CF_3C(O)Me$, and $CF_3C(O)Ph$, all exhibit stretching frequencies $\nu(CO) \geq 1717\ cm^{-1}$, whereas those ketones that did not react, for example, Me_2CO, Et_2CO, and MeC(O)CHCHPh, have $\nu(CO) \leq 1712\ cm^{-1}$. It was noted that such a reactivity pattern correlates with the hydration ability of a ketone RC(O)R′ to give $RR'C(OH)_2$. Although the reactions between $[\eta^4\text{-}Me_4taa]TiO$ and ketones may be regarded as formal [2 + 2] cycloadditions, in view of the nucleophilic character of the [Ti=O] oxo moiety, it was proposed that nucleophilic addition of the [Ti=O] oxo moiety to the carbonyl carbon atom of the ketone could be a possible initial step in the cycloaddition reaction.

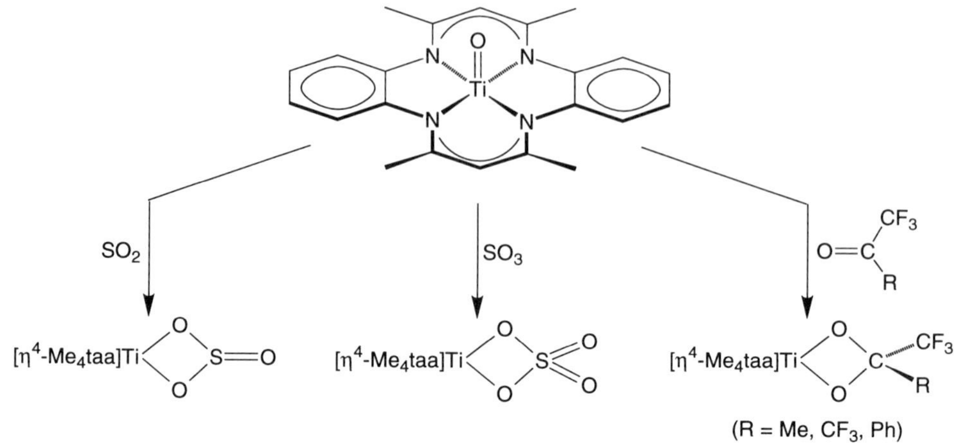

SO$_2$ → $[\eta^4\text{-Me}_4\text{taa}]\text{Ti}(\kappa^2\text{-O}_2\text{SO})$

SO$_3$ → $[\eta^4\text{-Me}_4\text{taa}]\text{Ti}(\kappa^2\text{-O}_2\text{SO}_2)$

O=C(CF$_3$)R → $[\eta^4\text{-Me}_4\text{taa}]\text{Ti}[\kappa^2\text{-O}_2\text{C}(\text{CF}_3)\text{R}]$

(R = Me, CF$_3$, Ph)

Scheme 4

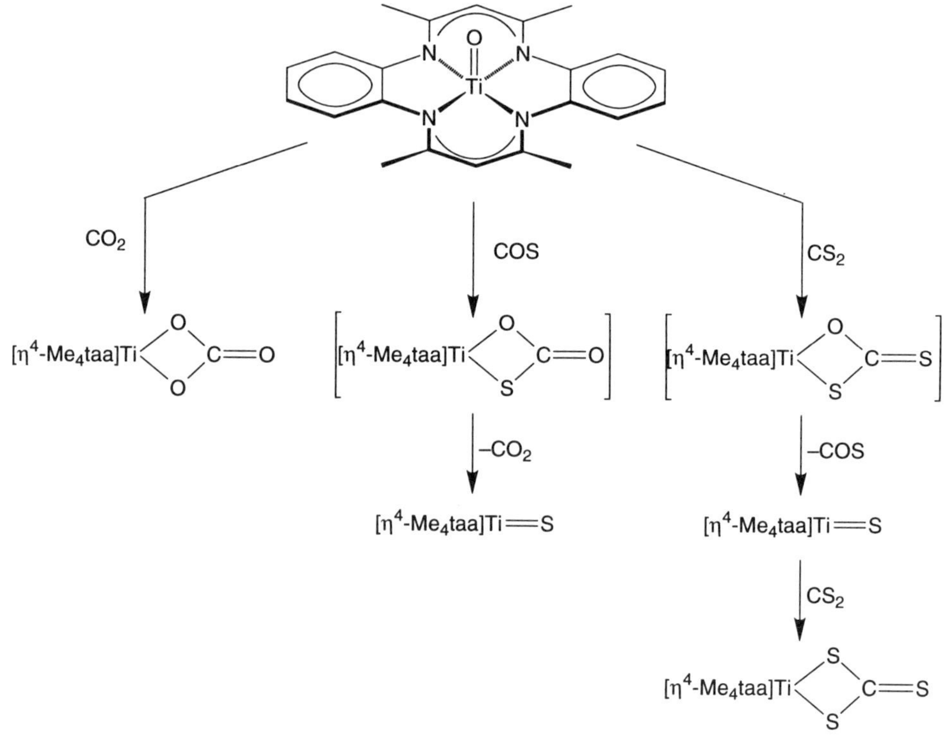

CO$_2$ → $[\eta^4\text{-Me}_4\text{taa}]\text{Ti}(\kappa^2\text{-O}_2\text{CO})$

COS → $[\eta^4\text{-Me}_4\text{taa}]\text{Ti}(\kappa^2\text{-OSCO})$ → –CO$_2$ → $[\eta^4\text{-Me}_4\text{taa}]\text{Ti}{=}\text{S}$

CS$_2$ → $[\eta^4\text{-Me}_4\text{taa}]\text{Ti}(\kappa^2\text{-OSCS})$ → –COS → $[\eta^4\text{-Me}_4\text{taa}]\text{Ti}{=}\text{S}$ → CS$_2$ → $[\eta^4\text{-Me}_4\text{taa}]\text{Ti}(\kappa^2\text{-S}_2\text{CS})$

Scheme 5

The complex $[\eta^4\text{-Me}_4\text{taa}]\text{TiO}$ undergoes a reversible cycloaddition reaction with CO_2 in which it forms a carbonate derivative $[\eta^4\text{-Me}_4\text{taa}]\text{Ti}(\eta^2\text{-O}_2\text{CO})$, as illustrated in Scheme 5. Interestingly, the reaction for the more electron-rich octamethyl derivative $[\eta^4\text{-Me}_8\text{taa}]\text{TiO}$ is shifted more in favor of the carbonate derivative. The reversibility of such cycloaddition reactions is also manifested by the reaction between $[\eta^4\text{-Me}_4\text{taa}]\text{TiO}$ and COS to give the sulfido derivative $[\eta^4\text{-Me}_8\text{taa}]\text{TiS}$. Thus, the formation of $[\eta^4\text{-Me}_8\text{taa}]\text{TiS}$ is proposed to occur via a sequence that involves initial formation of $[\eta^4\text{-Me}_4\text{taa}]\text{Ti}(\eta^2\text{-OSCO})$, followed by retrocycloaddition with elimination of CO_2. Moreover, $[\eta^4\text{-Me}_4\text{taa}]\text{TiO}$ reacts with CS_2 to give the thiocarbonate complex $[\eta^4\text{-Me}_4\text{taa}]\text{Ti}(\eta^2\text{-S}_2\text{CS})$, a transformation that involves initial formation of the sulfido complex $[\eta^4\text{-Me}_4\text{taa}]\text{TiS}$ followed by subsequent reaction with excess CS_2.

The oxo complex $[\eta^4\text{-Me}_4\text{taa}]\text{TiO}$ is also capable of ring-opening epoxides and cleaving anhydrides, as illustrated in Scheme 6. As with the reactions between $[\eta^4\text{-Me}_4\text{taa}]\text{TiO}$ and activated ketones, it has been suggested that the reactions with anhydrides proceed via nucleophilic attack of the titanium oxo group at the carbonyl group of the anhydride.

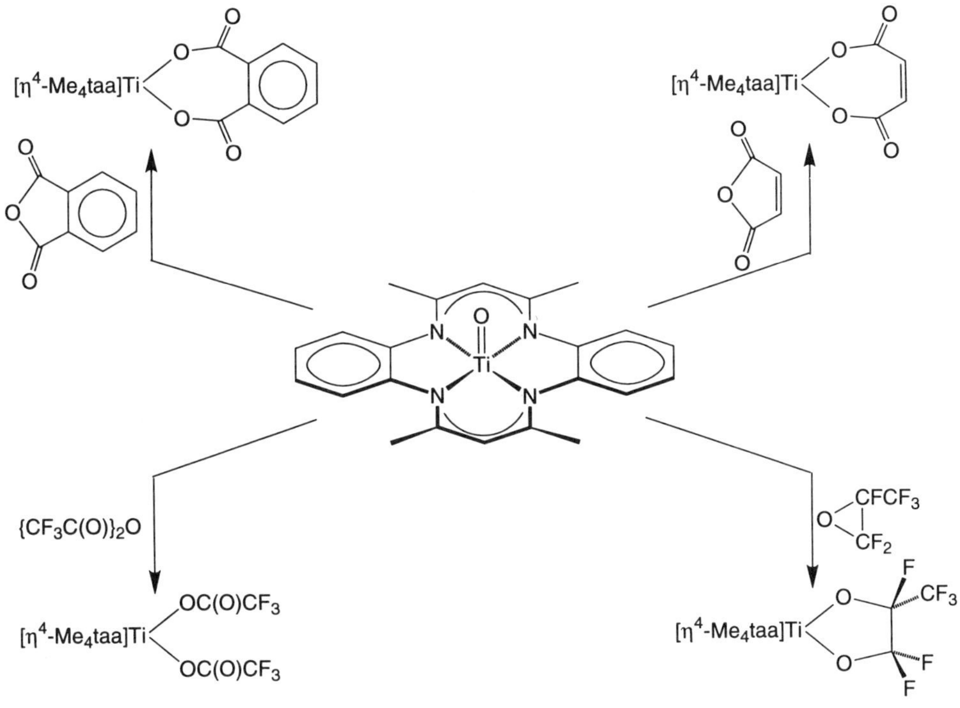

$[\eta^4\text{-Me}_4\text{taa}]\text{Ti}$ $[\eta^4\text{-Me}_4\text{taa}]\text{Ti}$ $\{CF_3C(O)\}_2O$ $CFCF_3$ CF_2 $OC(O)CF_3$ $OC(O)CF_3$ $[\eta^4\text{-Me}_4\text{taa}]\text{Ti}$ CF_3 F F F $[\eta^4\text{-Me}_4\text{taa}]\text{Ti}$

Scheme 6

Although the macrocyclic ligand stabilizes the titanyl moiety with respect to dimerization, the oxo ligand has indeed been shown to act as a bridge to other metal centers in forming heterobinuclear complexes, such as those illustrated in Scheme 7. The bonding in the majority of these complexes is best represented as a Ti=O → M interaction. For example, the Ti—O—Fe interaction in $\{[\eta^4\text{-}Me_4taa]Ti{-}O{-}Fe(salen)\}^+$ is asymmetric, with a Ti—O bond length of 1.701(6) Å and a Fe—O bond length of 2.036(6) Å. The Ti—O bond length is only slightly longer than that in $[\eta^4\text{-}Me_4taa]TiO$, indicating that the Ti—O bond retains much of its multiple-bond character in this complex. A bridging oxo complex is also obtained from the reaction of $[\eta^4\text{-}Me_4taa]TiO$ with $Cp^*W(O)_2Cl$ to give $[\eta^4\text{-}Me_4taa]Ti(Cl)(\mu\text{-}O)W(O)_2Cp^*$. However, in this example, a chloride ligand is also transferred to the titanium, so that the Ti—O interaction is best represented as a single bond.

In addition to tetradentate macrocycles, tridentate macrocycles have also been used to support the [Ti=O] moiety. The first structurally characterized titanyl complex of a tridentate macrocyclic ligand, namely, $[\eta^3\text{-}Me_3tacn]Ti(O)(NCS)_2$, was prepared by oxidation of the Ti^{III} derivative $[\eta^3\text{-}Me_3tacn]Ti(NCS)_3$ using

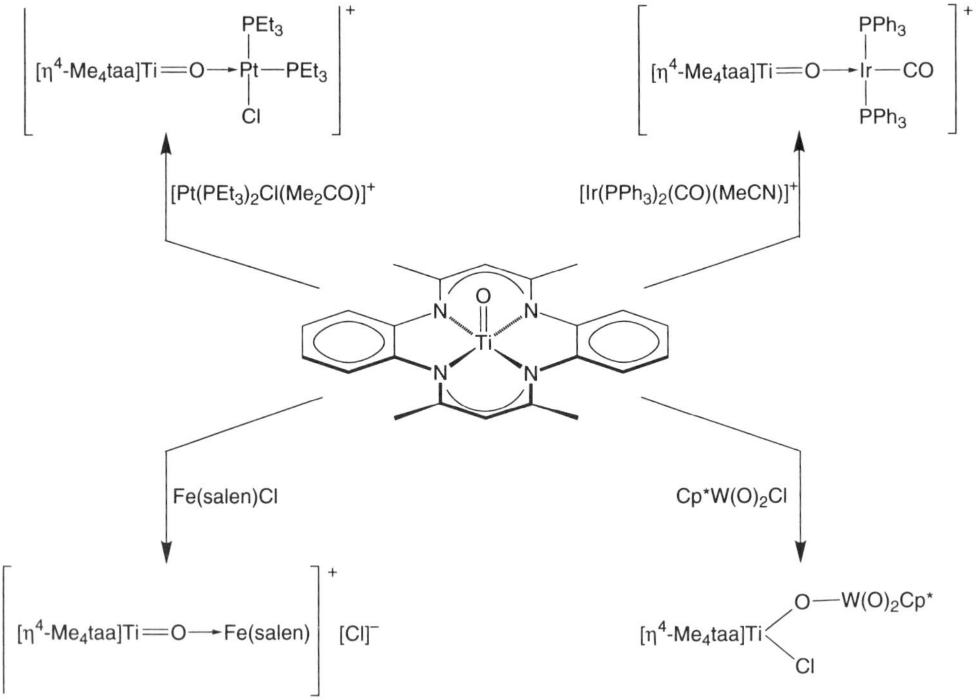

$[[\eta^4\text{-}Me_4taa]Ti{=}O{\rightarrow}Pt(PEt_3)_2Cl]^+$ ← $[Pt(PEt_3)_2Cl(Me_2CO)]^+$

$[[\eta^4\text{-}Me_4taa]Ti{=}O{\rightarrow}Ir(PPh_3)_2(CO)]^+$ ← $[Ir(PPh_3)_2(CO)(MeCN)]^+$

$[[\eta^4\text{-}Me_4taa]Ti{=}O{\rightarrow}Fe(salen)]^+[Cl]^-$ ← Fe(salen)Cl

$[\eta^4\text{-}Me_4taa]Ti(Cl)(O{-}W(O)_2Cp^*)$ ← $Cp^*W(O)_2Cl$

Scheme 7

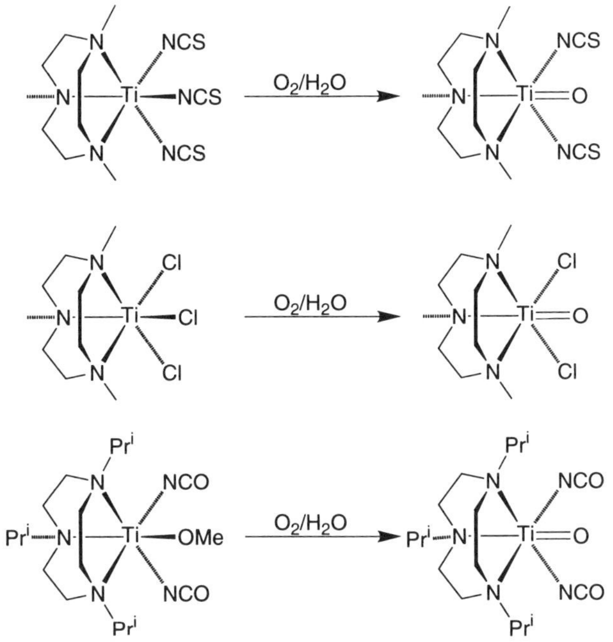

Scheme 8

O_2 in the presence of H_2O (Scheme 8) (90). Related terminal oxo derivatives have been similarly prepared. For example, colorless [η^3-Me_3tacn]Ti(O)Cl_2 and [η^3-Pr^i_3tacn]Ti(O)$(NCO)_2$ are obtained by the reactions of [η^3-Me_3tacn]$TiCl_3$ and [η^3-Pr^i_3tacn]Ti$(NCO)_2$(OMe), respectively, with air.

The oxo complexes [η^3-Me_3tacn]Ti(O)Cl_2 and [η^3-Me_3tacn]Ti(O)$(NCO)_2$ have been proposed to play a role in influencing the crystal structures of several derivatives. For example, the trichloride [η^3-Me_3tacn]$TiCl_3$ has been obtained as blue crystals by the reaction of Me_3tacn with $TiCl_3$ in MeCN at room temperature under anaerobic conditions, while under reflux, green crystals were obtained. It was originally concluded that the green crystals were compositionally disordered with the oxo impurity [η^3-Me_3tacn]Ti(O)Cl_2, since the green form exhibited an absorption at 938 cm^{-1} in the IR spectrum, which was absent in the blue form. However, when isolated in pure form, it was discovered that the oxo complex [η^3-Me_3tacn]Ti(O)Cl_2 (which does exhibit an absorption at 933 cm^{-1} in the IR spectrum) is *colorless*, and so it alone cannot be responsible for the green color. It was, therefore, subsequently proposed that the green form of [η^3-Me_3tacn]$TiCl_3$ is actually a *ternary* mixture of [η^3-Me_3tacn]$TiCl_3$, [η^3-Me_3tacn]Ti(O)Cl_2, and orange-red [η^3-Me_3tacn]Ti(η^2-O_2)Cl_2. However, the level of impurities was insufficient to exhibit a marked effect on the X-ray structure of the green form. In addition to [η^3-Me_3tacn]$TiCl_3$, the peroxo complex [η^3-Me_3tacn]Ti(η^2-O_2)$(NCO)_2$ also suffers from compositional disorder. Specifically, vibrational spectroscopy and electrochemistry have demonstrated that crystals of [η^3-Me_3tacn]Ti(η^2-O_2)$(NCO)_2$ are contaminated with the oxo derivative [η^3-Me_3tacn]Ti(O)$(NCO)_2$. The effect of the compositional disorder is to shorten the O—O bond distance as determined by X-ray crystallography to 1.348(5) Å, compared to the range 1.42–1.48 Å cited for other titanium peroxo complexes. Other than the anomalous O—O bond length, it was emphasized that it would have been impossible to establish the presence of the disorder by X-ray crystallography alone. Nevertheless, although X-ray crystallography alone did not allow the compositional disorder to be detected for [η^3-Me_3tacn]Ti(η^2-O_2)$(NCO)_2$, it was identified for the related derivative [η^3-Me_3tacn]Ti(η^2-O_2)Cl_2, which was contaminated with an appreciable amount (~50%) of oxo impurity, [η^3-Me_3tacn]Ti(O)Cl_2.

ii. Titanocene Derivatives. The permethyltitanocene oxo complexes Cp^*_2Ti(O)(NC_5H_4R) (R = H, Ph) have been obtained by Andersen and coworkers (91) from the reactions of either Cp^*_2Ti or Cp^*_2Ti(η^2-C_2H_4) with N_2O in the presence of the appropriate pyridine (Scheme 9). Pyridine is essential to stabilizing the permethyltitanocene oxo moiety and in its absence Bottomley et al. (92) had previously obtained a dinuclear species $[Cp^*Ti]_2$(μ-η^5, η^1-$C_5Me_4CH_2$)(μ-O)$_2$ from the reaction of Cp^*_2Ti with N_2O. The complex Cp^*_2Ti(O)(NC_5H_5) is characterized by a ν(Ti=O) stretch at 852 cm^{-1} in the IR spectrum, shifting to 818 cm^{-1} upon ^{18}O isotopic substitution. The structure of

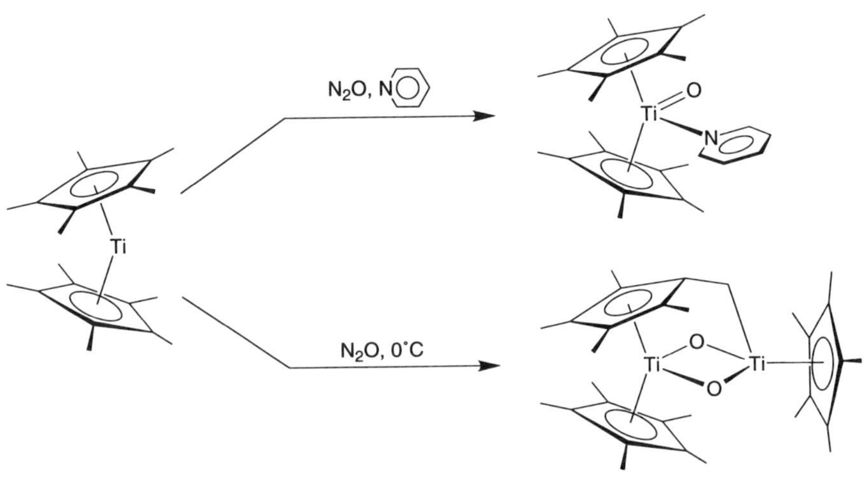

Scheme 9

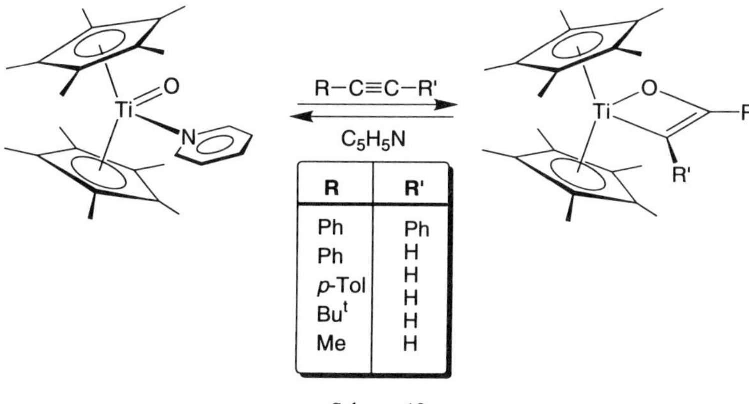

Scheme 10

the phenylpyridine analogue $Cp_2^*Ti(O)(NC_5H_4Ph)$ has been determined by X-ray diffraction [$d(Ti{=}O)$ = 1.665(3) Å].

The oxo complex $Cp_2^*Ti(O)(NC_5H_5)$ undergoes [2 + 2] cycloaddition reactions with acetylenes to give $Cp_2^*Ti[\eta^2\text{-}OC(R)C(R)]$ (Scheme 10) (93). Interestingly, the cycloaddition reactions are reversible and addition of excess pyridine regenerates the oxo derivative $Cp_2^*Ti(O)(NC_5H_5)$. Furthermore, the products derived from terminal acetylenes are converted to their hydroxy-acetylide tautomers at 45–75°C via a mechanism that involves initial retrocyclization generating $[Cp_2^*Ti(O)]$ followed by addition of the acetylinic C—H bond to the [Ti=O] moiety (Scheme 11). $Cp_2^*Ti(O)(NC_5H_5)$ also undergoes a [2 + 2] cycloaddition reaction with allene to give $Cp_2^*Ti[\eta^2\text{-}OC({=}CH_2)CH_2]$ (Scheme 12) (94). The latter complex is unstable in solution and decomposes to an enolate derivative $Cp^*(\eta^5, \eta^1\text{-}C_5Me_4CH_2)Ti[OC(CH_2)Me]$. Furthermore, $Cp_2^*Ti[\eta^2\text{-}OC({=}CH_2)CH_2]$ reacts with H_2 to give the enolate–hydride complex $Cp_2^*TiH[OC(CH_2)Me]$.

iii. Miscellaneous Derivatives. Some anionic terminal titanium oxo complexes have been prepared, although the degree of multiple-bond character in these complexes is questionable. For example, the tris(carbonato) titanium oxo complex $[(\eta^2\text{-}OCO_2)_3TiO]^{4-}$ has been isolated as the guanidinium salt $[C(NH_2)_3]_4[(\eta^2\text{-}OCO_2)_3TiO]$ by the reaction between $TiCl_4$ and $KHCO_3$ in the presence of $[C(NH_2)_3]_2(CO_3)$ (95). The complex $[Et_4N]_2[Ti(O)Cl_4]$ has been synthesized by the reactions of either $Ti_2OCl_6 \cdot 4(MeCN)$ or $Ti_4O_6Cl_8 \cdot 7(dioxane)$ with $[Et_4N]Cl$ (96) and is characterized by $\nu(Ti{=}O)$ = 975 cm^{-1} and $d(Ti{=}O)$ = 1.79 Å (97). However, the multiple-bond character of this complex should be regarded as circumspect, especially in view of the fact that the Ti=O bond is exceptionally long (see Tables I and V).

Scheme 11

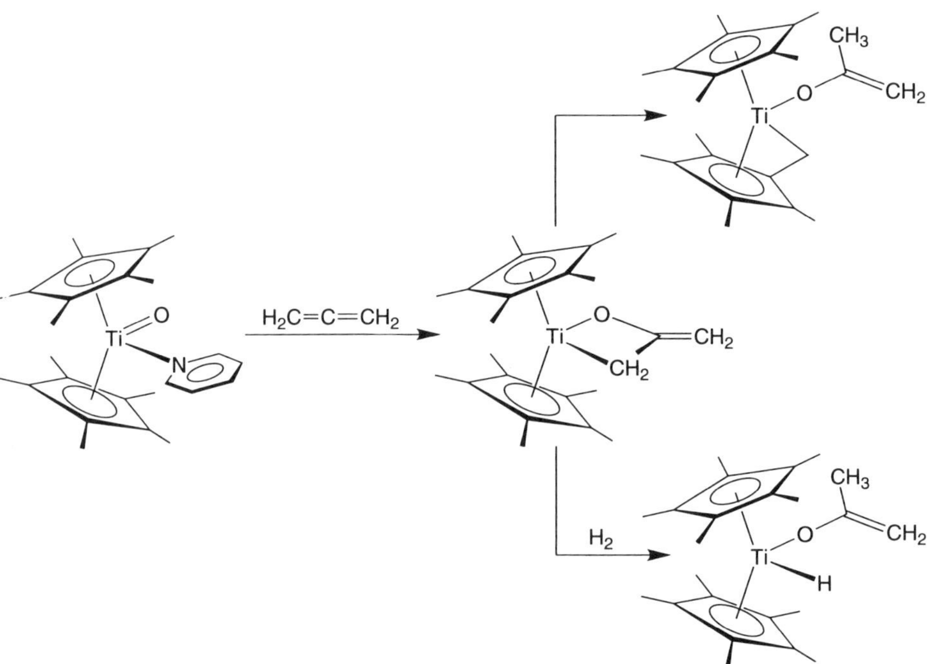

Scheme 12

b. Titanium Sulfido Complexes

i. Macrocyclic Derivatives. The most highly studied terminal titanium sulfido complex is the macrocyclic derivative [η^4-Me_4taa]TiS, prepared by the reaction of [η^4-Me_4taa]$TiCl_2$ with H_2S in the presence of Et_3N (Scheme 13) (86, 88). The sulfido complex is water sensitive and readily converted to the oxo derivative in the presence of traces of water (Scheme 14). In many regards, the reactivity of [η^4-Me_4taa]TiS parallels that of its oxo analogue. For example, the bis(triflate) complex [η^4-Me_4taa]Ti(OSO_2CF_3)$_2$ is obtained upon reaction of [η^4-Me_4taa]TiS with $Me_3SiOSO_2CF_3$, presumably via the intermediate {[η^4-Me_4taa]Ti($SSiMe_3$)(OSO_2CF_3)}.

The complex [η^4-Me_4taa]TiS also undergoes overall [2 + 2] cycloaddition reactions with the activated ketones $CF_3C(O)Me$ and $(CF_3)_2CO$ to give four-membered metallacycles (Scheme 14). Likewise, [η^4-Me_4taa]TiS reacts with CS_2 to give [η^4-Me_4taa]Ti(η^2-S_2CS), but does not react with CO_2 under comparable conditions.

The porphyrin sulfido complex (TTP)TiS has been prepared by the reaction of (TTP)Ti(η^2-Ph_2C_2) with PPh_3S, and is characterized by ν(Ti=S) at 572 cm^{-1} (81). In contrast, the corresponding reaction of (TTP)Ti(η^2-Ph_2C_2) with elemental sulfur gives the disulfido complex (TTP)Ti(η^2-S_2), which can be independently prepared by the reaction of (TTP)TiS with S_8. Finally, the titanium disulfido complex (OEP)Ti(η^2-S_2) may be converted to the terminal sulfido complex (OEP)TiS by using (TTP)Ti(η^2-Ph_2C_2) to abstract one of the sulfur atoms (85).

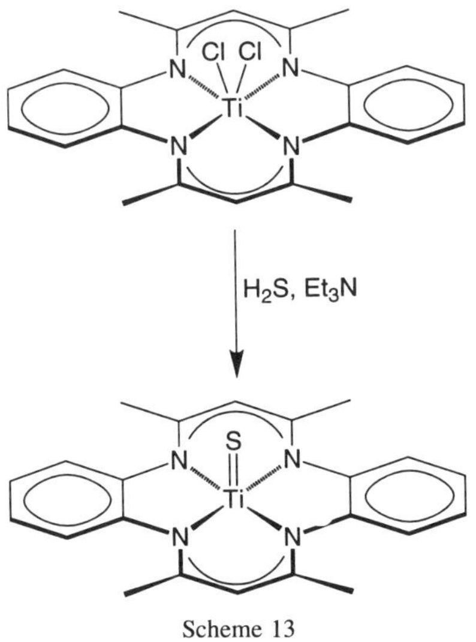

Scheme 13

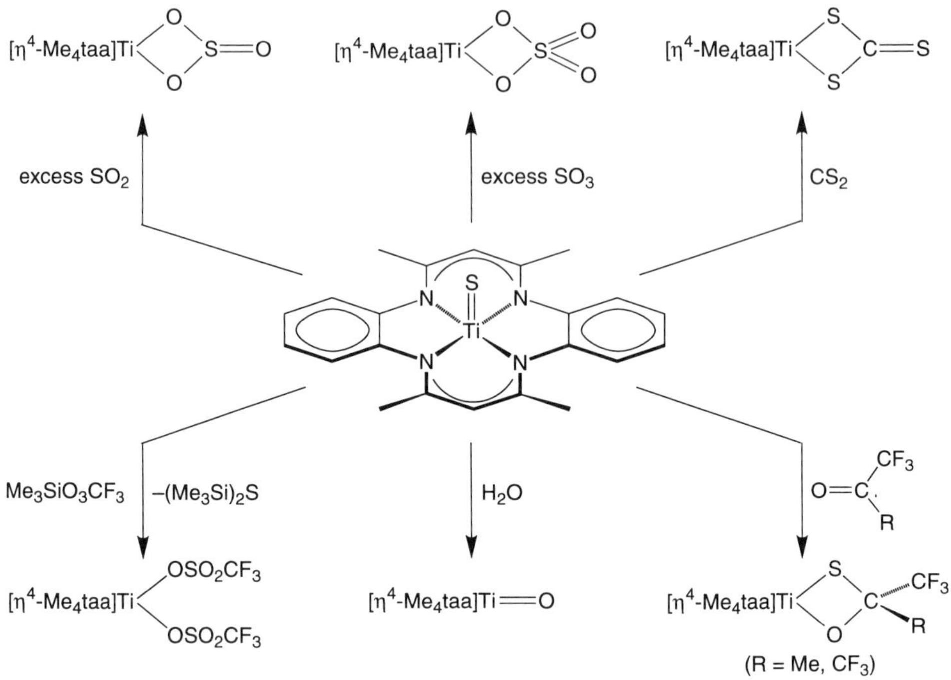

Scheme 14

ii. ***Miscellaneous Derivatives.*** The complexes $[Et_4N]_2[Ti(S)Cl_4]$ and $[Ph_4N]_2[Ti(S)Cl_4]$ have been synthesized by the reactions of $TiSCl_2$ with Et_4NCl and Ph_4PCl, respectively (98). The structure of $[Et_4N]_2[Ti(S)Cl_4]$ has been determined by X-ray diffraction, and the square pyramidal anion is characterized by a Ti=S bond length of 2.11(2) Å and ν(Ti=S) of 529 cm^{-1}. As with the oxo analogue, the degree of multiple-bond character in these complexes is uncertain. More recently, $Na_2[CpTi(\mu\text{-}S)(S)]_2$ has been obtained by the reaction of $Cp_2Ti(SH)_2$ with NaH, and is characterized by Ti=S bond lengths of 2.202(1) and 2.187(1) Å and ν(Ti=S) of 466 cm^{-1} (99).

C. Titanium Selenido Complexes. The selenido complex (TTP)TiSe has been prepared by the reaction of (TTP)Ti(η^2-Ph_2C_2) with Ph_3PSe (81). As with the sulfido analogue, (TTP)TiSe is not obtained by the reaction of (TTP)Ti(η^2-Ph_2C_2) with elemental selenium, which gives the diselenido complex (TTP)(η^2-Se_2) instead. The complex (TTP)TiSe is characterized by an IR absorption ν(Ti=Se) at 465 cm^{-1}, compared to 572 cm^{-1} for ν(Ti=S) of the sulfido derivative. Electrochemical studies indicate that the selenido complex (TTP)-TiSe is 0.13 V easier to oxidize than its sulfido analogue (81).

d. Titanium Tellurido Complexes. Terminal tellurido complexes of titanium have yet to be isolated and structurally characterized. Nevertheless, the terminal tellurido complex [$Cp_2^*Ti{=}Te$] has been considered as a possible intermediate in the mercury promoted conversion of $Cp_2^*Ti(\eta^2\text{-}Te_2)$ to $[Cp_2^*Ti]_2(\mu\text{-}Te)$; however, attempts to trap [$Cp_2^*Ti{=}Te$] with Lewis bases were unsuccessful (45c, 100).

2. *Zirconium Chalcogenido Complexes*

In contrast to titanium, zirconium is known to form multiple bonds to each of the chalcogens; oxygen, sulfur, selenium, and tellurium. The variety of supporting ligand complements for zirconium, however, is not as great as that observed for titanium and is almost exclusively restricted to cyclopentadienyl derivatives. In particular, as will be described in more detail below, the presence of bulky substituents on the cyclopentadienyl rings are required to isolate monomeric derivatives in preference to dimeric and trimeric species, for example, $[Cp_2Zr(\mu\text{-}O)]_3$ (101), $[Cp_2Zr(\mu\text{-}S)]_2$ (102–104), $[Cp_2^{Bu^t}Zr(\mu\text{-}S)]_2$ (105-107), $[Cp_2^{Bu^t}Zr(\mu\text{-}Se)]_2$ (105–108), and $[Cp_2^{Bu^t}Zr(\mu\text{-}Te)]_2$ (106a, 109–111).

Zirconium Oxo Complexes

*i. **Zirconocene Derivatives.*** Terminal zirconium oxo complexes have been isolated only in recent years. Indeed, the elusive nature of such complexes was emphasized by Holm and co-workers in 1990 (112). The majority of studies in this area has been concerned with pentamethylcyclopentadienyl derivatives, with one of the earliest discussions considering the potential of $Cp_2^*Zr(OH)_2$ and $Cp_2^*Zr(OH)Cl$ as precursors to a terminal zirconium oxo complex (113). However, the only evidence that such complexes may be a source of an oxo complex was provided by the observation of the $[Cp_2^*ZrO]^+$ fragment in the mass spectrum of $Cp_2^*Zr(OH)_2$. The first substantial evidence for the presence of a terminal zirconium oxo complex in a chemical system was described by Bergman and co-workers (114), who reported mechanistic studies indicating that the zirconium oxo species [$Cp_2^*Zr{=}O$] is generated as a reactive intermediate by both (a) elimination of benzene from $Cp_2^*Zr(OH)(Ph)$ and (b) deprotonation of $Cp_2^*Zr(OH)(OSO_2CF_3)$, as illustrated in Scheme 15. For example, *in situ* generated [$Cp_2^*Zr{=}O$] could be trapped by nitriles and diphenylacetylene to yield the oxametallacycles $Cp_2^*Zr[\eta^2\text{-}OC(R){=}NC(R){=}N]$ (R = Ph, Bu^t) and $Cp_2^*Zr[\eta^2\text{-}OC(Ph){=}C(Ph)]$, respectively.

Bergman and co-worker (115) also explored the possibility of generating the less-substituted [$Cp_2Zr{=}O$] intermediate by (a) [4 + 2] retrocycloaddition of the azaoxametallacycle $Cp_2Zr[\eta^2\text{-}N(Ar)C(Me){=}C(Ph)CH(Ar')O]$ (115), and (b) imido–oxo exchange between $Cp_2Zr(NBu^t)(thf)$ and certain carbonyl compounds, such as $Ph_2C{=}O$, $Ph_2C{=}C{=}O$, $Bu^tN{=}C{=}O$, or $CpCo(CO)_2$ (116),

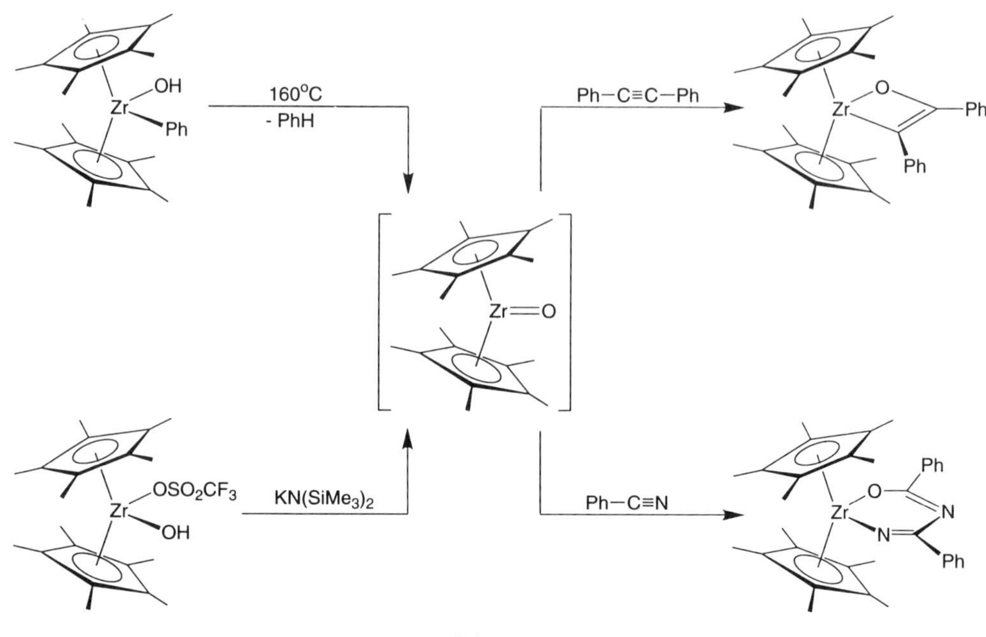

Scheme 15

as illustrated in Scheme 16. Carbonyl compounds with α-H substituents (e.g., $Bu^tC(O)CH_3$ and Pr^i_2CO), however, do not partake in imido–oxo exchange, but rather yield enolate derivatives (Scheme 17). Interestingly, the fate of the presumed $[Cp_2Zr{=}O]$ intermediate in the above reactions (Scheme 16) is dependent on the method of generation and it has been suggested that only the [4 + 2] retrocycloaddition of the azaoxometallacycle $Cp_2Zr[\eta^2\text{-N(Ar)C(Me)=C(Ph)CH(Ar')O}]$ actually yields free $[Cp_2Zr{=}O]$. For instance, whereas $Cp_2Zr(NBu^t)(thf)$ reacts with $Ph_2C{=}O$ to give $[Cp_2ZrO]_n$ as a mixture of oligomers (of which the trimer $[Cp_2Zr(\mu\text{-O})]_3$ is a major component), only the mixed (μ-oxo)(μ-imido) dimer $[Cp_2Zr]_2(\mu\text{-O})(\mu\text{-NBu}^t)$ is obtained from the reaction with metal carbonyls, for example, $CpCo(CO)_2$ (Scheme 16). Furthermore, the reaction of $Cp_2Zr(NBu^t)(thf)$ with $Ph_2C{=}C{=}O$ yields neither $[Cp_2ZrO]_n$ nor $[Cp_2Zr]_2(\mu\text{-O})(\mu\text{-NBu}^t)$, but rather gives $[Cp_2Zr]_2(\mu\text{-O})(\mu\text{-}\eta^2\text{-}O_2C{=}CPh_2)$. In contrast, although thermolysis of $Cp_2Zr[\eta^2\text{-N(Ar)C(Me)=C(Ph)CH(Ar')O}]$ in the absence of traps also gives oligomeric $[Cp_2ZrO]_n$, the presumed monomeric oxo species $[Cp_2ZrO]$ may be trapped by Cp_2ZrMe_2 giving $[Cp_2ZrMe]_2(\mu\text{-O})$.

The first terminal zirconium oxo complexes to be isolated used pyridine as a stabilizing ligand, in a fashion analogous to that described by Bergman for $Cp^*_2Zr(S)(NC_5H_5)$ (see Section IV.B.2) and by Andersen for $Cp^*_2Ti(O)$-

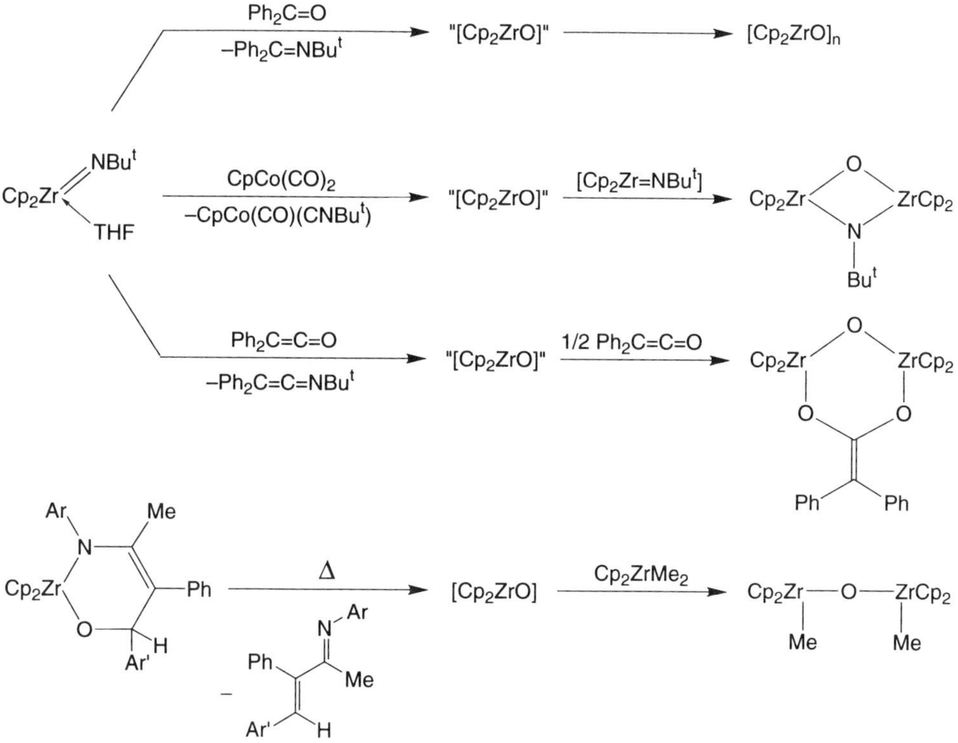

Scheme 16

(NC_5H_5) (see Section IV.A.1). Thus, the zirconium oxo complexes $Cp^R_2Zr(O)(NC_5H_5)$ ($Cp^R = Cp^*$, Cp^{Et*}) were synthesized by the reactions of $Cp^R_2Zr(CO)_2$ with N_2O and pyridine at 80°C (Scheme 18), with the structure of the latter having been determined by X-ray diffraction (Fig. 18) (32, 54, 117). As noted in Section III.A, the zirconium–oxo interactions in $Cp^R_2Zr(O)(NC_5H_5)$ have lit-

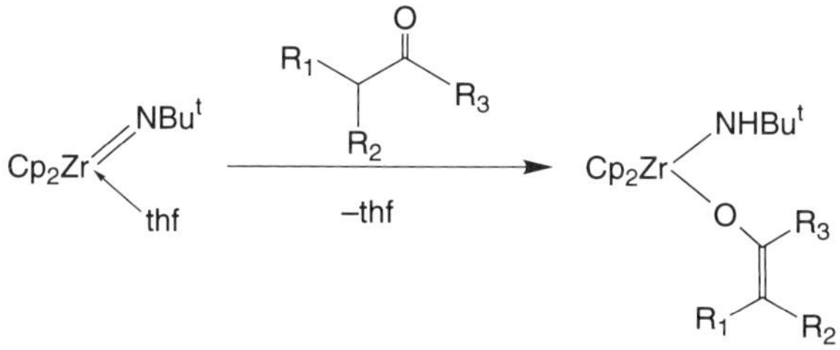

Scheme 17

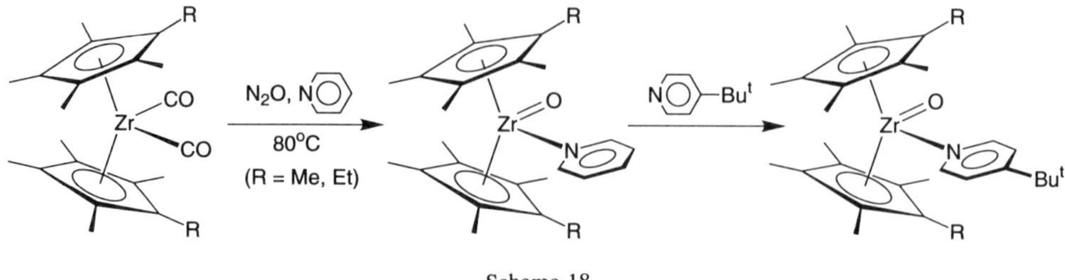

Scheme 18

tle triple-bond $\bar{Zr}\equiv\overset{+}{O}$ character due to both (a) the 18-electron nature of the zirconium centers, and (b) the commonly used MO description for bent metallocene derivatives. The Zr=O bond length and ν(Zr=O) IR absorption data for the above oxo complexes and related Group 4 (IV B) derivatives are listed in Tables VI and VII.

Pyridine is critical to the stability of the terminal oxo zirconocene moiety. In its absence, a species tentatively identified as the bridging oxo complex $[Cp^*Zr]_2(\mu\text{-}\eta^5,\eta^1\text{-}C_5Me_4CH_2)(\mu\text{-}O)_2$ is obtained from the reaction of $Cp_2^*Zr(CO)_2$ with N_2O; see Section IV.A.1 for the titanium analogue $[Cp^*Ti]_2(\mu\text{-}\eta^5,\eta^1\text{-}C_5Me_4CH_2)(\mu\text{-}O)_2$. Indeed, pyridine dissociation (and also rotation about the Zr—N bond) from $Cp_2^*Zr(O)(NC_5H_5)$ is slow on the NMR time scale, such that the pyridine ligand is characterized by five inequivalent resonances in its ^1H NMR spectrum. Nevertheless, pyridine exchange does occur on the chemical time scale and $Cp_2^*Zr(O)(NC_5H_5)$ reacts with excess 4-*tert*-butylpyridine to give $Cp_2^*Zr(O)(NC_5H_4Bu^t)$ (Scheme 18).

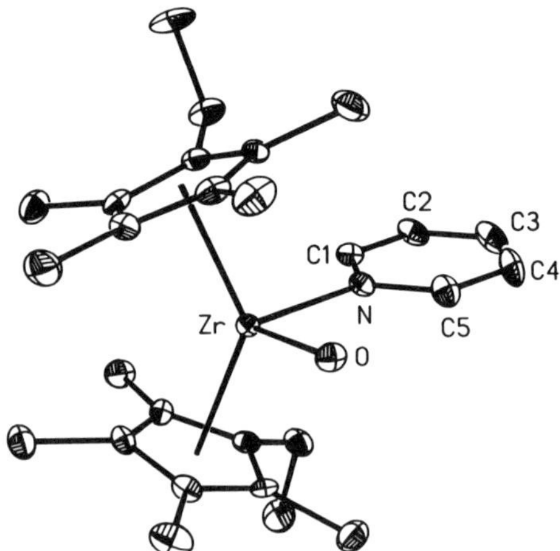

Figure 18. Molecular structure of $Cp_2^{Et*}Zr(O)(NC_5H_5)$. [Reprinted with permission from W. Howard and G. Parkin, *J. Am. Chem. Soc.*, *116*, 606 (1994). Copyright © 1994 American Chemical Society.]

TABLE VI
ν(M = O) Stretching Frequencies for Group 4 (IVB) Terminal Oxo Complexes

Complex	ν(M = O)/cm^{-1}	Reference
$Cp_2^*Ti(O)(NC_5H_5)$	852	*a*
$Cp_2^*Zr(O)(NC_5H_5)$	780	*b*
$Cp_2^*Zr(O)(NC_5H_4Bu^t)$	784	*b*
$Cp_2^{Et}*Zr(O)(NC_5H_5)$	773	*b*
$Cp_2^*Hf(O)(NC_5H_5)$	773	*c*
$Cp_2^{Et}*Hf(O)(NC_5H_5)$	770	*c*
$[\{\eta^5\text{-}\eta^1\text{-}\eta^5\text{-}\eta^1\text{-}Et_8C_4(C_5H_2N)_3(C_5H_3N)ZrO\}_2(\mu\text{-}K)_2]$	780	*d*
$[\{\eta^1\text{-}\eta^1\text{-}\eta^5\text{-}\eta^1\text{-}Et_7C_4(C_4H_2N)_3(CH_2CH_2C_5H_2N)ZrO\}_2(\mu\text{-}Li)_2]$	776	*e*
$[\{\eta^5\text{-}\eta^1\text{-}\eta^5\text{-}\eta^1\text{-}Et_8C_4(C_4H_2N)_3(3\text{-}CH_2{=}CHC_5H_2N)ZrO\}_2(\mu\text{-}Na)_2]$	785	*f*
$[cis\text{-}Et_8C_4(C_4H_2N)_2(C_5H_3N)(m\text{-}MeC_5H_2N)HfO]$	826	*f*

[a]M. R. Smith, III, P. T. Matsunaga, and R. A. Andersen, *J. Am. Chem. Soc.*, *115*, 7049 (1993).
[b]W. A. Howard, T. M. Trnka, M. Waters, and G. Parkin, *J. Organomet. Chem.*, *528*, 95, (1977).
[c]W. A. Howard and G. Parkin, *J. Organomet. Chem.*, *472*, C1 (1994).
[d]D. Jacoby, C. Floriani, A. Chiesi-Villa, and C. Rizzoli, *J. Am. Chem. Soc.*, *115*, 7025 (1993).
[e]D. Jacoby, S. Isoz, C. Floriani, A. Chiesi-Villa, and C. Rizzoli, *J. Am. Chem. Soc.*, *117*, 2805 (1995).
[f]D. Jacoby, S. Isoz, C. Floriani, A. Chiesi-Villa, and C. Rizzoli, *J. Am. Chem. Soc.*, *117*, 2793 (1995).

In addition to the essential stabilization of $[Cp_2^*Zr{=}O]$ provided by pyridine, the steric demands of the bulky peralkylatedcyclopentadienyl ligands are important for maintaining an unbridged structure. For example, rather than yielding a terminal oxo complex, the reaction of the unsubstituted cyclopentadienyl derivative $Cp_2Zr(CO)_2$ with N_2O in the presence of pyridine at 80°C gives the zirconocene oxo trimer $[Cp_2Zr(\mu\text{-}O)]_3$ (Scheme 19) (101).

TABLE VII
d(M=O) Bond Lengths for Group 4 (IVB) Terminal Oxo Complexes

Complex	*d*(M=O) (Å)	Reference
$Cp_2^*Ti(O)(NC_5H_4Ph)$	1.665(3)	*a*
$Cp_2^{Et}*Zr(O)(NC_5H_5)$	1.804(4)	*b*
$Cp_2^{Et}*Hf(O)(NC_5H_5)$	1.826(9)	*c*
$[\{\eta^5\text{-}\eta^1\text{-}\eta^5\text{-}\eta^1\text{-}Et_8C_4(C_4H_2N)_3(C_5H_3N)ZrO\}_2(\mu\text{-}K)_2]$	1.813(2)	*d*
$[\{\eta^1\text{-}\eta^1\text{-}\eta^5\text{-}\eta^1\text{-}Et_7C_4(C_4H_2N)_3(CH_2CH_2C_5H_2N)\}ZrO\}_2(\mu\text{-}Li)_2]$	1.838(7)	*e*
$[\{\eta^5\text{-}\eta^1\text{-}\eta^5\text{-}\eta^1\text{-}Et_8C_4(C_4H_2N)_3(3\text{-}EtC_5H_2N)\}ZrO\}_2(\mu\text{-}K)_2]$	1.804[2]	*f*

[a]M. R. Smith, III, P. T. Matsunaga, and R. A. Andersen, *J. Am. Chem. Soc.*, *115*, 7049 (1993).
[b]W. A. Howard and G. Parkin, *J. Am. Chem. Soc.*, *116*, 606 (1994).
[c]W. A. Howard and G. Parkin, *J. Organomet. Chem.*, *472*, C1 (1994).
[d]D. Jacoby, C. Floriani, A. Chiesi-Villa, and C. Rizzoli, *J. Am. Chem. Soc.*, *115*, 7025 (1993).
[e]D. Jacoby, S. Isoz, C. Floriani, A. Chiesi-Villa, and C. Rizzoli, *J. Am. Chem. Soc.*, *117*, 2805 (1995).
[f]D. Jacoby, S. Isoz, C. Floriani, A. Chiesi-Villa, and C. Rizzoli, *J. Am. Chem. Soc.*, *117*, 2793 (1995).

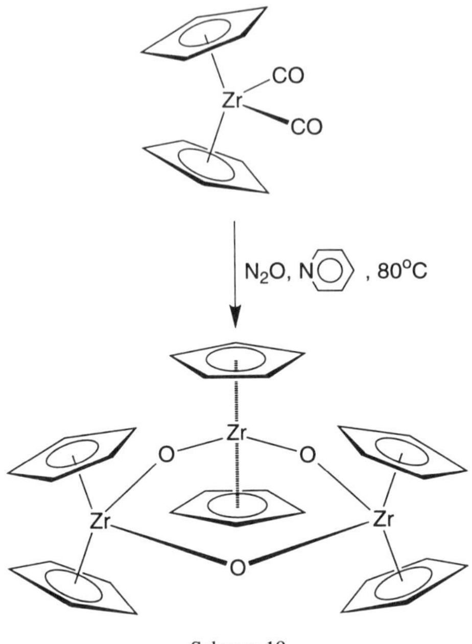

Scheme 19

The reactivity of the permethylzirconocene oxo complex $Cp_2^*Zr(O)(NC_5H_5)$ has been studied extensively. Consistent with X-ray diffraction (54) and computational (118) studies, the reactivity of $Cp_2^*Zr(O)(NC_5H_5)$ is indicative of $[\overset{+}{Zr}-\overset{-}{O}]$ character, as would be expected for an electrophilic d^0 zirconium center coupled to the highly basic, nucleophilic, oxo ligand. Thus, such dipolar character facilitates extensive reactivity of the zirconium oxo multiple bond, especially with polar ($X^{\delta+}-Y^{\delta-}$) substrates through formal 1,2 additions (Scheme 20). The X—Y bonds that partake in such reactions include $H^{\delta+}-Y^{\delta-}$ [Y = Cl; NH_2, NHPh, $N(Ph)NH_2$; OH, SH, SeH; $CH_2C(O)R$]; $X^{\delta+}-H^{\delta-}$ [X = SiH_2Ph], H—H, Me_3Si—Cl, and Me—I. For example, H_2O reacts rapidly and irreversibly with the [Zr=O] group of $Cp_2^*Zr(O)(NC_5H_5)$ to give the dihydroxide $Cp_2^*Zr(OH)_2$. This observation is in marked contrast to that for the related tungsten system, in which $[Cp_2^*W(OH)_2]$ is unstable with respect to Cp_2^*WO and H_2O (see Section IV.C) (51). The origin of the enhanced stability of $Cp_2^*Zr(OH)_2$ compared to that of $[Cp_2^*W(OH)_2]$ is proposed to be a consequence of the 16-electron configuration of the zirconium center in $Cp_2^*Zr(OH)_2$, which promotes oxygen-to-zirconium lone-pair donation and strengthens the Zr—OH interaction (119). An illustration of the particularly high reactivity of $Cp_2^*Zr(O)(NC_5H_5)$ is provided by its facile reactions with relatively nonpolar H—X bonds, such as Si—H and H—H. Thus, the Si—H

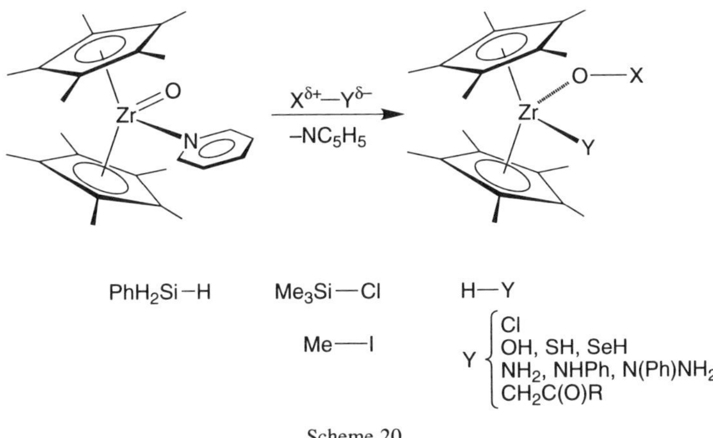

Scheme 20

bond of phenylsilane reacts with $Cp_2^*Zr(O)(NC_5H_5)$ at room temperature to give the siloxy-hydride derivative, $Cp_2^*Zr(H)(OSiH_2Ph)$, and with H_2 at 80°C to give the hydroxy hydride complex $[Cp_2^*Zr(H)](\mu\text{-}O)[Cp_2^*Zr(OH)]$. The [Zr=O] moiety is also capable of undergoing 1,2 additions with the activated C—H bonds of methyl ketones to give oxygen-bound enolate derivatives, $Cp_2^*Zr(OH)[\eta^2\text{-}OC(R)=CH_2]$. A possible mechanism for the formation of $Cp_2^*Zr(OH)][\eta^1\text{-}OC(R)=CH_2]$ has been proposed to involve initial coordination of the ketone to the zirconium center, followed by deprotonation via a six-membered ring (Scheme 21).

Cleavage of a C—H bond is also observed in the reaction of $Cp_2^*Zr(O)(NC_5H_5)$ with Bu^tI at room temperature to give the hydroxy-iodide derivative $Cp_2^*Zr(OH)I$, which is accompanied by elimination of isobutene (Scheme 22). By analogy with the reactions involving mcthyl ketones, a potential mechanism for the dehydrohalogenation involves a six-membered transition state via prior coordination to Bu^tI to the zirconium center. Dehydrohalogenation of Bu^tI, rather than 1,2 addition across [Zr=O], provides an interesting difference to the corresponding reaction with MeI which, as illustrated in Scheme 23, gives an alkoxy–iodide product $Cp_2^*Zr(OMe)I$. The formation of the methoxy iodide complex $Cp_2^*Zr(OMe)I$ also contrasts with the reactivity of the related tungstenocene oxo complex, Cp_2^*WO, which gives an oxo-alkyl derivative, $[Cp_2^*W(O)Me][I]$ (51). The difference in reactivities of the zirconium and tungsten systems has been attributed to the d^0 versus d^2 nature of the metal centers, with the d^0 Zr center of $Cp_2^*Zr(O)(NC_5H_5)$ unable to react in the nucleophilic fashion observed for the d^2 tungsten center in Cp_2^*WO.

Whereas methyl ketones are deprotonated by $Cp_2^*Zr(E)(NC_5H_5)$, aldehydes selectively undergo cycloaddition reactions to give six-membered oxametalla-

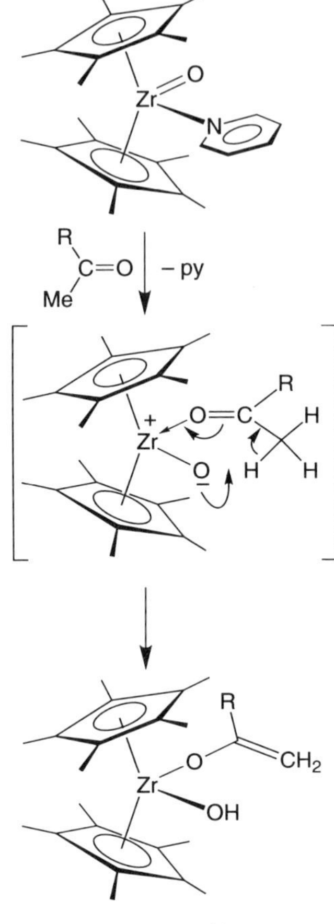

Scheme 21

cycles, $Cp_2^*Zr[\eta^2\text{-}OCH(R)OCH(R)O]$ (R = H, Pr^i, Bu^t) in which the six-membered rings exhibit an approximate chair conformation (Scheme 24). Likewise, $Cp_2^*Zr(O)(NC_5H_5)$ also rapidly undergoes a cycloaddition reaction with PhCN to give the six-membered metallacycle, $Cp_2^*Zr[\eta^2\text{-}OC(Ph)NC(Ph){=}N]$, a complex that was first synthesized by Bergman and co-workers (114) as a result of trapping *in situ* generated $[Cp_2^*Zr{=}O]$ with PhCN. Bergman has also trapped *in situ* generated $[Cp_2^*Zr{=}O]$ with alkynes; simple cycloaddition products $Cp_2^*Zr[\eta^2\text{-}OC(R){=}C(R)]$ are observed if the oxo species is generated at room temperature via deprotonation of $Cp_2^*Zr(OH)(OSO_2CF_3)$ (Scheme 15), but at the elevated temperatures required to generate $[Cp_2^*Zr{=}O]$ from $Cp_2^*Zr(OH)Ph$, subsequent rearrangement of $Cp_2^*Zr[\eta^2\text{-}OC(R){=}C(R)]$ occurs.

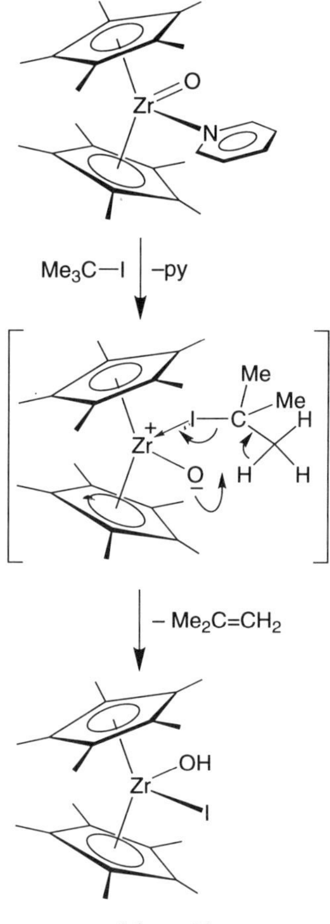

Scheme 22

*ii. **Macrocyclic Derivatives.*** Additional zirconyl derivatives have been described following the isolation of $Cp^R_2Zr(O)(NC_5H_5)$ (Cp^R = Cp*, Cp^{Et*}). Thus, Floriani and co-workers (120–122) reported the anionic oxo complexes $[\{\eta^5\text{-}\eta^1\text{-}\eta^5\text{-}\eta^1\text{-}Et_8C_4(C_4H_2N)_3(C_5H_3N)\}ZrO\}_2(\mu\text{-}K)_2]$, $[\{\eta^5\text{-}\eta^1\text{-}\eta^5\text{-}\eta^1\text{-}Et_8\text{-}C_4(C_4H_2N)_3(3\text{-}EtC_5H_2N)\}ZrO\}_2(\mu\text{-}K)_2]$, and $[\{\eta^1\text{-}\eta^1\text{-}\eta^5\text{-}\eta^1\text{-}Et_7C_4(C_4H_2N)_3\text{-}(CH_2CH_2C_5H_2N)\}ZrO\}_2(\mu\text{-}Li)_2]$, as illustrated in Scheme 25. For example, $[\{\eta^5\text{-}\eta^1\text{-}\eta^5\text{-}\eta^1\text{-}Et_8C_4(C_4H_2N)_3(C_5H_3N)\}ZrO\}_2(\mu\text{-}K)_2]$ is obtained from the porphyrinogen complex $[\{\eta^5\text{-}\eta^1\text{-}\eta^5\text{-}\eta^1\text{-}Et_8C_4(C_4H_2N)_4\}Zr(thf)]$ by the reaction with KH followed by CO (123, 124). The formation of the [Zr=O] moiety by this reaction is a rare example of the cleavage of the C—O bond of carbon monoxide (125). Interestingly, the reaction is also accompanied by the carbon atom of

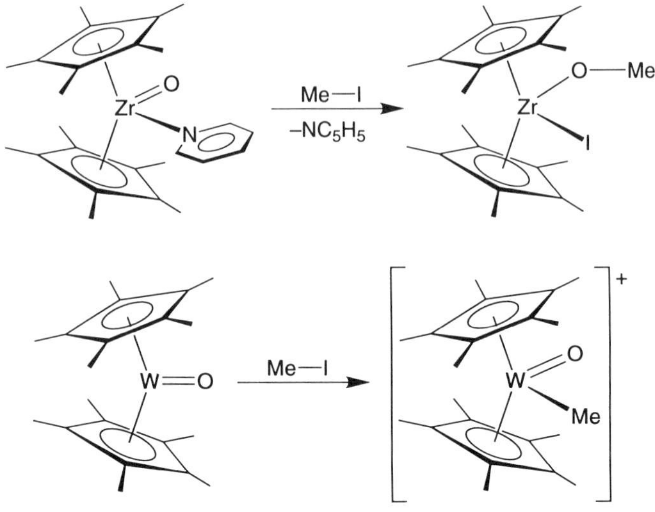

Scheme 23

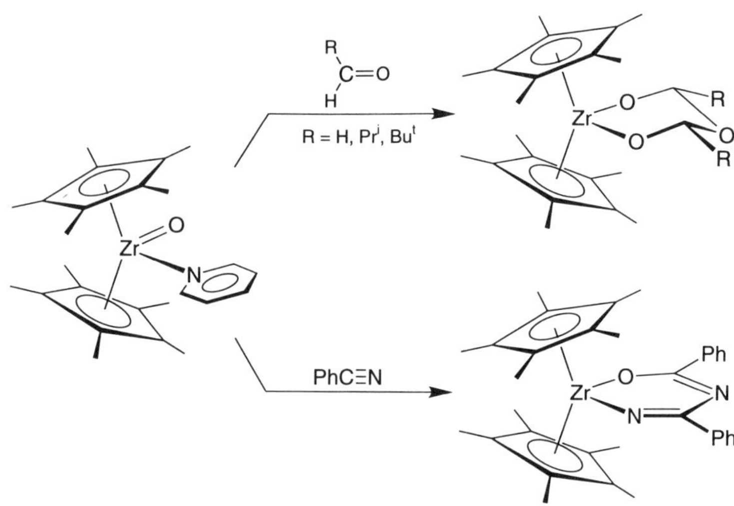

Scheme 24

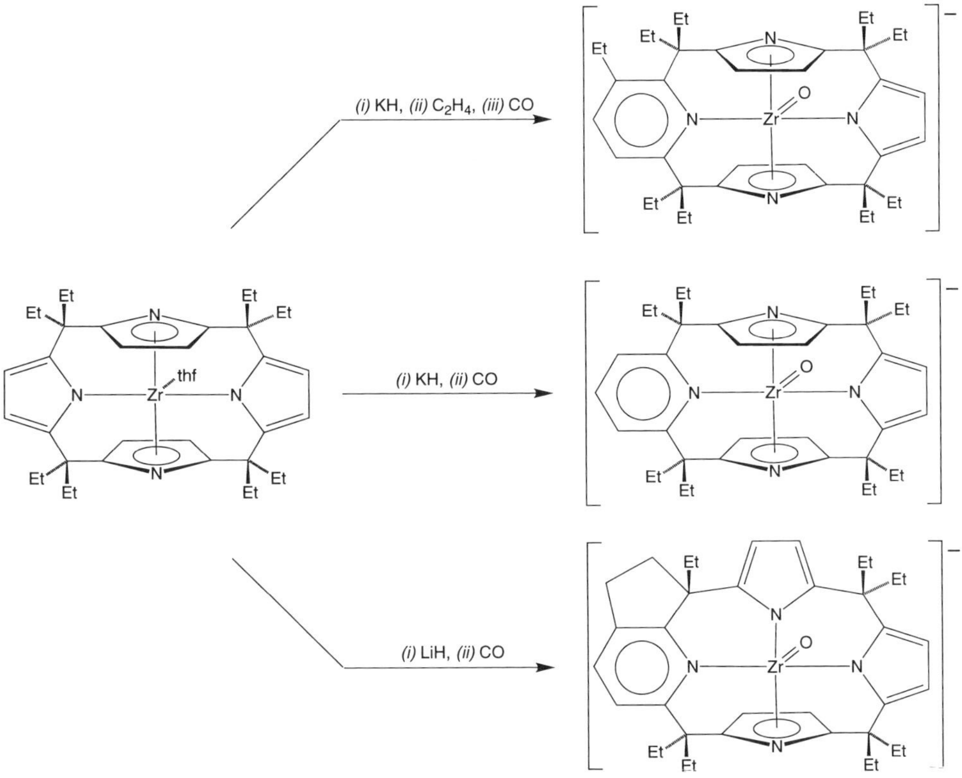

Scheme 25

CO being integrated into one of the five-membered pyrrole groups to form a pyridine nucleus. If the reaction is modified such that ethylene or acetylene is added prior to CO, an ethyl or vinyl moiety, respectively, is incorporated at the meta position of the pyridine fragment. Moreover, sequential reaction of $[\{\eta^5\text{-}\eta^1\text{-}\eta^5\text{-}\eta^1\text{-}Et_8C_4(C_4H_2N)_4\}Zr(thf)]$ with LiH and CO also results in homologation of the pyrrole fragment and formation of an anionic zirconium oxo complex $[\{\eta^1\text{-}\eta^1\text{-}\eta^5\text{-}\eta^1\text{-}Et_7C_4(C_4H_2N)_3(CH_2CH_2H_2N)\}ZrO\}_2(\mu\text{-}Li)_2]$, accompanied by functionalization of one of the porphyrinogen ethyl groups (Scheme 25) (126). The structures of these anionic zirconyl complexes have been determined by X-ray diffraction, demonstrating that the oxo ligands interact with the countercations, giving rise to dinuclear structures that are linked by four-membered $[M_2O_2]$ rings (M = K, Li). For example, the zirconium oxo moiety of the anion $[\{\eta^5\text{-}\eta^1\text{-}\eta^5\text{-}\eta^1\text{-}Et_8C_4(C_4H_2N)_3(C_5H_3N)\}ZrO]^-$ interacts with the K^+ counterion in a dinuclear arrangement via a four-membered $[K_2O_2]$ ring, with K—O bond lengths of 2.607(3) and 2.611(3) Å. Despite such interaction with the alkali metals, the Zr=O bond lengths and ν(Zr=O) vibrations are comparable to those of the neutral zirconocene derivatives summarized in Tables VI and VII. Finally, other terminal zirconium oxo complexes have also been reported, but definitive characterization is lacking (127).

b. Zirconium Sulfido Complexes. The first evidence for the generation of terminal zirconium sulfido complexes was reported by Tainturier et al. (105) who suggested that the redistribution of $[Cp_2^{Bu^t}Zr](\mu\text{-}S)_2[ZrCp_2]$ into

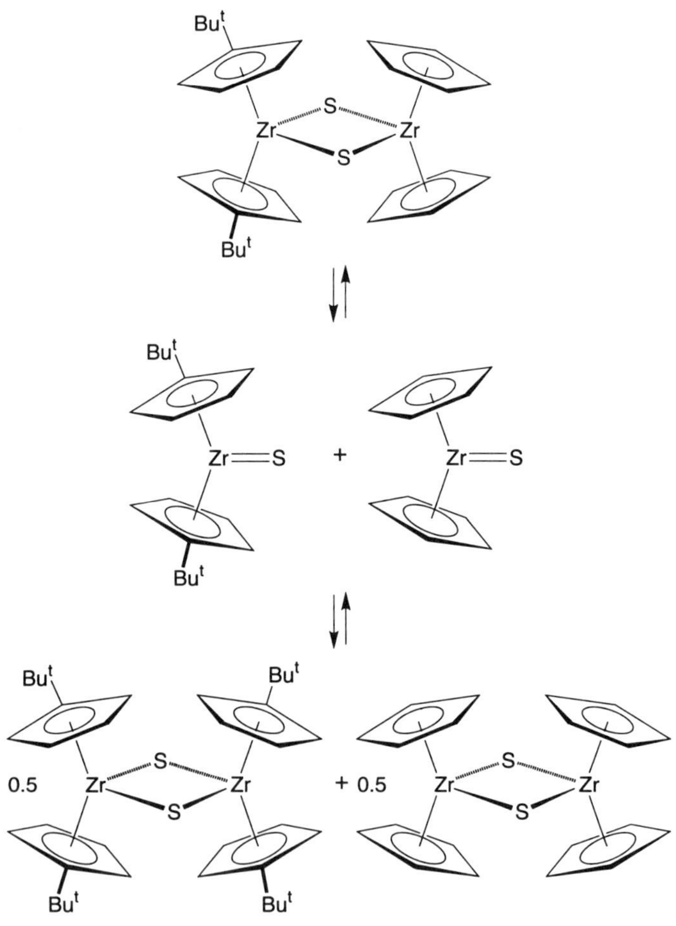

Scheme 26

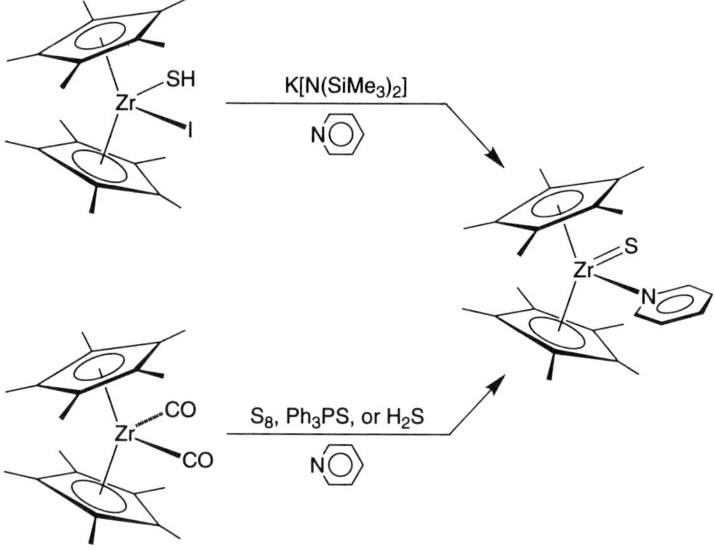

Scheme 27

$[Cp_2^{Bu^t}Zr(\mu\text{-}S)]_2$ and $[Cp_2Zr(\mu\text{-}S)]_2$ proceeded via dissociation into the intermediates $[Cp_2^{Bu^t}Zr{=}S]$ and $[Cp_2Zr{=}S]$ ($Cp^{Bu^t} = \eta^5\text{-}C_5H_4Bu^t$), as illustrated in Scheme 26. Bergman and co-workers (114) however, were the first to isolate terminal zirconium sulfido complexes as pyridine adducts, $Cp_2^*Zr(S)(NC_5H_4R)$ (R = H, Bu^t), by the dehydrohalogenation of $Cp_2^*Zr(SH)I$ with $KN(SiMe_3)_2$ in the presence of pyridine (Scheme 27). Subsequently, $Cp_2^*Zr(S)(NC_5H_5)$ (110) and $Cp_2^{Et*}Zr(S)(NC_5H_5)$ (32) have been synthesized via reaction of the respective dicarbonyl complex with 1 equiv of sulfur (or Ph_3PS) in the presence of pyridine (128). In the absence of pyridine, however, $Cp_2^*Zr(CO)_2$ reacts with excess sulfur to give sequentially $Cp_2^*Zr(\eta^2\text{-}S_2)(CO)$ and $Cp_2^*Zr(\eta^2\text{-}S_3)$ (Scheme 28) (129). The sulfido complex $Cp_2^*Zr(S)(NC_5H_5)$

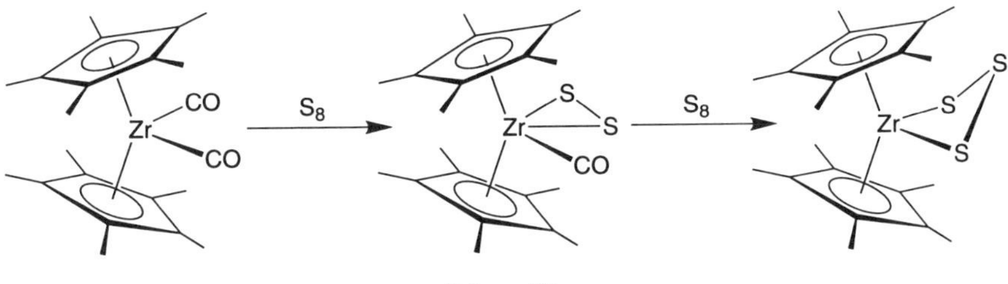

Scheme 28

has also been synthesized by the reaction of excess $Cp^*_2Zr(CO)_2$ with H_2S in the presence of pyridine (Scheme 27); however, in the presence of excess H_2S, the hydrosulfido complex $Cp^*_2Zr(SH)_2$ is obtained.

As with the zirconium oxo complexes described above, the importance of bulky cyclopentadienyl ligands in stabilizing the terminal chalcogenido moiety is indicated by the fact that bridging derivatives, for example, $[Cp_2Zr(\mu\text{-}S)]_2$ (130–132), $[Cp_2^{Bu^t}Zr(\mu\text{-}S)]_2$ (105, 133, 134), are obtained with less sterically demanding ligands.

The reactivity of the [Zr=S] moiety is summarized in Schemes 29 and 30. For example, $[Cp^*_2Zr{=}S]$ generated *in situ* reacts with PhCN to give a six-membered metallacycle (114a). The corresponding reaction with alkynes yields four-membered metallacycles, $Cp^*_2Zr[\eta^2\text{-}SC(R)C(R)]$. Interestingly, the formation of these four-membered metallacycles is reversible in the sense that $Cp^*_2Zr[\eta^2\text{-}SC(R)C(R)]$ reacts with *tert*-butylpyridine to give the terminal sulfido complex $Cp^*_2Zr(S)(NC_5H_4Bu^t)$. Bergman has interpreted this observation as indicating that the [Zr=S] moiety is more stable with respect to cycloaddition of alkynes than is the corresponding [Zr=O] moiety, and has noted that such behavior is unexpected in view of the high oxophilicity commonly associated with zirconium.

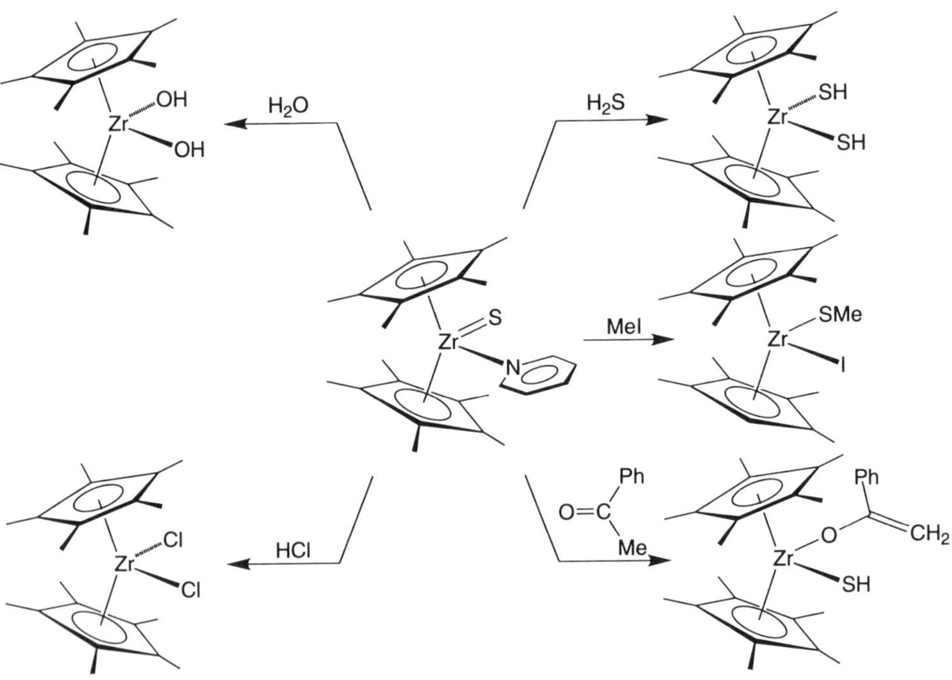

Scheme 29

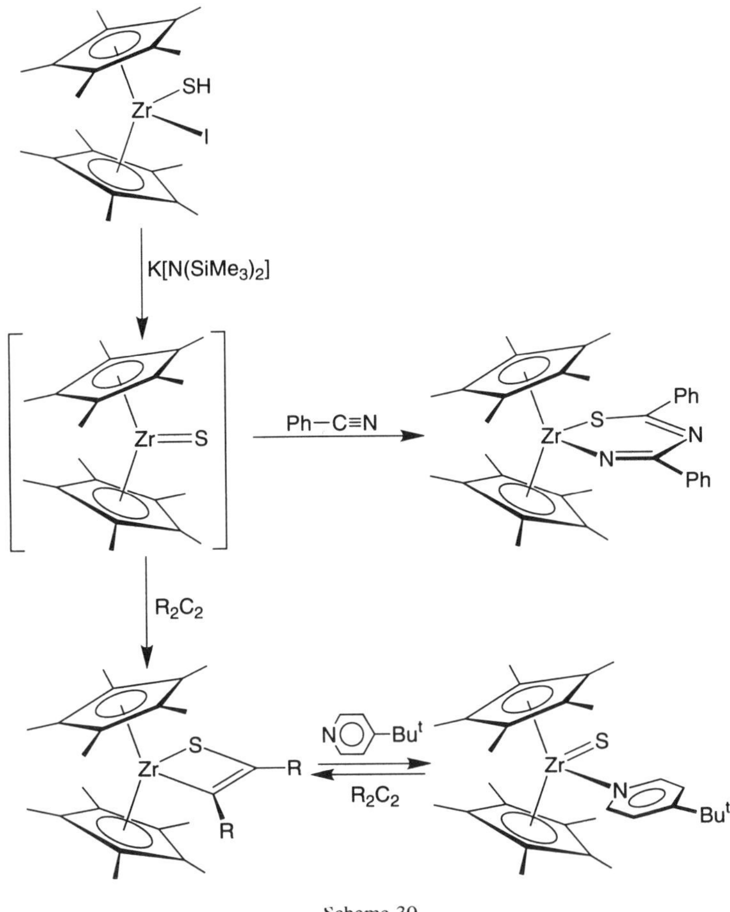

Scheme 30

C. Zirconium Selenido Complexes. The zirconocene selenido complexes $Cp_2^RZr(Se)(NC_5H_5)$ ($Cp^R = Cp^*$, Cp^{Et*}) have been synthesized by the reactions of $Cp_2^RZr(CO)_2$ with 1 equiv of selenium in the presence of pyridine (Scheme 31) (32, 128). In the absence of pyridine, $Cp_2^*Zr(CO)_2$ reacts with selenium to give sequentially $Cp_2^*Zr(Se)(CO)$ (135), $Cp_2^*Zr(\eta^2\text{-}Se_2)(CO)$, and $Cp_2^*Zr(\eta^2\text{-}Se_3)$ (129). Interestingly, this sequence may be reversed using $Cp_2^*Zr(CO)_2$ as a chalcogen trap, and thus $Cp_2^*Zr(\eta^2\text{-}Se_3)$ reacts with $Cp_2^*Zr(CO)_2$ to give a mixture of $Cp_2^*Zr(\eta^2\text{-}Se_2)CO$ and $Cp_2^*Zr(Se)CO$. As with the sulfido derivative described above, the selenido complex $Cp_2^*Zr(Se)(NC_5H_5)$ may be generated upon reaction of H_2Se with *excess* $Cp_2^*Zr(CO)_2$ in the presence of pyridine, but in the presence of excess H_2Se, Cp_2^*

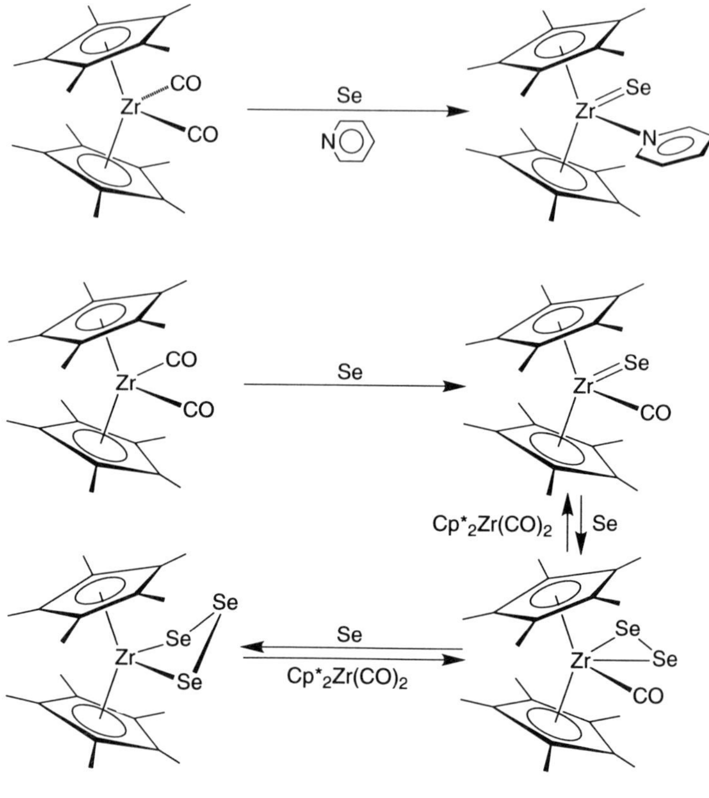

Scheme 31

Zr(SeH)$_2$ (136) is obtained preferentially. The reactivity of Cp$_2^*$Zr(Se)-(NC$_5$H$_5$) is summarized in Scheme 32.

The selenido carbonyl complex Cp$_2^*$Zr(Se)(CO) is particularly noteworthy since it is a unique example of a d^0 zirconium carbonyl complex which also contains a metal–ligand multiple bond. In fact, even $d^{>0}$ chalcogenido carbonyl complexes are quite rare, as described in Section IV.C. The molecular structure of Cp$_2^*$Zr(Se)CO has been determined by X-ray diffraction (Fig. 19) and the Zr=Se bond length of 2.478(2) Å is comparable to that in Cp$_2^{Et*}$Zr-(Se)(NC$_5$H$_5$) [2.480(1) Å].

Due to the d^0 nature of Cp$_2^*$Zr(Se)CO, π back-bonding does not stabilize the Zr—CO interaction significantly, and so it may be regarded as a nonclassical carbonyl complex (137, 138). The nonclassical nature of Cp$_2^*$Zr(Se)CO, and also the dichalcogenido complexes Cp$_2^*$Zr(η^2-E$_2$)CO, is indicated by the observation that (a) the Zr—CO and C—O bonds are marginally longer and

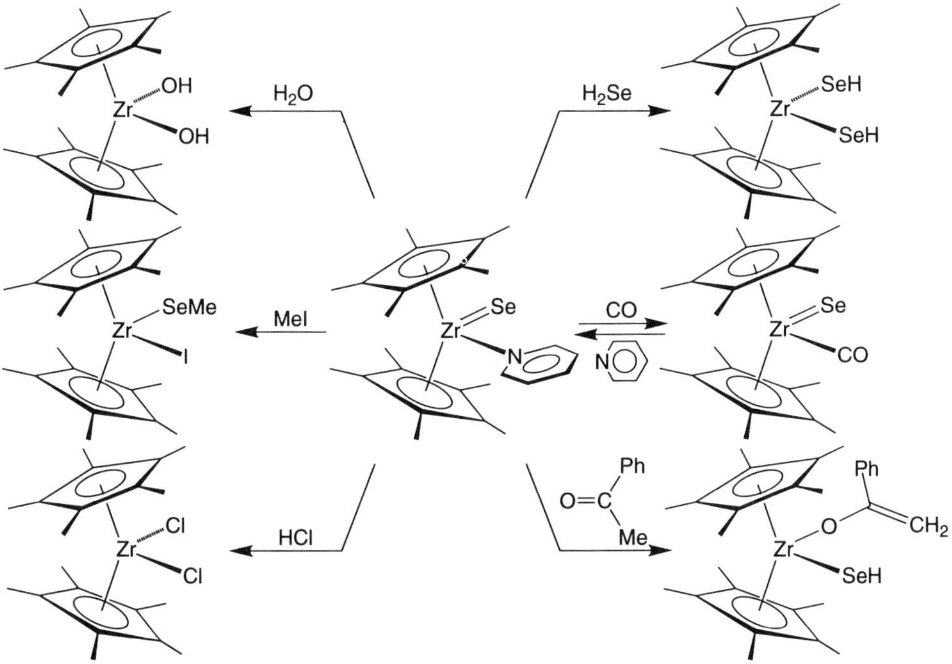

Scheme 32

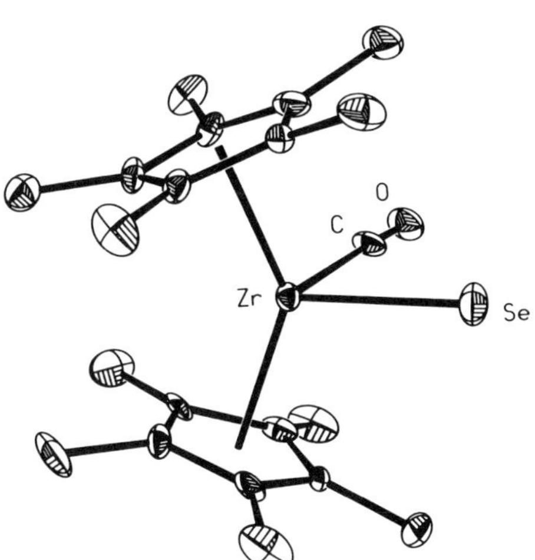

Figure 19. Molecular structure of $Cp_2^*Zr(Se)(CO)$. [Reprinted with permission from W. A. Howard, T. M. Trnka, and G. Parkin, *Organometallics*, *14*, 4037 (1995). Copyright © 1995 American Chemical Society.]

TABLE VIII
Metrical and IR Data for Structurally Characterized Neutral Permethylzirconocene Carbonyl Derivatives[a]

Complex	d(Zr—CO) (Å)	d(C—O) (Å)	ν(CO) (cm^{-1})
	$(J_{51_V}$ 2.233(15)	1.116(19)	2037
^{125}Te			
$Cp_2^*Zr(\eta^2\text{-}S_2)CO$	2.261(6)	1.126(8)	2057
$Cp_2^*Zn(\eta^2\text{-}Se_2)CO$	2.253(13)	1.128(16)	2037
$Cp_2^*Zr(\eta^2\text{-}Te_2)CO$	2.241[7]	1.123[9]	2006
$Cp_2^*Zr(CO)_2$	2.145(9)	1.16(1)	1945, 1852

[a]Taken from (135).

shorter, respectively, than the corresponding values for the related d^2 derivative $Cp_2^*Zr(CO)_2$ (Table VIII), and (b) the ν(CO) stretching frequencies are higher than those for $Cp_2^*Zr(CO)_2$ (139). However, since the ν(CO) stretching frequency for $Cp_2^*Zr(Se)CO$ is lower than that of free CO (2143 cm^{-1}), some degree of electron donation into the π^* CO orbital must occur, possibly via interaction with the [Zr=Se] or [$Zr(\eta^2\text{-}E_2)$] moieties. For example, Berry and co-workers (138f) suggested that the direct donation of electron density from silicon to the CO π^* orbital is responsible for the anomalously low value (1797 cm^{-1}) for the CO stretching frequency in $Cp_2Zr(\eta^2\text{-}Me_2Si{=}NBu^t)CO$. Furthermore, the reduced ν(CO) stretching frequency (2044 cm^{-1}) for $Cp_2^*ZrH_2(CO)$, compared to that of free CO, has been attributed to electron donation from a filled metal-hydride bonding orbital into an in-plane π^* CO orbital (138b, 140). The lability of the CO ligand in $Cp_2^*Zr(Se)CO$ is clearly indicated by its rapid displacement with pyridine to give $Cp_2^*Zr(Se)(NC_5H_5)$ (Scheme 32). The exchange is reversible, however, and exposure of a benzene solution of $Cp_2^*Zr(Se)(NC_5H_5)$ to CO (1 atm) results in approximately 10% conversion to $Cp_2^*Zr(Se)(CO)$ (135).

d. Zirconium Tellurido Complexes. Terminal tellurido complexes of zirconium were first described by Arnold and co-worker (141a) in 1992 (141) Specifically, Arnold reported that addition of dmpe to the tris(trimethylsilyl)silyltellurolate complex $(dmpe)Zr(TeR)_4$ [$R = Si(SiMe_3)_3$] induces elimination of R_2Te with the concomitant formation of the terminal tellurido complex $(dmpe)_2Zr(TeR)_2(Te)$ (Scheme 33). The tellurido complex is characterized by a ^{125}Te NMR signal at δ − 706 ppm and a Zr=Te bond length of 2.650(1) Å.

The zirconocene tellurido complexes $Cp_2^*Zr(Te)(NC_5H_5)$ and $Cp_2^{Et*}Zr(Te)(NC_5H_5)$ ($Cp^R = Cp^*, Cp^{Et*}$) have been synthesized in an analogous fashion to their sulfido and selenido counterparts by the reactions of $Cp_2^RZr(CO)_2$ with 1 equiv of tellurium in the presence of pyridine (Scheme 34) (32, 128). In

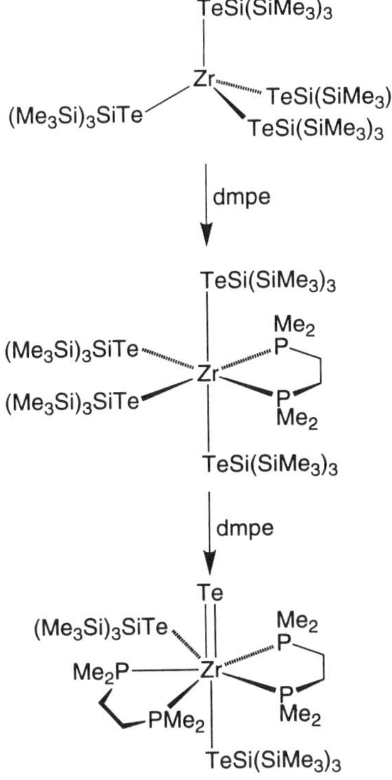

Scheme 33

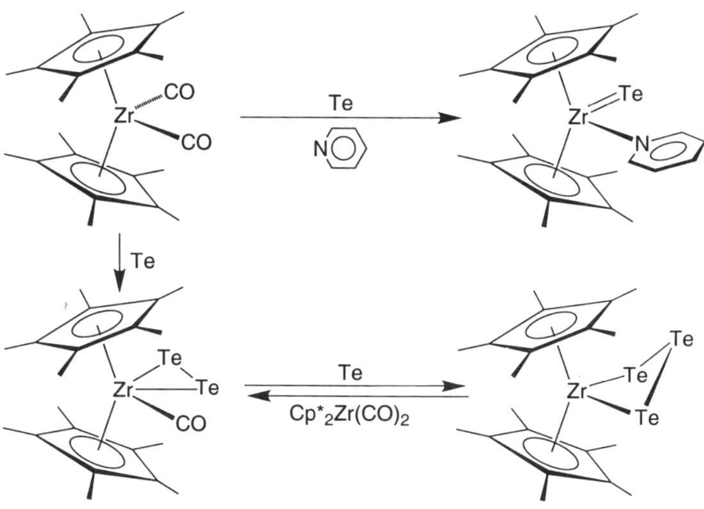

Scheme 34

the absence of pyridine, $Cp_2^*Zr(CO)_2$ reacts with tellurium (in an analogous fashion to selenium) to give sequentially $Cp_2^*Zr(\eta^2\text{-}Te_2)(CO)$, and $Cp_2^*Zr(\eta^2\text{-}Te_3)$ (129). The final step of this sequence may be reversed such that $Cp_2^*Zr(\eta^2\text{-}Te_3)$ reacts with $Cp_2^*Zr(CO)_2$ to give $Cp_2^*Zr(\eta^2\text{-}Te_2)CO$. The structure of $Cp_2^{Et*}Zr(Te)(NC_5H_5)$ has been determined by X-ray diffraction and the Zr=Te bond length of 2.729(1) Å is slightly longer than that in $(dmpe)_2Zr(TeR)_2(Te)$ [2.650(1) Å].

The importance of the bulky peralkylated cyclopentadienyl ligands in stabilizing these zirconocene terminal tellurido complexes is demonstrated by the fact that the corresponding reaction of $Cp_2^{Bu^t}Zr(CO)_2$ with tellurium in the presence of pyridine gives the bridged complex $[Cp_2^{Bu^t}Zr(\mu\text{-}Te)]_2$, rather than $Cp_2^{Bu^t}Zr(Te)(NC_5H_5)$ (32b). The inability to isolate a terminal tellurido complex in this system is precedented by Arnold's (142) report that the *tert*-butylpyridine induced elimination of $Te(SiPh_3)_2$ from $Cp_2^{Bu^t}Zr(TeSiPh_3)_2$ also yields $[Cp_2^{Bu^t}Zr(\mu\text{-}Te)]_2$ (142). On the basis of kinetic studies, however, Arnold suggested that the terminal tellurido complex $Cp_2^{Bu^t}Zr(Te)(NC_5H_4Bu^t)$ is an intermediate in the transformation.

The reactivity of $Cp_2^*Zr(Te)(NC_5H_5)$ is summarized in Scheme 35. As with

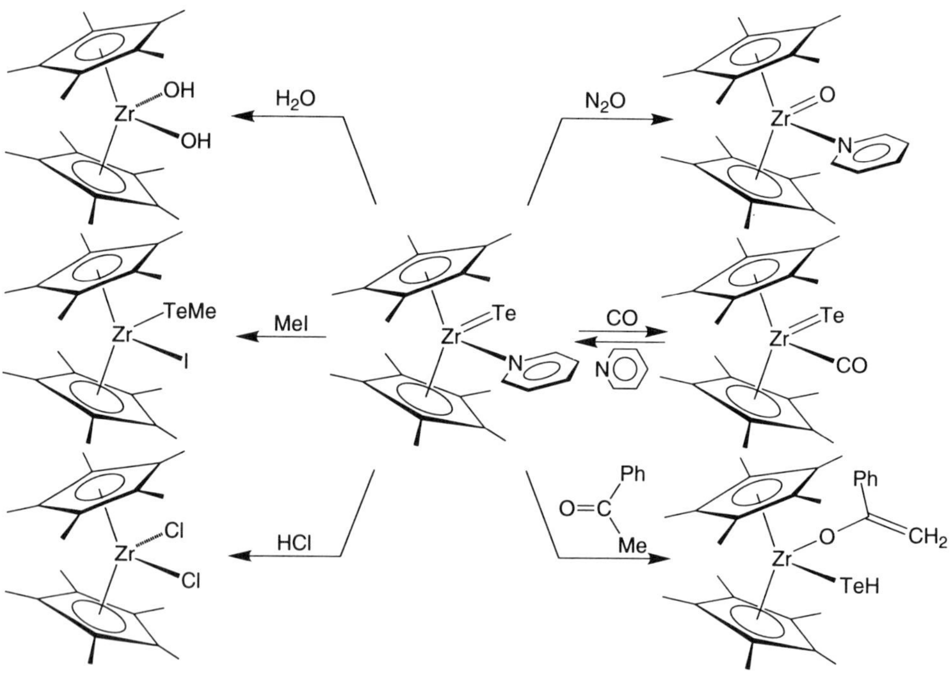

Scheme 35

its selenido counterpart, the pyridine ligand of $Cp_2^*Zr(Te)(NC_5H_5)$ is displaced by CO at room temperature. However, the tellurido–carbonyl $Cp_2^*Zr(Te)(CO)$ species has only been spectroscopically observed as an intermediate in the reaction, which ultimately yields a mixture of $Cp_2^*Zr(CO)_2$ and $Cp_2^*Zr(\eta^2\text{-}Te_2)(CO)$ (129). The displacement of the pyridine ligands of the selenido and tellurido complexes, $Cp_2^*Zr(Se)(NC_5H_5)$ and $Cp_2^*Zr(Te)(NC_5H_5)$, by CO thus provides a contrast with the oxo and sulfido complexes, $Cp_2^*Zr(O)(NC_5H_5)$ and $Cp_2^*Zr(S)(NC_5H_5)$, which are stable in the presence of CO at room temperature. Finally, it is noteworthy that the tellurido complex $Cp_2^*Zr(Te)(NC_5H_5)$ reacts instantaneously with N_2O at room temperature to give the oxo derivative, $Cp_2^*Zr(Te)(NC_5H_5)$.

e. Comparison of Zr=E Double and Zr—E Single-Bond Lengths. The structural characterization of the series of terminal chalcogenido complexes $Cp_2^{Et*}Zr(E)(NC_5H_5)$ (E = O, S, Se, Te) has allowed for comparison of Zr=E double bonds and Zr—E single bonds within a structurally related $(\eta^5\text{-}C_5R_5)_2Zr(EX)Y$ system. Thus, the data summarized in Table IX reveal that, as expected, each Zr=E double bond is substantially shorter than the corresponding Zr—E single bond, the difference ranging from 0.19 Å for oxygen to 0.15 Å for tellurium, corresponding to a 9.6 and 5.0% decrease, respectively.

3. *Hafnium Chalcogenido Complexes*

Terminal chalcogenido complexes of hafnium have been little studied in comparison to their zirconium counterparts, with Arnold and co-workers (141) reporting the first structurally determined terminal tellurido complex $(dmpe)_2Hf(TeR)_2(Te)$ [R = $Si(SiMe_3)_3$], characterized by a ^{125}Te NMR signal at δ −701 ppm and a Hf=Te bond length of 2.637(2) Å. Subsequently, the entire series of $Cp_2^RHf(E)(NC_5H_5)$ (Cp^R = Cp*, Cp^{Et*}; E = O, S, Se, Te) complexes has been synthesized (143). Thus, the sulfido, selenido, and tellurido derivatives are readily prepared in an analogous fashion to the zirconium com-

TABLE IX
Comparison of Zr=E Double and Zr—E Single Bond Lengths[a]

	d(Zr=E) (Å)	d(Zr—E) (Å)	[d(Zr—E) − d(Zr=E)] (Å)	% Decrease
O	1.804(4)	1.996[32]	0.192	9.6
S	2.334(2)	2.520[11]	0.186	7.4
Se	2.480(1)	2.658[10]	0.178	6.7
Te	2.729(1)	2.874[14]	0.145	5.0

[a]Taken from (54).

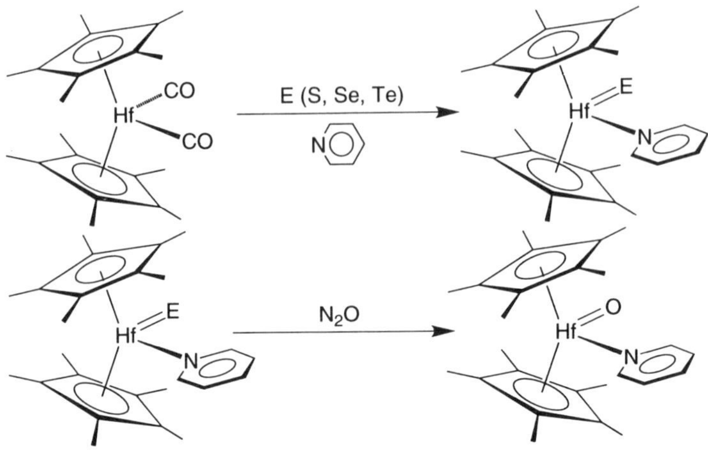

Scheme 36

plexes, by reaction of the dicarbonyl with the elemental chalcogen in the presence of pyridine (Scheme 36). However, the hafnium oxo complexes are not available via treatment of the dicarbonyl with N_2O in the presence of pyridine, which instead results in significant decomposition at the temperatures required to effect reaction. Nevertheless, the hafnium oxo complexes may be synthesized by the reactions of the tellurido derivatives $Cp_2^RHf(Te)(NC_5H_5)$ with N_2O at room temperature. The structures of the chalcogenido complexes Cp_2^{Et*}-$Hf(E)(NC_5H_5)$ (E = O, S, Se, Te) have been determined by X-ray diffraction and the Hf=E bond length data are summarized in Table X; for comparison, the bond lengths of the zirconium analogues are listed in Table IX.

Floriani and co-workers (121) synthesized the hafnium oxo complex $\{\eta^5$-η^1-η^1-η^1-$Et_8C_4(C_4H_2N)_2(C_5H_3N)(m$-$MeC_5H_2N)\}HfO$ by the reaction of the methyl derivative $\{\eta^5$-η^1-η^5-η^1-$Et_8C_4(C_4H_2N)_3(C_5H_3N)\}HfMe$ with CO (Scheme 37).

TABLE X
Hf=E Bond Lengths in $Cp_2^{Et}*M(E)(NC_5H_5)^a$

	d(Hf=E) (Å)
O	1.826(9)
S	2.311(3)
Se	2.467(1)
Te	2.716(1)

[a]Taken from (143).

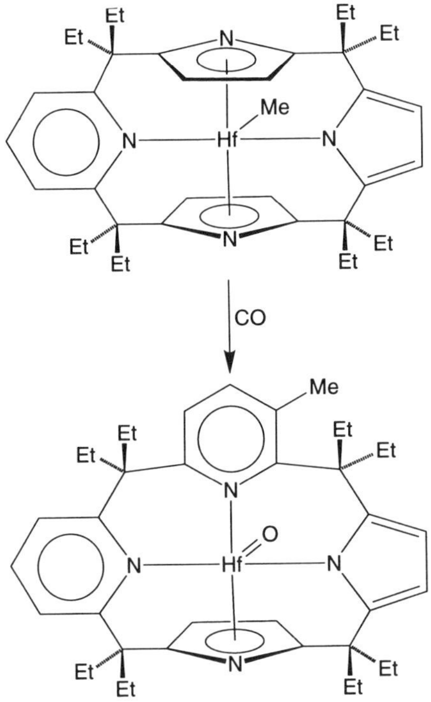

Scheme 37

The oxo complex has not been structurally determined by X-ray diffraction, but is characterized by a ν(Hf=O) stretch at 826 cm^{-1} in the IR spectrum.

C. Terminal Chalcogenido Complexes of Group 5 (V B): V, Nb, Ta

1. *Vanadium Chalcogenido Complexes*

Examination of Figs. 4–7 and 16 indicates that vanadium forms a substantial number of terminal chalcogenido complexes. The vast majority of these complexes, however, are V^{IV} oxo derivatives, which have been well reviewed (144).

a. Tris(amido)amine (Azatrane) Derivatives. The [η^4-$N(CH_2CH_2NSiMe_3)_3$] ligand has been used by Schrock and co-workers (145, 146) to generate the first complete series of terminal chalcogenido complexes of vanadium, [η^4-$N(CH_2CH_2NSiMe_3)_3$]VE (E = O, S, Se, Te), as illustrated in Scheme 38 (147). The oxo derivative [η^4-$N(CH_2CH_2NSiMe_3)_3$]VO, characterized by a

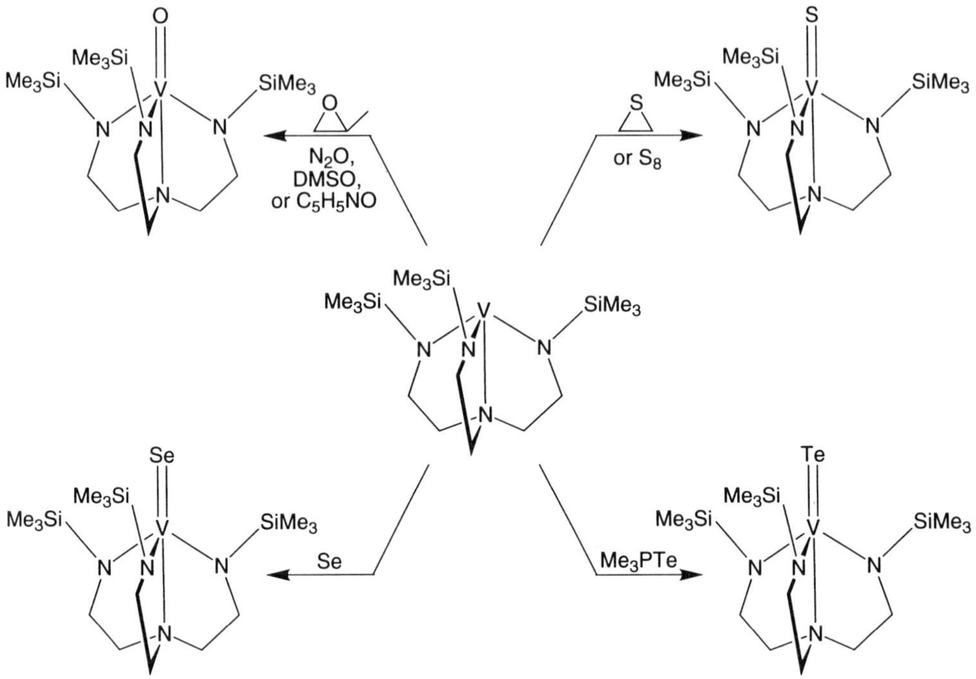

Scheme 38

ν(V=O) stretch of 996 cm^{-1}, may be obtained by the reaction between [η^4-$N(CH_2CH_2NSiMe_3)_3$]V and a variety of oxygen-atom transfer agents [e.g., N_2O, (DMSO), pyridine *N*-oxide, and epoxides]. The sulfido complex may be obtained by the reaction of [η^4-$N(CH_2CH_2NSiMe_3)_3$]V with either elemental sulfur or ethylene sulfide, while the selenido derivative is obtained by the corresponding reaction with selenium. The tellurido complex [η^4-$N(CH_2CH_2NSiMe_3)_3$]VTe is unstable and has only been spectroscopically identified in solution; specific evidence for the presence of the [V=Te] moiety is provided by a ^{51}V NMR signal at 1484 ppm, which exhibits satellites due to coupling to tellurium ($J_{^{51}V-^{125}Te}$ = 360 Hz). The ^{51}V NMR spectroscopic data for all [η^4-$N(CH_2CH_2NSiMe_3)_3$]VE derivatives are listed in Table XI, indicating that the resonances shift considerably to lower fields as the chalcogen becomes heavier. Such "inverse" relationships between metal NMR chemical shifts and ligand electronegativity are common for d^0 complexes (148).

The oxo complex [η^4-$N(CH_2CH_2NSiMe_3)_3$]VO is stable toward PMe_3, whereas the selenido ligand of [η^4-$N(CH_2CH_2NSiMe_3)_3$]VSe is abstracted by PMe_3 to regenerate [η^4-$N(CH_2CH_2NSiMe_3)_3$]V, and the tellurido derivative [η^4-

TABLE XI
^{51}V NMR Data for $[\eta^4\text{-}N(CH_2CH_2NSiMe_3)_3]VE$[a]

$[\eta^4\text{-}N(CH_2CH_2NSiMe_3)_3]VE$	$\delta(^{51}V)$ (ppm)
O	−173
S	621
Se	921
Te	1484

[a]Taken from (145).

$N(CH_2CH_2NSiMe_3)_3]VTe$ decomposes at room temperature. These observations nicely illustrate how the stability of the terminal metal chalcogenido [M=E] moiety decreases as the chalcogen becomes heavier.

Arnold has synthesized a related series of four-coordinate terminal chalcogenido complexes of vanadium supported by one chalcogenolate and two amido ligands (149). Thus, $[(Me_3Si)_2N]_2(RSe)VE$ and $[(Me_3Si)_2N]_2(RTe)VE$ (E = O, S, Se) have been obtained by addition of a chalcogen source to the three-coordinate precursors $[(Me_3Si)_2N]_2(Ph_3SiSe)V$ and $[(Me_3Si)_2N]_2(Ph_3SiTe)V$, as illustrated in Scheme 39. The structure of the selenido derivative

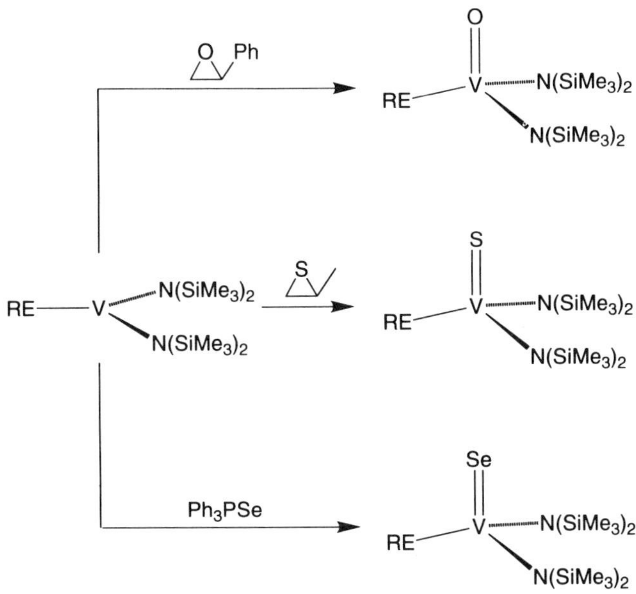

Scheme 39

$[(Me_3Si)_2N]_2[(Me_3Si)_3SiSe]VSe$ has been determined by X-ray diffraction and is characterized by a V=Se bond length of 2.1754(9) Å. As with the $[\eta^4$-$N(CH_2CH_2NSiMe_3)_3]VE$ complexes described above, the ^{51}V NMR signals for $[(Me_3Si)_2N]_2(RSe)VE$ and $[(Me_3Si)_2N]_2(RTe)VE$ shift to lower field in going from the oxo to selenido derivative. The $\nu(V{=}E)$ stretching frequencies for the oxo, sulfido, and selenido complexes, $[(Me_3Si)_2N]_2[(Me_3Si)_3SiSe]VE$, are 1012, 568, and 455 cm^{-1}, respectively.

b. Macrocyclic Derivatives. Porphyrin vanadium oxo complexes, (POR)VO, are well known, and may be prepared by reactions of the appropriate porphyrin with, for example, $V(O)(SO_4)_2$ or $(acac)_2VO$ (82). The vanadyl moiety in these complexes is typically characterized by an $\nu(V{=}O)$ IR absorption in the range 950–1050 cm^{-1} (82). The terminal sulfido and selenido analogues have also been obtained by the reactions $(POR)V(thf)_2$ (POR = OEP, TTP, TMTP) with S_8 and Cp_2TiSe_5, respectively (Scheme 40) (150). The $\nu(V{=}S)$ and $\nu(V{=}Se)$ IR absorptions for (POR)VS and (POR)VSe are observed in the ranges 555–561 cm^{-1} and 434–447 cm^{-1}, respectively. The V=E bond length data for some (POR)VE derivatives are summarized in Table XII.

Other macrocyclic chalcogenido complexes of vanadium include the tetramethyldibenzotetraza[14]annulene derivatives $[\eta^4$-$Me_4taa]VO$ and $[\eta^4$-Me_4-

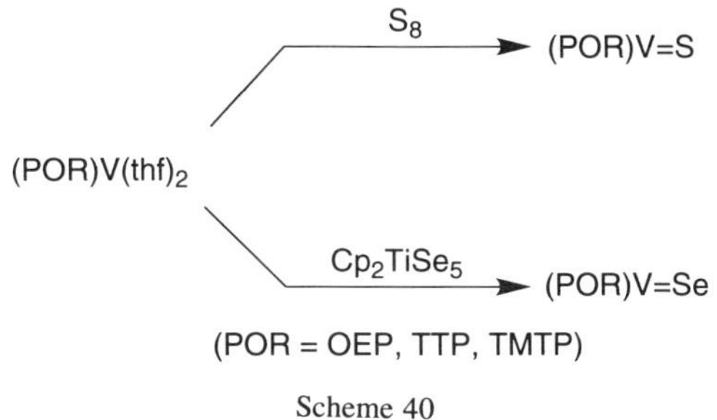

Scheme 40

TABLE XII
V=E Bond Lengths in (POR)VE[a]

	$d(V{=}O)$ (Å)	$d(V{=}S)$ (Å)	$d(V{=}Se)$ (Å)
(OEP)VE	1.620(2)	2.06(1)[a]	2.19(2)[b]
(DMOEP)VE	1.619(4)		
(DPEP)VE	1.62(1)[a]		

[a]Taken from (82 and 150).
[b]EXAFS data.

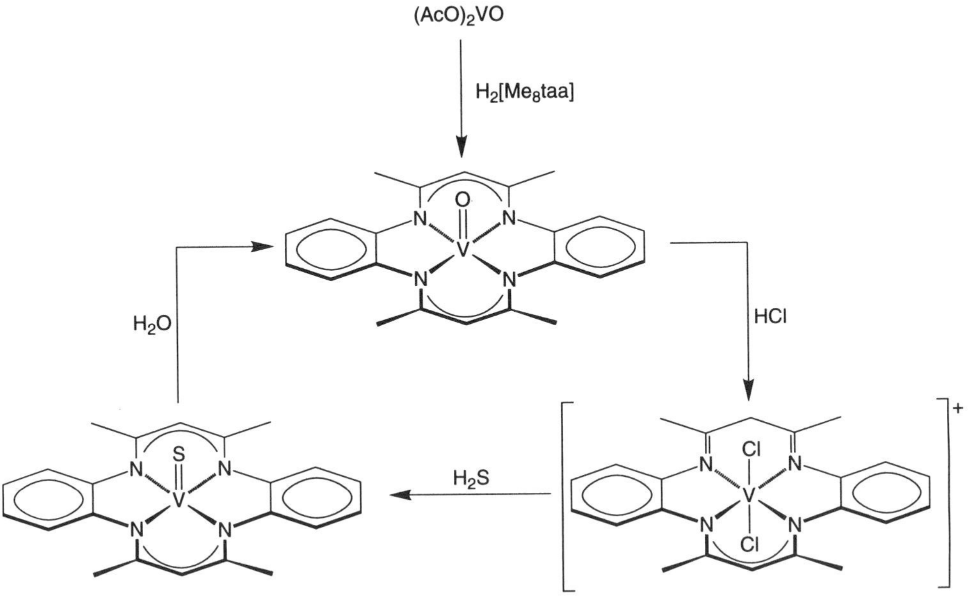

Scheme 41

taa]VS (151). The oxo complex was synthesized by reaction of $(AcO)_2VO$ with $[Me_4taa]H_2$, while the sulfido derivative $[\eta^4$-$Me_4taa]VS$ $[\nu(V{=}S) = 545\ cm^{-1}]$ was prepared by reaction of $[\eta^4$-$Me_4taa]VO$ with HCl followed by H_2S (Scheme 41). The sulfido complex is very unstable with respect to hydrolysis, thereby regenerating $[\eta^4$-$Me_4taa]VO$. The [V=O] moiety is characterized by an absorption at 965 cm^{-1} in the IR spectrum and a V=O bond length of 1.601(2) Å (151, 152). The oxo complex $[\eta^4$-$Me_4taa]VO$ reacts with $NaBPh_4$/*p*-TsOH to give $[\eta^4$-$Me_4taa]VOBPh_3$ and with $(Me_3Si)_2NH/NaBPh_4$ to give $\{[\eta^4$-$Me_4taa]V(OSiMe_3)\}[BPh_4]$ (153b). Finally, $[\eta^4$-$Me_4taa]VO$ may also be alkylated by Grignard reagents to give $[\eta^4$-$Me_4taa]VR$ (154a).

c. Vanadocene Derivatives. Andersen and co-workers (91) synthesized the vanadium oxo complex Cp^*_2VO by the reaction of Cp^*_2V with N_2O. The complex Cp^*_2VO is characterized by a $\nu(V{=}O)$ stretch at 855 cm^{-1} in the IR spectrum, shifting to 820 cm^{-1} upon ^{18}O substitution. Notably, the presence of pyridine is not required to stabilize the oxo derivative, in contrast to the titanium analogue $Cp^*_2Ti(O)(NC_5H_4R)$. However, prolonged reaction times between Cp^*_2V and N_2O result in formation of the cluster $[Cp^*V]_4(\mu$-$O)_6$ (155) in preference to the terminal oxo derivative Cp^*_2VO.

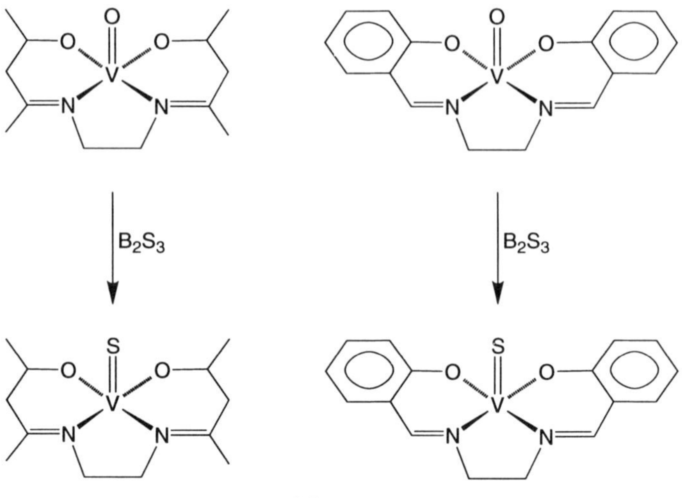

Scheme 42

d. Miscellaneous Derivatives. The first thiovanadyl [$V^{IV}S$] complexes, namely, (acen)VS and (salen)VS, were prepared by the reaction of the oxo derivatives with B_2S_3 (Scheme 42) (156). As a modification of this approach, Holm and co-workers (157) have demonstrated that $(Me_3Si)_2S$ is a convenient reagent to synthesize (acen)VS from (acen)VO. A variety of sulfido complexes of the type $(RO)_3V{=}S$ and $(RO)_2(RS)V{=}S$ have been prepared by several different methods, including substitution of the oxo ligand in $(RO)_3V{=}O$ using Lawesson's reagent $[RP(S)(\mu\text{-}S)]_2$, and oxidation of $V(OBu^t)_4$ with S_8 (Scheme (43) (158). The series of complexes has been characterized by ^{51}V NMR spec-

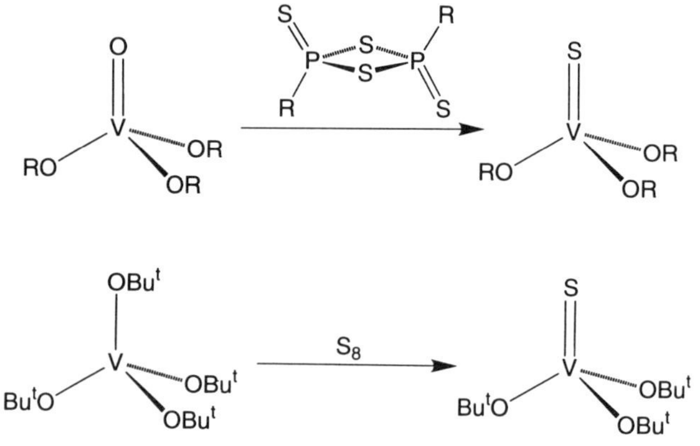

Scheme 43

TABLE XIII
V=S Bond Length and ν(V=S) IR Absorption Data for a Selection of Terminal Sulfido Complexes

Complex	*d*(V=S) (Å)	ν(V=S) (cm^{-1})	References
$(NH_4)_3[V(S)_4]$	2.15[2]	478	*a*
$Li_3[V(S)_4]\cdot 2TMEDA$	2.157(1)	477	*b*
(acen)VS	2.061(1)	556	*c, d*
(salen)VS		543	*d*
$[V(S)_2(\eta^2\text{-}S_2)(SPh)]^{2-}$	2.092(2), 2.099(2)		*e*
$[(PhS)_4V(S)]^{2-}$	2.078(2)	521	*f*

[a]Y. Do, E. D. Simhon, and R. H. Holm, *Inorg. Chem.*, *24*, 4635 (1985).
[b]S. C. Lee, J. Li, J. C. Mitchell, and R. H. Holm, *Inorg. Chem.*, *31*, 4333 (1992).
[c]M. Sato, K. M. Miller, J. H. Enemark, C. E. Strouse, and K. P. Callahan, *Inorg. Chem.*, *20*, 3571 (1981).
[d]K. P. Callahan, P. J. Durand, and P. H. Rieger, *J. Chem. Soc. Chem. Commun.*, 75 (1980). K. P. Callahan and P. J. Durand, *Inorg. Chem.*, *19*, 3211 (1980).
[e]J. K. Money, J. R. Nicholson, J. C. Huffman, and G. Christou, *Inorg. Chem.*, *25*, 4072 (1986).
[f]J. R. Nicholson, J. C. Huffman, D. M. Ho, and G. Christou, *Inorg. Chem.*, *26*, 3030 (1987).

troscopy. The V=S bond length and ν(V=S) IR absorption data for a selection of terminal sulfido complexes are listed in Table XIII.

Christou and co-workers (159) were the first to structurally characterize a terminal vanadium–selenido complex, namely, $[(\eta^2\text{-}SCH_2CH_2S)_2V(Se)]^{2-}$. For comparison, V=E bond length and ν(V=E) IR data for the series of oxo, sulfido, and selenido complexes $[(\eta^2\text{-}SCH_2CH_2S)_2V(E)]^{2-}$ are summarized in Table XIV.

TABLE XIV
Comparison of V=E Bond Length and ν(V=E) IR Absorption Data for $[(\eta^2\text{-}SCH_2CH_2S)_2V(E)]^{2-}$

Complex	*d*(V=E) (Å)	ν(V=E) (cm^{-1})	Reference
$[(\eta^2\text{-}SCH_2CH_2S)_2V(O)]^{2-}$	1.607(6)	948	*a*
	1.625(2)	930	*b*
$[(\eta^2\text{-}SCH_2CH_2S)_2V(S)]^{2-}$	2.098(2)	515	*a*
	2.087(1)	502	*b*
$[(\eta^2\text{-}SCH_2CH_2S)_2V(Se)]^{2-}$	2.196(3)	397	*c*

[a]D. Szeymies, B. Krebs, and G. Henkel, *Angew. Chem. Int. Ed. Engl.*, *23*, 804 (1984).
[b]J. K. Money, J. C. Huffman, and G. Christou, *Inorg. Chem.*, *24*, 3297 (1985). J. K. Money, J. C. Huffman and G. Christou, *Inorg. Chem.*, *27*, 507 (1988).
[c]J. R. Nicholson, J. C. Huffman, D. M. Ho, and G. Christou, *Inorg. Chem.*, *26*, 3030 (1987).

2. *Niobium Chalcogenido Complexes*

Terminal chalcogenido complexes of niobium are not as common as their vanadium counterparts, and the majority comprise oxo and sulfido derivatives, which have been well reviewed (160).

a. Niobocene Derivatives. A variety of niobocene oxo complexes has been synthesized. For example, Royo and co-workers (161) prepared a series of niobocene oxo chloride, $Cp^R_2Nb(O)Cl$, and oxo alkyl complexes, Cp^R_2-Nb(O)R, as illustrated in Scheme 44. Thus, the oxo chloride complexes are obtained by the reaction of the corresponding carbonyl derivative $Cp^R_2Nb(CO)Cl$ with O_2, and the alkyl derivatives $Cp^R_2Nb(O)R$ are obtained by subsequent alkylation with RLi. Interestingly, the reaction of the carbonyl–methyl derivative $(Cp^{SiMe_3})_2Nb(CO)Me$ with O_2 does not give the oxo complex, but rather gives the peroxo derivative $(Cp^{SiMe_3})_2Nb(\eta^2\text{-}O_2)Me$. The analogous chloro peroxo complex has been obtained by the reaction of the dichloride $(Cp^{SiMe_3})_2NbCl_2$ with H_2O_2. Nicholas and co-workers (162) have also described the synthesis of a niobocene oxo–alkyl derivative by decarbonylation of an η^2-CO_2 ligand; thus, heating $Cp^{Me}_2Nb(\eta^2\text{-}CO_2)CH_2SiMe_3$ at 60°C (or photolysis at −20°C) gives $Cp^{Me}_2Nb(O)CH_2SiMe_3$. A selection of $\nu(Nb{=}O)$ IR and $d(Nb{=}O)$ bond length data for niobocene oxo–alkyl and oxo–chloride derivatives is listed in Table XV.

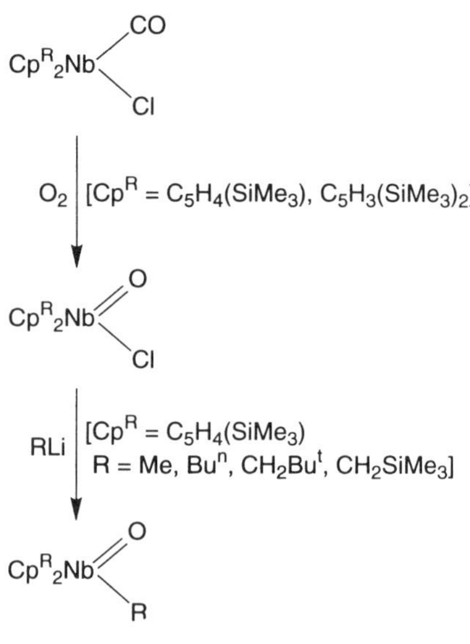

Scheme 44

TABLE XV
Nb=O Bond Length and ν(Nb=O) IR Absorption Data for Niobocene Oxo Complexes

Complex	ν(Nb=O) (cm^{-1})	d(Nb=O) (Å)	References
$Cp_2Nb(O)Me$	835		*a*
$Cp_2Nb(O)R$ (R = Bu, α-$CH_2C_5H_4N$, α-C_4H_3S)	865–870		*b*
$Cp_2Nb(O)[C_7H_5(CF_3)_2]$	980	1.63(3)	*c*
$Cp_2^{Me}Nb(O)CH_2SiMe_3$	837	1.741(3)	*d*
$(Cp^{SiMe_3})_2Nb(O)Me$	865	1.720(7)	*e*
$(Cp^{SiMe_3})_2Nb(O)Bu$	864		*e*
$(Cp^{SiMe_3})_2Nb(O)CH_2Bu^t$	857		*e*
$(Cp^{SiMe_3})_2Nb(O)CH_2SiMe_3$	865		*e*
$Cp_2Nb(O)Cl$	860 (867)	1.737(6)	*f, g, h*
$Cp_2^{Me}Nb(O)Cl$	853	1.732(11)	*i, j*
$(Cp^{SiMe_3})_2Nb(O)Cl$	865		*e*
$(Cp^{[SiMe_3]_2})_2Nb(O)Cl$	875		*e*

[a]A. R. Middleton and G. Wilkinson, *J. Chem. Soc. Dalton Trans.*, *1888* (1988).
[b]D. A. Lemenovskii, T. V. Baukova, V. A. Knizhnikov, E. G. Perevalova, and A. N. Nesmeyanov, *Dok. Chem.*, *226*, 65 (1973).
[c]R. Mercier, J. Douglade, J. Amaudrut, J. Sala-Pala, and J. E. Guerchias, *J. Organomet. Chem.*, *244*, 145 (1983).
[d]P.-F. Fu, M. A. Khan, and K. M. Nicholas, *Organometallics*, *10*, 382 (1991).
[e]A. Antiñolo, J. Martinez de llarduya, A. Otero, P. Royo, A. M. M. Lanfredi, and A. Tiripicchio, *J. Chem. Soc. Dalton Trans.*, 2685 (1988).
[f]P. M. Treichel and G. P. Werber, *J. Am. Chem. soc.*, *90*, 1753 (1968).
[g]D. A. Lemenovskii, T. V. Baukova, and V. P. Fedin, *J. Organomet. Chem.*, *132*, C14 (1977).
[h]A. L. Rheingold and J. B. Strong, *Acta Crystallogr. C47*, 1963 (1991).
[i]P.-F. Fu, M. A. Khan, and K. M. Nicholas, *Organometallics*, *11*, 2607 (1992).
[j]R. Broussier, H. Normand, and B. Gautheron, *J. Organomet. Chem.*, *155*, 347 (1978).

The disulfido–hydride complex of niobium $Cp_2^{Et*}Nb(\eta^2\text{-}S_2)H$ has been reported to convert to a mixture of $Cp_2^{Et*}Nb(S)SH$ and $Cp_2^{Et*}Nb(\eta^2\text{-}S_2)SH$ at 110°C (163). The sulfido complex $Cp_2^{Et*}Nb(S)SH$ is characterized by a ν(Nb=S) stretch of 437 cm^{-1}; it has also been structurally determined by X-ray diffraction, but with Nb—S bond lengths of 2.339(1) and 2.381(1) Å, the Nb=S and Nb—SH ligands are presumably disordered.

b. Macrocyclic Derivatives. Niobium(IV) and (V) oxo complexes of the types (POR)Nb(O) and (POR)Nb(O)X are well documented and have been reviewed (164). More recently, Floriana and co-workers (121) have synthesized the niobium oxo complex $\{\eta^5\text{-}\eta^1\text{-}\eta^1\text{-}\eta^1\text{-}Et_8C_4(C_4H_2N)_3(p\text{-}Me\text{-}C_5H_2N)\}Nb{=}O$ by the reaction of the methyl derivative $\{\eta^5\text{-}\eta^1\text{-}\eta^1\text{-}\eta^1\text{-}Et_8C_4(C_4H_2N)_4\}NbMe$ with CO (Scheme 45). As observed with the Zr and Hf porphyrinogen derivatives,

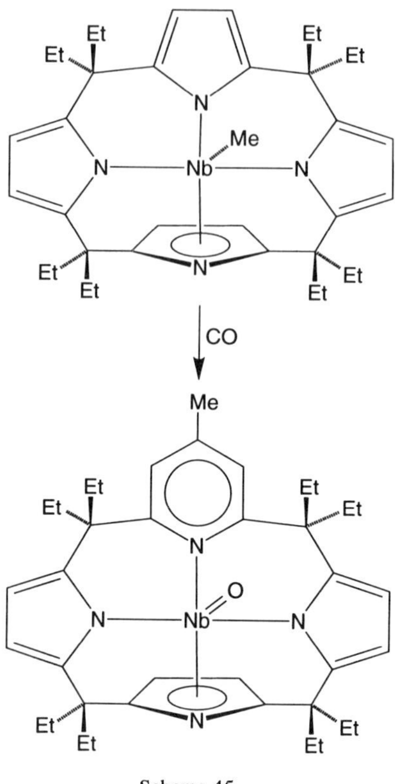

Scheme 45

the reaction involves cleavage of carbon monoxide and homologation of one of the pyrrole moieties.

c. **Miscellaneous Derivatives.** A series of oxo, sulfido, and selenido niobium complexes supported by the tropolonato ligand, $(\eta^2\text{-}C_7H_5O_2)_3Nb(E)$ (E = O, S, Se), have been prepared as illustrated in Scheme 46. Thus, the oxo complex is obtained by the reaction of $NbCl_5$ with tropolone, with the oxo ligand arising as a result of hydrolysis (165). The sulfido and selenido analogues have been prepared by the reactions of $(\eta^2\text{-}C_7H_5O_2)_3Nb(O)$ with $(Me_3Si)_2E$ (166), and the series of complexes $(\eta^2\text{-}C_7H_5O_2)_3Nb(E)$ (E = O, S, Se) are characterized by the data listed in Table XVI.

Gibson has reported the syntheses of the terminal chalcogenido complexes Cp*Nb(NAr)(PMe_3)(E) (E = S, Se, Te; Ar = 2,6-$Pr^i_2C_6H_3$) by the reactions of Cp*Nb(NAr)$(PMe_3)_2$ with elemental sulfur, selenium, and tellurium at room temperature (Scheme 47) (167). Significantly, Cp*Nb(NAr)(PMe_3)(Te) is the

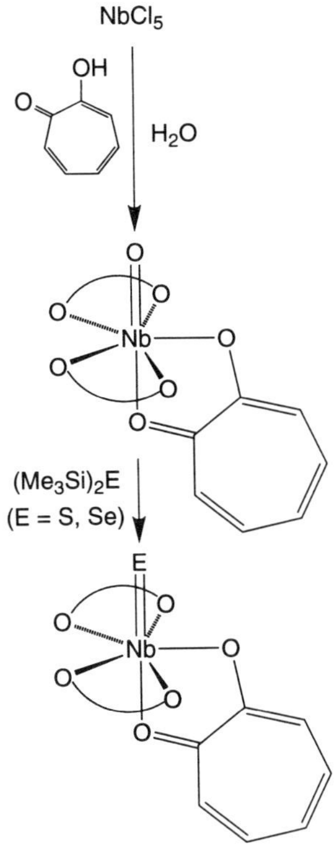

Scheme 46

sole terminal tellurido complex of niobium to have been isolated. The sulfido and selenido complexes are characterized by ν(Nb=S) and ν(Nb=Se) absorptions in the IR spectra at 465 and 335 cm^{-1}, respectively.

Gibson and co-workers (168–170) have also synthesized the oxo and sulfido complexes $(Me_3P)_3Nb(O)Cl_3$ and $(Me_3P)_3Nb(S)Cl_3$, which have been proposed

TABLE XVI

Nb=E Bond Length and ν(Nb=E) IR Data for $(\eta^2\text{-}C_7H_5O_2)_3Nb(E)$[a]

Complex	*d*(Nb=E) (Å)	ν(Nb=E) (cm^{-1})
$(\eta^2\text{-}C_7H_5O_2)_3Nb(O)$	1.712(14)	914
$(\eta^2\text{-}C_7H_5O_2)_3Nb(S)$		494
$(\eta^2\text{-}C_7H_5O_2)_3Nb(Se)$		350

[a]Taken from (166).

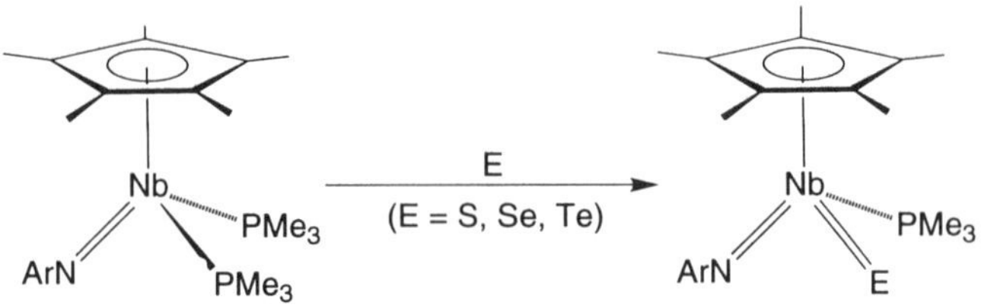

Scheme 47

as candidates for bond-stretch isomerism (Table XVII); however, the issue with respect to these complexes is yet to be fully resolved (75b).

Terminal sulfido derivatives of niobium are reasonably common (171, 172), as illustrated by the examples listed in Table XVIII. Interestingly, even though tetrathiovanadate derivatives $[V(S)_4]^{3-}$ are well-known, Holm and co-workers (173) have only recently synthesized soluble complexes of the niobium counterpart $[Nb(S)_4]^{3-}$ by the reaction of $Nb(OEt)_5$ with $(Me_3Si)_2S$ and LiOMe. Ibers and co-workers have prepared the sulfido and selenido complexes $K_3[Nb(E)_4]$ (E = S, Se) (174) and $Cs_3[Nb(Se)_4]$ (175) by the direct reaction of the elements at elevated temperatures.

Coucouvanis et al. (176) have described the formation of the terminal niobium sulfido complexes $[Nb(S)_2(SBu^t)_2]^-$ and $[Nb(S)(SBu^t)_4]^-$ as a result of C—S bond cleavage upon reaction of $NbCl_5$ with approximately 6 equiv of $NaSBu^t$ (176); in the presence of only 2 equiv, Christou has (172) noted that $[Nb(S)Cl_4]^-$ is obtained. Finally, $[(\eta^2\text{-}SCH_2CH_2S)_3Nb]^-$ undergoes an unusual isomerization when exposed to water, methanol, or phenol, to give the sulfido complex $[(\eta^2\text{-}SCH_2CH_2S)(\eta^3\text{-}SCH_2CH_2SCH_2CH_2S)Nb(S)]^-$ (177).

3. *Tantalum Chalcogenido Complexes*

a. Tantalocene Derivatives. By comparison to niobium, tantalum is known to form very few oxo complexes. The most extensive series of terminal

TABLE XVII

Nb=E Bond Length and ν(Nb=E) IR Data for $Nb(PMe_3)_3(E)Cl_3$[a]

$Nb(PMe_3)_3(E)Cl_3$	d(Nb=E) (Å)	ν(Nb=E) (cm^{-1})
Yellow-$Nb(PMe_3)_3(O)Cl_3$	1.781(6)	882
Green-$Nb(PMe_3)_3(O)Cl_3$	1.929(6)	871
Orange-$Nb(PMe_3)_3(S)Cl_3$	2.196(2)	455
Green-$Nb(PMe_3)_3(S)Cl_3$	2.296(1)	489

[a]Taken from (75b).

TABLE XVIII
MrNb=E and ν(Nb=E) IR Data for Niobium Sulfido and Selenido Complexes

Complex	d(Nb=E) (Å)	ν(Nb=E) (cm^{-1})	Reference
$Li_3[Nb(S)_4] \cdot 2TMEDA$	2.274(1)	462	*a*
$K_3[Nb(S)_4]$	2.241(8)–2.258(8)		*b*
$K_3[Nb(Se)_4]$	2.387(1)–2.403(1)		*b*
$(\eta^2\text{-}Et_2NCS_2)_3NbS$	2.164(3) and 2.112(3)	497	*c*
	2.168(1) and 2.122(1)	493	*d*
$[(\eta^2\text{-}SC_2H_4S)(\eta^3\text{-}SC_2H_4SC_2H_4S)Nb(S)]^-$	2.192(3)	487	*e*
$(\eta^2\text{-}Et_2NCS_2)_2(\eta^2\text{-}Et_2NCS)NbS$	2.18[1]		*f*
$(tht)Nb(S)Br_3$	2.09(8)		*g*
$Nb(S)Cl_3 \cdot SPPh_3$	2.114(4), 2.129(4)	536	*h*
$[Nb(S)_2(SBu^t)_2]^-$	2.184(3), 2.194(2)	492	*i*
$[Nb(S)(SBu^t)_4]^-$	2.194(2)	465	*i*
$[Nb(S)(SPh)_4]^-$	2.171(2)	525	*j*
$[Nb(S)Cl_4]^-$	2.085(5)	552	*k*

[a] S. C. Lee, J. Li, J. C. Mitchell, and R. H. Holm, *Inorg. Chem.*, *31*, 4333 (1992).
[b] M. Latroche and J. A. Ibers, *Inorg. Chem.*, *29*, 1503 (1990).
[c] M. G. B. Drew, D. A. Rice, and D. M. Williams, *J. Chem. Soc. Dalton Trans.*, 1821 (1985).
[d] Y. Do. and R. H. Holm, *Inorg. Chim. Acta*, *104*, 33 (1985).
[e] K. Tatsumi, Y. Sekiguchi, A. Nakamura, R. E. Cramer, and J. J. Rupp, *J. Am. Chem. Soc.*, *108*, 1358 (1986).
[f] P. F. Gilletti, D. A. Femec, F. I. Keen, and T. M. Brown, *Inorg. Chem.*, *31*, 4008 (1992).
[g] M. G. B. Drew, D. A. Rice, and D. M. Williams, *J. Chem. Soc. Dalton Trans.*, 2251 (1983).
[h] M. G. B. Drew and R. J. Hobson, *Inorg. Chim. Acta*, *72*, 233 (1983).
[i] D. Coucouvanis, S. Al-Ahmad, C. G. Kim, and S.-M. Koo, *Inorg. Chem.*, *31*, 2996 (1992).
[j] J. L. Seela, J. C. Huffman, and G. Christou, *Polyhedron*, *8*, 1797 (1989).
[k] U. Müller and P. Z. Klingelhöfer, *Anorg. Allg. Chem.*, *510*, 109 (1984).

chalcogenido complexes of tantalum is that of the permethyltantalocene system, and includes the alkyl and hydride derivatives $Cp_2^*Ta(E)Me$ and $Cp_2^*Ta(E)H$ (E = O, S, Se, Te).

i. ***$Cp_2^*Ta(E)H$.*** The oxo complex $Cp_2^*Ta(O)H$, the first oxo hydride complex and also the first member of this series to be isolated, was initially synthesized by the reaction of $Cp_2^*Ta(CH_2)H$ with H_2O. Subsequently, hydrolysis of other tantalocene derivatives has been used to prepare this complex (Scheme 48) (178, 179). More recently, $Cp_2^*Ta(O)H$ has been obtained by the reaction of $Cp_2^*TaCl(thf)$ with NaOH, while the corresponding reaction with N_2O gives the oxo–chloride derivative $Cp_2^*Ta(O)Cl$ [$\nu(Ta{=}O) = 850\ cm^{-1}$] (180, 181). The sulfido- and selenido–hydride analogues have been synthesized by the reactions of $Cp_2^*TaH_3$ with H_2S (182) and H_2Se (48), respectively, as illustrated in Scheme 49.

A convenient entry into terminal selenido complexes is also provided via the

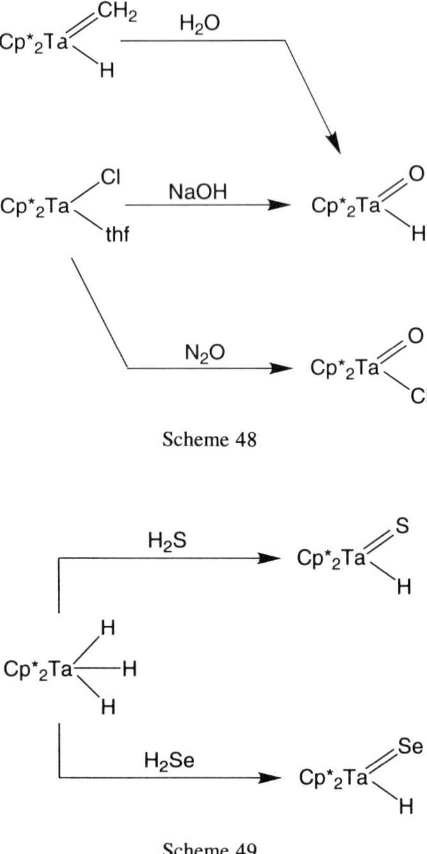

Scheme 48

Scheme 49

diselenido complex $Cp^*_2Ta(\eta^2\text{-}Se_2)H$, which is obtained by the reaction of $Cp^*_2TaH_3$ with elemental selenium at room temperature in the dark (Scheme 50) (53). In the presence of ambient light, however, $Cp^*_2Ta(\eta^2\text{-}Se_2)H$ is smoothly converted to the selenido hydroselenido complex $Cp^*_2Ta(Se)SeH$. Both $Cp^*_2Ta(\eta^2\text{-}Se_2)H$ and $Cp^*_2Ta(Se)SeH$ are converted to the selenido hydride complex $Cp^*_2Ta(Se)H$ upon reaction with PMe_3. The terminal tellurido complex $Cp^*_2Ta(Te)H$ is likewise prepared by the reaction of $Cp^*_2Ta(\eta^2\text{-}Te_2)H$ with mercury (to provide an effective driving force, presumably via formation of mercury telluride) in the presence of PMe_3 (Scheme 51) (183). An interesting contrast between the selenium and tellurium systems, however, is the observation that $Cp^*_2Ta(\eta^2\text{-}Te_2)H$ is stable with respect to $Cp^*_2Ta(Te)TeH$ under comparable conditions to which $Cp^*_2Ta(\eta^2\text{-}Se_2)H$ converts to $Cp^*_2Ta(Se)SeH$. The facile formation of the $Cp^*_2Ta(Se)SeH$ tautomer for the selenium system has been attrib-

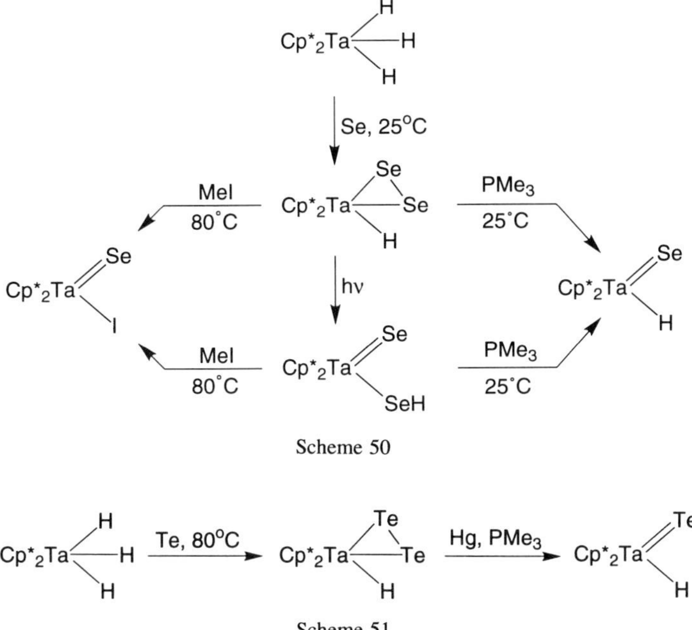

Scheme 50

Scheme 51

uted to (a) stronger Se—H versus Te—H bonds, and (b) the increased preference for the lighter element to partake in multiple bonding (35a, d).

In addition to the selenido hydride complex, the selenido–iodide derivative $Cp_2^*Ta(Se)I$ has been synthesized. Thus, $Cp_2^*Ta(Se)I$ is readily obtained from both $Cp_2^*Ta(\eta^2\text{-}Se_2)H$ and $Cp_2^*Ta(Se)SeH$ by reaction with MeI (Scheme 50). Interestingly, however, $Cp_2^*Ta(Se)I$ is *not* obtained as a product of the reaction of the selenido–hydride $Cp_2^*Ta(Se)H$ with MeI. In preference, the terminal selenido ligand is abstracted from the tantalum center giving the diiodide hydride complex $Cp_2^*TaHI_2$ (Scheme 52). Likewise, $Cp_2^*TaHI_2$ is also obtained from the reaction of $Cp_2^*Ta(Te)H$ with MeI.

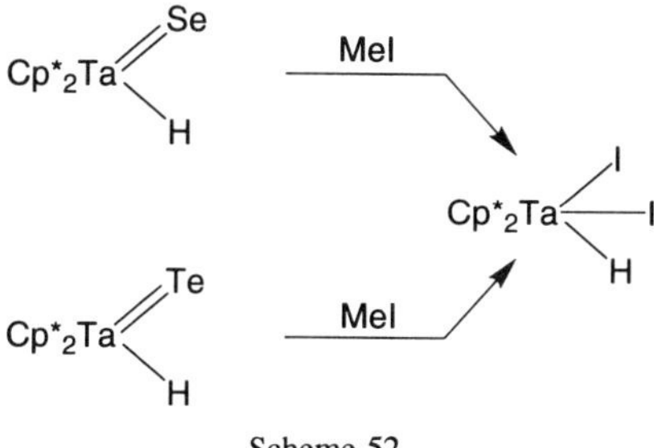

Scheme 52

TABLE XIX
Selected Data for $Cp_2^*Ta(E)H$[a]

Atom	d(M=E) (Å)	δ(Ta=H) (ppm)	δ(E) (ppm)	ν(Ta—H) (cm^{-1})	ν(Ta=E) (cm^{-1})
O	1.72(3)	7.45		1808	850
S	2.241(7)	8.05		1830	440
Se	2.329(2)	8.91	2153 (^{77}Se)	1844	
Te	2.588(2)	10.87	3085 (^{125}Te)	1837	

[a]Data taken from G. Parkin, A. van Asselt, D. J. Leahy, L. Whinnery, N. G. Hua, R. Quan, L. M. Henling, W. P. Schaefer, B. D. Santarsiero, and J. E. Bercaw, *Inorg. Chem.*, *31*, 82 (1992). J. H. Shin and G. Parkin, unpublished results. J. E. Nelson, G. Parkin, and J. E. Bercaw, *Organometallics*, *11*, 2181 (1992). J. H. Shin and G. Parkin, *Organometallics*, *14*, 1104 (1995). J. H. Shin, and G. Parkin, *Organometallics*, *13*, 2147 (1994).

The molecular structures of $Cp_2^*Ta(O)H$, $Cp_2^*Ta(Se)H$, and $Cp_2^*Ta(Te)H$ have been determined by X-ray diffraction. The Ta=O bond length in $Cp_2^*Ta(O)H$ is 1.72(3) Å; for comparison purposes, although there are no other terminal tantalum oxo complexes contained in the CSD, the structure of $[(Me_2CH)_2N]_3Ta{=}O$ has been cited as having a Ta=O bond length of 1.725(7) Å (184). The ν(Ta=O) stretching frequency of 850 cm^{-1} for $Cp_2^*Ta(O)H$, however, provides a good indication that the bond order is less than that in typical tantalum monooxo complexes, which commonly have ν(Ta=O) stretching frequencies in the range 905–935 cm^{-1} (185). Selected data for the series of complexes $Cp_2^*Ta(E)H$ are listed in Table XIX.

Relatively little reactivity has been described for the chalcogenido hydride complexes $Cp_2^*Ta(E)H$. Nevertheless, the oxo complex $Cp_2^*Ta(O)H$ has been observed to undergo rapid exchange of oxygen atoms with labeled H_2O, but without exchange of hydrogen atoms. Specifically, $Cp_2^*Ta(^{18}O)H$ reacts with $D_2^{16}O$ to give $Cp_2^*Ta(^{16}O)H$ and not $Cp_2^*Ta(^{16}O)D$ (Scheme 53). The absence of Ta–H/D exchange implies that α-hydrogen migration to oxygen and the formation of a hydroxy intermediate $[Cp_2^*Ta{-}OH]$ is not involved in the exchange process (51b). A related observation is that $Cp_2^*Ta(NPh)H$ reacts with D_2O to

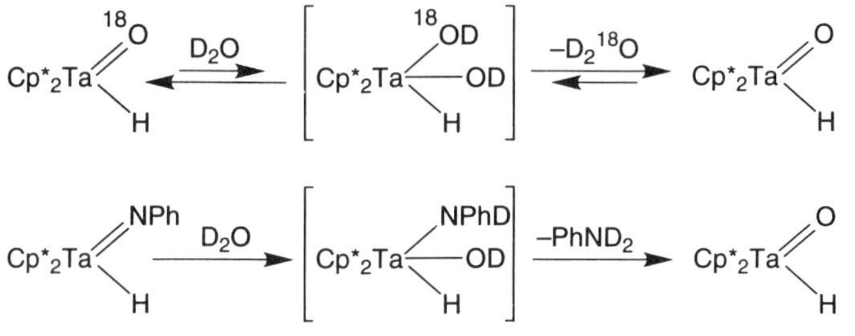

$Cp^*_2Ta(^{18}O)H \underset{}{\overset{D_2O}{\rightleftharpoons}} [Cp^*_2Ta(^{18}OD)(OD)H] \overset{-D_2{}^{18}O}{\rightleftharpoons} Cp^*_2Ta(O)H$

$Cp^*_2Ta(NPh)H \xrightarrow{D_2O} [Cp^*_2Ta(NPhD)(OD)H] \xrightarrow{-PhND_2} Cp^*_2Ta(O)H$

Scheme 53

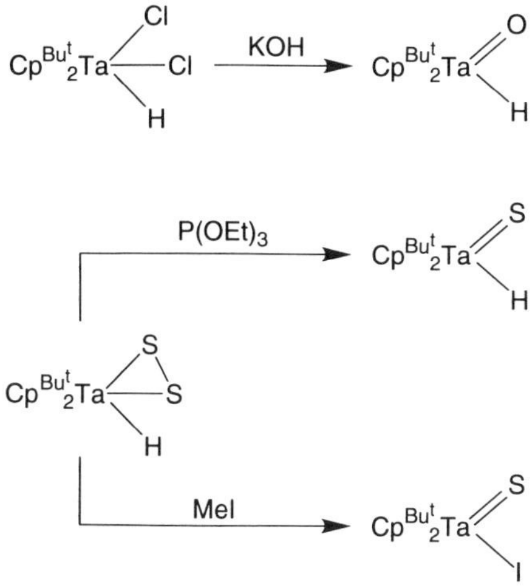

Scheme 54

give $Cp_2^*Ta(O)H$ and not $Cp_2^*Ta(O)D$, likewise implying that the mechanism does not involve α-hydrogen migration to nitrogen. Such observations are in accord with other studies indicating that migrations to heteroatoms have significantly higher activation barriers than to carbon (186).

In addition to the above permethyltantalocene complexes, $Cp_2^*Ta(E)H$, the less sterically demanding *tert*-butylcyclopentadienyl ligand has been used to afford terminal oxo and sulfido hydride complexes (Scheme 54). Thus, $Cp_2^{Bu^t}Ta(O)H$ [$\nu(Ta{=}O) = 850\ cm^{-1}$; $\nu(Ta{-}H) = 1819\ cm^{-1}$; $\delta(Ta{-}\underline{H}) = 6.97$ ppm] is obtained by the reaction of $Cp_2^{Bu^t}TaCl_2H$ with KOH (187), while $Cp_2^{Bu^t}Ta(S)H$ [$\nu(Ta{=}S) = 434\ cm^{-1}$; $\nu(Ta{-}H) = 1859\ cm^{-1}$; $\delta(Ta{-}\underline{H}) = 7.10$ ppm] has been prepared by the reaction of $Cp_2^{Bu^t}Ta(\eta^2\text{-}S_2)H$ with $P(OEt)_3$ (188). The sulfido iodide complex $Cp_2^{Bu^t}Ta(S)I$ [$\nu(Ta{=}S) = 434\ cm^{-1}$] has likewise been obtained from the reaction of $Cp_2^{Bu^t}Ta(\eta^2\text{-}S_2)H$ with MeI. The sulfido complex $Cp_2^{Bu^t}Ta(S)H$ undergoes a formal [2 + 2] addition with PhNCS to give $Cp_2^{Bu^t}Ta[\eta^2\text{-}SC(S)NPh]H$, and reacts with HCl to give $Cp_2^{Bu^t}TaCl_2H$ (Scheme 55) (187).

ii. $Cp_2^*Ta(E)R$. A general method of preparing chalcogenido methyl complexes involves the thermal isomerization of $Cp_2^*Ta(\eta^2\text{-}ECH_2)H$ derivatives, which are synthesized according to Scheme 56. For example, Bercaw and coworkers (182, 189, 190) prepared the formaldehyde and thioformaldehyde complexes $Cp_2^*Ta(\eta^2\text{-}OCH_2)H$ and $Cp_2^*Ta(\eta^2\text{-}SCH_2)H$ by the reactions of Cp_2^*-$Ta(CH_2)H$ and $Cp_2^*Ta(CCH_2)H$ with MeOH and MeSH, respectively. However, since MeSeH and MeTeH are not as readily available as the former reagents, different methods have been developed to synthesize the selenoformal-

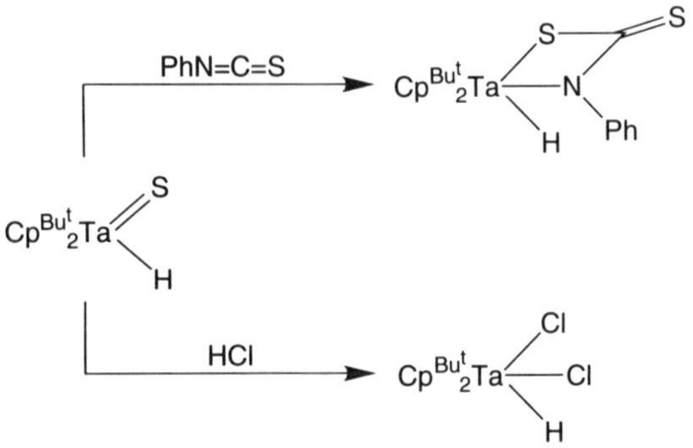

Scheme 55

dehyde and telluroformaldehyde complexes $Cp_2^*Ta(\eta^2\text{-}SeCH_2)H$ and $Cp_2^*Ta(\eta^2\text{-}TeCH_2)H$. Thus, $Cp_2^*Ta(\eta^2\text{-}TeCH_2)H$ is prepared by the addition of tellurium to the $[Ta{=}CH_2]$ double bond of $Cp_2^*Ta(CH_2)H$, using PMe_3 as a catalyst (191), while $Cp_2^*Ta(\eta^2\text{-}SeCH_2)H$ is obtained by functionalization of the terminal selenido ligand of $Cp_2^*Ta(Se)I$ by reaction with MeLi (53). The choice of alkylating agent is critical since, as described below, the Grignard reagent MeMgI gives exclusively the tantalum–methyl derivative $Cp_2^*Ta(Se)Me$ rather than $Cp_2^*Ta(\eta^2\text{-}SeCH_2)H$. Although the complexes $Cp_2^*Ta(\eta^2\text{-}ECH_2)H$ (E = O, S,

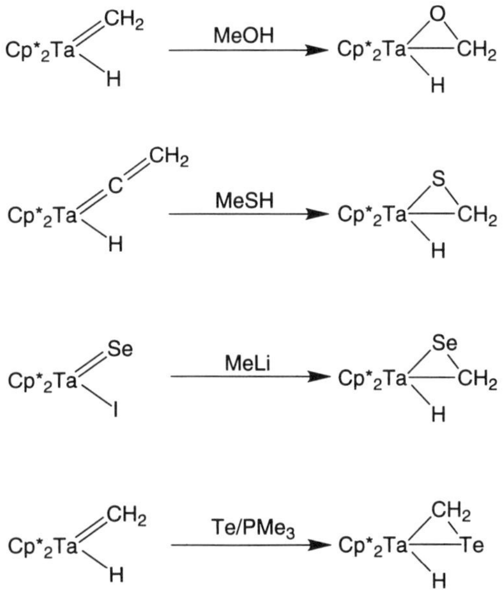

Scheme 56

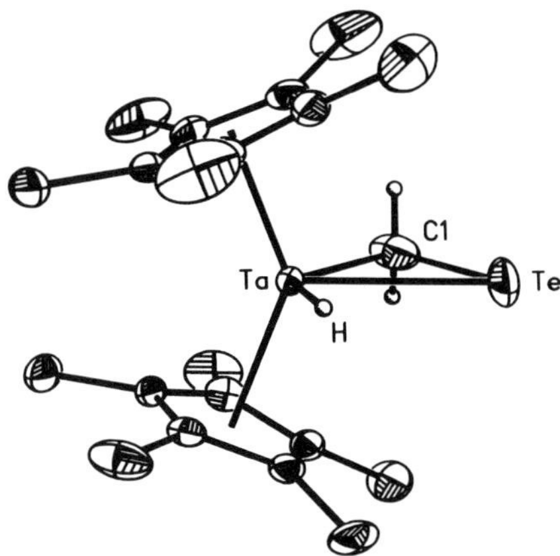

Figure 20. Molecular structure of $Cp_2^*Ta(\eta^2\text{-}CH_2Te)H$. [Reprinted with permission from J. H. Shin and G. Parkin, *Organometallics*, *13*, 2147 (1994). Copyright © 1994 American Chemical Society.]

Se, Te) are formally analogous, the structure of the telluroformaldehyde derivative (Fig. 20) is unique in that the $[\eta^2\text{-}CH_2Te]$ moiety adopts an orientation in which the CH_2 group is located in the lateral equatorial position, rather than in the central site as observed for the other derivatives $Cp_2^*Ta(\eta^2\text{-}ECH_2)H$ (E = O, S, Se).

Each of the complexes $Cp_2^*Ta(\eta^2\text{-}ECH_2)H$ (E = O, S, Se, Te) isomerize to the respective chalcogenido–methyl derivative at elevated temperatures (Scheme 57). For the oxygen, sulfur, and selenium complexes, the isomerizations are first order, with the reactions of the d_3-derivatives $Cp_2^*Ta(\eta^2\text{-}ECD_2)D$ being characterized by an inverse kinetic isotope effect ($k_H/k_D < 1$; see Table XX). The inverse kinetic isotope effects have been interpreted in terms of a mechanism involving a pre-equilibrium with $[Cp_2^*TaEMe]$, followed by rate-determining α-methyl elimination (Scheme 58). Consistent with a rate-determining step that involves cleavage of the E—Me bond, the rate constants for the isomerization increase in the order O < S < Se (Table XX). Interestingly, however, the tellurium derivative is anomalously slow (48), an observation that is proposed to be a result of the telluroformaldehyde complex adopting a different structural type to those of the other derivatives. Specifically, with the tellurium

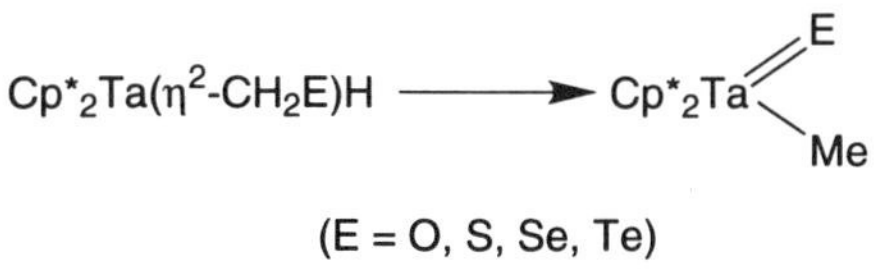

$$Cp^*_2Ta(\eta^2\text{-}CH_2E)H \longrightarrow Cp^*_2Ta(=E)(Me)$$

(E = O, S, Se, Te)

Scheme 57

TABLE XX

Rate Constants and Kinetic Isotope Effects for Isomerization of $Cp^*_2Ta(\eta^2\text{-}ECH_2)H$ to $Cp^*_2Ta(E)Me$

Atom	k/s^{-1} at 100°C[a]	k_H/k_D
O	$9.5(3) \times 10^{-8}$	0.46(3) [140°C][b]
S	$6.0(3) \times 10^{-6}$	0.72(3) [138°C][c]
Se	$1.5(1) \times 10^{-3}$	0.6(1) [110°C][d]
Te	$\sim 2 \times 10^{-4}$–2×10^{-5}[e]	

[a]J. H. Shin, Ph. D. thesis, "Syntheses, Structures and Reactivity of Permethyltantalocene and *tert*-butylmolybdenocene derivatives," Columbia University, New York, 1996.

[b]A. van Asselt, B. J. Burger, V. C. Gibson, and J. E. Bercaw, *J. Am. Chem. Soc.*, *108*, 5347 (1986).

[c]J. E. Nelson, G. Parkin, and J. E. Bercaw, *Organometallics*, *11*, 2181 (1992).

[d]J. H. Shin and G. Parkin, *Organometallics*, *14*, 1104 (1995).

[e]Observed pseudo-first-order rate constant varies as a function of concentration.

atom occupying the central equatorial position, the hydride and methylene moieties of $Cp^*_2Ta(\eta^2\text{-}TeCH_2)H$ do not adopt a suitable disposition to allow coupling. Thus, the mechanism proposed for the oxygen, sulfur, and selenium derivatives is not applicable to $Cp^*_2Ta(\eta^2\text{-}TeCH_2)H$ in the absence of additional steps, such as prior intramolecular rearrangement to bring the CH_2 and H groups into a cis disposition. However, several observations suggest that a prior rear-

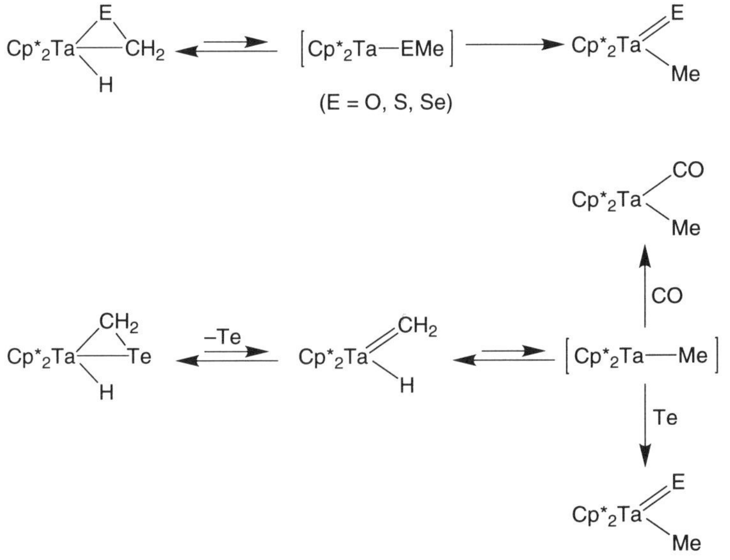

Scheme 58

rangement of the (η^2-$TeCH_2$) ligand is not involved in the isomerization. For example, the first-order rate constant for the isomerization varies with sample preparation, thereby suggesting that the isomerization is not a simple intramolecular rearrangement. Furthermore, the isomerization is catalyzed by PMe_3, and in the presence of CO the carbonyl–methyl derivative $Cp_2^*Ta(CO)Me$ is also obtained (192). The mechanism proposed for the isomerization (Scheme 58) comprises a sequence involving (a) dissociation of Te (which is catalyzed by PMe_3, see above) generating the methylene hydride complex $Cp_2^*Ta(CH_2)H$, (b) α-H migration to give [$Cp_2^*Ta{-}Me$], and (c) subsequent trapping with Te to give $Cp_2^*Ta(Te)Me$. Such a mechanism also accounts for the formation of the carbonyl–methyl derivative $Cp_2^*Ta(CO)Me$, which is independently known to be obtained by trapping of [$Cp_2^*Ta{-}Me$] with CO.

Relatively little reactivity of the above chalcogenido methyl complexes has been described, with the exception of the slow hydrogenation of $Cp_2^*Ta(S)Me$ to give $Cp_2^*Ta(S)H$ (182).

Permethyltantalocene chalcogenido alkyl complexes have also been obtained by other methods. For example, oxo alkyl complexes, $Cp_2^*Ta(O)R$, are available via the peroxo derivatives $Cp_2^*Ta(\eta^2\text{-}O_2)R$, which themselves are conveniently obtained by the reactions of [Cp_2^*TaR] precursors with O_2. Thus, $Cp_2^*Ta(CH_2)H$ reacts with O_2 to give a mixture of $Cp_2^*Ta(\eta^2\text{-}O_2)Me$ and $Cp_2^*Ta(O)Me$ (193), while $Cp_2^*Ta(CHPh)H$, $Cp_2^*Ta(\eta^2\text{-}C_6H_4)H$, $Cp_2^*Ta(\eta^2\text{-}C_2H_4)H$, and $Cp_2^*Ta(\eta^2\text{-}CH_2{=}CHMe)H$ react with O_2 to give $Cp_2^*Ta(\eta^2\text{-}O_2)R$ (R = CH_2Ph, Ph, Et, Pr), as illustrated in Scheme 59. These peroxo complexes are capable of oxidizing a limited number of substrates, including SO_2, PMe_3, and PPh_3, thereby generating the corresponding oxo–alkyl derivative $Cp_2^*Ta(O)R$ (Scheme 60). In the absence of a suitable substrate, the peroxo complexes $Cp_2^*Ta(\eta^2\text{-}O_2)R$ undergo an acid-catalyzed rearrangement to the oxo–alkoxo derivatives $Cp_2^*Ta(O)OR$.

Oxo alkyl complexes have also been synthesized by the reactions of [Cp_2^*TaR] precursors with epoxides, accompanied by the concomitant release of the olefin (194). For example, $Cp_2^*Ta(CH_2)H$ abstracts an oxygen atom from styrene oxide to give $Cp_2^*Ta(O)Me$, while $Cp_2^*Ta(CHPh)H$ abstracts an oxygen atom from ethylene oxide to give $Cp_2^*Ta(O)CH_2Ph$ (Scheme 61). Interestingly, mechanistic studies demonstrate that these oxo-transfer reactions do not proceed via four-membered metallaoxetane intermediates, with the consequence that the microscopic reverse, that is, olefin epoxidation, likewise does not involve such an intermediate for this specific system (195). Specifically, the potential intermediate metallaoxetane complexes $Cp_2^*Ta[\eta^2\text{-}OCH(R)CH_2]Me$ have been isolated (by the reaction of $Cp_2^*Ta(CH_2)Me$ with RCHO), and while the O-anti isomers eliminate olefin to give the oxo methyl complex $Cp_2^*Ta(O)Me$ (Scheme 62), the rates of these reactions are slower than the rate of oxygen-atom abstraction from epoxides by $Cp_2^*Ta(CH_2)H$. Consequently, metallaoxetanes cannot

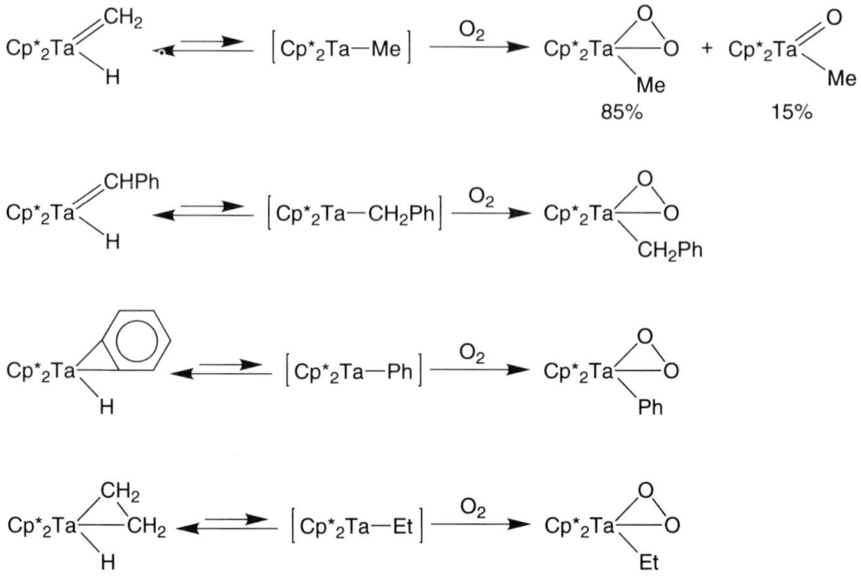

Scheme 59

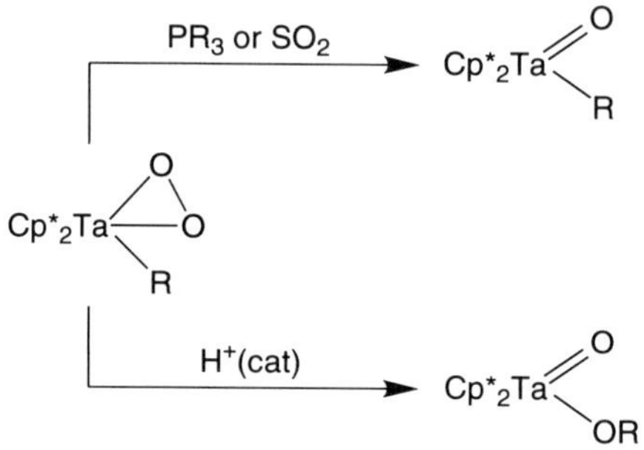

Scheme 60

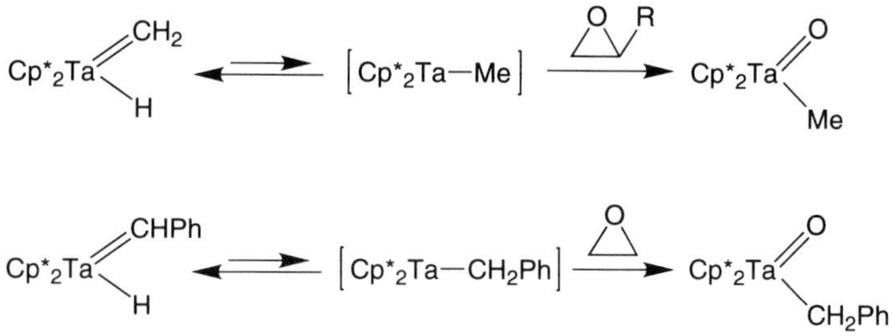

Scheme 61

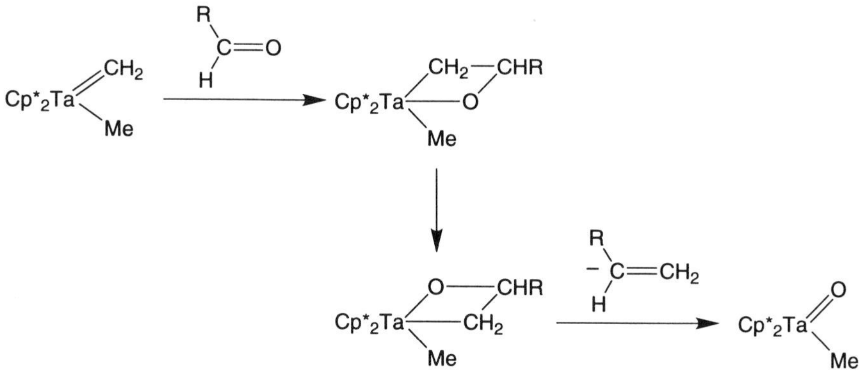

Scheme 62

be intermediates for the transformation, and the oxo transfer may proceed as illustrated in Scheme 63.

Related to the deoxygenation of epoxides, episulfides are also capable of transferring sulfur to [Cp_2^*TaR] intermediates, thereby generating sulfido alkyl complexes (196). For example, $Cp_2^*Ta(\eta^2\text{-}C_6H_4)H$ and $Cp_2^*Ta(\eta^2\text{-}CH_2CHPh)H$ react with ethylene sulfide to give $Cp_2^*Ta(S)Ph$ and $Cp_2^*Ta(S)CH_2CH_2Ph$, respectively (Scheme 64). Furthermore, sulfido alkyl complexes have also been prepared by isomerization of thioaldehyde derivatives, akin to the synthesis of $Cp_2^*Ta(S)Me$ from the thioformaldehyde complex $Cp_2^*Ta(\eta^2\text{-}SCH_2)H$ described above. For example, $Cp_2^*Ta(S)CH_2R$ (R = Ph, CH_2Ph, CH_2Bu^t) have been synthesized by isomerization of $Cp_2^*Ta(\eta^2\text{-}SCHR)H$ (Scheme 65). Deuterium labeling studies demonstrate that the migration of the [CHDCHDPh] group proceeds with greater than or equal to 85% retention of stereochemistry at carbon.

Several selenido–alkyl derivatives have been synthesized by functionalization of the selenido iodide complex $Cp_2^*Ta(Se)I$ with Grignard reagents, as illustrated in Scheme 66; for example, $Cp_2^*Ta(Se)Me$, $Cp_2^*Ta(Se)Bu$,

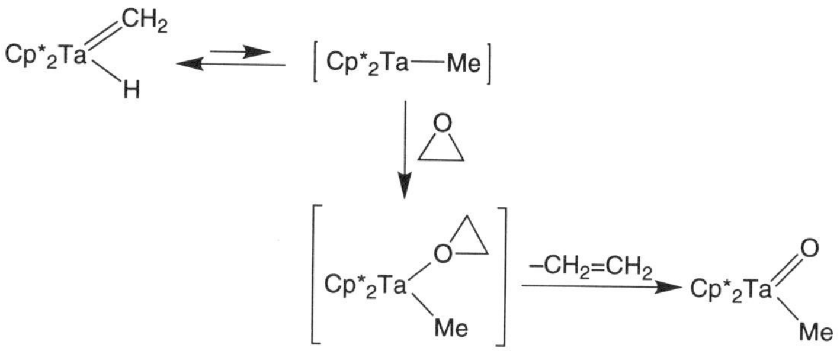

Scheme 63

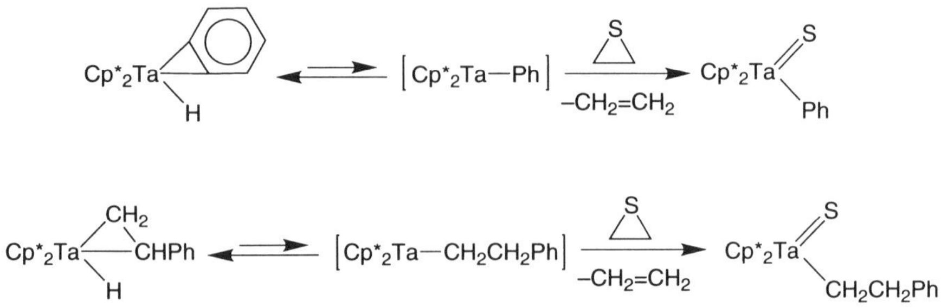

$Cp^*_2Ta(\eta^2\text{-}C_6H_4)H \rightleftharpoons [Cp^*_2Ta{-}Ph] \xrightarrow[-CH_2{=}CH_2]{\text{ethylene sulfide}} Cp^*_2Ta(=S)Ph$

$Cp^*_2Ta(CH_2CHPh)H \rightleftharpoons [Cp^*_2Ta{-}CH_2CH_2Ph] \xrightarrow[-CH_2{=}CH_2]{\text{ethylene sulfide}} Cp^*_2Ta(=S)CH_2CH_2Ph$

Scheme 64

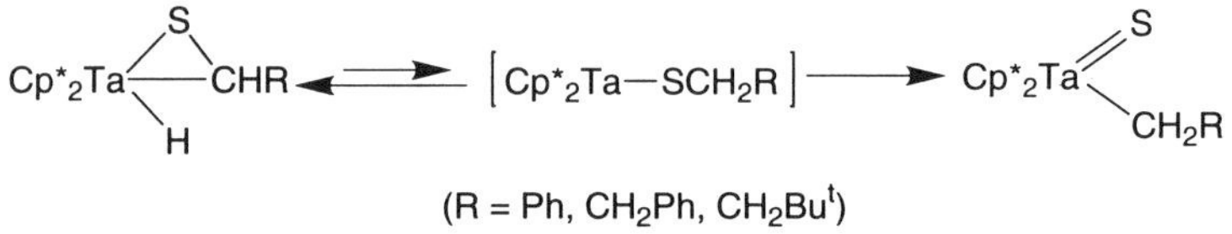

$Cp^*_2Ta(SCHR)H \rightleftharpoons [Cp^*_2Ta{-}SCH_2R] \longrightarrow Cp^*_2Ta(=S)CH_2R$

(R = Ph, CH_2Ph, CH_2Bu^t)

Scheme 65

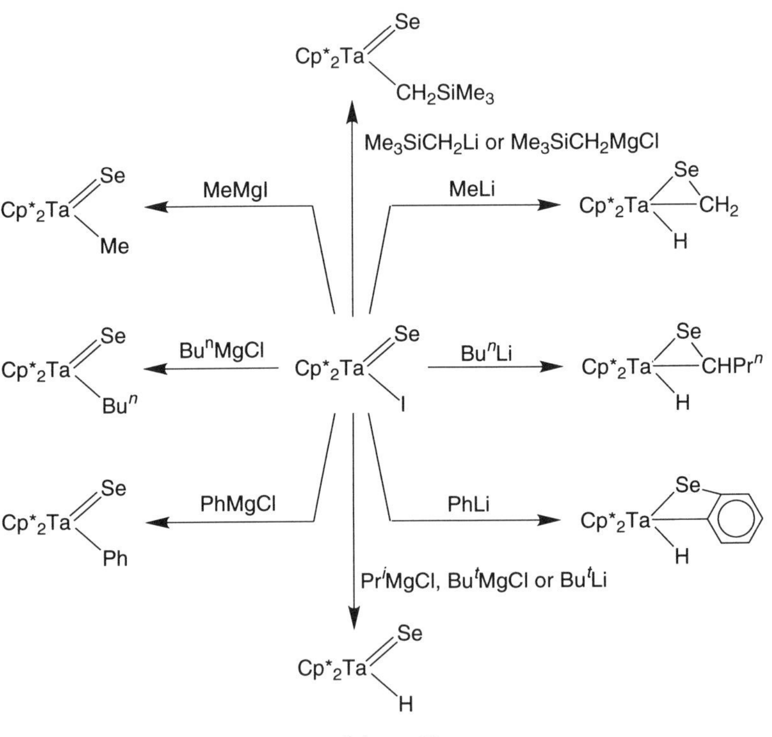

Scheme 66

$Cp_2^*Ta(Se)CH_2SiMe_3$, and $Cp_2^*Ta(Se)Ph$ have been prepared by such a method. The structures of $Cp_2^*Ta(Se)(CH_2SiMe_3)$ [d(Ta=Se) = 2.372(1) Å] and $Cp_2^*Ta(Se)Ph$ [d(Ta=Se) = 2.370(2) Å] have been determined by X-ray diffraction. The specific choice of alkylating agent is most important since, as indicated above, alkyllithium reagents (e.g., MeLi) functionalize the selenido ligand giving selenoaldehyde derivatives. Finally, an additional type of reactivity is illustrated by the reactions with the secondary and tertiary alkyl derivatives Pr^iMgCl, Bu^tMgCl, and Bu^tLi, which yield the selenido–hydride derivative, $Cp_2^*Ta(Se)H$, as a consequence of β-H elimination.

The ^{77}Se NMR data for a series of permethyltantalocene selenido complexes and some related derivatives with Ta—Se single bonds are compiled in Table XXI.

In addition to the above $Cp_2^*Ta(E)R$ complexes, oxo– and sulfido–methyl derivatives of the parent tantalocene system have been synthesized by Bergman (197). For example, $Cp_2Ta(CH_2)Me$ reacts with a variety of metal carbonyl complexes to give $Cp_2Ta(O)Me$ as a byproduct (197). For example, $Re_2(CO)_{10}$ reacts with $Cp_2Ta(CH_2)Me$ to give $Cp_2Ta(O)Me$ and $Cp_2Ta(C_3H_3)Re_2(CO)_9$, with the initial step of this reaction proposed to be metathesis between the Ta=CH_2 and the C=O moieties (Scheme 67). It is yet to be established, however, whether $Cp_2Ta(O)Me$ is actually a monomer or an oligomer. More re-

TABLE XXI
^{77}Se NMR Data for Permethyltantalocene Complexes[a]

Complex	δ_s (ppm)[b]	δ_d (ppm)[c]
$Cp_2^*Ta(\eta^2\text{-}SeCH_2)H$	−595	
$Cp_2^*Ta(\eta^2\text{-}SeCHPr)H$	−425	
$Cp_2^*Ta(\eta^2\text{-}SeC_6H_4)H$	−186	
$Cp_2^*Ta\eta^2\text{-}Se_2)H$	−408, 54	
$Cp_2^*Ta(SePh)H_2$	131	
$Cp_2^*Ta(Se)SeH$	−209	2363
$Cp_2^*Ta(Se)Ph$		2293
$Cp_2^*Ta(Se)H$		2153
$Cp_2^*Ta(Se)Bu$		2094
$Cp_2^*Ta(Se)Et$		2085
$Cp_2^*Ta(Se)CH_2SiMe_3$		2024
$Cp_2^*Ta(Se)Me$		1990

[a]Data were taken from J. H. Shin, Ph. D. Thesis, "Syntheses, Structures and Reactivity of Permethyltantalocene and *tert*-butylmolybdenocene derivatives," Columbia University, New York 1996.
[b]Chemical shifts for [Ta—Se] single-bonded moieties.
[c]Chemical shifts for [Ta=Se] double bonded moieties.

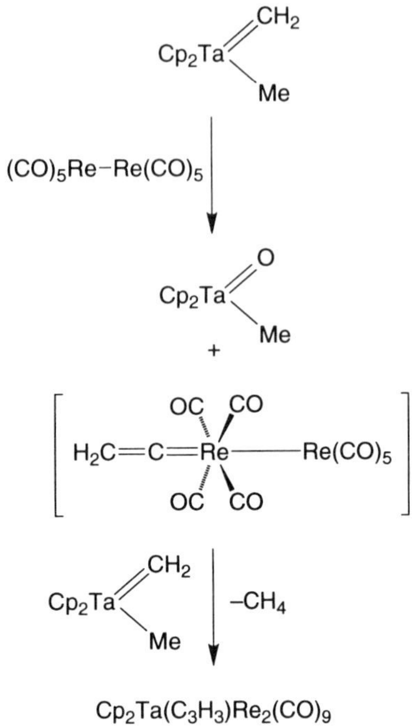

$Cp_2Ta(=CH_2)Me$

$(CO)_5Re-Re(CO)_5$

$Cp_2Ta(=O)Me$ + $[H_2C=C=Re(CO)_4-Re(CO)_5]$

$Cp_2Ta(=CH_2)Me$, $-CH_4$

$Cp_2Ta(C_3H_3)Re_2(CO)_9$

Scheme 67

cently, $Cp_2Ta(O)Me$ has been synthesized by the reaction of $[Cp_2TaMe]$, generated by either photolysis of $Cp_2Ta(\eta^2\text{-}C_2H_4)Me$ or thermolysis of $Cp_2Ta(PMe_3)Me$, with the epoxide $(MeCH)_2O$, as illustrated in Scheme 68 (198). Likewise, the corresponding thiirane yields the terminal sulfido complex $Cp_2Ta(S)Me$. The latter complex reacts with $(C_2H_4)S$ in the dark to give the disulfido complex $Cp_2Ta(\eta^2\text{-}S_2)Me$, as illustrated in Scheme 69; however, under photolysis conditions, isomeric $Cp_2Ta(S)SMe$ is obtained in preference. The sulfido complex $Cp_2Ta(S)Me$ may be regenerated upon treatment of $Cp_2Ta(\eta^2\text{-}S_2)Me$ with PR_3.

b. Tris(amido)amine (Azatrane) Derivatives. An azatrane ligand has been used to prepare the terminal oxo, selenido, and tellurido complexes $[\eta^4\text{-}N(CH_2CH_2NSiMe_3)_3]TaE$ (E = O, Se, Te). Thus, Schrock and co-workers (199) synthesized the oxo complex $[\eta^4\text{-}N(CH_2CH_2NSiMe_3)_3]TaO$ by the reaction of the phosphinidene complexes $[\eta^4\text{-}N(CH_2CH_2NSiMe_3)_3]TaPR$ (R = Ph, Cy, Bu^t) with aldehydes (Scheme 70), while Arnold and co-worker (200) synthesized the selenido and tellurido complexes $[\eta^4\text{-}N(CH_2CH_2NSiMe_3)_3]TaSe$

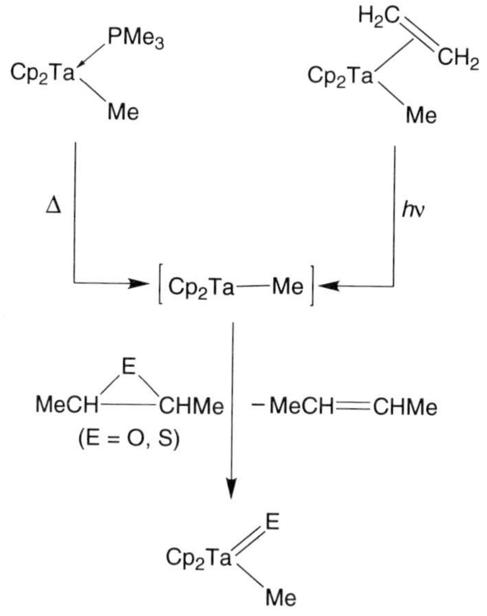

Scheme 68

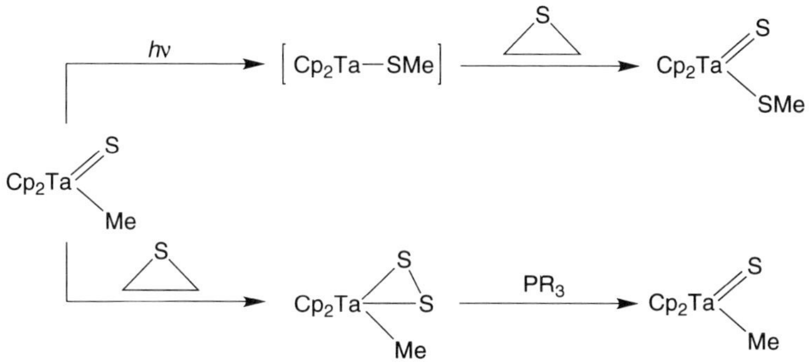

Scheme 69

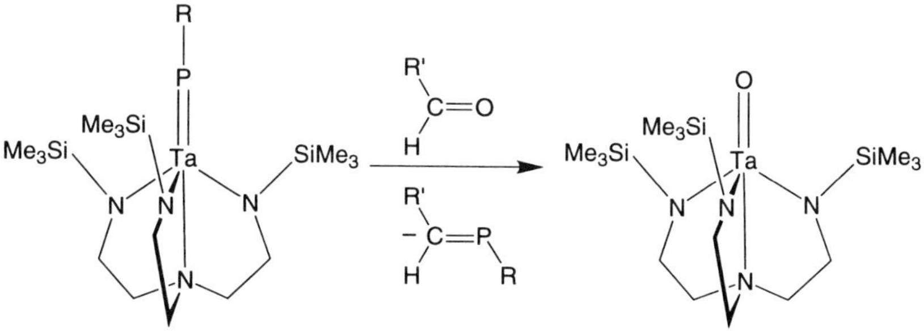

Scheme 70

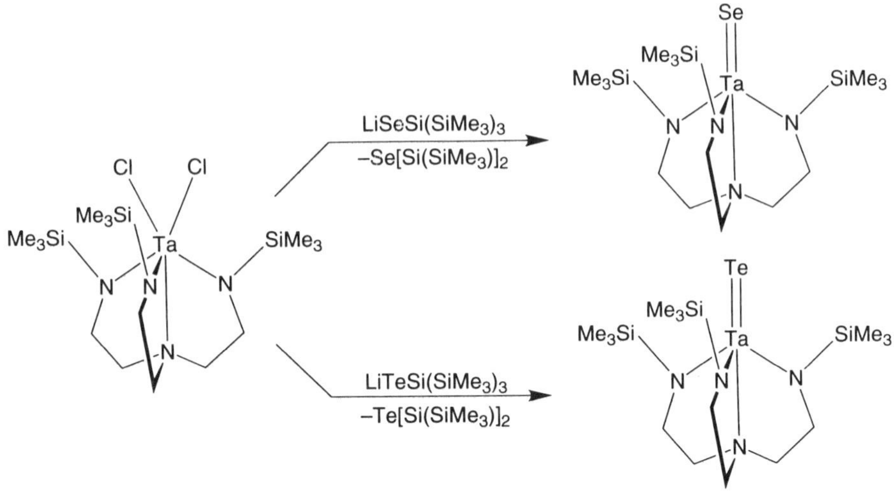

Scheme 71

and [η^4-N($CH_2CH_2NSiMe_3)_3$]TaTe by the reactions of [η^4-N($CH_2CH_2NSiMe_3)_3$]TaCl_2 with (thf)$_2$LiESi($SiMe_3)_2$ (Scheme 71). The latter reactions are proposed to occur via elimination of [($Me_3Si)_3Si]_2$E from a bis(chalcogenolate) intermediate [η^4-N($CH_2CH_2NSiMe_3)_3$]Ta[ESi($SiMe_3)_2]_2$. Selected data for [η^4-N($CH_2CH_2NSiMe_3)_3$]TaE (E = Se, Te) are summarized in Table XXII. The structures of [η^4-N($CH_2CH_2NSiMe_3)_3$]TaSe and [η^4-N($CH_2CH_2NSiMe_3)_3$]TaTe have been determined by X-ray diffraction and both exhibit a distorted trigonal bipyramidal geometry. However, the two complexes differ in the degree of interaction between tantalum and the axial nitrogen donor: The Ta—N bond length in the selenido derivative [2.349(2) Å] is significantly shorter than in the tellurido derivative [2.487(5) Å], an observation that has been attributed to the more electronegative selenido ligand promoting the trans dative interaction.

c. Miscellaneous Derivatives. Wolczanski and co-workers (201) demonstrated that the three coordinate Ta^{III} complex (silox)$_3$Ta (silox =Bu^t_3 SiO) reacts

TABLE XXII
Selected Data for [η^4-N($CH_2CH_2NSiMe_3)_3$]TaE[a]

	[η^4-N($CH_2CH_2NSiMe_3)_3$]TaSe	[η^4-N($CH_2CH_2NSiMe_3)_3$]TaTe
d(Ta=E) (Å)	2.330(1)	2.568(1)
d(Ta—N_{eq})	2.011(4)	2.000(5), 1.986(5), 1.982(5)
d(Ta—N_{ax})	2.349(2)	2.487(5)
δ ^{77}Se or ^{125}Te NMR (ppm)	1518	1681
ν(Ta=E) (cm^{-1})	507	460

[a]Taken from (200).

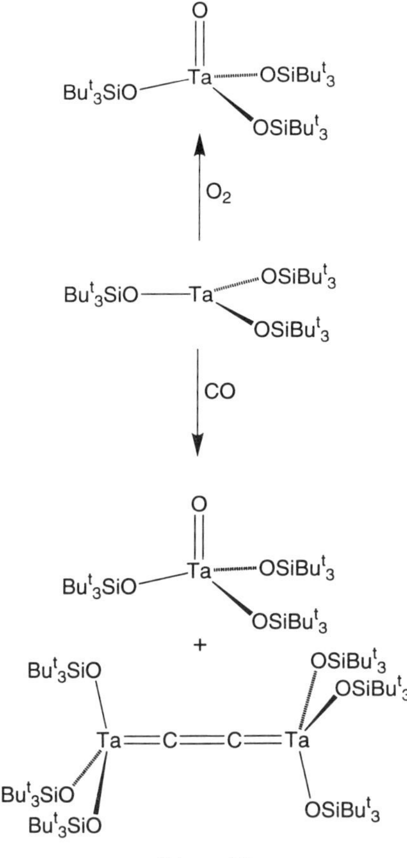

Scheme 72

rapidly with O_2 at room temperature to give the oxo derivative $(\text{silox})_3\text{TaO}$, identified by $\nu(\text{Ta=O}) = 905\ \text{cm}^{-1}$ (Scheme 72). More impressively, $(\text{silox})_3\text{Ta}$ is also capable of cleaving the carbon–oxygen bond of carbon monoxide to give the oxo and carbide complexes, $(\text{silox})_3\text{TaO}$ and $[(\text{silox})_3\text{Ta}]_2(\mu\text{-C}_2)$.

Other examples of molecular terminal sulfido complexes of tantalum include $(\eta^2\text{-PhSCH}_2\text{CH}_2\text{SPh})\text{TaCl}_3(\text{S})$ [$d(\text{Ta = S}) = 2.204(5)$ Å; $\nu(\text{Ta=S}) = 510$ and $516\ \text{cm}^{-1}$] (202) $(\eta^2\text{-Et}_2\text{NCS}_2)_3\text{TaS}$ [$d(\text{Ta=S}) = 2.181(1)$ Å (203); $\nu(\text{Ta=S}) = 479\ \text{cm}^{-1}$ (204, 205)], and $(\eta^2\text{-Et}_2\text{NCS}_2)_2(\eta^2\text{-Et}_2\text{NCS})\text{TaS}$ (206). Finally, as with its niobium counterpart, the tetrathiometallate $[\text{Ta(S)}_4]^{3-}$ has only recently been synthesized by the reaction of Ta(OEt)_5 with $(\text{Me}_3\text{Si})_2\text{S}$ and LiOMe, and the Ta=S bond length [2.279(2) Å] is very similar to that of the niobium derivaitve [2.274(1) Å] (173). The related derivatives $\text{K}_3[\text{Ta(E)}_4]$ (E = S, Se) (174) and $\text{Cs}_3[\text{Ta(Se)}_4]$ (207) have been prepared in the solid state by the direct reaction of the elements at elevated temperatures; the Ta=Se bond lengths in the latter derivative are in the range 2.369(4) to 2.397(6) Å.

D. Terminal Chalcogenido Complexes of Group 6 (VI B): Cr, Mo, W

1. Chromium Chalcogenido Complexes

Structurally characterized examples of terminal chalcogenido complexes of chromium are presently limited to oxo derivatives, some examples of which are illustrated in Fig. 21 (208, 209). Furthermore, chromium oxo alkyl complexes, for example, $Cp^*Cr(O)_2Me$ and $Cp^*Cr(O)Me_2$, have also recently been synthesized by the reactions of Cp^*CrMe_2py with O_2 and Me_3NO, respectively (Scheme 73) (210). High-valent chromium oxo complexes such as CrO_3 and CrO_2Cl_2 are useful oxidizing agents; indeed, the latter reagent is capable of functionalizing cyclohexane by a mechanism that is proposed to involve initial

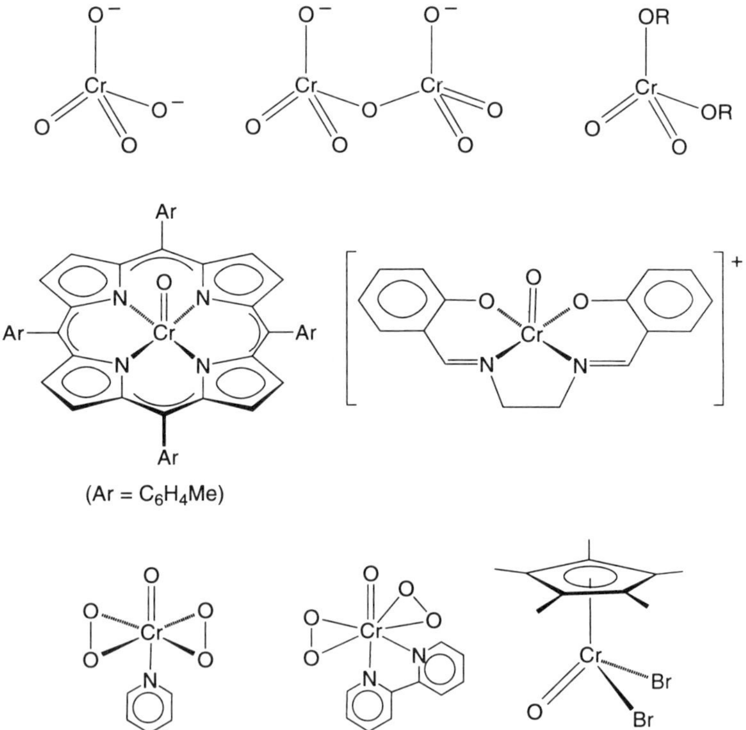

Figure 21. Representative examples of chromium oxo complexes.

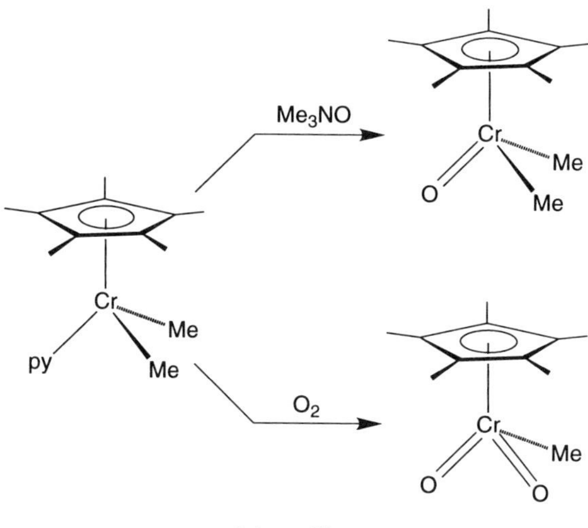

Scheme 73

hydrogen-atom transfer from cyclohexane to $Cr(O)_2Cl_2$ (211). Chromium oxo species, such as $Cr(O)_2(CO)_2$, have also been generated and spectroscopically identified upon photolysis of $Cr(CO)_6$ in O_2-doped argon matrices at low temperature (212, 213). Furthermore, $[Cr(O)_2(CO)_3]^-$ has been generated by the reaction of $[Cr(CO)_5]$ with O_2 in the gas phase (214).

2. *Molybdenum Chalcogenido Complexes*

Terminal chalcogenido complexes of molybdenum are highly prevalent, with this element possessing the largest number of structurally characterized examples. The majority of these studies, however, are concerned with oxo and sulfido derivatives (215), in part due to occurrence of [Mo=O] and [Mo=S] moieties in molybdoenzymes (13–15). For example, synthetic analogue systems designed to model aspects of oxo-transferase enzymes have been actively pursued by Holm and co-workers, as illustrated in Scheme 74 (13, 216, 217).

a. *Trans*-$Mo(PR_3)_4(E)_2$ Derivatives. The terminal chalcogenido complexes *trans*-$Mo(PMe_3)_4(E)_2$ (E = S, Se, Te) have been synthesized by using the electron-rich complexes $Mo(PMe_3)_6$ and $Mo(PMe_3)_4(\eta^2\text{-}CH_2PMe_2)H$ as precursors (Scheme 75) (218); see Section IV.C for their tungsten counterparts. For example, the selenido and tellurido complexes $Mo(PMe_3)_4(E)_2$ (E = Se, Te) are prepared by the reactions of $Mo(PMe_3)_6$ with the elemental chalcogens, while the sulfido derivative $Mo(PMe_3)_4(S)_2$ is obtained by the reaction of

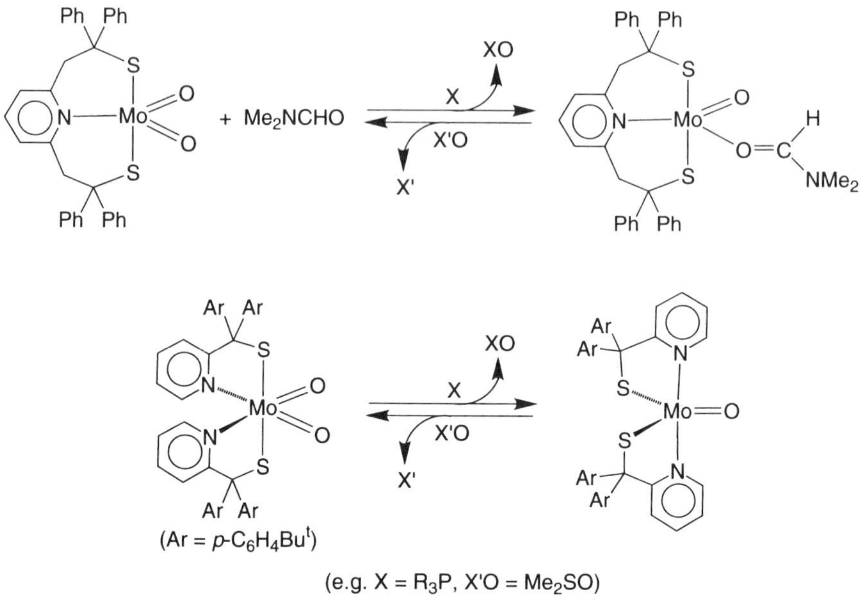

(Ar = p-$C_6H_4Bu^t$)

(e.g. X = R_3P, X'O = Me_2SO)

Scheme 74

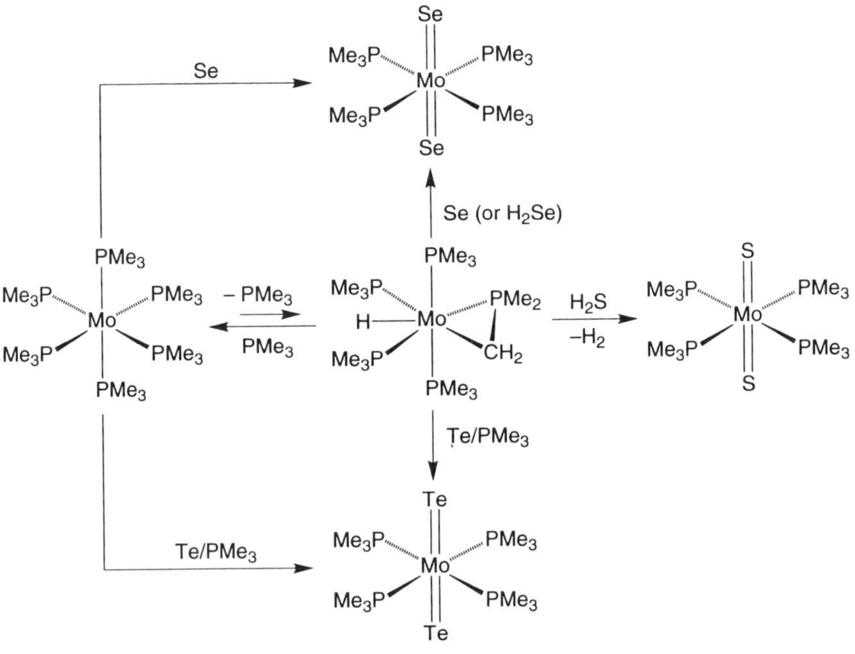

Scheme 75

TABLE XXIII
Selected NMR Spectroscopic Data for $Mo(PMe_3)_4(E)_2$[a]

Nucleus	$Mo(PMe_3)_4(S)_2$	$Mo(PMe_3)_4(Se)_2$	$Mo(PMe_3)_4(Te)_2$
^{31}P	−14.0 [s, $^1J_{Mo\text{-}P}$ = 150]	−16.0 [s, $^1J_{Mo\text{-}P}$ = 140]	−16.2 [s]
^{77}Se		1133 [s]	
^{125}Te			1507 [s]

[a]Taken from (218).

$Mo(PMe_3)_4(\eta^2\text{-}CH_2PMe_2)H$ with H_2S. Likewise, the selenido complex $Mo(PMe_3)_4(Se)_2$ is also obtained by the reaction of $Mo(PMe_3)_4(\eta^2\text{-}CH_2PMe_2)H$ with H_2Se. Interestingly, $Mo(PMe_3)_4(Te)_2$ is the first structurally characterized terminal tellurido complex of molybdenum. Thus, even though terminal chalcogenido derivatives such as $[Mo(O)_4]^{2-}$, $[Mo(S)_4]^{2-}$, and $[Mo(Se)_4]^{2-}$ are well known (1, 25–28), the tellurido analogue is yet to be isolated (219). Selected NMR spectroscopic data for $Mo(PMe_3)_4(E)_2$ (E = S, Se, Te) are presented in Table XXIII.

Very recently, Cotton et al. (220) synthesized two related series of terminal chalcogenido complexes, namely *trans*-$Mo(dppe)_2(E)_2$ and *trans*-Mo-$(dppee)_2(E)_2$ (E = O, S, Se, Te), which include the oxo derivative *trans*-$Mo(dppee)_2(O)_2$, as illustrated in Scheme 76 (221); the related anionic trans-dioxo complex $[Mo(O)_2(CN)_4]^{4-}$ is also known (58a).

The Mo=E bond lengths in $Mo(PMe_3)_4(E)_2$, $Mo(dppee)_2(E)_2$ and $Mo(dppe)_2(E)_2$ are summarized in Table XXIV. For comparison, the W=E bond lengths in $W(PMe_3)_4(E)_2$ (see Section IV.C) are also included, and are virtually identical to the corresponding values in $Mo(PMe_3)_4(E)_2$.

The 18-electron nature of the molybdenum centers in $Mo(PR_3)_4(E)_2$ dictate that the molybdenum–chalcogenido interactions are best represented as Mo=E double bonds, with little contribution from the triply bonded resonance structure $\overset{-}{Mo}\equiv\overset{+}{E}$. In this regard, although terminal sulfido complexes of molybdenum are common, derivatives with "pure" Mo=S double bonds are rare: nevertheless, one other example is *trans*-{syn-Me_8[16]aneS_4}$Mo(S)_2$ (see below) (222). The Mo=S bond lengths in the complexes with "pure" double bonds are longer than the values for complexes in which there is $\overset{-}{Mo}\equiv\overset{+}{S}$ triple bond character (Table XXV) (223). Likewise, the Mo=Se bond lengths in $Mo(PR_3)_4(Se)_2$ are longer than the other representative terminal selenido complexes listed in Table XXVI. The Mo=S bond length in $Mo(dppe)_2(O)(S)$, however, is abnormally long [2.415(7) Å] (224), an observation that has been suggested to be a result of the X-ray structure being performed on a crystal with the unusual composition $(dppe)_2Mo(O)(S)\cdot SO_2\cdot H_2SO_4\cdot PhMe\cdot EtOH$ (218). Specifically, due to the presence of H_2SO_4, the sulfido ligand may be protonated so that the derived Mo−S bond length is closer to that of a single bond. However, Cotton and coworker (220b) have also suggested that the anomalously long Mo=S bond

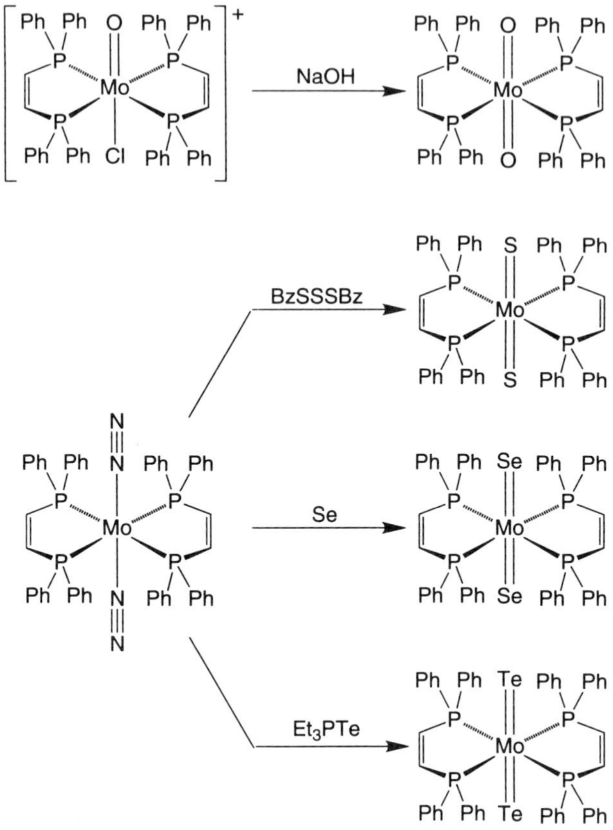

Scheme 76

TABLE XXIV

M=E Bond Lengths for $M(PR_3)_4(E)_2$

	d(M=E) (Å)			
	$Mo(PMe_3)_4(E)_2$[a]	$W(PMe_3)_4(E)_2$[b]	$Mo(dppee)_2(E)_2$[c]	$Mo(dppe)_2(E)_2$[c]
O			1.804(2)	
S	2.254(2)	2.252(3)	2.228(1)	2.236(1)
Se	2.383(2)	2.380(1)	2.356(2)	2.377(1)
Te	2.597(1)	2.596(1)	2.562(1)	

[a]V. J. Murphy and G. Parkin, *J. Am. Chem. Soc.*, *117*, 3522 (1995).
[b]D. Rabinovich and G. Parkin, *Inorg. Chem.*, *34*, 6341 (1995).
[c]F. A. Cotton and G. Schmid, *Inorg. Chem.*, *36*, 2267 (1997).

TABLE XXV
Representative Terminal Mo=S Bond Lengths in Mononuclear Complexes[a]

Complex	d(Mo=S) (Å)
$Mo(PMe_3)_4(S)_2$	2.254(2)
$\{syn\text{-}Me_8[16]aneS_4\}Mo(S)_2$	2.239(7)
$[\{syn\text{-}Me_8[16]aneS_4\}Mo(S)(SMe)][I]$	2.140(5)
$[\{syn\text{-}Me_8[16]aneS_4\}Mo(S)(F)][BF_4]$	2.118(3)
cis-$Mo(\eta^2\text{-}Et_2NO)_2(S)_2$	2.154(1)
cis-$Mo(\eta^2\text{-}C_5H_{10}NO)_2(S)_2$	2.145(2)
cis-$Mo(\eta^2\text{-}C_5H_{10}NO)_2(S)(O)$	2.106(5)
$[NH_4]_2[Mo(O)_2(S)_2]$	2.188(1)
$[(CH_2CH_2NH_2)_2][MoS_4]$	2.18
$[NEt_4]_2[(S_4)_2Mo(S)]$	2.128(1)
$[Tp^{Me_2}]Mo(\eta^2\text{-}S_2CNEt_2)(S)$	2.129(2)
$[\eta^3\text{-}\{(PPh_3P)AgS\}_3Cl]Mo(S)$	2.103(3)
$[PPh_4][(S_2CS_2)_2Mo(S)]$	2.126(3)
$[NEt_4][PPh_4][(S_2CS_2)_2Mo(S)]$	2.127(4)

[a]Taken from (218).

length in $(dppe)_2Mo(O)(S)$ could be due to disorder between oxo and sulfido ligands. In this regard, Cotton determined the molecular structure of the related complex $(dppee)_2Mo(O)(S)$, in which the oxo and sulfido ligands were disordered about an inversion center. Significantly, after successfully modeling the disorder, the derived Mo=O [1.843(6) Å] and Mo=S [2.244(3) Å] bond lengths are virtually identical to those in $Mo(dppee)_2(O)_2$ and $Mo(dppee)_2(S)_2$. It is, therefore, most likely that the anomalously long Mo=S bond length of 2.415(7) Å reported for $(dppe)_2Mo(O)(S)$ is erroneous.

Cotton and co-workers (220) carried out ab initio calculations on the model compounds *trans*-$Mo(PH_3)_4(E)_2$ and concluded that in each case the HOMO

TABLE XXVI
Representative Terminal Mo=Se Bond Lengths[a]

Complex	d(Mo=Se) (Å)
$Mo(PMe_3)_4(Se)_2$	2.383(2)
cis-$Mo(\eta^2\text{-}C_5H_{10}NO)_2(Se)(O)$	2.299(1)
$[Ph_4P]_2[Mo(Se)_4]$	2.293(1)
$[Ph_4P]_2[Mo(Se)(Se_4)_2]$	2.270(4)
$[Ph_4P]_2[Mo_2(Se)_2(\mu\text{-}Se)_2(\eta^2\text{-}Se_2)_2]$	2.237(9)

[a]Taken from (218).

corresponds to the chalcogen p_π lone-pair orbitals and not the d_{xy} nonbonding orbital that is predicted by Xα MO calculations (220). Consequently, whereas the lowest energy bands in *trans*-$ML_4(E)_2$ complexes, for example, *trans*-$W(PMe_3)_4(E)_2$ (225), have been previously assigned to a transition from a d_{xy} orbital (HOMO) to the metal–chalcogen π^* orbitals (lowest unoccupied molecular orbital, LUMO), Cotton reinterpreted such bands as a singlet–singlet transition from the nonbonding chalcogen (S, Se, Te) p_π lone-pair orbitals to the metal–chalcogen π^* orbitals. The electronic spectrum for the oxo complex *trans*-Mo(dppee)$_2$(O)$_2$, however, is quite distinct from those of its heavier congeners. The difference is due to the exceptional stabilization of the Mo—O σ and π bonding orbitals, accompanied by the significant destabilization of the antibonding counterparts. Thus, any transition to one of these Mo—O antibonding orbitals is expected to be of considerably higher energy, with the conclusion that the lowest energy bands correspond to the singlet and triplet components of a $d_{xy} \rightarrow d_{x^2-y^2}$ transition.

b. $Mo(PR_3)_2(L)(E)Cl_2$ Derivatives. A series of octahedral molybdenum oxo complexes with a meridonal arrangement of phosphine ligands, *mer*-$Mo(PR_3)_3(O)X_2$ (e.g., X = Cl, Br, I, NCO, NCS; PR_3 = PMe_3, PMe_2Ph, PEt_2Ph, PPr_2Ph, PBu_2Ph, $PMePh_2$, $PEtPh_2$, $PPrPh_2$, $MeC[CH_2PPh_2]_3$) have been synthesized by a variety of methods (226–231). One of these complexes, namely, *mer*-$Mo(PMe_2Ph)_3(O)Cl_2$, was isolated in blue and green forms with different Mo=O bond lengths, that is, so-called "bond-stretch isomers" (73, 74). Subsequent work, however, has demonstrated that the apparent observation of long Mo=O bond lengths is an artifact due to compositional disorder with isostructural *mer*-$Mo(PMe_2Ph)_3Cl_3$ (75).

Whereas the tungsten complexes $W(PR_3)_4Cl_2$ react with ethylene oxide to give oxo-olefin derivatives, $W(PR_3)_2(O)(\eta^2\text{-}C_2H_4)Cl_2$ (see Section IV.D.3), the principal products for the corresponding molybdenum system are $Mo(PR_3)_3(O)Cl_2$ (PR_3 = PMe_3, $PMePh_2$) (230, 232). Oxo olefin complexes $Mo(PR_3)_2(O)(\eta^2\text{-}C_2H_4)Cl_2$ may, however, be generated by treatment of $Mo(PR_3)_3(O)Cl_2$ with an excess of ethylene. Likewise, oxo carbonyl complexes $Mo(PR_3)_2(O)(CO)Cl_2$ may be generated by the reaction of $Mo(PR_3)_3(O)Cl_2$ with CO (Scheme 77); see Sections IV.B.3 and IV.C for further discussion of chalcogenido carbonyl complexes.

In contrast to the common occurrence of the oxo derivatives Mo-$(PR_3)_2(L)(O)Cl_2$, sulfido analogues are rare. Nevertheless, the related molybdenum sulfido complex $Mo(PMe_3)_3(S)Cl_2$ has been synthesized by the reaction of $Mo(PMe_3)_4Cl_2$ with ethylene sulfide; subsequent reaction with Mo-$(PMe_3)_4Cl_2$, however, gives the bridging sulfido complex $[(Me_3P)_3$-$MoCl](\mu\text{-}Cl)(\mu\text{-}S)[MoCl_2(PMe_3)_2]$, as illustrated in Scheme 78 (230, 233).

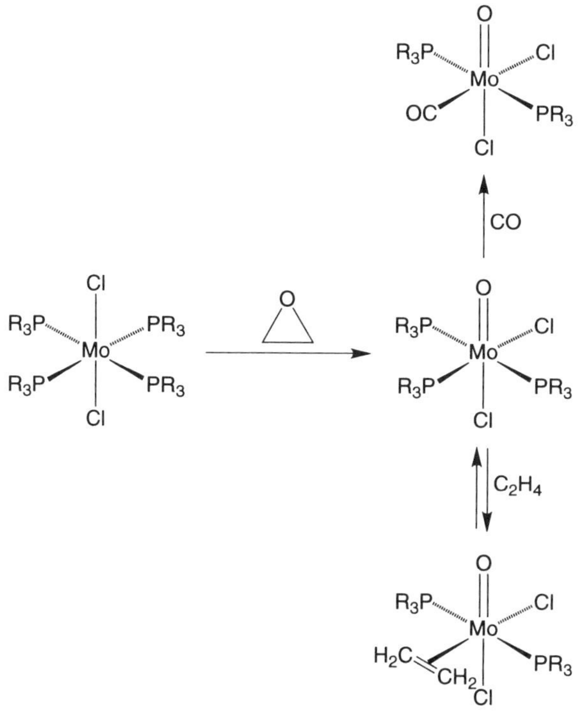

Scheme 77

c. **Macrocyclic Derivatives.** Closely related to the *trans*-chalcogenido complexes, $Mo(PR_3)_4(E)_2$, the 18-electron trans-sulfido complex $\{syn\text{-}Me_8\text{-}[16]aneS_4\}Mo(S)_2$ has been synthesized by the reaction of the dinitrogen complex $\{syn\text{-}Me_8[16]aneS_4\}Mo(N_2)_2$ with S_8 (Scheme 79) (222). The sulfido ligands in $\{syn\text{-}Me_8[16]aneS_4\}Mo(S)_2$ have been observed to exhibit enhanced nucleophilicity, and are alkylated by RX (RX = MeI, $PhCH_2Br$) to give $[\{syn\text{-}Me_8[16]aneS_4\}Mo(S)(SR)]^+$ (234, 235). Likewise, one of the sulfido ligands in $\{syn\text{-}Me_8[16]aneS_4\}Mo(S)_2$ may be protonated by HBF_4 in CH_2Cl_2 to give $\{[\{syn\text{-}Me_8[16]aneS_4\}Mo(S)]_2(\mu\text{-}S)\}^{2+}$, accompanied by evolution of H_2S (236). The reaction is proposed to proceed via the monosulfido intermediate $[\{syn\text{-}Me_8[16]aneS_4\}Mo(S)]^{2+}$. If the reaction is carried out in toluene, fluoride abstraction by $[\{syn\text{-}Me_8[16]aneS_4\}Mo(S)]^{2+}$ is observed, resulting in the formation of $[\{syn\text{-}Me_9[16]aneS_4\}Mo(S)(F)]^+$. Furthermore, protonation with CF_3SO_3H yields $[\{syn\text{-}Me_8[16]aneS_4\}Mo(S)(O_3SCF_3)]^+$.

Molybdenum porphyrin oxo complexes of the types (POR)Mo(O), $\{(POR)Mo(O)X\}^-$ (X = F, Cl, Br, NCS), (POR)Mo(O)X (X = Hal, OR),

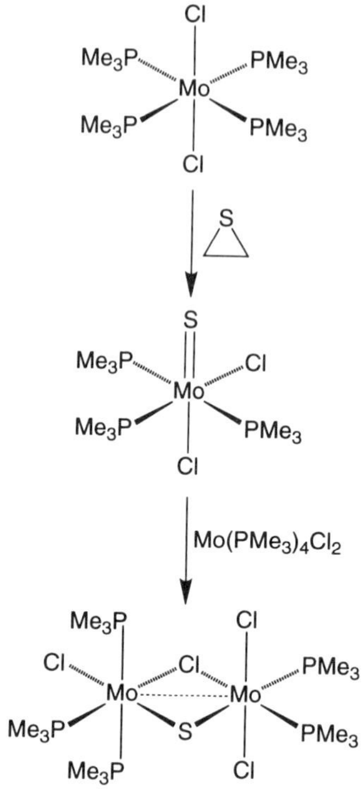

Scheme 78

and (POR)Mo(O)$_2$ are well known and have been reviewed (237); for instance, the molecular structure of (TTP)MoO [d(M=O) = 1.656(6) Å; ν(Mo=O) = 980 cm^{-1}] was determined in 1979 (238). In contrast, terminal sulfido and selenido complexes have only recently been synthesized by Woo and co-workers (239), as shown in Scheme 80. For example, (TTP)MoS may be synthesized by the reactions of (a) (TTP)Mo(η^2-Ph$_2$C$_2$) with S$_8$ or Cp$_2$TiS$_5$ and (b) (TTP)MoCl$_2$ (240) with Li$_2$S. Likewise, the terminal selenido analogue (TTP)MoSe may be prepared by the reaction of (TTP)MoCl$_2$ with Na$_2$Se. The sulfido complex (TTP)MoS is characterized by a ν(Mo=S) absorption at 542 cm^{-1} and a Mo=S bond length of 2.100(1) Å, considerably shorter than the value for Mo(PMe$_3$)$_4$(S)$_2$ [2.254(2) Å] described above. The sulfido and selenido complexes (TTP)MoE are also obtained by chalcogen-atom transfer from the corresponding tin chalcogenido complex (TTP)SnE to (TTP)Mo(η^2-Ph$_2$C$_2$); the chalcogen transfer is irreversible for the case of the sulfido derivative, but reversible for the selenido derivative (Scheme 80) (241). The chalcogenido

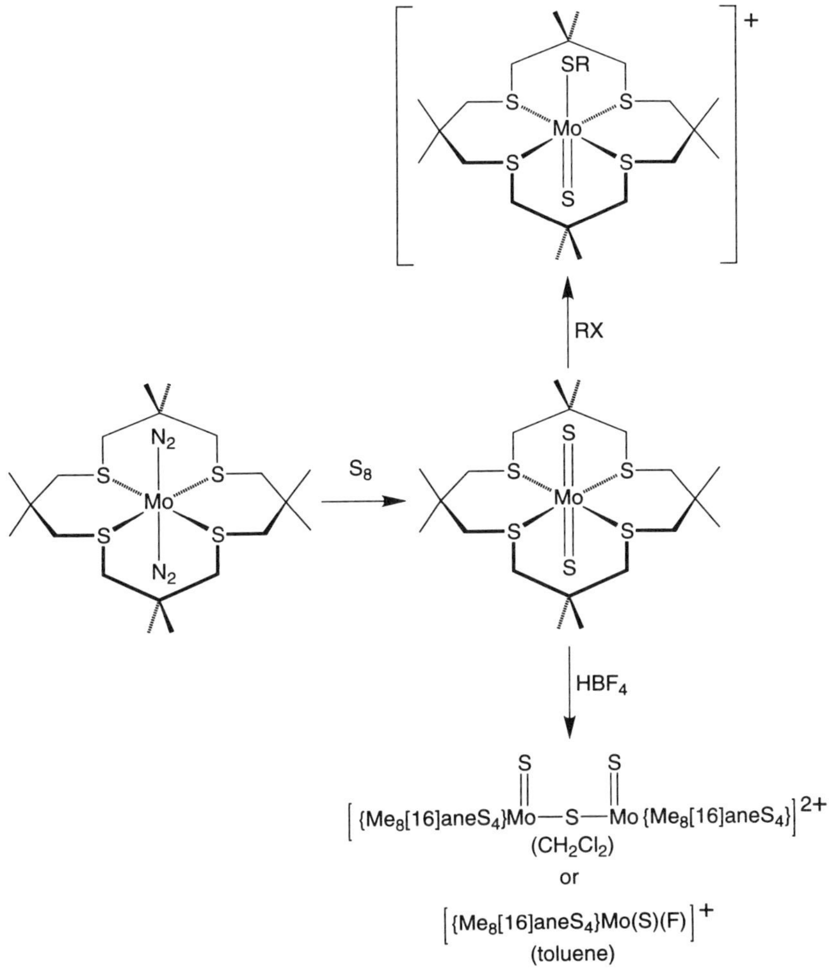

$[\{Me_8[16]aneS_4\}Mo(S)(SR)]^+$

RX

S_8

HBF$_4$

$[\{Me_8[16]aneS_4\}Mo(S)—S—Mo(S)\{Me_8[16]aneS_4\}]^{2+}$
(CH_2Cl_2)
or
$[\{Me_8[16]aneS_4\}Mo(S)(F)]^+$
(toluene)

Scheme 79

ligands in (TTP)MoE may also be abstracted by PPh_3 to give (TTP)Mo(PPh_3)$_2$ and Ph_3PE (239). In contrast, the corresponding molybdenum oxo complex (TTP)MoO is not reduced by PPh_3, presumably due to a very strong Mo=O bond. In line with such an observation, chalcogen abstraction from the selenido complex (TTP)MoSe is faster than from the sulfido analogue (TTP)MoS.

d. Tris(amide) Derivatives. Cummins (242) recently used the bulky amide ligand [(3,5-$Me_2C_6H_3$)NBu^t] to isolate a complete series of terminal chalcogen-

$$(TTP)MoCl_2 \xrightarrow{Li_2S \text{ or } Na_2Se} \underset{(E = S, Se)}{(TTP)Mo{=}E} \xleftarrow{S_8 \text{ or } Cp_2TiS_5} (TTP)Mo(\eta^2\text{-}Ph_2C_2)$$

$$(TTP)Mo(\eta^2\text{-}Ph_2C_2) + (TTP)Sn{=}S \longrightarrow (TTP)Mo{=}S + (TTP)Sn + Ph_2C_2$$

$$(TTP)Mo(\eta^2\text{-}Ph_2C_2) + (TTP)Sn{=}Se \rightleftharpoons (TTP)Mo{=}Se + (TTP)Sn + Ph_2C_2$$

$$(TTP)Mo{=}E + 3PPh_3 \xrightarrow[(E = S, Se)]{} (TTP)Mo(PPh_3)_2 + Ph_3PE$$

Scheme 80

ido complexes for molybdenum. The complexes $[(3,5\text{-}Me_2C_6H_3)NBu^t]_3MoE$ (E = O, S, Se, Te) were synthesized from the three-coordinate precursor, $[(3,5\text{-}Me_2C_6H_3)NBu^t]_3Mo$, as illustrated in Scheme 81. Interestingly, N_2O is not effective for the synthesis of the oxo derivative; thus, in marked contrast to other systems, the reaction of $[(3,5\text{-}Me_2C_6H_3)NBu^t]_3Mo$ with N_2O results in the unprecedented cleavage of the N—N bond, with the concomitant formation of $[(3,5\text{-}Me_2C_6H_3)NBu^t]_3MoN$ and $[(3,5\text{-}Me_2C_6H_3)NBu^t]_3MoNO$ (243). Each of the chalcogenido complexes $[(3,5\text{-}Me_2C_6H_3)NBu^t]_3MoE$ (E = O, S, Se, Te) have been structurally characterized by X-ray diffraction, as summarized in Table XXVII. All the derivatives $[(3,5\text{-}Me_2C_6H_3)NBu^t]_3MoE$ experience N—Bu^t homolysis upon heating to give the arylimido chalcogenido complexes $[(3,5\text{-}Me_2C_6H_3)NBu^t]_2Mo(E)[N(3,5\text{-}Me_2C_6H_3)]$, as shown in Scheme 81.

e. Molybdenocene Derivatives. The molybdenocene oxo complex Cp_2MO, first synthesized by Green et al. (244) in 1972 by the reaction of Cp_2MoCl_2 with NaOH, and also by reduction of Cp_2MoCl_2 with sodium amalgam in the presence of propylene oxide (245), is a classic example of a metal oxo complex supported only by organic ligands (Scheme 82). More recently, the $Cp_2^{Me}MoO$ (45a, 47) and $Cp_2^{Bu^t}MoO$ (48) derivatives have been prepared by analogous procedures. The permethylmolybdenocene complex Cp_2^*MoO has also been isolated, but only in small quantities as a side product in the synthesis of $Cp_2^*MoH_2$ by cocondensation of molybdenum atoms with Cp*H (246). Finally, an alternative method to prepare Cp_2MoO involves photolysis of the carbonate derivative $Cp_2Mo(\eta^2\text{-}O_2CO)$, a reaction that is also accompanied by elimination of CO_2 (Scheme 82) (247, 248). $Cp_2^{Me}MoO$ is also formed by irradiation of $Cp_2^{Me}MoH_2$ in the presence of H_2O or propylene oxide (45a).

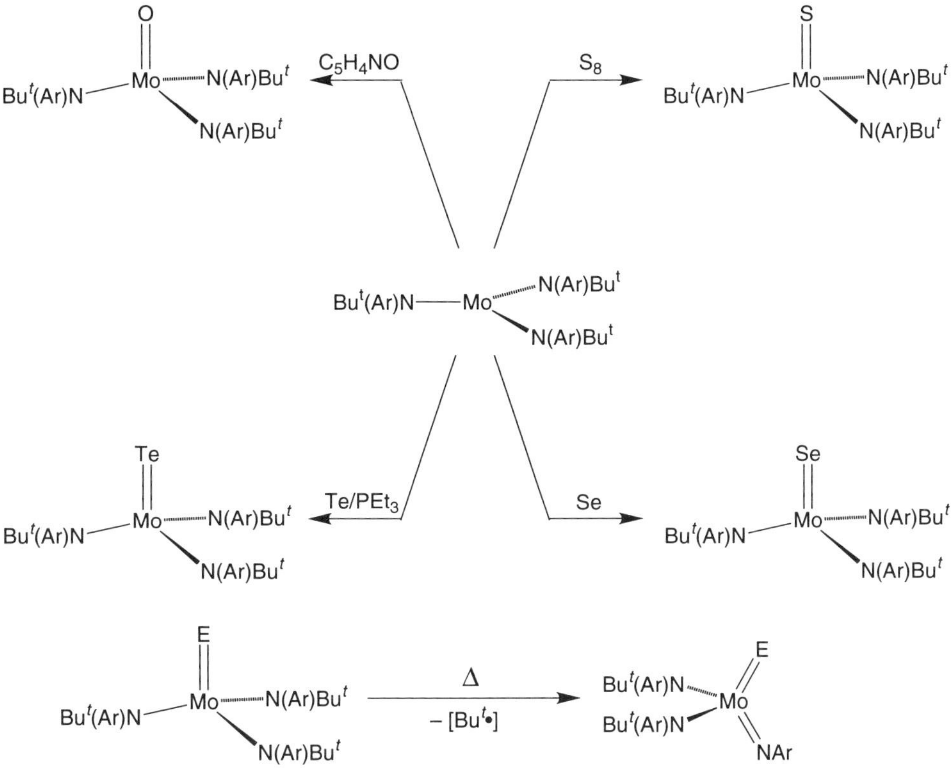

Scheme 81

The molecular structures of $Cp_2^{Me}MoO$ (47) and $Cp_2^{Bu^t}MoO$ (48) have been determined by X-ray diffraction, as illustrated in Fig. 22 for the latter complex. Bond length and IR data associated with the [Mo=O] moiety are summarized in Table XXVIII and, as noted in Section III.A, are indicative of Mo=O double-bond character, with little contribution from the triply bonded resonance structure $\overset{-}{Mo}\equiv\overset{+}{O}$.

TABLE XXVII

Mo=E Bond Lengths for $[(3,5\text{-}Me_2C_6H_3)NBu^t]_3MoE$[a]

	d(Mo=E) (Å)
O	1.706(2)
S	2.168(1)
Se	2.312(1)
Te	2.535(1)

[a]Taken from (242).

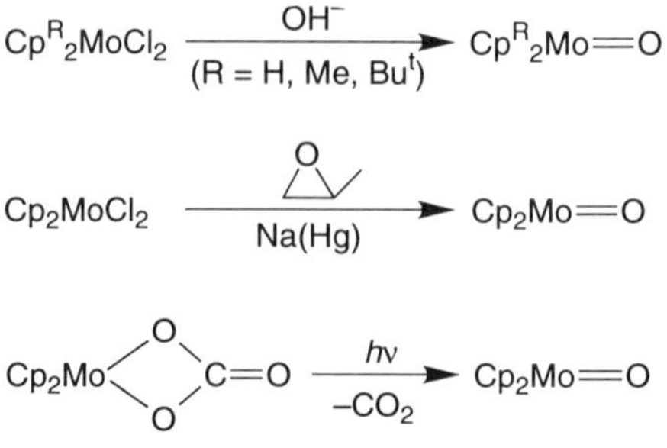

$$Cp^R_2MoCl_2 \xrightarrow[(R = H, Me, Bu^t)]{OH^-} Cp^R_2Mo{=}O$$

$$Cp_2MoCl_2 \xrightarrow[Na(Hg)]{\text{propylene oxide}} Cp_2Mo{=}O$$

$$Cp_2Mo(O_2C{=}O) \xrightarrow[-CO_2]{h\nu} Cp_2Mo{=}O$$

Scheme 82

Considering the preponderance of molybdenocene complexes, the reactivity of the terminal oxo derivatives has received remarkably little attention. Nevertheless, $Cp_2^{Bu^t}MoO$ does undergo formal 1,2-addition reactions with Me_3SiX reagents to give ultimately $Cp_2^{Bu^t}MoX_2$ derivatives (Scheme 83) (48), while Cp_2MoO undergoes a [2 + 2] cycloaddition reaction with PhNCO (Scheme 84) (249). The photochemistry of Cp_2MoO and $Cp_2^{Me}MoO$ has been investigated by Tyler and co-workers (45a), as summarized in Scheme 85. Thus, the initial species formed upon irradiation of Cp_2MoO is proposed to be $[Cp_2Mo]$, with the ultimate products arising from subsequent reaction of $[Cp_2Mo]$ with either itself or the photogenerated O_2. In the presence of dative trapping ligands (L), such as CO, PR_3, and olefins, Cp_2MoL is obtained, whereas Et_2S yields $Cp_2Mo(SEt)H$. A number of unidentified complexes, however, are also obtained during photolysis.

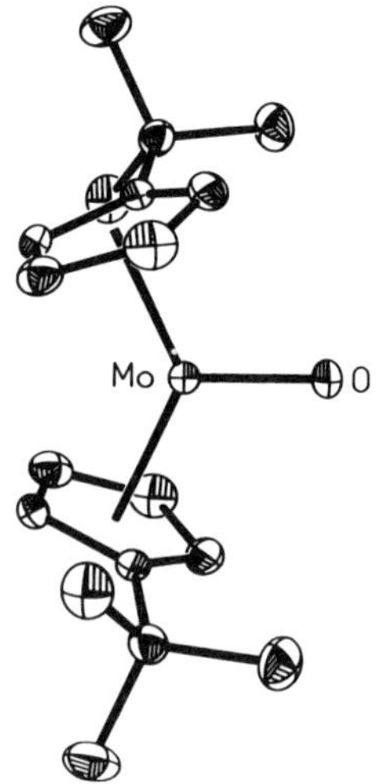

Figure 22. Molecular structure of $Cp_2^{Bu^t}MoO$.

TABLE XXVIII
Mo=O Bond Length and ν(Mo=O) IR Data for Molybdenocene Oxo Derivatives

	d(Mo=O) (Å)	ν(Mo=O) (cm^{-1})	References
Cp_2MoO		800 (793–868)	*a, b*
$Cp^{Me_2}MoO$	1.721(2)	827	*a, c*
$Cp_2^{Bu^t}MoO$	1.706(4)	838	*d*

[a]N. D. Silavwe, M. R. M. Bruce, C. E. Philbin, and D. R. Tyler, *Inorg. Chem.*, *27*, 4669 (1988).
[b]M. L. H. Green, A. H. Lynch, and M. G. Swanwick, *J. Chem. Soc. Dalton Trans.*, 1445 (1972).
[c]N. D. Silavwe, M. Y. Chiang, and D. R. Tyler, *Inorg. Chem.*, *24*, 4219 (1985).
[d]J. H. Shin, Ph. D. thesis, ''Syntheses, Structures and Reactivity of Permethyltantalocene and *tert*-butylmolybdenocene derivatives,'' Columbia University, New York, 1996.

In contrast to the terminal oxo derivatives Cp_2^RMoO, well-characterized examples of terminal sulfido, selenido, or tellurido analogues Cp_2^RMoE have not been isolated. Nevertheless, [Cp_2MoS] has been proposed as an intermediate in the formation of $Cp_2Mo[\eta^2$-SCN(Ar)NAr] by (a) the reaction of $Cp_2Mo(\eta^2-S_2)$ with PMe_3 in the presence of ArN=C=NAr (Ar = tolyl), and (b) the reaction

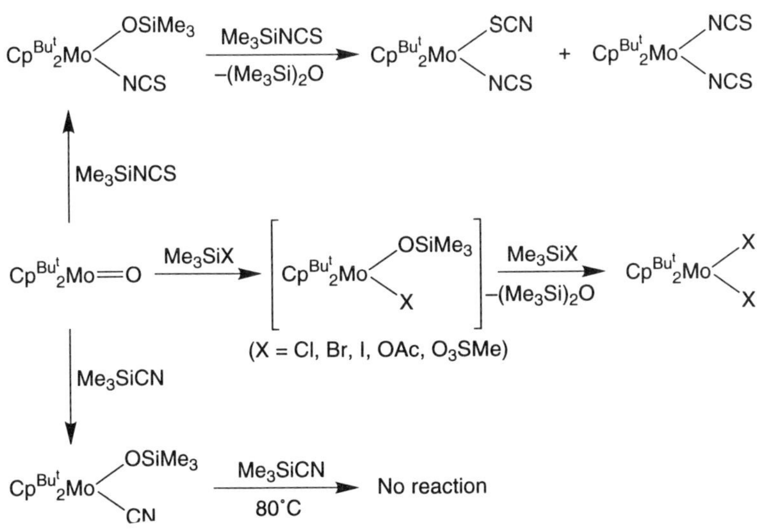

Scheme 83

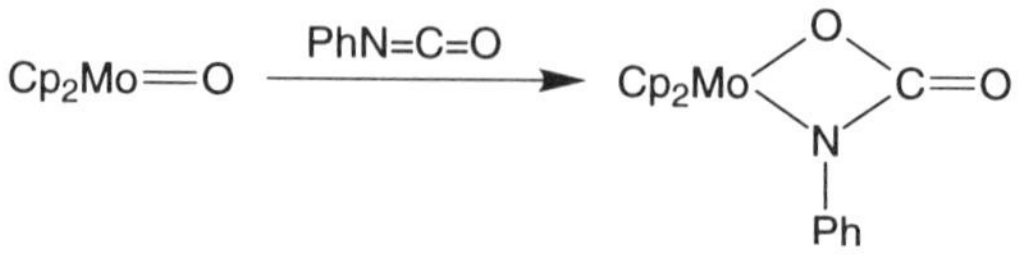

$Cp_2Mo{=}O \xrightarrow{PhN=C=O} Cp_2Mo[OC(=O)N(Ph)]$

Scheme 84

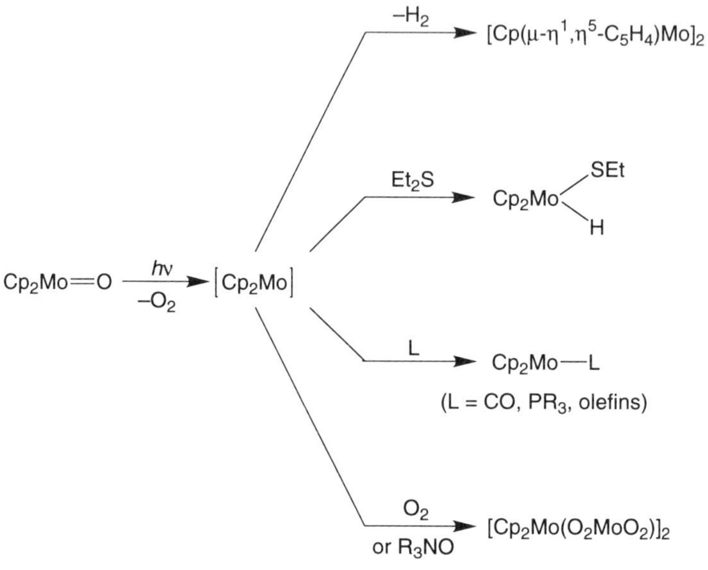

Scheme 85

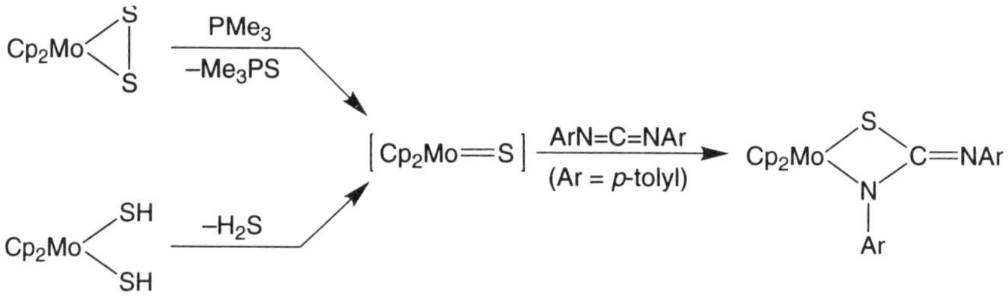

$Cp_2Mo(S_2) \xrightarrow[-Me_3PS]{PMe_3} [Cp_2Mo{=}S]$

$Cp_2Mo(SH)_2 \xrightarrow{-H_2S} [Cp_2Mo{=}S]$

$[Cp_2Mo{=}S] \xrightarrow[(Ar = p\text{-tolyl})]{ArN=C=NAr} Cp_2Mo[SC(=NAr)N(Ar)]$

Scheme 86

of $Cp_2Mo(SH)_2$ with ArN=C=NAr (Ar = tolyl), as illustrated in Scheme 86 (250, 251). Furthermore, only bridging complexes $[Cp_2^{Bu^t}Mo(\mu\text{-}E)]_2$ (E = S, Se, Te) have been obtained for the *tert*-butylcyclopentadienyl system, as illustrated in Scheme 87 (48).

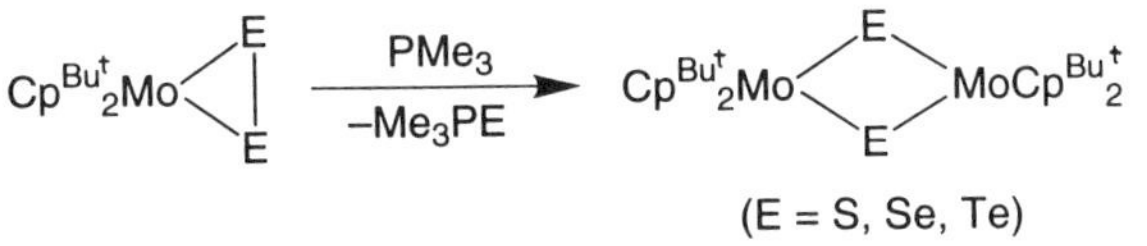

$Cp^{Bu^t}{}_2Mo(\eta^2\text{-}E_2) \xrightarrow[-Me_3PE]{PMe_3} Cp^{Bu^t}{}_2Mo(\mu\text{-}E)_2MoCp^{Bu^t}{}_2$

(E = S, Se, Te)

Scheme 87

3. *Tungsten Chalcogenido Complexes*

Of the third transition series, tungsten forms the most extensive compilation of terminal chalcogenido complexes, with oxo, sulfido, and selenido derivatives having been well reviewed (252). Therefore, particular emphasis in this section will be given to systems for which the tellurido complexes are known.

a. *Trans*-W(PMe$_3$)$_4$(E)$_2$ Derivatives. The electron-rich complex $W(PMe_3)_4(\eta^2\text{-}CH_2PMe_2)H$ (253) has been used as a precursor for a series of trans-sulfido, -selenido, and -tellurido complexes of tungsten, as illustrated in Scheme 88 (43, 254). Thus, highly reactive $W(PMe_3)_4(\eta^2\text{-}CH_2PMe_2)H$ is capable of abstracting chalcogen atoms from both H_2S and H_2Se to give

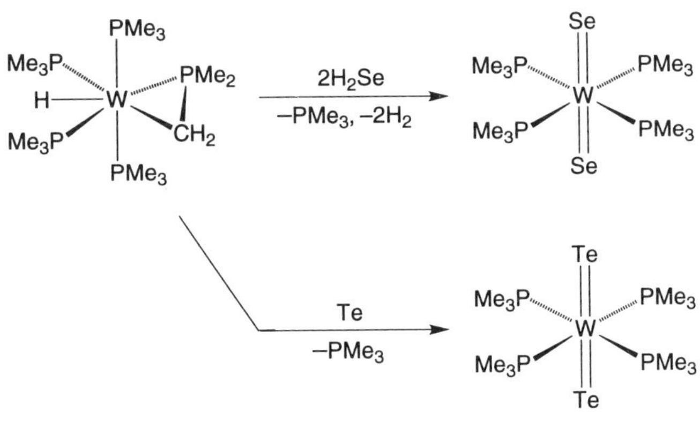

$W(PMe_3)_4(\eta^2\text{-}CH_2PMe_2)H \xrightarrow[-PMe_3,\ -2H_2]{2H_2S} trans\text{-}W(PMe_3)_4(S)_2$

$W(PMe_3)_4(\eta^2\text{-}CH_2PMe_2)H \xrightarrow[-PMe_3,\ -2H_2]{2H_2Se} trans\text{-}W(PMe_3)_4(Se)_2$

$W(PMe_3)_4(\eta^2\text{-}CH_2PMe_2)H \xrightarrow[-PMe_3]{Te} trans\text{-}W(PMe_3)_4(Te)_2$

Scheme 88

$W(PMe_3)_4(S)_2$ and $W(PMe_3)_4(Se)_2$, respectively, accompanied by elimination of H_2. In view of its inherent instability (255), H_2Te is not a viable reagent for the synthesis of the tellurido analogue. Nevertheless, $W(PMe_3)_4(Te)_2$, the first terminal tellurido complex of the transition metals, was successfully synthesized by the reaction of $W(PMe_3)_4(\eta^2\text{-}CH_2PMe_2)H$ with elemental tellurium. In this regard, it is worth noting that even though the sulfido and selenido complexes of tungsten $[W(S)_4]^{2-}$ and $[W(Se)_4]^{2-}$ are known, attempts to prepare the tellurido analogue $[W(Te)_4]^{2-}$ have so far been unsuccessful (256). However, Eichorn and co-workers (257) recently synthesized the tris(tellurido) complex $[Ph_4P]_2[W(O)(Te)_3]$ in low yield by the reaction of $W_2(O_2CC_3H_7)_4$ or WCl_4 with K_2Te_2 in the presence of $NH_2CH_2CH_2NH_2$ and Ph_4PBr; the source of the oxo ligand in $[W(O)(Te)_3]^{2-}$ is, however, unknown. Finally, even though the oxo analogue $W(PMe_3)_4(O)_2$ of the above sulfido, selenido, and tellurido complexes has not been isolated, the related carbonyl oxo complexes *trans*-$M(CO)_4(O)_2$ (M = Mo, W) have been identified as intermediates during the photolysis of $M(CO)_6$ in O_2-doped Ar or CH_4 matrices at 10 K (258–260); use of a chelating phosphine ligand has allowed the molybdenum oxo analogue, $Mo(dppee)_2(O)_2$, to be isolated (see Section IV.D.2.a).

The molecular structures of the terminal chalcogenido complexes $W(PMe_3)_4(E)_2$ (E = S, Se, Te) have been determined by X-ray diffraction, as illustrated in Fig. 23 for $W(PMe_3)_4(Te)_2$. The W=E double-bond lengths [d(W=S) = 2.253[3] Å, d(W=Se) = 2.380[1] Å, d(W=Te) = 2.596(1) Å] are considerably shorter than the corresponding typical W—E single bond values [d(W—S) = 2.39 Å, d(W—Se) = 2.45 Å, d(W—Te) = 2.82 Å]

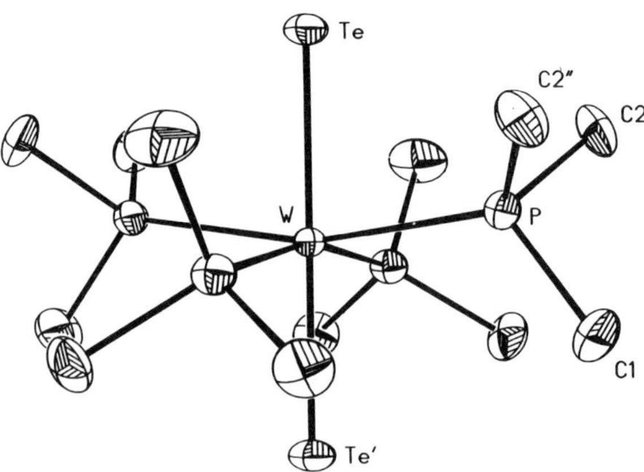

Figure 23. Molecular structure of $W(PMe_3)_4(Te)_2$. [Reprinted with permission from D. Rabinovich and G. Parkin, *Inorg. Chem.*, *34*, 6341 (1996). Copyright 1996 American Chemical Society.]

TABLE XXIX
Selected NMR Spectroscopic Data for $W(PMe_3)_4(E)_2$[a]

Nucleus	$W(PMe_3)_4(S)_2$	$W(PMe_3)_4(Se)_2$	$W(PMe_3)_4(Te)_2$
^{31}P	−44.3 [s, $^1J_{W-P}$ = 268]	−48.2 [s, $^1J_{W-P}$ = 255]	−51.2 [$^1J_{W-P}$ = 238 $^2J_{Te-P}$ = 17]
^{77}Se		803 [$^2J_{Se-P}$ = 13]	
^{125}Te			950 [$^1J_{W-Te}$ = 190 $^2J_{Te-P}$ = 17]

[a]Taken from (43).

(261), but are similar to the respective Mo=E bond lengths in $Mo(PMe_3)_4(E)_2$ (see Section IV.D.2).

Multinuclear NMR spectroscopies also furnish important characterizing information for $W(PMe_3)_4(E)_2$ in solution (Table XXIX), as illustrated for the tellurido derivative in Figs. 24 and 25. For example, the ^{31}P NMR signal at δ

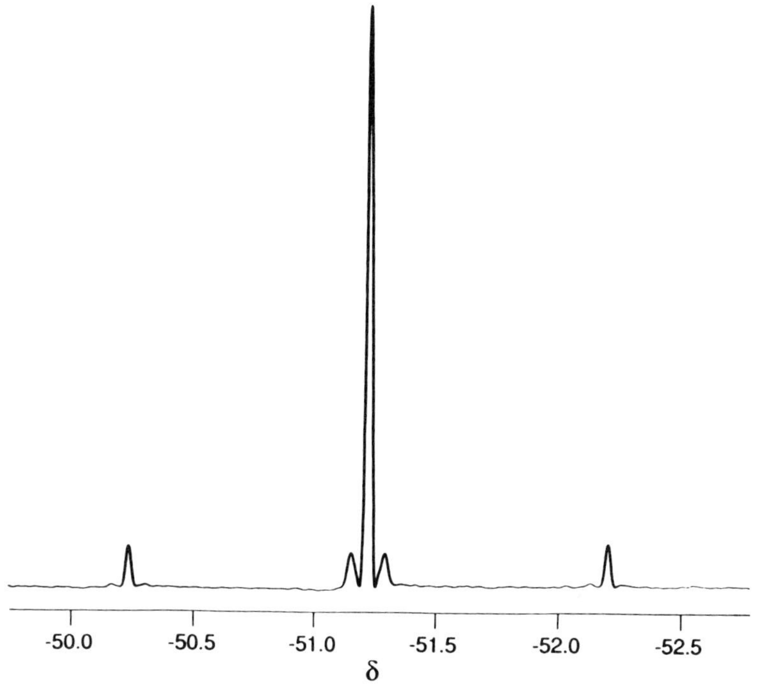

Figure 24. The $^{31}P\ \{^1H\}$ NMR spectrum of $W(PMe_3)_4(Te)_2$. [Reprinted with permission from D. Rabinovich and G. Parkin, *Inorg. Chem.*, *34*, 6341 (1996). Copyright 1996 American Chemical Society.]

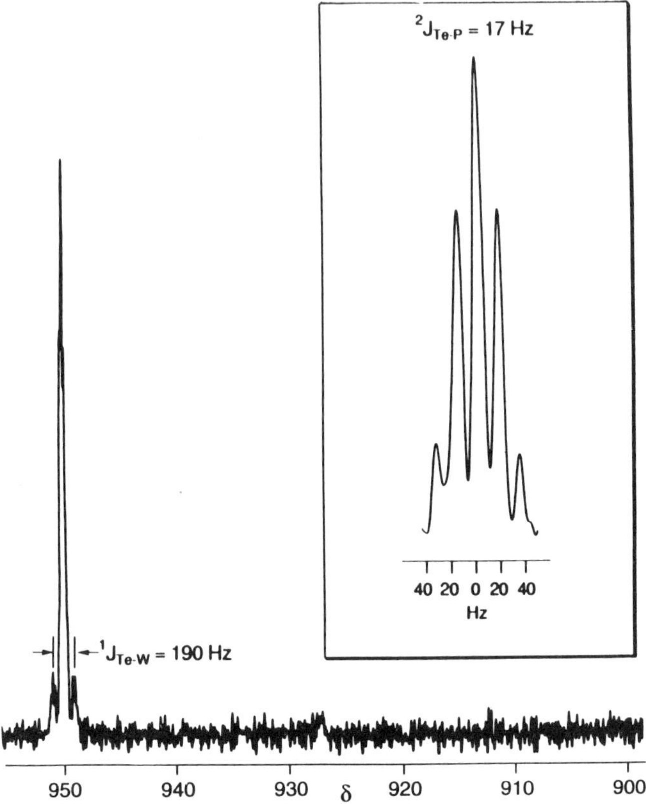

Figure 25. The ^{125}Te {^1H} NMR spectrum of $W(PMe_3)_4(Te)_2$. [Reprinted with permission from D. Rabinovich and G. Parkin, *Inorg. Chem.*, *34*, 6341 (1996). Copyright 1996 American Chemical Society.]

−51.2 ppm shows coupling to both tungsten ($^1J_{W-P}$ = 238 Hz) and tellurium ($^2J_{Te-P}$ = 17 Hz), and the relative intensities of these satellites is that predicted for a molecule of composition $W(PMe_3)_4(Te)_2$ [1.1 : 1]. Furthermore, $W(PMe_3)_4(Te)_2$ is also characterized by a ^{125}Te NMR signal at δ 950 ppm, which exhibits coupling to both tungsten ($^1J_{W-Te}$ = 190 Hz (183); W, $I = \frac{1}{2}$, 14.3%) and the four equivalent phosphorus nuclei of the PMe_3 groups (quintet, $^2J_{Te-P}$ = 17 Hz). For further discussion on ^{77}Se and ^{125}Te NMR studies, see Section V.

Thorp and co-workers (225) also studied the terminal chalcogenido complexes $W(PMe_3)_4(E)_2$ using electronic spectroscopy; see Section IV.D.2 for some related studies on molybdenum analogues.

An interesting feature of the reactions of $W(PMe_3)_4(\eta^2\text{-}CH_2PMe_2)H$ with H_2E (E = S, Se) is the facile elimination of H_2, an unusual reaction to be

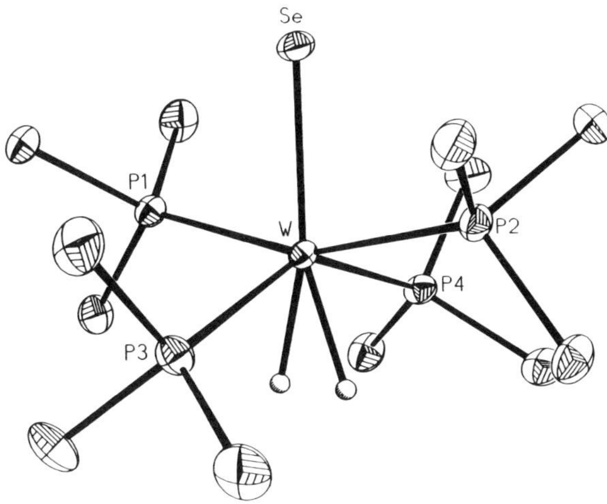

Figure 26. Molecular structure of $W(PMe_3)_4(Se)H_2$. [Reprinted with permission from D. Rabinovich and G. Parkin, *Inorg. Chem.*, *34*, 6341 (1996). Copyright 1996 American Chemical Society.]

achieved at a single metal center (262). The isolation of intermediates has provided information about the mechanism of the dehydrogenation. Thus, the hydrido hydrosulfido complex $W(PMe_3)_4H_2(SH)_2$ has been isolated from the reaction with H_2S, whereas the selenido–dihydride derivative $W(PMe_3)_4(Se)H_2$ has been isolated from the corresponding reaction with H_2Se. The structure of the latter complex has been determined by X-ray diffraction (Fig. 26) and is characterized by a W=Se bond length of 2.445(2) Å (see Section III.A).

Since the intermediates $W(PMe_3)_4H_2(SH)_2$ and $W(PMe_3)_4(Se)H_2$ differ in the number of equivalents of H_2E that have been added, they behave differently with regard to conversions to $W(PMe_3)_4(E)_2$. Thus, $W(PMe_3)_4H_2(SH)_2$ rapidly eliminates H_2 when dissolved to give $W(PMe_3)_4(S)_2$, whereas $W(PMe_3)_4(Se)H_2$ is stable under comparable conditions and requires addition of a second equivalent of H_2Se to give $W(PMe_3)_4(Se)_2$ (Scheme 89). Furthermore, addition of H_2S to $W(PMe_3)_4(Se)H_2$ also results in elimination of H_2 and formation of the mixed-chalcogenido complex $W(PMe_3)_4(S)(Se)$.

The general mechanism that has been suggested for the reactions of $W(PMe_3)_4(\eta^2\text{-}CH_2PMe_2)H$ with H_2E (E = S, Se) is illustrated in Scheme 90. The first stage involves the direct attack of H_2E at the W—C bond of $W(PMe_3)_4(\eta^2\text{-}CH_2PMe_2)H$ together with loss of PMe_3, generating the 16-electron intermediate $[W(PMe_3)_4H(EH)]$ (263), followed by rapid α-H elimination to give $W(PMe_3)_4(E)H_2$ (isolated for E = Se). The final steps are proposed to involve addition of a second equivalent of H_2E to $W(PMe_3)_4(E)H_2$ giving $W(PMe_3)_4H_2(EH)_2$ (isolated for E = S), which rapidly eliminates H_2 to give

Scheme 89

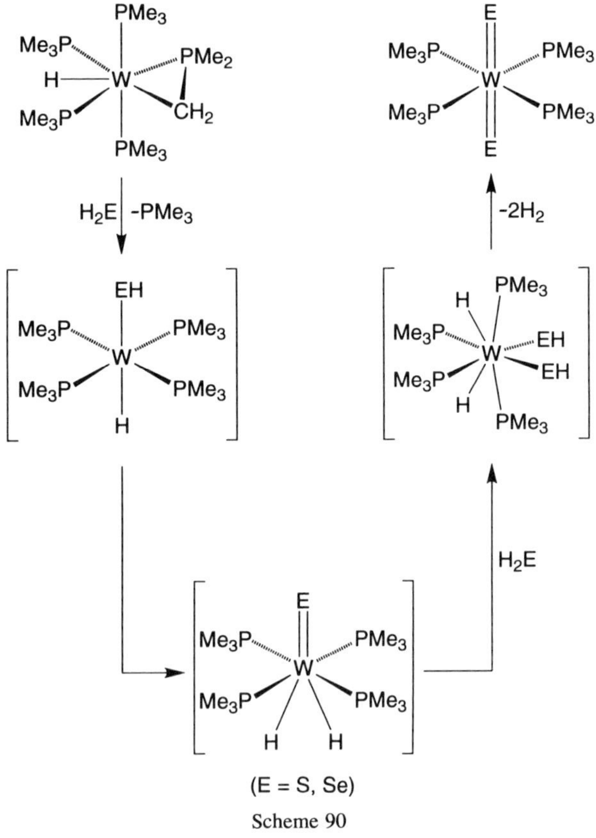

Scheme 90

$W(PMe_3)_4(E)_2$. Unfortunately, the precise details of the latter transformation are presently uncertain.

The formation of $W(PMe_3)_4(Te)_2$ is autocatalytic in PMe_3, and the mechanism for tellurium-atom transfer presumably involves the intermediacy of Me_3PTe (264). In this regard, R_3PTe derivatives have been used as tellurium atom transfer reagents in other systems (265, 266), and in some cases stable adducts have been isolated, for example, $W(CO)_5(TePBu^t_3)$ (267).

The complexes $W(PMe_3)_4(E)_2$ and $W(PMe_3)_4(Se)H_2$ [and also $W(PMe_3)_2(CNR)_2(E)_2$; see below] represent rare examples of structurally characterized 18-electron tungsten complexes containing terminal chalcogenido ligands. As such, the tungsten–chalcogenido interactions may be aptly represented as "pure" W=E double bonds, with little contribution from the triply bonded resonance from $\overset{-}{W}\equiv\overset{+}{E}$ (see Section III.A). Theoretical calculations on the model complexes $W(PH_3)_4(E)_2$ by Kaltsoyannis (268) and on $Mo(PH_3)_4(E)_2$ by Cotton and co-workers (220) are also in agreement with such a bonding scheme. In support of this formalism, the presence of "pure" W=S double bonds in $W(PMe_3)_2(L)_2(S)_2$ (L = PMe_3, CNR) is reflected by particularly low ν(W=S) stretching frequencies in the range 387–392 cm^{-1}. For reference, ν(W=S) stretching frequencies are typically observed in the range 450–570 cm^{-1} for complexes in which the contribution of the resonance form $\overset{-}{W}\equiv\overset{+}{S}$ would be expected to be significant (269).

b. *cis*-$W(PMe_3)_2(\eta^2$-OCHR) Derivatives. A variety of aldehydes (e.g., RCHO, R = H, Me, Ph) are capable of displacing two of the PMe_3 ligands in *trans*-$W(PMe_3)_4(E)_2$ (E = S, Se, Te) to give the η^2-aldehyde complexes $W(PMe_3)_2(E)_2(\eta^2$-OCHR), as illustrated in Scheme 91. The molecular structures of several of these derivatives have been determined by X-ray diffraction, and identify two common features, namely, (a) with C—O bond lengths within the aldehyde ligands in the range 1.363(22)–1.395(11) Å, the complexes are best described as W^{VI} metallaoxirane derivatives rather than W^{IV} aldehyde adducts, and (b) the complexes exhibit a cis disposition of chalcogenido ligands (see Section III.D).

The formation of $W(PMe_3)_2(E)_2(\eta^2$-OCHR) is reversible and the equilibrium constants K_E, $\Delta H°$, and $\Delta S°$ have been determined as a function of the chalcogen (Table XXX). From these values of $\Delta H°$ and $\Delta S°$, it is evident that the

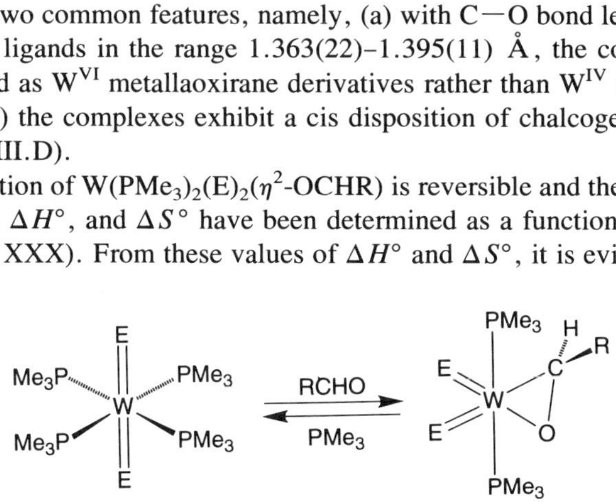

Scheme 91

TABLE XXX
Thermodynamic Data for the Equilibrium between $W(PMe_3)_4(E)_2$ and $W(PMe_3)_2(E)_2(\eta^2\text{-}OCHPh)$[a]

	K (30°C)	$\Delta H°$ (kcal mol^{-1})	$\Delta S°$ (eu)
S	8(3)	5(1)	20(4)
Se	$7.9(8) \times 10^{-2}$	8(2)	22(5)
Te	$4.0(9) \times 10^{-3}$	11(2)	25(5)

[a]Taken from (43).

formation of each of the aldehyde derivatives is driven entropically by dissociation of the trimethylphosphine ligands. It is also evident that equilibrium constants K_E vary dramatically as a function of the chalcogen, with $K_S >> K_{Se} > K_{Te}$, an observation that is due to enthalpic, rather than entropic, factors.

c. $W(PMe_3)_2(CNBu^t)_2(E)_2$ (E = S, Se) and $W(PMe_3)(CNBu^t)_4(\eta^2\text{-}Te_2)$. In contrast to the above reactions with aldehydes, the reactions of $W(PMe_3)_4(E)_2$ with alkyl isocyanides depend strongly on the nature of the chalcogen. Thus, the sulfido and selenido complexes $W(PMe_3)_4(E)_2$ react reversibly with alkyl isocyanides to substitute two PMe_3 ligands and give *trans,trans,trans*-$W(PMe_3)_2(CNR)_2(E)_2$, whereas the tellurido derivative $W(PMe_3)_4(Te)_2$ reacts irreversibly with Bu^tNC to give the η^2-ditellurido complex $W(PMe_3)(CNBu^t)_4(\eta^2\text{-}Te_2)$ (43), a result of the reductive coupling of the two tellurido ligands (Scheme 92) (270, 271). Similar results are also observed in the reactions of the aldehyde derivatives $W(PMe_3)_2(E)_2(\eta^2\text{-}OCHR)$ with Bu^tNC, where

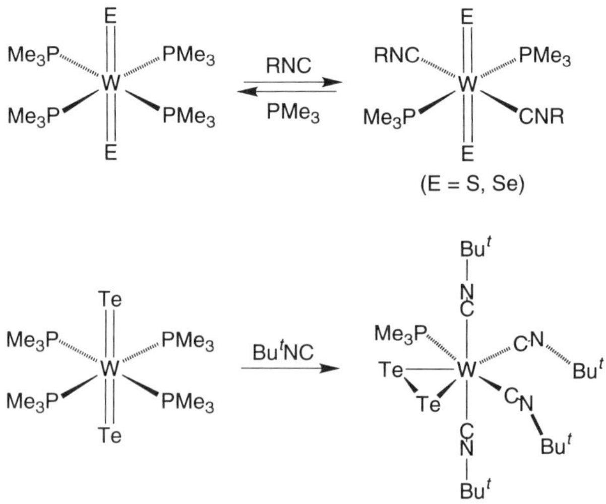

Scheme 92

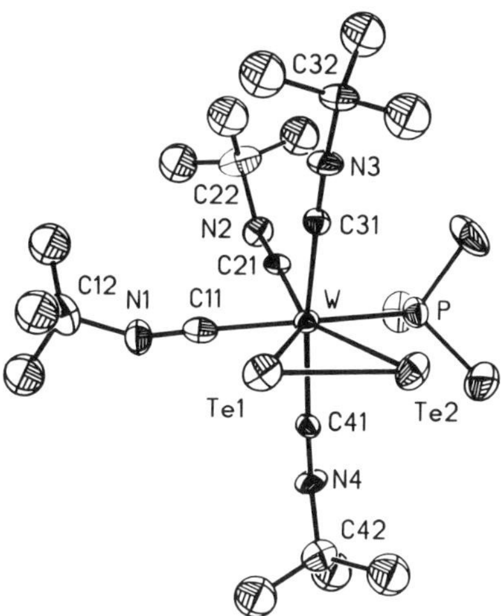

Figure 27. Molecular structure of $W(PMe_3)(CNBu^t)(\eta^2\text{-}Te_2)$. [Reprinted with permission from D. Rabinovich and G. Parkin, *Inorg. Chem.*, *34*, 6341 (1996). Copyright 1996 American Chemical Society.]

the sulfido and selenido complexes yield $W(PMe_3)_2(CNBu^t)_2(E)_2$ but the tellurido derivative results in coupling to give $W(PMe_3)(CNBu^t)_4(\eta^2\text{-}Te_2)$.

The molecular structure of $W(PMe_3)(CNBu^t)_4(\eta^2\text{-}Te_2)$ has been determined by X-ray diffraction (Fig. 27) and, as expected, the W—Te bond lengths [2.868(2) and 2.877(2) Å] are substantially longer than the corresponding multiple bonds in $W(PMe_3)_4(Te)_2$ [2.596(1) Å] and $W(PMe_3)_2(Te)_2(\eta^2\text{-}OCHR)$ [2.524(1) − 2.534(1) Å]. Interestingly, the X-ray structure also indicates that there are two types of Bu^tNC ligand present in the molecule, namely, linear and bent (272, 273). The observation of bent isocyanide ligands is consistent with extensive donation of electron density from the electron-rich metal center to the isocyanide ligand (π back-bonding) and is associated with a resonance structure of the type $M{=}C{=}\ddot{N}{-}R$ (**A**), in contrast to $\bar{M}{-}C{\equiv}\overset{+}{N}{-}R$ (**B**) for linear coordination.

$$M{=}C{=}\ddot{N}\backslash R \qquad \bar{M}{-}C{\equiv}\overset{+}{N}{-}R$$

A **B**

Furthermore, by comparison with PMe_3, the strong π-acceptor character of Bu^tNC was suggested to be the reason for its ability to induce coupling of the tellurido ligands: that is, π-acceptor ligands reduce electron density at a metal center and thereby stabilize a lower valence state.

The observation of coupling for the tellurido system, but not for the sulfido and selenido systems, is consistent with the notion that tellurium exhibits a reduced tendency to partake in multiple bonding. A related observation has also been noted for the permethyltantalocene system, as described in Section IV.C.3. However, although coupling of selenido and sulfido ligands was not observed for the $W(PMe_3)_4(E)_2$ system, it should be noted that the coupling of such ligands has been observed in other systems. For example, the reductive coupling of two sulfido ligands has been proposed as a possible sequence in (a) RSSR induced conversion of $[Mo(S)_4]^{2-}$ to $[(\eta^2\text{-}S_2)(S)Mo(\mu\text{-}S)]_2^{2-}$ (274a), (b) formation of $[(\eta^2\text{-}S_4)(S)M(\mu\text{-}S)]_2^{2-}$ by the reaction of $[M(S)_4]^{2-}$ (M = Mo, W) with S_8 (274b), (c) formation of $[W_2(S)_2(SH)(\mu\text{-}\eta^3\text{-}S_2)(\eta^2\text{-}S_2)_3]^-$ upon acidification of $[W(S)_4]^{2-}$ with HCl in MeCN (274c), (d) formation of $W(O)(\eta^2\text{-}S_2)_2(bpy)$ by reaction of $[W(S)_4]^{2-}$ with aqueous HCl in the presence of bpy (274d), (e) decomposition of $[Mo(O)_2(S)_2]^{2-}$ to $[(\eta^2\text{-}S_2)(O)Mo(\mu\text{-}S)_2Mo(O)(\eta^2\text{-}S_2)]^{2-}$ (274e), (f) synthesis of $(Bu^i_2NCS_2)_2Re(\mu\text{-}S)]_2$ by the reaction of $[Re(S)_4]^{3-}$ with $(Bu^i_2NCS_2)_2$, (274f), and (g) synthesis of $[(\eta^2\text{-}Bu^i_2NCS_2)_2V(\mu\text{-}\eta^2\text{-}S_2)]_2$ by the reaction of $[V(S)_4]^{3-}$ with $(Bu^i_2NCS_2)_2$ (274g). Most recently, Young and coworkers (11a, b) described an interesting example of a complex with an intramolecular Mo=S· · ·S interaction, namely, $[Tp^{Me_2}]Mo(O)(S)\{\eta^2\text{-}SP(S)Pr^i_2\}$, and suggested how such interactions may facilitate redox induced coupling and cleavage reactions. For example, oxidation of $[Tp^{Me_2}]Mo(O)(S)\{\eta^1\text{-}SP(S)Pr^i_2\}$ results in S—S coupling and the formation of $[Tp^{Me_2}]Mo(O)\{\eta^2\text{-}SP(SS)Pr^i_2\}^+$. Correspondingly, reduction of $[Tp^{Me_2}]Mo(O)\{\eta^2\text{-}SS(C_5H_4N)N,S\}$ is accompanied by cleavage of the S—S bond and formation of the sulfido–thiolate derivative $[Tp^{Me_2}]Mo(O)(S)\{S(C_5H_4N)\}^-$. Of relevance to these observations, Stiefel (274h) previously noted short S· · ·S contacts in dioxomolybdenum thiolate complexes of the type $[\eta^2\text{-}R_2NCR_2CMe_2S]_2Mo(O)_2$, and suggested that such "partial S—S bond formation" may also occur in molydoenzymes (274h).

By comparison to sulfido complexes, oxidative-cleavage and reductive-coupling reactions involving oxo ligands have been less well studied, but such transformations are proposed to be involved in (a) the photochemical conversion of $(TPP)Mo(\eta^2\text{-}O_2)_2$ to *cis*-$(TPP)Mo(O)_2$ (274i), and (b) the photoinduced decomposition of $[Mn(O)_4]^-$ (274j); also see Section IV.E.3.

While the reductive coupling of the two tellurido ligands and formation of $W(PMe_3)(CNBu^t)_4(\eta^2\text{-}Te_2)$ is irreversible, the oxidative cleavage of the η^2-ditellurido ligand has been observed for a closely related system. Specifically, elimination of H_2 from $W(PMe_3)_4(\eta^2\text{-}Te_2)H_2$ is accompanied by the oxidative cleavage of the ditellurido ligand and formation of the terminal tellurido complex $W(PMe_3)_4(Te)_2$, as illustrated in Scheme 93.

d. $W(PR_3)_2(L)(E)Cl_2$ Derivatives. Several examples of tungsten chalcogenido complexes of the type $W(PR_3)_2(L)(E)Cl_2$ are known, of which the oxo

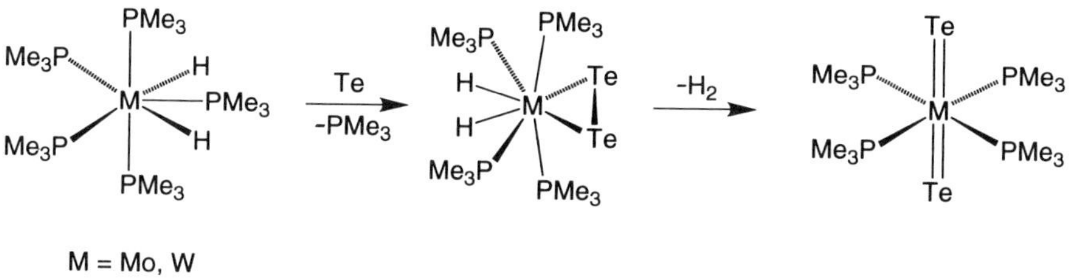

M = Mo, W

Scheme 93

complex *cis-mer*-$W(PMe_2Ph)_3(O)Cl_2$ is one of the earliest examples (275–277). More recently, Mayer described the syntheses of several related terminal oxo and sulfido complexes by the oxidative addition of E═X *multiple* bonds across low-valent tungsten centers. For example, cumulenes and heterocumulenes E═C═E′ (e.g., CO_2, COS) react with $W(PR_3)_4Cl_2$ to give $W(PR_3)_2$-(E)(CE′)Cl_2, as illustrated in Scheme 94 (PR_3 = $PMePh_2$, PMe_3) (230, 232, 278–280). Likewise, other multiply bonded substrates and epoxides react sim-

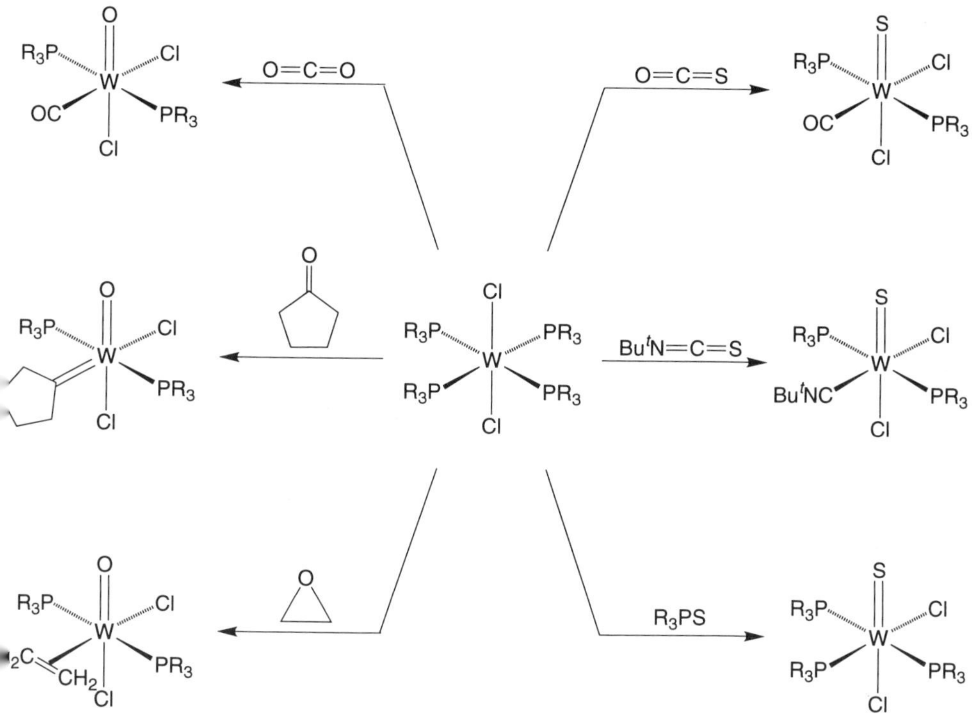

Scheme 94

ilarly. The carbonyl and ethylene complexes can also be obtained by substitution of one of the $PMePh_2$ ligands in $W(PMePh_2)_3(E)Cl_2$ by CO and C_2H_4, while the isocyanide complexes $W(PMePh_2)_2(E)(CNBu^t)Cl_2$ may be obtained similarly (279, 281).

As noted by Mayer (232), π-acid ligands (L = C_2H_4, CO, $CNBu^t$) in complexes of the type $W(PR_3)_2(E)(L)Cl_2$ adopt positions that maximize π back-bonding with the d^2 metal center. For a d^2 octahedral metal center with a single multiply bonded functionality (along z), the HOMO is a nonbonding d_{xy} orbital [cf. *trans*-$MoL_4(O)_2$; Section III.D], so that the π-acid ligand is required to be cis to the multiply bonded group. Furthermore, in order to maximize bonding, ligands with a single π-acceptor orbital, such as olefins and alkynes, are required to lie in a plane perpendicular to the multiply bonded moiety, as illustrated in Fig. 28.

Since the E=CE′ double bond of cumulenes is cleaved to generate both divalent (E) and dative (CE′) ligands, the overall result is a two-electron oxidation of the tungsten center. In contrast, $W(PMePh_2)_4Cl_2$ reacts with cyclopentanone to give ultimately the oxo–alkylidene complex $W(PMePh_2)_2Cl_2(O)[C(CH_2)_4]$ (282, a transformation that is a very interesting example of the "double" (i.e., a four electron) oxidative addition of a [C=O] moiety (Scheme 94). Furthermore, the [C=O] bond in cyclopentanone (~160 kcal mol^{-1}) and carbon dioxide (127 kcal mol^{-1}) represent the strongest bonds to be cleaved by a single metal center with both fragments remaining coordinated. The driving force for these reactions has been attributed to the strength of the tungsten–oxo bond, which has been estimated to be greater than 138 kcal mol^{-1}.

The formation of $W(PMePh_2)_2Cl_2(O)[C(CH_2)_4]$ proceeds via a bis(ketone) derivative $W(PMePh_2)_2Cl_2[\eta^2\text{-}OC(CH_2)_4]_2$, which subsequently converts to the oxo–alkylidene derivative $W(PMePh_2)_2Cl_2(O)[C(CH_2)_4]$ upon elimination of 1 equiv of ketone (283, 284). The related bis(acetone) complex $W(PMePh_2)_2Cl_2(\eta^2\text{-}OCMe_2)_2$ has also been synthesized and structurally char-

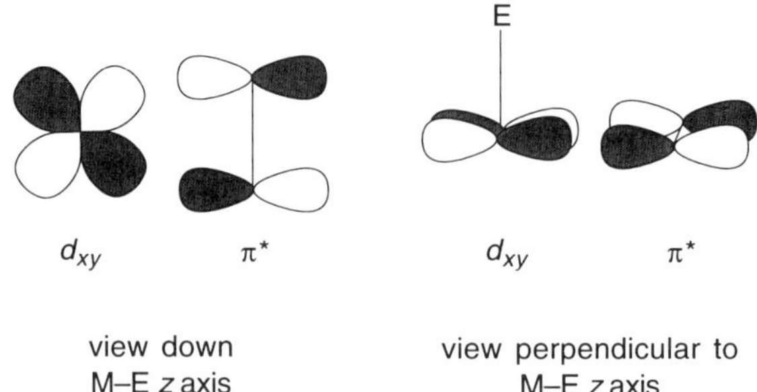

Figure 28. Preferred orientation of an olefin with respect to a multiply bonded [M=E] moiety in order to maximize π interactions.

acterized by X-ray diffraction; MeCHO and Bu^tCHO also form analogous $W(PMePh_2)_2Cl_2(\eta^2\text{-}OCHR)_2$ derivatives. However, these $W(PMePh_2)_2Cl_2(\eta^2\text{-}OCR_2)_2$ complexes do not convert to an oxo–alkylidene derivative, but rather decompose to $W(PMePh_2)_3(O)Cl_2$. In the case of the aldehyde derivatives, the olefins RCH=CHR are also observed as a byproduct. Likewise, the reactions of aromatic ketones [e.g., Ph_2CO and PhC(O)Me], with $W(PR_3)_4Cl_2$ yield $W(PR_3)_3(O)Cl_2$ and the olefins [$Ph_2C{=}CPh_2$ and Ph(Me)C=C(Me)Ph, respectively]; for these examples, however, the bis(ketone) complexes are not spectroscopically observed.

The chalcogenido complexes $W(PMePh_2)_2(E)(CO)Cl_2$ and $W(PMePh_2)_2(E)(\eta^2\text{-}C_2H_4)Cl_2$ (E = O, S) described above are particularly noteworthy since they are the first examples of oxo carbonyl, sulfido carbonyl, oxo olefin, and sulfido olefin complexes to be isolated (278, 279, 281). More recently, the oxo and sulfido carbonyl complexes $[Tp^{Me_2}]WX(O)(CO)$ (X = Br, I) (285), $[Tp^{Me_2}]W(O)_2(\mu\text{-}O)W(O)(CO)[Tp^{Me_2}]$ (286), $[Tp^{Me_2}]W(O)(CO)\{SP(S)Ph_2\}$ (287), and $[Tp^{Me_2}]W(S)(CO)\{SP(S)Ph_2\}$ (287) have been synthesized (Scheme 95), with accurate structures for the latter two having been determined by X-ray diffraction. Interest in these complexes derives from the fact that chalcogenido and carbonyl (or olefin) ligands typically bind to metal centers with different electronic characters. Thus, chalcogenido ligands are typically bound to metal centers with low d^n counts (in the present case d^2) and hence high oxidation states (1), where π-acceptor ligands such as CO and olefins require sufficiently high energy filled d orbitals to allow for π back-donation. More recently, the

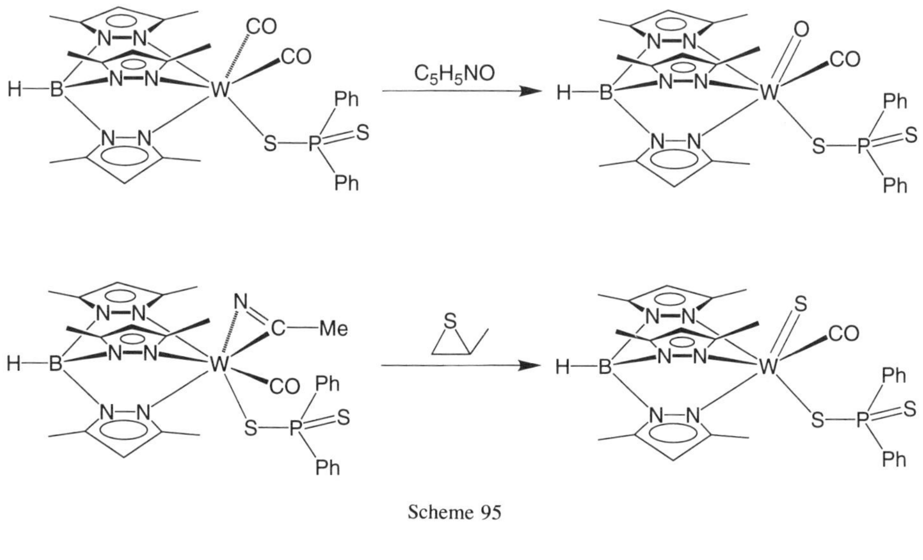

Scheme 95

TABLE XXXI
ν(CO) Data for Chalcogenido–Carbonyl and Related Tungsten Complexes

Complex	ν(CO) (cm^{-1})	Reference
$W(PMePh_2)_2(O)(CO)Cl_2$	2006	*a*
$W(PMePh_2)_2(S)(CO)Cl_2$	1986	*a*
$W(PMePh_2)_2(NPh)(CO)Cl_2$	1964	*a*
$W(PMePh_2)_2$(N-*p*-Tol)$(CO)Cl_2$	1964	*a*
$W(PMe_3)_2(O)(CO)Cl_2$	1995	*a*
$W(PMe_3)_2(NPh)(CO)Cl_2$	1946	*a*
$W(PMe_3)_2(CHPh)(CO)Cl_2$	1939	*a*
$[Tp^{Me_2}W(O)(CO)\{SP(S)Ph_2\}$	1985	*b*
$[Tp^{Me_2}]W(S)(CO)\{SP(S)Ph_2\}$	1960	*b*
$[Tp^{Me_2}]W(O)_2(\mu\text{-}O)W(O)(CO)[Tp^{Me_2}]$	1950	*c*
$[Tp^{Me_2}]WBr(O)(CO)$	1977	*d*
$[Tp^{Me_2}]WI(O)(CO)$	1975	*d*

[a] F.-M. Su, J. C. Bryan, S. Jang, and J. M. Mayer, *Polyhedron*, *8*, 1261 (1989).
[b] S. Thomas, E. R. T. Tiekink, and C. G. Young, *Organometallics*, *15*, 2428 (1996).
[c] C. G. Young, R. W. Gable, and M. F. Mackay, *Inorg. Chem.*, *29*, 1777 (1990).
[d] S. G. Feng, L. Luan, P. White, M. S. Brookhart, J. L. Templeton, C. G. Young, *Inorg. Chem.*, *30*, 2582 (1991).

first example of a d^0 chalcogenido carbonyl complex, namely, $Cp_2^*Zr(Se)(CO)$, has been isolated. In view of the d^0 nature of the metal center, π back-donation cannot occur to a first approximation, and so $Cp_2^*Zr(Se)(CO)$ is a particularly interesting example of a chalcogenido carbonyl complex (see Section IV.B.2).

Infrared studies on a series of carbonyl complexes, for example, $W(PR_3)_2(X)(CO)Cl_2$ (X = O, S, NR, CHR) identifies that the oxo ligand is the less electron donating of this group (see Table XXXI).

e. Tungstenocene Derivatives. Similar to Cp_2MoO, the tungstenocene oxo complex Cp_2WO has been synthesized by the reaction of Cp_2WCl_2 with NaOH (244). The oxo complex Cp_2WO is also obtained by irradiation of Cp_2WH_2 or thermolysis of $Cp_2W(Me)H$ in the presence of propylene oxide (Scheme 96) (245). More recently, the permethyltungstenocene analogue Cp_2^*WO has been prepared by the reaction of $Cp_2^*WCl_2$ with $KOH_{(aq)}$ and by the reaction of $Cp_2^*W(\eta^2\text{-}CH_2O)$ with H_2O (Scheme 97) (51). The oxo complex Cp_2^*WO is also obtained by thermolysis of $Cp_2^*W(OH)H$; since equimolar quantities of $Cp_2^*WH_2$ are also obtained, the proposed mechanism involves initial ligand redistribution giving $Cp_2^*WH_2$ and $Cp_2^*W(OH)_2$, followed by 1,2 elimination of H_2O (288). Finally, Cp_2^*WO has also been reported to be a trace product in the synthesis of $Cp_2^*WH_2$ by cocondensation of tungsten atoms with Cp*Li (246), and by photolysis of $Cp_2^*WH_2$ in the presence of water (47b).

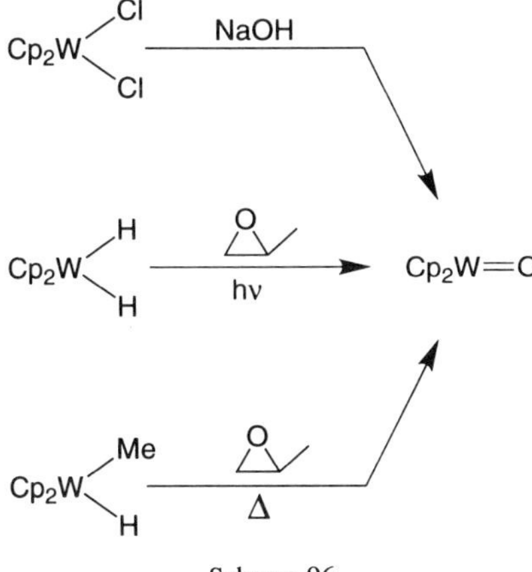

Scheme 96

The compounds Cp_2WO and Cp^*_2WO are 18-electron complexes and, as such, are better described as possessing a W=O double bond rather than a triple bond (see Section III.A). Thus, the ν(W=O) stretching frequencies of 799–879 and 860 cm^{-1}, respectively, are considerably lower than observed for other terminal tungsten oxo complexes; for example, a range of 922–1058 cm^{-1} is commonly observed for monooxo tungsten complexes (185). In support of the presence of comparatively weak tungsten–oxo interactions, Cp_2WO and Cp^*_2WO display particularly high reactivity associated with the [W=O] functionality. For example, Cp^*_2WO undergoes rapid exchange with labeled H_2^*O,

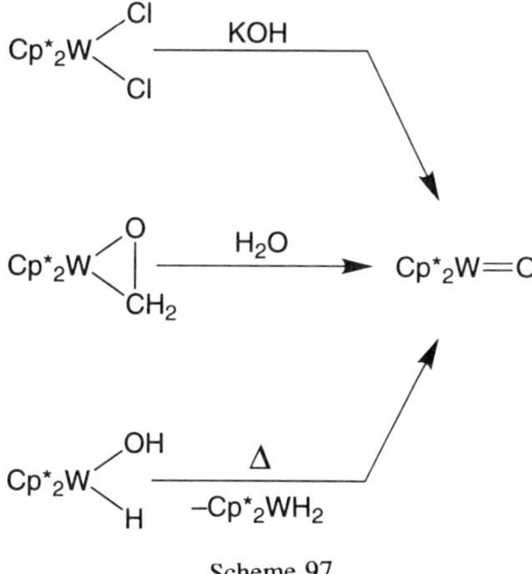

Scheme 97

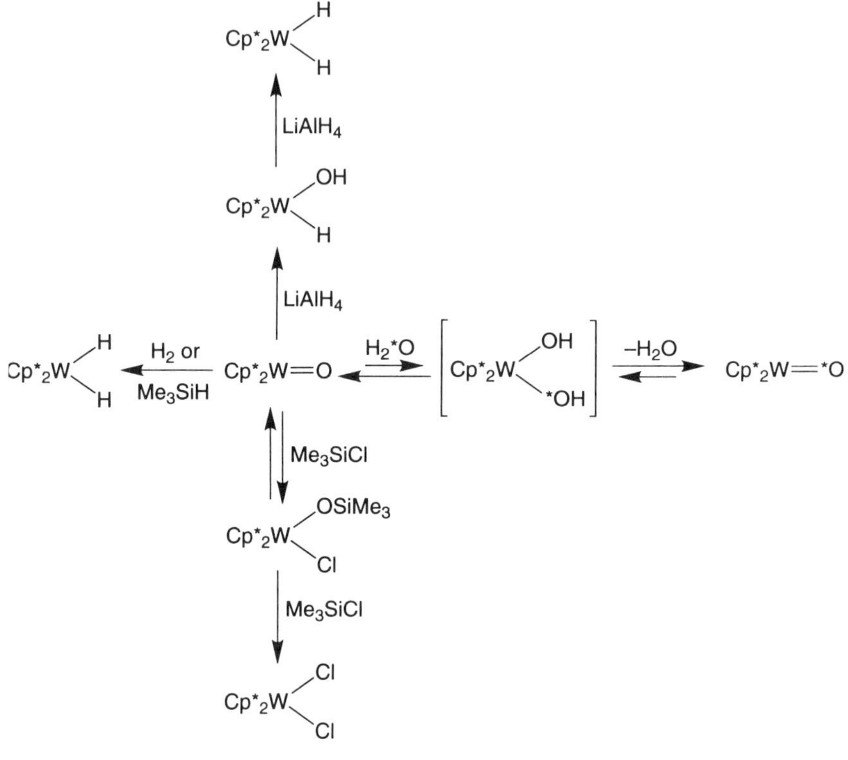

Scheme 98

presumably via the dihydroxide intermediate [$Cp_2^*W(OH)(^*OH)$] (Scheme 98) (51). Furthermore, this exchange is more rapid than observed for related ≤ 16-electron complexes; for example, (η^5-Cp*)(η^1-Cp*)$W(O)_2$ does not undergo facile exchange with labeled water.

Without implying any mechanistic interpretation, the formation of the dihydroxide intermediate [$Cp_2^*W(OH)_2$] may be considered to be the result of a formal 1,2 addition of the O—H bond across the W=O bond. In this regard, the reaction of Cp_2^*WO with Me_3SiCl proceeds in an analogous fashion via 1,2 addition of the Si—Cl bond across W=O to give $Cp_2^*W(OSiMe_3)Cl$. Subsequent reaction of $Cp_2^*W(OSiMe_3)Cl$ with excess Me_3SiCl gives $Cp_2^*WCl_2$ (Scheme 98). Under forcing conditions (~200°C), Cp_2^*WO may be hydrogenated to $Cp_2^*WH_2$ using either H_2 or Me_3SiH. The conversion can also be achieved under milder conditions using $LiAlH_4$, in which case the hydroxy hydride complex $Cp_2^*W(OH)H$ has been isolated as the initial product.

The [W=O] moiety of Cp_2^*WO exhibits nucleophilicity at two sites, namely,

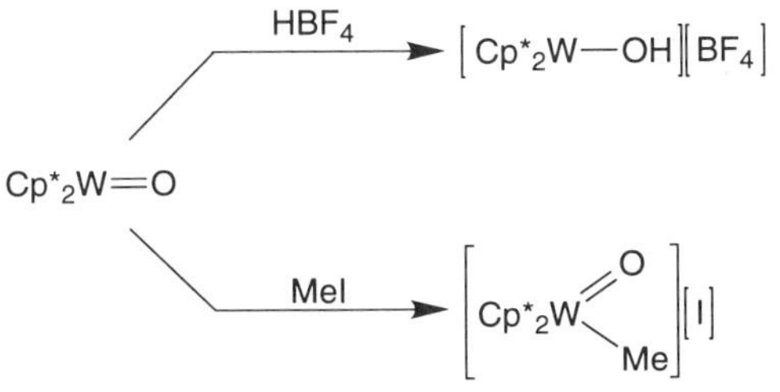

HBF_4 → $[Cp^*_2W{-}OH][BF_4]$

$Cp^*_2W{=}O$

MeI → $[Cp^*_2W(O)Me][I]$

Scheme 99

the oxo ligand by virtue of its lone pair, and the tungsten center by virtue of its d^2 configuration, as illustrated in Scheme 99. For example, the oxo ligand of Cp_2^*WO is protonated by HBF_4 to give $[Cp_2^*W(OH)][BF_4]$, whereas alkylation by MeI occurs at the tungsten center to give the oxo–methyl cation $[Cp_2^*W(O)Me][I]$, rather than $Cp_2^*W(OMe)I$. Such selectivity is in marked contrast to the reaction of the related d^0 zirconium complex, $Cp_2^*Zr(O)(NC_5H_5)$, in which the oxo ligand is the site for electrophilic attack, giving $Cp_2^*Zr(OMe)I$ (see Section IV.B.2). Cooper and co-workers (289) have also synthesized the unsubstituted tungstenocene oxo methyl complex $[Cp_2W(O)Me][PF_6]$ by photolysis of $[Cp_2W(NCMe)Me][PF_6]$ in the presence of O_2.

The oxo complex, Cp_2^*WO, is a useful precursor for other tungsten oxo derivatives (Scheme 100). For example, Cp_2^*WO reacts with $H_2O_{2(aq)}$ to give (η^5-

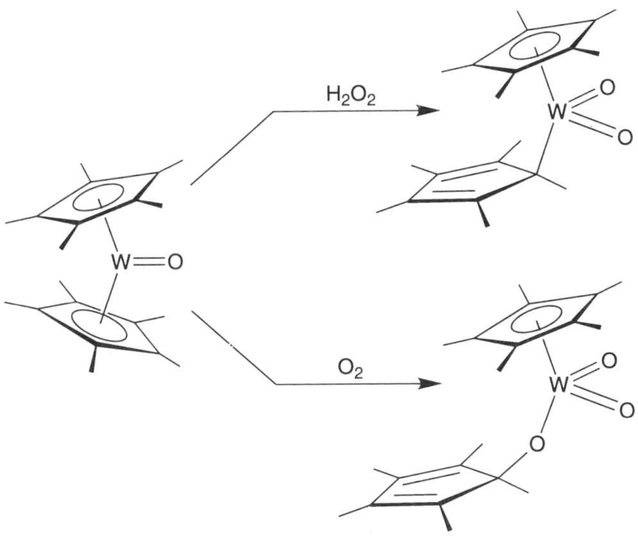

Scheme 100

$Cp^*)(\eta^1\text{-}Cp^*)W(O)_2$, the first example of a transition metal complex with a crystallographically characterized monohapto Cp* ligand (290). In contrast, the reaction of Cp_2^*WO with O_2 results in insertion of an oxygen atom into the [W—Cp*] moiety, thereby giving $Cp^*W(O)_2(OCp^*)$.

Geoffroy and co-workers (249) described additional chemistry of the parent tungstenocene oxo complex Cp_2WO, which also exhibits formal 1,2-addition reactions, as illustrated in Scheme 101. An interesting contrast between the Cp and Cp* systems, however, is provided by the observation that Cp_2WO reacts with MeOH to give $Cp_2W(OMe)_2$, whereas for the Cp* system, the reaction between $Cp_2^*WCl_2$ and NaOMe gives the formaldehyde complex $Cp_2^*W(\eta^2\text{-}OCH_2)$ and MeOH, rather than the bis(methoxide) complex $Cp_2^*W(OMe)_2$.

The oxo complex Cp_2WO has also been reported to undergo a variety of formal [2 + 2] cycloaddition reactions with unsaturated organic substrates, as illustrated in Scheme 102. Interestingly, although benzaldehyde and acetone do not react with Cp_2WO to give stable metallacycles, evidence for the generation of such species is provided by the observation of oxygen-atom exchange between $Cp_2W{=}^{17}O$ and R_2CO. The W=O moiety also undergoes [2 + 2] cycloaddition across the C=O bond of certain transition metal carbonyl complexes; for example, Cp_2WO reacts with $[Cp^{Me}Mn(CO)_2(NO)]^+$ to give $[Cp^{Me}Mn(CO)(NO)(\mu_2\text{-}\eta^3\text{-}CO_2)WCp_2]^+$ (Scheme 103). Interestingly, the corresponding reaction of $Cp_2W{=}^{17}O$ to give $[Cp^{Me}Mn(CO)(NO)(\mu_2\text{-}\eta^3\text{-}$

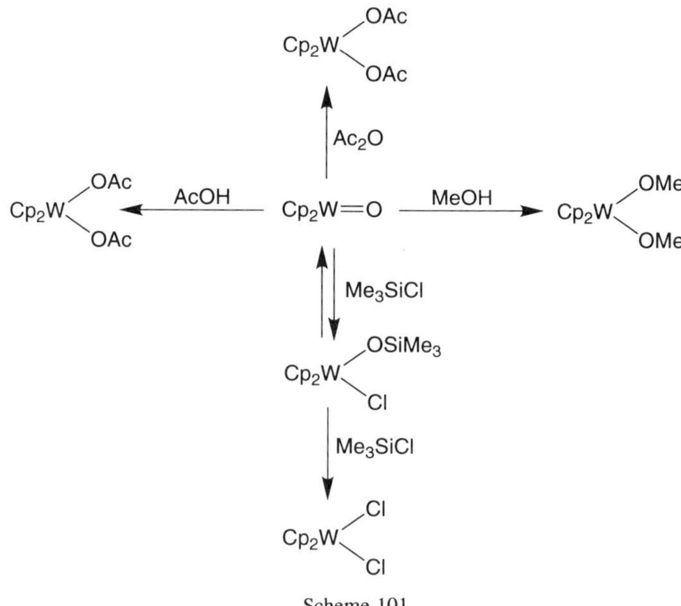

Scheme 101

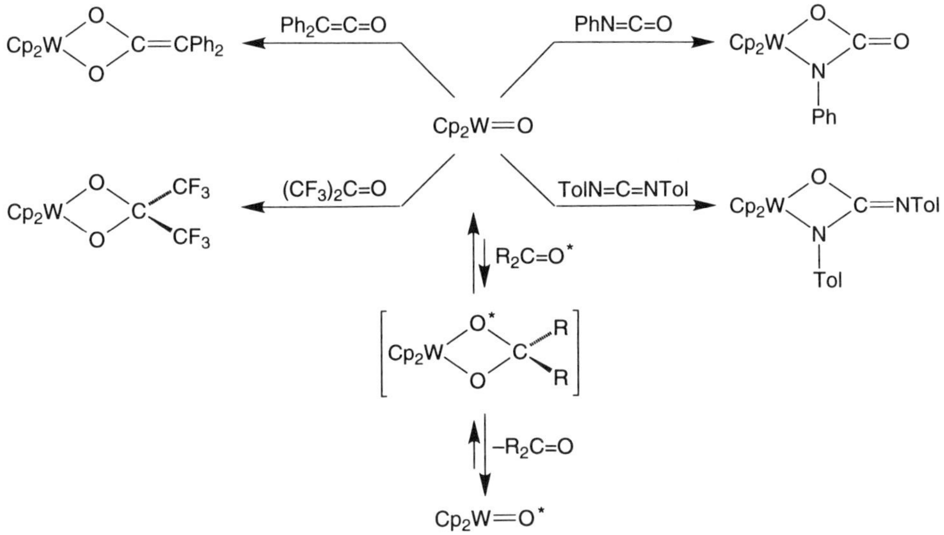

Scheme 102

$CO^{17}O)WCp_2]^+$ is accompanied by ^{17}O incorporation into a carbonyl ligand, that is $[Cp^{Me}Mn(C^{17}O)(NO)(CO_2)WCp_2]^+$. Although a possible explanation for ^{17}O incorporation into the carbonyl ligand is that the [2 + 2] addition of the [W=O] and [O=C] moieties is reversible, additional studies suggest that Cp_2WO is not released from the $[Cp^{Me}Mn(CO)(NO)(\mu_2\text{-}\eta^3\text{-}CO_2)WCp_2]^+$ product; an intramolecular mechanism has therefore been suggested to explain the exchange process (Scheme 103).

E. Terminal Chalcogenido Complexes of Group 7 (VII B): Mn, Tc, Re

1. *Manganese Chalcogenido Complexes*

Other than derivatives such as the well-known permanganate $[Mn(O)_4]^-$ anion (291), terminal chalcogenido complexes of manganese are extremely rare. Indeed, structurally characterized molecular terminal oxo complexes have only been obtained recently. Collins et al. (292, 293) were the first to isolate monomeric manganese(V) terminal oxo complexes, using both $[N_4]$ and $[N_2O_2]$ ligand-donor complements, as illustrated in Scheme 104. Likewise, O'Halloran and co-workers (294) used a closely related N_2O_2 ligand complement to prepare a terminal oxo complex by direct oxidation of a Mn^{III} derivative by O_2. The Mn=O bond length and ν(Mn=O) IR data for these complexes are summa-

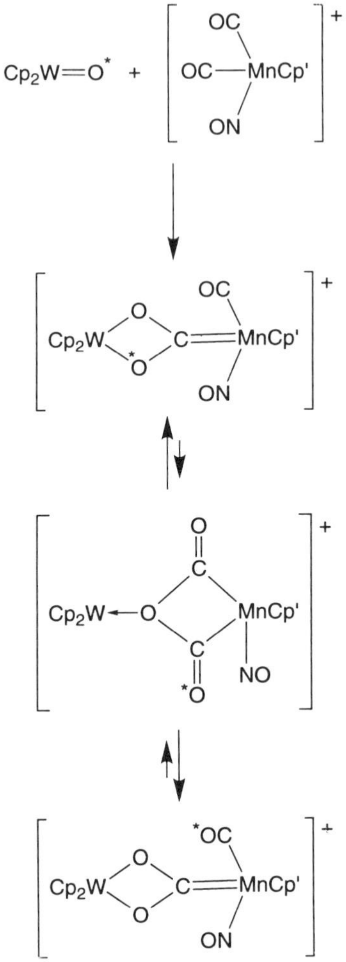

Scheme 103

rized in Table XXXII. Collins has also used ^{18}O labeling to provide a definitive assignment of the $\nu(Mn{=}O)$ vibration in the IR spectrum of $[\{\eta^4\text{-}(C_6H_2Cl_2)[NC(O)CMe_2NC(O)]_2(CEt_2)\}Mn(O)]^-$. The observed values [$\nu(Mn{=}^{16}O)$ = 979 cm^{-1} and $\nu(Mn{=}^{18}O)$ = 942 cm^{-1}] are, however, significantly higher than those reported for (TMP)Mn(O) (754 cm^{-1}) and (TTP)Mn(O) (754 cm^{-1}), which have been generated in situ (295). For comparison, $\nu(M{=}O)$ data for some related porphyrin–oxo derivatives are summarized in Table XXXIII.

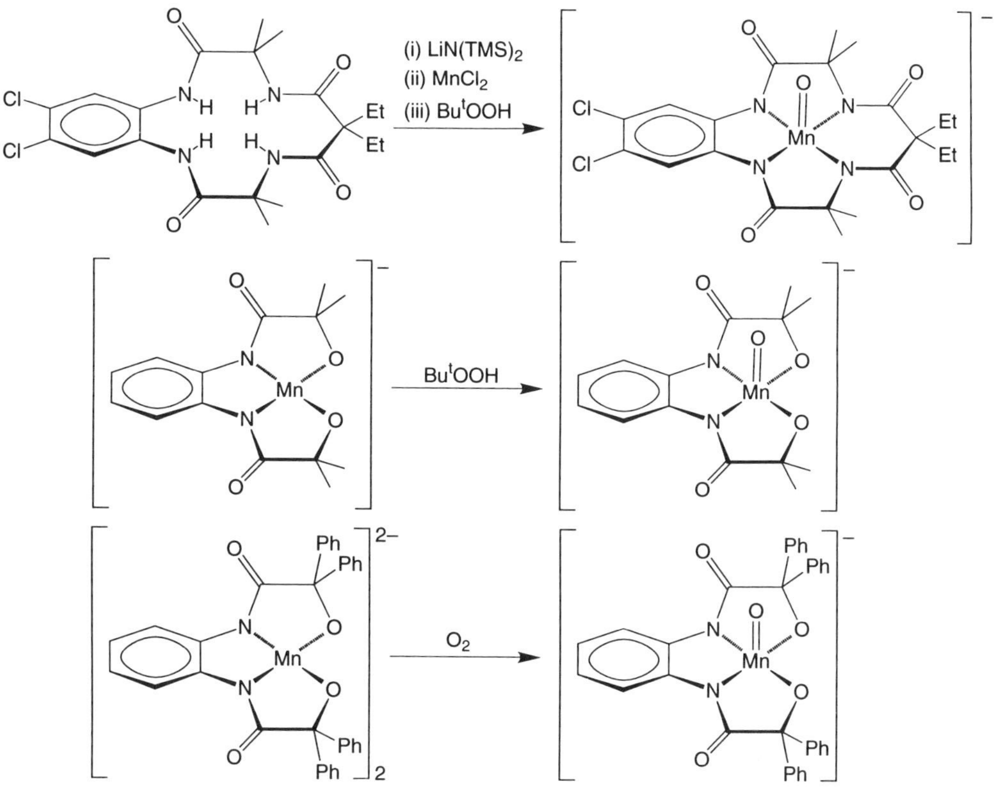

Scheme 104

TABLE XXXII

Mn=O Bond Length and ν(Mn=O) IR Data for Manganese Oxo Complexes

Complex	d(Mn=O) (Å)	ν(Mn=O) (cm^{-1})	Reference
$[\{\eta^4\text{-}(C_6H_2Cl_2)[NC(O)CMe_2NC(O)]_2(CEt_2)\}Mn(O)]^-$	1.555(4)	979	*a*
$[\{\eta^4\text{-}(C_6H_4)[NC(O)CMe_2O]_2\}Mn(O)]^-$	1.548(4)		*b*
$[\{\eta^4\text{-}(C_6H_4)[NC(O)CPh_2O]_2\}Mn(O)]^-$	1.558(4)		*c*

[a]T. J. Collins, R. D. Powell, C. Slebodnick, and E. S. Uffelman, *J. Am. Chem. Soc.*, *112*, 899 (1990).

[b]T. J. Collins and S. W. Gordon-Wylie, *J. Am. Che. Soc.*, *111*, 4511 (1989).

[c]F. M. MacDonnell, N. L. P. Fackler, C. Stern, and T. V. O'Halloran, *J. Am. Chem. Soc.*, *116*, 7431 (1994).

TABLE XXXIII
Resonance Raman $\nu(M{=}O)$ Data for Five- and Six-Coordinate Metalloporphyrin Derivatives[a]

	$\nu(M{=}O)$ (cm^{-1})	
	Five-Coordinate	Six-Coordinate
V^{IV}	1007	895
Cr^{IV}	1025	
Mn^{IV}	754	732
Fe^{IV}	843	

[a]Taken from (295).

A related cationic manganese oxo species [(TMP)MnO][CI] has been produced by the reaction of (TMP)MnCl with NaOCl (296a) and is characterized by a $\nu(Mn{=}O)$ absorption at 950 cm^{-1}, significantly higher in energy than that for the neutral complex (TMP)Mn(O) (754 cm^{-1}). The reagent NaOCl has also been used to generate proposed manganese(V) oxo complexes $[(salen)Mn(O)]^+$ (296b). These oxo complexes are suggested to be the active species responsible for catalytic epoxidation of olefins by (salen)Mn using NaOCl as the oxidant; however, the proposed $[(salen)Mn(O)]^+$ derivatives have not been isolated.

Manganese oxo species have also been generated by irradiation of carbonyl derivatives in O_2-doped Ar matrices at low temperature. For example, photolysis of $CpMn(CO)_3$ in the presence of O_2 yields a species proposed to be $[CpMn(O)_2]$ on the basis of the observed $\nu(Mn{=}O)$ values of 938 and 893 cm^{-1} (213, 297).

Finally, although structurally characterized terminal tellurido complexes of manganese are unknown, examples with multiply bonded *bridging* tellurido ligands have been synthesized, as illustrated in Scheme 105 (298).

2. *Technetium Chalcogenido Complexes*

By comparison with manganese and rhenium, the chemistry of technetium has been little studied since its natural abundance is very low due to the short half-lives of its isotopes; thus, terminal chalcogenido complexes of technetium are not common. Nevertheless, since technetium has become available in recent times as a product of nuclear fission, technetium chemistry has started to develop and, in fact, a significant number of terminal oxo complexes are now known (299, 300). For example, Davison has synthesized $[Tp]Tc(O)_3$ by (a) oxidation of $[Tp]Tc(O)Cl_2$ (300b, 301) with HNO_3 and (b) the reaction of Na[Tp] with $[NH_4][Tc(O)_4]$ (Scheme 106) (302). Likewise, Kläui and co-workers (303) prepared the $[\eta^3\text{-}CpCo\{P(OMe)_2(O)\}_3]Tc(O)_3$ counterpart. Interest-

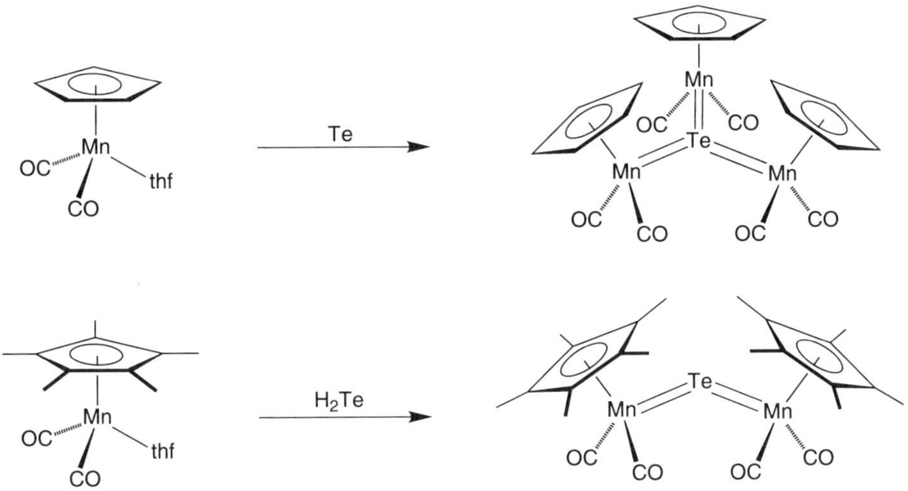

Te

H_2Te

Scheme 105

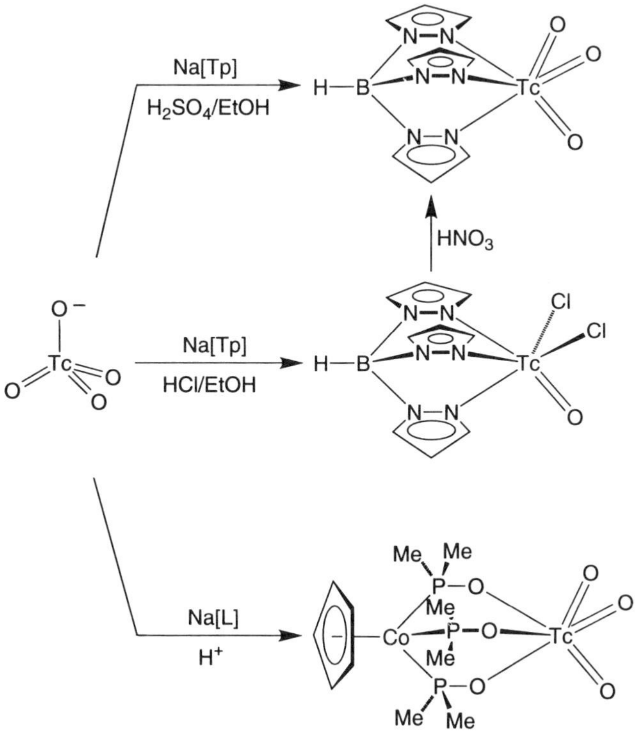

Na[Tp]

H_2SO_4/EtOH

HNO_3

Na[Tp]

HCl/EtOH

Na[L]

H^+

Scheme 106

ingly, $[Tp]Tc(O)_3$ reacts with ethylene to give $[Tp]Tc(O)(\eta^2\text{-}OCH_2CH_2O)$; in contrast, the rhenium complex $[Tp]Re(O)(\eta^2\text{-}OCH_2CH_2O)$ eliminates ethylene at elevated temperatures to give the trioxo complex $[Tp]Re(O)_3$ (302, 304), a difference that is in line with the ability of rhenium to stabilize the higher oxidation state (Scheme 107). Herrmann et al. (305) also reported the syntheses of alkyltechnetium oxo complexes by the reaction of Tc_2O_7 with Me_4Sn, as illustrated in Scheme 108.

One particular oxo complex that deserves some comment is the polymeric material $[Cp^*Tc(\mu\text{-}O)_3Tc]_n$ (Fig. 29), purported to be obtained by the reaction of $Cp^*Tc(CO)_3$ with H_2O_2 (306). The nature of $[Cp^*Tc(\mu\text{-}O)_3Tc]_n$ is most unusual considering that (a) the analogous reaction of $Cp^*Re(CO)_3$ with H_2O_2 yields the well-known trioxo complex $Cp^*Re(O)_3$ (see Section IV.E.3), and (c)

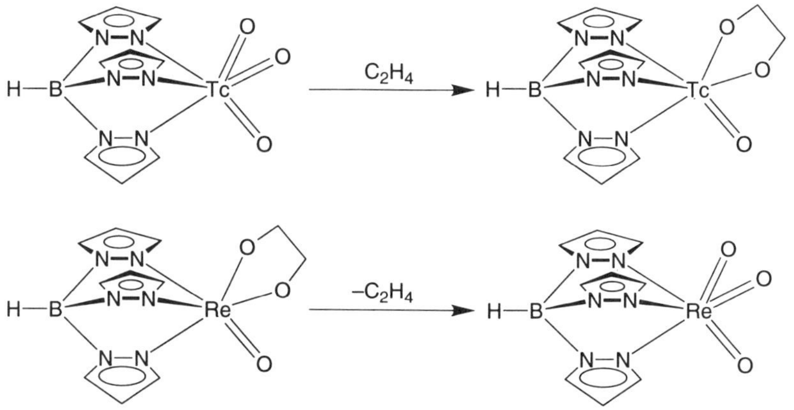

Scheme 107

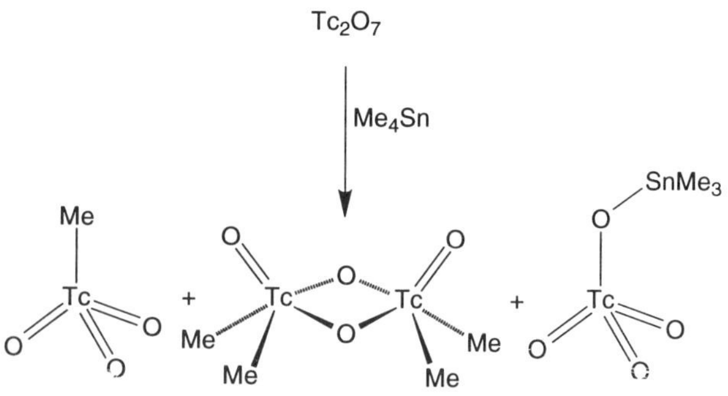

Scheme 108

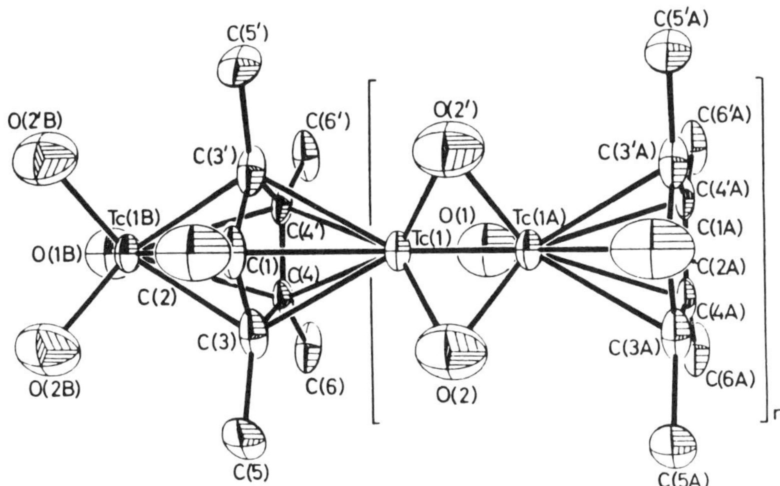

Figure 29. Originally proposed polymeric structure of $[Cp^*Tc(\mu\text{-}O)_3Tc]_n$. [Reproduced with permission from (306).]

the Tc—Tc separation of 1.867(4) Å is anomalously short; nevertheless, the structure of $[Cp^*Tc(\mu\text{-}O)_3Tc]_n$ has been rationalized theoretically (307). Herrmann et al. (308), however, questioned the characterization of polymeric $[Cp^*Tc(\mu\text{-}O)_3Tc]_n$ since its unit cell was similar to that of the rhenium trioxo complex $Cp^*Re(O)_3$. However, at the time that Herrmann questioned the nature of $[Cp^*Tc(\mu\text{-}O)_3Tc]_n$, the structure of $Cp^*Re(O)_3$ had not been solved. It is, therefore, significant that Cotton and co-workers (309) subsequently reexamined the crystallographic data for $Cp^*Re(O)_3$, identified the nature of the problem as a combination of twinning and disorder phenomena, and thereby obtained a satisfactory solution for its structure (Fig. 30); other workers have also reached a similar conclusion concerning the structure of $Cp^*Re(O)_3$ (310). With the structure of $Cp^*Re(O)_3$ having been successfully solved, Cotton duly noted that the polymeric structure alleged for "isostructural" $[Cp^*Tc(\mu\text{-}O)_3Tc]_n$ cannot be correct. However, he also pointed out that the stucture of $[Cp^*Tc(\mu\text{-}O)_3Tc]_n$, which had been refined with the Tc atoms at full occupancy, could *not* be refined successfully as the disordered trioxo complex $Cp^*Tc(O)_3$, with the Tc atoms at partial occupancies. This notion led Cotton to suggest that alleged polymeric $[Cp^*Tc(\mu\text{-}O)_3Tc]_n$ is actually the *rhenium* trioxo complex $Cp^*Re(O)_3$! In support of this suggestion, he cites that (a) the $\nu(Re{=}O)$ stretching vibrations in $Cp^*Re(O)_3$ are the same as reported for $[Cp^*Tc(\mu\text{-}O)_3Tc]_n$, and (b) the synthesis of $[Cp^*Tc(\mu\text{-}O)_3Tc]_n$ is irreproducible.

By comparison with the above terminal oxo complexes, the heavier chalcogenido derivatives of technetium are virtually unknown. Indeed, the first ter-

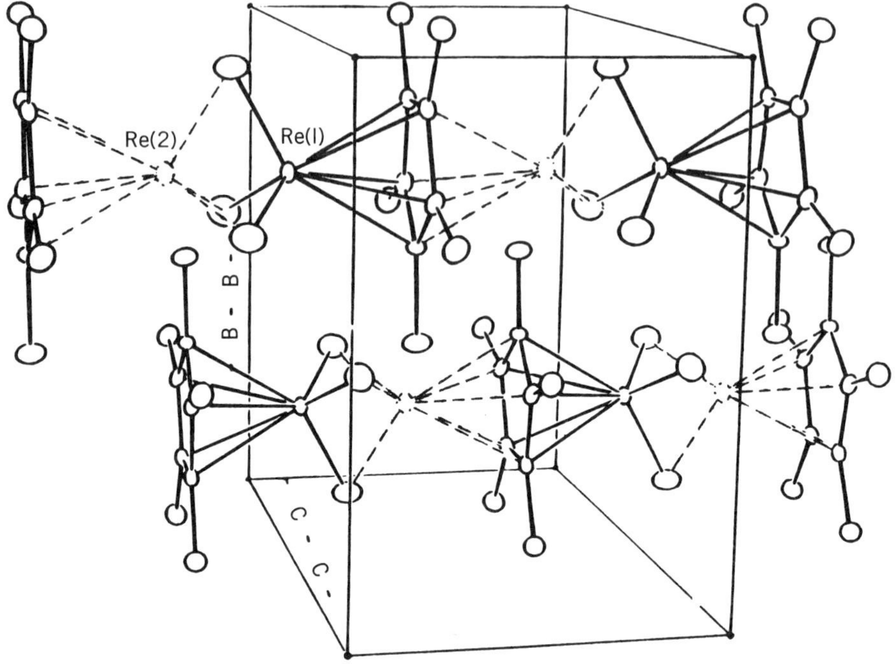

Figure 30. Molecular structure of $Cp^*Re(O)_3$ illustrating the nature of the disorder. [Reprinted with permission from A. R. Burrell, F. A. Cotton, L. M. Daniels, and V. Petricek, *Inorg. Chem.*, *34*, 4253 (1995). Copyright © 1995 American Chemical Society.] Note the similiarity with the originally proposed ordered structure for the polymeric techetium compound shown in Fig. 29.

minal technetium sulfido complex, $[Tp]Tc(S)Cl_2$, was only synthesized as recently as 1989 by Duatti et al. (311), from the reaction of the oxo derivative $[Tp]Tc(O)Cl_2$ with B_2S_3 (Scheme 109).

3. *Rhenium Chalcogenido Complexes*

Rhenium is known to form many terminal oxo complexes (312), a large number of which are organometallic. A selection of mononuclear oxo com-

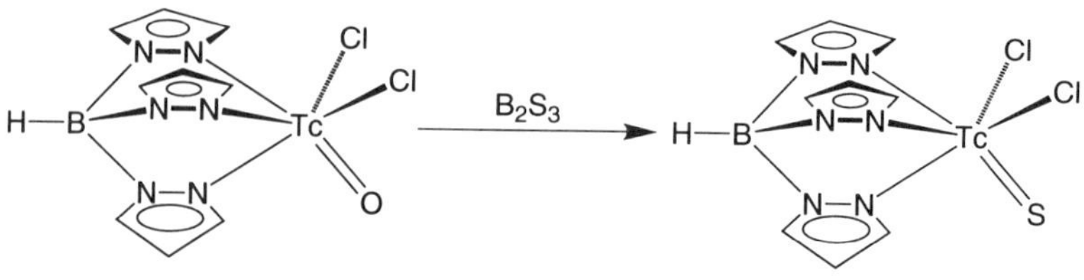

Scheme 109

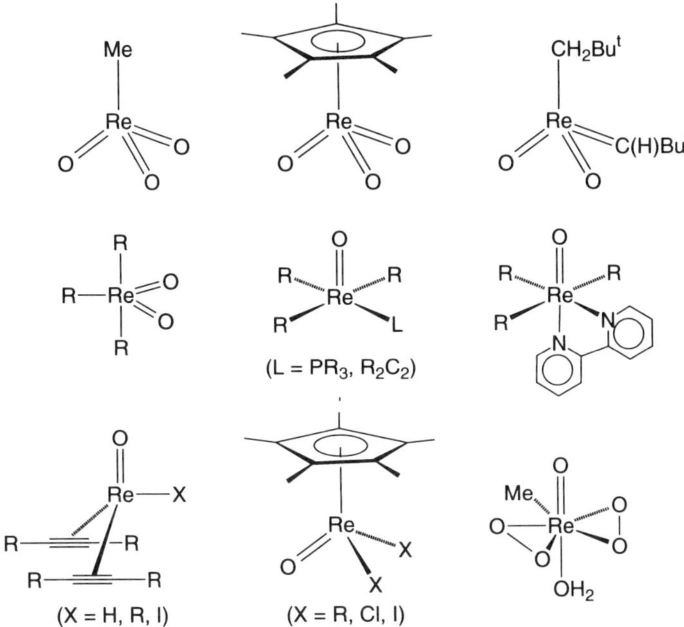

Figure 31. Representative examples of rhenium oxo complexes.

plexes is illustrated in Fig. 31 (313). Perhaps the most extensively studied of these complexes are the trioxo derivatives, for example, $MeRe(O)_3$ (314), $CpRe(O)_3$ (315), $(\eta^5\text{-}C_9H_7)Re(O)_3$ (316), and $Cp^*Re(O)_3$ (317, 318), which, as described by Herrmann and co-workers (25g, 319–322), exhibit diverse reactivities. The $[Tp]Re(O)_3$ (302, 304, 323) and $[\eta^3\text{-}(CpCo\{P(OMe)_2(O)\}_3]Re(O)_3$ (303) counterparts have also been synthesized. Photolysis of the oxalate complex $[Tp]Re(O)(\eta^2\text{-}C_2O_4)$ results in elimination of CO_2 and the formation of the mono-oxo species $\{[Tp]Re(O)\}$ as a reactive intermediate, as evidenced by its trapping reactions (Scheme 110) (304). The formation of $[Tp]Re(O)_3$ in the reaction with O_2 has been proposed to occur via the peroxo intermediate $\{[Tp]Re(O)(\eta^2\text{-}O_2)\}$. Interestingly, however, $\{[Tp]Re(O)(\eta^2\text{-}O_2)\}$ does not undergo a simple rearrangement to give $[Tp]Re(O)_3$ and a binuclear mechanism has been proposed. The barrier that prevents oxidative-cleavage within $[Re(\eta^2\text{-}O_2)]$ to give $[Re(O)_2]$ has been attributed to the transformation being symmetry forbidden.

Terminal rhenium sulfido derivatives are also reasonably well known, as illustrated by the examples listed in Table XXXIV. One of the earliest rhenium sulfido complexes, $[Re(S)_4]^-$, was obtained by the reaction of $[Re(O)_4]^-$ with H_2S (324a). Reactions of $[Re(S)_4]^-$ include (a) addition of alkynes in the presence of sulfur to give $[Re(S)\{SC(R)C(R)S\}_2]^-$, (b) addition of olefins in the

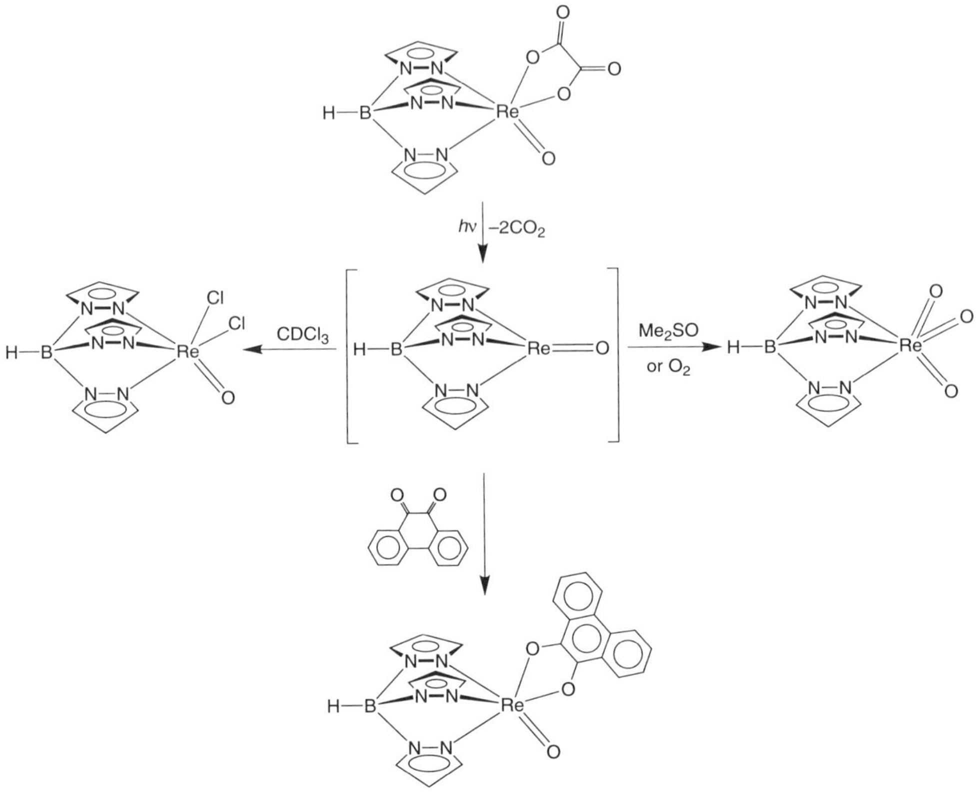

Scheme 110

presence of R_3NO to give $[Re(O)\{SC(R)_2C(R)_2S\}_2]^-$, and (c) addition of ethylene in the presence of sulfur to give $[Re(S)(S_4)(SCH_2CH_2S)]^-$ (324b). An analogy has been drawn between the latter reactions of $[Re(S)_4]^-$ with the osmylation reactions of $Os(O)_4$. The sulfido complex $[Re(S)_4]^-$ also reacts with disulfides $[R_2NC(S)S]_2$ to give $Re_2(\mu\text{-}S)_2(S_2CNR_2)_4$, a reaction that involves a formal reduction of the metal center upon addition of an oxidizing agent (324c).

The first nontetrahedral terminal sulfido complex $[(\eta^2\text{-}SCH_2CH_2S)_2Re(S)]^-$ was prepared by the reaction of $K_2[ReCl_6]$ with $HSCH_2CH_2SH$ in the presence of Et_3N; the terminal sulfido ligand was proposed to result from dealkylation of a dithiolate ligand (325, 326). Analogous to the technetium system, the rhenium sulfido complex $[Tp]Re(S)Cl_2$ was synthesized by the reaction of the oxo derivative $[Tp]Re(O)Cl_2$ with B_2S_3 (Scheme 111) (311). More recently, the alkyne derivatives $(\eta^2\text{-}C_2R_2)_2Re(S)I$ (R = Me, Et) have been obtained from $(\eta^2\text{-}C_2R_2)_2Re(O)I$ using a similar procedure (327). The structure of

TABLE XXXIV
Re=S and ν(Re=S) IR Data for Terminal Rhenium Sulfido Complexes

Complex	d(Re=S) (Å)	ν(Re=S) (cm^{-1})	Reference
$[Me_4N][Re(S)_4]$		500	*a*
$[Bu_4N][Re(S)_4]$	2.123[5]	484	*b*
$[Et_4N][Re(S)_4]$	2.125[4]		*c*
$[Bu_4N][Re(S)(S_4)_2]$	2.075(4)	528	*d*
		521	*e*
$[Pr_4N][Re(S)(S_4)_2]$	2.087(4)	526	*e*
$[Ph_4P][Re(S)(S_4)_2]$		523	*e*
$[Et_4P][Re(S)(S_4)(SCH_2CH_2S)]$	2.05		*f*
$[Pr_4N][Re(S)(S_4)(SSCMe_2S)]$	2.081(6)	511	*e*
$[(\eta^2\text{-}SCH_2CH_2S)_2Re(S)]^-$	2.104(2)	517	*g*
$(\eta^2\text{-}C_2Me_2)_2Re(S)I$	2.082(3)	531	*h*
$[Cp^{Me}Ru(dppe)SRe(S)_3$	2.115(3)–2.130(3)	490	*i*
$[Pr_4N]_2[Re(S)(\mu\text{-}S)_3(CuCl)_3Cl]$	2.10(1)	525	*j*

[a] A. Müller, E. Diemann, and V. V. K. Rao, *Chem. Ber.*, *103*, 2961 (1970).
[b] Y. Do, E. D. Simhon, and R. H. Holm, *Inorg. Chem.*, *24*, 4635 (1985).
[c] A. Müller, E. Krickemeyer, H. Bögge, M. Penk, and D. Rehder, *Chimia*, *40*, 50 (1986).
[d] F. A. Cotton, P. A. Kibala, and M. Matusz, *Polyhedron*, *7*, 83 (1988).
[e] A. Müller, M. Lemke, E. Krickemeyer, H. Bögge, and M. Penk, *Monats. Chem.*, *124*, 857 (1993).
[f] J. T. Goodman, S. Inomata, and T. B. Rauchfuss, *J. Am. Chem. Soc.*, *118*, 11674 (1996).
[g] P. J. Blower, J. R. Dilworth, J. P. Hutchinson, and J. A. Zubieta, *Inorg. Chim. Acta*, *65*, L225 (1982).
[h] S. K. Tahmassebi and J. M. Mayer, *Organometallics*, *14*, 1039 (1995).
[i] M. A. Massa, T. B. Rauchfuss, and S. R. Wilson, *Inorg. Chem.*, *30*, 4667 (1991).
[j] C. D. Scattergood, C. D. Garner, and W. Clegg, *Inorg. Chim. Acta*, *132*, 161 (1987).

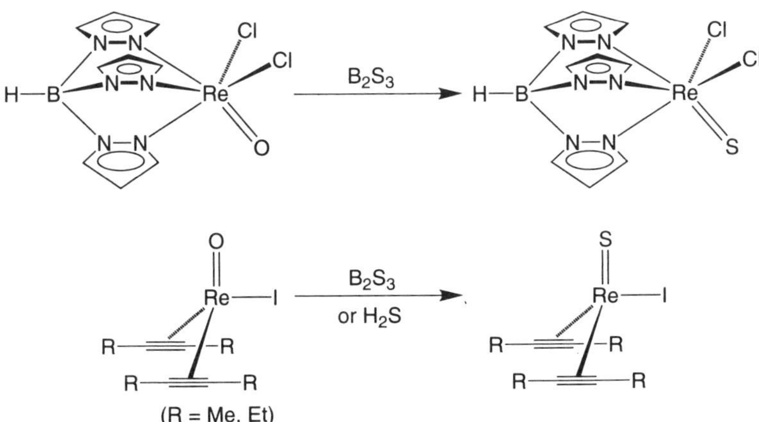

Scheme 111

(η^2-C_2Me_2)$_2$Re(S)I has been determined by X-ray diffraction and is characterized by a Re=S bond length of 2.082(3) Å; for comparison, the Re=O bond length in the oxo analogue is 1.697(3) Å. Furthermore, ν(Re=S) IR absorptions of 531 and 532 cm^{-1} for (η^2-C_2Me_2)$_2$Re(S)I and (η^2-C_2Et_2)$_2$Re(S)I, respectively, compare with a value of 971 cm^{-1} for ν(Re=O) in (η^2-C_2Et_2)$_2$Re(O)I. The oxo complex (η^2-C_2Et_2)$_2$Re(O)I is also slowly converted to the sulfido derivative upon treatment with H_2S, suggesting that the Re=O bond is no more than 46 kcal mol^{-1} stronger than the Re=S bond in this system.

F. Terminal Chalcogenido Complexes of Group 8 (VIII): Fe, Ru, Os

1. *Iron Chalcogenido Complexes*

Terminal chalcogenido complexes of iron are exceedingly rare, even though terminal oxo derivatives are of considerable current interest with respect to the mechanisms of action of cytochrome P450 and other heme containing enzymes such as peroxidases (1, 328, 329). In this regard, several iron–oxo porphyrin derivatives have been prepared, for example, by the reaction of an iron(II) porphyrin derivative with O_2 at low temperature in the presence of a nitrogen-donor Lewis base, as illustrated in Scheme 112 (330). The formation of (TPP)Fe(O)-(1-MeImH) is proposed to proceed via the peroxo-bridged dimer intermediate, [(TPP)Fe]$_2$(μ-O_2). However, none of these complexes has been structurally characterized by X-ray diffraction (331), although EXAFS studies on horseradish peroxidase and several model compounds indicate Fe—O bond lengths in the range 1.60–1.65 Å (332); for reference, the Fe—O bond length in the ferrate derivative K_2FeO_4 is 1.67[1] Å (333, 334). Despite such difficulties in obtaining diffraction data, the [Fe=O] moieties in the above complexes have

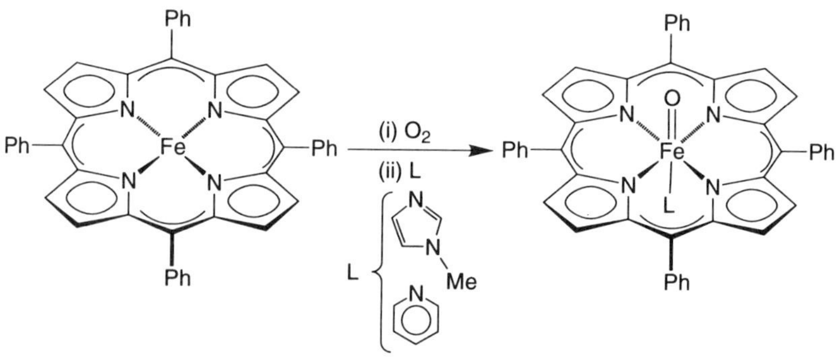

Scheme 112

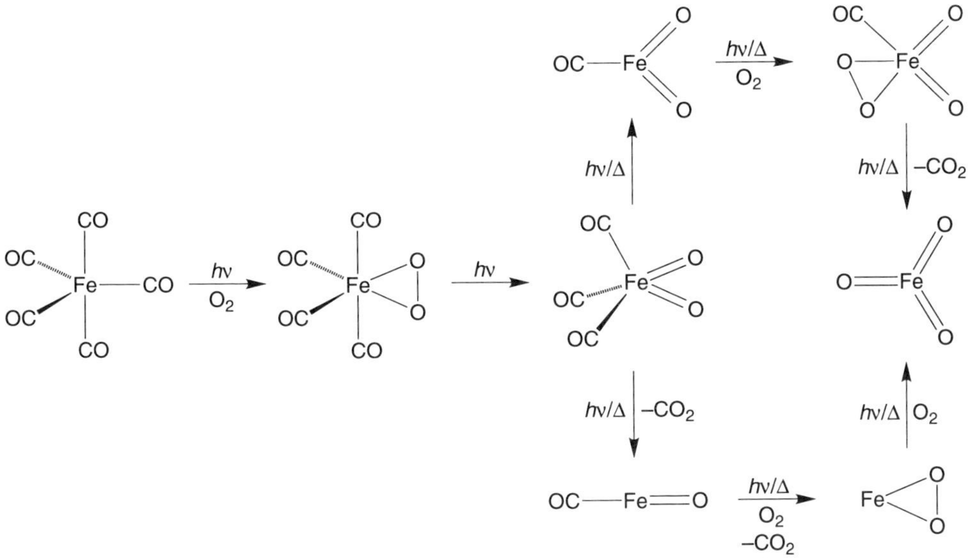

Scheme 113

been studied by a variety of spectroscopic techinques (1), such as resonance Raman spectroscopy (335).

The interaction of O_2 with an iron center has been studied using matrix isolation techniques to determine the nature of the products. Thus, UV photolysis of $Fe(CO)_5$ in an Ar matrix doped with O_2 at 20 K results in the formation of several terminal iron oxo species whose identities and interconversions have been proposed on the basis of IR spectroscopy, as summarized in Scheme 113 (213, 336). The final product of the oxidation is the D_{3h} iron(VI) oxo complex $Fe(O)_3$, and not O=Fe=O, as previously suggested.

2. *Ruthenium Chalcogenido Complexes*

Terminal chalcogenido complexes of ruthenium are presently limited to oxo derivatives (337), and a selection of mononuclear complexes is illustrated in Fig. 32 (63, 67–70, 338). Particular interest in ruthenium oxo complexes is associated with the oxidizing properties of the high-valent derivatives, for example, $Ru(O)_4$, toward organic substrates.

3. *Osmium Chalcogenido Complexes*

Osmium forms many terminal oxo complexes (339) of which $Os(O)_4$ is the most important, being a very useful mild oxidant (340, 341). Some illustrative

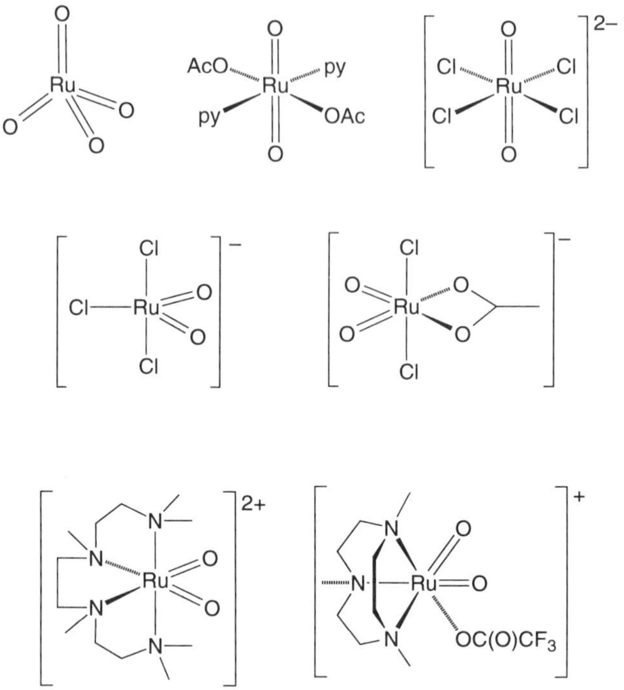

Figure 32. Representative examples of ruthenium oxo complexes.

examples of mononuclear osmium oxo complexes that have been synthesized are shown in Figs. 33 and 34, the latter containing organometallic derivatives (342, 343).

A terminal osmium sulfido complex $[(\eta^6\text{-}1,4\text{-MePr}^i\text{C}_6\text{H}_4)\text{OsS}]$ has been proposed to be an intermediate in the elimination of tetramethylfuran from $(\eta^6\text{-}1,4\text{-MePr}^i\text{C}_6\text{H}_4)\text{Os}(\eta^4\text{-SC}_3\text{Me}_3\text{COMe})$, ultimately yielding $[(\eta^6\text{-}1,4\text{-MePr}^i\text{C}_6\text{H}_4)\text{Os}]_3(\mu_3\text{-S})_2$ (344, 345).

G. Terminal Chalcogenido Complexes of Group 9 (VIII): Co, Rh, Ir

Trimesityliridium $(2,4,6\text{-Me}_3\text{C}_6\text{H}_2)_3\text{Ir}$ reacts cleanly with Me_3NO to give the first (and only) terminal oxo complex of iridium, namely, $(2,4,6\text{-Me}_3\text{C}_6\text{H}_2)_3\text{Ir(O)}$ (Scheme 114) (37). The oxo complex can also be obtained by the reaction of $(2,4,6\text{-Me}_3\text{C}_6\text{H}_2)_3\text{Ir}$ and $(2,4,6\text{-Me}_3\text{C}_6\text{H}_2)_4\text{Ir}$ with O_2, the latter reaction requiring heating to 60°C. The terminal iridium oxo moiety is characterized by $d(\text{Ir=O}) = 1.725(9)$ Å and $\nu(\text{Ir=O}) = 802\ \text{cm}^{-1}$. Interestingly, the iridium oxo complex may be reduced by aqueous $\text{Na}_2\text{S}_2\text{O}_4$, thereby regenerating $(2,4,6\text{-Me}_3\text{C}_6\text{H}_2)_3\text{Ir}$.

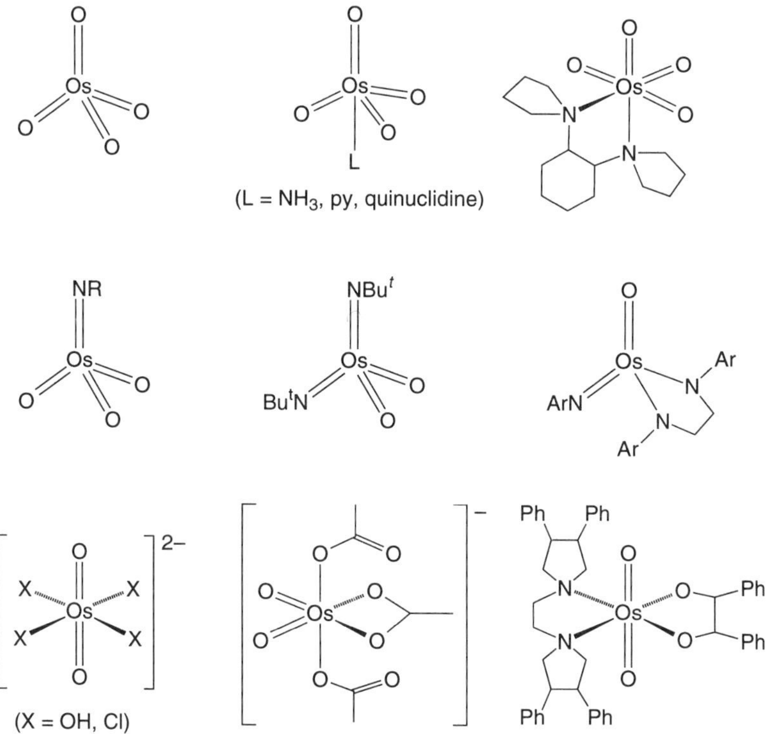

Figure 33. Representative examples of osmium oxo complexes.

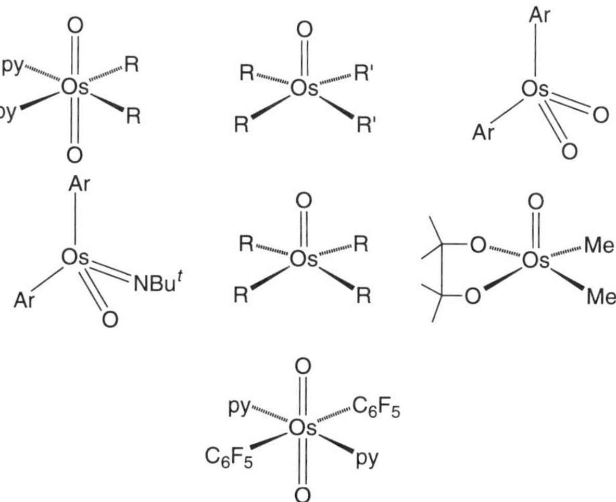

Figure 34. Representative examples of osmium oxo alkyl complexes.

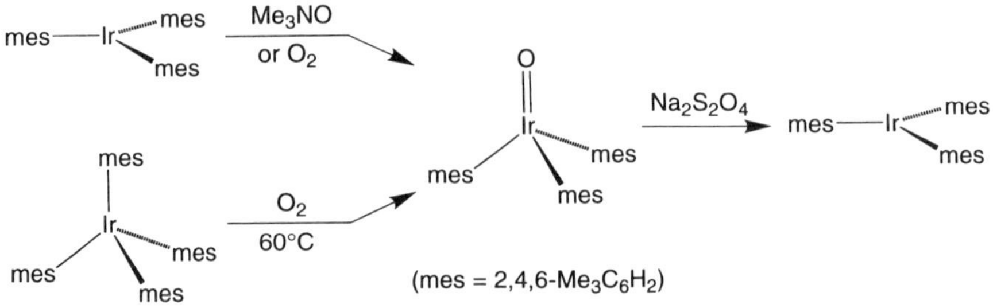

Me_3NO or O_2; O_2, 60°C; $Na_2S_2O_4$

(mes = 2,4,6-$Me_3C_6H_2$)

Scheme 114

Terminal rhodium oxo complexes have not been isolated. Nevertheless, a rhodium oxo species, $[Cp^*Rh(O)(OH_2)]$, has been proposed as the initial product of photolysis of $Cp^*Rh(CO_3)\cdot 2H_2O$, with subsequent reaction giving the hydroxy derivative $\{[Cp^*Rh]_2(\mu\text{-}OH)_3\}^{2+}$ (346).

H. Terminal Chalcogenido Complexes of Group 10 (VIII): Ni, Pd, Pt

As is evident from Figs. 4–7 and 16, structurally authenticated terminal chalcogenido complexes of Ni, Pd, and Pt are unknown. Mayer (38) attributed the absence of such complexes as a consequence of the necessarily high *d* electron count on late transition metal centers. Nevertheless, UV photolysis of $Ni(CO)_4$ in O_2-doped Ar matrices at low temperature (~20 K) has been proposed to generate $(CO)_2Ni{=}O$, in addition to $(CO)_2Ni(\eta^2\text{-}O_2)$ and $(CO)_3Ni(\eta^2\text{-}O_2)$ (347). Furthermore, room temperature generation of terminal platinum oxo species of the type $(dppp)(L)\overset{+}{Pt}{-}\overset{-}{O}$ has been invoked to rationalize (a) the dppp catalyzed exchange of the ^{13}C labeled bicarbonate complex $(dppp)Pt(\eta^2\text{-}O_2{}^{13}CO)$ with $^{12}CO_2$, (b) the formation of $(dppp)Pt(\eta^2\text{-}S_2CO)$ upon reaction of $(dppp)Pt(\eta^2\text{-}O_2CO)$ with CS_2, and (c) the formation of the phosphine oxide dpppO and $(dppp)_2Pt$ upon reaction of $(dppp)Pt(\eta^2\text{-}O_2CO)$ with dppp (Scheme 115) (348).

V. ^{77}Se AND ^{125}Te NMR SPECTROSCOPY OF TERMINAL CHALCOGENIDO COMPLEXES

Both ^{77}Se ($I = \frac{1}{2}$, 7.6%) and ^{125}Te ($I = \frac{1}{2}$, 7.0%) NMR spectroscopies (349) have been used extensively in the characterization of terminal selenido and tellurido complexes. A drawback of the technique, however, is that the ^{77}Se and ^{125}Te chemical shifts of terminal selenido and tellurido ligands span an enormous range, even for chemically similar compounds (see, e.g., Table XXI).

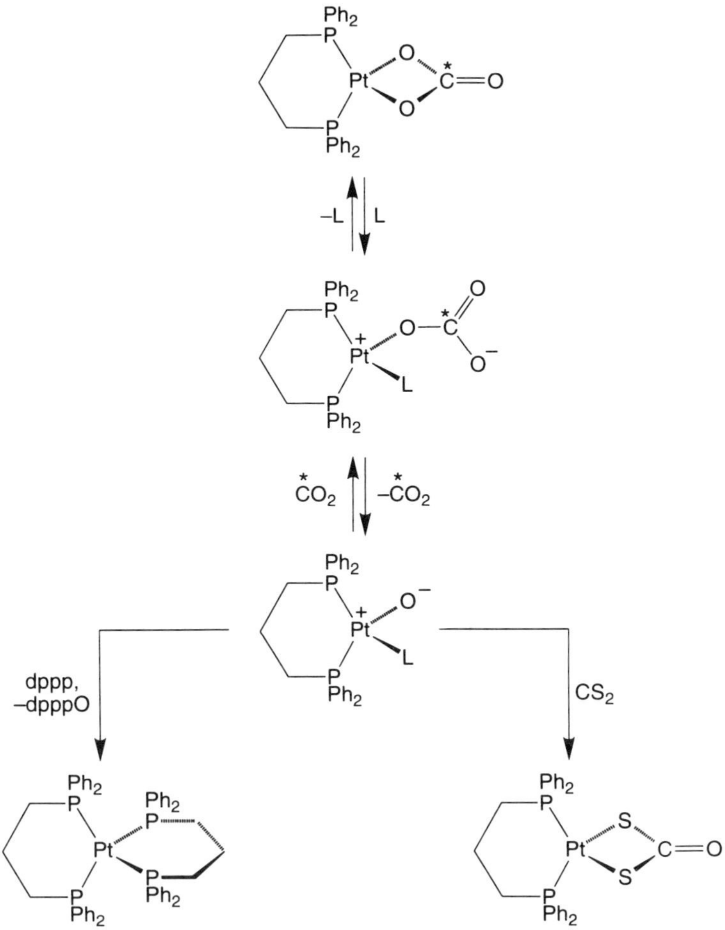

Scheme 115

Furthermore, it has been noted (43) that the chemical shifts of terminal chalcogenido ligands may occur in regions assigned to bridging or metal-bound chalcogenido ligands (350). As such, it is not prudent to use ^{77}Se and ^{125}Te NMR chemical shift data on their own to distinguish between terminal and bridging chalcogenido ligands. However, the observation of coupling to other nuclei may aid with ambiguities in assignment. The existence of an empirical linear correlation of ^{77}Se and ^{125}Te NMR chemical shifts for isostructural selenium and tellurium complexes, as illustrated in Fig. 35 (43, 351), is of some use in compound characterization. For example, if the ^{77}Se and ^{125}Te NMR chemical shifts of two related selenium and tellurium complexes correlate as illustrated in Fig. 34, it would support the notion that they are of the same

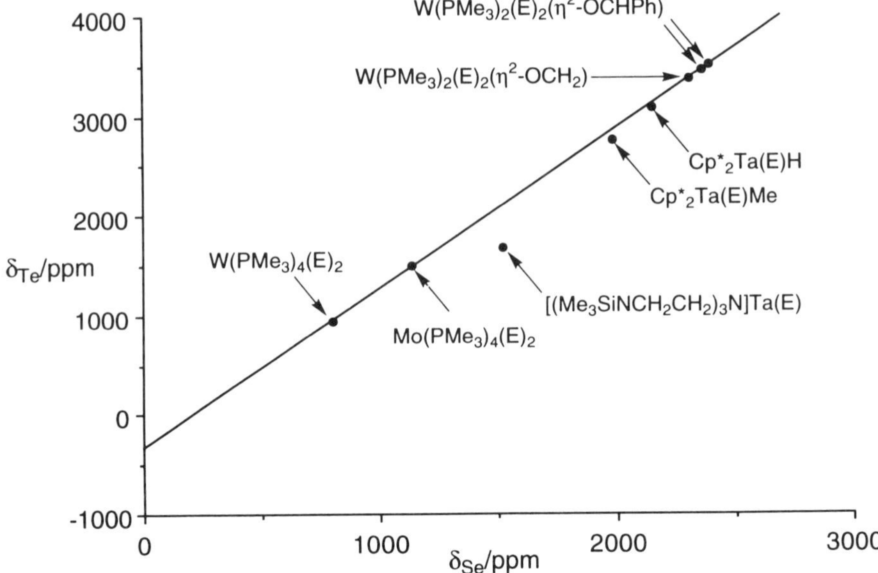

Figure 35. Correlation between ^{77}Se and ^{125}Te NMR chemical shifts for terminal selenido and tellurido complexes. [Reprinted with permission from D. Rabinovich and G. Parkin, *Inorg. Chem.*, *34*, 6341 (1996). Copyright © 1996 American Chemical Society.]

structural type. Finally, the correlation may also be used to predict the location of the ^{125}Te NMR chemical shift of a complex if the ^{77}Se NMR shift of the selenium analogue is known. Such information is useful because of the very large ^{125}Te NMR chemical shift range and the difficulty often encountered locating ^{125}Te NMR signals.

ABBREVIATIONS

Ac	Acetyl
acac	Acetylecetonate(1 −)
acen	*N,N′*-Ethylenebis(acetylacetoniminate)(2 −)
bpy	2,2′-Bipyridine
Bu^t	*t*-Butyl
Bz	Benzyl
CSD	Cambridge structural database
Cp*	η^5-C_5Me_5
Cp^{Et}*	η^5-C_5Me_4Et
Cp^{Me}	η^5-C_5H_4Me

Cp^R	A variously substituted cyclopentadienyl derivative
Cy	Cyclohexyl
DMSO	Dimethyl sulfoxide
dmp	2,9-Dimethyl-1-10-phenanthroline
dmpe	$Me_2PCH_2CH_2PMe_2$
dppe	$Ph_2PCH_2CH_2PPh_2$
dppee	$Ph_2PCH{=}CHPPh_2$
dppp	$Ph_2PCH_2CH_2CH_2PPh_2$
edta	Ethylenediaminetetraacetate(4−)
EHMO	Extended Hückel molecular orbital
Et	Ethyl
EXAFS	Extended X-ray absorption fine structure
HOMO	Highest occupied molecular orbital
IR	Infrared
LUMO	Lowest unoccupied molecular orbital
Me	Methyl
MeImH	*N*-Methylimidazole
Me_2OEP	α,γ-Dimethyl-α,γ-dihydrooctaethylporphyrinate(2−)
mes	2,4,6-$Me_3C_6H_2$
Me_3tacn	1,4,7-Trimethyl-1,4,7-triazacyclononane
$[Me_4taa]H_2$	Tetramethyldibenzotetraaza[14]annulene
$[Me_8taa]H_2$	Octamethyldibenzotetraaza[14]annulene
MO	Molecular orbital
MPOEP	*meso*-Phenyloctaethylporphyrinate(2−)
NMR	Nuclear magnetic resonance
OEP	Octaethylporphyrinate(2−)
PC	Phthalocycaninate(2−)
Ph	Phenyl
POR	A variously substituted porphyrinate(2−) derivative
Pr^i_3tacn	1,4,7-Triisopropyl-1,4,7-triazacyclononane
py	Pyridine
pz	Pyrazolyl
salen	1,2-bis(salicylideneamino)ethane(2−)
silox	*t*-Bu_3SiO
THF	Tetrahydrofuran
tht	Tetrahydrothiophene
TMP	Tetramesitylporphyrinate(2−)
TMTP	*meso*-Tetrakis(*m*-tolyl)porphyrinate(2−)
TMS	Trimethylsilyl
Tol	$C_6H_4CH_3$
Tp	Tris(pyrazolyl)hydroborate(1−), [η^3-HB(pz)$_3$]
Tp^{Me_2}	Tris(3,5-dimethylpyrazolyl)hydroborate(1−), [η^3-HB(3,5-Me_2pz)$_3$]

TPP	*meso*-Tetraphenylporphyrinate(2 −)
TTP	*meso*-Tetrakis(*p*-tolyl)porphyrinate(2 −)

ADDENDUM

Since this manuscript was submitted, there have been several advances with respect to the early first-row transition metals. For example, Bergman, Andersen and co-workers (352) have synthesized the terminal titanium sulfido complex $Cp_2^*Ti(S)(NC_5H_5)$ by reaction of $Cp_2^*Ti(\eta^2\text{-}C_2H_4)$ with $\frac{1}{8}$ equiv of S_8 in the presence of pyridine. The sulfido complex $Cp_2^*Ti(S)(NC_5H_5)$ is characterized by a Ti=S bond length of 2.217(1) Å. Interestingly, $Cp_2^*Ti(S)(NC_5H_5)$ reacts reversibly with H_2 to give $Cp_2^*Ti(SH)H$. Terminal selenido and tellurido complexes of titanium and vanadium supported by ligation of the macrocyclic octamethyldibenzotetraaza[14]annulene dianion, $[\eta^4\text{-}Me_8taa]M{=}E$ (M = Ti, V; E = Se, Te), have been synthesized by reaction of $[\eta^4\text{-}Me_8taa]MCl_2$ with $(Bu^tMe_2Si)_2E$, and have been structurally characterized by X-ray diffraction: d(Ti=Se) = 2.269(2) Å, d(Ti=Te) = 2.484(2) Å; d(V=Se) = 2.210(2) Å, d(V=Te) = 2.434[2] Å (353). The titanium selenido and tellurido complexes $[\eta^4\text{-}Me_8taa]Ti{=}E$ (E = Se, Te) react with both N_2O and O_2 to give the oxo derivative $[\eta^4\text{-}Me_8taa]Ti{=}O$, and with Me_3SiCl to give the dichloride $[\eta^4\text{-}Me_8taa]TiCl_2$. In addition, $[\eta^4\text{-}Me_8taa]Ti{=}Se$ reacts with elemental selenium to give the diselenido complex $[\eta^4\text{-}Me_8taa]Ti(\eta^2\text{-}Se_2)$. Finally, the bulky amide ligand, $[(3,5\text{-}Me_2C_6H_3)_2NAd]^-$ (Ad = adamantyl), has been used to prepare the terminal sulfido and selenido complexes of vanadium, $[(3,5\text{-}Me_2C_6H_3)_2(Ad)N]_3V{=}E$ (E = S, Se) (354).

ACKNOWLEDGMENTS

I sincerely wish to thank the students and postdoctoral researchers who carried out portions of the research described in this chapter, which was supported by the U.S. Department of Energy, Office of Basic Energy Sciences, and the National Science Foundation. I am also grateful to the students and postdoctoral researchers who provided critical comments during the preparation of this chapter. Professor James Mayer is also thanked for his valuable suggestions concerning the final manuscript.

REFERENCES

1. W. A. Nugent and J. M. Mayer, *Metal–Ligand Multiple Bonds*, Wiley-Interscience, New York, 1988.
2. For a more recent compilation covering certain aspects of metal–ligand multiple

bonding, see: C.-M. Che, and V. W. W. Yam, Eds., *Advances in Transition Metal Coordination Chemistry*, JAI Press, Greenwich, CT, 1996, Vol. 1.

3. Polonium, which is radioactive, has been little studied in this context and will not be considered in this chapter.
4. (a) V. M. Goldschmidt, *J. Chem. Soc. (London)*, 655 (1937). (b) V. M. Goldschmidt, *Naturwissenschaften*, *18*, 999 (1930). (c) K. A. Jensen, *Chem. Ind.*, *49*, 2016 (1964).
5. G. J. Leigh, Ed., *IUPAC Nomenclature of Inorganic Chemistry, Recomendations 1990*; Blackwell, London 1990.
6. (a) J. P. Collman, L. S. Hegedus, J. R. Norton, and R. G. Finke, *Principles and Applications of Organotransition Metal Chemistry*; University Science Books, Mill Valley, CA, 1987. (b) G. W. Parshall and S. D. Ittel, *Homogeneous Catalysis*, 2nd ed.; Wiley-Interscience: New York, 1992. (c) C. L. Hill, Ed., *Activation and Functionalization of Alkanes*, Wiley-Interscience: New York, 1989.
7. (a) B. C. Gates, J. R. Katzer, and G. C. A. Schuit, *Chemistry of Catalytic Processes*, McGraw-Hill, New York, 1979, Chapter 5, pp. 390–445. (b) J. G. Speight, *The Desulfurization of Heavy Oils and Residua*; Marcel-Dekker, New York, 1981. (c) M. Rakowski DuBois, *Chem. Rev.*, *89*, 1 (1989).
8. (a) R. J. Angelici, *Acc. Chem. Res.*, *21*, 387 (1988). (b) C. M. Friend, *Acc. Chem. Res.*, *21*, 394 (1988). (c) B. C. Wiegand and C. M. Friend, *Chem. Rev.*, *92*, 491 (1992). (d) R. A. Sánchez-Delgado, *J. Mol. Catal.*, *86*, 287 (1994).
9. (a) J. M. Thomas and K. I. Zamaraev, *Angew. Chem.*, *Int. Ed. Engl.*, *33*, 308 (1994). (b) V. N. Parmon, Z. R. Ismagilov, and M. A. Kerzhentsev, in *Perspectives in Catalysis*, J. M. Thomas, and K. I. Zamaraev, Eds., Blackwell Scientific Publications, Oxford, UK, 1992, pp. 337–357. (c) D. O'Sullivan, *Chem. Eng. News*, *59* (30), 40 (1981). (d) S. Li, and G. Lu, *New J. Chem.*, *16*, 517 (1992).
10. A. Müller, B. Krebs, Eds., *Sulfur, its Significance for Chemistry, for the Geo-, Bio-, and Cosmophere and Technology*, Elsevier Science Publishers, Amsterdam, The Netherlands, 1984, pp. 349–494.
11. (a) A. A. Eagle, L. J. Laughlin, C. G. Young, and E. R. T. Tielink, *J. Am. Chem. Soc.*, *114*, 9195 (1992) and references cited therein. (b) J. P. Hill, L. J. Laughlin, R. W. Gable, and C. G. Young, *Inorg. Chem.*, *35*, 3447 (1996). (b) G. L. Wilson, R. J. Greenwood, J. R. Pilbrow, J. T. Spence, and A. G. Wedd, *J. Am. Chem. Soc.*, *113*, 6803 (1991).
12. (a) J. Reedijk, Ed., *Bioinorganic Catalysis*, Marcel-Dekker, New York, 1993. (b) K. D. Karlin, *Science*, *261*, 701 (1993). (c) J. J. R. Fraústo da Silva, and R. J. P. Williams, *The Biological Chemistry of the Elements*, Clarendon Press: Oxford, UK, 1991.
13. (a) R. H. Holm, and J. M. Berg, *Acc. Chem. Res.*, *19*, 363 (1986). (b) A. G. Wedd, *Coord. Chem. Rev.*, *154*, 5 (1996).
14. (a) E. I. Stiefel, *Science*, *272*, 1599 (1996). (b) J. H. Enemark, and C. G. Young, *Adv. Inorg. Chem.*, *40*, 1 (1993). (c) M. A. Pietsch and M. B. Hall, *Inorg. Chem.*, *35*, 1273 (1996).
15. (a) A. A. Eagle, S. M. Harben, E. R. T. Tiekink, and C. G. Young, *J. Am.*

Chem. Soc., *116*, 9749 (1994). (b) C. G. Young, T. O. Kocaba, X. F. Yan, E. R. T. Tiekink, L. Wei, H. H. Murray, III, C. L. Coyle, and E. I. Stiefel, *Inorg. Chem.*, *33*, 6252 (1994).

16. U. Siemeling, *Angew. Chem. Int. Ed. Engl.*, *32*, 67 (1993).

17. (a) P. O'Brien, *Chemtronics*, *5*, 61 (1991). (b) M. L. Steigerwald and L. E. Brus, *Acc. Chem. Res.*, *23*, 183 (1990). (c) M. Bochmann and K. J. Webb, *Mater. Res. Soc. Symp. Proc.*, *204*, 149 (1991). (d) R. H. Bube, *Annu. Rev. Mater. Sci.*, *20*, 19 (1990). (e) T. Yokogawa, M. Ogura, and T. Kajiwara, *Appl. Phys. Lett.*, *50*, 1065 (1987). (f) A. C. Jones, *J. Cryst. Growth*, *129*, 728 (1993). (g) G. N. Pain, G. I. Christiansz, R. S. Dickson, G. B. Deacon, B. O. West, K. McGregor, and R. S. Rowe, *Polyhedron*, *9*, 921 (1990). (h) M. L. Steigerwald, *Mater. Res. Soc. Symp. Proc.*, *131*, 37 (1989).

18. (a) M. L. Steigerwald, T. Siegrist, and S. M. Stuczynski, *Inorg. Chem.*, *30*, 4940 (1991). (b) M. L. Steigerwald, T. Siegrist, S. M. Stuczynski, and Y.-U. Kwon, *J. Am. Chem. Soc.*, *114*, 3155 (1992). (c) M. L. Steigerwald, *Polyhedron*, *13*, 1245 (1994). (d) J. Arnold, J. M. Walker, K.-M. Yu, P. J. Bonasia, A. L. Seligson, and E. D. Bourret, *J. Cryst. Growth*, *124*, 647 (1992). (e) A. L. Seligson, P. J. Bonasia, J. Arnold, K.-M. Yu, J. M. Walker, and E. D. Bourret, *Mater. Res. Soc. Symp. Proc.*, *282*, 665 (1993).

19. For a recent review of transition metal–imido chemistry, see D. E. Wigley, *Progress in Inorganic Chemistry*, Vol. 42, Wiley-Interscience, 1994, pp. 239–482.

20. M. Scheer, *Angew. Chem. Int. Ed. Engl.*, *34*, 1997 (1995).

21. J. B. Bonanno, P. T. Wolczanski, and E. B. Lobkovsky, *J. Am. Chem. Soc.*, *116*, 11159 (1994).

22. C. E. Laplaza, W. M. Davis, and C. C. Cummins, *Angew. Chem. Int. Ed. Engl.*, *34*, 2042 (1995).

23. N. C. Zanett, R. R. Schrock, and W. D. Davis, *Angew. Chem. Int. Ed. Engl.*, *117*, 2044 (1995).

24. The terminal stibido complexes, $[(CO)_4M{\equiv}Sb]^-$ (M = Cr, Mo, W) and $[(CO)_3Fe{\equiv}Sb]^-$, however, have been generated in the gas phase. See F. P. Arnold, Jr., D. P. Ridge, and A. L. Rheingold, *J. Am. Chem. Soc.*, *117*, 4427 (1995).

25. (a) W. P. Griffith, *Coord. Chem. Rev.*, *5*, 459 (1970). (b) R. H. Holm, *Chem. Rev.*, *87*, 1401 (1987). (c) R. H. Holm and J. P. Donahue, *Polyhedron*, *12*, 571 (1993). (d) F. Bottomley and L. Sutin, *Adv. Organomet. Chem.*, *28*, 339 (1988). (e) H. Arzoumanian, *Bull. Soc. Chim. Belg.*, *100*, 717 (1991). (f) F. Bottomley, *Polyhedron*, *11*, 1707 (1992). (g) W. A. Herrmann, E. Herdtweck, M. Flöel, J. Kulpe, U. Küsthardt, and J. Okuda, *Polyhedron*, *6*, 1165 (1987). (h) R. C. Mehrotra and A. Singh, *Chem. Soc. Rev.*, *25*, 1 (1996).

26. (a) D. A. Rice, *Coord. Chem. Rev.*, *227*, 199 (1978). (b) H. Vahrenkamp, *Angew. Chem. Int. Ed. Engl.*, *14*, 322 (1975). (c) A. Müller, E. Diemann, R. Jostes, and H. Bögge, *Angew. Chem. Int. Ed. Engl.*, *20*, 934 (1981). (d) A. Müller, and E. Diemann, in *Comprehensive Coordination Chemistry*, G. Wilkinson, R. D. Gillard, and J. A. McCleverty, Eds., Pergamon, Oxford, UK, 1987, Chapter 16.1.

(e) A. Müller, *Polyhedron*, *5*, 323 (1986). (f) P. Kelly, *Chemistry in Britain*, 25 (1997).

27. L. C. Roof and J. W. Kolis, *Chem. Rev.*, *93*, 1037 (1993).

28. (a) A. Müller, and E. Diemann, *Adv. Inorg. Chem.*, *31*, 89 (1987). (d) J. Wachter, *Angew. Chem. Int. Ed. Engl.*, *28*, 1613 (1989). (e) M. G. Kanatzidis, *Comments Inorg. Chem.*, *10*, 161 (1990). (f) M. G. Kanatzidis, and S.-P. Huang, *Coord. Chem. Rev.*, *130*, 509 (1994). (g) J. W. Kolis, *Coord. Chem. Rev.*, *105*, 195 (1990). (h) P. Böttcher, *Angew. Chem. Int. Ed. Engl.*, *27*, 759 (1988). (i) M. A. Ansari, J. M. McConnachie, and J. A. Ibers, *Acc. Chem. Res.*, *26*, 574 (1993). (j) M. A. Ansari, and J. A. Ibers, *Coord. Chem. Rev.*, *100*, 223 (1990). (k) J. Beck, *Angew. Chem.*, *Int. Ed. Engl.*, *33*, 163 (1994). (l) M. G. Kanatzidis, and A. C. Sutorik, *Progress in Inorganic Chemistry*, Wiley-Interscience, New York, 1995, Vol. 43, pp. 151–265. (m) D. Fenske, J. Ohmer, J. Hachgenei, and K. Merzweiler, *Angew. Chem. Int. Ed. Engl.*, *27*, 1277 (1988).

29. (a) W. A. Herrmann, *Angew. Chem. Int. Ed. Engl.*, *25*, 56 (1986). (b) N. A. Compton, R. J. Errington, and N. C. Norman, *Adv. Organomet. Chem.*, *31*, 91 (1990). (c) C. E. Housecroft, in *Inorganometallic Chemistry*, T. P. Fehlner, Ed., Plenum, New York, 1992; Chapter 3, pp. 73–178. (d) D. Fenske, J. Ohmer, J. Hachgenei, and K. Merzweiler, *Angew. Chem. Int. Ed. Engl.*, *27*, 1277 (1988). (e) I. Dance, and K. Fisher, *Progress in Inorganic Chemistry*, Wiley-Interscience, New York, 1994, Vol. 41, pp. 637–803.

30. For a review of the factors that influence the bond angles in bridging oxo and sulfido complexes, see F. Bottomley and S.-K. Goh, *Polyhedron*, *15*, 3045 (1996).

31. T. M. Trnka, and G. Parkin, *Polyhedron*, *16*, 1031 (1997).

32. (a) W. A. Howard, M. Waters, and G. Parkin, *J. Am. Chem. Soc.*, *115*, 4917 (1993). (b) W. A. Howard, T. M. Trnka, M. Waters, and G. Parkin, *J. Organomet. Chem.*, *528*, 95 (1997).

33. J. Arnold, *Progress in Inorganic Chemistry*, Wiley-Interscience, New York, 1995, Vol. 43, pp. 353–417.

34. For example, approximately 300 complexes with transtion metal–tellurium bonds are listed in the present version of the Cambridge Structural Database.

35. For some leading articles on multiple bonding of the heavier main group elements, see (a) W. Kutzelnigg, *Angew. Chem. Int. Ed. Engl.*, *23*, 272 (1984). (b) L. E. Gusel'nikov and N. S. Nametkin, *Chem. Rev.*, *79*, 529 (1979). (c) P. Jutzi, *Angew. Chem. Int. Ed. Engl.*, *14*, 232 (1975). (d) N. C. Norman, *Polyhedron*, *12*, 2431 (1993). (e) M. W. Schmidt, P. N. Truong, and M. S. Gordon, *J. Am. Chem. Soc.*, *109*, 5217 (1987). (f) H. Jacobsen and T. Ziegler, *J. Am. Chem. Soc.*, *116*, 3667 (1994). (g) J. E. Huheey, E. A. Keiter, and R. L. Keiter, *Inorganic Chemistry: Principles of Structure and Reactivity*, 4th ed., Harper Collins, New York, 1993, Chapter 18, pp. 861–875. (h) H. Jacobsen and T. Ziegler, *Comments Inorg. Chem.*, *17*, 301 (1995). (i) J. Kapp, P. von Ragué Schleyer, M. Remko, *J. Am. Chem. Soc.*, *118*, 5745 (1996). (j) M. Dreiss, and H. Grützmacher, *Angew. Chem. Int. Ed. Engl.*, *35*, 828 (1996).

36. For example, the chemistry of the Si=Si double bond is substantially less developed than that of the C=C double bond. See (a) R. West, *Angew. Chem. Int. Ed. Engl.*, *26*, 1201 (1987). (b) R. West, *Pure Appl. Chem.*, *56*, 163 (1984). (c) T. Tsumuraya, S. A. Batcheller, and S. Masamune, *Angew. Chem. Int. Ed. Engl.*, *30*, 902 (1991).

37. R. S. Hay-Motherwell, G. Wilkinson, B. Hussain-Bates, and M. B. Hursthouse, *Polyhedron*, *12*, 2009 (1993).

38. J. M. Mayer, *Comments Inorg. Chem.*, *8*, 125 (1988).

39. For further discussion of the influence of filled-filled repulsions in transition metal chemistry, see K. G. Caulton, *New J. Chem.*, *18*, 25 (1994).

40. J. M. Mayer, *Inorg. Chem.*, *27*, 3899 (1988).

41. Mayer has noted, however, that M=O bond lengths do increase slightly in the presence of a second multiply bonded ligand (e.g., oxo, nitrido, imido).

42. For a theoretical discussion on the nature of the triple bond in Cl_4CrO, involving two normal π bonds and a dative O to Cr σ bond, see: A. K. Rappé and W. A. Goddard, III, *J. Am. Chem. Soc.*, *104*, 3287 (1982).

43. D. Rabinovich and G. Parkin, *Inorg. Chem.*, *34*, 6341 (1996).

44. (a) J. W. Lauher, and R. Hoffmann, *J. Am. Chem. Soc.*, *98*, 1729 (1976). (b) Z. Lin, and M. B. Hall, *Coord. Chem. Rev.*, *123*, 149 (1993).

45. (a) N. D. Silavwe, M. R. M. Bruce, C. E. Philbin, and D. R. Tyler, *Inorg. Chem.*, *27*, 4669 (1988). (b) A. J. Bridgeman, L. Davis, S. J. Dixon, J. C. Green, and I. N. Wright, *J. Chem. Soc. Dalton Trans.*, 1023 (1995). (c) J. M. Fischer, W. E. Piers, T. Ziegler, L. R. MacGillivray, and M. J. Zaworotko, *Chem. Eur. J.*, *2*, 1221 (1996).

46. For discussions on multiple bonding in related imido derivatives see (a) G. Parkin, A. van Asselt, D. J. Leahy, L. Whinnery, N. G. Hua, R. W. Quan, L. M. Henling, W. P. Schaefer, B. D. Santarsiero, and J. E. Bercaw, *Inorg. Chem.*, *31*, 82 (1992). (b) D. M. Antonelli, W. P. Schaefer, G. Parkin, and J. E. Bercaw, *J. Organomet. Chem.*, *462*, 213 (1993). (c) K. A. Jørgensen, *Inorg. Chem.*, *32*, 1521 (1993). (d) J. T. Anhaus, T. P. Kee, M. H. Schofield, and R. R. Schrock, *J. Am. Chem. Soc.*, *112*, 1642 (1990). (e) M. H. Schofield, T. P. Kee, J. T. Anhaus, R. R. Schrock, K. H. Johnson, and W. M. Davis, *Inorg. Chem.*, *30*, 3595 (1991). (f) D. S. Glueck, J. C. Green, R. I. Michelman, and I. N. Wright, *Organometallics*, *11*, 4221 (1992). (g) J. C. Green, M. L. H. Green, J. T. James, P. C. Konidaris, G. H. Maunder, and P. Mountford, *J. Chem. Soc. Chem. Commun.*, 1361 (1992).

47. (a) N. D. Silavwe, M. Y. Chiang, and D. R. Tyler, *Inorg. Chem.*, *24*, 4219 (1985). (b) M. Yoon and D. R. Tyler, *J. Chem. Soc. Chem. Commun.* 639 (1997).

48. J. H. Shin, and G. Parkin, unpublished results.

49. J. M. Mayer, *Inorg. Chem.*, *27*, 3899 (1988).

50. W. A. Nugent and J. M. Mayer, *Metal–Ligand Multiple Bonds*, Wiley-Interscience, New York, 1988, p. 117.

51. (a) G. Parkin and J. E. Bercaw, *J. Am. Chem. Soc.*, *111*, 391 (1989). (b) G. Parkin and J. E. Bercaw, *Polyhedron*, *7*, 2053 (1988).

52. W. A. Nugent, R. J. McKinney, R. V. Kasowski, and F. A. Van-Catledge, *Inorg. Chim. Acta*, *65*, L91 (1982).

53. J. H. Shin and G. Parkin, *Organometallics*, *14*, 1104 (1995).

54. W. Howard and G. Parkin, *J. Am. Chem. Soc.*, *116*, 606 (1994).

55. See, for example, the empirical Schomaker–Stevenson equation, $d(A - B) = r(A) + r(B) - c|\chi(A) - \chi(B)|$. (a) V. Schomaker and D. P. Stevenson, *J. Am. Chem. Soc.*, *63*, 37 (1941). (b) A. F. Wells, *J. Chem. Soc.*, 55 (1949). (c) A. F. Wells, *Structural Inorganic Chemistry*, 5th ed., Oxford University Press, London, 1984, pp. 287–291.

56. (a) M. T. Benson, T. R. Cundari, S. J. Lim, H. D. Nguyen, and K. Pierce-Beaver, *J. Am. Chem. Soc.*, *116*, 3955 (1994). (b) M. T. Benson, T. R. Cundari, Y. Li, and L. A. Strohecker, *Int. J. Quantum Chem.: Quantum Chem. Sym.*, *28*, 181 (1994). (c) M. S. Gordon and T. S. Cundari, *Coord. Chem. Rev.*, *147*, 87 (1996).

57. L. Deng and T. Ziegler, *Organometallics*, *15*, 3011 (1996).

58. Examples of d^2 dioxo complexes with a trans arrangement include $[Mo(O)_2(CN)_4]^{4-}$, $[Re(O)_2(py)_4]^+$, $[Ru(O)_2(OAc)_2(py)_2]$, and $[Os(O)_2(en)_2]^{2+}$. Furthermore, the d^1 derivative $[Re(O)_2(dmap)_4]^{2+}$ also exhibits a trans arrangement of oxo ligands (58a–d). (a) V. W. Day and J. L. Hoard, *J. Am. Chem. Soc.*, *90*, 3374 (1968). (b) C. S. Johnson, C. Mottley, J. T. Hupp, and G. D. Danzer, *Inorg. Chem.*, *31*, 5143 (1992). (c) J. C. Brewer, H. H. Thorp, K. M. Slagle, G. W. Brudvig, and H. B. Gray, *J. Am. Chem. Soc.*, *113*, 3171 (1991). (d) J. R. Winkler, and H. B. Gray, *Inorg. Chem.*, *24*, 346 (1985).

59. W. P. Griffith and T. D. Wickins, *J. Chem. Soc. (A)*, 400 (1968).

60. (a) D. M. P. Mingos, *J. Organomet. Chem.*, *179*, C29 (1979). (b) D. C. Brower, J. L. Templeton, and D. M. P. Mingos, *J. Am. Chem. Soc.*, *109*, 5203 (1987). (c) Z. Lin, and M. B. Hall, *Coord. Chem. Rev.*, *123*, 149 (1993). (d) K. Tatsumi, and R. Hoffmann, *Inorg. Chem.*, *19*, 2656 (1980). (e) I. Demachy, and Y. Jean, *New J. Chem.*, *19*, 763 (1995). (f) I. Demachy, and Y. Jean, *New J. Chem.*, *20*, 53 (1996). (g) I. Demachy, and Y. Jean, *Inorg. Chem.*, *35*, 5027 (1996).

61. For a recent review describing the electronic structures of metal–oxo complexes, see V. M. Miskowski, H. B. Gray, and M. D. Hopkins, *Adv. Transition Metal Coordination Chem.*, *1*, 159 (1996).

62. For some early rules concerning the stereochemistry of Mo^{VI} dioxo complexes, see R. J. Butcher, B. R. Penfold, and E. Sinn, *J. Chem. Soc. Dalton Trans.*, 668 (1979).

63. W. P. Griffith, J. M. Jolliffe, S. V. Ley, and D. J. Williams, *J. Chem. Soc. Chem. Commun.*, 1219 (1990).

64. R. L. Blackbourn, L. M. Jones, M. S. Ram, M. Sabat, and J. T. Hupp, *Inorg. Chem.*, *29*, 1791 (1990).

65. T. Behling, M. V. Capparelli, A. C. Skapski, and G. Wilkinson, *Polyhedron*, *1*, 840 (1982).

66. J. C. Dobson, K. J. Takeuchi, D. W. Pipes, D. A. Geselowitz, and T. J. Meyer, *Inorg. Chem.*, *25*, 2357 (1986).

67. W.-C. Cheng, W.-Y. Yu, K.-K. Cheung, and C.-M. Che, *J. Chem. Soc. Chem. Commun.*, 1063 (1994).

68. (a) W.-C. Cheng, W.-Y. Yu, C.-K. Li, and C.-M. Che, *J. Org. Chem.*, *60*, 6840 (1995). (b) C.-K. Li, C.-M. Che, W.-F. Tong, W.-T. Tang, K.-Y. Wong, and T.-F. Lai, *J. Chem. Soc. Dalton Trans.*, 2109 (1992).

69. (a) T. C. W. Mak, C.-M. Che, and K. Y. Wong, *J. Chem. Soc. Chem. Commun.*, 986 (1985). (b) C. M. Che, T. F. Lai, and K. Y. Wong, *Inorg. Chem.*, *26*, 2289 (1987).

70. (a) C. L. Bailey and R. S. Drago, *J. Chem. Soc. Chem. Commun.*, 179 (1987). (b) A. S. Goldstein, R. H. Beer, and R. S. Drago, *J. Am. Chem. Soc.*, *116*, 2424 (1994).

71. T. R. Cundari and R. S. Drago, *Inorg. Chem.*, *29*, 2303 (1990).

72. P. T. Wood, W. T. Pennington, J. W. Kolis, B. Wu, and C. J. O'Connor, *Inorg. Chem.*, *32*, 129 (1993).

73. (a) J. Chatt, L. Manojlovic-Muir, and K. W. Muir, *Chem. Commun.*, 655 (1971). (b) L. Manojlovic-Muir, and K. W. Muir, *J. Chem. Soc. Dalton Trans.*, 686 (1972). (c) L. Manojlovic-Muir, *J. Chem. Soc. (A)*, 2796 (1971). (d) L. Manojlovic-Muir and K. W. Muir, *J. Chem. Soc. Dalton Trans.*, 686 (1972). (e) B. L. Haymore, W. A. Goddard, III, and J. N. Allison, *Proc. Int. Conf. Coord. Chem.*, *23rd*, 535 (1984).

74. (a) Y. Jean, A. Lledos, J. K. Burdett, and R. Hoffmann, *J. Am. Chem. Soc.*, *110*, 4506 (1988). (b) Y. Jean, A. Lledos, J. K. Burdett, and R. Hoffmann, *J. Chem. Soc. Chem. Commun.*, 140 (1988).

75. (a) G. Parkin, *Acc. Chem. Res.*, *25*, 455 (1992). (b) G. Parkin, *Chem. Rev.*, *93*, 887 (1993). (c) K. Yoon, G. Parkin, and A. L. Rheingold, *J. Am. Chem. Soc.*, *113*, 1437 (1991). (d) K. Yoon, G. Parkin, and A. L. Rheingold, *J. Am. Chem. Soc.*, *114*, 2210 (1992).

76. For a brief review, see C. A. McAuliffe and D. S. Barratt, in *Comprehensive Coordination Chemistry*, G. Wilkinson, R. D. Gillard, and J. A. McCleverty, Eds.; Pergamon, Oxford, UK, 1987, Chapter 31.

77. P. Mountford and D. Swallow, *J. Chem. Soc. Chem. Commun.*, 2357 (1995).

78. M. Tsutsui, R. A. Velapoldi, K. Suzuki, and T. Koyano, *Angew. Chem. Int. Ed. Engl.*, *7*, 891 (1968).

79. J. W. Buchler, G. Eikelman, L. Puppe, K. Rohbock, H. H. Schneehage, and D. D. Weck, *Justus Liebigs Ann. Chem.*, *745*, 135 (1971).

80. P. Fournari, R. Guilard, M. Fontesse, J.-M. Latour, and J.-C. Marchon, *J. Organomet. Chem.*, *110*, 205 (1976).

81. L. K. Woo, J. A. Hays, V. G., Young, Jr., C. L. Day, C. Caron, F. D'Souza, and K. M. Kadish, *Inorg. Chem.*, *32*, 4186 (1993).

82. For a review of titanium and vanadium porphyrin oxo complexes, see R. Guilard, and C. Lecomte, *Coord. Chem. Rev.*, *65*, 87 (1985).

83. R. Guilard, J.-M. Latour, C. Lecomte, J.-C. Marchon, J. Protas, and D. Ripoll, *Inorg. Chem.*, *17*, 1228 (1978).

84. T. Malinski, D. Chang, J.-M. Latour, J.-C. Marchon, M. Gross, A. Giraudeau, and K. M. Kadish, *Inorg. Chem.*, *23*, 3947 (1984).

85. L. K. Woo, *Chem. Rev.*, *93*, 1125 (1993).

86. V. L. Goedken and J. A. Ladd, *J. Chem. Soc. Chem. Commun.*, 142 (1982).

87. C.-H. Yang, J. A. Ladd, and V. L. Goedken, *J. Coord. Chem.*, *19*, 235 (1988).

88. (a) C. E. Housmekerides, R. S. Pilato, G. L. Geoffroy, and A. L. Rheingold, *J. Chem. Soc. Chem. Commun.*, 563 (1991). (b) C. E. Housmekerides, D. L. Ramage, C. M. Kretz, J. T. Shontz, R. S. Pilato, G. L. Geoffroy, A. L. Rheingold, and B. S. Haggerty, *Inorg. Chem.*, *31*, 4453 (1992).

89. (a) C. E. Housmekerides, R. S. Pilato, G. L. Geoffroy, and A. L. Rheingold, *J. Chem. Soc. Chem. Commun.*, 563 (1991). (b) C. E. Housmekerides, D. L. Ramage, C. M. Kretz, J. T. Shontz, R. S. Pilato, G. L. Geoffroy, A. L. Rheingold, and B. S. Haggerty, *Inorg. Chem.*, *31*, 4453 (1992).

90. A. Bodner, P. Jeske, Wieghardt, K. Weyhermüller, E. Dubler, H. Schmalle, and B. Nuber, *Inorg. Chem.*, *31*, 3737 (1992).

91. M. R. Smith III, P. T. Matsunaga, and R. A. Andersen, *J. Am. Chem. Soc.*, *115*, 7049 (1993).

92. F. Bottomley, G. O. Egharevba, I. J. B. Lin, and P. S. White, *Organometallics*, *4*, 550 (1985).

93. J. L. Polse, R. A. Andersen, and R. G. Bergman, *J. Am. Chem. Soc.*, *117*, 5393 (1995).

94. D. J. Schwartz, M. R. Smith, III, and R. A. Andersen, *Organometallics*, *15*, 1446 (1996).

95. L. Peng-Ju, H., Sheng-Hua, H. Kun-Yao, W. Ru-Ji, and T. C. W. Mak, *Inorg. Chim. Acta*, *175*, 105 (1990).

96. A. Z. Feltz, *Anorg. Allg. Chem.*, *334*, 242 (1965).

97. W. Haase and H. Hoppe, *Acta Crystallogr.*, *B24*, 282 (1968).

98. (a) V. Krug, G. Koellner, U. Müller, *Z. Naturforsch.*,*43b*, 1501 (1988). (b) U. Müller and V. Krug, *Angew. Chem. Int. Ed. Engl.*, *27*, 293 (1988).

99. P. J. Lundmark, G. J. Kubas, and B. L. Scott, *Organometallics*, *15*, 3631 (1996).

100. J. M. Fischer, W. E. Piers, L. R. MacGillivray, and M. J. Zaworotko, *Inorg. Chem.*, *34*, 2499 (1995).

101. The complex $[Cp_2Zr(\mu\text{-}O)]_3$ was first synthesized by the reaction of $Cp_2Zr(CO)_2$ with CO_2. See G. Fachinetti, C. Floriani, A. Chiesi-Villa, and C. Guastini, *J. Am. Chem. Soc.*, *101*, 1767 (1979).

102. A. Shaver and J. M. McCall, *Organometallics*, *3*, 1823 (1984).

103. F. Bottomley, D. F. Drummond, G. O. Egharevba, and P. S. White, *Organometallics*, *5*, 1620 (1986).

104. E. Hey, M. F. Lappert, J. L. Atwood, and S. G. Bott, *J. Chem. Soc. Chem. Commun.*, 421 (1987).

105. G. Tainturier, M. Fahim, G. Trouvé-Bellan, and B. Gautheron, *J. Organomet. Chem.*, *376*, 321 (1989).

106. (a) G. Tainturier, M. Fahim, and B. Gautheron, *J. Organomet. Chem.*, *373*, 193 (1989). (b) M. Fahin and G. Tainturier, *J. Organomet. Chem.*, *301*, C45 (1986). (c) G. Tainturier, M. Fahim, and B. Gautheron, *J. Organomet. Chem.*, *362*, 311 (1989).

107. G. Erker, T. Mühlenbernd, R. Benn, A. Rufinska, G. Tainturier, and B. Gautheron, *Organometallics*, *5*, 1023 (1986).

108. (a) G. Tainturier, B. Gautheron, and S. Pouly, *Nouv. J. Chim.*, *10*, 625 (1986). (b) B. Gautheron, G. Tainturier, and S. Pouly, *J. Organomet. Chem.*, *268*, C56 (1984).

109. (a) G. Erker, T. Mühlenbernd, R. Nolte, J. L. Petersen, G. Tainturier, and B. Gautheron, *J. Organomet. Chem.*, *314*, C21 (1986). (b) G. Erker, R. Nolte, G. Tainturier, and A. Rheingold, *Organometallics*, *8*, 454 (1989).

110. Bottomley has investigated the reactions of $Cp_2^*Zr(CO)_2$ with stoichiometric quantities of H_2S and H_2Se in the absence of pyridine and reported the products to be $[Cp_2^*Zr(\mu\text{-}E)]_2$ (E = S,Se) (103). However, subsequent work has demonstrated that, for the case of H_2S, the isolated product is actually $[Cp_2^*Zr(SH)]_2(\mu\text{-}S)$ [W. A. Howard, and G. Parkin, *Organometallics*, *12*, 2363 (1993)].

111. In addition to the discussion concerning $[Cp_2^*Zr(\mu\text{-}S)]_2$ in (110), other syntheses of $[Cp_2^*Zr(\mu\text{-}S)]_2$ have been reported: (a) the reaction of $Cp_2^*Zr(CH{=}CH_2)_2$ with sulfur (111a) and (b) the reaction of $Cp_2^*ZrCl_2$ with "Li_2S" (111b). In neither of these examples, however, has the structure of $[Cp_2^*Zr(\mu\text{-}S)]_2$ been confirmed by X-ray diffraction. (a) R. Beckhaus and K.-H. Thiele, *Z. Anorg. Allg. Chem.*, *573*, 195 (1989). (b) R. Broussier, M. Rigoulet, R. Amardeil, G. Delmas, and B. Gautheron, *Phosphorus, Sulfur, Silicon*, *82*, 55 (1993).

112. E. W. Harlan and R. H. Holm, *J. Am. Chem. Soc.*, *112*, 186 (1990).

113. R. Bortolin, V. Patel, I. Munday, N. J. Taylor, and A. J. Carty, *J. Chem. Soc. Chem. Commun.*, 456 (1985).

114. (a) M. J. Carney, P. J. Walsh, F. J. Hollander, and R. G. Bergman, *Organometallics*, *11*, 761 (1992). (b) M. J. Carney, P. J. Walsh, and R. G. Bergman, *J. Am. Chem. Soc.*, *112*, 6426 (1990). (c) M. J. Carney, P. J. Walsh, F. J. Hollander, and R. G. Bergman, *J. Am. Chem. Soc.*, *111*, 8751 (1989).

115. (a) T. A. Hanna, A. M. Baranger, P. J. Walsh, and R. G. Bergman, *J. Am. Chem. Soc.*, *117*, 3292 (1995). (b) T. A. Hanna, A. M. Baranger, and R. G. Bergman, *J. Org. Chem.*, *61*, 4532 (1996).

116. (a) S. Y. Lee and R. G. Bergman, *J. Am. Chem. Soc.*, *118*, 6396 (1996). (b) P. J. Walsh, F. J. Hollander, and R. G. Bergman, *Organometallics*, *12*, 3705 (1993).

117. It is noteworthy that a terminal oxo complex is not obtained by the reaction of $Cp_2^*Zr(\eta^2\text{-}C_2Ph_2)$ with N_2O, which rather gives the oxametallacycle $Cp_2^*Zr[\eta^2\text{-}OC(Ph){=}C(Ph)]$. See (a) G. A. Vaughan, G. L. Hillhouse, and A. L. Rheingold, *J. Am. Chem. Soc.*, *112*, 7994 (1990). (b) G. A. Vaughan, C. D. Sofield, and G. L. Hillhouse, *J. Am. Chem. Soc.*, *111*, 5491 (1989). (c) G. A. Vaughan, G. L. Hillhouse, R. T. Lum, S. L. Buchwald, and A. L. Rheingold, *J. Am. Chem. Soc.*, *110*, 7215 (1988).

118. $\nu(Zr{=}E)$ for the heavier chalcogenido derivatives are believed to be less than 400 cm^{-1} but have not been identified with certainty. Cundari and co-workers (56) calculated $\nu(Zr{=}E)$ values for the series of hypothetical derivatives Cp_2ZrE: O, 945 cm^{-1}; S, 484 cm^{-1}; Se, 328 cm^{-1}; Te, 248 cm^{-1}.

119. In contrast, both $Cp_2^*W{=}O$ and $[Cp_2^*Zr{=}O]$ would be expected to have similar M=O bond orders, despite the fact that the metal centers are 18- and 16-electron, respectively, since the metallocene fragment is not capable of supporting a triply bonded ligand without compromising the metal–Cp* interactions. See Section III.A.

120. D. Jacoby, C. Floriani, A. Chiesi-Villa, and C. Rizzoli, *J. Am. Chem. Soc.*, *115*, 7025 (1993).

121. D. Jacoby, S. Isoz, C. Floriani, A. Chiesi-Villa, and C. Rizzoli, *J. Am. Chem. Soc.*, *117*, 2793 (1995).

122. D. Jacoby, S. Isoz, C. Floriani, A. Chiesi-Villa, and C. Rizzoli, *J. Am. Chem. Soc.*, *117*, 2805 (1995).

123. D. Jacoby, C. Floriani, A. Chiesi-Villa, and C. Rizzoli, *J. Am. Chem. Soc.*, *115*, 7025 (1993).

124. D. Jacoby, S. Isoz, C. Floriani, A. Chiesi-Villa, and C. Rizzoli, *J. Am. Chem. Soc.*, *117*, 2793 (1995).

125. For a further discussion of carbon monoxide cleavage by transition metals, see D. R. Neithamer, R. E. LaPointe, R. A. Wheeler, D. S. Richeson, G. D. Van Duyne, and P. T. Wolczanski, *J. Am. Chem. Soc.*, *111*, 9056 (1989).

126. D. Jacoby, S. Isoz, C. Floriani, A. Chiesi-Villa, and C. Rizzoli, *J. Am. Chem. Soc.*, *117*, 2805 (1995).

127. (a) M. M. Dawod, F. I. Khalili, and A. M. Seyam, *Synth. React. Inorg. Met.-Org. Chem.*, *24*, 663 (1994). (b) R. K. Agarwal, B. S. Tyagi, M. Srivastava, and A. K. Srivastava, *Thermochim. Acta*, *61*, 241 (1983).

128. The importance of using highly substitued cyclopentadienyl derivatives for isolating terminal chalcogenido complexes is further demonstrated by the fact that the reactions of $Cp_2^{Bu^t}Zr(CO)_2$ with elemental sulfur, selenium, and tellurium in the presence of pyridine give the bridged complexes $[Cp_2^{Bu^t}Zr(\mu\text{-}S)]_2$, $[Cp_2^{Bu^t}Zr(\mu\text{-}Se)]_2$, and $[Cp_2^{Bu^t}Zr(\mu\text{-}Te)]_2$. See (32b).

129. W. A. Howard, G. Parkin, and A. L. Rheingold, *Polyhedron*, *14*, 25 (1995).

130. A. Shaver and J. M. McCall, *Organometallics*, *3*, 1823 (1984).

131. F. Bottomley, D. F. Drummond, G. O. Egharevba, and P. S. White, *Organometallics*, *5*, 1620 (1986).

132. E. Hey, M. F. Lappert, J. L. Atwood, and S. G. Bott, *J. Chem. Soc. Chem. Commun.*, 421 (1987).

133. (a) G. Tainturier, M. Fahim, and B. Gautheron, *J. Organomet. Chem.*, *373*, 193 (1989). (b) M. Fahin and G. Tainturier, *J. Organomet. Chem.*, *301*, C45 (1986). (c) G. Tainturier, M. Fahim, and B. Gautheron, *J. Organomet. Chem.*, *362*, 311 (1989).

134. G. Erker, T. Mühlenbernd, R. Benn, A. Rufinska, G. Tainturier, and B. Gautheron, *Organometallics*, *5*, 1023 (1986).

135. W. A. Howard, T. M. Trnka, and G. Parkin, *Organometallics*, *14*, 4037 (1995).

136. W. A. Howard, and G. Parkin, unpublished results.

137. For leading recent references on nonclassical carbonyl complexes, see (a) P. K. Hurlburt, J. J. Rack, J. S. Luck, S. F. Dec, J. D. Webb, O. P. Anderson, and S. H. Strauss, *J. Am. Chem. Soc.*, *116*, 10003 (1994). (b) F. Aubke, and C. Wang, *Coord. Chem. Rev.*, *137*, 483 (1994). (c) L. Weber, *Angew. Chem. Int. Ed. Engl.*, *33*, 1077 (1994). (d) H. Willner, M. Bodenbinder, C. Wang, and F. Aubke, *J. Chem. Soc. Chem. Commun.*, 1189 (1994). (e) S. H. Strauss, *Chemtracts-Inorg. Chem.*, *10*, 77 (1997).

138. In addition to $Cp_2^*Zr(Se)CO$ and $Cp_2^*Zr(\eta^2\text{-}E_2)CO$ (E = S, Se, Te), other nonclassical carbonyl complexes of zirconium include $Cp_2^*ZrH_2(CO)$ (138a, b) $[Cp_2Zr\{\eta^2\text{-}CH(Me)(6\text{-ethylpyrid-2-yl})\text{-}C,N\}(CO)]^+$ (138c), $[Cp_2^*Zr(\eta^2\text{-}COCH_3)CO]^+$ (138d) $[Cp_2Zr(\eta^2\text{-}COCH_3)CO]^+$ (138d), $[Cp_2^*Zr(\eta^3\text{-}C_3H_5)CO]^+$ (138e), and $Cp_2Zr(\eta^2\text{-}Me_2SiNBu^t)CO$ (138f). (a) J. M. Manriquez, D. R. McAlister, R. D. Sanner, and J. E. Bercaw, *J. Am. Chem. Soc.*, *100*, 2716 (1978). (b) J. A. Marsella, C. J. Curtis, J. E. Bercaw, and K. G. Caulton, *J. Am. Chem. Soc.*, *102*, 7244 (1980). (c) A. S. Guram, D. C. Swenson, and R. F. Jordan, *J. Am. Chem. Soc.*, *114*, 8991 (1992). (d) Z. Guo, D. C. Swenson, A. S. Guram, and R. F. Jordan, *Organometallics*, *13*, 766 (1994). (e) D. M. Antonelli, E. B. Tjaden, and J. M. Stryker, *Organometallics*, *13*, 763 (1994). (f) L. J. Procopio, P. J. Carroll, and D. H. Berry, *Polyhedron*, *14*, 45 (1995).

139. D. J. Sikora, M. D. Rausch, R. D. Rogers, and J. L. Atwood, *J. Am. Chem. Soc.*, *103*, 1265 (1981).

140. (a) H. H. Brintzinger, *J. Organomet. Chem.*, *171*, 337 (1979). (b) S. H. Strauss, *Chemtracts—Inorg. Chem.*, *6*, 157 (1994).

141. (a) V. Christou and J. Arnold, *J. Am. Chem. Soc.*, *114*, 6240 (1992). (b) C. P. Gerlach, V. Christou, and J. Arnold, *Inorg. Chem.*, *35*, 2758 (1996).

142. D. E. Gindelberger and J. Arnold, *Organometallics*, *13*, 4462 (1994).

143. W. A. Howard and G. Parkin, *J. Organomet. Chem.*, *472*, C1 (1994).

144. L. V. Boas and J. C. Pessoa, in *Comprehensive Coordination Chemistry*, G. Wilkinson, R. D. Gillard, and J. A. McCleverty, Eds., Pergamon, Oxford, UK, 1987, Chapter 33.

145. C. C. Cummins, R. R. Schrock, and W. M. Davis, *Inorg. Chem.*, *33*, 1448 (1994).

146. The oxo complexes $[\eta^4\text{-}N(CH_2CH_2NMe)_3]VO$ [$d(V{=}O) = 1.599(6)$ Å, $\eta(V{=}O) = 962\ cm^{-1}$] (146a) and $[\eta^4\text{-}N(CH_2CH_2NC_6F_5)_3]VO$ are also known (146b). (a) W. Plass, and J. G. Verkade, *J. Am. Chem. Soc.*, *114*, 2275 (1992). (b) K. Nomura, R. R. Schrock, and W. M. Davis, *Inorg. Chem.*, *35*, 3695 (1996).

147. A series of vanadium complexes with *bridging* chalcogenido ligands is known, namely, $[V(CO)_3(dppe)]_2(\mu_2\text{-}E)$ (E = S, Se, Te). (a) J. Schiemann, P. Hübener, U. Behrens, and E. Weiss, *Angew. Chem. Int. Ed. Engl.*, *22*, 980 (1983). (b) N. Albrecht, P. Hübener, U. Behrens, and E. Weiss, *Chem. Ber.*, *118*, 4059 (1985).

148. W. Priebsch and D. Rehder, *Inorg. Chem.*, *24*, 3085 (1985).

149. C. P. Gerlach and J. Arnold, *Inorg. Chem.*, *35*, 5770 (1996).

150. (a) J. L. Poncet, R. Guilard, P. Friant, C. Goulon-Ginet, and J. Goulon, *Nouv. J. Chim.*, *8*, 583 (1984). (b) J. L. Poncet, R. Guilard, P. Friant, and J. Goulon, *Polyhedron*, *5*, 417 (1983).

151. (a) V. L. Goedken and J. A. Ladd, *J. Chem. Soc. Chem. Commun.*, 910 (1981). (b) C.-H. Yang, J. A. Ladd, and V. L. Goedken, *J. Coord. Chem.*, *18*, 317 (1988). (c) H. Schumann, *Z. Naturforsch.*, *50B*, 1494 (1995).

152. [η^4-Me_4taa]VO has also been reported in subsequent publications in which the [V=O] moiety is characterized by [ν(V=O) = 978 cm^{-1}, d(V=O) = 1.600(5) Å] (152a) and [ν(V=O) = 974 cm^{-1}, d(V=O) = 1.601(2) Å] (152b). (a) S. Lee, C. Floriani, A. Chiesi-Villa, and C. Guastini, *J. Chem. Soc. Dalton Trans.*, 145 (1989). (b) F. A. Cotton, J. Czuchajowska, and X. Feng, *Inorg. Chem.*, *30*, 349 (1991).

153. (a) V. L. Goedken and J. A. Ladd, *J. Chem. Soc. Chem. Commun.*, 910 (1981). (b) C.-H. Yang, J. A. Ladd, and V. L. Goedken, *J. Coord. Chem.*, *18*, 317 (1988).

154. [η^4-Me_4taa]VO has also been reported in subsequent publications in which the [V=O] moiety is characterized by [η(V=O) = 978 cm^{-1}, d(V=O) = 1.600(5) Å],(154a) and [ν(V=O) = 974 cm^{-1}, d(V=O) = 1.601(2) Å](154b). (a) S. Lee, C. Floriani, A. Chiesi-Villa, and C. Guastini, *J. Chem. Soc. Dalton Trans.*, 145 (1989). (b) F. A. Cotton, J. Czuchajowska, and X. Feng, *Inorg. Chem.*, *30*, 349 (1991).

155. (a) F. Bottomley, C. P. Magill, and B. Zhao, *Organometallics*, *9*, 1700 (1990). (b) F. Bottomley, C. P. Magill, and B. Zhao, *Organometallics*, *10*, 1946 (1991).

156. K. P. Callahan, P. J. Durand, and P. H. Rieger, *J. Chem. Soc. Chem. Commun.*, 75 (1980). (b) K. P. Callahan and P. J. Durand, *Inorg. Chem.*, *19*, 3211 (1980).

157. Y. Do, E. D. Simhon, and R. H. Holm, *Inorg. Chem.*, *22*, 3809 (1983).

158. F. Preuss and H. Noichl, *Z. Naturforsch.* ,*42B*, 121 (1987).

159. J. R. Nicholson, J. C. Huffman, D. M. Ho, and G. Christou, *Inorg. Chem.*, *26*, 3030 (1987).

160. L. G. Hubert-Pfalzgraf, M. Postel, and J. G. Riess, in *Comprehensive Coordination Chemistry*, G. Wilkinson, R. D. Gillard, and J. A. McCleverty, Eds., Pergamon, Oxford, UK, 1987, Chapter 34.

161. A. Antiñolo, J. Martinez de llarduya, A. Otero, P. Royo, A. M. M. Lanfredi, and A. Tiripicchio, *J. Chem. Soc. Dalton Trans.*, 2685 (1988).

162. P.-F. Fu, M. A. Khan, and K. M. Nicholas, *Organometallics*, *10*, 382 (1991).

163. H. Brunner, G. Gehart, W. Meier, J. Wachter, and B. Nuber, *J. Organomet. Chem.*, *454*, 117 (1993).

164. For a review of niobium and molybdenum porphyrin oxo chemistry, see Y. Matsuda and Y. Murakami, *Coord. Chem. Rev.*, *92*, 157 (1988).

165. E. L. Muetterties and C. M. Wright, *J. Am. Chem. Soc.*, *87*, 4706 (1965).

166. M. G. B. Drew, D. A. Rice, and D. M. Williams, *Inorg. Chim. Acta*, *118*, 165 (1986).

167. U. Siemeling and V. C. Gibson, *J. Chem. Soc. Chem. Commun.*, 1670 (1992).

168. V. C. Gibson, T. P. Kee, R. M. Sorrell, A. P. Bashall, and M. McPartlin, *Polyhedron*, *7*, 2221 (1988).

169. A. Bashall, V. C. Gibson, T. P. Kee, M. McPartlin, O. B. Robinson, and A. Shaw, *Angew. Chem. Int. Ed. Engl.*, *30*, 980 (1991).

170. V. C. Gibson and M. McPartlin, *J. Chem. Soc. Dalton Trans.*, 947 (1992).

171. Y. Do and R. H. Holm, *Inorg. Chim. Acta*, *104*, 33 (1985).

172. J. L. Seela, J. C. Huffman, and G. Christou, *Polyhedron*, *8*, 1797 (1989).

173. S. C. Lee, J. Li, J. C. Mitchell, and R. H. Holm, *Inorg. Chem.*, *31*, 4333 (1992).

174. M. Latroche and J. A. Ibers, *Inorg. Chem.*, *29*, 1503 (1990).

175. H. Yun, C. R. Randall, and J. A. Ibers, *J. Solid State Chem.*, *76*, 109 (1988).

176. D. Coucouvanis, S. Al-Ahmad, C. G. Kim, and S.-M. Koo, *Inorg. Chem.*, *31*, 2996 (1992).

177. K. Tatsumi, Y. Sekiguchi, A. Nakamura, R. E. Cramer, and J. J. Rupp, *J. Am. Chem. Soc.*, *108*, 1358 (1986).

178. A. van Asselt, B. J. Burger, V. C. Gibson, and J. E. Bercaw, *J. Am. Chem. Soc.*, *108*, 5347 (1986).

179. G. Parkin, A. van Asselt, D. J. Leahy, L. Whinnery, N. G. Hua, R. Quan, L. M. Henling, W. P. Schaefer, B. D. Santarsiero, and J. E. Bercaw, *Inorg. Chem.*, *31*, 82 (1992).

180. D. M. Antonelli, W. P. Schaefer, G. Parkin, and J. E. Bercaw, *J. Organomet. Chem.*, *462*, 213 (1993).

181. The mixed–ring oxo–chloride derivatives Cp*[η^5-$C_5H_4SiMe_3$]Ta(O)Cl [ν(Ta=O) = 901 cm^{-1}] and Cp*[η^5-$C_5H_3(SiMe_3)_2$]Ta(O)Cl have also been obtained by the reaction of the corresponding dichlorides with air; however, dry O_2 yields the dinuclear complexes, {Cp*[η^5-$C_5H_4SiMe_3$]$TaCl_2$}$_2$(μ-O) and {Cp*[η^5-$C_5H_3(SiMe_3)_2$]$TaCl_2$}$_2$(μ-O). See A. Castro, M. Gómez, P. Gómez-Sal, A. Manzanero, and P. Royo, *J. Organomet. Chem.*, *518*, 37 (1996).

182. J. E. Nelson, G. Parkin, and J. E. Bercaw, *Organometallics*, *11*, 2181 (1992).

183. J. H. Shin and G. Parkin, *Organometallics*, *13*, 2147 (1994).

184. W. A. Nugent and J. M. Mayer, *Metal–Ligand Multiple Bonds*, Wiley-Interscience, New York, 1988, p. 163.

185. W. A. Nugent and J. M. Mayer, *Metal–Ligand Multiple Bonds*, Wiley-Interscience, New York, 1988, p. 116.

186. G. Parkin, E. Bunel, B. J. Burger, M. S. Trimmer, A. van Asselt, and J. E. Bercaw, *J. Mol. Catal.*, *41*, 21 (1987).

187. H. Brunner, M. M. Kubicki, J.-C. Leblanc, C. Moise, F. Volpato, and J. Wachter, *J. Chem. Soc. Chem. Commun.*, 851 (1993).

188. H.-J. Bach, H. Brunner, J. Wachter, M. M. Kubicki, J.-C. Leblanc, and M. L. Ziegler, *Organometallics*, *11*, 1403 (1992).

189. A. van Asselt, B. J. Burger, V. C. Gibson, and J. E. Bercaw, *J. Am. Chem. Soc.*, *108*, 5347 (1986).

190. G. Parkin, E. Bunel, B. J. Burger, M. S. Trimmer, A. van Asselt, and J. E. Bercaw, *J. Mol. Catal.*, *41*, 21 (1987).

191. J. H. Shin and G. Parkin, *Organometallics*, *13*, 2147 (1994).

192. Note that the tellurido-methyl product $Cp_2^*Ta(Ta)Me$ does not react with CO to give $Cp_2^*Ta(CO)Me$ under similar conditions.

193. A. van Asselt, M. S. Trimmer, L. M. Henling, and J. E. Bercaw, *J. Am. Chem. Soc.*, *110*, 8254 (1988).

194. L. L. Whinnery, Jr., L. M. Henling, and J. E. Bercaw, *J. Am. Chem. Soc.*, *113*, 7575 (1991).

195. For a review of transition metal catalyzed epoxidations, see K. A. Jorgensen, *Chem. Rev.*, *89*, 431 (1989).

196. J. E. Nelson, G. Parkin, and J. E. Bercaw, *Organometallics*, *11*, 2181 (1992).

197. (a) G. Proulx and R. G. Bergman, *Science*, *259*, 661 (1993). (b) G. Proulx and R. G. Bergman, *J. Am. Chem. Soc.*, *115*, 9802 (1993). (c) G. Proulx and R. G. Bergman, *J. Am. Chem. Soc.*, *118*, 1981 (1996). (d) G. Proulx, F. J. Hollander, and R. G. Bergman, *Can. J. Chem.*, *73*, 1111 (1995).

198. (a) G. Proulx and R. G. Bergman, *Organometallics*, *15*, 133 (1996). (b) G. Proulx and R. G. Bergman, *J. Am. Chem. Soc.*, *116*, 7953 (1994).

199. C. C. Cummins, R. R. Schrock, and W. M. Davis, *Angew. Chem. Int. Ed. Engl.*, *32*, 756 (1993).

200. V. Christou and J. Arnold, *Angew. Chem. Int. Ed. Engl.*, *32*, 1450 (1993).

201. (a) R. E. LaPointe, P. T. Wolczanski, and J. F. Mitchell, *J. Am. Chem. Soc.*, *108*, 6382 (1986). (b) D. R. Neithamer, R. E. LaPointe, R. A. Wheeler, D. S. Richeson, G. D. Van Duyne, and P. T. Wolczanski, *J. Am. Chem. Soc.*, *111*, 9056 (1989).

202. M. G. B. Drew, D. A. Rice, and D. M. Williams, *J. Chem. Soc. Dalton Trans.*, 845 (1984).

203. E. J. Peterson, R. B. von Dreele, and T. M. Brown, *Inorg. Chem.*, *17*, 1410 (1978).

204. M. G. B. Drew, D. A. Rice, and D. M. Williams, *J. Chem. Soc. Dalton Trans.*, 1821 (1985).

205. The value of ν(Ta=S) = 950 cm^{-1} cited in (203) is incorrect. See (171).

206. P. F. Gilletti, D. A. Femec, F. I. Keen, and T. M. Brown, *Inorg. Chem.*, *31*, 4008 (1992).

207. H. Yun, C. R. Randall, and J. A. Ibers, *J. Solid State Chem.*, *76*, 109 (1988).

208. L. F. Larkworthy, K. B. Nolan, and P. O'Brien, in *Comprehensive Coordination Chemistry*, G. Wilkinson, R. D. Gillard, and J. A. McCleverty, Eds.; Pergamon Oxford, UK, 1987, Chapter 35.

209. (a) J. T. Groves, W. J. Kruper, Jr., R. C. Haushalter, and W. M. Butler, *Inorg. Chem.*, *21*, 1363 (1982). (b) T. L. Siddall, N. Miyaura, J. C. Huffman, and J. K. Kochi, *J. Chem. Soc. Chem. Commun.*, 1185 (1983). (c) D. B. Morse, T. B. Rauchfuss, and S. R. Wilson, *J. Am. Chem. Soc.*, *110*, 8234 (1988). (d) J. A. McGinnety, *Acta Crystallogr.*, *B28*, 2845 (1972).

210. S.-K. Noh, R. A. Heintz, B. S. Haggerty, A. L. Rheingold, and K. H. Theopold, *J. Am. Chem. Soc.*, *114*, 1892 (1992).

211. G. K. Cook and J. M. Mayer, *J. Am. Chem. Soc.*, *116*, 1855 (1994).

212. M. Poliakoff, K. P. Smith, J. J. Turner, and A. J. Wilkinson, *J. Chem. Soc. Dalton Trans.*, 651 (1982).

213. M. J. Almond, *J. Chem. Soc. Rev.*, *23*, 309 (1994).

214. K. Lane, L. Sallans, and R. R. Squires, *J. Am. Chem. Soc.*, *106*, 2719 (1984).

215. (a) C. D. Garner and J. M. Charnock, in *Comprehensive Coordination Chemistry*, G. Wilkinson, R. D. Gillard, and J. A. McCleverty, Eds., Pergamon, Oxford, UK, 1987, Chapter 36.4. (b) E. I. Stiefel, in *Comprehensive Coordination Chemistry*, G. Wilkinson, R. D. Gillard, J. A. McCleverty, Eds., Pergamon: Oxford, UK, 1987, Chapter 36.5. (c) C. D. Garner, in *Comprehensive Coordination Chemistry*, G. Wilkinson, R. D. Gillard, and J. A. McCleverty, Eds., Pergamon, Oxford, UK, 1987, Chapter 36.6.

216. R. H. Holm, *Coord. Chem. Rev.*, *100*, 183 (1990).

217. (a) B. E. Schultz and R. H. Holm, *Inorg. Chem.*, *32*, 4244 (1993). (b) B. E. Schultz, R. Hille, and R. H. Holm, *J. Am. Chem. Soc.*, *117*, 827 (1995). (c) B. E. Schultz, S. F. Gheller, M. C. Muetterties, M. J. Scott, and R. H. Holm, *J. Am. Chem. Soc.*, *115*, 2714 (1993).

218. V. J. Murphy and G. Parkin, *J. Am. Chem. Soc.*, *117*, 3522 (1995).

219. For example, whereas $[Mo(Se)_4]^{2-}$ has been prepared by the reactions of $Mo(CO)_6$ with polyselenides, the corresponding reaction of $Mo(CO)_6$ with $[Te_4]^{2-}$ gives only $(CO)_4Mo(\eta^2\text{-}Te_4)$. See L. C. Roof, W. T. Pennington, and J. W. Kolis, *J. Am. Chem. Soc.*, *112*, 8172 (1990).

220. (a) F. A. Cotton and X. Feng, *Inorg. Chem.*, *35*, 4921 (1996). (b) F. A. Cotton, and G. Schmid, *Inorg. Chem.*, *36*, 2267 (1997).

221. Some related Mo^{IV} oxo complexes include *trans*-[Mo(dppe)(O)(OMe)][OTf] (221a) and *trans*-$[Mo(dppe)(O)(OH)][BF_4]$ (221b) (a) T. Adachi, D. L. Hughes, S. K. Ibrahim, S. Okamoto, C. J. Pickett, N. Yabanouchi, and T. Yoshida, *J. Chem. Soc. Chem. Commun.*, 1081 (1995). (b) M. R. Churchill and F. Rotella, *J. Inorg. Chem.*, *17*, 668 (1978).

222. T. Yoshida, T. Adachi, K. Matsumura, K. Kawazu, and K. Baba, *Chem. Lett.*, 1067 (1991).

223. For a compilation of Mo=S bond lengths in dinuclear complexes, see J. L. Dulebohn, T. C. Stamatakos, D. L. Ward, and D. G. Nocera, *Polyhedron*, *10*, 2813 (1991). Representative examples include $[Cp'Mo(\mu\text{-}S)(S)]_2$ [2.135(2) Å], (223a) $[Cp^*Mo(\mu\text{-}S)(S)]_2$ [2.144(2) Å], (223b) $[(\eta^2\text{-}S_2)Mo(S)(\mu\text{-}S)][(\eta^2\text{-}S_4)Mo(S)(\mu\text{-}S)]^{2-}$ [2.108(2) Å; 2.12(2) Å], (223b) $[Mo_2Cu_5S_8(S_2CNMe_2)_3]^{2-}$ [1.942(7) Å; 2.123(4) Å] (223c), $\{[Tp^{Me_2}]Mo(S)(\mu\text{-}S)_2Mo(O)(\mu\text{-}OH)\}_2$ [2.124(2) Å] (223d), $[Tp^{Me_2}]Mo(S)(\mu\text{-}S)_2Mo(S)(S_2CNEt_2)$ [2.122(5) Å; 2.094(5) Å] (223d), $[(\eta^2\text{-}S_4)Mo(O)(\mu\text{-}S)_2Mo(S)_2]^{2-}$ [2.182(7) Å; 2.135(7) Å] (223e), (a) M. R. DuBois, D. L. DuBois, M. C. vanDerveer, and R. C. Haltiwanger, *Inorg. Chem.*, *20*, 3064 (1981). (b) W. Clegg, G. Christou, C. D. Garner, and G. M. Sheldrick, *Inorg. Chem.*, *20*, 1562 (1981). (c) X. Lei, Z. Huang, Q. Liu, M. Hong, and H.

Liu, *Inorg. Chem.*, *28*, 4302 (1989). (d) S. A. Roberts, C. G. Young, W. E. Cleland, Jr., K. Yamanouchi, R. B. Ortega, and J. H. Enemark, *Inorg. Chem.*, *27*, 2647 (1988). (e) D. Coucouvanis and S.-M. Koo, *Inorg. Chem.*, *28*, 2 (1989).

224. I.-P. Lorenz, G. Walter, and W. Hiller, *Chem. Ber.*, *123*, 979 (1990).
225. J. A. Paradis, D. W. Wertz, and H. H. Thorp, *J. Am. Chem. Soc.*, *115*, 5308 (1993).
226. A. V. Butcher and J. Chatt, *J. Chem. Soc. (A)*, 2652 (1970).
227. F. A. Cotton, M. P. Diebold, and W. J. Roth, *Inorg. Chem.*, *26*, 2848 (1987).
228. L. K. Atkinson, A. H. Mawby, and D. C. Smith, *Chem. Commun.*, 1399 (1970).
229. E. Carmona, A. Galindo, L. Sanchez, A. J. Nielson, and G. Wilkinson, *Polyhedron*, *3*, 347 (1984).
230. K. A. Hall and J. M. Mayer, *J. Am. Chem. Soc.*, *114*, 10402 (1992).
231. K. A. Hall and J. M. Mayer, *Inorg. Chem.*, *34*, 1145 (1995).
232. J. M. Mayer, *Adv. Transition Metal Coord. Chem.*, *1*, 105 (1996).
233. K. A. Hall and J. M. Mayer, *Inorg. Chem.*, *33*, 3289 (1994).
234. T. Yoshida, T. Adachi, K. Matsumura, and K. Baba, *Chem. Lett.*, 2447 (1992).
235. Alkylation of $[Ph_4P]_2[W(S)_4]$ by EtBr to give $[Ph_4P][WS_3(SEt)]$ has recently been reported. See P. M. Boorman, M. Wang, and M. Parvez, *J. Chem. Soc. Chem. Commun.*, 999 (1995).
236. T. Yoshida, T. Adachi, K. Matsumura, and K. Baba, *Angew. Chem. Int. Ed. Engl.*, *32*, 1621 (1993).
237. For a review of niobium and molybdenum porphyrin oxo chemistry, see Y. Matsuda, and Y. Murakami, *Coord. Chem. Rev.*, *92*, 157 (1988).
238. T. Diebold, B. Chevrier, and R. Weiss, *Inorg. Chem.*, *19*, 1193 (1979).
239. L. M. Berreau, V. G. Young, Jr., and L. K. Woo, *Inorg. Chem.*, *34*, 3485 (1995).
240. L. M. Berreau, J. A. Hays, V. G. Young, and L. K. Woo, *Inorg. Chem.*, *33*, 105 (1994).
241. For a review of such atom-transfer reactions see L. K. Woo, *Chem. Rev.*, *93*, 1125 (1993).
242. C. C. Cummins, personal communication.
243. C. E. Laplaza, A. L. Odom, W. M. Davis, C. C. Cummins, and J. D. Protasiewicz, *J. Am. Chem. Soc.*, *117*, 4999 (1995).
244. M. L. H. Green, A. H. Lynch, and M. G. Swanwick, *J. Chem. Soc. Dalton Trans.*, 1445 (1972).
245. M. Berry, S. G. Davies, and M. L. H. Green, *J. Chem. Soc. Chem. Commun.*, 99 (1978).
246. F. G. N. Cloke, J. P. Day, J. C. Green, C. P. Morley, and A. C. Swain, *J. Chem. Soc. Dalton Trans.*, 789 (1991).
247. M. Henary, W. C. Kaska, and J. I. Zink, *Inorg. Chem.*, *30*, 1674 (1991).
248. The complex $Cp_2Mo(\eta^2\text{-}O_2CO)$ is obtained by irradiation of Cp_2MoH_2 in the presence of CO_2. See K. A. Belmore, R. A. Vanderpool, J.-C. Tsai, M. A. Khan, and K. M. Nicholas, *J. Am. Chem. Soc.*, *110*, 2004 (1988).

249. R. S. Pilato, C. E. Housmekerides, P. Jernakoff, D. Rubin, G. L. Geoffroy, and A. L. Rheingold, *Organometallics*, *9*, 2333 (1990).

250. R. S. Pilato, K. A. Eriksen, E. I. Stiefel, and A. L. Rheingold, *Inorg. Chem.*, *32*, 3799 (1993).

251. It is noteworthy that $Cp_2Mo(SH)_2$ is stable with respect to elimination of H_2S and formation of Cp_2MoS. Such an observation contrasts markedly with the instability of $Cp^R_2M(OH)_2$ (M = Mo, W) species with respect to Cp^R_2MO.

252. Z. Dori, in *Comprehensive Coordination Chemistry*, G. Wilkinson, R. D. Gillard, J. A. McCleverty, Eds., Pergamon, Oxford, UK, 1987, Chapter 37.

253. (a) V. C. Gibson, C. E. Graimann, P. M. Hare, M. L. H. Green, J. A. Bandy, P. D. Grebenik, and K. Prout, *J. Chem. Soc. Dalton Trans.*, 2025 (1985). (b) M. L. H. Green, G. Parkin, M. Chen, and K. Prout, *J. Chem. Soc. Dalton Trans.*, 2227 (1986).

254. (a) D. Rabinovich and G. Parkin, *J. Am. Chem. Soc.*, *113*, 5904 (1991). (b) D. Rabinovich and G. Parkin, *J. Am. Chem. Soc.*, *113*, 9421 (1991). (c) D. Rabinovich and G. Parkin, *J. Am. Chem. Soc.*, *115*, 9822 (1993). (d) D. Rabinovich and G. Parkin, *Inorg. Chem.*, *33*, 2313 (1994).

255. Hydrogen telluride must be handled in the dark below 0°C to avoid decomposition. See N. N. Greenwood and A. Earnshaw, *Chemistry of the Elements*, Pergamon, New York, 1986, pp. 899–900.

256. (a) K. E. Howard, T. B. Rauchfuss, and S. R. Wilson, *Inorg. Chem.*, *27*, 1710 (1988). (b) L. C. Roof, W. T. Pennington, and J. W. Kolis, *J. Am. Chem. Soc.*, *112*, 8172 (1990).

257. D. R. Gardner, J. C. Fettinger, and B. W. Eichorn, *Angew. Chem. Int. Ed. Engl.*, *33*, 1859 (1994).

258. J. A. Crayston, M. J. Almond, A. J. Downs, M. Poliakoff, and J. J. Turner, *Inorg. Chem.*, *23*, 3051 (1984).

259. In contrast, the corresponding reaction of $Cr(CO)_6$ generates an intermediate that has been identified as four-coordinate tetrahedral $Cr(CO)_2(O)_2$. See M. Poliakoff, K. P. Smith, J. J. Turner, and A. J. Wilkinson, *J. Chem. Soc. Dalton Trans.*, 651 (1982).

260. For related studies on photooxidation of $M(CO)_6$, see (213) and (a) M. J. Almond, A. J. Downs, and R. N. Perutz, *Inorg. Chem.*, *24*, 275 (1985). (b) M. J. Almond and M. Hahne, *J. Chem. Soc. Dalton Trans.*, 2255 (1988). (c) M. J. Almond, J. A. Crayston, A. J. Downs, M. Poliakoff, and J. J. Turner, *Inorg. Chem.*, *25*, 19 (1986). (d) M. J. Almond and A. J. Downs, *J. Chem. Soc. Dalton Trans.*, 809 (1988).

261. The W—E bond lengths listed are the mean values obtained from a search of the Cambridge Structural Database (Version 5.09, April 1995) for bonds between tungsten and a two-coordinate chalcogen. The range of W—E bond lengths are $d(W{-}S) = 2.03 - 2.72$ Å, $d(W{-}Se) = 2.32 - 2.70$ Å, and $d(W{-}Te) = 2.67 - 2.88$ Å.

262. However, the formation of bridging sulfido derivatives accompanied by elimination of H_2 has been observed for the reactions of some dinuclear late transition

metal complexes with H_2S. See (a) C.-L. Lee, G. Besenyei, B. R. James, D. A. Neslon, and M. A. Lilga, *J. Chem. Soc. Chem. Commun.*, 1175 (1985). (b) G. Besenyei, C.-L. Lee, J. Gulinski, S. J. Rettig, B. R. James, D. A. Nelson, and M. A. Lilga, *Inorg. Chem.*, *26*, 3622 (1987). (c) A. F. Barnabas, D. Sallin, and B. R. James, *Can. J. Chem.*, *67*, 2009 (1989). (d) R. McDonald and M. Cowie, *Inorg. Chem.*, *32*, 1671 (1993). (e) T. Y. H. Wong, A. F. Barnabas, D. Sallin, and B. R. James, *Inorg. Chem.*, *34*, 2278 (1995).

263. It is noteworthy that closely related "16"-electron aryloxy complexes of molybdenum, $Mo(PMe_3)_4H(OAr)$ ($Ar = C_6H_3Pr^i_2$, $C_6H_2Me_3$), have recently been isolated and structurally characterized. See T. Hascall, V. J. Murphy, and G. Parkin, *Organometallics*, *15*, 3910 (1996).

264. Tertiary phosphines react with elemental tellurium to give phosphine tellurides R_3PTe. See (a) R. A. Zingaro, B. H. Steeves, and K. Irgolic, *J. Organomet. Chem.*, *4*, 320 (1965). (b) M. L. Steigerwald, and C. R. Sprinkle, *Organometallics*, *7*, 245 (1988).

265. (a) D. J. Berg, C. J. Burns, R. A. Andersen, and A. Zalkin, *Organometallics*, *8*, 1865 (1989). (b) S. M. Stuczynski, Y.-U. Kwon, and M. L. Steigerwald, *J. Organomet. Chem.*, *449*, 167 (1993). (c) J. G. Brennan, T. Siegrist, S. M. Stuczynski, and M. L. Steigerwald, *J. Am. Chem. Soc.*, *111*, 9240 (1989). (d) J. G. Brennan, R. A. Andersen, and A. Zalkin, *Inorg. Chem.*, *25*, 1761 (1986). (e) K. McGregor, G. B. Deacon, R. S. Dickson, G. D. Fallon, R. S. Rowe, and B. O. West, *J. Chem. Soc. Chem. Commun.*, 1293 (1990). (f) W. E. Piers, L. R. MacGillivray, and M. Zaworotko, *Organometallics*, *12*, 4723 (1993). (g) I. P. Beletskaya, A. Z. Voskoboynikov, A. K. Shestakova, and H. Schumann, *J. Organomet. Chem.*, *463*, C1 (1993). (h) M. L. Steigerwald and C. E. Rice, *J. Am. Chem. Soc.*, *110*, 4228 (1988).

266. For other examples of chalcogen transfer using R_3PE, see (a) R. D. Baechler, M. Stack, K. Stevenson, and V. Vanvalkenburgh, *Phosphorus, Sulfur, Silicon*, *48*, 49 (1990). (b) K. A. Hall and J. M. Mayer, *Inorg. Chem.*, *33*, 3289 (1994).

267. N. Kuhn, H. Schumann, and G. Wolmershäuser, *J. Chem. Soc. Chem. Commun.*, 1595 (1985).

268. N. Kaltsoyannis, *J. Chem. Soc. Dalton Trans.*, 1391 (1994).

269. M. H. Chisholm, J. C. Huffman, and J. W. Pasterczyk, *Polyhedron*, *6*, 1551 (1987) and references cited therein.

270. The term "reductive coupling" is used here to indicate changes in formal oxidation state at the metal center, and not at the ligands. (a) C. Elschenbroich and A. Salzer, *Organometallics: A Concise Introduction*, 2nd ed., VCH, New York, 1992, pp. 412–414. (b) R. H. Crabtree, *The Organometallic Chemistry of the Transition Metals*, 2nd ed., Wiley, New York, 1994, Chapter 6.

271. For some other examples of coupling and cleavage reactions, see (a) E. M. Carnahan, J. D. Protasiewicz, and S. J. Lippard, *Acc. Chem. Res.*, *26*, 90 (1993). (b) A. Mayr, and C. M. Bastos, *Progress in Inorganic Chemistry*, Wiley-Interscience, New York, Vol. 40, 1992, pp. 1–98. (c) R. Hoffmann, C. N. Wilker, S. J. Lippard, J. L. Templeton, and D. C. Brower, *J. Am. Chem. Soc.*, *105*, 146 (1983). (d) C. N. Wilker, R. Hoffmann, and O. Eisenstein, *Nouv. J. Chim.*, *7*,

535 (1983). (e) R. Hoffmann, C. N. Wilker, and O. Eisenstein, *J. Am. Chem. Soc.*, *104*, 632 (1982). (f) S. Nakamura and K. Morokuma, *Organometallics*, *7*, 1904 (1988). (g) F. Delbecq, *J. Organomet. Chem.*, *443*, 191 (1993).

272. Bent coordination of isocyanide ligands is not as common as linear coordination, although it does have precedence. See, for example, (a) E. M. Carnahan, R. L. Rardin, S. G. Bott, and S. J. Lippard, *Inorg. Chem.*, *31*, 5193 (1992). (b) E. M. Carnahan and S. J. Lippard, *J. Chem. Soc. Dalton Trans.*, 699 (1991). (c) J. Chatt, A. J. L. Pombeiro, R. L. Richards, G. H. D. Royston, K. W. Muir, and R. Walker, *J. Chem. Soc. Chem. Commun.*, 708 (1975). (d) A. M. Martins, M. J. Calhorda, C. C. Romão, C. Völkl, P. Kiprof, and A. C. Filippou, *J. Organomet. Chem.*, *423*, 367 (1992). (e) W. D. Jones, G. P. Foster, and J. M. Putinas, *Inorg. Chem.*, *26*, 2120 (1987). (f) J.-M. Bassett, D. E. Berry, G. K. Barker, M. Green, J. A. K. Howard, and F. G. A. Stone, *J. Chem. Soc. Dalton Trans.*, 1003 (1979). (g) K. W. Chiu, C. G. Howard, G. Wilkinson, A. M. R. Galas, and M. B. Hursthouse, *Polyhedron*, *1*, 803 (1982). (h) M. Green, J. A. K. Howard, M. Murray, J. L. Spencer, and F. G. A. Stone, *J. Chem. Soc. Dalton Trans.*, 1509 (1977).

273. For reviews on isocyanide complexes see (a) E. Singleton and H. E. Oosthuizen, *Adv. Organomet. Chem.*, *22*, 209 (1983). (b) Y. Yamamoto, *Coord. Chem. Rev.*, *32*, 193 (1980). (c) H. Werner, *Angew. Chem. Int. Ed. Engl.*, *29*, 1077 (1990).

274. (a) C. L. Coyle, M. A. Harmer, G. N. George, M. Daage, and E. I. Stiefel, *Inorg. Chem.*, *29*, 14 (1990). (b) S. A. Cohen and E. I. Stiefel, *Inorg. Chem.*, *24*, 4657 (1985). (c) F. Sécheresse, J. M. Manoli, and C. Potvin, *Inorg. Chem.*, *25*, 3967 (1986). (d) C. Simonnet-Jegat, N. Jourdan, F. Robert, C. Bois, and F. Sécheresse, *Inorg. Chim. Acta*, *216*, 201 (1994). (e) W. Rittner, A. Müller, A. Neumann, W. Bäther, and R. C. Sharma, *Angew. Chem. Int. Ed. Engl.*, *18*, 530 (1979). (f) L. Wei, T. R. Halbert, H. H. Murray, III, and E. I. Stiefel, *J. Am. Chem. Soc.*, *112*, 6431 (1990). (g) T. R. Halbert, L. L. Hutchings, R. Rhodes, and E. I. Stiefel, *J. Am. Chem. Soc.*, *108*, 6437 (1986). (h) J. M. Berg, D. J. Spira, K. O. Hodgson, A. E. Bruce, K. F. Miller, J. L. Corbin, and E. I. Stiefel, *Inorg. Chem.*, *23*, 3412 (1984). (i) H. Ledon, M. Bonnet, and J.-Y. Lallemand, *J. Chem. Soc. Chem. Commun.*, 702 (1979). (j) D. G. Lee, C. R. Moylan, T. Hayashi, and J. I. Brauman, *J. Am. Chem. Soc.*, *109*, 3003 (1987).

275. A. V. Butcher, J. Chatt, G. J. Leigh, and P. L. Richards, *J. Chem. Soc. Dalton Trans.*, 1064 (1972).

276. E. Carmona, L. Sánchez, M. L. Poveda, R. A. Jones, and J. G. Hefner, *Polyhedron*, *2*, 797 (1983).

277. K. Yoon, G. Parkin, D. L. Hughes, and G. J. Leigh, *J. Chem. Soc. Dalton Trans.*, 769 (1992).

278. J. C. Bryan, S. J. Geib, A. L. Rheingold, and J. M. Mayer, *J. Am. Chem. Soc.*, *109*, 2826 (1987).

279. F.-M. Su, J. C. Bryan, S. Jang, and J. M. Mayer, *Polyhedron*, *8*, 1261 (1989).

280. J. C. Bryan and J. M. Mayer, *J. Am. Chem. Soc.*, *112*, 2298 (1990).

281. F.-M. Su, C. Cooper, S. J. Geib, A. L. Rheingold, and J. M. Mayer, *J. Am. Chem. Soc.*, *108*, 3545 (1986).

282. J. C. Bryan and J. M. Mayer, *J. Am. Chem. Soc.*, *109*, 7213 (1987).

283. J. C. Bryan and J. M. Mayer, *J. Am. Chem. Soc.*, *112*, 2298 (1990).

284. For a MO analysis of this cleavage reaction, see (271g).

285. S. G. Feng, L. Luan, P. White, M. S. Brookhart, J. L. Templeton, and C. G. Young, *Inorg. Chem.*, *30*, 2582 (1991).

286. C. G. Young, R. W. Gable, and M. F. Mackay, *Inorg. Chem.*, *29*, 1777 (1990).

287. S. Thomas, E. R. T. Tiekink, and C. G. Young, *Organometallics*, *15*, 2428 (1996).

288. G. Parkin and J. E. Bercaw, *Organometallics*, *8*, 1172 (1989).

289. P. Jernakoff, J. R. Fox, J. C. Hayes, S. Lee, B. M. Foxman, and N. J. Cooper, *Organometallics*, *14*, 4493 (1995).

290. G. Parkin, R. E. Marsh, W. P. Schaefer, and J. E. Bercaw, *Inorg. Chem.*, *27*, 3262 (1988).

291. B. Chiswell, E. D. McKenzie, and L. F. Lindoy, in *Comprehensive Coordination Chemistry*, G. Wilkinson, R. D. Gillard, and J. A. McCleverty, Eds., Pergamon, Oxford, UK, 1987, Chapter 41.

292. T. J. Collins, R. D. Powell, C. Slebodnick, and E. S. Uffelman, *J. Am. Chem. Soc.*, *112*, 899 (1990).

293. T. J. Collins and S. W. Gordon-Wylie, *J. Am. Chem. Soc.*, *111*, 4511 (1989).

294. F. M. MacDonnell, N. L. P. Fackler, C. Stern, and T. V. O'Halloran, *J. Am. Chem. Soc.*, *116*, 7431 (1994).

295. R. S. Czernuszewicz, Y. O. Su, M. K. Stern, K. A. Macor, D. Kim, J. D. Groves, and T. G. Spiro, *J. Am. Chem. Soc.*, *110*, 4158 (1988).

296. (a) O. Bortolini and B. Meunier, *J. Chem. Soc. Chem. Commun.*, 1364 (1983). (b) P. J. Pospisil, D. H. Carsten, and E. N. Jacobsen, *Chem. Eur. J.*, *2*, 974 (1996).

297. M. J. Almond, R. W. Atkins, and R. H. Orrin, *J. Chem. Soc. Dalton Trans.*, 311 (1994).

298. (a) W. A. Herrmann, C. Hecht, M. L. Ziegler, and B. Balbach, *J. Chem. Soc. Chem. Commun.*, 686 (1984). (b) M. Herberhold, D. Reiner, and D. Neugebauer, *Angew. Chem. Int. Ed. Engl.*, *22*, 59 (1983).

299. For a review of structurally characterized technetium complexes, including oxo derivatives, see M. Melník, and J. E. van Lier, *Coord. Chem. Rev.*, *77*, 275(1987).

300. For other comparisons of Tc=O bond lengths, see (a) A. Duatti, A. Marchi, L. Magon, E. Deutsch, V. Bertolasi, and G. Gilli, *Inorg. Chem.*, *26*, 2182 (1987). (b) R. W. Thomas, G. W. Estes, R. C. Elder, and E. Deutsch, *J. Am. Chem. Soc.*, *101*, 4581 (1979).

301. R. W. Thomas, A. Davison, H. S. Trop, and E. Deutsch, *Inorg. Chem.*, *19*, 2840 (1980).

302. J. A. Thomas and A. Davison, *Inorg. Chim. Acta*, *190*, 231 (1991).

303. H. J. Banberry, W. Hussain, I. G. Evans, T. A. Hamor, C. J. Jones, J. A. McCleverty, H.-J. Schulte, B. Engles, and W. Kläui, *Polyhedron*, *9*, 2549 (1990).

304. S. N. Brown and J. M. Mayer, *Inorg. Chem.*, *31*, 4091 (1992).

305. W. A. Herrmann, R. Alberto, P. Kiprof, and F. Baumgärtner, *Angew. Chem. Int. Ed. Engl.*, *29*, 189 (1990).

306. B. Kanellakopulos, B. Nuber, K. Raptis, and M. L. Ziegler, *Angew. Chem. Int. Ed. Engl.*, *28*, 1055 (1989).

307. A. W. E. Chan, R. Hoffmann, and S. Alvarez, *Inorg. Chem.*, *30*, 1086 (1991).

308. See footnote 10 of (305).

309. A. K. Burrell, F. A. Cotton, L. M. Daniels, and V. Petricek, *Inorg. Chem.*, *34*, 4253 (1995).

310. N. Masciocchi, P. Cairati, F. Saiano, and A. Sironi, *Inorg. Chem.*, *35*, 4060 (1996).

311. A. Duatti, F. Tisato, F. Refosco, U. Mazzi, and M. Nicolini, *Inorg. Chem.*, *28*, 4564 (1989).

312. K. A. Conner and R. A. Walton, in *Comprehensive Coordination Chemistry*, G. Wilkinson, R. D. Gillard, and J. A. McCleverty, Eds., Pergamon, Oxford, UK, 1987, Chapter 43.

313. (a) D. M. Hoffman, in *Comprehensive Organometallic Chemistry II*, E. W. Abel, F. G. A. Stone, G. Wilkinson, Eds., Pergamon, New York, 1995, Vol. 6, Chapter 10. (b) C. C. Romao, in *Encyclopedia of Inorganic Chemistry*, R. B. King, Ed., Wiley, New York, 1994, Vol. 6, pp. 3437–3465.

314. I. R. Beattie and P. J. Jones, *Inorg. Chem.*, *18*, 2318 (1979).

315. F. E. Kühn, W. A. Herrmann, R. Hahn, M. Elison, J. Blümel, and E. Herdtweck, *Organometallics*, *13*, 1601 (1994).

316. W. A. Herrmann, F. E. Kühn, and C. C. Romao, *J. Organomet. Chem.*, *489*, C56 (1995).

317. (a) A. H. Klahn-Oliva and D. Sutton, *Organometallics*, *3*, 1313 (1984). (b) W. A. Herrmann, R. Serrano, and H. Bock, *Angew. Chem. Int. Ed. Engl.*, *23*, 383 (1984).

318. For the $(\eta^5\text{-}C_5Me_4Et)Re(O)_3$ derivative, see J. Okuda, E. Herdtweck, and W. A. Herrmann, *Inorg. Chem.*, *27*, 1254 (1988).

319. (a) W. A. Herrmann, *J. Mol. Catal.*, *41*, 109 (1987). (b) W. A. Herrmann, *J. Organomet. Chem.*, *300*, 111 (1986). (c) W. A. Herrmann, *Angew. Chem. Int. Ed. Engl Ed.Engl.*, *27*, 1297 (1988). (d) W. A. Herrmann, *J. Organomet. Chem.*, *382*, 1 (1990).

320. (a) J. Takacs, P. Kiprof, J. Riede, and W. A. Herrmann, *Organometallics*, *9*, 782 (1990). (b) W. A. Herrmann, W. Wagner, U. N. Flessner, U. Volkhardt, and H. Komber, *Angew. Chem. Int. Ed. Engl.*, *30*, 1636 (1991). (c) W. A. Herrmann, R. W. Fischer, and D. W. Marz, *Angew. Chem. Int. Ed. Engl.*, *30*, 1638 (1991). (d) W. A. Herrmann and M. Wang, *Angew. Chem. Int. Ed. Engl.*, *30*, 1641 (1991).

321. W. A. Herrmann, R. W. Fischer, M. U. Rauch, and W. Scherer, *J. Mol. Catal.*, *86*, 243 (1994).

322. (a) W. A. Herrmann and R. W. Fischer, *J. Am. Chem. Soc.*, *117*, 3223 (1995). (b) W. A. Herrmann, W. Scherer, R. W. Fischer, J. Blümel, M. Kleine, W.

Mertin, R. Gruehn, J. Mink, H. Boysen, C. C. Wilson, R. M. Ibberson, L. Bachmann, and M. Mattner, *J. Am. Chem. Soc.*, *117*, 3231 (1995). (c) H. S. Genin, K. A. Lawler, R. Hoffmann, W. A. Herrmann, R. W. Fischer, and W. Scherer, *J. Am. Chem. Soc.*, *117*, 3244 (1995).

323. (a) I. A. Degnan, W. A. Herrmann, and E. Herdtweck, *Chem. Ber.*, *123*, 1347 (1990). (b) I. A. Degnan, J. Behm, M. R. Cook, and W. A. Herrmann, *Inorg. Chem.*, *30*, 2165 (1991).

324. (a) A. Müller, E. Diemann, and V. V. K. Rao, *Chem. Ber.*, *103*, 2961 (1970). (b) J. T. Goodman, S. Inomata, and T. B. Rauchfuss, *J. Am. Chem. Soc.*, *118*, 11674 (1996). (c) H. H. Murray, L. Wei, S. E. Sherman, M. A. Greaney, K. A. Eriksen, B. Carstensen, T. R. Halbert, and E. I. Stiefel, *Inorg. Chem.*, *34*, 841 (1995).

325. P. J. Blower, J. R. Dilworth, J. P. Hutchinson, and J. A. Zubieta, *Inorg. Chim. Acta*, *65*, L225 (1982).

326. The oxo analogue, $[(\eta^2\text{-}SCH_2CH_2S)_2Re(O)]^-$, has also been synthesized by reaction of $Na[Re(O)_4]$ with $HSCH_2CH_2SH$. See A. Davison, C. Orvig, H. S. Trop, M. Sohn, B. dePamphilis, and A. G. Jones, *Inorg. Chem.*, *19*, 1988 (1978).

327. S. K. Tahmassebi and J. M. Mayer, *Organometallics*, *14*, 1039 (1995).

328. F. P. Guengerich and T. L. Macdonald, *Acc. Chem. Res.*, *17*, 9 (1984).

329. S. M. Nelson, in *Comprehensive Coordination Chemistry*, G. Wilkinson, R. D. Gillard, and J. A. McCleverty, Eds., Pergamon, Oxford, UK, 1987, Chapter 44.2.

330. (a) D.-H. Chin, A. L. Balch, and G. N. La Mar, *J. Am. Chem. Soc.*, *102*, 1446 (1980). (b) G. Simonneaux, W. F. Scholz, C. A. Reed, and G. Lang, *Biochim. Biophys. Acta*, *716*, 1 (1982).

331. The X-ray structure of an iron(IV) oxo porphyrin complex has been cited in footnote 28 of M. Schappacher, R. Weiss, R. Montiel-Montoya, and A. Trautwein, *J. Am. Chem. Soc.*, *107*, 3736 (1985). However, the sample was contaminated with an Fe^{III} bromide derivative, so that the derived Fe=O bond length [1.604(19) Å] should be regarded with caution [see (75)].

332. (a) J. E. Penner-Hahn, K. S. Eble, T. J. McMurry, M. Renner, A. L. Balch, J. T. Groves, J. H. Dawson, and K. O. Hodgson, *J. Am. Chem. Soc.*, *108*, 7819 (1986). (b) J. E. Penner-Hahn, T. J. McMurry, M. Renner, L. Latos-Grazynsky, K. S. Eble, I. M. Davis, A. L. Balch, J. T. Groves, J. H. Dawson, and K. O. Hodgson, *J. Biol. Chem.*, *258*, 12761 (1983).

333. M. L. Hoppe, E. O. Schlemper, and R. K. Murmann, *Acta Crystallogr.*, *B38*, 2237 (1982).

334. For a recent theoretical calculation on $[FeO_4]^{2-}$, see R. J. Deeth, *J. Chem. Soc. Faraday Trans.*, *89*, 3745 (1993).

335. (a) K. Czarnecki, S. Nimri, Z. Gross, L. M. Proniewicz, and J. R. Kincaid, *J. Am. Chem. Soc.*, *118*, 2929 (1996), and references cited therein. (b) M. Schappacher, G. Chottard, and R. Weiss, *J. Chem. Soc. Chem. Commun.*, 93 (1986).

336. (a) M. Fanfarillo, H. E. Cribb, A. J. Downs, T. M. Greene, and M. J. Almond, *Inorg. Chem.*, *31*, 2962 (1992). (b) M. Fanfarillo, A. J. Downs, T. M. Greene, and M. J. Almond, *Inorg. Chem.*, *31*, 2973 (1992).

337. M. Schröder and T. A. Stephenson, in *Comprehensive Coordination Chemistry*, G. Wilkinson, R. D. Gillard, and J. A. McCleverty, Eds., Pergamon, Oxford, UK, 1987, Chapter 45.

338. (a) S. Perrier, T. C. Lau, and J. K. Kochi, *Inorg. Chem.*, *29*, 4190 (1990). (b) S. Perrier and J. K. Kochi, *Inorg. Chem.*, *27*, 4165 (1988).

339. W. P. Griffith, in *Comprehensive Coordination Chemistry*, G. Wilkinson, R. D. Gillard, and J. A. McCleverty, Eds., Pergamon, Oxford, UK, 1987, Chapter 46.

340. M. Schröder, *Chem. Rev.*, *80*, 187 (1980).

341. For some recent theoretical calculations concerning the mechanisms of dihydroxylation of olefins, see (a) A. Veldkamp and G. Frenking, *J. Am. Chem. Soc.*, *116*, 4937 (1994). (b) P.-O. Norrby, H. C. Kolb, and K. B. Sharpless, *Organometallics*, *13*, 344 (1994).

342. (a) W. A. Nugent, R. L. Harlow, and R. J. McKinney, *J. Am. Chem. Soc.*, *101*, 7265 (1979). (b) M. H. Schofield, T. P. Kee, J. T. Anhaus, R. R. Schrock, K. H. Johnson, and W. M. Davis, *Inorg. Chem.*, *30*, 3595 (1991). (c) W. P. Griffith, A. C. Skapski, K. A. Woode, and M. J. Wright, *Inorg. Chim. Acta*, *31*, L413 (1978). (d) E. J. Corey, S. Sarshar, M. D. Azimioara, R. C. Newbold, and M. C. Noe, *J. Am. Chem. Soc.*, *118*, 7851 (1996). (e) M. Nakajima, K. Tomioka, Y. Iitaka, and K. Koga, *Tetrahedron*, *49*, 10793 (1993).

343. (a) P. Stavropoulos, P. G. Edwards, T. Behling, G. Wilkinson, M. Motevalli, and M. B. Hursthouse, *J. Chem. Soc. Dalton Trans.*, 169 (1987). (b) W. A. Herrmann, S. J. Eder, and P. Kiprof, *J. Organomet. Chem.*, *412*, 407 (1991). (c) W. A. Herrmann, S. J. Eder, and P. Kiprof, *J. Organomet. Chem.*, *413*, 27 (1991). (d) C. J. Longley, P. D. Savage, G. Wilkinson, B. Hussain, and M. B. Hursthouse, *Polyhedron*, *7*, 1079 (1988). (e) B. S. McGilligan, J. Arnold, G. Wilkinson, B. Hussain-Bates, and M. B. Hursthouse, *J. Chem. Soc. Dalton Trans.*, 2465 (1990). (f) W. A. Herrmann, S. J. Eder, and W. Scherer, *J. Organomet. Chem.*, *454*, 257 (1993). (g) R. W. Marshman, W. S. Bigham, S. R. Wilson, and P. A. Shapley, *Organometallics*, *9*, 1341 (1990). (h) S. J. Eder, W. A. Herrmann, and P. J. Kiprof, *J. Organomet. Chem.*, *428*, 409 (1992). (i) W. A. Herrmann, S. J. Eder, P. Kiprof, and P. Watzlowik, *J. Organomet. Chem.*, *428*, 187 (1992).

344. Q. Feng, H. Krautscheid, T. B. Rauchfuss, A. E. Skaugset, and A. Venturelli, *Organometallics*, *14*, 297 (1995).

345. Likewise, only bridging chalcogenido and polyselenido complexes have so far been obtained for the pentamethylcyclopentadienyl osmium system. See M. Herberhold, G.-X. Jin, L. M. Liable-Sands, and A. L. Rheingold, *J. Organomet. Chem.*, *519*, 223 (1996).

346. M. Henary, W. C. Kaska, and J. I. Zink, *Inorg. Chem.*, *28*, 2995 (1989).

347. A. J. Downs, T. M. Greene, and C. M. Gordon, *Inorg. Chem.*, *34*, 6191 (1995).

348. M. A. Andrews, G. L. Gould, and E. J. Voss, *Inorg. Chem.*, *35*, 5740 (1996).

349. (a) H. C. E. McFarlane and W. McFarlane, in *Multinuclear NMR*, J. Mason, Ed., Plenum, New York, 1987; Chapter 15, pp. 417–435. (b) N. P. Luthra and J. D. Odom, in *The Chemistry of Organic Selenium and Tellurium Compounds*,

S. Patai, Ed., Wiley, New York, 1987; Vol. 1, Chapter 6, pp. 189–241. (c) H. C. E. McFarlane and W. McFarlane, in *NMR of Newly Accessible Nuclei*, P. Laszlo, Ed., Academic, New York, 1983, Vol. 2, Chapter 10, pp. 275–299. (d) D. H. O'Brien, K. J. Irgolic, and C.-K. Huang, in *Proceedings of the Fourth International Conference on the Organic Chemistry of Selenium and Tellurium*, F. J. Berry and W. R. McWhinnie, Eds., The University of Aston, Birmingham, UK, 1983; pp. 468–491. (e) C. Rodger, N. Sheppard, C. McFarlane, and W. McFarlane, in *NMR and the Periodic Table*, R. K. Harris and B. E. Mann, Eds., Academic, New York, 1978, Chapter 12, pp. 383–419.

350. ^{77}Se chemical shifts in the ranges δ 1000–2500 and δ 600–1000 ppm have been reported for terminal and bridging selenido ligands, respectively, in a variety of anionic polyselenide complexes of tungsten. See, for example, (a) R. W. M. Wardle, C. H. Mahler, C.-N. Chau, and J. A. Ibers, *Inorg. Chem.*, *27*, 2790 (1988). (b) R. W. M. Wardle, S. Bhaduri, C.-N. Chau, and J. A. Ibers, *Inorg. Chem.*, *27*, 1747 (1988). (c) Y.-J. Lu, M. A. Ansari, and J. A. Ibers, *Inorg. Chem.*, *28*, 4049 (1989). (d) M. A. Ansari, C.-N. Chau, C. H. Mahler, and J. A. Ibers, *Inorg. Chem.*, *28*, 650 (1989). (e) C. C. Christuk, M. A. Ansari, and J. A. Ibers, *Inorg. Chem.*, *31*, 4365 (1992). (f) M. A. Ansari, C. H. Mahler, and J. A. Ibers, *Inorg. Chem.*, *28*, 2669 (1989). (g) C. C. Christuk and J. A. Ibers, *Inorg. Chem.*, *32*, 5105 (1993).

351. For other studies concerned with correlation of ^{77}Se and ^{125}Te NMR chemical shift data, see (a) H. C. E. McFarlane and W. McFarlane, *J. Chem. Soc. Dalton Trans.*, 2416 (1973). (b) M. Baiwir, G. Llabrès, L. Christiaens, M. Evers, and J.-L. Piette, *Magn. Reson. Chem.*, *25*, 129 (1987). (a) D. H. O'Brien, N. Dereu, C.-K. Huang, and K. J. Irgolic, *Organometallics*, *2*, 305 (1983). (c) M. Baiwir, in *Proceedings of the Fourth International Conference on the Organic Chemistry of Selenium and Tellurium*, F. J. Berry and W. R. McWhinnie, Eds., The University of Aston, Birmingham, UK, 1983, pp. 406–467. (d) M. Baiwir, G. Llabrès, A. Luxen, L. Christiaens, and J.-L. Piette, *Org. Magn. Reson.*, *22*, 312 (1984). (e) M. Björgvinsson and G. J. Schrobilgen, *Inorg. Chem.*, *30*, 2540 (1991). (f) T. Drakenberg, F. Fringuelli, S. Gronowitz, A.-B. Hörnfeldt, I. Johnson, and A. Taticchi, *Chem. Scr.*, *10*, 139 (1976). (g) T. Drakenberg, A.-B. Hörnfeldt, S. Gronowitz, J.-M. Talbot, and J.-L. Piette, *Chem. Scr.*, *13*, 152 (1979). (h) G. A. Kalabin, R. B. Valeev, and D. F. Kushnarev, *J. Org. Chem. USSR*, *17*, 830 (1981) (translation from *Zh. Org. Khim.*, *17*, 947 (1981). (i) E. A. V. Ebsworth, D. W. H. Rankin, and S. Cradock, *Structural Methods in Inorganic Chemistry*, 2nd ed., CRC Press, Boca Raton, FL, 1991.

352. Z. K. Sweeney, J. L. Polse, R. A. Andersen, R. G. Bergman, and M. G. Kubinec, *J. Am. Chem. Soc.*, *119*, 4543 (1997).

353. J. L. Kisko, T. Hascall, and G. Parkin, *J. Am. Chem. Soc.* *119*, 7611 (1997).

354. K. B. P. Ruppa, N. Desmangles, S. Gambarotta, G. Yap, and A. L. Rheingold, *Inorg. Chem.*, *36*, 1194 (1997).

Coordination Chemistry of Azacryptands

JANE NELSON*, VICKIE MCKEE, and GRACE MORGAN

Chemistry Department
Queen's University Belfast BT9 5AG

CONTENTS

*Postal address is Open University, Milton Keynes, MK7 6AA

Progress in Inorganic Chemistry, Vol. 47, Edited by Kenneth D. Karlin.
ISBN 0-471-24039-7 © 1998 John Wiley & Sons, Inc.

I. INTRODUCTION

It is now over 25 years since the strategy of cryptand complexation was developed by Lehn and co-workers (1). The idea behind this strategy was that replacement of the first solvation shell of a cation by the surrogate ''solvation'' shell of a three-dimensional organic ligand, within which the cation is concealed, would materially alter the properties of the cation. From this concept derives the term cryptand (from the Greek word kryptos meaning hidden) (Fig. 1).

This approach has contributed much to coordination chemistry and will continue to do so. The early cryptands, derived from diazapolyxocrown ethers by the addition of a third polyether strand, generated an invaluable range of applications in the areas of cation recognition, transport, and catalysis (2–4). The ligands have shown an unrivaled ability to selectively recognize particular alkali metal cations (3, 5), permitting their transport through membranes owing to the hydrophic nature (6) of the exterior of the cryptand within which the polar guest lies concealed. The polyether cryptands surpass crown ethers in their capacity to form stable and selective inclusion complexes (called cryptates) and have an enhanced capacity to solubilize polar salts in nonpolar solvents leading to applications (7–10) in phase-transfer catalysis and anion activation as well as in cation transport and selectivity. More elaborate versions of these cryptands are currently being developed on the supramolecular scale aimed at specific applications, for example, in nonlinear optics, liquid crystalline materials, and elec-

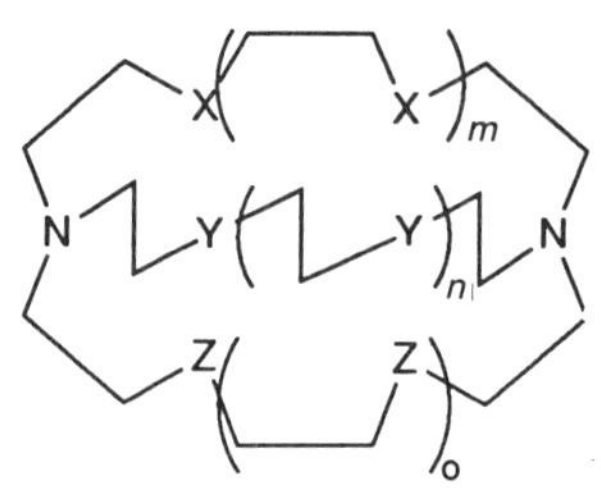

X, Y, Z = Donors

Figure 1. A cryptand (schematic).

tronic components (11). However, our concern in this chapter is mainly with host–guest interactions involving ionic guests.

What is vital to the importance of polyether cryptands is their propensity to coordinate Group I (IA) cations. The value of these O-donor cryptand hosts in such situations is that selectivity for a particular Group 1 (IA) cation, together with strong complexation, can be assured by using cryptands of dimensions appropriate to the needs of the target cation. The development of effective ligands for Group I (IA) cations gave impetus to the study of the coordination chemistry of this previously neglected group of elements, which had until then been regarded from the coordination chemists point of view as "Cinderella elements." Eventually, this work led to the synthesis and recognition of alkalide complexes involving Group I (IA) anions (12).

The polyether cryptands can accommodate other main group cationic guests such as Pb^{2+}, Cd^{2+}, and some lanthanides (13, 14), but in general, complexation of transition ion and heavy main group cations is not most efficiently achieved with neutral ether–O-donor ligands. The N donors, whether sp^2 N or sp^3 N, are much better equipped to coordinate transition ions and many main Group 3–14 (IIIB–IVA) cations (15). Several valuable or potentially toxic target cations such as Cu^+, Ag^+, Tl^+, and Hg^{2+} are expected to be efficiently sequestered by N-donor cryptands (azacryptands), leading to potential applications, for example, in hydrometallurgy or detoxification.

Azacryptands can be readily protonated, which enables them to encapsulate anions, an objective not readily achieved with the analogous O-donor hosts. The capacity of many of these ligands to encapsulate pairs of cations likewise enables the resultant cryptates to act as metalloreceptors, hosting bridging anions in "cascade" fashion (Fig. 2) (16).

With both sets of (oxo- or aza-) cryptands, there arises the possibility of

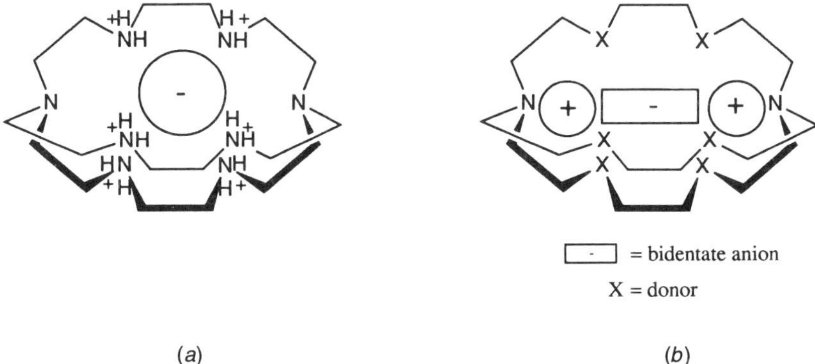

Figure 2. Anion receptors: (*a*) polyammonium type and (*b*) cascade type.

imposing on a cationic guest a coordination geometry other than its preferred (i.e., square based for many transition ions) donor disposition. It is believed that such "entactic state" geometries (17), which lie intermediate between the preferences of a pair of oxidation states, and thus exhibit intermediate energies, favor a redox catalytic role for the encapsulated transition series cation (Fig. 3).

Rapid electron transfer can be assured within a rigid cage where geometry is predetermined by the ligand; the cation is thus prevented from relaxing toward its preferred geometry in either redox state. The combination of kinetic stability toward decomplexation, and coordination geometry invariant upon redox change, makes these cryptates potentially valuable as redox reagents. Many cations with redox behavior appropriate to catalysis of important biochemical oxygen-handling processes are strongly coordinated by azacryptands, which encourages attempts at functional modeling with these biologically credible hosts.

There are further specialist applications better suited to azacryptand than oxocryptand hosts: for example, for photoactive cations, the higher extinction coefficients associated with sp^2 N-donor ligands represent an advantage. This statement is particularly true of lanthanide fluorophores. Also, because of the different synthetic strategies used for azacryptands, these ligands are often more

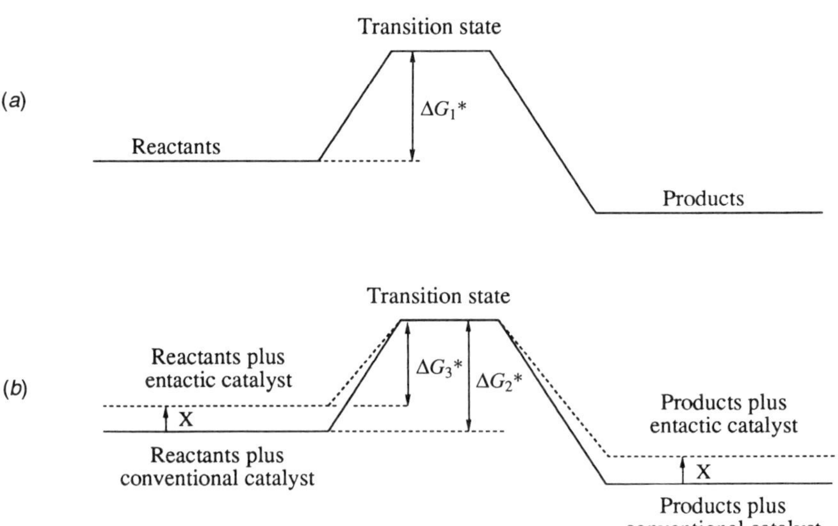

Figure 3. Energetic consequences of entactic state geometry: (*a*) Energy level diagram showing free energy of activation of an uncatalyzed reaction. (*b*) Energy level diagram showing energy paths of a normal catalyst with reduced free energy ΔG_2^* and for an entactic catalyst ΔG_3^*. The energy profile shows that the entactic catalyst site is energized relative to a normal catalyst by the strain energy, X, imposed on the catalytic group.

easily adapted to suit a particular purpose. For example, substitution aimed at enhancing lipophilicity or aqueous solubility, or attachment of particular functional groups, is more easily achieved with azacryptands, particularly where they incorporate an aromatic spacer unit. So they are potentially useful candidates for fluorescence labeling, achievable through attachment to monoclonal antibodies. The kinetic stability of cryptates, whether oxo- or aza-, is of course an essential property in respect of imaging processes, whatever (radioisotopic, fluorescence, or magnetic resonance) visualization technique is used.

II. COVERAGE AND SCOPE

There are many examples of macrobicycles intermediate between all-azacryptands and the original N-bridgehead polyether cryptands, so boundaries are hard to define. Several excellent recent reviews (18) cover the application of cryptand chemistry across the oxo/aza spectrum, and extend the range of cavity-containing systems beyond the macrobicycles and macrotricycles which are the main topic of this chapter. For our purposes, we will classify as azacryptands molecules in which the majority of potential donors are nitrogen atoms. By cryptand, we will understand macrobicyclic hosts that are sufficiently flexible to collapse around the ionic guest "hiding" it from the external environment. (The macrobicyclic criterion implies ring size greater than eight membered.) We will concentrate mainly on cryptands with nitrogen or carbon bridgehead atoms.

Following discussion of the various synthetic approaches that have been used and the short- or long-term target outcomes leading to potential applications of the chemistry, we will describe individual systems in more detail. This discussion will be organized roughly by ligand size, starting with small mononucleating cages and progressing to the larger potentially di- or polynucleating macrobicycles.

III. SYNTHESIS

A. Strapping, Capping, and Coupling

The synthetic strategy adopted in the first successful cryptand syntheses (19) and developed by subsequent groups (20) involved strapping an already-made macrocycle using a diacid chloride (e.g., that of diglycollic acid) under high dilution conditions (21) to reduce competition with intermolecular reaction processes that could lead to undesired, for example, polymerized byproducts. Because the synthesis of diacid chlorides is not easy and generation of both pre-

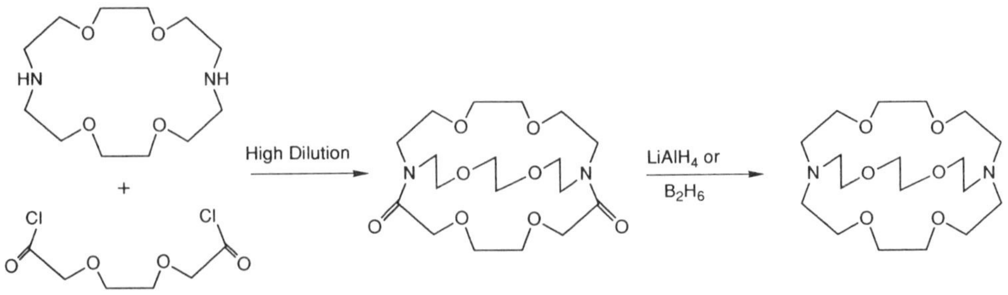

Figure 4. Synthetic scheme for polyether cryptands (Ts = tosyl).

cursor macrocycle and final macrobicycle requires two stages (acid chloride addition to form lactam and its subsequent reduction with hydride), polyether cryptands tend to be more expensive than crown ethers, which has somewhat limited their widespread application. Although other synthetic routes exist for polyamine macrobicycles (azacryptands), these can be made by the analogous stepwise macrocyclic strapping route (22, 23) if the appropriate acid chlorides are obtainable (Figs. 4 and 5).

Other methods of strapping an existing azamacrocycle to generate a cryptand

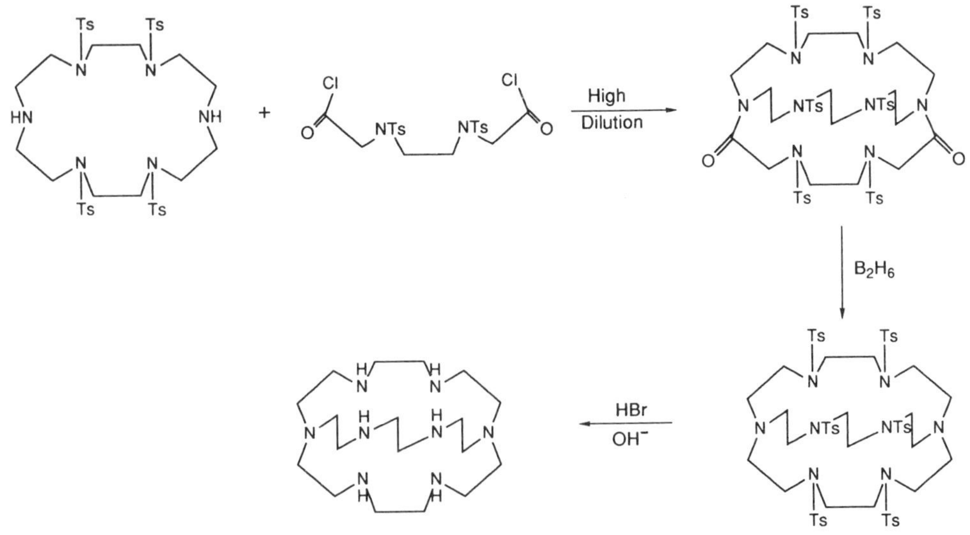

Figure 5. Synthetic scheme for aminocryptands.

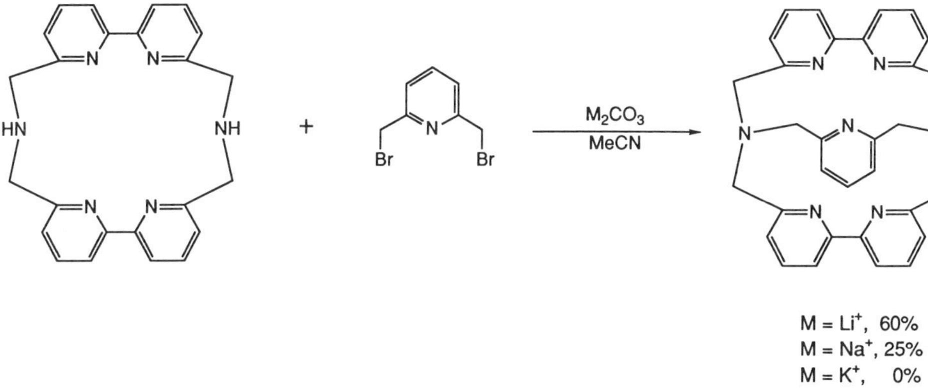

Figure 6. Template strapping of trans-macrocyclic NH functions with dibromoalkane.

have been used; for example, the condensation of a dibromoalkane with a pair of transmacrocyclic NH functions using a Group I (IA) metal carbonate to remove the HBr consequently formed (24). Template effects can influence the outcome of these reactions, favoring incorporation of the cation that best matches the eventual cavity size (Fig. 6).

The direct alkali-cation promoted macrocyclization route to crown ethers involving stoichiometric addition of dihalide to acyclic diamine, which was suggested on the basis of mechanistic studies (25), has proved successful also for cryptands (26) (Fig. 7). Once more it is demonstrated that template effects influence the yield of product. These methods have been successful even for nonidentically stranded cryptands.

Where three identical strands are desired, other direct methods can be applied, for example, the use of NH_3 as capping agent under conditions of high temperature and pressure (26), which obviates the need for high dilution; a form of "tripod–tripod capping". Such routes (Figs. 8 and 9), using dihalogeno compounds in the presence of Na_2CO_3 can generate good yields of symmetric cryptands, although in small quantity, limited by the capacity of the reaction vessel. The high-pressure route can also be used successfully in strapping mode (27).

A barrel-ended trisbpy-derived cryptand has been generated by Vogtle and co-worker (28) using direct alkali cation macrocyclization and 1,3,5-methylamino aromatic caps. Such direct tripod–tripod coupling and capping routes, although statistically disfavored, do succeed in generating good yields of product in some cases, particularly where a substituted benzene ring forms the basis of one of the tripods (29). Intermolecular condensation generally competes with intramolecular condensation, so low cyclization yields are to be expected. Bulky

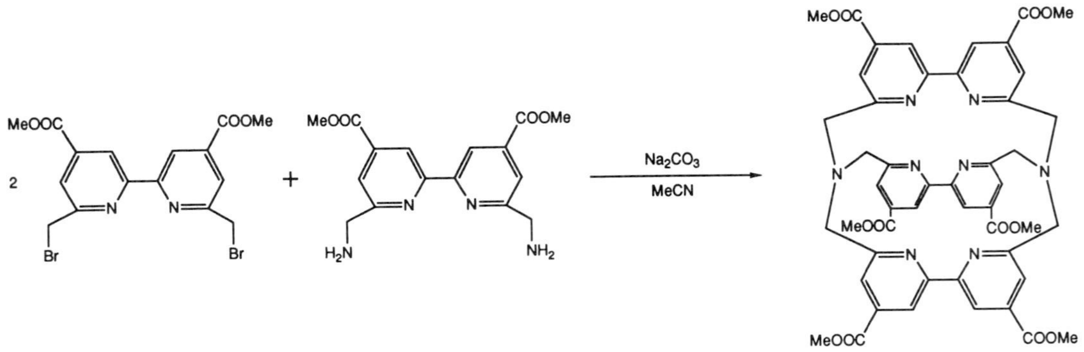

Figure 7. Direct alkali-cation macrobicyclization route.

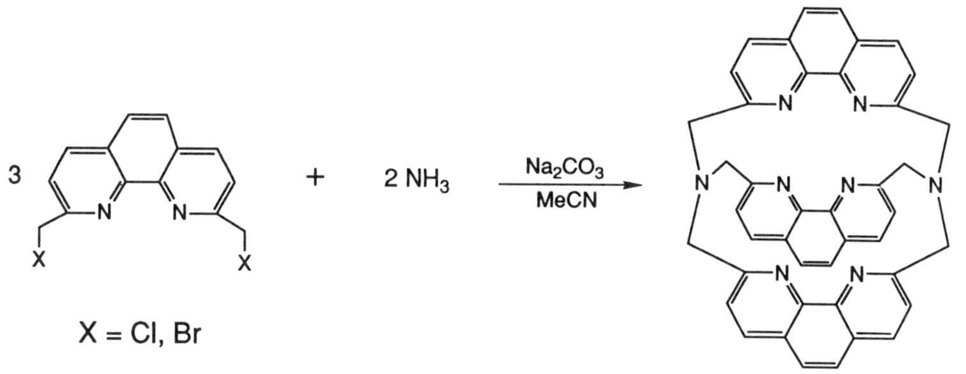

Figure 8. Tripod–tripod capping using NH_3.

tosyl substituents disfavor intermolecular reaction, so the use of a tosylamide salt improves yields in tripod coupling. Because of this, the Lehn group chose condensation of N-tosylated triamide with mesylate as their preferred route to polyazamacrobicycles. Several trispropylene-capped azacryptands (Figs. 10 and 11) have been prepared using the condensation of tosylamide with mesylate in the presence of Cs^+ without recourse to high-dilution conditions (29) [Fig. 10(*a*)]. The tripod coupling route is also appropriate to generation of laterally

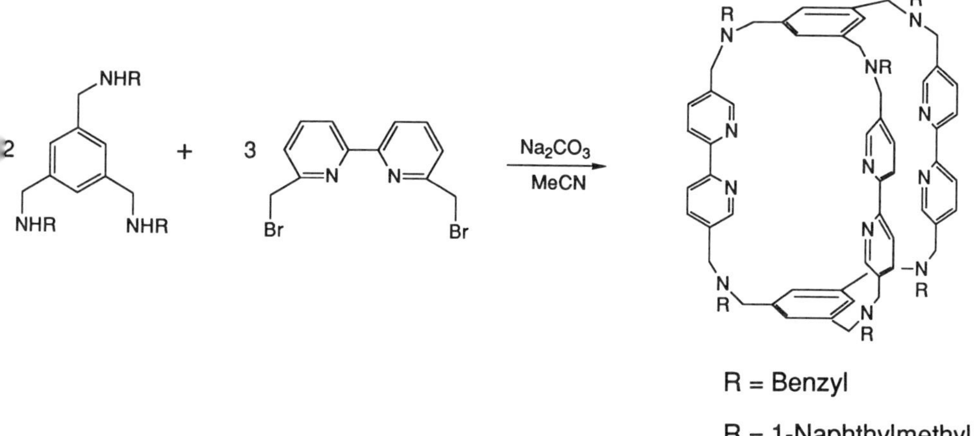

Figure 9. Synthesis of barrel-ended trisbipy cryptand.

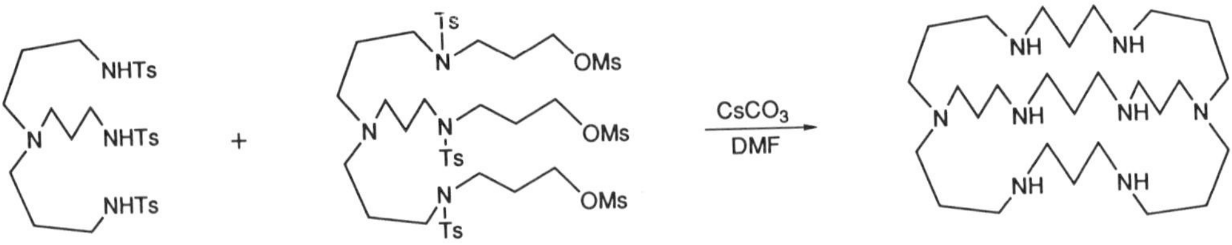

(*a*)

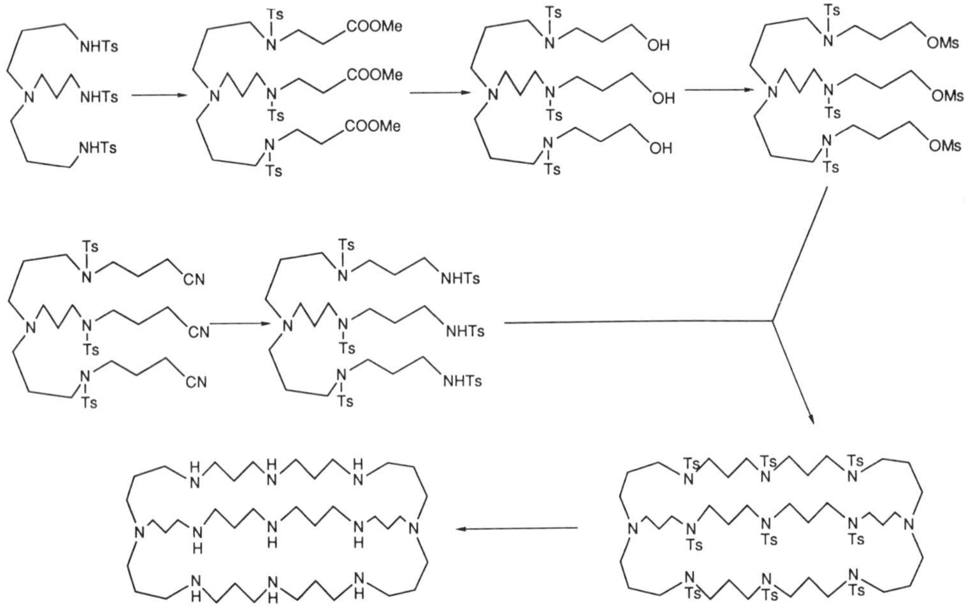

(*b*)

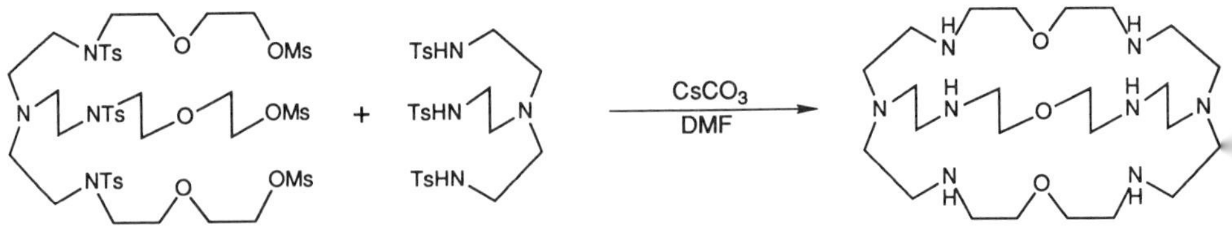

(*c*)

Figure 10. Tripod coupling route to azamacrobicycles: (*a*) C3-bistrpn, (*b*) NH-bistrpn, and (*c*) O-bistren (Ms = mesyl).

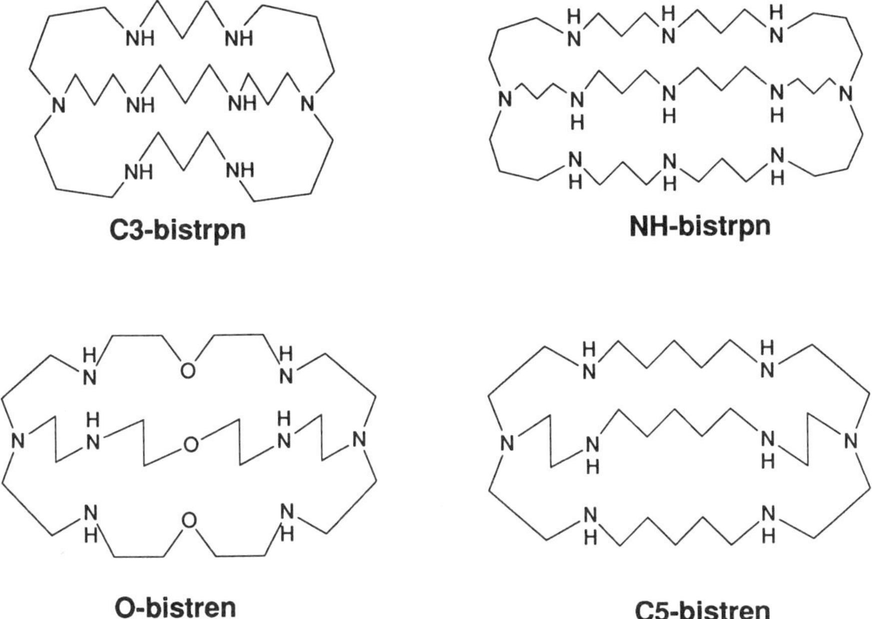

Figure 11. Azacryptands: O-bistren, C3-bistrpn, C5-bistren, and NH-bistrpn.

unsymmetric (i.e., differently capped) cryptands and has been successfully used for this purpose (Fig. 12) (30, 31).

Double tripod capping has been applied by Newkome et al. (32, 33) to synthesize symmetric pyridine-containing oxa-azacryptands (Fig. 13) in low (2–5%) yield.

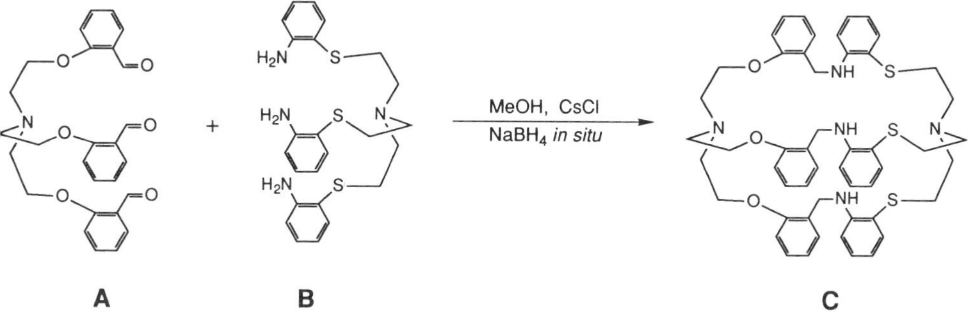

Figure 12. Tripod coupling route to axially unsymmetric azacryptands.

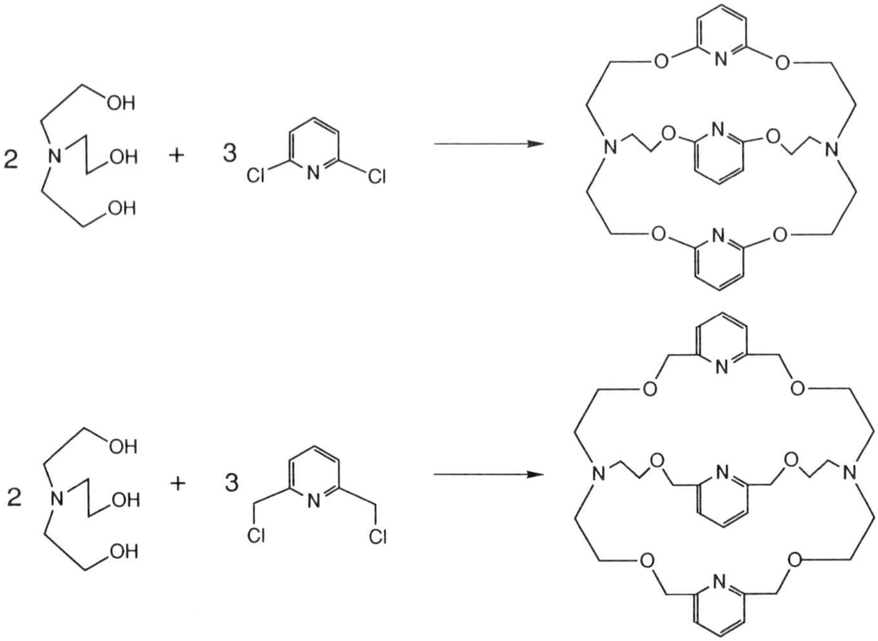

Figure 13. Synthetic route to pyridinophane cryptands.

B. Template Reactions

Highly effective use has been made of the template effect by Sargeson and co-workers (34–38) to generate azacryptates in high yield from simple precursors. The octaazacryptand "sep" (short for "sepulchrate") (Fig. 14) can be obtained in better than 95% yield by condensation of trisethylenediamine cobalt(III) with formaldehyde and ammonia (35, 38); a template capping procedure. Reaction procedure is simple and yields are high; tens of grams of cryptate can easily be prepared in a one-pot reaction in this way. The mechanism, of course, is not so simple; a complex multistage route (Fig. 14) involves attack by NH_3 on the initially formed transient imino C=N, deriving from condensation of HCHO with the amine NH_2 function. Transient intermediates can be withdrawn and characterized in the early stages of the reaction (35, 39), illustrating the mechanism of reaction.

The main drawback to application of this remarkable reaction is the requirement that the chemistry take place on a kinetically inert template cation. Much of the early work was done on a Co^{III} cation, which once incorporated, is difficult to remove from the crypt. Even when reduced to Co^{II}, which is not normally considered kinetically inert, the cation proves hard to remove from the N-bridgehead cages (35).

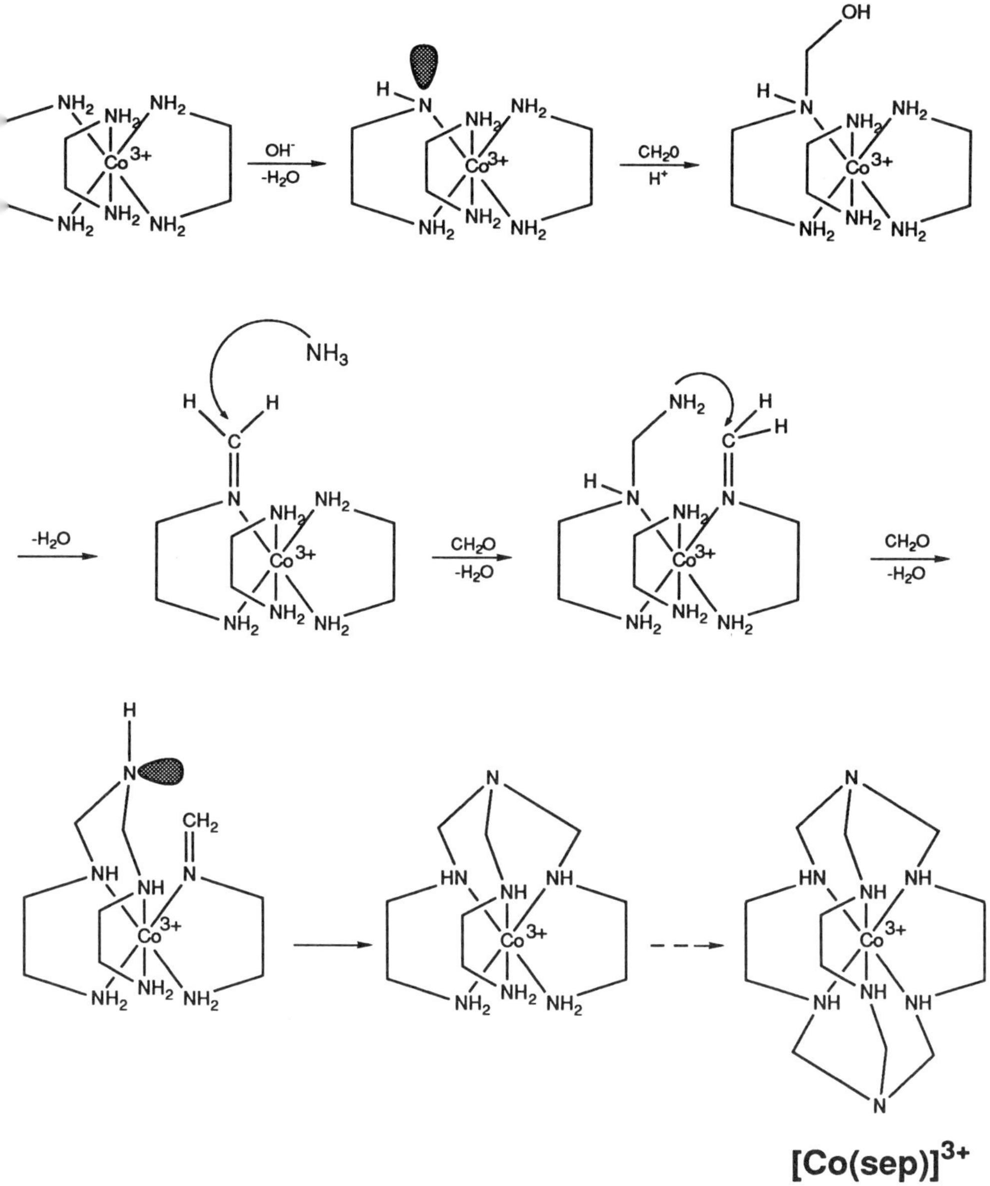

$[Co(sep)]^{3+}$

Figure 14. Template condensation leading to the N-bridgehead cryptand, sep.

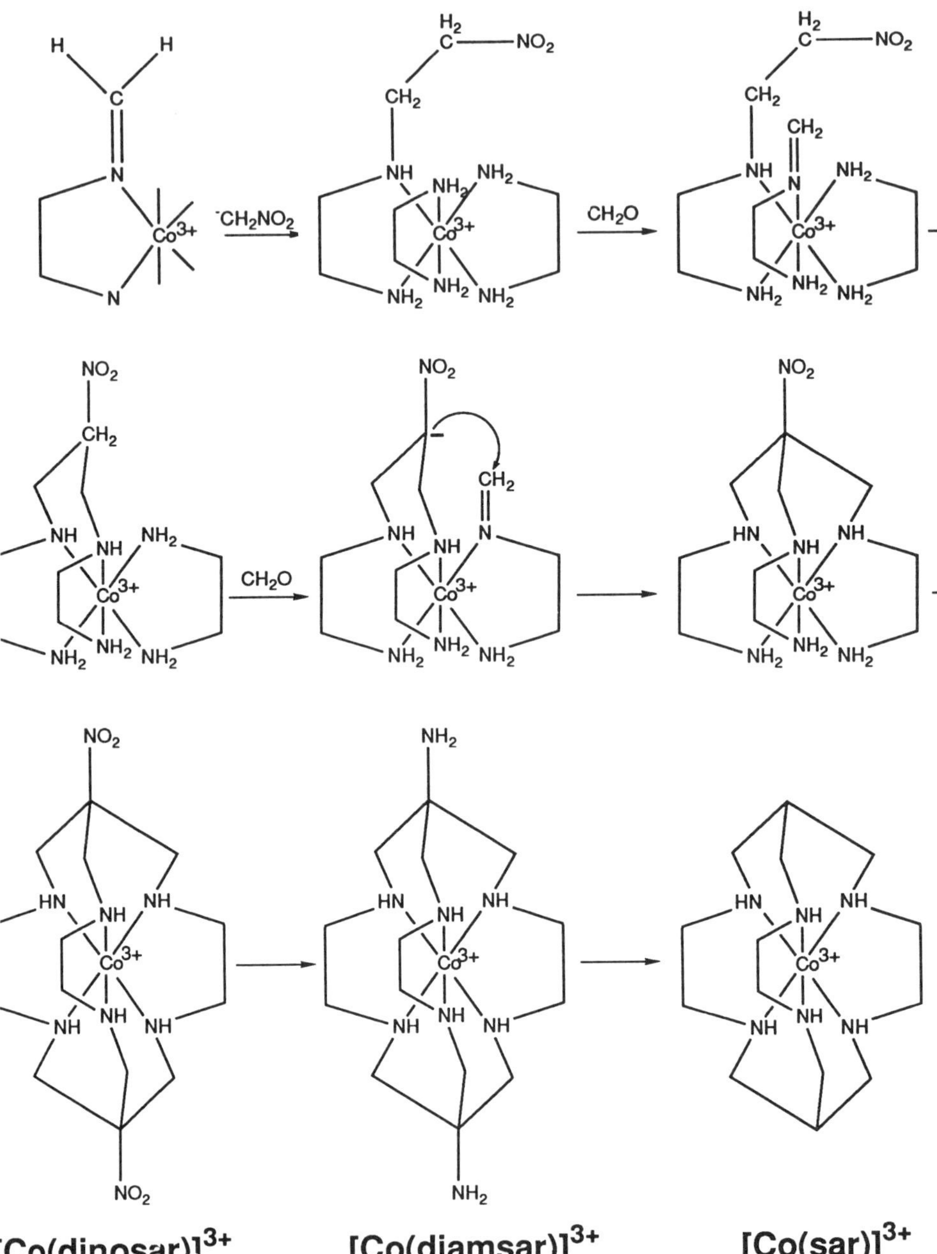

Figure 15. Template condensation leading to C-bridgehead cryptands.

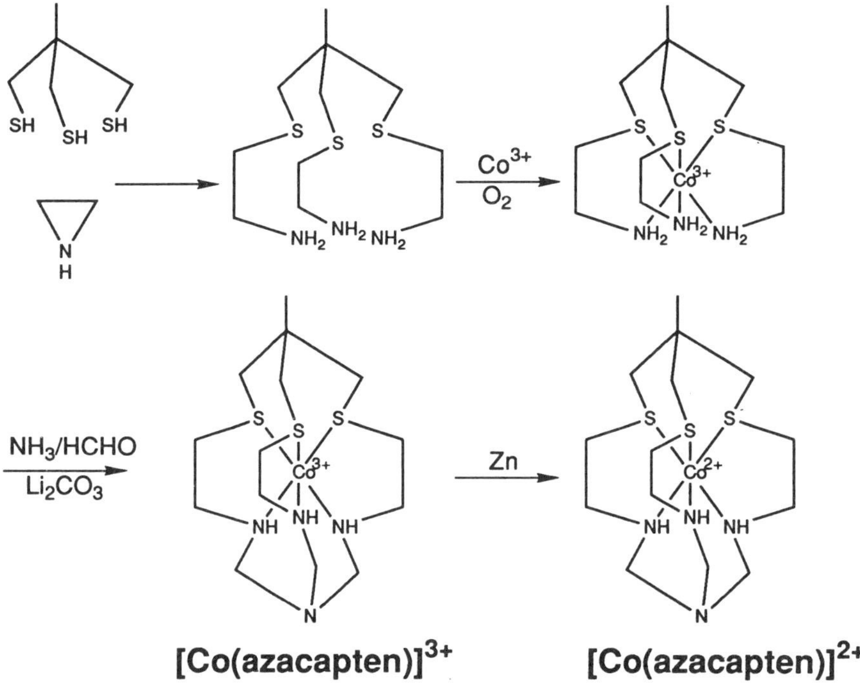

Figure 16. Template condensation starting with podate reagents.

Carbon-bridgehead cage synthesis can be achieved by a similar route (Fig. 15) (39, 40), exploiting the reactivity of the carbon atom in nitromethane. This strategy has been further developed to include tripod capping of a range of coordinated podand ligands on a Co^{III} template, making accessible a number of unsymmetrically capped (39) cages, including those with two different facially coordinating donor sets, for example, one N_3, and a second S_3, set of donors (Fig. 16) (41).

In addition, the strategy has been applied in "reverse" mode using bistriamine capped podates as reagents, linking the podand ends by means of condensation involving HCHO and nitromethane [or other reactive carbon function (42, 43)] to complete the strands that constitute the cage (Fig. 17). Sequential condensation of two different aldehydes has been used in ring closure steps, using the second aldehyde to replace nitromethane, both in tripod capping (42) or reverse mode bis podate condensation. Tri- (44, 45) or tetra- cyclic (46) (some rings so small as to barely to qualify for the prefix macro) azacryptates can be formed in these reactions. Although not yet obtained in the metal-free form that facilitates synthesis of a range of cryptates, it is anticipated that these small macrotricyclic or macrotetracyclic cages (Fig. 18) will induce properties

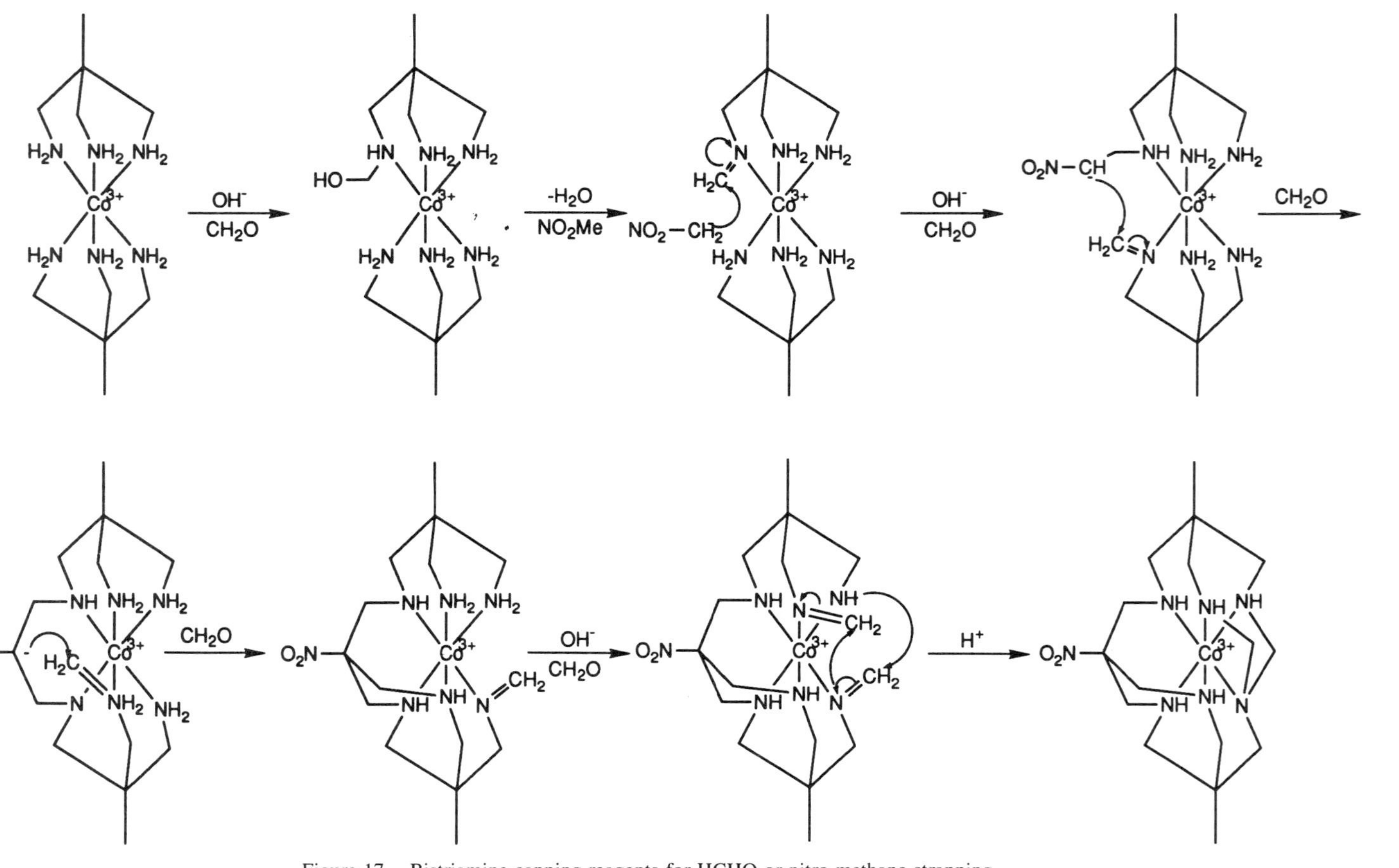

Figure 17. Bistriamine capping reagents for HCHO or nitro methane strapping.

in the guest cation markedly different from those found with sep- (Fig. 14) or sar- (Fig. 15) type cryptates.

Where sequential condensation of paraformaldehyde and propanal with $[Cobis\text{-}tame]^{3+}$ is used (47), the product (Figs. 18 and 19) is the simple "expanded" cage macrobicycle **I** with three 1,3-propanediamine-derived straps; when propanal alone is used, however, the small macrotricycle **II** (Fig. 19) is obtained (45), with remarkable stereospecificity, given the various possibilities presented within the complex reaction mechanism.

Kinetically inert cations such as Pt^{IV} (48), Rh^{III} (49, 50), and Ir^{III} (50) have been used in trisethylenediamine capping reactions to extend the range of cations that can be encrypted using this strategy. However, many cations of interest are kinetically labile and thus unsuitable for use as templates in these reactions, which generally require forcing conditions. In order to generate their cryptates, the metal-free form of the cryptand is required. Cations can be removed from the C-bridgehead cage ligands (51), albeit with some difficulty, via competition with protons, thus allowing complexation of kinetically labile

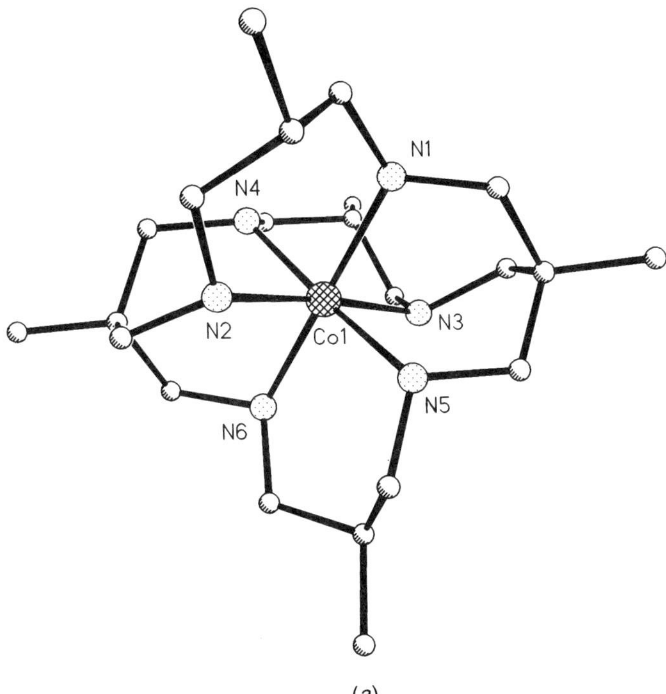

Figure 18. X-ray structure of macrobicyclic and macrotricyclic template condensation products (45, 47): (*a*) $Co[\mathbf{I}]^{3+}$ and (*b*) $Co[\mathbf{II}]^{3+}$.

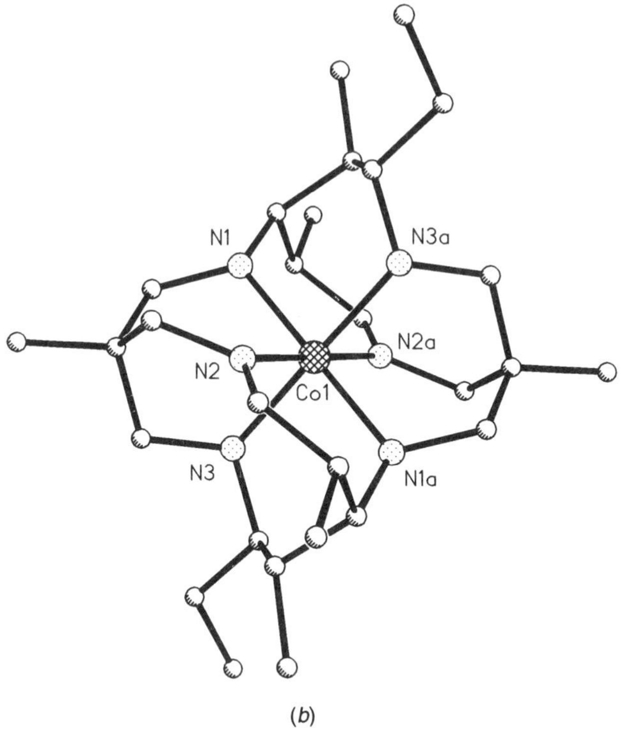

Figure 18. (*Continued.*)

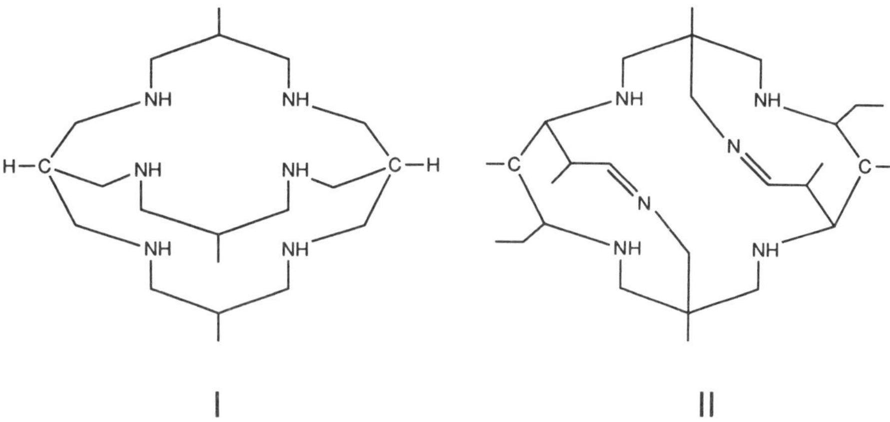

Figure 19. Expanded macrobicycle **I** and macrotricycle **II.**

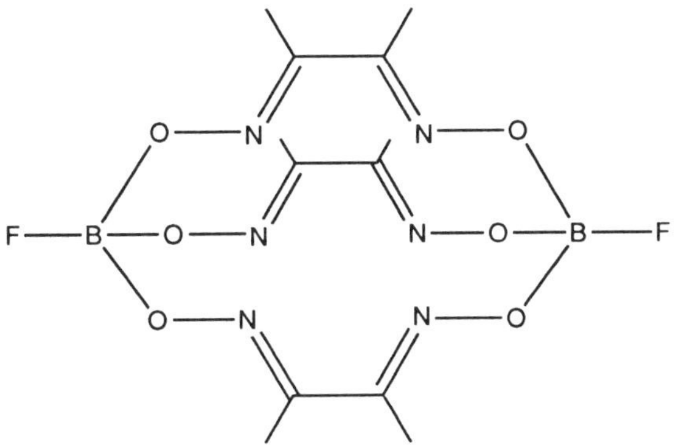

Figure 20. A boron-capped cage.

target cations (52–56) by direct reaction with the protonated form of the ligands. An alternative approach is to remove the encapsulated inert template ion (57) with a strong ligand such as CN^-. However, there is no report of either strategy having been successfully applied on a large scale, so these valuable receptors are not yet available for industrial processes.

In the early years of cryptand chemistry, a series of sp^2 N-donor boron oxime ester cryptates was synthesized on a range of transition cation templates, via hydrogen halide elimination between trisoxime complexes and boron trihalides (58–60). Although the products are superficially similar to the sep and sar cages (Fig. 20), this chemistry has not been exploited to the same extent, presumably because of susceptibility of the ligands to hydrolytic degradation. Voloshin and co-workers (61) have more recently undertaken Fe^{II} template syntheses of such boron-bridgehead systems, concentrating on solid-state characterization techniques such as Mössbauer spectroscopic and structural studies (Fig. 21). Given the striking success of trispyrazolylborate podands (62) in modeling copper and iron biosites, it will presumably not be long before trispyrazolylborate-capped azacryptands are synthesized, to revitalize boron-bridgehead cryptand chemistry.

C. [2 + 3] Schiff-Base Condensation

It is clear from the preceding discussion that easier routes to metal-free cryptands are a desirable objective. A recent and very productive discovery (63) has been the realization that the Schiff-base condensation reaction can be exploited to generate free cryptands from triamines and dicarbonyls in [2 + 3] condensation mode. We have shown that this effective high-yield tripod-coupling–capping route to azacryptands proceeds in high yield under mild conditions, without the need for high dilution techniques. By using a one-pot procedure (63).

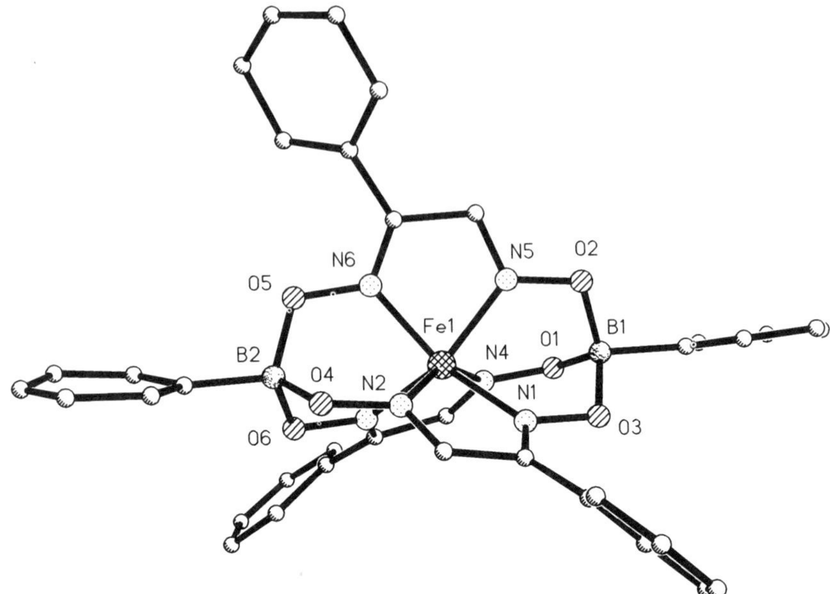

Figure 21. X-ray structure (61) of iron(II) complex of boron-bridgehead cage.

we have obtained high yields of several symmetric hexa-Schiff-base cryptands (Fig. 22), despite the apparent need for six simultaneous condensations. Yields are sometimes improved by the use of a template, although this may well be due to cryptate insolubility assisting isolation of the product; it is not clear that the template ion is a prerequisite for reaction. Many of the metal-free Schiff-base ligands show surprising long-term stability in the solid state when pure.

This efficient tripod-coupling–capping reaction has by now been used for generation of cryptands derived from tris(aminopropyl)amine (trpn) (64, 65) caps as well as tris(2-aminoethyl)amine (tren), and those incorporating spacer units ranging from a two-carbon link in imBT to a 9-carbon link in the diphenylmethane-linked cryptand DPMimBT (66) or as many as 16-carbon in the link in some carocryptands (Fig. 23) (67). In only one case so far has a diketone been successfully substituted for the dialdehyde normally required for efficient [2 + 3] condensation; this cryptand product is formed only under acidic conditions, which suggests that the proton may act as template or as catalyst (68). (Attempts to replace dialdehydes with diketones in the [2 + 3] Schiff-base condensation reaction normally meet with failure.) Schiff-base condensation can be used to couple a pair of large tripodal triamines and trialdehydes on a Cs^+ template, in this way generating unusually large laterally unsymmetric crypt-

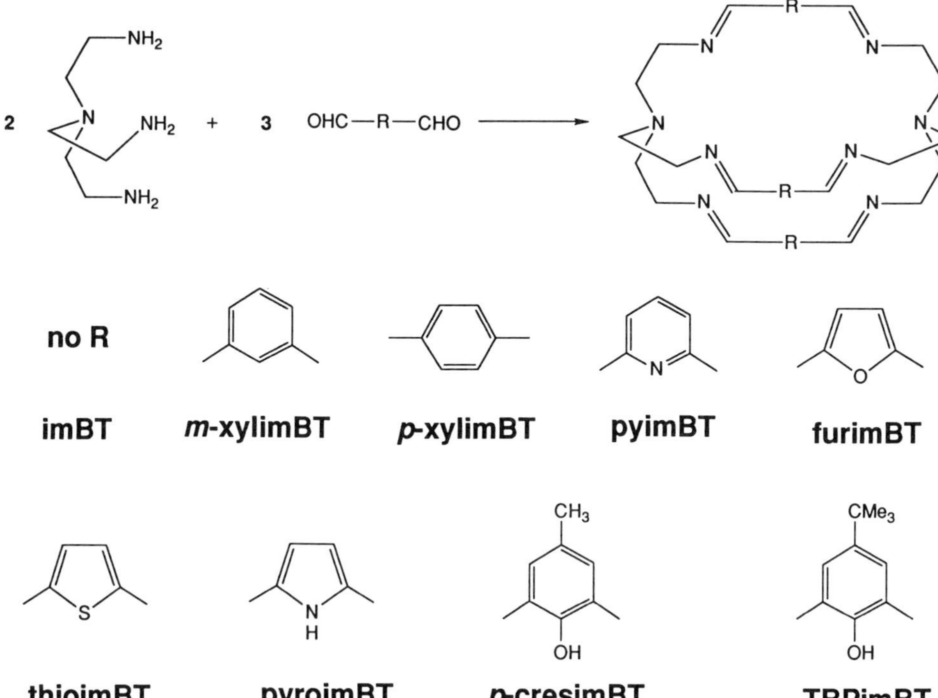

Figure 22. Schiff-base condensation route to azacryptands.

ands (30, 31). The tripodal amines–aldehydes utilized for these coupling reactions have also been successfully used in Cs^+-templated tripod capping syntheses to generate laterally symmetric cryptands (69, 70).

Once the three-dimensional skeleton is assembled, the Schiff-base cryptands can be readily reduced with $NaBH_4$ or $LiAlH_4$ to generate the analogous octaaminocryptands (69–72) which, being more chemically robust than their hexa-Schiff-base precursors, have many advantages as ligands for both cation and anion encapsulation. The large cryptands such as **D** and **E** (Fig. 24) generated in their imino-form by [1 + 1] condensation of trialdehyde with triamine (69, 70) are normally reduced in situ to the polyamine form.

D. Amidocryptands and Small Azacages

A similar [2 + 3] condensation strategy has been used by the Raymond group (73, 74) to generate amide cryptands (Fig. 25), both in metal-free mode

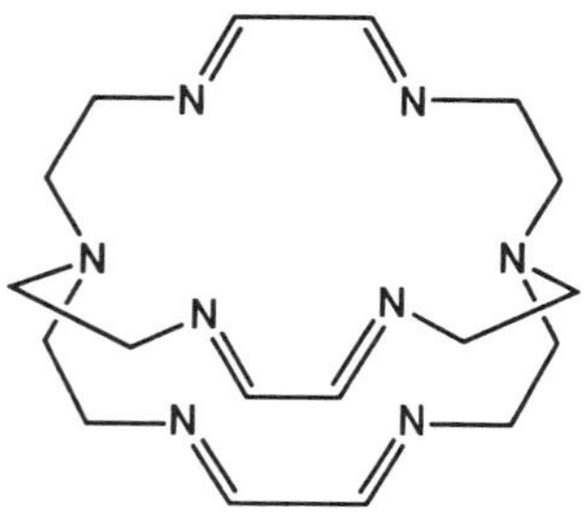

imBT

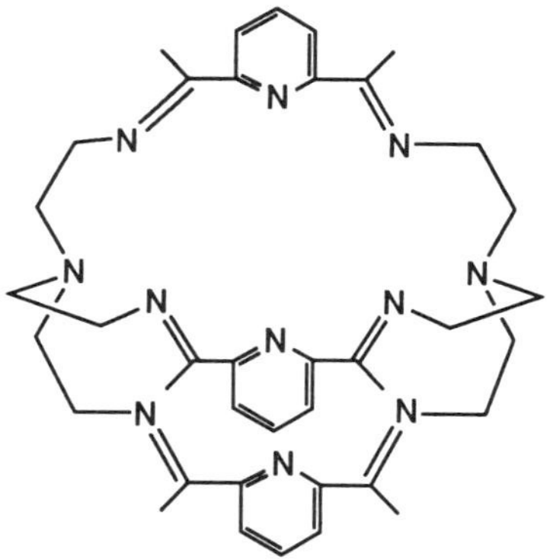

MepyimBT

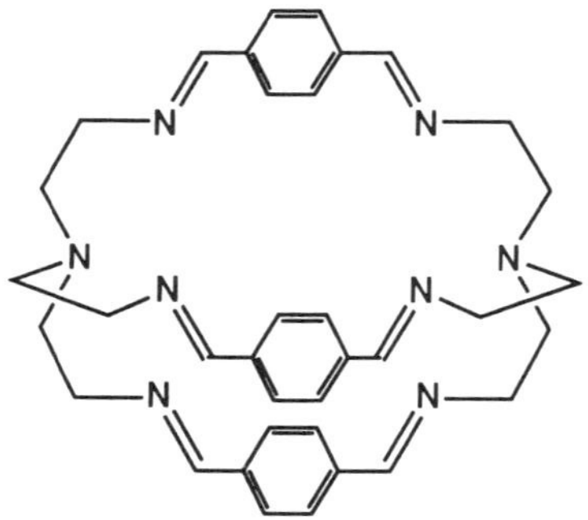

***p*-xylimBT**

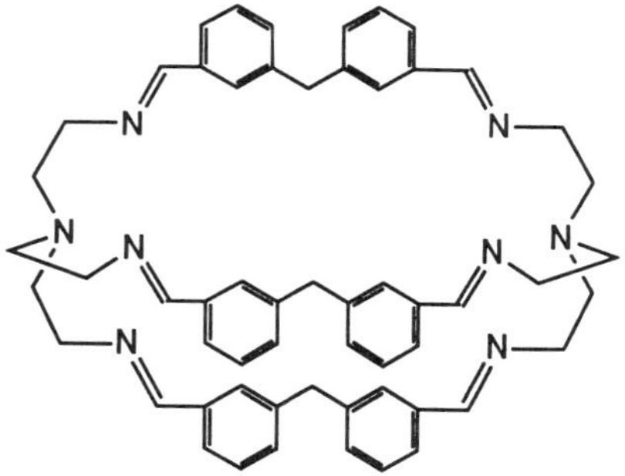

DPMimBT

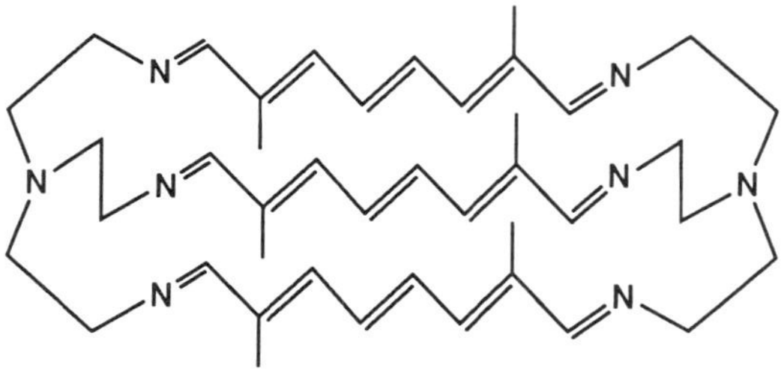

a carocryptand

Figure 23. Some Schiff-base cryptands.

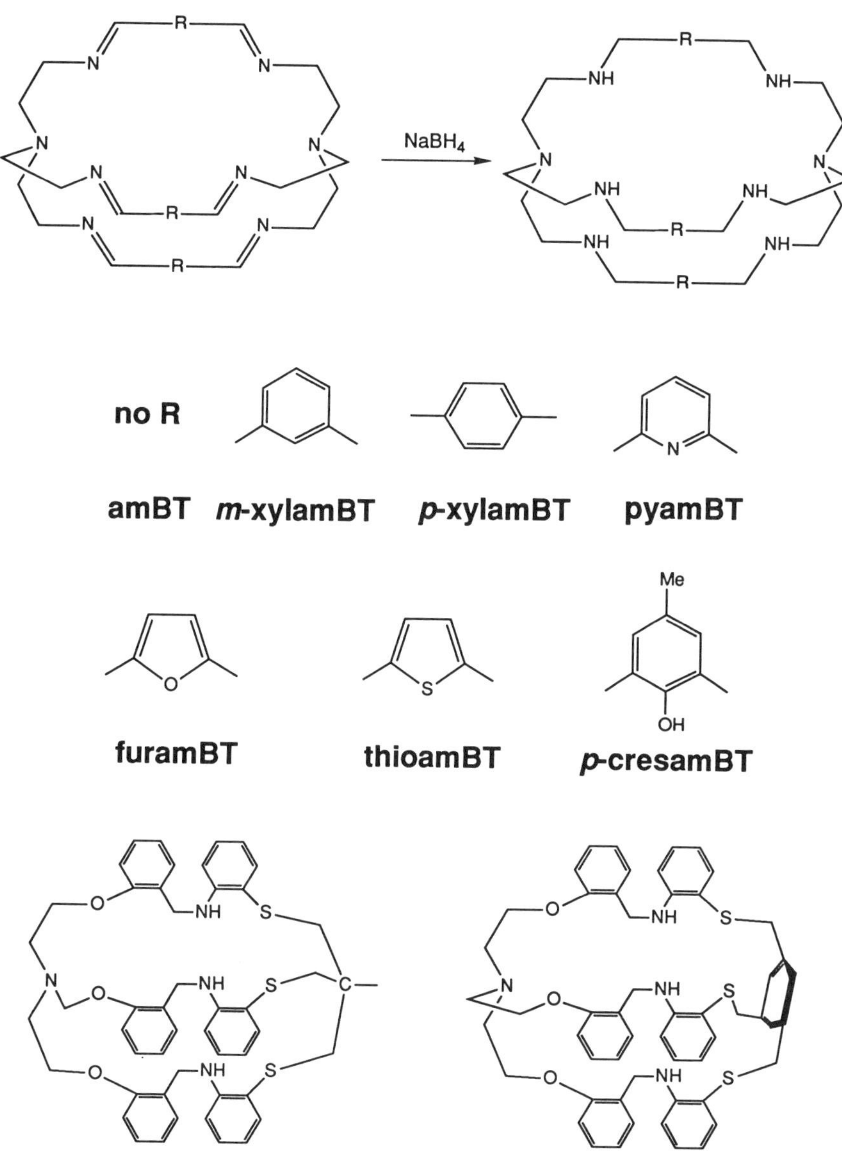

Figure 24. Aminocryptand ligands derived from hexa-Schiff-bases by borohydride reduction.

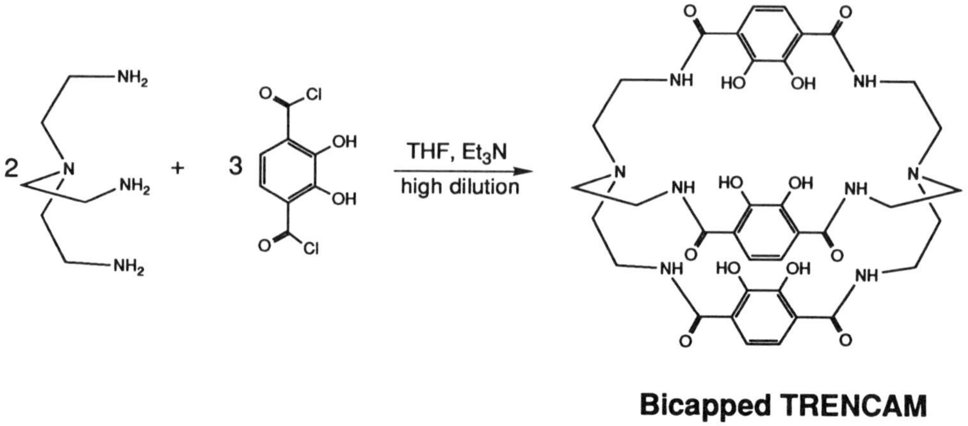

Bicapped TRENCAM

Figure 25. Amidocryptand synthesis: bicapped TRENCAM.

and via template synthesis. In these cryptates, the role of the amide is not coordinating but structural, providing the link between precursor catecholate chelators and triamine caps.

To efficiently complex lithium (75, 76) (i.e., to compete with large hydration or other solvation energies) closely fitting macrobicyclic ligands, with a high degree of preorganization are needed. The macrocycles upon which the relevant azacages are based are 12-membered N_4 macrocycles, of which an opposite (trans-disposed) pair of amine functions is then linked (77) via methylene chains constituting a pair of three-carbon or two-carbon units, often supporting an apical coordinating O or NR donor (Fig. 26). Many of these receptors are also efficient ligands for protons (78, 79) functioning as proton sponges (80).

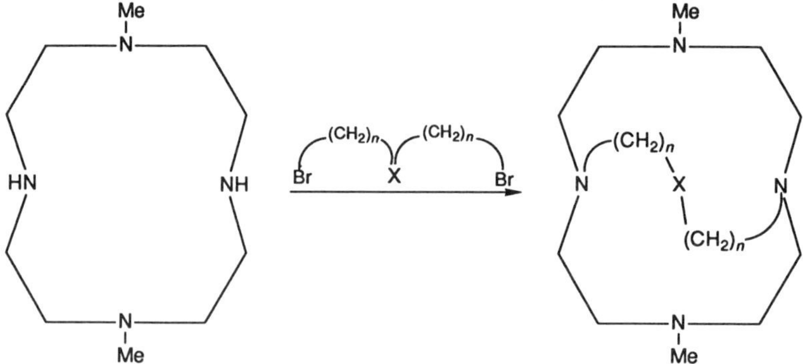

Figure 26. Small azacages synthesized by strapping a trans N_4 macrocycle.

The cages are synthesized by nontemplate methods; starting from *trans*-dimethyl 12-ane N_4, and strapping via the trans-NH functions in the usual way via a $Y{-}(CH_2)_n{-}X{-}(CH_2)_n{-}Y$ unit where Y = OTs, OMs, COCl and X (if present) = O, NH, or NMe. Where diacylchlorides are used diborane (B_2H_6) is the reducing agent; the ligands are often conveniently isolated in protonated form (78, 79, 81).

IV. TARGET APPLICATIONS

Some potentially useful applications of azacryptands arise directly from their coordination properties. The relatively high thermodynamic and kinetic stabilities deriving from the "cryptate effect" (5) (where stability constants, K, can be as much as 10^5 times as large as in analogous macrocyclic complexes) can lead to applications in detoxification or environmental remediation via sequestration of harmful cations; in medicine via transport of cations for imaging purposes; or again, in hydrometallurgy, for extraction of precious metals from mine waste water or other low-concentration aqueous sources. The cryptate effect makes azacryptands potentially valuable candidates for incorporation in ion-selective electrode membranes responsive to transition or heavy metal cations, in the same way that polyether cryptands are used for Gp 1 cations (82). As azacryptand chromoionophores are developed, sensor technology that relies on spectroscopic response will benefit from combination of the cryptate effect and the electronic spectral properties of nitrogen-donor cryptands. Thus the use of azacryptand chromophores (83) for fluoroimmunoassay or possible in vivo fluorescent marking relies on the protective function (3, 84, 85) of the cryptand host to reduce contact with solvent or other groups likely to cause deactivation of the photoactive cation by energy transfer. In this way, the long-lived excited states that generate intense luminescence can be assured.

So the particular properties that derive from the encrypted environment account for the choice of macrobicyclic ligands for many existing applications and for other potentially useful ones. For example, when a counterion is denied (through encryptation) interaction with its congugate ion or solvent, it can be expected to become more reactive. Strong anion activation occurs, permitting reactions more typical of the gas phase than solution, due to a tiny concentration of "naked anions." Applications (7, 8, 86–90), so far restricted to polyether cryptands, involving phase-transfer catalysis, solubilization of inorganic anions, or separation of anionic and cationic charge, rely on the hydrophobic nature of the cryptand exterior to carry the polar reagent into the nonpolar phase where reaction is possible under anhydrous conditions, and under a mild temperature regime that reduces competing side reactions and conserves energy. When these applications are extended to azacryptand carriers, the new environment should ensure a different and potentially useful set of transport and reactivity patterns in the numerous organic reactions which involve transformation of anions.

The transition ion cryptates of ligands such as sar and sep, described earlier, are suitable for use as redox reagents (91, 92). Here the fast and reversible nature of the electron transfer ensured by the rigid ligand system, which restricts alteration of geometry on change of oxidation state, is relevant to their proposed function in photoelectrochemical energy storage devices (92–94).

Where transport across membranes is required, it is not desirable to have very strong binding in the membrane-soluble cryptate (95, 96) as this would preclude availability of the free cation for release at the remote interface following diffusion across the membrane. The relatively low stability constants of some, for example, Group 1 (IA), azacryptates may not be a serious disadvantage here as transport is most efficient where stability constants are of intermediate strength (95, 97) (log $K \approx 5$). A related possibility, for less strongly binding and more labile systems is the use of cryptand host as carrier, for example in transport of the Li^+ used in medicine (98).

On protonation, the aminocryptand hosts are well suited to encapsulation of anionic guests. Anion complexation, although relatively neglected in the past, is now a rapidly developing research area (99–101). It is currently stimulated by concern about the possible adverse environmental effects of accumulation in surface waters of common soluble oxoanions such as NO_3^- and PO_4^{3-}. There are also local concentrations associated with industry or mining, of anions of admitted toxicity, such as oxoanions of chromium, selenium, and arsenic, which need to be removed from aquifers. The aim in complexing these anions is not activation but sequestration to assist in construction of monitoring, purification, or detoxification devices.

On the other hand, interest in modeling biological anion receptors is driven by the objective of mimicking their function, and thus achieving low-energy metal-assisted anion reactivity as, for example, in phosphatases or hydrolases. For such applications, more sophisticated preorganization is necessary to assure, first, complexation of the anionic guest to form an assembly, which then in turn acts as host for biological substrates (96, 102, 103). This situation is probably more readily achieved with macrocycles than macrobicycles; in one study of adenosine triphosphate (ATP) hydrolysis, no inherent catalytic advantage deriving from the use of macrobicyclic rather than macrocyclic host was seen (104).

Preorganization via self-assembly is the goal of those working with amphiphilic cryptands (105, 106) (i.e., macrobicycles with hydrophilic cryptate heads and hydrophobic tails). Most cationic surfactants have a charge of +1 on the head group; the charged cryptate heads repel each other strongly, while the hydrophobic tails aggregate to form micelles (Fig. 27). So far there has been

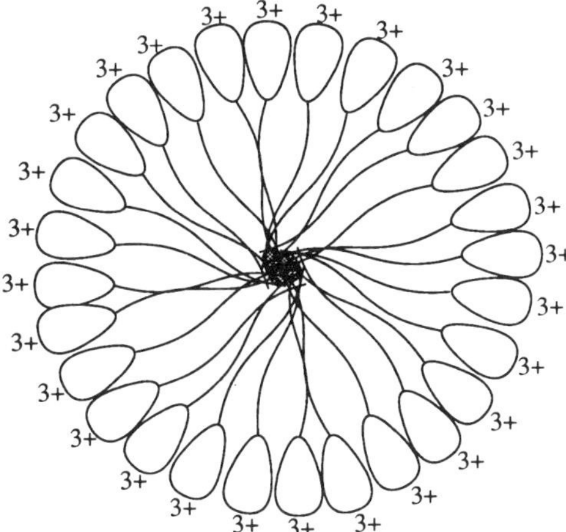

Figure 27. Representation of a micelle.

little exploration of the potentially very powerful group of detergents with head group charge as high as +3. Azacryptands are potentially valuable precursors for this new class of surfactant as the N donors are likely to accommodate and retain cations with such high charge levels (106).

V. INDIVIDUAL CRYPTAND SYSTEMS

A. Small Azacages for Li^+ and H^+

The impetus to design cages at the smaller cavity limit derives from the need to encapsulate and monitor biologically significant small cations such as Li^+ and H^+. Small metal-free azacages, like azacrowns, are water-soluble bases, which makes them suitable receptors for monitoring the concentration of these cations in biological fluids. The suspect toxicity of many existing crown ether or oxocryptand receptors means that the development of alternative receptors is a matter of lively interest. The series of small azacages developed for use as Li^+ receptors (77) is illustrated in Fig. 28.

Some of these cages function as "fast proton sponges", that is, very strong bases (stronger than OH^- in aqueous solution), which cannot be deprotonated in water (78–80). In contrast to the similarly sized polyether cryptand 1,1,1 (107), which although strongly basic, exhibits slow protonation, their proton-exchange kinetics are fast, which together with high basicity, represents an unusual and potentially valuable combination of properties. The molecular con-

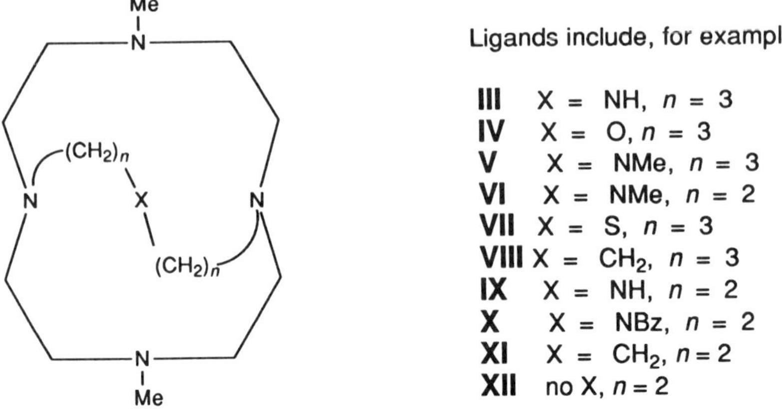

Figure 28. Small azacages **III–XII.**

formation that generates this situation by enforcing a network of relatively short hydrogen-bond interactions to stabilize the NH^+ cation, results from a delicate balance between electrostatic attractions and repulsions (108a). Minor alteration of the bridging link between trans nitrogens markedly affects the basicity of the cryptand; for example, the ligand **XII** with a C_4 all-methylene link is a proton sponge, but that with the analogous C_5 link **XI** has at least three orders of magnitude lower basicity (see Section F, Table VII). The cryptand **VIII** with a C_7 all-methylene link has only moderately high basicity, while the analogous cryptand **III** or **IV** with NH or O in place of this central methylene is a proton sponge; methylating the central NH reduces the basicity to measurable levels. This sensitivity of basicity to molecular conformation is the consequence of disruption of a hydrogen-bond network that allows the most stable NH^+ entities to form as many as three short $NH^+ \cdots N$ contacts (108b). We are thus ensured high stability in the monoprotonated state together with the possibility of fast exchange, which arises because no one hydrogen bond is particularly strong, and because the NH^+ site, lying near the surface of the cryptand, is accessible to the solution environment (Fig. 29). The basicity of the diprotonated cryptand is more normal because of the absence of any unusual hydrogen-bond stabilization; the addition of the second proton destroys the H-bond framework of the monoprotonated species. The third protonation constant is often too low to be measurable. This low basicity is considered to be in consequence of strong electrostatic repulsion between the three positive charges confined within the small, rigid cavity.

Recently, a series of propylene-strapped macrobicycles and macrotricycles have been made using the ditosylate of propane 1,3 diol to strap cyclen (109). The macrobicycles act as proton sponges while the tight macrotricyclic cages

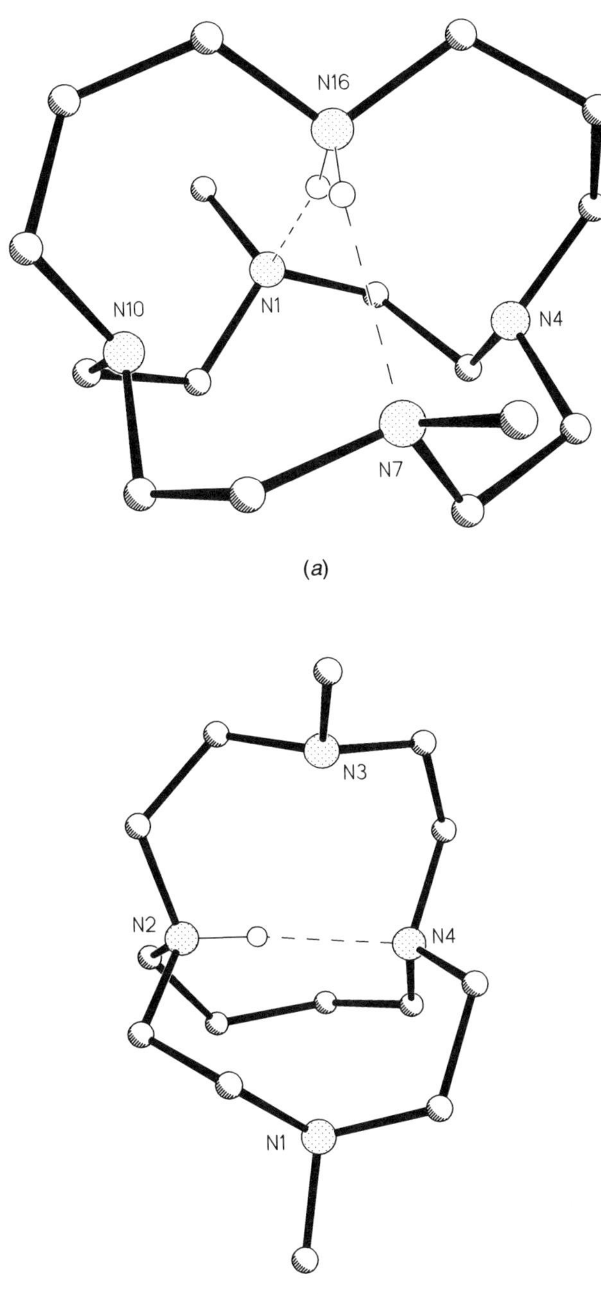

Figure 29. X-ray structures of proton sponges (78, 109): (*a*) [**III**H]$^+$ and (*b*) [**XII**H]$^+$.

christened "adamanzanes" suffer extreme kinetic inertness, which makes their basicity impossible to measure.

Where basicity low enough to be measurable occurs within these small receptors, it is possible to complex Li^+ ions at a reasonably accessible pH because of reduced competition with protonation, compared with the highly basic "proton sponge" cryptands. When the series of receptors **III–X** based on 12-ring N_4 macrocycles with a $(CH_2)_n{-}X{-}(CH_2)_n$ ($n = 2, 3$) strap added between trans nitrogen functions was tested (77, 110–112) for Li^+/Na^+ complexation, none were shown to encapsulate cations with radius above 0.85 Å, ensuring the desired selectivity for Li^+ against Na^+. The axial NR donor is necessary for Li^+ encapsulation, in comparison to H^+ encapsulation (77, 110) which does not depend specifically on the nature of the axial ligand. The structural preorganization of the five-coordinate site is crucial to effective complexation of Li^+ by azacryptands (Fig. 30) as this process appears entropically driven (110) (see Section V.F, Table VI), in contrast to the situation with the small polyether cryptand 2,1,1, where sizeable enthalpic stabilization, as demonstrated by a complexation enthalpy of -21.8 kJ mol^{-1} in aqueous solution, exists (113). Study of the thermodynamics of Li^+ complexation by these small azacages shows that the entropic driving force for complexation in aqueous solution arises from replacement of the solvation shell of Li^+ (aq) by the azacryptand ligand

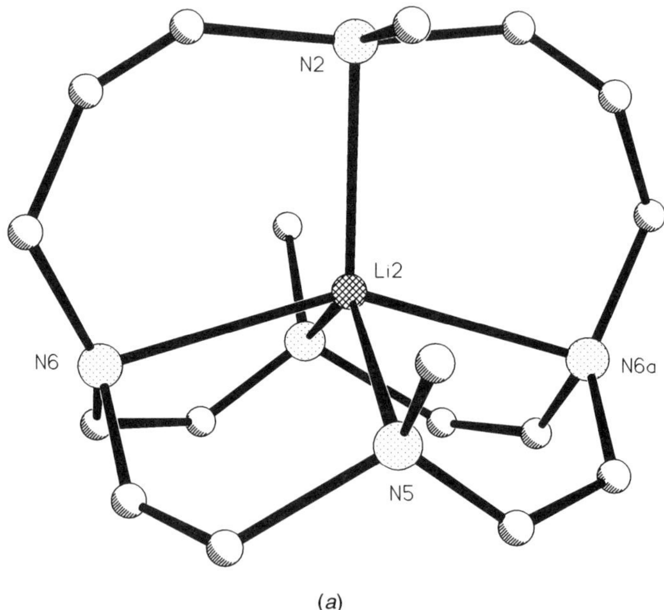

Figure 30. Lithium cryptates of small azacages (76, 112, 114): Structures of (*a*) Li[**V**]$^+$, (*b*) Li[**VI**]$^+$, and (*c*) Li[**IX**]$^+$.

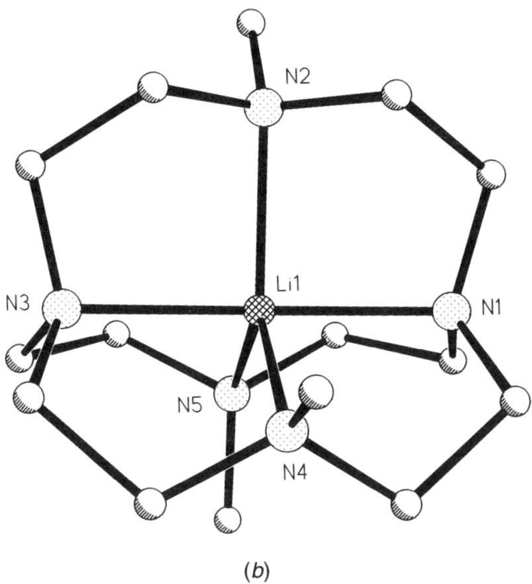

(*b*)

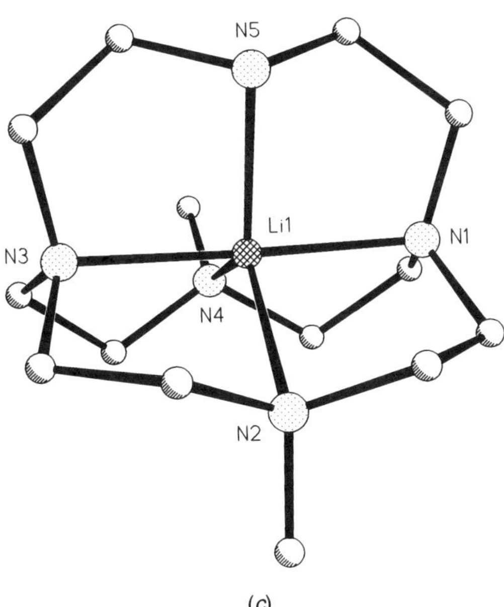

(*c*)

Figure 30. (*Continued.*)

(110). The approximate thermoneutrality of the complexation process arises from the smaller bond energy of the $Li-N_{amine}$ versus $Li-O_{water}$ coordinate bond, so no enthalpic advantage derives from exchange of solvate water for the azaligand. From this series, the tris N—Me analogue of 111 (**VI**) fits Li^+ best (114) and generates the largest stability constant so far reported for lithium complexes of azamacrocycles or azamacrobicycles in aqueous solution (log K = 5.5; see Section F, (Table VI)).

The ^7Li NMR of all the Li cryptates in the series show (77, 110–112, 114) two noncoalescing peaks due to aquated and cryptated Li^+, testifying to an exchange slow on the NMR time scale. The ^7Li and ^{23}Na NMR spectra also helped to characterize the first azacryptand alkalide (115), the sodide of the above-mentioned Li-selective cryptate $[Li\mathbf{VI}]^+$. Good thermal stability is indicated by the absence of decomplexation or irreversible decomposition below 65°C. If the syntheses and thermal stability can be extended to the highly reducing azacage electrides, the potential redox applications could be important.

Because the ionic radii of the later first transition series cations fulfill the cavity size criterion, some, for example, $(CH_2)_3-S-(CH_2)_3$ or $(CH_2)_3-NH-(CH_2)_3$ strapped cages **VII** and **III** are able to complex transition ions (116) (Fig. 31) such as Co^{2+}, Ni^{2+}, Cu^{2+}, or Zn^{2+}. In some cases, this

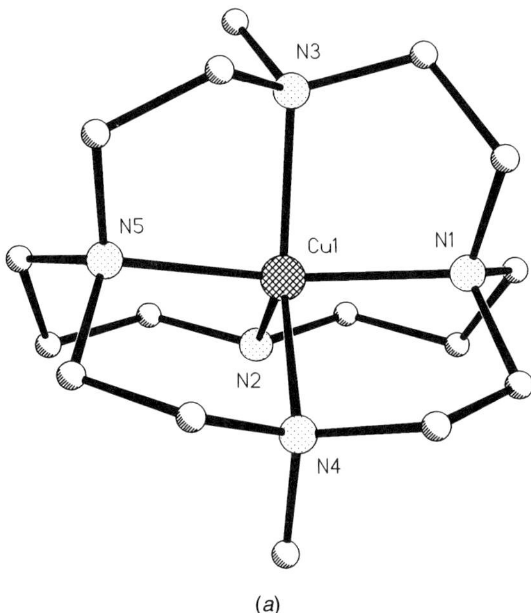

Figure 31. X-ray structures of copper and zinc cryptates of small azacages (77, 117, 119): (*a*) $Cu[\mathbf{III}]^{2+}$, (*b*) $Cu[\mathbf{IX}]^{2+}$, and (*c*) $Zn[\mathbf{XIII}]^{2+}$.

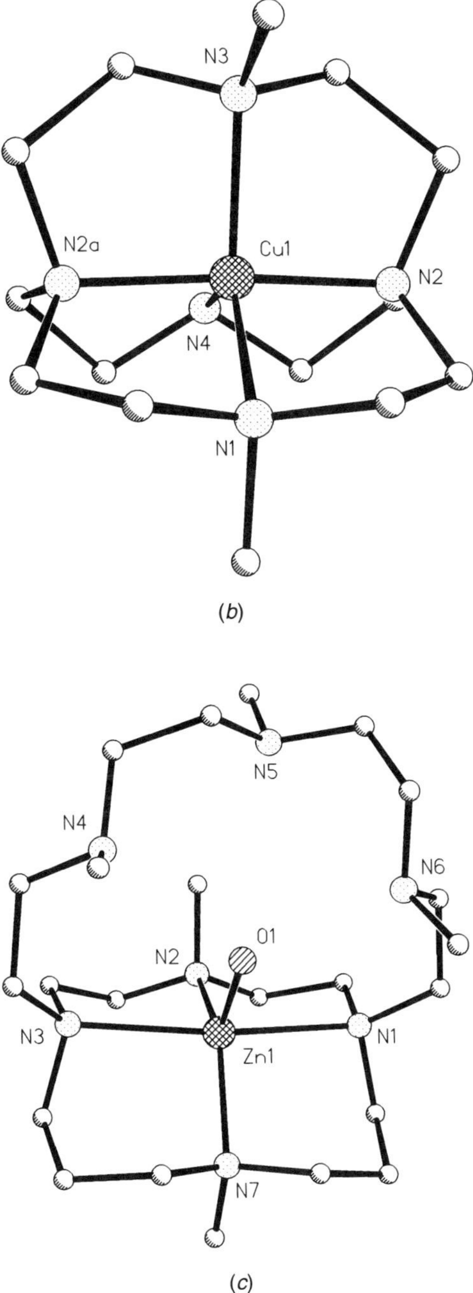

Figure 31. (*Continued.*)

can happen even when the cryptand is too basic to allow competition of Li^+ with H^+ (116, 117). The consequence of the good fit of these transition series cations in the small azacryptand cavity is high stability toward decomplexation, one zinc complex, for example, being unaffected by strongly acidic conditions over a period of months (118). The ligand **XIII,** studied as host for transition

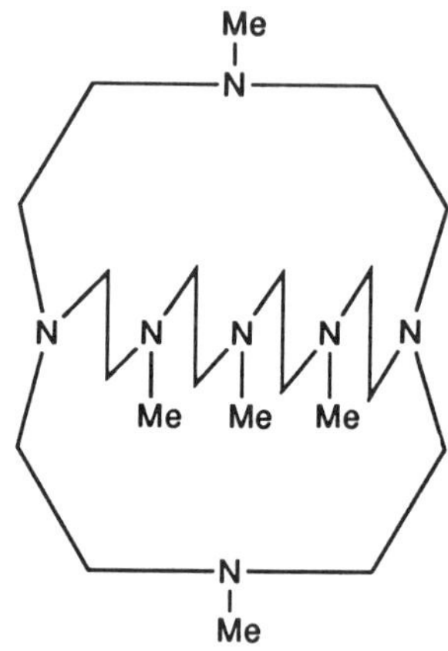

ions, has a long 11-membered N_3 strand, which generates properties intermediate between those of azamacrocycles and cryptands; although seven N donors are available for coordination, there is evidence that no more than four are in fact involved [Fig. 31(*c*)]. No cryptate effect is observed in solution complexation studies of this ligand (119) and an X-ray structure obtained for the Zn^{2+} complex [Fig. 31(*c*)] fails to show more than partial inclusion of the cation in the crypt.

B. Templated Mononucleating Polyazacages

The strategy of using transition cations to assemble azamacrocycles from simple precursors originated in the early 1960s, in the elegant work of Curtis (120a), Busch (120b,c) and co-workers. They utilized, as template ions, cations such as Ni^{II} or Cu^{II} preferring square planar coordination, which predisposes toward delivery of a macrocyclic product. The use of template cations that prefer octahedral coordination predisposes toward a different topological outcome. This second strategy, as applied by Sargeson (34–38) using the octahedrally coordinating Co^{III} cation as template, has been very successful in generating hexaamino cage complexes.

In the Curtis reaction (Fig. 32), the amino group, activated by reason of its coordination, readily condenses with a carbonyl group (e.g., formaldehyde or acetone) forming a macrocycle with two imine and two secondary amine groups. In synthesizing their azacage ligands, the Australian (34–38) group exploit a similar reaction strategy (Figs. 14 and 15.)

The apparent overall simplicity of the cryptate formation reaction conceals (35, 39, 40) a series of intermediate stages controlled by a Co^{III} template ion that is subsequently retained by the cage ligand. Initially, template-assisted formaldehyde–amine condensation generates imine bonds, which are at the same time protected from protonation by their coordination to Co^{III} and activated toward attack by ammonia, resulting in capping of the $[Co(en)_3]^{3+}$ complex at both ends, generating the kinetically inert Co^{III} cryptate.

The strong retention of the cation in the crypt makes for reversible electrochemistry (34–36, 91, 92). The kinetic and thermodynamic consequences of the cryptate effect enhance the suitability of the guest cations for redox or photophysical applications deriving from stabilization, within the cage ligands, of unusually high redox or electronically excited states (53, 48, 121, 122).

The carbon–bridgehead sar series, made available by extension of the capping strategy to reactive carbon atoms such as that in nitromethane (Fig. 15), is synthetically more adaptable than the sep series, and much of the more recent chemistry has been carried out using substituted C-bridgehead cages. The nitro function on the carbon cap is readily reduced to an amino group, generating amino-substituted cages such as diamsar. These are easily protonated whenever Lewis acid cations are used and lead to overall cryptate charges as high as +4 or +5. These cryptates are consequently strong oxidizing agents.

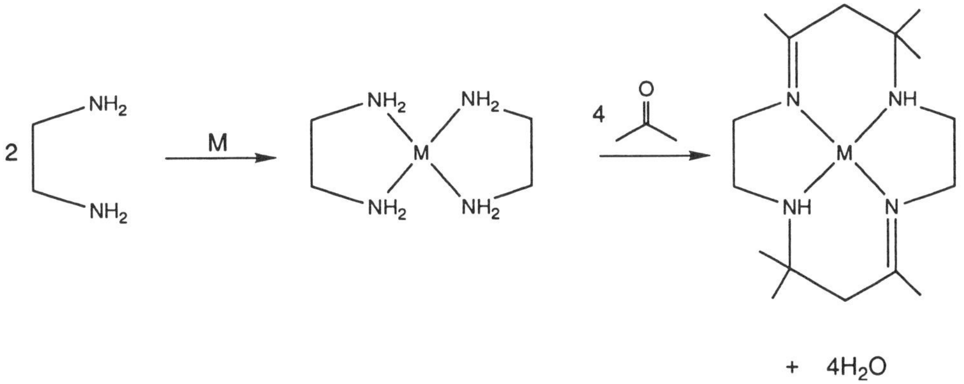

2 NH_2 NH_2 —M→ NH_2 NH_2 M NH_2 NH_2 —4 (acetone)→ N NH M NH N + $4H_2O$

Figure 32. Curtis reaction.

The substitution of donors, X, other than N has been achieved by a further extension of the strategy (41): a podand ligand with the X donors incorporated is first synthesized by template methods on the kinetically labile Co^{2+} ion, which is converted to Co^{3+} by aerobic oxidation. Capping of the kinetically inert Co^{III} podate can now be achieved, for example, with the NH_3/HCHO or $MeNO_2$/HCHO combinations using the techniques developed for sep and sar. By this route, a series of unsymmetric crypates (Fig. 33(*a* and *b*)] with mixed-donor sets, for example, N_x/X_{6-x} (X = S) has been obtained. The consequences of sequential replacement of N donors by S donors have been thoroughly analyzed, with reference to redox, magnetic, and electronic spectral properties (123–127). Unsymmetric cages with differently substituted caps, for example, Me and CO_2 substituents on opposite bridgehead carbons, the latter cap also bearing one amide oxygen [Fig. 33(*c*)], have likewise been generated (128) by applying this capping strategy with bifunctional methylene compounds such as diethyl malonate. These monoester-substituted oxosar and azaoxosar cages (Fig. 34) are anionic because of the deprotonation of the amide nitrogen, which makes the Co^{II} cryptates powerful reductants; the ester functionality offering the possibility of further derivatization to build larger macromolecules such as biological conjugates for imaging purposes.

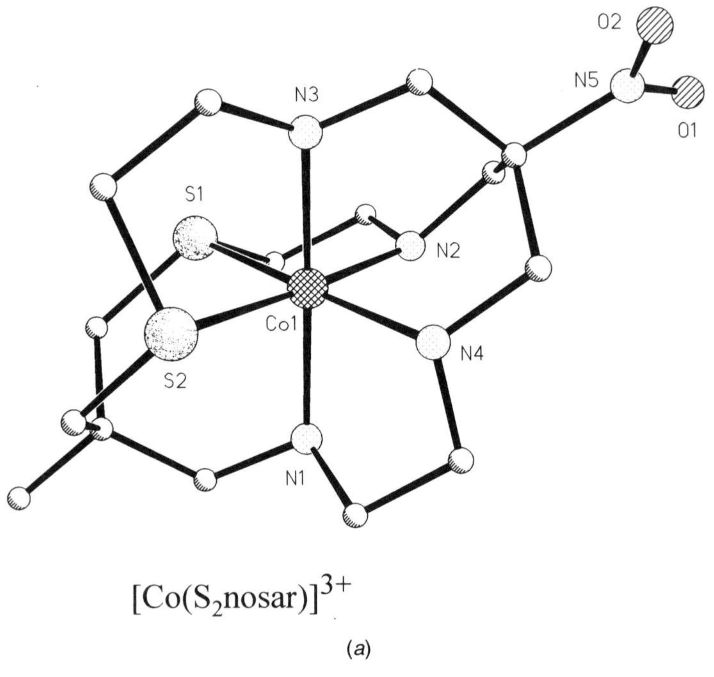

Figure 33. X-ray structures (43, 57, 124, 125) of some unsymmetric Co^{III}azasar derivatives. (*a*) $Co[S_2nosar]^{3+}$, (*b*) $Co[S_1amsar]^{3+}$, and (*c*) $Co[carboxyoxosar]^{3+}$.

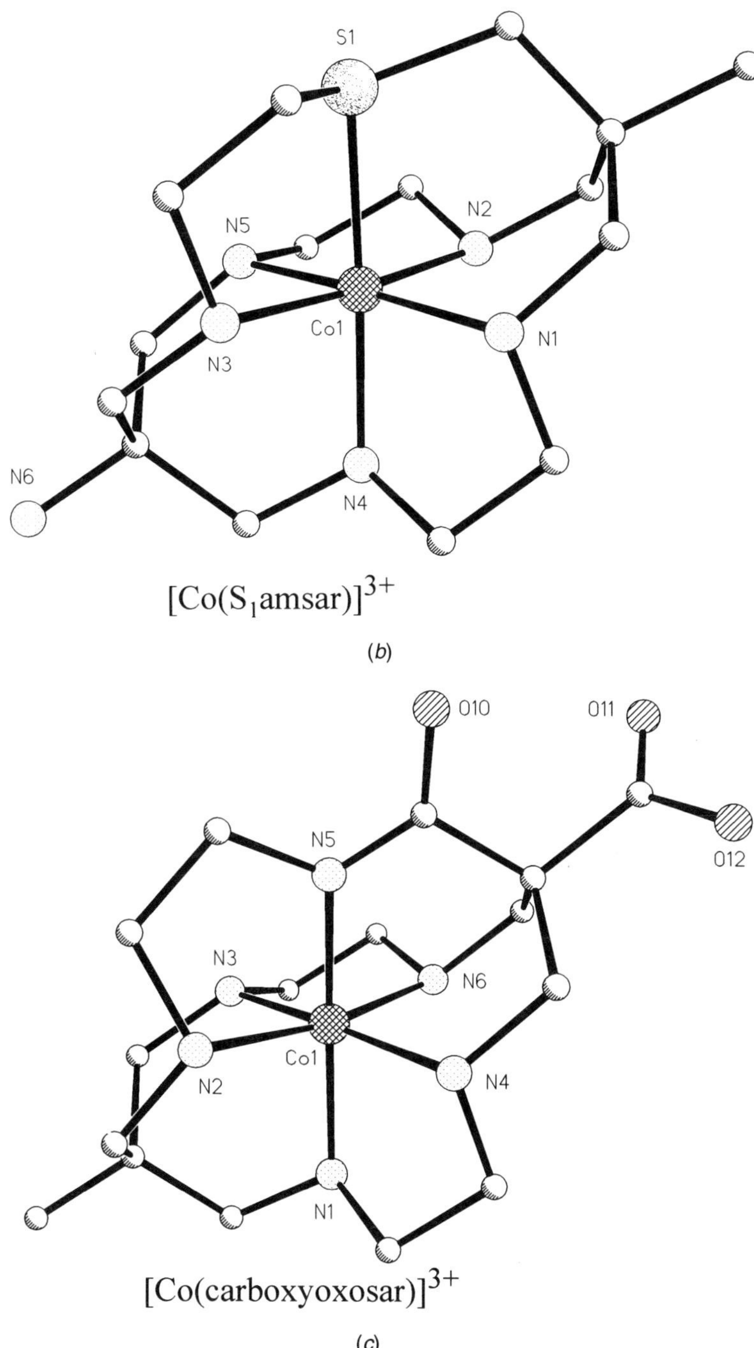

$[Co(S_1amsar)]^{3+}$

(b)

$[Co(carboxyoxosar)]^{3+}$

(c)

Figure 33. *(Continued.)*

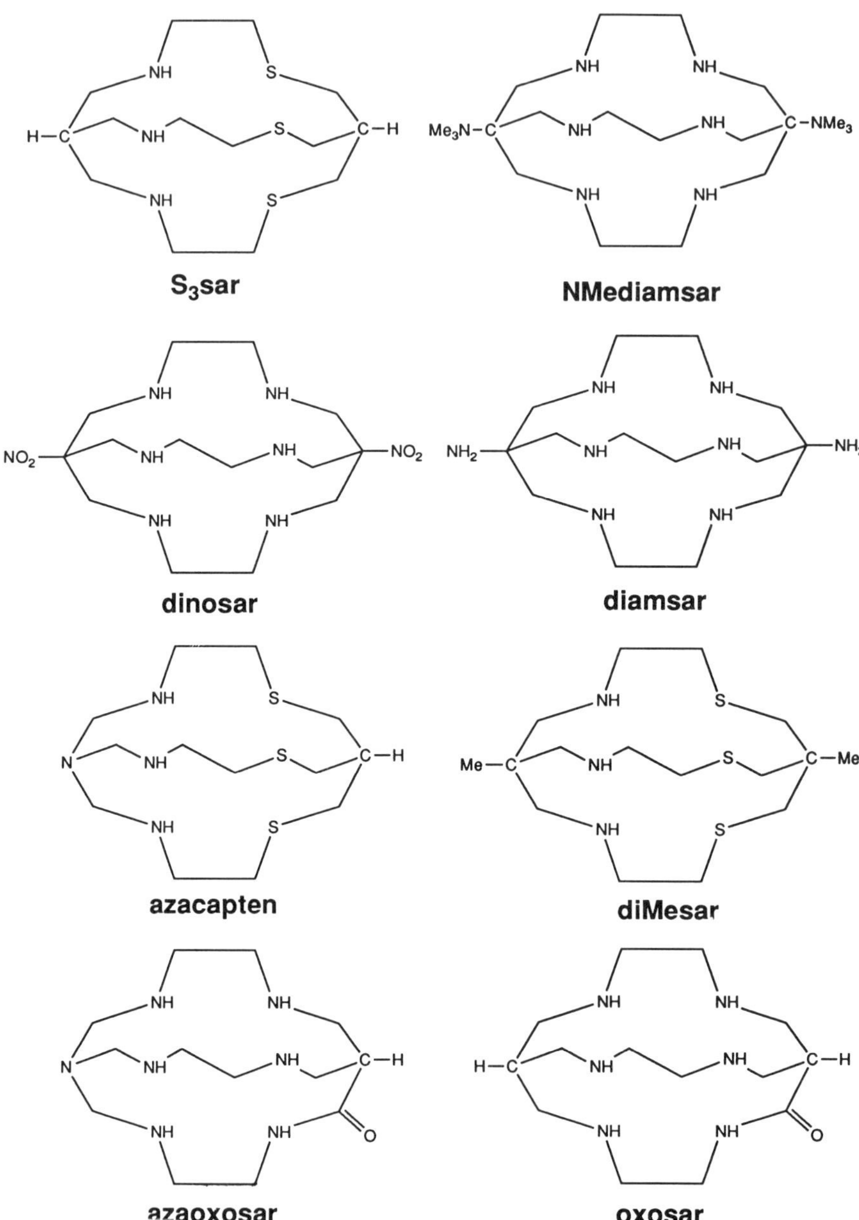

Figure 34. Some symmetric and unsymmetric cages, made by reactions described by Figs. 14–16.

Following the development of these syntheses, many azacryptates of kinetically inert cations other than Co^{III}, for example, Rh^{III} (49), Pt^{IV} (46, 129), and Ir^{III} (50) have been structurally characterized and their photophysical (where appropriate) and redox properties investigated. The kinetically inert Cr^{III} ion, however, did not prove suitable for use as template ion. The influence of cavity size on the lifetime of metal-centred emission was compared for Rh^{III} cryptates of the six-atom stranded cage sar as against the expanded seven-atom stranded "expanded" cage homologue **I,** (Fig. 19), showing a strongly enhanced phosphorescence for the latter. These expanded cage cryptates also show very different redox behavior from the smaller sar-based cages, stabilizing larger cations such as Co^{II} (130) without losing the high kinetic stability or relatively fast electron exchange rates characteristic of the smaller cages. The ability (51) to decomplex template cations from the carbon-bridgehead cages has enabled access to cryptates of nonkinetically inert but physically important cations; for example, redox reagents such as Ni^{III} (56), V^{IV} (53, 131), Ru^{II} (132–137), or the photophysically active Cr^{III} (54, 138). (In the latter case, the kinetically labile Cr^{II} ion is first encapsulated followed by aerobic oxidation.) The cage environment is crucial to protection of these highly charged cations from hydrolytic oligomerization, as it is in protecting photochemically excited states from deactivation by inhibiting vibrational and rotational modes of nonradiative decay. Both Ni^{III} and Ru^{III} are rapid oxidants; however, despite the favorable self-exchange kinetics of the Ru^{III}/Ru^{II} system deriving from the similar cage dimensions evident in both redox states (136), the Ru^{n+}/sar system cannot be used for redox purposes in solution because of the rapid ligand autoxidation that attends the oxidation stage and generates imine bonds (Fig. 35) in the sar cage via oxidative dehydrogenation (132). Successive oxidations lead to generation (135) of hexaimine species by this mechanism.

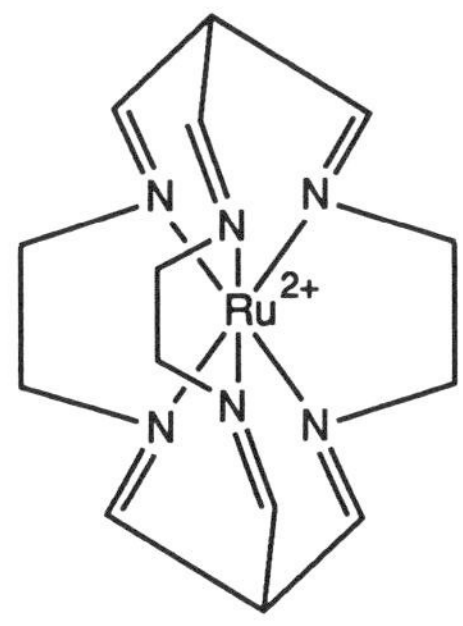

$[Ru(hexaimsar)]^{2+}$

Figure 35. Oxidative dehydrogenation of Ru^{II}sar cages.

The cobalt cryptates that show relatively slow exchange rates have found use, together with $[Ru(bpy)_3]^{2+}$ sensitizers, in photoreduction of water or other photoreducible substrates (93, 139–142): the wide range of available ligands, including the protonable diamsar series (Fig. 36), is valuable as it permits tuning of both redox potential (143, 144) and electron-transfer rates (145) for different applications. The Ni^{III} cryptates are potentially useful as powerful oxidants, which are nevertheless stable in dilute aqueous solution; these cryptates show reversible electrochemical behavior and self-exchange kinetics around 1000 times faster than their $Co^{III/II}$ analogues. This finding may be rationalized on the basis of structural data (56, 146), which shows Ni—N bond lengths for $Ni(diamsarH_2)^{4+}$ intermediate between the ideal bond lengths for Ni^{II} and Ni^{III}. For $Mn(sar)^{2+}$, on the other hand, it has been shown (147) using γ-active ^{56}Mn, that exchange with $[Mn(H_2O)_6]^{2+}$ is slow; at most 8% exchanging over about 18 h. In this case, there is also a slowish electron self-exchange rate due to the relatively large size of the Mn cation in both oxidation states; the poor fit is thought to hinder access to the transition state geometry. There is also some lack of reversibility arising from a deprotonation-linked disproportionation of Mn^{III}.

An extensive study of the magnetic properties of first transition series cryp-

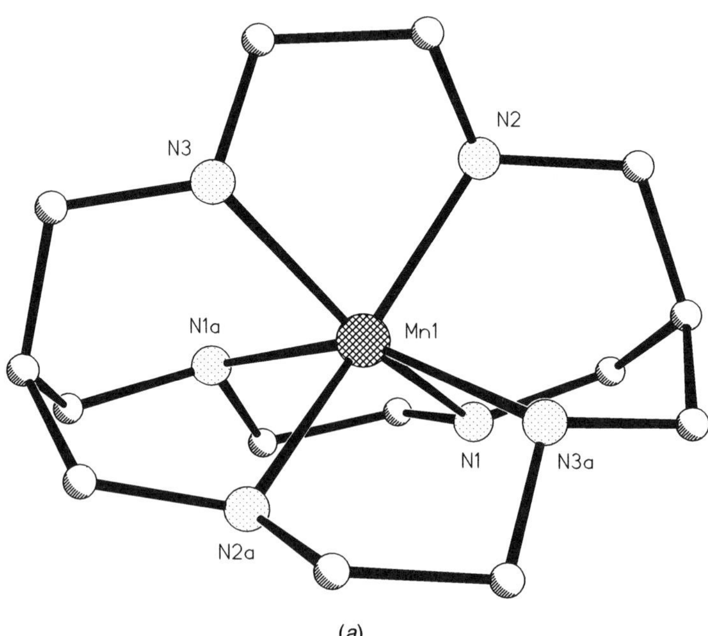

(*a*)

Figure 36. X-ray structures of Mn^{III}, Mn^{II}, and Ni^{II} cryptands (52, 56, 147): (*a*) $Mn[sar]^{3+}$, (*b*) $Mn[diamsarH_2]^{4+}$, and (*c*) $Ni[diamsarH_2]^{4+}$.

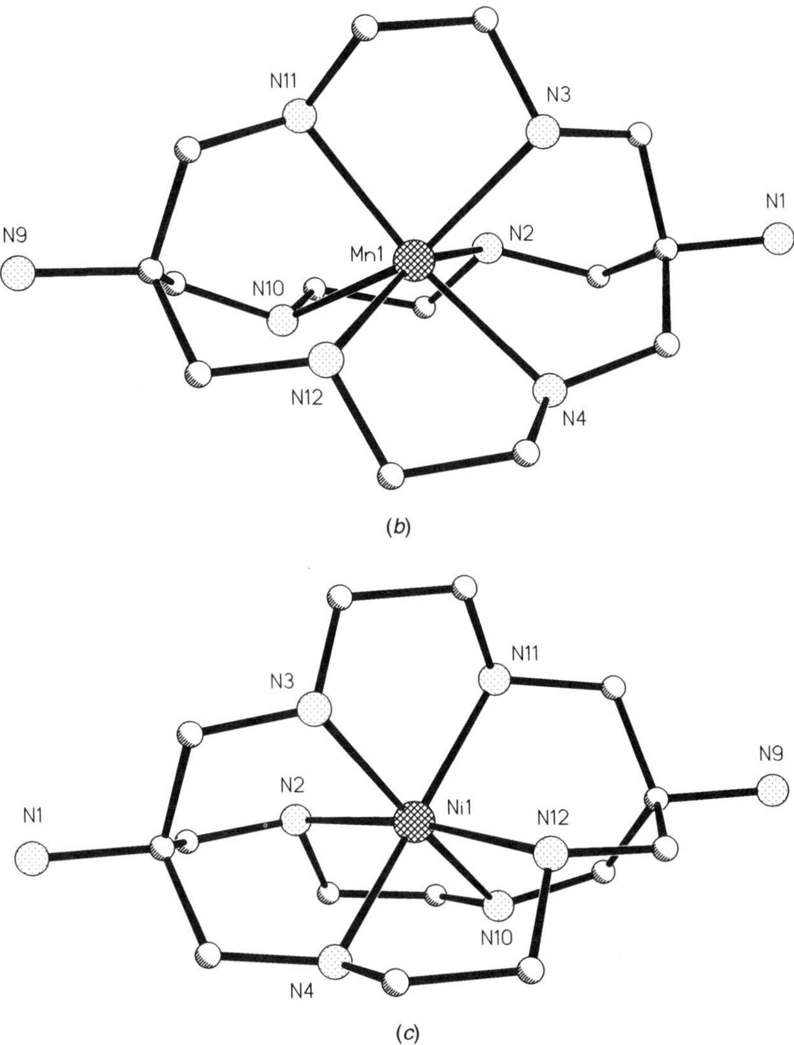

Figure 36. (*Continued.*)

tates of these C-bridgehead cage ligands has been carried out (148) and shows low-spin ground states for Fe^{III}, Co^{III}, and Ni^{III}, but high-spin ground state for Mn^{III}. The M^{II} cryptates are generally high spin, with the exception of the Fe^{II} series, members of which tend to exhibit ${}^6A$–${}^1T$ spin crossover behavior, which has been studied both in solution and in the solid state (149, 150). Thiaaza cryptates of Co^{II} such as $[Coazacapten]^{2+}$, (Fig. 37) are unusual in exhibiting low spin or intermediate moments (148, 151) at room temperature.

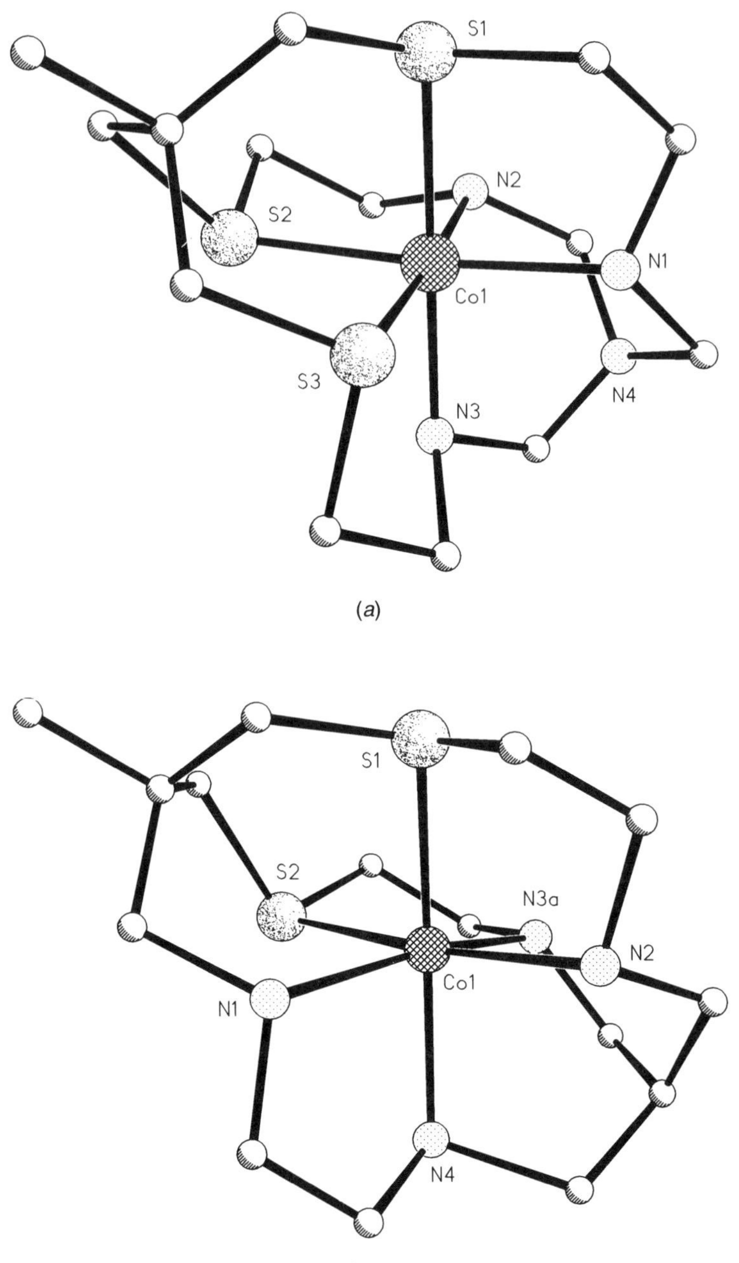

Figure 37. X-ray structures of thiaazacryptates (41, 151): (*a*) Co[azacapten]$^{3+}$ and (*b*) Co[S_2Mesar]$^{2+}$.

In our remaining discussion of these mononucleating cage ligands, the main emphasis will lie on conformational studies, leaving the reader to consult review articles (36, 37, 40) for more comprehensive treatment of the important body of work on other aspects such as electron-transfer kinetics.

C. Cryptand Conformations

It is common to describe cryptand ligands as imposing geometry on the encapsulated metals, although the coordination geometry is not absolutely fixed. The geometry of the cryptate is restrained within certain limits rather than constrained to one particular set of values. For a generic azacryptand, composed of two triamine caps and three linking spacers (Fig. 38), three major sources of flexibility can be identified. These sources are the length of the strands in the cap, the length and rigidity of the linking strands, and the helicity or twist of the crypt (rotation of one cap with respect to the other). In addition, the combination of a number of small deviations from ideal bond lengths and angles along a strand can be significant, especially if the strand is relatively long.

The simplest measure of the helicity of the ligand is the twist angle φ, the rotation between the two sets of cap N atoms [Fig. 39(*a*)]. In mononuclear cryptates a twist angle of 60° is required for octahedral geometry at the metal ion while a twist of 0° corresponds to D_{3h} (trigonal prismatic) symmetry. In practice, since the twist also affects the distance *d* between the N_3 planes, and is influenced by the length and rigidity of the linker, most real examples represent compromise between the preferences of cryptand and metal ion. In binuclear or polynuclear systems, a preferred metal–metal distance or M—X—M bridge size may also need to be accommodated, and φ can provide some fine tuning of the overall length of the cryptand. Increasing the helicity of the cage reduces *d* and, hence, the overall length of the crypt.

Where the bridgehead atom X is a tertiary amine nitrogen, there are two conformational possibilities (Fig. 40); the lone pair can be directed outward, away from the cavity (exo) or it can be directed inward, possibly coordinating to encapsulated metals (endo). In the case of the shortest (e.g., tame derived) caps, the N_3 bite only permits the cap to bind with fac geometry and the bridgehead geometry must always be exo, since otherwise the three amine nitrogen donors cannot get close enough to bind to a single metal ion. In the cases of

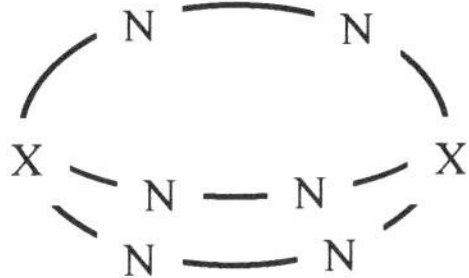

Figure 38. Azacryptand (schematic).

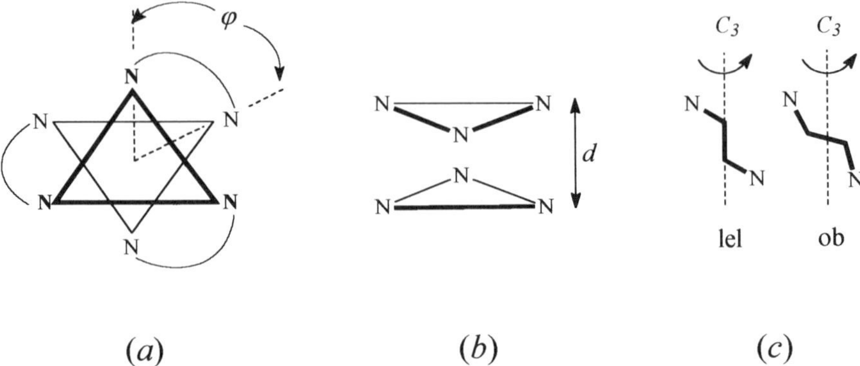

Figure 39. Parameters describing conformation in small azacryptands: (*a*) twist angle φ, (*b*) distance, d, between N_3 planes, and (*c*) ob and lel disposition of methylene spacers.

tren and trpn, for the same reason, the reverse usually applies; the bridgeheads are endo and the lone pairs may interact with encapsulated metal ions. Tren and trpn also differ from one another in the preferred size of the N_3N_{br} pyramid, and therefore in the preferred M—N distances and geometry as well as the orientation of the lone pairs on the nitrogen donors, as will be seen later.

The most obvious property of the spacers is to control the distance between the two sets of donors derived from the caps (although the linkers may also carry donor atoms). This distance is primarily a function of the total length of the spacer, but there are important fine adjustments depending on the flexibility of the spacer and the helicity of the cage.

The largest set of closely comparable azacryptands–cryptates comprises the sar and sep family developed by Sargeson and co-workers (51, 152). The cryptands are fully saturated and, at first glance, appear as if they should possess considerable conformational lability. There is NMR evidence for this in solution but in the solid state most examples seem to be relatively conformationally rigid. The very similar structures of [diamsar], [(diamsarH_3](NO_3)$_3$, and

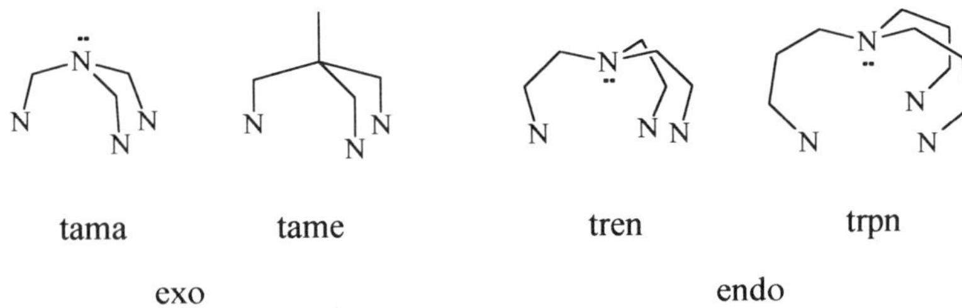

Figure 40. Exo and endo conformations in cryptand caps.

[diamsarH.Mg]$(NO_3)_3$ (the latter two are isomorphous) suggests a large degree of preorganization for binding (51) (Fig. 41). The amine donors are frequently involved in extensive hydrogen-bonding networks.

Each of the amine donors is potentially chiral but the theoretical range of isomers is not generally observed, most examples being (*RRRRRR*) or (*SSSSSS*)

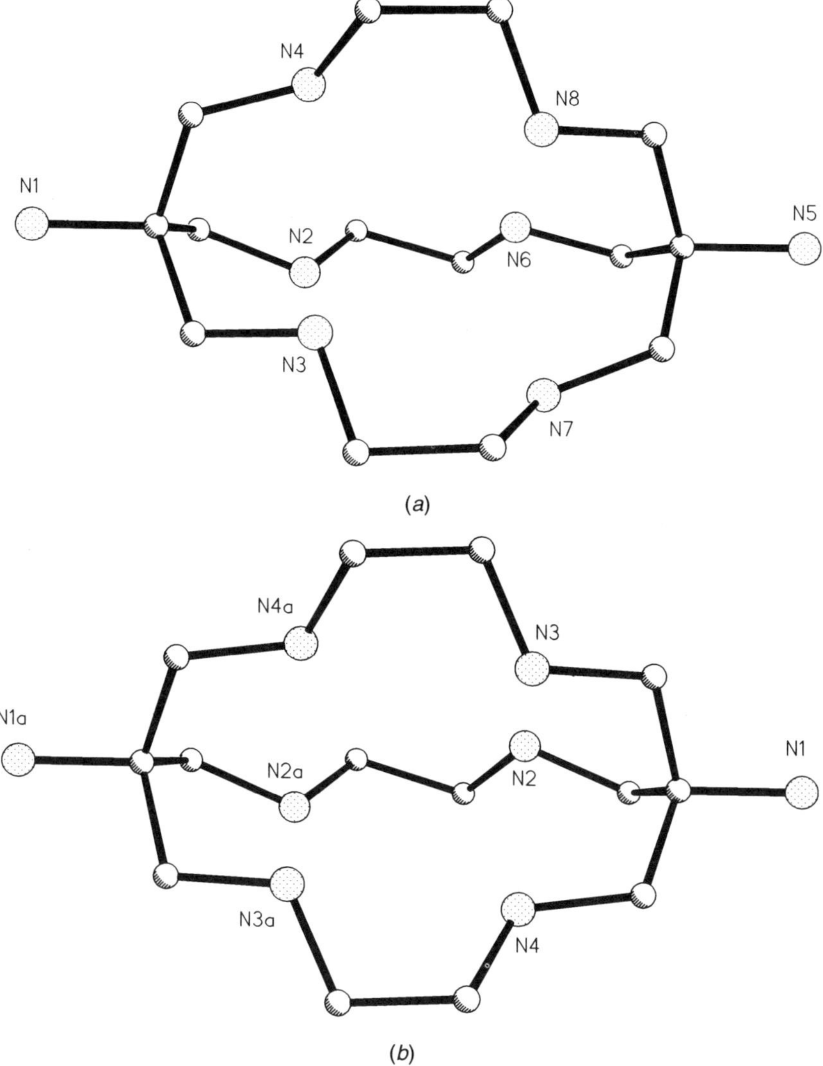

Figure 41. Structures of diamsar derivatives: (*a*) (diamsar) (*b*) $(diamsarH_3)^{3+}$, (*c*) $(diamsarH.Mg)^{3+}$, and (*d*) $[diamsarH_2Ni]^{4+}$; (*b*) and (*c*) are isomorphous.

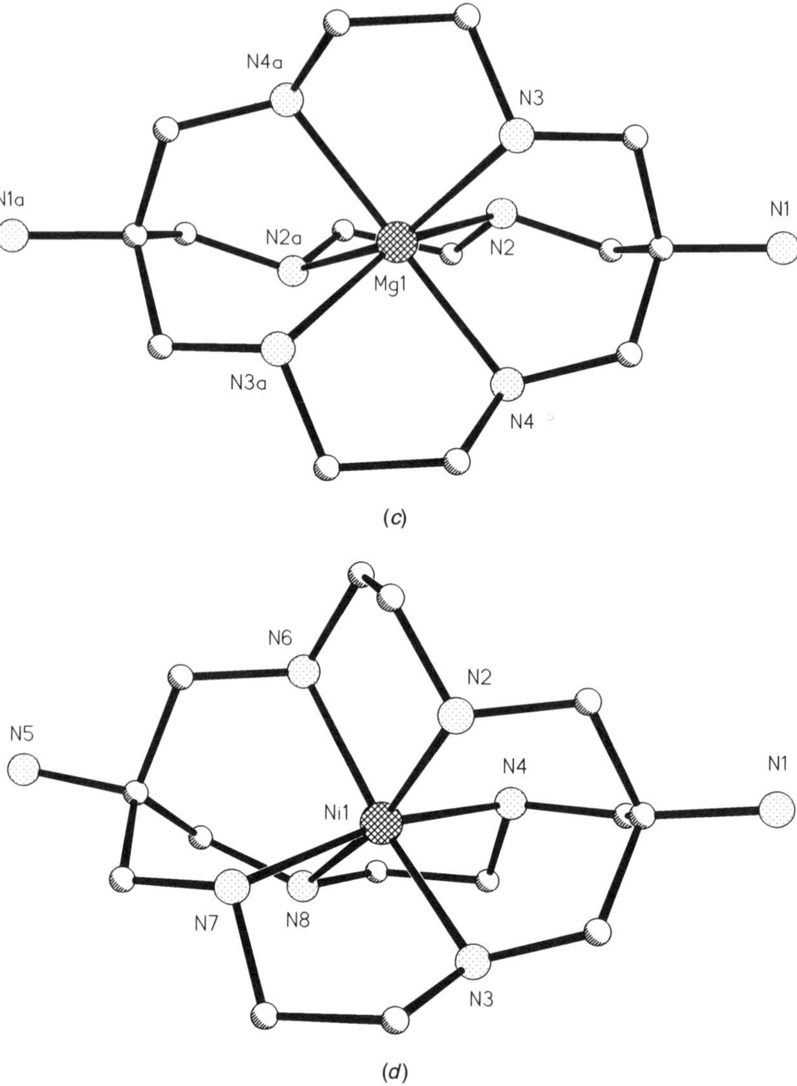

Figure 41. (*Continued*.)

(51). In this series, where the lateral spacers are $-CH_2CH_2-$ groups, it is convenient to describe geometry in terms of the ob/lel conformation of the spacers (Fig. 39) as well as the twist angle. For sep and sar cages, the lel_3 conformation is most common. Increasing the number of ob strands reduces the size of the cavity, by compression along the (real or pseudo-) C_3 axis, but also allows better octahedral geometry. Configurations with one or more ob strand are therefore favored by smaller ions or those with marked stereochemical preferences; an example of this is the lel_2ob conformation seen in $[Ni(diamsarH_2)]^{4+}$ (153).

Two sets of theoretical studies on cobalt cryptands, based on ligand field models (154) and on strain energy minimization as a function of M—N distance (155) gave good agreement with each other and with the experimental observations. The agreement of the two very different theoretical models was interpreted by Comba (155) as implying that the electronic ground state in these complexes is a result of the structural environment, rather than vice versa. Calculated conformations for the sep and sar systems were quite similar, although the N bridgehead in the sep cap permits flatter geometry than the C bridgehead in sar. In general, it seems that variation of the substituent at the C-bridgehead had little effect on the geometry although a few exceptions have been noted. For example, methylation of the pendant amine in diamsar to form NMediamsar (152) results in a change in the stereochemistry of the Co^{III} complex from lel_3 to ob_3. This change does not involve much alteration of the M—N distances but, significantly, in the ob_3 structure the cobalt geometry is much closer to octahedral.

Methylation of three of the secondary amines yields (only) the two *fac* conformers (lel_3 and ob_3) with Co^{III}, of which the lel_3 appears more stable but the overall structure of the complex is not much different from the unsubstituted analogues (145). Further methylation has more drastic effects and the three-fold symmetry of the cryptand is destabilized by Me–Me interactions. Both the free ligands and their Cu^{II} and Ni^{II} complexes take up a conformation reminiscent of Busch's lacunar complexes, with metal ions bound to only four amines in what is essentially a strapped macrocyclic arrangement (Fig. 42 on page 214) (156). The square planar geometry at the metals requires that the four nitrogen donor set is (RSRS) not (RRRR).

D. Schiff-Base Cryptands

1. Average Valence Dicopper

No binuclear complexes of the sep or sar type have been reported to date. The underlying reason for this is not the length of the en spacer but the steric constraints imposed by the small, for example, tame-type cap. The amine can bind facially to an octahedral metal ion but this leaves the metal ion substantially out of the N_3 planes and occupying the center of the cavity. Longer triamines tren or trpn as cap generally take up the endo configuration where the N lone pair is directed into the cavity and the N_3 triangle is large enough to permit trigonal geometry at a suitable metal (copper), because the metal can closely approach the N_3 plane and there is more space in the center of the cavity.

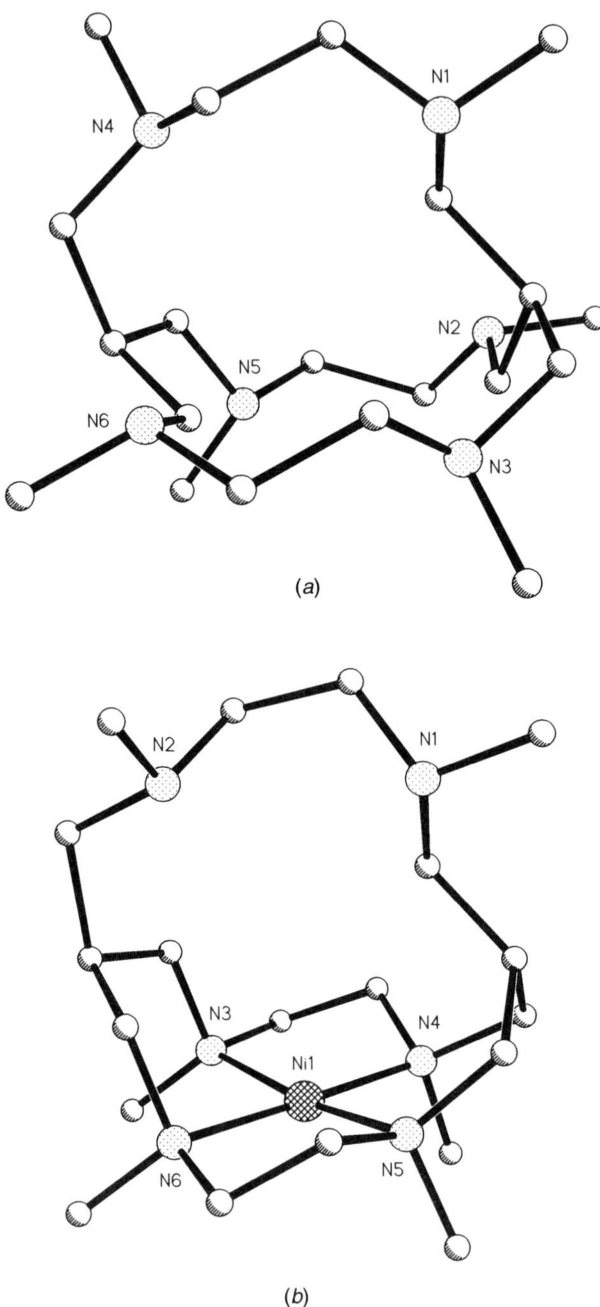

Figure 42. Site geometry in a methylated cryptand (*a*) (Me_6sar) and its Ni(II) cryptate (*b*).

The series of small aza cages generated by the Schiff-base [2 + 3] condensation method from tren and the two-carbon-spaced precursor dialdehyde, have demonstrated the capacity to accommodate a pair of cations, that is, copper ions, held in close proximity (157) by the constraints of the ligand. They can also act as mononucleating ligands [Fig. 43 (*a* and *b*)] for transition and main group cations (including Cu^{II} in some cases [Fig. 43(*b*)]) (158), but the demonstrated ability to form dinuclear cryptates within these two-carbon-spaced ligands is so far restricted to copper and silver. Within the hexaimino cages imBT and imbistrpn (Fig. 44) the (+I, +I) state of dicopper is the most stable redox state, although the cryptates readily undergo one-electron oxidation with mild oxidizing agents such as Ag^+ to generate $[Cu_2L]^{3+}$ salts. In the analogous sp^3-N octaamino derivative amBT (all Schiff-base C=N functions hydrogenated) the (+I, +I) oxidation state is not stable, presumably because of the "harder" nature of the donors, and attempts to generate it result only in disproportionation and precipitation of copper metal. In this case, only the $[Cu_2L]^{3+}$ cryptate has been obtained (159). Electron spin resonance spectroscopy demonstrates complete delocalization (160) of the unpaired electron over both copper nuclei in all three $[Cu_2L]^{3+}$ salts. The retention of the seven-line near-isotropic spectrum (Fig. 45) to temperatures as low as 4 K leaves no option

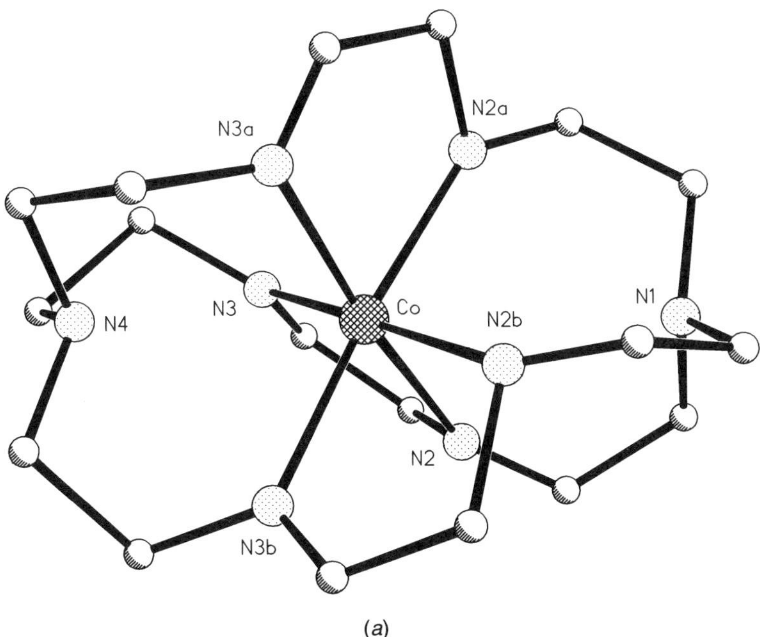

Figure 43. Structures of Schiff-base azacryptates (158, 161): (*a*) $Co[imBT]^{2+}$ and (*b*) $Cu[imBT]^{2+}$.

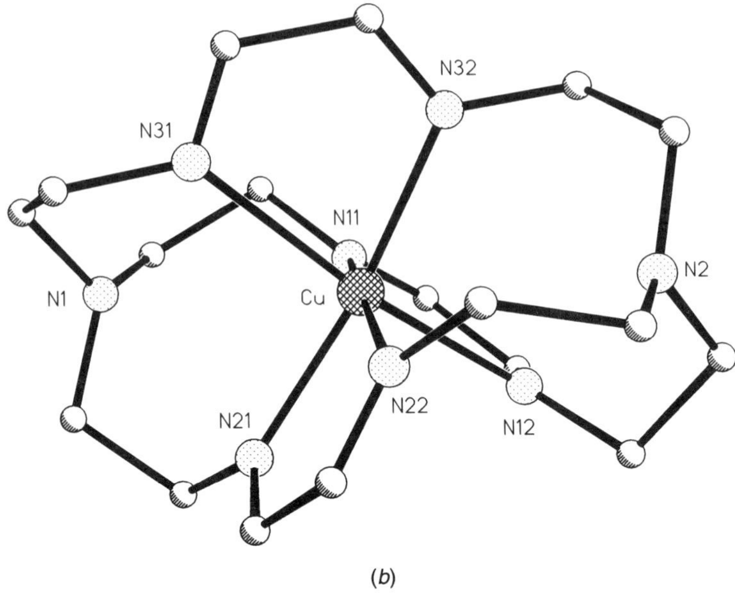

(*b*)

Figure 43. (*Continued*.)

but to classify the Cu redox state as <u>average</u> (di$Cu^{1.5}$) rather than mixed ($Cu^{I}Cu^{II}$) valence. At the approximate 2.4 Å distance, which the imBT (Fig. 46) and amBT (Fig. 47) cryptands impose, the necessarily efficient overlap of Cu d_z^2 orbitals along the threefold cryptand axis ensures the existence of a one-electron copper–copper bond in the $[Cu_2L]^{3+}$ cryptates. In the analogous trpn-capped imbistrpn system (Fig. 48), the dicopper (1.5) cryptate has an internuclear Cu· · ·Cu distance over one-half of an angstrom shorter, at 2.419(1) Å,

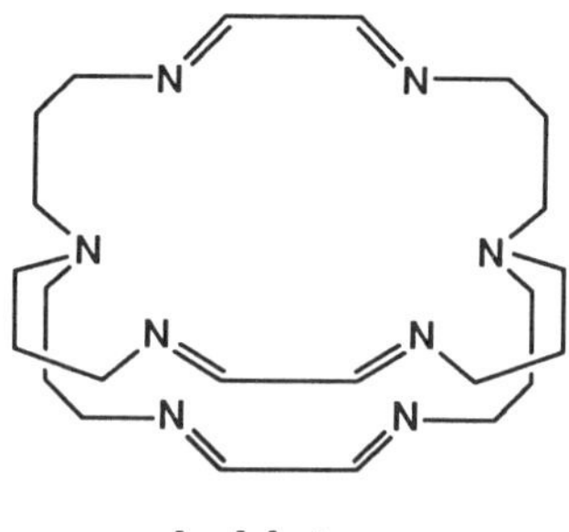

Figure 44. Trispropylene-capped cryptand, imbistrpn.

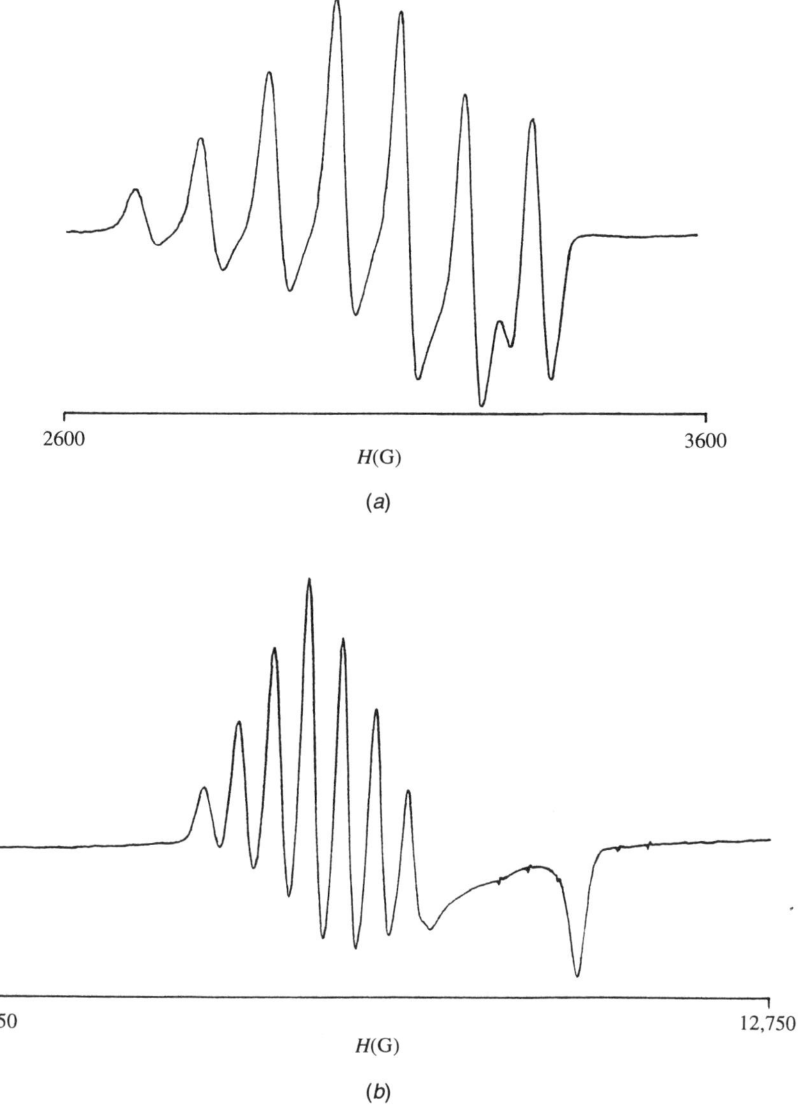

Figure 45. The ESR spectra of $Cu_2[imBT]^{3+}$ in frozen DMF solution (160): (*a*) *X*-band at 4 K and (*b*) *Q*-band at 80 K. In the *X*-band spectrum, the $g_{\parallel}$ signal overlaps the $g_{\perp}$. In the higher frequency *Q*-band spectrum, $g_{\parallel}$ and $g_{\perp}$ are clearly resolved. The seven-line pattern results from coupling of unpaired spin with both spin $\frac{3}{2}$ copper nuclei. [Reprinted with permission from The Royal Society of Chemistry. C. Harding, J. Nelson, M. C. R. Symons, and J. Wyatt, *J. Chem. Soc. Chem. Commun.*, 2499 (1994).]

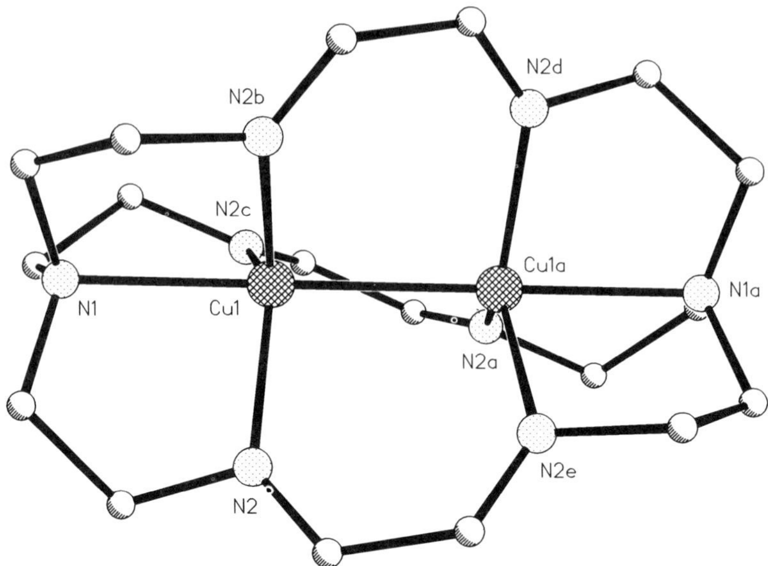

Figure 46. Dicopper(I) azacryptate: crystal structure (157) of $Cu_2[imBT]^{2+}$. Selected distances (Å): Cu—Cu 2.45, Cu—N_{im} 2.00, Cu—N_{br} 2.22; Cu—N_3 plane 0.18.

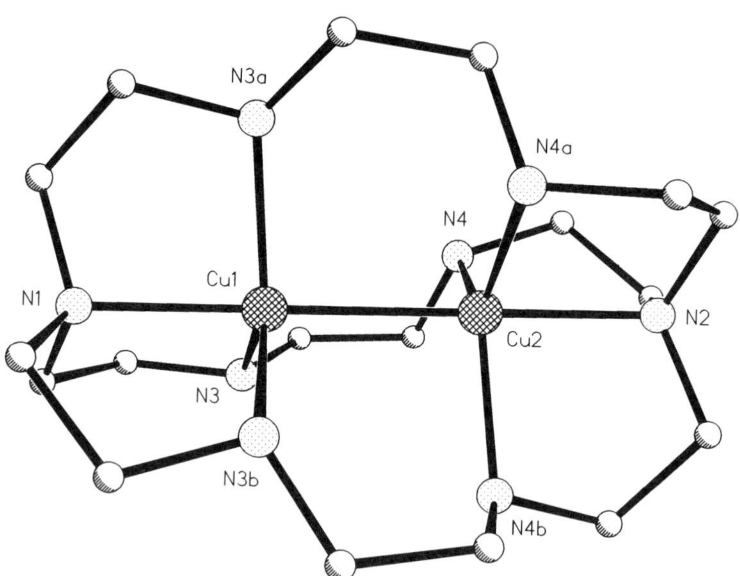

Figure 47. Average-valence dicopper (1.5) cryptate: X-ray structure (161) of $Cu_2[amBT]^{3+}$: Selected distances (Å): Cu—Cu 2.42, Cu—N 2.06, Cu—N_3 plane 0.12, Cu—N_{br} 2.05.

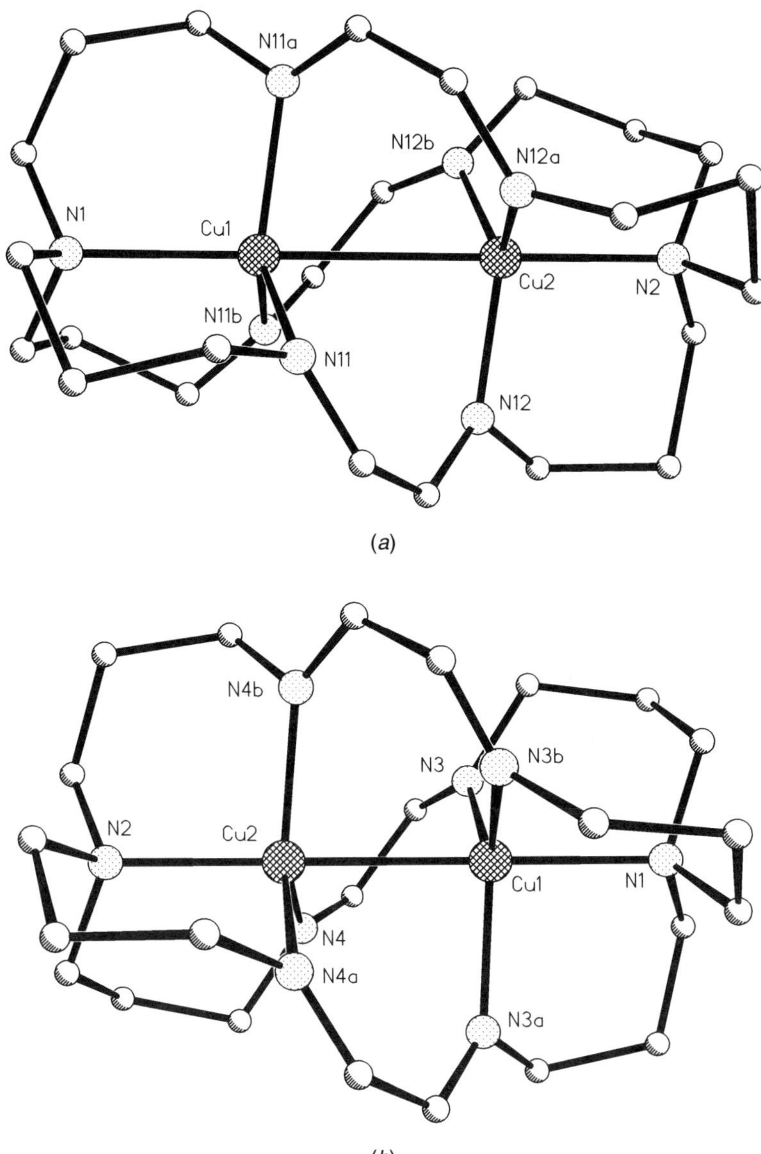

Figure 48. Comparison of the X-ray structures (161, 162, 163) of (*a*) Cu_2(I)[imbistrpn]$^{2+}$ and (*b*) Cu_2(1.5)[imbistrpn]$^{3+}$. Selected distances (Å): (*a*) Cu—Cu 2.93; Cu—N_{im} 2.04; Cu—N_3 plane −0.24 Cu—N_{br} 2.20; (*b*) Cu—Cu 2.42; Cu—N_{im} 2.02; Cu—N_3 plane −0.13 Cu—N_{br} 2.06.

than in the dicopper(I) analogue (161, 163). The larger trpn cap allows the noninteracting Cu^I cations to relax back into the N_3 imino plane, while in the analogous average-valence system the copper cations are pulled toward one another by the interaction of the one-electron bond. This sizeable reduction of internuclear distance on dicopper(I) oxidation confirms the presence of bonding interaction in the average-valent dicopper (+1.5) state.

Spectroscopic signatures of this unprecedented one-electron Cu—Cu bond, so far demonstrated, include intense and complex near-infrared (IR) absorption tailing to long wavelength together with absorption in the near-ultraviolet (UV); MCD spectra show features at matching frequencies (162). One characteristic of the MCD spectrum (Fig. 49) is the appearance of an intense feature in the 1000–1200-nm region, displaying vibrational fine structure of around 200 cm^{-1}

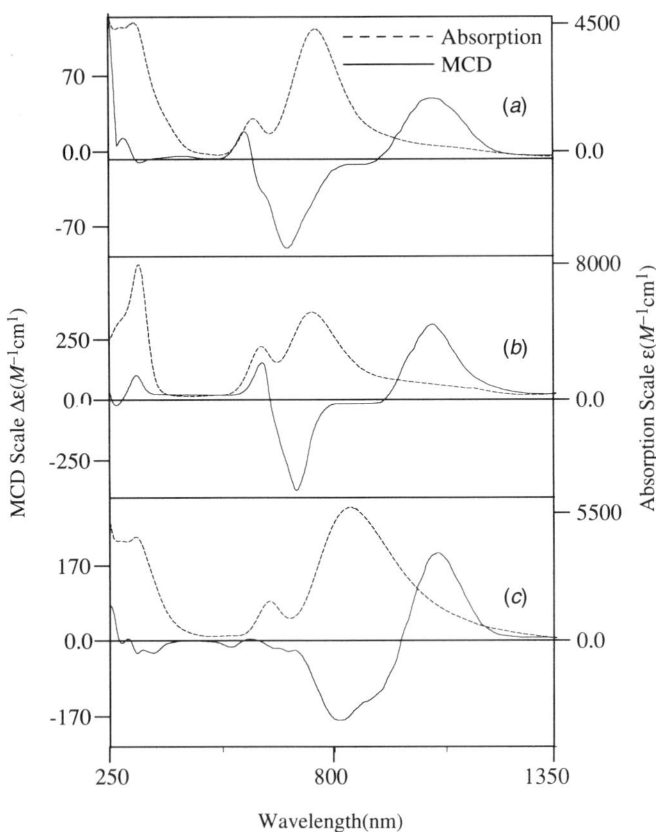

Figure 49. The MCD and electronic absorption spectra (162) of average-valence dicopper cryptates (*a*) $[Cu_2imBT]^{3+}$, (*b*) $[Cu_2amBT]^{3+}$, and (*c*) $[Cu_2imbistrpn]^{3+}$. Note the intensity of the 1100-nm band in MCD and its extensive vibrational structure in all three compounds.

($\pm$50) cm^{-1}, which corresponds to a weak broad shoulder in absorption. Resonance Raman studies (162, 163) demonstrate (Fig. 50) linkage of the near IR absorption to a Cu—Cu stretching mode around 250–290 cm^{-1}. Some of these signatures can be recognized in the now structurally characterized (164–166) Cu_A electron-transfer site of cytochrome oxidase (Fig. 51), which incorporates

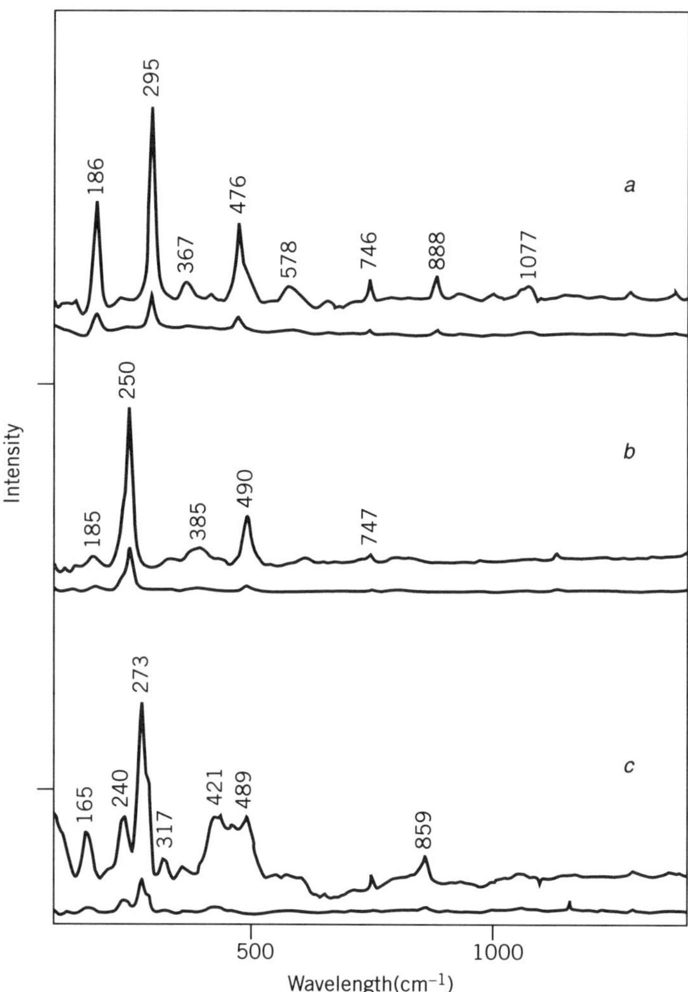

Figure 50. Resonance Raman spectra (162) (m*M* aqueous solution) of average valence dicopper cryptates (exciting frequency 840 nm) (*a*) $[Cu_2imBT]^{3+}$, (*b*) $[Cu_2imbistrpn]^{3+}$, and (*c*) $[Cu_2amBT]^{3+}$. The bands below 300 cm^{-1} are observed only when the spectrum is excited with $\lambda > 750$ nm.

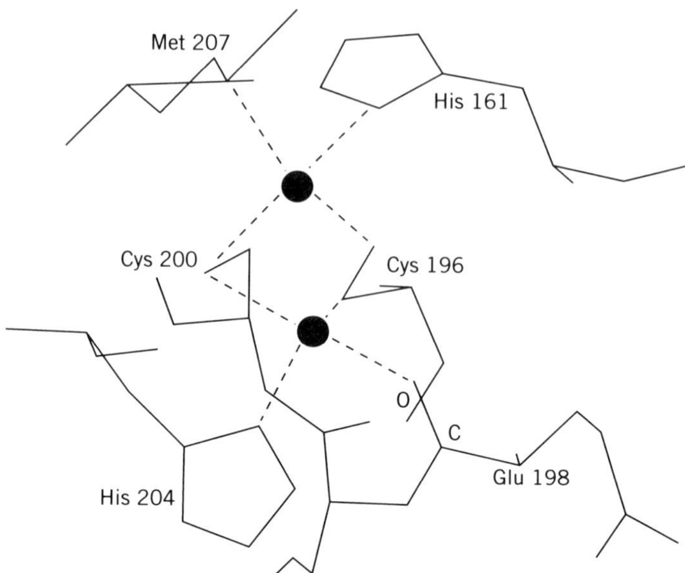

Figure 51. Schematic representation of the CuA coodination site in cytochrome oxidase, showing a [$2Cu{-}2S_\gamma$] structure. Solid circles show copper atom positions. Broken lines indicate coordination bonds. The peptide carbonyl groupings of Glu^{198} is illustrated with a bar marked with C and O.

average-valence dicopper supported by a pair of cysteinyl S^- bridges. The original controversy (167, 168) over whether copper atoms in the dicopper site were held together by direct Cu—Cu bonding or via cysteinyl bridges has thus been resolved by a neat natural compromise; as the crystal structure shows, both are involved! Calorimetric and thermodynamic studies are planned (169), which may establish the extent to which bonding of copper ions within the cryptates is cooperative; in the natural electron-transfer site such cooperativity may control the multielectron-transfer function.

2. *Potentially Binucleating Azacryptands*

Cryptands using tren and trpn allow the metal ion close to the base of the N_3N_{br} pyramid and therefore, in spite of not being much longer than sar or sep, can bind two metal ions. To date, the only binuclear azacryptates characterized with en or α-diimine spacers are the dicopper(I) or average valence dicopper(1.5) complexes discussed above. This finding is not so much a consequence of the small cavity size as of the enforced trigonal-based geometry for the metal ions. The geometry of a number of these systems is summarized in Table I.

TABLE I
Selected Dimensions of Small Azacryptands

Cryptand/Cryptate	Reference	M−N	$X_{br}-X_{br}^{a}$	$M-X_{br}$	en (NCCN) Torsion[d]	N−N	$M^{b}_{out\text{-}of\text{-}plane}$	d^{c}	$X_{br}-N_3$ Plane	Twist φ	M−M
sar	51		6.22		[1]54.0	3.04		2.56	1.79	9.3	
diamsar	51		6.38		56.6	3.01		2.71	1.84	25.3	
Zn^{II}/sar	51	2.17	6.27	3.14	52.8	2.99	1.31	2.63	1.82	30.8	
Co^{II}/sep	35	2.16	6.04	3.02	60.0	3.04	1.27	2.54	1.75	42.4	
amBT	172		6.37		65.4	4.14		2.98	1.69	15.1	
PbamBT(ob)	171	2.74	5.71	2.85	60.6	4.29	1.14	2.30	1.71	43.6	
CdamBT(ob)	170	2.49	5.66	2.83	65.7	3.92	1.03	2.06	1.80	53.5	
CdamBT(lel)	170	2.52	5.55	2.78	70.6	4.02	0.98	1.96	1.80	54.9	
$Cu_2^{1.5}$amBT(OAc)	159	2.06	6.48	2.06	73.4	3.55	0.12	2.62	1.92	52.1	2.37
$Cu_2^{1.5}$amBT(NO_3)	159	2.07	6.50	2.07	80.27	3.57	0.14	2.64	1.93	50.8	2.36
$Cu_2^{1.5}$amBT(ClO_4)	159	2.06	6.54	2.05	66.5	3.57	0.13	2.68	1.93	43.6	2.42
imBT[c]	172		6.86		169.5	4.19		3.33	1.76	24.5	
$[CuimBT]^{2+}$	161	2.15	6.48	3.24	3.3	3.20	1.08	2.18	2.15	48.8	
$[CoimBT]^{2+}$	158	2.16	6.16	3.08	7.4	3.26	1.06	2.11	2.02	56.6	
$[GdimBT(H_2O)_2]^{3+}$	227	2.58	5.96	2.45	5.8	3.80	1.24	2.59	1.74	15.8	
$Cu_2^{I}[imBT]^{2+}$	157	2.00	6.88	2.22	16.3	3.45	0.18	2.81	2.04	25.3	2.45
$Cu_2^{I}[imbistrpn]^{2+}$	161	2.04	7.32	2.20	28.1	3.50	−0.24	2.45	2.43	65.0	2.93
$Cu_2^{1.5}(imbistrpn]^{3+}$	162	2.02	6.53	2.06	32.5	3.49	−0.13	2.16	2.18	56.2	2.42

[a]Bridgehead atom = X.
[b]Positive values indicate displacement from the N_3 plane toward the center of the cryptand, negative values indicate displacement toward the bridgehead nitrogen, X_{br}.
[c]Figure 39(*b*). One strand is twisted in the opposite sense to the others—cryptand is $\lambda\delta\delta$ rather than $\delta\delta\delta$ or $\lambda\lambda\lambda$.
[d]Figure 39(*c*).

The average valence systems might be expected to show the shortest Cu· · ·Cu distances as a result of the bond between the metal ions (157, 163). This hypothesis is certainly justified in the imbistrpn system where there is a marked reduction in both Cu—Cu and overall cage length on going from $diCu^{I}$ to average valence. The smaller imBT and amBT cages appear to enforce a short Cu· · ·Cu distance, irrespective of whether a pair of Cu^{I} or $Cu^{1.5}$ cations is accommodated.

The C_2 linked tren-capped cages are also well suited to accommodation of single cation guests. The aminocryptand (amBT), on account of its endo configuration, is only slightly longer than sep or sar analogues bridgehead to bridgehead but the metal-binding site is considerably larger as judged by the dimensions of the N_3 triangle (Table I).

Calculations (170) suggest that amBT should form its most stable mononuclear complexes with the larger cations Hg^{2+}, Cd^{2+}, and Pb^{2+}. X-ray structures of two Cd cations have been reported (170), $[Cd(amBT)](BF_4)_2$ has the lel_3 conformation and $[Cd(amBT)](OAc)_2$ is ob_3 (see Section V.J); however, NMR data were interpreted as lel_3 for both complexes.

Although in sep and sar, a change from lel to ob results in a compression of the cage along the trigonal axis, this is not the case in amBT. The $N_{br}—N_{br}$ distance is slightly longer and the N_3 triangle slightly smaller in ob_3. The Cd—N distances are shorter in the lel_3 complex but there is no marked difference in the angles about Cd, the main constraint seems to be in the bite of the en linkages, leading to intrastrand N—Cd—N angles of about 70°, similar to those found in Cd complexes of en. The Pb^{II} complex (171), though of less regular geometry, is ob_3, as is the free ligand (172). As with the Sargeson cryptands, amBT displays a high degree of preorganization for metal binding. The most striking difference between amBT and its mononuclear cryptates is in the twist angle (φ); this is a consequence of the preference of bound metals for geometry closer to octahedral than to trigonal prismatic and the increased twist results in a shorter $N_{br}—N_{br}$ distance.

The related cryptand imBT introduces more rigid α-diimine spacers and, by virtue of the change from sp^3 to sp^2 at nitrogen, introduces major changes in the preferred C—N—C geometry and the orientation of the coordinating lone pairs. In the free hexaimine ligand, the imine groups are tangentially disposed with respect to the C_3 cylinder of the cryptand so that the lone pairs are not directed into the cavity. In addition, the imine groups on each strand are trans rather than cis. These are general features of the series of cryptands with diimine spacers and the reorganization required for binding might be expected to result in lower stability constants then in the equivalent amine series. The mono imBT complexes structurally characterized to date tend to be those of smaller cations than for amBT together with cations of metals with more preference for the softer imine donors. As expected, the torsion angle of the spacer is small,

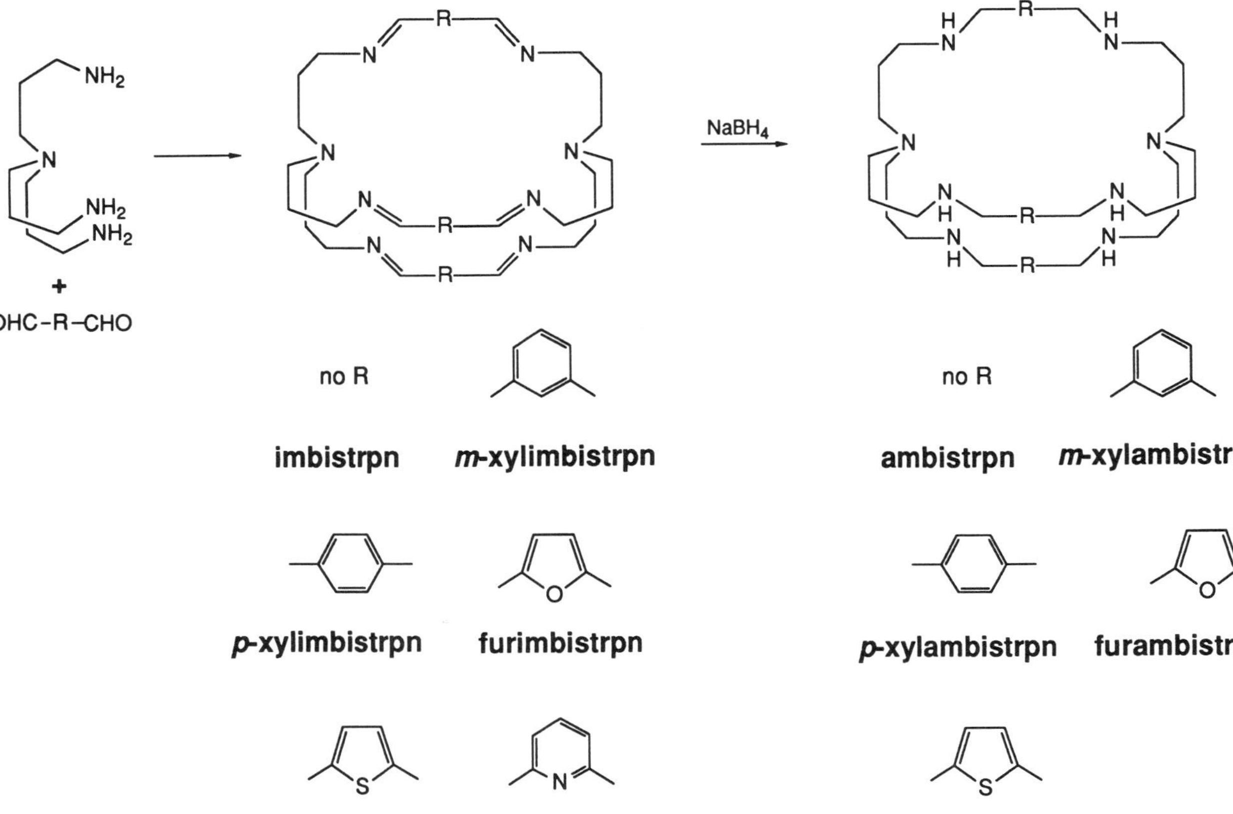

Figure 52. Imbistrpn series of Schiff-base cryptands and reduced ambistrpn derivatives.

though the twist angle is closer to 60° than in amBT. The exception to this is the irregular geometry of the eight-coordinate Gd complex (see Section V.J, Fig. 82) where the strands are opened up sufficiently to allow two water molecules into the coordination sphere.

The small cages imBT, amBT, and imbistrpn have 8 or 10 atoms per cryptand strand, and are intermediate in cavity size between Sargeson cages such as sep or sar with 6 atoms per strand and azacryptands with 5- or 6-atom links between tren-capped ends and thus 11 or 12 atoms in each strand. These larger cryptands (Figs. 11, 22, and 24) have a pair of well-separated N_4 coordinating sites together with the more flexible tren-derived caps, and are thus potentially dinucleating for all except the largest cations.

Aminocryptands such as O-bistren, C5-bistren, and NH-bistrpn or C3-bistrpn (Fig. 11) can be made by multistage reactions involving tripod coupling of the appropriate tritosyl and triol or trimesyl reagents (29). Another route successfully applied to generation of, particularly, aromatic-spaced aminocryptands is Schiff-base condensation of dicarbonyls with triamines tren (Fig. 22) or trpn (Fig. 52), followed by borohydride reduction (71, 173). Both routes have been used to generate the small cryptand amBT (172, 174). The latter route is much simpler synthetically, but possibly because of the greater flexibility of aliphatic dialdehydes, does not succeed (72, 173) in the synthesis of cryptands with long aliphatic links between the tren capped ends.

a. Hexa-Schiff-Base Cryptands. The hexaiminocryptands imBT ↔ *p*-xylimBT show good coordination properties for copper as Cu^I (175, 176). Dicopper cryptates can be made by direct reaction with free ligand, where available, or more commonly in better yield by template synthesis on a Cu^I salt. Crystal structures show that the Cu^I is normally coordinated by the tren-derived caps in approximate trigonal pyramidal geometry (Fig. 53), with the Cu cation held just outside (i.e., away from the bridgehead) the imino N_3 plane, although in some cases [Fig. 53(*c*)] (177) the Cu^I ion is pulled well outside this plane by the attraction of donors in the linker strand.

Most of these Cu^I_2 cryptates are inert toward atmospheric oxidation; cyclic voltammetric studies show irreversible or occasionally quasireversible oxidation waves indicating structural or chemical change on oxidation (179). Only in two cases besides imBT, that is, pyimBT and pyroimBT, both of which have potential N donors in the spacer link, does electrochemical oxidation indicate the presence of even transient mixed-valent forms of the copper cryptate. In the case of pyimBT, as broadening of the cyclic voltammogram suggests, the corresponding oxidized cryptate is unstable; we were unable to isolate the dicopper(I/II)$[pyimBT]^{3+}$ cryptate in a pure form in the solid state (177). The one-electron oxidized product of $[Cu_2pyroimBT]^{2+}$ which is isolable as a solid complex (180) behaves as localized mixed valent, due to nonidentical coordination sites deriving from a necessarily unsymmetric coordination of the single deprotonated pyrrole N donor; in the two-electron oxidized Cu_2^{II} cryptate where three deprotonated pyrrole N donors are involved, similarly unsymmetric coordination sites exist, as illustrated by the crystal structure (Fig. 54).

Evidence of steric strain is furnished by the facile ring-opening reactions of the dicopper(II) cryptates of the Schiff-base ligands, and to a lesser extent by a destabilizing loss of planarity in the conjugated bis-iminoaromatic assembly in some dicopper(I) structures, particularly *m*-xylyl linked hosts. Here, in both tren- (181) and trpn-capped (182) cryptates (Fig. 55), the bisiminophenyl assembly has had to twist out of plane to accommodate the pair of Cu^I cations, presumably because of hindrance arising from projection of the α-aromatic CH into the cavity. Even so, the Cu^I ions lie within van der Waals contact distance of the α-aromatic CH. The copper cations are separated by 4.23 and 4.44 Å for tren- and trpn-capped hosts, respectively; the dihedral twist of imino C=N vectors in each strand is 36.3 and 45.5°.

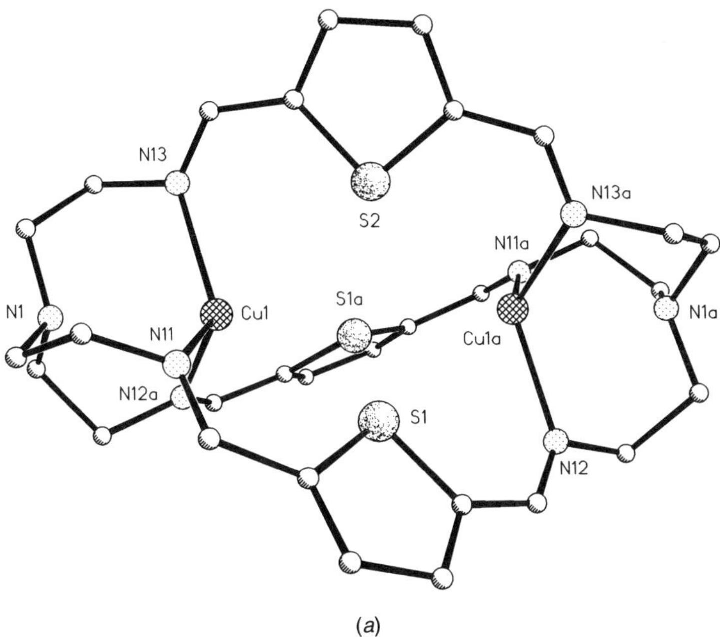

Figure 53. X-ray structures (176–178) of dicopper(I) azacryptates: (*a*) $[Cu_2thioimBT]^{2+}$, (*b*) $[Cu_2furimBT]^{2+}$, and (*c*) $[Cu_2pyimBT]^{2+}$. Selected internuclear distances (Å): (*a*) Cu—Cu 4.73; Cu—S_{th} 3.20; Cu—N_{br} 2.56; Cu—N_{im} 2.24: Cu—N_3 plane 0.5 and (*b*) Cu—Cu 4.20; Cu—O 3.18–3.28; Cu—N_{br} 2.39; Cu—N_{im} 2.00; Cu—N_3 plane 0.34. (*c*) Cu—Cu 3.04; Cu—N_{py} 2.64, 3.06, 3.01; Cu—N_{im} 2.20 Cu—N_{br} 2.75; Cu—N_3 plane 0.63.

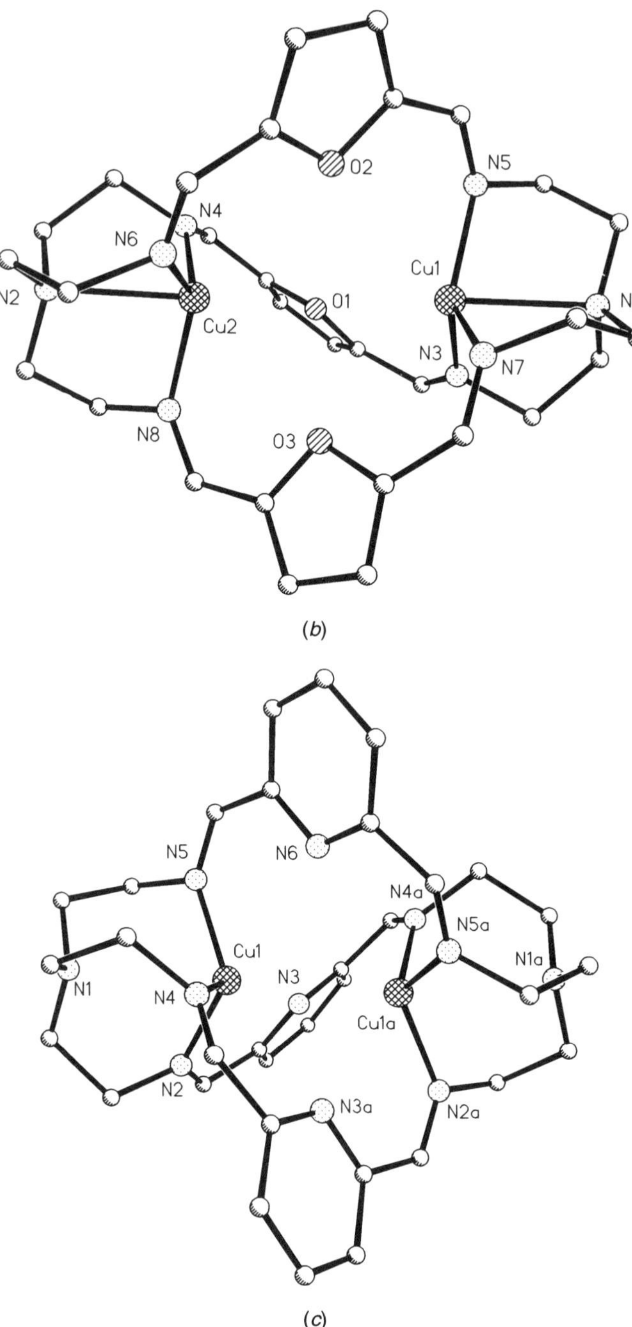

Figure 53. (*Continued*.)

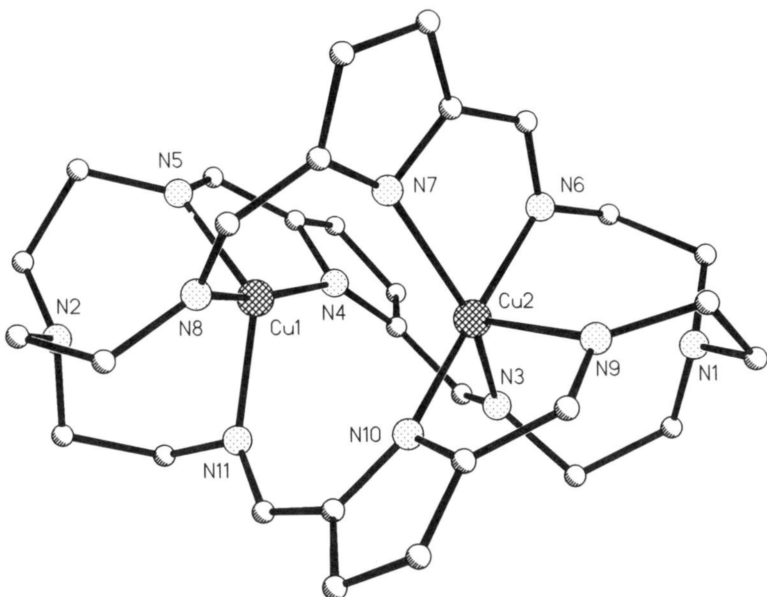

Figure 54. Structure of dicopper(II) cryptate of triply deprotonated pyrrole-based cryptand (180): Cu_2[pyroimBT-3H]$^+$. Selected internuclear distances (Å): Cu—Cu 3.01: Cu(1)—N_{py} 2.01; Cu(1)—N_{br} 2.76 Cu(1)—N_{im} 2.06, 2.03, 1.98; Cu(1)—N_{im} 0.71 Cu(2)—N_{py} 2.03, 2.15; Cu(2)—N_{br} 3.15; Cu(2)—N_{im} 2.00, 2.04, 2.21; Cu(1)—N_{im} 0.98.

Dicopper(II) cryptates of some of the Schiff-base ligands can be isolated as μ-hydroxo dimers following direct reaction of free ligand with hydrated Cu^{II} salt. This behavior is reminiscent of that of the dicopper(II) O-bistren cryptate, which shows a very high affinity for hydroxo anion as bridging ligand (see Section V.F), Table VII). The rigid steric constraints of the hexaimino cage enforce a linear Cu—O(H)—Cu geometry, which leads, (Fig. 56) to efficient antiferromagnetic interaction between Cu^{II} paramagnets generating ESR silence (175, 176).

The dicopper(II) cryptate of the pyrrole-linked cage pyroimBT, however, makes no use of hydroxo bridges, preferring instead the strong coordination of the three deprotonated monocoordinating pyrrole donors. The ESR spectra are of normal intensity but show no hyperfine coupling pattern, and consist of a single broadened $g \approx 2$ main band signal and a weak seven-line hyperfine split $\Delta M_s = 2$ half-band, which arises from weak interaction between the two different Cu^{II} sites that are revealed in the crystal structure (180). The generation of an anionic ligand via deprotonation of the NH function from the spacer link is valuable in that it reduces the tendency to metal-assisted hydrolytic ring-opening attack on the imine bonds. which occurs readily with neutral hexa-

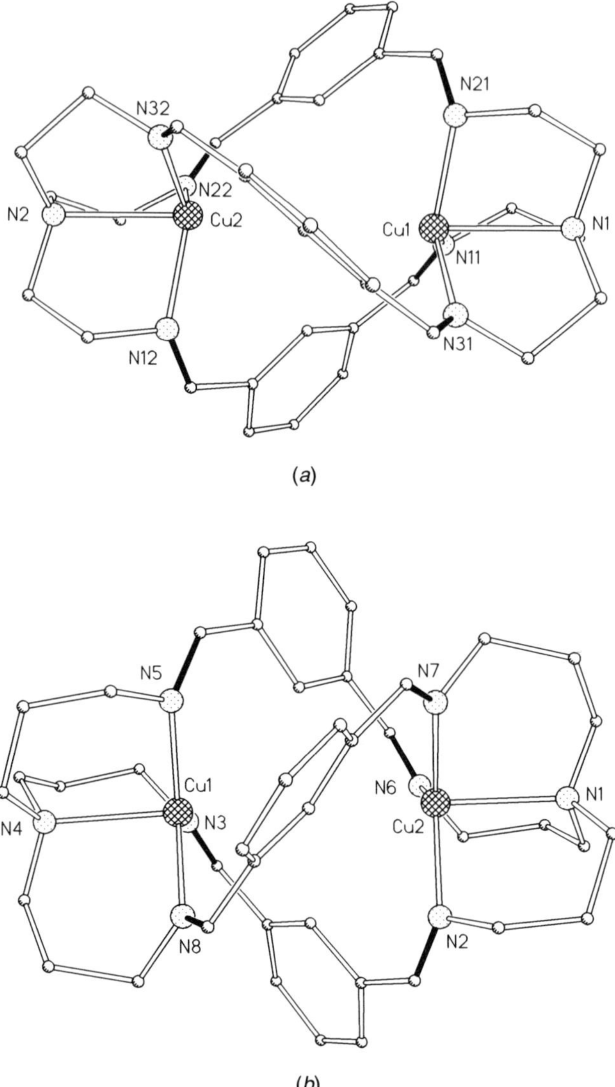

Figure 55. Structures of *m*-xylyl spaced cryptands (181, 182) to illustrate deviation from planarity of diiminoxylyl moiety: (*a*) $Cu_2[xylimBT]^{2+}$ and (*b*) $Cu_2[xylimbistrpn]^{2+}$. Selected dimensions (Å): (*a*) Cu—Cu 4.23; Cu—$\underline{C}H_{ar}$ 3.00–3.35; Cu—N_{br} 2.33; Cu—N_{im} 1.99: Cu—N_3 plane 0.29 and (*b*) Cu—Cu 4.44 Cu—$\underline{C}H_{ar}$ 3.29; Cu—N_{br} 2.28; Cu—N_{im} 2.09: Cu—N_3 plane 0.06 N=C(CCC)C=N. Torsion angles (*a*) 36.3° and (*b*) 45.5°.

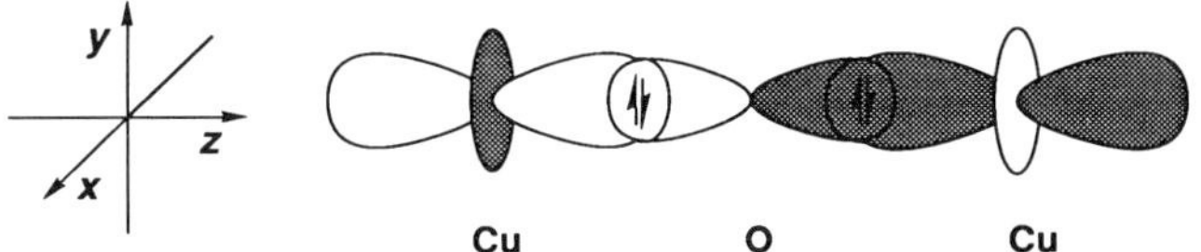

Figure 56. Orbital overlap of Cu^{II} d_z^2 and O $2p_z$, colinearly disposed.

Schiff-base cryptands. Another proton-ionizable azacryptand (**XIV**) (Fig. 57), deriving from [2 + 3] condensation of 3,5-diformyl pyrazoledicarbaldehyde with tren, generates both di- and tetracopper(II) cryptates on treatment with the appropriate stoichiometric amount of Cu^{II} salt (183).

Solution complexation studies (see Section F, Table II) reported (184) for some of the Schiff-base cryptates show that in the case of xylimBT the equilibria are not sufficiently stable over time to establish quantitative complexation parameters for the relatively strong Lewis acid M^{II} cations. This finding reinforces our synthetic experience: $[Cu_2OH]^{3+}$ and pairs of other M^{2+} cations are associated with metal-assisted hydrolytic attack on iminocryptand C=N functions, particularly in xylimBT, generating (Fig. 58) ring-opened derivatives such as [**XV**] (175).

The M^{I} cations such as Ag^+, which have only weak Lewis acidity, can be safely accommodated and tenaciously retained by the soft sp^2 N-donor Schiff-base receptors (175, 176, 178). For disilver cryptates, the range of Ag· · ·Ag distances (185) revealed within the very similar (as regards coordination site)

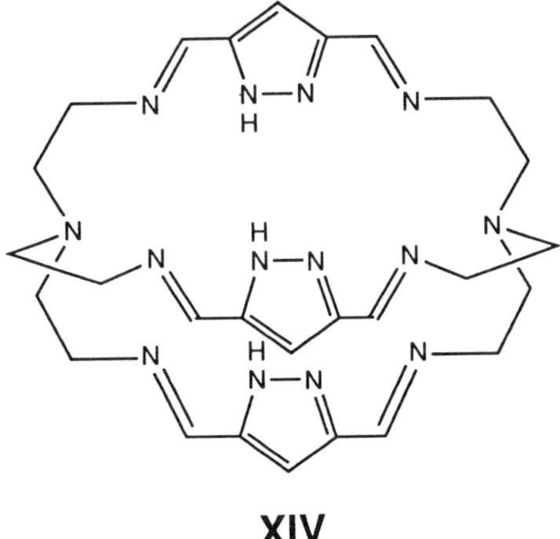

Figure 57. Pyrazolate cryptand **XIV**.

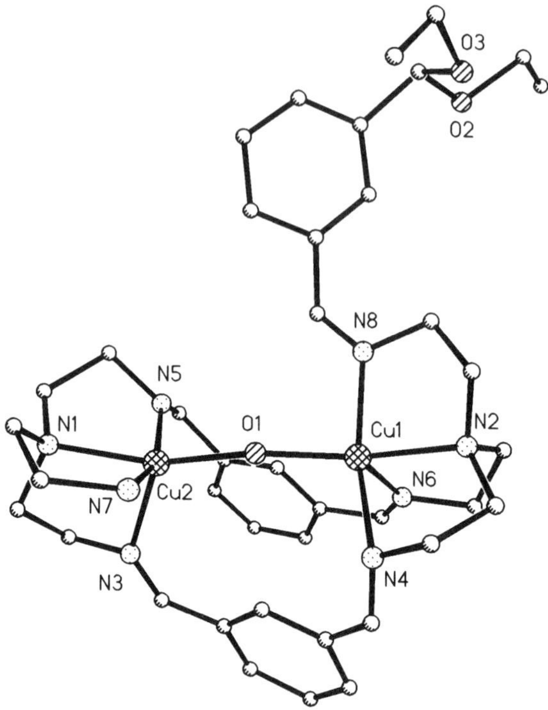

Figure 58. Ring-opened derivative. **XV**, of xylimBT: crystal structure (175) of $[Cu_2OH\mathbf{XV}]^{3+}$

N
N
N
N
H_2N
OEt
OEt
N
N

XV

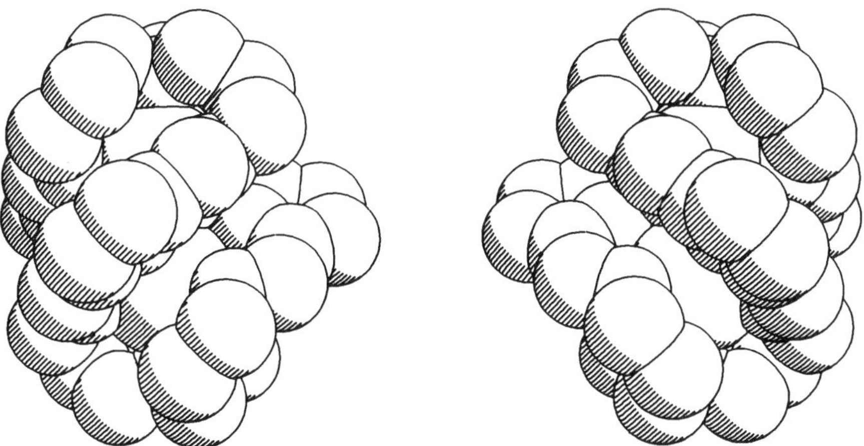

Figure 59. The two chiral forms of $Ag_2[furimbistrpn]^{2+}$ (185), showing the triple helical conformation.

cryptand hosts, both tren and trpn capped, is intriguing. Internuclear distance appears more responsive to the nature of the noncoordinating bridging link than to the size of the methylene caps. Differing helicity in the strands (these azacryptates adopt a triple helical conformation: Fig. 59) allows the Ag^+ cations to approach each other to the same extent unless hindered by steric barriers that some, for example, *p*-xylyl or thiopheno- spacer groups, may present. In the lower limit, achievable with furan-spaced cryptands (Fig. 60), the Ag· · ·Ag distance is close to that in elemental silver, but in some of the other hosts, considerably longer Ag· · ·Ag internuclear distances are observed.

The idea that Ag· · ·Ag "bonding" may be involved in the more closely spaced distances has received support from thermodynamic measurements, which reveal positive enthalpic cooperativity (see Section V.F, Table III) for the addition of a second Ag^+ cation to the monosilver cryptate (184).

Monocationic guests that may be encapsulated without danger of hydrolytic attack (particularly where donor atoms exist in the linker units) are the Group 1 (IA) and 2 (IIA) alkali or alkaline earth cations. The host cavity size is too small in the furimBT–pyimBT series to accommodate two alkali metal cations, except for lithium, where no strong complexation is observed (184). For the Gp 1 series, the only cryptates isolated in the solid state were the monosodium salts. These cryptates adopt, in one or all strands, the cis–trans conformation of C=N imino groups (with respect to the C—X heteroatom bond), which is typical of the free ligand conformation. The ^{23}Na NMR shows that these sodium

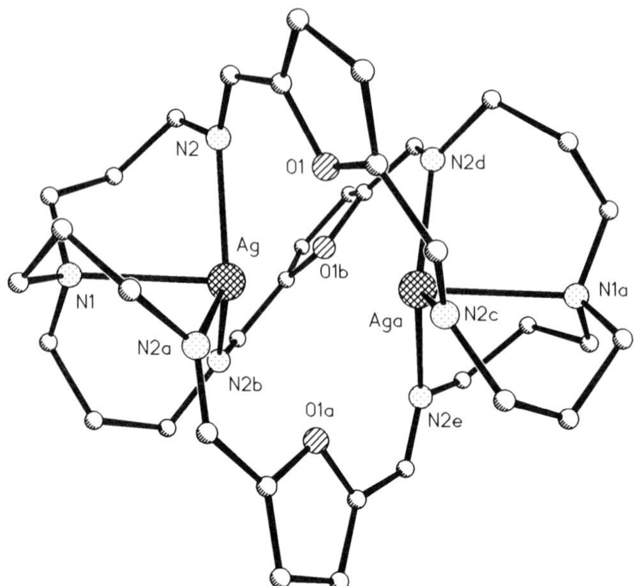

Figure 60. Structure of Ag_2[furimbistrpn]$^{2+}$: selected internuclear distances (Å): Ag—Ag 3.05; Ag—N_3 plane 0.23; Ag—N_{im} 2.34; Ag—O 2.91; Ag—N_{br} 2.51.

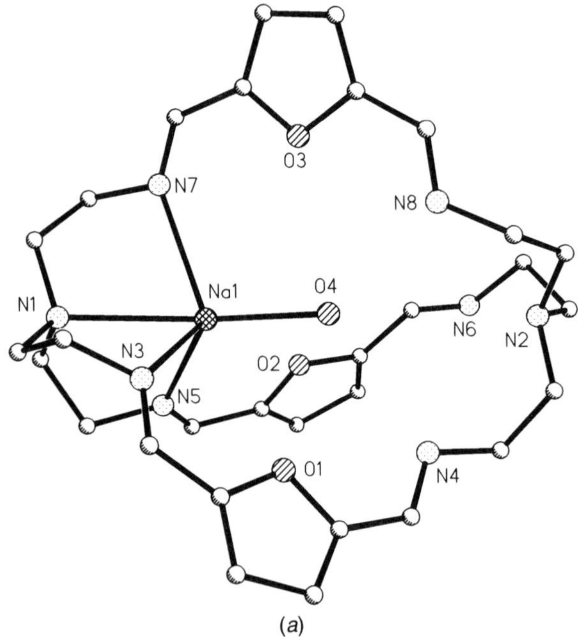

Figure 61. X-ray structures of monosodium cryptates (186): (*a*) Na[furimBT]$^+$ and (*b*) Na[pyimBT]$^+$. Selected internuclear distances (Å): (*a*) Na—N_{im} 2.57; Na—N_{br} 2.65; Na—O_{fur} 3.14—3.66; Na—O_w 2.24; O_w—N_{im} 2.83, 3.00; C=N(COC)N=C angle in non-hydrogen-bonded strand 170°. (*b*) Na—N_{im} 2.42; Na—N_{br} 3.12; Na—N_{py} 2.66; C=N(COC)N=C angle$_{av}$ 151.8°.

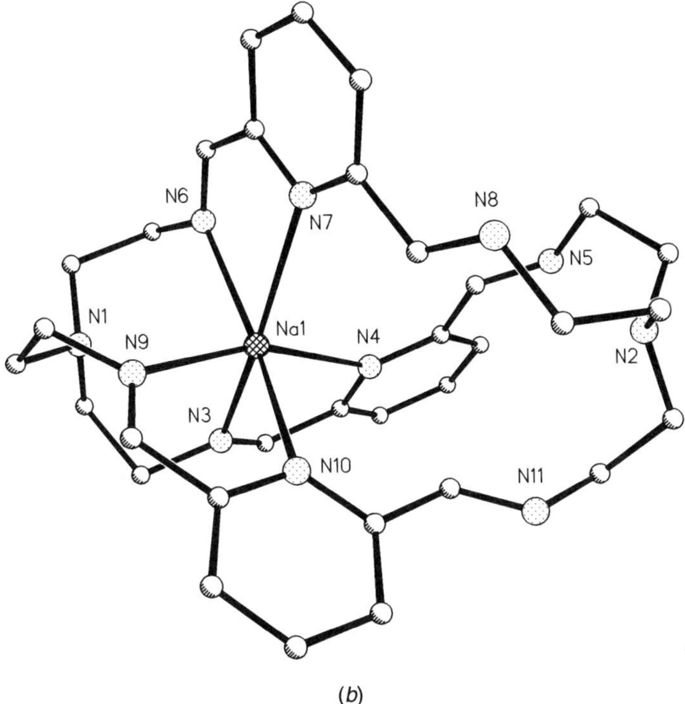

Figure 61. (*Continued.*)

cryptates (Fig. 61) are kinetically stabilized toward decomplexation (186) even though (see Section V.F, Table II) thermodynamics of cryptate formation are relatively weak. None of the other Gp 1 cations show important coordination of these cryptands.

b. Octaamino Cryptands. The octaamino ligands obtained by borohydride reduction of hexa-Schiff-base cryptands are chemically resistant to hydrolytic degradation and represent satisfactory hosts for Lewis acid cations, either singly or in pairs. This fact is demonstrated by the isolation of a wide range of stable dicopper(II) octaamino-cryptates, bridged and unbridged (175, 176, 178). At pH values close to neutral some of these are obtained in the μ-hydroxo $[Cu_2(OH)L]^{3+}$ form, as predicted by speciation experiments (187), and others as unbridged dicopper(II) salts (177, 180), but where the OH^- bridge is present, it can be replaced by various anionic bridges, allowing a series of "cascade" complexes to be identified (175–178, 188). In this cascade complexation, a dinuclear assembly with a pair of free coordination sites hosts a dicoordinating bridge of appropriate dimension, which then links the pair of cations; when the cations are paramagnetic the bridging link mediates their magnetic exchange interaction. Variable temperature magnetic susceptibility studies demonstrate the effect of the unusually large (approaching linear in most cases) bridging angle on mediation of the interaction between Cu^{II} paramagnets in cascade complexation of bridging anions such as OH^-, imidazolate, N_3^-, and NCO^- (175, 176, 189). Although the isoelectronic azide and cyanate anions generate isomorphous complexes (Fig. 62), their magnetic properties are not identical, as will be seen.

Where the Cu—O(H)—Cu angle is linear (Fig. 63) (190), efficient antiferromagnetic interaction is assured by overlap of d_z^2 magnetic orbitals with O_{2pz} of the bridge as illustrated for the more rigid hexaimino cryptate hosts (Fig. 56). However, not all the dicopper μ-hydroxo cryptates of the octaamino series show such strong interaction, suggesting that in some cases operation of steric constraints, such as repulsion between aromatic CH or thiopheno-S lone pair and bridging OH^- may dominate. A bent Cu—O(H)—Cu assembly or a long Cu—OH· · ·OH_2—Cu bridge then ensues generating only moderate antifer-

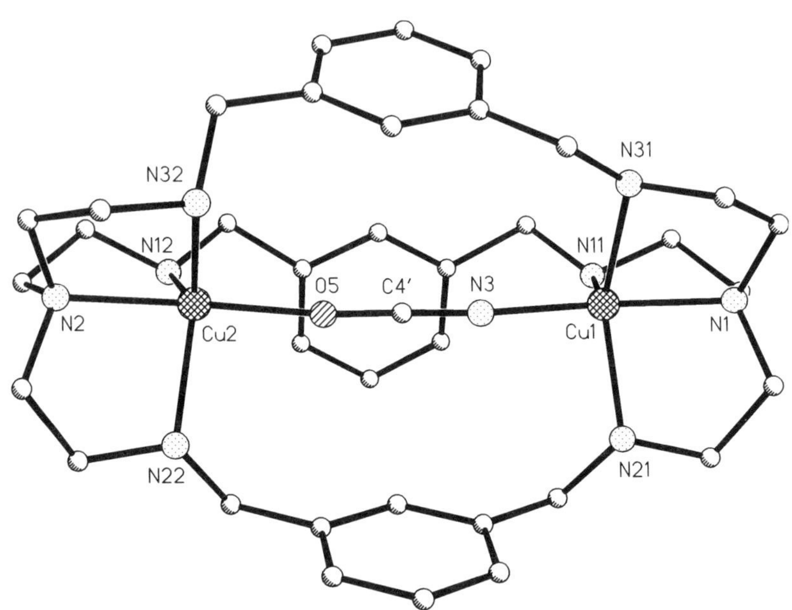

Figure 62. Cascade complexes of dicopper cryptate with pseudohalides: Structure of Cu_2[xylamBT(X)] X = NCO^-; complex with X = N_3^- is isomorphous (189). Selected dimensions (Å): Cu—N, O 1.95; Cu—N_{im} 2.12; Cu—N_{br} 2.05; Cu—NN (—NC or —NO) angle 170.5°.

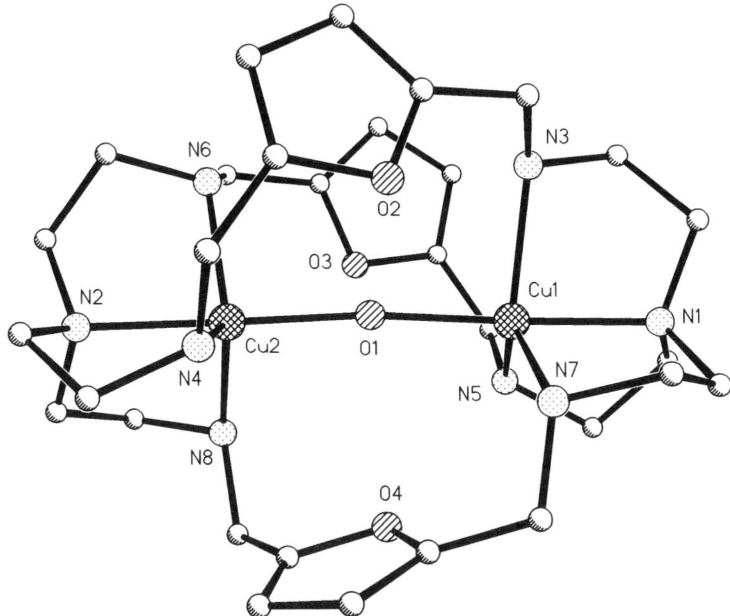

Figure 63. Linear hydroxo-bridged dicopper cryptate: Structure of $Cu_2OH[furamBT]^{3+}$ (190). Selected internuclear distances (Å): Cu· · ·Cu 3.90; Cu—N_3 plane 0.22; Cu—N_{br} 2.07; Cu—N_{am} 2.15; Cu—O—Cu angle 174°.

romagnetic interaction; in the first case via a more normal, approximately tetrahedral, Cu—O—Cu angle. This behavior is observed (175, 178, 188) for dicopper(II) μ-hydroxo cryptates of xylamBT, thioamBT, and *p*-xylamBT.

A μ-1,3,-carbonato-bridged complex of xylamBT, formed via CO_2 absorption on standing in air in basic solution, has been structurally characterized (191) and shows (Fig. 64) relatively large Cu—O—C angles of 134–154°. Because of the large bridging angles, this cryptate might be expected to exhibit moderately strong antiferromagnetic interaction, but magnetic suceptibility studies have yet to be reported.

Magnetic interaction mediated via imidazolate shows less variation from normal values, as a Cu—N· · ·N—Cu angle close to linear is the norm in such situations. Moderate antiferromagnetic interaction is the rule in this case, accompanied by ESR spectra (Fig. 65), which testify to thermal population of an excited triplet state (175, 176, 178, 188). Imidazolate-bridged dicopper(II) cryptates of the easily generated aminocryptand, xylamBT, have been exploited in experiments aimed at synthesis of a heterobinuclear Cu· · ·Zn assembly capable of superoxide dismutation (192). Both dicopper(II) and copper–zinc μ-imidazolate (Fig. 66) cryptates exhibit catalytic activity toward dismutation of superoxide anion, as supplied enzymatically from the hypoxanthine–xanthine oxidase reaction. The catalytic activity is significantly lower than in natural SOD, but of the same order as the best synthetic SOD mimics.

The versatile azido ligand, capable of terminal coordination and/or bridging in 1,1 or 1,3 mode, shows a different behavior. The steric constraints of the cryptand enforce the otherwise unknown colinear geometry on the $N_{br}-M-NNN-M-N_{br}$ assembly, which gives rise to unusual and characteristic spectroscopic signatures for the encapsulated azide anion. For example, the IR $\nu_{as}(N_3^-)$ absorption appears at anomalously high wavenumber, close to 2200 cm^{-1}, in a variety of bimetallic cryptate hosts [diiron, dinickel, dicobalt (188) as well as dicopper (175, 176, 178, 188, 189)]. For linearly bridged μ-azido dicopper, ligand-to-metal charge transfer (LMCT) electronic absorption appears at abnormally low (~400–450 nm) energy, and a complex and unusual X-band triplet-type ESR spectrum is seen [Fig. 67(*a*)]. This spectrum appears to derive from the circumstance where the microwave quantum is of the same order as the zero-field splitting, because the *Q*-band spectrum (193) (Fig. 67(*b*)] is much simpler.

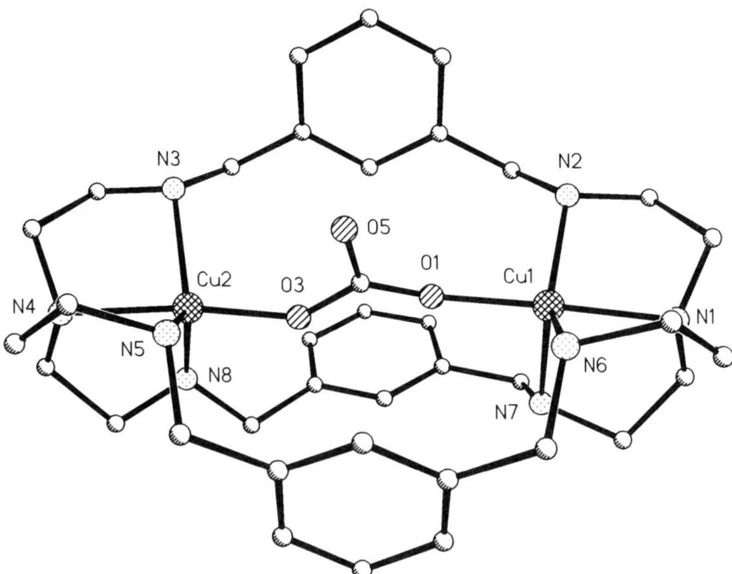

Figure 64. Carbonato-bridged cascade cryptate: Structure of $Cu_2(\mu CO_3)[xylamBT]^{3+}$ (182). Selected dimensions (Å): Cu· · ·Cu 5.85; Cu—O 1.85; CuOC angle 154.6, 134.8°.

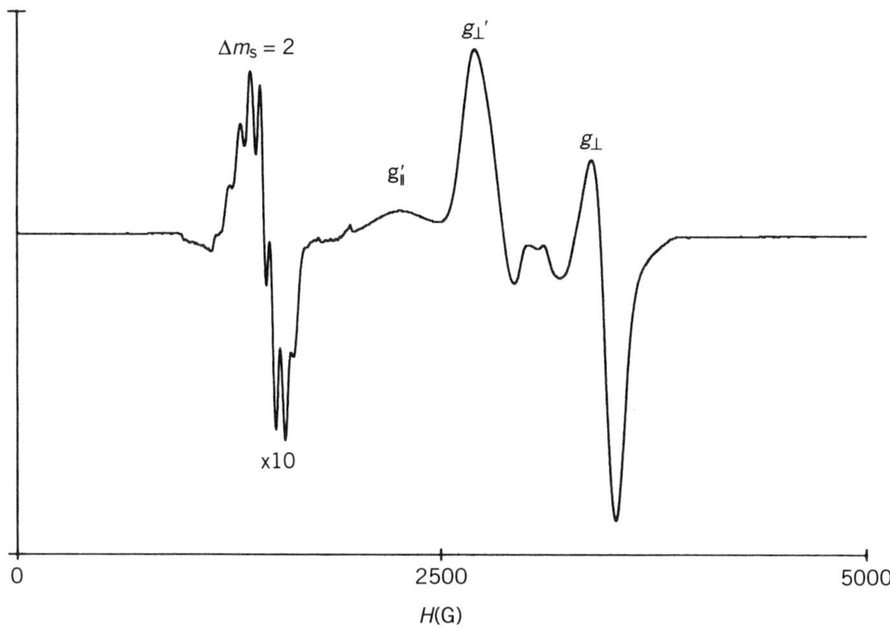

Figure 65. *X*-band ESR spectrum of Cu_2(imidazolate)[xylamBT]$^{3+}$ in frozen DMF solution at 80 K (175). Because of zero-field effects the signals are split into two, for example, $g_\perp$ and $g'_\perp$ components. Another unusual characteristic of triplet-state spectra is the observation of the normally forbidden $\Delta m_s = 2$ transition, split into seven lines by coupling to both spin $\frac{3}{2}$ copper nuclei, with one-half of the normal hyperfine coupling.

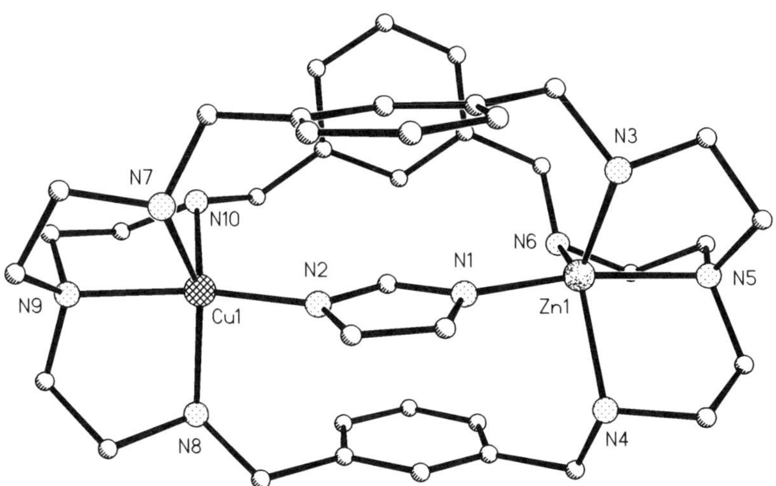

Figure 66. Heterobinuclear copper–zinc cryptate: Structure (192) of Cu—Zn(imidazolate)[xylamBT]$^{3+}$.

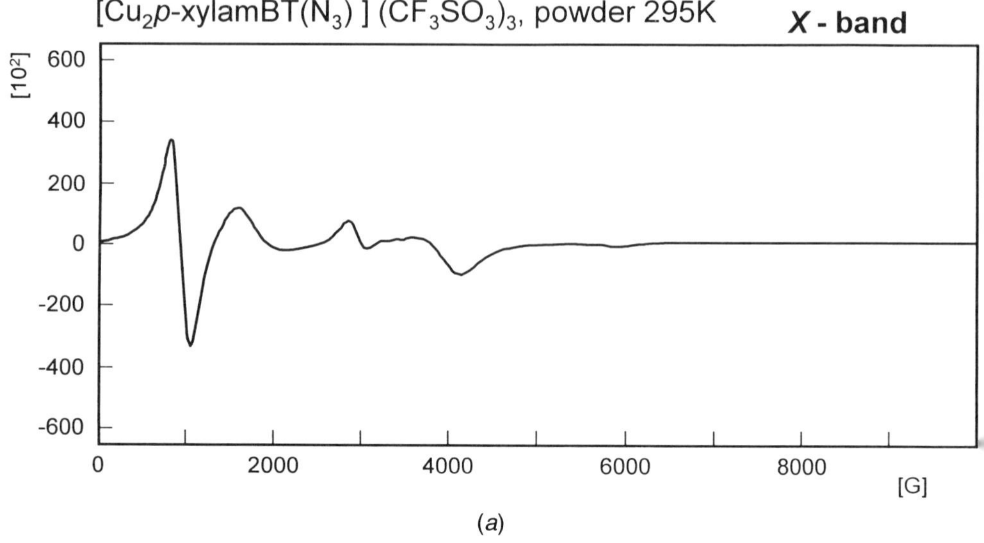

(*a*)

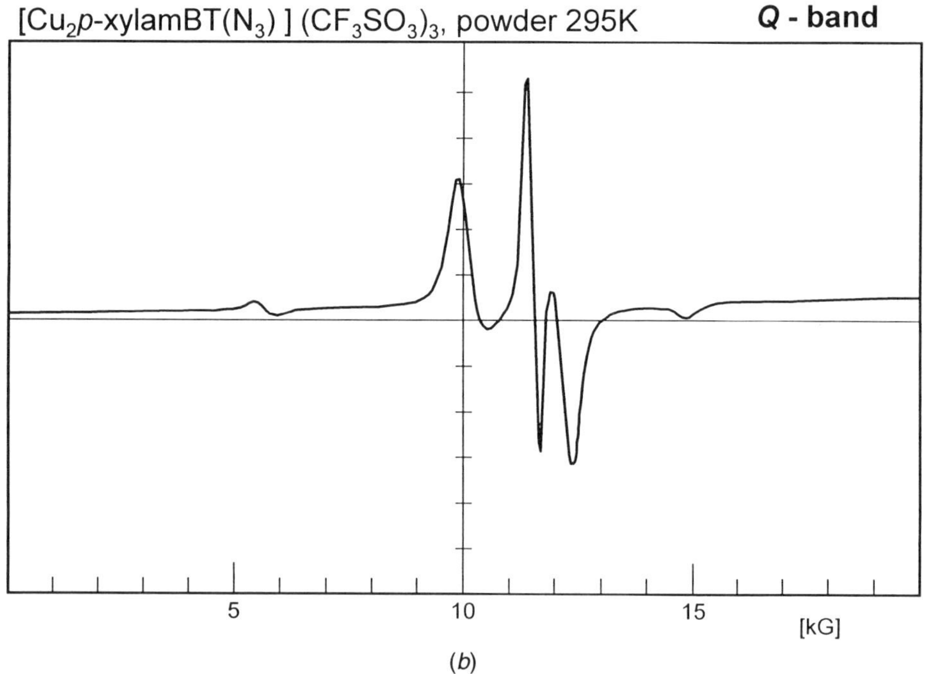

(*b*)

Figure 67. ESR spectra (295 K) of linear azido-bridged dicopper cryptates (193): $Cu_2(N_3)$[*p*-xylamBT]$^{3+}$. (*a*) *X*-band polycrystalline spectrum; (*b*) *Q*-band polycrystalline spectrum. The *X*-band polycrystalline specrum cannot be easily analyzed into $g_{\parallel}$ and $g_{\perp}$ components, which are recognizable in the *Q*-band spectrum.

Magnetic susceptibility measurements (175, 176, 178, 188, 189) confirm that where these unusual azide spectral signatures are evident, the triplet state is somewhat stabilized with respect to the singlet, giving weak ferromagnetic interaction within the dicopper assembly. As mentioned earlier, the isoelectronic NCO^- bridge behaves differently, in mediating a weak but definite antiferromagnetic interaction between the Cu^{II} paramagnets (Fig. 68). Finally, where no bridging ligand intervenes between the paramagnets (as for pyimBT and pyroimBT dicopper(II) cryptates), no significant magnetic interaction of either sense is observed despite close approach of copper nuclei as shown in Fig. 54.

Comparison of the magnetic properties of the Schiff-base derived octaamino dicopper(II) cryptates of the series furimBT to xylimBT and the analogous O-bistren and C5-bistren dicopper(II) cryptates should be interesting, but magnetic and spectroscopic information is as yet only partially available (194, 195) for the latter systems. Structural data are available, however, and the X-ray crystallographic structure determination of [Cu_2OH O-bistren] (196) (Fig. 69) shows a large but not linear Cu—O—Cu angle. No magnetic properties are reported for this green, μ-hydroxo, form of the dicopper(II) cryptate, but a blue form (194) has a room temperature magnetic moment of 2.00 BM/Cu^{2+} ion, in line with the observation of a normal intensity ESR spectrum for that complex,

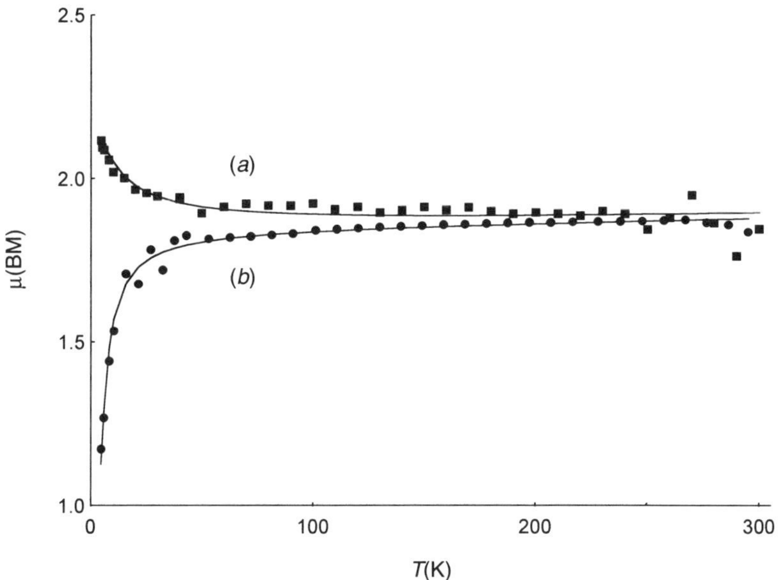

Figure 68. Magnetic interaction in pseudohalide bridged dicopper(II) cascade complexes (189): $Cu_2(X)[xylimBT]^{3+}$. (*a*) $=N_3^-$ and (b) X$=NCO^-$. The increase of moment, μ, at low temperature in (*a*) is due to ferromagnetic and the decrease in (*b*) to antiferromagnetic interaction.

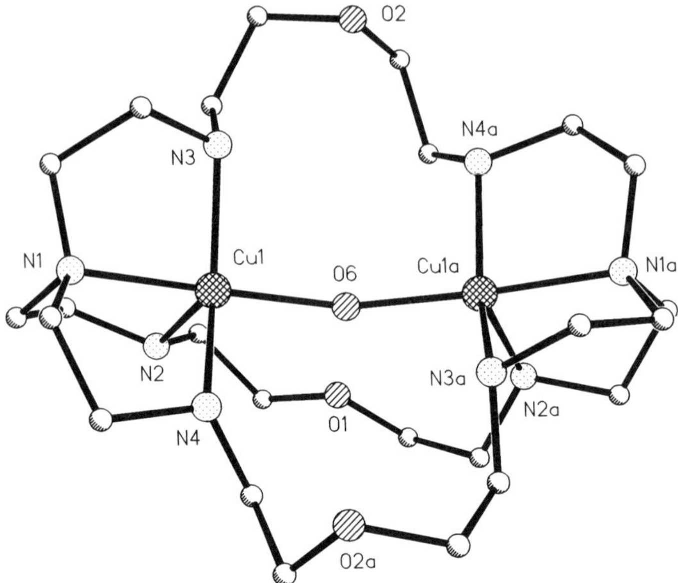

Figure 69. Hydroxo-bridged dicopper(II) cryptate of O-bistren: X-ray structure (196) of $Cu_2(OH)[O\text{-bistren}]^{3+}$. Selected dimensions (Å): Cu· · ·Cu 3.77; Cu—O 193; Cu—O—Cu 155.5°.

suggesting that this blue form of the complex is noninteracting, presumably because unbridged.

3. *Proton-Ionizable Schiff-Base Cryptands*

As we have seen, the easily made series of Schiff-base cryptands do not generally show sufficient hydrolytic stability to represent suitable hosts for M^{2+} or M^{3+} cations that have strong Lewis acid properties. Such Lewis acids can be accommodated, without danger of hydrolytic attack on the host, within aminocryptands or the other class of hydrolytically stable macrobicycles; the proton-ionizable, potentially anionic hosts containing phenol (Fig. 70) (197), pyrrole (180), or other acidic hydrogen functions (183) in the linker strands. The phenolate cryptands (Fig. 70) in particular, demonstrate good complexation properties for Lewis acid guests (Fig. 71b).

Cryptands such as cresimBT and TBPimBT (Fig. 70) have excellent capacity for encapsulation of M^{n+} cations of lanthanoid, Group 13 (IIIA) or transition series. The hosts are interesting in that they are pH sensitive. Under neutral

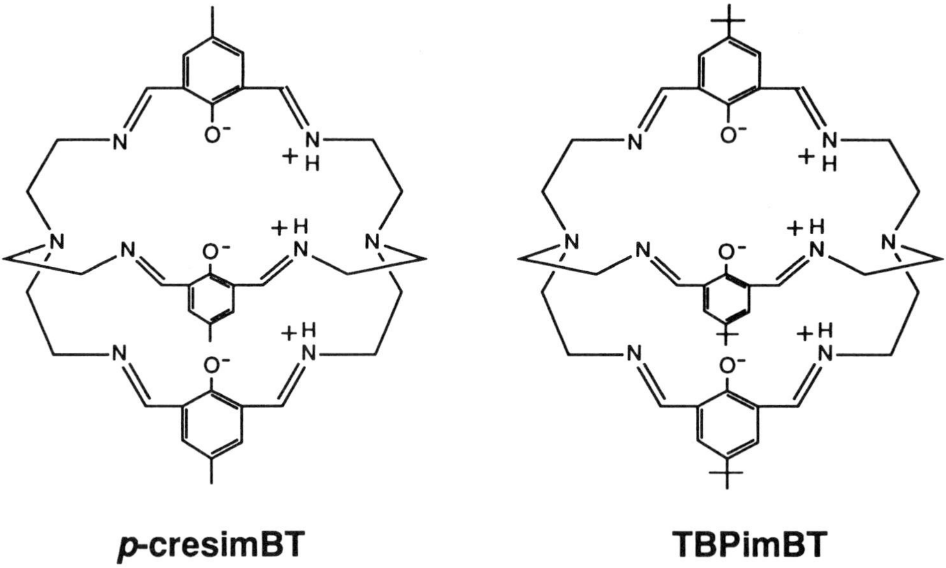

***p*-cresimBT** **TBPimBT**

Figure 70. Proton transfer in phenolate cryptands.

conditions, one M^{n+} cation only is accommodated (Fig. 72) (198), with proton transfer of three phenolate protons to the imine nitrogens at the other end of the cavity. This metal ion is often disordered between two possible sites that differ mostly in the positions of the imine nitrogens without requiring gross change in the shape of the cryptate cation. Under basic conditions, for example, use of acetate salt in transmetalation, a second M^{2+} cation can be accommodated within the crypt, presenting the opportunity for controlled synthesis of heterobinuclear cryptates (Fig. 73, see p. 247) (197, 199). One interesting facet of the coordination chemistry of these phenol-based cryptand hosts relates to the redox state in which transition series cationic guests are encountered. A comparison of iron and manganese monocryptates is instructive in this respect: the +III oxidation state cryptate is isolated starting either from Fe^{2+} or Fe^{3+} salts, unless careful inert-atmosphere conditions (197) are utilized, whereas the +II oxidation state cryptate is the invariable product of reaction with either Mn^{2+} or Mn^{3+} salts. The difference has been rationalized (200) in terms of the well-defined geometric requirements of Mn^{3+}, which are unattainable in the cryptand host.

In addition, dimanganese(II) and diiron(II) complexes of both ligands and a heterobinuclear $Fe^{II}Co^{II}$ cryptate (Fig. 73), synthesized under helium protection, have been obtained (64, 197) but diiron(III) cryptates have not. This finding may be rationalized on the basis of the greater repulsion expected between

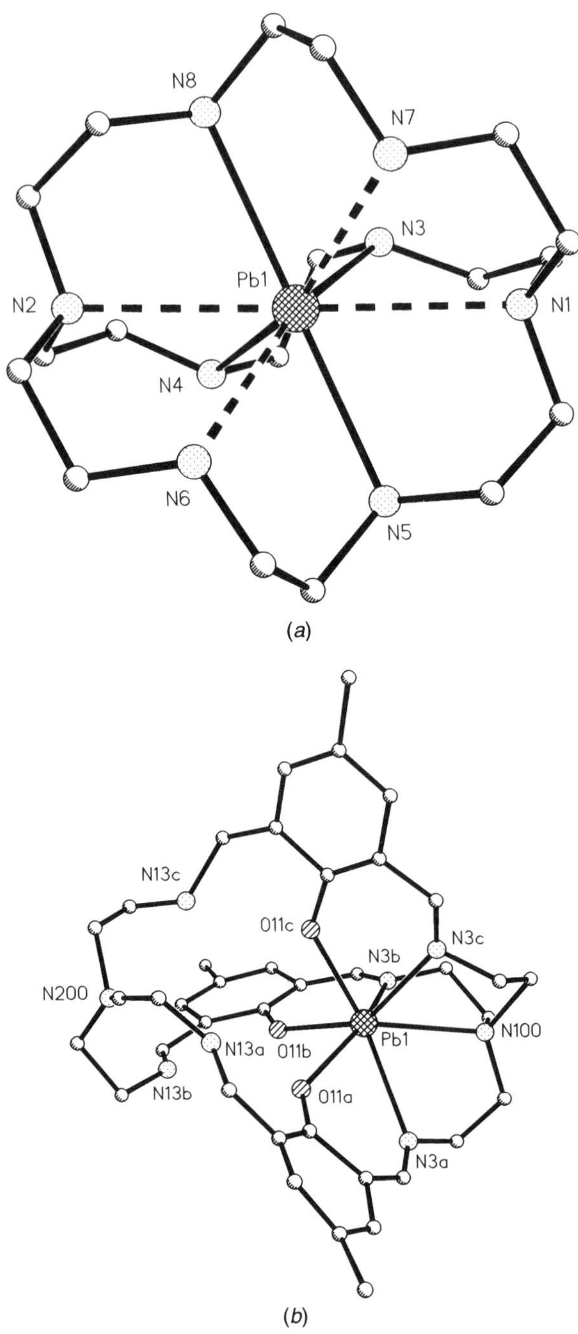

Figure 71. Structure of lead cryptates (171, 198): (*a*) Pb[amBT]$^{2+}$ and (*b*) Pb[cresimBT]$^{2+}$. Selected dimensions (Å): (*a*) Pb—N—2.69–2.74, · · · 2.81–2.87. (*b*) Pb—N 2.69; Pb—O_{Ph} 2.52 hydrogen-bond distance O· · ·NH^+ 2.74.

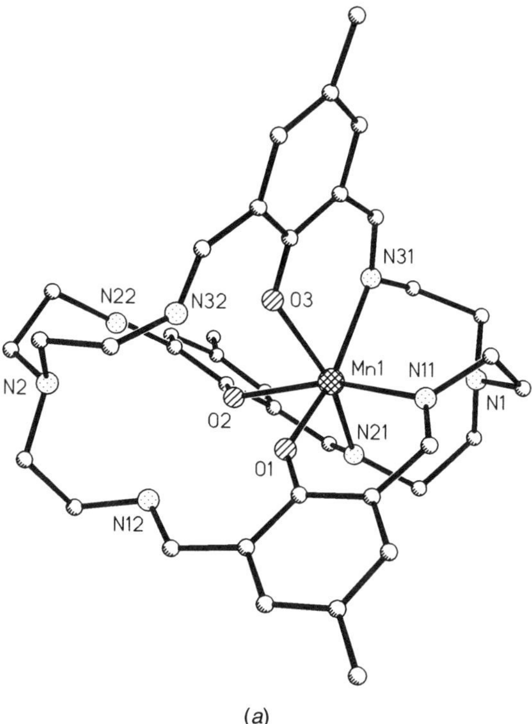

Figure 72. Mononuclear phenolate cryptates (200): (*a*) Mn[cresimBT]$^{2+}$ and (b) Fe[cresimBT]$^{3+}$.

closely spaced +III state cations. Only with the more flexible octaamino cryptand cresamBT are dinuclear cryptates of triply charged cations, such as Gd^{3+}, obtained (202). These cryptates have so far not been structurally characterized, but may show similar structures to that illustrated for the dizinc analogue in Fig. 74. The dimanganese(II) cryptates of cresimBT and TBPimBT show complex and interesting ESR spectra (64, 203) (Fig. 75) testifying to the presence of μ-OPh$^-$ mediated interaction between the paramagnetic sites, possibly via an excited state mechanism. However, in the series of homo- and heterobinuclear cresimBT $M^{II}M'^{II}$ cryptates, whose magnetic susceptibility (197) behavior has been studied as a function of temperature, there is little evidence for appreciable ground-state interaction.

The rare earth complexes of these phenolate cryptands are kinetically inert, as demonstrated chemically by the absence of precipitation in the presence of aqueous hydroxide or phosphate over days or weeks, and spectroscopically by the sharpness of the ^1H NMR spectra of diamagnetic cryptates; also by the

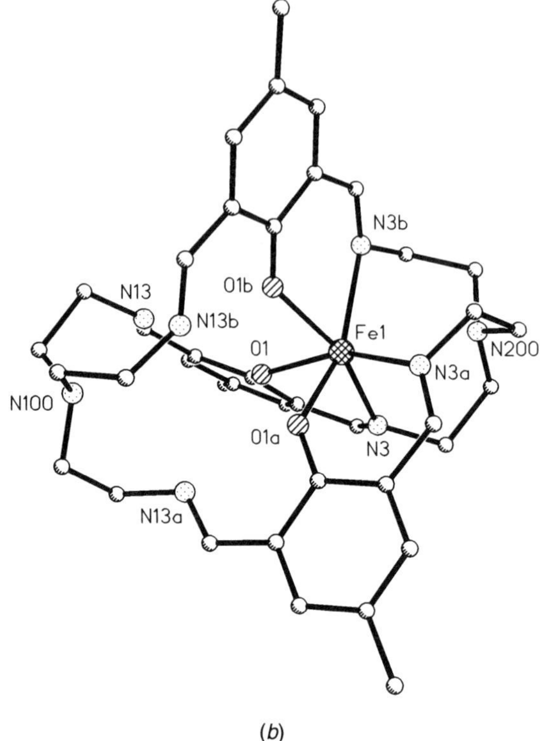

(*b*)

Figure 72. (*Continued.*)

appearance, where appropriate, of two nonexchanging signals for magnetically active encapsulated nuclei (204) (e.g., ^{45}Sc). This absence of kinetic lability, together with X-ray crystallographic characterization (Fig. 76, see p. 249) which, in more than one case, shows strong coordination of water, makes the Gd^{3+} cryptates attractive candidates for use as magnetic resonance imaging (MRI) agents. Relaxation rates around 8–9 s^{-1}, comparable with those of existing relaxation agents, have been observed (205) using partly nonaqueous solvent systems (DMSO/H_2O mixtures). However, achievement of the aqueous solubility needed for clinical application may require the use of the more soluble octaamino cryptands, the introduction of solubilizing substituents on the aromatic ring, or both.

Another potential application (particularly of the Schiff-base phenol-linked cryptands that incorporate hexaimino chromophores) derives from the long-lived fluorescence of Eu^{3+} and Tb^{3+}. The rapidly developing technique of fluoroim-

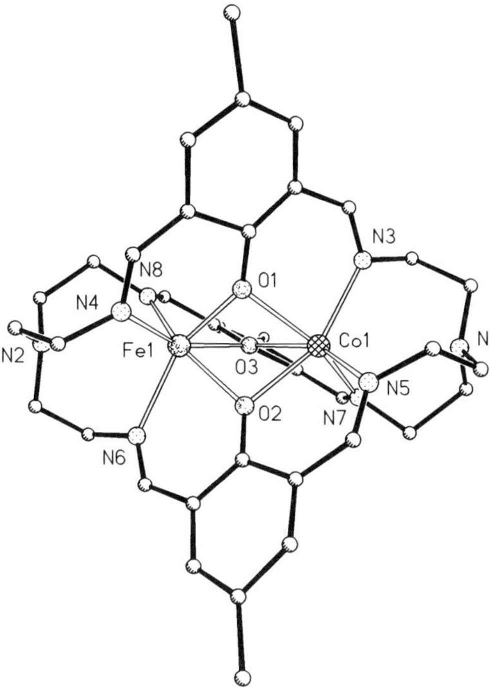

Figure 73. Heterobinuclear phenolate cryptates (197): X-ray structure of CoFe[cresimBT-3H]$^{+}$; for an alternative interpretation see (201).

munoassay (206, 207) relies on efficient Eu^{3+} or Tb^{3+} fluorescence to report on the concentration of target metabolites covalently attached to the lanthanide chelate. Excited states of $Eu(aq)^{3+}$ or $Tb(aq)^{3+}$ can be easily deactivated in solution by energy transfer to solvent vibrational modes such as ν_{OH} or ν_{NH}, whereas in a cage host the cations are sterically protected against this deactivation mechanism, even in aqueous medium, if kinetic stability toward decomplexation is sufficiently high. The monoeuropium cryptate of cresimBT is apparently deactivated through the $O\cdots H{-}N^{+}$ hydrogen-bond system (204) (Fig. 70) and fluoresces only weakly, but the typical Eu fluorescence is intensely observed in the heterobinuclear Eu-Zn complex (202), which lacks this deactivation route. The most intense fluoresence so far observed in the phenol-linked azacryptates is achieved with the diterbium cryptate of the aminophenolate ligand cresamBT, where neither charge-transfer nor energy-transfer mechanisms are available for deactivation.

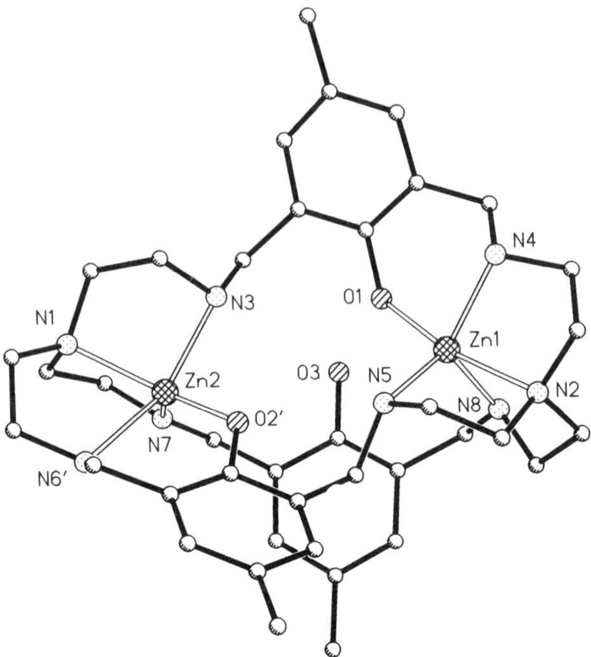

Figure 74. Dizinc cryptate of octaaminophenolate cryptand: X-ray structure (202) of Zn_2[cre imBT]$^{2+}$; the Zn(2) site is at 30% occupancy.

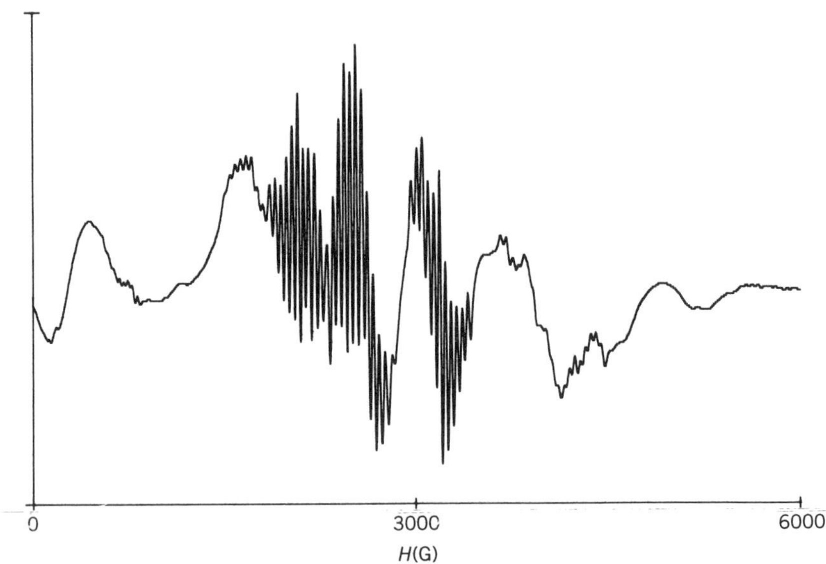

Figure 75. *X*-band ESR spectrum of a dimanganese phenolate cryptate (64): Mn_2[TBPimBT]$^+$ (DMSO/MeOH glass, 80 K). Zero-field splitting generates six main features as well as a number of forbidden transitions. Hyperfine splitting, well resolved on the central features, consists of 11 lines arising from coupling to both spin $\frac{5}{2}$ manganese nuclei, with one-half of the normal coupling constant.

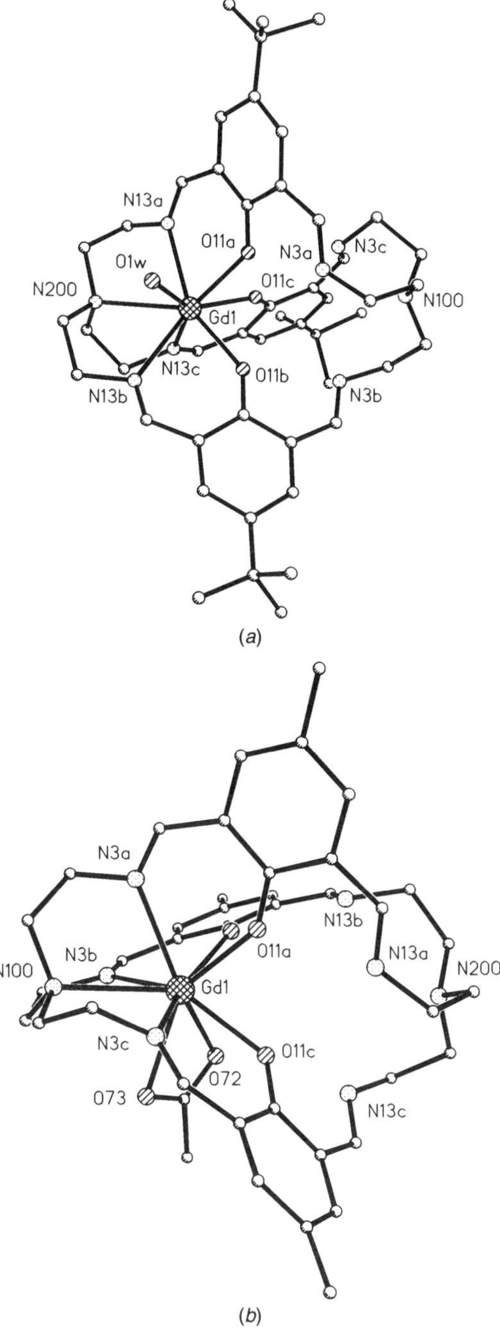

Figure 76. Gadolinium cryptates of phenolate ligands; X-ray structures (204) of (*a*) $Gd(H_2O)[$*TBPimBT*$]^{3+}$ and (*b*) $Gd(OAc)[cresimBT]^{2+}$.

E. Other Azacryptand Hosts for Photoactive Cations

All-azacryptands such as trisbpy and trisphen (208) are ideal hosts for photophysically active cations (11). The combination of the powerful complexing properties of the cryptand framework incorporating six sp^2 N donors, with the fluorescence deriving from energy transfer involving three efficient and closely linked chromophores confers valuable photoactivity on complexes of such ligands. Cryptates of ligands with strongly absorbing chromophores such as bpy or phen are designed to exploit the antenna effect that leads, via energy transfer of absorbed frequencies, to enhanced luminescent emission (206, 207). The Eu^{3+} and Tb^{3+} trisbpy cryptates synthesized by Lehn and co-workers (Fig. 77) (24, 83) have reasonably good quantum yield and absorption efficiency (209, 210), especially in the solid state (211, 212), and it is the combination of these two factors that determines the overall efficiency of light energy conversion. Ligands **XVI** and **XVII** incorporating other bis heterocyclic functions (213–216) were also designed to make photoluminescent complexes. However, the most effective cryptand ligands for this purpose are unsymmetric: **XVIII** and **XIX,** having one chelating *N*-oxide strand (217–219) and two bpy strands. The *N*-oxide function is a better donor for lanthanide cations than the heterocyclic sp^2 N donor and, consequently, a more effective sensitizer for the Eu^{3+} or Tb^{3+} emission (Fig. 78). However, when more than one such chelating *N*-oxide pair is incorporated in a cryptand such as **XX,** steric constraints prevent efficient coordination of the target lanthanide cation and lead to less effective luminescence (220). Comparable or even greater luminescence efficiency can be achieved using functionalized macrocyclic hosts such as the bpy-substituted calix-4*t*-Bu arene (221), dipendant bpy macrocycles **XXI** (222) or, best of all, those with a pair of pendant bpy *N*-oxide **XXII** (223) chelators. The improvement results from increased flexibility that permits the establishment of stronger M—L bonds and at the same time, by improving the access of ligand donors, reduces competition with water (which quenches the lanthanide fluorescence). In the case of *N*-oxide pendants, the more effective energy transfer possible with the pair of N → O chelates contributes to the enhanced fluorescence yield. The most favorable disposition of N → O chelates may eventually be achieved using specially designed macrotri- or tetracyclic ligands.

Any cryptate that proves efficient in terms of long-lived fluorescence in aqueous solution may have important future applications in diagnostic medicine (207), for example for in vivo fluorescence marking when attached to monoclonal antibodies.

Complexes of Ru^{3+} with sp^2 N donors have for many years been used in energy-conversion systems with the aim of converting light energy to electrochemical potential, for example, photoproduction of hydrogen. The most fa-

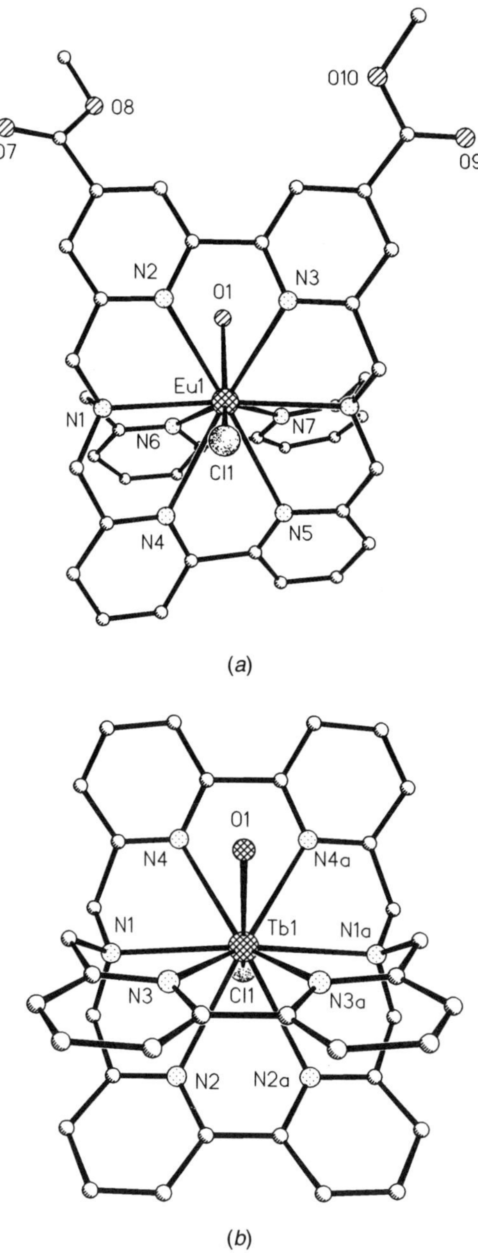

Figure 77. Azacryptand hosts for photoactive cations: X-ray structures (214, 213) of (*a*) Eu[diacetatotrisbpy]H_2O.Cl] and (*b*) [Tb[trisbpy]H_2O.Cl]$^{2+}$.

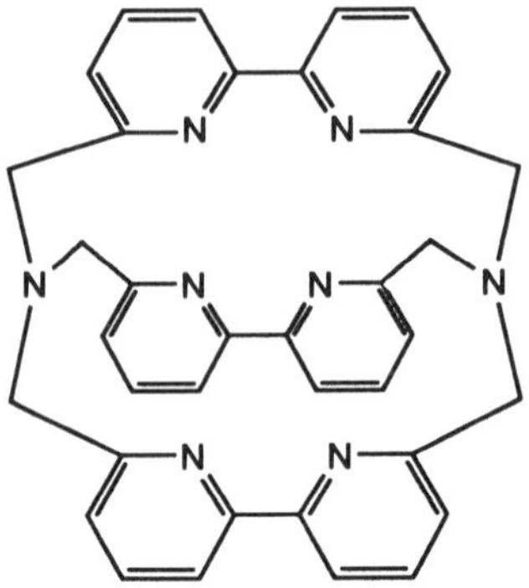

trisbpy 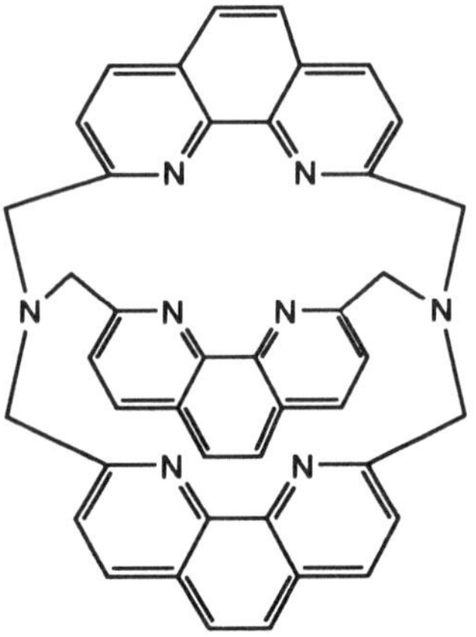 **trisphen**

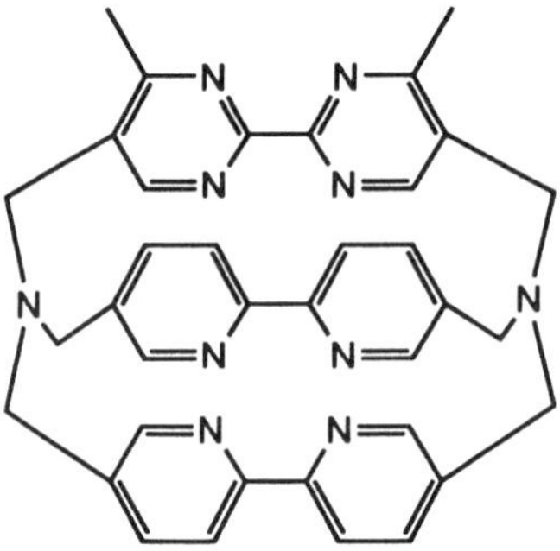

XVI 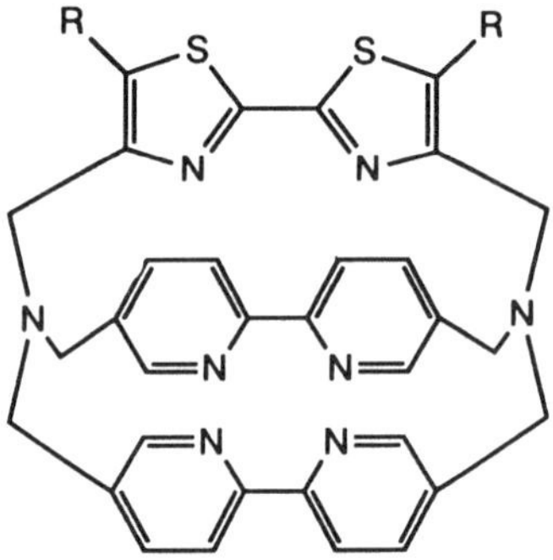 **XVII**

miliar systems utilize three bpy chelators to provide the necessary octahedral coordination site for ruthenium; however, a common limiting factor arises from photodissociation of the bpy ligands. One strategy proposed for overcoming photodissociation is to incorporate the set of bpy ligands in a preorganized trisbpy "cage." In one Ru^{2+} host (**XXIII**) (Fig. 79 on page 256) a 1,3,5-trisubstituted aromatic function is used as "barrel" end to the cryptand cage (224). This strategy increases photostability by 1000-fold, and at the same time, in consequence of protection of the photoexcited state from nonradiative deactivation, results in a slower radiative decay process, thereby doubling the lifetime of the triplet excited state (225).

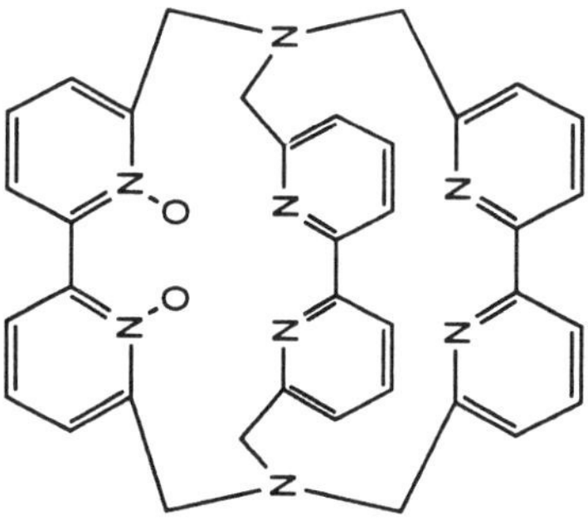

XX

XIX

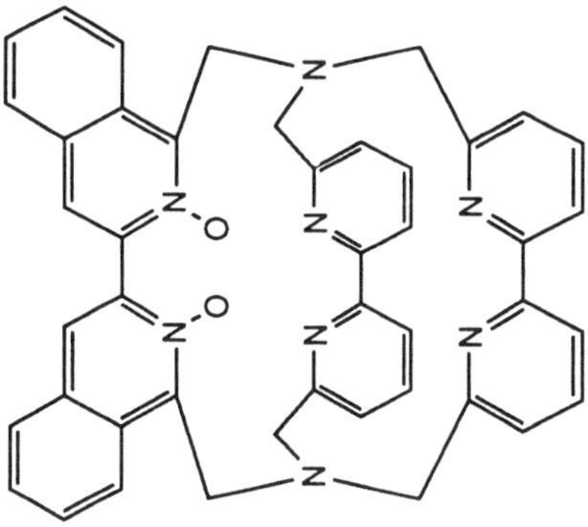

XVIII

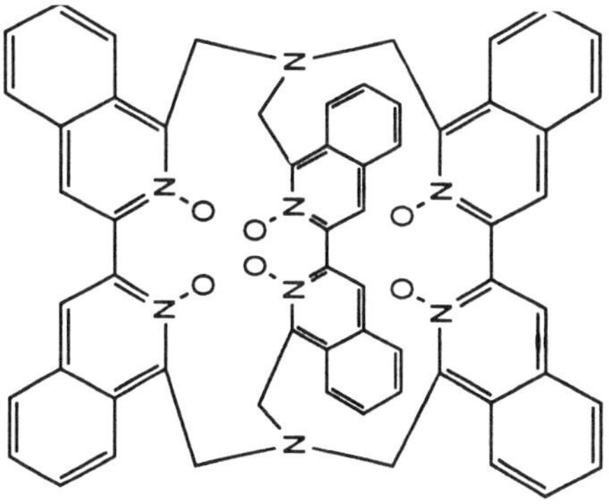

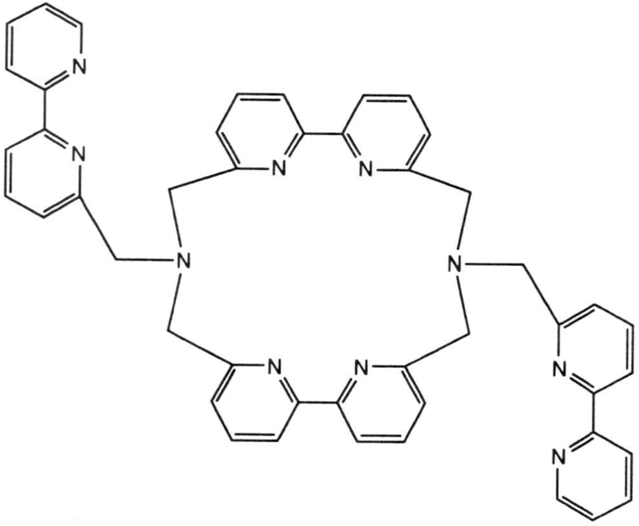

XXII

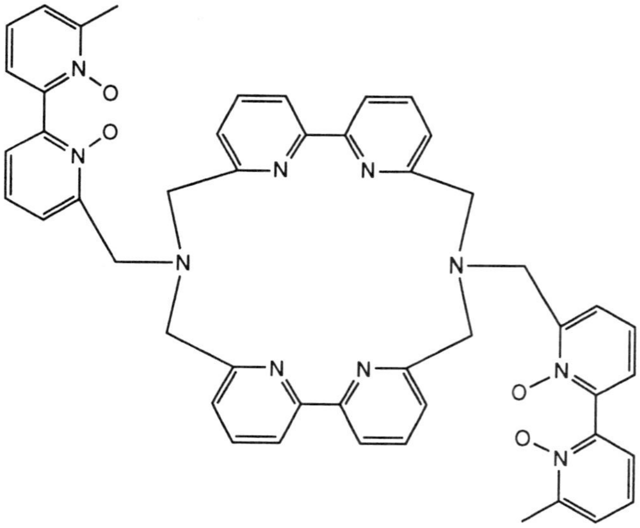

XXI

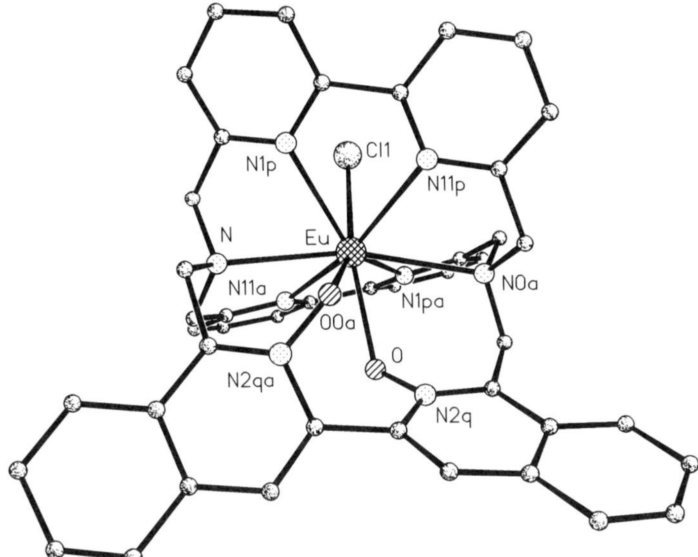

Figure 78. Heterocyclic N-oxide cryptand host for photoactive cations: X-ray structure (223) of $Eu(bpy)_2$(**XIX**)$^{3+}$

The Schiff-base strategy has been adapted by Lehn and co-workers (226) to generate a potentially binucleating trisbpy hexa-Schiff base **XXIV** [Fig. 80(*a*) on page 257] of potential use as Ru^{2+} receptor. However, reports to date of cryptate formation are confined to nonsterically demanding guest cations such as Ag^{I} and Cu^{I}. In consequence of the high donicity of the ligand and low preferred coordination number of the guests, as many as three cations can be encapsulated in a linear array [Fig. 80(*b*) on page 257].

F. Solution Complexation Studies

While much of the structural and other data described so far relates to the solid state, good solution complexation data are needed if the suitability of potential receptors for most applications is to be monitored satisfactorily. Both deprotonable Schiff-base and octaamino cryptands have potential value as detoxification agents for environmental or biomedical purposes, on account of their ability to sequester toxic heavy metal cations (171, 198, 227), such as Pb^{2+}, Cd^{2+}, Tl^{3+}/Tl^{+}, and Hg^{2+}. There are indications not only of high thermodynamic stability but, from NMR measurements, of slow decomplexation

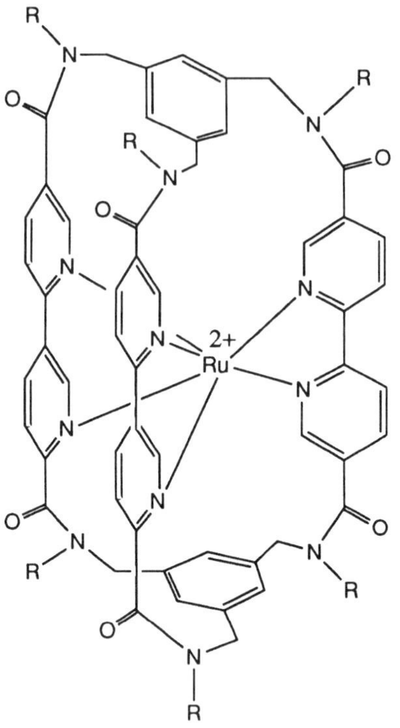

XXIII

Figure 79. Photoactive ruthenium cryptate: **XXIII.**

kinetics in solution. Couplings between magnetically active isotopes of the encapsulated cation and ligand protons are clearly resolved in some cases. Also, where direct observation of the magnetic nucleus is possible, two separate non-coalescing signals are observed corresponding to noninterconverting cryptated and solvated environments. For some biomedical applications, unusually high kinetic stability can compensate for lower thermodynamic stability, if other factors, for example, reduced levels of ligand toxicity, are favorable. Slow decomplexation means that the potentially toxic cation is not present in any significant amount in the free aqueous state to become bioavailable.

In the hexa-Schiff-base series, as anticipated, Group 1 (IA) cations are poorly coordinated by azacryptated hosts: formation constants measured in acetonitrile (Table II) lie in the low range: less than 4 for Na^+ and less than 7 for Group 2 (IIA) cations. One potentially useful exception occurs with Mg^{2+}, where on

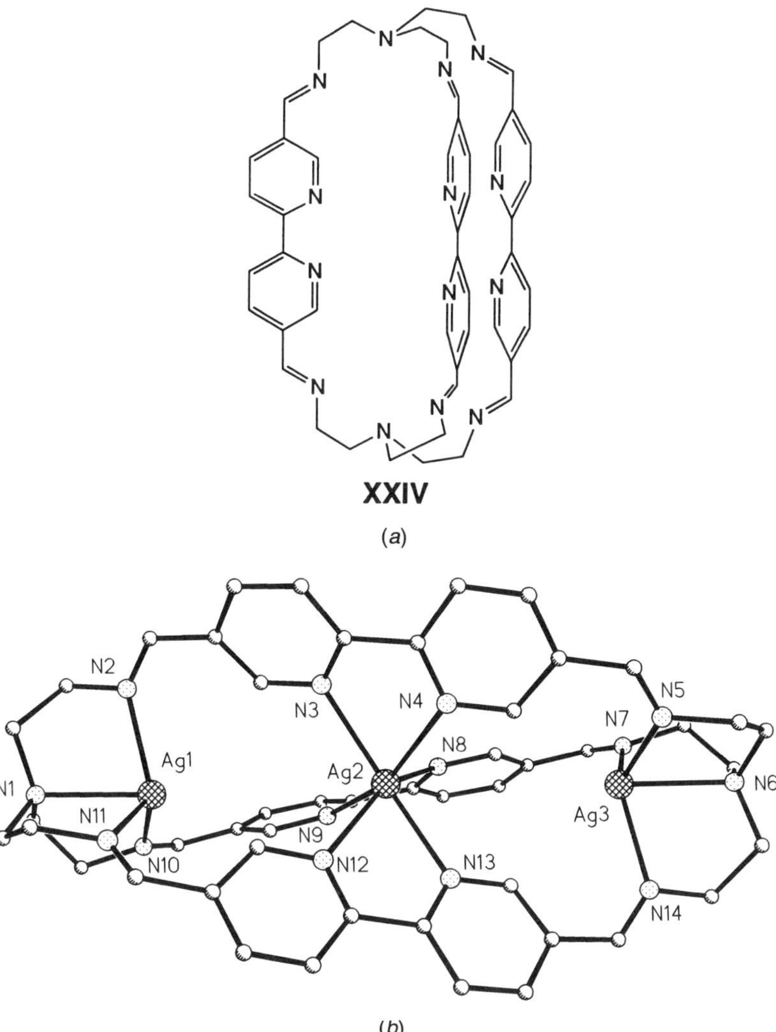

Figure 80. (*a*) **XXIV** and (*b*) X-ray structure (226) of a trisilver Schiff-base bpycryptate: Ag_3[**XXIV**]$^{3+}$.

account of the smaller size of the Mg^{II} cation, dinuclear cryptates are formed in preference to the mononuclear complexes favored by the other Group 2 (IIA) cations, leading to efficient sequestration of magnesium and an effective selectivity (184) for Mg^{2+} over Ca^{2+}, an unusual situation (Fig. 81). Calorimetric measurements show that in general, Group 2 (IIA) cations are entropically,

TABLE II

Formation Constants as Log $\beta_{11}(= [ML]^{n+}/[M^{n+}].\ [L])$ of Selected Cations with Schiff-Base Cryptands Compared with Cryptands 222 in Acetonitrile [at 25°C, I = 0.01 M (Et_4NClO_4)] and 333 in Aqueous Medium

Ligand	Li^+	Na^+	K^+	Rb^+	Cs^+	Mg^{2+}	Ca^{2+}	Sr^{2+}	Ba^{2+}	Co^{2+}	Ni^{2+}	Cu^{2+}	Cu^+	Zn^{2+}	Ag^+	Cd^{2+}
furimBT[a]	3.91	2.47	2.6	2.1	3.3	11.1[b]	6.64	5.3	5.8	6.2	5.42	*c*	7.1 (12.6[d])	5.39	7.2 (13.3)	4.13
pyimBT[a]	2.36	3.0	*e*	*e*	*e*	12.6[b]	6.26	6.01	6.22	7.2	6.24	*c*	4.77 (8.6[d])	5.4	6.0	6.50
xylimBT[a]	*e*	*e*	*e*	*e*	*e*	4.26 (7.5)[d]	4.5 (6.29)[d]	4.2 (7.86)[d]	*e*	*f*	*f*	7.2 (12.4)[d]	8.65[b]	*f*	4.8 (8.77)[d]	6.5 (12.8)
222	6.9[g]	10.68[h]	11.3[g]	9.65[g]	4.5[g]		10.5[i]		>9[g]					>9.5[j]	8.9[g]	
333	<2[k]	2.7[l]	5.4[k]	5.7[l]	5.9[l]	<2[k]	<2[k]	<2[k]	—	—	—	—	—	—	—	—

[a]Reference (184).

[b]Only dinuclear cryptates exist; formation constant log β_{12} (= $[M_2L]^{2n+}/[M^{n+}]^2.\ [L]$ listed.

[c]Irreproducible results.

[d]Dinuclear cryptates exist with the formation constants log β_{12} as indicated in parentheses.

[e]Not detected.

[f]Ligand hydrolysis possible on the time scale needed for measurement of stability constant.

[g]Reference (288).

[h]Reference (289).

[i]Reference (290).

[j]Reference (291): (I = 0.5 M Me_4NClO_4).

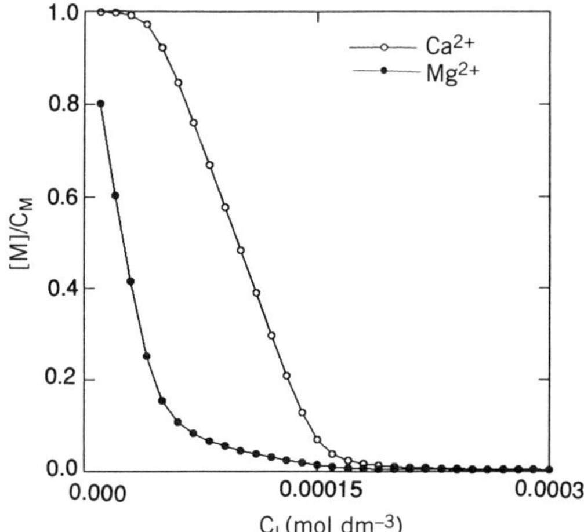

Figure 81. Calculated percentages of free Mg^{2+} and free Ca^{2+} as function of ligand [furimBT] concentration (184) ($C_M = 10^{-4}$ mol dm^{-3}). At an equivalent ligand concentration (10^{-4} mol dm^{-3}) percentages of free Mg^{2+} and Ca^{2+} are, respectively, 4.5 and 48. [Reproduced by permission from the Royal Society of Chemistry. R. A. Abidi, M. G. B. Drew, F. Arnaud-Neu, S. Lahely, M.-J. Schwing-Weil, D. J. Marrs, and J. Nelson, *J. Chem. Soc. Perkin Trans. 2*, 2747 (1996).]

rather than enthalpically (Table III), stabilized within the hexaimino azacryptand hosts, which explains their lower stability constants and also the absence of any size-dependent cation selectivity. In contrast, polyether cryptands show marked selectivity (5) based on complementarity of cation size and cryptand dimension; at least where there is a good steric match as in 222, 221, or 211, although selectivity is lacking in the longer, 11-membered strand cryptand 333, which represents a better steric comparison for the 1,5-linked hexa-Schiff-base ligands.

In the octaamino series, the small cryptand amBT shows considerable promise as host, for single large main group cations versus transition series cations, on account of the good complementarity of fit in the former case. Indeed, this good fit, and the kinetic stability that accompanies it, appears to protect the analogous hexa-Schiff-base cryptand (158) (imBT) from hydrolytic degradation, so that it may also be considered a potentially useful aqueous sequestering agent for some cations. Even the strong Lewis acid Gd^{III}, for example, can be encapsulated, giving a cryptate (Fig. 82) of potential value in MRI (227). Systematic study of complexation thermodynamics of these systems is in progress (187).

TABLE III
Thermodynamic Parameters (184) of Complexation of Silver and Alkaline Earth Cations by furimBT and pyimBT in Acetonitrile (I = 0.01 M Et_4NClO_4])[a]

Ligands	Cations	Species	$-\Delta G_{xy}$	$-\Delta H_{xy}$	$T\Delta S_{xy}$	ΔS_{xy}
furimBT	Mg^{2+}	M_2L	63.3	18	46	150
	Ca^{2+}	ML	37.8	11	27	92
	Sr^{2+}	ML	30.2	7	37	120
	Ba^{2+}	ML	33.1	1	32	110
	Ag^{+}	ML	41.0	18	23	79
		M_2L	75.8	49	27	90
pyimBT	Mg^{2+}	M_2L	71.8	−12	85	280
	Ca^{2+}	ML	35.68	32	4	12
	Sr^{2+}	ML	34.26	−30	65	210
	Ba^{2+}	ML	35.45	25	11	40
	Ag^{+}	ML	34.2	21	14	47

[a]In units of kilojoules per mole (kJ mol^{-1}) and ΔH and ΔG and joules per kelvin (JK^{-1} mol^{-1}) for (ΔS).

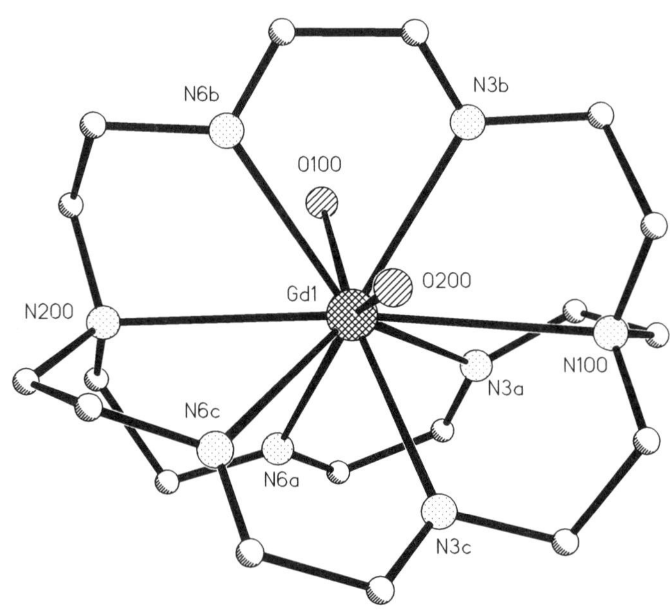

Figure 82. X-ray structure (227) of $[GdimBT.(H_2O)_2]^{3+}$.

Potentiometric study (Table IV) of the protonation constants of the octa-amino ligands shows a basicity pattern for xylamBT quite similar to C5-bistren (187, 191). The earlier study of xylamBT protonation (191) using KNO_3 supporting electrolyte generated lower values of protonation constants, which may derive from differential anion complexation, that is, NO_3^- versus ClO_4^-. As recent results (Section V.K) show that perchlorate cannot automatically be considered an innocent bystander where protonated aminocryptand hosts are concerned, it is likely that many of the "constants" in Table IV are anion dependent. (We have now embarked on a program to investigate anion dependency and to remeasure the protonation constants with a bulky supporting electrolyte such as tosylate.) The cryptands with heteroatoms in the linker units, pyamBT and furamBT, seem appreciably less basic than the other four ligands. There is no evidence, as there is with the small azacages (Section V.A and Section V.F, Table VI), of basicity strikingly reducing on successive protonation, as would suggest particularly strong repulsion within the cryptand host. The ligand amBT is to some extent an intermediate case; although basicity in the first protonation step (see Section V.F., Table VI) is noticeably higher than the 1,5-linked azacryptands, by the fourth protonation step it has reduced to the level of the sixth successive protonation for furamBT or O-bistren and C3-bistrpn. (Protonation of amBT at levels above 4 requires much higher acidity as the protonation plot shows a distinct discontinuity at that stage (228)). An interesting pairwise sequence of protonation constants appears for pyamBT and O-bistren; relatively smaller differences are seen between first and second, third and fourth, fifth and sixth protonation constants than between second and third or fourth and fifth. It seems that in these protonated cryptands the third proton is experiencing repulsion from the first or second proton, whereas the site of the second protonation is well removed from the first.

Thermodynamic data for complexation of cations by the polyamino cryptands xylamBT, *p*-xylamBT, pyamBT, furamBT, and for cascade complexation of anions by their dinuclear cryptates are compared in Tables V and VII with that available for transition ion complexation by the analogous O-bistren and C5-bistren cryptands (229–232). The stability constants, measured potentiometrically, for formation of dicopper cryptates of xylamBT and furamBT, are much higher than for any polyether cryptand but 2–3 log units lower than for O- or C5-bistren. That of pyamBT, however, represents the highest aqueous stability constant so far reported for a dicopper(II) cryptate (187, 195, 230, 231). These stability constants are much greater than with the analogous Schiff-base cryptands, especially allowing for the difference in solvent systems used. This fact is at least partly due to the better preorganization of the amine ligands, referred to earlier. Study of the octaamino cryptand hosts xylamBT, furamBT, and pyamBT (187) with a range of cations shows that Irving–Williams behavior

TABLE IV

Logarithm of Successive Protonation Constants (187) of Polyamine Cryptands at 25°C, in Water[a]

	Protonation Stage (n)	pyamBT	furamBT	p-xylamBT	xylamBT	xylamBT[b]	C5-bistren[c]	C3-bistrpn[d]	O-bistren[e]
log K_n	1	9.47	9.14	9.6	10.13	9.92	10.35	10.10	9.99
	2	8.78	8.81	9.00	9.14	9.26	9.88	10.45	9.02
	3	7.80	8.02	8.62	8.70	8.75	8.87	9.40	7.98
	4	7.11	6.66	7.4	7.2	7.67	8.38	8.65	7.20
	5	6.24	5.93	6.7	6.92	7.16	8.14	7.00	6.40
	6	5.24	5.78	6.52	6.74	6.59	7.72	6.75	5.67

[a] $I = 0.1\ M\ Et_4NClO_4$ unless otherwise stated.

[b] Reference (191): ($I = 0.1\ M\ KNO_3$).

[c] Reference (230): ($I = 0.1\ M\ NaClO_4$).

[d] Reference (282): ($I = 0.1\ M$ NaOTs); Two further protonation steps observed: $n = 7$: 4.95; $n = 8$: 4.15.

[e] Reference (195): ($I = 0.1\ M\ NaClO_4$).

TABLE V

Formation Constants of Mono- and Dinuclear Azacryptates (187) with Selected Transition Ion, Main Group, and Lanthanide Cations

log β_{ij}	Ligand	Cu^{II}	Co^{II}	Ni^{II}	Zn^{II}	Pb^{II}	Cd^{II}	Ag^{I}	Fe^{IIb}	Eu^{III}	Gd^{III}
ML	furamBT	17.20	7.8	6.5	8.92	7.8	10.62	*c*		*c*	*c*
M_2L	furamBT	25.38		9.8	16.10	13.27	15.50	*d*		*d*	7.9
ML	pyamBT	20.93	12.04	8.99	12.8	11.9	14.08			5.18	*c*
M_2L	pyamBT	33.07	17.75	15.2	21.20	16.37	19.40				
ML	xylamBT		7.53	5.50	*c*		9.7	*c*		*c*	*c*
M_2L	xylamBT	25.86			*c*		15.3	18.78			
ML	*p*-xylamBT	*c*	7.0		*c*					*c*	*c*
ML	xylamBT[e]	16.79	9.81								
M_2L	xylamBT[e]	26.20	13.56								
ML	C_5-bistren[f]	15.39									
M_2L	C_5-bistren[f]	28.75									
ML	O-bistren[g]	16.54	11.20	11.70	11.86				3.62[h,b]		
M_2L	O-bistren[g]	29.21	16.80	18.5	18.22				4.14[i,b]		

[a]Given as log β_{ij}; solvent: water/0.1 *M* Et_4NClO_4, $T = 25°C$ unless otherwise stated.
[b]Reference (292): deprotonated and protonated cryptates.
[c]Protonated mononuclear cryptates $[MH_nL]^{(n+2)+}$ only.
[d]Protonated binuclear cryptates $[M_2H_nL]^{(n+4)+}$ only.
[e]Reference (191) ($I = 0.1$ *M* KNO_3).
[f]Reference (232) ($I = 0.1$ *M* $NaClO_4$).
[g]Reference (195) ($I = 0.1$ *M* $NaClO_4$).
[h]The log *K* for the process: $Fe^{2+} + L = [Fe_2[H - 1]L]^{3+} + H^+$ ($I = 0.1$ *M* KCl).
[i]The log *K* for the process: $Fe^{2+} + [LH_2]^{2+} = [FeH_2L]^{4+}$ ($I = 0.1$ *M* KCl).

is only roughly observed in the log K_s series (viz; the order $Eu^{3+} \approx Gd^{3+} < Ni^{2+} < Co^{2+} < Cu^{2+} > Zn^{2+}$ applies, with maximum stability at Cu^{2+}). This exception to Irving–Williams behavior at Ni^{2+}, whose pyamBT–xylamBT cryptates are over 10 times less stable than their Co^{2+} analogues, does not occur with O-bistren, another trigonal-type azacryptand. Deviation from Irving–Williams behavior suggests that coordination geometry in aqueous solution is not regular octahedral in all cases; it may indeed vary for a particular cryptand across the first transition series. This deviation leads to good selectivity for Cu^{2+} against Ni^{2+}; selectivity for Cu^{2+} against Zn^{2+} is also at a useful level.

Presumably it is coordination of pyridine N donors from the linker unit that is responsible for the enhanced stability of pyamBT complexes over the other azacryptates. A similar effect is noticed in the hexa-Schiff-base series (184) (Table II). In this series, pyimBT is the strongest ligand for transition ions, although in this more rigid host the complexation of transition ions tends to mono- rather than binuclear, and the coordination number is more likely to be 6, although its geometry is not necessarily regular octahedral. Indeed, with the Schiff-base cryptand the reversal of the Irving–Williams order once more ob-

TABLE VI

Successive Protonation Constants[b] and Formation Constants (log β_{11}) for Cryptates of Small Azacryptand Ligands **III–XIII** and amBT Compared with 222 and 211 and Analogous NMe Derivatives[b]

Ligand	References	Co^{2+}	Ni^{2+}	Cu^{2+}	Zn^{2+}	Ag^{+}	Cd^{2+}	Li^{+}	H^{+}[a]
$N_6O_2$222NMe[c]	16	5.2	5.7	12.5	6.8	13.0	10.7	3.5[d]	9.68 [9.36] {5.65} (2.26)
$N_4O_2$211NMe[c]	16	9.9	10.0	16.0	11.2	12.7	12.4	3.8	11.18 [9.75] {2.42}
$N_4O_4$222NMe[c]	16	4.9	5.1	12.7	6.0		12.0	2.4	10.01 [8.92] {2.75}
XIII[e]	119			16.02[f]	9.36		14.22[f]	g	
III[h]	117							g	>14; [8.41] {<2}
IV[h]	79								>14 [11.21]
V[h]	110							3.2[i]	11.83[j] [9.53][j] {3.43}[j]
VI[h]	114							5.5	11.8 [10.0]
VII[h]	116			18.2					11.91 [8.78]
VIII[h]	108							g	12.0 [7.86]
IX[h]	76							4.8	12.48 [9.05]
X[h]	293							3.0	11.8 [8.3]
XI[h]	81								11.53 [6.94]
XII[h]	78								>14 [7.8]
222[k]	5	<2.5	<3.5	6.81	<2.5	9.85	7.1	<2.0[l]	10.00 [7.53]
amBT[m]	170, 228						18.3		11.18 [9.43] {7.59} (5.78)
211[c]	16	<4.7	<4.5	7.8	>5.3	9.82[m]	>5.5	6.98[l]	10.64 [7.85]

[a]The log K_1 unbracketed, first protonation constant; log K_2, [ ] second protonation constant; log K_3, { } third protonation constant; log K_4, 0 fourth protonation constant.

[b]In water, $I = 0.1\ M$ R_4NX, 25°C unless otherwise stated.

[c]Reference (16): ($I = 0.1\ M$ R_4NX, 25°C).

[d]95% alcohol/water.

[e]Reference (119): (0.15 M $NaClO_4$, 25°C).

[f]Monoprotonated species also have significant stability.

[g]Not measurable.

[h]Reference (77): ($I = 0.15\ M$ NaCl).

[i]Reference (75): $-\Delta H = 2.7$ kJ mol^{-1}; $\Delta S = 52.3$ JK^{-1}.

[j]Reference (114); $-\Delta H = 54.4$ [42.7] {13.0} kJ mol^{-1}; $T\Delta S = 13.2$ [11.7] {6.6} kJ mol^{-1}.

[k]Reference (5): [$I = 0.05\ M$ Me_4NBr, 25°C].

[l]Reference (294): ($I = 0.05\ M$ NR_4X).

[m]Reference (170): ($I = 0.1\ M$ KNO_3, 25°C).

served for the Co^{II}/Ni^{II} pair must again derive from trigonal distortion from coordination geometry.

Speciation plots for 2 : 1 Cu^{2+}/L ratio (Fig. 83) show the major species present at neutral pH with furamBT (Fig. 83b) is the μ-hydroxo cryptate, while a higher pH is needed for formation of this derivative with pyamBT (Fig. 83a), in line with log Q values (Table VII) of 6.91 for $[Cu_2OHpyamBT]^{3+}$ and 8.12 for $[Cu_2OHfuramBT]^{3+}$. Measurements of the cascade complexation of dicopper cryptates of xylamBT with other dicoordinating anions in buffered (pH 8) aqueous solution have been carried out (233) (Table VII) and log Q values show peak selectivity for azide. This finding is interpreted as resulting from a situation that represents the best stereochemical match of anion bite with Cu· · ·Cu separation; the dicopper cryptate host is considered to recognize the length of

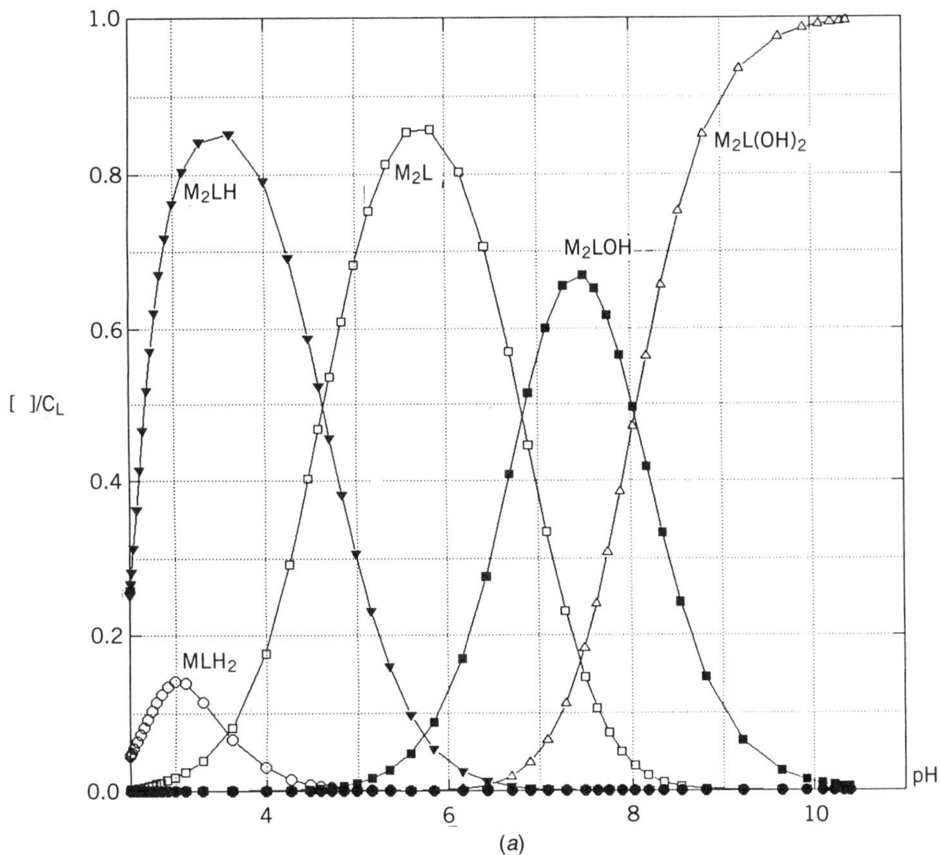

Figure 83. Speciation plots for Cu^{II} aminocryptate systems in water at 25°C at 2 : 1 M : L ratio (187): (*a*) L = pyamBT and (*b*) L = furamBT.

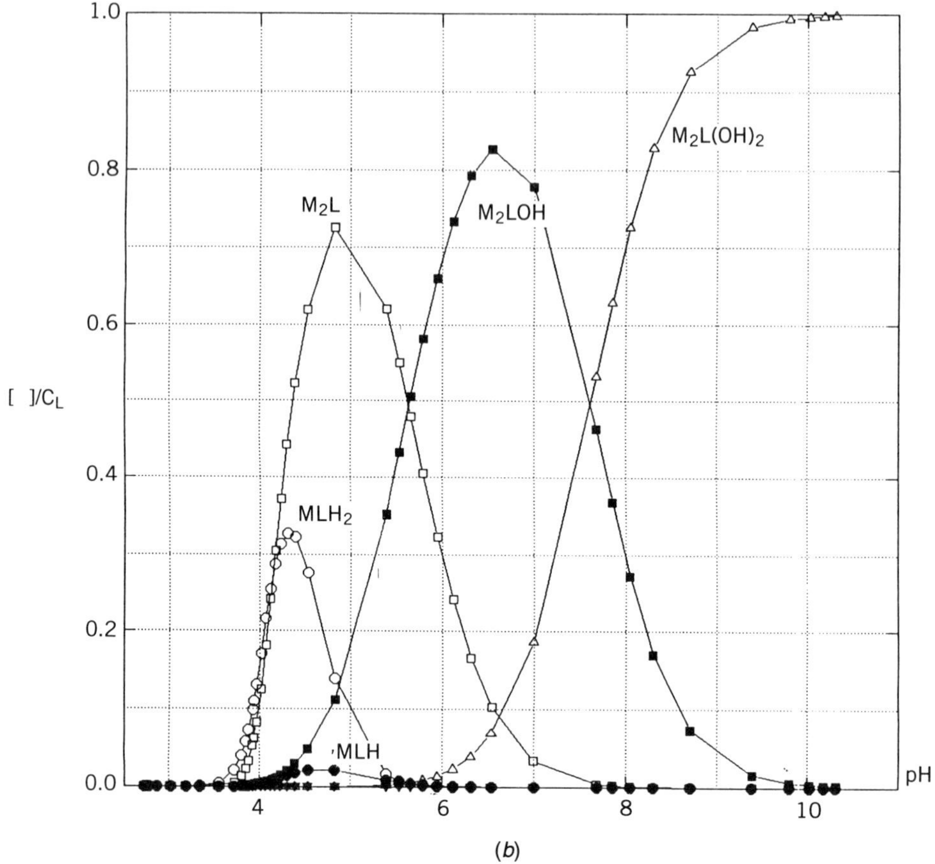

Figure 83. (*Continued.*)

the anion, rather than its coordinating ability or overall size. No quantitative results were reported in this article for cascade complexation of hydroxide with xylamBT, although the μ-hydroxo derivative is dominant in aqueous solution at pH 8, where the measurements of anion complexation were made, and the complex has, of course, been isolated in the solid state (175). In our work, problems with precipitation prevented evaluation of log Q for $[Cu_2\text{-}OHxylamBT]^{3+}$. Cascade complexation studies, using as host the more flexible dicopper(II) cryptates of O-bistren and C5-bistren, show particularly strong affinity for hydroxide anion with O-bistren. The dicopper(II) cryptate of C5-bistren was studied (230) to evaluate the contribution of hydrogen-bond stabilization to OH^- affinity in the O-bistren analogue, and the comparison shows (Table VII) an enhanced stability for the μ-hydroxo anion in the O-bistren ver-

TABLE VII

Cascade Complexes: log Q (= $[M_2LX]^{(4-n+}/[M_2L^{4+}]$. $[X^{n-}]$) of Host Complexes $[M_2L]^{4+}$ with Various Anions, in Water at 25°C

Host Complex	F^-	Cl^-	Br^-	I^-	N_3^-	NCO^-	NCS^-	SO_4^{2-}	NO_3^-	HCO_3^-	HCO_2^-	$MeCO_2^-$	OH^-
$[Cu_2O\text{-bistren}]^{4+}$	4.71[a,b] 4.5[d]	2.11[a] 3.55[c]		3.2[a]									11.56[c]
$[Cu_2C5\text{-bistren}]^{4+}$	3.3[e]												6.19[e]
$[Fe_2O\text{-bistren H}]^{3+}$								1.6[f,g] 2.6[f,h]					
$[Cu_2xylamBT]^{4+}$[i]	[j]	[j]	[j]	[j]	4.78	4.6	2.95	3.26	2.70	4.56	3.32	2.97	[k]

[a]Reference (229): (I = 1.00 M $NaClO_4$) corrected for bulk electrolyte effects relative to perchlorate.
[b]Reference (229): $NaF/NaClO_4$ to total ionic strength 1M.
[c]Reference (231): (I = 0.1 M $NaClO_4$).
[d]Reference (230): [I = 0.01 M: (0.09 M $NaClO_4$ + 0.01M NaF)].
[e]Reference (230): (I = 0.1 M $NaClO_4$).
[f]Reference (292): (I = 0.1 M KCl, 25°C).
[g]Reference (292): FeH_2L host.
[h]Reference (292): Fe_2OHL host.
[i]Reference (233): (0.1 M CF_3SO_3H/morpholine buffer; pH 8).
[j]Not studied.
[k]Reference (187): I = 0.1 M NEt_4ClO_4, precipitation during iteration prevented evaluation. The $[Cu_2OHL]^{3+}$ formation has been studied in related systems; log Q = 6.95 for pyamBT; 8.12 for furamBT.

sus the C5-bistren cryptate host, arising from hydrogen bonding of coordinated OH to the inbuilt ether–oxygen acceptor. Our results for the $[Cu_2furamBT]^{4+}$ host indicate a similar enhancement of stability over cryptates which lack an appropriately sited hydrogen-bond acceptor. However, it is not clear that this rationale translates validly to the hexaimino series where both furimBT and xylimBT are isolated as dicopper(II)hydroxo-bridged cryptates (175, 176) at the same ($\approx$ neutral) pH.

Cascade complexation of halide anions has been studied with O-bistren and C5-bistren dicopper(II) cryptates; fluoride being the most strongly retained of the halide series (230, 231), but much more weakly (3–7 log units) than the similarly sized hydroxo anion. No cascade complexation studies of halides with xylamBT cryptates are available to reveal which system performs best with these anions.

In conclusion, stability constants measured to date (184, 187) for the azacryptand series show values intermediate between the best macrocyclic hosts and the unsymmetrically stranded "designer" macrobicyclic O_xN_2 hosts such as 221 or 211 (5). With improvement of azacryptand ligand design aimed at matching cavity size and target cation radius, the selectivity of azacryptand ligands should be capable of enhancement, enabling them, for some purposes, to compete with polyether cryptands. As the fit of the guest cation in the azacryptand cavity improves (Fig. 84), it is noticed that impressive stability constants are indeed achieved. The value of log K (at 18.3) for Cd^{2+} with amBT (170) in water, for example, is well over double that for the analogous polyether cryptand in the same solvent (Table VI), and about 50% greater than for the analogous di- or tetra- NMe-substituted analogues (18, 113, 234). In the hexaimino series, although cadmium, unlike other +II oxidation state cations, shows no tendency to hydrolytic degradation of imino bonds, the formation constants of monocadmium cryptates (Table II) are low. With the octaamino cryptates, however, the formation constants of dicadmium cryptates are sizeable (Table V). The pyamBT ligand once again shows the greatest affinity, while those for the monocadmium cryptates of these larger ligands are several log units smaller than for amBT, on account of the poorer fit.

G. Oxygen Uptake Studies

Martell and co-workers (173, 235) studied the oxygen affinities of Co^{2+} and Cu^{+} in the dinucleating cryptates O-bistren and C5-bistren, compared with the octaamino derivatives of the Schiff-base cryptands furamBT, pyroamBT, pyamBT, and *p*-xylamBT. The dicobalt complex of O-bistren is an excellent oxygen carrier, which can reversibly oxygenate and deoxygenate without appreciable degradation. However, the well-documented difficulty in synthesizing

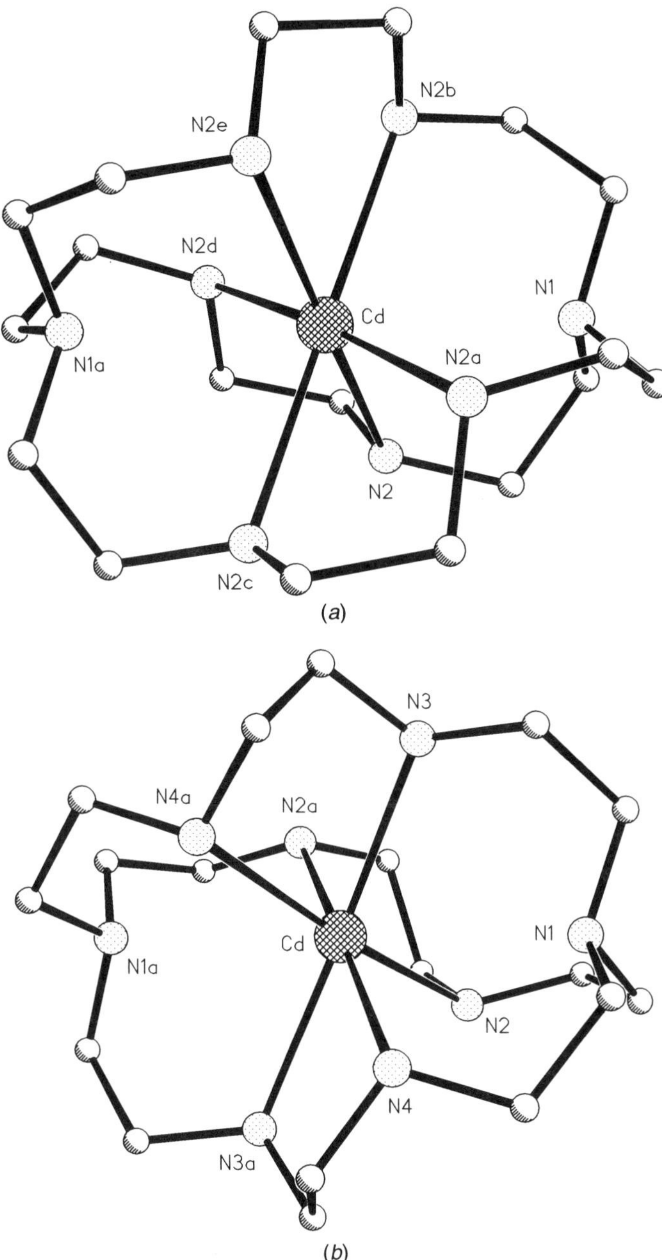

Figure 84. Cadmium cryptates of amBT:Crystal structures (170) of (*a*) Cd[amBT]$(BF_4)_2$ lel and (*b*) Cd[amBT]$(OAc)_2$ ob.

O-bistren directed attention to the analogous and easily prepared Schiff-base-derived octaaminocryptates as possible alternatives. Disappointingly, although some of these dicobalt octaaminocryptates absorbed oxygen, their use was limited by degradation reactions (probably of the oxidative dehydrogenation type), and none showed the long-term stability that characterizes the dicobalt O-bistren system. Two of the dicobalt cryptates, for example, those of furamBT and *p*-xylamBT, in fact fail to absorb oxygen at all, which emphasizes the importance of kinetic factors in aerobic oxidation processes. A similar kinetic effect explains the lack of reactivity of dicopper(I) hexa-Schiff-base cryptates (179) toward O_2; good access to the encapsulated cation appears to be a prerequisite for aerobic oxidation or oxygenation reactions. The more favorable kinetics for oxygen uptake with the analogous dicobalt(II) complex of the macrocyclic ligand O-bisdien illustrate the access requirement; however, the dicobalt complex of this ligand has thermodynamic stability higher than ideal for a carrier.

The advantage of the O-bistren dicobalt(II) carrier (Fig. 85) is that its oxygenation constant is three orders of magnitude lower (235, 236) than the values observed for other dicobalt(II) polyamine chelates, allowing O_2 uptake to be reversed at a temperature low enough to avoid ligand degradation. Only above 90°C are there indications that the system slowly converts to the inert Co_2^{III} cryptate. The exploitation of this promising system is currently inhibited by expense arising from the difficulty of synthesis of the carrier, so the development of similar but more easily synthesized azacryptand hosts is a priority objective.

The exclusion, on kinetic and thermodynamic grounds, of some otherwise promising dicobalt cryptates from function as oxygen carriers justifies the strategy of the Busch group (237), which is to engineer a rigid vault (described as a lacuna) over the redox-active mono- or binuclear site to ensure the necessary combination of access and steric protection for efficient O_2 transport. The Busch cyclidene macrobicycles (Fig. 86) are aza-boxes rather than cryptands, which facilitates reversible O_2 uptake for cobalt and iron carriers. Because of their rigidity and the essential incorporation within the ligand of a "permanent void" for O_2 accommodation, these hosts are not strictly classifiable as cryptands, so we will not describe the work in detail here, but refer those interested in this elegant chemistry to one of the number of recent comprehensive reviews (237–239).

H. Artificial Siderophores

There is increasing interest in using anionic receptors, such as catecholate, to transport relatively highly charged cations of biomedical interest. Iron detoxification is one reason for the interest; another is the complexation of Group 13

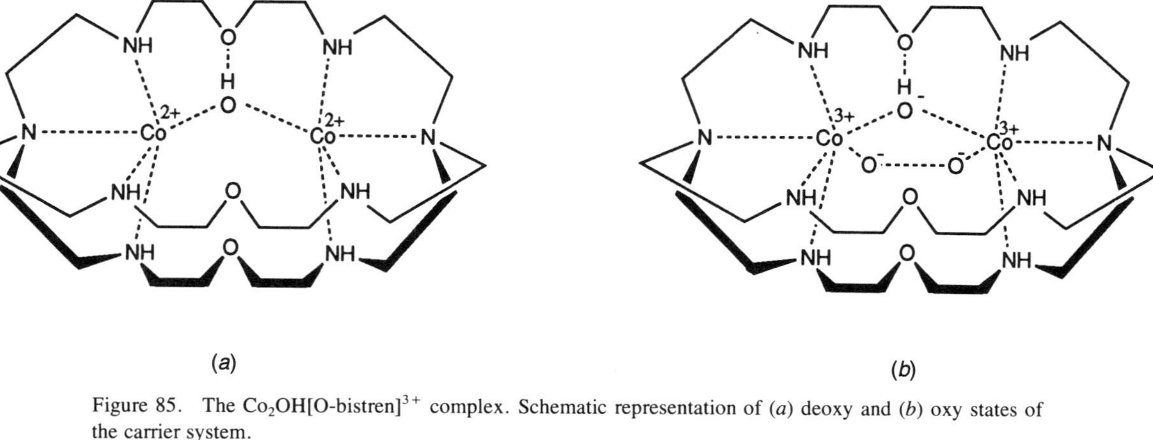

Figure 85. The $Co_2OH[O\text{-bistren}]^{3+}$ complex. Schematic representation of (*a*) deoxy and (*b*) oxy states of the carrier system.

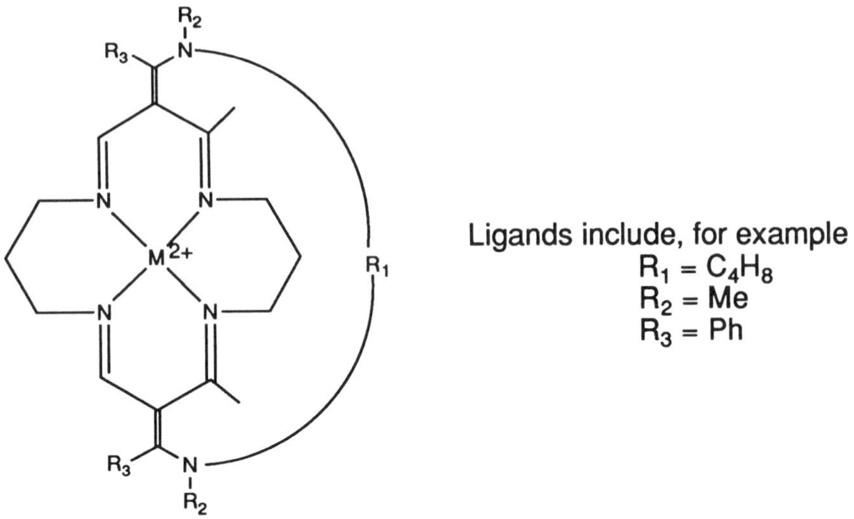

Ligands include, for example
$R_1 = C_4H_8$
$R_2 = Me$
$R_3 = Ph$

Figure 86. Cobalt dioxygen receptors with an inbuilt permanent void.

(IIIA) or lanthanoid cations for use as radiotherapeutic or radioisotopic imaging agents. Over 10 years ago the value of using the macrobicyclic strategy to ensure kinetic stabilization was appreciated by Vogtle and co-workers (240, 241) and by Raymond and co-workers (242, 243), who designed amido cryptand Fe^{3+} receptors such as bicapped-TRENCAM (Fig. 87) to incorporate the necessary tris catecholate function in a macrobicyclic framework. The Fe^{III} cryptate shows an unusual trigonal prismatic coordination geometry [Fig. 88(*a*)] (243, 245) deriving from the presence of strong hydrogen bonds, which maintain the planarity of the catecholamide functions. The best yields were obtained when the host was assembled on an Fe^{3+} template, but it could be also made metal-free, [Fig. 88(*b*)], which allowed the evaluation (244) of the stability constant of the Fe^{III} cryptate (log $K = 10^{43}$); however, complexation by this cryptand proved to be no stronger than in podand triscatecholate analogues such as MECAM (246), and indeed was weaker than in the natural triscatecholate podand siderophore, enterobactin (247). The absence of any observed cryptate effect has been attributed (248) to lack of preorganization of coordination sites in the cryptand, which in turn derives from a divergent organization of the catecholate donors in the metal-free state. The attempt to mimic naturally occurring siderophores was extended by Pierre et al. (249), who have incorporated the 2,2′-dihydroxobiphenyl function into an amido cryptand (Fig. 87) **XXV** and infer, on the basis of competition studies with desferrioxamine, a high stability constant for Fe^{III} within **XXV.**

bicapped TRENCAM

MECAM

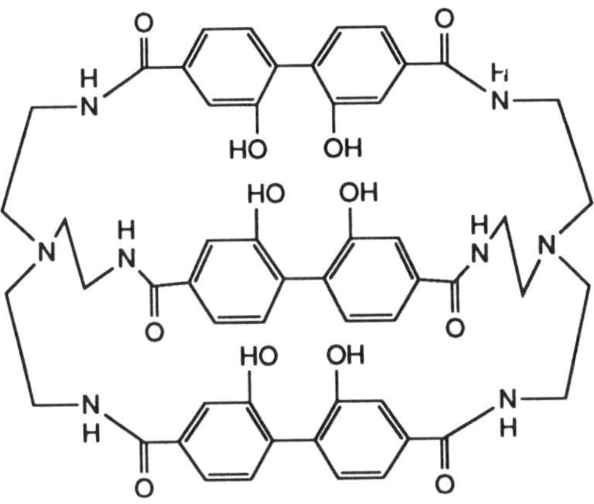

XXV

Figure 87. Podand and cryptand receptors for Fe^{III}: MECAM and bicapped TRENCAM and **XXV**.

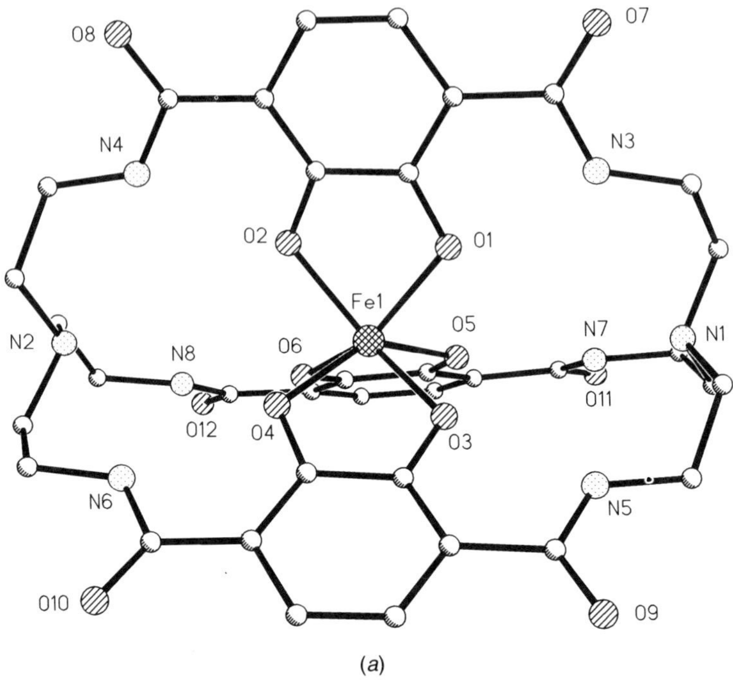

(a)

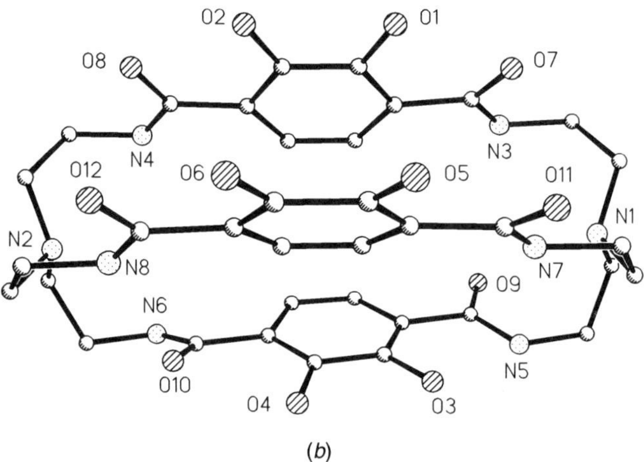

(b)

Figure 88. X-ray structure (244, 245) of Fe^{III} TRENCAM and the metal-free cryptand, showing convergent and divergent conformations.

The hydroxamate group $^{-}ON{-}C{=}O$ represents a valuable chelating group for transition ions of biological interest, particularly Fe^{3+}, and has been incorporated in a cryptand framework by Martell and Sun (Fig. 89) (250, 251). This trishydroxamate cryptand **XXVI,** like TRENCAM, failed to show any marked cryptate effect (252) presumably owing to lack of preorganization. Chana and Hider (253) used a different synthetic route to prepare more flexible N-bridgehead hydroxamate cryptands such as **XXVII,** which they hope may achieve the selective complexation of target M^{3+} cations for biomedical purposes. They believe that their more versatile synthesis will facilitate any necessary functionalization.

A small urea cryptand **XXVIII** (Fig. 89) has been synthesized by an interesting tripod–tripod coupling reaction in moderate ($\approx 22\%$) yield. Disappoint-

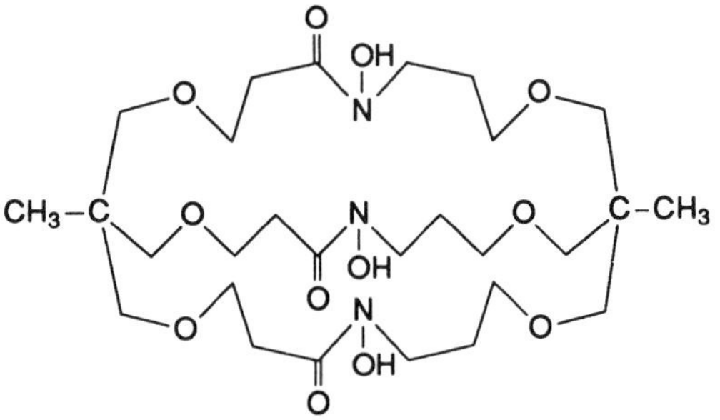

XXVI

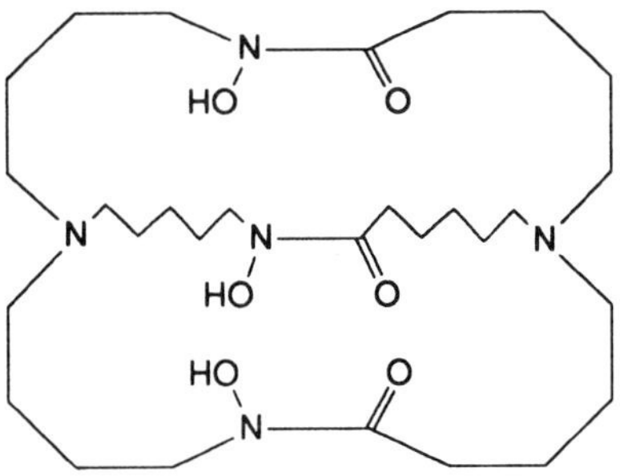

XXVII

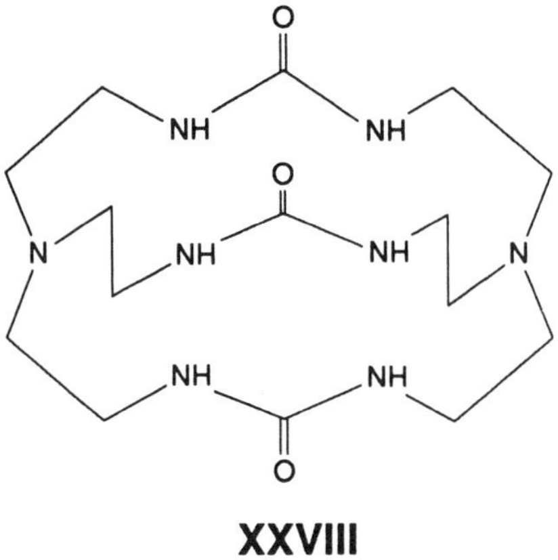

XXVIII

Figure 89. Cryptand receptors (241).

ingly, this cryptand is a weak cation binder, although a moderately strong two-proton binder showing cooperatively (254). The cooperativity is believed to arise because hydrogen bonding of the first NH^+ ensures, via appropriate orientation of the C=O dipoles, preorganization of the second protonation site. Both first and second protonation constants lie close to 8.0, with the second protonation constant marginally exceeding the first, in a manner unprecedented for macrobicyclic proton receptors. Proton-exchange rates, like those of the small aza cages studied by Bencini et al. (78) (Section V.A), are fast.

Medical interest also stimulated the synthesis of the trispyrazolate cryptand **XIV** by the Schiff-base condensation route (183) followed by BH_4^- reduction to the aminocryptand. Although this protonionizable cryptand has been reported to accommodate as many as four Zn^{2+} or Cu^{2+} cations, as yet no quantitative complexation studies nor investigations of sequestration of M^{3+} cations such as Fe^{3+} have been reported. It is indeed unlikely that an all-N-donor cryptand, even an anionic one like **XIV,** could approach the very large stability constants achievable with TRENCAM; the low values (Table V) recorded for Fe^{2+} complexation with O-bistren suggest that N-donor ligands are inefficient hosts for this cation. However, the quantitative complexation parameters for hard cations, Mn^{2+} and Fe^{3+} with the $N_3O_3^-$ cryptands such as cresimBT, which have the potential to act as siderophores, will be of interest, when available (169).

I. Oligobicycles

As mentioned earlier (Section III.B), the Schiff-base condensation route was used by Bharadwaj and co-workers (30, 31, 70) in the Cs^+ templated synthesis of several oligocyclic azacryptands (Fig. 24, Structures D and E). The first cryptand with disulfide linker groups (69, 255) made by this route (**XXXI**), (Fig. 90) coordinates Cu^{2+} in both its hexa-Schiff-base and octaamino form, as well as Ni^{2+} in the latter case. Other larger hosts, generated by capping of dialdehydes with tren or of diamines with trialdehyde podands (Fig. 12) (70) also demonstrate affinity for transition ions assumed to be coordinated in the N_4 tren-derived site or N_6 tris-*o*-diaminobenzene site, although no X-ray crystallographic structural evidence exists to confirm inclusion of the cation within the cryptand. Crystallographic evidence has, however, been obtained for encapsulation of a single Cu^{2+} ion coordinated by the tren-derived site in a smaller cryptand (Fig. 91) (256).

Lindoy's group synthesized two types of large cryptand (Fig. 92) whose conformational preferences appear to inhibit cation coordination. A large tripyridine cryptand (**XXIX**) with trisphenol–ether derived caps (Fig. 92), synthesized by a multistage capping strategy (257), shows no coordination properties toward cations, nor does the analogue **XXX,** which lacks the pyridine

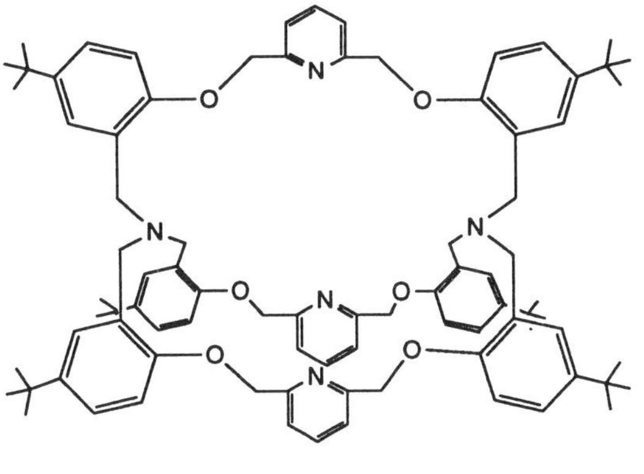

XXIX

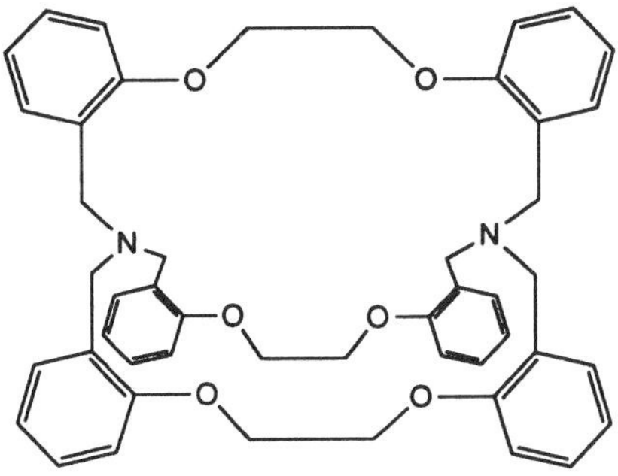

XXX

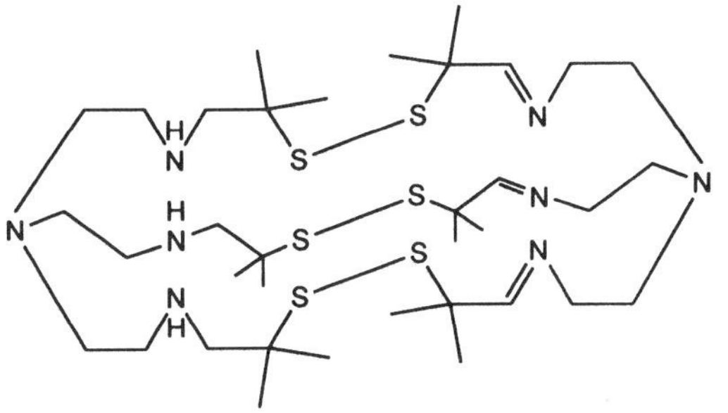

XXXI

Figure 90. Disulfide **XXXI** and phenol ether-derived azacryptands **XXIX** and **XXX**.

functionality (258). The topological rigidity of these cages, relative to aliphatic cryptands like O-bistren, restricts them to the exo–exo bridgehead conformation that disfavors cation encapsulation. The anion coordinating properties of the protonated derivatives of these large hosts should prove interesting. Compound **XXIX** crystallizes out of solution with an encapsulated benzene solvate molecule (Fig. 92) (257), which suggests the possibility of encapsulating a planar anion of similar size.

Although some of the larger cryptands can coordinate one central cation via their set of three chelating groups, and indeed, the template action of Cs^+ is responsible in some cases for their high-yield assembly, they are verging on

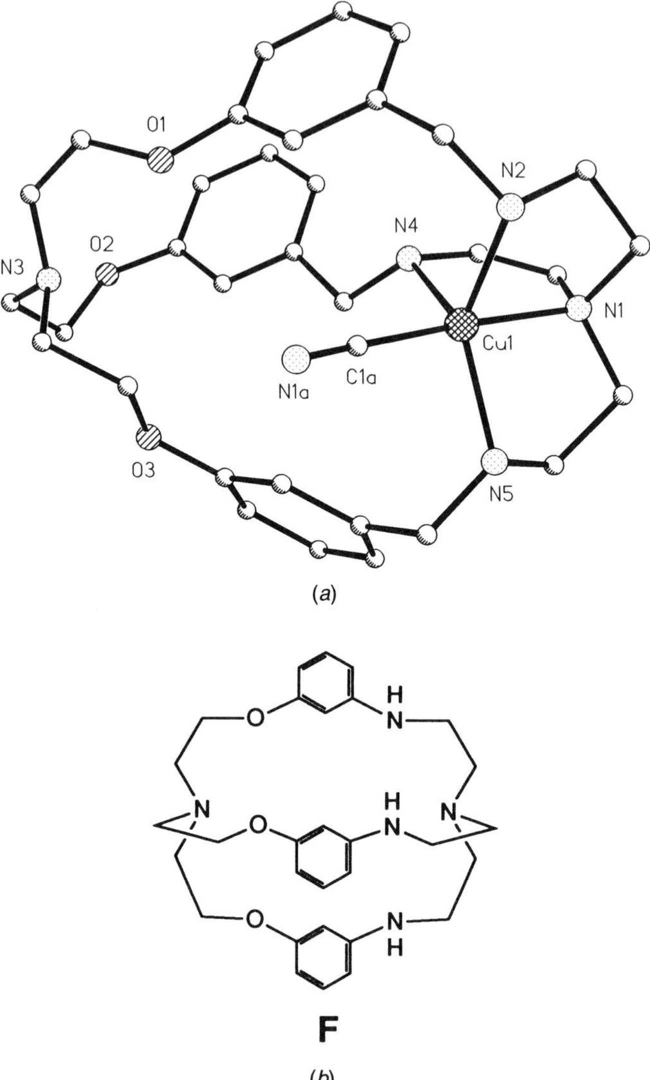

Figure 91. (*a*) X-ray structure (256) of $[Cu(\mathbf{F})CN]^+$ and (*b*) **F**.

molecular rather than ionic receptors because of their complexity and large "cavity" size. Ambitious objectives for application rest upon supramolecular assemblies of this type. Among these are the possibility of generating molecular magnets, molecular wires, switches, and photoionic devices, as well as cooperative binding of substrates leading to catalysts.

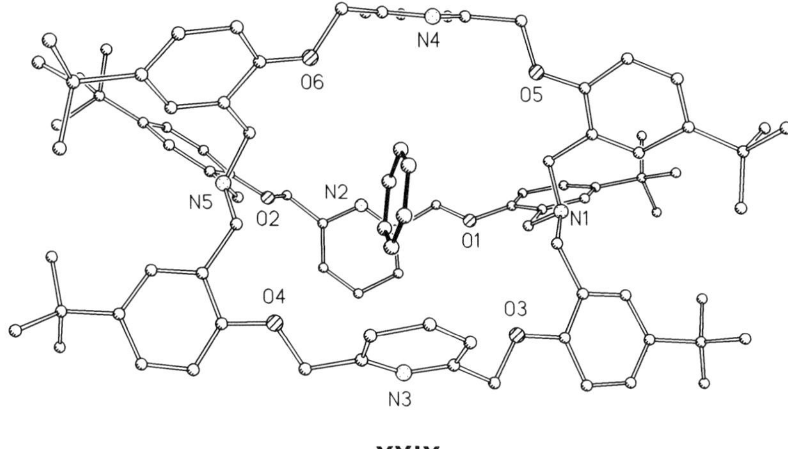

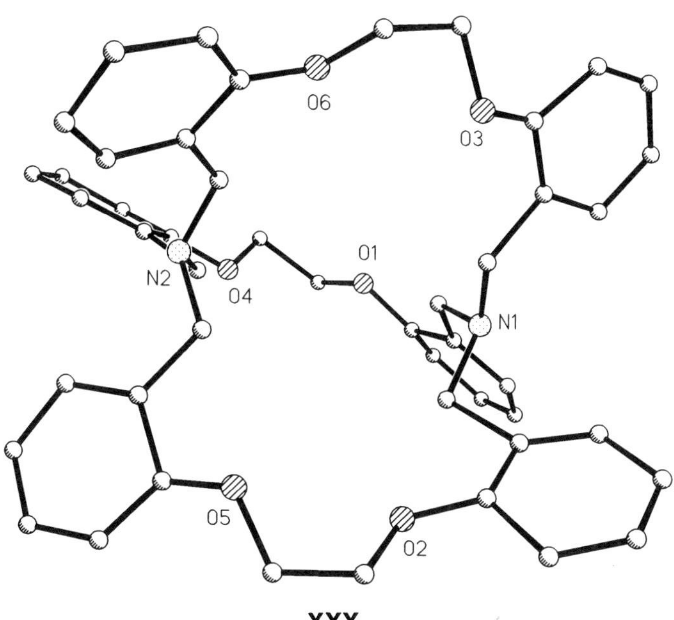

Figure 92. X-ray structures (257, 258) of (*a*) **XXIX** and (*b*) **XXX**.

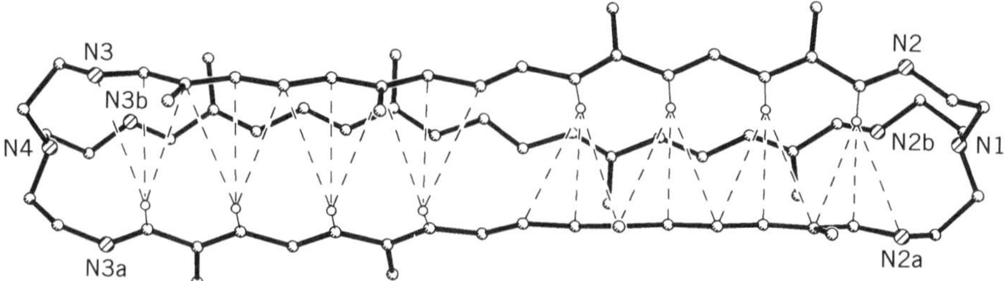

Figure 93. X-ray structure of carocryptand **XXXI** (67) showing interstrand interactions.

These perspectives have been imaginatively set out by Lehn (11) who has used cryptands to move toward their achievement in some fields, for example, in light-converting energy transfer (Section V.E) and molecular electronics. The triscarotenoid azacryptands (67), (Fig. 93) assembled by the [2 + 3] Schiff-base condensation route, have the ability to bind Cu^+ ions in the N_4 caps. As carotenoids have been shown (29) to behave as molecular wires, systems like these dinuclear carocryptates can be taken to represent a three-wire molecular cable polarized at each end; the analogy of carocryptates with rhodopsin is stressed (67).

An azacryptand with azobenzene photochemistry (Fig. 94) and, therefore, capable of acting as a molecular switch, has been synthesized in gram amounts

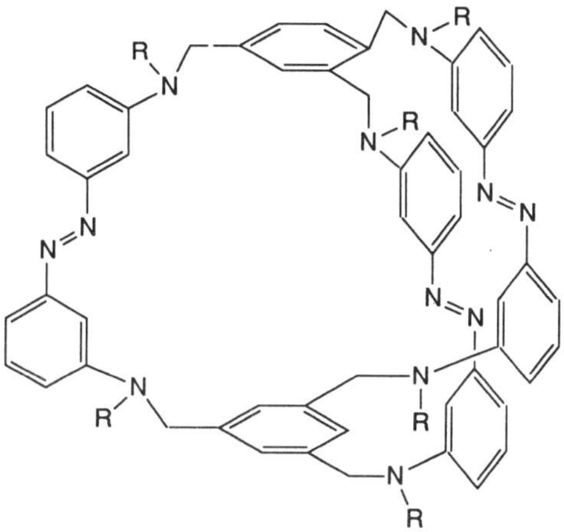

Figure 94. A photoswitchable cryptand.

from readily available starting materials by Vogtle and co-workers (260). The fact that it is capable of photoisomerization despite the steric constraints existing within the cryptand shows that an inversion mechanism must be involved.

An interesting redox-active metallocene cryptand has been generated by the useful [2 + 3] Schiff-base condensation route (261). This novel pentanuclear entity contains two encapsulated copper ions and three photo- or electroactive metallocene units (Fig. 95 on page 282). It suggests the possibility of monitoring the encapsulation of guests via redox change in the metallocene segment of the cryptand, and of eventual development of redox-sensitive sensors for target cations or anions. Another possibility is the synthesis of optically pure cryptand receptors from available chiral arene tricarbonyl chromium building blocks for use in enantioselective recognition of chiral molecules.

Another potentially useful development in this area results from the electrogeneration of an anionic trisbpy-cryptand (262). When $[Natrisbpy]^+$ is electrolytically reduced, a deep blue air-sensitive solid, the neutral [Natrisbpy] complex, precipitates in which the electron appears to be delocalized over the cryptand skeleton. The assembly has been named "cryptatium" to signify its resemblance to an expanded metal atom, rather than an electride (Fig. 96 on page 283). As such, it is expected to have unusual conductivity and redox behavior. The redox properties leading to cryptatium generation in solution have been investigated for the series La^{3+}, Ca^{2+}, Na^+ (263).

J. Conformations and Preorganization in the Schiff-Base Derived Series

We have seen that preorganization in small azacage ligands is important to their role as receptors for single small transition series or main group cations. Equally, in the more flexible hydroxamate or catecholate ligands, it is the lack of preorganization that explains the absence of a cryptate effect.

When we examine the hexaimine cryptands from this point of view, we note that the solid-state structures of the free imine ligands are not as closely similar to the complexes as seen in the Sargeson systems. For imBT, imbistrpn and generally for free (natural) hexaimine cryptands, the ligands are less helical than the complexes and have the imine bonds aligned tangentially to the C_3 cylinder of the ligand (Fig. 97). In many cases, it is possible to suggest some interactions between the strands across the cavity, which may be responsible for the conformation. Drew et al. (264) have compared the crystallographic structures of furimBT, xylimBT, and pyimBT with the conformations calculated from molecular mechanics. In each case a weak but significant attractive H–π interaction between imine protons on one strand and the π system of the next could be identified and this is thought to be the reason for the striking similarity of the three structures. The individual cryptand structures are further modified by interactions between the ring systems (either attractive or repulsive), but the dom-

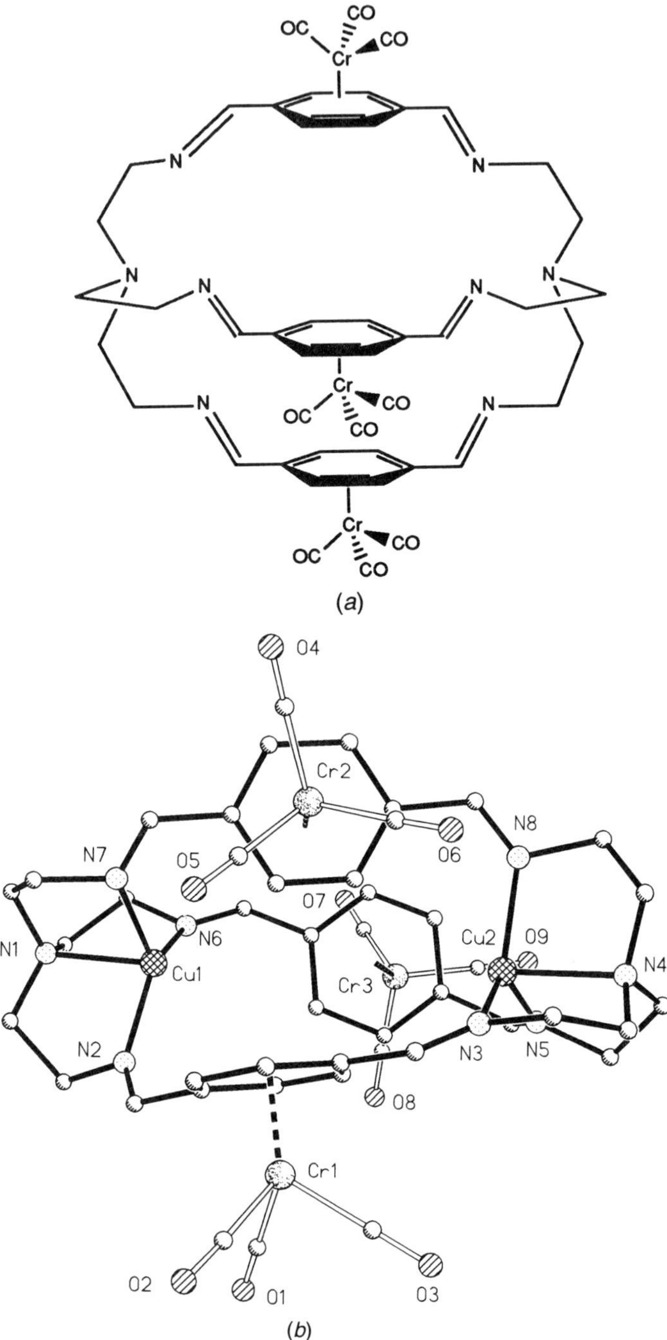

Figure 95. (*a*) Scheme and (*b*) X-ray structure of a Schiff-base metallocene cryptate (261).

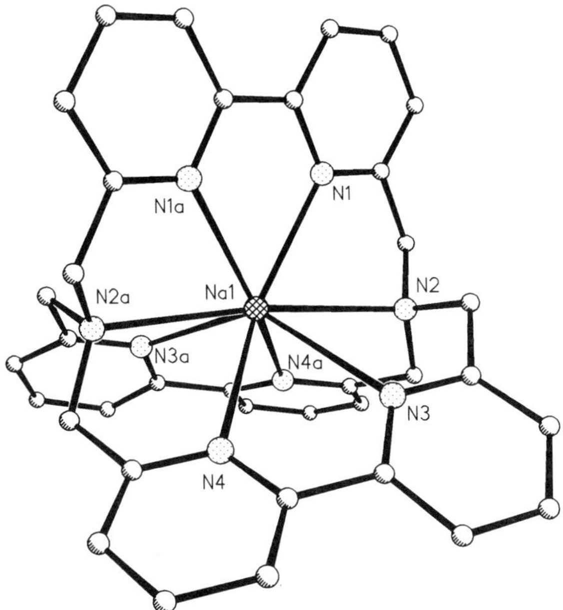

Figure 96. X-ray structure of neutral [Natrisbpy] (262): cryptatium.

inant factor is the imine CH–π attraction. In the absence of any particular stronger interaction [e.g., phenyl CH–π in *p*-xylimBT (264, 265)], this arrangement might be expected to be general in all aromatic hexaimino cryptands and to contribute to the rearrangement costs of complexation. Binding one or more metal ions in a hexaimino cryptand requires significant reorganization; the N_3 triangle contracts, and the imine nitrogen rotates so that the lone pair can be directed into the cavity. It is this requirement that imparts chirality to the system and goes some way toward explaining why binuclear systems are generally helical even though there is no requirement for both N_3 sets to bind the same metal ion. Sometimes there is a small but significant dihedral angle between the imine groups along a strand, suggesting that steric factors are competing with the preferred planar geometry of the conjugated systems.

The ^{1}H NMR spectra can be used to monitor the conformational change between cryptand and cryptate in solution, where the guest cation is diamagnetic. It is interesting to note (67) that the carocryptands (and here the small Schiff-base cage imBT can be considered to constitute the smallest number of this series) show a "frozen conformation" *d,t,t,d* ^{1}H NMR spectral pattern for the methylene cap spectrum at ambient temperature (67, 202), in contrast to free cryptands with 1,5-linked aromatic spacer units that are more conformationally mobile (264, 266). On the other hand, the dicopper(I) carocryptates show a freely rotating methylene ^{1}H NMR spectrum, in marked contrast to the

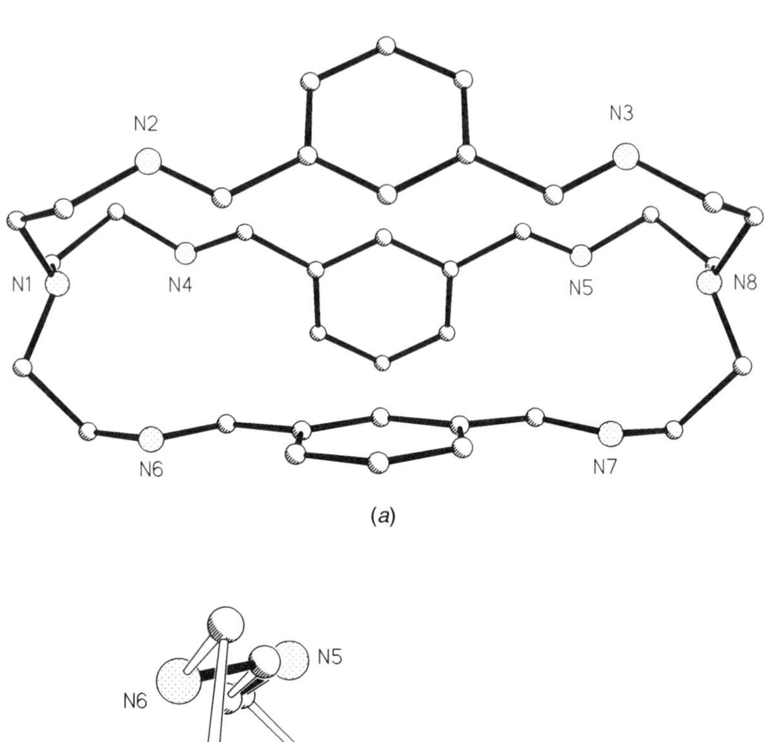

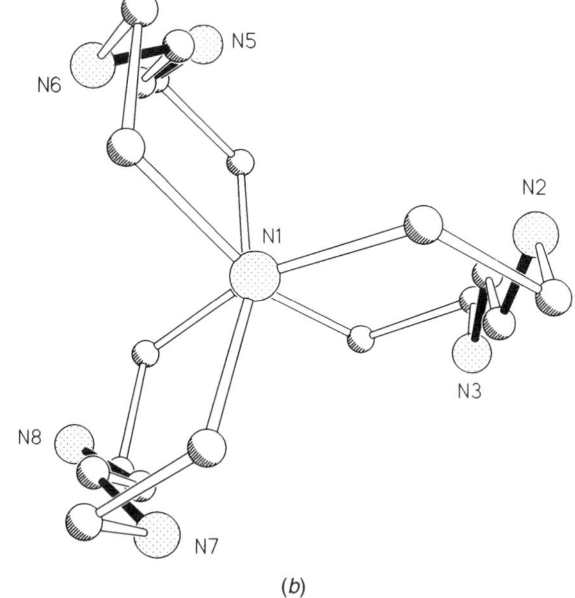

Figure 97. Some metal-free Schiff-base cryptands, showing divergent trans, trans conformations: X-ray structures (267, 268, 66) of (*a*) xylimBT, (*b*) imBT (end on view), (*c*) furimBT, and (*d*) DPMimBT

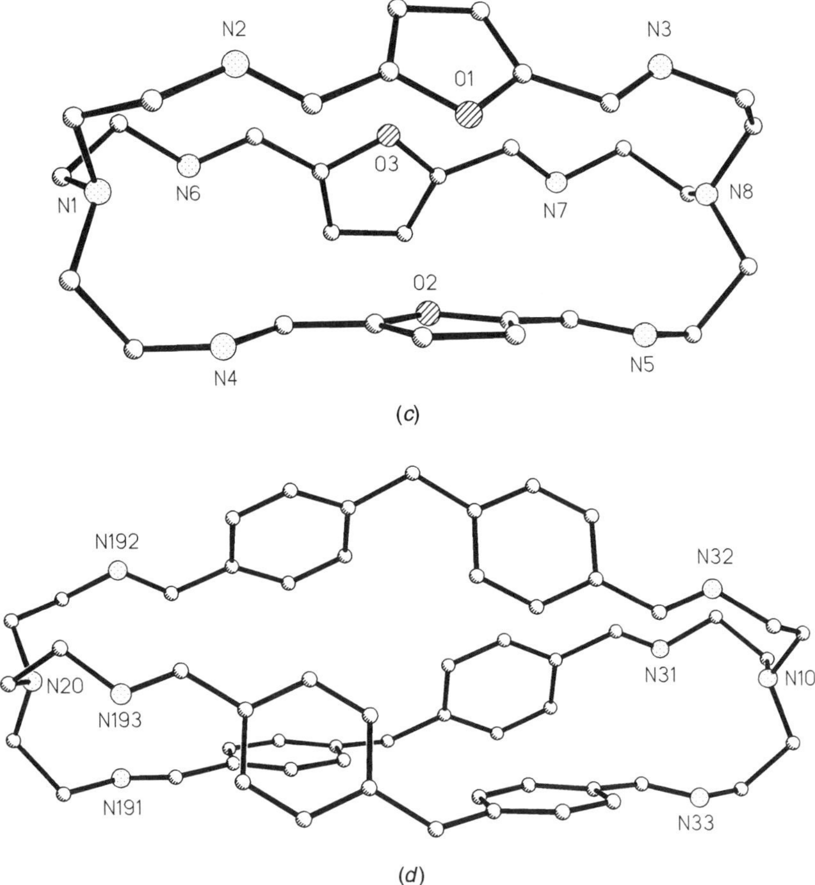

Figure 97. (*Continued.*)

dicopper cryptates of the 1,5-aromatic-linked Schiff-base cryptands, where the methylene protons are in frozen conformation at all temperatures in the acetonitrile solvent fluid range. In larger *p*-xylyl- (188) and diphenylmethane-spaced (66) tren-capped cryptands, both the dicopper(I) complex and free ligand (265, 66) are conformationally mobile. The conformational rigidity of the free carocryptands arises from hydrophobic π–π interactions that align the polyunsaturated linker chains in the divergent hexaimine conformation. Conformational preference in the aromatic-linked Schiff-base cryptands (267, 268), on the other hand, is affected by mainly edge-to-face aromatic π–π interactions, as shown by ^{1}H NMR spectra of the free cryptand xylimBT. These π–π interactions give rise, in the divergent conformation of that cryptand, to an anomalous position of ^1H NMR resonance for the unique aromatic proton α to both imino substituents, deriving from the cumulative ring current of the three aromatic rings. The large (5 ppm) coordination shift for this unique aromatic proton demonstrates that the ring currents have dramatically opposite effects in free ligand and in cryptate. The close approach of aromatic rings in the convergent form presumably contributes to steric strain, and strain energy including that arising from consequent twisting out of plane of the bisimino–xylyl assembly (Fig. 55) destabilizes the cryptate, explaining both the approximate thermoneutrality (184) of the dicopper(I) complex and the nonexistence of the monocopper(I) form.

The crystal structure of the free carocryptand (Fig. 93) shows relatively close contact (3.9–4.0 Å) between strands in the polyunsaturated region suggesting attractive interaction in this region, with some opening out at the ends. When the guest cation is accommodated, adoption of the convergent conformation disrupts the hydrophobic interaction, giving more freedom to individual cryptand strands.

Conformation in the protonated bisiminomethylpyridine cryptate, $[MepyimBTH_2]^{2+}$ [Fig. 98b], apart from closer approach of strands in the methylene caps deriving from hydrogen bonding, is similar to that in the unmethylated cryptand pyimBT (Fig. 98a). This observation suggests that there is no steric reason for the general failure to isolate cryptands from the [2 + 3] Schiff-base condensation of triamines with diketones, so that kinetic factors

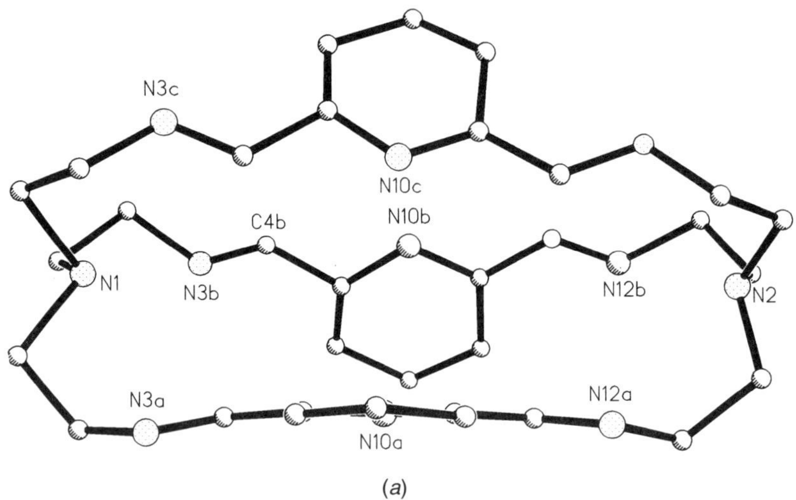

Figure 98. Comparison of conformations of methylated and unmethylated (264, 68) PyimBT: X-ray structures of (*a*) pyimBT and (*b*) $MepyimBT.H_2^{2+}$.

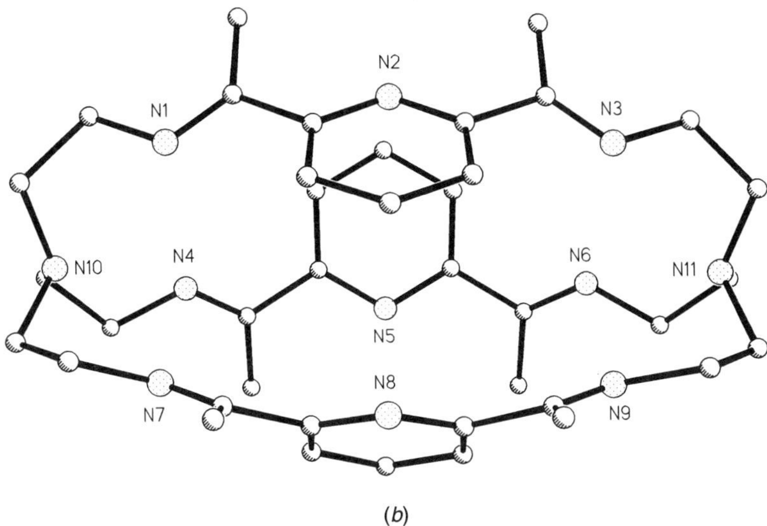

(*b*)

Figure 98. (*Continued.*)

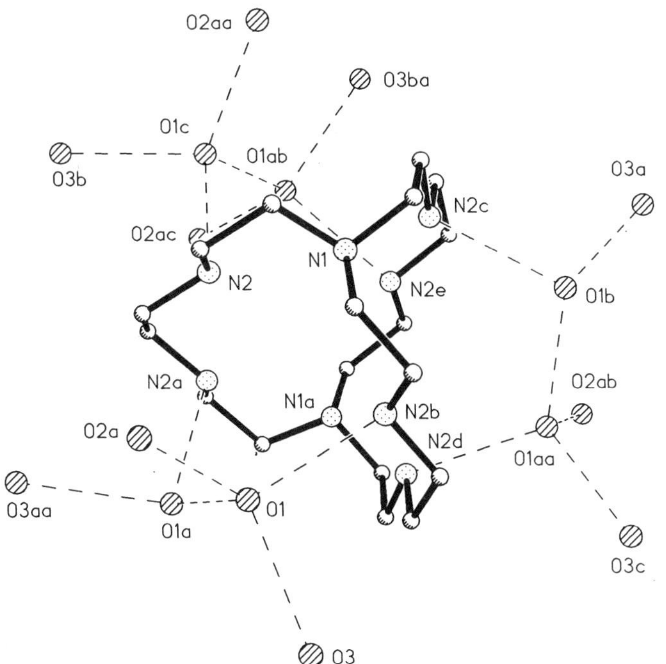

Figure 99. Structure of amBT (172) showing associated water molecules and hydrogen bonding.

must be considered responsible for the perceived difference between dialdehydes and diketones in this reaction.

In many free cryptand or cationic ligands, particularly amine systems, hydrogen bonding is important in deciding the details of the solid-state conformation, which presumably is also true of solution conformations. These cages frequently crystallize with many molecules of water solvate and extensive hydrogen-bond networks are built up (64, 286). Typical is amBT.8H_2O, in which each NH is hydrogen bonded to a water molecule (172) (Fig. 99) and hence to the rest of the network. An extreme example is the complex *tentatively* formulated as (H_8 *p*-xylambistrpn)$_2$ $(ClO_4)_{15}OH \cdot 16H_2O$ (64) where the trpn cap-N is exo, allowing inter- and intramolecular π interactions between phenyl groups but also permitting extensive hydrogen bonding involving all the secondary amines (but interestingly, not the exo bridgehead), the water molecules and the perchlorate anions (Fig. 100). One perchlorate anion sits in the cleft between each pair of cryptand strands. The anions are not directly bonded to the crypt

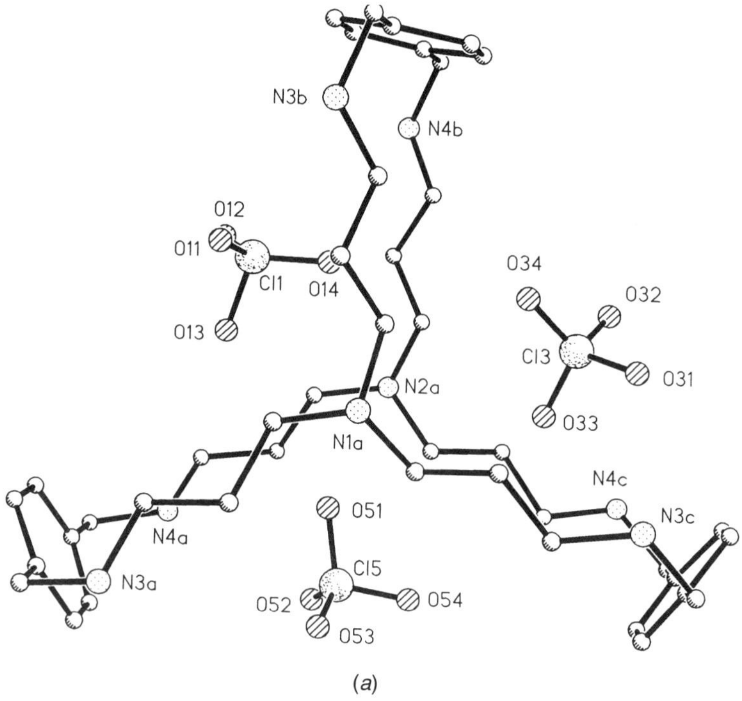

Figure 100. Structure of protonated *p*-xylambistrpn (64) showing water hydrogen bonded to anions and to NH^+: (*a*) side-on view showing perchlorate anions associated with one cryptand, (*b*) side-on view showing associated water molecules and hydrogen-bonded anions, and (*c*) view of unit cell showing hydrophobic and hydrophilic channels.

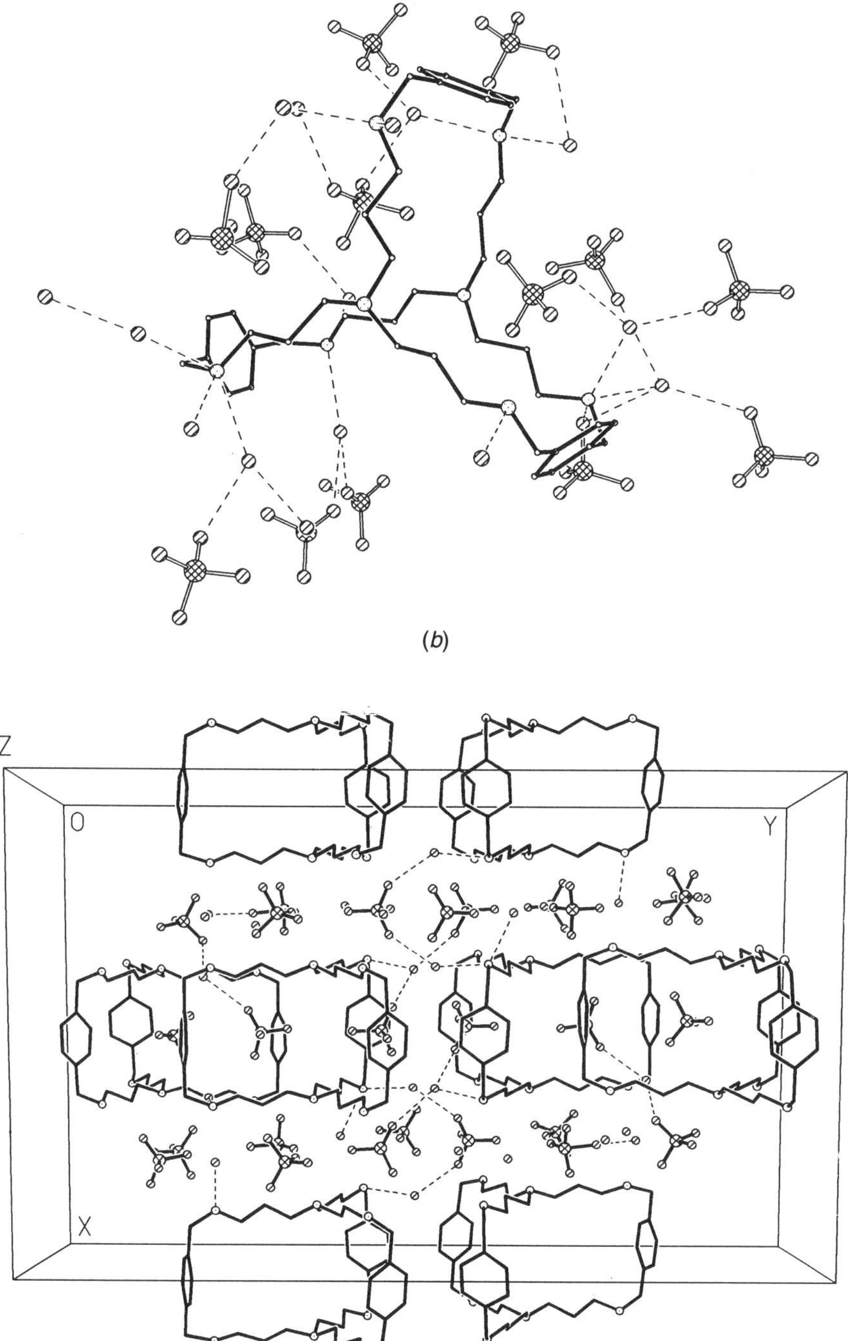

Figure 100. (*Continued*.)

but are linked via amine–water–perchlorate hydrogen bonds. A similar motif is seen in protonated furamBT, where the encapsulated perchlorate ion is coordinated to protonated amines both directly, and via water molecules, as will be seen in Section V.

K. Cryptands for Anion Complexation

Anion complexation is without doubt a growing branch of chemistry (99–101) on account of potentially important applications in transport, selectivity, and catalysis, which match those, by now well developed, for cation complexation. Likewise, in biology, the field of anion complexation must be considered of equal significance to that of cation complexation, given that each positively charged center is associated with one or more negatively charged anionic center and remembering that anions have to be located, transferred across membrances, and chemically transformed in the same way as cations (269); indeed, the majority of enzyme substrates are anions rather than cations (270). Complexation of anions is, however, a relatively challenging task on account of their larger size and higher free energies of solvation, which reduces their chance of successful competition with cations for the coordination site.

Leaving aside the "cascade" complexes discussed earlier, polyammonium cryptands are among the more promising of hosts for anions, including the oxoanions that are the subject of current environmental concern. The first cryptand, synthesized by Park and Simmons in 1968 (19) was isolated as a hydrochloride salt. X-ray crystallographic structure determination later (271) showed the chloride anion to be contained within the cavity. The small tetrahedrally symmetric oxo–azamacrotricyclic hosts **XXXII–XXXIV** when protonated or quaternized (272, 273) encapsulate halide anions, as does the similar macrob-

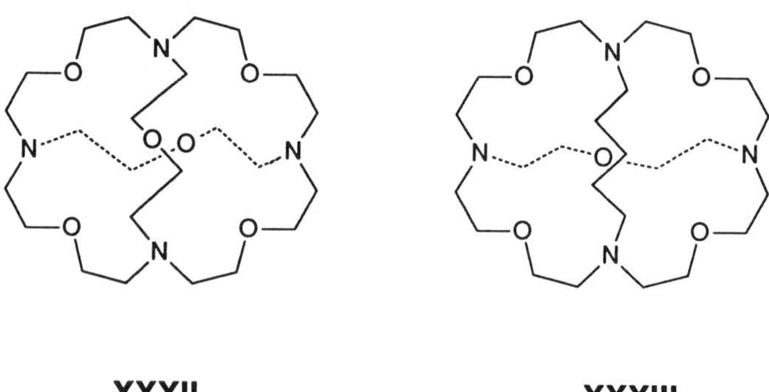

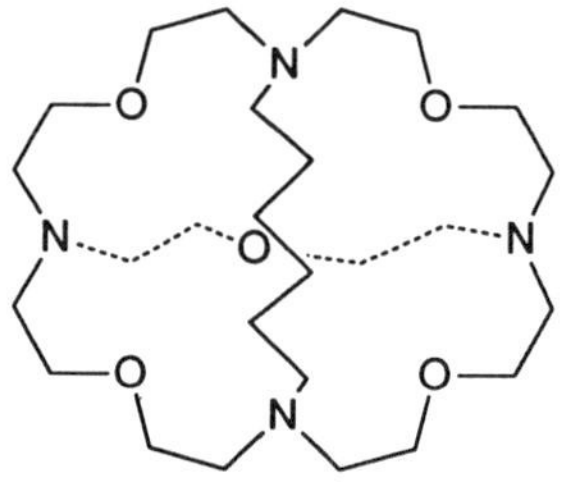

XXXIV 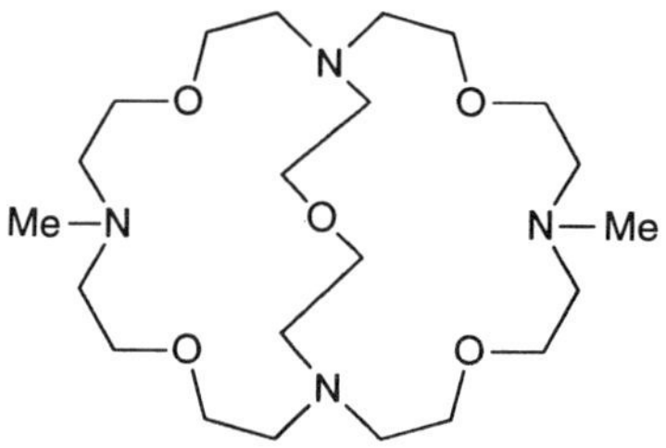 **XXXV**

icycle **XXXV** and the all-azamacrotricycles **XXXVI** and **XXXVII** (274), which appear to act as "soft" anion receptors favoring more polarizable guests such as iodide or azide over the less polarizable chloride. Complexation has been both structurally (Fig. 101) and quantitatively (Table VIII) demonstrated for these systems (275–279).

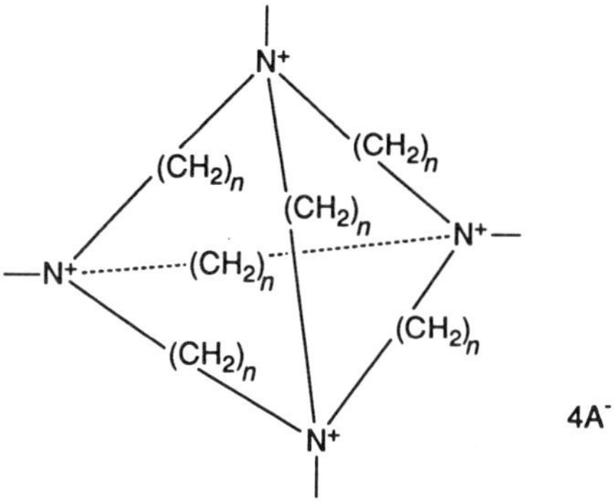

XXXVI $n = 6$

XXXVII $n = 8$

Macrobicyclic azacryptand hosts such as O-bistren are easily protonated and, in their hexaprotonated form, bind halides (Fig. 102) such as chloride and bromide in the center of the cryptand cavity. The smaller anion fluoride lies unsymmetrically in the ellipsoidal cavity, which, however, is ideally suited to accommodation of the linear triatomic anion N_3^- (280), as shown by X-ray

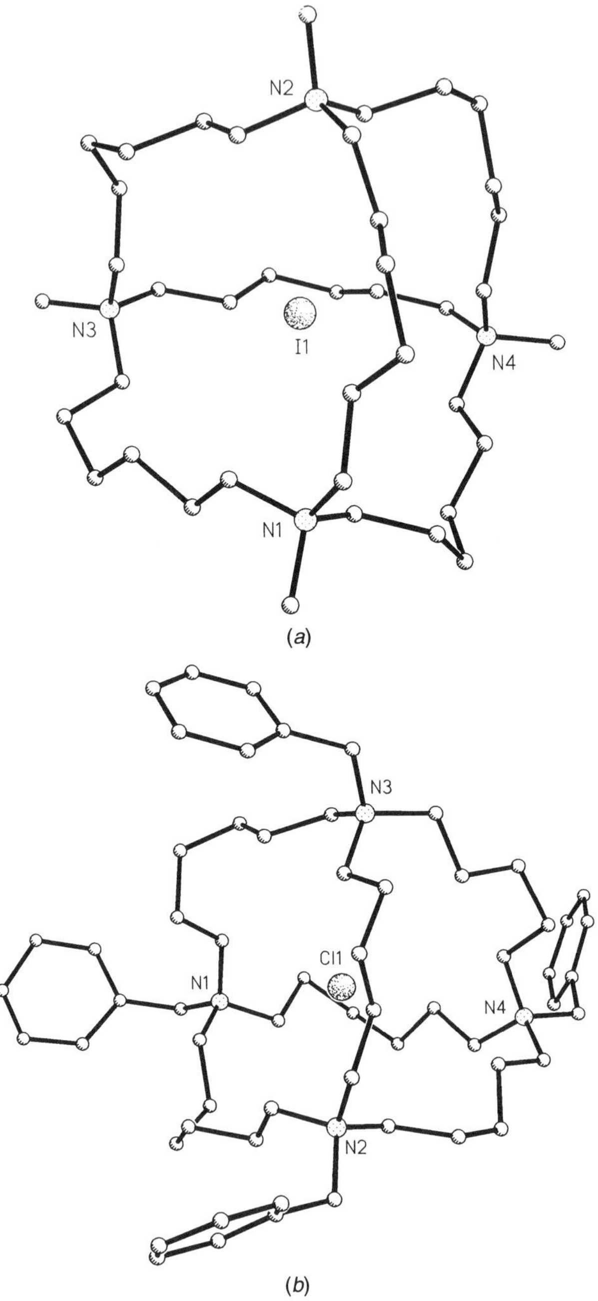

Figure 101. Macrotricyclic hosts for anions: Structures of (276, 277) (*a*) **[XXXVI.I]**$^-$ and (*b*) **[XL.Cl]**$^-$.

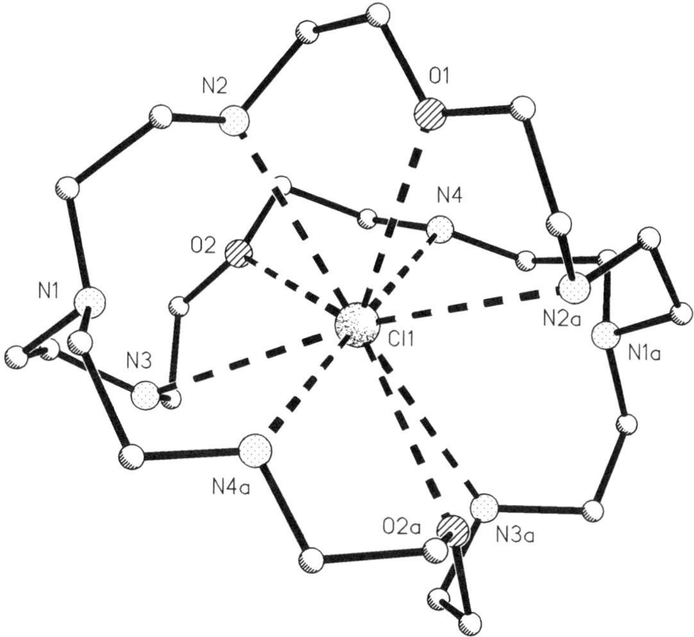

Figure 102. X-ray structure (280) of $[O\text{-bistrenH}_6]^{6+}$ chloride cryptate.

crystallographic structure determination (Fig. 103). Potentiometric methods have been used to monitor (280, 281) the complexation of other anions, including oxoanions. The stability constants thus obtained for anion complexation by O-bistren are relatively high, particularly for highly charged species like phosphate-derived anions (Table VIII.) The azacryptand C3-bistrpn (282) can accommodate eight protons, explaining stability constants with sulfate or oxalate a couple of orders of magnitude greater than with the tren-capped analogue. It has been shown (283) that polyamines with C_2 links between N donors are less basic than those with C_3 links, which accounts for the better complexation of anions by trpn- capped cryptands.

The 1,3,5-trisubstituted benzene capped ''barrel-end'' cryptands (284) **XXXIX** and **XXXVIII** with propylene or $(CH_2)_2{-}O{-}(CH_2)_2$ spacers form stable complexes in their hexaprotonated form with Cl^-, NO_3^-, and N_3^- monocharged anions and stronger complexes with dinegative anions. In aqueous solution, there is NMR evidence for encapsulation of the nitrate ion within the cryptand cavity of the propylene bridged cryptand **XXXIX,** although X-ray crystallographic structure determination shows that all six nitrate anions are located outside the cryptand cavity.

TABLE VIII

Formation Constants log Q (= $[LH_nX]^{(n-m)+}/[LH_n]^{n+}.\ [X^{m-}]$) of Anion Cryptates

Anion Host	HF_2^-	F^-	Cl^-	Br^-	I^-	Oxalate	Sulfate	Azide	HPO_4^{2-}	ATP^{4-}	$P_2O_7^{2-}$	HCO_3^-	NO_3^-
(**XXXII**H_4)$^{4+}$ [a]		[b]	>4	<1	<1								[b]
(**XXXIII**H_4)$^{4+}$ [a]		[b]	>4.5	1.55	[b]								[b]
(**XXXV**H_4)$^{4+}$ [a]		[b]	1.7	<1.0	[b]								[b]
(C_7, C_7, $C_7N_2H_2$)$^{2+}$ [a]		[b]	0.7	<1.0	[b]								[b]
(**XXXII**Me_4)$^{4+}$ [c]		—	1.0	1.8	[b]								
(**XXXVI**)$^{4+}$ [c]		—	1.3	2.45	2.3				2.54	2.46		1.76	
(**XXXVII**)$^{4+}$ [d]		—	<0.5	2.45	2.4			1.24	0.32	1.97			
O-bistren$H_6$$^{6+}$ [e]	6.4 [f]	4.1	3.0	2.60	2.15	4.95	4.9	4.3	5.5	8.0	10.3	2.30	2.80
C3-bistrpn$H_8$$^{8+}$ [g]			2.40	2.95	3.40	6.55	7.45						
C5-bistren$H_6$$^{6+}$	5.2 [f]												
[**XXXVIIIH**]$^{6+}$ and			In range			In range	In range					In range	
[**XXXIX**]$^{6+}$ [h]			2.5–4			5.0–6.5	2.5–4						2.5–4
amBT.$H_5$$^{+}$ [i]		>8.8 [f]	>1.2 [k]										
amBT.$H_6$$^{+}$ [i]		11.2 [j]											

[a] Reference (272): water, pH 1.5 HNO_3. T = 22°C.

[b] The ^{13}C NMR evidence (272) shows no complex formed.

[c] Reference (278): (I = 0.1 M NaOTs/NEt_4OTs).

[d] The 1 M trisfluoride pH 8.6.

[e] Reference (280): 0.1 M NaOTs [additional values for acetate (3.1) and ADP (5.85) anions].

[f] Reference (230): total ionic strength 0.1 M ($NaClO_4$/NaF).

[g] Reference (282): 0.1 M NaOTs, 25°C (additional value for malonate: 4.00).

[h] Reference (284): NMR method.

[i] Reference (228).

[j] (I = 0.1 M KNO_3, 25°C).

[k] (I = 0.1 M KCl, 25°C).

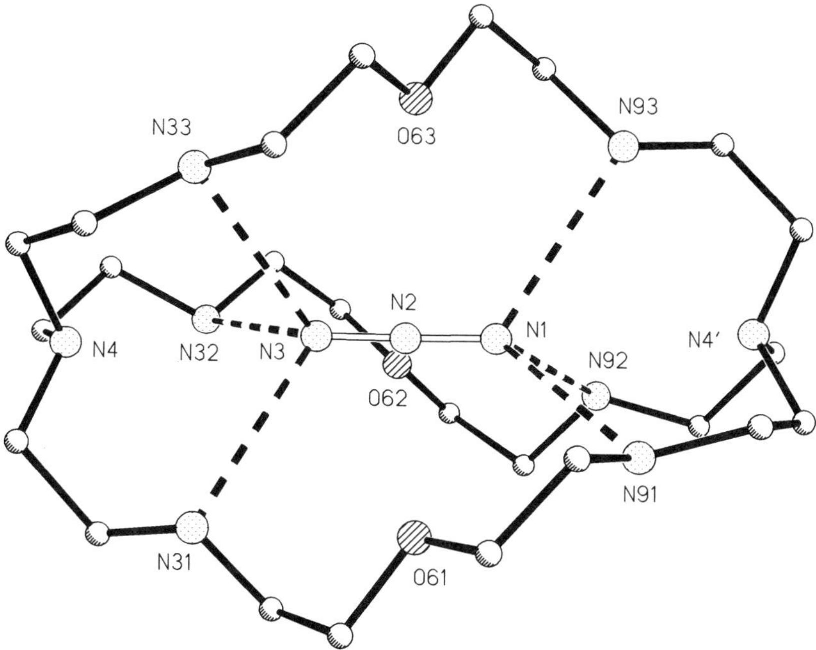

Figure 103. Polyatomic anion cryptates: crystal structure of $[\text{O-bistrenH}_6]^{6+}$ azide (280).

NH
R
NH
NH
HN
R
HN
R
NH

XXXVIII R = $-CH_2CH_2OCH_2CH_2-$

XXXIX R = $(CH_2)_3$

The diphenylmethane-spaced aminocryptand DPMamBT (285) acts as receptor for dicarboxylic anions when protonated, and displays chain-length dependent structural selectivity toward the guest dicarboxylates (Fig. 104). Structural characterization of the most strongly bound (terephthalate) anion cryptate shows that the strong selectivity for this anion against other aliphatic dicarbox-

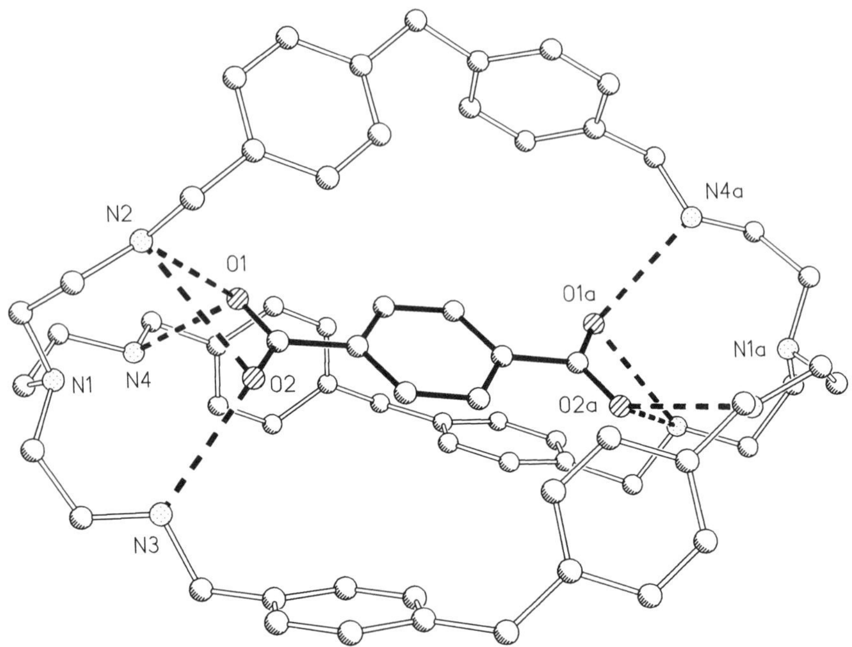

Figure 104. X-ray structure (285) of [DPMamBTH$_6$.terephthalate]$^{4+}$.

ylates is partly explained by hydrogen-bonding complementarity and partly by hydrophobic effects. This rare example is one of a structurally characterized polyatomic anion cryptate, another being the azide cryptate of O-bistren (Fig. 103) (280), where the anionic guest is seen to fit perfectly into the cryptand cavity.

Excellent complementarity of fit has also been demonstrated (174) for the fluoride anion encapsulated within the protonated small cryptand [H$_6$amBT]$^{6+}$, in comparison with the fluoride cryptate of larger macrobicycles such as O-bistren, where complementarity of fit is poor. Within [H$_6$amBT]$^{6+}$ the N—H· · ·F distances are 281 Å, representing a typical strong N—F hydrogen bond. Exceptionally large stability constants for formation of protonated amBT fluoride cryptates and excellent selectivity against chloride have been recently reported by Smith and co-workers (228) (Table VIII).

Figure 105. Oxoanion and perfluoroanion cryptates (286): (*a*) [furamBT.H$_6$.perchlorate.9H$_2$O]$^{5+}$ and (*b*) [pyamBT.H$_6$.hexafluosilicate]$^{4+}$. Note the three hydrogen-bonded water molecules linking amine and perchlorate in (*a*).

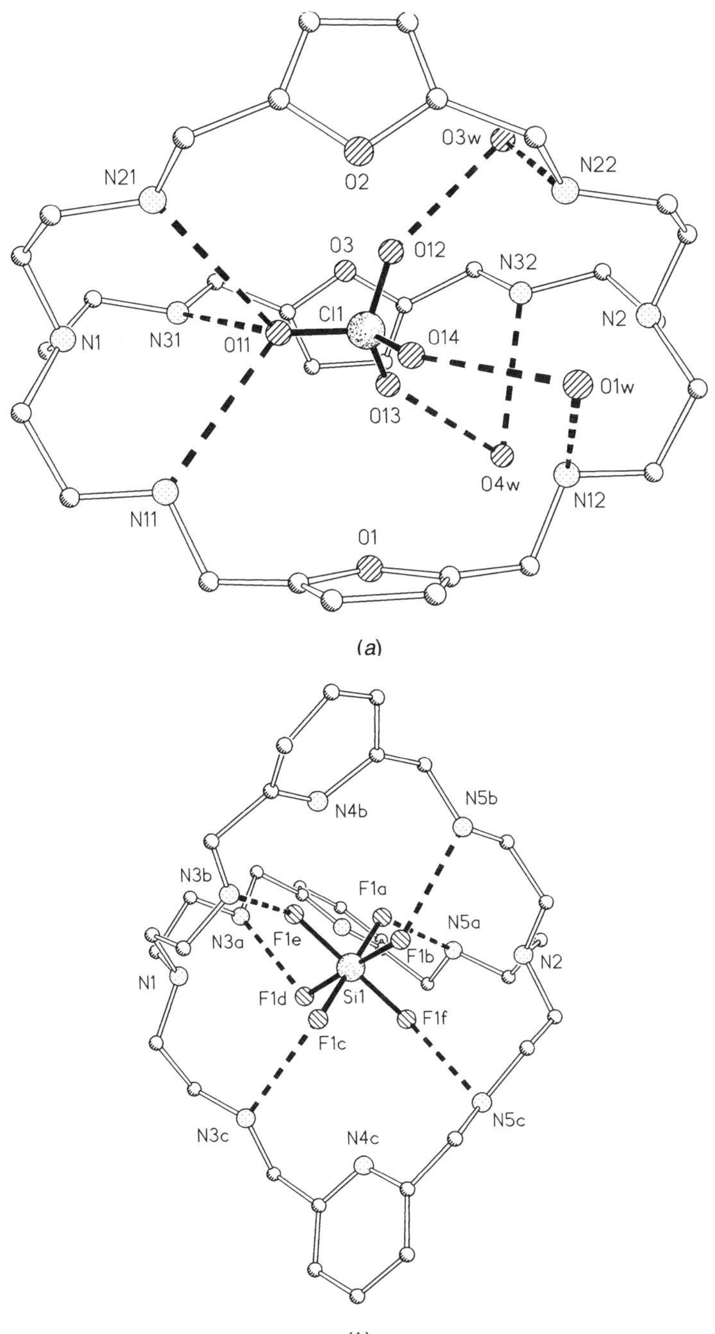
O3w
N22
N21
O2
O3
O12
N32
Cl1
N2
N1
N31
O11
O14
O1w
O13
O4w
N12
N11
O1
(a)
N4b
N5b
N3b
F1a
N5a
N3a
F1e
F1b
N2
N1
F1d
Si1
F1f
F1c
N3c
N4c
N5c
(b)

For complex inorganic ions such as oxoanions and perhaloanions the protonated aminocrypts deriving from 1,5-linked hexa-Schiff-bases have an appropriately complementary conformation, with many potentially convergently directed hydrogen-bonding sites. We have recently demonstrated by X-ray crystallographic structure determination (Fig. 105) (286) the encapsulation of $[ClO_4]^-$ and $[SiF_6]^{2-}$ within such hosts. These are the first structures of oxoanion or perfluoroanion cryptates to be determined, and show the importance of hydrogen-bonding complementarity, which ensures selectivity for the octahedral $[SiF_6]^{2-}$ guest over the less well-matched tetrahedral ion, $[BF_4]^-$. It is noticeable that in all the structurally characterized anion cryptates, the hydrogen-bonding network (285, 286), which stabilizes anion binding, includes associated water as well as endogeneous NH^+ sites. This observation is in accord with modeling studies (287) on other cryptand systems.

VI. CONCLUSION

This chapter represents a brief and subjective overview of azacryptate chemistry; a topical and exciting area where the numerous potential applications justify the continuing search for new synthetic methods and new systems. Encapsulation of single cations generates effects that follow from principles well developed in polyether cryptand chemistry, but the consequences of constraining a pair of cations, bridged or unbridged, within an azacryptand host are only now in the course of definition and systematization. Equally, the consequences of anion encapsulation, chemical or electrochemical, are as yet not fully understood. We hope that this chapter will help bring coherence to the results of the various research groups who have worked separately and with different objectives in the field of azacryptand complexation. It is clear that these valuable host molecules will continue to be exploited, in both cation and anion coordination chemistry, over future decades, and in such context this attempt at formalization of common principles may be of value to future workers in this segment of supramolecular chemistry.

ACKNOWLEDGMENTS

We are grateful to Professor S. M. Nelson for introducing us to the exciting and unpredictable world of Schiff-base chemistry: ''chemistry with a life of its own . . . ''. Over the last 10 years, many postgraduates and undergraduate students at Queens University have made significant contributions to the development of the Schiff-base derived cryptands, notably Debbie Marrs, Josie Hunter, Lu Qin, Noreen Martin, Yann Dussart,

Beatrice Maubert, and Joanne Coyle. Many of the crystal structures were obtained by Dr. M. G. B. Drew at Reading, whose expertise proved invaluable in the initial and subsequent demonstration of the success of the [2 + 3] Schiff-base condensation strategy. Other collaborators (too numerous to mention individually) have contributed significantly to the understanding of spectroscopy and bonding of the cryptate systems obtained by this route; we are particularly indebted to the groups of Professors Schwing and Arnaud-Neu at Strasbourg for solution complexation measurements; Dr. Harding at the Open University for magnetic studies; Professor Thomson at University of East Anglia and Professor McGarvey at Queens University for spectroscopy.

ABBREVIATIONS

111

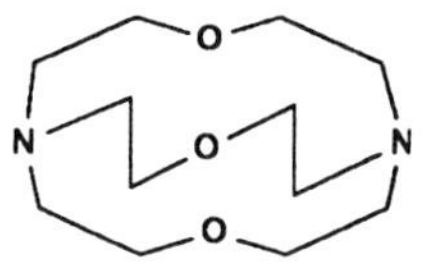

222

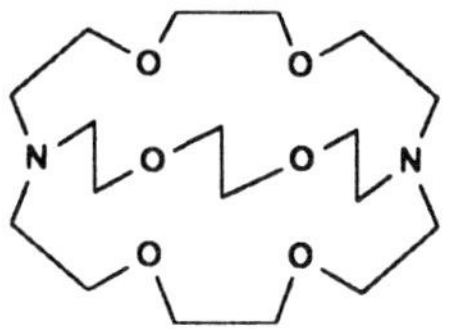

221

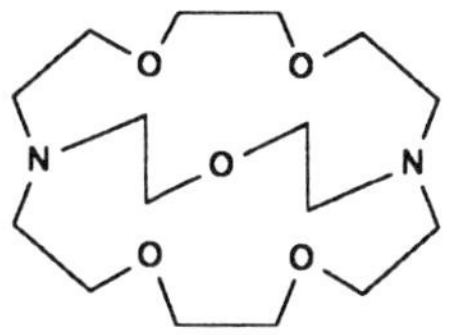

211

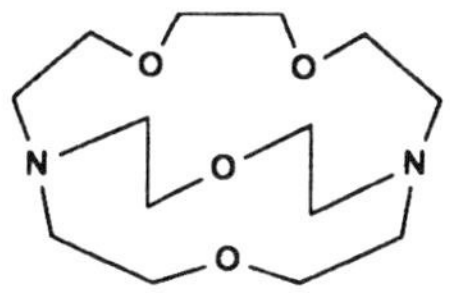

333

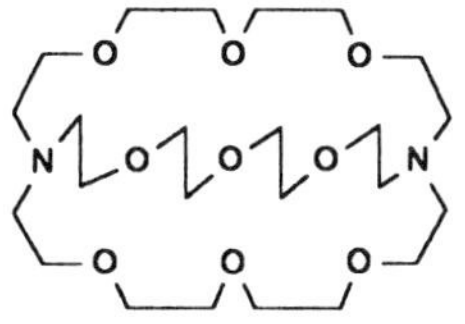

$C_7C_7C_7N_2$

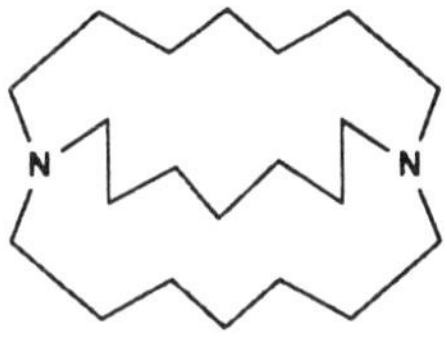

$N_4O_2$211NMe

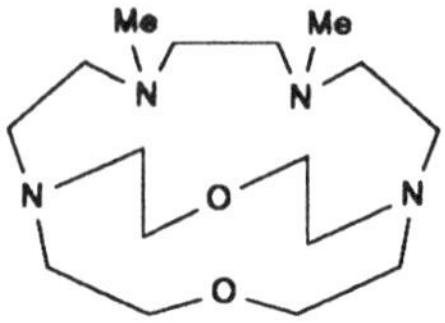

$N_4O_4$222NMe

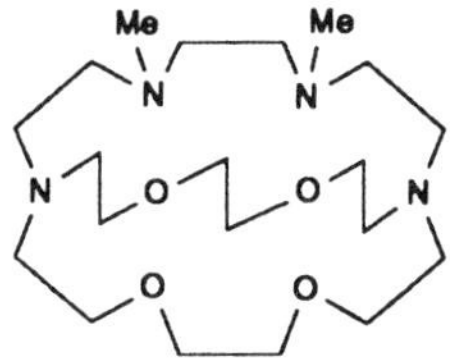

$N_6O_2$222NMe

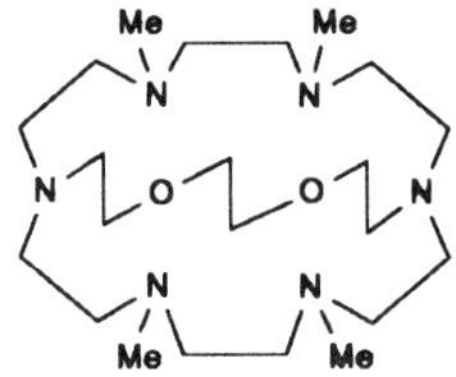

amBT

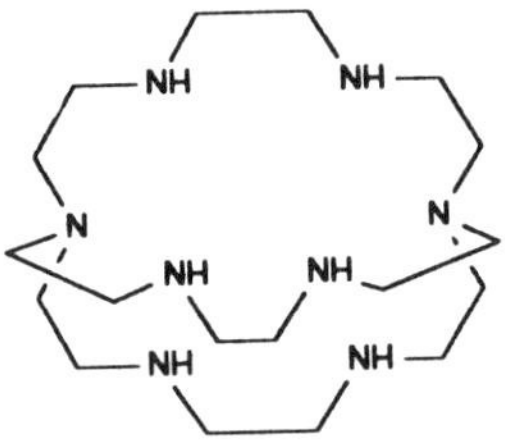

ATP Adenosine triphosphate

C3-bistrpn

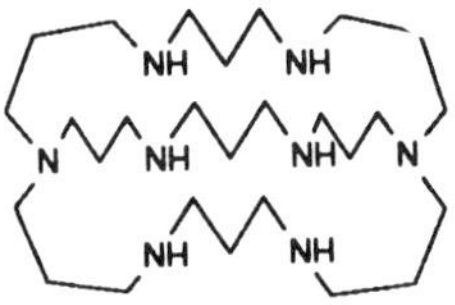

C5-bistren	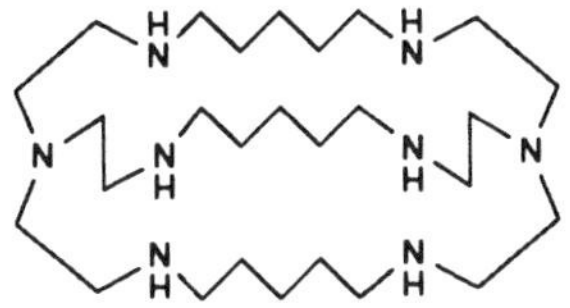
cyclen	12-aneN_4
diamsar	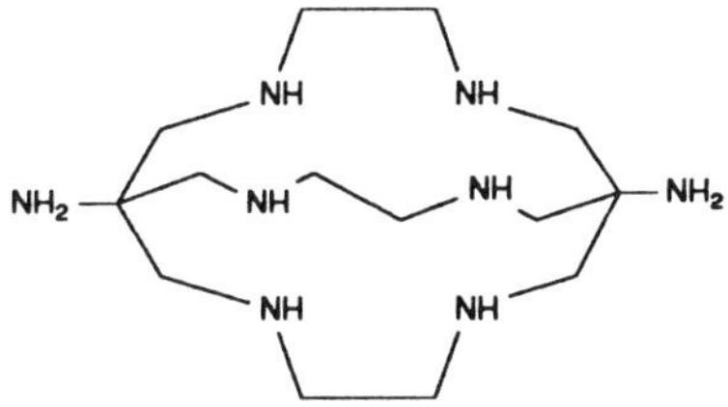
DMF	*N*,*N*-Dimethylformamide
DMSO	Dimethyl sulfoxide
en	Ethylenediamine
ESR	Electron spin resonance
furamBT	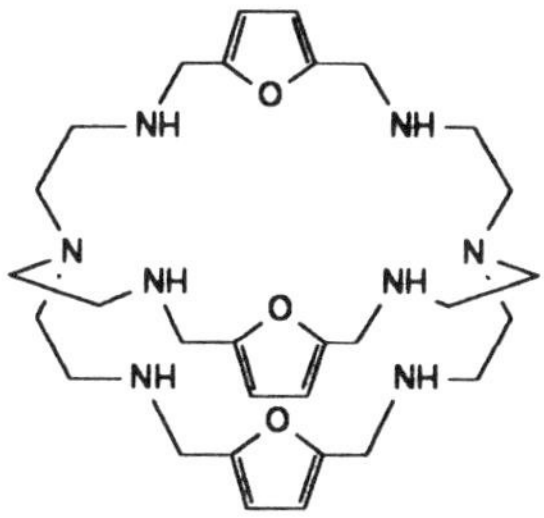
furimBT	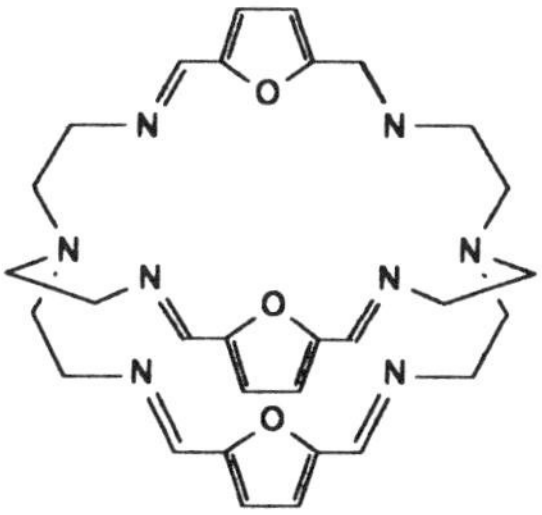
imbistrpn	

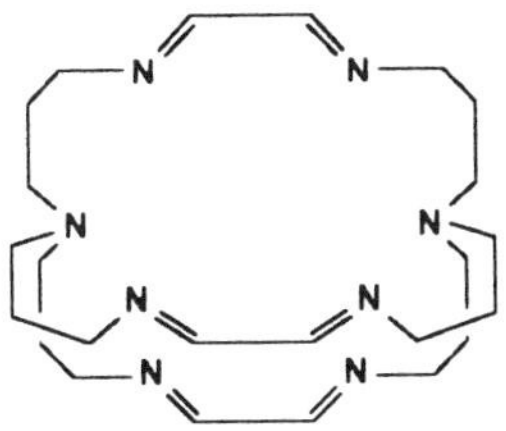

imBT

IR	Infrared
LMCT	Ligand-to-metal charge transfer
lel	Parallel
MRI	Magnetic resonance imaging
Ms	Mesyl
MCD	Magnetic circular dichroism
NMR	Nuclear magnetic resonance

O-bistren

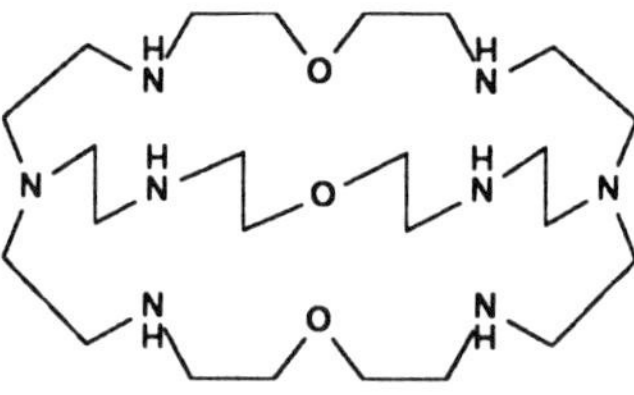

ob	Oblique

pyamBT

pyimBT

sar	Sarcophogine
sep	Sepulchrate
SOD	Superoxide dismutase
tame	Trisaminomethylmethane
tama	Tris(aminomethyl)amine
THF	Tetrahydrofuran
tren	Tris(2-aminoethyl)amine
trpn	Tris(aminopropyl)amine
trisbpy	Trisbipyridyl
trisphen	Trisphenanthroline
Ts	Tosyl (*p*-toluene sulfonate)
UV	Ultraviolet

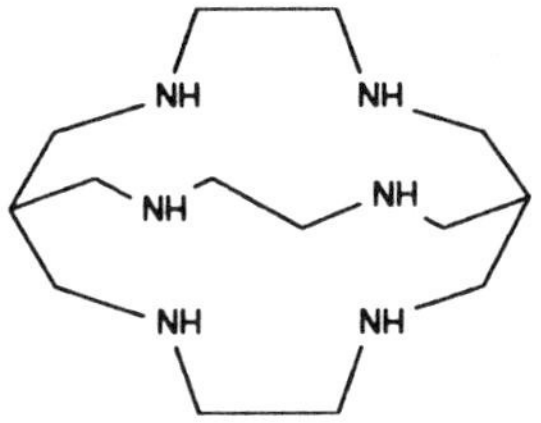

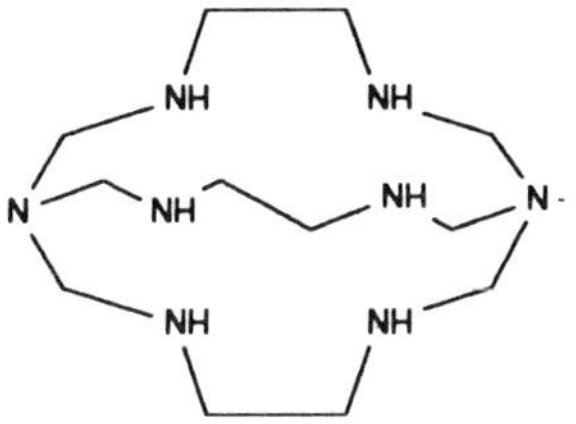

xylamBT

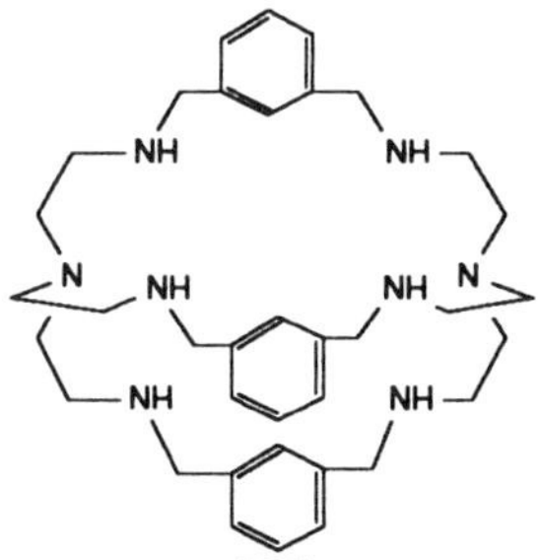

***p*-xylamBT**

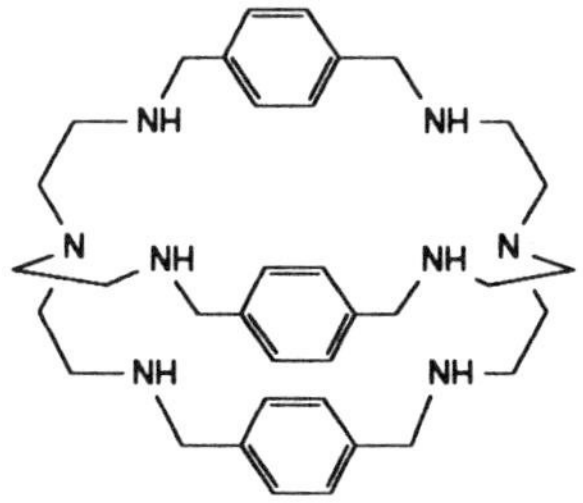

xylimBT

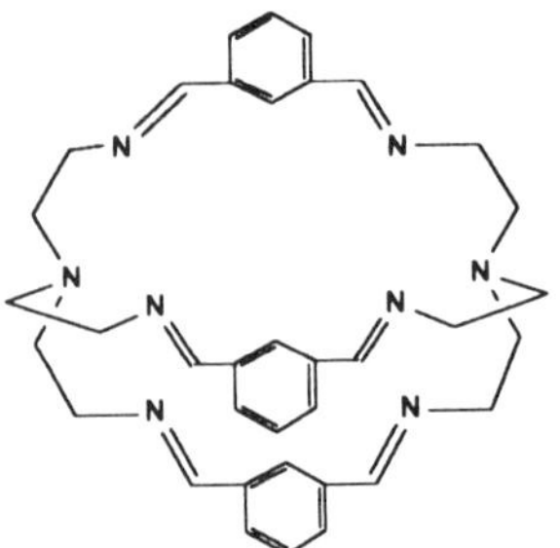

REFERENCES

1. B. Dietrich, J.-M. Lehn, and J.-P. Sauvage, *Tetrahedron Lett.*, 2885 (1969).
2. J.-M. Lehn, *Pure Appl. Chem.*, *49*, 857 (1977).
3. J.-M. Lehn, *Acc. Chem. Res.*, *11*, 49 (1978).
4. J.-M. Lehn, *Pure Appl. Chem.*, *50*, 871 (1978).
5. J.-M. Lehn and J.-P. Sauvage, *J. Am. Chem. Soc.*, *97*, 6700 (1975).
6. B. Dietrich, J.-M. Lehn, and J. P. Sauvage, *J. Chem. Soc. Chem. Commun.*, 15 (1973).
7. F. Montanari, D. Landini, and F. Rolla, *Top. Curr. Chem.*, *101*, 147 (1982).
8. S. Quici, P. L. Anelli, H. Molinari, and T. Beringhelli, *Pure Appl. Chem.*, *58*, 1503 (1986).
9. J.-M. Lehn, *Pure Appl. Chem.*, *51*, 979 (1979).
10. J.-M. Lehn, *Science*, *227*, 849 (1985) and references cited therein.
11. J.-M. Lehn, *Angew. Chem. Int. Ed. Engl.*, *29*, 1304 (1990).
12. J. L. Dye, *Progress in Inorganic Chemistry*, Wiley-Interscience, New York, 1984, Vol. 32, p. 327.
13. (a) F. Arnaud-Neu, B. Speiss, and M.-J. Schwing-Weill, *Helv. Chim. Acta*, *60*, 2633 (1977). (b) M. C. Almassio, F. Arnaud-Neu, and M.-J. Schwing-Weill, *Helv. Chim. Acta*, *66*, 1296 (1983).
14. G. Anderegg, *Helv. Chim. Acta*, *64*, 1790 (1981).
15. J.-M. Lehn and F. Montavon, *Helv. Chim. Acta*, *61*, 67 (1978).
16. J.-M. Lehn, *Pure Appl. Chem.*, *50*, 2441, (1980) and references cited therein.
17. B. Vallee and R. J. P. Williams, *Proc. Natl. Acad. Sci.*, *59*, 498 (1968).
18. For complete coverage of these topics, see J. L. Atwood, J.-M. Lehn, Eds., *Comprehensive Supramolecular Chemistry:* Pergamon, New York, 1996. For cryptands, see also B. Dietrich, in *Inclusion Compounds*, J. L. Atwood, J. E. Davies, and D. D. MacNicol, Eds. Academic, New York, 1984; Y. Inoue and G. W. Gokel, Eds., *Cation binding by Macrocycles*, Dekker, New York, 1990.
19. H. E. Simmons and C. H. Park, *J. Am. Chem. Soc.*, *90*, 2431 (1968).

20. B. Dietrich, J.-M. Lehn, J. P. Sauvage, and J. Blanzat, *Tetrahedron*, *29*, 1629 (1973).
21. H. Stetter and J. Marx, *Liebigs Ann.*, *607*, 59 (1957).
22. J.-M. Lehn and F. Montavon, *Helv. Chim. Acta*, *59*, 1566 (1976).
23. J.-M. Lehn and F. Montavon, *Tetrahedron Lett.*, *44*, 4557 (1972).
24. J.-C. Rodriguez, B. Alpha, D. Plancherel, and J.-M. Lehn, *Helv. Chim. Acta*, *67*, 2264 (1984).
25. S. Kulstad and L. A. Malmsten, *Tetrahedron*, *36*, 521 (1980).
26. B. Alpha, E. Anklam, R. Deschenaux, J.-M. Lehn, and M. Pietraszkiewicz, *Helv. Chim. Acta*, *71*, 1042 (1988).
27. M. Pietraszkiewicz, P. Santanski, and J. Jurczak, *Tetrahedron*, *40*, 2971 (1984).
28. F. Ebemeyer and F. Vogtle, *Chem. Ber.*, *122*, 1725 (1989).
29. B. Dietrich, M. W. Hosseini, J. M. Lehn, and R. B. Sessions, *Helv. Chim. Acta*, *68*, 289 (1985).
30. K. G. Raghunathan and P. K. Bharadwaj, *Tetrahedron Lett.*, *33*, 7581 (1992).
31. P. Ghosh, R. Shukla, D. K. Chand, and P. K. Bharadwaj, *Tetrahedron*, *51*, 3265 (1995).
32. G. R. Newkome, V. Majestic, F. Fronczek, and J. I. Atwood, *J. Am. Chem. Soc.*, *101*, 1047 (1979).
33. G. R. Newkome, V. K. Majestic, and F. R. Fronczek, *Tetrahedron Lett.*, *22*, 3035 (1981).
34. I. I. Creaser, J. MacB. Harrowfield, A. J. Herlt, A. M. Sargeson, M. R. Snow, and J. Springborg, *J. Am. Chem. Soc.*, *99*, 3181 (1977).
35. I. I. Creaser, R. J. Geue, J. MacB. Harrowfield, A. J. Herlt, A. M. Sargeson, M. R. Snow, and J. Springborg, *J. Am. Chem. Soc.*, *104*, 6016 (1982).
36. A. M. Sargeson, *Chem. Br.*, *15*, 23 (1979).
37. A. M. Sargeson, *Pure Appl. Chem.*, *50*, 905 (1978).
38. J. MacB. Harrowfield, A. J. Herlt, and A. M. Sargeson, *Inorg. Syntheses*, *20*, 85 (1980).
39. R. J. Geue, T. W. Hambley, J. M. Harrowfield, A. M. Sargeson, and M. R. Snow, *J. Am. Chem. Soc.*, *106*, 5478 (1984).
40. A. M. Sargeson, *Pure Appl. Chem.*, *56*, 1603 (1984).
41. L. R. Gahan, T. W. Hambley, A. M. Sargeson, and M. R. Snow, *Inorg. Chem.*, *21*, 2699 (1982).
42. A. Hohn, R. J. Geue, and A. M. Sargeson, *J. Chem. Soc. Chem. Commun.*, 1473 (1990).
43. R. J. Geue, W. R. Petri, A. M. Sargeson, and M. R. Snow, *Aust. J. Chem.*, *45*, 1681 (1992).
44. G. J. Gainsford, R. J. Geue, and A. M. Sargeson, *J. Chem. Soc. Chem. Commun.*, 233 (1982).
45. K. N. Brown, R. J. Geue, T. W. Hambley, A. M. Sargeson, and A. C. Willis, *J. Chem. Soc. Chem. Commun.*, 567 (1996).

46. P. Comba, J. M. Engelhardt, J. M. Harrowfield, G. A. Lawrance, L. L. Martin, A. M. Sargeson, and A. H. White, *Inorg. Chem. 24*, 1607 (1985).
47. R. J. Geue, A. Hohn, S. F. Ralph, A. M. Sargeson, and A. C. Willis, *J. Chem. Soc. Chem. Commun.*, 1513 (1994).
48. H. A. Boucher, G. A. Lawrance, P. A. Lay, A. M. Sargeson, A. M. Bond, D. F. Sangster, and J. C. Sullivan, *J. Am. Chem. Soc.*, 4652 (1983).
49. R. J. Geue, M. B. McDonnell, A. W. H. Mau, A. M. Sargeson, and A. C. Willis, *J. Chem. Soc. Chem. Commun.*, 667 (1994).
50. J. MacB. Harrowfield, A. J. Herlt, P. A. Lay, A. M. Sargeson, A. M. Bond, W. H. Mulac, and J. C. Sullivan, *J. Am. Chem. Soc.*, *105*, 5503 (1983).
51. G. A. Bottomley, I. J. Clark, I. I. Creaser, L. M. Engelhardt, R. J. Geue, K.' S. Hagen, J. M. Harrowfield, G. A. Lawrance, P. A. Lay, A. M. Sargeson, A. J. See, B. W. Skelton, A. H. White, and F. R. Wilner, *Austr. J. Chem.*, *47*, 143 (1994).
52. I. I. Creaser, L. M. Engelhardt, J. M. Harrowfield, A. M. Sargeson, B. W. Skelton, and A. W. White, *Aust. J. Chem.*, *46*, 465 (1993).
53. D. P. Comba, L. M. Engelhardt, J. M. Harrowfield, G. A. Lawrance, L. L. Martin, A. M. Sargeson, and A. H. White, *J. Chem. Soc. Chem. Commun.*, 174 (1985).
54. P. Comba, I. I. Creaser, L. R. Gahan, J. M. Harrowfield, G. A. Lawrance, L. L. Martin, A. W. H. Mau, A. M. Sargeson, W. H. F. Sasse, and M. R. Snow, *Inorg. Chem.*, *25*, 384 (1986).
55. J. I. Bruce, L. R. Gahan, T. M. Hambley, and R. Stranger, *J. Chem. Soc. Chem. Commun.*, 702 (1993).
56. I. J. Clark, I. I. Creaser, L. M. Engelhardt, J. M. Harrowfield, E. R. Krausz, G. M. Moran, A. M. Sargeson, and A. H. White, *Aust. J. Chem.*, *46*, 111 (1993).
57. L. R. Gahan, T. M. Donleavy, and T. W. Hambley, *Inorg. Chem.*, *29*, 1451 (1990).
58. D. R. Boston and N. J. Rose, *J. Am. Chem. Soc.*, *90*, 6859 (1968).
59. J. E. Parks, B. E. Wagner, and R. H. Holm, *Inorg. Chem.*, *10*, 2472 (1972).
60. S. C. Jackels and N. J. Rose, *Inorg. Chem.*, *12*, 1232 (1973).
61. V. E. Zavodnik, V. K. Belsky, Y. Z. Voloshin, and O. A. Varzstskii, *J. Coord. Chem.*, *28*, 97 (1993) and references cited therein.
62. N. Kitajima and W. Tolman, *Progress in Inorganic Chemistry*, Wiley-Interscience, New York, Vol. 43, 1995, p. 419.
63. J. Nelson and D. McDowell, *Tetrahedron Lett.*, 385 (1988).
64. G. Morgan, Ph.D. Thesis, ''Cryptand and Podand Receptors for Anions and Calions,'' Open University, Milton Kaynes. 1995.
65. G. Morgan, V. Mc Kee, and J. Nelson, *Inorg. Chem.*, *24*, 4427 (1994) and references cited therein.
66. J. Jazwinski, J.-M. Lehn, D. Lilienbaum, R. Ziessel, J. Guilheim, and C. Pascard, *J. Chem. Soc. Chem. Commun.*, 1691 (1987).

67. J.-M. Lehn, P. Vigneron, I. Bkouche-Waksman, J. Guilheim, and C. Pascard, *Helv. Chim. Acta*, *75*, 1069 (1992).

68. H. Adams, N. A. Bailey, D. E. Fenton, C. Fukahara, and M. Kanesato, *Supramol. Chem.*, *2*, 325 (1993).

69. K. Ragunathan and P. K. Bharadwaj, *Polyhedron*, *14*, 693 (1995).

70. K. G. Raghunathan, R. Shukla, S. Misra, and P. K. Bharadwaj, *Tetrahedron Lett.*, *34*, 5631 (1993).

71. M. G. B. Drew, D. McDowell, and J. Nelson, *Polyhedron*, *7*, 2229 (1988).

72. D. Chen and A. E. Marteli, *Tetrahedron*, *47*, 6895 (1991).

73. K. N. Raymond, T. J. McMurry, and T. M. Garrett, *Pure Appl. Chem.*, *60*, 545 (1988).

74. K. N. Raymond and T. M. Garrett, *Pure Appl. Chem.*, *60*, 1807 (1988).

75. A. Bencini, A. Bianchi, M. Ciampolini, E. Garcia-Espana, P. Dapporto, M. Micheloni, P. Paoli, J. A. Ramirez, and B. Valtancoli, *J. Chem. Soc. Chem. Commun.*, 702 (1989).

76. A. Bencini, A. Bianchi, A. Borselli, S. Chimichi, M. Ciampolini, P. Dapporto, M. Micheloni, N. Nardi, P. Paoli, and B. Valtancoli, *Inorg. Chem.*, *29*, 3282 (1990).

77. M. Caimpolini, N. Nardi, B. Valtancoli, and M. Micheloni, *Coord. Chem. Rev.*, *120*, 223 (1992).

78. A. Bencini, A. Bianchi, C. Bazzicalupi, M. Ciampolini, P. Dapporto, V. Fusi, M. Micheloni, N. Nardi, P. Paoli, and B. Valtancoli, *J. Chem. Soc. Perkin Trans.*, *2*, 115 (1993).

79. A. Bencini, A. Bianchi, M. Ciampolini, N. Nardi, B. Valtancoli, S. Mangani, E. Garcia-Espana, and J. A. Ramirez, *J. Chem. Soc. Perkin Trans. 2*, 1131 (1989).

80. T. Saupe, C. Krieger, and H. A. Staab, *Angew. Chem.*, *Int. Ed. Engl.*, *25*, 451 (1986).

81. A. Bencini, A. Bianchi, C. Bazzicalupi, M. Ciampolini, P. Dapporto, V. Fusi, M. Micheloni, N. Nardi, P. Paoli, and B. Valtancoli, *J. Chem. Soc. Perkin Trans.*, *2*, 715 (1993).

82. K. Kimura and T. Shono, in *Ion-selective electrodes*, *4*, E. Pungor, Ed., Akademai Klado, Budapest, 1985.

83. B. Alpha, J.-M. Lehn, and G. Mathis, *Angew. Chim. Int. Ed. Engl.*, *26*, 266 (1987).

84. N. Sabbatini, M. Guardigli, I. Manet, R. Ungaro, A. Casnati, R. Ziessel, G. Ulrich, Z. Asfari, and J. M. Lehn, *Pure Appl. Chem.*, *67*, 135 (1995).

85. N. Sabbatini, S. Perathroner, V. Balzani, B. Alpha, and J.-M. Lehn, in *Supramolecular Photochemistry*, Reidel, Dordrecht, The Netherlands, 1987.

86. D. Landini, F. Montenari, and F. Rolla, *Synthesis*, *223* (1978).

87. D. Clement, F. Damm, and J.-M. Lehn, *Heterocycles*, *5*, 3903 (1977).

88. M. Cinquini, F. Montenari, and P. Tundo, *J. Chem. Soc. Chem. Commun.*, 393 (1975).

89. J. M. Lehn, *Pure Appl. Chem.*, *52*, 2303 (1980).
90. J. M. Lehn and P. G. Potvin, in *Progress in Macrocyclic Chemistry*, Wiley, New York, 1987, Chapter 4, Vol. 3, p. 167.
91. J. R. Pladziewicz, M. A. Accola, P. Osvath, and A. M. Sargeson, *Inorg. Chem.*, *32*, 2525 (1993).
92. P. A. Lay, J. Lydon, A. W. H. Mau, P. Osvath, A. M. Sargeson, and W. H. F. Sasse, *Aust. J. Chem.*, *46*, 641 (1993).
93. P. A. Lay, A. W. H. Mau, W. H. F. Sasse, I. I. Creaser, L. R. Gahan, and A. M. Sargeson, *Inorg. Chem.*, *22*, 2347 (1983).
94. A. M. Sargeson, *Pure Appl. Chem.*, *58*, 1511 (1986).
95. J.-M. Lehn, in *Physical Chemistry of Transmembrane Ion Motions*, G. Spach, Ed., Elsevier, Amsterdam, The Netherlands, 1983, p. 181.
96. J.-M. Lehn, *Ann. N.Y. Acad. Sci.*, *471*, 41 (1986).
97. J. D. Lamb, J. J. Christensen, J. L. Oscarson, B. L. Nielsen, B. W. Assay, and R. M. Izatt, *J. Am. Chem. Soc.*, *102*, 6820 (1980).
98. J. M. Lehn, *Neurosci. Res. Prog. Bull.*, *14*, 133 (1976).
99. B. Dietrich, *Pure Appl. Chem.*, *65*, 1457 (1993).
100. L. Pierre and P. Baret, *Bull. Soc. Chim. Fr.*, 367 (1983).
101. P. J. Beer, *J. Chem. Soc. Chem. Commun.*, 689 (1996).
102. F. P. Schmidtchen, A. Gleich, and A. Schummer, *Pure Appl. Chem.*, *61*, 1535 (1989).
103. M. P. Mertes and K. B. Mertes, *Acc. Chem. Res.*, *23*, 413 (1990).
104. M. W. Hosseini, J.-M. Lehn, L. Maggiora, K. Bowman-Mertes, and M. M. Mertes, *J. Am. Chem. Soc.*, *109*, 537 (1987).
105. C. A. Behm, I. I. Creaser, B. K. Korybutdaskiewicz, R. J. Geue, A. M. Sargeson, and G. W. Walker, *J. Chem. Soc. Chem. Commun.*, 1844 (1993).
106. C. A. Behm, P. F. L. Boreham, I. I. Creaser, B. K. Korybutdaskiewicz, D. J. Maddelena, A. M. Sargeson, and G. M. Snowdon, *Aust. J. Chem.*, *48*, 1009 (1995).
107. H. C. Brugge, D. Carboo, K. van Denten, A. Knochel, J. Kopf, and W. Dreissig, *J. Am. Chem. Soc.*, *108*, 107 (1986).
108. (a) A. Bencini, A. Bianchi, A. Borselli, M. Ciampolini, M. Micheloni, P. Paoli, and V. Valtancoli, *J. Chem. Soc. Perkin 2*, 209 (1990). (b) M. Ciampolini, M. Micheloni, P. Orioli, F. Vizza, S. Mangani, and F. Zanobini, *Gazz. Chim. Ital.*, *116*, 189 (1986).
109. (a) J. Springborg, P. Kofod, C. E. Olsen, H. Toftlund, and I. Sotofte, *Acta Chem. Scand.*, *49*, 547 (1995). (b) J. Springborg, C. E. Olsen, and I. Sotofte, *Acta Chem. Scand.*, *49*, 555 (1995). (c) J. Springborg, U. Pretzmann, and C. E. Olsen, *Acta Chem. Scand.*, *50*, 294 (1996).
110. A. Bencini, A. Bianchi, A. Borselli, M. Ciampolini, E. Garcia-Espana, P. Dapporto, M. Micheloni, P. Paoli, J. A. Ramirez, and B. Valtancoli, *Inorg. Chem.*, *28*, 4279 (1989).

111. A. Bencini, A. Bianchi, M. Ciampolini, E. Garcia-Espana, P. Dapporto, M. Micheloni, P. Paoli, J. A. Ramirez, and B. Valtancoli, *J. Chem. Soc. Chem. Commun.*, 701 (1989).
112. A. Bencini, A. Bianchi, A. Borselli, M. Ciampolini, M. Micheloni, N. Nardi, P. Paoli, B. Valtancoli, S. Chimichi, and P. Dapporto, *J. Chem. Soc. Chem. Commun.*, 174 (1990).
113. R. M. Izatt, J. S. Bradshaw, S. A. Nielsen, J. P. Lamb, J. Christensen, and D. Sen, *Chem. Rev.*, *85*, 271 (1985).
114. A. Bencini, A. Bianchi, S. Chimichi, M. Ciampolini, P. Dapporto, E. Garcia-Espana, M. Micheloni, N. Nardi, P. Paoli, and B. Valtancoli, *Inorg. Chem.*, *30*, 3687 (1991).
115. J. L. Eglin, E. P. Jackson, K. L. Moeggenborg, J. L. Dye, A. Bencini, and M. Micheloni, *J. Incl. Phenon. Mol. Recogn.*, *12*, 263 (1992).
116. A. Bianchi, E. Garcia-Espana, M. Micheloni, N. Nardi, and F. Vizza, *Inorg. Chem.*, *25*, 4379 (1986).
117. M. Ciampolini, M. Micheloni, F. Vizza, S. Chimichi, P. Dapporto, and F. Zanobini, *J. Chem. Soc. Dalton Trans.*, 505 (1986).
118. M. Micheloni, *J. Coord. Chem.*, *18*, 3 (1988).
119. A. Bencini, A. Bianchi, P. Dapporto, V. Fusi, E. Garcia-Espana, M. Micheloni, P. Paoletti, P. Paoli, A. Rodriguez, and B. Valtancoli, *Inorg. Chem.*, *32*, 2753 (1993).
120. (a) N. F. Curtis, *Coord. Chem. Rev.*, *3*, 3 (1968). (b) D. H. Busch, *Rec. Chem. Prog.*, *25*, 107 (1964). (c) M. C. Thompson and D. H. Busch, *J. Am. Chem. Soc.*, *84*, 1762 (1962).
121. T. Ramasami, J. F. Endicott, and G. R. Brubaker, *J. Phys. Chem.*, *87*, 5057 (1983).
122. V. Balzani, N. Sabbatini, and F. Scandola, *Chem. Rev.*, *86*, 319 (1986).
123. T. M. Donleavy, L. R. Gahan, T. W. Hambley, G. R. Hanson, K. L. McMahon, and R. Stranger, *Inorg. Chem.*, *33*, 5131 (1994).
124. J. I. Bruce, L. R. Gahan, T. W. Hambley, and R. Stranger, *Inorg. Chem.*, *32*, 5997 (1993).
125. T. M. Donleavy, L. R. Gahan, T. W. Hambley, and R. Stranger, *Inorg. Chem.*, *31*, 4376 (1992).
126. L. R. Gahan, G. A. Lawrance, and A. M. Sargeson, *Inorg. Chem.*, *23*, 4369 (1984).
127. R. V. Dubs, L. R. Gahan, and A. M. Sargeson, *Inorg. Chem.*, *22*, 2523 (1983).
128. R. J. Geue, W. R. Petri, and A. M. Sargeson, and M. R. Snow, *Aus. J. Chem.*, *45*, 1681 (1992).
129. K. S. Hagen, P. A. Lay, and A. M. Sargeson, *Inorg. Chem.*, *27*, 3424 (1988).
130. P. Osvath and A. M. Sargeson, *J. Chem. Soc. Chem. Commun.*, 40 (1993).
131. P. Comba and A. M. Sargeson, *Aust. J. Chem.*, *39*, 1029 (1986).
132. P. Bernhard and A. M. Sargeson, *J. Chem. Soc. Chem. Comm.*, 1516 (1985).

133. P. Bernhard and A. M. Sargeson, *Inorg. Chem.*, *26*, 4122 (1987).
134. P. Bernhard and A. M. Sargeson, and F. C. Anson, *Inorg. Chem.*, *27*, 2754 (1988).
135. P. Bernhard and A. M. Sargeson, *J. Am. Chem. Soc.*, *111*, 597 (1989).
136. P. Bernhard, H. B. Burgi, A. Raselli, and A. M. Sargeson, *Inorg. Chem.*, *28*, 3234 (1989).
137. P. Bernhardt, D. J. Bull, W. T. Robinson, and A. M. Sargeson, *Austr. J. Chem.*, *43*, 1241 (1992).
138. P. Comba, A. W. H. Mau, and A. M. Sargeson, *J. Phys. Chem.*, *89*, 394 (1985).
139. I. I. Creaser, L. R. Gahan, R. J. Geue, A. Laukonis, P. A. Lay, J. D. Lydon, M. G. McCarthy, A. W. H. Mau, A. M. Sargeson, and W. H. F. Sasse, *Inorg. Chem.*, *24*, 2671 (1985).
140. A. Launikonis, P. A. Lay, A. W. H. Mau, A. M. Sargeson, and W. H. F. Sasse, *Aust. J. Chem.*, *39*, 1053 (1986).
141. A. W. H. Mau, W. H. F. Sasse, I. I. Creaser, and A. M. Sargeson, *Nouv. J. Chem.*, *10*, 589 (1986).
142. I. I. Creaser, A. Hammershoi, A. Launikonis, A. W. H. Mau, and A. M. Sargeson, *Photochem. Photobiol.*, *49*, 19 (1989).
143. R. J. Geue, M. G. McCarthy, and A. M. Sargeson, *J. Am. Chem. Soc.*, *106*, 8282 (1984).
144. A. Hammershoi, G. A. Lawrance, and A. M. Sargeson, *Aust. J. Chem.*, *39*, 2183 (1986).
145. R. J. Geue, A. J. Hendry, and A. M. Sargeson, *J. Chem. Soc. Chem. Commun.*, 1646 (1989).
146. L. M. Engelhardt, J. M. Harrowfield, A. M. Sargeson, and A. H. White, *Aust. J. Chem.*, *46*, 127 (1993).
147. P. A. Anderson, I. I. Creaser, C. Dean, J. M. Harrowfield, E. Horn, L. L. Martin, A. M. Sargeson, M. R. Snow, and E. R. T. Tiekink, *Aust. J. Chem.*, *46*, 449 (1993).
148. L. L. Martin, R. L. Martin, K. S. Murray, and A. M. Sargeson, *Inorg. Chem.*, *29*, 1387 (1990).
149. L. L. Martin, K. S. Hagen, A. Hauser, R. L. Martin, and A. M. Sargeson, *J. Chem. Soc. Chem. Commun.*, 1313 (1988).
150. L. L. Martin, R. L. Martin, and A. M. Sargeson, *Polyhedron*, *13*, 1969 (1994).
151. T. M. Donleavy, L. R. Gahan, and T. W. Hambley, *Inorg. Chem.*, *33*, 2668 (1994).
152. P. V. Bernhardt, A. M. T. Bygott, R. J. Geue, A. J. Hendry, B. R. Korybut-Daskiewicz, P. A. Lay, J. R. Pladziewicz, A. M. Sargeson, and A. C. Willis, *Inorg. Chem.*, *33*, 4553 (1994).
153. L. M. Engelhardt, J. M. Harrowfield, A. M. Sargeson, and A. H. White, *Aust. J. Chem.*, *46*, 127 (1993).
154. P. Comba, A. M. Sargeson, L. M. Engelhardt, J. M. Harrowfield, A. H. White, E. Horn, and M. R. Snow, *Inorg. Chem.*, *24*, 2325 (1985).

155. P. Comba, *Inorg. Chem.*, *28*, 426 (1989).
156. P. W. Bernhardt, J. M. Harrowfield, D. C. R. Hockless, and A. M. Sargeson, *Inorg. Chem.*, *33*, 5659 (1994).
157. C. Harding, V. McKee, and J. Nelson, *J. Am. Chem. Soc.*, *113*, 9684 (1991).
158. J. Hunter, J. Nelson, C. Harding, M. McCann, and V. McKee, *J. Chem. Soc. Chem. Commun.*, 1148 (1990).
159. M. E. Barr, P. H. Smith, W. E. Antholine, and B. Spencer, *J. Chem. Soc. Chem. Commun.*, 1649 (1993).
160. C. Harding, J. Nelson, M. C. R. Symons, and J. Wyatt, *J. Chem. Soc. Chem. Commun.*, 2499 (1994).
161. J. Coyle, V. McKee, and J. Nelson, *Biocoordination chemistry; inorganic compounds with framework structures*, Karrebaeksminde, Denmark, Sept. 1996.
162. J. A. Farrar,V. McKee, A. H. R. al-Obaidi, J. J. McGarvey, J. Nelson, and A. J. Thomson, *Inorg. Chem.*, *34*, 1302 (1995).
163. A. Al-obaidi, G. Baranovich, C. Coates, J. Coyle, J. J. McGarvey, and J. Nelson, submitted to Inorganic Chemistry.
164. S. Iwata, C. Ostermeir, B. Ludwig, and H. Michel, *Nature* (London), *37*, 660 (1995).
165. T. Tsukihara, H. Aoyama, E. Yamashita, T. Tomizaka, H. Yamaguchi, K. Shinzawa-Itoh, R. Nakashima, R. Yaono, and S. Yoshikawa, *Science*, *269*, 1069 (1995).
166. S. E. Wallace Williams, C. A. James, S. Davies, M. Saraste, P. Lapplainen, J. Vanderoost, M. Fabian, G. Palmer, and W. H. Woodruff, *J. Am. Chem. Soc.*, *118*, 3986 (1996).
167. N. Blackburn, M. E. Barr, W. H. Woodruff, and J. van der Cost, S. de Vries, *Biochemistry*, *33*, 10401 (1994).
168. H. Bertagnolli and W. Kaim, *Angew. Chem. Int. Ed. Engl.*, *34*, 771 (1995).
169. J. Nelson and F. Arnaud-Neu to be published.
170. J. A. Thompson, M. E. Barr, D. K. Ford, L. A. Silks, B. J. McCormick, and P. H. Smith, *Inorg. Chem.*, *35*, 2025 (1996).
171. N. Martin, J. Nelson, and V. McKee, *Inorg. Chim. Acta*, *218*, 5 (1994).
172. P. H. Smith, M. E. Barr, J. R. Brainard, D. K. Ford, H. Frieser, S. Muralidharar, S. D. Reilly, R. R. Ryan, L. A. Selles, and W. H. Yu, *J. Org. Chem.*, *58*, 7939 (1993).
173. D. Chen, R. J. Motekaitis, I. Murase, and A. E. Martell, *Tetrahedron*, *51*, 77 (1995).
174. B. Dietrich, J.-M. Lehn, J. Guilheim, and C. Pascard, *Tetrahedron Lett.*, *30*, 4125 (1989).
175. C. J. Harding, Q. Lu, J. F. Malone, D. J. Marrs, N. Martin, V. McKee, and J. Nelson, *J. Chem. Soc. Dalton Trans.*, 1739 (1995), and reference cited therein.
176. Q. Lu, J.-M. Latour, C. J. Harding, N. Martin, D. J. Marrs, V. McKee, and J. Nelson, *J. Chem. Soc. Dalton Trans.*, 1471 (1994) and references cited therein.
177. Q. Lu, C. Harding, V. McKee, and J. Nelson, *Inorg. Chim. Acta*, *211*, 195 (1993).

178. C. J. Harding, Q. Lu, D. J. Marrs, G. Morgan, M. G. B. Drew, O. Howarth, V. McKee, and J. Nelson, *J. Chem. Soc. Dalton Trans.*, 3021 (1996).
179. L. Qin, M. McCann, and J. Nelson, *J. Inorg. Biochem.*, *51*, 633 (1993).
180. Q. Lu, V. McKee, and J. Nelson, *J. Chem. Soc. Chem. Commun.*, 649 (1994).
181. V. McKee and J. Nelson to be published.
182. M. P. Ngwenya, J. Ribenspeis, and A. E. Martell, *J. Chem. Soc. Chem. Commun.*, 1207 (1990).
183. M. Kumar, V. J. Aran, and P. Navarro, *Tetrahedron Lett.*, *35*, 2161 (1995).
184. R. Abidi, M. G. B. Drew, F. Arnaud-Neu, S. Lahely, M.-J. Schwing-Weil, D. J. Marrs, and J. Nelson, *J. Chem. Soc. Perkin Trans. 2*, 2747 (1996).
185. G. Morgan, V. McKee, and J. Nelson, *Coord. Chem. Rev.* in preparation.
186. D. Marrs, J. Malone, V. McKee, and J. Nelson, *J. Chem. Soc. Chem. Commun.*, 383 (1992).
187. S. Fuangswasdi, D. E. A. Thesis, ''Complexation Metallique en Solution par des Cryptands Octaamines et un Cryptand derivé de Bases de Schiff,'' Strasbourg, France, 1995; F. Arnaud-Neu, S. Fuangswasdi, J. Nelson, in preparation.
188. M. G. B. Drew, J. Hunter, D. Marrs, J. Nelson, and C. Harding, *J. Chem. Soc. Dalton Trans.*, 3235 (1992).
189. C. Harding, F. Mabbs, E. MacInnes, V. McKee, and J. Nelson, *J. Chem. Soc. Dalton Trans.*, 3227 (1996).
190. Q. Lu, C. J. Harding, V. McKee, and J. Nelson, *J. Chem. Soc. Chem. Commun.*, 1768 (1993).
191. R. Menif, J. Reibenspeis, and A. E. Martell, *Inorg. Chem.*, *30*, 3446 (1991).
192. J. L. Pierre, P. Chautemps, S. Refaif, C. Beguin, A. Elmarzouki, G. Serratrice, and P. Rey, *J. Am. Chem. Soc.*, *117*, 1965 (1995).
193. E. J. MacInnes, F. Mabbs, C. J. Harding, and J. Nelson, work in progress.
194. J.-M. Lehn, S. H. Pine, E. Watanabe, and A. K. Willard, *J. Am. Chem. Soc.*, *99*, 6766 (1991).
195. R. Motekaitis, A. E. Martell, J.-M. Lehn, and E. I. Watanabe, *Inorg. Chem.*, *21*, 4253 (1982).
196. R. J. Motekaitis, P. R. Rudolf, A. E. Martell, and A. Clearfield, *Inorg. Chem.*, , *28*, 112 (1989).
197. M. D. Timken, R. A. Gagne, W. A. Marritt, D. N. Hendrickson, and E. Sinn, *Inorg. Chem.*, *24*, 4202 (1985).
198. M. G. B. Drew, O. W. Howarth, G. G. Morgan, and J. Nelson, *J. Chem. Soc. Dalton Trans.*, 3149 (1994).
199. N. Martin, V. McKee, and J. Nelson, to be published.
200. M. G. B. Drew, C. J. Harding, V. McKee, G. G. Morgan, and J. Nelson, *J. Chem. Soc. Chem. Commun.*, 1035 (1995).
201. R. E. Marsh and W. P. Schaefer, *Inorg. Chem.*, *25*, 3661 (1986).

202. N. Martin, PhD Thesis, ''Azacryptand Hosts for Transition, Main Group, and Lanthanide Cations,'' Queens University, Belfast, 1996.

203. P. Chakraborty and S. K. Chandra, *Polyhedron*, *13*, 683 (1994).

204. M. G. B. Drew, O. W. Howarth, C. J. Harding, N. Martin, and J. Nelson, *J. Chem. Soc. Chem. Commun.*, 903 (1995).

205. R. Ruloff, N. Martin, and J. Nelson, unpublished work.

206. N. Sabbatini, M. Guardigli, and J.-M. Lehn, *Coord. Chem. Rev.*, *123*, 201 (1993) and references cited therein.

207. G. Mathis, *Clinical Chem.*, *41*, 1391 (1995).

208. A. Caron, J. Guilheim, C. Riche, C. Pascard, B. Alpha, J. M. Lehn, and J. C. Rodriguez-Ubis, *Helv. Chim. Acta*, *68*, 1577 (1985).

209. B. Alpha, R. Ballardini, V. Balzane, J.-M. Lehn, S. Perathroner, and N. Sabbatini, *Photochem.*, *Photobiol.*, *52*, 299 (1990).

210. B. Alpha, V. Balzani, J.-M. Lehn, S. Perathoner, and N. Sabbatini, *Angew. Chem. Int. Ed. Engl.*, *26*, 1266 (1987).

211. G. Blasse, G. J. Dirksen, N. Sabbatini, S. Perathoner, J.-M. Lehn, and B. Alpha, *J. Phys. Chem.*, *92*, 2419 (1988).

212. G. Blasse,G. J. Dirksen, D. van der Voort, N. Sabbatini, S. Perathoner, J.-M. Lehn, and B. Alpha, *Chem. Phys. Lett.*, *146*, 347 (1988).

213. I. Bkouche-Waksman, J. Guilhem, C. Pascard, B. Alpha, R. Deschenaux, and J.-M. Lehn, *Helv. Chim. Acta*, *74*, 1163 (1991).

214. M. Cesario, J. Guilheim, C. Pascard, E. Anklam, J.-M. Lehn, and M. Pietraszkiewicz, *Helv. Chim. Acta*, *74*, 1157 (1991).

215. J.-M. Lehn and J. B. R. Devains, *Tetrahedron Lett.*, *30*, 2209 (1989).

216. J.-M. Lehn and J. B. R. Devains, *Helv. Chim. Acta*, *75*, 1221 (1992).

217. J.-M. Lehn and C. Roth, *Helv. Chim. Acta*, *74*, 572 (1991).

218. L. Prodi, M. Maestri, V. Balzani, J.-M. Lehn, and C. Roth, *Chem. Phys. Lett.*, *180*, 45 (1991).

219. J. M. Lehn, M. Pietraszkiewicz, and J. Karpuik, *Helv. Chim. Acta*, *73*, 106 (1990).

220. M. Pietraszkiewicz, J. Karpuik, and A. K. Rout, *Pure Appl. Chem.*, *3*, 563 (1993).

221. N. Sabbatini, M. Guardigli, A. Mecati, V. Balzani, R. Ungaro, E. Ghidini, A. Casnati, G. Ulrich, Z. Asfari, and A. Pochini, *J. Chem. Soc. Chem. Commun.*, 878 (1990).

222. V. Balzani, J.-M. Lehn, J. van de Loosdrecht, A. Mecati, N. Sabbatini, and R. Ziessel, *Angew. Chem. Int. Ed. Engl.*, *30*, 191 (1991).

223. C. Roth, J.-M. Lehn, J. Guilheim, and C. Pascard, *Helv. Chim. Acta*, *78*, 1895 (1995).

224. F. Barigelletti, L. de Cola, V. Balzani, P. Belser, A. von Zelewsky, F. Ebmeyer, F. Vogtle, and S. Grammenudi, *J. Am. Chem. Soc.*, *110*, 7210 (1988).

225. L. de Cola, F. Barigelletti, V. Balzani, P. Belser, A. von Zelewsky, F. Ebmeyer, F. Vogtle, and S. Grammenudi, *J. Am. Chem. Soc.*, *111*, 4642 (1989).

226. J. de Mendoza, E. Mesa, J.-C. Rodriguez-Ubis, P. Vazquez, F. Vogtle, J.-M. Lehn, D. Lilienbaum , and R. Ziessel, *Angew. Chem. Int. Ed. Engl.*, *30*, 1331 (1991).

227. B. Maubert, N. Martin, M. G. B. Drew, and J. Nelson, *J. Chem. Soc. Dalton Trans.*, to be submitted for publication.

228. S. D. Reilly, G. R. K. Khalsa, D. K. Ford, J. R. Brainard, B. P. Hay, and P. H. Smith, *Inorg. Chem.*, *34*, 569 (1995).

229. R. J. Motekaitis, A. E. Martell, and I. Murase, *Inorg. Chem.*, *25*, 938 (1986).

230. R. J. Motekaitis, A. E. Martell, I. Murase, J.-M. Lehn, and M. W. Hosseini, *Inorg. Chem.*, *27*, 3630 (1988).

231. R. J. Motekaitis, A. E. Martell, B. Dietrich, and J.-M. Lehn, *Inorg. Chem.*, *23*, 1588 (1984).

232. A. E. Martell and R. J. Motekaitis, in *Determination and Use of Stability Constants*, VCH, Weinheim, Germany, 1988.

233. L. Fabrizzi, P. Pallavicini, L. Parodi, and A. Taglietti, *Inorg. Chim. Acta*, *238*, 5 (1995).

234. R. M. Izatt, K. Pawlak, J. S. Bradshaw, and R. Bruening, *Chem. Rev.*, *91*, 1721 (1991).

235. R. J. Motekaitis and A. E. Martell, *J. Chem. Soc. Chem. Commun.*, 1020, (1988).

236. R. J. Motekaitis and A. E. Martell, *J. Am. Chem. Soc.*, *110*, 7715 (1988).

237. A. Sauermasarwa, L. D. Dickerson, N. Herron, and D. H. Busch, *Coord. Chem. Rev.*, *128*, 117 (1993).

238. D. H. Busch and N. W. Alcock, *Chem. Rev.*, *94*, 585 (1994).

239. C. Cairns and D. H. Busch, *Prog. Macr. Chem.*, *3*, 1 (1987).

240. W. Kiggen and F. Vogtle, *Angew. Chim. Int. Ed. Engl.*, *23*, 714 (1984).

241. W. Kiggen, F. Vogtle, S. Franken, and H. Puff, *Tetrahedron*, *42*, 1859 (1986).

242. T. J. McMurry, S. J. Rodgers, and K. N. Raymond, *J. Am. Chem. Soc.*, *109*, 3451 (1987).

243. T. J. McMurry, M. W. Hosseini, T. M. Garrett, F. E. Hahn, Z. E. Reyes, and K. N. Raymond, *J. Am. Chem. Soc.*, *109*, 7196 (1987).

244. T. M. Garrett, T. J. McMurry, M. W. Hosseini, Z. E. Reyes, F. E. Hahn, and K. N. Raymond, *J. Am. Chem. Soc.*, *113*, 2965 (1991).

245. T. P. Karpishin, T. D. P. Stack, and K. N. Raymond, *J. Am. Chem. Soc.*, *115*, 182 (1993).

246. S. J. Rodgers, C.-W. Lee, C. Y. Ng, and K. N. Raymond, *Inorg. Chem.*, *26*, 1622 (1987).

247. W. R. Harris, C. J. Carrano, and K. N. Raymond, *J. Am. Chem. Soc.*, *101*, 2213 (1979).

248. A. E. Martell, R. D. Hancock, and R. J. Motekaitis, *Coord. Chem. Rev.*, *133*, 39 (1994).

249. J.-L. Pierre, P. Baret, and G. Gellon, *Angew. Chem. Int. Ed. Eng.*, *30*, 85 (1991).

250. A. E. Martell and Y. Sun, *J. Am. Chem. Soc.*, *111*, 8023 (1989).
251. A. E. Martell and Y. Sun, *Tetrahedron*, *46*, 2725 (1990).
252. R. J. Motekaitis, Y. Sun, and A. E. Martell, *Inorg. Chem.*, *30*, 1554 (1991).
253. S. S. Chana and R. C. Hider, *Tetrahedron Lett.*, *35*, 9455 (1994).
254. P. G. Potvin and M. H. Wong, *Can. J. Chem.*, *66*, 2914 (1988).
255. K. G. Ragunathan and P. K. Bharadwaj, *Proc. Ind. Acd. Sc.*, *105*, 215 (1993).
256. D. K. Chand and P. K. Bharadwaj, *Inorg. Chem.*, *35*, 3380 (1996).
257. R. J. Jannsen, L. F. Lindoy, O. A. Matthews, G. V. Meehan, A. N. Sobolev, and A. H. White, *J. Chem. Soc. Chem. Commun.*, 735 (1995).
258. I. M. Atkinson, L. F. Lindoy, O. A. Matthews, G. V. Meehan, A. N. Sobolev, and A. H. White, *Aust. J. Chem.*, *47*, 1155 (1994).
259. S. I. Kujiyama, T. Lazrak, M. Blanchard-Desce, and J.-M. Lehn, *J. Chem. Soc. Chem. Commun.*, 1179 (1991).
260. H.-W. Losensky, H. Spelthann, A. Ehlen, F. Vogtle, and J. Bargon, *Angew. Chem. Int. Ed. Engl.*, *27*, 1189 (1988).
261. M.-T. Youinou, J. Suffert, and R. Ziessel, *Angew. Chem. Int. Ed. Engl.*, *21*, 775 (1992).
262. L. Echegoyen, A. de Cian, J. Fischer, and J.-M. Lehn, *Angew. Chem. Int. Ed. Engl.*, *30*, 838 (1991).
263. L. Echegoyen, E. Perez-Cordero, J.-B. R. de Vains, C. Roth, and J.-M. Lehn, *Inorg. Chem.*, *32*, 572 (1993).
264. M. G. B. Drew, V. Felix, V. McKee, G. Morgan, and J. Nelson, *Supramolecular Chem.*, *5*, 281 (1995).
265. M. G. B. Drew, D. Mc Dowell, J. Hunter, and J. Nelson, *J. Chem. Soc. Dalton Trans.*, *11* (1992).
266. D. McDowell, Ph.D. Thesis, "Macrocycles, Macrobicycles: A Study," Open University, Milton Keynes, 1990.
267. D. McDowell, V. McKee, J. Nelson, and W. T. Robinson, *Tetrahedron Lett.*, 7453 (1987).
268. D. McDowell, V. McKee, and J. Nelson, *Polyhedron*, 1143 (1989).
269. J. J. R. F. de Silva and R. J. P. Williams, *Struct. Bonding* (*Berlin*) *29*, 67 (1976).
270. L. G. Lange, III, J. F. Riordan, and B. L. Vallee, *Biochemistry*, *13*, 4361 (1974).
271. R. A. Bell, G. G. Christoph, F. R. Fronczek, and R. E. Marsh, *Science*, *190*, 151 (1975).
272. E. Graf and J.-M. Lehn, *J. Am. Chem. Soc.*, *98*, 6403 (1976).
273. E. Graf and J.-M. Lehn, *Helv. Chim. Acta*, *64*, 1040 (1981).
274. F. P. Schmidtchen, *Chem. Ber.*, *113*, 864 (1980).
275. B. Metz, J. M. Rosalsky, and R. Weiss, *J. Chem. Soc. Chem. Commun.*, 533 (1976).
276. F. P. Smidtchen and S. Muller, *J. Chem. Soc. Chem. Commun.*, 1115 (1984).

277. K. Ichikawa and A. M. A. Hossein, *J. Chem. Soc. Chem. Commun.*, 1721 (1996).
278. F. P. Smidtchen, *Chem. Ber.*, *114*, 597 (1981).
279. F. P. Scmidtchen, *Angew. Chim. Int. Ed. Engl.*, *16*, 720 (1977).
280. B. Dietrich, J. Guilheim, J.-M. Lehn, C. Pascard, and E. Sonveaux, *Helv. Chim. Acta*, *67*, 91 (184).
281. J.-M. Lehn, E. Sonveaux, and A. K. Willard, *J. Am. Chem. Soc.*, *100*, 4914 (1978).
282. M. W. Hosseini and J.-M. Lehn, *Helv. Chim. Acta*, *71*, 749 (1988).
283. B. Dietrich, M. W. Hosseini, J.-M. Lehn, and R. B. Sessions, *Helv. Chim. Acta*, *66*, 1262 (1983).
284. D. Heyer and J.-M. Lehn, *Tetrahedron Lett.*, *27*, 5869 (1989).
285. J.-M. Lehn, R. Meric, J.-P. Vigneron, I. Bkouche-Waksman, and C. Pascard, *J. Chem. Soc. Chem. Commun.*, *62* (1991).
286. G. G. Morgan, V. McKee, and J. Nelson, *J. Chem. Soc. Chem. Commun.*, 1649 (1995).
287. A. A. Varnek, G. Wipff, A. S. Glebov, and D. Feil, *J. Comput. Chem.*, *16*, 1 (1995).
288. B. Cox, J. G-Rosas, and H. Schneider, *J. Am. Chem. Soc.*, *103*, 1384 (1981).
289. H.-J. Buschmann, *Inorg. Chim. Acta*, *120*, 125 (1986).
290. B. Cox, J. Stroka, and H. Schneider, *Inorg. Chim. Acta*, *128*, 207 (1989).
291. J. Bessiere and A. F. Lejaille, *Anal. Lett.*, *12(A7)*, 753 (1979).
292. R. J. Motekaitis, A. E. Martell, and W. Benutely, *Inorg. Chim. Acta*, *205*, 23 (1993).
293. A. Bencini, A. Bianchi, M. Ciampolini, P. Dapporto, M. Micheloni, N. Nardi, P. Paoli, and B. Valtancoli, *J. Chem. Soc. Perkin Trans 2*, 181 (1992).
294. S. Lincoln and T. Rodopoulos, *Inorg. Chim. Acta*, *190*, 233 (1991).

Polyoxometalate Complexes in Organic Oxidation Chemistry

RONNY NEUMANN

Casali Institute of Applied Chemistry
Graduate School of Applied Science
The Hebrew University of Jerusalem
Jerusalem 91904, Israel

CONTENTS

Progress in Inorganic Chemistry, Vol. 47, Edited by Kenneth D. Karlin.
ISBN 0-471-24039-7 © 1998 John Wiley & Sons, Inc.

I. INTRODUCTION

A. General Background

The original discovery in 1826 by Berzelius of the first heteropoly compound (1) at first did not bring about considerable research into heteropolyanion chemistry, probably because the compound's structure remained unknown. In fact, the modern age of heteropolyanion chemistry essentially began only one century later, in 1934, with the milestone X-ray structure determination by Keggin (2) of the structure of the hydrates of 12-tungstophosphoric acid ($H_3PW_{12}O_{40}$). Aptly, the name Keggin is now synonymous with the structure of the $[PW_{12}O_{40}]^{3-}$ anion and a rather large family of isostructural compounds. During the following 50 years of research, heteropolyanion chemistry and that of the related isopoly compounds centered mostly around the preparation, structure, properties, and analytical chemistry applications. This earlier pre-1983 interest in the field of heteropolyanion or polyoxometalate chemistry was thoroughly summarized in Pope's (3) comprehensive book with strong emphasis in fact being placed on synthesis, structure, and properties. Pope's monograph was probably illuminating to many scientists, for since its publication there has been a significant and much increased research interest in many aspects of polyoxometalate chemistry. Many possible applications of polyoxometalates became apparent. This diverse research has been described in various rather recent reviews that dealing with different aspects of polymetalate chemistry, including general overviews (4), structure (5), photochemistry and photocatalysis (6), clinical chemistry (7), medicinal applications (antiviral agents) (8), solid-state applications (electronic conductors) (9), and electrocatalysis (10).

The use of polyoxometalates as catalysts really started only approximately 20 years ago. The first push in this direction came when it was realized that the heteropoly*acids* of the Keggin structure, $H_nXM_{12}O_{40}$, where X is B, As, and Ge but mostly P or Si and M is W or Mo, are very strong Brønsted acids with typical pK_a values up to four units lower than sulfuric acid. This high Brønsted acidity was effectively used in many typical acid catalyzed reactions in fields ranging from synthesis of fine chemicals to manufacture of petrochemicals. A certainly nonexhaustive listing of reaction types includes Friedel–Crafts alkylation of aromatics with alkenes or alcohols; alkene hydration; hydrolysis; esterification and amidation, etherification; alkane and alkene isomerization; and polymerization, most notably of tetrahydrofuran (THF). In this context, a very important current discovery is that the insoluble $Cs_{2.5}H_{0.5}PW_{12}O_{40}$ is even a measurably stronger acid compared to $H_3PW_{12}O_{40}$. It is reasonable to assume that the use of heteropoly acids in acid catalysis will continue to be a focal point in the use of these compounds, especially in cases where "superacidity" is required (11). Inclusion of heteropolyacids in well-defined solid supports such as zeolites and pillared clays may impart additional shape selectivity, thereby further enhancing the scope of heteropolyacid catalysis.

In the framework of this chapter, we will not, however, dwell further on the use of heteropolyacids for acid catalysis. Instead, we will limit ourselves to the use of polyoxometalates as oxidation catalysts. This field has many aspects and may be subclassified in various ways. Thus, one can differentiate between homogeneous liquid-phase and heterogeneous gas-phase reactions. Different oxidants such as molecular oxygen, hydrogen peroxide, and others may be compared. Classification can also be based on the oxidative transformations carried out, and so forth. Various parts of these research efforts based on these classifications have been reviewed during the last decade. First, Matveev and coworkers (12) described some of their own early research carried out during the late 1970s to the early 1980s. Tsigdinos (13) also reviewed the earlier research. Somewhat later on, in 1987, Misono described some of the earlier work in heterogeneous catalysis (14) and later he (15) and others (16) expanded on the earlier work. Ishii and Ogana (17) compiled their work on the use of hydrogen peroxide and Hill et al. (8a, 18) described the use of transition metal substituted polyoxometalates. Recently, Kozhevnikov (19) discussed the use of heteropolyanions in fine chemical synthesis. In this chapter, we have endeavored to give a comprehensive outlook on the field in an attempt to interconnect the somewhat dissimilar research efforts. In order to do so, we have chosen to categorize polyoxometalate (POM) catalyzed oxidations by the principal mode of the catalytic reaction. Along this line of reasoning, one may distinguish between three major reaction modes. For the first reaction type, molecular oxygen is generally the oxidant. The catalytic cycle can be best described by the division of the reaction into two stages. First, the substrate is oxidized to yield the product and the catalyst is reduced. The reduced polyoxometalate catalyst is then reoxidized by molecular oxygen in the second and possibly separate stage completing the catalytic cycle, Eq. 1.

$$\begin{aligned} \text{Substrate} + \text{POM}_{\text{ox}} &\longrightarrow \text{product} + \text{POM}_{\text{red}} \\ \text{POM}_{\text{red}} + \text{O}_2 &\longrightarrow \text{POM}_{\text{ox}} \end{aligned} \tag{1}$$

In the liquid phase, a metal ion or redox mechanism is generally operating, whereas for reactions in the gas phase, a Mars–van Krevelen type mechanism is often invoked. Thus, the oxidation of the substrate may be viewed either as a dehydrogenation (electron transfer from the substrate to the catalyst) or an oxygenation, whereby an oxygen atom of the polyoxometalate is transferred to the substrate. Regeneration of the catalyst therefore implies either electron donation of oxygen to the catalyst with coformation of water or insertion of oxygen into the oxygen atom depleted polyoxometalate.

The second reaction type views the oxidation catalyzed by the polyoxometalate as an interaction with a primary oxidant, typically a peroxygen compound such as hydrogen peroxide. This interaction yields an activated catalyst intermediate, for example, a peroxo of hydroperoxo species that oxidizes the organic substrate, Eq. 2.

$$\text{Oxidant} + \text{POM}_{\text{ox}} \longrightarrow [\text{POM}{-}\text{Ox}]_{\text{activated}}$$

$$[\text{POM}{-}\text{Ox}]_{\text{activated}} + \text{Substrate} \longrightarrow \text{POM}_{\text{ox}} + \text{Product} \qquad (2)$$

Oxidation may take place via both heterolytic or homolytic cleavage of the peroxo bond to yield the product. In the formation of polyoxometalate–peroxo species, the polyoxometalate is similar to other simplier tungsten, molybdenum, or vanadium containing compounds, which react in an analogous manner.

Finally, one may consider polyoxometalates as inorganic ligands for non-tungsten and molybdenum transition metals (cobalt, manganese, ruthenium, etc.) These compounds are now often termed transition metal substituted polyoxometalates (TMSP). In mechanistic scenarios for such reactions, the catalytically active site is at the substituted transition metal center; the polyoxometalate functions as a ligand with a strong capacity for accepting electrons. In this last group of oxidation reactions, the actual reaction mechanism certainly varies as a function of the transition metal and oxidant, but can be conceived as taking place via a general intermediate "transiton metal–oxo" species.

Therefore, the use of polyoxometalates in oxidation has evolved along several lines of research taking advantage of the different oxidation pathways that are, mechanistically speaking, available. Several themes, however, are nevertheless common to the use of polyoxometalate compounds as oxidation catalysts. The first is their high thermal stability, generally to at least 350°C but often much higher under molecular oxygen. This characteristic has led to the obvious use of polyoxometalates in liquid-phase dioxygen mediated reactions, but also to their application in high-temperature gas-phase oxidations of hydrocarbons with dioxygen. Researchers in this field have also been intrigued by these compounds as well-defined transition metal oxides as opposed to often nonstoichiometric and/or amorphous transition metal oxides so commonly used in industrially important petrochemical oxidations. Second, the inorganic nature of the polyoxometalate coupled with the presence of high-valent molybdenum and tungsten has additionally led to the supposition that they are in fact intrinsically stable to oxidation. For liquid-phase catalysis, this argument bodes strongly for the use of transition metal substituted polyoxometalates as alternatives for organometallic compounds highly susceptible to self-oxidation. Some have even suggested that TMSPs may be considered "inorganic porphyrins." Third, heteropolyanions usually assemble under acidic conditions from simpler oxides. This leads to the general assumption that these compounds are thermodynamically stable at conditions where they are formed, that is, at acidic to neutral pH. The exact range of solvolytic stability to pH varies from compound to compound. In general, polyoxometalates are not stable to base (e.g., at 1 *M* NaOH) and decompose to simple oxides. In addition, polyoxometalates once formed are more stable to solvolysis in organic solvents as compared to aqueous media. Solubility of the polyoxometalates is also an important consideration when discussing catalysis. Heteropolyacids are soluble in water and most protic and aprotic polar solvents. In general, replacing the proton with a larger cation from alkali metal to quaternary ammonium cation will decrease solubility in water and polar solvents and increase solubility in apolar solvents. Thus, good control of solubility is possible. Finally, it is most important to recognize that the overwhelming majority of catalytic applications use polyoxometalates based on the Keggin structure. In the relatively few cases where this is not so, the catalysts are often derived from the relevant Keggin compound.

B. The Keggin Compounds and Derivatives

In general, heteropolyanions can be described by the general formula $[X_xM_mO_y]^{q-}$ ($x \leq m$) where X is defined as the heteroatom and M represents the addenda atoms. The Keggin heteropolyanion (α isomer) (Fig. 1) represents

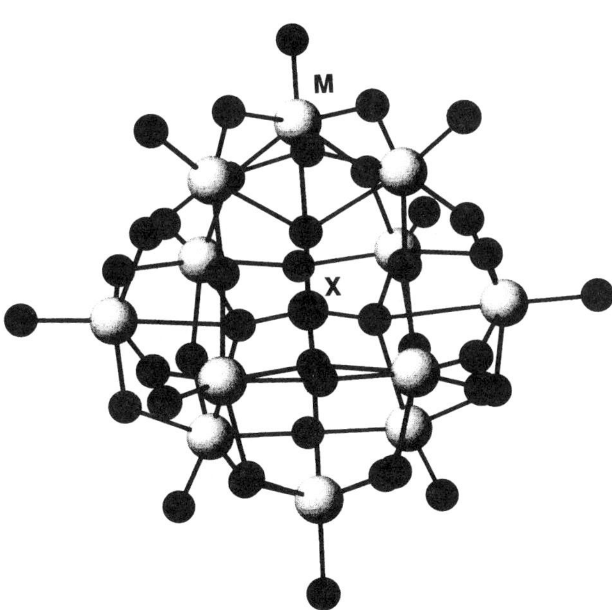

Figure 1. The α-isomer of the Keggin structure, $[XM_{12}O_{40}]^{q-}$.

the major subclass. The basic compound is described by the general formula, $[XM_{12}O_{40}]^{q-}$. The Keggin structure has an approximate T_d symmetry based on a central XO_4 tetrahedron surrounded by 12 MO_6 octahedra arranged in four groups of three edge-shaped octahedra, M_3O_{13}. The central heteroatom in the XO_4 tetrahedra are usually of the main group elements, B, Ge, As, Si, and P, but may also be first-row transition metals such as Co and Zn. Phosphorous and silicon are the most common heteroatoms. In the primary compounds, the addenda are either tungsten or molybdenum. One may distinguish between four kinds of oxygen atoms, 4 internal oxygens connecting the heteroatom to the addenda, 12 edge-sharing oxygens, 12 corner-sharing oxygens connecting M_3O_{13} units, and 12 terminal oxygens. The four different types of oxygen atoms lead to four peaks in a distinctive IR spectrum. Rotation by 60° of one M_3O_{13} triad leads to the less stable β isomer, which, however, has not been catalytically significant.

The simple Keggin heteropolyanions have been used in two of the three basic reaction types described in Section I.A. The first use has been with peroxygen oxidants, mainly hydrogen peroxide, where metal–peroxo intermediates are the catalytically active species. These reactions are typical for those found for molybdenum and tungsten oxides (20). The second use has been in reactions that proceed according the mechanism outlined in Eq. 1. Electrochemical measurements show that the molybdenum-based Keggin compounds, most notably $[PMo_{12}O_{40}]^{3-}$, have higher oxidation potential compared to the analogous tungsten compounds. The relatively easy reduction of the molybdenum species allows it to accept electrons from a substrate to form heteropoly "blues" and "browns" of equivalent structure. The redox potential is such that reoxidation of the reduced species with molecular oxygen is thermodynamically possible although relatively slow. Thus, usually the reoxidation step is rate determining. Another way to carry out the reaction according to Eq. 1 is to photoactivate the tungstate $[PW_{12}O_{40}]^{3-}$. This photoactivation leads to an excited Keggin compound, $[PW_{12}O_{40}]^{3-*}$, of high oxidation potential capable, for example, of oxidative dehydrogenation of hydrocarbons. Photocatalysis will not be discussed further in the framework of this chapter.

Early on, it was realized that the oxidation potential of the Keggin molybdate compounds could be improved by partial replacement of the molybdenum addenda by vanadium leading to a "mixed"-addenda Keggin compound $[XM_{12-x}M'_xO_{40}]^{q-}$, where X is usually P or Si and M and M′ are Mo, W, and V. The compound $[PMo_{10}V_2O_{40}]^{5-}$ seems to have optimal characteristics as concerns the combination of oxidation potential ($\sim$0.7 V), ease of reoxidation, and stability. Higher stoichiometries of V/Mo tend to significantly destabilize to a heteropolyanion. The $[PW_6Mo_6O_{40}]^{3-}$ anion has also been shown to be effective in redox-type reactions. An important point, and one that is often missed in the literature, is that the redox mechanism in fact is not operating in all oxidations reporting the use of vanadium containing molybdates. There are cases, where reactions are simply vanadium centered oxidations and the polyoxometalate is a ligand to the vanadium center. In much of the literature, the distinction between the two has not been made. There are many methods that may be used to differentiate between these mechanisms but comparisons of activity of vanadium-containing polyoxometalates with very different redox potentials such as $[PMo_{10}V_2O_{40}]^{5-}$ and $[PW_{10}V_2O_{40}]^{5-}$ is simple and often effective.

An important family of Keggin polyoxometalate derivatives are compounds with transition metal substitution. These compounds, at least per definition, react at singular metal centers with a polyoxometalate ligand as described in section I.A. Synthesis and characterization of transition metal substituted heteropolyoxometalates with lower valent transition metals such as cobalt, manganese, copper, or iron, as the heteroatom has been mentioned in the literature (21). These compounds, however, have received relatively little attention as catalysts because of the inaccessibility of the transition metal to coordination by a substrate. Transition metal substitution at the polyoxometalate surface is more interesting. These compounds may be prepared by pH controlled formation of lacunary or ''defect'' Keggin structures. Over 20 years ago, it was shown (22) that removal of one M=O unit yields a lacunary compound, $[XM_{11}O_{39}]^{q-}$, with C_s symmetry and the following structure, Fig. 2. The vacancy brought about by the loss of the addenda atom creates a structure that may be viewed as a pentadentate ligand. Upon addition of a transition metal cation, it is enclosed by the polyoxometalate ligand creating the monosubstituted transition metal compounds, Fig. 2. The octahedral coordination sphere is usually completed by water or other liable ligands. It is worthwhile commenting that the TMSPs are with relatively few exceptions always polyoxotungstates due to the lower stability of the polyoxomolybdates.

Lacunary anions of the Keggin structure with three vacancies are also well known. Depending on which addenda atoms have been removed, two isomers termed A- and $[B\text{-}XM_9O_{34}]^{q-}$ with C_{3v} symmetry are formed, Fig. 3. The vacancies in these truncated lacunary compounds may be filled with first-row transition metals forming $[TM_3PW_9O_{37}]^{q-}$. Alternatively, two fragments may assemble directly forming the well-known Wells–Dawson structure $[P_2W_{18}O_{62}]^{6-}$. Of more interest, from a catalyst point of view, these fragments may link with the incorporation of transition metal ion linkers that lead to what we call ''sandwich''-type polyoxometalates, Fig. 4. Use of this type of polyoxometalate is a more recent and less researched development. However, from our initial research these ''sandwich'' compounds would seem to have important advantages compared to other transition metal substituted polyoxometalates in oxidation analysis. This advantage includes greater solvolytic stability toward hydrogen peroxide and possible molecular oxygen activation as discussed below. Similar

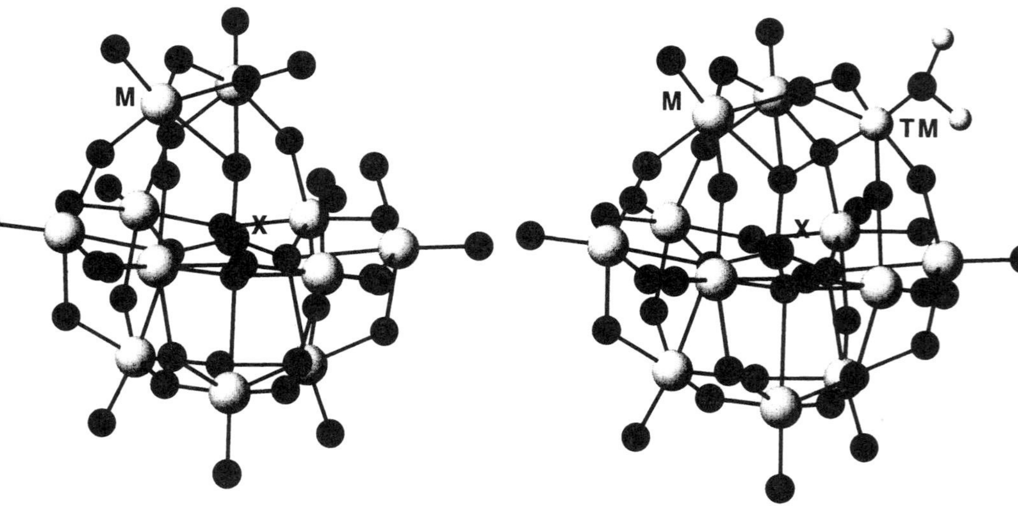

Figure 2. The monovacant Keggin structure, $[XM_{11}O_{39}]^{q-}$ and the transition metal substituted Keggin compound. $[TM(H_2O)XM_{11}O_{39}]^{q-}$, where TM equals one or more transition metals.

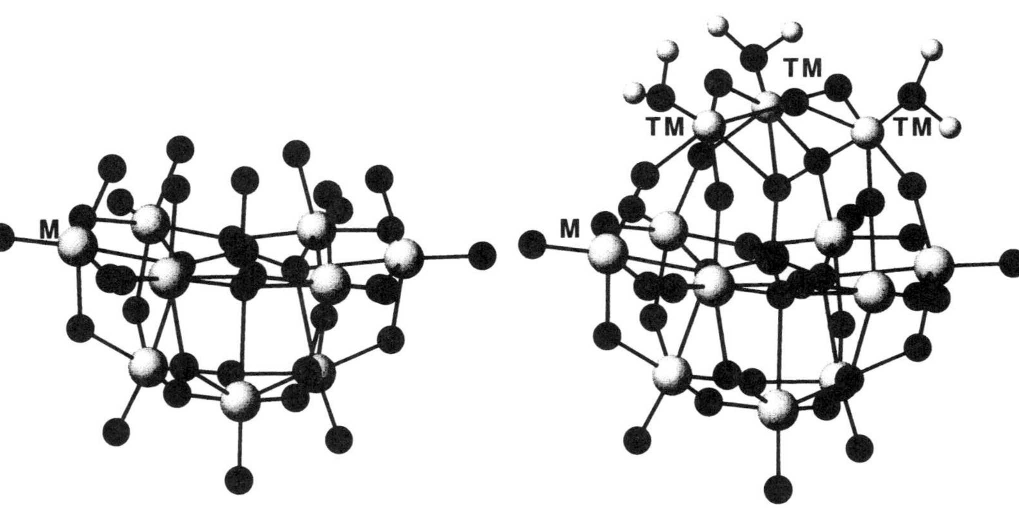

Figure 3. The trivacant Keggin structure. $[B\text{-}XM_9O_{34}]^{q-}$ and the transition metal substituted Keggin compound. $[TM_3XM_{11}O_{37}]^{q-}$.

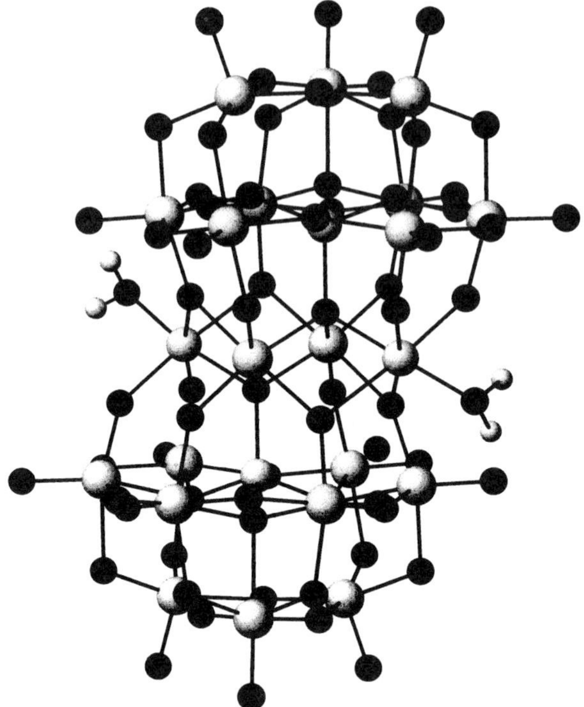

Figure 4. "Sandwich"-type polyoxometalates.

to what is known with the Keggin compounds, lacunary structures from the Wells–Dawson compounds with both one and three vacancies may be formed and filled with transition metals. Although these compounds are aesthetically very pleasing, they have shown little advantage in oxidation catalysis compared to the analogous Keggin compounds.

II. THE POLYOXOMETALATE REDUCTION–SUBSTRATE OXIDATION REACTION MODE: OXYGEN AS PRIMARY OXIDANT

A. Reactions with Both a Metal Catalyst and a Polyoxoanion

One of the first reports on polyoxometalates in oxidation catalysis was the use of the $H_{3+x}PMo_{12-x}V_xO_{40}$ (x = 1–6) heteropolyacid as cocatalyst in the

Pd^{II} catalyzed oxidative hydration of ethylene, the so-called Wacker process, Eq. 3 (e.g., for $x = 2$).

$$CH_2{=}CH_2 + Pd^{II} + H_2O \longrightarrow MeCHO + Pd^0 + 2H^+$$

$$Pd^0 + H_5PMo_{10}V_2^VO_{40} + 2H^+ \longrightarrow Pd^{II} + H_7PMo_{10}V_2^{IV}O_{40}$$

$$H_7PMo_{10}V_2^{IV}O_{40} + \tfrac{1}{2}O_2 \longrightarrow H_5PM_{10}V_2^VO_{40} + H_2O \qquad (3)$$

In this work, Matveev and co-workers (12, 23) at Novosibirsk showed that in principle the polyoxometalate could replace the corrosive $CuCl_2$ as the reoxidant of the reduced primary catalyst $Pd^{II}SO_4$, and the reaction could be carried out in two stages: catalyst hydration and catalyst recycle. The absence of chloride would importantly also eliminate troublesome chlorinated side products. On the positive side, the results found were a conceptual breakthrough in the use of polyoxometalates as oxidation catalysts. However, careful examination of the results shows that reactions required very large amounts of heteropolyacid and, in the reaction medium used, acidic water, the reoxidation step was very slow. The oxidation of Pd^0 to Pd^{II} was believed to be mainly by free VO_2^+ for $x \geq 3$, because of the instability of the heteropolyanion and by a undisassociated more stable heteropolyacid, where $x \leq 2$. The reoxidation with molecular oxygen takes place only in the coordination sphere of $H_{3+x}PMo_{12-x}V_xO_{40}$; free VO^{2+} is not reoxidized (24). These latter findings have been staple in this field and led to the preferred use of $H_5PMo_{10}V_2O_{40}$ in many applications.

Within a short time period, the original work of the Novosibirsk group in ethylene oxidation was followed up with similar research on oxidation of terminal alkenes such as propene, butene, and other higher alkenes to methylketones, Eq. 4 (25).

$$\text{R-CH}_2\text{CH=CH}_2 + H_2O \xrightarrow[{[PV_2Mo_{10}O_{40}]^{5-}}]{Pd^{II},\ O_2} \text{R-CH}_2\text{C(=O)CH}_3 \qquad (4)$$

Unfortunately, much of this work is not translated and is difficult to acquire. The $H_3PMo_6W_6O_{40}$ heteropolyacid was described by Izumi and co-workers (26) as an alternative for the vanadium substituted phosphomolybdates in the ketonization of 1-butene. A similar reaction was also carried out in diglyme–acetic acid to yield vinyl acetate from ethylene (27). Again, large quantities of polyoxometalate were required and reoxidation was very slow. Additional palladium, catalyzed reactions using the phosphomolybdovanadates as reoxidants included the acetylation of benzene and its derivatives (28). Only a small change in reduction conditions leads to oxidative coupling of benzene and other aromatics, Eq. 5 (29). Turnover numbers are low in both cases.

$$\xrightarrow[\text{[PV}_2\text{Mo}_{10}\text{O}_{40}]^{5-}]{\text{Pd}^{II}, \text{O}_2, \text{HOAc}}$$ (5)

Use of more electron-rich substrates such as durene (1,2,4,5-tetramethylbenzene) or diphenyloxide led to more efficient reactions, of about 90% conversion, and oligomerization to tetramers, Eq. 6 (30).

$$\xrightarrow[\text{[PV}_2\text{Mo}_{10}\text{O}_{40}]^{5-}, 90^\circ\text{C}]{\text{Pd}^{II}, \text{O}_2, \text{HOAc}}$$ (6)

Heck-type reactions could also be realized with ethylene and benzene as well as with heteroaromatic compounds such as thiophene and furan (31). Oxidative dehydrogenation of secondary alcohols to ketones was also possible using Pd^{II} and $H_5PMo_{10}V^V_2O_{40}$ (32). Despite all these early advances in a relatively short period of time, the low turnover rates and large amount of heteropolyanion catalyst required, deterred further reported research and the use of palladium and $H_5PMo_{10}V^V_2O_{40}$ was dormant for almost a decade.

In the 1990s, the oxidative hydration of ethylene was significantly improved by Grate's group at Catalytica (33). Their important finding was that the original system could be easily improved by addition of small amounts, 5–25 m*M*, of chloride anions to the solution. This result is most probably based on the fact that the $Pd^{II}Cl_2/Pd^0$ redox couple is better suited to the heteropolyacid system than the $Pd^{II}SO_4/Pd^0$ couple. They furthermore stated that their present optimized system is viable for industrialization. A Slovak group also adopted the Russian system for oxidation of 1-octene and 2-octanone in order to overcome problems of chlorinated side products in the classic system, but the lifetime of the catalytic system was short (34). Several groups have investigated the use of Pd^{II} and $H_5PMo_{10}V^V_2O_{40}$ for oxidation of 1-butene to 2-butanone both using the usual liquid-phase procedure as well as by using supported catalysts in heterogeneous gas-phase reactions (35). In general, the reports are not encouraging; both slow rates and catalyst deactivation have been reported. Similarly, attempts at carrying Wacker-type oxidations with cyclohexene yielded mixtures containing the Wacker product, cyclohexanone, and the autooxidation product,

2-cyclohexen-1-one (36). An interesting report on the physicochemical characteristics of the heteropolyanion Wacker system using the oxidation of 1-decene has also been carried out using electron microscopy at the reaction interface (37). Finally, it is worth mentioning that there have been some reports on the use of vanadium substituted Keggin compounds with Pd^{II} for the catalytic oxidation of CO to CO_2 relevant to environmental chemistry (38).

The early success in the use of $H_{3+x}PMo_{12-x}V_x^{V}O_{40}$ as cocatalyst in the palladium catalyzed reactions led to attempts to expand the concept to further redox couples. One interesting attempt was to apply the method to the Tl^{III}/Tl^{I} couple. Thallium(III) is capable of unique synthetic transformations but reactions are stoichiometric. A catalytic process would certainly be desirable. The Novosibirsk (12) group attempted such a scheme in the presence of bromide to reduce the Tl^{III}/Tl^{I} redox potential. In all cases, the reactions were not selective, and at best only two turnovers even at 8-atm O_2 were obtained (39). Similar attempts in ruthenium and iridium catalysis were equally discouraging (40). We might add that over the years our group has also attempted to use these and other redox couples in such oxidation chemistry. These attempts have all been unsuccessful and one may conclude that the use of $H_{3+x}PMo_{12-x}V_xO_{40}$ as cocatalyst in metal catalyzed reactions will probably be limited mostly to the Pd^{II}/Pd^0 redox couple. A word of caution is in order here, for even under Catalytica's best reported conditions the reaction is about 0.3 *M* in $H_5PMo_{10}V_2O_{40}$, which translates to a very significant, about 600 g/L^{-1}, catalyst.

B. Homogeneous Systems

The use of $H_{3+x}PMo_{12-x}V_xO_{40}$ heteropolyacids to oxidize substrates in a liquid-phase single-catalyst system also began about 15 years ago. As in the palladium-catalyzed reactions, if we take into account the oxidation potential of $H_{3+x}PMo_{12-x}V_xO_{40}$, it seemed possible that Br^-/Br_2 and I^-/I_2 systems could also be of interest. In this way an acetic acid–water solution containing iodide and $H_8PMo_6V_6O_{40}$ under propene and oxygen pressure yielded the acetate of 1,2-propanediol. The reaction proceeded by oxidation of hydrogen iodide to iodine, which in the presence of propene and water formed the iodohydrin. The later reacted with acetic acid to form the product, recycling the hydrogen iodide. A similar concept was used in the bromination of arenes. First, hydrogen bromide was oxidized to molecular bromine, in an acetic acid–water–sulfuric acid solution. The latter reacted with an organic substrate such as naphthalene and the catalyst was oxidized by oxygen, Eq. 7 (41).

$$\begin{aligned} 2HBr + H_5PMo_{10}V_2^{V}O_{40} &\longrightarrow Br_2 + H_7PMo_{10}V_2^{IV}O_{40} \\ Br_2 + ArH &\longrightarrow ArBr + HBr \\ H_7PMo_{10}V_2^{IV}O_{40} + \tfrac{1}{2}O_2 &\longrightarrow H_5PMo_{10}V_2^{V}O_{40} + H_2O \end{aligned} \tag{7}$$

The reaction required high oxygen pressures of 10–15 atm and rates and turnover frequencies (TOF) for bromide oxidation were very low ($\sim$2–5 mol Br_2 h^{-1}). Our interpretation of this result was that water was an unattractive and probably detrimental solvent for this type of redox chemistry. Along this line of thought, we proposed to carry out this chemistry in an organic media such as 1,2-dichloroethane. Solubilization of the catalyst in the organic phase was made possible by addition of polyethers, known complexing agents for heteropolyacids. This addition enabled the reaction described in Eq. 6 at only 0.2 atm of O_2 with a TOF of 50/h. An additional advantage was that highly selective para bromination of phenol was possible (42).

Another early interest in the catalytic chemistry of $H_{3+x}PMo_{12-x}V_xO_{40}$ was in the oxidation of sulfur-containing compounds of interest in the purification of industrial waste and natural gas. Among the first transformations contemplated one may note in order of required oxidation potential, the oxidation of hydrogen sulfide to elemental sulfur, sulfur dioxide to sulfur trioxide (sulfuric acid), mercaptans to disulfides and sulfides to sulfoxides and sulfones, Eq. 8 (43).

$$8H_2S + 4O_2 \longrightarrow S_8 + 8H_2O$$

$$SO_2 + 1/2O_2 \longrightarrow SO_3 \xrightarrow{H_2O} H_2SO_4$$

$$2RSH + 1/2O_2 \longrightarrow RSSR + H_2O$$

$$RCH_2SCH_2R' \xrightarrow{O_2} R'CH_2S(\rightarrow O)CH_2R \longrightarrow R'CH_2S(\rightarrow O)_2CH_2R \qquad (8)$$

This chemistry was again carried out in water or, in the case of sulfides, in acetone. Selectivities were quite excellent. In recent years, Hill and co-workers (44) continued the investigation of the oxidation chemistry of sulfur compounds. In one work, they oxidized hydrogen sulfide to sulfur using the redox-type mechanism. In this case, certain tungstates are of sufficient redox potential to allow the reactions to proceed. Reactions for the most stable catalyst (Pressler's compound, $[NaPW_{30}O_{110}]^{14-}$) are very selective ($>$99.5%), but unfortunately turnover frequencies are only about 7 per day. Hill's group has also investigated the oxidation of tetrahydrothiophene to the sulfoxide as a model for mustard gas oxidation, which is useful for its disposal. In the first work, an alkoxy substituted vanadotungstate was used with *tert*-butylhydroperoxide (TBHP) as oxidant (45). This system, however, seems to be operating by the usual alkylperoxo–vanadium (tungsten) heterolytic mechanism. In a subsequent investigation, $H_{3+x}PMo_{12-x}V_nO_{40}$ (mostly $x = 2$, homogeneous and supported catalyst) was used for the sulfide oxidation (46). This system is unique in that it is very selective without significant overoxidation to the sulfone and the catalyst reoxidation is by *tert*-butylhydroperoxide instead of molecular oxygen. In other words, instead of formation of the typical inorganic alkylperoxo intermediate, which may have been expected with TBHP, the reaction appears to proceed by the general two-step mechanism, Eq. 1.

The use of $H_{3+x}PMo_{12-x}V_xO_{40}$, in reactions that can be described by the general two-step mechanism, was expanded into oxidation but not oxygenation of organic substrates, Eq. 9 (SH_2 = substrate, S = product).

$$SH_4 + POM_{ox} \longrightarrow S + 2H^+ + POM_{red}$$

$$2H^+ + POM_{red} + \tfrac{1}{2}O_2 \longrightarrow H_2O + POM_{ox} \qquad (9)$$

The first manifestation of this concept was the oxidative dehydrogenation of cyclic dienes such as dihydroanthracene and α-terpinene with $H_5PMo_{10}V_2O_{40}$ (47). For exo–endo dienic substrates such as limonene and 4-vinylcyclohexene, the dehydrogenation was preceded by isomerization of the exo double bond to an endo ring position, Eq. 10.

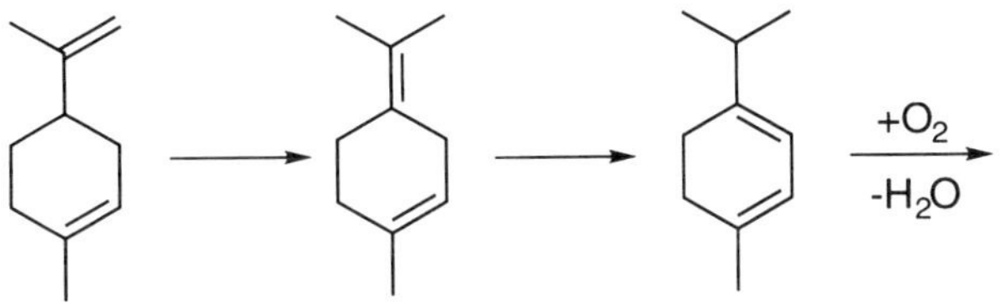

$+O_2$ / $-H_2O$ (10)

This research was continued using carbon supported catalysts for the oxidative dehydrogenation of alcohols and amines (48). These reactions worked most effectively for benzylic derivatives; secondary alcohols and amines reacted more sluggishly while primary derivatives were generally unreactive. In the case of the benzyl alcohols, benzaldehyde was the only product, Eq. 11.

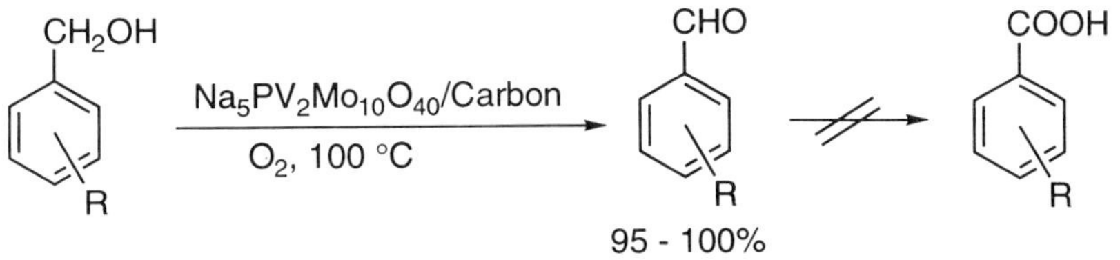

CH_2OH (R) — $Na_5PV_2Mo_{10}O_{40}$/Carbon, O_2, 100 °C → CHO (R), 95 - 100% ↛ COOH (R)

(11)

Note that the autooxidation of the aldehyde to the carboxylic acid was strongly inhibited. For benzyl amines the original dehydrogenation product was the unstable imine [ArCH=NH], which in the presence of even a very small amount of water yielded the carbonyl compound, ArCHO, which in the presence of the unreacted amine gave the Schiff base ArCH=NH—CH_2Ar, as the initial product in about 20 min. The reaction then continued slowly (24 h) via disassociation of the Schiff base to yield the carbonyl compound as final product, Eq. 12.

CH_2NH_2 (R) $\xrightarrow[O_2,\ 100\ °C]{Na_5PV_2Mo_{10}O_{40}/Carbon}$ CH=NHCH$_2$ (R, R) 20 min $\longrightarrow$ CHO (R) 24 hr

(12)

Ishii and co-workers (49) later showed that the amine dehydrogenation could be carried out with $(NH_4)_5H_4PMo_6V_6O_{40}$ and performed the reaction with a few additional substrates. Similar to the case with amines, the oxidative dehydrogenation of hydrazo compounds has also been studied. Thus, azodicarboxamide is formed from hydrazocarboxamide and dioxygen using a heteropolyanion as catalyst (50).

Another reaction of interest that was studied by several groups including our own is the oxidation of phenols. For the oxidation of 2,5,6-trimethylphenol in acetic acid (51), the main product was 2,5,6-trimethylbenzoquinone with the coupled phenol as byproduct, Eq. 13. Increasing amounts of water led to increased amounts of coupled products. Reactions in alcohol on the other hand gave the monomeric benzoquinone as the sole product (52). In a similar reaction using trimethoxyphenol as substrate, the substituted *p*-benzoquinone was the major product (53). Again, using alcohol as solvent 2,6-substituted phenols were oxidatively dimerized to diphenoquinones as sole products (51) Eq. 14. Less activated phenols did not react in this system (mild temperature and atmospheric pressure).

OH $\longrightarrow$ O, O + HO—, —OH

(13)

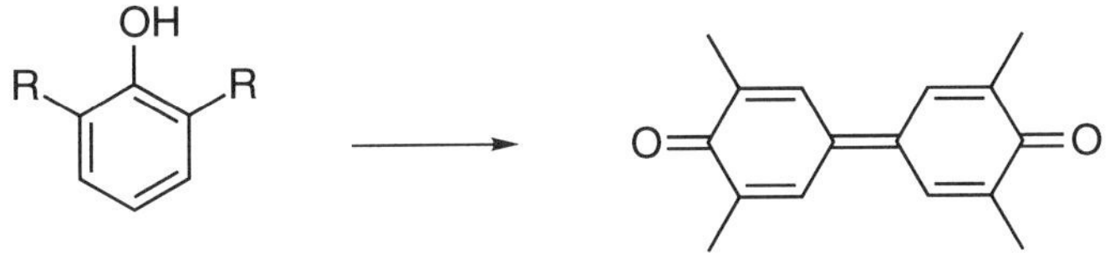

(14)

An interesting extension of this work is the oxidation of 2-methyl-1-naphthol to 2-methyl-1,4-naphthaquinone (vitamin K_3, menadione) in fairly high selectivities, at about 83% and atmospheric dioxygen (54). This work could lead to a new environmentally favorable process to replace the stoichiometric CrO_3 oxidation of 2-methylnaphthalene used today. Interestingly, the same authors have claimed in a patent that the more difficult compound to oxidize 2-methylnaphthalene, can be oxidized with $H_{3-x}PMo_{12-x}V_xO_{40}$ to 2-methyl-1,4-naphthaquinone at elevated temperatures (140°C) and dioxygen pressures of 8 atom at 78% conversion and 82% selectivity (55). Our results using the described procedure were much poorer. We observed that product isolation and catalyst recycle were difficult. Another phenol oxidation of the same type under more extreme conditions is reported in a recent Japanese patent (56), where phenol was oxidized at about 100 atm of air to *p*-benzoquinone at 17% conversion and 52% selectivity. An improvement in the results, mainly selectivity, could lead to an alternative process for the preparation of hydroquinone from phenol replacing the more expensive hydrogen peroxide as oxidant. Note that under similar reaction conditions the same researchers reported that even phenol can be obtained from benzene at a 3.3% yield (57). Other oxidation reactions have been carried out using mixed-addenda heteropolyanions containing vanadium. However, we believe that these other reactions do not proceed by the two-stage mechanism. Rather, they are often vanadium-centered reactions and as such will be described in the following sections.

An interesting application of the two-stage reaction scheme has been proposed by Weinstock et al. (58). They observed from the literature that while substituted phenols were reactive species, aliphatic alcohols were quite inert. Based on this observation (48, 51, 52) they proposed a process whereby wood (cellulose and lignin) could be reacted or bleached to remove the phenolic lignin by selective reaction of the lignin with a polyoxometalate, Eq. 15.

$$\text{Cellulose} + \text{LigninH}_2 + \text{POM}_{ox} \xrightarrow{N_2} \text{Cellulose}\downarrow + \text{Lignin} + 2\text{H}^+ + \text{POM}_{red}$$

$$\text{Lig} + 2\text{H}^+ + \text{POM}_{red} \xrightarrow{O_2} CO_2 + H_2O + \text{POM}_{ox}$$

(15)

In the first stage, the wood pulp is reacted anaerobically with an aqueous solution of a polyoxometalate, with the complex $[PV_2Mo_{10}O_{40}]^{5-}$ being the most effective. This reaction leads to bleaching or depolymerization of lignin as has been observed using a model compound, Eq. 16.

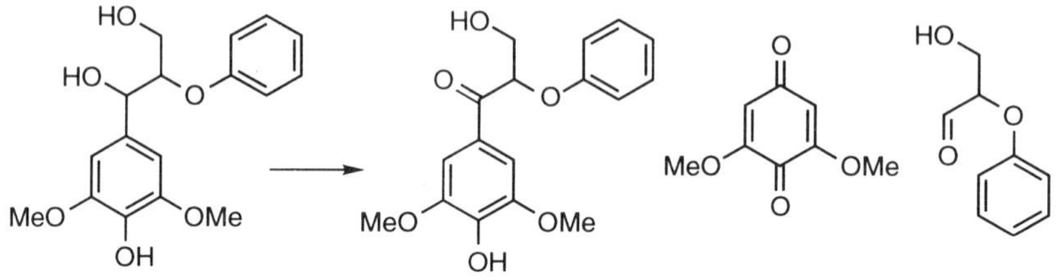

(16)

After separation of the undamaged and bleached cellulose and concentration of the depolymerized lignin, the aqueous phase is thermally treated under dioxygen to bring about the polyoxometalate reoxidation and oxidative mineralization (CO_2 and H_2O) of the depolymerized lignin. The mineralization works best with $K_5SiVW_{11}O_{40}$. This proposed process is conceptually very elegant, however, as presented here this is essentially a multicycle stoichiometric polyoxometalate reaction wherein the polyoxometalate is recycled for each pass. By considering the tremendous size of the paper industry, it is difficult to imagine that such a process could be carried out on a large scale due to the amount of polyoxometalate required even assuming there were no losses. Also, a polyoxometalate must be found that is highly effective for both stages.

The specific mechanism of the generalized reaction scheme described in Eqs. 1 and 9 is a matter of very significant interest and is most probably dependent on the reaction conditions including the specific transformation under consideration and the solvent. For the general case, it is now well accepted that the reduction step is via separate electron and proton transfer from the substrate to the catalyst, although in certain cases arguments may be put forth for hydrogen-atom transfer. Often, in the steady state of the reaction the color of the catalyst is blue pointing clearly to a reduced POM catalyst or catalyst–substrate complex as a stable or quasistable intermediate. The catalyst reduction can be observed by UV–vis and/or ESR. The catalyst reoxidation reaction is usually the rate-determining step and much research effort has and still needs to be invested for a complete understanding of the mechanism of the catalyst reoxidation. Recent patents have suggested that the reoxidation reaction may be accelerated by addition of transition metals (59). Matveev and co-workers (60) naturally first studied the reaction mechanism in aqueous media that was the common solvent in their reactions. Initially, it was found that for the reduced form of the catalyst, $H_{3+x}PMo_{12-x}V_xO_{40}$, the catalyst configuration in solution depended on the amount of vanadium substitution. For $x \geq 3$, the vanadium was disassociated from the heteropolyanion as a free VO^{2+} cation, whereas for $x \leq 2$, the vanadium was within the polyoxoanion framework. At the same time, with the use of ESR Pope et al. (61) showed that the electrons were at the vanadium centers in $H_5PMo_{10}V_2O_{40(red)}$. Using almost exclusively kinetic-based studies in the model oxidation of 2-propanol, Matveev and co-workers (62) postulated that the reaction is a radical chain process with participation of a hydroxy radical as the intermediate responsible for the alcohol oxidation. For the hexa-substituted phosphomolybdovanadate, a different pathway suggesting an intermediate superoxide–heteropolyanion complex that formed hydrogen peroxide was presented. Later, again from kinetic measurements only, it was concluded, that the four-electron reduction of molecular oxygen coordinated by the heteropolyanion is the fastest catalyst reoxidation pathway (63). These reactions are thought to occur by dynamic formation in water of reduced heteropoly species containing four V^{IV} atoms regardless of what the nominal number of vanadium atoms in the original polyoxoanion (64). The major problem in these studies was that most conclusions were heavily based on kinetic measurements and there was little or no spectroscopic support for the conclusions.

Our subsequent research in nonaqueous media used the α-terpinene to *p*-cymene oxidative dehydrogenation as the model reaction (65). We found, with the use of kinetic and several spectroscopic techniques (UV–vis, ESR, NMR, IR), that the reaction mechanism involves two steps. First, there was a fast formation of a catalyst–substrate complex based on electron transfer. After the completion of the catalyst reduction and substrate dehydrogenation, the last step was catalyst reoxidation via a proposed (not observed) μ-peroxo intermediate. In this step, the molecular oxidant undergoes a four-electron reduction to water with 2 equiv of $H_5PMo_{10}V_2O_{40}$ catalyst. There is no intermediacy of the two-electron reduced hydrogen peroxide. In addition, it was shown that the $H_5PMo_{10}V_2O_{40}$ tends to scavange free radicals. Thus, for example, tetralin autooxidation was completely inhibited by the catalyst. We are quite certain that free radicals are of little consequence in the liquid-phase oxidative dehydrogenations. Note that this reaction is more difficult to carry out in the absence of proton countercations. This finding has been attributed to a more favorable oxidation potential of the acid form. However, at least in some cases, acid cocatalysis cannot be ruled out. Most recently, Hill and Duncan (66) used ^{17}O NMR spectroscopy as a probe for reduced catalyst reoxidation. A four-electron reduction of molecular oxygen without intermediacy of H_2O_2 appears to be confirmed. However, their results showing negligible ^{17}O incorporation into the polyoxoanion seem to indicate an outer-sphere molecular oxygen reduction rather than dioxygen reduction through a μ-peroxo intermediate. A generalized scheme for the possible mechanism of the two-step reaction is presented in Fig. 5. Certainly, more research aimed at further in depth understanding of the

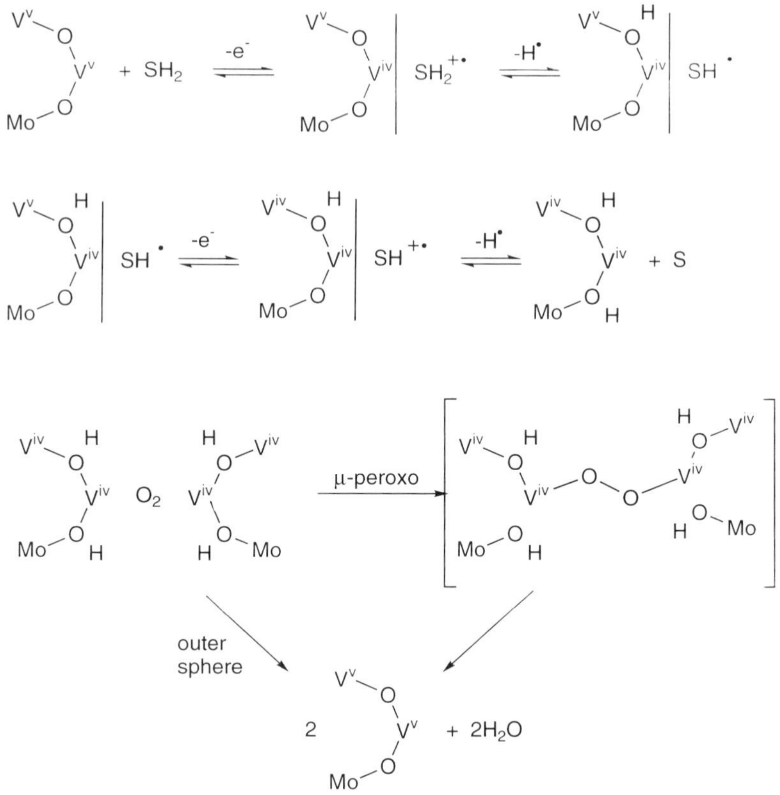

Figure 5. Mechanistic scheme for the two-step redox reaction.

reaction mechanism would be most desirable, especially with additional organic substrates. Some work has been carried out in reoxidation of reduced polyoxotungstates formed in photochemical reactions (67). Work by Hill and Duncan (60) indicated that in these reactions the two-electron reduction of molecular oxygen to hydrogen peroxides occurs via an outer-sphere reaction.

Finally, it is worth pointing out that oxidation of sulfur-containing compounds have received considerable attention as concerns reaction kinetics. First, Kuznetsova et al. (68) and later Hill and co-workers (44) have similarly investigated H_2S oxidation. A thorough kinetic investigation has been carried out in the thioether oxidation with TBHP rather than with molecular oxygen, (44). In contrast to the more common systems, where molecular oxygen is the primary oxidant, here the rate-limiting step was the catalyst reduction, that is, electron transfer between the sulfide and $H_5PMo_{10}V_2O_{40}$.

C. Heterogeneous Catalysis

If we consider that with few exceptions catalytic oxidation reactions in the gas phase are all Mars–van Krevelen type reactions (69), it is no surprise that the $H_{3+x}PMo_{12-x}V_xO_{40}$ family of compounds (usually $x = 0, 1, 2$) has attracted attention as a potentially interesting catalytic species for gas-phase heterogeneous catalysis. Over the last 20 years, this especially has led to very extensive industrial activity. Thus, much of the research has been described in the patent literature and only relatively few reports in the open literature are available. In this section, we will first describe the trends in the use of polyoxometalates for heterogeneous oxidation taken from analysis and perusal of the patent literature. The major interest in the use of heteropolyacid-based oxidation catalysis has been to develop a process for methacrylic acid. Methacrylic acid is still most commonly made by the acetone cyanohydrin process that requires toxic HCN. This process produces very large amounts of ammonium sulfate that has lost its appeal as a fertilizer and therefore the process has become waste intensive. There are several newer processes for methacrylic acid, the common denominator of which is formation of methacrolein as the key intermediate. The later has been made by condensation of propionaldehyde and formaldehyde or by oxidation of isobutene or *tert*-butyl alcohol. As compared to the acrolein oxidation to acrylic acid, the analogous oxidation of methacrolein to methacrylic acid is more complicated. The methacrylic acid can be obtained by aerobic oxidation of methacrolein, supposedly with $H_{3-y}Cs_yPMo_{12-x}V_xO_{40}$ ($0 < x < 2$; $2 < y < 3$) as catalyst (70). The overall conversion of methacrolein is said to be 70–90% with 80–85% selectivity to methacrylic acid.

Since isobutene has become a highly required and somewhat scarce feedstock for oxygenated octane boosters such as methyl- and ethyl-*tert*-butyl ethers, alternative oxidative routes with polyoxometalates to methacrylic acid are called for and have been investigated. The first method is by the well-known hydroformylation of propene to give isobutyraldehyde, which can then be oxidatively dehydrogenated directly to methacrylic acid or via isobutyric acid and/or methacrolein, Eq. 17.

CO,H_2 — CHO — CHO — COOH — COOH (17)

The final approach taken for a new process for methacrylic acid has been by oxidation of isobutane. While this reaction is certainly the most difficult from a chemical and catalytic point of view, requiring both selective oxygenation to form the carboxylic acid and oxidative dehydrogenation to form the double bond, it is the reaction that makes the most "economic sense." Along these lines, selective alkane oxidation in general is definitely a high-priority goal in heterogeneous catalytic oxidation research. Polyoxometalates have recently been among the compounds tested for such catalytic oxidation of methane, ethane, propane, isobutane, pentane, and so forth.

The oxidation of methacrolein to methacrylic acid, which is catalyzed by $H_{3+x}PMo_{12-x}V_xO_{40}$, is the most developed and most understood use of polyoxometalates in heterogeneous oxidation. The process is carried out by passing mixtures of methacrolein, steam, and oxygen over a fixed bed reactor at elevated temperatures. The best results that have been reported refer to a 90% conversion and 86% selectivity at 280°C (71). It is thought that both the oxidative and acidic properties of the catalysts interplay in the reaction. The most effective catalysts appear to have one or two vanadium atoms in the Keggin structure. Partial substitution of the protons by alkali metal or ammonium cations have been shown to increase the stability of the catalysts without reducing activity. The effects of Cs, V substitution, and temperature on conversion and selectivity have been very nicely summarized (72). Catalysts $H_{3-y}Cs_yPMo_{12}O_{40}$ showed that the relative activity was $HCs_2PMo_{12-x}V_xO_{40} \sim H_3PMo_{12-x}V_xO_{40} > H_2CsPMo_{12-x}V_xO_{40}$ although slightly higher temperatures were required in the presence of Cs as a countercation. Although reactivity increased as a function of temperature, selectivity decreased; CO and CO_2 were the major byproducts and acetic acid was the lesser side product. In a series of $H_{3-x}PMo_{12-x}V_xO_{40}$ catalysts, it was observed that the monovanadate was the superior catalyst in terms of reactivity and selectivity. In the $H_3PMo_{12}O_{40}$ catalyzed reaction, Misono et al. (73) showed that the reaction was of a "redox" type by stopping and starting the flow of oxygen. They used an ^{18}O exchange and a pulse reactor attached to a mass spectrometer to suggest the following reaction mechanism, Eq. 18.

$$CH_2{=}C(CH_3)CHO + \text{"MoO"} \underset{}{\overset{H^+}{\rightleftharpoons}} CH_2{=}C(CH_3)CH(OMo)(OMo) \xrightarrow[2)\ +O_2]{1)\ -e^-,\ -H^+} CH_2{=}C(CH_3)COOH \qquad (18)$$

In the first fast step, the molybdate reacts with the aldehyde functionality to form a molybdate diester. This reaction is thought to be acid catalyzed and explains the ^{18}O exchange found between the molybdate and methacrolein. The second rate-determining step is the redox reaction where the methacrylic acid is formed and the reduced catalysts is then regenerated with molecular oxygen.

Some early work by others showed a correlation between the surface and counterions (74) and the critical necessity of acid functionality (75). A more recent article discussed the question of reaction selectivity to methacrylic acid. Thus, measurement of the reduction potential of $[XMo_{12}O_{40}]^{q-}$ revealed that both the experimental and theoretical relative reducibility and catalytic selectivity were correlated for X as As ~ P > Ge > Si (76). All these findings have been thoroughly summarized by Misono et al. (77) in a recent review of the subject.

In the context of the mechanism of the methacrolein oxidation, it is very worthwhile mentioning unique features of heterogeneous catalysis using heteropolyacids as catalysts. There features are (a) a result of the secondary structure of crystals of Keggin acids and refers to the molecular crystal that includes the water molecules of the hydrate and (b) a result of mobility of electrons and protons in the bulk heteropolycompounds. Catalytic reactions have been divided into three groups (78). First, there are ordinary surface-type reactions where the reaction takes place on the two-dimensional surface of the solid. In this reaction type, activity is related to oxidizing ability. The second reaction mode is bulk-type I or reactions in a pseudoliquid phase. In these reactions, the reactants are adsorbed into the molecular crystal. Adsorption was shown to take place by substitution of water of hydration and/or by insertion causing expansion between polyanion units. This type of reation is most common for acid-catalyzed reactions. The third reaction type is termed bulk-type II. In these reactions, the initial redox interaction at the surface between the catalyst and substrate leads to formation of protons and electrons at the catalyst surface. Instead of remaining at the surface (surface-type reaction) it is possible for the electrons and protons to migrate throughout the bulk. Thereby the entire catalyst bulk participates in the reaction even though the initial redox reaction is at the surface. The methacrolein oxidation to methacrylic acid has been classified as a surface-type oxidation (79), whereas the isobutyric acid oxidative dehydrogenation described below has been found to be bulk-type II reaction. Differentiation between the two mechanisms is based on proportionality of rate-to-surface area. In the surface-type reactions, the rate is proportional to the surface area, whereas in the latter bulk-type II case, the rate is independent of the surface area (80).

Since this review (78) there has been almost no research published in this area besides the works of Sumimoto and Mitsubishi Rayon who have patented some new catalytic preparations. Other noteworthy research in this area has been the study of Misono and co-workers (81) on the oxidation of acetaldehyde to acetic acid using various alkali metal substituted heteropolyanions. Again, as in the case of methacrolein, it was found that both the acidic (weak) and oxidative functionality was required and the redox step was rate determining. Similar research on acrolein oxidation to acrylic acid has been carried out in both Poland (82) and Russia (83). The Russian group also measured the heat of adsorption to acrolein to the heteropolysurface. No new conclusions were drawn as to the catalytic oxidation of methacrolein.

Oxidative dehydrogenation of isobutyric acid via hydroformylation of propene is another potential route to methacrylic acid that has attracted interest concerning the use of heteropolyanion as catalyst. In fact, the $PMo_{12-x}V_xO_{40}$ series of compounds is uniquely active and selective in this oxydehydrogenation reaction compared to other molybdenum, vanadium, phosphorous, and other transition metal oxide combinations (84). This result is in contrast to oxidation of propionic acid, propene, and 1-butene, which are not effectively or selectively oxidized using heteropolyacids. The favorable role of the α-methyl group is notable (85). Selectivities of approximately 75% are often reported at fairly high conversions. The major byproducts are formation of acetone and propene in addition to CO and CO_2. As is the case of the methacrolein oxidation, the balance between the acidic and oxidative properties of the catalyst is important. Excessive acidity compared to weak oxidation properties bring about excessive formation of propene and CO, which was shown by using tungstates in place of molybdates. On the basis of these results, Akimoto et al. (86) proposed a reaction mechanism based on a carbocation intermediate, Eq. 19.

+ CO ← H^+ — [COOH]ads — $-H^{\cdot}$ → [* COOH]ads

$-H^{\cdot}$ → COOH ; O^{2-} → O + CO

(19)

After adsorption of the isobutyric acid, acid catalysis brings about the nondesired formation of propene and carbon monoxide. On the other hand, disassociation of a hydrogen atom leads to an activated intermediate adsorbed species that can again loose a hydrogen atom leading the methacrylic acid product. The twice hydrogen reduced catalyst is reoxidized with molecular oxygen in the typical Mars–van Krevelen manner. Acetone was proposed to be formed by interaction of the activated intermediate species with lattice oxygen before the second dehydrogenation step. In this way, the acetone and methacrolein/propene ratio is a measure of the oxidative versus acidic pathway. Subsequent extensive and thorough research by this group concentrated on effects of countercations, heteroatoms, addenda atoms, and temperature (87). Among the further conclusions that were drawn, one may point out the positive effect of countercations with high oxidation potentials, the effect of temperature on the rate-determining step, and the best activity and selectivity of $[PMo_{11}VO_{40}]^{4-}$ in the molybdate series.

Although the selectivity–activity results in the oxydehydrogenation of isobutyric acid would be sufficient for industrialization, the catalytic activity is lost at an unacceptable rate (88). Much of the research in the last 10 years has been devoted to solving this problem. Numerous patents have been given on new catalyst preparations, formulations, and regeneration, but apparently the problem has not been solved. In the recent literature, various catalyst supports have also been investigated (89). Over the last year or so, several papers have been published in attempts to understand and solve the deactivation problem. In a trinational effort (90) the mechanism of thermal decomposition of phosphomolybdate salts and its effects on the catalytic performance has been studied. With potassium as the countercation, excessive heating led to structural collapse and loss of activity. On the other hand, when ammonium salts were used, heating had to preferential formation of a molybdenum-salified heteropolycompound and an increase in catalytic activity. The most recent research in this area (91) concludes that the removal of the hydration water creates the bare Keggin ions that have high catalytic activity. After removal of the water of hydration, which is required for activity, there is a series of solid-state reactions that lead to detrimental formation of MoO_3. The deactivated catalyst may be reorganized under sufficient partial pressure of water to the initial polyoxometalate provided the components are not spatially separated and formation of crystalline MoO_3 is avoided. Indeed, it has been shown by others that use of the correct partial pressure of water in the catalyst can revert deactivation of the catalyst (92). These latest results give hope to the idea that isobutyric acid oxydehydrogenation to methacrylic acid using heteropoly acids may still be an attainable industrial goal.

Although most research interest as per oxidation of isobutyraldehyde goes through isobutyric acid, it is possible to oxidize isobutyraldehyde to methacrolein with heteropolyacids, although this pathway appears to be much inferior at this time (93). The pathway from isobutane to methacrylic is the most attractive due to the low cost of the raw material, simplicity of a one-step process, and the minimization of environmental concerns. There have been numerous patents given for this route including the use of C_4 LPG mixtures (94). Literature coverage has been more scant. Mizuno et al. (95) first found that $C_{2.5}H_{0.5}PMo_{12}O_{40}$ compounds with some slight cation substitution, especially Ni^{II}, yielded methacrolein and methacrylic acid at a combined selectivity of 33% at a conversion of 24%. The major byproducts were carbon monoxide and carbon dioxide and the minor byproduct was acetic acid. Trifiro, Hecquet, and co-workers (96) received similar results by addition of Fe^{III}. Selectivities of about 50% at up to 12% conversions were observed with no observation of structural decomposition.

Studies on the use of heteropolyanions in the oxidation of alkanes other than isobutane have also been described in recent years. In oxidation of butane and pentane, it was found that phosphomolybdates with vanadium incorporation either in the anion or as a countercation were effective catalysts. It was concluded that the vanadium was the active site in both cases (97). Indeed, an interesting point that often comes up in the use of molybdovanadates is the position of the vanadium atom within the heteropoly framework or as a VO^{2+} (VO_2^+) countercation. This point has recently been discussed by Hervé and coworkers (98). In the oxidation of pentane with the $PMo_{12-x}V_xO_{40}$ series, it was found that maleic anhydride was the only product (yields were very low), whereas for vanadyl pyrophosphate both maleic and phthalic anhydrides were formed (99). The oxidation of propane to acrylic acid is also a possible transformation of industrial importance. The $[PMo_{12-x}V_xO_{40}]^{q-}$ type cataylysts (100) at 12% conversion gave selectivities of 15% to acetic acid and 24% to acrylic acid. Partial iron substitution as the countercation along with cesium gave acrylic acid in a relatively high 13% yield (101). Antimony stabilized potassium phosphomolybdates have been used for dehydrogenation of ethane to ethene at about 5% conversion and 50% selectivity (102). There has also been some interest in methane oxidation to formaldehyde (103) also with N_2O as oxidant (104). In the presence of small amounts of carbon tetrachloride, formaldehyde was also formed but methyl chloride was the major product (105). At Sun chemicals Lyons, Ellis, and co-workers (106) also oxidized alkanes; however, they used transition metal substituted polyoxometalates to form alcohols and ketones at the most substituted position as products instead of oxygenation–oxydehydrogenation products.

Other oxidative heterogeneous polyoxometalate reactions that have been studied include methanol to formaldehyde oxidative dehydrogenation (107) and furan formation from butadiene (108). As a final note, it is important to point out an often overlooked work where the $[PMo_{12}O_{40}]^{3-}$ polyoxoanion was shown to donate a lattice oxygen to triphenylphosphine at room temperature. The intermediate $PMo^{VI}_{12-2x}Mo^{V}_{2x}O_{40-x}$ compounds for $x = 1, 2,$ or 3 could be isolated and characterized by X-ray diffraction (XRD), IR, and X-ray photoelectron spectroscopy (XPS) showing no structural decomposition (109). This result shows that a Mars–van Krevelen mechanism is possible even at room temperature with the polyoxometalate compounds. Once more knowledge is gained on the structural stability of heteropoly anions at high temperatures and ways discovered to control the stability, it is conceivable that new significant inroads will be made on the use of polyoxometalates in heterogeneous oxidation catalysis.

III. THE POLYOXOMETALATE OXIDATION-SUBSTRATE REACTION MODE: PEROXYGEN COMPOUNDS AS PRIMARY OXIDANTS

Since all polyoxometalates have high valent d^0 tungsten, molybdenum, and sometimes vanadium and niobium addenda atoms, one may conclude that polyoxometalates should have catalytic properties common to these classes of compounds. Indeed, a main use of these high valent d^0 compounds in catalysis is in the activation of hydrogen peroxide, alkyl hydroperoxides, and, to a lesser degree, peracids. Interaction of these transition metal centers with these peroxides brings about formation of inorganic peroxo, hydroperoxo, alkyl peroxo, or acyl peroxo intermediates. Reaction proceeds by a homolytic (radical type) or heterolytic (electrophilic type) pathway leading to oxygenation of the organic substrate. Usually, the selective electrophilic addition of a polarized (δ^+) peroxo oxygen to a nucleophilic substrate, such as alkene or sulfide, leading to epoxides or sulfoxides and sulfones, is the desired first step. Alternatively, homolytic cleavage of the peroxo bond may lead to oxidation of alcohols and alkanes. Beginning in 1984, Ishii and co-workers (110) oxidized a large spectrum of substrates with the environmentally attractive oxidant, hydrogen peroxide. The basic catalytic procedure involved mixing the organic substrate, the hexadecylpridinium quaternary ammonium salt of $PM_{12}O_{40}$ (M = Mo or W) dissolved in a chlorohydrocarbon and aqueous 30–35% hydrogen peroxide. In this way, allylic alcohols (110) and alkenes were epoxidized, Eq. 20 (111).

$$R_1\text{–CH(X)–CH=CH–}R_2 \xrightarrow{30\%\ H_2O_2} R_1\text{–CH(X)–}\underset{\text{epoxide}}{\text{CH–CH}}\text{–}R_2 \qquad R_1, R_2 = \text{H, alkyl, aryl};\ X = \text{H, OH} \tag{20}$$

Use of *tert*-butyl alcohol in a one-phase system somewhat changed the product selectivity. In the alkene epoxidation, the reaction continued by hydrolysis to yield the vicinal diol and under more elevated temperatures oxidative bond cleavage to carboxylic acids was possible, Eq. 21.

$$R_1\text{–CH}_2\text{–CH=CH–}R_2 \longrightarrow R_1\text{–CH}_2\text{–}\underset{\text{epoxide}}{\text{CH–CH}}\text{–}R_2 \longrightarrow R_1\text{–CH}_2\text{–CH(OH)–CH(OH)–}R_2 \longrightarrow R_1\text{–CH}_2\text{–COOH} + R_2\text{–COOH} \tag{21}$$

In the homogeneous systems, ketonization of secondary alcohols was effective, whereas in two-phase systems, *tert*-butylhydroperoxide was required for alcohol oxidation (112). The paper published in 1988 (111) describing the oxidation of alkenes would prove over the following 5–7 years to be a source of much discussion concerning the identity of the catalytically active compound. Ishii's group noticed at the time that the presence of phosphorus versus silicon as heteroatom was preferred but did not elaborate on this point. At the time, they also compared two catalytic procedures, one where hydrogen peroxide was added to a chloroform solution of tris-hexadecylpyridinium phosphododecatungstate and one where hexadecylpyridinium chloride, hydrogen peroxide, and $H_3PW_{12}O_{40}$ were mixed prior to reaction to form a catalytic active species. The study showed similar activity for both procedures but different elementary analyses for an isolated catalyst. The catalyst prepared according to the first procedure was thought to retain the Keggin structure. Ishii and others continued to describe other oxidative transformations using the tris-hexadecylpyridinium phosphododecatungstate or molybdate as catalyst. This study included oxidation of alkynes to ketones, α,β-dicarbonyl compounds, and acids, Eq. 22 (113) the synthesis of α-hydroxy ketones from alkenes (114) and both α-hydroxy ketones and α,β-diketones from vicinal diols, Eq. 23 (115).

COOH

COOH

(22)

OH OH O

R_1 R_2 R_1 R_2 R_1 R_2 R_1 R_2

OH O O

(23)

The epoxidation of α,β-unsaturated carboxylic acids (116) the lactonization in *tert*-butanol of α,ω-diols to lactones such as 1,4-butanediol to γ-butyrolactone (117) the formation of N-oxides from amines (118) and oxidation of sulfides to sulfoxides and sulfones were also reported (119). In a slightly different two-phase system, tris-tetrabutylammonium salts of the polyoxoanions were used to oxidize cyclohexene to 1,2-cyclohexanediol, adipaldehyde, and adipic acid (120). The heteropolyanion could also be immobilized as the quaternary ammonium salt of poly(4-vinylpyridine) and used to oxidize 1,4-butanediol to γ-butyrolactone (121).

Similar reactions were carried out using slightly different homogeneous catalyst systems. In this way, 3,4,5-trimethoxytoluene was oxidized to 2,3-dimethoxy-5-methyl-*p*-benzoquinone in acetic or formic acid with H_2O_2 using $H_xXM_{12}O_{40}$ (X = Si or P; M = Mo or W) as catalyst, Eq. 24 (122).

MeO OMe OMe → O OMe OMe O (24)

Analogously, 2,3-dimethyl-*p*-benzoquinone was prepared from 2,3-dimethylphenol (123) 2,3,5-Trimethyl-*p*-benzoquinone, the vitamin E precursor, was prepared by oxidation of 1,2,4-trimethylbenzene, Eq. 25 (124) or 2,3,5- or 2,3,6-trimethylphenol (125).

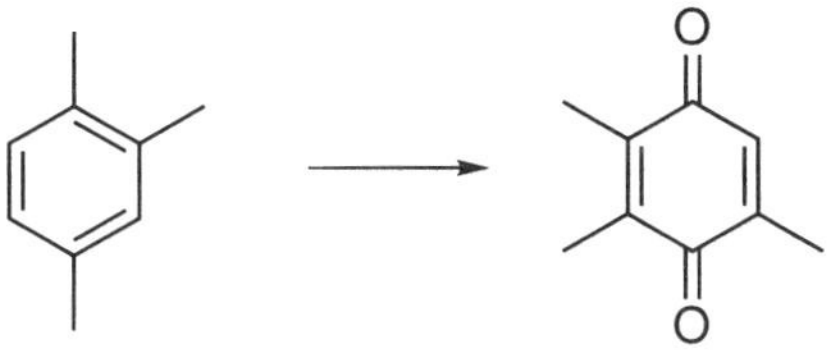

→ O O (25)

Cyclopentene was oxidatively cleaved to glutyraldehyde in tributyl phosphate again with H_2O_2 using $H_3PM_{12}O_{40}$ (M = Mo or W) as catalyst (126). Cyclohexene was similarly oxidized (127). Ethylbenzene was oxidized to acetophenone and 1-phenylethanol in low yield by carrying out the oxidation reaction with the heteropolyacids, $H_3PM_{12}O_{40}$ (M = Mo or W), in a single-phase system using acetonitrile as solvent (128). We have shown that partial substitution of vanadium as an addenda atom in the Keggin heteropolyacid leads to a significantly more potent catalyst for alkyl aromatic side chain oxidation with a series of compounds, $H_yXM_{12-x}V_xO_{40}$ (X = Si or P; M = Mo or W; x = 1, 2) using acetic acid (129). Ketones, acetates, and carboxylic acids are the common products, Eq. 26.

(26)

An important point in Reaction 26 is that although there is vanadium substitution in the addenda, the reaction is not of the "redox"-type since tungstates and molybdates react equally well. Another research paper well worth mentioning in the context of these tungstate and molybdate mediated reactions, is the preparation of and use of the triperoxo niobium compound $[SiW_9(NbO_2)_3O_{37}]^{7-}$, prepared from the niobium capped silicotungstate (130). This compound is a catalyst for epoxidation and hydrolysis of allylic alcohols as well as maleic acid, but is inactive for simple alkenes.

As indicated above, there has been much contention about the identity of the true catalyst in the reactions using aqueous hydrogen peroxide as oxidant and Keggin-type compounds as catalysts. Most likely, the identity of the catalyst depends on (a) whether the reaction is being carried out in a one- or two-phase system, and (b) the oxidizability of the organic substrates. First, concerning two-phase systems, van Bekkum from the start (120) observed that the less stable lacunary Keggin compounds were more reactive than complete Keggin analogues. As stated above, Ishii also found that when silicon was used as a heteroatom it was highly inactive for alkene epoxidation compared to phosphorous but this finding was not explained (111). Although Ishii hinted that the Keggin compounds were perhaps not completely stable under all conditions, it was Brégault and co-workers (131) who were the first to suggest and then convincingly prove that the heteropolyanion in the Ishii system for alkene epoxidation was only a precursor to the true catalyst, $\{PO_4[MO(O_2)_2])^{3-}$. This complex is the so-called Venturello compound, which was synthesized and used in similar catalytic oxidation reactions over 10 years ago (132). Note that no silicon analogue of the Venturello compound is known, easily explaining the lack of activity of $[SiM_{12}O_{40}]^{4-}$. The "culprit" for the lack of solvolytic stability is the presence of the <u>aqueous</u> hydrogen peroxide phase. Isolation of the compound and solution spectroscopic studies nicely support both of these results. Since then Brégault and co-workers (133), Griffith (134), and Hill (135) have performed additional thorough studies with speciation both by isolation and spectroscopy of several intermediate compounds in such biphasic systems. It is quite clear that of all the possible "real" catalysts in this system the Venturello complex is the most active. Ishii's work on the oxidation of sulfides in the biphasic media shows, however, that a Keggin-peroxo intermediate may still be catalytically active for the tungsten but not for the molybdenum analogues provided the oxidation is of active substrates such as sulfides (118). Investigation by ^{31}P NMR shows that the tris-hexadecylpyridinium phosphododecatungstate can initially be stable under reaction conditions, that is aqueous hydrogen peroxide. Furthermore, comparison of reaction selectivities for sulfide oxidation shows the proposed Keggin peroxo compound yields sulfoxide in high selectivity, whereas conditions proven to lead to the Venturello compound yield only sulfone under identical conditions.

By combining the research studies for the biphasic systems, it would seem clear that the Keggin heteropolyanion is inherently unstable to aqueous hydrogen peroxide. The lacunary Keggin compounds, $[PM_{11}O_{39}]^{7-}$, are more reactive because they are even less stable. For reaction to occur, the organic substrate and the Keggin catalyst precursor compound must be well mixed with aqueous hydrogen peroxide at the organic–water interface. The requirement for good contact between phases also explains why surfactant-like quaternary ammonium cations are superior to nonsurfactant tetraalkyl ammonium salts; the former tend to be at the interface and are inherently more active, whereas the later are likely to be in the bulk organic phase and are less active (136). The oxidant formed, $\{PO_4[MO(O_2)_2]\}^{3-}$, is a strong oxidant catalytically active via a heterolytic mechanism and capable of highly efficient alkene epoxidation. Keggin peroxo intermediates can probably be initially formed in biphasic reactions. These oxidants are weak and are capable of oxidizing easier to oxidize compounds such as sulfides. The fact that the niobium Keggin compound with a known peroxo intermediate can catalytically activate hydrogen peroxide to the easier allylic alcohol epoxidation but not to the more difficult alkene epoxidation also supports the notion of different oxidizing species for different Keggin heteropolyanions and different substrates.

The results found for the typical Ishii two-phase system may not hold for monophasic reaction media. For example, in our research on the oxidation of alkyl aromatics with 30% hydrogen peroxide in acetic acid and acetonitrile solvents using the acidic vanadium substituted polyoxometalates instead of the quaternary substituted compound mentioned above, there appears to be ample evidence including postreaction isolation of the Keggin compound in quantitative amounts and *in situ* spectroscopic studies that the polyoxometalates are stable under reaction conditions (129). It is not a priori clear why the heteropolyanions appear to be more solvolytically stable in one-phase systems with polar solvents. It is important to note here that the reaction is vanadium centered, since in its absence there is little reaction. As mentioned above, the reaction does not proceed via the "redox" mechanism. Since vanadium is generally considered an addenda atom and has a peroxide chemistry related to that of molybdenum and tungsten, we have categorized the reaction here, although a good point could be made for its classification in the reaction of transition metal substituted types to be discussed in Section IV. It would also be interesting to follow up on the solvolytic stability of Keggin anions to aqueous hydrogen peroxide in other one-phase systems using solvents such as *tert*-butanol and tributyl phosphate among others already used in literature examples.

Other oxidants besides hydrogen peroxide have been contemplated for use in the oxidation using complete Keggin anions. We have already mentioned the use of *tert*-butylhydroperoxide in the oxidation of secondary alcohols to ketones (110). This oxidant has also been used in the oxidation of 2,4,6-trisubstituted phenols (137). Both the 4-*tert*-butylperoxy-2,5-cyclodien-1-ones and *p*-benzoquinones were formed depending on the substituents on the para position. For 2,4-disubstituted phenolic compounds, dimers were formed. A very interesting oxidant is of course molecular oxygen. In the presence of aldehydes, it is well known that peracids can be formed and can be used to epoxidize alkenes and for the Baeyer–Villiger oxidation of ketones to esters. It has been shown that in the presence of Keggin-type heteropolyanions, preferably containing vanadium as addenda and isobutyraldehyde as peracid precursor, effective catalysis for both the alkene epoxidation and Baeyer–Villiger oxidation can be observed (138). The catalytic effect is probably mostly in the peracid generation step but catalysis of the substrate oxygenation could not be ruled out. We have just recently completed an extensive investigation of the use of vanadium containing Keggin compounds, aldehydes, and dioxygen for the oxidation of alkanes (139). It is clear that multiple reaction pathways are possible.

Finally, it has been observed that vicinal diols (140) and ketones (141) can be oxidatively cleaved using only dioxygen and vanadium containing heteropolyacids. For example, 1-phenyl-2-propanone can be cleaved to benzaldehyde (benzoic acid) and acetic acid ostensibly through the α,β-diketone intermediate 1-phenyl-1,2-propane dione, Eq. 27.

CHO(COOH)

(27)

Similarly, cycloalkanones can be cleaved to ketoacids and diacids. The reaction is certainly vanadium centered and is apparently furthermore also acid catalyzed.

One may summarize the use of Keggin heteropolyanions as a subset of other d^0 containing compounds by concluding that the majority of research has been carried out using aqueous hydrogen peroxide as oxidant. The Keggin compound tends to be solvolytically unstable under these conditions, the most stable and catalytically active compound eventually formed being that first prepared by Venturello. The actual oxidizing species in the reaction depends on the solvent (mono- or biphasic) system, the order of addition of the reaction components, and the relative oxidizability of the substrate. Less research has been carried out using other oxidants such as *tert*-butyl hydroperoxide, dioxygen and aldehyde, and dioxygen alone. Further research in these latter areas is certainly warranted. Another promising avenue of research is the use of polyoxometalates in well-defined solids. In such an example, polyoxometalates were attached in the gallery of a pillared layered double hydroxides (LDH) and used for shape selective epoxidation of alkenes with aqueous hydrogen peroxide (142). 2-Hexene was more reactive than cyclohexene although the opposite is true in liquid-phase catalysis.

IV. TRANSITION METAL SUBSTITUTED POLYOXOMETALATES

The first use of transition metal substituted polyoxometalates for oxidation in organic chemistry was with the transition metal as heteroatom (143). Most notably, it was recognized by electrochemical measurements for the compounds with cobalt and cerium heteroatoms, $[Co^{III}W_{12}O_{40}]^{5-}$ and $[Ce^{IV}Mo_{12}O_{42}]^{8-}$, which showed high oxidation potentials. These high potentials were sufficient for the noncatalytic outer-sphere oxidation of *p*-xylene and *p*-methoxytoluene to the corresponding aldehydes, tolualdehyde, and anisaldehyde, as the main product. Unfortunately, no catalytic cycle could be generated because the reduced Co^{II} and Ce^{III} species could not be recycled easily with an attractive oxidant such as molecular oxygen. It is now clear that an approach where the substituted transiton metal is amenable to coordination by an organic substrate or oxidant, as in substitution in lacunary structures, can lead to a much more interesting catalytic chemistry. Katsoulis and Pope (144) first showed the importance of the accessibility of the substituted transition metal. In the first example, they showed that oxygen could be reversibly bound to a manganese containing Keggin compound, $[Mn^{II}(H_2O)XW_{11}O_{39}]^{6-}$ (X = Ge and Si), dissolved in an organic phase. The second report was of oxidation of a substituted Keggin species to the Cr^V-oxo analogue (145). These two papers reported properties indigenous to some well-known properties of metalloporphyrins, that is, dioxygen binding and formation of high-valent oxo species. Although transition substituted polyoxometalates are certainly not metalloporphyrins, these findings and the fact that the polyoxometalate ligand may function as a multielectron accepting ligand, as is known for the porphyrin ligand, led Hill and Brown (146) at the same time to use manganese, iron, cobalt, and copper substituted Keggin compounds as catalysts for alkene epoxidation with iodosobenzene. This paper was not important because a new method of epoxidation was discovered but because iodosobenzene was a oxygen donor to TMSP compounds used in a manner comparable to those known for iron and manganese porphyrins. In the latter case, it was known that addition of iodosobenzene yield high-valent intermediates via the "shunt" pathway leading to epoxidation of alkenes. This result was also observed for the TMSP compounds leading eventually to the phrase "inorganic porphyrins." The point of comparability was obvious for the manganese case; however, copper and especially cobalt substitued Keggin anions were also catalytically active even though there is no analogous metalloporphyrin reactivity. Thus, the terminology of inorganic porphyrin for TMSP compounds is certainly arguable as concerns various elements of reactivity but its use must be placed in context of the specific property being evaluated. In our opinion, the terminology inorganic porphyrin is too far-reaching although in certain cases the catalytic chemistry has been demonstrated to be similar.

In reviewing the ensuing reports on the use of transition metal substituted polyoxometallates, it is important to divide the various oxidants used in three groups. The first are single oxygen donors such as iodosobenzene, pentafluoroiodosobenzene, periodate, and hypochlorite. In general, these oxidants can be viewed as single oxygen donors to transition metal centers, although as we shall see below, certain exceptions are known. The second group consists of the peroxides such as hydrogen peroxide, *tert*-butylhydroperoxide, and potassium monopersulfate. These oxidants may be considered to have dual reactivity to TMSP compounds. On the one hand, interaction with the low-valent transition metal can lead to oxygen donation but also homolysis of the peroxide bond leading to nonconstructive decomposition to dioxygen and water and radical, for example, Haber-Weiss and Fenton-type chemistry is also possible. Alternatively, peroxides may react with the high-valent tungsten or molybdenum atoms of the polyoxometalate "ligand" and lead to heterolytic peroxide chemistry with high-valent d^0 atoms as described in the section above. Finally, the last group deals with the use of molecular oxygen as oxidant. Here the triplet molecular oxygen ground state leads to the well-known metal catalyzed autooxidation mechanisms that are often the basis of reactivity. In certain cases, other possibilities are available and will be discussed.

The use of iodosobenzene and its more reactive pentafluoro analogue were used by Hill et al. (147) in research subsequent to the original alkene epoxidation, leading also to oxidation of alkanes to alcohols and ketones with manganese and cobalt substituted Keggin species $[M(H_2O)PW_{11}O_{39}]^{5-}$. The iron and chromium counterparts, however, showed no catalysis. Mansuy et al. (148) further studied the use of transition metal substituted Keggin and Wells-Dawson (149) compounds. In particular, they more thoroughly compared the reactivity of the TMSP compounds with those of metalloporphyrins in both alkene epoxidation and alkane oxidation. The manganese substituted polyoxometalates were the most active of all the TMSP compounds (Mn $\gg$ Fe $>$ Co $\gg$ Cu $\sim$ Ni $\gg$ no substitution) and often, but not always, compared well to the activity of manganese porphyrins. From the stereoselectivities and regioselectivities observed in probe substrates such as limonene, *cis*-stilbene, naphthalene, and others, it was suggested, but not proven, that a reactive intermediate ($Mn^V = O$) similar to that contemplated in the manganese porphyrin systems could be postulated. More recently, Liu et al. (150) in China used various mono- and trisubstituted polyoxometalates in similar epoxidation of alkenes with iodosobenzene.

As in the case of metalloporphyrins, the chromium analogue is much less reactive than the manganese, iron, and ruthenium derivatives. This finding has allowed the isolation of the high-valent $Cr^{V}{=}O$ compound (145) and its use in stoichiometric reactions (151). Interestingly and a priori rather unexpectedly, the reaction of the $Cr^{V}{=}O$ compound with organic substrates shows a preponderance of allylic epoxidation products in the oxidation of cyclohexane in addition to evidence for a radical intermediate in other oxidation reactions. This and other lines of evidence led to formulation of radical-type transition state rather than a concerted one.

The different reactivity of the $Cr^{V}{=}O$ polyoxometalate to alkenes (substitution products) compared to the manganese(III) compound with iodosobenzene (addition products) leave the mechanism of oxidation in the later case unsolved. The $Cr^{V}{=}O$ compound was also prepared electrochemically and investigated (152). Attempts at sustained electrocatalytic alcohol oxidation to aldehyde yielded poor results. Another very notable investigation concerning the use of iodosobenzene as oxidant deals with the assembly (synthesis) of the $[Co^{II}PW_{11}O_{39}]^{5-}$ compound from its component fragments, Co^{2+}, $H_2PO_4^-$ and WO_4^{2-} under simultaneous turnover conditions in the epoxidation of *cis*-stilbene (153). It was claimed that if fragmented during reaction, this catalyst could repair itself. Unfortunately, for the catalysts commonly experiencing destruction and/or deactivation, say in the presence of hydrogen peroxide, the resulting degradation products are not equivalent to those necessary for de novo synthesis.

Other iodosobenzene-type oxygen donors have been used rather sparingly in polyoxometalate-catalyzed oxidation reactions. For example, due to the general intrinsic instability of polyoxometalates to basic conditions, the use of inexpensive sodium hypochlorite as oxidant has not been significant. In one publication, a copper substituted Wells-Dawson compound was shown to catalyze the oxidation of alkenes, however, the reaction was not selective (154). In addition to epoxides, chlorohydrins and allylic oxidation products were formed. Most recently, *p*-cyano-*N*,*N*-dimethylaniline-*N*-oxide and other *N*-oxides were used as oxygen donors in a tetracobalt substituted ''sandwich'' polyoxometalate catalyzed epoxidation of alkenes (155). The cobalt substituted Keggin compound had similar activity, but surprisingly, the manganese, iron, and nickel analogues were much less reactive. From mostly kinetic studies, it was concluded that the active catalytic intermediate was a relatively rare $Co^{IV}{=}O$ species. This conclusion may prove to be controversial considering that the corresponding manganese compound, $Mn^{IV}{=}O$ or $Mn^{V}{=}O$ is perhaps not formed, as the latter is an inferior catalyst with *N*-oxides by two orders of magnitude.

An important step forward in the use of transition metal substituted polyoxometalates was the incorporation of second-row transition metal ruthenium into the polyoxometalate structure. Neumann and Abu-Gnim (156) first synthesized the $[Ru^{III}(H_2O)SiW_{11}O_{39}]^{5-}$ compound and used it to catalyze the oxidation of alkenes and alkanes. With iodosobenzene as oxidant for alkenes, epoxidation and carbon–carbon bond cleavage was predominant. For alkanes, both alcohols and ketones were formed. Sodium periodate proved to be a milder oxidant and therefore also more selective. Especially noteworthy was the oxidative cleavage of styrene to yield benzaldehyde in almost quantitative yield. Peroxide oxidants were also used and will be discussed below. The $[Ru^{III}(H_2O)SiW_{11}O_{39}]^{5-}$ periodate cleavage of styrene was further investigated both using kinetic and spectroscopic tools (157). From the reaction kinetics including a Hammett plot, ^{18}O isotope effects, UV–vis, and ESR spectra a reaction mechanism including a metallocyclooxetane intermediate, which is transformed to a Ru^{V} cyclic diester in the rate-determining step requiring water, was proposed. The diester is rearranged via carbon–carbon bond cleavage to form the aldehyde product. There has been some concern raised by others (158) that the $[Ru^{III}(H_2O)SiW_{11}O_{39}]^{5-}$ is not a pure compound. Apparently, we have found that the impurities ($\sim 3\%$) are due to the presence of small amounts of a compound where the ruthenium is ligated with hydroxy and possibly chloride ligands instead of water. In any case, there is no evidence of ex-framework ruthenium nor do these impurities affect the catalytic activity. Since these publications, Rong and Pope (159) synthesized the analogous $[Ru^{III}(H_2O)PW_{11}O_{39}]^{4-}$ but concentrated their studies on preparation and definition of the coordination chemistry. They found clear evidence for ligation of organic substrates to the ruthenium center. Ligation of sulfoxides to the ruthenium center allowed electrocatalytic formation of sulfone using a graphite electrode. The ligation of the ruthenium center with nitrogen-containing ligands such as NO and N_2 has also been reported (160). The periodate system was also nicely adapted in an electrochemical two-phase system where iodate formed from spent periodate is reoxidized at a lead oxide electrode (161). Highly selective and efficient synthesis of aldehydes by carbon–carbon bond cleavage was possible. Others also used these ruthenium substituted Keggin compounds for oxidation of alkanes to alcohols and ketones using sodium hypochlorite and *tert*-butylhydroperoxide as oxidant (162). Similarly, alcohols and aldehydes have been oxidized to carboxylic acid with potassium chlorate as oxidant (163).

Although the above mentioned oxygen-atom donors are certainly interesting from a conceptual and mechanistic point of view, they are prohibitively expensive for most practical applications. Since polyoxometalates have been touted as being potentially significant for industrial applications because of their inherent stability to self-oxidation, it goes without saying that this line of thought should be continued in the use of inexpensive oxidants such as peroxides and molecular oxygen. If we first consider peroxides, essentially three compounds have been studied; potassium persulfate as a triple salt (Oxone), TBHP, and hydrogen peroxide. Of the three, hydrogen peroxide is the cheapest and most environmentally benign. Early on, TBHP in an anhydrous form was used to oxidize alkanes such as cyclohexane to a 2 : 1–1 : 1 mixture of cyclohexanol and cyclohexanone using $[M^{II}(H_2O)PW_{11}O_{39}]^{5-}$ as catalysts (164). The order of reactivity for M = Mn, Cu, Co, and Fe varied according to the solvent used; benzene being preferred over 1,2-dichloroethane and acetonitrile. Turnover numbers of up to 100 were observed. In a similar reaction, the $[Ru^{III}SiW_{11}O_{39}]^{5-}$ Keggin compound was catalytically effective using TBHP and Oxone (156). In similar oxidation reactions of alkenes, epoxide formation and carbon–carbon bond cleavage were the prominent result. In both alkane and alkene oxidation, the persulfate was the more effective oxidant. More recently, noble metal (Ru^{III}, Pd^{II}, and Pt^{II}) and also Mn^{II} substituted "sandwich"-type polyoxometalates $[(WZnTM_2)(ZnW_9O_{19})_2]^{q-}$ were also used for alkane oxidation with 70% TBHP as oxidant (165). These polyoxometalates are about one order of magnitude more efficient than the transition metal substituted Keggin compounds. Interestingly, the alcohol/ketone ratio found for the oxidation of cyclohexane and the tertiary/secondary ratio found in the oxidation of adamantane are different for Ru and Mn versus Pd and Pt. These differences may possibly be attributed to different mechanistic pathways in each case, but this has not been investigated. Similar oxidation with TBHP of alkenes was not selective.

Oxidation with hydrogen peroxide using TMSP compounds is an attractive goal. Already, early research showed that addition of a low-valent transition metal into a Keggin structure caused either high rates of hydrogen peroxide decomposition and/or solvolytic destabilization to the aqueous peroxide media (120). The formation of cationic metal species separated from the polyoxometalate framework additionally severely reduced reaction selectivity, for example, in the epoxidation of 1-octene (135). Thus, literature reports discussing the use of TMSP compounds in which the question of catalyst decomposition is not addressed must be taken with a grain of salt and investigated further. Quite often it may be possible that the TMSP compound, commonly of the Keggin structure, claimed as catalyst is only a precursor to a decomposed compound that is the true catalytic species. Kuznetsova et al. (166) divided the Keggin TMSP compounds into three groups; those rapidly decomposing hydrogen peroxide (Co^{II}, Cu^{II}, and Ru^{III}), those with moderate decomposition rates (Mn^{II}, Fe^{III}, and Cr^{III}) and those with little or no decomposition (Ni^{II} and Ti^{IV}). Lanthanide-containing polyoxometalates of the formula $[LnW_{10}O_{36}]^{q-}$ for Ln = Ce^{IV}, Nd^{III}, and Sm^{III} (Weakley type) decomposed hydrogen peroxide at high rates with cooxidation of cyclohexanol to cyclohexanone, probably by intermediate radicals formed in the H_2O_2 decomposition (167). This work was extended to the oxidation of various alcohols to the corresponding carbonyl products using $[Ce^{IV}W_{10}C_{36}]^{8-}$ as catalyst (168). Good yields were obtained at high molar excesses of hydrogen peroxide but it is not clear if the polyoxometalate is stable (169). The epoxidation of stilbene with H_2O_2 and TBHP has also been investigated with TMSP compounds of the Keggin structure (170). Finally, TMSP compounds with a gallium heteroatom have been prepared and used for substrates such as allylic alcohols and maleic acid (171). No specific information is given as to H_2O_2 decomposition or catalyst stability.

The goal, therefore, in this area of research of hydrogen peroxide activation was to find a TMSP compound that is stable to aqueous hydrogen peroxide and decomposes the oxidant at a minimal rate. This goal has been apparently met by using transition metal substituted "sandwich"-type polyoxometalates. One active compound is the tetra iron substituted $[(Fe^{II})_4(PW_9O_{34})_2]^{10-}$ polyoxometalate, which was shown to be catalytically active in a monophasic reaction media using acetonitrile as solvent (172). Reasonable turnover numbers were reported for the epoxidation of alkenes with only moderate nonproductive decomposition of hydrogen peroxide. Unfortunately, beyond this preliminary communication there has yet to appear a more in depth study. The second sandwich-type compound, which was successfully applied, was the $[(WZnMn_2^{II})(ZnW_9O_{19})_2]^{12-}$ complex developed by Neumann and Gara (173). Among this class of compounds, the manganese derivative is uniquely active. Reactions were carried out in biphasic systems, preferably 1,2-dichloroethane–water. Tri- and tetramanganese substituted compounds are inactive. Most importantly, at low temperatures highly selective epoxidation could be carried out on cyclohexene, which is normally highly susceptible to allylic oxidation. Nonproductive decomposition of hydrogen peroxide at low temperatures was minimal. In a further report, an in depth kinetic and spectroscopic study was carried out (174). This investigation showed that the catalyst was stable under turnover conditions and tens of thousands of turnovers could be attained with little H_2O_2 decomposition. In addition, a peroxo intermediate was clearly identified by IR spectra. The mechanism of the reaction was further investigated by preparation and use of the rhodium derivative, $[(WZnRh_2^{III})(ZnW_9O_{19})_2]^{10-}$. The catalytic activity of the rhodium compound was similar to that of the manganese compound (175). Since the oxo chemistry of Rh^{III} and Mn^{II} differ significantly and in both cases a peroxo intermediate is observable by IR, whereas for nontransition metal substituted compounds there is no such peroxo species, a mechanistic scheme to explain the catalytic activity was given as follows, Fig. 6. Ini-

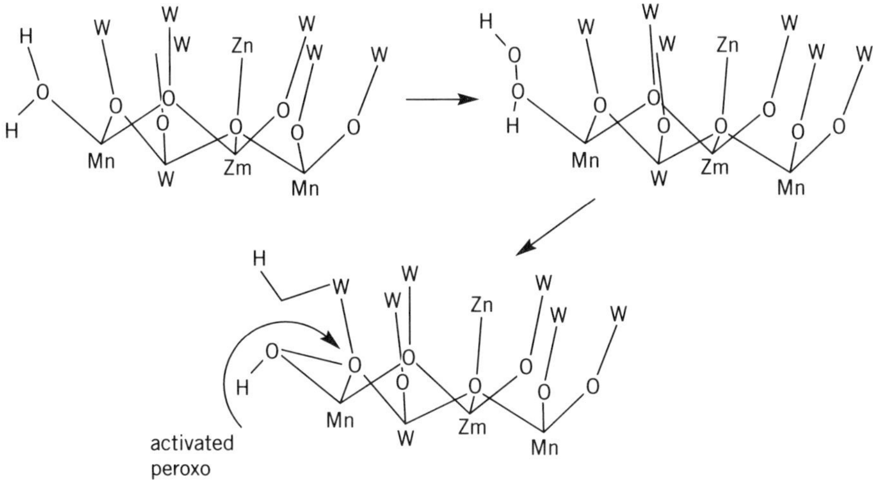

Figure 6. Mechanistic scheme for the activation of hydrogen peroxide with sandwich-type compounds, $[WZnMn^{II}_2)(ZnW_9O_{19})_2]^{12-}$.

tially, the hydrogen peroxide is ligated to the labile sixth coordination site of the transition metal. This reaction is followed by formation of a tungsten peroxo intermediate at the proximal tungsten position. This peroxo complex is highly active due to polarization of the peroxo oxygen atom by three neighboring metal atoms. The use of the $[(WZnMn^{II}_2)(ZnW_9O_{19})_2]^{12-}$ polyoxometalate as catalyst for hydrogen peroxide activation has been extended to additional substrates besides the original alkenes (176). These substrates include alcohols, allylic alcohols, dienes, diols, and sulfides. Special attention has been paid to questions of stereo- and regioselectivity in multifunctional compounds such as geraniol and limonene. One of the pitfalls of these catalytic systems is that although high turnover conditions can be easily achieved, reaction conversions and yields are not always so high, especially for substrates with low reactivity such as terminal alkenes. This problem can be corrected by continuous addition of hydrogen peroxide. Another problem that is intrisinc to these reactions is the use of organic solvents that reduce the environmental attractiveness for the use of hydrogen peroxide. We have shown that the a functionalized silica catalytic assembly can be prepared containing a chemically attached polyoxometalate with controllable hydrophobicity of the surface, Fig. 7. In this way, one can dispense with the use of an organic solvent and reactions can be carried out by mixing aqueous hydrogen peroxide and the organic solvent with a solid catalyst particle, which is easily recoverable (177). Catalytic activity is essentially the same as in the traditional biphasic liquid–liquid reaction medium.

An important research goal in all oxidation chemistry is concerned with the use of molecular oxygen as the primary oxidant. In the liquid phase, this chem-

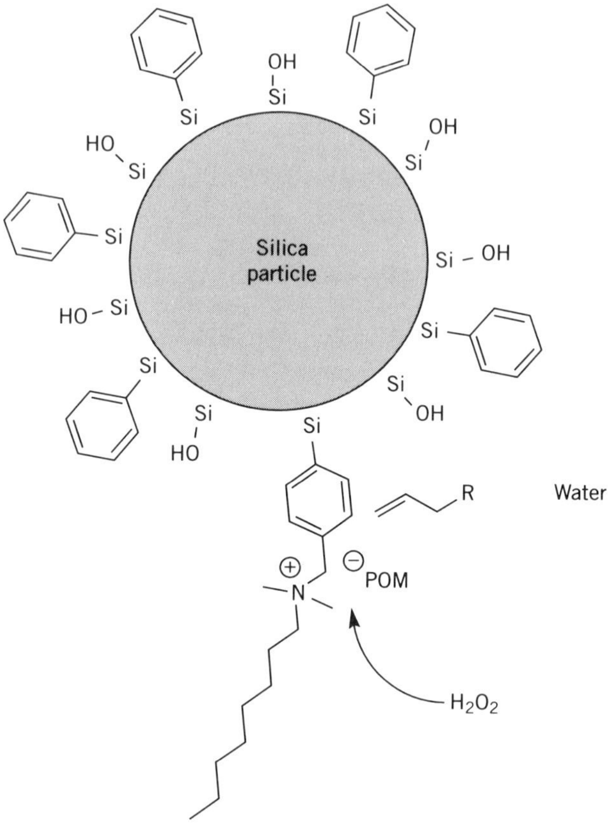

Figure 7. A cartoon of a functionalized silica particle for reactions in water.

istry is often dominated by autooxidation pathways, which in certain important cases can be utilized for preparation of basic chemicals of industrial importance. In fine chemical synthesis, these autooxidation pathways often lead to nonselective reactions. One way to circumvent this problem is to add a reducing agent (two electrons) as is often observed in nature. In the laboratory, the reducing agent may be added in various forms. One popular method is to add aldehydes that form peracids *in situ*. This method is in fact from the formalistic point of view, which is also a two-electron reduced dioxygen or peroxo intermediate. As concerns TMSP compounds, dioxygen has been used to oxidize inorganic substrates such as hydrogen sulfide to elemental sulfur and nitric oxide to nitric and nitrous acid (178). Another use that has been reported is that of a simple autooxidation catalyst. In this way, trisubstituted Keggin compounds of the formula $[M_3(H_2O)_3PW_9O_{37}]^{6-}$ ($M = Fe^{III}$ and Cr^{III}) and $[Fe_2M(H_2O)_3PW_9O_{37}]^{7-}$ ($M = Ni^{II}$, Co^{II}, Mn^{II}, and Zn^{II}) were used in the autooxidation of alkanes such as propane and isobutane to acetone and *tert*-butanols (179). Mizuno et al. (180) later synthesized and used the similar $[Fe_2^{III}Ni^{II}(OAc)_3PW_9O_{37}]^{10-}$ compound to oxidize alkanes such as adamantane, cyclohexane, ethylbenzene, and decane. The reaction products (alcohol and ketone) and regioselectivities are typical for metal-catalyzed autooxidations. Others have also oxidized cyclohexane using more common monosubstituted Keggin-type TMSPs (181). In this same vein, the oxidation of *p*-xylene to terephthalic acid under 60-atm air using TMSP compounds, most specifically $[SiRu^{III}W_{11}O_{39}]^{5-}$, has been patented (182). Other substrate types could also be oxidized, most notably activated phenols were oxidized to quinones and dimers (183).

The transition metal catalyzed autooxidation could also be utilized by viewing the transition metal substituted polyoxometalates as bifunctional catalysts. In such a scenario, the transition metal center can catalyze autooxidation of a first substrate, for example, aldehyde, alkane, or alkene. This autooxidation yields a peracid, alkyl hydroperoxide, or allylic hydroperoxide, respectively, as the initial product. The peroxo complexes can then be further activated, preferably by the tungsten or molybdenum center of the polyoxometalate "ligand," Eq. 28.

$$RCH_2C(H)R'' + \mathbf{TM}^{n+}[XM_{11}O_{39}] \longrightarrow RCH_2\dot{C}R'' + \mathbf{TM}^{(n-1)+}[XM_{11}O_{39}]$$

$$RCH_2\dot{C}R'' + O_2 \longrightarrow RCH_2C(OO\cdot)R'' \xrightarrow{RH} RCH_2\dot{C}R'' + RCH_2C(OOH)R''$$

$$RCH_2C(OOH)R'' + CH_2{=}CHR' \xrightarrow{TM[\mathbf{XM_{11}O_{39}}]} \text{epoxide}(R') + RCH_2C(OH)R''$$

R" = oxygen of carbonyl, alkyl, or aryl

(28)

This latter activation is required for alkyl and allylic hydroperoxides but not for peracids, as concerns the epoxidation of alkenes. These methods were applied first for epoxidation of alkenes in the presence of isobutyraldehyde and molecular oxygen using $[Co^{II}PW_{11}O_{39}]^{5-}$ as catalyst (184). In another system, the epoxidation of alkenes (1-octene and cyclohexene) was kinetically coordinated with the autooxidation of cumene to cumene hydroperoxide and cyclohexene to cyclohexene hydroperoxide using molybdenum but not tungsten-based TMSP compounds, such as $[Co^{II}PMo_{11}O_{39}]^{5-}$ and $[Ru^{III}PMo_{11}O_{39}]^{4-}$ (185). Instead of forming a two-electron reduced peroxide via an autooxidation pathway, a reducing agent or electrons may be directly added to the reaction system in the form of sodium borohydride or hydrogen in the presence of a platinum colloid, H_2/Pt among others (186). In the presence of Pt/H_2, effective reductive dioxygen activation is possible with $[Mn^{II}SiW_{11}O_{39}]^{6-}$ but turnover numbers were very low. In the presence of borohydride, the reaction was more efficient with both the manganese and copper substituted silicotungstates. The reaction, however, probably proceeded by a type of hydroboration reaction through intermediate formation of alkylboranes. The primary products are alcohols with anti-Markovnikov orientation toward addition to the double bonds.

An important but very elusive goal in the chemistry of ground-state molecular oxygen is its activation in the absence of reducing agents and autooxidation pathways. In our group, we have used the ruthenium substituted sandwich compound $\{[(WZnRu_2(H_2O)(OH)][(ZnW_9O_{19})_2]\}^{11-}$ to activate molecular oxygen, as shown in the hydroxylation of adamantane (187). This reaction takes place at the tertiary carbon–hydrogen bond only and several lines of evidence indicated that the reaction was not a transition metal catalyzed autooxidation. Since then, we have intensively investigated this subject and have come to the firm conclusion that this specific ruthenium-substituted polyoxometalate is acting as an inorganic dioxygenase (188).

The use of transition metal substituted polyoxometalates promises to remain a dynamic research topic in the future. The large variability in structures and substitution patterns available make the potential for this class of compounds great for future developments in oxidation chemistry. Certainly, new applications in oxidative catalysts in organic chemistry could be developed. Especially interesting will be the studies concerning mechanism. In this area, kinetic measurements can generally be carried out in the usual way. However, spectroscopic measurements have been greatly hindered due to the difficulty in obtaining NMR measurements, especially in TMSP compounds containing paramagnetic centers and in organic solvents. This difficulty in sharp contrast to studies using metalloorganic catalysts, where NMR measurements have been of great utility in mechanistic research.

V. OTHER USES OF POLYOXOMETALATES IN CATALYSIS

In the sections above, we have hopefully outlined and described the majority of the research discussed in the literature in the three major areas of oxidation catalysis with polyoxometalates. There are, of course, other areas where polyoxometalates are of interest as catalysts. The first, already mentioned at the outset, is in the area of acid catalysis. In this area, the petrochemical industry is always on the lookout for new acid catalysis with specific and selective catalytic properties. Pillared clays containing heteropolyacids may lead to interesting results. The $Cs_{2.5}H_{0.5}PW_{12}O_{40}$ heteropolyacid may also prove to have advantageous activity for a variety of acid-catalyzed reaction such as alkylation and transalkylation, isomerization, and so on (189). More research needs to be performed using combined oxidative and acidic catalysis. Interesting synergetic affects may be waiting to be reaped. Another petrochemical concern is desulferization of oils. Generally, this is a reductive process termed hydrosulfurization. It has recently been shown that heteropolyanions may be active in this reaction (190). Alternatively, one could contemplate oxidative desulfurization via formation of more polar sulfoxides and sulfones, which could then be removed by adsorption (191).

Although generally reduced polyoxometalates, heteropoly blues, formed in a redox cycle are reoxidized by molecular oxygen, one can contemplate the use of reduced polyoxometalates as reducing agents for organic compounds. One such example is the use of a photoreduced tungstate for the reduction of nitroaromatics to aminoaromatics (192). Another thought along these lines is the use of mixed-addenda polyoxoanions in the presence of palladium chloride for the reductive carbonylation of nitrobenzene to the *N*-phenyl carbamate (193). Another approach to use of polyoxometalates in reduction chemistry has been to use iridium complexes supported on electron-rich polyoxometalates, for example, $[(1,5\text{-cod})Ir \cdot P_2W_{15}Nb_3O_{62}]^{8-}$ (cod = 1,5-cyclooctadiene as catalysts for alkene hydrogenation. Originally, it was thought that the hydrogenation is catalyzed by the discrete iridium compound. Later, it was shown that catalysis is in fact by iridium colloids (194). Such compounds and other similar ruthenium and rhodium analogues have been shown to oxidize cyclohexene with molecular oxygen in what is quite likely an autooxidation reaction (195).

At this point, we must reiterate that there is significant and important chemistry being carried out in the fields of photocatalysis that might prove invaluable in the field of waste decomposition, such as treatment of water pollutants; electrocatalysis, where there might be interesting developments in the use of polyoxometalates supported on electrodes. Finally, there is a significant research effort into the chemistry of a large variety of peroxotungstates (not reviewed here) and other similar compounds designed for application with hydrogen peroxide.

VI. A FINAL WORD AND A VIEW TO THE FUTURE

In this chapter, my own personal views have frequently been given on the research described in the literature. My interpretations of the oxidation–polyoxometalate research may therefore sometimes be at odds with different acceptable views of others in the field. I can only hope that such differing interpretations will stimulate new research and lead to further exciting advances. Now a comment concerning future research directions. In the Babylonian Tal-mud there are two interpretations about forecasting the future. One view is that "the gift of prophecy has been denied to prophets and bestowed upon scholars," whereas the opposing view holds that "the gift of prophecy . . . has been bestowed upon fools." For fear of being considered a fool rather than a scholar, I will say that I can hardly predict what the future will bring in any field including that of oxidation catalysis with polyoxometalates. Any outlook I may present would be more of a statement of personal taste. In this vein, it would be fair to say that catalysis by polyoxometalates is still in its infancy. Many reactions are still to be discovered and our understanding of even the known catalytic reactions is clearly very limited. Thus, personally I would like to see new applications taking advantage especially of the intrinsic stability and many compositional variations available. On the other hand, much more fundamental research into what is already known would certainly be desired.

ACKNOWLEDGMENT

I gratefully acknowledge the Basic Research Foundation, administered by the Israel Academy of Sciences and Humanities, for support.

ABBREVIATIONS

Ac	Acetate
cod	1,5-Cyclooctadiene
ESR	Electron spin resonance
IR	Infrared
LDH	Pillared layered double hydroxides
LPG	Liquid petroleum gas
NMR	Nuclear magnetic resonance
POM	Polyoxometalate
TBHP	*tert*-Butylhydroperoxide
THF	Tetrahydrofuran
TMSP	Transition metal substituted polyoxometalate
TOF	Turnover frequency
UV–vis	Ultraviolet–visible
XPS	X-ray photoelectron spectroscopy
XRD	X-ray diffraction

REFERENCES

1. J. Berzelius, *Pogg. Ann.*, *6*, 369, 380 (1826).
2. J. F. Keggin, *Proc. R. Soc.*, *A144*, 75 (1934).

3. M. T. Pope, *Heteropoly and Isopoly Oxymetallates*; Springer-Verlag, New York, 1983.
4. M. T. Pope and A. Müller, *Angew. Chem. Int. Eng. Ed.*, *30*, 34 (1991). M. T. Pope and A. Müller, *Polyoxometalates: From Platonic Solids to Anti-Retroviral Activity;* Kluwer, The Netherlands, 1993. Y. Izume, K. Urabe, and M. Onaka, *Zeolite, Clay and Heteropoly Acid in Organic Reactions*, VCH, New York, 1992, p. 99.
5. V. W. Day and W. G. Klemperer, *Science*, *228*, 533 (1985). M. T. Pope, *Progress in Inorganic Chemistry*, Wiley-Interscience, New York, 1991, Vol. 39, p. 181. J. Zubieta and Q. Chen, *Coord. Chem. Rev.*, *114*, 107 (1992). K. Isobe and A. Yagasaki, *Acc. Chem. Res.*, *26*, 524 (1993). Y. Jeannin, G. Hervé, and A. Proust, *Inorg. Chim. Acta*, *198*, 319 (1992).
6. E. Papaconstantinou, *Chem. Soc. Rev.*, *18*, 1 (1989). C. L. Hill and C. M. Prosser-McCartha, in *Photosensitization and Photocatalysis Using Inorganic and Organometallic Complexes*, K. Kalyanasundaram and M. Grätzel, Eds., Kluwer, The Netherlands, 1993, p. 307.
7. E. N. Semenovskaya, *J. Anal. Chem. USSR*, *41*, 1339 (1986).
8. C. L. Hill, G.-S. Kim, and C. M. Prosser-McCartha, *Mol. Eng.*, *3*, 263 (1993). B. Krebs, *NATO ASI Ser. Ser. C*, *459*, 359 (1995).
9. E. Coronado and C. Gomez-Garcia, *J. Mol. Eng.*, *3*, 171 (1992).
10. J. E. Toth and F. C. Anson, *J. Am. Chem. Soc.*, *111*, 2444 (1989).
11. M. Misono and T. Okuhara, *Chemtech*, *23*, 23 (1993).
12. K. I. Matveev and I. V. Kozhevnikov, *Kinet. Catal.*, *21*, 855 (1980). I. V. Kozhevnikov and K. I. Matveev, *Russ. Chem. Rev.*, *51*, 1075 (1982). I. V. Kozhevnikov and K. I. Matveev, *Appl. Catal.*, *5*, 135 (1983).
13. G. A. Tsigdinos, *Top. Curr. Chem.*, *26*, 1 (1978).
14. M. Misono, *Catal. Rev.-Sci. Eng.*, *29*, 269 (1987).
15. N. Mizuno and M. Misono, *J. Mol. Catal.*, *86*, 319 (1994). T. Okuhara, N. Mizuno, and M. Misono, *Adv. Catal.*, *41*, 113 (1996).
16. R. J. J. Jansen, H. M. Vanveldhuizen, M. A. Schwegler, and H. van Beffum, *Rec. Trav. Chim. Pays-Bas*, *113*, 115 (1994). Y. Ono, in *Perspectives in Catalysis*, J. M. Thomas and K. Zamaraev, Eds., Blackwell, Oxford, 1992, p. 431.
17. Y. Ishii and M. Ogawa, in *Reviews on Heteroatom Chemistry*, S. Oae, Ed., MYU, Tokyo, 1990.
18. C. L. Hill, D. C. Duncan, and M. K. Harrup, *Comments Inorg. Chem.*, *14*, 367 (1993). C. L. Hill and C. M. Prosser-McCartha, *Coord. Chem. Rev.*, *143*, 407 (1995).
19. I. V. Kozhevnikov, *Catal. Rev.-Sci. Eng.*, *37*, 311 (1995). I. V. Kozhevnikov, *Russ. Chem. Rev.*, *62*, 473 (1993).
20. R. A. Sheldon, in *Aspects of Homogenous Catalysis Vol. 4*, R. Ugo, Ed., Riedel, The Netherlands, 1981, p. 1.
21. L. C. W. Baker and T. P. McCutcheon, *J. Am. Chem. Soc.*, *78*, 4505 (1956). L. C. W. Baker, G. A. Gallagher, and T. P. McCutcheon, *J. Am. Chem. Soc.*, *75*, 2493 (1953).

22. F. Zonnevijlle, C. M. Tourné, and G. F. Tourné, *Inorg. Chem.*, *21*, 2242 (1982). F. Zonnevijlle, C. M. Tourné, and G. F. Tourné, *Inorg. Chem.*, *22*, 1198 (1983). T. J. R. Weakley and S. A. Malik, *J. Inorg. Nucl. Chem.*, *29*, 2935 (1967). S. A. Malik and T. J. R. Weakley, *J. Chem. Soc. (A)*, 2647 (1968). C. M. Tourné and G. F. Tourné, *J. Bull. Soc. Chim. Fr.*, 1124 (1969).

23. K. I. Matveev, *Kinet. Catal.*, *18*, 716 (1977). K. I. Matveev and I. V. Kozhevnikov, *Kinet. Catal.*, *21*, 855 (1980).

24. K. I. Matveev, E. G. Zhizhina, M. B. Shitova, and L. I. Kuznetsova, *Kinet. Catal.*, *18*, 380 (1977). L. I. Kutsenova and K. I. Matveev, *React. Kinet. Catal. Lett.*, *3*, 305 (1975). V. M. Berdnikov, L. I. Kutsenova, K. I. Matveev, N. P. Kirik, and E. N. Yurchenko, *Sov. J. Coord. Chem.*, *5*, 39 (1979).

25. M. Cihova, M. Hrushovsky, and J. Vojtko, *J. Petrochem.*, *19*, 97 (1979).

26. K. Urabe, F. Kimura, and Y. Izumi, *VII Int. Cong. Catal.*, C14 (1980).

27. I. V. Kozhevnikov, V. E. Tarabanko, K. I. Matveev, and V. D. Vardanyan, *React. Kinet. Catal. Lett.*, *7*, 297 (1977).

28. L. N. Rachkovskaya, K. I. Matveev, G. N. Il'inich, and N. K. Eremenko, *Kinet. Catal.*, *18*, 854 (1977).

29. H. U. Mennenga, A. I. Rudenkov, K. I. Matveev, and I. V. Kozhevnikov, *React. Kinet. Catal. Lett.*, *5*, 401 (1976). V. E. Tarabanko, I. V. Kozhevnikov, and K. I. Matveev, *React. Kinet. Catal. Lett.*, *8*, 77 (1978).

30. L. N. Rachkovskaya, K. I. Matveev, A. I. Rudenkov, and G. U. Mennenga, *React. Kinet. Catal. Lett.*, *6*, 73 (1977).

31. V. E. Tarabanko, I. V. Kozhevnikov, and K. I. Matveev, *Kinet. Catal.*, *19*, 1160 (1978).

32. I. V. Kozhevnikov, V. E. Tarrabanko, and K. I. Matveev, *Kinet. Catal.*, *21*, 947 (1980).

33. J. R. Grate, D. R. Mamm, and S. Mohajan, *Mol. Eng.*, *3*, 205 (1993). J. R. Grate, D. R. Mamm, and S. Mohajan, in *Polyoxometalates: From Platonic Solids to Anti-Retroviral Activity*, M. T. Pope and A. Müller, Eds., Kluwer, The Netherlands, 1993, p. 27.

34. M. Cihova, J. Vojtko, and M. Hrusovsky, *Ropa Uhlie*, *28*, 297 (1986); *Chem. Abs.* *107*, 6740 (1986).

35. G. Centi, S. Perathoner, and G. Stella, *Stud. Surf. Sci. Catal.*, *88*, 393 (1994). G. van der Lans, M. Makkee, and J. J. F. Sholten, *J. Catal.*, *154*, 187 (1995). A. W. Stobe-Kreemers, R. B. Dielis, M. Makkee, and J. J. F. Scholten, *J. Catal.*, *152*, 175 (1995). H. Ishii, M. Tsuzuki, Masanori, and M. Kanzawa, Japanese Patent, 07 149 685, 1995; *Chem. Abstr.* *123*, 256163t.

36. M. Misono, T. Okuhara, and H. Soeda, Japanese Patent, 06 091 170, 1994; H. Soeda, T. Okuhara, and M. Misono, *Nippon Kagaku, Kaishi*, 917 (1993); *Chem Abstr.* *121*, 157194f.

37. C. Mathieu, E. Monfkier, and E. Y. Barbaux, *Spectra Anal.*, *23*, 23 (1994).

38. S. N. Pavlova, R. I. Maksimovskaya, and L. I. Kutsenova, *Kinet. Catal.*, *32*, 410 (1991). E. G. Zhizhina, L. I. Kutsenova, and K. I. Matveev, *Kinet. Catal.*,

29, 135 (1988). V. A. Golodov and B. S. Dzhumakaeva, *J. Mol. Catal.*, *35*, 309 (1986). N. G. Grunzinskaya, B. S. Dzhumakaeva, and V. A. Golodov, *Kinet. Catal.*, *36*, 191 (1995).

39. T. A. Gorodetskaya, I. V. Kozhevnikov, and K. I. Matveev, *React. Kinet. Catal. Lett.*, *16*, 17 (1981).
40. L. N. Arzamaskova, A. V. Romanenko, and Y. I. Yermakov, *Kinet. Catal.*, *21*, 1068 (1980).
41. T. A. Gorodetskaya, I. V. Kozhevnikov, and K. I. Matveev, *Kinet. Catal.*, *23*, 992 (1982).
42. R. Neumann and I. Assael, *J. Chem. Soc. Chem. Commun.*, 1285 (1988).
43. I. V. Kozhevnikov, V. I. Simagina, G. V. Varnakova, and K. I. Matveev, *Kinet. Catal.*, *20*, 506 (1979). B. S. Dzhumakaeva and V. A. Golodov, *J. Mol. Catal.*, *35*, 303 (1986).
44. M. K. Harrup and C. L. Hill, *Inorg. Chem.*, *33*, 5448 (1994). M. K. Harrup and C. L. Hill, *J. Mol. Catal. A: Chem.*, *106*, 57 (1996).
45. Y. Hou and C. L. Hill, *J. Am. Chem. Soc.*, *115*, 11823 (1993).
46. R. D. Gall, M. Faraj, and C. L. Hill, *Inorg. Chem.*, *33*, 5015 (1994). R. D. Gall, C. L. Hill, and J. E. Walker, *J. Catal.*, *159*, 473 (1996).
47. R. Neumann and M. Lissel, *J. Org. Chem.*, *54*, 4607 (1989).
48. R. Neumann and M. Levin, *J. Org. Chem.*, *56*, 5707 (1991).
49. K. Nakayama, M. Hamamoto, Y. Nishiyama, and Y. Ishii, *Chem. Lett.*, 1699 (1993).
50. A. Kaszonyi, M. Hronec, and M. Harustiak, *J. Mol. Catal.*, 80, L13 (1993).
51. O. A. Kholdeeva, A. V. Golovin, and I. V. Kozhevnikov, *React. Kinet. Catal. Lett.*, *46*, 107 (1992). O. A. Kholdeeva, A. V. Golovin, R. A. Maksimovskaya, and I. V. Kozhevnikov, *J. Mol. Catal.*, *75*, 235 (1992).
52. M. Lissel and H. Jansen van de Wal, R. Neumann, *Tetrahedron Lett.*, *33*, 1795 (1992).
53. H. Orita, M. Shimizu, T. Hayakawa, and K. Takehira, K., *React. Kinet. Catal. Lett.*, *44*, 209 (1991).
54. K. I. Matveev, E. G. Zhizhina, and V. F. Odyakov, *React. Kinet. Catal. Lett.*, *55*, 47 (1995).
55. T. A. Gorodetskaya, I. V. Kozhevnikov, and K. I. Matveev, USSR Patent 1 121 255, 1984; *Chem. Abstr. 102*, 203754w.
56. F. Matsuda and K. Inoe, K. Kato, Japanese Patent, 07 069 968, 1995; *Chem. Abstr. 123*, 9152s.
57. F. Matsuda, K. Inoe, K. Kato, Japanese Patent, 06 116 187, 1994; *Chem. Abstr. 121*, 157280f.
58. I. A. Weinstock, R. H. Attala, R. S. Reiner, M. A. Moen, K. E. Hammel, C. J. Houtman, and C. L. Hill, *New J. Chem.*, *20*, 269 (1996).
59. H. Niwa, M. Oguri, and M. Watanabe, Japanese Patent, 05 221 645-7, 1993, *Chem. Abstr. 120*, 87788f, *120*, 87789g, *120*, 87790a.

60. L. I. Kuznetsova, E. N. Yurchenko, R. I. Maksimovskaya, and K. I. Matveev, *Sov. J. Coord. Chem.*, *2*, 67 (1975). L. I. Kuznetsova, E. N. Yurchenko, R. I. Maksimovskaya, N. I. Kirik, and K. I. Matveev, *Sov. J. Coord. Chem.*, *3*, 39 (1977).

61. M. T. Pope, S. E. O'Donnell, and R. A. Prados, *J. Chem. Soc. Chem. Commun.*, 22 (1975).

62. I. V. Kozhevnikov, S. M. Kulikov, V. E. Tarabanko, and K. I. Matveev, *Dokl. Akad. Nauk SSSR (Eng. Trans.)*, *240*, 497 (1978).

63. L. I. Kuznetsova, V. M. Berdnikov, and K. I. Matveev, *React. Kinet. Catal. Lett.*, *17*, 401 (1981).

64. L. I. Kuznetsova, R. I. Maksimovskaya, and K. I. Matveev, *Inorg. Chim. Acta*, *121*, 137 (1986).

65. R. Neumann and M. Levin, *J. Am. Chem. Soc.*, *114*, 7278 (1992).

66. C. L. Hill and D. C. Duncan, *J. Am. Chem. Soc.*, *119*, 243 (1997).

67. A. Hiskia and E. Papaconstantinou, *Inorg. Chem.*, *31*, 163 (1992).

68. L. I. Kuznetsova, Yu. V. Chernushova, and R. I. Maksimovskaya, *Inorg. Chim. Acta*, *167*, 223 (1990).

69. C. N. Satterfield, *Heterogeneous Catalysis in Practice*, McGraw-Hill, New York, 1980.

70. M. Misono and N. Nojiri, *Appl. Catal.*, *64*, 1 (1990).

71. M. Misono, *Proceedings of the Climax 4th International Conference, Chemistry and Uses of Molybdenum*, H. F. Barry and P. C. H. Mitchell, Eds., Climax Molybdenum Co., Ann Arbor, MI, 1982, p. 289.

72. S. Nakamura and H. Ichihashi, in *Proceedings of the 7th International Congress on Catalysis, Tokyo*, T. Seiyama and K. Tanabe, Eds., Elsevier, Amsterdam, The Netherlands, 1981, p. 755.

73. M. Misono, K. Sakata, Y. Yoneda, and W. Y. Lee, in *Proceedings of the 7th International Congress on Catalysis, Tokyo*, T. Seiyama and K. Tanabe, Eds., Elsevier, Amsterdam, The Netherlands, 1981, p. 1047. Y. Konishi, K. Sakata, M. Misono, and Y. Yoneda, *J. Catal.*, *77*, 169 (1982).

74. K. Eguchi, I. Aso, N. Yamazoe, and T. Seiyama, *Chem. Lett.*, 1345 (1979).

75. M. Ai, *Appl. Catal.*, *4*, 245 (1982).

76. K. Eguchi, T. Seiyama, N. Yamazoe, S. Katsuki, and H. Taketa, *J. Catal.*, *111*, 336 (1988).

77. M. Misono, T. Okuhara, and N. Mizuno, *Stud. Surf. Sci. Catal.*, *44*, 267 (1989).

78. M. Misono, *Catal. Rev.-Sci. Eng.*, *30*, 339 (1988).

79. N. Mizuno, T. Watanbe, H. Mori, and M. Misono, *J. Catal.*, *123*, 157 (1990).

80. N. Mizuno, K. Katamura, M. Misono, and Y. Yoneda, *J. Catal.*, *83*, 384 (1983). N. Mizuno and M. Misono, *J. Phys. Chem.*, *93*, 3334 (1989). N. Mizuno, T. Watanbe, and M. Misono, *J. Phys. Chem.*, *94*, 890 (1990).

81. H. Mori, N. Mizuno, and M. Misono, *J. Catal.*, *131*, 133 (1991).

82. K. Bruckman, J. Haber, E. Lalik, and E. M. Serwicka, *Catal. Lett.*, *1*, 35 (1988).

83. G. Y. Popova and T. V. Andrushkevich, *Kinet. Katal.*, *35*, 135 (1994). T. V. Andrushkevich, V. M. Bondareva, G. Y. Popova, and Y. D. Pankratiev, *React. Kinet. Catal. Lett.*, *52*, 73 (1994).

84. M. Ai, *Polyhedron*, *5*, 103 (1986).

85. M. Otake and T. Onoda, in *Proceedings of the 7th International Congress in Catalysis*, *Tokyo*, T.Seiyama and K. Tanabe, Eds., Elsevier, Amsterdam, The Netherlands, 1981, p. 780.

86. M. Akimoto, Y. Tsuchida, K. Sato, and E. Echigoya, *J. Catal.*, *72*, 83 (1981).

87. M. Akimoto, K. Shima, H. Ikeda, and E. Echigoya, *J. Catal.*, *86*, 173 (1984). M. Akimoto, H. Ikeda, A. Okabe, and E. Echigoya, *J. Catal.*, *89*, 196 (1984).

88. O. Watzenberger, G. Emig, and D. T. Lynch, *J. Catal.*, *124*, 247 (1990).

89. M. J. Bartoli, L. Monceaux, E. Bordes, G. Hecquet, and P. Courtine, *Stud. Surf. Sci. Catal.*, *72*, 81 (1992).

90. S. Albonetti, F. Cavani, M. Koutyrev, and F. Trifiro, *Stud. Surf. Sci. Catal.*, *82*, 471 (1993). S. Albonetti, F. Cavani, F. Trifiro, M. Gazzano, M. Koutyrev, F. C. Aissi, A. Aboukais, and M. Guelton, *J. Catal.*, *146*, 491 (1994).

91. Th. Ilkenhans, B. Herzog, Th. Braun, and R. Schögl, *J. Catal.*, *153*, 275 (1995).

92. L. Weismantel, J. Stoeckel, and G. Emig, *Appl. Catal. A*, *137*, 129 (1996).

93. E. Mueller-Erlwein and J. Guba, *Chem.-Ing.-Tech.*, *60*, 1072 (1988).

94. H. Krieger and L. S. Kirch, U.S. Patent, 4 260 822, 1981; *Chem. Abstr. 93*, 221287w. H. Imai, T. Yamaguchi, and M. Sugiyama, Japanense Patent, 63 145 249, 1988; *Chem. Abstr. 110*, 76262x. S. Yamamatsu and T. Yamaguchi, World Patent, 9014325, 1990; *Chem. Abstr. 114*, 123263t. K. Nagai, Y. Nagaoka, H. Sato, and M. Ohsu, Eur. Patent 418 657, 1990; *Chem. Abstr. 115*, 50483y. T. Ushikubo, Takashi, Japanese Patent, 06 172 250, 1994; *Chem.. Abstr. 121*, 301574z. E. Bielmeier, T. Haeberle, H.-J. Siegert, and W. Gruber, German Patent, 4 240 085, 1994; *Chem. Abstr. 122*, 56780j. A. Okusako and T. Ui, K. Nagai, Koichi, Japanese Patent, 08 012 606, 1994; *Chem. Abstr. 124*, 235544y. E. Bordes, L. Tessier, and M. Gubelmann-Bonneau, Eur. Patent, 683 153, 1995; *Chem. Abstr. 124*, 149211f.

95. N. Mizuno, M. Tateishi, and M. Iwamoto, *J. Chem. Soc. Chem. Commun.*, 1411 (1994).

96. F. Cavani, E. Etienne, M. Favaro, A. Galli, F. Trifiro, and G. Hecquet, *Catal. Lett.*, *32*, 215 (1995).

97. T. Bergier, K. Brueckman, and J. Haber, *Recl. Trav. Chim. Pays-Bas*, *113*, 475 (1994).

98. E. Gadot, C. Marchal, M. Fournier, A. Tezé, and G. Hervé, *Top. Mol. Organ. Eng.*, *10*, 315 (1994).

99. G. Centi, J. Lopez-Nieto, C. Iapalucci, K. Bruecjman, and E. M. Serwicka, *Appl. Catal.*, *46*, 197 (1989).

100. T. Jinbo, Y. Kogure, H. Io, T. Muraguchi, and O. Kinkai, Japanese Patent, 06 218 286, 1994; *Chem. Abstr. 122*, 134129v. W. Ueda and Y. Suzuki, *Chem. Lett.*, 541 (1995).

101. N. Mizuno, M. Tateishi, and M. Iwamoto, *Appl. Catal.*, *A*, *128*, L165 (1995).
102. A. Albonetti, F. Cavani, F. Trifiro, and M. Koutyrev, *Catal. Lett.*, *30*, 253 (1995).
103. J. B. Moffat, *Chem. Eng. Commun.*, *83*, 9 (1989).
104. S. Kasztelan and J. B. Moffat, *J. Catal.*, *106*, 512 (1987).
105. S. Ahmed and J. B. Moffat, *Catal. Lett.*, *1*, 141 (1988).
106. S. N. Shaikh, P. E. Ellis, and J. E. Lyons, US Patent 5 334 780, 1994; *Chem. Abstr. 121*, 255246u. P. E. Ellis and J. E. Lyons, US Patent 4 898 989, 1990; *Chem. Abstr. 114*, 23420y.
107. K. Bruckman, J. M. Tatibouet, M. Che, E. Serwicka, and J. Haber, *J. Catal.*, *139*, 455 (1993). M. A. Banares, H. Hu, and I. E. Wachs, *J. Catal.*, *155*, 249 (1995).
108. M. Ai, *J. Catal.*, *67*, 110 (1981). M. Ai, T. Tsai, and A. Ozaki, *Bull. Chem. Soc. Jpn.*, *53*, 2647 (1980).
109. I. Kawafune, *Chem. Lett.*, 1503 (1986). I. Kawafune, *Chem. Lett.*, 185 (1989). I. Kawafune and G. Matsubayashi, *Bull. Chem. Soc. Jpn.*, *68*, 838 (1995).
110. Y. Matoba, Y. Ishii, and M. Ogawa, *Synth. Commun.*, *14*, 865 (1984).
111. Y. Ishii, K. Yamawaki, T. Ura, H. Yamada, T. Yoshida, and M. Ogawa, *J. Org. Chem.*, *53*, 3587 (1988).
112. K. Yamawaki, T. Yoshida, H. Nishihara, Y. Ishii, and M. Ogawa, *Synth. Commun.*, *16*, 537 (1986).
113. F. P. Ballistreri, S. Failla, E. Spina, and G. A. Tamaselli, *J. Org. Chem.*, *54*, 947 (1989). Y. Ishii and Y. Sakata, *J. Org. Chem.*, *56*, 5545 (1990).
114. Y. Sakata, Y. Katayama, and Y. Ishii, *Chem. Lett.*, 671 (1992).
115. Y. Sakata and Y. Ishii, *J. Org. Chem.*, *56*, 6233 (1991). T. Iwahama, S. Sakaguchi, Y. Nishiyama, and Y. Ishii, *Tetrahedron Lett.*, *36*, 1523 (1995).
116. T. Oguchi, Y. Sakata, N. Takeuchi, K. Kaneda, Y. Ishii, and M. Ogawa, *Chem. Lett.*, 2053 (1989).
117. Y. Ishii, K. Yamawaki, T. Yoshida, and M. Ogawa, *J. Org. Chem.*, *53*, 5589 (1988).
118. S. Sakaue, Y. Sakata, Y. Nishiyama, and Y. Ishii, *Chem. Lett.*, 289 (1992).
119. Y. Ishii, H. Tanaka, and Y. Nishiyama, *Chem. Lett.*, 1 (1994).
120. M. Schwegler, M. Floor, and H. van Bekkum, *Tetrahedron Lett.*, *29*, 823 (1988).
121. K. Nomiya, H. Murasaki, and M. Miwa, *Polyhedron*, *5*, 1031 (1986). M. Ogawa, H. Tanaka, and Y. Ishii, *Sekiyu Gakkaishi*, *36*, 27 (1993).
122. H. Orita, M. Shimizu, T. Haykawa, and K. Takehira, *React. Kinet. Catal. Lett.*, *44*, 209 (1991).
123. M. Shimizu, K. Takehira, H. Orita, and T. Hayakawa, Japanese Patent, 02 019 336, 1990; *Chem. Abstr. 113*, 23372p.
124. L. A. Petrov, N. P. Lobanova, V. L. Volkov, G. S. Zakharova, I. P. Kolenko, and L. Yu. Buldakova, *Izv. Akad. Nauk SSSR, Ser. Khim.*, 1967 (1989).
125. M. Shimizu, H. Orita, T. Hayakawa, and K. Takehira, *Tetrahedron Lett.*, *30*, 471 (1989).

126. H. Furukawa, T. Nakamura, H. Inagaki, E. Nishikawa, C. Imai, and M. Misono, *Chem. Lett.*, 877 (1988). K. Y. Lee, K. Itoh, M. Hashimoto, N. Mizuno, and M. Misono, *Stud. Surf. Sci. Catal.*, *82*, 561 (1994).

127. N. Mizuno, S. Yokota, I. Miyazaki, and M. Misono, *Nippon Kagaku Kaishi*, 1066 (1991).

128. D. Attanasio, D. Orru', and L. Suber, *J. Mol. Catal.*, *57*, L1 (1989).

129. R. Neumann and M. de la Vega, *J. Mol. Catal.*, *84*, 93 (1993).

130. M. W. Droege and R. G. Finke, *J. Mol. Catal.*, 323 (1991).

131. L. Salles, C. Aubry, F. Robert, G. Chottard, R. Thouvenot, H. Ledon, and J.-M. Brégault, *New J. Chem.*, *17*, 367 (1993). C. Aubry, G. Chottard, N. Platzer, J.-M. Brégault, R. Thouvenot, F. Chauveau, C. Huet, and H. Ledon, *Inorg. Chem.*, *30*, 4409 (1991).

132. C. Venturello, E. Alneri, and M. Ricci, *J. Org. Chem.*, *48*, 3831 (1983). C. Venturello, R. D'Aloiso, J. C. Bart, and M. Ricci, *J. Mol. Catal.*, *32*, 107 (1985).

133. L. Salles, C. Aubry, R. Thouvenot, F. Robert, C. Dorémieux-Morin, G. Chottard, H. Ledon, Y. Jeannin, and J.-M. Brégault, *Inorg. Chem.*, *33*, 871 (1994).

134. A. C. Dengel, W. P. Griffith, and B. C. Parkin, *J. Chem. Soc. Dalton Trans.*, 2683 (1993). A. J. Bailey, W. P. Griffith, and B. C. Parkin, *J. Chem. Soc. Dalton Trans.*, 1833 (1995).

135. D. C. Duncan, R. C. Chambers, E. Hecht, and C. L. Hill, *J. Am. Chem. Soc.*, *117*, 681 (1995).

136. R. Neumann and A. M. Khenkin, *J. Org. Chem.*, *59*, 7577 (1994).

137. M. Shimizu, H. Orita, T. Hayakawa, Y. Watanabe, and K. Takehira, *Bull. Chem. Soc. Jpn.*, *64*, 2583 (1991).

138. M. Hamamoto, K. Nakayama, Y. Nishiyama, and Y. Ishii, *J. Org. Chem.*, *58*, 6421 (1993). S. Teshigahara and Y. Kano, Japanes Patent, 05 213 916, 1993; *Chem. Abstr. 120*, 217242m.

139. R. Neumann and A. M. Khenkin, *Inorg. Chem.*, accepted for publication.

140. J.-M. Brégault, B. El Ali, J. Mercier, J. Martin, and C. Martin, *C. R. Acad. Sci. II*, *309*, 459 (1989).

141. B. El Ali, J.-M. Brégault, J. Martin, and C. Martin, *New J. Chem.*, *13*, 173 (1989). B. El Ali, J.-M. Brégault, J. Mercier, J. Martin, C. Martin, and O. Convert, *J. Chem. Soc. Chem. Commun.*, 825 (1989). A. Atlamsani, M. Ziyad, and J.-M. Brégault, *J. Chim. Phys. Phys.-Chim. Biol.*, *92*, 1344 (1995).

142. T. Tatsumi, K. Yamamoto, H. Tajima, and H. Tominaga, *Chem. Lett.*, 815 (1992). T. Tatsumi and K. Yamamoto, *Trans. Mater. Res. Soc. Jpn.*, *15A*, 141 (1994).

143. L. Eberson, *J. Am. Chem. Soc.*, *105*, 3192 (1983). L. Eberson and L.-G. Wistrand, *Acta Chem. Scand.*, *B34*, 349 (1980). L. Eberson and L. Jönsson, *Acta Chem. Scand.*, *B40*, 79 (1986). U. Bietti, E. Baciocchi, and J. B. F. N. Engberts, *Chem. Commun.*, 1307 (1996).

144. D. E. Katsoulis and M. T. Pope, *J. Am. Chem. Soc.*, *106*, 2737 (1984).

145. D. E. Katsoulis and M. T. Pope, *J. Chem. Soc. Chem. Commun.*, 1186 (1986).

146. C. L. Hill and R. B. Brown, *J. Am. Chem. Soc.*, *108*, 536 (1986).

147. C. L. Hill, R. B. Brown, and R. F. Renneke, *Prep. Am. Chem. Soc. Div. Pet. Chem.*, *32*, 205 (1987). C. L. Hill, R. F. Renneke, M. K. Faraj, and R. B. Brown, in *The Role of Oxygen in Chemistry and Biochemistry*, W. Ando and Y. Moro-oko, Eds., Elsevier, Amsterdam, The Netherlands, 1988, p. 185.

148. D. Mansuy, J.-F. Bartoli, P. Battioni, D. K. Lyon, R. G. Finke, *J. Am. Chem. Soc.*, *113*, 7222 (1991).

149. D. K. Lyon, W. K. Miller, T. Novet, P. J. Domaille, E. Evitt, D. C. Johnson, and R. G. Finke, *J. Am. Chem. Soc.*, *113*, 7209 (1991).

150. J. F. Liu, Q. H. Yang, and M. X. Li, *Chin. Chem. Lett.*, *6*, 155 (1995). M. X. Li, J. F. Liu, Y. Wang, Q. H. Yang, and Y. G. Chen, *Chin. Chem. Lett.*, *6*, 153 (1995). J. F. Liu, X. P. Zhan, F. K. Wang, G. P. Li, and J. P. Wang, *Chem. Res. Chin. Univ.*, *11*, 1 (1995). J. F. Kiu, B. L. Zhou, F. Q. Wang, and Q. X. Fang, *Chem. Res. Chin. Univ.*, *9*, 357 (1993). J. F. Liu, L. Meng, W. Y. Zhao, J. C. Li, and B. L. Zhao, *Huaxue Xuebao*, *53*, 46 (1995).

151. A. M. Khenkin and C. L. Hill, *J. Am. Chem. Soc.*, *115*, 8178 (1993).

152. C. Rong and F. C. Anson, *Inorg. Chem.*, *33*, 1064 (1994).

153. C. L. Hill and X. Zhang, *Nature (London)*, *373*, 324 (1995).

154. C. L. Hill, M. S. Weeks, A. M. Khenkin, and Y. Hou, *Prep. Am. Chem. Soc. Div. Pet. Chem.*, 1093 (1992).

155. X. Zhang, K. Sasaki, and C. L. Hill, *J. Am. Chem. Soc.*, *118*, 4809 (1996).

156. R. Neumann and C. Abu-Gnim, *J. Chem. Soc. Chem. Commun.*, 1324 (1989).

157. R. Neumann and C. Abu-Gnim, *J. Am. Chem. Soc.*, *112*, 6025 (1990).

158. W. J. Randall, T. J. R. Weakley, and R. G. Finke, *Inorg. Chem.*, *32*, 1068 (1993).

159. C. Rong and M. T. Pope, *J. Am. Chem. Soc.*, *114*, 2932 (1992).

160. K. Filipek, *Inorg. Chim. Acta*, *231*, 237 (1995).

161. E. Steckhan and C. Kandzia, *Synlett*, 139 (1992).

162. M. Bressan, A. Morvillo, and G. Romanello, *J. Mol. Catal.*, *77*, 283 (1992).

163. L. I. Kuznetsova, V. A. Likholobov, and L. G. Detusheva, *Kinet. Catal.*, *34*, 914 (1993).

164. M. Faraj and C. L. Hill, *J. Chem. Soc. Chem. Commun.*, 1487 (1987).

165. R. Neumann and A. M. Khenkin, *Inorg. Chem.*, *34*, 5759 (1995).

166. N. I. Kuznetsova, L. G. Detusheva, L. I. Kuznetsova, M. A. Fedetov, and V. A. Likholobov, *Kinet. Catal.*, *33*, 415 (1992).

167. R. Shiozaki, H. Goto, and Y. Kera, *Bull. Chem. Soc. Jpn.*, *66*, 2790 (1993).

168. R. Shiozaki, H. Kominami, and Y. Kera, *Synth. Commun.*, , *26*, 1663 (1996).

169. R. Shiozaki, E. Nishio, M. Morimoto, H. Kominami, M. Maekawa, and Y. Kera, *Appl. Spectrosc.*, *50*, 541 (1996).

170. O. A. Kholdeeva, G. M. Maksimov, M. A. Fedetov, and V. A. Grigoriev, *React. Kinet. Catal. Lett.*, *53*, 331 (1994).

171. J. F. Liu, Q. H. Yang, S. D. Chen, and M. X. Li, *Chin. Chem. Lett.*, *6*, 159 (1995).

172. A. M. Khenkin and C. L. Hill, *Mendeleev Commun.*, 140 (1993).

173. R. Neumann and M. Gara, *J. Am. Chem. Soc.*, *116*, 5509 (1994).

174. R. Neumann and M. Gara, *J. Am. Chem. Soc.*, *117*, 5066 (1995).

175. R. Neumann and A. M. Khenkin, *J. Mol. Catal.*, *114*, 169 (1996).

176. R. Neumann and D. Juwiler, *Tetrahedron*, *47*, 8781 (1996). R. Neumann, A. M. Khenkin, D. Juwiler, H. Miller, and M. Gara, *J. Mol. Catal.*, *117*, 169 (1997).

177. R. Neumann and H. Miller, *J. Chem. Soc. Chem. Commun.*, 2277 (1995).

178. N. I. Kuznetsova and E. N. Yurchenko, *React. Kinet. Catal. Lett.*, *39*, 399 (1989). M. A. Fedtov, O. M. Il'inich, L. I. Kuznetsova, G. L. Semin, Y. S. Vetchinova, and K. I. Zamaraev, *Catal. Lett.*, *6*, 417 (1990). L. I. Kuznetsova, M. A. Fedtov, and E. N. Yurchenko, *React. Kinet. Catal. Lett.*, *41*, 330 (1990). E. N. Yurchenko, T. D. Gutsel, and L. I. Kuznetsova, *Sov. J. Coord. Chem.*, *18*, 939 (1992).

179. J. E. Lyons, P. E. Ellis, and V. A. Durante, *Stud. Surf. Sci. Catal.*, *67*, 99 (1991). P. E. Ellis and J. E. Lyons, US Patent, US 4.898,989, 1990.

180. N. Mizuno, T. Hirose, M. Tateishi, and M. Iwamoto, *J. Mol. Catal.*, *88*, L125 (1994). N. Mizuno, M. Tateishi, T. Hirose, and M. Iwamoto, *Chem. Lett.*, 2137 (1993).

181. D. Qin, G. Wang, and Y. Wu, *Stud. Surf. Sci. Catal.*, *82*, 603 (1994).

182. J. Mizuho, Y. Asahi, and Y. Sasaki, Japanese Patent, 08 053 391, 1996; *Chem. Abstr.* *124*, 342862j.

183. D. E. Katsoulis and M. T. Pope, *J. Chem. Soc. Dalton Trans.*, 1483 (1989).

184. N. Mizuno, T. Hirose, M. Tateishi, and M. Iwamoto, *Chem. Lett.*, 1839 (1993). N. Mizuno, M. Tateishi, T. Hirose, and M. Iwamoto, *Chem. Lett.*, 1985 (1993). N. Mizuno, T. Hirose, M. Tateishi, and M. Iwamoto, *Stud. Surf. Sci. Catal.*, *82*, 593 (1994).

185. R. Neumann and M. Dahan, *J. Chem. Soc. Chem. Commun.*, 171 (1995).

186. R. Neumann and M. Levin, *Stud. Surf. Sci. Catal.*, *66*, 121 (1991).

187. R. Neumann, A. M. Khenkin, and M. Dahan, *Angew. Chem. Int. Eng. Ed.*, *34*, 1587 (1995).

188. R. Neumann and M. Dahan, *Nature (London)*, *388*, 353 (1997).

189. T. Okuhara, T. Nishimura, and M. Misono, *Chem. Lett.*, 155 (1995).

190. A. Spojakina and N. G. Kostova, *Stud. Surf. Sci. Catal.*, *88*, 651 (1994). A. Spojakina, N. G. Kostova, and T. Kh. Shokhireva, *Kinet. Catal.*, *35*, 924 (1994).

191. F. M. Collins, A. R. Lucy, D. J. H. Smith, Eur. Patent Appl., 482 841, 1991; *Chem. Abstr.* *117*, 216065g.

192. B. Amadelli, G. Varani, A. Moldotti, and V. Carassiti, *J. Mol. Catal.*, *59*, L9 (1990).

193. Y. Izumi, Y. Satoh, H. Kondoh, and K. Urabe, *J. Mol. Catal.*, *72*, 37 (1992).

194. R. G. Finke, D. K. Lyon, K. Nomiya, S. Sur, and N. Mizuno, *Inorg. Chem.*, *29*, 1784 (1990). R. G. Finke, *Top. Mol. Organ. Eng.*, *10*, 267 (1994). M. Pohl, D. K. Lyon, N. Mizuno, K. Noyima, and R. G. Finke, *Inorg. Chem.*, *34*, 1413 (1995).

195. N. Mizuno, D. K. Lyon, and R. G. Finke, *J. Catal.*, *128*, 84 (1991).

Metal Phosphonate Chemistry

ABRAHAM CLEARFIELD

Texas A&M University
College Station, TX

CONTENTS

Progress in Inorganic Chemistry, Vol. 47, Edited by Kenneth D. Karlin.
ISBN 0-471-24039-7 © 1998 John Wiley & Sons, Inc.

I. INTRODUCTION

Metal phosphonate chemistry is of relatively recent origin. In the mid 1970s, Yamanaka and co-workers (1, 2) carried out a reaction between a layered zirconium phosphonate, γ-$Zr(PO_4)(H_2PO_4)$, and ethylene oxide to form a phosphate ester, Eq. 1.

$$Zr(PO_4)(H_2PO_4) + H_2C\overset{O}{\frown}CH_2 \longrightarrow Zr(PO_4)[O_2P(OCH_2CH_2OH)_2] \quad (1)$$

Shortly thereafter, Alberti et al. (3) prepared zirconium phosphonate and phosphate ester derivatives by direct reaction of Zr^{IV} with phosphonic acids, RPO_3H_2 (R = alkyl or aryl group) and phosphoric acid esters, H_2O_3POR. These compounds were also layered and it was surmised that the structures of these compounds were derivatives of the parent zirconium phosphate α-$Zr(HPO_4)_2 \cdot H_2O$. We shall provide more details later, but for now we wish to point out why the field grew so rapidly. Here we take a page from a recent review article (4) in which it was observed "that these compounds have two endearing properties that have stimulated extensive exploration of their chemistry." First, they are made at low temperatures in which the bonding within the reactant molecules is retained. Thus, the phosphonate or phosphate ester moiety may bear functional groups that do not disturb the bonding to the inorganic portion of the layers. The second attractive feature is their potential for supramolecular assembly. Once the basic structures of the phosphonates are known they may be manipulated to design structures for specific purposes. For example, one can intersperse purely inorganic groups such as HPO_4^{2-} and HPO_3^{2-} in between the organic groups on the layer. By combining such groups with bis(phosphonic acids), it is possible to create three-dimensional porous compounds. Alternatively, one may build up structures layer by layer to produce highly stable thin films with a variety of structure types.

The ability to design structures with specific properties is a very attractive feature for the materials chemist. Consequently, there has been an exponential growth of research in the field. Compounds exhibiting behavior as ion exchangers, sorbents, sensors, proton conductors, nonlinear optical materials, photochemically active materials, catalysts, and hosts for intercalation of a broad spectrum of guests have been prepared. We shall attempt in what follows to enlighten the reader in all aspects of this research.

II. ZIRCONIUM PHOSPHATES

Very often the layered phosphonate has a structure very close to that of a wholly inorganic phosphate. This finding is particularly true for the Group 4

(IVB) metal phosphonates. There are two main types of zirconium phosphates termed α-zirconium phosphate, $Zr(HPO_4)_2 \cdot H_2O$, and γ-zirconium phosphate, $Zr(PO_4)(H_2PO_4) \cdot 2H_2O$. For short, these compounds will be referred to as α-ZrP and γ-ZrP.

The α-ZrP compound was first prepared by refluxing zirconium phosphate gels in 10–12 *M* phosphoric acid (5). The gels had been examined fairly extensively in the 1950s as ion exchangers for possible use in nuclear reactor water purification (6, 7). Single crystals were formed hydrothermally in 10 *M* H_3PO_4 at 170°C. The structure of the layers (8) is illustrated in Fig. 1. The crystals are monoclinic, space group $P2_1/n$ with $a = 9.060(2)$, $b = 5.297(1)$, $c = 15.414(3)$, $\beta = 101.71(2)°$. The layers consist of metal atoms lying slightly above and below the mean plane and bridged by phosphate groups from above and below. Three oxygen atoms of each phosphate group are bonded to three different zirconium atoms, which form a distorted equilateral triangle. Each zirconium atom is thus octahedrally coordinated by six oxygen atoms from six different phosphate groups. The fourth oxygen bonds to a proton and points into

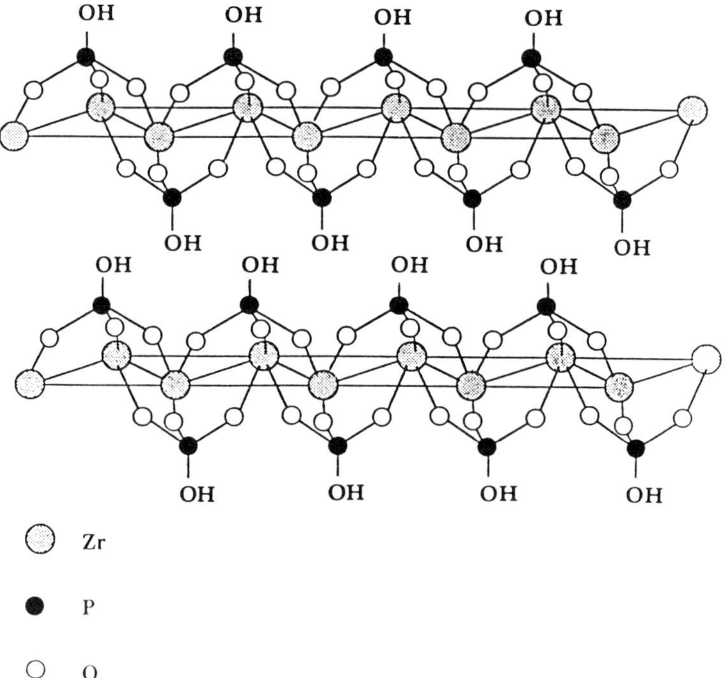

Figure 1. Schematic view of a portion of the layers in α-zirconium phosphate, $Zr(HPO_4)_2 \cdot H_2O$. The water molecule resides in the spaces created by three adjacent OH groups.

the interlayer space. The layers are staggered such that a phosphorus atom on one layer lines up with a zirconium atom in the adjacent layer. A water molecule resides in the interlayer space, hydrogen bonded to monohydrogen groups all from the same layer, alternately, at about $+\frac{1}{4}c$ and $-\frac{1}{4}c$. There are no hydrogen bonds between layers, so that only van der Waals forces hold the layers together. Similar metal phosphates are formed by Ti, Hf, Si, Ge, Sn, and Pb (9b) and they also behave as ion exchangers. One-half of the protons are exchanged at a pH 2–3 and the remainder at several pH units higher (9a,b, 10).

Although γ-ZrP was discovered many years ago (11), its structure was not uncovered until recently because of the difficulty in obtaining single crystals. It was known from ^{31}P NMR solid state spectra that the two phosphate groups in γ-ZrP are different and are assigned as an orthophosphate and a dihydrogen phosphate (12). The structure of the layers was determined for α-titanium phosphate (13) and later a complete structure including the water molecules was obtained for γ-ZrP (14) using powder data. A polyhedral representation is given in Fig. 2. The crystals are monoclinic, a = 5.3825(2), b = 6.6337(1), c = 12.4102(4) Å, β = 98.678(2)°, space group $P2_1$, and Z = 2. The structure indeed contains alternating orthophosphate groups and dihydrogen phosphate groups as predicted from the NMR spectra. The zirconium atom is octahedrally coordinated by four oxygen atoms from four different PO_4^{3-} and two from oxygens from two different $H_2PO_4^-$ groups. Two of the oxygens of the orthophosphate group bridge zirconium atoms in the a-axis direction and the other two do so in the b direction. These groups are alternately located 0.65 Å above and below the mean plane of the layers. The dihydrogen phosphate groups reside on the outer periphery of the layer utilizing two of its oxygen atoms to bridge across zirconium atoms in the a direction. The two hydroxyl groups extend into the interlamellar space. The location of the water molecules and the hydrogen-bonding scheme is shown in Fig. 2. Both protons are replaceable through ion exchange reactions (9b, 15).

The γ phase is usually prepared by refluxing a solution of a soluble zirconium salt in phosphoric acid containing high concentrations of ammonium or sodium phosphates (10). This reaction produces a half-exchanged phase $Zr(PO_4)(MHPO_4)\cdot H_2O$, where M = NH_4^+ or Na^+. The acid form is obtained by washing out the cations with dilute acid. The structure of the ammonium phase has recently been solved (16). The crystals were prepared by refluxing a mixture of $ZrOCl_2\cdot 8H_2O$ and $NH_4H_2PO_4$ in 2 M HCl containing 2 mol of H_3PO_4 per mole of Zr with addition of HF. The oven dried (60°C) solid was anhydrous. Its structure was solved from X-ray powder data. The NH_4^+ group does not bond to a single phosphate group but is equally shared by four such groups. The same is true for the fully ammonium ion exchanged α-phase, $Zr(NH_4PO_4)_2\cdot H_2O$ (17). In the α phase, each ammonium ion is within a hydrogen-bond distance of four PO^- groups and in turn each PO^- group is sur-

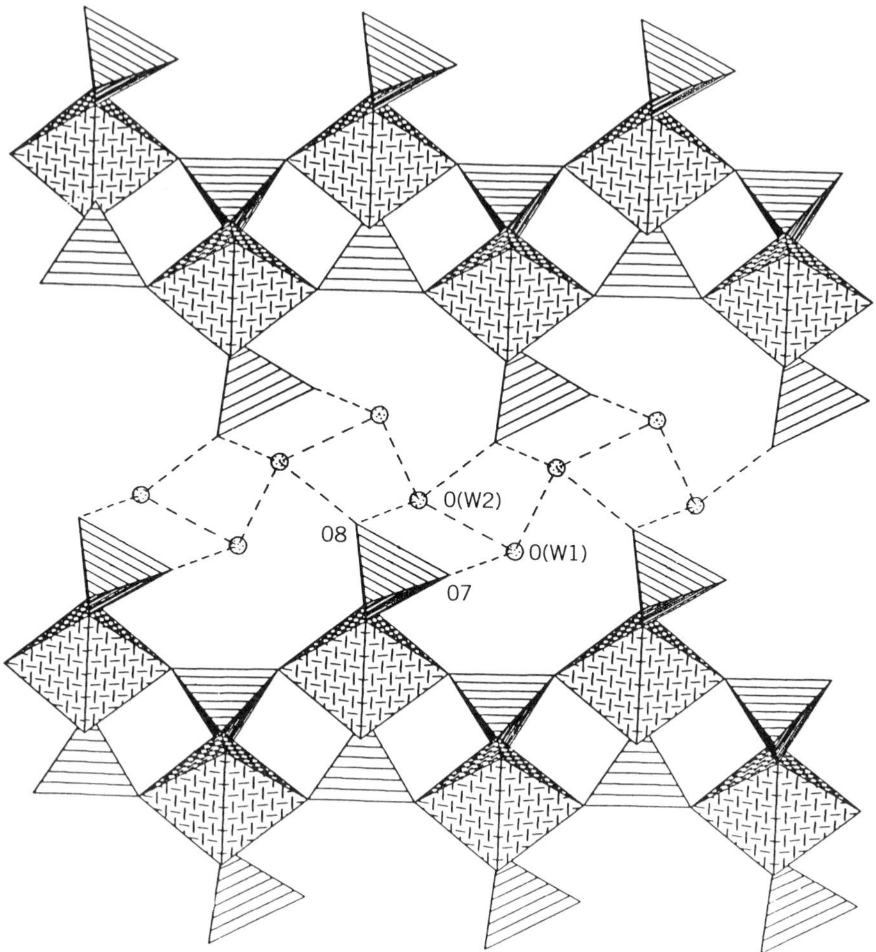

Figure 2. Polyhedral representation of the structure of γ-ZrP down the *a* axis. Hydrogen bonds involving the water molecules and hydroxyl groups are marked. The *c* axis is vertical and the *b* axis is horizontal.

rounded by four NH_4^+ ions. The ammonium ions are situated slightly above and below the plane at the midpoint between adjacent layers. In the γ phase, the ammonium ions are also disposed about the plane at $z = \frac{1}{2}$, halfway between two and -POOH groups in the same layer, forming hydrogen-bond chains in the *b*-axis direction. Since the phosphate groups from adjacent layers are shifted along *b*, the NH_4^+ groups also interact with the oxygen atoms in the adjacent layer (NH—O = 3.0 Å). The remaining hydroxyl proton may then be shared equally by the two phosphate groups in adjacent layers. This distance between

PO^- oxygen atoms in adjacent layers is 2.48 Å. The interlayer distance in the half-exchanged NH_4^+ phase is 11.25 Å as compared to 12.2 Å for the parent dihydrate γ-ZrP. Additionally, we see that the twofold axis of the dihydrogen phosphate groups is perpendicular to the layers in the ammonium salt, whereas in γ-ZrP (Fig. 2), these groups are tilted away from the perpendicular to achieve maximum hydrogen bonding.

III. LAYERED ZIRCONIUM PHOSPHONATES

A. Zirconium Alkyl and Arylphosphonates

One of the first phosphonates synthesized was the phenylphosphonate, $Zr(O_3Ph)_2$ (3). It was obtained in microcrystalline form with only a dozen or so reflections. Consequently, its structure was not known until recently. Alberti et al. (3) prepared a sample by a hydrothermal process in the presence of dilute HF. After about 30 days, a product was obtained for which 35–40 reflections were visible in the X-ray powder pattern. Surprisingly, this degree of crystallinity was sufficient for an "ab initio" structure solution. A projection of the structure as viewed down the *b* axis is shown in Fig. 3 (18).

The crystals belong to space group *C2/c* with $a = 9.0985(5)$, $b = 5.4154(3)$, $c = 30.235$ Å, and $\beta = 101.333(5)°$. The in-plane *a* and *b* dimensions are very close to those corresponding dimensions in α-ZrP. However, the phenyl groups are tilted relative to the layers by 30°. The tilt arises from the fact that in this space group the Zr atoms are constrained to lie in the plane, whereas in

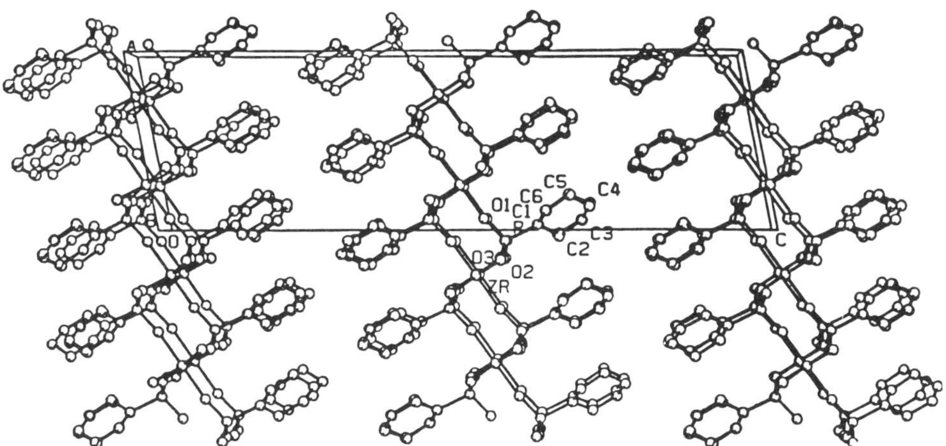

Figure 3. Projection of the structure of zirconium phenylphosphonate down the *b* axis.

α-ZrP, these atoms are slightly above and below a mean plane through the layer. The interlayer distance of this compound is 14.82 Å.

The phenylphosphonate may be considered as the prototype compound of the phosphonate derivatives of the Group (IVB) metals. All of the layered alkyl and aryl derivatives of this group prepared by direct precipitation have the α-layered structure. The in-layer dimensions of the parent α-ZrP are approximately 9.06 × 5.3 Å^2, or about 48 Å^2. Within this area are two phosphate groups bonded to the layers (8). Any alkyl or aryl group whose lateral area does not exceed 24 Å^2 can fit onto an α-ZrP type layer. However, as we shall see, bulkier groups may form new layer types or three-dimensional materials.

Whether a particular compound has the α-ZrP structure can be shown indirectly from a density measurement coupled with a knowledge of the basal spacing of the layers (19, 20). In Fig. 4, we have reproduced a plot of interlayer space as a function of the molecular weight of the compound divided by its density. The compounds include phosphates and phosphites of Group 4 (IVB) metals as well as phosphonates. Division of the abscissa value by the corresponding interlayer distance and twice Avagadro's number yields values close to 24 Å^2.

We stated previously that addition of a soluble zirconium salt (or for that matter, any of the group (IVB) metals) to dilute phosphoric acid yields an amorphous gel. The gels can be converted into products of increasing crystallinity by refluxing in successively stronger H_3PO_4 solutions (9a, 21). Another technique utilized to improve crystallinity is to reflux in the presence of HF (22). The HF forms the complex hexafluorozirconate anion that is stable at room temperature. However, heating to above 60°C releases Zr^{4+} as shown in Eq. 2.

$$ZrF_6^{2-} \longrightarrow Zr^{4+} + 6F^- \qquad (2)$$

The slowly released zirconium ion forms microcrystalline products by a homogeneous precipitation process.

Similar behavior is exhibited by the Group 4 (IVB) phosphonates. Rapid precipitation from dilute solutions produce amorphous materials with a fine particle size. Refluxing in excess acid improves the crystallinity somewhat but even in the presence of a 20-fold excess of HF, only imperfectly crystalline materials are obtained. Improved crystallinity may be obtained by use of hydrothermal methods, but the practitioner should be aware that this procedure may result in structural changes as will be demonstrated later. In general, the amorphous compounds are more reactive, less stable thermally to alkaline hydrolysis, and have higher surface areas. Dines et al. (19) used zirconium methylphosphonate as a prototype, and by different treatments, prepared a series of products with particle sizes ranging from less than 100 Å to 0.3 μm. The compound with the

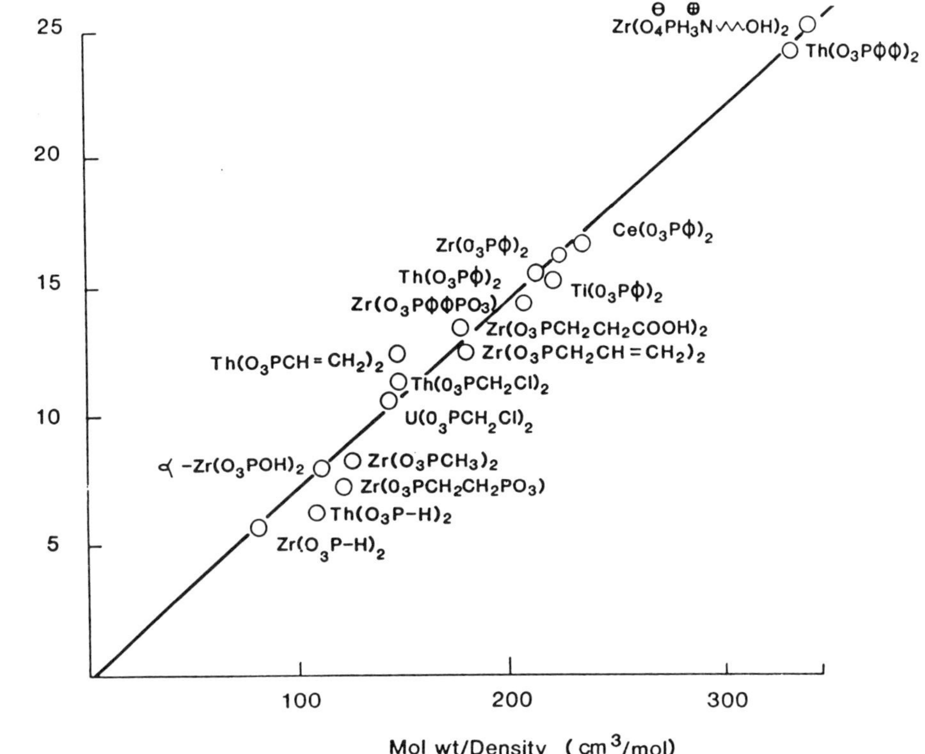

Figure 4. Molecular volume versus interlayer spacing, revealing that the various compounds share a common site area (~24 Å^2 from the slope) and are essentially isostructural. [Reprinted with permission from M. D. Dines, P. M. DiGiacomo, K. P. Callahan, P. C. Griffith, R. H. Cane, and R. E. Cooksey, in *Chemically Modified Surfaces in Catalysis and Electrocatalysis*, J. S. Miller, Ed., American Chemical Society Symposium Series, Vol. 192, Washington, DC, 1982, Chapter 12. Copyright © 1982 American Chemical Society.]

largest particle size was the most crystalline and yielded a surface area of about 16 $m^2\ g^{-1}$ while the sample with the smallest particle size and least crystalline X-ray diffraction (XRD) pattern possessed a surface area of about 600 $m^-\ g^{-1}$. Measurement of their densities showed that they differed by no more than 2%, indicating that differences in porosity were not a factor in the surface area determinations.

A systematic examination of *n*-alkyl derivatives of Group (IVB) metals has not been carried out. However, many different kinds of compounds have been prepared showing the versatility of the reaction. A partial list of compounds prepared by Dines and Di Giacomo (23) is given in Table I. A complementary list (Table II) has also been supplied by Alberti and Costantino (20). We note

TABLE I
Compounds Prepared by Dines and DiGiacomo (23)

Compound	d_{002} (Å)	SA (m^2 g^{-1})
$Zr(O_3PCH_2Cl)_2$	10.05	26
$Zr(O_3PH)_2$	5.61	108
$Zr(O_3Ph_5)_2$	15.0	180–220
$Zr(O_3POCH_2CH_2CN)_2$	13.2	
$Zr(O_3PCH_2CO_2H)_2$	11.1	
$Zr(O_3PCH_2CH_2CO_2H)_2$	12.8	
$Zr[O_3P(CH_2)_3CO_2H]_2$	14.8	
$Zr[O_3P(CH_2)_4CO_2H]_2$	16.8	
$Zr[O_3P(CH_2)_5CO_2H]_2$	19.0	
$Zr(O_3PCH_2CH{=}CH_2)_2$	12.6	90
$Zr(O_3PCH{=}CH_2)_2$	10.6	2
$Zr(O_3PCH_2S_3SEt)_2$	14.6	
$Zr(O_3AsPh)_2$	14.3	
$Zr[O_3P(CH_2)_2SH]_2$	15.5	44
$Zr[O_3P(CH_2)_{10}PO_3]_2$	17.3	
$Zr(O_3AsC_6H_4Cl)_2$	17.7	
$Zr(O_3PCH_2OPh_5)_2$	19.0	

that the compounds shown in these tables contain a variety of functional groups. In fact, this is a hallmark of metal phosphonate chemistry. By attaching functional groups to the organic moiety, different types of chemical properties may be imparted to the resultant compounds. For example, intercalation of amines is one such property. The carboxylic acid derivatives readily intercalate amines,

TABLE II
Formula, Density, and Interlayer Distance of Some Organic Derivatives of Zirconium Phosphate Having α- or γ-Layered Structure

Formula	Density (gm cm^{-3})	Interlayer Distance (Å)	Reference
α-$Zr(HOCH_2PO_3)_2 \cdot H_2O$	2.20	10.1	3
α-$Zr(HOCH_2PO_3)_2$	2.30	9.2	3
α-$Zr(HOOCCH_2PO_3)_2$	2.31	11.1	24
α-$Zr(PhPO_3)_2$	1.89	14.7	3
α-$Zr(EtOPO_3)_2$	1.92	11.7	3
α-$Zr(BuOPO_3)_2$	1.66	15.9	25
α-$Zr(C_{12}H_{25}OPO_3)_2$	1.28	32.7	25
α-$Zr(C_{14}H_{21}OPO_3)_2$[a]	1.34	20.7	25
γ-$Zr(HOCH_2CH_2OPO_3)_2 \cdot H_2O$		18.4	2
γ-$Zr\{[HOCH_2CH(Me)OPO_3](HPO_4)\}$		21.4	26
γ-$Zr(PhOPO_3)(HPO_4) + 2H_2O$		16.4	27

[a]Zirconium bis(monooctylphenylphosphate).

whereas the derivatives containing alcohol functions do so with difficulty. The α-$Zr(O_3PCH_2OH)_2$ compound requires prolonged contact with neat amines at 60° to affect intercalation, whereas $Zr(O_3PCH_2CO_2H)_2$ may be loaded with 0.1 *M* methanol solutions of *n*-alkylamine (20).

Often the most difficult part of the synthesis is the preparation of the phosphonic acid, since in many cases these acids with the required R- groups are not readily available. One of the most versatile methods for the formation of carbon phosphorus bonds involves the reaction of an ester of trivalent phosphorus with an alkylhalide, Eq. 3.

$$P(OR)_3 + R'X \longrightarrow (OR)_2\overset{\overset{\displaystyle O}{\|}}{P}R' + RX \qquad (3)$$

This reaction was discovered by Michaelis and Kaehne (28) and later was explored extensively by Arbuzov (29). Temperatures of 150–200°C are required depending on the R′ group. It is reasonably well established that the Michaelis–Arbuzov reaction occurs in two stages via an ionic phosphonium intermediate accompanied by valency expansion of the phosphorus, Eq. 4.

$$P(OR)_3 + R'X \longrightarrow [R'\text{—}P^+(OR)_3] \longrightarrow (OR)_2\overset{\overset{\displaystyle O}{\|}}{P}R' + RX \qquad (4)$$

The ester is then hydrolyzed in strong hydrochloric acid. In the event that R′ is an aryl group, then a catalyst, $NiCl_2$ or $PdCl_2$, is required to promote the reaction.

Because no systematic study of *n*-methylene derivatives has been carried out, it is necessary to infer the structures of the alkyl zirconium phosphonate derivatives by indirect procedures. We have seen that density measurements indeed indicate that the *n*-alkyl derivatives all have an α-ZrP-type layer. Furthermore, it is well known from studies of *n*-alkylamine intercalation that a double layer of amine forms with the general formula $Zr(HPO_4)_2 \cdot 2RNH_2$ (30–34). For most amines, the monohydrogen phosphate protons reside on the amino groups. A plot of interlayer distance versus the number of carbon atoms in the chain yields a straight line with a slope of 2–2.1 Å. If the *n*-alkyl chains were oriented in a direction perpendicular to the mean plane of the layer, then each addition of a carbon atom should increase the interlayer spacing by 1.26 Å. This value assumes an all-trans–trans conformation of the chains. The actual value observed confirms that double stack of chains is present with the chains inclined to the mean plane at about 55–60° (32).

A number of zirconium bis(carboxymethylenephosphonates) are listed in Table I. The change in interlayer spacing for each added carbon atom is very

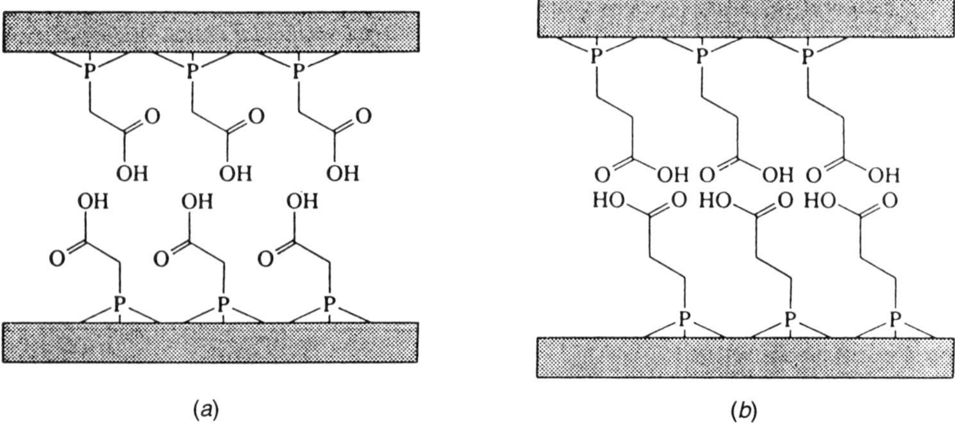

Figure 5. Possible arrangement of the alkyl carboxy chains in compounds with (*a*) even and (*b*) odd number of carbon atoms [207, Copyright © 1992 with kind permission from Elsevier Science S.A., P.O. Box 564, 1001, Lausanne, Switzerland.] Layered metal phosphonate and covalently pillared diphosphonates.

close to 2.0 Å. Thus, it is very likely that these alkyl chains assume the same conformation as the amine intercalates. Alberti (119) presented an idealized representation of zirconium bis(carboxymethanephosphonate), $Zr(O_3PCH_2CO_2H)_2$ reproduced in Fig. 5a. These modeled structures may be compared to that of zirconium chloromethylphosphonate, $Zr(O_3PCH_2Cl)_2$ (35). The original precipitate was treated hydrothermally in dilute HF at 130°C for 24 h to improve the crystallinity. The structure of the microcrystals was then solved ab initio from X-ray powder data. The crystals are monoclinic with a = 9.3402(7), b = 5.3926(3), c = 21.374 Å, β = 107.892(5)°, space group $P2_1/c$, and Z = 4. The inlayer dimensions are very similar to those of α-ZrP and the interlayer spacing is cos 17.892° × 21.374/2 = 10.17 Å. The structure as viewed down the b axis is shown in Fig. 6. An interesting feature of this compound is the disorder of the chlorine atoms. The disordered positions have nearly the same x and z parameters but differ in the y parameter by about $\frac{1}{2}b$. Apparently, the phosphonate ligand may bond with the chlorine atom pointing in the positive or negative direction but having done so a certain order must follow. The two chlorine atoms are independent, Cl1(A) being at approximately 0.2, 0, 0.2, and Cl2(A) at 0.6, 0.5, 0.2. However, the glide related Cl1(B) atom at 0.1, 0.5, 0.2 would only be 2.6 Å from, Cl1(B). Thus, if we have Cl1 in the A position, then Cl2 has to be in the B position and vice versa.

The powder X-ray method utilized to solve the structure of the chloromethylphosphonate and the zirconium phenylphosphonate has been a valuable tool that we have used to good effect as will be demonstrated later. However,

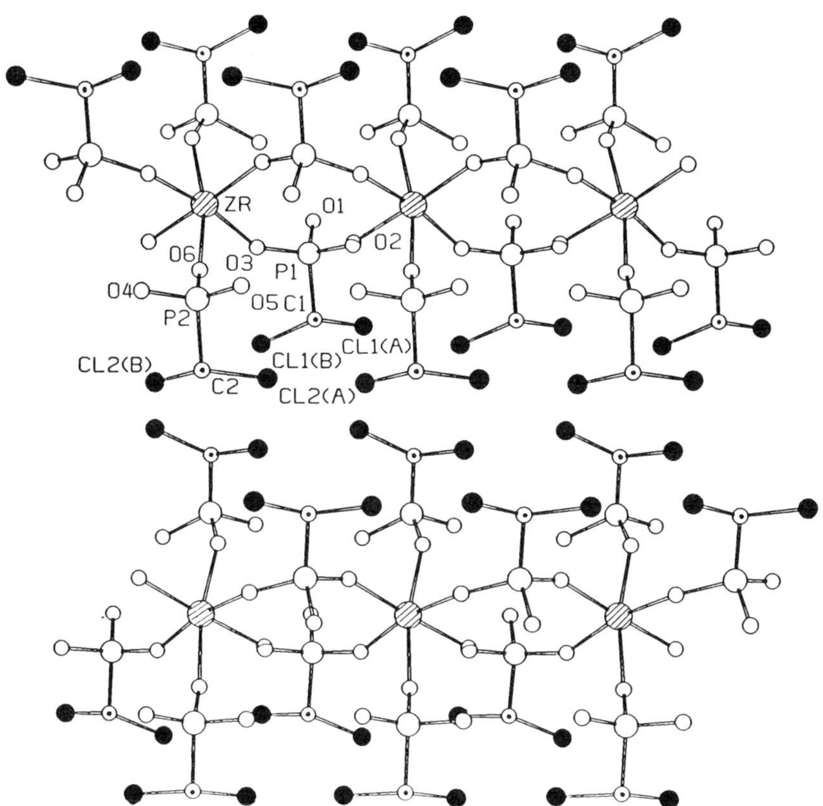

Figure 6. Representation of the structure of α-zirconium chloromethylphosphonate, $Zr(O_3PCH_2Cl)_2$.

before this technique was generally available, Thompson and co-workers (36) demonstrated the value of NMR in structure elucidation. They mounted the sample of zirconium bis(carboxymethanephosphonate) on thin microscope slides. The majority of the crystallites lay with the layers flat against the slide. Thus, they were able to orient the thin film either parallel or perpendicular to the direction of the magnetic field. From the changes in line shape as a function of the orientation of the sample they were able to deduce the orientation of the carbonyl group relative to the layer normal. The P—C bond was determined to lie nearly perpendicular to the ZrP layers and the carboxylic acid group has a P—C—C—O dihedral angle of 90 $\pm$ 15°. This structure rationalizes a lower reactivity for *n*- odd relative to *n*- even for $Zr[O_3P(CH_2)_nCO_2H]_2$ compounds. For *n* odd the metal phosphonate layers act to retard the reactions of the carboxyl group (37) relative to the *n*-even compounds in which the carbonyl groups would point away from the layers. This positioning of the carboxyl groups with n even allows for strong hydrogen bonding relative to n-odd carboxylates (38).

We note in passing that the structure deduced from NMR is at odds with the picture developed by Alberti and Costantino (20) illustrated in Fig. 5. In order to achieve an interlayer spacing of 11.1 Å, the carboxyl group is not parallel to the layers but is in a more perpendicular orientation.

The structural picture that emerges from the foregoing presentation is that the α-type layered compounds have an inorganic layer backbone resembling that of α-ZrP with pendant organic groups residing in the interlamellar space. There is no interpretation of the organic groups so that the forces holding the layers together are either van der Waals in nature or, if hydroxyl or carboxyl end groups are present, hydrogen bonds between layers may form. These forces can be very strong, preventing interpretation of the layers by groups of similar polarity. For example, stirring zirconium phenyphosphonate in benzene for several days results in no swelling or imbibement of the solvent. Similarly, n-alkyl derivatives form emulsions with hydrocarbon solvents only with difficulty. These compounds are also thermally stable. The phenylphosphonate decomposes in air in the neighborhood of 400°C and n-alkyl derivatives decompose about 80°C lower (36). This stability apparently stems from the tightness with which the organic groups are packed together, preventing oxygen from penetrating between the groups. Since the organic groups are anchored to the layers, they are not volatile. The ultimate in resistance to thermal decomposition was exhibited by $Zr(O_3PCH_2CH_2PO_3)$. A thermogravimetric experiment showed no loss in weight to 700°C (39).

The behavior of phosphonate derivatives containing functional groups differs strikingly from that described for the all hydrocarbon types. Those with acidic functional groups such as SO_3H, CO_2H, and even OH readily intercalate amines and participate in ion exchange reactions. In fact, the reactivity of metal phosphonates largely depends on the nature of the functional groups appended to the aryl or alkyl moiety. The reactivity of metal phosphonates can be further increased by spacing large groups by smaller ones to create void spaces thereby increasing the porosity of the compound. The synthesis and properties of the mixed derivatives are described in Section III.B.

B. Mixed-Component Phases

The ability to prepare layered compounds with two or more organic pendant groups has some appealing features. As was pointed out, if one group is small and the other large, then microporosity can be built into the structure. This feature allows access to reactive groups within the bulk of the solid. It also may allow incorporation of bulky groups, those having cross-sectional areas greater than 24 Å, onto the layers. Very special types of complexing agents or redox couples may be fixed to the layers. In fact, the ability to understand and control the processes that result in mixed-component phases will allow the synthetic chemist great latitude in preparing a variety of compounds with desirable features.

As was pointed out by Dines et al. (19) in preparing a multicomponent system, there is no guarantee that the ratio of components in the product will be the same as that in the reactant mix. In fact, this is rarely the case. If the relative rates of reaction with the metal of the two components is very different, this difference will be reflected in the amounts incorporated in the product. Incompatible groups may yield phase segregated products. In a model system in which one R group was H (O_3PH) and the other group was Ph in a 3 : 1 ratio and a 4 : 1 ratio of phosphorus to zirconium, a poorly crystalline product was isolated after 2-h reflux that exhibited a *d* spacing of 11.3 Å. The solid contained a 1 : 1 ratio of the two components. This *d* spacing was interpreted as being the approximate average of the interlayer spacings of zirconium phosphite, 5.6 Å, and zirconium phenylphosphonate, 14.8 or 10.2 Å. The difference in the observed and calculated values is the result of the low level of crystallinity of the product. The infrared (IR) spectrum contained a broad P—H band at 2455 cm^{-1} as well as bands at 695 and 750 cm^{-1}, characteristic of the phenyl ring. On redissolving the 11.3 Å product in dilute HF and recrystallizing the solid by partial evaporation, the new product gave an X-ray pattern with prominent reflections at 15, 10.6, and 5.6 Å. This pattern was interpreted on the basis that three phases were now present: $Zr(O_3PPh)_2$, 15 Å; $Zr(O_3PH)_2$, 5.6 Å; and the mixed-component phase $Zr(O_3PPh)(O_3PH)$. The reduction in interlayer spacing from 11.3 to 10.6 Å was attributed to the more ordered nature of the recrystallized product.

Alberti et al. (40) prepared mixed-component derivatives utilizing dilute HF solutions. Three types of derivatives were prepared and characterized:

1. $Zr(HPO_4)_x(R'PO_3)_{2-x}$ (R′ = —H; —Ph: $—CH_2CH_2CO_2H$)
2. $Zr(RPO_3)_x(R'PO_3)_{2-x}$ (R = (R = $—CH_2CH_2CO_2H$; R′ = $—CH_2OH$)
3. $Zr(HPO_3)_x(R'PO_3)_{2-x}$ (R′ = (R′ = —Ph; $—CH_2OH$; $—CH_2CO_2H$; $—CH_2CH_2CO_2H$)

In a typical reaction, the zirconyl chloride is dissolved in dilute HF (F/Zr = 10–20) and a mixture of the two phosphorus compounds added to it. No precipitation occurs because of the excess HF present. On heating to 60—100°C a gradual precipitation sets in. The total concentration of phosphorus in solution ranged from 1.5 to 6.8 *M* with high ratios of phosphorus to zirconium. It was found that the system is discontinuous, not all values of *x* are possible. Three types of products were obtained all with the α-type layered structure.

1. Pure $Zr(RPO_3)$, with small amounts of $R'PO_3$ groups present in solid solution.
2. Pure $Zr(R'PO_3)$ with small amounts of RPO_3 groups in solid solution.
3. The $Zr(RPO_3)_x(R'PO_3)_{2-x}$ complex, with RPO_3 or $R'PO_3$ groups as solid solutions with x between 0.8 and 1.3 depending on the type of R groups used.

Because the mixed derivatives were prepared by the HF method, fairly well developed X-ray diffraction powder patterns were obtained. Normally, the first reflection represents the interlayer spacing when a single ligand was used in the preparation. With the mixed-ligand products, this first reflection was always much larger than expected. For example, for the compound $Zr(HPO_4)_{1.15}(O_3PPh_5)_{0.85}\cdot 0.84H_2O$, the first reflection or (001) was recorded with a d spacing of 24.9 Å, relative intensity 40, and the (002) reflection at d = 12.4 Å was observed with intensity 100. These results were interpreted on the basis of an interstratified or staged structure in which one layer is predominantly populated by one of the ligands with a small amount of the second and a separate layer consists mainly of the second ligand with a small amount of the first randomly distributed in this layer. A schematic representation of such an alternating layer structure is shown in Fig. 7. For the compound in question, the phosphonate layer should have a 14.8-Å interlayer spacing while the phosphate layer should be somewhat greater than 7.6 Å, based on the assumption that all the water resides in this layer. In α-ZrP, the 7.6-Å spacing is on the basis of 1 mol of water, whereas in the interstratified compound a total of 0.84 mol was observed. This value would correspond to 1.68 mol of water residing in the phosphate layer. Thus, about 2 Å needs to be allowed for the extra water, so an (001) reflection at 24.4 Å is to be expected. This value is reasonably close to the observed one. A relatively complete interpretation of the powder pattern for a similarly staged compound, $Zr(HPO_4)_{0.85}(HPO_3)_{1.15}$, has been carried out (41), which confirmed its interstratified nature.

Additional studies on mixed-component system were conducted by Wang et al. (42). In agreement with Alberti's results, it was found that the product obtained by the slow thermal decomposition of ZrF_6^{2-} in a solution of phosphoric and phenylphosphonic acid yielded predominantly the interstratified compound with a 22.5 Å interlayer spacing. This compound contained just 0.5 mol of water and, hence, yielded the expected interlayer spacing. On exposure to a moist atmosphere for 17 h the interlayer spacing expanded to the value observed by Alberti, 24.5 Å. The composition of the solid was $Zr(HPO_4)_{1.2}(O_3PPh)_{0.8}\cdot 0.5H_2O$. However, as we shall see the product was not phase pure but contained small amounts of α-ZrP (d = 7.6 Å) and $Zr(O_3PPh)_2$ (d = 14.8 Å). A solid-state MAS NMR of the ^{31}P nucleus gave peaks for two resonances, −5.31 and −20.2 ppm. Pure zirconium phenylphosphonate yields a peak at

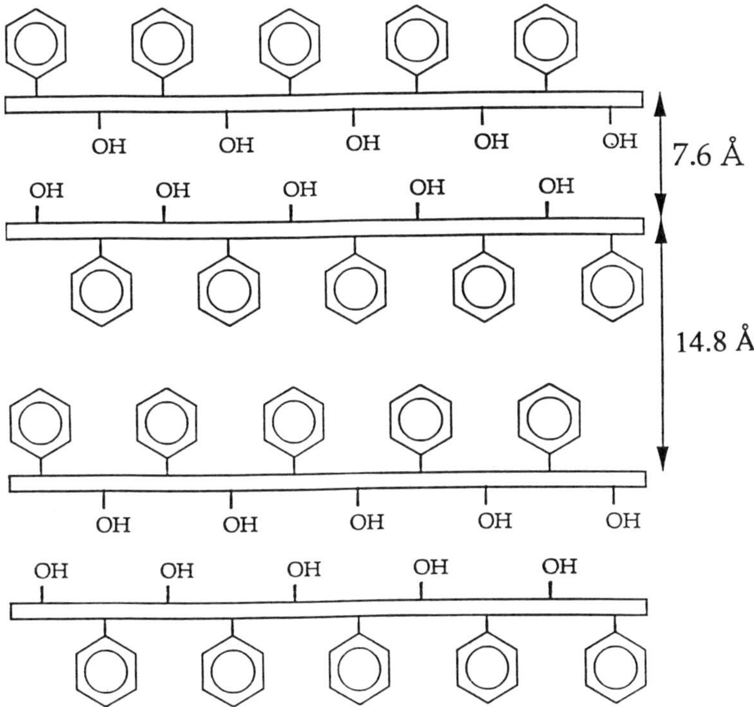

Figure 7. Schematic representation of the staged zirconium phosphate phenylphosphonate, $Zr(HPO_4)(O_3Ph)$.

−5.3 ppm, whereas pure α-ZrP gave a single peak at −18.7 ppm (12). Thus, the spectra reveal that the two phosphorus-containing groups retain almost the same environment in this staged compound as they possessed in their pure compounds. Based on this evidence and on what follows, the staged or interstratified compound was formulated as shown in Fig. 7.

Amines are readily intercalated into α-zirconium phosphate with an expansion of the interlayer distance. For example, intercalation of butylamine into α-ZrP causes the *d* spacing to increase from 7.6 to 18.6 Å. This large expansion is due to the formation of an amine bilayer inclined at an angle of approximately 55° relative to the layers (32). In order to show that a compound in which the two ligands are concentrated in separate layers had indeed been prepared, the solid was slurried in a solution of butylamine. The X-ray powder pattern for the intercalated phase revealed the presence of several phases. The major one is represented by reflections at 34.7, 17.3, and 11.4 Å, which are the first three (001) reflections representing the staged butylamine intercalate $Zr(C_4N_9NH_2 \cdot HPO_4)_{1.2}(O_3PPh)_{0.8}$. If a bilayer of amine forms in the phosphate layer, then this layer expands to 18.6 Å. The sum of this spacing and 14.8 Å for the phenylphosphonate layer yields an expected value of 33.4 Å. There are actually seven orders of this basal spacing present in the diffractogram. Consistent with this structure, assignment of the ^{31}P NMR spectra shows the resonance for the phosphate group is shifted downfield (from −20.2 to −18.7 ppm) and broadened upon intercalation of amine, whereas the phenylphosphonate resonance (−5.3 ppm) remains unaffected.

There is also a low-intensity reflection at 18.6 Å resulting from amine intercalation into α-ZrP. In fact, a small reflection was present at 7.6 Å (2θ = 11.65°) in the original X-ray pattern, which represents the interlayer separation for α-ZrP. In addition, the original diffraction patterns contained a small reflection from zirconium phenylphosphonate. While it is somewhat hazardous to infer too much from these results because of the presence of small amounts of these impurity phases, it is highly likely that the picture of the staged compound that emerges is not too different from that shown in Fig. 7. Judging from the X-ray pattern, the amount of α-ZrP and zirconium phenylphosphonate present in the mixture is about the same and in any case small. Therefore, given the formula, we can assume with some degree of certainty, that the ratio of phosphate to phenylphosphonate is close to that in the analyzed product. If any significant level of phenylphosphonate was present in the phosphate layer, it would have to expand beyond 7.6 Å. Thus, any excess of phosphate groups beyond 1 mol could easily be accommodated within the phenylphosphonate layer. This formulation is also in accord with the shift of the phosphate ^{31}P NMR peak downfield as is the general case of amine intercalation into α-ZrP (43).

The Alberti reaction involving mixtures of phosphorus acid and phenylphosphonic acid with ZrF_6^{2-} (F/Zr = 20) was repeated by Clearfield et al. (42a,b). The reaction was carried out over an extended period of time which reduced the volume in half. A mixture of products resulted that were identified based upon the *d* spacings in the XRD pattern. The most abundant phase (60%) had a *d* spacing of 25.5 Å as well as higher order reflections of this spacing. On dehydration, the *d* spacing of 21 Å (Alberti's result was 21.2 Å) represents the interstratified or staged compound similar to that obtained with the phosphate (Fig. 11). The expected value for such a layered derivative is 20.4 Å (14.8 Å + 5.6 Å). Interestingly, this phase did not rehydrate when exposed to moisture. This lack of rehydration is apparently due to the hydrophobic nature of the interlayer groups. The second most abundant phase (~25%) had a basal spacing of 10.3 Å and the third phase was identified with a 14.7-Å spacing. This latter phase is probably $Zr(O_3PPh)_{2-x}(HPO_3)_x$, where x has a value of 0.1 or less.

Repetition of this experiment, but with increased levels of HF, led to mixtures containing largely zirconium phosphite (5.6-Å interlayer spacing), some

$Zr(O_3PPh)_2$ containing small amounts of phosphite, and a more highly staged product with a d spacing of 30.4 Å. This phase could be a highly hydrated form containing a repeat distance involving two phosphite layers and one phenylphosphonate layer or a mixed-phosphonate–phosphite layer of lower d spacing and three phosphite layers.

In order to sort out the different types of species that are possible, a schematic representation of the possibilities was developed [42a,b] as shown in Fig. 8 in which ligands of dissimilar size occur in the same compound. Types A and B contain equal numbers of the two ligands. In a type A derivative, there is interpenetration of the large ligands that can only occur when the larger ligand abutts a small ligand in the opposite layer. This arrangement is required because the spacing along the layers if 5.3 Å (8) and the ligands are of the order of 3.3 Å in diameter so that no interpretation is possible. Actually, the layers in α-ZrP are staggered so that adjacent layers are shifted from each other by about 3 Å. Thus, it would appear that only a type C arrangement, where there is an excess of the small ligand, can the layers approach each other by interpenetration. The more common form is represented by the type B arrangement where the larger ligand determines the interlayer spacing.

In types E, F, and G, the ligands are arranged in an ordered fashion. We

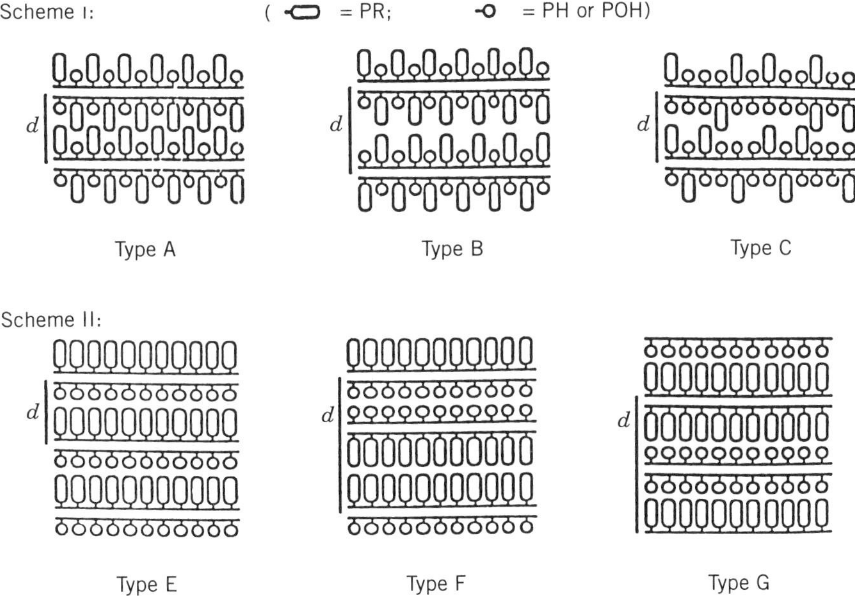

Figure 8. Cartoon representation of the arrangement of two ligands, one large and one small, on the α-ZrP-type layers.

have already demonstrated the existence of the interstratified arrangement of type F. Type G would also lead to the same repeat interlayer spacing. However, intercalation of amines or other molecules should occur equally in the type G layers and we have shown that this is not the case. Type E requires that the *d* spacing be one-half of the sum of the interlayer spacings of the two pure ligand compounds. The 10.3-Å phase observed in the phenylphosphonate phosphite compound may be of this type.

In order to prepare single phases rather than mixtures, a larger number of preparative procedures were tried (42b). The reactions can be classified into two types, thermal and hydrothermal. A series of thermal reactions using mixtures of phosphoric and phenylphosphoric acid were carried out at F/Zr = 22 in the same fashion as described earlier. The ratio $PhPO_3H_2:Zr$ was held constant at 0.82. If the ratio of H_3PO_4 to phenylphosphonic acid was greater than 50, the staged compound as depicted in Fig. 7, was obtained. Given the general formula $Zr(O_3PPh)_x(HPO_4)_{2-x} \cdot nH_2O$, x was in the range 0.7–0.95. This variation depended on the ratio of the two phosphorus compounds in the reactant mix and to some extent upon the ratio of phenylphosphonic acid to Zr. We note that in every case the phosphate–Zr ratio was greater than 1, strengthening the hypothesis that the phosphate layer contained only HPO_4^{2-} groups and the remaining phosphate groups were randomly dispersed within the phenylphosphonate layers. In these compounds, x was in the range 0.5–1.2. When the ratio $H_3PO_4/PhPO_3H_2$ fell below 50 but was greater than 30, a mixture of the staged product and a 14.8-Å phase was obtained. Finally, as the ratio fell below 30, only the 14.8-Å phase was obtained. In three such preparations, the formula was close to $Zr(O_3PPh)_{1.4}(HPO_4)_{0.6} \cdot 0.47H_2O$. In all of these preparations, the amount of phenylphosphonate incorporated was much greater than the amount of phosphate leading to a type B arrangement. The water must then fit into the spaces between the small phosphate groups surrounded by the layer phenyl groups.

A series of reactions using mixtures of phosphorous and phenylphosphonic acids were run but without the presence of HF. The ratio $PhPO_3H_2/Zr$ was kept at 1 and the amount of phosphorous acid varied in relation to the phosphoric acid from 1 : 1 mole ratio to 10 : 1. A precipitate formed as soon as the zirconium solution was added to the acid mixture. Figure 9(*a*) shows the X-ray pattern for the product formed when the acid ratios are 1. Refluxing the solid in its mother liquor improved the crystallinity somewhat, but the first reflection remained at 15.2 Å. Analysis gave $Zr(O_3PPh)_{1.04}(HPO_3)_{0.94} \cdot 2.7H_2O$. When the phosphorous/phosphonic acid ratio was 2 : 1, the interlayer spacing decreased to 11.8

Figure 9. X-ray difraction patterns of zirconium phenylphosphonate phosphites: (*a*) the 15-Å phase formed at 90–100°C reflux; (*b*) a similar phase to (*a*) but having an excess of phosphite over phenylphosphonate; and (*c*) Compound (B) heated in concentrated sulfuric acid at 65°C.

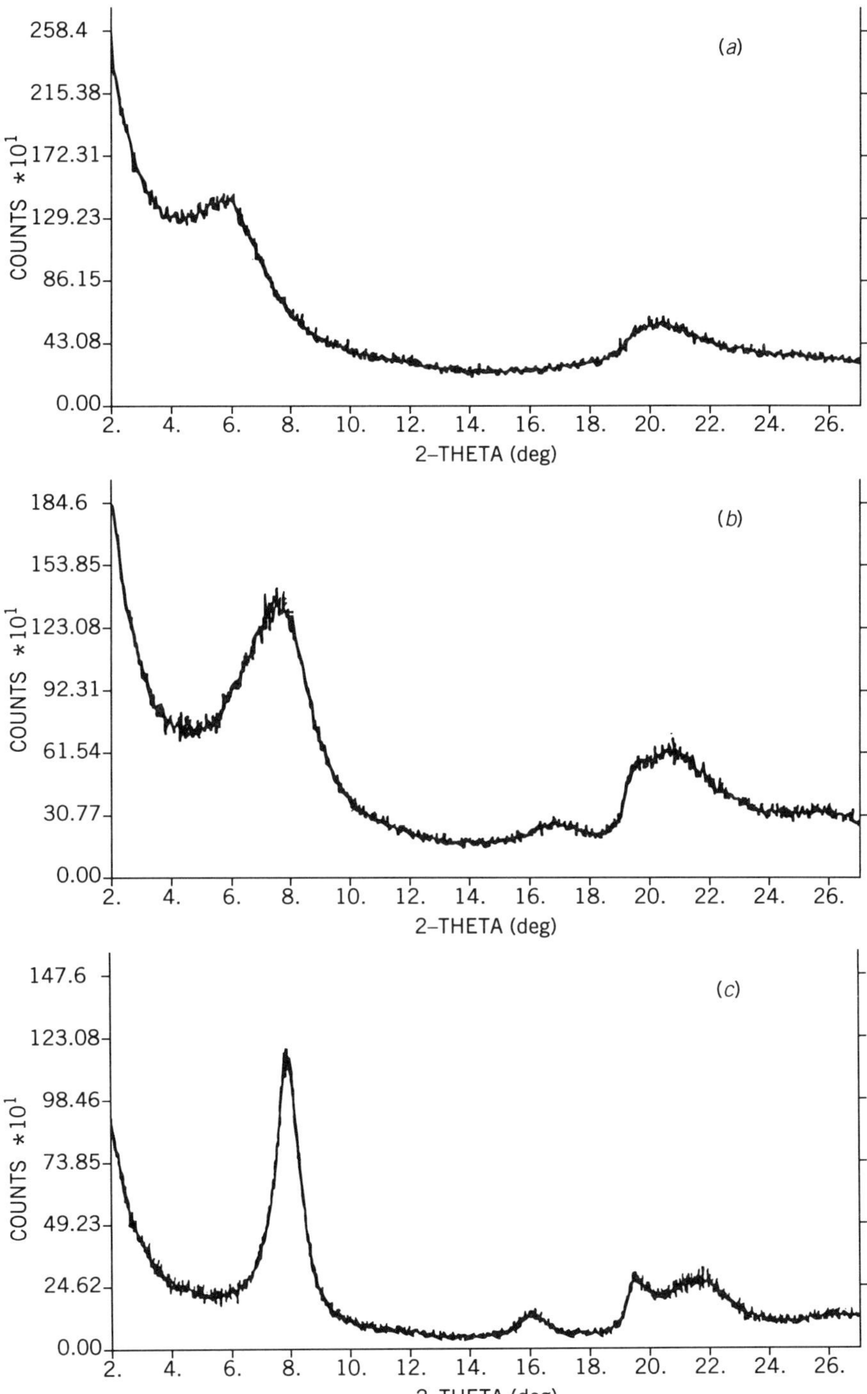
(a)
COUNTS *10^1
258.4
215.38
172.31
129.23
86.15
43.08
0.00
2. 4. 6. 8. 10. 12. 14. 16. 18. 20. 22. 24. 26.
2–THETA (deg)
(b)
COUNTS *10^1
184.6
153.85
123.08
92.31
61.54
30.77
0.00
2. 4. 6. 8. 10. 12. 14. 16. 18. 20. 22. 24. 26.
2–THETA (deg)
(c)
COUNTS *10^1
147.6
123.08
98.46
73.85
49.23
24.62
0.00
2. 4. 6. 8. 10. 12. 14. 16. 18. 20. 22. 24. 26.
2–THETA (deg)

[Fig. 9(*b*)] and at a ratio of 6:1 the value was 10.8 Å [Fig. 9(*c*)]. Furthermore, as the ratio of $H_3PO_3/PhPO_3H_2$ increased so did the crystallinity [Fig. 9(*c*)]. The foregoing results may be rationalized on the basis of the following reasoning. Without the presence of HF, the reaction is kinetically controlled with 100% precipitation of the zirconium. The product contained slightly more phenylphosphonate than phosphite requiring a type B arrangement. The layers are ill formed even after heating in the mother liquor for 2 days. Thus, the increased layer spacing of 15.2 versus 14.8 Å for well-crystallized products must arise from the disorder and consequent larger amount of water incorporation. Increasing the amount of H_3PO_3 relative to phenylphosphonic acid to 3:1 gives products in which the amount of phosphite incorporated onto the layers slightly exceeds the amount of phenylphosphonate incorporated. Some interpenetation may now occur as in the type A arrangement (d = 11.8–11.2 Å). Further increase in the amount of H_3PO_3 used leads to a type C structure with close approach of the layers. Finally, at the lowest basal spacings compositions approaching $Zr(O_3PPh)_{0.5}(HPO_3)_{1.5}$ are obtained. The foregoing explanation is further substantiated by the IR and NMR spectra. The former contain C—H stretching bands at 3078 and 3058 cm^{-1}, a P—H stretch at 2452 cm^{-1}, PO_3 vibrations at 1016 and 1150 cm^{-1}, and the phenyl out-of-plane vibrations at 692 and 750 cm^{-1} with a third peak at 726 cm^{-1}. These latter three bands serve as the signature bands for the singly monosubstituted phenyl group. The large increase in the intensity of the P—H stretching band as the ratio of H_3PO_3/PhPOH increases signifies the greater incorporation of phosphite groups into the product. As confirmation, the ^{31}P solid state NMR peak for O_3PH at −16.3 ppm increases relative to the phenylphosphonate peak at −5 ppm in harmony with the IR spectra. The increase in crystallinity may signal a more ordered structure of the type E.

Insufficient results have been obtained in the hydrothermal reactions to present more than general trends. In the temperature range of 100–150°C and $H_3PO_3/PhPO_3H_2$ = 1–3 the 15 Å phase was obtained. Increasing the temperature and the time tends to yield higher levels of the staged compound with $d \cong$ 22–25 Å. At higher levels of added H_3PO_3, a more highly staged product $d \cong 30$ Å was obtained. More work both with and without HF additions is required to determine with certainty the nature and control of these hydrothermal reactions.

C. Porous Pillared Bisphosphonates

An important advance in metal phosphonate chemistry was made by Dines et al. (19) using terminal bisphosphonic acids to cross-link the layers. Both alkyl and aryl bisphosphonates were prepared. The interlayer spacings for these compounds as reported by Dines are given in Table III. A schematic represen-

TABLE III
Interlayer Distances Found for Cross-Linking Groups

Compound	*d*-Spacing from XRD (Å)
$Zr(O_3PCH_2CH_2PO_3)$	7.8
$Zr(O_3PCH_2CH_2CH_2PO_3)$	No reflection observed
$Zr[O_3P(CH_2)_{10}PO_3]$	17.2
$Zr(O_3PC_6H_4PO_3)$	9.6
$Zr[O_3P(C_6H_4)_2PO_3]$	13.9
$Zr[O_3P(C_6H_4)_3PO_3]$	18.5
$Zr[O_3PCH_2(C_6H_4)CH_2PO_3]$	10.8

tation of the zirconium biphenylenebis(phosphonate) is shown in Fig. 10(*a*). This representation makes it clear that there should be no micropores present because of the closeness of the phenylene pillars. Therefore, to create the desired microporosity mixed-component derivatives in which a small group such as H, OH, Me was interposed between the larger pillaring groups were prepared as illustrated in Fig. 10(*b*). The lateral pore dimension is dictated by the density of pillars whereas the pore height is determined by the length of the pillar. Dines et al. (19) prepared a series of phosphite bisphosphonates based on the biphenyl pillar. They used surface area measurements based on the one point N_2 Brunauer–Emmet–Teller (BET) method as a means of verifying pillaring. The idea is that there would be very little change in the external surface area from sample to sample. Therefore, any increase in surface area would stem from the increased microporosity. Indeed the particle size was constant (0.06 μm in diameter) and the surface area increased as the amount of phosphite incorporated increased, as shown in Fig. 11. For example, at 50 mol percent the surface area is 481 $m^2\ g^{-1}$ versus 316 $m^2\ g^{-1}$ for the 100% product. To show that microporosity was indeed responsible for the increase in surface area, the sample containing 30 mol % diphenyl pillars was allowed to sorb nonane, whereupon its measured surface area dropped precipitously. The pore size was estimated to be 8.3 × 5 Å but no pore size measurements were actually carried out. One of the questions left unanswered by Dines et al. (19) was the high surface area of $Zr(O_3PPhPhPO_3)$, 316 $m^2\ g^{-1}$. This compound should have no micropores and the particle size is too large to create such a large surface area.

In an effort to provide information on several questions left unanswered by the Dines group, additional studies were undertaken in our laboratory (45). Two types of reactions were carried out, those for which no HF was included and reactions run in the presence of HF. The phosphonic acid employed was 4,4′-biphenylene bis(phosphonic acid), $H_2O_3PC_6H_4C_6H_4PO_3H_2$, either alone or mixed with phosphorous or phosphoric acid and in some cases ethylphosphate. In each case, the zirconium solution was added to the acid. In the reactions

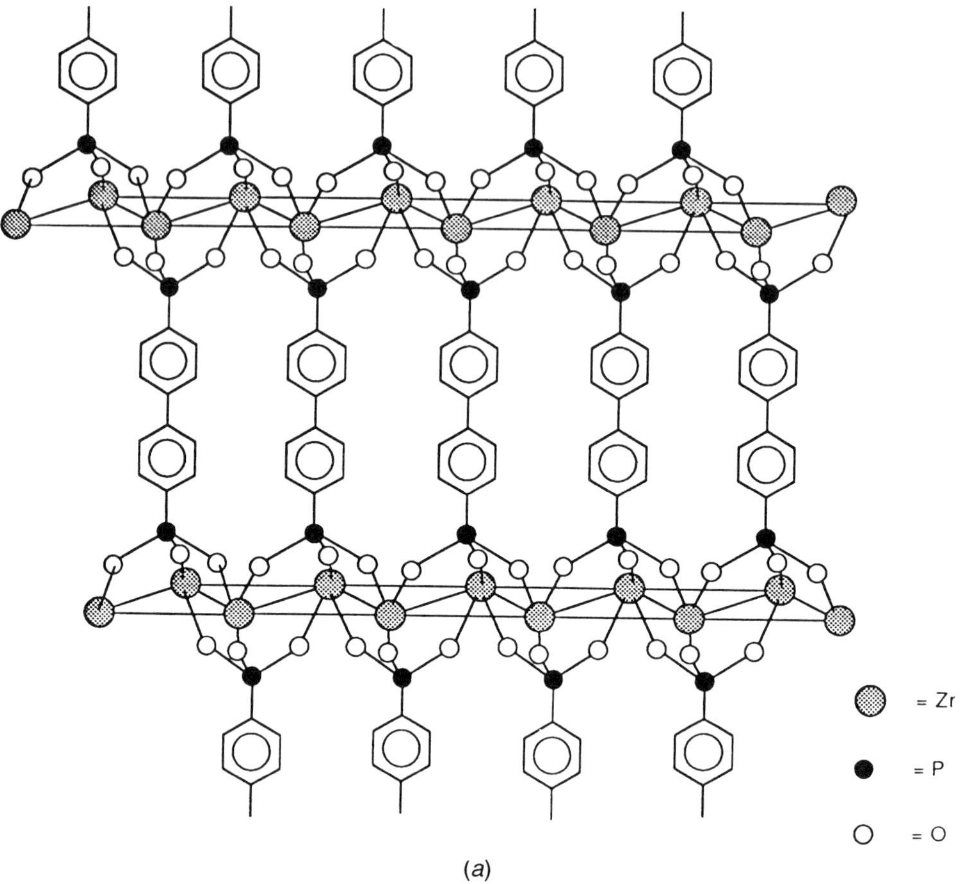

Figure 10. Idealized conception of a portion of the structure of $Zr(O_3PC_6H_4C_6H_4PO_3)$ indicating that no microporosity is to be expected (*a*) and idealized conception of a portion of the structure modified by inclusion of a small spacer moeity, HPO_4, producing micropores (*b*).

containing no HF, the precipitate that formed initially was refluxed for 18 h. The HF containing reactions were carried out at 60–70°C in an open container. These products were much more crystalline than those obtained by the reflux method (39, 44). The situation as far as surface area and pore size are concerned is complex and depends on the solvent used in the reaction as detailed below. The X-ray powder patterns of these preparations show a major peak at 13.8 Å with a second order of this reflection at 6.9 Å. These reflections are a clear indication of an interlayer spacing of this magnitude. This value is very close to what is expected from the length of the biphenyl group (7.05 Å) and the

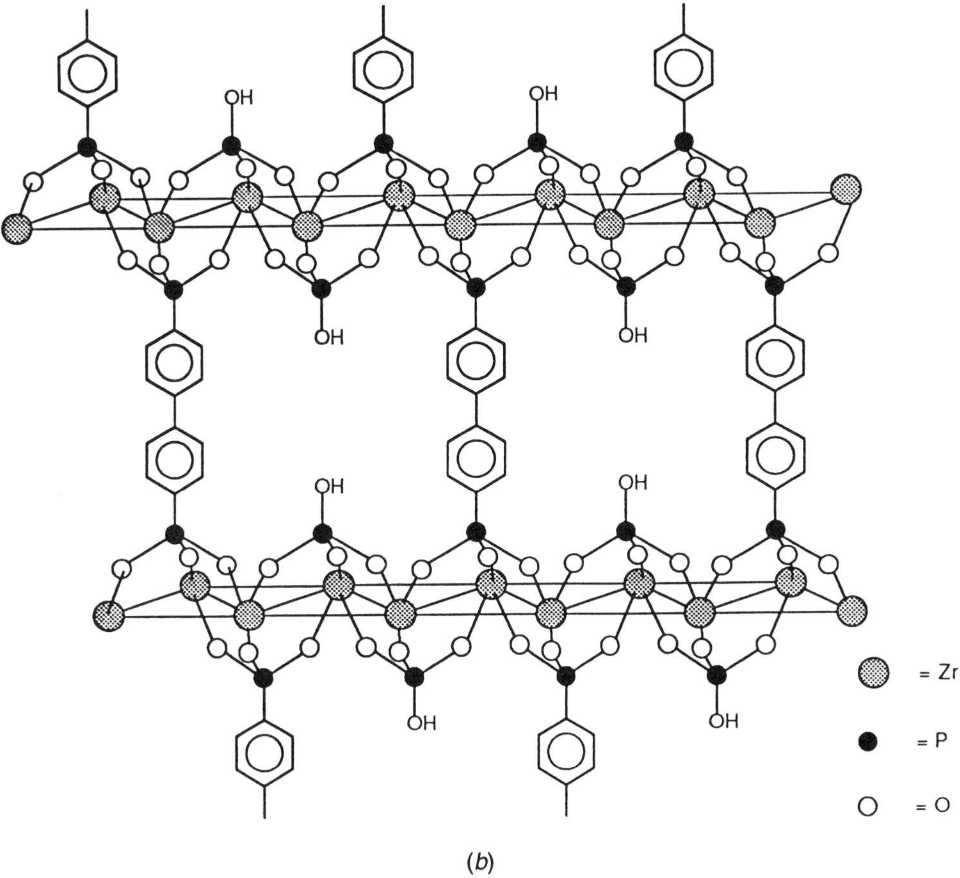

(*b*)

Figure 10. (*Continued.*)

thickness of the inorganic layer. The latter value is derived from the layer thickness of α-ZrP, 6.3 aN (8), augmented by the difference in bond lengths of P—C (1.80 Å) and the P—O bond length (1.55 Å). From this analysis, it would be expected that the surface area would depend on the particle size distribution. However, the surface areas were much larger than could be accounted for on the basis of particle size (45). The compounds prepared in dimethyl sulfoxide (DMSO) appeared to be grouped into two categories. Those that were prepared in DMSO diluted with benzene had relatively low surface areas (133–213 m^2 g^{-1}) and no micropores. The average pore radius was in the range 48–90 Å. In contrast, when the synthesis was carried out in concentrated DMSO solutions, the surface areas ranged from 270–397 m^2 g^{-1} and there was a significant

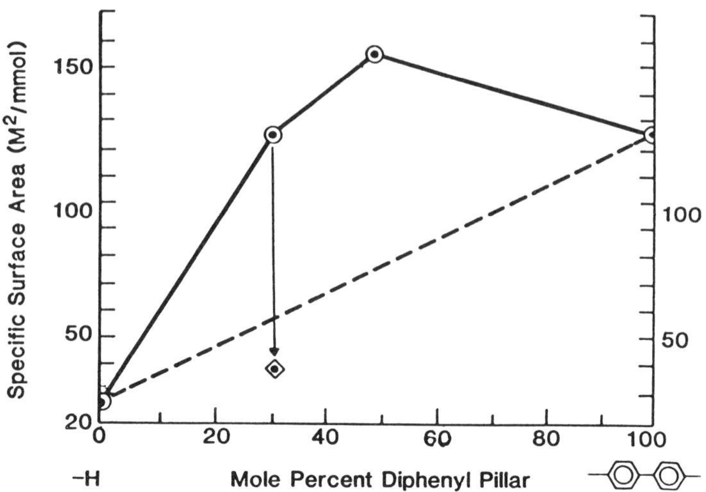

Figure 11. Variation of the specific surface area of mixed-component pillared compounds, whose end members are the hydride and the diphenyl bisphosphonate. The point denoted with a diamond corresponds to the area measured after nonane treatment to block the pore. [Reprinted with permission from M. D. Dines, P. M. DiGiacomo, K. P. Callahan, P. C. Griffith, R. H. Cane, and R. E. Cooksey, in *Chemically Modified Surfaces in Catalysis and Electrocatalysis*, J. S. Miller, Ed., American Chemical Society Symposium Series, Vol. 192, Washington, DC, 1982, Chapter 12. Copyright © 1982 American Chemical Society.]

contribution from micropores (25–50% of the surface area). The average pore size for these products was in the range 11–15 Å. Finally, the products obtained from the alcohol–water mixtures were high in surface area and had intermediate pore sizes (average 24–30 Å). These preparations were carried out in considerably more dilute solutions than any of the DMSO preparations because of the low solubility of the bisphosphonic acid in this medium. Yet the textural properties were intermediate to those prepared in diluted and concentrated DMSO solutions. Thus, it appears that the solvent plays a significant role in determining the textural properties of the end product. Typical BET N_2 sorption–desorption curves for two types of these compounds are shown in Fig. 12.

The ^{31}P MAS NMR spectrum of the zirconium biphenylenebis(phosphonate) contains a major peak at −4.8 ppm ascribed to the phosphorus atoms that belong to the phosphonate bonded to zirconium. A second weak resonance is also present at +4.2 ppm and is attributed to the unbonded end of a surface located phosphonic acid group. By way of comparison, the ^{31}P spectrum for zirconium phenylphosphonate, $Zr(O_3PPh)_2$, contains a single peak at −5.1 ppm. In this

Figure 12. The BET N_2 sorption–desorption curves of $Zr(O_3PC_{12}H_8PO_3)$, (*a*) Surface area (SA) 213 $m^2\ g^{-1}$. Micropores account for 10 $m^2\ g^{-1}$ and mesopores 203 $m^2\ g^{-1}$ of the SA; (*b*) Surface area 308 $m^2\ g^{-1}$, micropores account for 135 $m^2\ g^{-1}$, mesopores for 173 $m^2\ g^{-1}$. Volume is ×100.

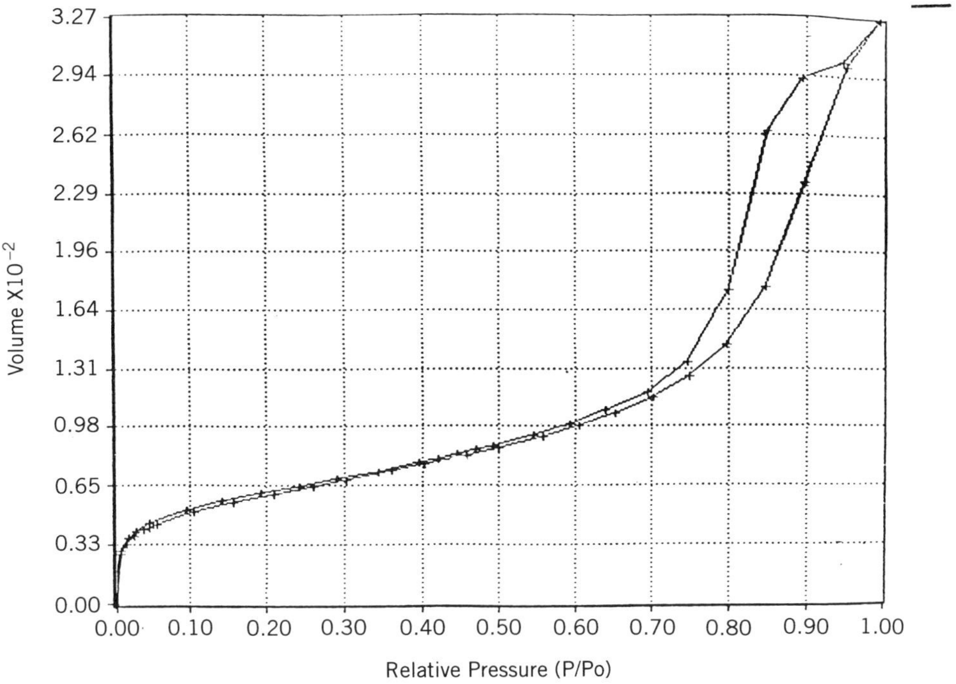

3.27
2.94
2.62
2.29
1.96
1.64
1.31
0.98
0.65
0.33
0.00
Volume X10^{-2}
0.00
0.10
0.20
0.30
0.40
0.50
0.60
0.70
0.80
0.90
1.00
Relative Pressure (P/Po)

(*a*)

1.23
1.10
0.98
0.86
0.74
0.61
0.49
0.37
0.25
0.12
0.00
Volume (X10^{-2})
0.00
0.10
0.20
0.30
0.40
0.50
0.60
0.70
0.80
0.90
1.00
Relative Pressure (P/Po)

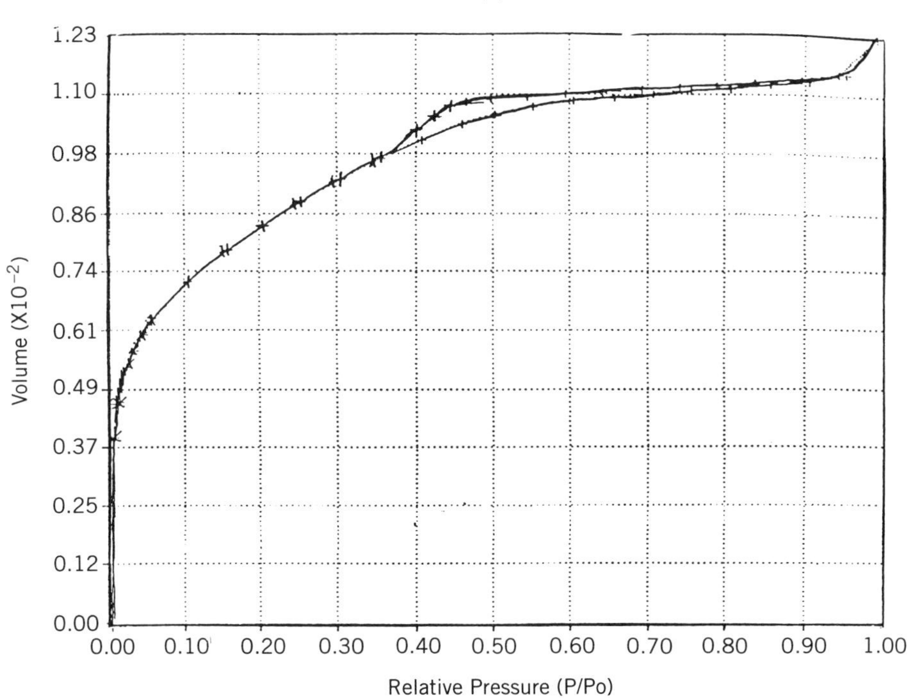

(*b*)

compound, all of the phosphonate groups are bonded to zirconium atoms giving rise to the single resonance. The free acid $PhPO_3H_2$ contains a single peak at +15 ppm and biphenylenebisphosphonic acid yields a single resonance at +15.3 ppm (relative to H_3PO_4 at 0 ppm). Thus, the observed weak resonance at +4.8 ppm is not due to free acid, but rather, we believe to diphosphonate groups, one end of which is bound to zirconium but the other is not.

In order to account for the high surface area and the pore structure, we proposed a model that is shown schematically in Fig. 13. The pores result from the coming together of layers of unequal size. Thus, one layer may terminate while another bonded to it keeps growing laterally and then binds to other layers leaving a gap. Variations on this procedure are readily envisioned to give pores of different sizes from large mesopores, which may be strictly between particles to micropores and pores of intermediate size. The solvents used and the concentration of reagent in these solvents mediate the type of porous structure obtained. By studying these effects, it may be possible to control the pore size to fit reactions of different sized molecules. In this connection, we have sulfonated these porous materials in fuming sulfuric acid. The resultant products contain about 3% sulfur, which amounts to sulfonation of about 25% of the rings. Thus, we interpret this result to mean that only those groups lining the pores are sulfonated. The pores are now highly hydrophilic and extreme measures are required to obtain dry products. These sulfonates may exhibit interesting catalytic behavior.

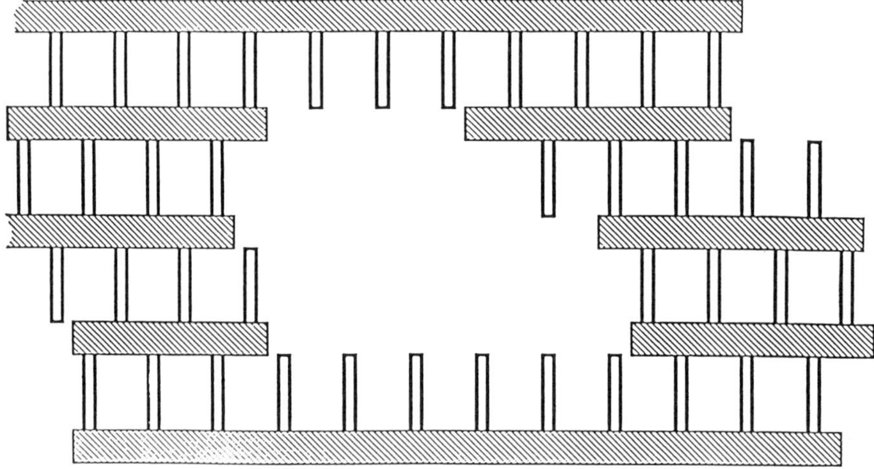

Figure 13. Conceptual model of pore formation in zirconium biphenyldiphosphonate. The double vertical lines represent biphenyl groups and the horizontal striped bars represent the inorganic PO_3ZrO_3P layers. The biphenyl groups protruding into the pore have free PO_3H_2 groups signified by connecting the two parallel vertical lines.

There are several major differences between the HF prepared compounds and the refluxed products. With the more crystalline HF treated products, the range of compositions obtained from mixed derivatives containing phosphate or phosphite was limited (44–46). The bisphosphonic acid was always present in greater amount for these reactions. With more amorphous materials (no HF present), the full range of compositions, that is, a complete solid solution, could be prepared. These compounds generally had higher surface areas than the HF treated products, but there was no regular pore structure. Pores ranging in radius from 5 to 100 Å were obtained. In contrast, the HF treated products had a bimodal pure distribution, one range of pore sizes being in the micropore region, and the other centered in the mesopore region (47).

The compounds described above contain a mixture of micropores and mesopores, the ratio depending on the synthetic method. However, Alberti et al. (48) used a clever tactic to obtain a pillared microporous α-type zirconium phosphonate with a very narrow pore size distribution. By placing methyl groups on the ring positions alpha to the phosphonate groups, they created pillars too large to fit adjacent to each other at a distance of 5.3 Å required by the α-type layer structure. For this purpose 3,3′,5,5′-tetramethylbiphenylene bis(phosphonic acid) (**1**) was synthesized

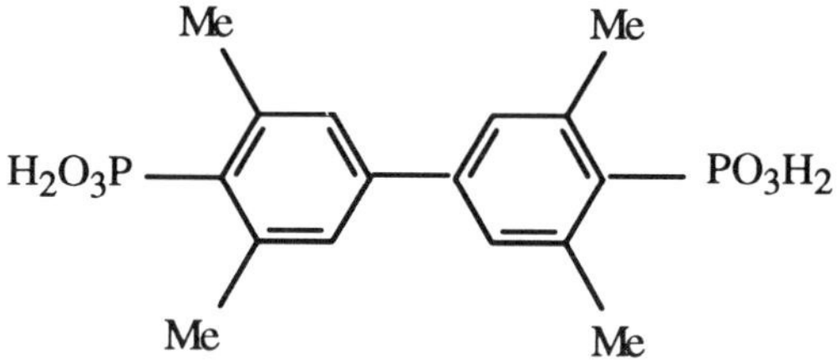

Me Me

H_2O_3P — — PO_3H_2

Me Me

1

A mixture of **1** and H_3PO_3 in DMSO was reacted with $ZrOCl_2$ in HF as before. The phosphite served to separate the pillars so that they were never adjacent to each other. Otherwise, the methyl groups would sterically hinder each other. It requires a 3 : 1 ratio of phosphite/phosphonate to avoid steric interactions. Thus, the composition of the product was $Zr(HPO_3)_{1.34}(O_3PRO_3P)_{0.33}$ and a computer drawn model of the structure is shown in Fig. 14. The interlayer spacing is 14.5 Å and a narrow pore size distribution centered at about 6 Å was observed. The specific surface area as determined from BET analysis was 375 $m^2\ g^{-1}$.

Derouane and Jullien-Lardot (49) prepared a series of n-alkyldiphosphonate pillared zirconium compounds in methanol containing HF. The reactants were treated hydrothermally for 3–175 h but no temperature of treatment was given. The products had surface areas in the range of 132–314 $m^2\ g^{-1}$ for the octyl-

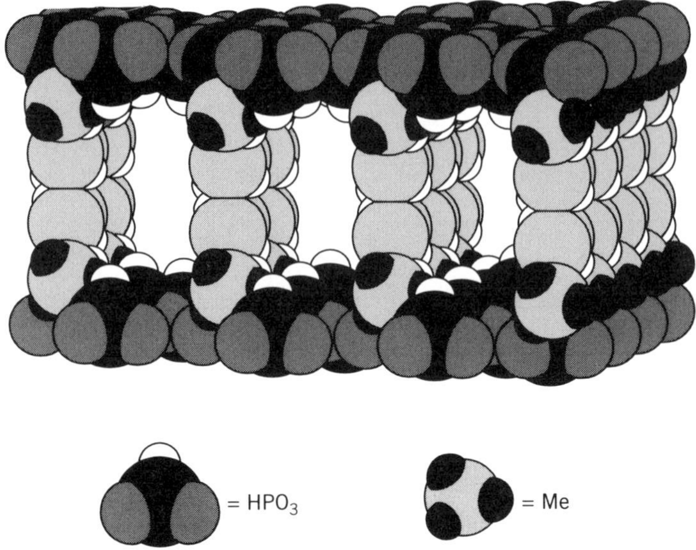

Figure 14. Computer-generated model of the pillared microporous compound with the idealized composition $Zr(HPO_4)_{1.34}(O_3PRPO_3)_{0.33}$, where R = Compound **1**. [Reproduced with permission from (48).]

diphosphonate and a constant mean pore diameter of 38 Å. One other point needs to be emphasized. Elemental analysis almost always showed a P/Zr greater than 2 (50) and this value was attributed to the fact that over the high observed surface area only one of the acid groups of the bis(phosphonic acid) was bonded to the layer and the other was free. Thus, the formula had to be modified to take this fact into account. For example, if P/Zr = 2.2, then the formula is $Zr(O_3PRPO_3)_{0.9}(O_3PRPO_3H_2)_{0.2} \cdot nH_2O$. This excess of phosphate and organic could also be observed by the increased weight loss measured by thermogravimeric analysis (TGA) (44, 46). Alberti et al. (50) arrived at the same formulation and carried out titrations of the free protons and found correspondence between the number of free groups determined by titration and elemental analysis. In order to account for the pore dimensions, surface area, and free groups, we proposed a structure as illustrated in Fig. 13. The pores are thought to arise from gaps in the layers that grow at different rates, depending on the availability of reactants. Alberti et al. (48) suggest the hypothesis that DMSO acts as a template around which the layers grow. This hypothesis seems reasonable from our results. Dilution of the DMSO changes the nature of the micelles accounting for the change in pore size. Also, the very large amount of solvent trapped in the pores also accords with this idea. Derouane and Jullien-Lardot (49) proposed a somewhat similar but more restricted structural model to account for the pores formed in alkyldiphosphonates and the increased layer spacing.

Additional information on porous materials will be presented in Section VI dealing with γ-zirconium phosphonate derivatives and in Section X.F.3 on naturally porous materials.

IV. FUNCTIONALIZED ZIRCONIUM PHOSPHONATES

A. Polyether Derivatives

1. Interaction of Zirconium Polyether Compounds with Alkali Thiocyanates

The first organic derivatives of zirconium were prepared by Yamanaka (1, 2) by treating γ-ZrP with ethylene oxide. The reaction was formulated as shown in Eq. 5 in the belief that γ-ZrP is a monohydrogen phosphate, $Zr(HPO_4)_2 \cdot 2H_2O$.

$$Zr(HPO_4)_2 + 2H_2C\overset{O}{\frown}CH_2 \longrightarrow Zr(O_3POCH_2CH_2OH)_2 \qquad (5)$$

The correct formulation based upon the now known structure of the γ-phase is represented by Eq. 1. This compound readily sorbed 1 mol of water to attain an interlayer spacing of 18.4 Å. A similar reaction was carried out with propylene oxide (51), but only one-half of the groups were esterified. This behavior results from the steric interference of the methyl groups on interaction with adjacent hydroxyl groups on the same phosphate group. Evidence for diester formation with ethylene oxide was deduced from NMR data (52).

Yamanaka (2) was unable to obtain similar derivatives of α-ZrP. This negative result stemmed from the fact that ethylene oxide could not diffuse between the more tightly spaced α-type layers. However, reaction should occur on the surface, and this was in fact found to be the case. Reactions were carried out with α-ZrP samples of differing crystallinities (53). The less crystalline the sample the higher the surface area (54). With the least crystalline samples (surface areas $\geq 90\ m^2\ g^{-1}$) swelling of the layers occurred and the level of ester formation approached 2 mol mol^{-1} of α-ZrP. Crystalline samples with surface areas below 26 $m^2\ g^{-1}$ did not swell and therefore were esterified only on the outer surfaces, retaining the α-ZrP bulk structure. Extended reactions produced polymers as large as hexamers on the surface. With samples of intermediate crystallinities, having surface areas of 26–90 $m^2\ g^{-1}$, swelling of the outer layers to 13 Å was observed as polyether chains accumulated on these layers.

Apparently, the increased hydrophilic character of the outer layer as esterification proceeds allows swelling to occur. This swelling in turn results in the next to the surface layer becoming esterified. The process continues to the extent that, sufficient layers are converted to be seen by X-ray diffraction, while the core of the particle retains the α-ZrP structure.

Polyether derivatives of α-ZrP could be prepared by a direct method. Polyethylene oxides (PEO) of different degrees of polymerization were converted into PEO phosphates by reaction with $POCl_2$ followed by a hydrolysis in aqueous acidic solutions (53). The phosphates were then isolated as their barium salts (Eq. 6).

$$\begin{array}{ccc} H(OCH_2CH_2)_nOH + POCl_3 \longrightarrow & H(OCH_2CH_2)_nOPOCl_2 & \\ & \big\downarrow H_2O \mid H^+ & \\ H(OCH_2CH_2)_nOPO_3Ba \xleftarrow{Ba^{2+}} & H(OCH_2CH_2)_nOPO_3H_2 & \end{array} \qquad (6)$$

Prior to reaction with $ZrOCl_2$ the barium salts were converted to the soluble acid state by treatment with dilute sulfuric acid. Reaction of the PEO phosphate with a soluble zirconium salt yielded the desired product. Reactions conducted at room temperature yielded stoichiometric products of composition $Zr[O_3PO(CH_2CH_2O)_nH]\cdot xH_2O$. Reactions carried out at reflux temperature yielded products that had undergone a certain amount of ester cleavage by the excess acid present in the solution. These products were of the type $Zr[O_3PO(CH_2CH_2O)_aH]_x[[HPO_4]_{2-x}\cdot nH_2O$. Because of the hydrophilic nature of the PEO chains the layers imbibe water with consequent increase of the interlayer spacings as shown in Table IV by the increase in interlayer spacing when wet.

The PEO products, prepared by the direct method, have the α-ZrP-type layers, which were shown by comparison with products of similar chain length that had been synthesized from the reaction of ethylene oxide with both α- and γ-ZrP. A plot of the basal spacings of the zirconium polyethylene oxide phosphates in the dehydrated condition is shown in Fig. 15 (55).

The fully extended trans–trans conformation of the polyether chains would require an increase of 3.5 Å per monomer unit or a 7-Å increase in interlayer spacing for every integral increase in the value of n because of the bilayer arrangement. We observe less than one-half of this value. Polyethylene oxide has a helical structure in the solid state (56), the polymer chain having seven monomer units in two turns of the helix with a repeat distance of 19.3 Å. Such a helix would not fit within the α-ZrP layers and still occupy 24 $Å^2$ of layer surface (the area required for each phosphate group). One possible arrangement would require that the chains be inclined at an angle to the layers and overlap. Another, perhaps more satisfactory, conformation is the cis–cis orientation of

TABLE IV
Polyethylene Glycol Esters of Zirconium Phosphate $Zr[O_3PO(CH_2CH_2O)_nR]_x [HPO_4]_{2-x}$

		(EA)[a]		(x)		Intelayer Spacing (Å)		
n	R	%C	%H	From EA	From TGA	Anhydrous	Wet	TGA[b] *T* (°C)
1	Me	17.91	3.48	1.99	2.00	14.9	15.3	200
2	H	20.87	3.47	2.00	2.01	14.9	17.6	197
3	H				2.00	17.8	21.3	190
3	Et	21.20	4.40	1.33	1.30	20.8	25.3	160
4	H	22.87	4.30	1.52	1.58	18.8		160
4	H	30.33	5.05	2.01	2.01	20.8	28.5	160
9	H			1.13		25.3	148	148
9	H	40.50	6.75	2.02	2.01	39.3	>44	147
12	Me				1.00	39.3	>44	142
13	H	42.25	7.20	1.98	1.98	>44	>44	140
22	H				1.97	>44	>44	135

[a]Elemental analysis = EA.
[b]Temperature at which decomposition begins.

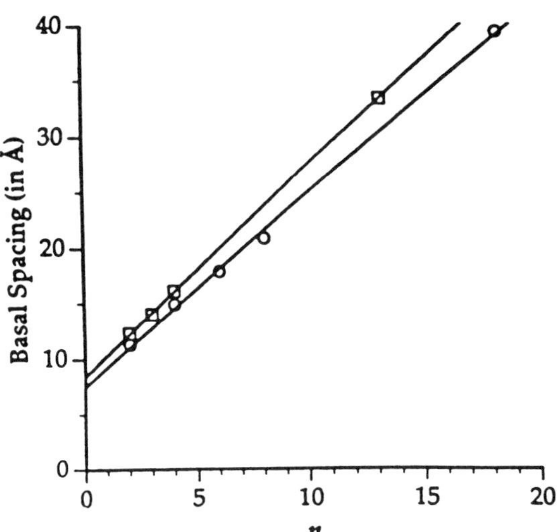

Figure 15. Plot of basal spacing of zirconium polyethylene oxide phosphates versus *n*, the degree of polymerization of the PEO: unbridged monophosphates; ○; phosphate bridged structures, □.

the chains. In this arrangement, each monomer is 2.5 Å from oxygen to oxygen. The basal spacing data is plotted in Fig. 15. The slope of the line in which n is plotted against the interlayer spacing of the dry zirconium polyether ester is 1.77 Å per monomer unit. If we assume the all-cis conformation, the chains would then have to be inclined at 45° to the layers. Extrapolation of the line $n = 0$ yields a value of 7.6 Å, which is just the interlayer spacing of the layers is α-ZrP. Therefore, we assume that the chains do not overlap and the cis conformation is the preferred one. This angle of tilt is considerably greater than the one observed for n-alkylamine intercalates of α-ZrP, about 35° from the perpendicular to the layers (31).

Diphosphates of the types $H_2O_3PO(CH_2CH_2O)_nPO_3H_2$ were prepared and reacted with Zr^{IV} to produce pillared or cross-linked layers. The diesters were synthesized in essentially the same way as the monoesters but with the use of excess $POCl_3$. Unlike the monophosphate derivatives of Zr, the cross-linked products precipitate readily from the reaction solution in high yields and when dried, yield free-flowing powders. Table V summarizes the data for this class compounds. Since the chains cannot overlap in this case, we consider only the cis conformation as reasonable. This information is summarized in the plot of interlayer spacing as a function of chain length in Fig. 15. The slope of the line is 1.92 Å per monomer requiring an angle of tilt of 39.5° from the perpendicular to the planes. Extrapolation of the line to $n = 0$ yields a value of 8.2 Å. This value is considerably larger than the 7.6-Å basal spacing of α-ZrP. Furthermore, with these cross-linked layers there are no O—H end groups to consider, so the layer thickness without the hydrogens in α-ZrP is 6.6 Å. To account for the 1.6-Å increase we assume that the layers are oriented such that a phosphate group in one layer is lined up so that it is above and below phosphate groups in adjacent layers. In α-ZrP, the sequence in adjacent layers is P over Zr over P, which allows for close packing of the layers.

TABLE V

Polyethylene Glycol Esters of Zirconium Bisphosphate $Zr[O_3PO(CH_2CH_2O)_nPO_3]_x\ [HPO_4]_{2-2x}$

	Elemental	Analysis	From	From	Interlayer Spacing (Å)		TGA[a]
n	%C	%H	EA	TGA	Anhydrous	Wet	T (°C)
2	13.56	2.17	0.96	0.96	12.3	15.5	246
3	15.16	3.14	0.83	0.87	14.0	16.1	234
4	21.04	3.70	0.97	1.00	16.1	17.2	220
9	22.81	3.96	0.73	0.71	20.1[b]	21.3	195
13	37.97	6.32	1.02	1.00	33.3	>44	167
22	35.56	6.24	0.83	0.87	>44	>44	150

[a]Temperature of decomposition of organic groups as determined by thermogravimetric analysis.
[b]This lower than expected interlayer spacing may be due to the lower organic content.

Although the polyether chains are anchored at both ends, there is still some swelling of the layers with uptake of water. This swelling could occur in a simple fashion by the layers shifting so that the chains do not tilt as much. For example, if we subtract 8.2 from 16.1 Å (n = 3, wet) and divide by 3, the value per monomer is 2.6 Å, which could mean the chains are now perpendicular to the layers. For n = 4, this value is 2.25 Å per monomer. However, the expanded value for n = 2, 15.5 Å is too large to be explained this way and may indicate some breaking of the cross-links with further swelling.

1. *Interaction of Zirconium Polyether Compounds with Alkali Thiocyanates*

Polyethers are known to complex alkali thiocyanates forming ion conducting films (57). The uptake of NaSCN and LiSCN from 0.1 *M* methanol solutions was determined and the results are collected in Table VI. Two types of reactions were observed, ion exchange and neutral salt complexing. In the case of NaSCN, about one-half of the sodium ion is exchanged in the monophosphates and one-half is complexed. These reactions are accompanied by extensive hydrolysis. With the diphosphates, much less salt is taken up and very little is complexed, the predominant reaction being ion exchange. Apparently tying down the chains at both ends does not follow them to incorporate ions within the coiled chain as it does for free polyether chains. The LiSCN yielded similar results to those obtained with NaSCN but a higher proportion of Li^+ was exchanged.

The synthesis of solvent-free metal salt complexes of polyethylene oxides

TABLE VI
Reaction of Polyethylene Glycol Esters of Zirconium Phosphate with Electrolytes

Sample		NaSCN			LiSCN		
n	x	mmol SCN / mol ZrP	mmol Na / mol ZrP	Δx^a	mmol SCN / mol ZrP	mmol Li / mol ZrP	Δx
			Monophosphates				
4	2.00	141	309	−0.29	75	114	−0.12
9	2.00	344	796	−0.53	158	306	−0.42
13	2.00	424	1005	−0.84	203	724	−0.78
22	2.00	654	1250	−0.86	331	938	−0.81
			Diphosphates				
4	1.00	36	52	0	28	46	0
9	0.64	57	264	−0.03	53	224	−0.04
13	0.90	93	566	−0.31	82	456	−0.23
22	0.90	194	887	−0.38	111	570	−0.30

[a]The parameter Δx is the change in moles of polyetherphosphate on the layers resulting from hydrolysis during the complexation reaction.

prompted detailed electrical measurements with the thought that these materials might prove to be useful electrolytes, in a hydrous environment, for high-energy density batteries (58–60). Conductivity measurements were performed on the polyethylene oxide zirconium phosphate–NaSCN complex ($x = 22$, $n = 2.0$) between room temperature and 100°C (55). Conductivities of the order 10^{-6} Ω^{-1} cm^{-1} at room temperature were obtained, having an increment by order of magnitude with an increase in temperature of about 100°C. The data conformed to an Arrhenius regime, where σ versus $1/T$ is a straight line with $E_a = 0.22$ eV. The conductivity is associated with the motion of cations in the solid phase formed by chains of polyethylene oxide–NaSCN helical complex. The formation of cation vacancies within the PEO helix is responsible for the conductivity at temperatures below 60°C (60). A higher conductivity is observed at temperatures greater than 60°C, due to the melting of the PEO. At this temperature, it becomes slightly soluble in the stoichiometric complex, creating more of the vacancies needed for conduction. In contrast, it appears that no melting of PEO occurs in the zirconium PEO–phosphate. Even so, it is surprising that the conductivity data conforms to an Arrhenius-type dependence on temperature with an activation energy of 0.22 eV because there are two conductors present: Na^+ in the ion-exchange sites and Na^+ in the PEO complex. The conductivity of sodium ion phases of α-zirconium phosphate are of the order of 10^{-6}–10^{-8} Ω^{-1} at 200°C (61, 62). Activation energies for these phases are of the order 0.7 eV. Thus, the conductivities in the range of 25–100°C would be several orders of magnitude lower for $Zr(NaPO_4)_2$. Therefore, the Na^+ ions in the exchange sites can contribute very little to the overall conductivity. Given the changed environment in the present case one cannot be certain of this point.

B. Sulfonic Acid Derivatives of α-Zirconium Phosphate Structure

1. *Preparation and Structure*

DiGiacomo and Dines (63) prepared the first sulfonic acid derivatives of zirconium phosphonates. The reactions used to prepare the substituted phosphonic acids are illustrated below.

$$(EtO)_2P(=O)H + NaH \longrightarrow (EtO)_2P(=O)ONa + H_2 \quad (7)$$

$$(EtO)_2PONa + \text{(1,3-propane sultone, cyclic } -CH_2CH_2CH_2-S(=O)_2-O-) \longrightarrow (EtO)_2P(=O)(CH_2)_3SO_3Na \quad (8)$$

$$(EtO)_2P(=O)(CH_2)_3SO_3Na \xrightarrow{HBr} H_2O_3P(CH_2)_3SO_3H$$

In aromatic compounds, a sulfonation reaction was used rather than a sulfone followed by an Arbuzov reaction.

$$PhCH_2CH_2Br + H_2SO_4{\cdot}SO_3 \longrightarrow HO_3S-C_6H_4-CH_2CH_2Br \quad (9)$$

$$NaO_3S-C_6H_4-CH_2CH_2Br + P(OEt)_3 \longrightarrow NaO_3S-C_6H_4-CH_2CH_2P(OEt)_3 \xrightarrow{HBr} HO_3S-C_6H_4-CH_2CH_2PO_3H \quad (10)$$

Reaction of the phosphonic acids either with or without added HF yielded the desired zirconium compounds. Not surprisingly, these compounds were shown to behave as ion exchangers and acid catalysts. However, their behavior depended on the degree of crystallinity of the preparations not unlike that of zirconium phosphates (9a, 21). Subsequently, it was shown (64) that aromatic phosphonate derivatives of zirconium could be sulfonated in fuming sulfuric acid directly. Two methods were used for the preparation, either the zirconium phosphonate was allowed to react with fuming sulfuric acid or the phenylphosphonic acid was first sulfonated (55) and then allowed to react with an added zirconium salt. The extent of sulfonation depends on the strength of the fuming acid used and the time and temperature of exposure to the acid. For $Zr(O_3PPh)_2$, complete sulfonation was attained in 28% fuming sulfuric acid in 1 h at 70°C. To recover the solid, water is added dropwise to convert the excess SO_3 to H_2SO_4 and then a slight excess is added to convert the mixture to a gel. Addition of methanol results in precipitation of the sulfonated product. This step is necessary in order to prevent dispersion of the product in aqueous solution. If too much water is added to the gel, a clear dispersion of the zirconium bis(sulfophenylphosphonate), $Zr(O_3PC_6H_4SO_3H)_2$ occurs. This solid is almost impossible to recover by filtration or centrifugation. However, if the number of sulfonic acid groups is reduced either by reducing the amount of sulfonation or by spacing the phenyl rings with alkyl groups or phosphite groups, then the product can be recovered from aqueous dispersions by filtration or centrifugation. Addition of methanol aids both of these recovery processes.

The X-ray powder pattern of the sulfonated zirconium phenylphosphonate consists of a series of (00ℓ) reflections (64, 65). Thus, it is not possible to determine the structure of this compound from such a small amount of X-ray data. However, based upon the fact that the thoroughly dried product has an interlayer spacing of 16.1 Å, it was deduced that sulfonation occurred in the ring position meta to the phosphonate group (64). This deduction was made before the correct structure of zirconium phenylphosphonate was known. However, a computer simulated model, shown in Fig. 16, confirms the meta substitution. A similar model based upon para substitution requires at least a 2–3-Å larger basal spacing.

A number of pillared derivatives of the type shown in Fig. 10(*b*) have also been sulfonated (54). These compounds cannot swell and, therefore, are recovered by filtration. It is possible to sulfonate all the rings even when the biphenyl content approaches a high value (close to 1). This statement stands in contrast to our earlier assertion that in the mesoporous biphenyl pillared compounds only the phenyl rings bounding the pores were sulfonated. The degree of sulfonation depends on the conditions used. Use of higher temperatures and stronger fuming sulfuric acid is able to accomplish sulfonation of the biphenyl rings even when adjacent to or surrounded by other biphenyl groups (5.3-Å

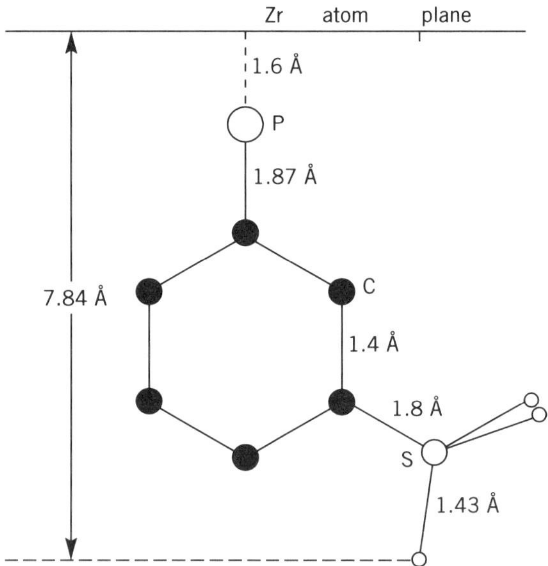

Figure 16. Schematic representation of sulfophenyl groups bonded to an α layer. It is assumed that the C—C—S and C—S—O angles are 120° and 109°, respectively. [Reprinted from *Reactive Polym.*, 13 (1987) C.-Y. Yang and A. Clearfield, "The Preparation and Ion-Exchange Properties of Zirconium Sulphophosphonates." Copyright © 1987 with kind permission of Elsevier Science-NL. Sars Burgerharttstraat-25, 1055 KV Amsterdam, The Netherlands.]

apart). The closeness of the phenyl rings in adjacent biphenyl groups requires them to be rotated in opposite directions to each other to avoid steric crowding.

2. *Zirconium Sulfophosphonates as Proton Conductors*

The swelling and complete dispersion of $Zr(O_3PC_6H_4SO_3H)_2$ in water results from hydration of the protons to form a highly or completely ionized polyelectrolyte consisting of negatively charged phenylsulfonate ions bonded to the inorganic layers and freely floating hydronium ions. Such a structure might be expected to behave as a proton conductor. Alberti et al. (66) prepared $Zr(O_3PC_6H_4SO_3H)_{0.73}(O_3PCH_2OH)_{1.27} \cdot nH_2O$. The reason for choosing a mixed derivative was to avoid the difficulty of isolating a more highly sulfonated compound. The interlayer spacing of this compound in the anhydrous condition was estimated to be 14.6 Å on the assumption that the sulfonic groups of one layer were opposite the CH_2OH groups of an adjacent layer. The observed value of 14.2 Å was interpreted to indicate some interpenetration of the layers. The proton conductivity of this compound was strongly dependent on the relative humidity, as shown in Fig. 17. The activation energies varied from about 13 kcal mol^{-1} at a relative humidity (RH) of 20% to less than 5 kcal mol^{-1} at greater than 80% RH. At the same time the preexponential factor decrease from $\log A = 8$ at 20% RH to $\log A = 4$ above 80% RH. Subsequently, the conductivity of two similar derivatives $Zr(O_3PC_6H_4SO_3H)_{0.85}(O_3Et)_{1.15}$ and $Zr(O_3PC_6H_4SO_3H)_{0.97}(O_3PCH_2OH)_{1.03}$ was determined in the anhydrous state at temperatures in the range 100–200°C (67). The best conductivity for the ethyl derivative at 180°C was $\sigma = 10^{-5}\ \Omega^{-1}\ cm^{-1}$ while in a wet nitrogen atmosphere it was of the order of $10^{-3}\ \Omega^{-1}\ cm^{-1}$.

The high proton conductivities exhibited by the sulfophenylphosphonate derivatives at temperatures up to 180°C recommends these materials for use in electrochemical devices. Accordingly, it was of some interest to determine the conductivity of more highly sulfonated products because it is the sulfonic acid groups that are largely responsible for the proton mobility. We therefore measured the conductivity of two additional compounds, $Zr(O_3PC_6H_4SO_3H)_2$ (EWS-3-89) and $Ti(O_3PPh)_{0.15}(O_3PC_6H_4SO_3H)_{1.60}(HPO_4)_{0.25}$ (EWS-4-1) (65). The conductivities as a function of RH are given in Table VII. It is seen that the conductivity values are somewhat higher than those found for the mixed derivatives containing alkyl groups. In fact, the room temperature value for the titanium compound at 85% RH is among the highest recorded for a solid proton conductor.

3. *Ion Exchange Properties of Zirconium Sulfophosphonates*

The zirconium sulfophenylphosphonate phosphates of general composition $Zr(O_3PC_6H_4SO_3H)_x(HPO_4)_{2-x}$ contain two ion exchange sites, the hydroxyl

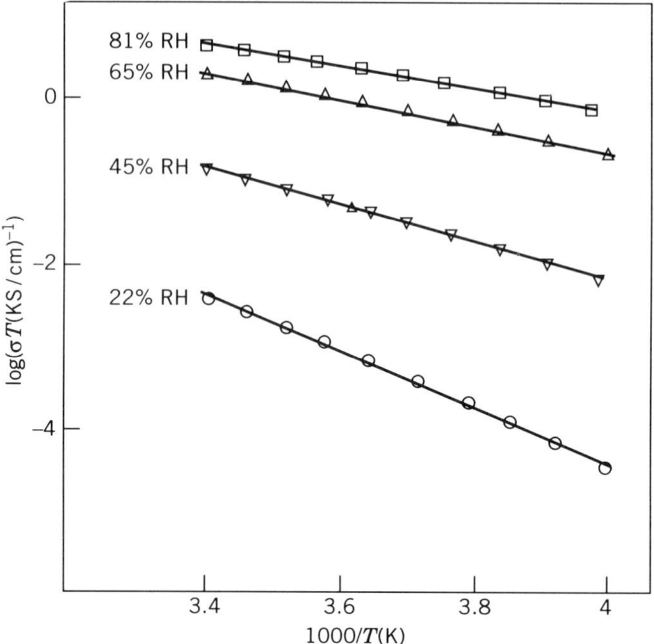

Figure 17. Arrhenius plots for $Zr(O_3PC_6H_4SO_3H)_{0.73}(O_3^-PCH_2OH)_{1.27} \cdot nH_2O$ at different relative humidities. [Reprinted from *Solid State Ionics*, *50*, G. Alberti, M. Caiciola, U. Constantino, A. Peraio, and E. Montonere, 315, Copyright © 1992 with kind permission of Elsevier Science-NL, Sara Burgerhartstrat 25, 1055 KV Amsterdam, The Netherlands.]

proton and the sulfonic acid protons (68a,b). It is to be expected that the more acidic sulfonic acid protons are more readily displaced by cations. In fact, the two types of protons can be distinguished by means of their titration curves using a 0.1 and 0.01 *M* supporting electrolyte (NaCl). The exchanger samples had compositions for which $x = 0.43$ (sample MY-VI-2) and $x = 0.767$ (MY-VI-95). The titration curves exhibit an initial flat portion at low pH that repre-

TABLE VII

Conductivity (Ω^{-1} cm^{-1}) at Room Temperature as a Function of Relative Humidity for Zirconium and Titanium Sulfophenylphosphonates

	Relative Humidity (RH %)				
Sample	20(3)[a]	30(3)	50(4)	65(4)	85(5)
EWS-3-89 (Zr)	5×10^{-6}	2×10^{-4}	1.1×10^{-3}	7.8×10^{-3}	2.1×10^{-2}
EWS-4-1 (Ti)	4×10^{-5}	3×10^{-3}	1.2×10^{-2}	7.2×10^{-2}	1.3×10^{-1}

[a]The number in parentheses is the estimated error in the humidity measurement.

sents neutralization of the sulfonic acid protons. The two titration curves, carried out for sample MY-VI-95, coincide perfectly at pH values below 4. This coincidence demonstrates that virtually all the strongly acidic protons are replaced by Na^+ even at the lower (0.01 M) (68a) sodium ion concentration. At higher pH, the curve for 0.1M NaCl falls below the curve for 0.01M NaCl. For weakly acidic protons, an increase in sodium ion concentration should bring about an increase in the extent of exchange and, hence, an increase in the hydrogen ion concentration. According to the formula for MY-VI-95 the sulfonic acid protons amount to 1.96 mequivalents per gram (1.96 meq g^{-1} of exchanger. The two titration curves separate at close to this value (1.8 meq g^{-1}) and so they represent an excellent method by which to distinguish between the two types of protons. We observe that the titration curve for sample MY-VI-2 rises in pH at an earlier point as expected because of its lower sulfonic acid content.

The higher the charge and the larger the size of the unhydrated ion, the greater is the affinity for the ion. Relevant thermodynamic data are given in Table VIII. We note that for the univalent ions the entropy is negative, whereas for the divalent ions it is positive. Thus, the alkali metal–hydrogen ion reactions are largely enthalpy driven, whereas those for alkaline earth exchange are entropy driven.

In general, the greater the density of the sulfonic acid groups attached to the layers the greater is the affinity for a given species as measured by the distribution coefficient, K_d, (defined below).

$$K_d = \frac{\text{Concentration of ion in the exchanger}}{\text{Concentration of ion in solution at equilibrium}} \times \frac{\text{Volume of Solution}}{\text{mass of exchanger}} \tag{11}$$

Evidence for this assertion is given in Table IX. We note that the K_d values for sample MY-VI-95 are higher than for the sample containing less sulfonic acid groups (MY-VI-2). Both samples exhibit much higher K_d values than either

TABLE VIII
Thermodynamic Data for $MZ^+—H^+$ Exchanger MY-VI-2 at 25°C

Metal Ion	K_c	$\Delta G°$ (kJ mol^{-1})	$\Delta H°$ (kJ mol^{-1})	$\Delta S°$ (J K mol^{-1})
Na^+	0.898	0.27	−8.3	−29
Cs^+	50.0	−9.69	−15.9	−21
Mg^{2+}	3.58	−3.16	0.9	14
Ba^{2+}	91.2	−11.18	−2.7	28

TABLE IX
Distribution Coefficients for Alkali and Alkaline Earth Metal Ions on Exchangers MY-IV-95, MY-VI-2, Amorphous Zirconium Phosphate, and a Polystyrene Sulfonic Acid Resin AG 50W-X8 at 25°C

Ion	MY-IV-95[a]	K_d(mL g^{-1}) MY-VI-2[a]	Amorphous ZrP[a,b]	AG 50W-X8[c]
Li^+	110		7	33
Na^+	205		11	54
K^+	1,500	650	120	99
Cs^+	6,500		1,600	148
Mg^{2+}	21,000	9,800		790
Ca^{2+}	89,000	37,000		1,450
Ba^{2+}	400,000	190,000		5,000

[a]The parameter K_d at pH 2.00 and a metal loading of 0.1 meq g^{-1}.
[b]Calculated from selectivity coefficients given by L. Kullberg and A. Clearfield, *J. Phys. Chem.*, *85*, 1578 (1981).
[c]The parameter K_d in 0.1 *M* HNO_3. Data from F. W. E. Strelow, R. Rethemeyer, and C. J. C. Bothma, *Anal. Chem.*, *37*, 106 (1965).

amorphous zirconium phosphate or a typical strong acid ion exchange resin containing sulfonic acid groups (68a). The affinity of the layered sulfonates for coordinated metal species such as $[Co(NH_3)_6]^{2+}$ or $[Ru(bpy)_3]^{2+}$ is very large. The sulfonates are colloidally dispersed in aqueous media and on adding the coordination compound the layers come together trapping the species between the layers quantitatively with accompanying precipitation.

In contrast, the K_d values for bis(sulfophosphonates) are much lower than their two dimensional counterparts (69). Because the phosphonate pillars are rigidly held in place, they more closely resemble the highly cross-linked sulfonic acid ion exchange resins. However, the phosphonate compounds exhibit ion sieving ability. For example, large complexes such as $[Ru(bpy)_3]^{2+}$ may be completely rejected in favor of other species that can fit into the cavities. By systematic control of the pore size it should be possible to create a series of ion sieves.

C. Zirconium Aminophosphonates and Polyiminephosphonates

Amino derivatives of the type $Zr(O_3PRNH_2)_2$ are readily prepared by the reaction of the corresponding phosphonic acid, $H_2O_3PRNH_2$, where R is an alkyl or aryl group, with a soluble zirconium salt. In this process, the amino groups are protonated because zirconium salts are generally highly acidic. In the protonated state, the layered amino derivatives are colloidally dispersed. In this state, the amino derivatives should behave as anion exchangers in acid solutions. Rosenthal and Caruso (70) prepared $Zr[O_3P(CH_2)_3NH_3 \cdot Cl]_2$ with an interlayer spacing of 15.3 Å. This compound did not exhibit anion exchange behavior, which was attributed to a highly crowded interlayer gallery. Therefore, they prepared a mixed-derivative $Zr[O_3P(CH_2)_3NH_3^+Cl^-]_{0.2}(O_3PMe)_{1.8}$ (interlayer spacing, 11.3 Å). This compound did behave as an anion exchanger. When the hydrochloride salt was stirred in NaOH, the free base had a basal spacing of 10.4 Å, indicating that the compound did not have a staged structure. This compound complexed Cu^{2+} as its sulfate salt whereas $Zr[O_3P(CH_2)_3NH_2]_2$ did not.

The behavior described above for the propylamino derivative may be typical for alkylamines but we prepared the monoamine $Zr(O_3PCH_2CH_2NH_2)_2$ and found, in agreement with Rosenthal, that it dispersed only with difficulty. It had to be thoroughly wet, in the protonated form, and subjected to sonication before colloidal dispersion took place. In contrast, the analogous phosphate, $Zr(O_3POCH_2CH_2NH_2)_2$, was difficult to isolate because it formed a clear gel containing up to 95% water. The gel could be broken by rotovaping to concentrate the solid and then adding acetone or other water-soluble organic. This tendency to form gels was general for the phosphates but $Zr[O_3POCH_2CH_2N^+(Me)_2Cl^-]_2$ exhibit facile anion exchange as a swollen-layered compound.

A series of polyimines that readily exfoliated in the protonated state in aqueous media was synthesized (55, 71). Table X provides data for some of these compounds. The phosphonates were prepared by refluxing a mixture of chloromethylphosphonic acid with ethyleneimines in aqueous NaOH. The products were then combined with zirconyl chloride solutions, generally at temperatures from 50 to 100°C, to yield layered compounds in which a portion of the

TABLE X
Representative Interlayer Spacings of Zirconium Polyimine Phosphonates or Phosphates of General Composition $Zr(O_3PR)_x(O_3PR')_{2-x}$

R	R′	X	Basal Spacing (Å)
$-CH_2NHCH_2CH_2NH_2$		2	17.1
$-CH_2NHCH_2CH_2NH_2$	OH	1.7–1.8	16.3
$-CH_2NHCH_2CH_2NH_2$	OH	0.63	14.5
$CH_2(NHCH_2CH_2)_2NH_2$		2	21.8
$CH_2(NHCH_2CH_2)_3NH_2$		2	27.6
$O-CH_2CH_2NH_2$		2	12.7
$O-CH_2CH_2NHCH_2CH_2NH_2$		2	21.0
$-CH_2NHCH_2CH_2NHCH_2PO_3$[a]		1	14.3
$-CH_2NHCH_2CH_2NHCH_2PO_3$[a]	OH	0.5	Amorphous
$-CH_2NH(CH_2CH_2NH)_2CH_2PO_3$		1	17.7
$-CH_2NH(CH_2CH_2NH)_3CH_2PO_3$		1	19.2

[a]The diphosphonates conform to the general formula $Zr(O_2P-R)_x(O_3PR')_{2-2x}$.

amino groups is protonated. The extent of protonation can be determined by analysis for Cl^-. The general reactions are shown in Eqs. 12 and 13.

$$ClCH_2PO_3H_2 + NH_2(CH_2CH_2NH)_nH \xrightarrow{NaOH} H_2O_3PCH_2(NHCH_2CH_2)_nNH_2 \quad (12)$$

$$ZrOCl_2 + 2H_2O_3PCH_2(NHCH_2CH_2)NH_2 \xrightarrow{H_2O} Zr[O_3PCH_2(NHCH_2CH_2)_nNH_2]_2 \cdot xHCl \cdot yH_2O \quad (13)$$

The interlayer spacings listed in Table X are highly dependent on the water content and the degree of protonation. A high level of protonation of the amino groups results in colloidal dispersion of the layers, presumably as single exfoliated layers. In this condition, the aminophosphonates readily undergo anion exchange and complex formation.

Complex and polyvalent anions not only exchange but become encapsulated and form insoluble zirconium polyimine complexes. Among the species so precipitated are $[Fe(CN)_6]^{3-}$, $[Fe(CN)_6]^{4-}$, $[PtCl_4]^{2-}$, and a variety of heteropolyacid anions, including $[PW_{12}O_{40}]^{3-}$, $[PV_2W_{10}O_{40}]^{5-}$, and $[SiW_{12}O_{40}]^{4-}$. The increase in basal spacings of the complex ion containing polyimines were obtained from X-ray powder patterns. For the octahedral and square planar complexes, this increase is of the order 1.8–2 Å. This small increase is possible only if the complexes partially protrude into the space between the amino chains. Slightly more than one-half of a mole of the hexacyanoferrate(II) ion and 0.68 mol of hexacyanoferrate(III) ion were sequestered by $Zr(O_3PCH_2NHCH_2CH_2NH_2)_{1.75}(HPO_4)_{0.25} \cdot 2HCl$ but a full mole of $[PtCl_4]^{2-}$ per mole of the exchanger was taken up. In this latter complex, one of the chlorine atoms was replaced in the square planar coordination sphere by an amino group as shown by UV–vis spectra and elemental analysis.

The cross-linked polyimines behave as solid anion exchangers in acid solution. However, in neutral solution they become complexing agents. In the protonated form, they take up small amounts of Cu^{2+} presumably by cation exchange with protons. In neutral solution, the uptake is much greater yielding deep blue complexes. Much of the chemistry of the polyimines and amino derivatives remains unexplored. Previously, Dines et al. (19) reported the preparation of zirconium(ethylpyridyl)phosphonate, $Zr(O_3PCH_2CH_2C_5H_4N)_2$. About 0.2 mol of Pd could be complexed by the pyridine groups, and this complex behaved as an active hydrogenation catalyst. Maya (72) prepared $Zr(OH)_{0.64-}(O_3PCH_2CH_2NH_3NO_3)_{1.68+}$, which he showed to exhibit anion-exchange properties, but no details were given. Thus, our understanding of these interesting compounds is in its infancy and much remains to be done to understand their complexing dual behavior as agent and ion exchanger.

V. SELF-ASSEMBLY AND LAYER-BY-LAYER ASSEMBLY OF THIN FILMS

A. Layer-by-Layer Constructs

Before discussing additional types of phosphonate compounds, we will describe interesting procedures employed to produce layer-by-layer constructs or thin films with interesting properties. Thin films of inorganic materials have great practical value in many applications such as optical coatings and microelectronics. However, by employing phosphonate chemistry, films may be synthesized sequentially layer by layer at room temperature. Excellent reviews on this subject are available (37, 73). Previously, multilayer thin films were prepared by the Langmuir–Blodgett (LB) method (74, 75). An amphipilic molecule with a polar head group and a nonpolar tail is spread on the surface of water. These monolayers are then compressed by a movable barrier into a close packed array and transferred to a flat solid surface. The array is held together only by van der Waals forces and is therefore unstable. An alternative strategy, in which covalent forces are utilized to bind the layers, produces stable films (76). Mallouk and co-workers (77, 78) pointed out the similarity between LB films and zirconium alkylphosphonates [Fig. 18(*b*)] and devised a scheme to prepare thin films based on this similarity. A substrate such as a gold film or silica surface is primed by binding one end of a long-chain molecule to the surface [Fig. 18(*a*)]. The other end contains a phosphonic acid group. This surface is then dipped into a Zr^{IV} solution, which then binds to the free phosphonic acid end of the monolayer. That the metal is actually bonded is evidenced by its not being removed on washing and subsequent adsorption of an α, ω-bis(phosphonic acid). The process is repeated as many times as desired. Characterization by ellipsometry and angle resolved X-ray photoelectron spectroscopy (XPS) demonstrated that the thickness of each deposited layer was close to that of the layer spacing in the analogous microcrystalline solid. A plot of interlayer spacing as a function of the number of layers of $Zr[O_3P(CH_2)_{10}PO_3]$ (78) is shown in Fig. 19.

Time-resolved ellipsometry shows that monolayer growth is substantially complete in 10 min of exposure to the alkane bis(phosphonic acid) solution. The exposed acid groups also adsorb Zr^{4+} rapidly and in turn the Zr terminated surface also bonds the phosphonic acid rapidly (77, 79). However, much longer exposure times are required in order to achieve compact monolayers that form well-ordered multilayer films. This point is illustrated in ellisometric data for growth of the α, ω-octyl bis(phosphonate) of zirconium. With long 4-h immersions the change in thickness for each new layer was constant at 13.4 Å. This matches well the interlayer spacing of the bulk compound, 13.8 Å. For short immersion times, the films are much thinner than expected at 12.8 Å per layer after about 10 layers (79). The thickening of layers in the lateral direction with

surface —— $X\text{-}(CH_2)_n\text{-}PO_3H_2$ (1) → PO_3^{2-} / PO_3^{2-} —— $ZrOCl_2$ (2) → PO_3 / PO_3 Zr —— $H_2O_3PC_{10}H_{20}PO_3H_2$ (3) →

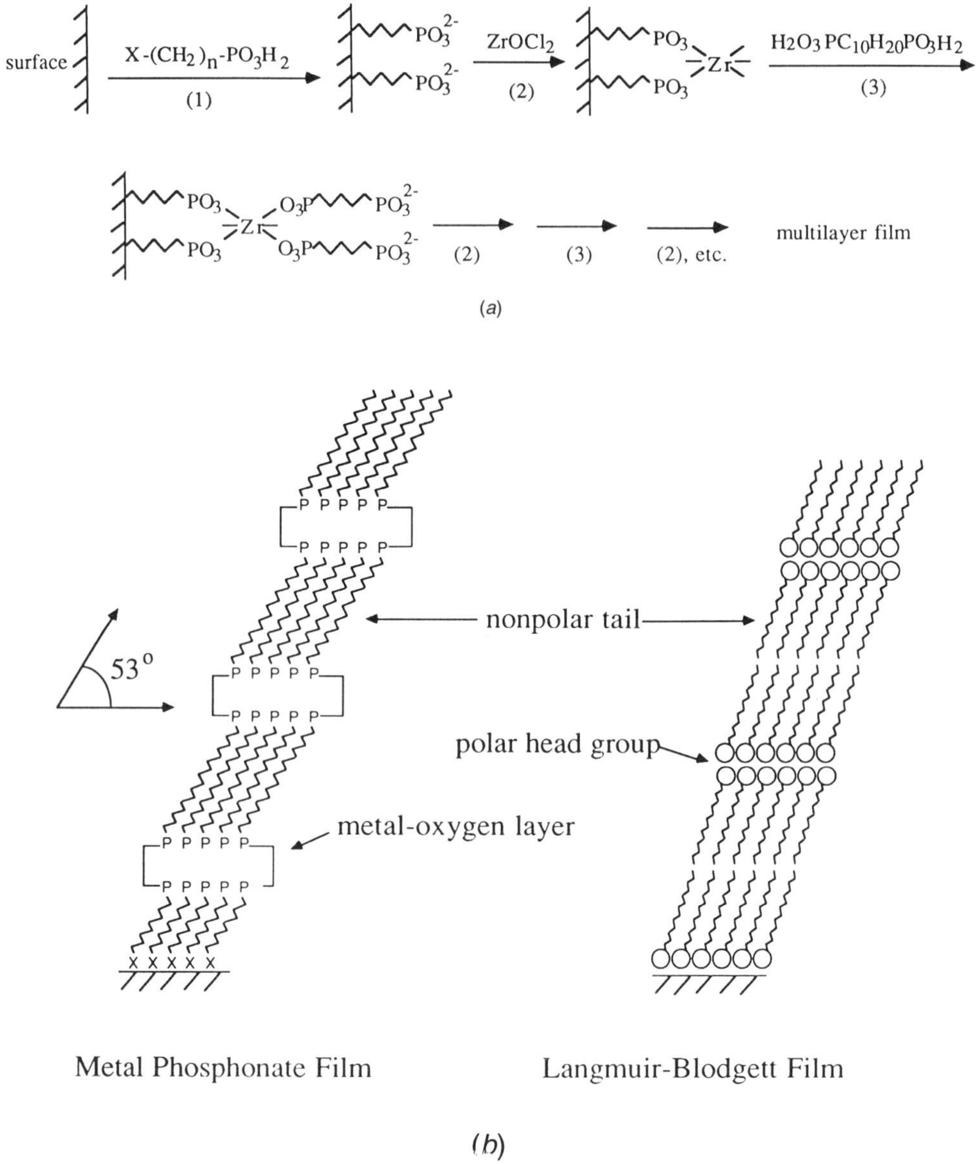

PO_3 / PO_3 Zr O_3P PO_3^{2-} / O_3P PO_3^{2-} → (2) → (3) → (2), etc. multilayer film

(*a*)

(*b*)

Figure 18. Scheme for step-by-step assembly of zirconium phosphonate thin films according to Mallouk et al. (*a*) and structural analogy between such films and Langmuir–Blodgett films (*b*). [Reprinted with permission from H. Lee, J. Kepley, H. G. Hong, and T. E. Mallouk, *J. Am. Chem. Soc.*, *110*, 618 (1988). Copyright © 1988 American Chemical Society.]

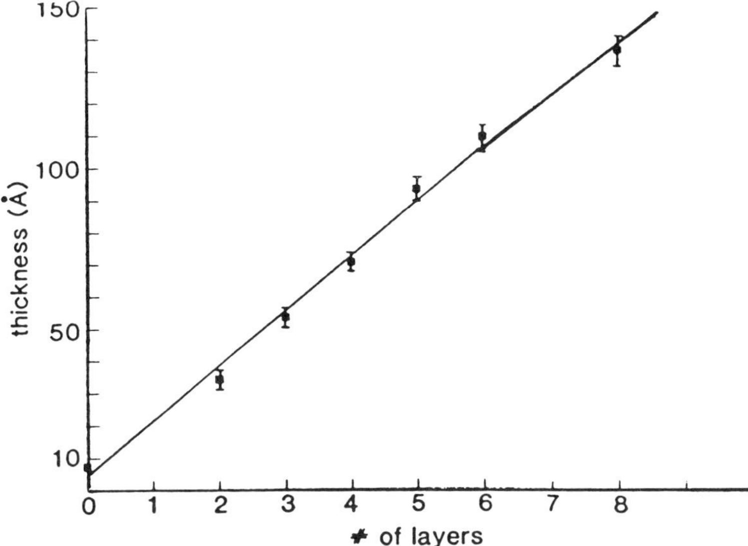

Figure 19. Film thickness versus the number of layers for 1,10-decanebis(phosphonate) multilayers on a silicon surface. [Reprinted with permission from H. Lee, J. Kepley, H. G. Hong, and T. E. Mallouk, *J. Am. Chem. Soc.*, *110*, 618 (1988). Copyright © 1988 American Chemical Society.]

time may well result from the same mechanism that allows ester interchange reactions. We have seen that in γ-ZrP such reactions are facile and also reversible. The same is true for α-ZrP (80). The surface of α-ZrP can be functionalized by contact with alkylphosphates or phosphonic acids. The interior of the crystals remain unreacted only because it is inaccessible to the ligand. Once the layers are spread apart as by an intercalation reaction, then the interior phosphate groups also exchange. The angle of inclination to the layer of long-chain *n*-alkylamines intercalated into α-ZrP increases as more amine is loaded between the layers (81). Thus, the lower layer thickness on short immersion time may result from positions on the surface Zr layer unoccupied by the phosphonate groups. With prolonged immersion, the phosphonate ligands can move laterally along the surface to maximize the van der Waals attractions by filling all available surface positions.

Ellipsometry requires a knowledge of the index of refraction of the film for interpretation of film thickness. The index is an unknown quantity but is assumed to be equal to that of the corresponding microcrystalline solid. Grazing angle X-ray diffraction, which is capable of assessing layer thickness and uniformity of the multilayer films, does not depend on a knowledge of the index of refraction of the films. This technique together with ellipsometry was used by Page and co-workers (82) to examine Hf/1, 10-decanediylbis(phosphonate), or Hf-DPB multilayers. Hafnium was used in place of zirconium in order to enhance the diffraction from the metal layers. It was found that the layer thickness varied from sample to sample from about 15–21 Å per layer as compared to 16.7 Å for the bulk solid. Estimates of layer density from grazing angle XRD indicated that the films with lower interlayer distances were about 75% as dense as the theoretical bulk density of the corresponding microcrystalline solid. The index of refraction of such thin films is lower than that of the bulk material (1.48–1.50 vs. 1.54). Thus, the use of bulk indexes of refraction can lead to substantial errors in ellipsometric measurements of metal phosphonate films.

To account for the different layer thicknesses, it was proposed that the films have a domain structure as depicted in Fig. 20 (82). Within each layer, there exist "domains" that correspond to regions of particular alkyl chain density, tilt angle and orientation, and metal–phosphonate binding motif, all of which are presumably dictated by lateral hafnium density on the functionalized substrate surface. Such domains would have different thicknesses to account for the variation in film thickness. The domain boundaries as shown in Fig. 20 are disordered and could encompass void space and solvent to account for the lower layer density. In a sense, this domain model has some features in common with the bulk or microcrystalline cross-linked α-type zirconium phosphonates described in Section III.C. In these latter materials, it is proposed that the layers

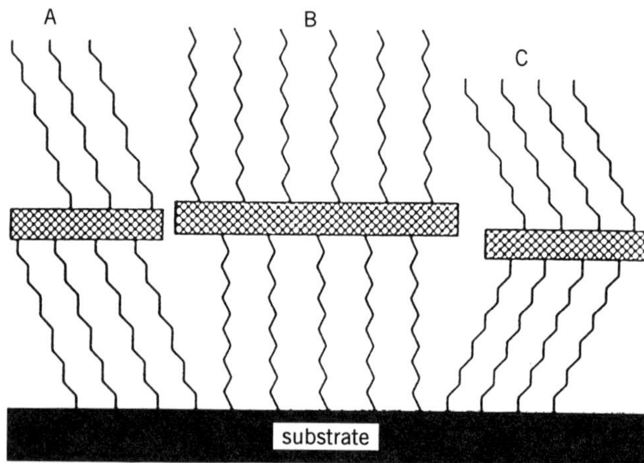

Figure 20. Schematic of the proposed domain structure. Lines represent the bis(phosphonate) alkyl chains; cross-hatched areas represent the inorganic hafnium–phosphonate layers. Lateral density of bis(phosphonate) chains, chain tilt angle, and layer thickness vary from one domain to the next. [A. C. Zeppenfield, S. L. Fiddler, W. K. Ham, B. J. Klopfenstein, and C. J. Page, *J. Am. Chem. Soc.*, *116*, 9158 (1994). Copyright © 1994 American Chemical Society.]

have different lateral dimensions creating void spaces. In the films, a single layer actually consists of a group of layers in the lateral direction with voids between them. If domains A and B in Fig. 20 continue to grow in the directions shown the void space between them will continue to increase. However, domain C may close in on domain B and form a single domain.

An important point to consider is the nature of the underlying layer or primer layer, that is, the layer attached to the substrate. In the case discussed above, the primer layer consisted of Hf bonded to a silica wafer. The arrangement of the Hf atoms on the wafer surface is most assuredly different than their arrangement in α-HfP or Zr in α-ZrP (83). Therefore, the bonding to the phosphonate may have an arrangement that is neither the α nor the silica type. This aspect of the layer growth needs closer study. The same is true for thiols on gold surfaces. The degree of order and packing density in the films was probed by IR spectroscopy (84–87). A monolayer of 11-mercapto-1-undecanol (MVD) on a gold film was prepared by Frey et al. (84). Treatment with $POCl_2$ in ethanol yielded a mixture of phosphate mono- and diesters that bonded Zr^{4+} to initiate the self-assembly of Zr/1, 10-decanediylbis(phosphonate) layers. A second film was prepared on a germanium wafer. These films were examined with a combination of attenuated total reflection–Fourier transform infrared (ATR–FTIR) and polarization modulation–Four transform infrared (PM–FTIR) spectroscopies. The primer layer on germanium was a phosphate amide formed by phosphorylation of 4-aminobutyl or 3-aminopropylsilane. In both types of films, it was found, by observation of the CH stretching regions, that the band positions indicated that the alkyl chains were not as conformationally ordered or as tightly packed as in crystalline *n*-decane (84).

Examination of the two primer layers showed that the one on the gold surface was much more organized than the primer layer on the germanium crystal. Therefore it was concluded that the conformational disorder (deviation from a close-packed all-trans conformation) was due to the phosphonate–Zr–phosphonate control of the lateral spacings in the films. Similar results were reported by Page and co-workers (85). A thiophosphonic acid anchor layer on gold was more ordered for a C_{12} alkyl chain length than for a shorter C_4 chain length. Multilayers grown on the two primer layers were not affected by the degree of order of the primer layer. However, the overall degree of alkyl chain order was greater for longer chain lengths. A film grown with $Zr[O_3P(CH_2)_6PO_3]$ showed essentially liquid-like disorder.

Katz and co-workers (86, 87) at AT&T examined films grown on silicon and gold substrates. Their IR results were similar to those of Frey et al. (84). If the substrate is properly etched, a very ordered initial layer will be deposited. For a poorly prepared surface, the intensity of the symmetric and antisymmetric CH_2 stretches was reduced and the bands were broadened and shifted to higher frequencies, indicating more liquid-like behavior. However, whether the initial layer was well ordered or not, deterioration of the order occurred upon formation of the zirconium phosphonate layers. X-ray photoelectron spectroscopy examination of each step in the film formation showed that the washing step, after addition of Zr, removed considerable amounts of the metal species. This result suggests that some of the Zr is loosely bound and perhaps longer aging times may be required. Bent et al. (87) concluded that the degree of order in the films is closely tied to the match of the lattice spacings of the organic layer to the metal ion and that Zr may not be well matched to the preferred hydrocarbon crystal packing. We shall return to this point later.

Talham and co-workers (88) prepared zirconium phosphonate films by a LB technique. The surface of a silicon wafer was coated with a monolayer of octadecyltrichlorosilane (OTS). Then a single layer of octadecylphosphonic acid was transferred from an LB trough to the surface of the OTS–silicon wafer held in the trough. This procedure produced a monolayer of the phosphonic acid in a tail-to-tail arrangement with the OTS on the silicon wafer. The wafer was then removed from the trough and dipped into a beaker containing a 5-mM solution of Zr^{4+} to self-assemble the zirconium at the organic template. To complete the layer, the wafer was rinsed in water and then replaced into the LB trough where a new octadecylphosphonic acid film was compressed and transferred to the substrate. The film was built up by repetition of this three step process (Fig. 21). The position of the asymmetric methylene (ν_a CH_2) band at 2918 cm^{-1} with a full width at half-maximum (fwhm) of 20 cm^{-1} indicated that in this case an all-trans, close-packed template formed. The shape and position of this band remained unchanged as the multilayer film was built up. Ellipsometry and X-ray diffraction were in agreement as to the thickness of the bilayer (51–52 Å).

Subsequently, tilt angles of the octadecyl chains were measured by polarized ATR–FTIR (89). In the initial primer layer, the chains are tilted at a 31° angle to the surface normal. This angle is about the same tilt of the phenyl rings in $Zr(O_3PPh)_2$ (18) and of the chains in α-ZrP amine intecalates (81). However, different tilt angles were obtained for the capping layer of octadecylphosphonate. If the capping layer was applied by the LB technique, the tilt angle was only 5°. This result was attributed to the fact that on compressing the ODP in the trough the chains are almost perpendicular to the water surface and are transferred to the zirconated template layer intact. In contrast, if the self-assembly method is used to bind the capping layer, the tilt angle was 22°. This angle is closer to that observed in the bulk zirconium phosphonates (90).

A self-assembled Zr/1, 10-decanediylbis(phosphonate) multilayer, similar to those prepared by Mallouk and co-workers (78, 79), was grown on a zirconated LB template layer. The IR spectra indicated that these layers are more poorly ordered because the LB layer binds a high density of Zr atoms. The DDPB is able to bond to more Zr atoms in the template than in the next monolayer of

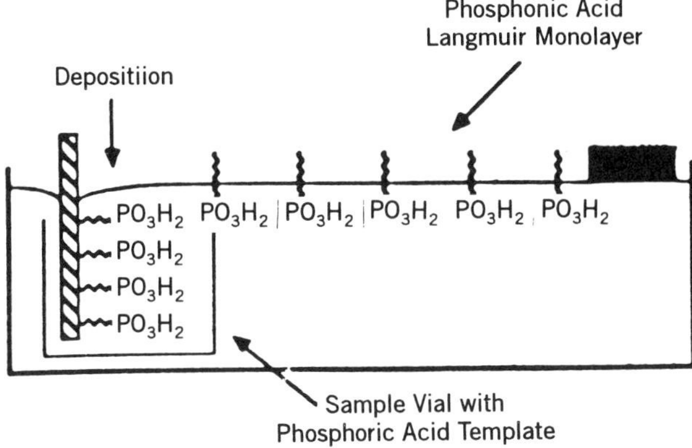

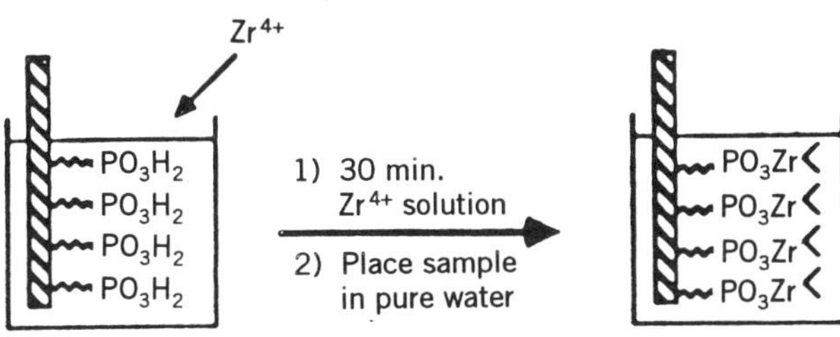

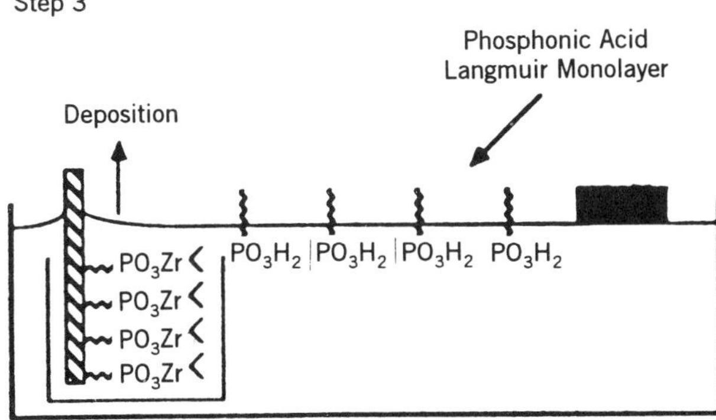

Figure 21. Scheme for deposition procedure for the preparation of zirconium octadecylphosphonate bilayers. [Reprinted with permission from H. Byrd, J. K. Pike, and D. R. Talhem, *Chem. Mater.*, *5*, 709 (1993). Copyright © 1993 American Chemical Society.]

zirconium creating a mismatch. After four such layers, the mismatch is overcome and the organization in each layer remains the same.

The films described in the foregoing paragraphs are centrosymmetric. Putvinski et al. (91) devised a method for depositing layers with polar order. 11-Hydroxyundecanethiol was deposited as a monolayer on a gold substrate. The hydroxyl groups were then phosphorylated with an acetonitrile solution of $POCl_3$ containing triethylamine to yield the corresponding phosphonic acid. The acid was in turn treated with Zr^{4+} that in turn was treated with 11-hydroxyundecyl phosphonic acid. The multilayers so built up are polar in the sense that in one layer the Zr is bonded to a PO_4 group and to a PO_3 group in the next layer.

A similar procedure was utilized to fix azo dyes into repeating noncentrosymmetric layers (92, 93). These films exhibited nonlinear optical behavior comparable in magnitude to that of $LiNbO_3$ and were stable to 150°C. Subsequently, films were prepared from a variety of nonlinear chromophores (93, 94). The intensities of the second harmonic generated waves were proportional to the square of the number of layers (94). Figure 22 shows an idealized schematic structure of such a dye–zirconium multilayer.

The zirconium–phosphonate bonding in the layer-by-layer thin films is thought to yield the same arrangement as in the layers of α-zirconium phosphate. The area available to the phosphonate is 24 Å^2 (8). This requirement is met for most alkyl and linear aryl chains. However, the AT&T group have prepared films from phosphonated porphyrins (95) and other ligands (96) whose cross-sectional areas exceed 24 Å^2. The XPS analysis of such films shows that the amount of zirconium in the film is larger than expected for an α-ZrP-type structure (89). In Section VII, we shall show that in the preparation of bulk zirconium phosphates with ligands exceeding the required 24 Å^2, new structure-types form.

Phosphonic acids, as we shall see later, can form compounds with divalent and trivalent elements. Therefore, it is not surprising that layer-by-layer thin films can be prepared from divalent elements (79). However, these films were prepared in aqueous ethanol solutions since the solubility of many divalent phosphonates increases as the pH is lowered. Film formation is very rapid, of the order of 10 min at room temperature, as determined by time resolved mass measurements made with a piezoelectric quartz crystal microbalance. A variety of physical and chemical methods were utilized to show that the structure and stoichiometry of these films resemble closely those of the analogous bulk solids (79, 97). A particularly interesting procedure was to show that a multilayer manganese octadecylphosphonate film prepared by Talham's LB film method had similar magnetic properties as the bulk manganese alkylphosphonate compounds of the general formula $Mn(O_3PR)\cdot H_2O$ (98). The electron paramagnetic resonance (EPR) studies of the LB film (50 bilayers) gave evidence for antiferromagnetic exchange in a two-dimensional inorganic extended lattice. The g values are characteristic of Mn^{2+} in a nearly cubic field (1.99–2.00) and the

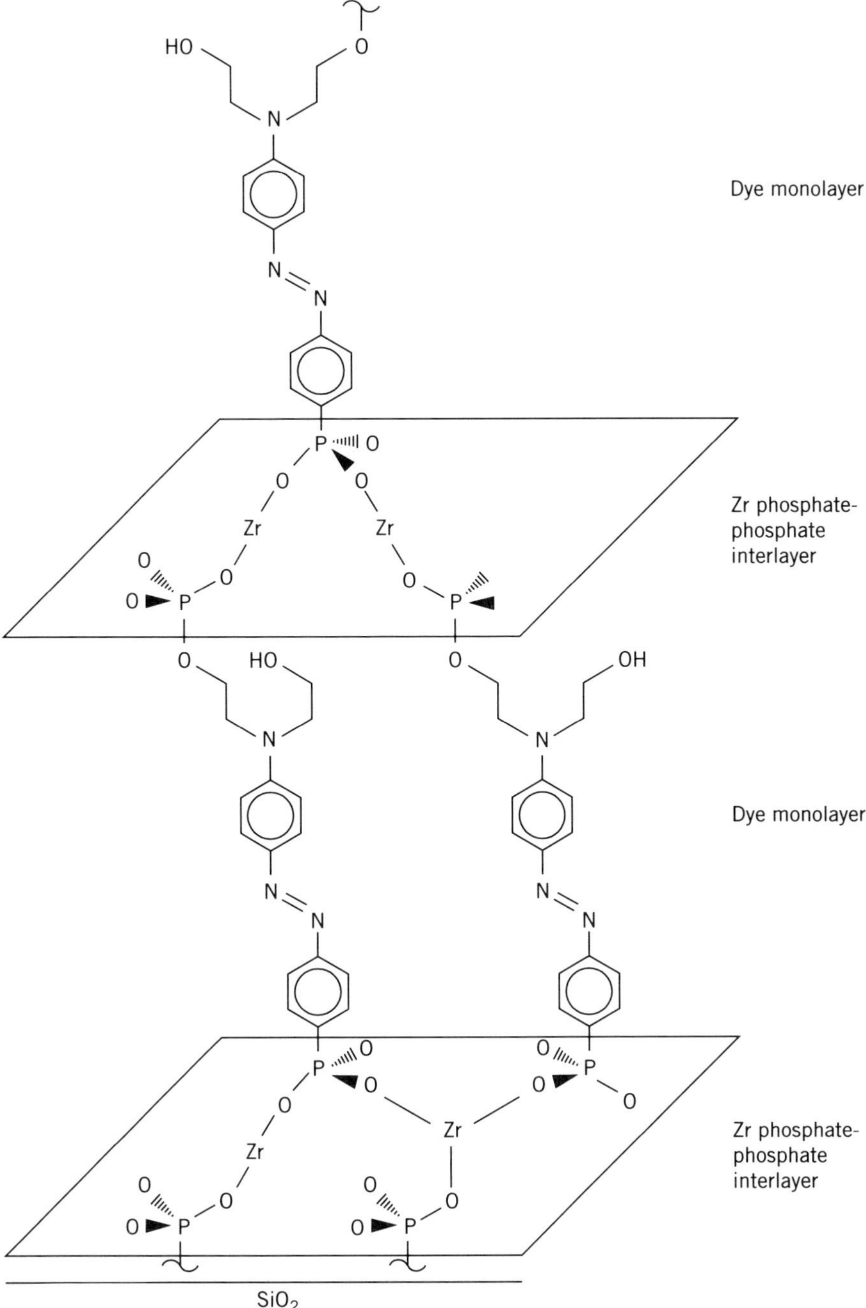

Figure 22. Idealized schematic structure of dye multilayer film based on the structure of α-ZrP. [Reproduced with permission from (92).]

measured antiferromagnetic exchange constant, J/K of -2.8 K, was in excellent agreement with the J/K for the bulk compound.

B. Special Methods of Self-Assembly

A triple decker film containing Y^{3+}, Zr^{4+}, and Hf^{4+} in separate layers was prepared by Mallouk and co-workers (99) in the usual sequential manner. Angle-resolved XPS was utilized to depth profile the film and showed that the three different metals lay in approximately planar sheets separated by the length of the phosphonate groups.

Feng and Bein (100) were able to nucleate and grow a zinc phosphate molecular sieve on the surface of a gold substrate covered with a phosphonic acid. Mercapto-1-undecanol (MUD) was sorbed onto the gold surface, phosphorylated, and converted to a phosphonic acid as before. Several layers of a zirconium bisphosphonate film were then grown on the functionalized gold substrate. The zeolite-X analogue of zinc phosphate was prepared from a mixture of NaOH, H_3PO_4, $Zn(NO_3)_2$, and the template 1,4-diazabicyclo[2.2.2]octane. This mixture forms a milky suspension that settles rapidly and crystallizes on standing (5 h at 7°C). When the gold–phosphonate substrate was dipped into a beaker of the settled milk, the zinc phosphate crystallized on the phosphonate surface. The majority of the crystals (~90%) grew with their basal surfaces oriented parallel to one of the (111) planes of the zeolite as observed by scanning electron microscopy. These materials may find application as sensors by control of access to the surface to molecules preferred by the zeolite and as a means of orienting molecules for nonlinear optical applications.

Recent work on thin-film growth has involved adsorption of oppositely charged polyelectrolytes (101). Mallouk and co-workers (102) extended this technique to layering of structurally well-defined, two-dimensional colloidal polyanions and polymeric cations. The α-ZrP layers may be completely exfoliated by intercalating a small amine [$MeNH_2$, $CH_3CH_2CH_2NH_2$, $(Me)_4NH^+$, etc.) between the layers, diluting and sonicating the suspended solid (103). When such a suspension is placed in contact with an amine-modified gold or silicon surface, the surface–NH_3^+ groups displace the loosely held amine associated with one side of the ZrP layer. This procedure electrostatically attaches the α-ZrP layer to the gold or silicon surface. A number of polymeric or oligomeric cations were then attached to the exposed side of the α-ZrP layer by an ion exchange reaction. Among the ions utilized were the aluminum Keggin ion $[Al_{13}O_4(OH)_{24}(H_2O)_{12}]^{7+}$, polyallylamine hydrochloride (PAH), and cytochrome *c*.

One advantage of this procedure is that both negatively and positively charged layers can be utilized and almost any positively or negatively charged species intercalates between layers of opposite charge. General methods for exfoliating layered materials have been presented (104).

We have already described the preparation of zirconium and titanium phosphonates functionalized with sulfonic acid or amino groups. These compounds exfoliate spontaneously to yield negatively and positively charged single layers, respectively. This result was shown by self-assembling alternate sulfonate and amine layers by merely mixing the two as separately exfoliated species. A schematic drawing of one of the resultant products is shown in Fig. 23 (65). The solid product formed immediately. The interlayer spacing for the ethylpyridinium–phenylsulfonate complex shown in Fig. 23 was 18.4 Å. The sum of the individual half-layer distances is about 16.4 Å leaving 2 Å for the positioning of the proton between the amino and sulfonic acid layers. As shown, the 2 Å is the distance from the pyridinium ring to the base of the phenyl ring. Two strong peaks were observed in the solid-state ^{31}P NMR spectrum, one at −5.5 ppm representing the sulfophosphonate groups and the other at 6.38 ppm for the aminophosphonate groups. The proton conductivity for a number of these mixed or interstratified complexes was determined and the results are collected in Table XI. It should be obvious to the reader that this procedure together with the layer-by-layer technique of Mallouk is capable of preparing an infinite variety of new materials.

C. Applications of Self-Assembled Films

1. *Self-Assembled Films as Insulators*

A series of multilayer films of zirconium alkylbis(phosphonates) were grown on silicon (*p*-type) and gold substrates. The chain length varied from 2 to 10 carbon atoms (105). Films as thin as five layers were pinhole free for the longer chains ($n = 6$, 8, and 10). The dielectric constant of all the films was 4.0 ± 0.2 and breakdown fields were on the order of 6×10^6 V cm^{-1}. Films were also prepared from 4,4′-biphenylbis(phosphonic acid). These films were too

TABLE XI

Room Temperature Conductivity (σ, Ω^{-1} cm^{-1}) at Various Relative Humidities

	Relative Humidity (RH%)				
Sample	20%	30%	50%	65%	85%
EWS-3-83[a]	1×10^{-7}	3×10^{-6}	6×10^{-5}	4.0×10^{-4}	1.5×10^{-3}
EWS-3-90[b]	$<1 \times 10^{-8}$	1×10^{-7}	9×10^{-6}	2.3×10^{-4}	8.7×10^{-4}
EWS-3-91[c]	$<1 \times 10^{-8}$	4×10^{-7}	1×10^{-5}	4.6×10^{-4}	7.2×10^{-4}
EWS-4-2[d]	$<1 \times 10^{-8}$	6×10^{-7}	7×10^{-4}	8.2×10^{-4}	2.4×10^{-4}

[a] $[Ti(HPO_4)_{0.14}(O_3PPh)_{0.81}(O_3PC_6H_4SO_3H)_{1.05}]_{0.51}[Zr(O_3PCH_2CH_2C_5H_4N)_2]_{0.49}$.
[b] $[Zr(HPO_4)_{0.12}(O_3PC_6H_4SO_3H)_{1.88}]_{0.59}[Th(O_3PCH_2CH_2C_5H_4N)_2]_{0.41}$.
[c] $[Zr(HPO_4)_{0.12}(O_3PC_6H_4SO_3H)_{1.88}]_{0.5}[Th(O_3PCH_2CH_2C_5H_4N)_2]_{0.5}$.
[d] $[Ti(HPO_4)_{0.25}(O_3PPh)_{0.12}O_3PPhSO_3)_{1.63}]_{0.55}[Th(O_3PCH_2CH_2C_5H_4N)_2]_{0.45}$.

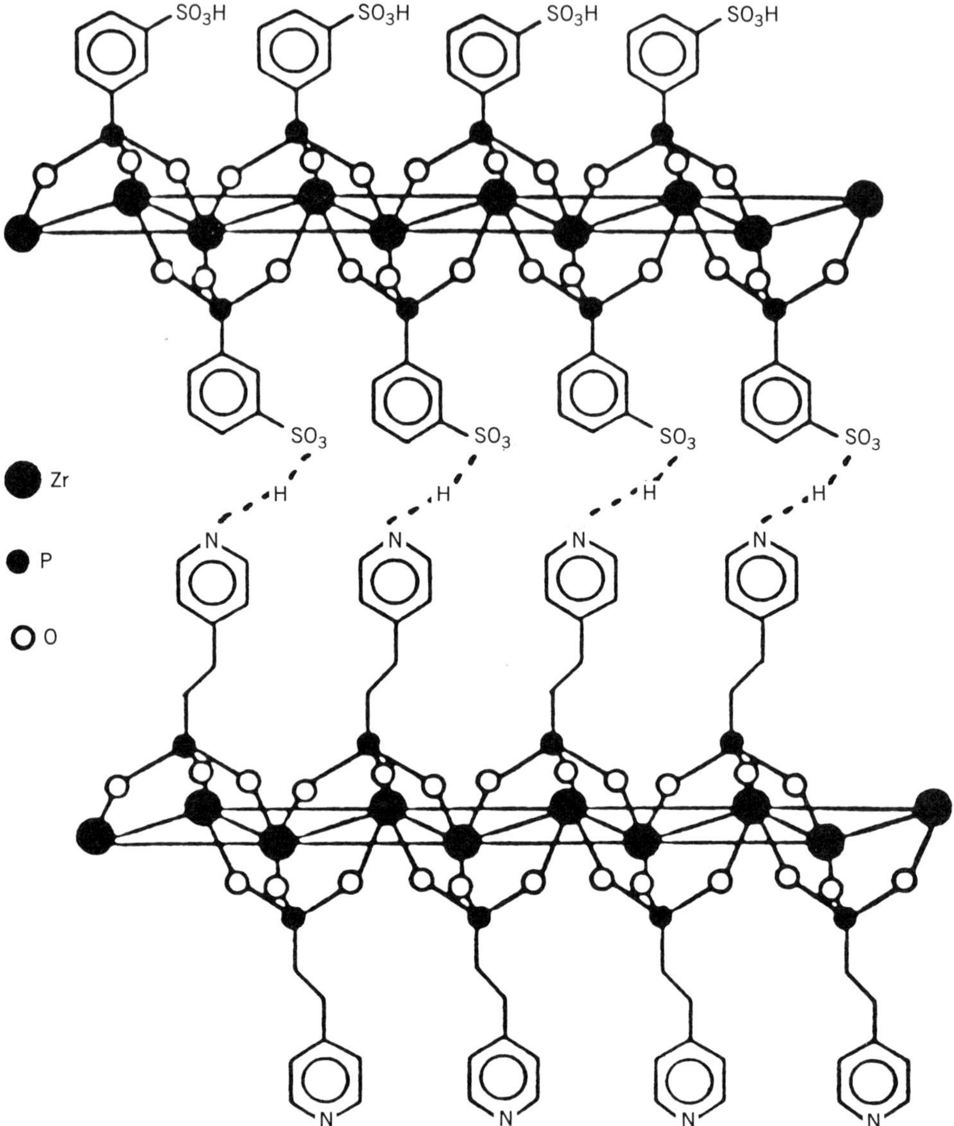

Figure 23. Schematic drawing of the interstratified compound $[Zr(O_3PC_6H_4SO_3)_2]^{2-}$ $Zr(O_3PCH_2CH_2C_5H_4NH)_2]^{2+}$.

porous to allow pinhole-free evaporated metal contacts (106). In fact, interfacial capacitance measurements conducted in aqueous electrolytes showed that the films were permeable to water and chloride ions (107). However, with nonwetting liquids (e.g., mercury) the films behaved as good insulators with dielectric constants in the range of 3–6. Also, ionic conductivity in these films was observed possibly due to free ions in the film. We mentioned in Section III.C that bulk zirconium bis(phosphonates), whether made with alkyl or arylphosphonic acids, yield porous products as shown in Fig. 13. Protons may be present as a result of unreacted $-PO_3H_2$ groups, which ionize in the presence of water-forming hydronium ion.

A single layer of self-assembled film of zirconium 1,10-decanediylbis(phosphonate) was grown over a 7-Å thick primer layer on a gold electrode. The phosphonate layer was 17-Å thick but was still sufficient to block electron transfer between the gold electrode and a ferricyanide ion solution (77). The distance dependence of electron tunneling examined by using increased numbers of layers of zirconium 1,2-ethanediylbis(phosphonate) (ZEDP). These layers are only 8-Å thick and are the smallest that can be made. The rate constant for ferrocyanide oxidation with progressively thicker films was determined. It was found that the rate constant varied exponentially with distance with a damping constant of 0.4 $Å^{-1}$. Similar measurements were conducted with ferrocene groups at the outer surface of the film (108) with the same results. The very low value of the tunneling constant was attributed to defects in the films.

2. *Nonlinear Optical Thin Films*

In addition to preparing multilayered films from the ligand shown in Fig. 22, similar films were prepared from a variety of polar molecules by the AT&T scientists (91, 92, 94). The three-step adsorption sequence devised by this group ensures the polar orientation of these molecules in the film. Unlike LB films that can be easily disrupted the covalent bonding locks the polar molecules into place making any type of rearrangement difficult. This strong bonding is evidenced by their stability to 150°C, a relatively high temperature for organic NLO media. The order parameters for these films are comparable to those reported for poled polymers. The films did not lose activity when a nonpolar molecule was inserted along with the polar ligands during the deposition sequence.

3. *Miscellaneous Applications*

We must necessarily defer discussions of certain optical effects and electron-transfer reactions of the LB-type thin films as well as their role in sensing applications until we have described additional metal phosphonate types. Here we discuss some recent innovative uses of the films. Feldheim and Mallouk (109) prepared C_{10} and C_{16} alkane phosphonate films on gold substrates of Fe^{III} and Fe^{II}. The Fe^{3+} film was used to initiate the polymerization of adsorbed pyrrole vapor. The film thickness was found to increase by 5 Å per layer of Fe^{3+}, presumably due to formation of polypyrrole on contact with Fe^{3+} in the process. It is suggested that these and similar films are potentially useful for catalysis, electrocatalysis, magnetic applications, and optical waveguides.

An interesting use of the Fe^{3+} polymerization of pyrrole procedure was used to create an ohmic contact for an ultrasmall capacitor (110). A layer-by-layer film was built up on a gold substrate in which insulating layers of exfoliated α-ZrP alternated with PAH. The films terminated in cationic PAH, were grown to the 40–100-Å level. The film assembly was then allowed to sorb gold particles (2.5 nm) from a citrate stabilized suspension. Additional layers were then grown by reversing the layer sequence, that is, by sorbing PAH followed by α-ZrF to produce a layer sequence (α-ZrP/PAH)/gold nanoparticles/(PAH/α-ZrP) film. Such films are termed MINIM for metal–insulator–(gold) nanocluster–insulator–metal. The film serves as a capacitor and ohmic contact was made as shown in Fig. 24. The $FeCl_3$ was introduced into the film followed by pyrrole vapor, which then polymerized on the top surface of the film. Excess $FeCl_3$ was washed out to obtain the MINIM heterostructure and ohmic contacts added as shown in the figure. Devices such as these display electronic properties characteristic of ultrasmall capacitance tunnel junctions ($\sim 10^{-18}$ farads). A high impedance plateau centered at 0 V was observed in the current–voltage curve. The voltage range of the gap can be tuned by changing the thickness of the insulating films. Such devices are predicted to display single electron-transfer current steps in the I–V curve and these characteristics of the devices are under continuing investigation.

We shall now return to a discussion of additional classes of metal phosphonate compounds and involve LB-type phosphonate films as appropriate to the applications of these new compound types.

VI. γ-ZIRCONIUM PHOSPHONATES

We have already described the structure of γ-ZrP in Section II and the preparation of γ-type organic phosphates by Yamanaka (2). These phosphate derivatives were prepared by ester interchange between γ-ZrP and the organic phosphates. Alberti et al. (111, 112) conducted many trials to obtain the direct preparation of γ-type phosphonates and concluded that this was not possible. However, they found that the topotactic replacement of the dihydrogen phosphate groups by O_2PRR' groups is a general reaction (112, 113). With methyl

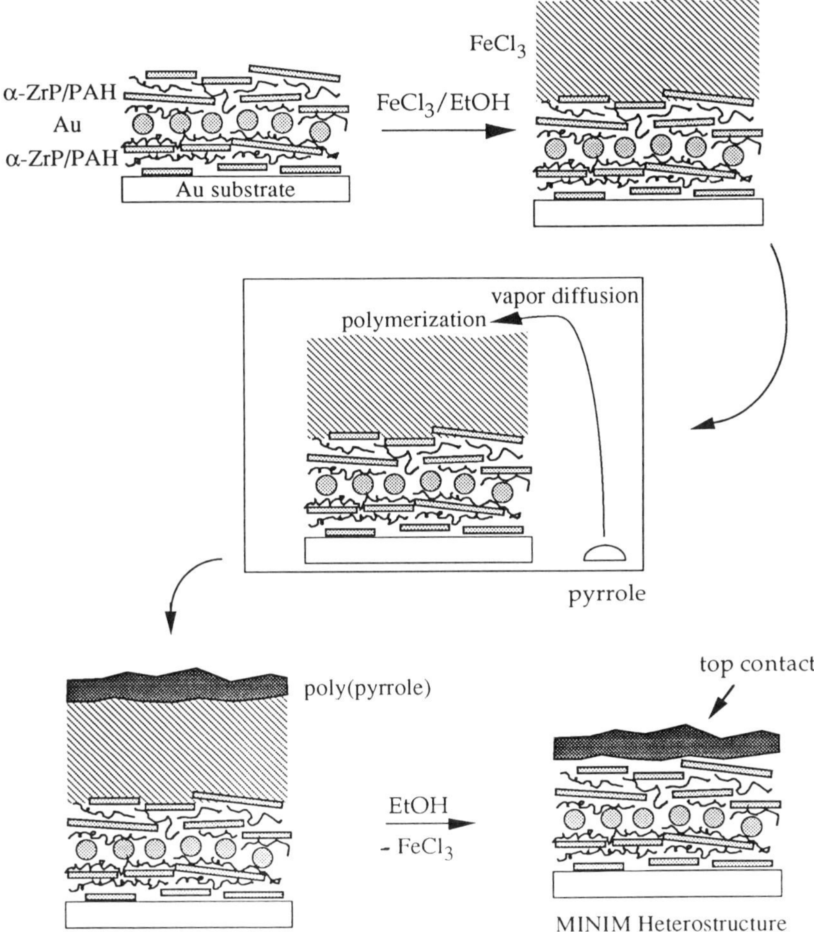

Figure 24. Illustration of the polymerization of pyrrole to form a top metal contact in MINIM devices. [Reprinted with permission from D. L. Feldheim, K. C. Grolar, M. J. Natan, and T. E. Mallouk, *J. Am. Chem. Soc.*, *118*, 7640 (1996). Copyright © 1996 American Chemical Society.]

and propylphosphonic acids complete replacement of the dihydrogen phosphate group was obtained but with the bulkier cychohexylphosphonic acid only two-thirds replacement occurred (112). The reaction may be formulated as in Eq. 14 for complete replacement or as in Eq. 15 for partial replacement of the $O_2P(OH)_2$ groups.

$$Zr[PO_4][O_2P(OH)_2] + MePO_3H_2 \longrightarrow Zr[PO_4][HO_3PMe] + H_3PO_4 \quad (14)$$

TABLE XII

Compositions and Interlayer Distances of Some Derivative Compounds of γ-Zirconium Phosphate

Acid Employed	Composition	Interlayer Distance (Å)	Reference
H_3PO_3	$ZrPO_4(O_2PHOH) \cdot 2H_2O$	12.2	111a
H_3PO_2	$ZrPO_4(O_2PH_2) \cdot H_2O$	8.8	117
H_2O_3PMe	$ZrPO_4(O_2POHMe \cdot 2H_2O$	12.8	112
$H_2O_3PC_3H_7$	$ZrPO_4(O_2POHPr) \cdot 1.2H_2O$	15.1	112
$HO_2P(Me)_2$	$ZrPO_4(H_2PO_4)_{0.33}(O_2P(Me)_2)_{0.67} \cdot H_2O$	10.3	117
$H_2O_3PC_6H_5$	$ZrPO_4(H_2PO_4)_{0.33}(O_2POHPh_5)_{0.67} \cdot 2H_2O$	15.4	111b
$H_2O_3P(C_6H_{11})$	$ZrPO_4(H_2PO_4)_{0.33}(O_2POHC_6H_{11})_{0.67} \cdot H_2O$	16.9	112
H_2O_2PPh	$ZrPO_4(O_2PHPh)$	15.1	[a]

[a]S. Murcia-Mascaro's, Ph.D. Thesis, "Synthesis, characterization and reactivity of pellared layered composites of the phosphate and phosphate–phosphonates of zirconium of type-gamma," University Perugia, Italy, 1993.

$$Zr[PO_4][O_2P(OH)_2] + HO_2PRR' \longrightarrow Zr[PO_4][O_2P(OH)_2]_{1-x}[O_2PRR']_x + xH_3PO_4 \quad (15)$$

The completeness of the reaction or the value of x depends on steric factors. For a series of γ-derivatives, the in-plane a and b unit cell dimensions are very close to those for $\gamma \cdot \cdot \cdot$ZrP. Only the c parameter, measuring the interlayer spacing, increased with increasing size of the ingoing group. Thus, the reaction is a topotactic replacement of H_2PO_4 groups with the phosphonate groups, leaving the γ-layer essentially unaltered. On this basis, Alberti used computer models to show whether groups will sterically interfere with each other. The models show that n-alkylphosphonates can be accommodated but the presence of a second alkyl group, even if both are methyl groups, cannot. Table XII lists most of the derivatives prepared by the Alberti group. It is seen that phosphonic acids with bisalkyl (methyl) groups, the rigid phenyl group, and the bulky cyclohexyl group all yield γ derivatives with $x = 0.67$. Thus, the steric constraints are alleviated by retaining one-third of the smaller H_2PO_4 groups.

α-ZrP is the more stable form of zirconium phosphate. Consequently, γ phases often revert to α-ZrP under severe chemical treatment such as hydrothermal conditions or in strong acids. Therefore, it is not surprising that some ester interchange reactions led to α-type phases (113). Measured heats of reaction indicate that the formation of α phases takes place by a dissolution–reprecipitation mechanism whereas topotactic ester interchange reactions are probably diffusion controlled. Therefore, to minimize the formation of α phases, the temperature should be kept low, dilute solutions of phosphonic acids should be used and contact times kept to a minimum (113).

1. *Pillared γ-Type Zirconium Phosphonates*

Alberti et al. (114–115) also prepared pillared γ-zirconium phosphates bis(phosphonates) by topotactic reactions employing diphosphonic acids. An aryl compound of composition $Zr(PO_4)(H_2PO_4)_{0.56}(HO_3PC_6H_4PO_3H)_{0.22}$-$\cdot 2.2H_2O$ and an alkylderivative $Zr(PO_4)(H_2PO_4)_{0.18}(HO_3PC_4H_8PO_3H)_{0.41}\cdot 1.3H_2O$ were prepared by ester interchange reactions at 80°C. The reactions may be formulated in the following two ways:

$$Zr(PO_4)(H_2PO_4)\cdot 2H_2O + \frac{x}{2} R(PO_3H_2)_2 \xrightarrow{(n-2)H_2O}$$

$$Zr(PO_4)(H_2PO_4)_{1-x}(HO_3P{-}R{-}PO_3H)\frac{x}{2} \cdot nH_2O + xH_3PO_4 \quad (16)$$

$$Zr(PO_4)(H_2PO_4)\cdot 2H_2O + xR(PO_3H_2)_2 \xrightarrow{(n-2)H_2O}$$

$$Zr(PO_4)(H_2PO_4)_{1-x}(HO_3P{-}R{-}PO_3H_2)_x \cdot nH_2O + xH_3PO_4 \quad (17)$$

The first reaction indicates that cross-linkage of the layers has occurred (Eq. 16). The second reaction represents noncross-linkage such that only one of the PO_3H_2 groups bonds to zirconium leaving a PO_3H_2 pendant group in the interlamellar space. That cross-linkage had occurred was shown from ion exchange experiments. It has been determined (116) that below pH 7 γ-ZrP exchanges just 1 mol of protons for alkali metal ions. If the reaction of Eq. 16 takes place, then the ion exchange capacity of the system should remain unchanged because both of the phosphonate protons are exchangeable below pH 7. In contrast, all three phosphonate protons of Reaction 17 are titrable, so the exchange capacity should increase as x increases. In fact, the ion exchange capacity was found to remain unchanged confirming that cross-linking had occurred.

The layers of γ-ZrP are more rigid than those of α-ZrP and therefore, in the γ system, a broad range of solid solutions, or degrees of substitution of diphosphonates for $[H_2PO_4]^-$ is possible. The formulas given above for the monophenyl and butyl derivatives represent the maximum amount of substitution but several derivatives with lesser amounts of substituted organic phosphonate were readily obtained. Although some of these derivatives contained small numbers of cross-links, none of them were microporous. By using a longer rigid cross-linking agent, such as a biphenyl group, microporous derivatives were obtained (117). The cross-linked biphenylene derivatives had interlayer spacings of 16–16.3 Å. However, if the amount of bis(phosphonate) incorporated was less than 25% of the total $[H_2PO_4]^{2-}$ the derivative was unstable to elevated temperature (200°C) dehydration. Disproportionation to yield γ-ZrP and a more highly cross-linked phase formed. Since the microporosity should increase with decrease in the number of pillars, the maximum in microporosity should occur at slightly more than 25% pillaring where a single stable phase is obtained. This was found to be the case. The maximum surface area was 320 $m^2\ g^{-1}$ and a Horvath–Kawazoe differential pore volume plot gave a narrow micropore size distribution centered about 5.8 Å. Computer generated models of the γ-zirconium biphenylenebis(phosphonate) are shown in Fig. 25.

Ester interchange reactions with phosphinates, O_2PR_2, have been carried out with R = H or Me (118). The full extent of reactions of this type have not been exploited. Compounds with crown ethers as pendant groups have also been obtained (119, 120) in the γ system. This chemistry is too new to describe the behavior of these compounds but they are sure to exhibit interesting supramolecular chemistry. Crown ether derivatives of a different type are discussed in Section VII.

2. *Proton Conduction of γ-Type Sulfophosphonates*

In Section IV.B.2 we have seen that sulfophosphonates having α-type layered structures are excellent proton conductors. Therefore, it was of some interest to see whether a similar range of favorable conductivities could be obtained with γ-type derivatives. In this connection, Alberti et al. (121) prepared a number of sulfophenyl and sulfobenzylphosphonic acids and incorporated them into γ-zirconium phosphate by ester interchange reactions. Because a high con-

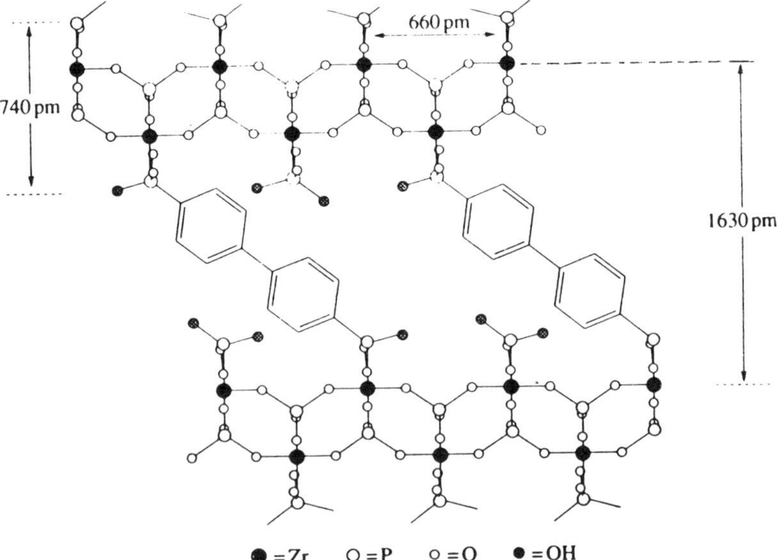

Figure 25. Schematic drawing of two layers of γ-ZrP covalently pillared by 4,4′-biphenyl bis(phosphonate) groups.

tent of ammonium fluoride was used in the synthesis of γ-ZrP, the product also contained fluoride ion bonded to zirconium. We shall have more to say about such compounds later. The γ-ZrP was formulated as. $Zr[PO_4][O_2P(OH)_2]_{0.8}$-$[F \cdot H_2O]_{0.2} \cdot 2H_2O$. It was found that the fluoride ion was not affected by the exchange reactions that were carried out at 80°C. The highest loading of sulfonic acid was 0.46 mol using 3-sulfophenylphosphonic acid. This compound contained 2.2 mol of water at 11% RH and 6.6 mol at 90% RH. Conductivities were run at 100°C as a function of RH and the results were found to converge to a common conductivity at 90% RH. The highest value obtained was 0.05 Ω^{-1} cm^{-1} at 100°C and 95% RH. This value is somewhat less than achieved with α-$Zr(O_3PC_6H_4SO_3H)_2$ but still quite good.

VII. ADDITIONAL ZIRCONIUM PHOSPHONATE STRUCTURE TYPES

A. *N*-(Phosphonomethyl)iminodiacetic Acid Compounds

The metal atoms in α-ZrP are arranged in almost equisided parallelograms about 5.3 Å on a side. The phosphate group sits nearly in the center of the equilateral triangles formed by the two halves of the parallelogram and bonds to the three metal atoms at the corners of the triangle. This area is about 24 Å and will be slightly smaller or larger depending on the length of the metal–oxygen bond. If the organic group of the phosphonic acid is smaller than the area required for α-type bonding, a layered compound with the α-type layer will form. However, if the organic group is larger, compound formation will still take place because of the strong bonding tendency of the phosphonate oxygens, but a new structure will form. This new structure type is illustrated for the zirconium PMIDA derivatives, that is, the methyliminodiacetic acid phosphonate ion,

$$O_3P-CH_2-N\begin{matrix}\diagup CH_2CO_2H \\ \diagdown CH_2CO_2H\end{matrix}$$

This ligand is too large to fit in the space provided by an α-type layer. Therefore, on refluxing a solution of zirconyl chloride with the phosphonic acid of PMIDA a linear chain compound formed (122). In order to dissolve the phosphonic acid, it was necessary to add ammonia. Consequently, both ammonium ion and chloride ion were incorporated between the chains with NH_4^+ groups near the carboxyls. The formula of this compound is $(NH_4)Zr\{H_3[(O_3PCH_2$-$NH(CH_2COO)_2]_2\}F_2 \cdot 3H_2O \cdot NH_4Cl$. Only one of the phosphonic acid protons is replaced leaving two oxygen atoms per phosphonate group for bonding. These oxygen atoms bridge across zirconium atoms on both sides to form the chain. Interesting ion exchange and adsorption reactions of this compound are under study.

In attempts to preserve an α-type layer, mixtures of H_3PO_4 and PMIDA were refluxed with a solution of $[ZrF_6]^{2-}$. A compound of composition

$$Zr_2(PO_4)[O_3PCH_2N(CH_2CO_2H)_2]\left[O_3PCH_2N\begin{matrix} \diagup CH_2CO_2^- \\ \diagdown CH_2CO_2H \end{matrix}\right]\cdot 2H_2O$$

was obtained when the phosphoric acid/PMIDA acid ratio was 1 : 1. The crystal structure of this PMIDA derivative was solved using a combination of X-ray data obtained from a rotating anode source and synchrotron radiation. This compound has a new type of layer arrangement (123). The structure has features of both α- and γ-ZrP with a ratio of Zr/P = 2 : 3. In order to achieve this ratio, one of the protons from a PMIDA acetate group is missing. The Zr atoms are tetrahedrally disposed about the PO_4 groups as in γ-ZrP. However, the phosphonate groups bridge to three Zr atoms as in α-ZrP. Each Zr is octahedrally coordinated with the coordination sphere comprised of three phosphonate oxygen atoms from three different groups, two phosphate oxgyen atoms from two PO_4 groups, and a water molecule. A schematic representation is shown in Fig. 26. The organic groups project into the interlayer space and participate in an intricate network of hydrogen bonding through the carboxyl groups. There is a hydrogen bond [2.82(3) Å] between a carboxyl group in one layer with the oxygen of a second carboxyl group in the same layer as shown by the dashed line in Fig. 26. Additionally, there is a short O—O distance, 2.51(1) Å, between O8 of one carboxyl group to O8′ of an adjacent layer, thus binding the layers together. There is only one proton for both carboxyl groups. Either it is randomly distributed or equidistant from the two oxygen atoms.

A second solid was obtained from a reactant mixture in which the H_3PO_4/PMIDA ratio was increased to 4. The XRD powder pattern was similar to the first compound but with broadened peaks. This solid was also similar in its IR and solid state NMR spectra to the PMIDA derivative described above except that the spectra revealed the presence of $[HPO_4]^{2-}$ groups. These results coupled with elemental analysis led us to conclude (123) that structurally the two PMIDA compounds were similar except that increasing the H_3PO_4/PMIDA ratio resulted in replacement of some of the PMIDA groups by monohydrogen phosphate groups. Our results were in accord with the formula

$$Zr_2(PO_4)(HPO_4)_{0\cdot5}\left[O_3PCH_2N\begin{matrix} \diagup CH_2CO_2^- \\ \diagdown CH_2CO_2H \end{matrix}\right]\left[O_3PCH_2N\begin{matrix} \diagup CH_2CO_2H \\ \diagdown CH_2CO_2H \end{matrix}\right]_{0\cdot5}\cdot 2H_2O$$

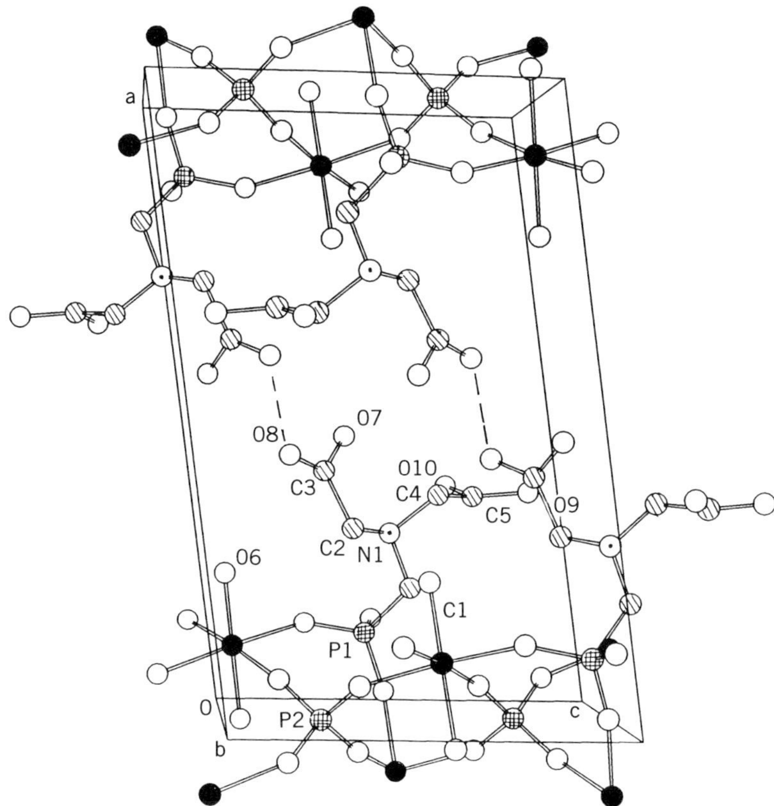

Figure 26. Projection of the structure of $Zr_2(PO_4)(HPMIDA)(PMIDA)(H_2O)_2$ down the *b* axis showing the layer arrangements and hydrogen bonding (dashed lines).

A third solid was obtained by increasing the H_3PO_4/PMID ratio to 6 or above. In this case, the X-ray and spectroscopic data indicated an α-ZrP-type layered compound, of approximate composition $Zr[O_3PCH_2N(CO_2H)_{0.5}]$-$(HPO_4)_{0.5} \cdot H_2O$ had formed. Amine intercalation reactions were in accord with the stoichiometry proposed for these compounds. Thus, we conclude that the bonding of phosphonate groups to Zr is a sufficient driving force to develop new structure types to accommodate organic groups too large to fit the space provided by the α-type layer. However, of the structures presented so far the α-phase is the most stable and, given the opportunity, this is the layer type that will result. The PMIDA compounds exhibit interesting metal complexion behavior above pH 4.

B. Alkylviologen Compounds

Another example illustrative of these structural principles is provided by the formation of a zirconium phosphonate based on the bis(*N*, *N*′-diethylviologen phosphonic acid), $H_2O_3PCH_2CH_2N(C_5H_4)_2NCH_2CH_2PO_3H_2$ (**2**) ligand. Vermeulen and Thompson (124) prepared the zirconium salt of compound **2** with the expected composition $Zr(O_3PCH_2CH_2{-}viologen{-}CH_2CH_2PO_3)X_2$ (X = Cl, Br, or I). These layered compounds exhibit photoinduced charge transfer forming a charge separated state that is long lived and relatively stable in air (125, 126). As prepared, the compounds are poorly crystalline. However, by heating the reactants in a strong HF solution beginning at 55°C and raising the temperature 1°C day^{-1} for 30 days small crystals were obtained. The structure of this compound was solved from X-ray powder data and the results are illustrated in Fig. 27 (127). The composition was determined to be $Zr_2(O_3PCH_2CH_2{-}viologen{-}CH_2CH_2PO_3)F_6 \cdot 2H_2O$. The compound is layered but the layers consist of double chains of ZrF_3O_3 octahedra with the oxygen atoms originating from the phosphonate groups. Two of the oxygen atoms bridge the octahedra parallel to the chains while the third oxygen links two such chains as shown in Fig. 27. These double chains are then connected to each other by the phosphonoviologen groups as shown in the figure. Water molecules reside in the space between the phosphonate groups. The layers are stacked perpendicular to the *c* axis. Note that, as shall be shown later, the viologen groups could fit into an α-ZrP type layer. They do not apparently do so because then the positively charged viologen nitrogen atoms would strongly repel each other. In the chain structure, the Zr atoms are approximately 6.6-Å apart as opposed to 5.3 Å in α-ZrP. This larger distance reduces the repulsion and permits water molecules to enter, further shielding the charge. The photochemical behavior of this compound is discussed in Section VII.D.

Just as the PMIDA ligand formed different phases when phosphate ion was included as a reactant so too with the viologen (**2**) ligand. We shall refer to the bonded form of **2** as PV. Two different phases were obtained as mixed-Zr phosphate–phosphonates (126). One of these phases had a composition $Zr(HPO_4)(PV)_{0.5}X$, where X = halide ion and showed only 00ℓ reflections (d_{001} = 18 Å) in its X-ray powder pattern. A schematic representation of its structure is shown in Fig. 28(*a*). The second phase was more crystalline since it was prepared in HF and had basal spacing of 13.5 Å. Hydrothermal treatment of this phase in a teflon lined pressure vessel at 190–200°C with dilute HF gave a significantly more crystalline sample. Its crystal structure was solved ab initio from powder diffraction data (128). This compound contains well-defined pores as shown in Figs. 28(*a*) and 29. Its composition is $Zr_2(PO_4)(PV)F_3 \cdot 3H_2O$ with the water and halide ions situated in the tunnels. Two of the three halides are bonded to Zr

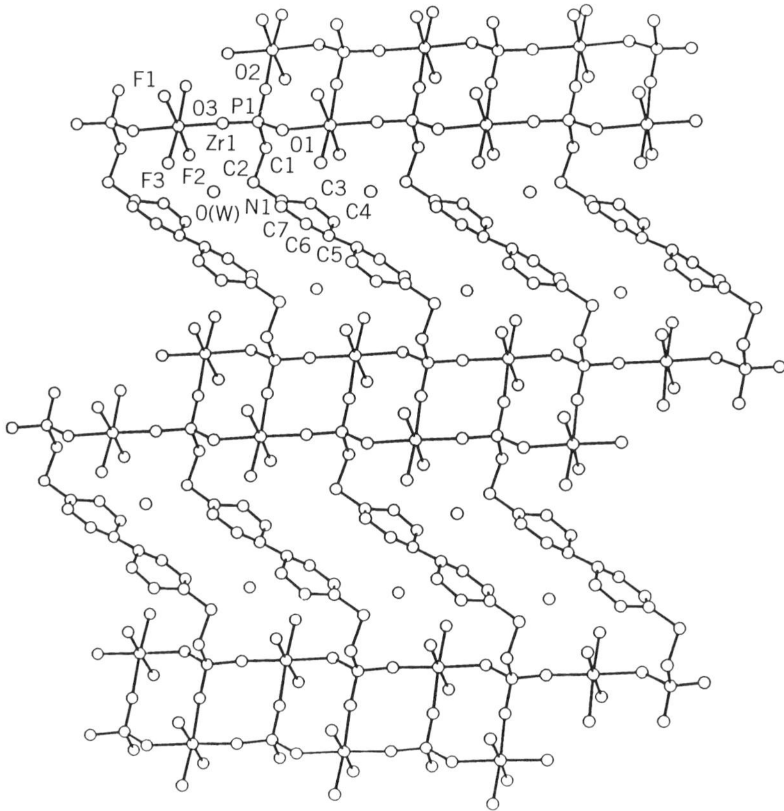

Figure 27. Portion of the structure of $Zr_2(O_3PCH_2CH_2\text{—viologen—}CH_2CH_2PO_3)F_6 \cdot 2H_2O$ as viewed along the *b* axis. The *a* axis is horizontal and the *c* axis vertical.

The structure of $Zr_2(PO_4)(PV)F_3 \cdot 3H_2O$ is closely related to that of the PMIDA derivative, $Zr_2(PO_4)(PMIDAH_2)PMIDAH \cdot 2H_2O$, shown in Fig. 26. The stoichiometry is similar if we recognize that the bis(phosphonate) PV is equivalent to the two PMIDA groups and the halide ions are necessary to counterbalance the positive charge of the PV group. On the other hand, the PMIDA compound balances its charge by juggling the carboxyl protons. The Zr atoms are octahedrally coordinated by two phosphate oxygens, three oxygen atoms from the phosphonate groups in a facial geometry and a fluoride ion that points into the tunnels. In the corresponding PMIDA structure this site is filled by a water molecule. The viologen phosphonate molecules bridge the inorganic layers and form a criss-cross stack as shown in Fig. 28(*b*). This type of stacking reduces the electrostatic repulsions of adjacent viologen groups. The closest

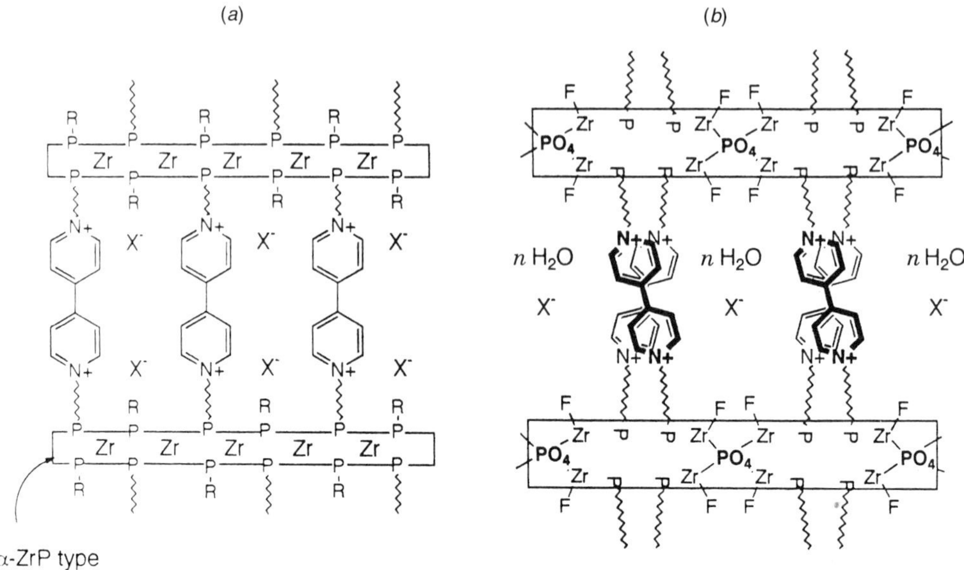

Figure 28. Schematic representation of (*a*) $Zr(O_3PR)PV$ and (*b*) $Zr_2(PO_4)PVX_3$. Here R = alkyl and OH. Lone P atoms represent $-PO_3$ groups bound to three different Zr atoms.

face-to-face contact between viologen groups in the *c*-axis direction is 4.6 Å. The closest viologen–viologen contact across the pore (*b*-axis direction) is about 8 Å and the basal spcaing is 13.6 Å leading to a fairly large pore. The positioning of the F^- and three water molecules in the cavity is shown in Fig. 29 (see color insert). This fluoride ion is exchangeable.

Thompson and co-workers (128, 129) prepared catalysts from the porous zirconium viologen compound by incorporation of active metal species such as platinum and palladium. This was accomplished by exchange of anionic metal halides, $[MX_4]^{2-}$, for the fluoride ions in the cavity. Treatment of the exchanged solid with hydrogen gas reduces the salts to colloidal metal particles. In the process, the viologen groups are also reduced taking on a blue or purple color. Exposure of the reduced material to air results in bleaching of the solid with production of hydrogen peroxide. If streams of H_2 and O_2 are passed through an aqueous suspension of the porous metal containing material H_2O_2 is produced. Mixed platinum–palladium catalysts worked best. This catalytic reaction is reminiscent of the reduction of O_2 by colloidal copper (130) or silver (131) metal dispersed on the surface of zirconium phosphate. As a stream of air was passed over these materials, the metal particles were oxidized and diffused back into the zirconium phosphate while the protons present at the exchange sites diffused to the surface forming either H_2O_2 or H_2O.

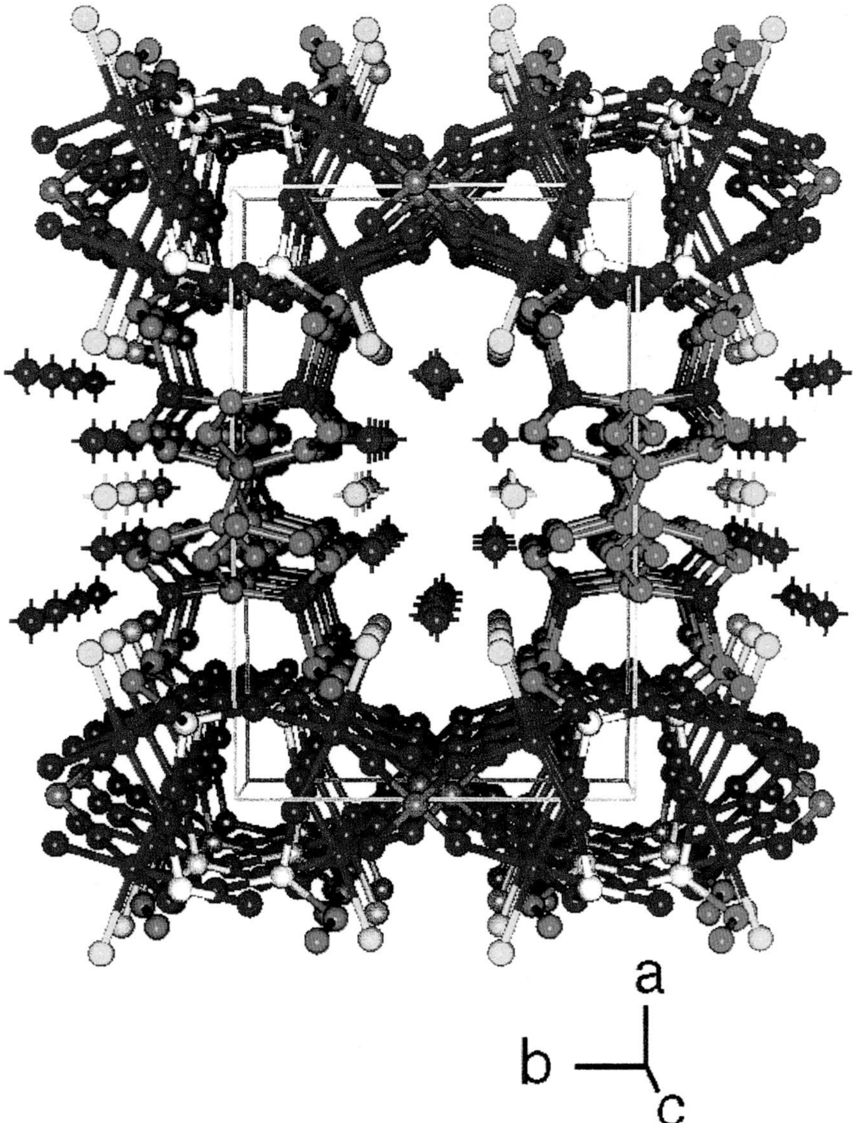

Figure 29. Structure of $Zr_2(PO_4)PVF_3$ viewed down the c-axis (shown in perspective). Phosphate phosphorus atoms are shown in orange, PV phosphorus atoms are gray, Zr atoms purple, F atoms are yellow, and the layer connecting ethyl viologens have carbon atoms in green and N in purple.

C. Crown Ether Containing Compounds

A series of zirconium crown ether phosphonates were prepared, not by ester interchange reactions as described for γ-ZrP, but by direct reaction between Zr(IV) and crown ether phosphonic acids (132). The phosphonic acids were prepared by a Mannich-type reaction (133) as shown in Scheme I. The reactions are similar to those described for the PMIDA derivatives. Only the interlayer spacings were evident from the X-ray powder patterns but the similarity of the ^{31}P NMR spectra to those obtained for the PMIDA derivatives allowed the structure types to be identified. These are represented in Fig. 30. Compound **5** was prepared from a mixture of the 1-aza-15-crown-5-*N*-methylene phosphonic

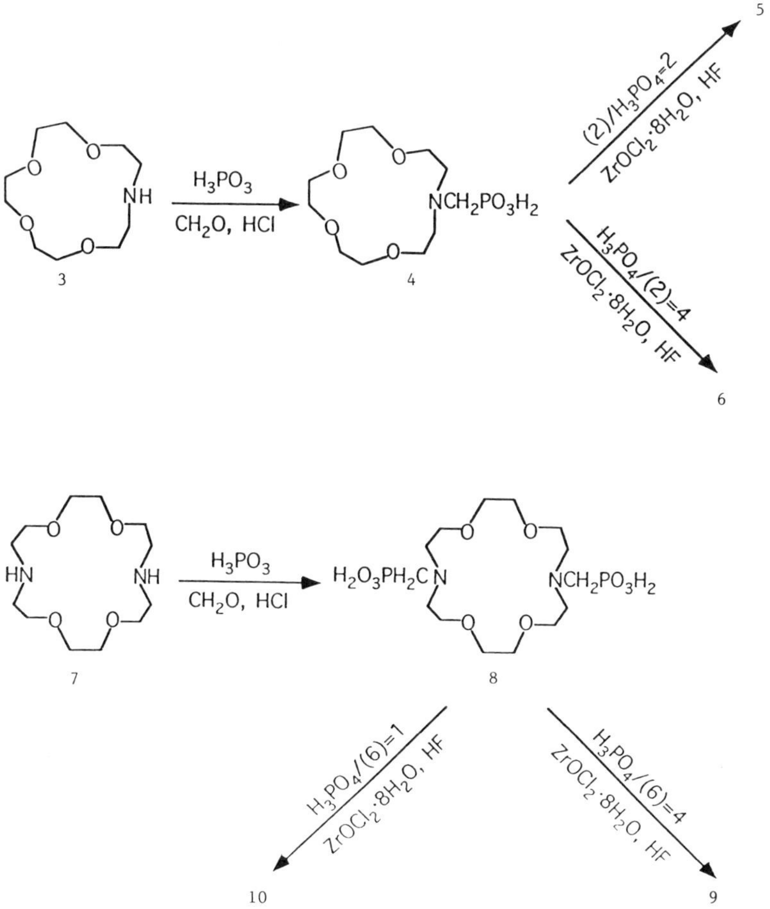

Scheme 1

acid (**4**), H_3PO_4, $ZrOCl_2 \cdot 8H_2O$, and HF. The ratio of the phosphonic acid to H_3PO_4 was 2 and the formula, deduced from elemental analysis and the NMR data is $Zr_2(PO_4)(O_3PCH_2NHC_{10}H_{20}O_4)_{1.17}(HPO_4)_{0.83}F_2Cl_{0.17} \cdot 1.69H_2O$. A schematic of compound (**5**) is given in Fig. 30(*a*). The ratio of crown ether to monohydrogenphosphate is greater than one, requiring some of the pendant crowns in one layer to lie opposite a crown ether in an adjacent layer. This arrangement accounts for the observed 20-Å basal spacing. Increasing the amount of phosphoric acid in the reaction mixture allows more HPO_4^{2-} groups to be incorporated on the layers resulting in a 13-Å interlayer spacing with close approach of a crown ether moiety in one layer to a phosphate group in an adjacent layer (compound **6**). These layered crown ether derivatives swell in water allowing easy sequestration of ions.

In similar reactions, a diaza crown ether [compound **7**, Scheme I] was con-

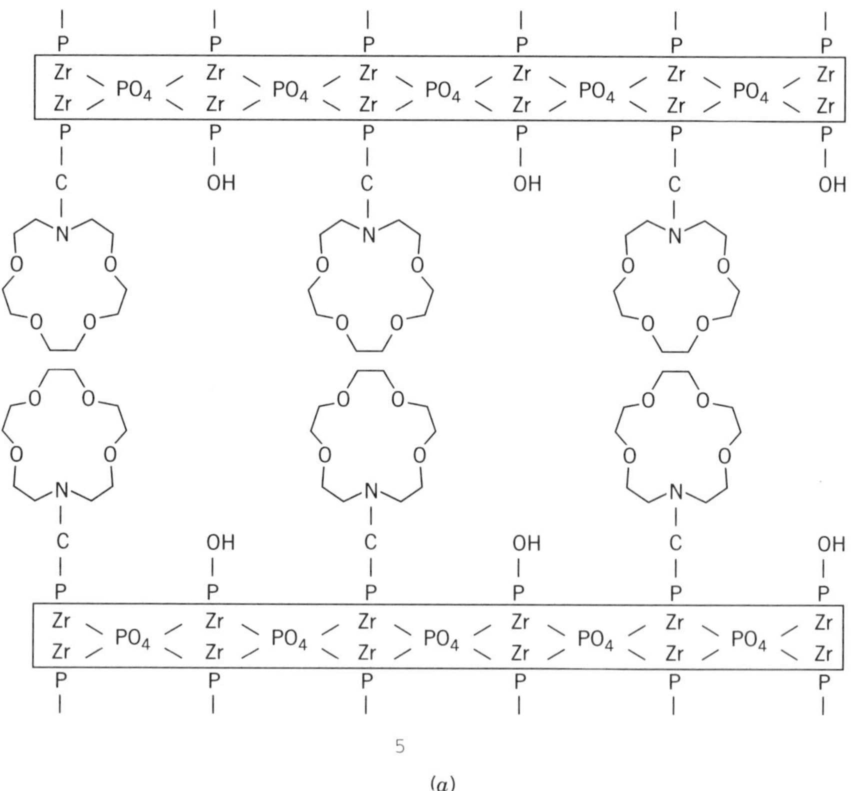

(*a*)

Figure 30. Schematic representation of (*a*) layered zirconium phosphate 1-aza-15-crown-5 phosphonate with a 20-Å interlayer spacing and (*b*) the cross-linked bisphosphonate.

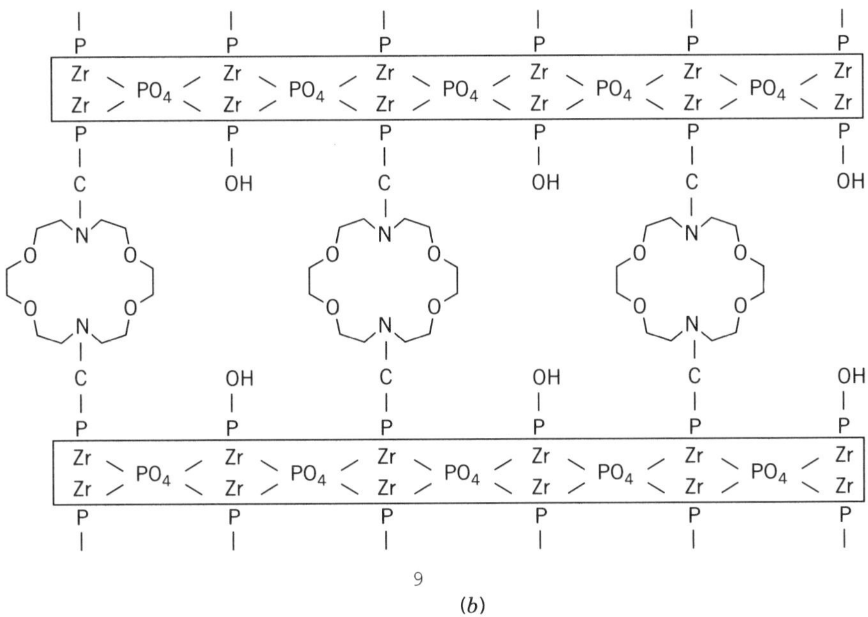

(*b*)

Figure 30. (*Continued.*)

verted to a bisphosphonic acid (compound **8**), to produce cross-linked compounds [Fig. 30(*b*), compound (**9**)]. They are the first examples of cross-linked or pillared crown ether α-zirconium phosphate-type derivatives. As microporous immobilized crown ether arrays they may exhibit synergistic effects coupled with the HPO_4^{2-} ion exchange sites, or exhibit restricted conformations because of the bridge structure.

VIII. SOME INTERESTING INTERLAYER REACTIONS

A. Preparation of Ester and Amide Phosphonates

In some instances, it is not possible to prepare a particular zirconium phosphonate in a direct way. Such a case in point is the direct synthesis of ester or amide phosphonates in the presence of acids, particularly HF, because of the accompanying hydrolysis. Omission of HF leads only to amorphous or poorly crystalline products. Thompson developed a method (134) to obtain such products from a preformed, highly crystalline carboxylic phosphonate, as shown in the following reactions:

$$Zr(O_3PCH_2CH_2CO_2H)_2 \longrightarrow Zr(O_3PCH_2CH_2COCl)_2 \longrightarrow Zr(O_3PCH_2CH_2COXR)_2 \qquad (18)$$

X = O, NH
R = alkyl, α, ω-alkyl

The zirconium carboxylic phosphonate could be prepared in highly crystalline form and was converted to the acid chloride with $SOCl_2$. The amide, $Zr(O_3PCH_2CH_2CONH_2)_2$, was prepared by passing ammonia vapors over the acid chloride. Similarly, the esters or alkylamides were prepared by heating mixtures of the acid chloride and alcohols or *n*-alkylamines. By using diamines, cross-linked diamides were formed. A large number of derivatives were prepared in this way.

A second amidation reaction was discovered by Kijima et al. (135). α, ω-Alkyldiamines were intercalated into $Zr(O_3PCH_2CH_2COOH)_2$, which upon heating to about 200°C condensed to form cross-linked diamides. In the case of ethylenediamine, the original carbonyl stretching vibration of the carboxyl group at 1730 cm^{-1} was shifted to 1680 cm^{-1} upon intercalation of the amine. This shift was caused by removal of the carboxyl proton in order to protonate the amino groups. Upon heating to 210°C, two new peaks at 1655 and 1560 cm^{-1}, indicative of amide formation, replaced the one at 1680 cm^{-1}. Similar results were obtained with other diamines.

B. Interlayer Polymerization Reactions

Thin films of polyacetylenes have shown promise for use in ultrafast signal processing because of their unusual nonlinear optical properties (136). Attempts have also been made to use diacetylene containing LB films as ultrathin negative resists for X-ray or scanning probe microlithography (137). Cao and Mallouk (138) were able to polymerize diacetylenes between the layers of divalent acetylene phosphates. In order to prevent undue polymerization of the acetylene during preparation of the required ligand, the diacetylene had to be prepared under mild conditions. We have seen that the Arbuzov reaction required temperatures of around 180°C but the phosphate ester $H_2O_3POCH_2CH_2{-}C{\equiv}C{-}C{\equiv}C{-}CH_2CH_2OPO_3H_2$ (7) was obtained at much lower temperatures as shown in Eq. 9.

$$HOCH_2CH_2C{\equiv}CH \xrightarrow[CuCl,\ O_2]{NH_4OH} (HOCH_2CH_2C{\equiv}C)_2 \xrightarrow{POCl_3}$$
$$(Cl_2\overset{\overset{\displaystyle O}{\|}}{P}OCH_2CH_2C{\equiv}C)_2 \xrightarrow{H_2O} (H_2O_3POCH_2CH_2C{\equiv}C)_2 \qquad (19)$$

Compound **7** was kept in solution with the desired metal salt by pH adjustment and then precipitated by slow diffusion of NH_3 into the solution. This procedure ensured a relatively good crystalline product without the necessity of refluxing. The polymerization of the diacetylenes is a radical addition reaction that yields high polymers only if the monomer units are aligned properly and if polymerization induces minimal strain in the host lattice (4). For these requirements to be met, the monomer chains need to be coplanar or nearly so. Geometric considerations also dictate that the separation between monomer head groups be between 4.7 and 5.2 Å. These conditions were met by the Mn, Zn, and Mg derivatives but not by Cd and Ca. Photolysis of the trivalent derivatives of Y, La, and Sm yielded only short-chain oligomers but the crystal structures of the three divalent ions (see Section X) constrained the R groups to be coplanar and the head groups to have a 4.8-Å separation. The diacetylenes were therefore prealigned in the correct way so that a higher degree of polymerization was obtained. However, the desired level of polymerization was not obtained probably due to the covalent bonding of the diacetylene groups to the layers acting to restrain their mobility. It was suggested that a more flexible linkage may overcome this problem.

Polyaniline (PANI) is an electronically conducting polymer that has exhibited a variety of different forms (139). Several attempts to form PANI in the interlamellar space of layered compounds have been reported (140). Rosenthal et al. (141) prepared copper exchanged α-ZrP, $CuZr(PO_4)_2$, added it to neat aniline and kept the stirred slurry at 60°C for 4 days. A pale violet solid was obtained that gave IR and electronic spectra for the nonconductive emeraldine form of PANI. The X-ray powder pattern indicated no layer expansion so that the reaction took place only on the surface. In contrast, the interaction between aniline (0.1 *M* aqueous solution) and $Zr[O_3P(CH_2)_3CO_2H]_{0.75}(HPO_4)_{1.25}$ occurred in the interlayer space. Initially, the aniline was intercalated as shown by a lattice expansion to 14.5 Å, an expansion of 4.5 Å. Upon polymerization, this value decreased to 13 Å indicating that the polymer rings lie parallel to the layers. The PANI resembled the emeraldine phase but also contained IR bands not present in that form of PANI.

C. Epitaxial Growth of Semiconductor Particles

In metals and semiconductors the valence electrons exhibit spatial delocalization to various degrees. Consequently, nanometer size clusters may have only partial band structure development (142) and therefore exhibit unique properties that are size dependent. These so-called semiconductor "quantum dots" or nanoclusters have optical spectra that can be wavelength tuned by varying their size (143). Therefore many different methods of limiting the growth of semiconductor crystallites such as CdS have been explored (144). The interlayer spaces of metal phosphonate layered compounds provide a well-ordered matrix in which to prepare nanoclusters (4). The technique is to use derivatives that contain sulfonic acid or carboxylic acid groups in which the Cd^{2+} or other suitable ions Pb^{2+}, Zn^{2+} lie midway between the layers. Exposure of the ion exchanged phosphonates to gaseous H_2Se or H_2S results in slow conversion of the metal ion to the chalcogenide, Eq. 19 (145).

$$Zr(O_3PCH_2CH_2O_2^-)_2Cd^{2+} \xrightarrow{H_2Se} Zr(O_3PCH_2CH_2CO_2H)_2/CdSe \qquad (19)$$

The CdSe particles 40–50 Å in diameter having a cubic zincblende structure form. The crystallites grow with the [111] zone axis parallel to the pseudo-sixfold axis (146) of the inorganic ZrO_3PO^- portion of the layer (8). Although the host phosphonate lattice is destroyed, the crystallites remain aligned. Electron diffraction patterns taken perpendicular to the host layers show only the (111) and (220) symmetry equivalent reflections of CdSe as if the reflections were from a single crystal with sixfold symmetry. Apparently, in the crystal growth process the host layers buckle but hold the crystallites to their initial orientation.

Lead sulfide particles, 30–40 Å in diameter within the interior of the phosphonate layers and 100–400 Å along the outer edges, were also prepared. These results indicate that in the interlamellar space the growth of the particles is controlled by the restriction of mass transport but along the edges the growth is less restricted.

D. Optical Properties and Electron-Transfer Reactions

The rates of intermolecular electron-transfer reactions vary exponentially with the distance between the electron donor and acceptor. Therefore, many different approaches to control the spatial arrangement of donor and acceptor molecules have been tried (147). Metal phosphonates, both in bulk and as thin films, allow microcontrol over the arrangement and/or segregation of donor and acceptor. In addition, they are generally transparent in the visible portion of the spectrum allowing the electron-transfer (ET) reactions to be studied spectroscopically. The first report of an ET reaction carried out in a metal phosphonate matrix was carried out by Clearfield and co-workers (148). They incorporated $[Ru(bpy)_3]^{2+}$ between the layers of $Zr(HPO_4)(O_3PC_6H_4SO_3H)$. Recall that this sulfonated product spontaneously exfoliates in water. Upon addition of $[Ru(bpy)_3]^{2+}$, the layers come together encapsulating the ruthenium bipyridyl cation. The interlayer spacing increased from 16 to 19.4 Å. This small increase for such a large complex ion indicates that the bipyridyl rings interdigitate between phenyl rings and approach closely to the phosphate groups. Diffuse reflectance spectroscopy was used to obtain electronic spectra for this solid complex. Both the π to π^*

band observed at 285 nm for an aqueous $[Ru(bpy)_3]^{2+}$ solution and the metal-to-ligand charge transfer (MLCT) band were strongly red shifted. Similar red shifts were observed for $[Ru(bpy)_3]^{2+}$ in other environments (147) and attributed to chemical, structural, and electrochemical characteristics of the $[Ru(bpy)_3]^{2+}$ complex itself. None of these explanations applied to the sulfophenyl matrix. The red shift was attributed to the interactions of the aromatic rings in the phosphonate with those in the ruthenium complex. In a subsequent study, methylviologen (MV^{2+}) and $[Ru(bpy)_3]^{2+}$ were simultaneously encapsulated between the layers of the sulfophosphonate (149). While the microenvironment of the sulfophosphonate was found to restrain the movement of the large ions through the interlayer space, diffusional processes led to dynamic quenching. A model combining diffusional quenching and sphere of action quenching was invoked to account for the observed quenching of $[Ru(bpy)_3]^{2+}$ by MV^{2+}.

In a continuation of this study, it was planned to separate the $[Ru(bpy)_3]^{2+}$ and MV^{2+} into separate layers. For this purpose, the staged compound illustrated in Fig. 7 was to be sulfonated. In water, the sulfophenyl layer would swell to incorporate and fix $[Ru(bpy)_3]^{2+}$ between these layers. Methylviologen or another suitable amine would then be intercalated between the phosphate layers. However, on sulfonation of the phenyl rings, it was found that the pendent groups redistributed themselves to make identical layers containing a random distribution of the two ligands (150). However, this separation of donor and acceptor was achieved by using a layer-by-layer self-assembled film (151). High-surface area α-ZrP was prepared using a previously published method (152). Exfoliation of this material gave single sheets of about 50-Å diameter (~7 Å thick). The α-ZrP lamellae were affixed to silica particles of about 300-Å diameter. Monolayer films of methylviologen affixed to a polystyrene polymer and the ruthenium bipyridyl species affixed to polyethylene were grown on alternate layers of the zirconium phosphate. Electron-transfer quenching of the luminescence of the ruthenium complex was extremely fast. However, when the solution-phase electron donor was added, the charge separated state was intercepted before ET could occur and a relatively long-lived (~30 μs) charge-separated state was obtained.

Photophysical studies have been carried out in which the donor and acceptors were made into bis(phosphonic) acids. These acids were in turn used to make thin films by the self-assembly methods previously described. For example, a viologen bis(phosphonate) layer was prepared from **8** and Zr and these layers alternated with those prepared from porphyrin tetraphosphonic acid.

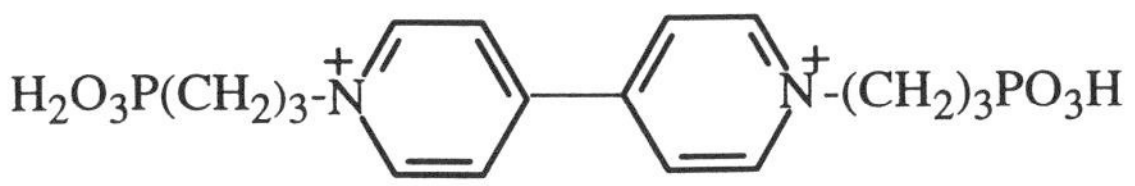

$$H_2O_3P(CH_2)_3\text{-}\overset{+}{N}C_5H_4\text{—}C_5H_4\overset{+}{N}\text{-}(CH_2)_3PO_3H$$

8

The porphyrin acted as the electron donor and the MV^{2+} as the acceptor (95). The fluorescence observed for this film was much less than that observed for the porphyrin above. This quenching was attributed to ET across the zirconium phosphate layer from the photoexcited porphyrin to the viologens in the adjacent layer. In a second prepared film, the viologen and porphyrin were separated by a layer containing $O_3P(CH_2)_3{-}C_6H_4(CH_2)_3PO_3$ as spacers. In this case, no quenching of the porphyrin fluorescence was observed.

Vermeulen and Thompson (124) prepared zirconium phosphonates from *N*,*N*′-diethyl-4,4′-bipyridinium dihalides, and $Zr(SO_4)_2 \cdot 4H_2O$ abbreviated as ZrPV(X), X = Cl^-, Br^-, or I^-. These compounds undergo photoinduced charge transfer producing a charge separated long-lived state. Dialkyl viologens can behave as one- or two-electron acceptors. The one-electron reduced product is a blue radical action that is oxidized rapidly by molecular oxygen and other radical scavengers (153). Spectroscopic studies were interpreted on the basis that the photoproduct was the blue radical cation, produced in the interlayer region of the zirconium phosphonate. The electron was donated by the halide, which was believed held in the layer depression position normally occupied by the water molecule in α-$Zr(HPO_4)_2 \cdot H_2O$ (8). The stability of the photoreduced state was attributed to trapping of the oxidized halogen and the inability of oxygen to diffuse into the interior of the phosphonate.

The structure of ZrPV(X) has already been described in Section VII.B and Fig. 27. The pure ZrPV(F) compound of composition $Zr_2(O_3PCH_2CH_2NC_5H_4C_5H_4NCH_2CH_2PO_3)F_6 \cdot 2H_2O$ did not undergo photoreduction. This finding is in keeping with the fact that photoexcited viologen cannot oxidize fluoride ion. Analysis showed that all the ZrPV(F) preparations contain Cl^- or Br^-, either through the presence of the 18-Å phase or replacement of some of the fluoride by chloride. Thus, the non-fluorine halogen content is sufficient to effect the photoreduction. Another possibility is that without hydrothermal treatment some of the product micropores that could imbibe Cl^- given the propensity of the cross-linked products to create pores (Section III.C).

More detailed mechanistic studies were conducted on both the bulk ethylviologen phosphonate and on similar thin films (125) grown on fused silica substrates. Photoreduction of the viologen on these thin films was very efficient with a quantum yield of 0.15. The thin films were very sensitive to air, but in the absence of air the blue color was found to persist indefinitely. These results probably stem from defects in the films that allow oxygen to diffuse into the interior as opposed to the denser bulk structure. Photoaction spectra show no halide dependence for photoreduction but the rates are in the order Cl > Br > I. To account for these observations, it was proposed that the primary photoprocess involved excitation of the viologen followed by oxidation of the halide to X· (125). Deuterium isotope effects (K_H/K_D = 3) suggest that the oxidized halide ions abstract hydrogen atoms from the methylene groups of ZrPV(Cl).

The NMR data were presented to show that structural rearrangements took place leading to the final charge separated state. The formation of HCl and subsequent structural rearrangements are irreversible. In contrast, the ZrPV(Br) compound shows a high degree of reversibility in the photoreduction. When a film of ZrPV(Br) was photolyzed in a closed system to a high level of viologen reduction and the UV light was removed, the blue color gradually faded. Similar experiments were carried out under dynamic vacuum and the sample color was again found to fade but about 40% of the viologen was still in the reduced state (37). A possible explanation for these results considers that some Br forms Br_2 that might escape into the gas phase resulting in the permanent reduction. However, Br or even Br_2^- may participate in a back-electron-transfer reaction before forming volatile bromine accounting for the reversibility.

Because of the nonporosity of ZrPV(X) and its effect upon exclusion of oxygen, it was of interest to prepare porous versions of this compound. The strategy was to include either H_3PO_4 or H_3PO_3 in the reaction mixture in the hopes of obtaining a mixed-ligand product in which the large bis(phosphonate) groups are spaced by the smaller phosphate or phosphite groups. In general, the reactant mix containing HF was refluxed for 7 days yielding three separate products, none of which yielded X-ray powder patterns that matched those of the original ZrPV(Cl) compound (126). The interlayer spacing for the phosphate-containing derivative prepared in HF was 13.6 Å and that prepared in HCl was about 18.5 Å. We have already described the structure of the 13.6-Å phase in Section VII.B as pictured in Fig. 28(*b*). In the 18.5-Å phase, NMR spectra showed that the phosphate groups are protonated and the charge of the MV^{2+} is compensated by halide ions as shown in Fig. 28(*a*). The phosphite-containing derivative also had an approximate 18-Å basal spacing and its solid state MAS NMR spectrum indicated the presence of only phosphonate (+5.5 ppm) and phosphite PH(−9.5 ppm) chemical shifts. Thus, it was considered to have a structure similar to that of the phosphate–phosphonate of Fig. 28(*b*). These porous compounds underwent photoreduction of the viologen at a faster rate than ZrPV(X) but only under anaerobic conditions. However, the mechanism of photoreduction was considered to be similar to that of ZrPV(X).

IX. VANADIUM PHOSPHONATES

In a recent review, Khan and Zubieta (154) described a large body of oxovanadium and oxomolybdenum phosphonates. Much of this review dealt with polynuclear clusters that we will touch on in Section IX.E. In this section, we will describe the oxovanadium phosphonates that are analogues of the zirconium-type compounds discussed to this point.

A. Synthesis and Structure of Oxovanadium Phosphonates

Vanadyl phosphonates are prepared by addition of V_2O_5 to a hot alcohol solution of the desired phosphonic acid (155, 156). A small amount of aqueous mineral acid is added as a catalyst for the reaction. The alcohol not only acts as solvent but also reduces V^V to V^{IV}. The alcohol is incorporated between the layers as an intercalant but it can be removed by extensive water washing. The general reaction in benzyl alcohol is as shown in Eq. 21.

$$V_2O_5 + 2RPO_3H_2 + 3BzOH \xrightarrow[85^\circ]{H^+} 2VO(RPO_3)\cdot H_2O\cdot BzOH + PhCHO + H_2O \qquad (21)$$

Removal of the alcohol by water washing yields a dihydrate, $VO(O_3PR)\cdot 2H_2O$.

Subsequently, single crystals of the phenylphosphonate $VO(O_3PPh)\cdot H_2O$ were grown hydrothermally in water at 200°C. The structure consists of layers of corner-sharing VO_6 octahedra and PO_3C tetrahedra (157). The phenyl groups extend into the interlamellar space above and below the plane as shown in Fig. 31(*a*). The repeat distance perpendicular to the layers is 14.14 Å as compared to 14.82 Å in $Zr(O_3PPh)_2$ (18). Within the oxide portion of the layer the octahedra share an axial oxygen with a neighboring vanadium to form —V=O—V=O chains that run in the *b*-axis direction. These chains have alternating short, 1.610(9) Å, and long, 2.14(1) Å, VO bonds. The phosphonate groups bridge across two vanadiums in the same chain and a single metal atom in an adjacent chain as shown in Fig. 31(*b*). A molecule of water completes the coordination sphere of the VO_6 octahedron.

The dihydrate derived from washing the alcohol intercalate has a structure resembling that of newberyite. This difference is evident in the unit cell dimensions for the several compounds collected in Table XIII. The monohydrate is monoclinic, space group *C2/c*, whereas the dihydrate is orthorhombic with the *b* axis as the interlayer distnace. Indirect evidence from magnetic and thermogravimetric data suggest that the dihydrate has a structure similar to that of newberyite, $MgHPO_4\cdot 3H_2O$. The layer structure of newberyite is shown in Fig. 32 (158). The dihydrate has both water molecules coordinated to vanadium, in positions trans and cis to the vanadyl oxygen atom. This positioning of oxygen atoms eliminates the —V=O—V— chains that are present in the monohydrate. The octahedral coordination of the metal atoms is completed by bonding from oxygen atoms of three different phosphonate groups. This bonding positions the phenyl groups far enough apart that the phenyl groups from adjacent layers interpenetrate, accounting for the small interlayer distance. Upon exposure to alcohol, reintercalation occurs up to a limit of 1 mol mol^{-1} of

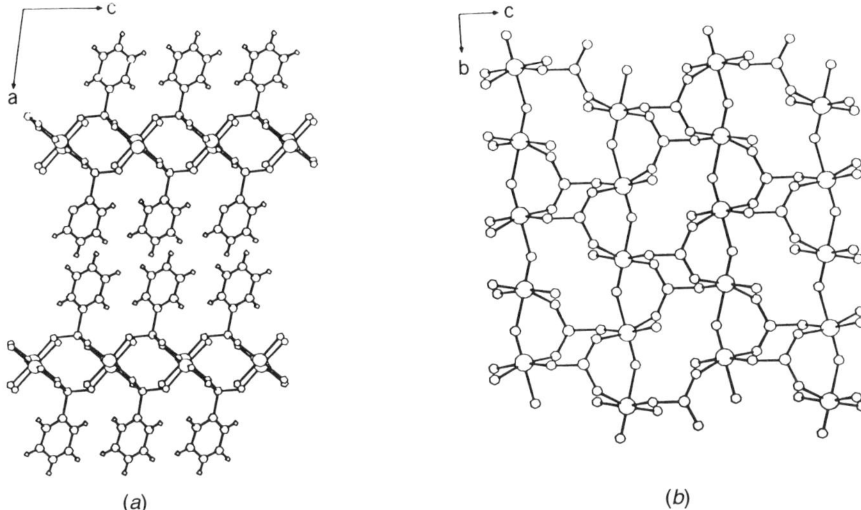

Figure 31. (*a*) View of the *ac* plane of $VO(O_3PPh) \cdot H_2O$ showing the vanadium/phosphorus oxide layer separated by a bilayer of phenyl groups and (*b*) a view of the *bc* plane illustrating the —V—O—V— chains of the corner sharing VO_6 octahedra. Phenyl groups have been omitted for clarity. [Reprinted with permission from G. Huan, A. L. Jacobson, J. W. Johnson, and E. W. Corcoran, Jr., *Chem. Mater.*, *2*, 91 (1990). Copyright © American Chemical Society.]

phosphonate. This upper limit of alcohol uptake indicates that the intercalated alcohol molecules are associated with specific sites between the layers.

A series of vanadyl alkylphosphonates of general composition $VO(O_3PC_nH_{2n+1}) \cdot H_2O \cdot BzOH$ was prepared by the alcohol reflux method (156). The interlayer spacings were plotted as a function of alkyl chain length both with the benzyl alcohol and as monohydrates. In the process of benzyl alcohol removal the interlayer spacings are reduced by about 6.3 Å. The slope of the alcohol free compounds is 1.05 Å per CH_2 unit. An all-trans polymethylene chain has a repeat distance of 1.27 Å per CH unit. For a bilayer, the slope would be 2.54 Å. Since the slope is less than one-half of this value, the alkyl groups from one layer must interpenetrate those of the adjacent layer. In addition, the chains must be inclined to the layers at an angle given by $\sin^{-1}$ 1.05/127° or 56° in order to account for the observed interlayer spacings.

We have seen that vanadyl phosphonates prepared hydrothermally in water differ from those synthesized by refluxing in alcohol. To further elucidate the structural relationships, a series of preparations was carried out with alkyl phosphonic acids and V_2O_3 in aqueous media, hydrothermally at 200°C (159). The products are all in oxidation state IV, oxidation of the V^{III} presumably being effected by dissolved oxygen. Two types of compounds were obtained. In the

TABLE XIII
Synthesis Method, Composition, and Unit Cell Dimensions of Vanadyl Phosphonates

Compound	United Cell Dimensions (Å)	Structure Type	Preparatory Method	Reference
$VO(O_3PPh) \cdot H_2O$	$a = 28.50(3)$ $b = 7.18(2)$ $c = 9.42(2)$ $\beta = 97.1(2)°$	H[a]	$V_2O_3 + PhPO_3H_2$ Hydrothermal in water 200°C, 4 days	157
$VO(O_3PPh) \cdot H_2O \cdot EtOH$	$a = 9.96$ $b = 12.06$ $c = 9.69$	N[b]	V_2O_5 + Phosphonic acid Reflux in 95% alcohol 6 days	155
$VO(O_3PPh) \cdot 2H_2O$	$a = 10.03$ $b = 9.69$ $c = 9.77$	N	Wash ethanol compound with water at 60°C Expose to moisture	155
$VO(O_3PC_nH_{2n+1}) \cdot 1.5H_2O$ $n = 1–3$	$a = 17.281(5)$, $b = 7.499(6)$ $c = 9.415(4)$, $\beta = 106.04°$	V[c]	Hydrothermal as for type H	159
$VO(O_3PC_nH_{2n+1}) \cdot H_2O$	$n > 3$	H	Hydrothermal as for $VO(O_3PPh_5)$	159

[a]Structure as described in (157) (—V=O—V=O chains).
[b]Newberyite structure.
[c]The $VO(HOPO_3) \cdot 0.5H_2O$ structure (160).

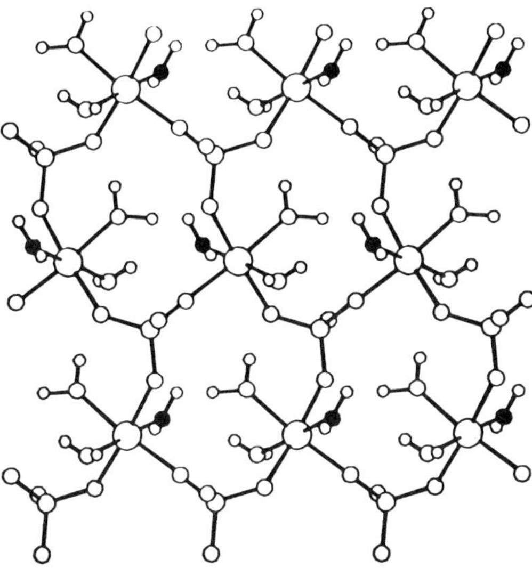

Figure 32. View of the layer structure of newberyite, $MgHPO_4 \cdot 3H_2O$. In $VO(O_3PPh) \cdot 2H_2O$, V replaces Mg in newberyite, a phenyl group replaces the hydroxyl group, and the water molecule (stippled) is replaced by the vanadyl oxygen. [Reprinted with permission from G. Huan, A. L. Jacobson, J. W. Johnson, and E. W. Corcoran, Jr., *Chem. Mater.*, 2, 91 (1990). Copyright © American Chemical Society.]

series of composition $VO(O_3PC_nH_{2n+1}) \cdot yH_2O$, compounds with n greater than 3, $y = 1$ have structures identical to those prepared in benzyl alcohol for which the alcohol has been removed by water washing (type H in Table XIII). Compounds for which $n = 1$–3 have $y = 1.5$ and a different layer type. Poorly developed single crystals were grown for $VO(O_3PMe) \cdot 1.5H_2O$, which gave an apparent monoclinic unit cell (Table XIII) in space group $P2_1/a$. The structure is very similar to that of $VO(HOPO_3) \cdot 0.5H_2O$ (160) with the methyl groups replacing the hydroxyl groups. The extra water molecule in the methylphosphonate is thought to occupy interlayer sites. The layer is built up of face-shared $V_2O_8(H_2O)$ dimers connected through corners by PO_3C tetrahedra. The interlayer spacing is 8.30(2) Å. The differences in the two structure types (H and V) are evident in their TGA curves, interlayer spacings, and magnetic properties (see below).

B. Intercalation Reactions

We have already seen that the synthesis of vanadyl phosphonates in alcohol leads to intercalation of 1 mol of alcohol (156). The vanadyl phosphonates synthesized this way may have the newberyite structure (Table XIII) and it was shown that a series of *n*-alkanols containing 2–10 carbon atoms yield intercalation derivatives whose interlayer spacings lie on a straight line with slope 1.1 Å per CH_2 unit. Closer inspection of the plot revealed that there is strong odd–even effect (155). However, the alcohol intercalation line parallels that for $VO(O_3PC_nH_{2n+1})\cdot H_2O$. In the newberyite structure the water molecule is coordinated to V in an equitorial position, whereas the alcohol and vanadyl oxygen occupy the axial positions. Consequently, the alkanol chains lie parallel to the phosphonate alkyl chains. This positioning of the chains is supported by the closeness of the slopes in the plots of interlayer spacing versus chain length, 1.05 Å per CH_2 unit for the alkyl chains and 1.1 Å per CH_2 for the alkanol chains.

The vanadyl phosphonates exhibit molecular recognition effects in their alcohol intercalation reactions (156). This selectivity can be controlled by the nature of the organic group in the phosphonic acid incorporated into the vanadyl phosphonate. Adsorption of alkanols by $VO(O_3PPh)\cdot 2H_2O$ requires several hours, but the same alcohols are intercalated rapidly by vanadyl alkylphosphonates. This difference in sorption rates is attributed to the greater flexibility of the alkyl chains. Vanadyl hexylphosphonate, $VO(O_3PC_6H_{13})\cdot H_2O$, intercalates the primary alcohols *n*-butanol and isobutyl alcohol but not *sec*-butanol. In general, primary alcohols can be separated from secondary and tertiary alcohols with high selectivity using the vanadyl alkylphosphonates (156).

Recently, alkanediols, $HO(CH_2)_nOH$ (n = 2–5), have been intercalated into layered vanadyl phenylphosphonate, $VO(O_3PPh)\cdot 2H_2O$, and vanadyl ethylphosphonate, $VO(O_3PEt)\cdot 1.5H_2O$. These compounds were prepared either by heating the preformed phosphonate in the desired alkanediol or by preparation in the alkanediol (161). Interlayer spacings of the alkanediols are listed in Table XIV. One mole of diol was intercalated in $VO(O_3PPh)\cdot 2H_2O$. In the C_4 and C_5 derivatives 1 mol of water was also present but the C_2 and C_3 intercalates were water free. In the C_2 and C_3 cases, thermal gravimetric analysis shows that 0.5 mol of diol is removed at a time, indicating that they exist in two

TABLE XIV
Interlayer Spacings of Alkanediols Intercalated into $VO(O_3PPh) + 2H_2O$ and $VO(O_3P\,Et)\cdot 1.5\,H_2O$

Intercalated Alkanediol	d(Å)[a]	d(Å)[b]
$OH(CH_2)_2OH{=}C_2$	10.74	8.80
$OH(CH_2)_3OH{=}C_3$	12.26	9.50
$OH(CH_2)_4OH{=}C_4$	14.24	11.27
$OH(CH_2)_5OH{=}C_5$	15.20	11.68

[a]R = Ph (d = 9.82 Å)
[b]R = Et (d = 9.71 Å)

different sites. The host molecule has the newberysite structure with one water molecule in the equatorial position and one in the axial position. The axial position water molecule is the less tightly bonded, being replaced by alcohol in the $VO(O_3PR) \cdot H_2O \cdot R'OH$ type compounds (156, 161). It is suggested that in the case of the C_2 and C_3 diol intercalates of the phenylphosphonate that both sites are occupied by the diol molecules.

In the case of the C_2 and C_3 intercalates of $VO(O_3Et) \cdot 1.5H_2O$, anhydrous products, containing 1 mol of alkandiol, were obtained (162). Although the diols were removed in one rather than two steps, Gendraud et al. (162) used the similarity of IR spectra to conclude that the ethylphosphonate intercalates had the same structure as the phenylphosphonate. For this to occur, they state that the structure of the ethylphosphonate would have to rearrange. This finding implies that the structure of the ethylphosphonate host is type V, although they never state that this is the case. The type V structure has been shown to form hydrothermally but Gendraud et al. (162) prepared it by the reflux method. However, they concluded from extended X-ray absorption fine structure (EXAFS) spectra that the vanadyl ethylphosphonate appears to contain $V_2O_8(H_2O)$ dimers.

The composition of the C_4, C_5 intercalated phenylphosphonates are $VO(O_3PPh) \cdot H_2O \cdot HO(CH_2)_nOH$. The TGA results show that the diol is lost at 130–200°C followed by the water molecule. This result leads to the conclusion that the diol bonds to vanadium in the axial position. In the vanadylethylphosphonate only 0.5 mol of diol is intercalated indicating that the diols bridge across the layers.

Several vanadyl organophosphonates were exposed to amines either as a liquid or vapor (163). All these reactions resulted in formation of $VOHPO_4 \cdot 0.5H_2O$ by scission of the P—C bond followed by intercalation of the amines. On very long exposure to amines (~ 1 month), oxidation to the +V state occurs presumably through the agency of dissolved oxygen to form $VOPO_4 \cdot 2H_2O$ followed by intercalation of 2 mol of amine.

Torgerson and Nocera (164) demonstrated that intercalation reactions of vanadyl phosphonates can be used to control optical properties. 2-Naphthalenemethanol, $C_{10}H_7CH_2OH$, was intercalated into a series of vanadyl alkylphosphonates in which the carbon chain ranged from one to eight carbons in length. A plot of interlayer spacing as a function of the number of carbon atoms in the chain yielded a straight line with slope of 1.01 Å per CH_2 unit. The optical spectra (UV–vis and luminescence) of the naphthalene chromophore was directly affected by the change in interlayer spacing. For the methylphosphonate an intense absorption manifold is observed with maxima at 220 and 280 nm. In addition, a highly structured fluorescence band of naphthalene, with λ_{max} = 370 nm, was obtained. Two prominent changes in the spectra were obtained on increasing the carbon chain by one CH_2 group. The luminescence spectrum was now that of naphthalene excimers and not individual monomers. Excimer luminescence results form the overlap of the π systems of a ground state and an electronically excited naphthalene that requires a cofacial arrangement of the naphthalenes. The absorption spectrum indicated that this cofacial arrangement is "preorganized" within the interlayer region. The interdigitation of the naphthalene rings in the ethylphosphonate establishes a high degree of interlayer order in which the naphthalene rings were cofacially organized and overlapping. This order is presumably present for the methylphosphonate but the decreased interlayer spacing results in greater overlap of the naphthalene rings eliminating the spectral features of the cofacially overlapping rings. As the alkyl chains become too long, the spectra of isolated naphthalene chromophores is once again obtained.

A second way to control the organization of the naphthalene groups is to attach them directly to the layers. Accordingly, a series of compounds of composition $VO(O_3PC_{10}H_7) \cdot H_2O \cdot ROH$ (R = methyl, ethyl, butyl, hexyl, and octyl) were prepared. The interlayer spacings are now controlled by the alcohol intercalant. The excimer emission at $\lambda \cong 410$ nm was only present when alcohol was absent. In these systems, the interlayer spacing is too large to permit overlay of the naphthalene rings.

Tongerson and Nocera (164) conclude that the vanadyl phosphonates allow incremental tuning of an organized system in an incremental manner and is not confined to optical properties. Modification of the phosphonate by introduction of functional groups, metal-binding sites, paramagnetic centers, and unsaturated olefins provides a framework for materials design. A start has been made in this direction by incorporation of the dimolybdenum core as $[Mo_2(MeCN)_8]^{4+}$ between the layers of $VO(O_3PCH_2CH_2CO_2NH_4^+)$ (165).

C. Magnetic Properties

We have already described results for the preparation of alkylphosphonates by hydrothermal methods (159) leading to structure type V for $n = 1$–3 and type H for $n > 3$, where n equals the number of carbon atoms in the alkyl chain (Table XIII). These differences in structure types are reflected in the magnetic properties of these compounds. The magnetic susceptibility for the larger alkyl chain compounds are similar and could be fit to the Curie–Weiss expression

$$\chi = \chi_0 + C/(T + \theta) \qquad (22)$$

for temperatures greater than 40 K. Below 40 K, weak antiferromagnetic coupling is observed, resulting in departure from Curie–Weiss behavior. Compounds prepared by the benzyl alcohol method also exhibited similar magnetic behavior.

For compounds with $n = 1$–3, the magnetic susceptibility as a function of temperature displays a maximum at a temperature below 100 K, characteristic of antiferromagnetically exchange coupled dimers (159). The susceptibility data could be fit with a Bleaney–Bowers expression for an isolated dimer model containing two $S = \frac{1}{2}$ cations with isotropic g tensor (166). Similar magnetic susceptibility behavior at temperatures below 100 K was found for two additional compounds, vanadyl phosphite, $VO(HPO_3)\cdot H_2O$ and the vanadyl butylphosphonate hydrates with 1 or 1.5 H_2O (166).

Huan et al. (167) attempted to prepare a continuous solid solution of two phosphonate ligands of the type $VO(O_3PPh)_{1-y}(O_3PMe)_y\cdot 2H_2O$. The reactions were carried out hydrothermally as before. The reactant ratios were selected according to Eq. (23) with increments of $m = 0.1$.

$$2.5(1 - m)PhPO_3H_2 + 2.5m\ MePO_3H_2 + V_2O_5 \longrightarrow$$
$$VO(O_3PPh)_{1-y}(O_3PMe)_y\cdot xH_2O \qquad (23)$$

A continuous series of compositions was not obtained. Rather, compounds with $y = 0.5$ and 0.75 were the only ones that formed in this system. Both compounds exhibited magnetic susceptibility behavior of the $VO(HPO_4)\cdot 0.5H_2O$ or $VO(O_3PMe)\cdot H_2O$ types that we have already described. This magnetic data corroborated the X-ray powder data indicative of layer structures. The intense reflections were all of the 00ℓ type with interlayer spacings of 11.2 Å for the $y = 0.5$ compound and 19.4 Å for the $y = 0.75$ compound.

The layer spacing for the compound with $y = 0.5$ is the sum of the layer spacings of the individual pure methyl and phenylphosphonates divided by 2. On this basis, Huan et al. (167) indicated two possible structures, types A and E in Fig. 8. A decision between these two structures could not be made with certainty from the available data.

The interlayer spacing of 19.4 Å for the $y = 0.75$ compound is close to the sum of the value for the methylphosphonate and the $y = 0.5$ compounds (8.3 + 11.2 Å). Again, two structures as shown in Fig. 33(*a* and *b*) are possible. The Huan et al. (167) suggest that the structure in (*b*) is more likely because it maximizes the use of space. The fact that the 002 reflection is stronger than 001 also favors the structure *B* as was demonstrated for the interstratified zirconium compound $Zr(O_3PPh)(HPO_4)$ shown in Fig. 8.

Nocera and co-workers (168) found an interesting correlation between the magnetic properties of layered vanadyl phosphonates and the Hammett σ function. A series of vanadyl phosphonates of the type $VO(O_3PC_6H_4X)\cdot nH_2O$ with $X = NO_2$, F for $n = 1$ and Cl, Me for $n = 1.5$. The nitro compounds behave as simple paramagnets exhibiting Curie–Weiss behavior with a Weiss constant of about 0.1 K. In contrast, the compounds with X = H and F, $n = 1$ show a

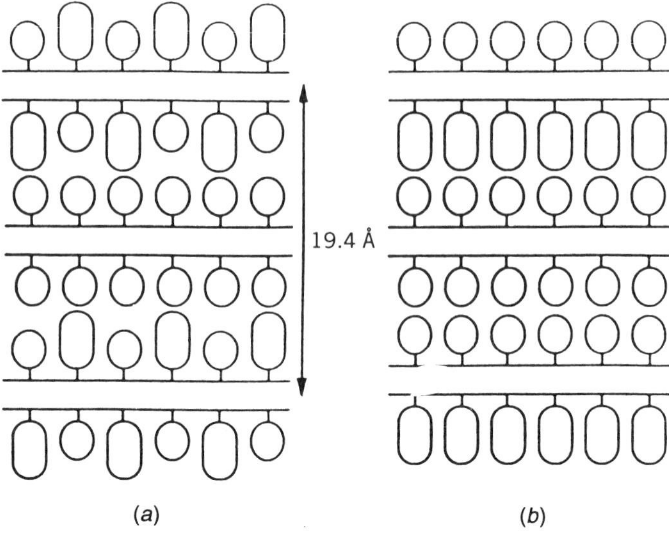

Figure 33. Structural Models for $VO(O_3PPh)_{0.25}(O_3PMe)_{0.15}$. [Reprinted with permission from G. Huan, A. J. Jacobson, J. W. Johnson, and D. P. Goshorm, *Chem. Mater.*, *4*, 661 (1992). Copyright © 1992 American Chemical Society.]

maximum in the χ versus T curve at about 5 K and then turn downward revealing antiferromagnetic couplings between the V^{IV} centers. All three compounds have the same layer structure in which the paramagnetic vanadium centers are joined via chair-like $V(OPO)_2V$ links. These chair subunits are then stacked together resulting in $V{=}O\cdot\cdot\cdot O$ chains, as already described (157). A similar layer structure was found in the alkali metal $VOPO_4 \cdot nH_2O$ type compounds (169). It is felt that the dominant magnetic coupling elements in these phosphonates is thought to be the chair-like exchange pathways while the $V{=}O{-}V{=}O$ chains play a less significant role. In these three compounds, the coupling strength is in the order Ph > C_6H_4F > $C_6H_4NO_2$. This order is opposite to their Hammett σ terms (σ_p) of 0, 0.15, 0.81, respectively. In these systems, where the $O{-}P{-}O$ pathway dominates exchange interactions between V centers, the magnetic couplings respond to variations in the σ values that, in turn, reflect the electronic environment at the P atom. Weaker coupling with increase in σ appears reasonable based on frontier orbital grounds. As the substituents become more electron withdrawing, the energy mismatch between phosphonate orbital energies and vanadyl *d* levels widens leading to decreased interaction and greater coupling (168).

The compounds with X = Cl or Me_3 for n = 1.5 have the $VO(HOPO_3)\cdot 0.5H_2O$ structure type (Table XIII). These layers contain vanadyl centers interacting directly through μ^2-bridging oxygens of $V(\mu^2\text{-}O)_2$ dimers. It is believed that the change in the layer type is derived from the steric bulk of the substituents. The maxima in the χ versus T plots are above 50 K. The higher $T(\chi_{max})$ in these systems is consistent with the stronger coupling ability of the μ-oxo dimer pathway. Consequently, the magnetic coupling in these compounds show little sensitivity to variation of the ring substituents.

D. Relation to VPO Catalysis

Vanadium phosphates are excellent catalysts for the oxidation of C_4 hydrocarbons to maleic anhydride (170). Vanadyl(IV) pyrophosphate, $(VO)_2P_2O_7$, is thought to be the active phase of the commercially used catalysts (171). The pyrophosphate is usually made by heating $VO(HPO_4)\cdot 0.5H_2O$. We have seen in Section IX.C that vanadyl alkylphosphonates from methyl to butyl have the same layer structure as vanadyl monohydrogen phosphate. Indeed, $(VO)_2P_2O_7$ was formed (166) upon thermal decomposition of the alkylphosphonates. Vanadyl phosphite, $VO(HPO_3)$, behaved similarly. Recall that the phosphite was similar in its magnetic properties to the small-chain phosphonates indicating the presence of V_2O_8 dimers in this compound also. In fact, a superior catalyst was obtained from vanadyl phosphite attributed to the low temperature at which it converts to the pyrophosphate.

E. Additional Structures

A new layer type was obtained from combination of V_2O_3 with the bis(phosphonic acid) $CH_2(PO_3H_2)_2$ (172). The compound $(VO)_2[CH_2(PO_3)_2]\cdot 4H_2O$ is layered with a = 12.805(4), b = 10,592(3), c = 15.037(5) Å, space group *Pbca* (orthorhombic) (172). There are two types of vanadiums in the structure. V1 structure is coordinated by one terminal oxygen atom, three water oxygen atoms, and two oxygen atoms from chelation by a phosphonate group (Fig. 34(a)]. V2 is coordinated by one terminal oxygen, one water molecule, two oxygen atoms from separate $[O_3PCH_2PO_3]^{4-}$ groups, and it is chelated by a phosphonate group. Each methylene bis(phosphonate) group chelates both V1 and V2 atoms forming six-membered rings as shown in Fig. 34(a). The remaining oxygen atoms act as monodentate ligands to two additional V2 atoms to link the structure in two dimensions. The relationship of the layers to each other is shown in Fig. 34(b). This compound followed the Curie–Weiss behavior characteristic of the larger chain alkylphosphonates.

A new series of vanadyl, vanadium(V), methylphosphonates of the general composition $M(VO_2)_3(O_3Me)_2$, M = NH_4, K, Rb, or Tl, was recently synthesized by a hydrothermal procedure at 150–190°C (173). These new compounds are based on the hexagonal tungsten oxide (HTO) motif that contains triangular

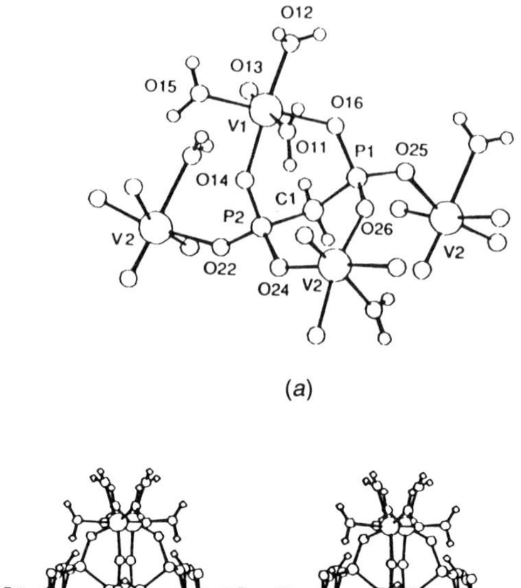

(a)

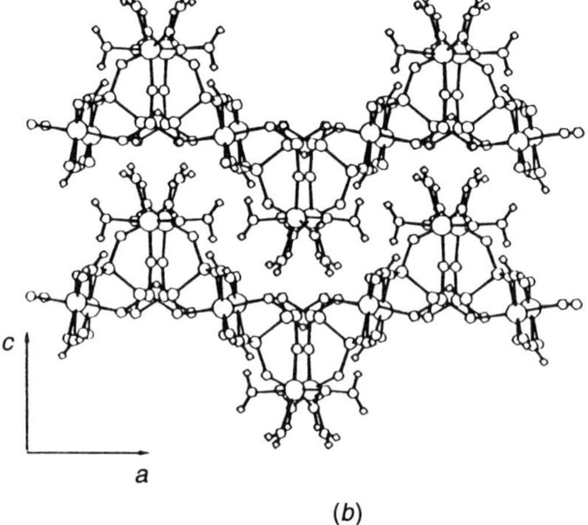

(b)

Figure 34. The coordination of the $[O_3PCH_2PO_3]^{4-}$ group and V1 atom in $(VO)_2[(O_3P)_2CH_2]\cdot 4H_2O$ (*a*) and the arrangement of the layers (*b*). Large circles represent V atoms. [Reproduced with permission from (172).]

and hexagonal voids. The potassium phase is hexagonal (trigonal) $a = 7.139(3)$, $c = 19.109(5)$ Å, and space group *R*32. The compounds are built up from corner sharing VO_6 octahedra and phosphonate tetrahedra. The vanadium atoms are shifted in the [110] direction creating distorted octahedra with two long, two medium, and two short V—O bonds. Each VO_6 unit shares four of its vertices with similar neighbors. This distortion allows linkage through short (1.66 Å) and long (2.17 Å) V—O bonds of the type V=O—V′. The P and C atoms sit on a threefold axes forming three equivalent P—O bonds each of which bridges to a vanadium atom. Thus, the equatorial oxygens are all vanadyl oxygen atoms and the axial oxygens are from the phosphonate groups. Consequently, the layers of octahedra are capped by phosphonate tetrahedra. The layers have an ABC repeat motif in the *z* direction and are staggered such that every methylphosphonate group points toward a six-ring window in the adjacent layer. The M^+ ion resides in the void spaces created by the layer arrangement.

The diversity of vanadium phosphonate chemistry is shown by the recent synthesis of four additional compounds based on $CH_2(PO_3H_2)_2$ (174). The methods of preparation and pertinent data are given in Table XV. Illustrations of the structures of two of the compounds are presented as Fig. 35(*a* and *b*). It is seen in Table XV that small changes in the ratio of reactants or temperature of preparation lead to the different structures. This sensitivity to the nature of the vanadium source and oxidation state and reaction conditions is responsible for the very large number and variety of vanadium phosphonates. The structural versatility of the vanadium organophosphonate system and the bonding and coordination modes are discussed in (174) and the previously mentioned review (154).

Another way to alter structure is to use the directing influence of a template. The hydrothermal reaction involving a mixture of $RbVO_3$, $PhPO_3H_2$, $EtNH_2Cl$, and $(Me)_2NH_2Cl$ in water was heated at 160°C for 4 days to yield green platelets (175). A layered compound of composition $[(Et)_2NH_2][(Me)_2NH_2]$-$[V_4O_4(OH)_2(PhPO_3)_4]$ was obtained. A representation of the structure is shown in Fig. 36. In this compound, the layer consists of corner sharing VO_5 square pyramids capped on one side only by the phosphonate groups. The V=O groups are directed away from the phenyl groups. The vanadyl groups in the adjacent layer are directed toward those of the neighboring layer creating a hydrophilic interlayer in which the amino cations reside. This positioning of the vanadyl groups then creates a hydrophobic region filled with phenyl groups from two adjacent layers. Discrete $(V_2O_2OH)^{3+}$ binuclear units within the layers are bridged by phosphonate groups to form the layer motif. Many other metal phosphonates exhibit separation of hydrophobic and hydrophilic regions. For example, the staging of the $Zr(O_3PPh)(HPO_4)$ compound (Section III.B) has been cited as resulting from the desire of like groups to segregate in this way. Other examples will be described in succeeding sections.

The type of product obtainable in the V—O—P system, where the phosphorus atom represents phosphonate rather than phosphate or phosphite is critically dependent on the reaction conditions. Changes in factors such as temperature, reactant ratios, pH, source of oxovanadium, and templating reagents can lead to new and unexpected results. We shall close this section by describing a few more examples. The reaction of VCl_4, propylenebis(phosponic acid), piperazine, and water in the mole ratio 1:1.5:3:945 hydrothermally at 200°C for 96 h yielded green rod-like crystals of $[H_2N(C_2H_4)_2NH_2][(VO)_2$-$(O_3PCH_2CH_2CH_2PO_3H)_2]$ (176). The vanadium coordination is square pyram-

TABLE XV

Structural Features of Several Phases Formed in the Vanadium $CH_2(PO_3H)_2$ System

Preparation Method	Formula	Unit Cell Dimensions (Å)	Space Group	Structure
(1) VCl_4 + MDPA[a] + CsCl + H_2O in mole ratio 1:0.80:7.39:1104, 24 h r.t.: Blue-green blocks	$Cs[VO(HO_3PCH_2PO_3H)_2(H_2O)]$	$a = 10.991(2)$ $b = 10.161(2)$ $c = 7.445(1)$	$C2$	Discrete Cs^+ and $[VO(HO_3PCH_2PO_3H)_2(H_2O)]$ ions MDPA chelates V to yield mononuclear species.
(2) $CsVO_3$ + MDPA + H_2O Ratio 1.00: 2.46:1450 Autoclave 48 h, 200°C Blue needles	$Cs[VO(HO_3PCH_2PO_3)]$	$a = 10.212(2)$ $b = 10.556(2)$ $c = 14.699(3)$ $\beta = 94.57(2)°$	$C2$	Infinite chains negatively charged $[VO(HO_3PCH_2PO_3)]^-$ with Cs^+ between the chains [Fig. 35(*a*)]
(3) VCl_4 + CsCl + MDPA + H_2O. Mole ratio 1.00:9.68:1.02:994 Autoclave 48 h, 270°C Green plates	$Cs[(VO)_2(O_3PCH_2PO_3)_2(H_2O)_2]$	$a = 9.724(2)$ $b = 8.136(2)$ $c = 10.268(2)$ $\beta = 94.57(2)$*	$C2/m$	Cs^+ within channels of two-dimensional V—O—MDP network. V^{IV}/V^{III} trinuclear unit of VO_6 corner sharing octahedral [Fig. 35(b)]
(4) VCl_4 + CsCl + MDPA + H_2O	$V(HO_3PCH_2PO_3)(H_2O)$	$a = 5.341(1)$ $b = 11.516(2)$ $c = 10.558(2)$ $\beta = 99.89(1)°$	$P2_1/n$	$V^{III}O_6$ octahedra linked by MDP into a three-dimensional lattice

[a] Methylene diphosphonic acid, $CH_2(PO_3H)_2$.

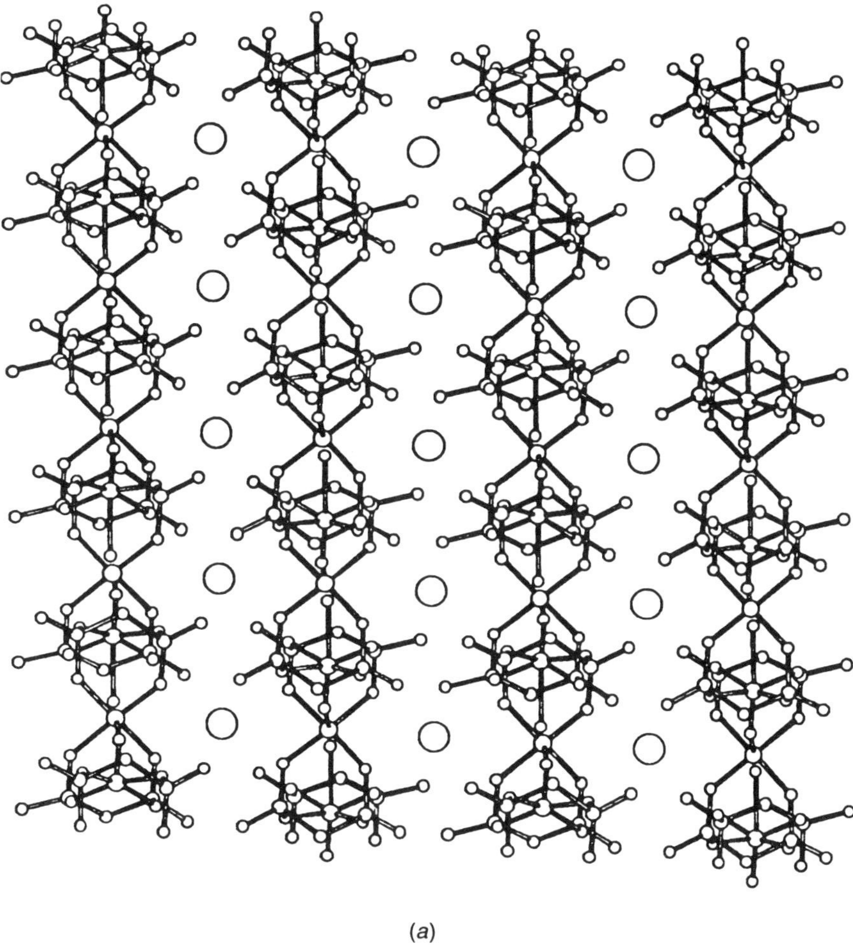

Figure 35. (*a*) A view of the anionic chains of $Cs[VO(HO_3PCH_2PO_3]$ with the Cs^+ between the chains and (*b*) a view of the structure of $Cs[(VO)_2V(O_3PCH_2PO_3)_2(H_2O)_2$ illustrating the intralayer cavities occupied by Cs^+. [Reprinted with permission from M. I. Khan, Y.-S. Lee, C. J. O'Connor, R. C. Haushalter, and J. Zubieta, *J. Am. Chem. Soc.*, *116*, 4525 (1994). Copyright © 1994 American Chemical Society.]

idal. Four of the oxygen atoms come from four adjacent phosphonate groups. Each *P*1 type phosphonate group bridges across three vanadiums and the fourth phosphorus site bonds the methylene chain terminating in the *P*2 site. This phosphorus contributes one oxygen to the vanadium square pyramid in an adjacent row. The other two oxygen atoms are present as P=O and P—OH groups.

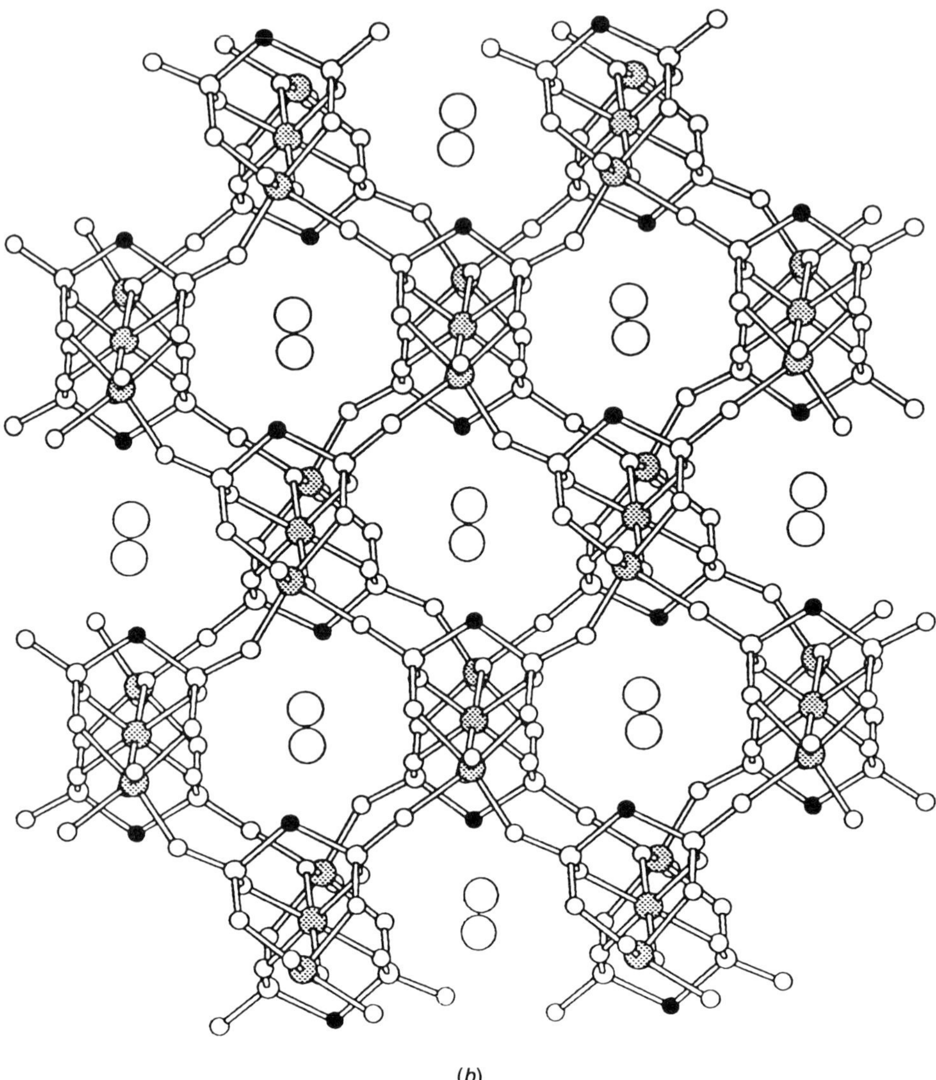

(*b*)

Figure 35. (*Continued.*)

The motif, shown in Fig. 37, forms infinite ribbons four rows thick in which the phosphonate tetrahedra have their chains alternately up and down as we move along the ribbon direction. This alternation of chains creates a ladder-like or stepwise arrangement as shown in Fig. 37. The POH groups at the edges of the ribbons serve to interconnect the adjacent ribbons through hydrogen bond-

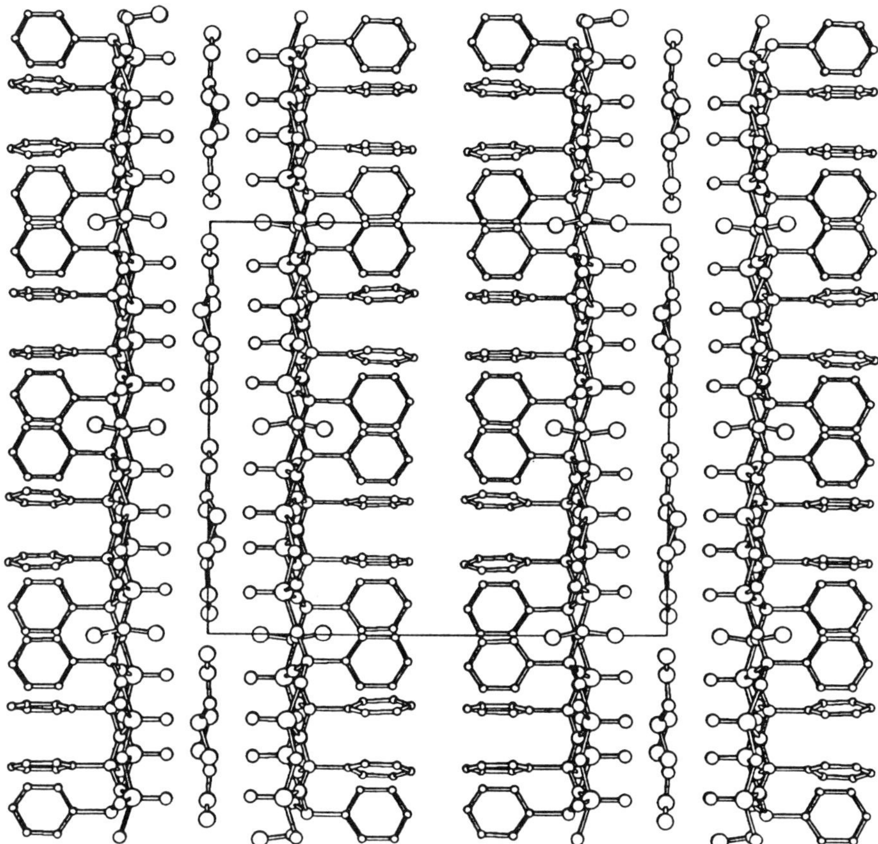

Figure 36. View of the *ac* plane of $[(Et)_2NH_2][(Me)_2NH_2][V_4O_4(OH)_2(PhPO_3)_4]$ (1), showing the layer of $[(Et)_2NH_2]^+$ cations intercalated between inorganic V/P/O layers, which are in turn separated by bilayers of phenyl groups. The positions of the $[(Me)_2NH_2]^+$ are also highlighted and shown to penetrate the V/P/O layers.

ing. The stacking of the layers creates cavities in which the piperazinium cations reside.

Substitution of *N,N′*-piperezinebis(methylenephosphonic acid) as the source of phosphonate groups produced an entirely different type of layered compound (176). The inorganic layers consist of corner-sharing vanadium octahedra and phosphonate tetrahedra cross-linked by $[-CH_2NH(C_2H_4)_2NHCH_2-]^{2+}$ groups. Its composition is $[VO(H_2O)\{O_3PCH_2NH(C_2H_4)_2NHCH_2PO_3\}$. We note that charge balance is achieved by protonation of the piperazine nitrogens unlike the polyimine bridged zirconium phosphonates that require the presence of anions for charge balance (Section IV.C).

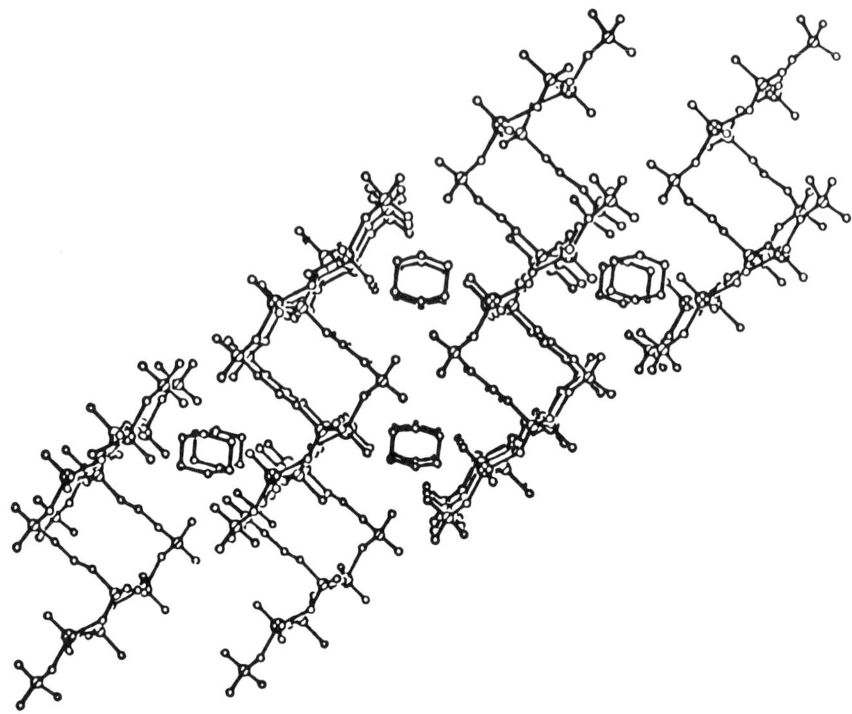

Figure 37. View of the structure of $[H_2N(C_2H_4)_2NH_2][(VO)_2(O_3P(CH_2)_3PO_3H)_2]$ parallel to the *a* axis showing the stair–step motif consisting of infinite ribbons four polyhedra in width and the $[H_2N(C_2H_4)NH_2]^{2+}$ cations in the interlayer cavities. [Reprinted with permission from V. Soghomonian, R. Diaz, R. C. Haushalter, C. J. O'Connor, and J. Zubieta, *Inorg. Chem.*, *34*, 4460 (1995). Copyright © 1995 American Chemical Society.]

An interesting polyoxovanadium phosphonate anion was prepared hydrothermally from a mixture of V_2O_3, V_2O_5, and $MePO_3H$ using $(Me)_4NOH$ as template (177). The compound consists of discrete $(Me)_4N^+$ cations and $[H_6(VO_2)_{16}(O_3PMe)_8]^{8-}$ anions. The anions are composed of VO_5 square pyramids and $MePO_3$ tetrahedra with μ_2 and μ_3 linkages. The six protons are bonded to phosphonate oxygens requiring the vanadyl groups to supply only 2+ in charge. Thus, 14 of the vanadiums are in the oxidation state IV and 2 are in oxidation state V. The bonding motif is shown in Fig. 38 and four such groupings are arranged about a fourfold axis connected by the terminal phosphonate groups to form the ion shown in Fig. 38. One tetramethylammonium ion resides in the center of the cavity. Similar ethyl and phenylphosphonate derivatives have also been prepared.

A major research project pursued by Zubieta and co-workers (178) was to

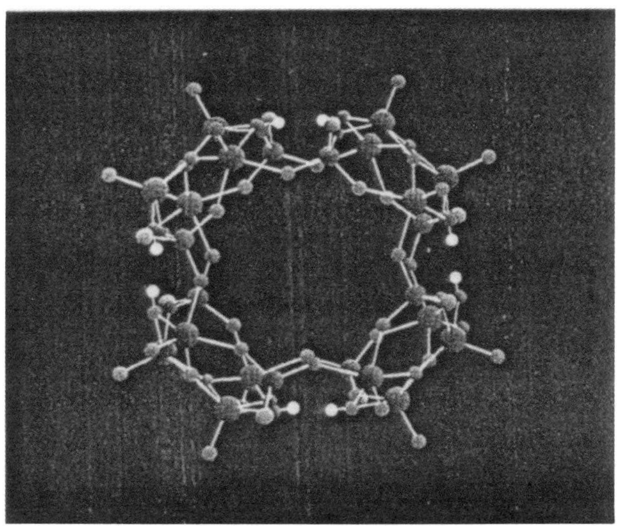

Figure 38. View of the $[H_6(VO_2)_{16}(O_3PMe)_8]^{8-}$ anion down the fourfold axis: V green, P, blue; O, red; C, gray; H, white. [Reproduced with permission from (177).]

isolate simple low molecular weight units to use as building blocks to prepare higher oligomers. Hence we describe the formation of two such simple units. The $[(VOCl)_2(PhPO_3H_2)_2(H_2O)_2]$ complex was obtained by using $(Ph)_4P[VO_2Cl_2]$ as a starting source of vanadium, as shown in Eq. 24.

$$2(Ph)_4P[VO_2Cl_2] + 2PhPO_3H_2 + RCH_2OH \longrightarrow$$
$$[(VOCl)_2(HO_3PPh)_2(H_2O)_2] + 2RCHO + 2(Ph)_4PCl \tag{24}$$

The vanadium compound cocrystallizes with the tetraphenyl phosphonium chloride. The structure consists of discrete binuclear vanadyl groups of square pyramids bridged by two phosphonate groups. Each vanadium is coordinated to a terminal oxo group, and a chloride ion and water molecule. This grouping is centrosymmetric and is similar to the chair-like arrangement in $VO(PhPO_3) \cdot H_2O$.

The second unit was prepared from alcoholic solutions of $[(n\text{-Bu})_4N)]_3V_5O_{14}$ and (2-methylbenzyl)phosphonic acid at room temperature. Bright red diamagnetic crystals of the binuclear V^V species $[(n\text{-Bu})_4N]_2[V_2O_4(O_3PCH_2C_6H_4Me)_2]$ were obtained in 40% yield. Each V^V site bonds to two terminal oxo groups in a cis geometry and two oxygen atoms from the μ_2-bridging phosphonate ligands. This leaves one oxygen of each phosphonate group as P=O since no protons are bonded to these oxygens. This tetrahedrally coordinated vanadium dimer fragment is a rare species. Both species have been polymerized into large oligomers.

X. DIVALENT AND TRIVALENT METAL PHOSPHONATES

A. Divalent Transition Metal Phosphonates

Ten years after the discovery of zirconium phenyl and alkylphosphonates (3) the results of three research groups studying the preparation and behavior of divalent and trivalent metal phosphonates began to appear (179–181). Metal(II) phosphonates are soluble in acid solutions and therefore it is often possible to obtain single crystals by slow evaporation of the solutions. The crystal structures of the Mn (179) and Zn (180) compounds were determined from single crystals. The composition of the divalent phosphonate is generally $M(O_3PR)\cdot H_2O$. The crystals belong to space group $Pmn2_1$ and the unit cell dimensions for the Zn compound are $a = 5.634(2)$, $b = 14.339(5)$, $c = 4.833(2)$ Å, and $Z = 2$. The metal atoms are six coordinated. Since these compounds only contain one phosphonate group, in order to obtain six coordination, two of the phosphonate oxygens chelate the metal and at the same time bridge across adjacent metal atoms in the same row. These oxygens are therefore three coordinate. The third phosphonate oxygen bridges to an adjacent row creating the layer arrangement shown in Fig. 39. The sixth site is occupied by the water molecule. The chelation-bridging arrangement produces a kinked or crenelated layer as shown in Fig. 40. Consequently, the phenyl rings approach each other closely and this is evidenced by disorder of the rings in the refinement stage of the structure. Closer inspection of the data reveals the presence of additional X-ray reflections indicating a doubling of the unit cell along the c axis. Refinement in the larger cell shows the phenyl rings in one cell to be rotated 90° relative to those in adjacent unit cells (Fig. 40).

Mallouk and co-workers (179) prepared a series of Mg^{2+} alkyl derivatives C_1–C_{12}, and showed that the unit cells differ only in the b dimension and the X-ray patterns conform to $Pmn2_1$ symmetry. Thus, Mn, Zn, and Mg phosphonates and probably Ni^{2+} and Co^{2+} all conform to the same structure type. The $Cd(O_3PPh)\cdot H_2O$ complex has the same structure but $Cd(O_3PMe)\cdot H_2O$ crystallizes in space group $Pna2_1$ (182).

Copper phosphonate structures are an exception to the six-coordinate crenelated layer type just described. The composition is the same, $Cu(O_3PR)\cdot H_2O$, R = Me, Ph but the copper atoms are five coordinate (183). The coordination is distorted tetragonal pyramidal. The four nearly coplanar oxygen atoms are derived from one water and three phosphonate oxygen atoms. Fig. 41 A is a view of the layer omitting water coordinated to Cu and the organic group bonded

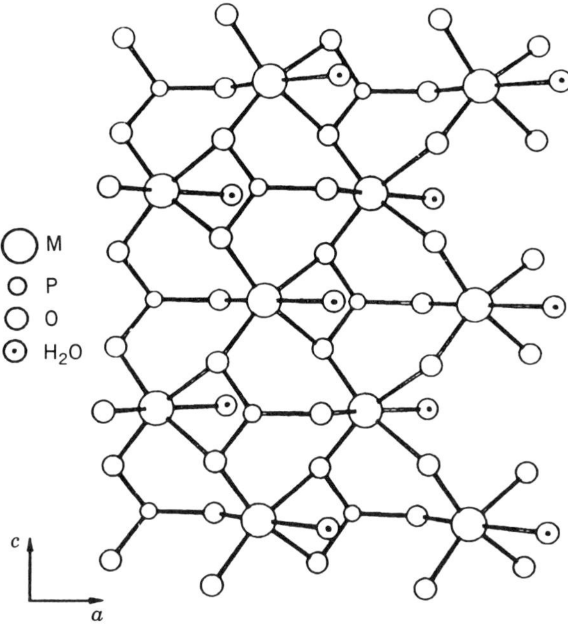

Figure 39. Representation of the layers of $M(O_3PR)\cdot H_2O$ compounds, M = Mn, Mg, Co, or Zn. [Reprinted with permission from G. Cao, H. Lee, V. M. Lynch, and T. E. Mallouk, *Solid State Ionics*, *26*, 63 (1988). Copyright © American Chemical Society.]

to P. The axial copper oxygen bond is 16% longer than the bonds in the equatorial plane, which range from 1.928(4) to 1.991(5) Å. The three phosphonate oxygen atoms all bond to copper atoms. One of them bridges two Cu atoms that are 3.13 Å apart forming a short [1.973(4) Å] and a long bond, 2.32 Å. These same copper atoms are bridged by a single oxygen from another phosphonate group forming a four-membered parallelogram-shaped ring. Thus, this oxygen is in the equatorial position for one copper coordination sphere and at the same time occupies the apical position of an adjacent copper coordination sphere. The remaining two oxygen atoms of each phosphonate group then bridge across copper atoms in adjacent rows to form eight-membered rings with a chair conformation reminiscent of vanadyl phenylphosphonate. A second more circular eight-membered ring forms in rows parallel to the *b* axis in between the rows of parallelogram four-membered rings alternating with the eight-membered chair-type rings. The interlayer spacings are 8.495(4) Å for the methylphosphonate and 13.991(1) Å for the phenylphosphonate. In the latter compound, the phenyl rings are nearly at right angles to adjacent phenyl groups.

The $Cu(O_3PEt)$ compound was prepared hydrothermally and in the anhydrous condition had a structure different from that described for

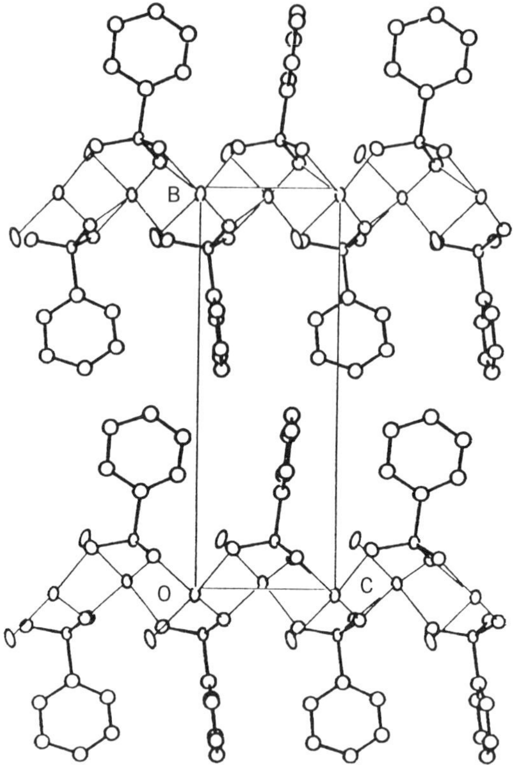

Figure 40. Layer structure of $Mn(O_3PPh)$ viewed down the *a* axis showing the pleated sheet nature of the layers and the arrangement of the phenyl rings. [Reprinted with permission from G. Cao, H. Lee, V. M. Lynch, and T. E. Mallouk, *Inorg. Chem.*, *27*, 2781 (1988). Copyright © 1988 American Chemical Society, G. Cao, H. Lee, V. M. Lynch, and T. E. Mallouk, *Solid State Ionics*, *26*, 63 (1988). Copyright © 1988 American Chemical Society.]

$Cu(O_3PMe) \cdot H_2O$ (184). Heating the ethylphosphonate to 180°C led to loss of the water and transformation to trigonal bipyramidal Cu coordination as shown in Fig. 41. It will be shown below that in $M(O_3PR) \cdot H_2O$ hydrates, dehydration is reversible and thought to be topotactic (185, 186). Only in the case of Ni was there evidence for a structural change on dehydration (185). The reason provided (184) for the rearrangement in the case of copper phosphonate is the unfavorable coordination that would result upon dehydration if it were topotactic. Either the copper would have a distorted tetrahedral coordination of trigonal pyramidal coordination, both highly unfavorable. However, it will be shown in Section X.D that zinc phosphonates also rearrange upon dehydration and/or on intercalation of amines. Therefore, caution must be exercised in interpreting

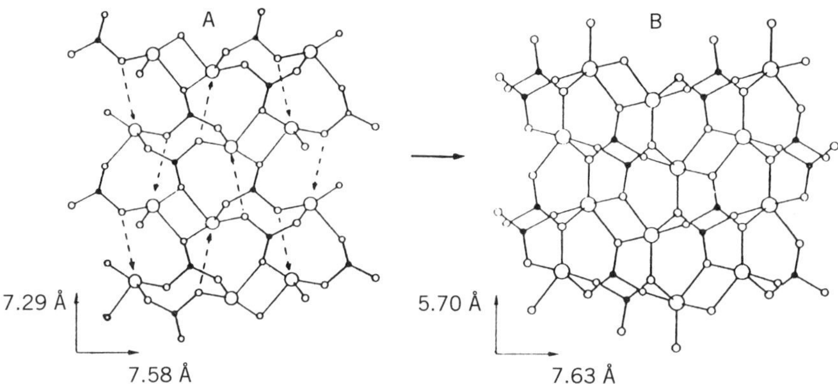

Figure 41. Schematic illustration of the structural transition of $Cu(O_3PEt)\cdot H_2O$ during dehydration. The view is perpendicular to the *a* axis. Ethyl groups have been omitted for clarity. [Reprinted with permission from J. Le Bideau, B. Bijoli, A. Jouanneaux, C. Payen, P. Polvadeau, and J. Rouxel, *Inorg. Chem.*, *32*, 4617 (1993). Copyright © 1993 American Chemical Society.]

reactions of these transition metal phosphonates. Additional aspects of copper phosphonates will be presented in Sections X.D.

It has been found that attempts to prepare Fe^{II} phosphonates always yield Fe^{III} compounds or mixtures contaminated by Fe^{III} phosphonates when prepared by the methods reported above (187). This result has been confirmed but the ethylphosphonate, $Fe(O_3PEt)\cdot H_2O$, was synthesized from FeOCl in acetone at 80°C (sealed tube) (188). This compound was found to have a structure similar to the one described above for Mn and Zn phosphonates. However, the unit cell is monoclinic, $a = 4.856(8)$, $b = 10.33(1)$, $c = 5.744(3)$ Å, $\beta = 91.0(1)°$, and space group *Pn*. We note that the in-plane unit cell dimensions are similar to those of the Mn and Zn orthorhombic parameters. The same situation exists in the copper system where $Cu(O_3PMe\cdot H_2O$ is monoclinic and $Cu(O_3PPh)\cdot H_2O$ is orthorhombic with both compounds having similar in-plant unit cell dimensions. The orthorhombic cell is better able to accommodate the bulkier phenyl groups. In the present case, of iron(II) ethylphosphonate, the ethyl groups are disordered, a condition thought to arise from static disorder, rather than dynamic. This situation is the same as the one observed with the Mn and Zn phenylphosphonates (179, 180). However, it was shown (179) that weak reflections exist, which indicated the true unit cell is monoclinc, but insufficient weak reflections were present to either refine the structures on the larger unit cell or to determine with certainty the monoclinic parameters.

The plot of $1/\chi$ versus temperature for iron phosphonate indicated a Curie–Weiss behavior in the temperature range 180–300 K with a Curie constant of $C = 3.9$ emu K^{-1} mol^{-1} and $\mu_{eff} = 5.5$ μ_B. These values are consistent with high-spin Fe^{2+} (188). Deviation from Curie–Weiss behavior occurs below 180 K and a broad maximum in χ occurs near 35 K, characteristic of low-dimensional antiferromagnetism. At still lower temperature, weak ferromagnetic ordering sets in.

Another interesting iron(II) compound, $Fe(HO_3PC_2H_4CO_2H)$ was described by the French group (189). The layered structure was found to resemble that of α-ZrP (8). Two of the phosphonate oxygen atoms and one carboxylate oxygen are used for bonding. Two phosphonate groups bridge across Fe atoms forming infinite chains along the orthorhombic *a* axis. These oxygens occupy the equatorial positions of the octahedral coordination sphere. The organic groups then serve to link the chains into layers along the [110] direction occupying the axial positions of the octahedra. The remaining carboxyl and phosphonic oxygens bond to the protons. The distance between O—P—O bridged iron atoms is 5.14 Å (Ar—Zr in α-ZrP is 5.3 Å) and the interlayer distance 7.4 Å is close to that in α-ZrP, 7.6 Å. This iron complex exhibits Curie–Weiss susceptibility down to 5 K and has a magnetic moment of 5.35 μB. It was prepared in a sealed tube reaction in acetone at 80°C.

We have seen that if the bulkiness of the phosphonate group exceeds the area on the layer formed with straight-chain alkyl phosphonates, new structures able to accommodate the bulky ligand are formed. This finding is also true for the divalent metals. Le Bideau et al. (190) prepared tertiary butylphosphonates of Co, Mn, Zn, and Cu hydrothermally at 150°C. The composition of the products of Co, Mn, and Zn were the same as before, $M[O_3PC(Me)_3]\cdot H_2O$ but the layer structure was entirely different. The crystal structure of the cobalt compound was determined from single-crystal data. There are three independent Co atoms. The Co1 has an octahedral environment being bonded to three phosphonate oxygen atoms. These cobalt octahedra form dimers by edge sharing. The remaining two cobalt atoms are tetrahedrally coordinated by phosphonate oxygens. One of the cobalt atoms, Co3, forms pairs, also by edge sharing. Both IR and X-ray data were used to show that the Mn and Zn compound have the same structure as the Co compound. However, the Cu compound showed a different stoichiometry, $Cu_{1.75}O_{0.75}[O_3P(CMe)_3]\cdot H_2O$. The higher amount of Cu relative to phosphonate decreases the number of bulky groups to be accommodated.

B. Alkaline Earth Phosphonates

In Section X.A it was mentioned that Mg^{2+} forms layered phosphonates of the manganese(II), *Pmn2*-type structure. There are two exceptions to this statement. If the phosphonate has multiple ligating sites, a molecular coordination-type compound may form rather than an extended two-dimensional structure. Examples are provided by 2-aminoethylphosphonic acid (191) and the calcium and copper complexes of *N*-phosphonomethylglycine (192). Additionally, if the ligand cannot fit into the 28-Å^2 space on the layer, a coordination compound or new structure type may form as indicated for the copper tertiary butyl derivative. This new structure formation was the case when benzhydrylphosphonic acid was reacted with Mg^{2+} (179b). The magnesium was present as $[Mg(H_2O)_6]^+$ in the compound $Mg[HO_3PCH(Ph)_2]_2 \cdot 8H_2O$ interlaced between rows of monoprotonated benzhydrylphosphonic acids with the PO_3H groups hydrogen bonded to each other and to $[Mg(H_2O)_6]^{2+}$. This arrangement creates hydrophilic and hydrophobic regions in the layer stacking.

Two types of calcium phosphonates were prepared by Mallouk and co-workers (193). One of them had the familiar composition $Ca(O_3P(Me) \cdot H_2O$ but a new structure type. It is monoclinic, with a space group $P2_1/c$, $a = 8.8562(13)$, $b = 6.6961(10)$, $c = 8.1020(10)$ Å, $\beta = 96.910(11)°$, and $Z = 4$. The Ca atoms are coordinated by six oxygen atoms of average bond length (2.40 Å) and by a seventh at 2.71 Å forming a distorted pentagonal bipyramid coordination polyhedron. The connectivity within the layers is more complex, phosphonate oxygen forming bonds to three Ca atoms, one of which is the long bond. Another phosphonate oxygen bonds to the same Ca to which the first oxygen formed the long bond and is thus chelating this calcium atom. At the same time, this oxygen bridges to another Ca forming one-half of a four-membered ring. The oxygen forming the long bond also forms one-half of a diamond shaped four-membered ring in the opposite direction of the long bond. The remaining coordination sites are filled by the third phosphonate oxygen and water.

Compounds of the type $Ca(O_3PC_nH_{2n+1}) \cdot H_2O$ for $n = 1$–5, all have the same C–oxygen–phosphorus–water layer network (193). This conclusion is supported by IR spectra and the smooth progression of interlayer spacings with chain length size. This higher coordination number of 7 rather than 6 is made possible by the larger size of Ca^{2+} ($y = 0.99$ Å) over the size of the first-row transition elements. For $n = 6$–18, the composition changed to $Ca(HO_3PC_nH_{2n+1})_2$, $n \geqq 6$. These larger alkyl chain compounds were made in water–alcohol mixtures in order to dissolve the phosphonic acid. However, this was shown not to be the driving force of the structural change. Interestingly, since these compounds have a 1:2 Ca/P stoichiometry, they adopt the α-ZrP layer structure. The similarity is shown in Fig. 42.

Under normal aqueous synthetic conditions, refluxing a mixture of Ba^{2+} or Pb^{2+} with a stoichiometric (1 : 1) or excess (12 : 1) phenylphosphonic acid/metal ratio gives the product $M(HO_3PPh)_2$. The metal atoms are eight coordinate with a distorted dodecahedral geometry (194). Both phosphonate groups chelate the metal atoms and one of the chelating oxygens of each group coordinates to an adjacent metal atom in the *c*-axis direction. The final two coordination sites are taken by the remaining phosphonate oxygen that bridges across the zigzag chains

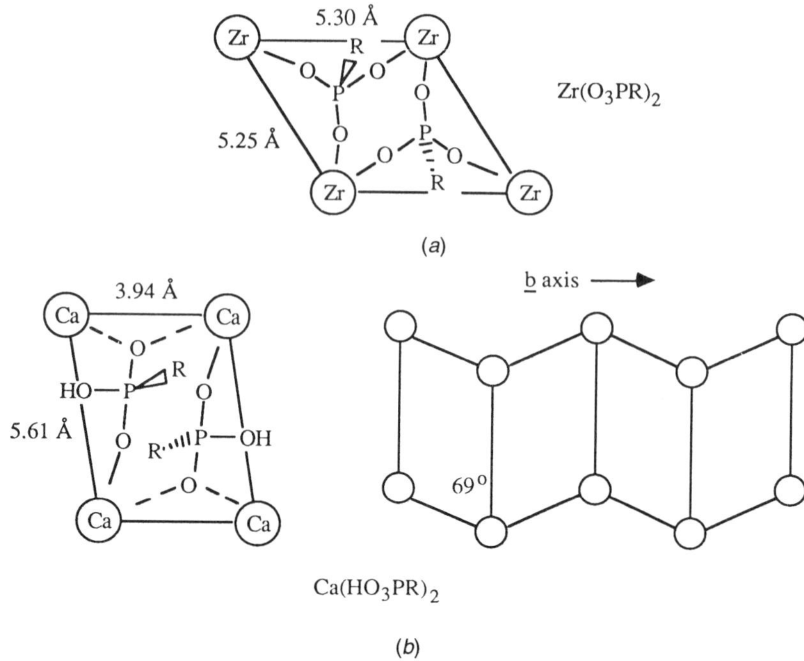

Figure 42. (*a*) Idealized structure of an α-ZrP layer (top view) showing the pseudohexagonal in plane unit cell (8) and (*b*) schematic drawings of the related metal phosphonate structures $Zr(O_3PR)_2$ and $Ca(HO_3PR)_2$. [Reproduced with permission from ref. 4).

in the *b*-axis direction. The phenyl rings are oriented roughly as the phenyl rings in $Zr(O_3PPh)_2$. In fact, the two structures crystallize in the same space group C^2/c and in the plane dimensions are similar [5.546(4), 8.495(4) Å for Ba and 5.415(3), 9.0985(5) Å for Zr].

C. Trivalent Metal Phosphonates

Mallouk and co-workers (193) prepared lanthanide methylphosphonates of composition $LnH(O_3PMe)_2$, Ln = La, Sm, or Ce, in aqueous or mixed-alcohol–aqueous media. Improvement in the crystallinity of the products was achieved by holding the precipitate in its mother liquor at 70°C for several days. The solubility of the lanthanide phosphonates is in general greater than that of the Groups 4 (IV.B) and 14 (IV.A) phosphonates but less than that of the corresponding divalent compounds. Poorly developed single crystals of $LaH(O_3PMe)_2$ were obtained, giving a triclinic unit cell with a = 5.398(7), b = 8.168(18), c = 10.162(19) Å, α = 73.76(16)°, β = 83.89(13)°, γ

=73.50(14)°, and $Z = 2$. The structure could not be determined because of the thinness of the crystal flakes and twinning.

Subsequently, $LaH(O_3PPh)$ and $LaH(O_3PCH_2Ph) \cdot 2H_2O$ were prepared and their structures solved (195). Indeed, the former compound is triclinic, space group $P\bar{1}$ with $a = 8.410(3)$, $b = 15.696(7)$, $c = 5.636(1)$ Å, $\alpha = 90.24(4)°$, $\beta = 108.99(1)°$, $\gamma = 85.89(4)°$, and $Z = 2$. The crystals were grown hydrothermally in 0.4 *M* HCl. The structure is layered as expected, in which the La atoms are eight-coordinated. The layer structure is very similar to that of $Ba(HO_3PPh)_2$ (194). Once again it was found that the phenyl groups were disordered and that the true unit cell required doubling the *c* axis. In this larger cell, there is no disorder. In any one row in the *a*-axis direction the rings lie parallel to each other but in the *c*-axis direction are they are canted at an angle of 59° to the rings in adjacent rows.

Lanthanum benzylphosphonate dihydrate crystallized in the orthorhombic system with $a = 10.801(2)$, $b = 10.301(2)$, $c = 33.246(8)$ Å, $Z = 8$, and space group *Pbcn*. In this compound, the La atom is also eight-coordinate distorted dodecahedral. However, this comopund contains two water molecules coordinated to La. Therefore the two phosphonate groups need supply one six oxygens rather than eight to the metal coordination sphere. This coordination is accomplished by having one phosphonate group chelate the La and then have one of these chelating oxygens bridge to an adjacnet La atom. The remaining phosphonate group then bridges across two adjacent La atoms in the *b*-axis direction. The third oxygen does not bond to any metal atoms and apparently remains bonded to the proton. The interlayer spacing is 16.62 Å. It is interesting to observe that single crystals were grown by slow evaporation of a mixture of $LaCl_3$ and benzylphosphonic acid. In contrast, crystals of $LaH(O_3PPh)_2$ were prepared hydrothermally in an acid mother liquor. A different stoichiometry is obtained in alkaline solution. The solid has the composition of $Ln_2(O_3PPh)_3$ and forms at pH values above 8. Acidification of the compound converts it to $LaH(O_3PR)_2$, Eq. 25.

$$2Ln_2(O_3PPh)_3 + 3HCl \longrightarrow 3LnH(O_3PPh) + LnCl_3 \qquad (25)$$

The reverse reaction, conversion of $LnH(O_3PPh)$ to the base formed compound $Ln_2(O_3PPh)_3$, did not take place.

A series of iron phosphonates was obtained from FeOCl as the source of iron (196). The $FeH(HO_3PPh)_4$ compound was prepared in a Pyrex tube reaction with excess phenylphosphonic acid and dichloromethane as solvent. The crystals are triclinic, $a = 14.968(9)$, $b = 5.36(1)$, $c = 8.678(6)$ Å, $\alpha = 88.5(1)°$, $\beta = 86.41(6)°$, $\gamma = 89.6(1)°$, and $Z = 1$. The structure is layered in which the Fe atoms are coplanar. The iron atoms are octahedrally coordinated by oxygen atoms from six different phosphonate groups. Four of the oxygen atoms come from phosphonate groups bridging two Fe atoms to form chains parallel to the *b* axis. The remaining two octahedral positions are filled by two phosphonate groups through only one of their oxygens. The chains are then linked to adjacent chains through hydrogen bonds to form layers. The P—C bonds point the phenyl rings toward adjacent sheets in the *a*-axis direction. Only van der Waals forces are operative between sheets.

The iron tetraphosphonate forms when the ratio of phosphonic acid to Fe is equal to or greater than 4. However, when this ratio is equal to 2, a series of three compounds, all with the same composition $FeH(O_3PPh)_2 \cdot H_2O$, was obtained. The three phases were labeled α, β, γ, and they are obtained one from the other depending on the length of the reaction time (Eq. 26).

$$FeOCl + 2PhPO_3H_2 \xrightarrow{t_1} \alpha \xrightarrow{t_2} \beta \xrightarrow{t_3} \gamma \tag{26}$$

In addition, the tetraphenylphosphonate converts to the α phase upon heating in a mixed-aqueous organic solvent. These iron diphosphonate phases were examined by IR, Mössbauer, and thermal analysis methods from which it was concluded that they have an α-ZrP-type structure. The same was found to hold for the methylphosphonate, $FeH(O_3PMe)_2$ (181).

D. Amine Intercalation Reactions and Layer Restructuring

We have previously discussed the molecular recognition reaction discovered by Johnson et al. (156) in the vanadyl phosphonates prepared in benzyl alcohol. On heating, the benzyl alcohol leaves creating a pocket into which straight-chain primary alcohols can fit but secondary and tertiary alcohols are excluded. The transition metal M^{II} phosphonates of composition $M(O_3PR) \cdot H_2O$ have the water molecule coordinated to the metal atom (Section X.A). Some of these compounds can be dehydrated and remain in the same space group and unit cell as the initial hydrated phase (185). For example, the powder pattern of the dehydrated $Zn(O_3PMe)$ phase could be indexed on the basis of the unit cell dimensions (space group $Pmn2_1$) $a = 5.42(1)$, $b = 6.94(1)$, $c = 5.22(1)$ Å. Similar unit cells hold for the Mg and Co phases. Thus, it was concluded that the loss of water was topotactic and it was also shown to be reversible. The dehydrated phase was exposed to gaseous amines for several days. Shape selectivity was evidenced in that only *n*-butyl and isobutylamine were intercalated while *sec*-butyl and *tert*-butylamines were not. *n*-Alkyl amines with one-to-eight carbon atoms were sorbed on both Co and Zn methylphosphonates to yield a series of intercalates of composition $M(O_3PMe)(C_nH_{2n+1}NH_2)$, M = Co or Zn. For Co compounds with $n = 3$–5, the slope of the line obtained by plotting the interlayer distances as a function of the number of carbon atoms in the alkyl chains is 1.155 Å per CH_2 unit. However, both lines extrapolate to the same point at $n = 0$, the interlayer distance required for $Co(O_3PMe)(NH_3)$. These results were interpreted on the basis that the amine chains are interdigitated for the smaller chain amines but for the larger amines a double layer with an angle of tilt of 31.4° with respect to the layer plane was proposed.

Zinc methylphosphonate behaved differently. All the amine compounds with n even fall on one straight line and those with n odd fall on a line with a different slope. From the slopes of the lines it was concluded that the even numbered amines pack as a double layer, whereas those with n odd pack as interdigitated chains. In the zinc phenylphosphonate system, no amines were intercalated (186). Computer modeling showed that the bulky phenyl rings effectively blocked the amines from the vacated metal site. However, ammonia was taken up rapidly to fill this position on the metal.

Subsequently, it was found that amines in the liquid state do intercalate into zinc phenylphosphonate (197). A pronounced odd–even effect was observed but all the data could be plotted on a single line for which the average slope was 1.24 Å per CH_2 unit. A full mole of amine was taken up in each case for $n = 3–9$. An interesting observation was the fact that the zinc phenylphosphonate hydrate, $Zn(O_3PPh) \cdot H_2O$, has an interlayer spacing of 14.34 Å (180) but the interlayer spacing for the propylamine intercalate is less than this value based on its powder pattern. The methylphosphonates of Mg, Co, and Zn show a decrease of 1.8 Å in interlayer distance upon loss of the coordinated water. This decrease in interlayer distance results from the nesting of the methyl groups of an adjacent layer toward the space vacated by the loss of water. Intercalation of amines results in an expansion of the layer spacing. Furthermore, 2-methylbutylamine and 2-aminoheptane were also intercalated into zinc phenylphosphonate even though computer models showed that this was not possible without some rearrangement of the structure.

Since the zinc phenylphosphonate amine intercalates were fairly crystalline, it was possible to solve their structures from the X-ray powder patterns (198). Structure solutions were obtained for the propyl, butyl, and pentylamine intercalates. Indeed it was found that an interesting rearrangement takes place. The zinc becomes tetrahedrally coordinated by three phosphonate oxygens from three different groups and the amino nitrogen. In the process, the structure changes from orthorhombic $Pmn2_1$ to monoclinic $P2_1/c$. The layers become relatively flat, moving the phenyl rings further apart so that the amino groups can coordinate to the metal as shown in Fig. 43 and the connectivity within the layers is shown in Fig. 44(*c*). The magnitude of the interlayer spacing is controlled by the degree of flattening of the layers and the extension of the phenyl rings. Because the propylamine chains do not extend beyond the phenyl rings into the interlamellar space, it is the extension of the phenyl rings into this space that dictates the 13.7-Å interlayer spacing. The butylamine intercalate has an interlayer spacing of 14.4 Å. In this case, the methyl protons increase the repul-

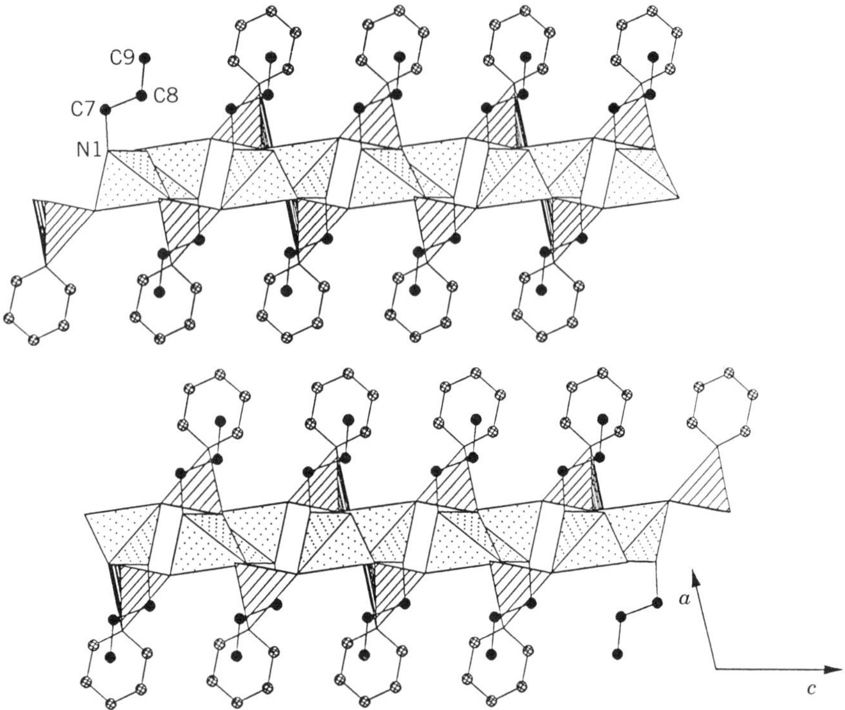

Figure 43. Projection of the zinc phenylphosphonate propylamine complex, $Zn(O_3PPh)(NH_2Pr)$, down the *b* axis showing that the propylamine forms part of the zinc coordination polyhedron and does not extend beyond the phenyl group into the interlayer space.

sion between adjacent layers leading to a slightly increased interlayer distance. Only beginning with the pentylamine intercalate does the carbon chain clearly extend beyond the phenyl ring and now there is a regular odd–even effect in the succeeding interlayer spacings as additional carbon atoms are added to the intercalating amine (197).

In a subsequent study, it was shown that mixed derivatives of the zinc phosphitephenylphosphonate, $Zn(O_3PH)_x(O_3PPh)_{1-x} \cdot H_2O$ could be prepared (199). Compounds with $x = 0.26–0.60$ were obtained. In order to obtain single-phase solid solutions, it was necessary to carry out the reactions pH $\sim$ 11 to achieve rapid precipitation of the mixed-ligand phases. Kinetic control of the precipitation was necessary to ensure that a homogeneous solid solution phase was attained. A combination of X-ray powder data, IR, and solid state was NMR spectra were utilized to show that the layer structure of these mixed-ligand phases was the same as that of the $x = 0$ compound but with a random distribution of the two ligands. Further, the IR spectra of the amine intercalates of

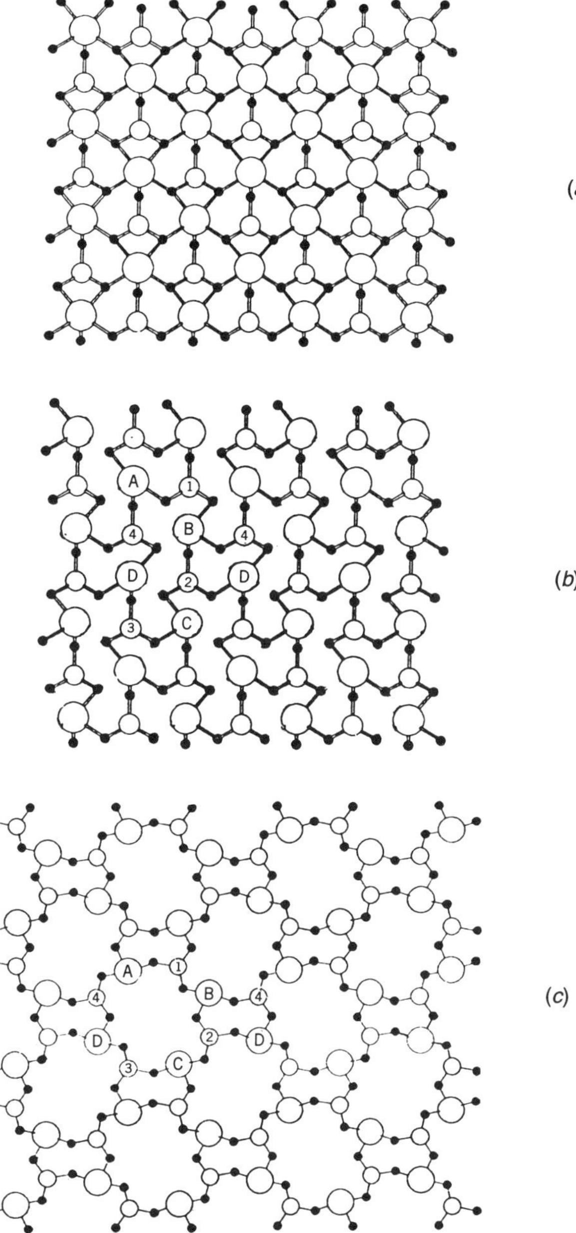

Figure 44. Schematic representation of the rearrangement occurring during dehydration or amine intercalation processes in $Zn(O_3PMe) \cdot H_2O$. (*a*) Layer of $Zn(O_3PR) \cdot H_2O$ showing the bridging and chelation of Zn atoms (large circles). (*b*) Suggested pathway that involves bond breaking such that each phosphonate oxygen bonds to only one Zn atom and (*c*) expansion of the in plane cell dimensions to produce the layer of the dehydrated or amine intercalated phases. [Reproduced from (201) in permission.]

the dehydrated phases were identical in their major bands with the corresponding zinc phenylphosphonates. Thus, the same layer rearrangement must have taken place with the mixed derivatives upon amine intercalation, even though the presence of small phosphite groups on the layers created space for the amines to contact the metal. A full mole of amine was intercalated in these mixed-ligand phases as for the stoichiometric phenylphosphonate.

The question was raised (198) as to whether other zinc phosphonates and, in particular, zinc methylphosphonate also undergo layer rearrangement upon amine intercalation. This question is important in terms of shape selectivity and the use of these layered phosphonates as sensors (see Section XI). However, based on observations of the IR spectrum of $Zn(O_3PC_2H_4NH_2)$ (200) as compared to that of the butylamine intercalate of $Zn(O_3PMe)$, it was suspected the dehydration of the hydrated zinc phosphonates keyed the rearrangement rather than the actual intercalation reaction. Bujoli and co-workers (201) used a combination of IR, EXAFS, and X-ray absorption near edge structure (XANES) to show that $Zn(O_3PMe) \cdot H_2O$ on dehydration converts from the distorted octahedral zinc coordination to the four-coordinate layer type found in the amine intercalates (Fig. 44*a–c*). However, Clearfield and co-workers (198) found that the same results are achieved whether the intercalation of amines occurs with the hydrated or dehydrated phase. Thus, liquid amines are able to trigger the same type of rearrangement as the dehydration reaction. A rearrangement pathway has been proposed as shown in Fig. 44. Upon removal of the water from the zinc phosphonate layer [Fig. 44(*a*)] it is only necessary to break the Zn—O bonds as indicated in Fig. 44(*b*). In the layer structure of the hydrate, each phosphonate group forms five bonds to zinc. Two of these bonds must break in such a way that no phosphonate oxygen bridges to two zinc atoms. The connectivity is now Zn—O—P—O—Zn. Following bond breaking, the layers need to expand along the in-plane *b*- and *c*-axial directions (198, 201) to obtain the alternating 8 and 16 ring arrangement of the amine intercalate layer shown in Fig. 44(*c*). This transformation must be very facile as it was shown to be reversible upon removal of the amine and rehydration.

Cadmium methyl- and phenylphosphonates behaved much like their zinc counterparts upon intercalation (182) and it was suggested that the structure rearranged on dehydration and/or intercalation.

Amine intercalation reactions were examined for both $Cu(O_3PMe) \cdot H_2O$ and the copper phenylphosphonate (183, 202). Intercalation for the methylphosphonate occurred from the amines in the gas phase as well as liquid to yield the same products of composition $Cu(O_3PMe) \cdot RNH_2$. Surprisingly, the phenylphosphonate incorporated 2 mols of amines, yet the interlayer spacings were almost identical to those observed for the methylphosphonate. There were several indications that a layer rearrangement takes place not only on intercalation of amines but on dehydration. Dehydration is accompanied by a color change from bright blue to light green. For the methylphosphonate, dehydration results in an increase in the interlayer spacing from 8.5 to 9.7 Å (202). Contact with amines produces a bright blue color. A plot of interlayer spacing as a function of amine chain length for the methylphosphonate host yielded a straight line with a slope of 2.01, indicating formation of a bilayer of amines with a tilt angle of 53° with respect to the layer plane. Extrapolation of the line to $n = 0$ yields a value of about 7.01 Å. This value does not correspond to the interlayer spacing of $Cu(O_3PMe) \cdot H_2O$ or the corresponding anhydrous compound. We recall that copper ethylphosphonate rearranges to a trigonal bipyramidyl coordination for the metal on dehydration (184). In the process, the rearrangement of the layer results in one-dimension shrinking from 7.29 to 5.70 Å, while the other in-plane dimension hardly changes. However, the interlayer spacing increases from 9.92 to 10.78 Å. This increase (0.96 Å) is somewhat lower than that observed for the methylphosphonate, 1.2 Å, but is a good indication that a similar rearrangement takes place with the methyl derivative.

Additional copper phosphonate phases were prepared by hydrothermal methods (203). An anhydrous phase of the methylphosphonate results by treatment of the monohydrate at 200°C. This phase has been labeled α-$Cu(O_3PMe)$. The structure of α-$Cu(O_3PMe)$ is thought to be similar to the trigonal bipyramidal ethyl analogue as discussed above. However, on further treatment of the α-phase at 180°C hydrothermally for 15 days, the β phase was obtained together with a small amount (5%) of $Cu_3O(O_3PMe)_2 \cdot 2H_2O$. The β phase has a tubular structure as shown in Fig. 45. The unit cell is hexagonal $a = 16.343(6)$, $c = 7.092(4)$ Å, $Z = 18$, and space group $R\bar{3}$. The Cu atoms have distorted tetragonal pyramidal coordination spheres with the apical bond longer by about 0.3–0.4 Å than the bonds in the pyramidal base. There are zigzag chains of Cu atoms running parallel to the *c* axis. These metal atoms are connected by phosphonate oxygens, one oxygen per 2 Cu, forming four-membered rings in the shape of distorted parallelograms. These chains are in turn linked together in the *ab* plane by O—P—O bridges to form the tunnels. The methyl groups project into the tunnel imparting hydrophobic character to the interior.

The $Cu_3O(O_3PMe)_2 \cdot 2H_2O$ complex is layered but the copper atoms have two different types of coordination. Two-thirds of the Cu atoms have the familiar tetragonal pyramidal coordination and the remainder have a distorted (4 + 2) octahedral coordination. One of the phosphonate groups forms Cu—O—P—O—Cu bridging to three different copper atoms. The other phosphonate group has one oxygen bonded to three Cu atoms and the other two bonded to two Cu atoms each. Methyl groups project into the interlamellar space.

The magnetic behaviors of the methyl, ethyl, and phenyl monohydrates are similar. They are paramagnetic with room temperature magnetic moments of 2.0–2.2 μ_B per Cu atom. Below 20 K the χT product continuously decreases

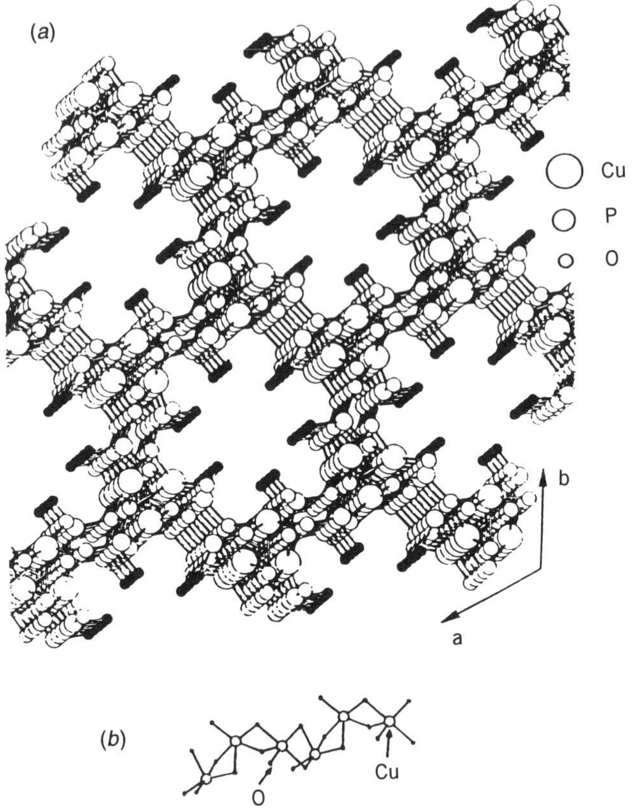

Figure 45. (*a*) Perspective representation of β-$Cu(O_3PMe)$ viewed down the *c* axis illustrating the tubular structure. Carbon atoms are in black and the large circles are Cu atoms. (*b*) Representation of copper chain along the *c* axis. [Reprinted with permission from J. Le Bideau, C. Payen, P. Palvadeau, and B. Bijoli, *Inorg. Chem.*, *33*, 4885, 1994. Copyright © 1994 American Chemical Society.]

revealing weak antiferromagnetic interactions between Cu^{II} ions. The room temperature magnetic moment for $Cu_3O(MePO_3)_2 \cdot 2H_2O$ is equal to 2.1 μ_β per Cu atom. The $\chi_m T$ product then continually decreases until 70 K, where it reaches a plateau indicating a lowering of the magnetic moment. At very low temperatures the χT product becomes zero showing that an S=O antiferromagnetic ground state has been attained (203).

Very recently it has been shown that small changes in the preparative conditions lead to the formation of new copper phosphonates (204). For example, a mixture of copper sulfate (2 mmol), (2-aminoethyl) phosphonic acid (1 mmol), and NaOH (2 mmol) in 15 mL of water treated hydrothermally at 110°C for 4 days yielded $Cu_{2.5}(OH)_2(O_3PCH_2CH_2NH_3)(SO_4)$. Substitution of $Cu(NO_3)_2$ for the sulfate under the same preparative conditions led to the formation of $Cu_2(OH)(O_3PCH_2CH_2NH_2)(NO_3) \cdot H_2O$. The former compound is layered with Cu in three different environments. Two of the Cu atoms have distorted 4 + 1 + 1 elongated tetragonal octahedral geometry (longer axial bonds and shorter equatorial bonds). The third Cu atom has a distorted square pyramidal coordination. Each of the hydroxides is linked to three copper atoms and a proton. The layer is built up from a complex network of edge-sharing copper–oxygen polyhedra propagating in the *a* and *c*-axial directions. The layers are held together by van der Waals forces and hydrogen bonds between the NH_3^+ ends of the alkyl chains and oxygens of the sulfate group that do not coordinate to copper atoms.

The nitrate compound is also layered with one Cu in sixfold distorted elongated tetragonal geometry and the other in distorted square pyramidal geometry with the amino nitrogen as one of coordinating atoms. The hydroxide ion is bonded to three Cu atoms in this layer as in the sulfate compound. The octahedra and square pyramids share edges forming zigzag chains parallel to the *b* axis. These chains are connected to each other in the *a*-axis direction by O—P—O bridges as well as $OPCH_2CH_2N$ links. The nitrate groups reside between the layers and probably hydrogen bonds to the layers through water and the amino group. The nitrate group is exchangeable by other anions. This behavior and the structure may be compared to the hydroxyphosphates, $M_2^{II}(OH)(PO_4)$ (205) in which a divalent phosphonate group replaces a trivalent phosphate group necessitating incorporation of an anion between the layers.

E. Pillared Layered Compounds of Zn and Cu

In Section III.C, we described efforts by several groups to prepare microporous zirconium bis(phosphonates). One difficulty with describing the nature of these compounds is the lack of definitive crystal structure data owing to the very low solubility of this class of compounds. Consequently, in only a few cases were structure solutions possible (Section VII.B) and these were not of the type conceived by Dines et al. (19) (Figs. 27 and 29). It was therefore felt that divalent metal phosphonates, being soluble in dilute acid, may more readily form sufficiently crystalline products for X-ray analysis, which proved to be the case.

Zinc 1,4-phenylenebis(phosphonate) hydrate, $Zn_2(O_3PC_6H_4PO_3) \cdot 2H_2O$, was prepared from stoichiometric amounts of $ZnCl_2$ and the bisphosphonic acid in water by heating the mixture at 60°C for 7 days (206). The crystal structure was solved using powder X-ray data. The inorganic layer portion is almost identical to that in zinc phenylphosphonate (180). These layers are bridged by the phenylene groups as shown in Fig. 46. At least they were refined in those

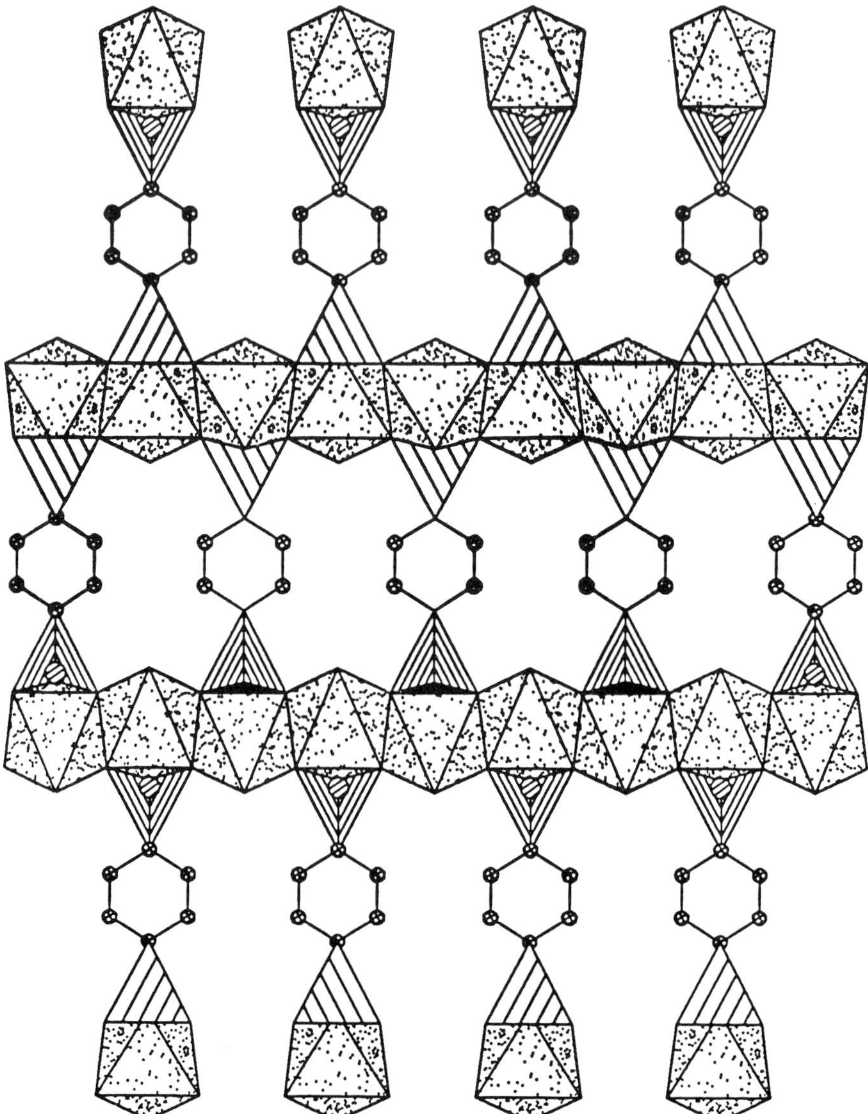

Figure 46. Projection of the structure of zinc phenylenebis(phosphonate) down the *b* axis showing the bridging nature of the diphosphonate.

positions. However, the distance from the center of a phenyl ring to that of an adjacent ring in the *c*-axis direction is 5.65 Å. This distance is close enough to lead to severe nonbonded hydrogen–hydrogen interactions. However, symmetry restrictions of space group *Pnnm* do not allow rotation out of the *ac* plane. Attempts at refinement in a lower symmetry space group did not converge. Another possibility is the doubling of the *c* dimension as was the case for $Zn(O_3PPh) \cdot H_2O$. However, the powder pattern did not allow determination of this possibility.

A protonated zinc, 4,4′-biphenylylenebis(phosphonate), $Zn_2\{[O_3P\text{-}(C_6H_4)_2PO_3H]$ was prepared at pH 1.6 by heating the zinc–phosphonic acid mixture at 90°C for 7 days (206). A new structure-type formed because of the presence of the protons. The structure is not layered in the traditional sense but is better described as consisting of a double chain structure running parallel to the *b*-axis direction [Fig. 47(*a*)]. These chains are then linked to similar double chains in the *c*-axis direction by the phenylyl groups as shown in Fig. 47(*b*) to form layers. These layers are then hydrogen bonded to each other into a three-dimensional structure. The rings are not coplanar but are rotated 52° from each other. The Zn coordination is tetrahedral. It will be interesting to see whether this compound can intercalate amines or exchange its protons for cations.

Reaction of copper salts with 1,4-phenylenebis(phosphonic acid), $H_2O_3PC_6H_4PO_3H_2$, yielded the layered cross-linked analogue of copper phenylphosphonate (207). The metal atoms retained the same distorted square pyramidal geometry, where four of the coordination sites are occupied by phosphonate oxygens and the remaining site is filled by a water molecule. The phenyl rings serve to cross-link the CuO_3P layers into a three-dimensional structure. A new layered structure was obtained when 4,4′-biphenylenebis(phosphonic acid) was reacted with a soluble copper salt. The reaction was carried out hydrothermally at 150°C for 12 days to achieve a high degree of crystallinity. The $Cu[HO_3P(C_6H_4)_2PO_3H]$ complex is triclinic, *P1*, with $a = 4.856(2)$, $b = 14.225(5)$, $c = 4.788(2)$ Å, $\alpha = 97.85(1)°$, $\beta = 110.14(1)°$, $\gamma = 89.38(1)°$, and $Z = 1$. The structure consists of linear chains of Cu bridged by centrosymmetrically related phosphonate groups utilizing two of their oxygen atoms. This binding mode leads to square planar geometry for the copper atoms. The biphenylene groups serve to link the chains into slab-like layers. The third oxygen of the phosphonate groups is protonated and is involved in linking the adjacent layers via hydrogen bonds. These hydroxyl oxygens also interact weakly (Cu---O = 3.14 Å) with the Cu atoms of the adjacent chain imparting a pseudo-square pyramidal character to the copper coordination.

In the zinc and copper phenylene and biphenylene compounds, the rings are closely spaced so that no microporosity is evident. Introduction of phosphite or phosphate spacer groups has been possible (208) increasing the surface area. The X-ray powder patterns of the mixed-ligand compounds indicates that they

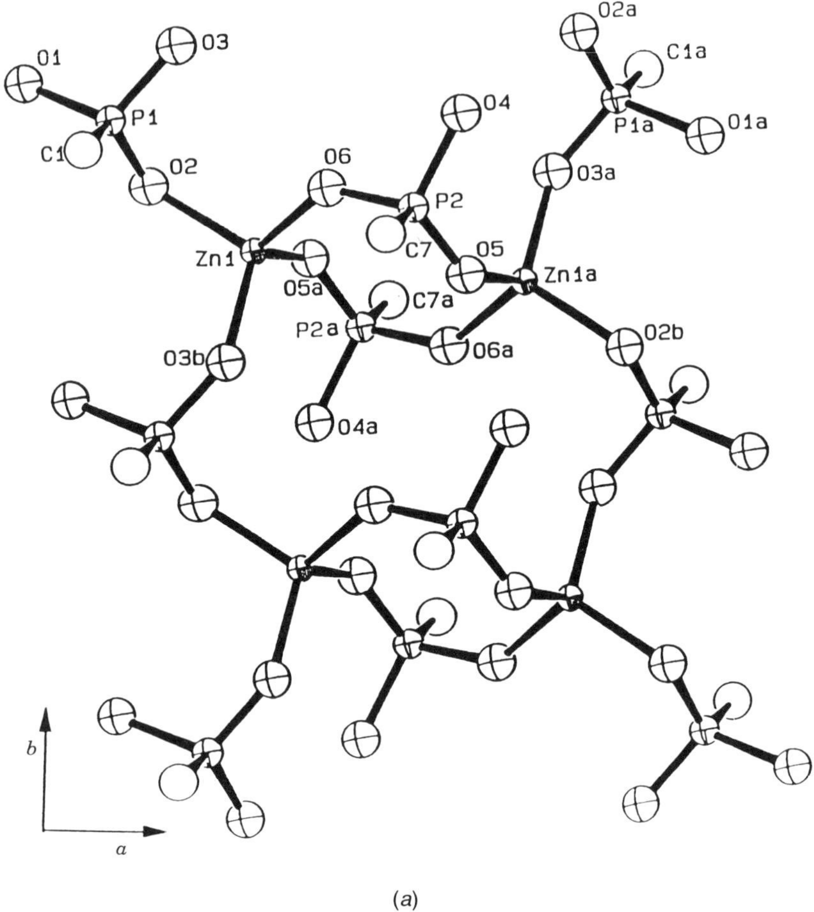

(*a*)

Figure 47. (*a*) Representation of the Zn coordination in Zn biphenylylene bis(phosphonate) and the mode of bridging of the $-PO_3$ groups to form double chains and (*b*) the structure as viewed down the *b* axis. Dashed lines indicate hydrogen bonds.

have unit cells similar to those of the unsubstituted bis(phosphonates). However, the crystallinity of these mixed-ligand derivatives is poor so that structure solutions have not yet been obtained. Amine intercalation behavior and ion exchange of the protonated derivatives is under examination.

F. Functionalized and Porous Phosphonates

A number of divalent (Zn or Cu) metal phosphonates have been prepared in which the organic group was functionalized. By varying synthetic procedures

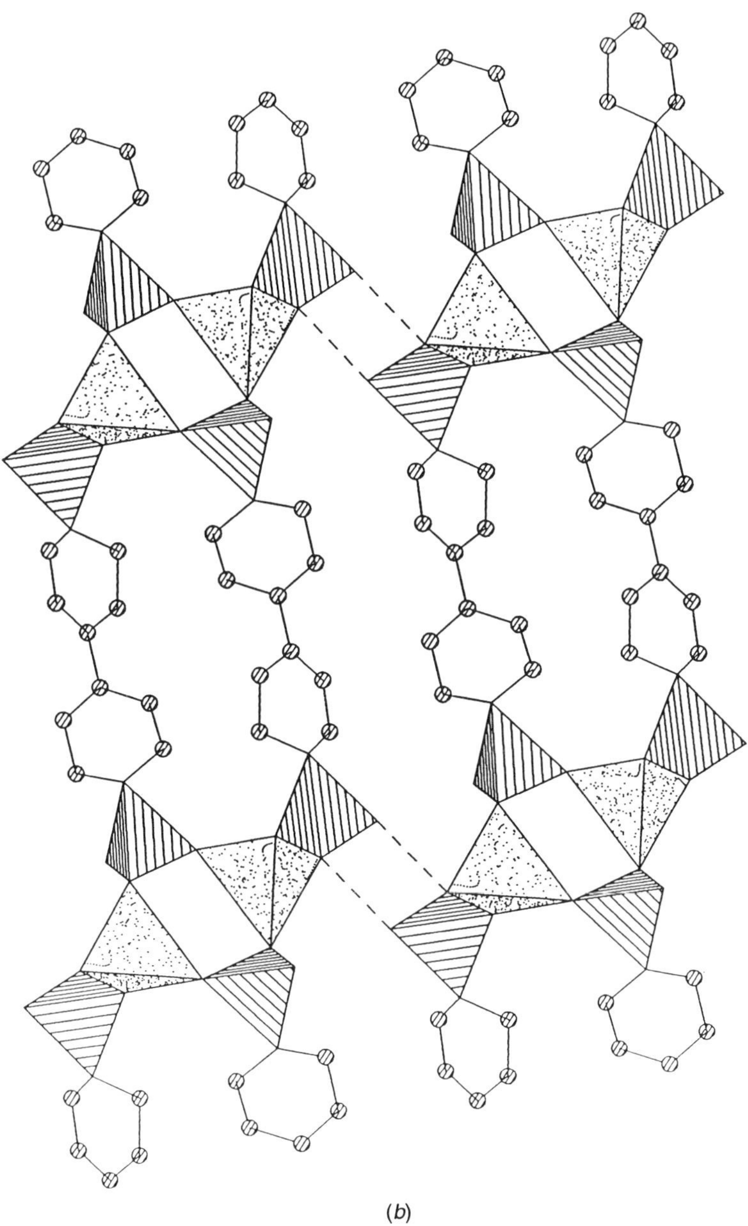

(b)

Figure 47. (*Continued*.)

and the nature of the functional groups several new structure types and some interesting chemistry resulted.

1. *Halogenated Phosphonate Derivatives*

Zinc chloromethyl phosphonate, $Zn(O_3PCH_2Cl) \cdot H_2O$ was prepared using $Zn(CH_3COO)_2$ and $H_2O_3PCH_2Cl$ at 70°C followed by slow evaporation (209). The layer structure was found to be the same as that presented earlier for $Zn(O_3PPh) \cdot H_2O$ except that the chloromethyl derivative is monoclinic not orthorhombic. The in-plane unit cell dimensions are very nearly the same as those of the zinc phenylphosphonate with the chloromethyl groups occupying the interlayer space. A similar bromo derivative was prepared in a mixture of 5% water in acetone (210) and found to have the same structure as the chloromethyl compound. In these halo derivatives, only van der Waals forces hold the layers together.

When $ZnCl_2$ was used as the source of Zn and the pH was increased with NaOH, a new derivative of composition $Zn_2Cl(O_3PCH_2Cl)(HO_3PCH_2Cl) \cdot 3H_2O$ was obtained (209). This compound is also layered but the layers are built up of both tetrahedrally and octahedrally coordinated zinc atoms. One Cl and three phosphonate oxygens constitute the atoms bonded to Zn in the tetrahedra. The octahedrally coordinated zinc atoms occur in clusters of two, bridged by phosphonate oxygens, to form four-membered rings. Each of these zinc atoms is then linked to a tetrahedrally coordinated Zn through phosphonate O—P—O bridges. The coordination sphere is completed by three water molecules. These groupings lead to 12- and 14-membered rings in addition to the four-membered rings. One-half of the chlorine atoms of the chloromethyl groups are within van der Waals distances from those in adjacent layers and one-half are at larger distances creating gaps between the layers.

A third component of composition $Zn(O_3PCH_2Cl) \cdot$urea was obtained when urea was substituted for NaOH in the preceding synthesis. The zinc atom in this structure is tetrahedrally coordinated, being bonded to three phosphate oxygens and a urea oxygen (209). All the oxygens are two-coordinate participating in O—P—O bridging of the metal atoms. Two Zn atoms combine with two phosphonate groups to form eight-membered rings. These zinc atoms form a zigzag chain parallel to the *a*-axis direction. The third phosphonate oxygen then links the eight-membered rings together forming double-stranded chains. The chains are then linked to adjacent chains through hydrogen bonding involving the urea NH_2 groups.

2. *Carboxyl Derivatives*

Drumel et al. (210) prepared $Zn[O_3P(CH_2)_2CO_2H] \cdot H_2O$ and $Zn(O_3PCH_2CH_2NH_3^+NO_3^-) \cdot H_2O$ in a water (5%)–acetone medium as well as the bromomethyl phosphonate. All three compounds had structures, as shown by IR and X-ray data, that correlated with the standard $Zn(O_3PR) \cdot H_2O$ structure (179, 180). The same conclusion was drawn for the corresponding copper compounds in that they have the standard five-coordinate Cu layers (183).

Attempts were made to carry out the amidation of the carboxyl group in $Zn[O_3P(CH_2)_2CO_2H] \cdot H_2O$ without first forming an acetyl chloride (Section VIII.A). Reaction of aniline with zinc carboxyethylphosphonate at 60°C for 5 days yielded the amine intercalate with one-half of a mole of aniline incorporated. However, when the same reaction was carried out in water, the product was $Zn_2[O_3PCH_2CH_2CO_2)]_2 \cdot 3H_2O$. Under more severe conditions, 140°C for a week in an autoclave using 3 mol of aniline, the product was $Zn_3(O_3PC_2H_4CO_2)_2$. However, if instead of starting with the already formed zinc carboxyethylphosphonate, zinc nitrate, 2-carboxyethylphosphonic acid and an excess of aniline were heated at 140°C for 1 week in a sealed tube, the product was the desired amide $Zn(O_3PCH_2CH_2CONHPh)$.

The trihydrate was identified as the compound prepared earlier by Cao et al. (145). This structure is built up from $Zn(O_3PCH_2CH_2CO_2)$ layers in which trihydrated zinc ions have displaced two carboxylate protons and are coordinated by the two carboxylate groups from adjacent layers. The key to the preparation of the amide was the use of excess base. Three moles of aniline were required to neutralize the carboxyethylphosphonic acid. The excess aniline not used for neutralization can then form the amide. The $Zn(O_3PCH_2CH_2CONHC_6H_4Br)$ complex was also prepared by this direct method. A similar amidation reaction carried out with the Mn compound yielded $Mn(O_3PCH_2CH_2CONHC_6H_5) \cdot H_2O$. This compound is thought to have the normal $Mn(O_3PR) \cdot H_2O$ structure. Unit cell dimensions for the zinc and copper compounds are presented in (210).

The compound $Zn_3(O_3PC_2H_4CO_2)_2$ has an interesting structure. It is monoclinic, $P2_1/c$ with $a = 8.126(1)$, $b = 9.237(1)$, $c = 8.587(2)$ Å, $\beta = 106.26(3)°$ (210). Two of the three zinc atoms are tetrahedrally coordinated by three phosphonate oxygens and one carboxyl oxygen. The remaining zinc is six-coordinate bonding with two oxygens of two different carboxyl groups and four phosphonate oxygens. The four-coordinate Zn atoms occur in pairs bridged by single phosphonate oxygens to form four-membered rings. These rings are then flanked on either side by six-membered rings in the *c*-axis direction and eight-membered rings in the *b*-axis direction to form the layers. The layers are bridged by the phosphonate group bonding is one layer and its carboxyl group bonding zinc atoms in the adjacent layer. Yet another zinc carboxyethylphosphonate $Zn(O_3PC_2H_4COOH) \cdot 1.5H_2O$, was prepared (210) from zinc nitrate, carboxyethylphosphonic acid, and excess NaOH in a sealed tube at 110°C for 2 weeks. It is orthorhombic, *Pccn*, $a = 9.885(1)$, $b = 10.020(1)$, $c = 16.438(3)$ Å, and $Z = 8$. In this compound, Zn is tetrahedrally coordinated by three phosphonate oxygens and one carboxyl oxygen. The other carboxyl oxygen atom is protonated. The layers are built up of alternating 8- and 16-membered rings in the *a*- and *b*-axis direction forming layers. The layers are connected to each other through a phosphonate group in one layer bridging an adjacent layer through its carboxyl group to form a three-dimensional structure.

3. *Porous Compounds*

Recent research has produced a rash of compounds that contain micropores. Those dealing with divalent and trivalent metals will be presented in this section, and others especially based upon uranyl ion will be presented in Section XII.B. One compound that we have already described is β-Cu(O_3Me). A compound very much like that one was prepared from Zn($O_3PC_2H_4NH_2$). This compound was formed from $Zn(NO_3)_2$ and $H_2O_3PC_2N_4NH_2$ together with added sodium benzoate in a hydrothermal pressure vessel at 180° for 60 h (200). The Zn atoms are coordinated by three phosphonate oxygens from three different phosphonate groups and by an amino nitrogen of a fourth phosphonate group. In the *bc* plane this arrangement forms 16-membered rings constructed by corner sharing of four ZnO_3N tetrahedra and four PO_3C tetrahedra. The continuous linkage within the sheet results in eight-membered rings alternating with the 16-membered rings. The sheets are linked together via O—P—C—C—N links in the *a*-axis direction. The sheets are perfectly aligned in the *a*-axis direction to create a tunnel of the type shown in Fig. 45 of approximate dimensions 3.6 × 5.3 Å with the walls lined with NH_2 groups.

Two porous aluminum methylphosphonates have been prepared (211a,b). Both compounds were prepared hydrothermally from mixtures of pseudoboehmite and methylphosphonic acid in 48h. The α-phase (211b) was obtained by carrying out the reaction at 220°C, whereas the β-phase required a temperature of 160°C. The compositions were $Al_2(O_3PMe)_3 \cdot nH_2O$ with $n = 1$ for the β phase while the α phase has a variable water content ($n = 0$–1.5). The structure of the β phase has been solved from single-crystal data (211c). It is trigonal, space group *R3c*, with hexagonal all dimensions $a = 24.650(2)$, $c = 25.299(5)$ Å, and $Z = 18$. The framework contains octahedrally coordinated Al sharing corners with six $[MePO_3]^-$ tetrahedra. The phosphonate tetrahedra in turn share corners with three tetrahedrally coordinated Al. This peculiar connectivity forms 18 ring tunnels running parallel to the *c* axis. The methyl groups line the tunnels cutting down the effective diameter to about 5.8 Å. This compound, represented as AlMepO-β was able to sorb 2,2-dimethylpropane a molecule that has a 6.2-Å molecular diameter.

The α phase is also trigonal with $a = 13.9949(13)$ and $c = 8.5311(16)$ Å. The framework of this compound also consists of the same distribution of Al octahedra and tetrahedra. Unidimensional 18-ring channels run parallel to the *c* axis but the openings are in the form of an equilateral triangle 7 Å on a side. As with the β phase the channels are lined with methyl groups. Thermal treatment at 500°C under vacuum did not alter the framework structure. This compound also sorbed 2,2-dimethylpropane.

Another way to achieve microporosity is to use a diphosphonic acid with a variable length to the chains separating the two acid groups. We have previously described the use of *N*,*N*′-piperazinebis(methylenephosphonic acid) in the vanadyl phosphonate system (Section IX.E). The use of this ligand was extended to prepare manganese and cobalt derivatives (212). A mixture of $(Et_4N)_2MnCl_4$, piperazine phosphonic acid, $(Et)_4NCl \cdot H_2O$, and water in a ratio of 1 : 1 : 5 : 300 adjusted to pH 5 with 40% aqueous $(Bu_4N)OH$ was treated hydrothermally at 160°C for 63 h to yield $[Mn\{O_3PCH_2NH(C_2H_4)_2NHCH_2PO_3\}] \cdot H_2O$. The corresponding cobalt compound was prepared in a similar fashion. The metal atoms are tetrahedrally coordinated by oxygens from four different phosphonate groups. Each phosphonate bridges across two metal atoms leaving a P=O moiety at each phosphorus. Two metal tetrahedra corner share with two phosphonate groups to form the familiar chair shaped eight-membered ring. These metal dimer units are connected to neighboring metal sites through the organodiphosphonate groups resulting in the arrangement seen in Fig. 48. The water molecule resides within the cavity. In the [111] direction the cavity is formed by 44-membered rings and has dimensions of 4.7 × 18.0 Å. The magnetic properties are of the same type as described for other metal phosphonates.

We shall close this section with a seemingly unrelated topic. The Mn(porphyrin)Cl complexes are active homogeneous catalysts for many oxidation reactions with H_2O_2 as oxidizing agents in the presence of nitrogen base cocatalysts (213). However, recovery of the porphyrin is difficult. Immobili-

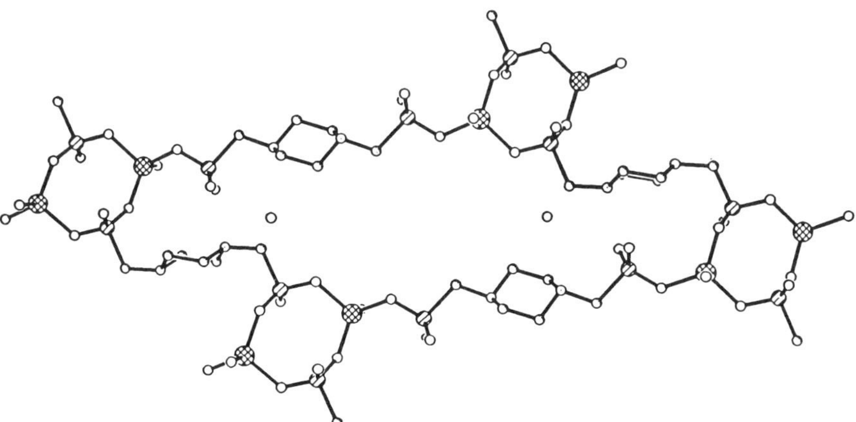

Figure 48. A view of $Mn[O_3PCH_2NH(C_2H_4)_2NHCH_2PO_3] \cdot H_2O$ down the [111] direction showing the large ellipsoidal 44-membered ring. [Reproduced with permission from (212).]

zation of the porphyrin complexes on an insoluble support would allow for easy separation and reuse, provided the activity of the catalyst was not severely restricted by its immobilization. Bujoli and co-workers (214) prepared phosphonate derivatives of the required porphyrin by two routes. Scheme 1 consisted of using p-$(EtO)_2OPC_6H_4CH_2O$ as the source of benzaldehyde to prepare the

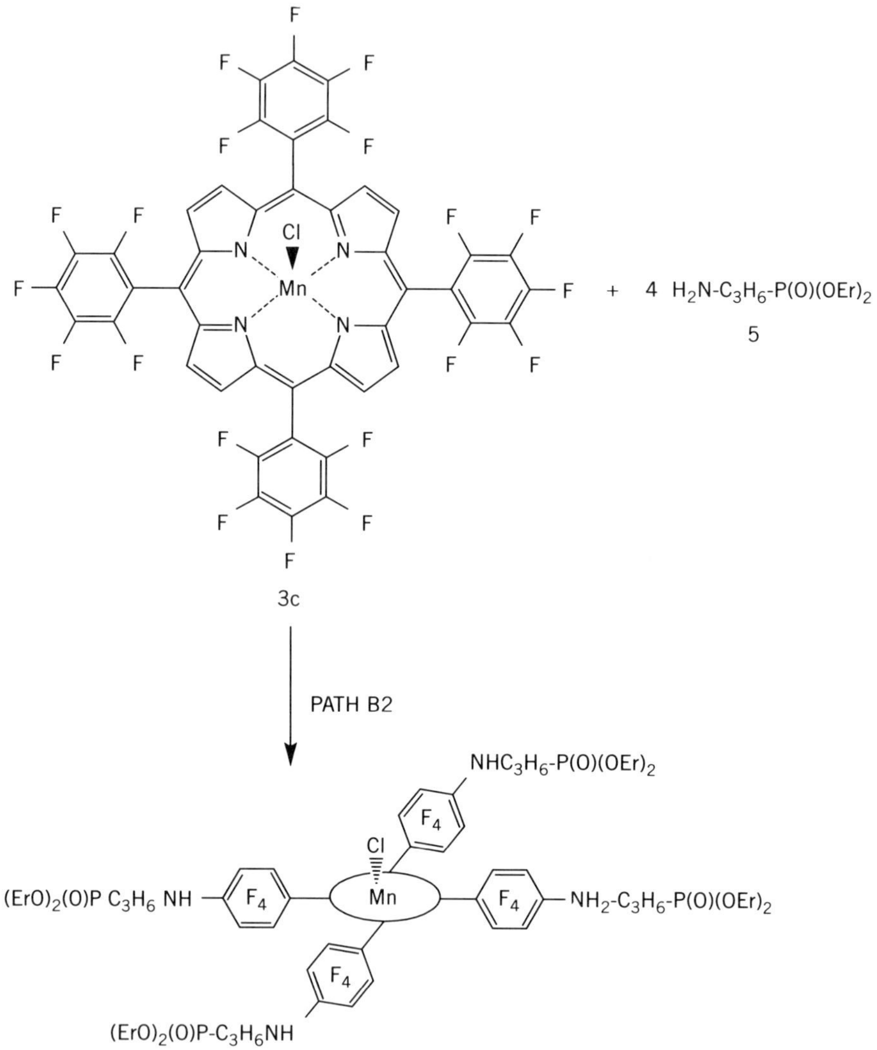

3c + 4 H_2N-C_3H_6-$P(O)(OEr)_2$ (5)

PATH B2

3d

Scheme 2

tetraphosphonated porphyrin or by reacting 1 mol of the phosphite, 3 mol of benzaldehyde and 4 mol of pyrrole to obtain a monophosphonated porphyrin. The latter compound formed only soluble products with zinc. This result is not unexpected as the very high bulkiness of the porphyrin makes it difficult to form extended layers. Schemes 2 and 3 are shown below. The tetraphosphonated porphyrins made by both schemes were then treated, either with zinc nitrate in methanol or $ZnCl_2$ in water. The insoluble products had a Zn/porphyrin ratio of 3 : 1 with retention of Mn in the porphyrin. The porphyrin complexes that had phenyl groups as side chains had low surface areas (3–6 $m^2\ g^{-1}$), whereas the porphyrin complexes with tetrafluorophenylaminopropyl side groups exhibited surface areas of 260–300 $m^2\ g^{-1}$ with an average 17-Å pore diameter. All the catalysts were active but after 24-h reaction time the high surface area catalysts gave the higher yields. Of more than passing interest is the large pore structure observed with the large spacer group, which may develop into a protocol for synthesis of microporous and mesoporous solids.

Because of the amorphous nature of the porphyrin phosphonate, critical structural information is lacking. Bujoli and co-workers (215) made a first attempt to utilize ^{31}P solid state MAS NMR to supply part of the missing information. They observed that there are three modes of bonding of the phosphonate group; for example, in zinc phosphonates (a) the three oxygen atoms can bond to one zinc atom each, a connectivity designated as (111), (b) one of the oxygen atoms bridges two Zn atoms while the others bond to one Zn each (112), (c) two bridging oxygens are present with a connectivity of (122). It was found that the isotropic chemical shifts move downfield as the connectivity increases,

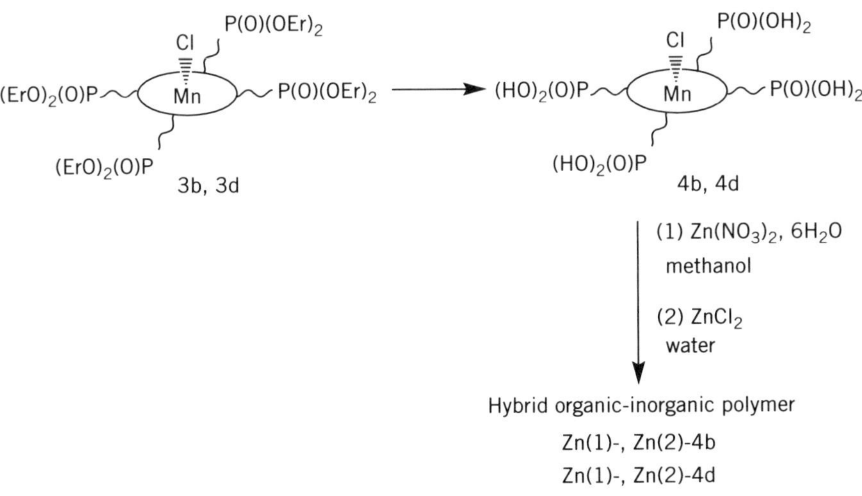

Scheme 3

corresponding to an increase in the paramagnetic contribution to the nuclear shielding. By analysis of the chemical shift tensors it is hoped that definite identification of the various bonding features can be ascertained. This correlation of chemical shift with connectivity is an interesting development to watch.

XI. SENSORS

Alberti et al. (216, 217) were able to devise hydrogen sensors based on zirconium phosphate and sulfophosphonate proton conductors. A layer of titanium hydride (TiH_x) was formed on the surface of a small disk of titanium metal by heating at 600°C in the presence of H_2. The hydride was then covered by a thin layer of exfoliated α-ZrP or $Zr(O_3PC_6H_4SO_3H)_x(HPO_4)_{2-x}$. A second electrode was added by deposition of a thin layer of platinum over the protonic conductor. The electronmotive force (E) of this cell is given by Eq. 27.

$$E = -0.029 \log (P_{H_2}) + C \tag{27}$$

where P_{H_2} is the pressure exerted by H_2 and C is a constant. In the presence of oxygen a mixed potential,

$$E = (RT/aF) \ln (P_{O_2})^{1/4}(P_{H_2})^{-1/2} + C \tag{28}$$

is measured, where P_{O_2} is the pressure of oxygen and F is Faraday's constant. If the pressure of H_2 is held constant by use of a zirconium hydride electrode, the cell may be used as an oxygen sensor. Sensors for detection of NO_x and SO_x are also underdevelopment by this group.

Mallouk and co-workers examined a variety of metal phosphonates as sensors for ammonia in the form of thin films. These include $Zn(O_3PPh)$ (218) and copper octanediylbis(phosphonate) (219) thin films grown on gold electrodes of a quartz crystal utilizing biphenylenebis(phosphonic acid) (220). The corresponding copper compound $Cu_2(O_3PC_6H_4C_6H_4PO_3)$ is bright green and anhydrous. The copper biphenylene compound described in Section X.E is hydrated and bright blue. The structure of the green copper compound could not be determined because of its poor crystallinity (208). The films were well ordered and with ultra-thin films of the order of 70 Å the response was rapid with the ammonia intercalation reaction 90% complete in 90 s. Intercalation occurs at two sites on the copper atoms, one of which is irreversible at room temperature. Larger Lewis bases are excluded from the ammonia binding sites.

A similar self-assembled thin film of the layered copper octanediylbis(phosphonate) was used together with a QCM to design a CO_2 sensor (221). This film intercalates amines to the coordinatively unsaturated Cu ions. By selecting amines with a functional tail group that reacts with CO_2 a binding site for that molecule is created. Amines are regularly used in industry to remove CO_2 from gas streams by the following reaction:

$$RNH_2 + CO_2 \rightleftharpoons RN(H)\text{—}C(=O)\text{—}OH \qquad (29)$$

Carbon dioxide also combines with alcohols to form carbonates (Eq. 30).

$$ROH + CO_2 \rightleftharpoons ROC(=O)\text{—}OH \qquad (30)$$

Based on this understanding, the molecules intercalated were 3-aminopropanol, 3-aminopropyl(methyl)dihydroxysilane, and *p*-xylylenediamine. Response time was 3–4 min. A good short treatment of phosphonate sensors has been published (222).

XII. MOLYBDENYL AND URANIUM PHOSPHONATES

A. Molybdenyl Compounds

A molybdenyl phenylphosphonate, $MoO_2(O_3PPh)\cdot H_2O$, was prepared hydrothermally at 150°C from an aqueous mixture of MoO_3 and phenylphosphonic acid (223). The structure, solved from powder data, is characterized by double molybdenyl phosphate-type chains similar to those in MoO_2HPO_4 (223). The chains run parallel to the *a* axis and consist of MoO_6 octahedra formed from three phosphonate oxygens, two oxygens of the molybdenyl group (OMoO angle is 116°) and a water molecule. Within single strands of the chains the Mo atoms are bridged by O—P—O groups with the third oxygen of the phosphonate group bridging the two chains together. These double-stranded chains arrange themselves such that the water molecules in one double-strand hydrogen bond to water and phosphonate oxygens in an adjacent double-stranded chain. This positioning has the effect of placing all the phenyl groups on the outer periphery of a set of two double chains creating alternating hydrophobic and hydrophilic regions within the crystal.

The molybdenyl phenylphosphonate intercalates amines slowly, about 20 h being required to saturation. Up to a point the reactions are reversible and occur by replacement of H_2O coordinated to Mo. Longer immersion results in rearrangement of the structure in an as yet unknown way without reversibility. Polar molecules other than amines can be intercalated but these reactions have not been explored in any detail.

In the presence of CsOH or RbOH, a mixture of MoO_3 plus methylphosphonic acid form the corresponding Mo^{VI} compounds of composition $M_2(MoO_3)O_3PMe$. The reactions were carried out hydrothermally at 180°C for 48 h (224). These compounds are layered, the layers being of the tungsten oxide type. The MoO_6 octahedra have three short and three long bonds. Two of the MoO_3 oxygen atoms form the equatorial plane of the octahedra and share corners with adjacent octahedra. Three near neighbor Mo atoms are then bridged by the phosphonate moiety. The sixth position is occupied by the remaining MoO_3 oxygen in a short [1.698(7) Å] molybdenyl type bond as it is not connected to any other atoms. All the methyl groups are on one side of the layer and the molybdenyl oxygens on the other side. The Cs(Rb) ions reside between layers in two different sites one in which it is 6-coordinate and another with 12-coordinate geometry.

B. Uranium(IV) and Uranium(VI) Phosphonates

The uranium atom occurs in phosphonate compounds in two oxidation states; as a U^{IV} ion which is in many ways similar to the Zr^{IV} cation, and more often in the form of the linear uranyl cation UO_2^{2+}. Since solid uranium^{IV} and uranyl phosphonates with extended structures are rarely obtained in the single-crystal form, to date, relatively few compounds belonging to this group have been structurally characterized. The structure of $U(O_3PCH_2CH_2CH_2SO_3H)_2$ was assumed to be layered (225) with an interlayer distance of 29.4 Å as well as that of $U(O_3PCH_2Cl)_2$ with an interlayer distance of approximately 10.7 Å. Both compounds were considered to be isostructural with α-ZrP. This conclusion was derived for an entire family of metal(IV) phosphonates based on the linear dependence between the interlayer distance and the molecular volume with a slope equivalent to 24 Å^2 per a molecule, which coincides well with the known molecular area in the layer of α-ZrP.

As a part of the recent increased progress in structure determination from X-ray diffraction powder data, the structure of a layered $U(O_3PPh)_2$ has recently been determined (225). The compound crystallizes in the monoclinic space group *C2/m*, with a = 9.4559 Å, b = 5.6769 Å, c = 14.9687 Å, and b = 96.539°. Its layers are parallel to the *ab* plane, and the phenyl rings are tilted 10° away from the *c* axis.

The area of solid uranyl phosphonates with extended structures has progressed only very recently, but the few known examples of this family of compounds exhibit an extraordinarily rich variety of structure types. This results partly from the directional anisotropy of the linear uranyl ion in which the two uranyl oxygen atoms occupy the axial positions, and in which 4–6 substituents (most often 5) are found in the equatorial plane. This anisotropy often causes the uranyl ion to act as a bridging block between the chelating phosphonate groups, which subsequently results in low-dimensional structures.

The $UO_2(O_3PCH_2OH) \cdot 5H_2O$ complex crystallizes (226) in the monoclinic space group $P2_1/c$ with a = 7.004 Å, b = 8.579 Å, c = 16.754 Å, and β = 90.65°, and forms linear chains propagating along the a axis. All phosphonate oxygen atoms are involved in coordination to uranium atoms. The space in between the linear chains is filled with five water molecules per a uranium atom, which form a rich network of interconnecting hydrogen bonds between themselves and the phosphonate chains.

The $UO_2(O_3PCH_2Cl)$ complex also crystallizes in the $P2_1/c$ space group (227) with a = 6.74212(2) Å, b = 7.11507(2) Å, c = 13.1490(1) Å, and β = 98.808(1)°, but the structure forms layers parallel to the ab plane. The methylchloride groups fill the interlayer space, and the extremely small interlayer distance, 6.50 Å is partly due to the small size of the organic group and partly to the absence of solvent molecules in the interlayer space.

The $UO_2(O_3PPh) \cdot 0.7H_2O$ complex forms extended molecular tubes (228) with six uranium atoms and six phosphorus atoms forming the tube's perimeter and with the phenyl groups located on the outside and pointing away from each tube (Fig. 49). The tubes are not connected by covalent bonds, but they are rather held together by weak van der Waals interactions. The distance between the opposite uranium atoms in each tube, 12.2 Å, can be considered the diameter of a tube, and the distance between the adjacent tubes spaced by the phenyl groups is 21.827 Å. The water molecules are located inside the tubes. Interestingly enough, the coordination environment of the uranyl ion is virtually the same as that in $UO_2(O_3PCH_2Cl)$, and the structure of the inorganic tubular wall in $UO_2(O_3PPh) \cdot 0.7H_2O$ is virtually the same as the structure of the inorganic layer in $UO_2(O_3PCH_2Cl)$. The structure of another uranyl phenylphosphonate, $(UO_2)_3(HO_3PPh)_2(O_3PPh)_2 \cdot H_2O$, is also formed by unconnected tubes (229) with the phenyl rings protruding on the outside of each tube. However, unlike in the above described case, the shape of the inorganic frame of each tube is almost rectangular with dimensions 7 × 6.5 Å. The structure of the inorganic frame is rather complex, and it includes three independent uranyl groups and four independent phenylphosphonate groups, two of which are protonated. Remarkably, this 50 atom structure was solved "ab initio" from X-ray powder data.

Very recently, three structurally very similar linear-chain uranyl bis(hydrogenphenylphosphonates) have been synthesized and characterized. In all of these compounds, the adjacent uranyl ions are bridged by pairs of phenylphosphonate groups. A relatively unstable $UO_2(HO_3PPh)_2 \cdot 2EtOH$ contains rarely seen octahedral uranium(VI) atoms (230) with the uranyl oxygen

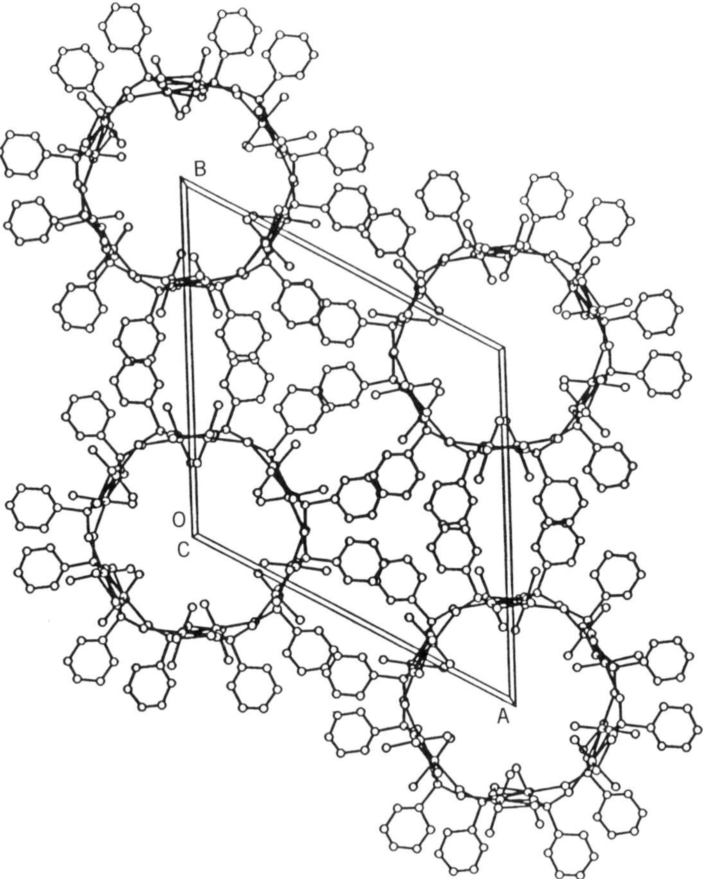

Figure 49. Projection of the structure of uranyl phenylphosphonate down the *a* axis showing the one dimensional pores about 12 Å in diameter separated by the phenyl rings. Uranyl oxygens line the inner walls of the tunnels.

atoms in the apices and with four oxygen atoms of the four bridging phenylphosphonate groups in the equatorial plane. In each phosphonate group, two oxygen atoms are used for bridging of the adjacent uranium atoms, while the third atom is protonated. The phenyl rings are arranged in two rows standing perpendicularly to the linear chain and on the opposite sides of the chain. The phenyl rings of the neighboring chains overlap each other so as to create pseudolayers spaced by the intercalated ethanol molecules. This compound is very unstable in air due to the loss of its solvent molecules.

The $[UO_2(HO_3PPh)_2(H_2O)]_2 \cdot 8H_2O$ (α-UPP) complex also forms linear chains (230) in which each uranyl ion is coordinated by four different bridging phenylphosphonate groups. However, in α-UPP, an additional water molecule coordinates the uranium atom in the equatorial plane to make the geometry of the uranium atom pentagonal bipyramidal. The phenyl groups are also arranged in two rows perpendicular to the direction of the phosphonate chains, but unlike in the ethanol intercalate, the two rows of phenyl rings are tilted 33° from each other. When viewed along the direction of the propagating chain, its shape resembles a boat (Fig. 50, α-UPP). The adjacent chains turn their phenyl sides (the boats' rims) together, and their inorganic parts (the boats' bottoms) together

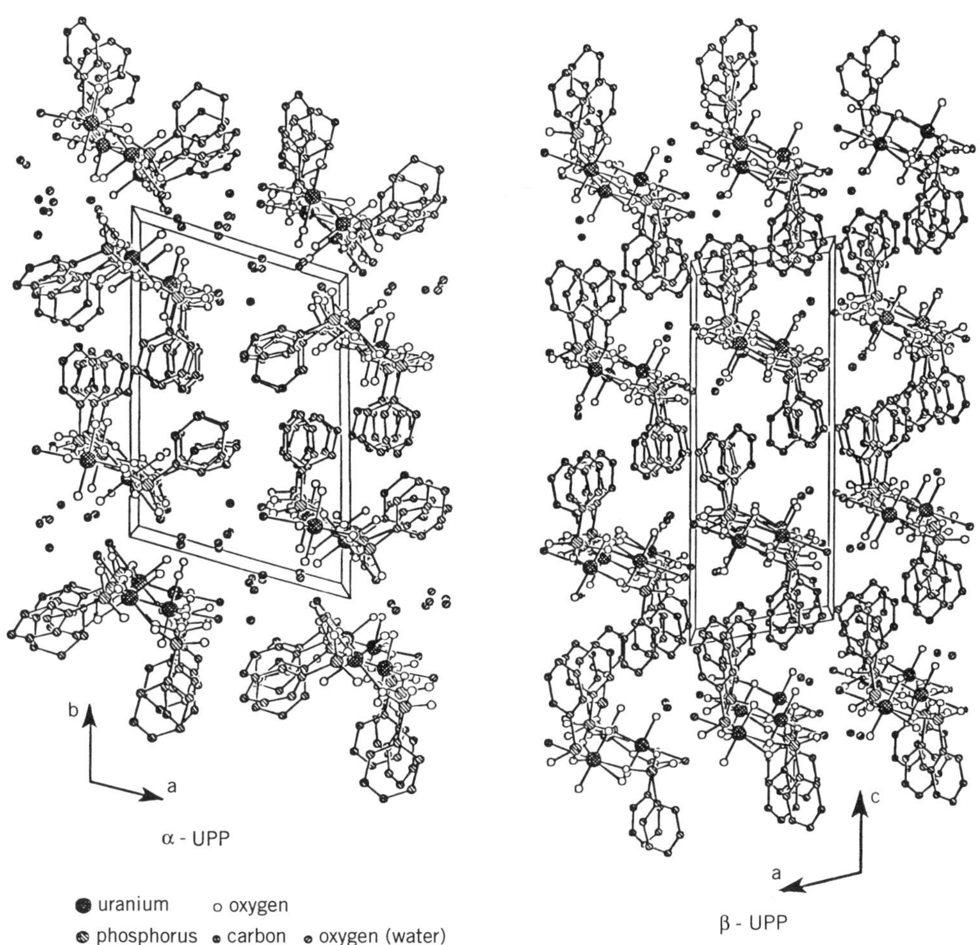

Figure 50. View of the unit cells of α-UP and β-UPP parallel to the direction of the linear chains. The α-phase has a boat conformation along the chains and the β-phase the chair-like conformation.

to form alternating hydrophobic and hydrophilic planes parallel to the *ac* crystallographic plane.

The coordination environment around the uranium atom in another linear uranyl bis(hydrogenphenylphosphonate) (231) $UO_2(HO_3PPh)_2(H_2O) \cdot 2H_2O$ (β-UPP) is virtually the same as that in α-UPP, but when viewed along the direction of the propagating uranyl phosphonate chains, the two rows of phenyl rings point opposite from each other to form a chair-like shape (Fig. 50, β-UPP). The adjacent uranyl phenylphosphonate chains interlock each other to form a more compact structure than that of α-UPP. The α-UPP transforms at room temperature and moderate humidity into β-UPP. A suggested pathway for the transformation consists of a breaking of a $U{-}O_{eq}$ bond, followed by the rotation of the phenylphosphonate group by 120° about the other bridging and unbroken $U{-}O_{eq}$ bond. The rotation is accompanied by a deprotonation of the protonated phosphonate oxygen and by a protonation of the oxygen atom previously anchored at the uranium atom. The newly formed chair-like uranylphosphonate chains then shift and rearrange to actually form the lattice consistent with that of β-UPP. The powder patterns, ^{31}P MAS solid state NMR spectra, the electronic absorption spectra and TGA curves of the β-UPP formed by transformation from α-UPP and of that grown from single crystals are virtually identical.

Interestingly, the two uranyl compounds α- and β-UPP with virtually the same coordination environment around the main chromophore, the uranyl ion, showed a dramatically different luminescence behavior (231). While β-UPP exhibited a very strong luminescence at room temperature, α-UPP did not luminesce under the same conditions. Preliminary studies have indicated that the quenching of luminescence in α-UPP is of a vibrational nature since both electron and energy transfer could be excluded. Further explorations of this interesting phenomenon are still in progress.

XIII. CONCLUSIONS

Metal phosphonate chemistry is a relatively new field of study originating in the second one-half of the 1970s decade. From its modest beginning, it has continued to grow in many different directions. Initially, the research was centered around zirconium compounds but has now been extended to a variety of elements with oxidation states from +2 to +6. More recently complex oxometalate phosphonates have been prepared as well as mixed-metal compounds. The field has further been greatly enlarged by the use of templates and hydrothermal methods of synthesis. As a result an astonishing array of different and novel structure types have been prepared.

It is evident then that phosphonic acids are highly versatile ligands. Furthermore, because they can be prepared with a variety of functional groups they are ideal starting materials for molecular design. In this chapter, we have provided examples of porous phosphonates that can behave as sorbents and ion exchangers, others that are proton conductors, are photochemically active and catalytically active. In addition, the combination of metal species with phosphonate ligands lend themselves to self assembly and layer by layer construction. All of these properties recommend them for the design of materials with specific properties. Several practitioners are already engaged in such endeavors and this aspect of the field is sure to grow as more scientists become aware of the possibilities. Thus, I predict a great deal of future activity in metal phosphonate chemistry and the benefits that will arise from these endeavors are only limited by the energy and creativity of the practitioners.

ABBREVIATIONS

ATR–FTIR	Attenuated total reflection Fourier transform infrared spectroscopy
BET	Brunauer–Emmett–Teller
bpy	2,2-bipyridyl
DMSO	Dimethyl sulfoxide
EPR	Electron paramagnetic resonance
ET	Electron transfer
EXAFS	Extended X-ray absorption fine structure
fwhm	Full width at half-maximum
HTO	Hexagonal tungsten oxide
IR	Infrared
LB	Langmuir–Blodgett
MAS NMR	Magic angle spinning nuclear magnetic resonance
MepO	Methylphosphonate
MINIM	Metal–insulator–nanocluster–insulator–metal
MLCT	Metal-to-ligand charge transfer
MUD	Mercapto-1-undecanol
NLO	Nonlinear optical
ODP	Octadecylphosphonate
OTS	Octadecyltrichlorosilane
PAH	Polyallylamine hydrochloride
PANI	Polyanaline
PEO	Polyethyleneoxide
Ph	Phenyl
FM–FTIR	Polarization modulation Fourier transform infrared spectroscopy

PMIDA	*N*-(Phosphonomethyl)iminodiacetic acid
QCM	Quartz crystal microbalance
RH%	Percent relative humidity
SA	Surface area
TGA	Thermogravimetric analysis
XANES	X-ray absorption near edge structure
XPS	X-ray photoelectron spectroscopy
XRD	X-ray diffraction
ZEDP	Zirconium ethanediylbis(phosphonate)

ACKNOWLEDGMENTS

Much of my own work in metal phosphonate chemistry has been supported by the National Science Foundation, at first in the Chemistry Division and, more recently, by the Division of Materials Research, for which I am sincerely grateful. I have also received steady and unfailing support from the Robert A. Welch Foundation, Grant No. 673A, which has provided that extra incentive to accomplish more. Finally, my sincere thanks to those unsung heroes, my students and postdocs, without whose efforts nothing would be accomplished. A special thanks to Daniel Grohol for helping to edit the entire manuscript.

REFERENCES

1. S. Yamanaka and M. Koizumi, Clays, *Clay Miner*, *23*, 477 (1975).
2. S. Yamanaka, *Inorg. Chem.*,*15*, 2811 (1976).
3. G. Alberti, U. Costantino, S. Allulli, and N. Tomassini, *J. Inorg. Nucl. Chem.*, *40*, 1113 (1978).
4. G. Cao, H. G. Hong, and T. E. Mallouk, *Acc. Chem. Res.*, *25*, 420 (1992).
5. A. Clearfield and J. A. Stynes, *J. Inorg. Nucl. Chem.*, *26*, 117 (1964).
6. K. A. Kraus and H. O. Phillips, *J. Am. Chem. Soc.*, *78*, 644 (1956); K. A. Kraus, H. O. Phillips, T. A. Carlson, and J. S. Johnson, *Proceedings of the 2nd U.N. Conf. Peaceful Uses of Atomic Energy*, Geneva 1958, Vol. 28, p. 3.
7. C. B. Amphlett, *Proceedings of the 2nd U.N. Conference Peaceful Uses of Atomic Energy*, Geneva 1958, Vol. 28, p. 3.
8. A. Clearfield and G. D. Smith, *Inorg. Chem.*, *8*, 431 (1969); J. M. Troup and A. Clearfield, *Inorg. Chem.*, *16*, 3311 (1977).
9. (a) A. Clearfield, in *Inorganic Ion Exchange Materials*, A. Clearfield, Ed., CRC Press, Boca Raton, FL, 1982, Chapter 1; (b) G. Alberti, in *Inorganic Ion Exchange Materials*, A. Clearfield, Ed., CRC Press, Boca Raton, FL, 1982, Chapter 2.
10. G. Alberti, *Accts. Chem. Res.*, *11*, 163 (1978).

11. A. Clearfield, R. H. Blessing, and J. A. Stynes, *J. Inorg. Nucl. Chem.*, *30*, 2249 (1968).
12. N. J. Clayden, *J. Chem. Soc. Dalton Trans.*, 1877 (1987).
13. A. N. Christensen, E. K. Andersen, I. G. Andersen, G. Alberti, M. Nielsen, and M. S. Lehmann, *Acta Chem. Scand.*, *44*, 865 (1990).
14. D. M. Poojary, B. Shpeizer, and A. Clearfield, *J. Chem. Soc. Dalton Trans.*, 111 (1995).
15. A. Clearfield and J. M. Garces, *J. Inorg. Nucl. Chem.*, *41*, 903 (1979).
16. D. M. Poojary, B. Zhang, Y. Dong, G. Z. Peng, and A. Clearfield, *J. Phys. Chem.*, *98*, 13616 (1994).
17. A. Clearfield and J. M. Troup, *J. Phys. Chem.*, *77*, 243 (1973).
18. D. M. Poojary, H.-L. Hu, F. L. Campbell, III, and A. Clearfield, *Acta Crystallgr. Sect. B*, *49*, 996 (1993).
19. M. D. Dines, P. M. DiGiacomo, K. P. Callahan, P. C. Griffith, R. H. Lane, and R. E. Cooksey, in *Chemically Modified Surfaces in Catalysis and Electrocatalysis*, J. S. Miller, Ed., American Chemical Society Symposium Series, Washington, DC, 1982, Vol. 192, Chapter 12.
20. G. Alberti and U. Costantino, in *Intercalation Chemistry*, M. S. Whittingham and A. J. Jacobson, Eds., Academic, New York, 1982, p. 147.
21. A. Clearfield, A. Oskarsson, and C. Oskarsson, *Ion Exch. Membr.*, *1*, 91, (1972).
22. G. Alberti and E. J. Torracca, *J. Inorg. Nucl. Chem.*, *30*, 317 (1968).
23. M. B. Dines and P. M. Di Giacomo, *Inorg. Chem.*, *20*, 92, 1981.
24. G. Alberti, U. Costantino, and M. L. Luciani Giovagnotti, *J. Chromatog.*, *180*, 45 (1979).
25. L. Maya, *Inorg. Nucl. Chem. Lett.*, *15*, 207 (1979).
26. S. Yamanaka, M. Tsujimoto, and M. Tanaka, *J. Inorg. Nucl. Chem.*, *41*, 615 (1979).
27. S. Yamanaka and M. Hattori, *Chem. Lett.*, 1073 (1979).
28. A. Michaelis and R. Kaehne, *Chem. Ber.*, *31*, 1048 (1898).
29. A. E. Arbuzov, *J. Russ. Phys. Chem. Soc.*, *38*, 687 (1906).
30. E. Michel and A. Z. Weiss, *Z. Naturforsch.*, *20B*, 1307 (1965); *B22*, 1100 (1967).
31. A. Clearfield and R. M. Tindwa, *J. Inorg. Nucl. Chem.*, *41*, 871 (1979).
32. R. M. Tindwa, D. K. Ellis, G.-Z. Peng, and A. Clearfield, *J. Chem. Soc. Faraday Trans.*, *81*, 545 (1985).
33. G. Alberti and U. Costantino, *J. Mol. Catal.*, *27*, 235 (1984).
34. U. Costantino, in *Inorganic Ion Exchange Materials*, A. Clearfield, Ed., CRC Press, Boca Raton, FL, 1982, Chapter 3.
35. D. M. Poojary, C. Bhardwaj, and A. Clearfield, *J. Mater. Chem.*, *5*, 171 (1995).
36. D. A. Burwell, K. G. Valentine, J. H. Timmermans, and M. E. Thompson, *J. Am. Chem. Soc.*, *114*, 4144 (1992): D. A. Burwell, K. G. Valentine, and M. E. Thompson, *J. Magn. Reson.*, *97*, 498 (1992).

37. M. E. Thompson, *Chem. Mater.*, *6*, 1168 (1994).
38. G. Alberti, U. Costantino, M. Casciola, R. Vivani, and A. Peraio, *Solid State Ionics*, *46*, 61 (1991).
39. C.-H. Yang, Ph.D. Thesis, ''Organo-Derivatives of Zirconium Phosphate,'' Texas A&M University, December 1986.
40. G. Alberti, U. Costantino, J. Kornyei, and M. L. Luciani Giovagnotti, *React. Polym.*, *4*, 1 (1985).
41. G. Alberti, U. Costantino, and G. Perego, *J. Solid State Chem.*, *63*, 455 (1986).
42. (a) J. D. Wang, A. Clearfield, and G.-Z. Peng, *Mater. Chem. Phys.*, *35*, 208 (1993); (b) A. Clearfield, J. D. Wang, Y. Tian, E. Stein, and C. Bhardwaj, *J. Solid State Chem.*, *117*, 275 (1995).
43. D. J. MacLachlan and K. R. Morgan, *J. Phys. Chem.*, *96*, 3458 (1992).
44. F. L. Campbell, III, Ph.D. Thesis, ''reparation and Characterization of Organic Derivatives of Zirconium Phosphate,'' Texas A&M University, 1990.
45. A. Clearfield, J. D. Wang, Y. Tian, F. L. Campbell, III, and G. Z. Peng, *Materials Synthesis and Characterization*, D. Perry, Ed., Plenum, New York, in press.
46. P. Bellinghausen, M. S Thesis, ''The Synthesis and Characterization of Zirconium p-phenylbis (phosphonate) Phosphate and other Zirconium Arylbis (phosphonates) for the Application of Ion Exchange,'' Texas A&M University, Dec. 1995.
47. A. Clearfield, in *Design of New Materials*, D. L. Cocke and A. Clearfield, Eds., Plenum, New York, 1987, p. 130.
48. G. Alberti, U. Costantino, F. Marmottini, R. Vivani, and P. Zappelli, *Angew. Chem. Int. Engl. Ed* , *32*, 1357 (1993).
49. E. G. Derouane and V. Jullien-Lardot, *Stud. Surf. Sci. Catal.*, *83*, 11 (1994).
50. G. Alberti, U. Costantino, R. Vivani, and P. Zappelli, in *Synthesis/Characterization and Novel Applications of Molecular Sieve Materials*, R. L. Bedard, T. Bein, M. E. Davis, J. Garces, V. A. Maroni, and G. D. Stucky, Eds., Materials Research Society, Pittsburg, PA, 1991, p. 105.
51. S. Yamanaka and M. Tsujimoto, *J. Inorg. Nucl. Chem.*, *24*, 1773 (1985).
52. H. Nakano, T. Ohno, and S. Yamanaka, *Chem. Lett.*, *9* (1994).
53. C. Y. Ortiz-Avila and A. Clearfield, *Inorg. Chem.*, *24*, 1773 (1985).
54. A. Clearfield and J. Berman, *J. Inorg. Nucl. Chem.*, *43*, 2141 (1981).
55. A. Clearfield and C. Y. Ortiz-Avila, in *Supramolecular Chemistry*, T. Bein, Ed., American Chemical Society Symposium Series, 499, Washington, DC, 1992. Chapter 14.
56. Y. Takahashi and H. Tadokoro, *Macromolecules*, *6*, 672 (1973).
57. D. E. Fenton, J. M. Parker, and P. V. Wright, *Polymers*, *14*, 589 (1973).
58. J. R. Owen, S. C. Lloyd-Williams, G. Lagos, P. C. Spurdens, and B. C. Steele, in *Lithium Nonaqueous Battery Electrochemistry*, E. B. Yeager et al., Eds. Electrochemical Society, Pennington, NJ, 1980, p. 293.

59. M. B. Armand, J. M. Chabagno, and J. M. Duclot, in *Fast Ion Transport in Solids*, P. Vashista, J. N. Mundy, and G. K. Shenoy, Eds., North-Holland, New York, 1979, p. 131.
60. J. B. Boyce, L. C. DeJohnghe, and R. A. Huggins, *Solid State Ionics*, North-Holland, New York, 1986, pp. 253–325.
61. S. Yamanaka, *J. Inorg. Nucl. Chem.*, *42*, 717 (1980).
62. P. Jerus and A. Clearfield, *Solid State Ionics*, *6*, 79 (1982).
63. P. M. Di Giacomo and M. B. Dines, *Polyhedron*, *1*, 61 (1982).
64. C.-Y. Yang and A. Clearfield, *Reactive Polym.*, *5*, 13 (1987).
65. E. W. Stein, Sr., A. Clearfield, and M. A. Subramanian, *Solid State Ionics*, *83*, 113 (1996).
66. G. Alberti, M. Casciola, U. Costantino, A. Peraio, and E. Montoneri, *Solid State Ionics*, *50*, 315 (1992).
67. G. Alberti, M. Casciola, R. Palombari, and A. Peraio, *Solid State Ionics*, *58*, 339 (1992).
68. (a) L. H. Kullberg and A. Clearfield, *Solv. Extr. Ion Exch.*, *7*, 527 (1989); (b) *8*, 187 (1990).
69. A. Clearfield, in *Industrial Environmental Chemistry*, D. T. Sawyer and A. E. Martell, Eds., Plenum, New York, 1992, p. 289.
70. G. L. Rosenthal and J. Caruso, *Inorg. Chem.*, *31*, 3104 (1992).
71. C. Y. Ortiz-Avila, C. Bhardwaj, and A. Clearfield, *Inorg. Chem.*, *33*, 2499 (1994).
72. L. Maya, *J. Inorg. Nucl. Chem.*, *43*, 3104 (1992).
73. T. E. Mallouk, H.-N. Kim, P. J. Ollivier, and S. W. Keller, in *Comprehensive Supramolecular Chemistry*, G. Alberti and T. Bein, Eds., Pergamon, Tarrytown, NY, 1996, Vol. 7, p. 189.
74. I. Langmuir, *J. Am. Chem. Soc.*, *39*, 1848 (1917); K. A. Blogett, *J. Am. Chem. Soc.*, *57*, 1007 (1935).
75. N. J. Geddes, M. C. Jurich, J. D. Swalen, R. Tweig, and J. F. Rabollt, *J. Chem. Phys.*, *94*, 1603 (1991).
76. A. Ullman, *Adv. Mater.*, *2*, 573 (1990); L. Netzer and J. Sagiv, *J. Am. Chem. Soc.*, *105*, 674 (1983).
77. H. Lee, L. J. Kepley, T. E. Mallouk, H. G. Hong, and S. Akhter, *J. Phys. Chem.*, *92*, 2597 (1988).
78. H. Lee, L. J. Kepley, H. G. Hong, and T. E. Mallouk, *J. Am. Chem. Soc.*, *110*, 618 (1988).
79. H. C. Yang, K. Aoki, H.-G. Hong, D. D. Sackett, M. F. Arendt, S.-L. Yau, C. M. Bell, and T. E. Mallouk, *J. Am. Chem. Soc.*, *115*, 11855 (1993).
80. C. Y. Ortiz-Avila and A. Clearfield, *J. Chem. Soc. Dalton Trans.*, 1617 (1984).
81. A. Clearfield and R. M. Tindwa, *J. Inorg. Nucl. Chem.*, *41*, 871 (1979).
82. A. C. Zeppenfeld, S. L. Fiddler, W. K. Ham, B. J. Klopfenstein, and C. J. Page, *J. Am. Chem. Soc.*, *116*, 9158 (1994).

83. H.-G. Hong, D. D. Sackett, and T. E. Mallouk, *Chem. Mater.*, *3*, 521 (1991).
84. B. L. Frey, D. G. Hanken, and R. M. Corn, *Langmuir*, *9*, 1815 (1993).
85. J. T. O'Brien, A. C. Zeppenfeld, G. L. Richmond, and C. J. Page, *Langmuir*, *10*, 4657 (1994).
86. M. L. Schilling, H. E. Katz, S. M. Stein, S. F. Shane, W. L. Wilson, S. Buratto, S. B. Ungashe, G. N. Taylor, T. M. Putvinski, and C. E. D. Chidsey, *Langmuir*, *9*, 2156 (1993).
87. S. F. Bent, M. L. Schilling, W. L. Wilson, H. E. Katz, and A. L. Harris, *Chem. Mater.*, *6*, 122 (1994).
88. H. Byrd, J. K. Pike, and D. R. Talham, *Chem. Mater.*, *5*, 709 (1993).
89. H. Byrd, S. Whipps, J. K. Pike, J. Ma, S. E. Nagles, and D. R. Talham, *J. Am. Chem. Soc.*, *116*, 295 (1994).
90. A. Clearfield, Comments, *Inorg. Chem.*, *10*, 89 (1990).
91. T. M. Putvinski, M. L. Shilling, H. E. Katz, C. E. D. Chidsey, A. M. Mujsce, and A. B. Emerson, *Langmuir*, *6*, 1567 (1990).
92. H. E. Katz, G. Scheller, T. M. Putvinski, M. L. Schilling, W. L. Wilson, and C. E. D. Chidsey, *Science*, *254*, 1485 (1991).
93. H. E. Katz, M. L. Schilling, S. Ungashe, T. M. Putvinski, and C. E. D. Chidsey, in *Supramolecular Chemistry*, T. Bein, Ed., American Chemical Society Symposium Series 499, Washington, DC, 1992. p. 24.
94. H. E. Katz, W. L. Wilson, and G. Scheller, *J. Am. Chem. Soc.*, *116*, 6636 (1994).
95. S. B. Ungashe, W. L. Wilson, H. E. Katz, G. R. Scheller, and T. M. Putvinski, *Chem. Mater.*, *114*, 8717 (1992).
96. H. E. Katz, M. L. Schilling, C. E. D. Chidsey, T. M. Putvinski, and R. S. Hutton, *Chem. Mater.*, *3*, 699 (1991).
97. H. Byrd, J. K. Pike, and D. R. Talham, *J. Am. Chem. Soc.*, *116*, 7903 (1994); C. T. Siep, H. Byrd, and D. R. Talham, *Inorg. Chem.*, *35*, 3479 (1996).
98. S. G. Carling, P. Day, D. Visser, and R. K. Kremer, *J. Solid State Chem.*, *106*, 111 (1993).
99. S. Akhter, H. Lee, H.-G. Hong, T. E. Mallouk, and J. M. White, *Vac. Sci. Technol.*, *A7*, 1608 (1989).
100. S. Feng and T. Bein, *Nature(London)*, *368*, 834 (1994); *Science*, *265*, 1839 (1994).
101. Y. Lvov, F. Essler, and G. Decker, *J. Phys. Chem.*, *97*, 13773 (1993).
102. S. W. Keller, H.-N. Kim, and T. E. Mallouk, *J. Am. Chem. Soc.*, *116*, 8817 (1994).
103. G. Alberti, M. Casciola, and U. Costantino, *J. Colloid Interface Chem.*, *107*, 256 (1985).
104. A. J. Jacobson, in *Materials Science Forum*, Vol. 152-153, 1, (1994), J. Rouxel, M. Tournoux, and R. Brec, Eds., Trans Tech Publications, Switzerland.
105. L. J. Kepley, D. D. Sackett, C. M. Bell, and T. E. Mallouk, *Thin Solid Films*, *208*, 132 (1992).

106. H. E. Katz and M. L. Schilling, *Chem. Mater.*, *5*, 1162 (1993).
107. M. L. Schilling, H. E. Katz, S. M. Stein, S. F. Shane, W. L. Wilson, S. B. Ungashe, G. N. Taylor, T. M. Putvinski, and C. E. D. Chidsey, *Langmuir* ,(1994).
108. H.-G. Hong and T. E. Mallouk, *Langmuir*, *7*, 2362 (1991).
109. D. L. Feldheim and T. E. Mallouk, *J. Chem. Commun.*, 2591 (1996).
110. D. L. Feldheim, K. C. Grabar, M. J. Natan, and T. E. Mallouk, *J. Am. Chem. Soc.*, *118*, 7640 (1996).
111. G. Alberti, U. Costantino, R. Vivani, and R. K. Biswas, *React. Polym.*, *17*, 245 (1992).
112. G. Alberti, R. Vivani, R. K. Biswas, and S. Murcia-Mascaros, *React. Polym.*, *19*, 1 (1993).
113. G. Alberti, M. Casciola, and R. Vivani, *Inorg. Chem.*, *32*, 4600 (1993).
114. G. Alberti, S. Murcia-Mascaro's, and R. Vivani, in *Soft Chemistry Routes to New Materials, Materials Science Forum*, **152–153,** 87 (1994) Trans Tech Publications, Switzerland.
115. G. Alberti, S. Murcia-Mascaro's, and R. Vivani, *Mater. Chem. Phys.*, *35*, 187 (1993).
116. A. Clearfield and J. M. Garces, *J. Inorg. Nucl. Chem.*, *41*, 879 (1979).
117. G. Alberti, F. Marmottini, S. Murcia-Mascaro's, and R. Vivani, *Angew. Chem. Int. Eng. Ed.*, *33*, 1594 (1994).
118. G. Alberti, M. Casciola, and R. K. Biswas, *Inorg. Chim. Acta*, *201*, 207 (1992).
119. G. Alberti, in *Comprehensive Supamolecular Chemistry*, G. Alberti and T. Bein, Eds., Pergamon, New York, 1996, Vol. 7, p. 151 ff.
120. E. Brunet, M. Huelva, and J. Rodriguez-Uhis, *Tetrahedron Lett.*, *35*, 8697 (1994).
121. G. Alberti, L. Boccali, M. Casciola, L. Massinelli, and E. Montoneri, *Solid State Ionics*, *84*, 97 (1996).
122. B. Zhang, D. M. Poojary, and A. Clearfield, *Inorg. Chem.*, submitted for publication.
123. B. Zhang, D. M. Poojary, A. Clearfield, and G. Z. Peng, *Chem. Mater.*, *8*, 1333 (1996).
124. L. A. Vermeulen and M. E. Thompson, *Nature(London)*, *358*, 656 (1992).
125. L. A. Vermeulen, J. L. Snover, L. S. Sapochak, and M. E. Thompson, *J. Am. Chem. Soc.*, *115*, 11767 (1993).
126. L. A. Vermeulen and M. E. Thompson, *Chem. Mater.*, *6*, 77 (1994).
127. D. M. Poojary, L. A. Vermeulen, E. Vincenci, A. Clearfield, and M. E. Thompson, *Chem. Mater.*, *6*, 1845 (1994).
128. H. Byrd, A. Clearfield, D. M. Poojary, K. P. Reis, and M. E. Thompson, *Chem. Mater.*, *8*, 2239 (1996).
129. K. P. Reis, V. K. Joshi, and M. E. Thompson, *J. Catal.*, *161*, 62 (1996).
130. A. Clearfield and S. P. Pack, *J. Catal.*, *51*, 431 (1978).
131. S. Cheng and A. Clearfield, *J. Catal.*, *94*, 455 (1985).

132. B. Zhang and A. Clearfield, *J. Am. Chem. Soc.*, *119*, 2751 (1997).

133. M. Tazaki, K. Nita, M. Takagi, and K. Ueno, *Chem. Lett.*, 571 (1982).

134. D. A. Burwell and M. E. Thompson, *Chem. Mater.*, *3*, 730 (1991); *3*, 14 (1991).

135. T. Kijima, Y. Kawagoe, K. Mihara, and M. Machida, *J. Chem. Soc. Dalton Trans.*, 3827 (1993); T. Kijima, S. Watanabe, and M. Machida, *Inorg. Chem.*, *33*, 2586 (1994).

136. G. M. Carter, M. K. Thakur, J. V. Hryniewicz, Y. J. Chen, and S. E. Meyer, in *Crystallographically Ordered Polymers*, D. J. Sandman, Ed., ACS Symposium Series, 337; American Chemical Society, Washington, DC, 1987, p. 168.

137. J. B. Lando, in *Polyacetylenes*, D. Bloor and R. Chance, Eds.; Martinus Nijhoff, Dordrecht, The Netherlands, 1985, p. 363.

138. G. Cao and T. E. Mallouk, *J. Solid State Chem.*, *94*, 59 (1991).

139. J.-C. Chiang and A. G. Mac Diarmid, *Synth. Metals*, *13*, 193 (1986); A. J. Epstein and A. G. Mac Diarmid, *Molec. Cryst. Liq. Cryst.*, *160*, 165 (1988).

140. M. G. Kanatzidis, C.-G. Wu, H. O. Marcy, D. C. DeGroot, J. L. Schindler, C. R. Kennewurf, M. Benz, and E. LeGoff, in *Supramolecular Architecture*, T. Bein, Ed., ACS Symposium Series 499, American Chemical Society, Washington, DC, 1991.

141. G. L. Rosenthal, J. Caruso, and S. G. Stone, *Polyhedron*, *13*, 1311 (1994).

142. R. Rossetti, S. Nakahana, and L. E. Brus, *J. Chem. Phys.*, *79*, 1086 (1983).

143. M. L. Steigewald and M. L. Brus, *Annu. Rev. Mater. Sci.*, *19*, 471 (1989).

144. G. A. Ozin, A. Kuperman, and A. Stein, *Angew. Chem. Int. Ed. Engl.*, *28*, 359 (1989); G. D. Stucky and J. E. McDougall, *Science*, *247*, 669 (1990); P. A. Bianconi, J. Lin, and A. R. Stzelecki, *Nature(London)*, *349*, 315 (1991).

145. G. Cao, L. K. Rabenberg, C. Nunn, and T. E. Mallouk, *Chem. Mater.*, *3*, 149 (1991).

146. A. Clearfield and G. D. Smith, *J. Colloid Interface Sci.*, *28*, 325 (1968).

147. S. Abdo, P. Canesson, M. Cruz, J. J. Fripiat, and H. Van Damme, *J. Phys. Chem.*, *85*, 797 (1981); R. A. Della Guardin and J. K. Thomas, *J. Phys. Chem.*, *87*, 990 (1983); P. K. Ghosh and A. Bard, *J. Phys. Chem.*, *88*, 5519 (1984); D. P. Vliers, D. Collin, R. A. Schoonheydt, and F. C. De Schrijver, *Langmuir*, *2*, 165 (1986); C. Miller and M. Gratzel, *J. Phys. Chem.*, *95*, 5225 (1991); C. E. D. Chidsey, *Science*, *251*, 919 (1991).

148. J. L. Colon, C.-Y. Yang, A. Clearfield, and C. R. Martin, *J. Phys. Chem.*, *92*, 5777 (1988).

149. J. L. Colon, C.-Y. Yang, A. Clearfield, and C. R. Martin, *J. Phys. Chem.*, *94*, 874 (1990).

150. D. Grohol and A. Clearfield, unpublished results.

151. S. W. Keller, S. A. Johnson, E. S. Brigham, and T. E. Mallouk, *J. Am. Chem. Soc.*, *117*, 12879 (1995); D. M. Kaschak and T. E. Mallouk, *J. Am. Chem. Soc.*, *118*, 4222 (1996).

152. A. Clearfield and J. Berman, *J. Inorg. Nucl. Chem.*, *43*, 2141 (1981).

153. H. Miyata, Y. Sugahara, K. Kuroda, and C. Kato, *J. Chem. Soc. Faraday Trans.*, *84*, 2677 (1988); T. Nakamato, K. Kuroda, and C. Kato, *J. Chem. Soc. Chem. Commun.*, 1144 (1989).
154. M. I. Khan and J. Zubieta, *Progress in Inorganic Chemistry*, Wiley-Interscience, New York, Vol. 43, 1995.
155. J. W. Johnson, A. J. Jacobson, J. F. Brody, and J. T. Lewandowski, *Inorg. Chem.*, *23*, 3842 (1984).
156. J. W. Johnson, A. J. Jacobson, W. M. Butler, S. E. Rosenthal, J. F. Brody, and J. T. Lewandowski, *J. Am. Chem. Soc.*, *111*, 381 (1989).
157. G. Huan, A. L. Jacobson, J. W. Johnson, and E. W. Corcoran, Jr., *Chem. Mater.*, *2*, 91 (1990).
158. D. J. Sutor, *Acta Crystallogr.*, *23*, 418 (1967); F. Abbonna, R. Boistelle, and R. Haser, *Acta Crystallogr.*, *B35*, 2514 (1979).
159. G. Huan, J. W. Johnson, J. F. Brody, D. P. Goshorn, and A. J. Jacobson, *Mater. Chem. Phys.*, *35*, 199 (1993).
160. J. W. Johnson, D. C. Johnston, A. J. Jacobson, and J. F. Leonowicz, *J. Am. Chem. Soc.*, *106*, 8123 (1994); C. C. Torardi and J. C. Calabrese, *Inorg. Chem.*, *23*, 1308 (1984); M. E. Leonowicz, J. W. Johnson, J. F. Brody, H. F. Shannon, Jr., and J. M. Newsome, *J. Solid State Chem.*, *56*, 370 (1985).
161. P. Gendraud, M. E. de Roy, and J. P. Besse, *J. Solid State Chem.*, *106*, 577 (1993).
162. P. Gendraud, L. Bigey, M. E. de Roy, and J. P. Besse, *Chem. Mater.*, *9*, 539 (1997).
163. P. Gendraud, M. E. de Roy, and J. P. Besse, *Inorg. Chem.*, *35*, 6108 (1996).
164. M. R. Torgerson and D. G. Nocera, *J. Am. Chem. Soc.*, *118*, 8739 (1996).
165. Y.-G. K. Shin, E. A. Saori, M. R. Torgerson, and D. G. Nocera, in *Selectivity in Catalysis*, ACS Symposium Series 517, M. E. Davis and S. L. Suib, Eds., American Chemical Society, Washington, DC, 1993.
166. V. V. Guliants, J. B. Benziger, and S. Sundaresan, *Chem. Mater.*, *7*, 1493 (1995).
167. G. Huan, A. J. Jacobson, J. W. Johnson, and D. P. Goshorn, *Chem. Mater.*, *4*, 661 (1992).
168. J. LeBideau, D. Papoutsakis, J. E. Jackson, and D. G. Nocera, *J. Am. Chem. Soc.*, *119*, 1313 (1997).
169. D. Papoutoakis, J. E. Jackson, and D. G. Nocera, *Inorg. Chem.*, *35*, 800 (1996).
170. G. Centi, F. Trifiro, J. R. Elner, and V. M. Franchetti, *Chem. Rev.*, *88*, 55 (1988).
171. G. Centi, *Catal. Today*, *16*, 5 (1993).
172. G. Huan, J. W. Johnson, and A. J. Jacobson, *J. Solid State Chem.*, *89*, 220 (1990).
173. W. T. A. Harrison,L. L. Dussack, and A. J. Jacobson, *Inorg. Chem.*, *35*, 1461 (1996).
174. G. Bonavia, R. C. Haushalter, C. J. O'Connor, and J. Zubieta, *Inorg. Chem.*, *35*, 5603 (1996).

175. M. I. Khan, Y.-S. Lee, C. J. O'Connor, R. C. Haushalter, and J. Zubieta, *J. Am. Chem. Soc.*, *116*, 4525 (1994).
176. V. Soghomonian, R. Diaz, R. C. Haushalter, C. J. O'Connor, and J. Zubieta, *Inorg. Chem.*, *34*, 4460 (1995).
177. G. Huan, A. J. Jacobson, and V. W. Day, *Angew. Chem. Int. Engl. Ed.*, *30*, 422 (1991).
178. Q. Chen, J. Salta, and J. Zubieta, *Inorg. Chem.*, *32*, 4485 (1993).
179. (a) G. Cao, H. Lee, V. M. Lynch, and T. E. Mallouk, *Inorg. Chem.*, *27*, 2781 (1988); (b) *Solid State Ionics*, *26*, 63 (1988).
180. K. Martin, P. J. Squattrito, and A. Clearfield, *Inorg. Chim. Acta*, *155*, 7 (1989).
181. P. Palvadeau, M. Quegnec, J. P. Venien, B. Bujoli, and J. Villieras, *Mat. Res. Bull.*, *23*, 1561 (1988).
182. G. Cao, V. M. Lynch, and L. N. Yacullo, *Chem. Mater.*, *5*, 100 (1993).
183. Y. Zhang and A. Clearfield, *Inorg. Chem.*, *31*, 2821 (1992).
184. J. LeBideau, B. Bujoli, A. Jouanneaux, C. Payen, P. Palvadeau, and J. Rouxel, *Inorg. Chem.*, *32*, 4617 (1993).
185. G. CaO and T. E. Mallouk, *Inorg. Chem.*, *30*, 1434 (1991).
186. K. J. Frink, R.-C. Wang, J. L. Colon, and A. Clearfield, *Inorg. Chem.*, *30*, 1438 (1991).
187. D. Cunningham, P. J. D. Hennely, and T. Deeney, *Inorg. Chim. Acta*, *37*, 95 (1979).
188. B. Bujoli, O. Pena, P. Palvadeau, J. LeBideau, C. Payen, and J. Rouxel, *Chem. Mater.*, *5*, 583 (1993).
189. B. Bujoli, A. Courilleau, P. Palvadeau, and J. Rouxel, *Eur. J. Solid State Inorg. Chem.*, *29*, 171 (1992).
190. J. LeBideau, A. Jounanneaux, C. Payen, and B. Bujoli, *J. Mater. Chem.*, *4*, 1319 (1994).
191. A. Schier, S. Glamper, and G. Muller, *Inorg. Chim. Acta*, *177*, 179 (1990).
192. P. R. Rudolf, E. T. Clark, A. E. Martell, and A. Clearfield, *Acta Crystallogr.*, *C44*, 796 (1985); *Inorg. Chim. Acta*, *164*, 59 (1989).
193. G. Cao, V. M. Lynch, J. S. Swinnea, and T. E. Mallouk, *Inorg. Chem.*, *29*, 2112 (1990).
194. D. M. Poojary, B. Zhang, A. Cabeza, M. A. G. Aranda, S. Bruque, and A. Clearfield, *J. Mater. Chem.*, *6*, 639 (1996).
195. R.-C. Wang, Y. Zhang, H.-L. Hu, R. R. Frausto, and A. Clearfield, *Chem. Mater.*, *4*, 864 (1992).
196. B. Bujoli, P. Palvadeau, and J. Rouxel, *Chem. Mater.*, *2*, 582 (1990).
197. Y. Zhang, K. J. Scott, and A. Clearfield, *J. Mater. Chem.*, *5*, 315 (1995).
198. D. M. Poojary and A. Clearfield, *J. Am. Chem. Soc.*, *117*, 11278 (1995).
199. K. J. Scott, Y. Zhang, R.-C. Wang, and A. Clearfield, *Che. Mater.*, *1*, 1095 (1995).

200. S. Drumel, P. Janvier, D. Deniaud, and B. Bujoli, *Chem. Commun.*, 1051 (1995).
201. S. Drumel, P. Janvier, M. Bujoli-Doeuff, and B. Bujoli, *J. Mater. Chem.*, *6*, 1843 (1996).
202. Y. Zhang, K. J. Scott, and A. Clearfield, *Chem. Mater.*, *5*, 495 (1993).
203. J. LeBideau, C. Payen, P. Palvadeau, and B. Bujoli, *Inorg. Chem.*, *33*, 4885 (1994).
204. S. Drumel, P. Janvier, M. Bujoli-Doeuff, and B. Bujoli, *Inorg. Chem.*, *35*, 5786 (1996).
205. W. T. A. Harrison, J. T. Vaughey, L. L. Dussak, A. J. Jacobson, T. E. Martin, and G. D. Stucky, *J. Solid State Chem.*, *114*, 151 (1995).
206. D. M. Poojary, B. Zhang, P. Bellinghausen, and A. Clearfield, *Inorg. Chem.*, *35*, 5254 (1996).
207. D. M. Poojary, B. Zhang, P. Bellinghausen, and A. Clearfield, *Inorg. Chem.*, *35*, 4942 (1996).
208. B. Zhang and A. Clearfield, unpublished results.
209. C. Bhardwaj, H.-L. Hu, and A. Clearfield, *Inorg. Chem.*, *32*, 4294 (1993).
210. S. Drumel, P. Janvier, P. Barboux, M. Bujoli-Doeuff, and B. Bujoli, *Inorg. Chem.*, *34*, 148 (1995).
211. (a) K. Maeda, Y. Kiyozumi, and F. Mizukami, *Angew. Chem. Int. Ed. Engl.*, *33*, 2335 (1994); (b) K. Maeda, J. Akimoto, Y. Kiyozumi, and F. Mizukami, *Angew. Chem. Int. Ed. Engl.*, *34*, 1199 (1995); (c) *J. Chem. Soc. Chem. Commun.*, 1033 (1995).
212. R. LaDucam, D. Rose, J. R. D. DeBord, R. C. Haushalter, C. J. O'Connor, and J. Zubieta, *J. Solid State Chem.*, *123*, 408 (1996).
213. P. Battioni, J. P. Renaud, J. F. Bartoli, M. Reina-Artiles, M. Fort, and D. Mansuy, *J. Am. Chem. Soc.*, *110*, 8462 (1988).
214. D. Deniaud, B. Schollorn, D. Mansuy, J. Rousel, P. Battioni, and B. Bujoli, *Chem. Mater.*, *7*, 995 (1995).
215. D. Massiot, S. Drumel, P. Janvier, M. Bujoli-Doeuff, and B. Bujoli, *Chem. Mater.*, *9*, 6 (1997).
216. G. Alberti and R. Palombari, *Solid State Ionics*, *35*, 153 (1989).
217. G. Alberti, M. Casiola, and R. Palombari, *Solid State Ionics*, *52*, 291 (1992).
218. L. C. Brousseau, III, K. Aoki, M. E. Garcia, G. Cao, and T. E. Mallouk, *NATO ASI Series C, V.400 Multifunctional Mesoporous Inorganic Solids*, C. Sequeira and M. J. Hudson, Eds., NATO ASI Series Kluwer Academic, Dordrecht, The Netherlands, 1993, p. 225.
219. K. Aoki, L. C. Brousseau, III, and T. E. Mallouk, *Sensors and Actuators*, *14*, 703 (1993).
220. L. C. Brousseau, III, and T. E. Mallouk, *Anal. Chem.*, *69*, 679 (1997).
221. L. C. Brousseau, III, D. Aurentz, A. J. Benesi, and T. E. Mallouk, *Anal. Chem.*, *69*, 688 (1997).
222. L. C. Brousseau, III, K. Aoka, H. C. Yang, and T. E. Mallouk, in *Interfacial*

Design and Chemical Sensing, ACS Symposium Series 561, T. E. Mallouk and D. J. Harrison, Eds., American Chemical Society, Washington, DC, 1994, p. 60.

223. D. M. Poojary, Y. Zhang, B. Zhang, and A. Clearfield, *Chem. Mater.*, *7*, 822 (1995).
224. W. T. A. Harrison, L. L. Dussack, and A. J. Jacobson, *Inorg. Chem.*, *34*, 4774 (1995).
225. A. Cabeza, M. A. G. Aranda, F. M. Cantero, D. Lozano, M. Martinez-Lara, and S. Bruque, *J. Solid State Chem.*, *121*, 181 (1996).
226. P. A. Brittel, M. Wozniak, J. C. Boivin, G. Nowogrocki, and G. Thomas, *Acta Crystallogr.*, *C 42*, 1502 (1986).
227. D. M. Poojary, D. Grohol, and A. Clearfield, *J. Phys. Chem. Solids*, *56*, 1383 (1995).
228. D. M. Poojary, D. Grohol, and A. Clearfield, *Angew. Chem. Int. Ed. Engl.*, *34*, 1508 (1995).
229. D. M. Poojary, A. Cabeza, M. A. G. Aranda, S. Bruque, and A. Clearfield, *Inorg. Chem.*, *35*, 1468 (1996).
230. D. Grohol, M. A. Subramanian, D. M. Poojary, and A. Clearfield, *Inorg. Chem.*, *35*, 5264 (1996).
231. D. Grohol and A. Clearfield, *J. Am. Chem. Soc.*, *119*, 4662 (1997).

Oxidation of Hydrazine in Aqueous Solution

DAVID M. STANBURY

Department of Chemistry
Auburn University
Auburn, AL 36849

CONTENTS

Progress in Inorganic Chemistry, Vol. 47, Edited by Kenneth D. Karlin.
ISBN 0-471-24039-7 © 1998 John Wiley & Sons, Inc.

I. INTRODUCTION

The literature on the chemistry of hydrazine (N_2H_4) goes back to 1887 when it was first prepared by Curtius (1). Its subsequent studies have been summarized by Audrieth and Ogg in 1951 (2), and again by Schmidt in 1984 (3). The latter volume is over 1000 pages in length and cites more than 4000 references. Although many of the reports on hydrazine chemistry are focused on its technical applications, there is a rich tradition of studying its fundamental reactivity in aqueous solution. The books by Aldrieth and Ogg and by Schmidt review the early history in the study of these reactions, which was largely oriented toward establishing the factors that would promote HN_3 formation in reactions where hydrazine was oxidized. This early work was central in the development of ideas regarding the distinction between one- and two-electron oxidation reactions, since only two-electron oxidants led to HN_3 production. In more recent years, the interest in using hydrazine to synthesize HN_3 has waned, but fundamental studies of the oxidation reactions of hydrazine have continued to play a central role in the development of mechanistic chemistry.

A critical point in the development of the reaction mechanisms was Higginson's 1957 review (4). In this review, Higginson surveyed the literature and presented good evidence for a general treatment of the mechanisms of hydrazine oxidation. We briefly outline his position.

The reactions can be classified according to the following limiting stoichiometries:

$$N_2H_4 \longrightarrow N_2 + 4H^+ + 4e^- \tag{1}$$

$$N_2H_4 \longrightarrow \tfrac{1}{2}N_2 + NH_3 + H^+ + 1e^- \tag{2}$$

$$N_2H_4 \longrightarrow \tfrac{1}{2}NH_3 + \tfrac{1}{2}HN_3 + 2H^+ + 2e^- \tag{3}$$

In favorable cases, the net four-equivalent oxidation can be achieved exclusively, but the net one- and two-equivalent oxidations always occur with some contribution from the four-equivalent process. It was also recognized that the oxidants could be classified as one- or two-equivalent oxidants in that one-equivalent oxidants would oxidize N_2H_4 to "N_2H_3" in the first step, whereas two-equivalent oxidants would bypass this radical and lead to N_2H_2 directly.

The following mechanism was inferred:

The one-equivalent oxidants

$$N_2H_4 \longrightarrow N_2H_3 + H^+ + e^- \qquad (4)$$

$$2N_2H_3 \longrightarrow N_4H_6 \longrightarrow 2NH_3 + N_2 \qquad (5)$$

$$N_2H_3 \longrightarrow N_2H_2 + H^+ + e^- \qquad (6)$$

$$2N_2H_2 \longrightarrow N_2 + N_2H_4 \qquad (7)$$

The two-equivalent oxidants

$$N_2H_4 \longrightarrow N_2H_2 + 2H^+ + 2e^- \qquad (8)$$

$$2N_2H_2 \longrightarrow N_2 + N_2H_4 \qquad (9)$$

The formation of HN_3 was assigned to an alternative decay pathway for N_2H_2 as in

$$2N_2H_2 \longrightarrow N_4H_4 \longrightarrow HN_3 + NH_3 \qquad (10)$$

The variable yields of HN_3 were then attributed to the subsequent oxidation of HN_3 to N_2 and to the oxidation of N_2H_2 to N_2. In general, HN_3 is produced in significant amounts only at high temperatures and in strongly acidic media, and will not be considered further in this chapter.

Thus, one-equivalent oxidants were seen to yield a mixture of net four-equivalent and one-equivalent products, while two-equivalent oxidants yield net four-equivalent products exclusively.

Higginson's mechanistic scenario has been a powerful organizing concept in the years since its publication. It has been cited in a large number of the research articles discussed in this chapter, and for many workers it is viewed almost as a law of chemistry that enables a clear criterion of mechanism. Descriptions of Higginson's mechanism can be found in *Comprehensive Chemical Kinetics* (5), Sykes' text on reaction mechanisms (6), Skyes' review on redox reaction mechanisms (7), Bottomley's review on hydrazine reactions (8), and Stedman's review on mechanisms of nitrogen compound reactions (9).

A potentially significant elaboration to Higginson's general mechanism was put forth in a 1972 paper by Brown and Higginson (10). This paper was written in response to a prior suggestion that reactions of one-electron oxidants with sulfite could be classified on the basis of the lability of the oxidant, with labile oxidants yielding $S_2O_6^{2-}$ and SO_4^{2-} and inert oxidants yielding only SO_4^{2-} (11).

Brown and Higginson examined a series of reactions of sulfite with one-electron oxidants with regard to this suggestion, and they extended the survey to analogous reactions of hydrazine. It was observed that significant yields of NH_3 were obtained with no simple correlation to the lability of the oxidant. A mechanism to account for this observation was proposed in which free N_2H_3 produced in an outer-sphere mechanism by inert oxidants normally should yield some NH_3 through the dimerization of N_2H_3, while labile oxidants react through inner-sphere mechanisms that can also yield NH_3.

Subsequent to Higginson's 1957 review, papers have been published on the homogeneous uncatalyzed oxidation of N_2H_4 by a wide variety of oxidants. These studies provide an opportunity to reassess the mechanistic proposals described above. There have also been major advances in our understanding of the intermediates proposed for these reactions, especially regarding the chemistries of the hydrazyl radical and diazene. It is now possible to assess proposed mechanisms much more critically.

Papers have also been published on autoxidations catalyzed by manganese (12a), pentacyanoferrate(II) (12b), cobalt(II)trisulfophthalocyanine, copper(II)-tetrasulfophthalocyanine (13), and by cobalt(II)tetrasulfophthalocyanine (14, 15). Studies of other catalytic reactions include the copper-catalyzed reactions with $S_2O_8^{2-}$ (16), $[Ni^{IV}(dmg)_3]^{2-}$ (dmg = dimethylglyoximato) (17, 18), and H_2O_2 (19-23); the silver(I)-catalyzed reactions with $P_2O_8^{4-}$ and with the ethylenebis(biguanide)silver(III) cation (24, 25); and the Mo^{VI}-catalyzed reaction with methylene blue (26). Heterogeneous reactions that have been studied include the oxidations by OsO_4 (27), ClO_3^- (catalyzed by OsO_4) (27), Ag_2O (28), and $BaCrO_4$ (29). Another class of oxidants that has been studied comprises enzymes, such as ribonucleotide reductase (30, 31) and photosystem II (32). All of these catalytic, heterogeneous, and enzyme studies are beyond the scope on this chapter.

A related area is the oxidation of hydrazine at electrode surfaces. Such reactions are electochemically irreversible and are highly dependent on the nature of the electrode surface. A few representative publications can be cited for the interested reader, but no attempt is made in this chapter to discuss this challenging subject (33–39).

In what follows, we present a summary of the relevant thermochemical properties of hydrazine, followed by a survey (comprehensive literature coverage through 1995 and extensive through 1996) of the characteristics of each of the homogeneous noncatalytic reactions, excluding the enzyme systems. These are organized with the main group oxidants first, according to the periodic table, and with the transition metal oxidants second. The transition metal oxidants are organized with the two-electron oxidants first, the inert one-electron oxidants second, and the labile one-electron oxidants third. We then introduce additional information regarding the nitrogenous intermediates, and finally organize the results so as to assess the viability of Higginson's classification, the general mechanism of Brown and Higginson, and the significance of other trends.

II. THERMOCHEMICAL PROPERTIES OF HYDRAZINE

Thermochemical data discussed and evaluated in this and later sections are summarized in Table I.

A. The pK_a of $N_2H_5^+$ and $N_2H_6^{2+}$

Hydrazine is a weak base, such that $N_2H_5^+$ has a pK_a of 7.98 at 25°C (40). Acid dissociation constants for $N_2H_5^+$ as a function of ionic strength and the corresponding enthalpy and entropy changes are critically summarized in the tables of Smith and co-workers (41). This pK_a has recently been recorded as a function of temperature up to 80°C (42).

A number of the reactions have been studied in strongly acidic media, such as those of $[Co(acac)_3]$, $[Co(ox)_3]^{3-}$, Co^{3+}, Ce^{4+}, Mn^{3+}, and so on. The use of such acidic conditions raises the question as to whether $N_2H_5^+$ or $N_2H_6^{2+}$ is the predominant species. The question is not trivial, since the degree of protonation can have a profound influence on the mechanism (e.g., is it a hydrogen-atom transfer reaction?) On the other hand, there are so many other uncertainties

TABLE I
Thermodynamic Data at 25°C[a]

Reaction	Value	Section
$N_2H_6^{2+} \rightleftharpoons N_2H_5^+ + H^+$	pK_a = −0.5	II.A
$N_2H_5^{+\cdot} \rightleftharpoons N_2H_4 + H^+$	pK_a = 7.98	II.A
$N_2(g) + 5H^+ + 4e^- \rightleftharpoons N_2H_5^+$	$E°$ = −0.23 V	II.B
$\frac{1}{2}N_2(g) + NH_4^+ + H^+ + e^- \rightleftharpoons N_2H_5^+$	$E°$ = −1.68 V	II.B
$\frac{1}{2}NH_4^+ + \frac{1}{2}HN_3 + \frac{1}{2}H^+ + 2e^- \rightleftharpoons N_2H_5^+$	$E°$ = 0.20 V	II.B
$N_2H_3^+ \rightleftharpoons N_2H_2 + H^+$	pK_a = −1.5	II.B
$N_2H_2 \rightleftharpoons N_2H^- + H^+$	pK_a = 34.3	IV.C
$N_2H_2 + 3H^+ + 2e^- \rightleftharpoons N_2H_5^+$	$E°$ = 0.63 V	II.B
$N_2H_4^+ \rightleftharpoons N_2H_3 + H^+$	pK_a = 7.2	IV.A
$N_2H_4^+ + e^- \rightleftharpoons N_2H_4$	$E°$ = 0.69 V	IV.B
$N_2H_2 + H^+ + e^- \rightleftharpoons N_2H_3$	$E°$ = −0.31 V	IV.B
$N_3H_4^+ \rightleftharpoons N_3H_3 + H^+$	pK_a = 4.95	IV.A
$N_3H_3 \rightleftharpoons N_3H_2^- + H^+$	pK_a = 11.37	IV.A

[a]All species in aqueous solution unless otherwise designated, with e^- representing NHE. $E°$ designates the standard reduction potential relative to NHE.

in these specific reactions that the issue may not be essential. Pollard and Nickless (43) mention $N_2H_6^{2+}$ and cite Higginson and Wright for a K_a of 6 *M*; Higginson and Wright determined this result at 60°C and 3.5 *M* ionic strength (44). Davies and Kustin (45) cite Yui (1941) for a K_a of 0.54 *M* at zero ionic strength and Schwarzenbach (1936) for a K_a of 7.7 *M* at 0.06 *M* ionic strength, both at 25°C. Bruhn et al. (46) cite Schwarzenbach (1936) for a K_a of 11.2 *M*. Bengtsson (47) cites work of Rosotti and Rosotti, but makes no quantitative statement about the behavior of $N_2H_6^{2+}$. Schmidt's book (3) cites a K_a of 0.54, which was obtained from a technical report. Smith and Martell (40) suggest a K_a of 7.9 at zero ionic strength and 20°C. Basak and Banerjea (48, 49) cite Banerjea and Singh as a reference for the pK_a of $N_2H_6^{2+}$ (50); the latter reported a K_a of 1.4 $\times\ 10^{-4}$ *M*, which disagrees strongly with the reports of Schwarzenbach, of Higginson and Wright, and of Yui. Note that Banerjea and Singh apparently were unaware of these prior studies. The weight of evidence supports a pK_a for $N_2H_6^{2+}$ near -0.5, which means that studies conducted in 1 *M* acid or stronger will have to contend with significant concentrations of both $N_2H_5^+$ and $N_2H_6^{2+}$.

B. Pertinent Reduction Potentials

Given the three limiting stoichiometries for oxidation of hydrazine, there are three pertinent overall reduction potentials in acid solution:

$$N_2(g) + 5H^+(aq) + 4e^- \longrightarrow N_2H_5^+(aq) \qquad E_4^\circ \qquad (11)$$

$$\tfrac{1}{2}N_2(g) + NH_4^+(aq) + H^+(aq) + e^- \longrightarrow N_2H_5^+(aq) \qquad E_1^\circ \qquad (12)$$

$$\tfrac{1}{2}NH_4^+(aq) + \tfrac{1}{2}HN_3(aq) + \tfrac{1}{2}H^+(aq) + 2e^- \longrightarrow N_2H_5^+(aq) \qquad E_2^\circ \qquad (13)$$

According to Bard, Parsons, and Jordan, (51) a standard potential of -0.23 V versus NHE at 25°C can be assigned to E_4°. Corresponding standard potentials for the other two half-cells can be calculated by use of the data in the NBS tables (52). These calculations lead to values of -1.68 V for E_1° and 0.20 V for E_2°. Note that despite the very favorable driving force for the net one-electron process (formation of NH_4^+ plus N_2), the results summarized below show that there are many cases where it does not occur for kinetic reasons. Note also that the unfavorable driving force for the net two-electron process makes it inaccessible for some of the weaker oxidants.

The species diazene, suggested to be an intermediate in many of the reactions of hydrazine, does not appear in the NBS tables or in Bard, Parsons, and Jordan's (51) book. Because of its fleeting existence, thermochemical studies are not very reliable. The most accurate estimates are probably those obtained from ab initio quantum calculations, which give a value of 58.6 kcal mol^{-1} for $\Delta_f G°$ for trans-N_2H_2 (53). McKee (54) estimates a hydration free energy of −9.6 kcal mol^{-1} for this molecule, which leads to a value of 205 kJ mol^{-1} for $\Delta_f G°$ for N_2H_2(aq). McKee also estimates that $N_2H_3^+$(aq) is a strong acid with pK_a = −1.5. Thus, from a mechanistic point of view, an important consideration is its two-electron standard reduction potential as defined by

$$N_2H_2(aq) + 3H^+(aq) + 2e^- \longrightarrow N_2H_5^+(aq) \tag{14}$$

The above result plus NBS data lead to a value of 0.63 V for this standard potential. It can be seen that diazene is energetically accessible with most of the oxidants reviewed below.

Data relating to the free radicals N_2H_3 and $N_2H_4^+$ are reviewed in Section IV.B.

III. SURVEY OF REACTION STOICHIOMETRY AND KINETICS STUDIES

A. Main Group Oxidants

Information describing the stoichiometry or kinetics of oxidation of hydrazine by main group compounds includes 32 oxidants, as listed in Table II.

The oxidation of hydrazine by O_3 in acidic media is unusual in producing NH_3, N_2, and NO_3^- in similar amounts (55). A mechanism was suggested in which diazene was formed by the oxidation of hydrazine; subsequent disproportionation of diazene would then lead to NH_3 plus HN_3, and then NO_3^- would be formed by the reaction of HN_3 with O_3. This mechanism is contradictory to the present view that diazene disproportionates to give N_2 + N_2H_4. A reason-

TABLE II

A List of Main Group Species Whose Oxidations of Hydrazine Are Discussed

O_3	IO_3^-	$P_2O_8^{4-}$	H
OCl^-	IO_4^-	H_3PO_5	OH
ClO_2^-	HSO_5^-	*N*-Chlorobenzamide	O^-
ClO_3^-	$H_2NOSO_3^-$	Chloramine-T	O_3^-
Br_2	HNO_2	Organic carbonyls	Cl_2^-
HOBr	HNO_3	Triplet 4-benzoylbenzoate	SO_4^-
BrO_3^-	NH_2Cl	Tl^{III}	HPO_4^-
I_2	NH_2Br	e_{aq}^-	H_2PO_4

able alternative explanation is that O_3 decomposes to produce OH radicals, which then attack hydrazine to produce NH_3 via N_2H_3 as is discussed in Section IV.A.

Little has been published since Higginson's review regarding the reaction of OCl^- with hydrazine, save for the comment that it is so rapid that it does not accumulate in the reaction of ClO_2^- with hydrazine (27).

Gas evolution was used to follow the reaction of $N_2H_5^+$ with excess ClO_2^- (27). The quantitative formation of N_2 was interpreted in terms of the following reaction:

$$N_2H_5^+ + ClO_2^- \longrightarrow N_2 + Cl^- + 2H_2O + H^+ \tag{15}$$

No quantitative results were presented regarding the kinetics of the reaction, although the results suggest a time scale of several minutes. Qualitative observations on the reaction of ClO_2 indicated that it is much slower.

Contrary to an early report, it is claimed that ClO_3^- does not react with $N_2H_5^+$ in the absence of a catalyst (27). This apparent contradiction is resolved by the dual requirements of high temperature and high acidity indicated for reaction in the early report (56).

The rate of reaction of Br_2 with hydrazine was investigated by electrolyzing a solution of Br^- to generate Br_2 (57). The reaction was initiated by rapid injection of a hydrazine solution, and its progress was monitored by amperometric detection of the Br_2. The pseudo-second-order rate constant was found to be strongly inhibited by acid for concentrations greater than 1 *M*, but it reached an apparent limiting value of $2 \times 10^5\ M^{-1}\ s^{-1}$ in dilute acid. These results seem to imply a rate-limiting reaction of Br_2 with $N_2H_5^+$. Rate constants obtained from chronopotentiometric experiments were some two orders of magnitude slower than those obtained from the amperometric method and were taken to refer to the reaction of Br_2 with N_2H_2. One should be skeptical of this interpretation since the method did not take into account the decomposition of N_2H_2, which is known to be rapid (see below) (58).

In a brief study of the rapid reaction of hydrazine with HOBr in dilute H_2SO_4, it was reported that 2 mol of HOBr were consumed per mole of N_2H_4 at 25°C, although it is not clear what analytical method was used to derive this result (59). We may infer that the reaction is

$$2HOBr + N_2H_4 \longrightarrow N_2 + 2Br^- + 2H_2O + 2H^+ \tag{16}$$

From the stoichiometric point of view, the oxidation of N_2H_4 by BrO_3^- is of great interest: apart from the special case of HNO_2, it is the only reaction discussed in this review that yields significant amounts of HN_3 at 25°C (59). For example, with excess N_2H_4 in 0.5 *M* H_2SO_4, somewhat more than 10% of the hydrazine consumed is converted into HN_3, and yields of NH_3 are comparable. Of course, N_2 is the predominant product. The kinetics and mechanism of the reaction are complex because of the important role played by the formation of Br_2 from the reaction of BrO_3^- with Br^-. An attempt was made to avoid this complication by using allyl alcohol to scavenge the HOBr intermediate, but only under conditions of 3–5 M H_2SO_4 did this really minimize the formation of Br^- (59). Allyl alcohol is also converted in substantial yields to propanol during these reactions, which is evidence that N_2H_2 is an intermediate.

The oxidation of N_2H_4 by I_2 has been studied many times and with conflicting opinions regarding the kinetics, as noted by Janovsky (60), by King et al. (61), and by Radhakrishnamurti et al. (62). There is a consensus that the stoichiometry is

$$2I_2 + N_2H_5^+ \longrightarrow N_2 + 4I^- + 5H^+ \quad (17)$$

The reaction rate is highly pH dependent, and is usually investigated at low pH in order to keep the rates low enough to be measured conveniently. King et al. (61) studied the kinetics of the reaction with excess N_2H_4 and I^-, and with $[H^+] = 0.04 - 0.2$ M and obtained the rate law

$$-\frac{d[I_2]_{tot}}{dt} = \frac{2k_1Q_a[I_2]_{tot}[N_2H_4]_{tot}}{[H^+](1 + Q_1[I^-]\left(1 + \frac{k_{-1}}{k_2}[I^-]\right)} \quad (18)$$

The following mechanism was inferred:

$$I_2 + N_2H_4 \rightleftharpoons N_2H_4I^+ + I^- \qquad k_1, k_{-1} \quad (19)$$

$$N_2H_4I^+ \longrightarrow \text{Products} \qquad k_2 \quad (20)$$

$$I_2 + I^- \rightleftharpoons I_3^- \qquad Q_1 \quad (21)$$

$$N_2H_5^+ \rightleftharpoons N_2H_4 + H^+ \qquad Q_a \quad (22)$$

The results led to a value of 4.4×10^7 M^{-1} s^{-1} for k_1. Gopalan and Karunakaran (63) reported an inverse dependence of the rate on [iodine]; unfortunately, they did not include data showing this peculiar effect and since it has not been observed by other workers its significance is questionable. A simple rate law was obtained by Sultan et al. (64), but they failed to investigate the dependence on $[I^-]$. Rao and Dalvi (65) used a continuous-flow method under second-order conditions and obtained a strictly inverse dependence on iodide concentration. Janovsky (60) used pulse radiolysis to generate I_2 in the presence of N_2H_4 and, hence, was able to work at higher pH. These experiments led to a value of $1.2 \times 10^7\ M^{-1}\ s^{-1}$ for k_1, which is in good agreement with the results of King et al. (61). Concurrently with the work of Janovsky, Radhakrishnamurti et al. (62) studied the reaction under rather acidic conditions ($[H^+] = 1.6 \times 10^{-3}$–$2\ M$) and obtained a more complex dependence on $[H^+]$. These workers overlooked the results of King et al. (61), but derived a similar mechanism that also contained a pathway for reaction of HOI with N_2H_4. Their value for k_1 at 35°C is in reasonable agreement with those indicated above. Janovsky (66) published a second paper on the reaction, in which a phosphate-dependent term was reported. This phosphate dependence was attributed to the formation of a reactive complex, $[N_2H_4 \cdot HPO_4]^{2-}$, although there was no direct evidence for such a species. Most recently, Liu and Margerum (67) reinvestigated this reaction, and found evidence for general base catalysis by CH_3COO^- and HPO_4^{2-}. Presumably, this accounts for the phosphate effect reported by Janovsky. Liu and Margerum (67) report no evidence for a HOI pathway, from which we may infer that prior reports of this pathway were due to unrecognized general base catalysis effects.

A mechanism similar to that suggested by King et al. (61) was previously suggested by Rottendorf and Sternhell (68), albeit without the support of kinetic data. These earlier workers obtained evidence for N_2H_2 as an intermediate on the basis of trapping experiments in which the I_2/N_2H_4 reaction led to the hydrogenation of a variety of unsaturated organic compounds. They suggested that N_2H_2 is formed from N_2H_3I by elimination of HI.

Rábai and Beck (69) inferred from their study of the reaction of IO_4^- with N_2H_4 that IO_3^- reacts with excess N_2H_4 according to the following stoichiometry:

$$2IO_3^- + 3N_2H_5^+ \longrightarrow 2I^- + 3N_2 + 6H_2O + 3H^+ \qquad (23)$$

Hasty (70) studied the kinetics of the reaction and found that it has an iodide-catalyzed pathway as well as an uncatalyzed pathway. The iodide-catalyzed mechanism involves rate-limiting formation of I_2 by the reaction of IO_3^- with I^-, followed by rapid reaction of I_2 with N_2H_4. For the uncatalyzed pathway the rate law is

$$-d[IO_3^-]/dt = k[IO_3^-][N_2H_5^+][H^+]^2 \qquad (24)$$

with $k = 1.5 \times 10^{-3}\ M^{-3}\ s^{-1}$. A mechanism was suggested in which IO_2^+, formed by double protonation, donates an oxygen atom to $N_2H_5^+$.

The oxidation of hydrazine by excess periodate apparently yields IO_3^- and N_2 (71). The quantitative conversion to N_2 was confirmed by volumetric determination of gas evolution and mass spectrometry (69). In a potentiometric titration study of N_2H_4 with IO_4^- under mildly acidic conditions and adding the IO_4^- to the solution of N_2H_4, an endpoint was observed at a 2.0 : 1 ratio of N_2H_4 to IO_4^-, under which conditions I_2 was observed as a product (72). Although no determination of NH_3 was performed, the result of the titration was interpreted in terms of the reaction

$$4IO_4^- + 8N_2H_4 + 4H^+ \longrightarrow \tfrac{3}{2}I_2 + 7.5N_2 + NH_3 + 16H_2O + HI \quad (25)$$

It is difficult to accept this result, since I_2 is known to react rapidly with N_2H_4. Under similar conditions Browne and Shetterly (73) obtained no evidence for NH_3 formation. In a thorough investigation, Rábai and Beck (69) found that the products could be I^-, I_2, or IO_3^-, depending on conditions, but that under no conditions were NH_3 or HNO_2 formed. Overall, the weight of evidence is that NH_3 is not a product of the oxidation of N_2H_4 by IO_4^-, and that with excess periodate the stoichiometry is

$$2IO_4^- + N_2H_5^+ \longrightarrow 2IO_3^- + N_2 + 2H_2O + H^+ \quad (26)$$

Under these conditions at a given pH the rate law is simply

$$-d[N_2H_5^+]dt = k[N_2H_5^+][IO_4^-] \quad (27)$$

with $k = 1.5\ M^{-1}\ s^{-1}$ (apparently at pH 2), but with excess hydrazine the rates become quite complex because of interference from the reactions of IO_3^- and I_2 (69). Note that the pH dependence in rate law 27 is not indicated, but it should be complex in view of the various species present in acidic solutions of IO_4^- (74). Presumably, the direct reaction of IO_4^- with $N_2H_5^+$ is a two-electron process, but the details of the mechanism are unclear.

It has been reported that this reaction generates N_2H_2 on the basis that it induces the hydrogenation of unsaturated compounds (75). It is unclear whether N_2H_2 is generated directly from IO_4^- or from one of its reduced states such as I_2.

The reaction of peroxomonosulfate with hydrazine has been studied in acidic perchlorate media (76). The stoichiometry was investigated by determining consumption ratios, supplemented with the qualitative observation of nitrogen evolution, and it is given as

$$2HSO_5^- + N_2H_5^+ \longrightarrow N_2 + 2SO_4^{2-} + 2H_2O + 3H^+ \quad (28)$$

Under conditions of a pseudo-first-order excess of hydrazine, the rates saturate with increasing hydrazine concentrations, consistent with the rate law

$$-\frac{d[\mathrm{HSO_5^-}]}{dt} = \frac{k_1K[\mathrm{N_2H_5^+}][\mathrm{HSO_5^-}]}{1 + K[\mathrm{N_2H_5^+}]} \tag{29}$$

where $k_1 = 2.8 \times 10^{-4}\ \mathrm{s^{-1}}$ and $K = 3.4 \times 10^{-4}\ M^{-1}$ at 30°C. The rates are independent of pH over the range from pH 4–0. Tests for possible catalysis by trace metal ions were negative. The authors suggested two possible mechanisms. One involves prior formation of a complex between the two reactants, while the other envisions isomerization of peroxomonosulfate as the rate-limiting step.

Hydroxylamine-*O*-sulfonic acid reacts with hydrazine according to

$$\mathrm{2H_2NOSO_3^- + N_2H_4 + 2OH^- \longrightarrow N_2 + 2NH_3 + 2SO_4^{2-} + 2H_2O} \tag{30}$$

and the rate law is first order in both $[N_2H_4]$ and $[H_2NOSO_3^-]$ between pH 10.7 and 13.7 with a rate constant of $0.010\ M^{-1}\ \mathrm{s^{-1}}$ (77). The implied mechanism involves N_2H_4 acting as a nucleophile as follows:

$$\mathrm{H_2NOSO_3^- + N_2H_4 \longrightarrow H_2NNHNH_2 + SO_4^{2-} + H^+} \tag{31}$$

$$\mathrm{H_2NNHNH_2 \longrightarrow N_2H_2 + NH_3} \tag{32}$$

$$\mathrm{2N_2H_2 \longrightarrow N_2H_4 + N_2} \tag{33}$$

Isotopic labeling studies with ^{15}N have been taken to demonstrate that triazane (H_2NNHNH_2) is formed and decomposes as in the mechanism given above (78).

On superficial examination, the reaction of hydrazine with nitrous acid appears quite unusual, but in fact it is worth careful consideration because of its potential to serve as a model for a wide range of two-electron oxidants. The reaction generates a variety of products including N_2, NH_3, N_2O, and HN_3. Hence, it does not fall easily into any mechanistic classification based on stoichiometry (79). A similar range of products was suggested in a study that used potentiometric titrations (80). As Perrott et al. (79) showed, a complicating factor is the follow-up reaction

$$\mathrm{HN_3 + HNO_2 \longrightarrow N_2 + N_2O + H_2O} \tag{34}$$

When this complication is taken into account, the following two reactions describe the basic process with their relative contributions being pH dependent:

$$\mathrm{HNO_2 + N_2H_5^+ \longrightarrow NH_3 + N_2O + H_2O + H^+} \tag{35}$$

$$\mathrm{HNO_2 + N_2H_5^+ \longrightarrow HN_3 + 2H_2O + H^+} \tag{36}$$

By conducting the reaction in excess hydrazine, the effects of the reaction of HNO_2 with HN_3 can be minimized, and in $HClO_4$ media a rate constant of 611 $M^{-2}\ s^{-1}$ is obtained for the rate law

$$-d[HNO_2]/dt = k[HNO_2][N_2H_5^+][H^+] \tag{37}$$

An intermediate having an absorption maximum at 225 nm is formed as HNO_2 is lost. This intermediate, formulated as a mixture of cis- and trans-$NH_2N{=}NOH$, decomposes by parallel pathways to give the two sets of products. The formation of the intermediate is proposed to have the following mechanism:

$$H^+ + HNO_2 \rightleftharpoons NO^+ + H_2O \tag{38}$$

$$NO^+ + N_2H_5^+ \longrightarrow NH_2NHNO + 2H^+ \tag{39}$$

$$NH_2NHNO \longrightarrow NH_2N{=}NOH \quad \text{(fast)} \tag{40}$$

In support of this mechanism, the reaction is catalyzed by Cl^-, Br^-, and SCN^-, as is typical of nitrosation reactions; the catalysis operates through the reaction of HNO_2 with X^- to form NOX, which can nitrosate hydrazine formally through an NO^+ transfer mechanism. Even though N_2H_2 is not an intermediate in this mechanism, it qualifies as a two-electron mechanism because it does not involve free radical intermediates.

Results obtained from an isotopic tracer study of the reaction with excess HNO_2 gave conclusive evidence against a mechanism involving direct formation of N_2H_2, and the results were taken to suggest the occurrence of nitrosation of NH_2NHNO or a cyclic isomer of HN_3 (81a). The suggestion of a cyclic isomer of NH_3 was subsequently withdrawn, as a result of an ^{15}N NMR investigation of this reaction (81b). Kinetic evidence for the former has recently been presented (82): Direct observation of the $NH_2N{=}NOH$ intermediate shows that it is susceptible to further nitrosation, which then leads to formation of N_2 and N_2O.

The report of NH_3 formation from oxidation by HNO_3 describes experiments conducted at very high acidity and temperatures (83a). The reaction appears to involve the pathway that generates HN_3, which is beyond the scope of this chapter. Essentially identical results were reported much earlier by Koltunov and co-workers (83b).

Oxidation of hydrazine by chloramine is of great importance as a parasitic reaction in the Raschig synthesis of hydrazine. Schmidt notes that it is catalyzed by metal ions, and thus gelatin is added to suppress this side reaction during the Raschig synthesis (84). The direct reaction has been studied in alkaline media by Delalu and Cohen-Adad (85), who find that the stoichiometry is given by

$$2NH_2Cl + N_2H_4 \longrightarrow N_2 + 2NH_4Cl \tag{41}$$

The rate law is

$$-d[NH_2Cl]/dt = (k_1 + k_2[H^+])[N_2H_4]_{tot}[NH_2Cl] \tag{42}$$

with $k_1 = 5 \times 10^{-3}\ M^{-1}\ s^{-1}$ and $k_2 = 1.95 \times 10^{10}\ M^{-2}\ s^{-1}$ at 25°C. A two-step mechanism was inferred, although no speculation was provided regarding the identity of the intermediates.

Consumption ratios ranging from 1 to 2 were obtained in a study of the alkaline oxidation of hydrazine by bromamine (86). A second-order rate constant of 40 $M^{-1}\ s^{-1}$ was obtained at pH 11.1.

An initial report on the reaction of $P_2O_8^{4-}$ with hydrazine in 1 M $HClO_4$ showed that the rate-limiting step is decomposition of $P_2O_8^{4-}$ to form PO_5^{3-}, and it is this latter species that attacks hydrazine (87). At higher pH, the decomposition is slower, which enables the direct reaction of $P_2O_8^{4-}$ with N_2H_4 to be studied (24). In acetate buffers with excess hydrazine, a consumption ratio of 2 ($= \Delta[P_2O_8^{4-}]/\Delta[N_2H_4]$) was obtained, indicating that the stoichiometry is

$$2H_nP_2O_8^{n-4} + N_2H_5^+ \longrightarrow 4H_2PO_4^- + N_2 + (2n - 3)H^+ \tag{43}$$

The rate law was determined under the same conditions (40°C, acetate buffer) and was found to be

$$-\frac{d[pdp]}{dt} = \frac{(k_1[H^+] + k_2K_3)[pdp][N_2H_5^+]}{[H^+] + K_3} \tag{44}$$

where K_3 ($= 4.4 \times 10^{-5}\ M$) is the acid dissociation constant for $H_2P_2O_8^{2-}$ and k_1 and k_2 take values of 2.25×10^{-3} and $7.5 \times 10^{-3}\ M^{-1}\ s^{-1}$. The authors made no suggestions as to the nature of the reaction intermediates. In principle, the reaction could have either a one- or two-electron mechanism.

The reaction of hydrazine with H_3PO_5 has been shown to be highly susceptible to catalysis by I^-, and the stoichiometry was investigated only for the iodide-catalyzed reaction (88). Bromide and chloride were also found to be very efficient catalysts. A rather complex rate law was obtained under conditions of no added halide, and it was interpreted (at least in part) in terms of an isomerization of normal four-coordinate H_3PO_5 to a five-coordinate form. One wonders whether this unusual result is due to halide contamination of the solutions; indeed, just such contamination was suggested (88) as the cause of irreproducibility in a prior study (89). The mechanism of the iodide-catalyzed reaction appears to involve rate-limiting oxidation of iodide by H_3PO_5, followed by rapid oxidation of N_2H_4 by iodine.

When the oxidation of N_2H_4 by *N*-chlorobenzamide (NCB) is conducted at 40°C with excess oxidant in HCl media the stoichiometry is

$$N_2H_4 + 2C_6H_5CONHCl \longrightarrow N_2 + 2C_6H_5CONH_2 + 2HCl \qquad (45)$$

The rates are independent of the concentration of hydrazine but depend directly on the concentrations of Cl^- and H^+, indicating rate-limiting conversion of NCB to Cl_2, which then attacks the hydrazine directly (90).

Hydrazine is oxidized by chloramine-T (CAT) to give N_2 as indicated by consumption ratios and the qualitative identification of *p*-toluenesulfanamide as a product (91). The inferred stoichiometry is

$$2\,CH_3-C_6H_4-SO_2-NHCl + N_2H_5^+ \rightarrow$$

$$2\,CH_3-C_6H_4-SO_2-NH_2 + N_2 + 2Cl^- + 3H^+ \qquad (46)$$

This reaction was investigated in acidic ethanolic mixtures in order to make the rates conveniently slow. At 35°C, the rate law is reported to be

$$-\frac{d[\mathrm{CAT}]}{dt} = \frac{k_1k_3[\mathrm{CAT}][\mathrm{H}^+][\mathrm{N_2H_5^+}]}{k_2 + k_3[\mathrm{N_2H_5^+}]} \qquad (47)$$

with values of $8.6 \times 10^{-4}\ s^{-1}$ and $3600\ M^{-1}$ claimed for k_1 and k_3/k_2. The dimension indicated for k_1 is incompatible with the rate law given, but insufficient data are available to rectify the error. The unusual limiting behavior in $[N_2H_5^+]$ was interpreted in terms of rate-limiting protonation of CAT. We suggest as an alternative hypothesis a mechanism analogous to that of *N*-chlorobenzamide (described above). Regrettably, the implied dependence on $[Cl^-]$ was not investigated.

Hydrazine is widely reported to react with aldehydes and ketones to yield hydrazones, which can be considered one-electron oxidized derivatives of hydrazine. Base-catalyzed decomposition of the hydrazones leads to formation of N_2 through the well-known Wolff–Kishner reduction, and thus an overall four-electron reaction is achieved. Despite the implied generality of these reactions, they can yield other products, and quantitative studies in aqueous solution are rather uncommon.

The reaction of hydrazine with the first member of the series, formaldehyde, does not give a hydrazone but rather tetraformal trisazine (92):

HN–N–NH / HN–N–NH (tetraformal trisazine ring structure)

Acetone reacts to give an equilibrium mixture of the hydrazone and dimethylketazine (93). The two equilibria can be controlled by adjusting the pH, as indicated:

$$N_2H_5^+ + (CH_3)_2CO \rightleftharpoons (CH_3)_2C{=}NNH_3^+ + H_2O$$

$$(CH_3)_2C{=}NNH_3^+ + (CH_3)_2CO \rightleftharpoons$$

$$(CH_3)_2C{=}NN{=}C(CH_3)_2 + H^+ + H_2O \qquad (49)$$

Values of 10.0 M^{-1} and 2.7×10^{-3} were measured for K_{48} and K_{49} at 15°C. Equilibration occurs on the time scale of some tens of minutes, which allowed kinetic measurements at the early date of 1929.

The reaction of hydrazine with *p*-chlorobenzaldehyde proceeds simply to form the corresponding hydrazone. The reaction mechanism has two steps as shown.

$$N_2H_4 + Cl{-}C_6H_4{-}CHO \rightleftharpoons Cl{-}C_6H_4{-}CH(OH){-}NH{-}NH_2 \qquad (50)$$

$$Cl{-}C_6H_4{-}CH(OH){-}NH{-}NH_2 \rightarrow Cl{-}C_6H_4{-}CH{=}N{-}NH_2 + H_2O \qquad (51)$$

The first step, carbinolhydrazine formation, is rapid and has an equilibrium constant of 14 M^{-1} (94). The next step, dehydration, has a rate constant of 1.7 $\times$ 10^{-2} s^{-1} and displays general acid catalysis (95).

By way of contrast, the reaction of hydrazine with trifluoroacetophenone stops at the carbinolhydrazine stage (96).

It was initially reported that the triplet excited state of 4-benzoylbenzoate (4-carboxybenzophenone) behaves quite unusually relative to most ground-state one-electron oxidants in that $N_2H_5^+$ but not N_2H_4 is oxidized (97). Further study revealed that the pH dependence of the overall photoredox quantum yield is more complex, having a maximum at pH 7.5 (98). A drastically revised mechanism was proposed in which $N_2H_5^+$ is relatively ineffective in deactivating the excited state. A third publication on this system presented flash photolysis evidence for the formation of radicals derived from the oxidant, determined their quantum yields, and used some of their properties to resolve certain mechanistic ambiguities; this final mechanism preserved the essential feature of an electron transfer to the excited state from N_2H_4 to generate the corresponding radicals (99). In none of the studies were the identity of the nitrogen-containing products determined, nor were consumption ratios determined.

Oxidation of N_2H_4 by Tl^{III} has been examined in various media, which is significant because of the tendency of Tl^{III} to bind anions. In all media, the stoichiometry is

$$2Tl^{III} + N_2H_4 \longrightarrow 2Tl^{I} + N_2 + 4H^+ \tag{52}$$

In 0.1–1.0 M $HClO_4$, the reaction is relatively fast (100), and it has the rate law

$$-\frac{d[Tl^{III}]}{dt} = \frac{2k[Tl^{III}][N_2H_4]_{tot}}{[H^+]} \tag{53}$$

The rate constant is 8.4 M^{-1} s^{-1} at 8°C. This rate law leads to a proton ambiguity in that the reactive species could be either $Tl(OH)^{2+}$ plus $N_2H_5^+$ or Tl^{3+} plus N_2H_4. In either case, a two-electron mechanism is proposed in which the first step leads to Tl^{I} plus N_2H_2. A one-electron mechanism is not ruled out by these data, although a two-electron mechanism seems more likely.

In acetate–perchlorate media, the rates are slower due to the formation of relatively unreactive acetato–Tl^{III} complexes (101, 102). The rate constants for all species $[Tl(OAc)_x]^{(3-x)+}$ for x = 0–4 show that increasing acetate substitution systematically decreases the rates.

Ethylenediaminetetracetic acid (edta) also inhibits the rates, although the Tl^{III}–edta complexes react directly with N_2H_4 (103). The complex pH dependence of the kinetics is attributed to varying degrees of protonation of coordinated edta.

A study of the reaction in chloride/sulfate media shows that chloride inhibits the rates, but it is difficult to interpret the results since the effects of binding of sulfate by Tl^{III} were not specifically considered (104). A similar effect was found in chloride/perchlorate media (100).

Reactions of several transient free radicals with hydrazine have been studied by use of pulse radiolysis. Hayon and Simic (105) determined a rate constant of $2.2 \times 10^8 M^{-1} s^{-1}$ for reaction of e_{aq}^- with $N_2H_5^+$ and cited prior evidence that H_2 is one of the ultimate products. They were unable to determine whether the first step in the mechanism involved formation of $H + N_2H_4$ or $H_2 + N_2H_3$, but if the latter is the case, then the reaction qualifies as an oxidation of N_2H_4. The former was conclusively identified to prevail in a recent reinvestigation of this reaction (106). On the other hand, hydrogen atoms do appear to act as an oxidant, reacting with N_2H_4 to yield $H_2 + N_2H_3$ with $k = 6 \times 10^7 M^{-1} s^{-1}$ (42); a comparison of the activation parametrs with those for the corresponding gas-phase reaction was used to infer a hydrogen-atom abstraction mechanism. The activation parameters for the reaction of hydrogen atoms with $N_2H_5^+$ indicate a more complex process, which was proposed to involve concerted formation of H^+, NH_2, and NH_3 (42). Hayon and Simic (105) also investigated the reactions of OH with $N_2H_5^+$ and N_2H_4 ($k = 1.0 \times 10^9$ and $1.4 \times 10^{10} M^{-1} s^{-1}$), and obtained spectrophotometric evidence for the formation of N_2H_3 or $N_2H_4^+$, depending on pH. The ultimate products of decay of these radicals are NH_3 and N_2, and the details of this decay process are discussed below (Section IV.A.). Ershov and Mikhailova (107) determined the rate of reaction of O^- with N_2H_4 ($k = 1.6 \times 10^9 M^{-1} s^{-1}$); they assumed N_2H_3 as the immediate product of this reaction, although evidence is lacking. These workers also measured the rate of reaction of O_3^- with N_2H_4 ($k = 1.2 \times 10^6 M^{-1} s^{-1}$), although they offered no suggestion as to the nature of the reaction products. The rate constant for the reaction of Cl_2^- with $N_2H_5^+$ was studied by spectrophotometric determination of the loss of Cl_2^- ($k = 8.0 \times 10^6 M^{-1} s^{-1}$); no information was obtained as to the nature of the products, but a hydrogen-atom-transfer mechanism of oxidation was inferred from a consideration of rate trends (108). Maruthamuthu and Neta (109) determined that SO_4^- reacts with N_2H_4 and $N_2H_5^+$ with rate constants of 8.1×10^8 and $2.1 \times 10^8 M^{-1} s^{-1}$, that HPO_4^- reacts with the same species with rate constants of 4.9×10^8 and $1.4 \times 10^8 M^{-1} s^{-1}$, and that H_2PO_4 reacts with $N_2H_5^+$ with a rate constant of $1.9 \times 10^8 M^{-1} s^{-1}$. A hydrogen-atom abstraction mechanism was assigned for all of these reactions on the basis of a LFER between the rate constants for the reactions of the sulfate and phosphate radicals.

B. Two-Electron Transition Metal Oxidants

This section summarizes the oxidation of hydrazine by the following two-electron transition metal oxidants: V^V, $[Cr^VO(L)_2]^-$, Cr^{VI}, Mo^{VI}, and Fe^{VI}.

Under certain circumstances the reaction of V^V with N_2H_4 yields only N_2, while under other circumstances a mixture of products is formed (47, 110). As is discussed below, the rate law implies a two-electron mechanism under conditions, where N_2 is the exclusive product.

Bengtsson (47) published a thorough study of the oxidation of N_2H_4 by V^V in perchloric acid media with excess N_2H_4. Although NH_3 is a product under certain conditions, N_2 is formed quantitatively under the conditions of the kinetics studies according to the stoichiometry

$$4VO_2^+ + N_2H_6^{2+} + 2H^+ \longrightarrow 4VO^{2+} + N_2 + 4H_2O \tag{54}$$

Ignoring the pH dependence, the rate law has the form

$$-\frac{d[V^V]}{dt} = \frac{k[V^V]^2[N_2H_4]_{tot}}{(1 + K[V^V])} \tag{55}$$

The possibility that V^{III} is formed as an intermediate was ruled out by showing that Ag^+ is not reduced during the reaction. The following mechanism was proposed:

$$V^V + N_2H_4 \rightleftharpoons VN_2H_4 \tag{56}$$

$$VN_2H_4 + V^V \longrightarrow 2V^{IV} + N_2H_2 \tag{57}$$

$$2V^V + N_2H_2 \longrightarrow 2V^{IV} + N_2 \quad \text{(fast)} \tag{58}$$

This mechanism bypasses the N_2H_3 stage of oxidation, going directly to N_2H_2, and thus it qualifies as a two-electron mechanism even though the individual metal ions are undergoing one-electron reductions. It is clear that under certain conditions the reaction can yield NH_3 and V^{III} (110), and hence the general mechanism must be more complex than given above.

A brief study of the effects of nta, edta, and dtpa on this reaction showed that these chelates inhibit the rate of the reaction (111). It was found that the edta and dtpa complexes are unreactive, whereas the nta complex reacts directly, albeit more slowly than free V^V.

Another purported two-electron oxidation is that by $[Cr^VO(L)_2]^-$, where HL = 2-ethyl-2-hydroxybutyrate (112). On the basis of a 2 : 1 $Cr^V/N_2H_5^+$ consumption ratio, a negative test for NH_4^+, and an analysis of the Cr-containing products, the reaction is

$$2Cr^V + N_2H_5^+ \longrightarrow 2Cr^{III} + N_2 \tag{59}$$

in mildly acidic media. Its rate law shows inhibition by HL, which implies that it is the monoligated complex that is reactive. The complex pH dependence was interpreted in terms of two parallel paths, one involving a direct redox reaction of $N_2H_5^+$ with Cr^V, and the other involving reversible formation of a N_2H_4 complex of Cr^V followed by an internal redox step. Both of these redox steps were assigned two-electron mechanisms to form Cr^{III} plus N_2H_2, but it is not clear that a sequence of one-electron steps can be ruled out. It was argued that the fate of N_2H_2 was to undergo further oxidation to N_2 by reaction with Cr^V. Its alternative, the disproportionation of N_2H_2, was ruled out on the basis that such a process would be too slow, but current information on this disproportionation shows that it is quite rapid (see below). Thermodynamic data for the $Cr^{V/IV/III}$ redox couples have recently been revised, but the impact of these revisions on the proposed mechanism remains to be assessed (113).

A large number of reports have been published on the oxidation of N_2H_4 by Cr^{VI}. Most of the reports agree that the stoichiometry is given by

$$4[HCrO_4]^- + 3N_2H_5^+ + 13H^+ + 8H_2O \longrightarrow 4[Cr(H_2O)_6]^{3+} + 3N_2 \quad (60)$$

in acidic media (114–118). A dissenting report was published on the basis of a potentiometric titration: When a solution of Cr^{VI} was gradually added into a solution of N_2H_4 in strong H_2SO_4 the endpoint corresponded to the above stoichiometry, but in the reverse case the endpoint was somewhat delayed, indicating the formation of some NH_3 (119). These workers cited similar observations by Cuy and Bray (120). In fact, Cuy and Bray reported just the opposite, that gradual addition of Cr^{VI} to the N_2H_4 solution led to deviations from the ideal stoichiometry. An explanation for these contradictory results is suggested by the work of Haight et al. (121), who found that the stoichiometry is affected by the presence of Mn^{2+} in sulfate media (116). Dinitrogen is formed quantitatively in the absence of Mn^{2+}, but in its presence there is a substantial yield of NH_3. Perhaps the other reports of NH_3 formation can be attributed to the catalytic effects of metal ion impurities.

A variety of rate laws have been reported for the reaction. Beck and Durham (114) studied the reaction in perchlorate media and reported the following complex rate law:

$$-d[Cr^{VI}]/dt = [N_2H_5^+][CrO_4H^-])(k_1[H^+] + k_2[CrO_4H^-]) \quad (61)$$

In a preliminary report, Ramanujam and Venkatasubramanian (115) in 1970 obtained simple kinetics in sulfate media consistent with only a first-order dependence on $[Cr^{VI}]$ and kinetic saturation with respect to $[N_2H_4]_{tot}$. In their full report on this reaction they demonstrated that the rates increase with $[H^+]$, although they did not obtain an analytical description of this effect (117). Haight et al. (121) examined the reaction in perchlorate media, confirmed the saturating dependence on $[N_2H_4]_{tot}$, and resolved the acid dependence. Their rate law is

$$-\frac{d[Cr^{VI}]}{dt} = \frac{k_1K_0[Cr^{VI}][N_2H_5^+][H^+]}{1 + 0.15[H^+] + K_0[N_2H_5^+]} \quad (62)$$

which was explained by the following mechanism:

$$HCrO_4^- + H^+ \rightleftharpoons H_2CrO_4 \qquad (K = 0.15\ M^{-1}) \qquad (63)$$

$$HCrO_4^- + N_2H_5^+ \rightleftharpoons N_2H_4CrO_3 + H_2O \qquad K_0 = 3.2\ M^{-1} \qquad (64)$$

$$H^+ + N_2H_4CrO_3 \longrightarrow \text{Products} \qquad k_1 = 6.8\ M^{-2}\ s^{-1} \qquad (65)$$

They found that the reaction is subject to general acid catalysis, and they concluded that the second-order dependence on $[Cr^{VI}]$ reported by Beck and Durham was due to $HCrO_4^-$ acting as an acid catalyst. Gupta et al. (122), also reported on the rate law, but they did not develop a simple expression for the pH dependence. Senent et al. (118) obtained a similar rate law, although they did not detect saturation in $[N_2H_4]$; this lacuna led to the proposal of rate-limiting formation of the complex between $N_2H_5^+$ and $HCrO_4^-$.

The effects of edta and Mn^{2+} on the reaction have been examined because of their potential to provide clues to details of the mechanism. Beck and Durham (123) were the first to report on the effect of edta, which includes an increase in the rates and the formation of an edta$-Cr^{III}$ complex. These workers interpreted the results as indicative of formation of a Cr^{VI}—edta—N_2H_4 complex, which then participated in an internal redox transformation to give a Cr^{IV}—edta complex. Ramanujam et al. (117) reported the spectral detection of Cr^{VI}—edta complexes. They also reported a remarkable dependence of the kinetics on [edta], with the rates showing two maxima as the edta concentration is increased. Finally, Beck et al. (124) demonstrated that the unusual results of Ramanujam et al. (117) were due to a failure to maintain constant pH. The Mn^{2+} ion has no effect in ClO_4^- media, but in SO_4^{2-} media it leads to the production of NH_3 and it decreases the rates when Cr^{VI} is in excess (116). The formation of NH_3 was attributed to the oxidation of N_2H_4 by Mn^{3+}, which was formed by the oxidation of Mn^{2+} by Cr^{IV}. The decrease in rate was attributed to the scavenging of Cr^{IV} by Mn^{2+}; in the absence of Mn^{2+}, Cr^{IV} would be converted to Cr^V either by disproportionation or by reaction with Cr^{VI}, and the Cr^V would then consume additional N_2H_4. In a subsequent publication, Haight et al. (121) used thermodynamic estimates to argue against the formation of Cr^V by reaction of Cr^{IV} with Cr^{VI}. These results allow the mechanism in the absence of Mn^{2+} and edta given above to be elaborated as follows:

$$H^+ + N_2H_4CrO_3 \longrightarrow N_2H_2 + Cr^{IV} \qquad k_1 \qquad (66)$$

$$2Cr^{IV} \longrightarrow Cr^{3+} + Cr^V \qquad \text{(fast)} \qquad (67)$$

$$Cr^V + N_2H_5^+ \longrightarrow Cr^{3+} + N_2H_2 + 3H^+ \qquad \text{(fast)} \qquad (68)$$

$$2N_2H_2 \longrightarrow N_2 + N_2H_4 \qquad \text{(fast)} \qquad (69)$$

The above appears reasonable, with the evidence for Cr^{IV} as an intermediate being rather strong. On the other hand, the evidence to rule out one-electron oxidation of N_2H_4 by Cr^{IV} seems to be limited to the fact that NH_3 is not produced under normal conditions. This conclusion seems not to be justified, since it is now known that many one-electron oxidants do not yield NH_3. Now that methods are available to prepare Cr^{IV} solutions, it should be possible to make decisive tests of certain aspects of the above mechanism (125). Data have also recently become available on the pH-dependent kinetics of disproportionation of Cr^{V} (126), but they do not appear to require a revision of the mechanism described above.

There is some dispute regarding the stoichiometry of the oxidation of N_2H_4 by Mo^{VI}. Huang and Spence (127) studied the reaction with excess N_2H_4 at 70°C and pH 1.6, and they obtained a negative test for NH_3. They used mass spectrometry to identify the N_2 as the gas evolved, and they used respirometry to demonstrate quantitative conversion of N_2H_4 to N_2. In a potentiometric titration study, it was reported that N_2 is the ultimate product, but additional inflections were observed and interpreted as corresponding to formation of $N_2H_4^+$, $N_2H_3^+$, and $N_2H_2^+$ (128). Similarly, incomprehensible results were reported in a subsequent potentiometric study (72). It would appear advisable to accept provisionally the simple results of Huang and Spence (127), that Mo^{VI} oxidizes N_2H_4 according to the stoichiometry

$$4Mo^{VI} + N_2H_4 \longrightarrow 2[Mo^{V}]_2 + N_2 + 4H^+ \qquad (70)$$

The two kinetic studies of this reaction agree that the rate law is

$$-\frac{d[Mo^{VI}]}{dt} = k[Mo^{VI}][N_2H_5^+] \qquad (71)$$

in acidic media, with one paper reporting data in phosphate buffer and the other in perchlorate media, the rate constant in phosphate media being 9.8×10^{-2} M^{-1} s^{-1} (127, 129). Huang and Spence demonstrated that N_2H_2 is an intermediate by showing that the reaction leads to the hydrogenation of fumaric acid. In both papers, the suggested mechanism is

$$Mo^{VI} + N_2H_5^+ \longrightarrow Mo^{IV} + N_2H_2 + 3H^+ \qquad (72)$$

$$2N_2H_2 \longrightarrow N_2 + N_2H_4 \qquad (73)$$

$$Mo^{VI} + Mo^{IV} \rightleftharpoons [Mo^{V}]_2 \qquad (74)$$

In fact, the above data do not rule out a one-electron mechanism. The observation that the reaction can induce vinyl polymerization has been taken in support of a one-electron mechanism (128).

The reaction of $[Fe^{VI}O_4]^{2-}$ appears to be quite straightforward (130). Between pH 7.5 and 11 it has the stoichiometry

$$[FeO_4]^{2-} + N_2H_4 \longrightarrow Fe^{II} + N_2 + 4OH^- \tag{75}$$

Tests for Fe^{III} showed that this oxidation state is bypassed during the reaction, while tests for olefin hydrogenation indicated that N_2H_2 is an intermediate. Over the pH range 8–10 the rate law was found to be

$$-d[Fe^{VI}]/dt = (k_0 + k_H[H^+])[FeO_4^{2-}][N_2H_4] \tag{76}$$

with $k_0 = 5.0 \times 10^3\ M^{-1}\ s^{-1}$ and $k_H = 5.6 \times 10^{12}\ M^{-2}\ s^{-1}$. These results led to the proposal that both N_2H_4 and $N_2H_5^+$ react with FeO_4^{2-} to give N_2H_2 plus Fe^{IV}. Since oxygen exchange on $[FeO_4]^{2-}$ is a slow process, the reaction with N_2H_4 must occur without formation of an inner-sphere $Fe^{VI}—N_2H_4$ intermediate. This stands in contrast with the proposed mechanism for the reaction of N_2H_4 with CrO_4^{2-}, where an inner-sphere $Cr^{VI}—N_2H_4$ intermediate is specifically suggested (121). It was argued that the reaction of $[FeO_4]^{2-}$ should have an initial two-electron step because of the lack of formation of NH_3. It is clear, however, that many one-electron oxidants do not yield NH_3 [e.g., $[IrCl_6]^{2-}$ and $[Fe(CN)_6]^{3-}$] and so this is not a valid mechanistic criterion. It was also argued that the intermediacy of N_2H_2 indicates the operation of a two-electron mechanism; again, there are many examples of one-electron oxidants that react via N_2H_2, and so this too is not diagnostic of a two-electron mechanism. In fact, from the present information it does not seem possible to rule out a one-electron mechanism, although a two-electron mechanism is also plausible.

C. Inert One-Electron Transition Metal Oxidants

Reactions of hydrazine with inert one-electron oxidants usually show rate laws with terms corresponding to the bimolecular reaction of N_2H_4 with the oxidant. For reactions where this is the case, the second-order rate constants are collected in Table III. A list of all inert one-electron oxidants in given in Table IV.

Brown and Higginson (10) found a consumption ratio of about 4 for the reaction of $[Mo(CN)_8]^{3-}$ between pH 2.4 and 4.4; above pH 6 they obtained consumption ratios near 5. In alkaline media, the excess consumption of Mo^V was attributed to its partial decomposition. In mildly acidic media, where decomposition is not a problem, the metal product is $[Mo(CN)_8]^{4-}$, and hence the

TABLE III
One-Electron Rate Constants for Outer-Sphere Oxidation of N_2H_4[a]

Oxidant	k, M^{-1} s^{-1}	Reference
$[Mo(CN)_8]^{3-}$	1.1×10^4	132
$[W(CN)_8]^{3-}$	1.65×10^2	132
$[Fe(CN)_6]^{3-}$	0.6	134
$[IrCl_6]^{2-}$	2×10^4	145

[a]Rate constants refer to the overall second-order reaction of N_2H_4 at 25°C.

reaction is

$$4[Mo(CN)_8]^{3-} + N_2H_5^+ \longrightarrow 4[Mo(CN)_8]^{4-} + N_2 + 5H^+ \quad (77)$$

A consumption ratio of 4 was also reported in a subsequent study, although at an unspecified pH (131). The first kinetic study of the reaction was conducted at 17°C and led to a rate law of the same form as that for the reaction of $[IrCl_6]^{2-}$ (131):

$$\text{Rate} = k_1K_a[N_2H_4]\text{tot}[Mo(CN)_8^{3-}]/([H^+] + K_a) \quad (78)$$

Notably, the derived acid dissociation constant for $N_2H_5^+$ is in agreement with other reports. The second kinetic study issued from the same research group and included temperature-dependent rates to permit extrapolation to 25°C (132). The proposed mechanism consisted of the following steps:

$$N_2H_5^+ \rightleftharpoons N_2H_4 + H^+ \qquad K_a \quad (79)$$

$$[Mo(CN)_8]^{3-} + N_2H_4 \longrightarrow [Mo(CN)_8]^{4-} + N_2H_4^+ \qquad k_1 \quad (80)$$

$$3[Mo(CN)_8]^{3-} + N_2H_4^+ \longrightarrow N_2 + \text{Products} \qquad \text{(fast)} \quad (81)$$

Dennis et al. (132) considered the k_1 step to correspond to an outer-sphere electron-transfer process. Their treatment of this in terms of Marcus theory is discussed in Section V.C.

Table IV
Inert One-Electron Oxidants Discussed

$[Mo(CN)_8]^{3-}$	$[Ru(bpy)_3]^{3+}$	$[IrBr_6]^{2-}$
$[W(CN)_8]^{3-}$	$[Co(acac)_3]$	$[Ir(OH_2)Cl_5]^-$
$[MnO_4]^-$	$[Co(ox)_3]^{3-}$	$[Ir(OH_2)_2Cl_4]$
$[Fe(CN)_6]^{3-}$	$[IrCl_6]^{2-}$	$[PtCl_6]^{2-}$

Stoichiometric data on the reaction of $[W(CN)_8]^{3-}$ with N_2H_4 are limited to a determination of the consumption ratio under unspecified conditions (131). The result indicated complete conversion to N_2, similarly to the reaction of $[Mo(CN)_8]^{3-}$. There are two reports on the kinetics of this reaction, both of which indicate a rate law analogous with that for $[Mo(CN)_8]^{3-}$ (131, 132). The two reports differ only in that one was conducted at 16°C and the other at 25°C.

Oxidation by $[MnO_4]^-$ has been studied both with regard to its kinetics and its stoichiometry, although these two aspects were studied under highly divergent conditions (119, 133). Thus, Brown and Higginson investigated the reaction at pH 12.6 in the presence of Ba^{2+}; under these conditions $BaMnO_4$ precipitates with a consumption ratio consistent with quantitative N_2 production (10). A possible interpretation is that the reaction has a two-electron mechanism to form N_2H_2 plus Mn^V; comproportionation of Mn^V with $[MnO_4]^-$ would then give $[MnO_4]^{2-}$, which would precipitate as its barium salt. An alternative is a one-electron mechanism giving N_2H_3 and $[MnO_4]^{2-}$ directly; further oxidation would then lead to N_2H_2 and hence to N_2. Costner and Ganapathisubramanian (133) omitted Ba^{2+} and used mildly acidic solutions, with the outcome that Mn^{2+} is produced and a consumption ratio indicative of NH_4^+ production is obtained. Prasad and Kumar (119) studied the reaction in the absence of Ba^{2+} in strongly acidic H_2SO_4 solutions and likewise obtained evidence for formation of NH_4^+. These results seem to imply that NH_4^+ arises by reaction of intermediate oxidation states of Mn that do not form when Ba^{2+} is present. An alternative interpretation is that NH_4^+ is only produced under acidic conditions. Thus, it is ambiguous whether the direct reaction of $[MnO_4]^-$ has a one- or two-electron mechanism, and it is also ambiguous whether NH_4^+ arises from the direct reaction of $[MnO_4]^-$ in acidic media or from one of the lower oxidation states.

The rate law for this reaction does little to resolve the mystery of its mechanism. Autocatalytic kinetics is observed in mildly acidic media; in a large excess of Mn^{2+} the rates are independent of $[N_2H_4]$ and second order with respect to $[MnO_4^-]$ (133). It seems clear that Mn^{2+} is involved in the autocatalysis, but a plausible mechanism has not yet been proposed.

The oxidation of hydrazine by $[Fe(CN)_6]^{3-}$ has been studied extensively. Brown and Higginson (10) obtained a consumption ratio of 4.025 ± 0.035 at pH 9.9 with excess $[Fe(CN)_6]^{3-}$ under anaerobic conditions. An important study of the reaction by Meehan et al. (134) showed that it is highly susceptible to Cu^{2+} catalysis and that the stoichiometry is altered by the presence of oxygen. Unfortunately, the details of the O_2 effect have not been disclosed. In the absence of oxygen, they obtained quantitative conversion to N_2 with a slight excess of $[Fe(CN)_6]^{3-}$ under alkaline conditions. Kumar and Prasad (72) report the formation of NH_3, N_2H_3, HN_3, H_2, N_2, and H_2O_2, depending on conditions, when the reaction is studied by conducting a potentiometric titration in oxygenated media; these peculiar results are difficult to understand, but they may be a consequence of hydrazine decomposition occurring catalytically at the electrode surfaces or unrecognized copper catalysis. In summary, the weight of evidence supports exclusive formation of N_2 when the reactions are performed under properly controlled conditions.

Minato et al. (135) noted that the reaction is slow in acidic media and much faster in alkaline media, being first order with respect to the concentrations of N_2H_4 and $[Fe(CN)_6]^{3-}$. The importance of copper catalysis was first reported by Veprek-Siska (136), and shortly thereafter by Meehan et al. (134). Meehan et al. (134) used edta to mask the effects of catalytic metal ions and obtained a rate constant of 0.6 M^{-1} s^{-1} in alkaline media. A significantly greater rate constant was reported by Dennis et al. (2.3 M^{-1} s^{-1}), which may be attributed to their failure to take precautions against copper catalysis. The significance of copper catalysis was further emphasized by Madlo et al. (137). Kinetic data as a function of methanol cosolvent were reported by Jindal et al. (138), although the results are questionable because precautions were not taken against copper catalysis.

Evidence for N_2H_2 as an intermediate comes from trapping studies. Thus, hydrogenation of olefins occurs during the oxidation of hydrazine by $[Fe(CN)_6]^{3-}$ (139, 140). Stoichiometry studies indicate that N_2H_2 is generated quantitatively (141).

The oxidation of N_2H_4 by $[Ru(bpy)_3]^{3+}$ is notable for being intensely chemiluminescent (142). Spectral analysis of the emitted light shows that it originates from an excited state of $[Ru(bpy)_3]^{2+}$. Stopped-flow traces of the decay of Ru^{III} in 0.1 N H_2SO_4 with excess N_2H_4 showed biphasic behavior, with a rapid consumption of about one-half of the initial Ru^{III} and a slower exponential loss of the remaining Ru^{III}. The pH dependence was not studied, nor was an explanation for the biphasic kinetics provided.

Kinetic studies of the reactions of $[Co(acac)_3]$ and $[Co(ox)_3]^{3-}$ with N_2H_4 have been reported with Co^{II} assumed as the product, although no data were provided regarding the stoichiometry or nature of the products (48, 49). In 0.5–1.2 M $HClO_4$ with excess hydrazine, the rate law for both oxidants reportedly is

$$-\frac{d[Co^{III}]}{dt} = [Co^{III}]\{k_H[H^+] + (k_R + k_{H,R}[H^+])[N_2H_6^{2+}]\} \quad (82)$$

In the proposed mechanism, the k_H term represents the acid-catalyzed decomposition of the complex, while the two other terms indicate the direct reaction of the complex with $N_2H_6^{2+}$. Protonation of the coordinated acac or ox ligand is proposed in explanation of the $k_{H,R}$ term. Both the reported rate law and proposed mechanism should be viewed with caution, since they were developed under the erroneous impression that $N_2H_6^{2+}$ is a rather weak acid (see above).

The oxidation of hydrazine by $[IrCl_6]^{2-}$ has led to considerable confusion because of disagreements among the various reports on this pivotal reaction. Brown and Higginson (10) obtained consumption ratios between 3.5 and 3.8 over the range from pH 1.6 to 6.5 when there was a modest excess of hydrazine. The defect from a consumption ratio of 4 was taken as evidence for formation of some NH_3 (10). Sengupta and Sen (143) examined the reaction at pH 1.3–0, and they obtained a consumption ratio of 3.7 with excess $[IrCl_6]^{2-}$ and a consumption ratio of 1.06 $\pm$ 0.1 with a large excess of hydrazine; they also obtained a positive qualitative test for formation of NH_3. Morris and Ritter (144) examined the stoichiometry with excess $[IrCl_6]^{2-}$ and obtained a consumption ratio of 3.81, a negative NH_3 test, and quantitative N_2 yield by mass spectrometry and Warburg respirometry. Unfortunately, these workers did not indicate the pH of these experiments. It can be concluded that consumption ratios somewhat less than 4 (in the range of 3.6) do not require the formation of NH_3. In the most recent report on the reaction, Stanbury (145) used ion chromatography to determine the yield of NH_3 and the consumption of N_2H_4 at pH 3.3 and 8.1 with a moderate excess of N_2H_4. Under these conditions, there was no detectable yield of NH_3, and the consumption ratio was 3.3 $\pm$ 0.5, with the large uncertainty reflecting the difficulty in measuring small changes in the concentration of N_2H_4 by ion chromatography. The evidence is clear that the results of Brown and Higginson do not imply the formation of NH_3. There remains a possibility as Sengupta and Sen reported that NH_3 is formed at very low pH with a large excess of N_2H_4, but these results are questionable. At very low pH the reaction is quite slow (hours), so that a significant portion of the Ir^{III} product becomes hydrolyzed. Moreover, NH_3 can be a contaminant in preparations of N_2H_4 salts and thus lead to spurious results in qualitative tests. Note that the rate law reported by Sengupta and Sen differs significantly from those reported by Stanbury and by Morris and Ritter. Presumably, the results of Sengupta and Sen were affected by catalysis by some unknown impurity. Overall, we consider that there is no strong evidence that NH_3 is produced in the reaction of $[IrCl_6]^{2-}$; the consistently reported defect from an ideal consumption ratio of 4 may be due to a minor contribution from the induced oxidation by O_2, as was found for the reaction with $[Fe(CN)_6]^{3-}$.

The first kinetic study of this reaction reported that the rate law saturates in $[N_2H_5^+]$, which is quite unexpected for a species such as $[IrCl_6]^{2-}$ (143). As we suggested above, this study was probably affected by some form of unrecognized catalysis and hence may be disregarded. Morris and Ritter (144) examined the kinetics in 0.1–1.0 *M* $HClO_4$ and found a rate law of the following form:

$$-\frac{d[IrCl_6^{2-}]}{dt} = \frac{k[IrCl_6^{2-}][N_2H_4]_{tot}}{1 + [H^+]/K_a} \tag{83}$$

The acid dependence was explained in terms of formation of unreactive $N_2H_5^+$, with N_2H_4 reacting directly with $[IrCl_6]^{2-}$. Inspection of the data reveals that the value of K_a is several orders of magnitude greater than the known dissociation constant of $N_2H_5^+$. Stanbury (145) examined the reaction from pH 1 to 10 and obtained a rate law of the same form but with a K_a value consistent with the acid dissociation constant of $N_2H_5^+$. Apparently, the data of Morris and Ritter (144) were also affected by some unknown form of catalysis. Morris and Ritter also reported kinetic data for the reactions of $[IrBr_6]^{2-}$, $[Ir(OH_2)Cl_5]^-$, and $[Ir(OH_2)_2Cl_4]$, although the pH dependence was not investigated for these oxidants.

The most reasonable mechanism consistent with the above results is

$$N_2H_5^+ \rightleftharpoons N_2H_4 + H^+ \qquad K_a \qquad (84)$$

$$[IrCl_6]^{2-} + N_2H_4 \longrightarrow [IrCl_6]^{3-} + N_2H_4^+ \qquad k \qquad (85)$$

$$N_2H_4^+ \rightleftharpoons N_2H_3 + H^+ \qquad \text{(fast)} \qquad (86)$$

$$[IrCl_6]^{2-} + N_2H_3 \longrightarrow [IrCl_6]^{3-} + N_2H_2 + H^+ \qquad \text{(fast)} \qquad (87)$$

$$2N_2H_2 \longrightarrow N_2 + N_2H_4 \qquad \text{(fast)} \qquad (88)$$

Further support for N_2H_2 as an intermediate comes from the observation that the reaction induces olefin hydrogenation (144). Since both $[IrCl_6]^{2-}$ and $[IrCl_6]^{3-}$ are substitution inert, it is evident that N_2H_4 does not enter into the coordination sphere of the metal ion during the reaction.

The oxidation of hydrazine by $[PtCl_6]^{2-}$ gives a large yield of NH_3, as inferred from a consumption ratio of 1.95 for $\Delta N_2H_4/\Delta Pt^{IV}$ at pH 4.7 (146). The rate law is

$$-\frac{d[PtCl_6^{2-}]}{dt} = \frac{k_1k_2[PtCl_6^{2-}][N_2H_5^+]}{k_{-1}[Cl^-] + k_2[N_2H_5^+]} \qquad (89)$$

which was interpreted in terms of a mechanism involving reversible loss of Cl^-, followed by rate-limiting reaction of $[PtCl_5]^-$ with $N_2H_5^+$. A value for k_1 (Cl^- loss) of about $5 \times 10^{-3}\ s^{-1}$ at 30°C, can be deduced from a figure of the data, which seems much too fast for $[PtCl_6]^{2-}$. A one-electron mechanism proceeding via Pt^{III} and $N_2H_4^+$ was suggested, but in view of the anomalous kinetics and the notorious sensitivity of Pt^{IV} substitution reactions to Pt^{II} catalysis it would be appropriate to await further investigation of this system.

D. Labile One-Electron Transition Metal Oxidants

As is indicated in Table V, there are a large number of labile transition metal oxidants whose reactions with hydrazine have been investigated.

TABLE V
Labile Transition Metal Species as Oxidants of Hydrazine

CrO_2^{2+}	$Fe^{III}edta$	$[Ni^{III}(NH_3)_n]^{3+}$
Mn^{3+}	Co^{3+}	$[Cu(NH_3)_4]^{2+}$
$[Mn^{III}(Cydta)(H_2O)]^-$	$[Co(CO_3)_3]^{3-}$	$AuCl_4^-$
$[Mn^{III}edta]^-$	$[Co(nta)(H_2O)_2]$	Ce^{4+}
$[Mn(P_2O_7)_3]^{9-}$	$[Co^{III}edta]^-$	Excited UO_2^{2+}
Fe^{III}	$[Ni^{III}(OH)]^{2+}$	

Classification of $[CrO_2]^{2+}$ [superoxochromium(III)] as a labile oxidant is not rigorously established, but it is suggested by general trends for superoxo complexes (147). The complex is formed from the reaction of Cr^{2+} with O_2, and it is reduced by hydrazine to $[Cr(H_2O)_6]^{3+}$ (46). Hydrogen peroxide is an expected product of the reaction, but technical difficulties hindered its detection. The nitrogeneous products were assumed to be N_2 and NH_3. When the reaction is conducted with a large excess of N_2H_4 in 0.01–0.05 *M* $HClO_4$, deviations characteristic of product inhibition are seen in the pseudo-first-order plots. These deviations are not seen in the presence of $HCrO_4^-$, where the rate law is

$$-\frac{d[CrO_2^{2+}]}{dt} = (k_0 + k[H^+][N_2H_5^+])[CrO_2^{2+}] \tag{90}$$

with values of $7 \times 10^{-4}\ s^{-1}$ and $58.1\ M^{-2}\ s^{-1}$ for k_0 and k. The k_0 term corresponds to the spontaneous decomposition of the complex. The following mechanism was advanced in explanation of the k term:

$$[CrO_2]^{2+} + H^+ \rightleftharpoons [CrO_2H]^{3+} \tag{91}$$

$$[CrO_2H]^{3+} + N_2H_5^+ \rightleftharpoons [CrO_2H]^{2+} + N_2H_5^{2+} \tag{92}$$

$$N_2H_5^{2+} \rightleftharpoons N_2H_4^+ + H^+ \tag{93}$$

$$N_2H_4^+ + [HCrO_4]^- \longrightarrow H_3CrO_4 + N_2H_2 \tag{94}$$

The original report gave the deprotonation of $N_2H_5^{2+}$ as an irreversible step, but its corrected form is shown above (148). Subsequent work has shown that the standard potential for the reduction of $[CrO_2]^{2+}$ to $[CrO_2H]^{2+}$ is 0.82–0.97 V, which is certainly adequate to achieve the oxidation of hydrazine (149, 150). Decomposition of $[CrO_2H]^{2+}$ to Cr^{3+} plus H_2O_2 has been found to occur with a half-life of about 15 min, which explains the ultimate formation of Cr^{3+} in the hydrazine reaction (150, 151). One difficulty with the above mechanism is that it requires a significant fraction of the hydrazyl radicals to be in the $N_2H_5^{2+}$ form in order for $[CrO_2H]^{2+}$ to act as an inhibitor. This would imply a pK_a of 1 or greater for $N_2H_5^{2+}$, while the evidence from ESR spectroscopy indicates that $N_2H_4^+$ is the predominant species in 1 M H_2SO_4(152, 153).

Single-electron oxidants are expected to yield either N_2 or a mixture of N_2 and NH_3, and among the reactions described since 1957 there is only one exception. This exception is the oxidation by Mn^{3+} in strong $HClO_4$ solutions, and here it seems clear that the data are not trustworthy (45). The results certainly are highly peculiar, in that it was reported that no gaseous products were generated. Although the products were not identified, a value of 1.05 was obtained for the consumption ratio, $\Delta Mn^{3+}/\Delta N_2H_4$. It was inferred that the species N_4H_6 was produced and that it did not decompose. However, in the same paper it was also reported that the oxidation of N_2H_4 by Ce^{4+} yielded no gaseous products, which is contrary to another report on this latter reaction where gas evolution was used to monitor the progress of the reaction (154). In a potentiometric study of the reaction of Mn^{3+} with hydrazine in 9 M H_2SO_4, breakpoints were observed that were attributed to the formation of N_2H_3, N_2H_2, and N_2H as stable species in solution (155). There is little information available on the stability of these species in 9 M H_2SO_4, but in view of their fleeting existence under most other conditions it seems quite improbable that the conclusions from this potentiometric study are correct. A possible explanation for the odd results is that the response of the electrodes is perturbed by surface-induced decomposition of N_2H_4.

Between 0.5 and 3.5 M H^+ in ClO_4^- media, this reaction has the following rate law (45):

$$-\frac{d[\mathrm{Mn^{III}}]}{dt} = \frac{[\mathrm{Mn^{III}}][\mathrm{N_2H_4}]_{\mathrm{tot}}(k_1[\mathrm{H^+}] + k_2K_{\mathrm{H}})}{[\mathrm{H^+}] + K_{\mathrm{H}}} \tag{95}$$

Here K_H is the acid dissociation constant of Mn^{3+} (= 0.93 M), and k_1 (= 5.2 $\times$ 10^2 M^{-1} s^{-1}) and k_2 (= 1.0 $\times$ 10^4 M^{-1} s^{-1}) refer to the reactions of $N_2H_5^+$ with Mn^{3+} and $[Mn(OH)]^{2+}$, respectively. A one-electron mechanism giving the radical $N_2H_4^+$ may be inferred in view of the inaccessibility of Mn^I. Davies and Kustin (45) suggest that this occurs through a hydrogen-atom transfer process.

A consumption ratio of 1.0 $\pm$ 0.1 was determined for the oxidation of hydrazine by $[Mn^{III}(Cydta)(H_2O)]^-$ between pH 2 and 5 (156), consistent with the stoichiometry

$$[\mathrm{Mn^{III}(Cydta)(H_2O)}]^- + \mathrm{N_2H_5^+} \longrightarrow [\mathrm{Mn^{II}(Cydta)(H_2O)}]^{2-} + \mathrm{NH_4^+} + \tfrac{1}{2}\mathrm{N_2} + \mathrm{H^+} \tag{96}$$

The rate law is

$$-\frac{d[\mathrm{Mn^{III}}]}{dt} = [\mathrm{Mn^{III}}][\mathrm{N_2H_4}]_{\mathrm{tot}}\left(k + \frac{k'}{[\mathrm{H^+}]}\right) \qquad (97)$$

with the k' term dominant, the individual rate constants being 700 M^{-1} s^{-1} and 0.9 s^{-1}. A rate-limiting reaction between $[Mn^{III}(Cydta)(H_2O)]^-$ and N_2H_4 is implied by the k' term, and by comparison with other reactions of $[Mn^{III}(Cydta)(H_2O)]^-$ an inner-sphere mechanism was deduced. The manganese oxidant is very robust, but it has a labile coordinated aqua ligand (pK_a = 8). A picture of the mechanism emerges that is very similar to that for the reaction of Fe^{III}edta with N_2H_4, with the exception that self-reaction of the $N_2H_4^+$ radicals to give NH_3 is dominant over their oxidation. It is possible that the difference is related to the acid dissociation of $N_2H_4^+$ (see below) and the differing pH ranges for the studies of the Fe^{III}edta and Mn^{III}Cydta reactions.

Reactions of $[Mn^{III}edta]^-$ and $[Mn(P_2O_7)_3]^{9-}$ with hydrazine have been investigated by Brown and Higginson (10) with respect to their stoichiometries. When the reaction of $[Mn^{III}edta]^-$ was investigated with excess Mn^{III} at pH 2.4 a consumption ratio of 1.1 was obtained, and larger ratios were obtained at higher pH. Yields of NH_3 were found in agreement with the consumption ratios. The formation of NH_3 is consonant with the behavior found for $[Mn^{III}Cydta]^-$. Consumption ratios increasing from 1.1 to 3.9 as the pH increased from 2 to 6 were found for the reaction of $[Mn(P_2O_7)_3]^{9-}$, and here too the yields of NH_3 were in agreement.

The oxidation of N_2H_4 by Fe^{III} was given special attention in Higginson's review because of its use in establishing the fate of the N_2H_3 radical (4). At issue was the question of whether it could disproportionate as in $2N_2H_3 \rightarrow N_2H_4 + N_2H_2$. A crucial component of the argument was the reversibility of the initial electron-transfer step. Subsequent to Higginson's review, Pollard and Nickless (43) described experiments that used more reliable analytical methods to determine the stoichiometry and kinetics. These experiments provided firm support for the reversibility of the first step by showing the inhibitory effects of Fe^{II} and the catalytic effects of Cu^{II}. Thus, at most 1.5% of the disproportionation of N_2H_3 led to N_2H_2 plus N_2H_4, the major disproportionation path being the formation of N_2 plus NH_3. Note that this and the prior studies on the oxidation by Fe^{III} were conducted with Cl^- or SO_4^{2-} media. In a more recent study in ClO_4^- media, it was found that the rates were irreproducible, depending on the specific stock solution of Fe^{III} (157). The effect was attributed to varying degrees of dimerization of Fe^{III} depending on the history of the solution, while SO_4^{2-} and Cl^- act to maintain the Fe^{III} in a monomeric state. Catalysis of this reaction by Cu^{2+} in sulfuric acid media is the subject of a very recent report (158).

Frank et al. (159) studied the kinetics of oxidation by Fe^{III} in acid media containing phenanthroline. They found that phenanthroline inhibited the reaction, and they attributed this effect to the formation of unreactive Fe^{III} complexes of phenanthroline. Note that the strong oxidant $[Fe(phen)_3]^{3+}$ does not form in such solutions. Their observation of ammonia as a product is consistent with a one-electron mechanism.

Minato et al. (135) found that the reaction of Fe^{III}edta with N_2H_4 in mildly alkaline media gives small but significant yields of NH_3, depending on conditions. The reaction is strongly retarded by Fe^{II}edta, but it is possible to drive the reaction to completion with good pseudo-first-order kinetics by the addition of phenanthroline as a Fe^{II} scavenger. Under these conditions, the reaction is first order with respect to both [Fe^{III}edta] and $[N_2H_4]$ with a rate constant of 660 $M^{-1}\ s^{-1}$ at 40°C. The kinetic inhibition by Fe^{II}edta is good evidence that the rate-limiting step is a reversible one-electron redox reaction. This result is consistent with the significant yields of NH_3 that were obtained, which imply the formation of N_2H_3 as an intermediate.

Oxidation of hydrazine by Co^{III} in acidic nitrate and sulfate media yields a mixture of N_2 and NH_3; consumption ratios in the range from 1 to 2 were obtained with excess hydrazine (160). Different rate laws were obtained in H_2SO_4 and HNO_3 media, as might be anticipated from the differing tendencies of these anions to bind Co^{III}. A saturating dependence on $[N_2H_5^+]$ was obtained in H_2SO_4 media, which was interpreted as arising from a rapid preequilibrium complexation between Co^{III} and $N_2H_5^+$, followed by rate-limiting internal redox decomposition. The free Co^{3+} ion was suggested as the predominant species because the rates were independent of sulfate concentration. This proposal should be taken with some skepticism in view of a prior demonstration that Co^{3+} binds SO_4^{2-} (161). In HNO_3 media, the reaction is first order with respect to $[N_2H_5^+]$ and $[Co(OH)^{2+}]$ with a rate constant of 0.18 $M^{-1}\ s^{-1}$ at 10°C, and is suggested to have an inner-sphere one-electron-transfer mechanism.

Brown and Higginson (10) measured consumption ratios for the reaction of N_2H_4 with $[Co(CO_3)_3]^{3-}$ from pH 7 to 9. They found ratios ranging from 1.9 to 3.1, depending on the ratio of the reactants. Since the ultimate metal ion product is Co^{II}, these ratios indicate variable yields of NH_3 and N_2. These workers also noted the formation of a purple intermediate, which they suggested is a $Co^{III}{-}N_2H_4{-}CO_3^{2-}$ complex. Tanner (162) has shown that the reaction is autocatalytic due to the formation of Co^{II}. The nonautocatalytic part of the rate law can be studied by the addition of edta to scavenge Co^{II}, and the result is

$$-d[Co^{III}]/dt = k[Co^{III}][N_2H_4]_{tot}/[HCO_3^-] \tag{98}$$

with a complex pH dependence. An inner-sphere mechanism with rate-limiting substitution was deduced from the inverse dependence on $[HCO_3^-]$. Tanner found that the purple intermediate is produced only in the autocatalytic pathway. It is not clear how the variable stoichiometry is apportioned between the noncatalytic and catalytic terms of the rate law.

Brown and Higginson (10) cited Thaker and Higginson's (163) results on the consumption ratio in the reaction of $[Co(nta)(H_2O)_2]$ with N_2H_4. These latter workers obtained a consumption ratio of 1.1–1.2 at pH 3.5. Their study of the kinetics showed that the reaction involved rate-limiting substitution of an aqua ligand in $[Co(nta)(OH)(OH_2)]^-$ by $N_2H_5^+$.

In a kinetic study of the reaction of N_2H_4 with $[Co^{III}edta]^-$, it was assumed that the process studied corresponds to a redox reaction, although no product analyses were performed, nor was the consumption ratio determined (164). In 0.5–1.0 *M* $HClO_4$ with excess hydrazine at 50–80°C the rate law is given as

$$-\frac{d[Co^{III}]}{dt} = [Co^{III}](k_R + k_{H,R}[H^+])[N_2H_5^+] \tag{99}$$

It is puzzling that Banerjee et al. (164) concluded $N_2H_5^+$ to be the predominant species, while the same workers indicated $N_2H_6^{2+}$ to be predominant at the same acid concentrations in their studies of the reactions of $[Co(acac)_3]$ and $[Co(ox)_3]^{3-}$. A further complication is the known acid–base chemistry of the Co^{III}edta complex.

A very brief report on the reaction of $[Ni^{III}(OH)]^{2+}$ with hydrazine describes it as a one-electron reaction of Ni^{III}, although the only evidence presented is that the Ni^{III} is consumed (165). This species, obtained by the dissolution of NiO(OH) in acid, is claimed to be formulated as $[Ni(OH)]^{2+}$. In dilute H_2SO_4 and with a pseudo-first-order excess of hydrazine, the rate law shows saturation in [hydrazine].

Lati and Meyerstein (166) used pulse radiolysis to generate $[Ni^{III}(NH_3)_n]^{3+}$, and they found that it reacts with N_2H_4 at pH 11 with a rate constant of $4 \times 10^6\ M^{-1}\ s^{-1}$. It may be assumed that this is a one-electron redox reaction, but the details are as yet unknown.

The reaction of $[Cu(NH_3)_4]^{2+}$ with hydrazine was investigated by Brown and Higginson (10) with respect to its stoichiometry. The experiments were conducted at pH 10 in 1.5 *M* NH_3 and thus the yield of NH_3 could not be determined. However, with a consumption ratio of 4.0 it appears as though N_2 is formed quantitatively.

In an early study of the reaction of $[AuCl_4]^-$ with N_2H_4, Kirk and Browne (167) found that metallic gold is formed at 100°C, along with HN_3 and NH_3; they noted that their observations were in agreement with two prior studies. In a more recent study, at 30°C, Sen Gupta and Basu (168) claimed that Au^I is formed; apparently, these workers were unaware of the prior studies and offered no suggestions as to why Au^0 was not formed. Sen Gupta and Basu also determined a consumption ratio of 2.0 and measured the yield of NH_4^+ in 0.5 *M* HCl, from which they assigned the stoichiometry as

$$Au^{III} + 2N_2H_5^+ \longrightarrow Au^I + 2NH_4^+ + N_2 + 2H^+ \qquad (100)$$

Their kinetic studies were conducted in 0.2–2 M HCl with excess hydrazine, and led to the following chloride-independent rate law:

$$-\frac{d[Au^{III}]}{dt} = \frac{k[Au^{III}][N_2H_4]_{tot}}{1 + [H^+]/K_a} \qquad (101)$$

Here, the K_a value corresponds to the acid dissociation of $HAuCl_4$, such that the rate-limiting step is the reaction of $[AuCl_4]^-$ with $N_2H_5^+$, which occurs with a rate constant of 2.4 M^{-1} s^{-1} at 25°C. The yield of NH_4^+ and the observation that the reaction induces the polymerization of acrylamide led to the proposal that Au^{III} is acting as a one-electron oxidant, forming Au^{II} as an intermediate. Skibsted's commentary on this work notes an apparent typographical error in the sign of $\Delta S^{\ddagger}$ and suggests that the intermediate could be a Au^{II}-radical species rather than the free Au^{II} and $N_2H_4^+$ (169).

Cerium (IV) has long been a popular oxidant for study, and it was included among Higginson's ammonia-generating group. More recent studies have used the reaction to generate the $N_2H_4^+$ radical for ESR experiments (152, 153). Mishra and Gupta (170) found that the consumption ratio was dependent on the concentration of H_2SO_4, with quantitative conversion to N_2 occurring only at very high acid concentrations. Morrow and Sheeres (154) found variable stoichiometry in 1 M $HClO_4$, depending on the relative concentrations of the reactants. Mishra and Gupta (170) used a potentiometric titration to probe the stoichiometry and observed inflections that were interpreted as indicating the formation of species such as N_4H_6, N_3H_5, N_2H_2, and N_3H_3 (72). This report should probably be disregarded since these species are not sufficiently persistent to lead to the reported behavior.

A kinetic study of the reaction in SO_4^{2-} media showed that the reaction is first order with respect to both Ce^{IV} and $[N_2H_4]_{tot}$ (170). Because of the formation of various $Ce^{IV}{-}SO_4^{2-}$ complexes and the acid dissociations of Ce^{IV} and $N_2H_5^+$, there is considerable ambiguity as to the mechanistic significance of the rate law. The rates are substantially faster in $HClO_4$ media and the form of the rate law is simpler (154):

$$-\frac{d[Ce^{IV}]}{dt} = \frac{k[Ce^{IV}][N_2H_4]_{tot}}{[H^+]} \qquad (102)$$

The rate constant is 6.7 s^{-1} at 25°C. The rate law was interpreted as indicating that $[Ce(OH)]^{3+}$ reacts with $N_2H_5^+$ in the rate-limiting step. A sequence of one-electron steps is indicated by the detection of $N_2H_4^+$ as an intermediate and the inaccessibility of Ce^{II}. An inner-sphere mechanism is possible because of the high lability of Ce^{IV}, although no intermediate complex was detected. The variable stoichiometry is best explained by competition between the self-reaction of $N_2H_4^+$ and its oxidation by Ce^{IV}.

The Ce^{IV} ion forms complexes with nta, edta, and dtpa, and the kinetics of their reactions with N_2H_4 indicates that they react directly with this substrate (171).

The photochemical oxidation of hydrazine by $[UO_2]^{2+}$ has been investigated in HNO_3 and $HClO_4$ media (172). Yields of NH_3 and U^{IV} and consumption of N_2H_4 were determined, allowing the idealized stoichiometry to be expressed as

$$[UO_2]^{2+} + 2N_2H_5^+ \longrightarrow U^{IV} + N_2 + 2NH_4^+ \qquad (103)$$

Significant deviations from this ideal (less NH_4^+) were observed at short irradiation times. Butler and Kemp (172) suggested a mechanism in which excited UO_2^{2+} accepts an electron from N_2H_4 to form U^V plus $N_2H_4^+$; the final products arise from subsequent disproportionation of both $N_2H_4^+$ and U^V.

IV. STUDIES OF REACTION INTERMEDIATES

A. N_2H_3, N_4H_6, N_3H_3, and Their Acid and Base Forms

Since Higginson's review, there are a number of important developments in our understanding of species derived from hydrazine. The radical $N_2H_4^+$ has been detected by ESR spectroscopy and its spectrum has been interpreted in terms of a planar radical (152, 153). Ab initio calculations at the G2 level support a C_{2h} structure that has "only a small deviation from planarity" (173). Hayon and Simic (105) used pulse radiolysis to generate N_2H_3 and $N_2H_4^+$ and detected them by UV spectroscopy. These workers obtained a pK_a of 7.1 ± 0.1 for $N_2H_4^+$ (105), a result that has been confirmed by Buxton and Stuart (106) (pK_a = 7.2 ± 0.1). Hayon and Simic also reported that these radicals decay with second-order kinetics, with $2k_{+,+} = 6 \times 10^8$ and $2k_{0,0} = 2.5 \times 10^9\ M^{-1}\ s^{-1}$, respectively, where the subscripts indicate the reactant ionic charges. These rate constants were revised by Inbar and Cohen (98) in order to account for the reaction of N_2H_3 with $N_2H_4^+$, such that $k_{+,+} = 2 \times 10^8$, $k_{+,0} = 3 \times 10^9$, and $k_{0,0} = 9.5 \times 10^8\ M^{-1}\ s^{-1}$. Hayon and Simic deduced that the radicals combine to form N_4H_6, which has no detectable UV spectral features, and that N_4H_6 undergoes pH-dependent first-order decay ($k = 850\ s^{-1}$ at pH 9.2) to form the absorbing species N_3H_3 plus NH_3 as in

$$N_4H_6 \longrightarrow N_3H_3 + NH_3 \qquad (104)$$

Buxton and Stuart (106) confirmed the general features of this process and, like Hayon and Simic (105), have not suggested a mechanism that accounts for its pH dependence.

Triazene (N_3H_3) was found to have a pH-dependent UV spectrum that led to an estimate of 7.0 ± 0.2 for the pK_a of $N_3H_4^+$ (105). This interpretation of the chemistry of N_3H_3 was revised by Sutherland (174), who studied its decay to form N_2 plus NH_3 from pH 1 to 13. He found that there are three species, $N_3H_4^+$, N_3H_3, and $N_3H_2^-$ (or $N_3H_4O^-$), and that they are related by pK_a values of 4.95 and 11.37. This revision has been confirmed by Buxton and Stuart (106). The decay of triazene shows general acid catalysis, but over the pH range from 1–13 with normal buffer concentrations the decay half-life of N_3H_3 is never greater than 70 s, which requires that the decay half-life of N_4H_6 be less than 70 s over this pH range.

Ab initio calculations have been reported on several isomers of gas-phase N_3H_3 (175). These calculations consistently predict that triazene ($NH_2N{=}NH$) is the stablest of the isomers, in agreement with the inferences regarding its structure by prior workers (174). It would be of interest to examine pathways for the decomposition of triazene in view of the 1–3 proton transfer that must take place.

Ershov and Mikhailova (107) recently reported that N_2H_3 reduces O_2 to O_2^- with a near-diffusion controlled rate constant, while $N_2H_4^+$ appears unreactive. This may account for the reported effect of O_2 in certain redox reactions of N_2H_4, such as that with $[Fe(CN)_6]^{3-}$ (134). It also permits us to test the mechanism proposed by Lim and Fagg (12) for the Mn^{2+} catalyzed autoxidation of N_2H_4. These workers proposed a chain mechanism with the reaction of N_2H_3 plus O_2 being one of the propagation steps. Their derived rate constant for this process, 13 M^{-1} s^{-1}, is totally inconsistent with the results of Ershov and Mikhailova, and it raises serious questions as to the validity of the proposed chain mechanism. Very recently, the reaction of O_2 with N_2H_3 was reinvestigated, and a rate constant of 3.8×10^8 M^{-1} s^{-1} at 20°C was determined for the electron-transfer process mentioned above (176).

Lim and Zhong (23) presented arguments that the results of Hayon and Simic (105) were seriously misinterpreted. One of the points made by Lim and Zhong is correct: namely, that Hayon and Simic were in error when they inferred that N_2H_2 is not formed in the radiolysis of N_2H_4. This inference was based on the assumption that N_2H_2 would decompose to form substantial yields of H_2, and such yields were not obtained. As Lim and Zhong point out, N_2H_2 does not decompose via this pathway, and thus low H_2 yields cannot be used to rule out N_2H_2. In other ways, however, Lim and Zhong's critique is in error. They dispute Hayon and Simic's (105) deductions about N_3H_3 without reference to Sutherland's (174) extensive studies of this species. Nor do they cite Inbar and Cohen's (98) revision to the rate constants for the radical decay. Most seriously, they infer that N_2H_3 reacts promptly with H_2O_2 so that the spectra reported by Hayon and Simic (their Fig. 3) refer to solutions in which 99% of the N_2H_3 is already depleted. Zhong and Lim based this deduction on a presumed initial N_2H_3 concentration of 20 m*M*, which is quite impossible. Such high radical concentrations are orders of magnitude greater than those attainable with the methods available to Hayon and Simic. In fact, the value of 20 m*M* refers to the concentration of N_2H_4 in the experiments. Except for the matter regarding the decomposition of N_2H_2 we may safely disregard Lim and Zhong's critique.

B. Reduction Potentials Involving $N_2H_4^+$ and N_2H_3

Minato et al. (135), in their study of the reaction of N_2H_4 with Fe^{III}edta at pH 6.7–9, suggested a potential of 0.0 V for the N_2H_3, H^+/N_2H_4 redox couple on the basis of the kinetic inhibition by Fe^{II}edta. Two factors that could affect this conclusion are the acid–base reactions of N_2H_3 and Fe^{III}edta. It is now known that $N_2H_4^+$ has a pK_a of 7.1 and that $[Fe^{III}edta(H_2O)]^-$ has a pK_a of 7.6 (177). These are within the range of pH used in the study of Minato et al. (135), and therefore they could account for the mild pH dependence seen for the ratio of k_{-1}/k_2, where k_{-1} refers to the reverse electron-transfer process and k_2 refers to oxidation of the radical to N_2H_2 by Fe^{III}edta. These pK_a values are close enough to the pH values at which the rate constants were determined that they should not significantly affect the estimate of the redox potential of the radical.

An upper limit to $E°$ of 0.94 V for the $N_2H_4^+/N_2H_4$ couple was estimated from the measured rate constant for electron transfer from N_2H_4 to $[Fe(CN)_6]^{3-}$ and the assumption of a diffusion-limited value for the reverse process (145). An attempt to use pulse radiolysis to measure the rate of reaction of $N_2H_4^+$ with $[Fe(CN)_6]^{4-}$ at pH 5 was unsuccessful, the inferred rate constant being less than $3 \times 10^6\ M^{-1}\ s^{-1}$. This low rate constant might seem to be inconsistent with the kinetic inhibition by $[Fe(CN)_6]^{4-}$ of the reaction of $[Fe(CN)_6]^{3-}$ with N_2H_4 reported by Jindal et al. (138), but Meehan et al. (134) have shown that such inhibition is an artifact arising from the precipitation of trace metal catalysts by the $[Fe(CN)_6]^{4-}$. By combining this upper limit to the rate constant for electron transfer from $[Fe(CN)_6]^{4-}$ to $N_2H_4^+$ with the observed forward rate constant, an upper limit of 0.73 V was obtained for $E°$ for the $N_2H_4^+/N_2H_4$ couple (145).

In a review on reduction potentials of free radicals (178), it was mentioned that Pearson (179) used a thermochemical cycle based on the ionization potential of N_2H_4 to estimate $E° = 0.65$ V for the $N_2H_4^+/N_2H_4$ couple. It was also noted that this estimate should be revised to 0.01 V because of a revision to the ionization potential of N_2H_4. Ritchie (180) used the same revised ionization potential with different assumptions regarding hydration energies to derive $E° = 0.70$ V. On the other hand, a determination of $\Delta_f H°$ (0) = 55.3 kcal mol^{-1}

for $N_2H_3(g)$ has recently been reported that can also be used in a thermochemical cycle (181). The calculation makes use of an estimate (by analogy with N_2H_4) of -9.9 kJ mol^{-1} as a heat capacity correction to 298 K, which gives $\Delta_f H°$ (298 K) = 221 kJ mol^{-1} for $N_2H_3(g)$. A calculated value of 56.5 cal K^{-1} mol^{-1} for $S°$ for $N_2H_3(g)$ (182) and ancillary thermochemical data from the NBS tables (52) can be used to estimate a value of 266 kJ mol^{-1} for $\Delta_f G°$ for $N_2H_3(g)$. An estimate of -7.5 kcal mol^{-1} for $\Delta G°_{\text{hyd}}$ for N_2H_3 by analogy with N_2H_4 and the pK_a of $N_2H_4^+$ then lead to estimates of $\Delta_f G° = 235$ kJ mol^{-1} for $N_2H_3(aq)$, $\Delta_f G° = 194$ kJ mol^{-1} for $N_2H_4^+(aq)$, and $E° = 0.69$ V for the $N_2H_4^+/N_2H_4$ couple, which is in fortuitous agreement with Pearson's early estimate.

We can reexamine the work of Minato et al. (135) on the reaction of N_2H_4 with $[Fe^{III}(edta)(H_2O)]^-$. They reported a rate constant of 3.3×10^{-2} M^{-1} s^{-1} for this reaction at 25°C. If we assume that this refers to the reversible formation of $[Fe^{II}edta]^{2-}$ plus $N_2H_4^+$, and if we assume an upper limit of 10^{10} M^{-1} s^{-1} for the reverse rate constant, then a lower limit of 3×10^{-12} can be calculated for the electron-transfer equilibrium constant. Since the Fe^{III}edta/Fe^{II}edta couple has $E° = 0.12$ V, this sets an upper limit of 0.8 V for $E°$ for the $N_2H_4^+/N_2H_4$ couple. This calculation is consistent with the other estimates discussed above, and it implies that we should not place too much weight on the original estimate for this potential by Minato et al. (135).

An estimate of $E°$ for the one-electron reduction of N_2H_2, as in

$$N_2H_2(aq) + H^+(aq) + e^- \longrightarrow N_2H_3(aq) \tag{105}$$

can be made by making use of the data given above for N_2H_3 and the data given in Section II.B for N_2H_2. The result, -0.31 V, indicates that N_2H_3 is a good one-electron reducing agent, capable of reducing O_2. This lends support to reaction mechanisms in which the oxidation of N_2H_3 is an essential step.

C. Diazene and Related Species

The chemistry of diazene (diimide, diimine, and N_2H_2) has been extensively explored since Higginson's review (4), with much of the interest focusing on its ability to reduce olefins stereospecifically (141, 183–185). Quantitative studies of its aqueous chemistry have been based on the use of stopped-flow methods to generate it through the acid-assisted decomposition of azodiformate, $[(NCO_2)_2]^{2-}$. In the first of these studies, the hydrogenation of azobenzene-4,4′-disulfonate was used as an indicator to determine the kinetics of the disproportionation of diazene (186). The process was found to have a second-order rate law, with $k = 2.0 \times 10^4$ M^{-1} s^{-1} from pH 3.4 to 7.0. It was then found that the UV spectrum of diazene could be detected and used to monitor the reactions of diazene directly (58). Analysis of the UV spectrum led to the conclusion that the species generated is the trans isomer, in agreement with ab initio predictions that this is the most stable isomer. The prior results for the decay kinetics were confirmed and extended to pH 1.5. The disproportionation activation parameters were determined ($\Delta H^{\ddagger} = 14$ kJ mol^{-1} and $\Delta S^{\ddagger} = -116$ J K^{-1} s^{-1}) and interpreted as implying that the decay mechanism involves a rapid solvent-catalyzed cis–trans equilibrium, followed by rate-limiting transfer of both hydrogen atoms from cis-N_2H_2 to trans-N_2H_2 through a pericyclic symmetry-allowed transition state. The inferred mechanism is as follows:

$$\text{trans-}N_2H_2 \rightleftharpoons \text{cis-}N_2H_2 \tag{106}$$

$$\textit{trans}\text{-}N_2H_2 + \textit{cis}\text{-}N_2H_2 \rightarrow \left[\text{cyclic } N_4H_4 \text{ transition state: } N{=}N \cdots H \cdots N(H) - N(H) \cdots H \cdots \right] \rightarrow N_2 + N_2H_4 \tag{107}$$

Ab initio studies have shown that the highly exothermic double-hydrogen atom transfer from cis-N_2H_2 to trans-N_2H_2 should have virtually no enthalpy barrier, in agreement with the experimental activation parameters (53). In contrast, the thermoneutral reaction of cis-N_2H_2 with N_2 is predicted to have a substantial enthalpy barrier (182). Ab initio studies have also led to the prediction that $N_2H_3^+$ is a rather strong acid, although its formation is sufficient to provide a pathway for rapid solvent-catalyzed cis–trans isomerization of N_2H_2 (54). The possibility of N_2H_2 acting as an acid has been considered by Bauer, whose estimates imply a pK_a of 34.3 (187). Thus reaction through N_2H^- appears highly unlikely.

Ab initio calculations have been reported for tetrazetidine (188). This species has the formula N_4H_4, as does the transition state for the decomposition of diazene. Tetrazetidine, however, has a cyclic N_4 structure. The calculations predict that it has an energy 33 kcal mol^{-1} greater than for two free trans-N_2H_2 molecules. Thus, tetrazetidine can play no role in the decomposition of diazene.

Another isomer of N_4H_4, trans-tetrazene ($H_2NN{=}NNH_2$), has been prepared by the acid-catalyzed decomposition of $(CH_3)_3Si{-}N{=}N{-}Si(CH_3)_4$ in diethyl ether (189). X-ray crystallography has confirmed the structure (190). This species is assumed to be formed by the acid-catalyzed dimerization of N_2H_2, and it decomposes in part as in

$$H_2N{-}N{=}N{-}NH_2 \longrightarrow NH_4^+ + N_3^- \tag{108}$$

The requirement for highly acidic conditions suggests that this may be the pathway for production of HN_3 in some of the two-electron reactions of hydrazine.

Early reports of aminonitrene ($H_2N{=}N$, 1,1-diazene) have been considered inconclusive (184). On the other hand, a more recent low-temperature matrix preparation of this species appears fairly conclusive (191). There is, as yet, no suggestion that this high-energy species plays a role in aqueous hydrazine chemistry.

Zhong and Lim (22) argued on the basis of a study of the copper-catalyzed reaction of H_2O_2 with N_2H_4 that N_2H_2 deprotonates with a pK_a of 11.5. This inference depends on a drastic revision of the pulse-radiolysis work of Hayon and Simic (105). As discussed above, this revision is based on faulty understanding of the pulse-radiolysis method, and hence little confidence should be placed on Zhong and Lim's derived pK_a value.

V. OVERALL TRENDS

A. Reappraisal of Higginson's Classification

The above discussion shows that the reduction products cannot be used to distinguish between oxidants having one- and two-electron mechanisms. There are oxidants (e.g., Mo^{VI}) that undergo a net one-electron reduction but are believed to have a two-electron mechanism, and others (e.g., $AuCl_4^-$) that undergo a net two-electron reduction but are believed to operate via one-electron steps. As the example of V^V shows, even if it can be established that the mechanism involves a one-electron reduction of the metal center, it is still possible for two such reductions to act cooperatively so as to achieve a direct oxidation of N_2H_4 to N_2H_2 without formation of any free radicals. The most meaningful distinction between one- and two-electron mechanisms would focus on hydrazine rather than the oxidant, and it would be based on whether the $N_2H_3/N_2H_4^+$ radical is formed. This is essentially the definition adopted by Higginson in his review (4).

Given this definition, we are now equipped to enquire whether Higginson's classification of oxidants according to the yield of NH_3 is still valid. According to Higginson, two-electron oxidants should not yield NH_3. The only clear exceptions to this rule are the reactions of $H_2NOSO_3^-$ and HNO_2, where some of the NH_3 arises from the nitrogen in the oxidant. Surprises are given by the reactions of $[PtCl_6]^{2-}$ and $[AuCl_4]^-$, which give NH_3 but would be expected to have two-electron mechanisms; in both cases the reactions have been proposed to have one-electron mechanisms, although the evidence is not very strong.

Also according to Higginson's classification, one-electron oxidants are expected to give either N_2 or a mixture of N_2 and NH_3. There are no clear exceptions to this rule. Higginson further noted that one-electron oxidants give significant yields of NH_3 only below pH 4, and he rationalized this effect by proposed pH dependencies for the rates of reactions of N_2H_4 with the oxidants, of N_2H_3 with itself, and of N_2H_3 with the oxidant. Exceptions to this rule include the reaction of $[AuCl_4]^-$, which gives high NH_3 yields at pH 4.7, and the reaction of $[Co(CO_3)_3]^{3-}$, which gives high yields of NH_3 at pH 7–9. The many peculiar features of the $[AuCl_4]^-$ reaction call for its reinvestigation. In the case of the $[Co(CO_3)_3]^{3-}$ reaction, its complex kinetics suggests the operation of an unusual mechanism.

In summary, the basic classification scheme of Higginson appears to be intact. Indeed, it is so entrenched that its basis in fact is difficult to assess, since for many workers the yield of NH_3 is considered sufficient evidence to assign the mechanism.

B. Stoichiometric Trends with One-Electron Oxidants

As mentioned above, the relative yields of N_2 and NH_3 in reactions of one-electron oxidants are pH dependent, and this has been attributed to the pH dependence of the kinetics of the reactions of $N_2H_5^+/N_2H_4$ and $N_2H_4^+/N_2H_3$. Despite Higginson's qualitative discussion of this effect in 1957 there has been no subsequent publication that has attempted to place it on quantitative grounds. Now that the pK_a of $N_2H_4^+$ and the pH dependence of its decay is known it should be possible to attempt such a study.

Brown and Higginson (10) considered the question of whether a mechanistic distinction could be made on the basis of lability, where labile oxidants would yield some NH_3 through inner-sphere mechanisms and inert oxidants would yield NH_3 through dimerization of N_2H_3. To this end they determined consumption ratios for reactions of the following oxidants: $[Cu(NH_3)_4]^{2+}$, $[Mn(P_2O_7)_3]^{9-}$, $[Mn(edta)]^-$, $[Co(CO_3)_3]^{3-}$, $[Co(nta)(H_2O)_2]$, $[Fe(CN)_6]^{3-}$, $[Mo(CN)_8]^{3-}$, $[IrCl_6]^{2-}$, and $[MnO_4]^-$ (in the presence of Ba^{2+}). They found that of the labile oxidants ($[Cu(NH_3)_4]^{2+}$, $[Mn(P_2O_7)_3]^{9-}$, $[Mn(edta)]^-$, $[Co(CO_3)_3]^{3-}$, and $[Co(nta)(H_2O)_2]$) all except $[Cu(NH_3)_4]^{2+}$ had consumption ratios less than 4, and hence they concluded that with the exception of $[Cu(NH_3)_4]^{2+}$ all yielded some NH_3. They also found that of the inert oxidants ($[Fe(CN)_6]^{3-}$, $[Mo(CN)_8]^{3-}$, $[IrCl_6]^{2-}$, and $[MnO_4]^-$) only $[IrCl_6]^{2-}$ led to a consumption ratio significantly less than 4, and thus that with this one exception all of the inert oxidants yielded N_2 exclusively. Brown and Higginson were persuaded on the basis of the reactions of $[Cu(NH_3)_4]^{2+}$ and $[IrCl_6]^{2-}$ that a simple mechanistic distinction in terms of lability could not be made.

From the above discussion, it is evident that Brown and Higginson (70) were unduly swayed by their mistaken belief that NH_3 is a product of the reaction of $[IrCl_6]^{2-}$. With this correction, the reactions that they considered lead to the conclusion that inert oxidants lead to the exclusive formation of N_2, while labile oxidants lead to the formation of some NH_3. The sole exception to this rule is the reaction of $[Cu(NH_3)_4]^{2+}$; its rate law has not been studied, but Brown and Higginson suggested that the reaction could actually have a two-electron mechanism with two Cu^{II} centers acting cooperatively as was suggested for the oxidation by V^{V}.

In addition to those reactions discussed above, the reactions of the following one-electron oxidants have been reported since Higginson's review: CrO_2^{2+}, $[W(CN)_8]^{3-}$, Mn^{3+}, Mn^{III}Cydta, Fe^{3+}, Fe^{III}edta, $[Ru(bpy)_3]^{3+}$, Co^{3+}, $[Co(acac)_3]$, $[Co(ox)_3]^{3-}$, Co^{III}edta, $[Ni^{III}(OH)]^{2+}$, $[Ni^{III}(NH_3)_n]^{3+}$, $[PtCl_6]^{2-}$, $[AuCl_4]^-$, Ce^{4+}, and $*UO_2^{2+}$. We need not consider the reports on the reactions of CrO_2^{2+}, $[Ru(bpy)_3]^{3+}$, $[Co(acac)_3]$, $[Co(ox)_3]^{3-}$, Co^{III}edta, $[Ni^{III}(OH)]^{2+}$, and $[Ni^{III}(NH_3)_n]^{3+}$, since pertinent stoichiometric studies were not performed. We can also exclude Mn^{3+} and $PtCl_6^{2-}$ because of the questionable validity of the results. Of those remaining, the labile oxidants (Mn^{III}Cydta, Fe^{3+}, Fe^{III}edta, Co^{3+}, $AuCl_4^-$, Ce^{4+}, and $*UO_2^{2-}$) all give significant yields of NH_3 while $[W(CN)_8]^{3-}$, an inert species, does not yield NH_3. Thus, subsequent to Higginson's review all of the studies that describe reactions of one-electron oxidants that were not considered in the Brown and Higginson paper are also consistent with the rule that NH_3 is formed only from labile oxidants.

C. Kinetic Trends with One-Electron Oxidants

As is discussed above, inert one-electron oxidants uniformly give N_2 as the sole product from oxidation of hydrazine. These oxidants also share a common rate law, which is bimolecular with respect to [oxidant] and $[N_2H_4]$. Moreover, they all have the same mechanism, in which $N_2H_5^+$ is unreactive and the rate-limiting step is electron transfer from N_2H_4 to form the $N_2H_4^+$ radical. It is reasonable to suggest that these reactions have outer-sphere mechanisms for their electron-transfer steps, and thus they should have rates that are in agreement with Marcus theory. Morris and Ritter (144) were the first to analyze their data from this perspective by plotting log k versus $E°$ for a series of oxidants. The significance of their analysis is unclear, however, since at least some of their data (the reaction of $[IrCl_6]^{2-}$) appear incorrect (see above). Stanbury (145) applied the Marcus cross-relationship to the reaction of N_2H_4 with $[IrCl_6]^{2-}$ to derive an upper limit of 0.3 M^{-1} s^{-1} for the $N_2H_4^+/N_2H_4$ self-exchange rate constant; only an upper limit for this rate constant could be derived because only an upper limit to $E°$ for the $N_2H_4^+/N_2H_4$ couple could be assigned. Dennis et al. (132) examined the reactions of N_2H_4 with a series of inert oxidants, including $[IrCl_6]^{2-}$, and derived similar conclusions. According to the G2 ab initio calculations of Pople and Curtiss (173) the N—N bond length contracts by 0.12 Å as N_2H_4 is oxidized to $N_2H_4^+$. This significant structural change is in qualitative agreement with the low derived self-exchange rate constant. It is interesting to compare these results with the self-exchange rate constant ($3 \times 10^{-3}\ M^{-1}\ s^{-1}$) determined for the analogous tetraisopropylhydrazine system in acetonitrile (192).

The central question raised by the above discussions is Why do labile one-electron oxidants yield NH_3 while inert one-electron oxidants do not? The kinetic inhibition by Fe^{2+} and Fe^{II}edta in the reactions of their corresponding oxidized forms is good evidence for N_2H_3 as an intermediate for these labile oxidants. The consistent results from a Marcus-type analysis of the reactions of the inert oxidants are good evidence that N_2H_3 is an intermediate in these reactions also. Since NH_3 is considered to derive from the dimerization of N_2H_3, the inert oxidants must be quite efficient in removing N_2H_3 from the system by further oxidation to N_2H_2. The oxidation of N_2H_3 by the labile oxidants must not be highly competitive with dimerization.

$$N_2H_4 \xrightarrow{-1e^-} N_2H_4^+ \begin{cases} \xrightarrow[\text{dimerization}]{\text{ox(labile)}} \frac{1}{2} N_4H_6 \longrightarrow \frac{1}{2} N_2 + NH_3 \\ \xrightarrow[\text{oxidation}]{\text{ox(inert and labile)}} N_2H_2 \longrightarrow \frac{1}{2} N_2 + \frac{1}{2} N_2H_4 \end{cases}$$

This situation could arise if the inert oxidants react rapidly with N_2H_3 through an outer-sphere mechanism, while the labile oxidants react more slowly because the rates are limited by substitution at the metal center. Pulse-radiolysis studies on the oxidation of $N_2H_4^+$ (or N_2H_3) to N_2H_2 are lacking but could provide a meaningful test of this proposal. Note that arguments such as these must be considered tentative, since most of the data upon which they are based come from experiments in which oxygen was not excluded. It is not possible at present to predict how the rapid reaction of O_2 with N_2H_3 affects the yields of NH_3.

D. Two-Electron Mechanisms

There has been modest progress on the front of establishing the mechanisms of two-electron oxidations. The reactions of I_2, $H_2NOSO_3^-$, HNO_2, and *p*-chlorobenzaldehyde are presently interpreted in terms of hydrazine acting as a nucleophile either through a displacement or addition process, and with subsequent elimination reactions leading to diazene or equivalent reactive intermediates, as in

$$RX + :N_2H_4 \longrightarrow R{-}NHNH_2 + HX \longrightarrow RH + N_2H_2 \qquad (109)$$

General acid–base catalysis appears to be a diagnostic feature of these reactions. Extensive studies have also been performed on two-electron transition metal oxidants, including V^V, Cr^{VI}, Mo^{VI}, and Fe^{VI}. It is disappointing to note that for none of these latter oxidants has a mechanism been proposed that gives enough detail to resolve the specific atom-transfer processes that must be occurring.

E. Potentiometric Titrations

Potentiometric titrations have been used to assess the mechanism of oxidation of N_2H_4 by Mn^{3+}, V^{V}, Mo^{VI}, Ce^{IV}, $[Fe(CN)_6]^{3-}$, and IO_4^- (72, 110, 128, 155). In each of these systems, the results obtained are in substantial disagreement with other reports or with reasonable expectations. It appears that potentiometric titrations do not give meaningful results in hydrazine reactions, although the reasons for its failure are not clear. One possibility is that the method is unreliable because the platinum electrodes generally used catalyze the decomposition of N_2H_4 and the H_2 so generated leads to spurious potentials. This effect has been discussed by Cuy and Bray (120). Reasonable results have been attained in the potentiometric titration of Cr^{VI} with N_2H_4, but in this case large concentrations of Cu^{2+} as a catalyst are required (193).

F. Summary

The general mechanistic scheme presented in 1957 by Higginson for oxidations of hydrazine has held up remarkably well. One should, however, be careful not to oversimplify the picture by making the suggestion that all one-electron oxidants give a full yield of NH_3. In fact, there are virtually no examples where a full yield is obtained. Moreover, there are many one-electron oxidants that yield essentially no NH_3 and, hence, cannot be distinguished from two-electron oxidants on the basis of stoichiometry. Less successful has been Brown and Higginson's (10) later (1972) assertion that there is no simple correlation between stoichiometry and oxidant lability. It now appears to be the rule that NH_3 is produced in significant yields only by labile oxidants.

At this point, it is useful to classify oxidants into three groups: (1) inert one-electron oxidants, (2) labile one-electron oxidants, and (3) two-electron oxidants. Inert one-electron oxidants give a full yield of N_2, and the rate constants agree with the cross-relationship of Marcus theory. The insignificant yield of NH_3 with inert oxidants can be traced to the thermodynamic ease of oxidation of N_2H_3, which leads to oxidation of this radical in preference to its dimerization. Labile one-electron oxidants may yield some NH_3, but no simple correlation has emerged for the rate constants. The requirement of a labile oxidant for a significant NH_3 yield implies an inner-sphere oxidation mechanism. This mechanism could lead to rate-limiting substitution in the oxidation of the N_2H_3 radicals that provides an opportunity for dimerization. Two-electron oxidants normally yield only N_2; while no overall kinetic scheme has yet emerged, there are signs of some generality for a two-step sequence of electrophilic attack at hydrazine followed by elimination to yield N_2H_2 as an intermediate.

Recent advances in the kinetic and thermochemical properties of the important intermediates N_2H_3, $N_2H_4^+$, and N_2H_2 should help in future studies of the reaction mechanisms.

ACKNOWLEDGMENTS

This research was supported by a grant from the NSF Fellow. During the preparation of much of this chapter, DMS was a Sloan Research Fellow. I would also like to acknowledge the helpful remarks of Professors Dale Margerum (Purdue University), Gil Haight (University of Washington), Tom Webb (Auburn University), and Geoff Stedman (University of Wales Swansea).

ABBREVIATIONS

acac	Acetylacetonate
bpy	2,2′-Bipyridine
CAT	Chloramine-T
Cydta	Cyclohexanediaminetetraacetate (ligand)
dmg	Dimethylglyoxime
dtpa	Diethylenetriaminepentaacetate (ligand)
edta	Ethylenediaminetetraacetate acid (ligand)
ESR	Electron spin resonance
LFER	Linear free energy relationship
NBS	National Bureau of Standards
NCB	*N*-Chlorobenzamide
NHE	Normal hydrogen electrode
nta	Nitrilotriacetate (ligand)
OAc	Acetate
ox	Oxalate
phen	1,10-Phenanthroline

REFERENCES

1. T. Curtius, *Berichte*, *20*, 1632 (1887).
2. L. F. Audrieth and B. A. Ogg, *The Chemistry of Hydrazine*, Wiley, New York, 1951, p. 244.
3. E. W. Schmidt, *Hydrazine and its Derivatives*, Wiley, New York, 1984, p. 1059.

4. W. C. E. Higginson, *Chem. Soc. Sp. Pub. No. 10*, 95 (1957).
5. T. J. Kemp, in *Comprehensive Chemical Kinetics*, C. H. Bamford and C. F. H. Tipper, Eds., Elsevier, New York, 1972, Vol. 7, pp. 274–509.
6. A. G. Sykes, *Kinetics of Inorganic Reactions*; Pergamon, New York, 1966, pp. 195–197.
7. A. G. Sykes, *Adv. Inorg. Chem.*, *10*, 153 (1967).
8. F. Bottomley, *Q. Rev.*, *24*, 617 (1970).
9. G. Stedman, *Adv. Inorg. Chem. Radiochem.*, *22*, 113 (1979).
10. A. Brown and W. C. E. Higginson, *J. Chem. Soc. Dalton Trans.*, 166 (1972).
11. J. Veprek-Siska, D. M. Wagnerova, and K. Eckschlager, *Coll. Czech. Chem. Commun.*, *31*, 1248 (1966).
12. (a) P. K. Lim and B. S. Fagg, *J. Phys. Chem.*, *88*, 1136 (1984).
12. (b) I. A. Funai, M. A. Blesa, and J. A. Olabe, *Polyhedron*, *8*, 419 (1989).
13. J. Gadooni and U. Onken, *Ber. Bunsenges. Phys. Chem.*, *90*, 154 (1986).
14. D. M. Wagnerova, E. Schwertnerova, and J. Veprek-Siska, *Coll. Czech. Chem. Commun.*, *38*, 756 (1973).
15. A. P. Hong and T.-C. Chen, *Environ. Sci. Technol.*, *27*, 2404 (1993).
16. A. P. Bhargava, R. Swaroop, and Y. K. Gupta, *J. Chem. Soc. (A)*, 2183 (1970).
17. S. Acharya, G. Neogi, R. K. Panda, and D. Ramaswamy, *Bull. Chem. Soc. Jpn.*, *56*, 2821 (1983).
18. S. Acharya, G. Neogi, R. K. Panda, and D. Ramaswamy, *J. Chem. Soc. Dalton Trans.*, 1477 (1984).
19. H. Erlenmeyer, C. Flierl, and H. Sigel, *Chimia*, *22*, 433 (1968).
20. H. Erlenmeyer, C. Flierl, and H. Sigel, *J. Am. Chem. Soc.*, *91*, 1065 (1969).
21. C. R. Wellman, J. R. Ward, and L. P. Kuhn, *J. Am. Chem. Soc.*, *98*, 1683 (1976).
22. Y. Zhong and P. K. Lim, *J. Am. Chem. Soc.*, *111*, 8398 (1989).
23. P. K. Lim and Y. Zhong, *J. Am. Chem. Soc.*, *111*, 8404 (1989).
24. A. K. Gupta, K. S. Gupta, and Y. K. Gupta, *Inorg. Chem.*, *24*, 3670 (1985).
25. R. Banerjee, A. Das, and S. Dasgupta, *J. Chem. Soc. Dalton Trans.*, 2271 (1990).
26. T. Huang and J. T. Spence, *J. Phys. Chem.*, *72*, 4573 (1968).
27. S. Sattar and K. Kustin, *Inorg. Chem.*, *30*, 1668 (1991).
28. R. J. Hodges and W. F. Pickering, *Aust. J. Chem.*, *19*, 981 (1966).
29. E. Baumgartner, M. A. Blesa, R. Larotonda, and A. J. G. Maroto, *J. Chem. Soc. Faraday Trans. 1*, *81*, 1113 (1995).
30. C. Gerez and M. Fontecave, *Biochemistry*, *31*, 780 (1992).
31. J.-Y. Han, J. C. Swarts, and A. G. Sykes, *Inorg. Chem.*, *35*, 4629 (1996).
32. U. Kebekus, J. Messinger, and G. Renger, *Biochemistry*, *34*, 6175 (1995).
33. S. Karp and L. Meites, *J. Am. Chem. Soc.*, *84*, 906 (1962).
34. G. Kokkinidis and P. D. Jannakoudakis, *J. Electroanal. Chem.*, *130*, 153 (1981).

35. T. Kodera, M. Honda, and H. Kita, *Electrochimica Acta*, *30*, 669 (1985).
36. C. Lin and A. B. Bocarsly, *J. Electroanal. Chem.*, *300*, 325 (1991).
37. M. D. Garcia, M. L. Marcos, and J. G. Velasco, *Electroanalysis*, *8*, 267 (1996).
38. U. Scharf and E. W. Grabner, *Electrochim. Acta*, *41*, 233 (1996).
39. J. Zhang, Y.-H. Tse, W. J. Pietro, and A. B. P. Lever, *J. Electroanal. Chem.*, *406*, 203 (1996).
40. R. M. Smith and A. E. Martell, *Critical Stability Constants*; Plenum, New York, 1976, Vol. 4, p. 43.
41. R. M. Smith, A. E. Martell, and R. J. Motekaitis, *NIST Critically Selected Stability Constants of Metal Complexes Database*, *2.0*, Gaithersburg, MD, 1995.
42. S. P. Mezyk, M. Tateishi, R. MacFarlane, and D. M. Bartels, *J. Chem. Soc. Faraday Trans.*, *92*, 2541 (1996).
43. F. H. Pollard and G. Nickless, *J. Chromatogr.*, *4*, 196 (1960).
44. W. C. E. Higginson and P. Wright, *J. Chem. Soc.*, 1551 (1955).
45. G. Davies and K. Kustin, *J. Phys. Chem.*, *73*, 2248 (1969).
46. S. L. Bruhn, A. Bakac, and J. H. Espenson, *Inorg. Chem.*, *25*, 535 (1986).
47. G. Bengtsson, *Acta Chem. Scand.*, *25*, 2989 (1971).
48. A. K. Basak and D. Banerjea, *Z. Anorg. Allg. Chem.*, *425*, 277 (1976).
49. A. K. Basak and D. Banerjea, *Ind. J. Chem.*, *17A*, 250 (1979).
50. D. Banerjea and I. P. Singh, *Z. Anorg. Allg. Chem.*, *349*, 213 (1967).
51. A. J. Bard, R. Parsons, and J. Jordan, Eds., *Standard Potentials in Aqueous Solution*, Marcel-Dekker, New York, 1985, p. 133.
52. D. D. Wagman, W. H. Evans, V. B. Parker, R. H. Schumm, I. Halow, S. M. Bailey, K. L. Churney, and R. L. Nuttall, *J. Phys. Chem. Ref. Data*, *11*, Suppl. No. 2 (1982).
53. M. L. McKee, M. E. Squillacote, and D. M. Stanbury, *J. Phys. Chem.*, *96*, 3266 (1992).
54. M. L. McKee, *J. Phys. Chem.*, *97*, 13608 (1993).
55. A. A. Grinberg, E. A. Shashukov, and N. N. Popova, *Russ. J. Inorg. Chem.*, *13*, 1055 (1968).
56. A. W. Browne and F. F. Shetterly, *J. Am. Chem. Soc.*, *30*, 54 (1908).
57. Q. W. Choi and B. B. Park, *J. Korean Chem. Soc.*, *19*, 403 (1975).
58. H. R. Tang and D. M. Stanbury, *Inorg. Chem.*, *33*, 1388 (1994).
59. Q. W. Choi and K. H. Chung, *Bull. Korean Chem. Soc.*, *7*, 462 (1986).
60. I. Janovsky, *J. Radioanal. Nucl. Chem. Lett.*, *103*, 31 (1986).
61. S. E. King, J. N. Cooper, and R. D. Crawford, *Inorg. Chem.*, *17*, 3306 (1978).
62. P. S. Radakrishnamurty, N. K. Rath, and R. K. Panda, *J. Chem. Soc. Dalton Trans.*, 1189 (1986).
63. R. Gopalan and J. Karunakaran, *Indian J. Chem.*, *19A*, 162 (1980).
64. S. M. Sultan, I. Z. Al-Zamil, A. M. Al-Hajjaji, S. A. Al-Tamrah, and A. M. Aziz Al Rahman, *J. Chem. Soc. Pak.*, *7*, 93 (1985).

65. T. S. Rao and S. P. Dalvi, *Proc. Indian Natl. Sci. Acad., Part A*, *56*, 153 (1990).
66. I. Janovsky, *J. Radioanal. Nucl. Chem. Lett.*, *128*, 433 (1988).
67. R. M. Liu and D. W. Margerum, *211th ACS National Meeting*, New Orleans, LA, 1996, INOR 191.
68. H. Rottendorf and S. Sternhell, *Aust. J. Chem.*, *16*, 647 (1963).
69. G. Rábai and M. T. Beck, *J. Chem. Soc. Dalton Trans.*, 919 (1984).
70. R. A. Hasty, *Mikrochimica Acta*, 925 (1973).
71. P. S. Verma and K. C. Grover, *Aust. J. Chem.*, *20*, 1533 (1967).
72. A. Kumar and R. K. Prasad, *J. Indian Chem. Soc.*, *51*, 366 (1974).
73. A. W. Browne and F. F. Shetterly, *J. Am. Chem. Soc.*, *31*, 221 (1909).
74. A. J. Bard, R. Parsons, and J. Jordan, Eds., *Standard Potentials in Aqueous Solution*, Marcel-Dekker, New York, 1985, p. 89.
75. J. M. Hoffman and R. H. Schlessinger, *Chem. Commun.*, 1245 (1971).
76. M. Sharma, P. V. S. Madnawat, D. S. N. Prasad, and K. S. Gupta, *Indian J. Chem.*, *32A*, 251 (1993).
77. W. E. Steinmetz, D. H. Robison, and M. N. Ackermann, *Inorg. Chem.*, *14*, 421 (1975).
78. E. Schmitz, R. Ohme, and G. Kozakiewicz, *Z. Anorg. Allg. Chem.*, *339*, 44 (1965).
79. J. R. Perrott, G. Stedman, and N. Uysal, *J. Chem. Soc. Dalton Trans.*, 2058 (1976).
80. R. K. Prasad and A. Kumar, *J. Indian Chem. Soc.*, *50*, 572 (1973).
81. (a) K. G. Phelan and G. Stedman, *J. Chem. Soc. Dalton Trans.*, 1603 (1982). (b) R. J. Gowland, K. R. Howes, and G. Stedman, *J. Chem. Soc. Dalton Trans.*, 797 (1992).
82. A. M. M. Doherty, K. R. Howes, G. Stedman, and M. Q. Naji, *J. Chem. Soc. Dalton Trans.*, 3103 (1995).
83. (a) D. G. Karraker, *Inorg. Chem.*, *24*, 4470 (1985). (b) V. S. Koltunov, V. A. Nikol'skii, and Yu. P. Agureev, *Kinet. Katal.*, *3*, 764 (1962).
84. E. W. Schmidt, *Hydrazine and Its Derivatives*, Wiley, New York, 1984, pp. 36–45.
85. H. Delalu and R. Cohen-Adad, *J. Chim. Phys.*, *76*, 465 (1979).
86. R. Risk-Ouaini, C. Tremblay-Goutaudier, and M. T. Cohen-Adad, *J. Chim. Phys.*, *90*, 579 (1993).
87. S. Kapoor and Y. K. Gupta, *J. Inorg. Nucl. Chem.*, *39*, 1019 (1977).
88. T. P. A. Dhas, D. K. Mishra, R. K. Mittal, and Y. K. Gupta, *Int. J. Chem. Kinet.*, *23*, 203 (1991).
89. P. Keshwani, A. K. Gupta, and Y. K. Gupta, *J. Indian Chem. Soc.*, *62*, 878 (1985).
90. B. S. Rawat and M. C. Agrawal, *Indian. J. Chem.*, *15A*, 713 (1977).
91. S. Jha, L. Bhatt, P. D. Sharma, and Y. K. Gupta, *Ind. J. Chem.*, *24A*, 531 (1985).

92. E. W. Schmidt, *Hydrazine and Its Derivatives*, Wiley, New York, 1984, p. 313.
93. E. C. Gilbert, *J. Am. Chem. Soc.*, *51*, 3394 (1929).
94. E. G. Sander and W. P. Jencks, *J. Am. Chem. Soc.*, *90*, 6154 (1968).
95. J. M. Sayer, M. Peskin, and W. P. Jencks, *J. Am. Chem. Soc.*, *95*, 4277 (1973).
96. C. D. Ritchie, *J. Am. Chem. Soc.*, *106*, 7187 (1984).
97. S. Ojanpera, A. Parole, and S. G. Cohen, *J. Am. Chem. Soc.*, *96*, 7379 (1974).
98. S. Inbar and S. G. Cohen, *J. Am. Chem. Soc.*, *100*, 4490 (1978).
99. S. Inbar, H. Linschitz, and S. G. Cohen, *J. Am. Chem. Soc.*, *103*, 7323 (1981).
100. B. M. Thakuria and Y. K. Gupta, *J. Chem. Soc. Dalton Trans.*, 2541 (1975).
101. K. S. Gupta, *J. Inorg. Nucl. Chem.*, *39*, 2093 (1977).
102. K. S. Gupta and Y. K. Gupta, *Indian J. Chem.*, *11*, 1285 (1973).
103. V. V. Vekshin, N. I. Pechurova, and V. I. Spitsyn, *Russ. J. Inorg. Chem.*, *21*, 530 (1976).
104. V. S. Srinivasan and N. Venkatasubramanian, *Indian J. Chem.*, *15A*, 115 (1977).
105. E. Hayon and M. Simic, *J. Am. Chem. Soc.*, *94*, 42 (1972).
106. G. V. Buxton and C. R. Stuart, *J. Chem. Soc. Faraday Trans.*, *92*, 1519 (1996).
107. B. G. Ershov and T. L. Mikhailova, *Bull. Acad. Sci. USSR; Div. Chem. Sci.*, *40*, 288 (1991).
108. K. Hasegawa and P. Neta, *J. Phys. Chem.*, *82*, 854 (1978).
109. P. Maruthamuthu and P. Neta, *J. Phys. Chem.*, *82*, 710 (1978).
110. R. K. Prasad and A. Kumar, *J. Indian Chem. Soc.*, *49*, 819 (1972).
111. M. S. Stuklova, N. I. Pechurova, and V. I. Spitsyn, *Bull. Acad. Sci. USSR, Ser. Chem.*, *28*, 1791 (1979).
112. V. S. Srinivasan and E. S. Gould, *Inorg. Chem.*, *20*, 3176 (1981).
113. R. N. Bose, B. Fonkeng, G. Barr-David, R. P. Farrell, R. J. Judd, P. A. Lay, and D. F. Sangster, *J. Am. Chem. Soc.*, *118*, 7139 (1996).
114. M. T. Beck and D. A. Durham, *J. Inorg. Nucl. Chem.*, *32*, 1971 (1970).
115. V. M. S. Ramanujam and N. Venkatasubramanian, *Indian J. Chem.*, *8*, 948 (1970).
116. G. P. Haight, T. J. Huang, and B. Z. Shakhashiri, *J. Inorg. Nucl. Chem.*, *33*, 2169 (1971).
117. V. M. S. Ramanujam, S. Sundaram, and N. Venkatasubramanian, *Inorg. Chim. Acta*, *13*, 133 (1975).
118. S. Senent, L. Ferrari, and A. Arranz, *Rev. Rou. Chim.*, *23*, 179 (1978).
119. R. K. Prasad and A. Kumar, *J. Indian Chem.*, *50*, 612 (1973).
120. E. J. Cuy and W. C. Bray, *J. Am. Chem. Soc.*, *46*, 1786 (1924).
121. G. P. Haight, T. J. Huang, and H. Platt, *J. Am. Chem. Soc.*, *96*, 3137 (1974).
122. K. K. S. Gupta, S. S. Gupta, and H. R. Chatterjee, *J. Inorg. Nucl. Chem.*, *38*, 549 (1976).
123. M. T. Beck and D. A. Durham, *J. Inorg. Nucl. Chem.*, *33*, 461 (1971).

124. M. T. Beck, D. Durham, and G. Rábai, *Inorg. Chim. Acta*, *18*, L17 (1976).
125. S. L. Scott, A. Bakac, and J. H. Espenson, *J. Am. Chem. Soc.*, *114*, 4205 (1992).
126. G. V. Buxton and F. Djouider, *J. Chem. Soc. Faraday Trans.*, *92*, 4173 (1996).
127. T. Huang and J. T. Spence, *J. Phys. Chem.*, *72*, 4198 (1968).
128. H. C. Mishra and R. N. P. Sinha, *Indian J. Chem.*, *9*, 1300 (1971).
129. M. F. Tikhonov and V. S. Koltunov, *Russ. J. Phys. Chem.*, *53*, 1143 (1979).
130. M. D. Johnson and B. J. Hornstein, *Inorg. Chim. Acta, 225*, 145 (1994).
131. J. G. Leipoldt, L. D. C. Bok, A. J. van Wyk, and C. R. Dennis, *React. Kinet. Catal. Lett.*, *6*, 467 (1977).
132. C. R. Dennis, A. J. Van Wyk, S. S. Basson, and J. G. Leipoldt, *Inorg. Chem.*, *26*, 270 (1987).
133. T. G. Costner and N. Ganapathisubramanian, *Inorg. Chem.*, *28*, 3620 (1989).
134. E. J. Meehan, I. M. Kolthoff, and K. Mitsuhashi, *Suomen Kem.*, *B 42*, 159 (1969).
135. H. Minato, E. J. Meehan, I. M. Kolthoff, and C. Auerbach, *J. Am. Chem. Soc.*, *81*, 6168 (1959).
136. J. Veprek-Siska, *Disc. Faraday Soc.*, *46*, 184 (1968).
137. K. Madlo, A. Hasnedl, and J. Veprek-Siska, *Coll. Czech. Chem. Commun.*, *41*, 7 (1976).
138. V. K. Jindal, M. C. Agrawal, and S. P. Mushran, *Z. Naturforsch.*, *25b*, 188 (1970).
139. S. Hünig, H.-R. Müller, and W. Thier, *Tetrahedron Lett.*, 353 (1961).
140. S. Hünig, and H. R. Müller, *Angew. Chem. Int. Ed. Engl.*, *1*, 213 (1962).
141. S. Hünig, H. R. Müller, and W. Thier, *Angew. Chem. Int. Ed. Engl.*, *4*, 271 (1965).
142. F. E. Lytle and D. M. Hercules, *Photochem. Photobiol.*, *13*, 123 (1971).
143. K. K. Sengupta and P. K. Sen, *Inorg. Chem.*, *18*, 979 (1979).
144. D. F. C. Morris and T. J. Ritter, *J. Chem. Soc. Dalton Trans.*, 216 (1980).
145. D. M. Stanbury, *Inorg. Chem.*, *23*, 2879 (1984).
146. K. K. S. Gupta, P. K. Sen, and S. S. Gupta, *Inorg. Chem.*, *16*, 1396 (1977).
147. A. Bakac, T.-J. Won, and J. H. Espenson, *Inorg. Chem.*, *35*, 2171 (1996).
148. A. Bakac, personal communication.
149. J. H. Espenson, A. Bakac, and J. Janni, *J. Am. Chem. Soc.*, *116*, 3436 (1994).
150. C. Kang and F. C. Anson, *Inorg. Chem.*, *33*, 2624 (1994).
151. W.-D. Wang, A. Bakac, and J. H. Espenson, *Inorg. Chem.*, *32*, 5034 (1993).
152. H. R. Falle, *Can. J. Chem.*, *46*, 1703 (1968).
153. J. Q. Adams and J. R. Thomas, *J. Chem. Phys.*, *39*, 1904 (1963).
154. J. I. Morrow and G. W. Sheeres, *Inorg. Chem.*, *11*, 2605 (1972).
155. H. C. Mishra and L. N. Choubey, *J. Indian Chem. Soc.*, *61*, 98 (1979).
156. P. Arselli and E. Mentasti, *J. Chem. Soc. Dalton Trans.*, 689 (1983).
157. S. S. Gupta and Y. K. Gupta, *J. Chem. Soc.*, *Dalton Trans.*, 547 (1983).
158. D.-P. Cheng and S.-J. Xia, *Chinese J. Chem.*, *14*, 48 (1996)

159. M. S. Frank, A. K. Ramaiah, and P. V. K. Rao, *Indian J. Chem., 18A*, 369 (1979).
160. K. Jijee and M. Santappa, *Proc. Ind. Acad. Sci., 69A*, 117 (1969).
161. G. Hargreaves and L. H. Sutcliffe, *Trans. Faraday Soc., 51*, 786 (1955).
162. S. P. Tanner, *Inorg. Chem., 17*, 600 (1978).
163. M. A. Thacker and W. C. E. Higginson, *J. Chem. Soc. Dalton Trans.*, 704 (1975).
164. A. K. Banerjee, A. K. Basak, and D. Banerjea, *Ind. J. Chem., 18A*, 332 (1979).
165. T. R. Reddy, V. Jagannadham, and G. S. S. Murthy, *Indian J. Chem., 25A*, 1120 (1986).
166. J. Lati and D. Meyerstein, *Inorg. Chem., 11*, 2393 (1972).
167. R. E. Kirk and A. W. Browne, *J. Am. Chem. Soc., 50*, 337 (1928).
168. K. K. Sen Gupta and B. Basu, *Transition Met. Chem., 8*, 3 (1983).
169. L. H. Skibsted, *Adv. Inorg. Bioinorg. Mech., 4*, 137 (1986).
170. S. K. Mishra and Y. K. Gupta, *J. Chem. Soc. (A)*, 2918 (1970).
171. M. S. Stuklova and N. I. Pechurova, *Russ. J. Inorg. Chem., 27*, 799 (1982).
172. K. R. Butter and T. J. Kemp, *J. Chem. Soc. Dalton Trans.*, 923 (1984).
173. J. A. Pople and L. A. Curtiss, *J. Chem. Phys., 95*, 4385 (1991).
174. J. W. Sutherland, *J. Phys. Chem., 83*, 789 (1979).
175. D. H. Magers, E. A. Salter, R. J. Bartlett, C. Salter, B. A. Hess Jr., and L. J. Schaad, *J. Am. Chem. Soc., 110*, 3435 (1988).
176. G. V. Buxton and C. R. Stuart, *J. Chem. Soc., Faraday Trans., 93*, 1535 (1997).
177. C. Bull, G. J. McClune, and J. A. Fee, *J. Am. Chem. Soc., 105*, 5290 (1983).
178. D. M. Stanbury, *Adv. Inorg. Chem., 33*, 69 (1989).
179. R. G. Pearson, *J. Am. Chem. Soc., 108*, 6109 (1986).
180. C. D. Ritchie, in *Nucleophilicity*, J. M. Harris and S. P. McManus, Eds., ACS, Washington, DC, 1987; Advances in Chemistry Series, Vol. 215, pp. 169–179.
181. B. Ruscic and J. Berkowitz, *J. Chem. Phys., 95*, 4378 (1991).
182. M. L. McKee, and D. M. Stanbury, *J. Am. Chem. Soc., 114*, 3214 (1992).
183. C. E. Miller, *J. Chem. Educ., 42*, 254 (1965).
184. R. A. Back, *Rev. Chem. Intermed., 5*, 293 (1984).
185. D. J. Pasto and R. T. Taylor, *Org. React. 40*, 91 (1991).
186. D. M. Stanbury, *Inorg. Chem., 30*, 1293 (1991).
187. N. Bauer, *J. Phys. Chem., 64*, 833 (1969).
188. G. Ritter, G. Häfelinger, E. Lüddecke, and H. Rau, *J. Am. Chem. Soc., 111*, 4627 (1989).
189. N. Wiberg, H.-W. Häring, and S. K. Vasisht, *Z. Naturforsch., 34B*, 356 (1979).
190. M. Veith and G. Schlemmer, *Z. Anorg. Allg. Chem., 494*, 7 (1982).
191. A. P. Sylwester and P. B. Dervan, *J. Am. Chem. Soc., 106*, 4648 (1984).
192. S. F. Nelsen, R. F. Ismagilov, L.-J. Chen, J. L. Brandt, X. Chen, and J. R. Pladziewicz, *J. Am. Chem. Soc., 118*, 1555 (1996).
193. S. Syamsunder and T. K. S. Murthy, *Indian J. Chem., 11*, 669 (1973).

Metal Ion Reconstituted Hybrid Hemoglobins

B. VENKATESH and P. T. MANOHARAN

Department of Chemistry
Regional Sophisticated Instrumentation Centre
Indian Institute of Technology
Madras 600 036, India

J. M. RIFKIND

Molecular Dynamics Section
Laboratory of Cellular and Institute on Aging
Molecular Biology, National Institute of Health,
Baltimore, MD

CONTENTS

Progress in Inorganic Chemistry, Vol. 47, Edited by Kenneth D. Karlin.
ISBN 0-471-24039-7 © 1998 John Wiley & Sons, Inc.

I. INTRODUCTION

Hemoglobins are essential to the life of all vertebrates. All hemoglobins carry the same prosthetic heme group, iron(II)protoporphyrin IX, associated with the polypeptide chains (2α and 2β), which contain between 141 and 153 amino acid residues. The three-dimensional structure of hemoglobin (Fig. 1), was first established by Perutz (1). The literature covering all aspects of this important protein is exhaustive. The books written by Dickerson and Geis (2), Antonini and Brunori (3), and Cantor and Schimmel (4) provide extensive details on the structural and functional aspects of hemoglobin. Several excellent reviews by Baldwin and Chothia (5), Perutz (1), and Rifkind (6, 7) have appeared in the past on the physical aspects and structure–function relationships in hemoglobin. Even though the hemoglobin literature extends back to the early twentieth century, a new expanded literature is continually appearing. Academic Press in the series *Methods in Enzymology* came out with an entire volume on hemoglobin in (1981) (Vol. 76) and has just come out with an updated two volumes (Vols. 231 and 232) on hemoglobin in 1994. Vol. 231 deals with biochemical and analytical methods and Vol. 232 deals with biophysical methods.

In spite of considerable efforts devoted to hemoglobin research, particularly during the past two decades, the detailed molecular basis for cooperative oxygenation of hemoglobin is not fully understood. The excellent work by Perutz (8), in his proposed stereochemical mechanism, is based on the X-ray structures of crystalline deoxyhemoglobin and oxyhemoglobin. While this model has been the basis for much of the research on hemoglobin, it does not account for perturbations in the solution state nor for conformational changes associated with partially oxygenated intermediates.

It has long been recognized that a full understanding of hemoglobin oxygen-

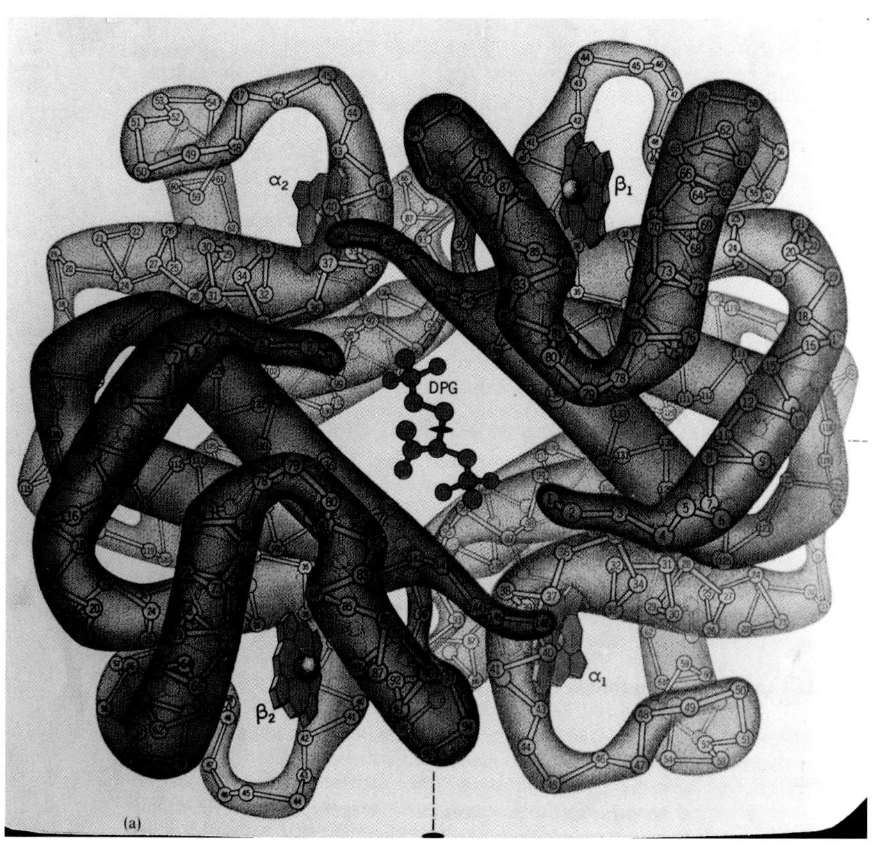

Figure 1. Pictorial representation of deoxy hemoglobin consisting of 2α and 2β poly peptide chains, with four hemes one on each chain. Here DPG binds in the centre cavity of hemoglobin molecule. (Reproduced with permission from reference 18)

ation requires knowledge of the structures of the partially oxygenated intermediates. However, because of the high cooperativity, it is extremely difficult to obtain this information from partially oxygenated hemoglobin samples. It is the resolution of this problem that has been the impetus for studies of hybrid hemoglobins. It was hoped that, by preparing hemoglobins with very different ligand reactivities for some of the chains, partially liganded stable intermediates could be prepared and studied, which would uncouple the individual steps for the cooperative binding of four ligands to hemoglobin. Now, it becomes necessary to define our use of hybrid. The ($\alpha_2\beta_2$) tetrameric form of hemoglobin, termed as HbA_0 of the mammalian protein, is said to be normal. A hybrid is defined as the one with modification introduced (a) either at the heme site or at the amino acid sites of the globin chain in a manner not normal to the actual properties of oxygen uptake by HbA_0, (b) by change of the oxidation state of iron, (c) by change of the metal ion, (d) by changing the prosthetic groups of porphyrin, (e) by attachment of groups, (f) by change of chains creating mutants, (g) by labels, or (h) by small molecules such as NO, CO, spin labels, and so on, at different places of the hemoglobin for monitoring purposes. In other words, any modification to some of the heme sites are termed hybrids; in addition, even if the heme site is not modified but the globins are modified, either by substituting the entire globin with that of another species or by the globins being chemically modified, the resulting hemoglobin will be called as hybrid hemoglobin. The classification given in Section II will provide a clear picture for defining hybrids.

The preparation of hybrid hemoglobins requires the removal of some of the hemes from hemoglobin and reconstitution with different hemes. The reconstitution work was initiated in 1926 by Hill and Holden (9) by preparing the first reconstituted form of oxyhemoglobin using proto-, meso-, deutero-, and hematohemes. They studied globin complexes involving copper, nickel, cobalt, zinc, and metal free porphyrins.

Starting in the 1960s, metal substitution in hemoglobin was again initiated and modern tools for probing the protein structure in solution were used to study these hybrids. In these studies, both thermodynamic and kinetic aspects of electron transfer and oxygenation were studied. Structural characterization was made using various spectroscopic methods including NMR, EPR, ENDOR, ESEEM, resonance Raman, fluorescence, UV–vis, IR, CD, and XRD. Much of the work on hybrid hemoglobins has been reviewed by Hoffman (10), Parkhurst (11), Scholler et al. (12), Cassoly (13, 14), Ikeda-Saito et al. (15), Bunn (16), Ascoli et al. (17), and Perrella and Rossi-Bernardi (18, 19). Most of these reviews, however, are almost 15 years old with the more recent ones primarily emphasizing methods of preparation.

Recently, extensive research has been directed toward hybrid hemoglobins in order to understand the control mechanism for ligand affinity in hemoglobin by analyzing the tertiary and quaternary structures and the changes induced by ligation. It has even been possible to obtain high-resolution X-ray data on several hybrid hemoglobins (20). An important advance in recent years is the ability to prepare many of the hemoglobin intermediates that could not previously be prepared because of the rapid dissociation of the tetramer to the dimer. This preparation is now accomplished by using cross-linked hemoglobins. Another important advance is the development of methods for gel electrophoresis at subzero temperatures. It has thus been possible to identify the relative concentration of hemoglobin intermediates produced during the course of ligand binding. By combining these two advances, results are beginning to focus on the actual intermediates associated with ligand binding. These studies should provide for significant progress in explaining the detailed mechanism for hemoglobin oxygenation.

This chapter provides an updated critical review of the literature on mammalian hybrid hemoglobins. We have reviewed the methods for preparation and characterization of both symmetric and asymmetric hybrids as well as those for the preparation of mono- and trisubstituted hemoglobins. Then, we have extensively reviewed the spectroscopic and other data, which measure functional properties of hemoglobin. In addition to highlighting the achievements in this field, we indicate future possibilities and potentials for the use of hybrids.

In the past, many specific mechanisms were proposed to account for cooperativity based on thermodynamic and kinetic experiments and crystallographic data (21–23). Adair (21) proposed a tetrameric model and considered a single conformation for a given ligation state, that is, the molecule HbX_4 is built up or broken down in stages.

$$\mathrm{Hb} \underset{k_1}{\overset{k_1'}{\rightleftharpoons}} \mathrm{HbX} \underset{k_2}{\overset{k_2'}{\rightleftharpoons}} \mathrm{HbX_2} \underset{k_3}{\overset{k_3'}{\rightleftharpoons}} \mathrm{HbX_3} \underset{k_4}{\overset{k_4'}{\rightleftharpoons}} \mathrm{HbX_4}$$

where X = ligand, k_i' and k_i (i = 1, 2, 3, 4) are association and dissociation rate constants, respectively with the ratios equal to the equilibrium constants. Without postulating any scheme for propagation of conformational changes, cooperativity is indicated by an increase in the k_i' values as oxygen is bound.

The Monod, Wyman, Changex (MCW) model (concerted model) proposed by Monod et al. (22) can be taken as two noncooperative Adair schemes coupled by conformational equilibria, that is, there is an equilibrium between the T and R conformations and the transition between them is a concerted process. According to Monod et al. (22), the interaction energy might be equally well interpreted as a stabilization energy between the unligated or ligated hemes. This model rules out the presence of any intermediate conformations for partially oxygenated hemoglobins.

The sequential Koshland, Nemethy, Filmer (KNF) model proposed by Koshland et al. (23) says that there is no equilibrium between T and R conformations in the absence of the ligand. The transition from the T to the R conformation is induced by the binding of the ligand. The first oxygen enters the molecule with more difficulty than the second one owing to the necessity of breaking up the preexisting "tight" partnership between the hemes; the second oxygen enters more easily since the interaction between hemes is already weakened by binding the first oxygen. This process continues as each oxygen is bound. This model predicts a crucial role for partially ligated hybrids.

When Perutz (24) examined the atomic models of oxy- and deoxyhemoglobin through XRD, he found a change in the quaternary structure. This change was due to the construction of the heme group such that it amplifies a small change in the atomic radius of the iron atom during the transition from the high- to the low-spin state on oxygen uptake. The rupture of each salt bridge and/or hydrogen bond removes one of the constraints holding the molecule in the deoxy or T conformation and tips the equilibrium between the two alternative quaternary structure R and T conformations. Perutz model shares many features with the MCW model.

Hence, it becomes evident that the two extreme quaternary states alone cannot explain the ligand-binding properties of hemoglobin. The next problem that arises is about the nature of other intermediate conformations between the extreme states and how these resulting intermediate conformations are switched on. Symmetric hybrid hemoglobin in which only two subunits of the same type of chain can bind ligands, is likely to provide adequate information on mono-, di-, and tri-ligated hybrids may also provide direct information on the partially ligated states. Since binding ligands to one subunit affects the other subunit, these changes often do correlate with the T → R quaternary conformational changes. In recent days, cross-linked hybrids have been given their due attention since they are expected to yield additional information due to this artificial or synthetic perturbation on normal hemoglobin. Slight modification of the above models are also found in the literature (25, 26).

II. CLASSIFICATION

Hybrid hemoglobins are classified into eight types.

1. Hybrids with reconstituted mixed-metal centers.
2. Valency hybrids.
3. Partially ligated hybrids.
4. Hybrids with different prosthetic groups.
5. Spin-labeled hybrids.
6. Semihemoglobin hybrids.
7. Mixed hybrids.
8. Chemically modified hybrids.

The complete range of hybrids and the experimental methods employed to study their physicochemical properties are displayed in Table I.

The above classification includes both symmetric and asymmetric hybrids. Asymmetric hybrids are usually in the cross-linked (XL) form. The cross-linking may be $\alpha\alpha$ or $\beta\beta$ depending on the nature of cross-linking reagent used. Some of the hybrids mentioned here are in unligated conformations but the corresponding ligated conformation have also been studied through various techniques. Here "des" represents deletion of one or more amino acids from the polypeptide chain. Subscript A, C, S, and so on denote the natural polypeptide chains from HbA, HbC, HbS, respectively.

Among the metal reconstituted hybrids, only the cobaltous ion, other than Fe can reversibly bind oxygen, whereas Ru^{II} can take up CO. But Co^{II} cannot take up CO, and Ru^{II} cannot take up oxygen. Other than these two metal ions, all others take up neither oxygen nor CO. It is worth mentioning here that the Fe^{III} ion in methemoglobin can be in high- or low-spin state depending on the nature of axial coordination. Pisciotta et al. (145) found that in some Hb molecules, two out of four iron centers are oxidized (similar to valency hybrid). Also note that the oxidation of iron in HbM (Boston) is at the α chain while it is at the β chain for HbM (Milwaukee) (146). It was later confirmed through XRD studies by Perutz et al. (147) that in HbM (Milwaukee) the glutamyl residue's E11 γ-carboxylate oxygen binds to the β iron as the sixth ligand facilitating the oxidation of the β-heme. The partially liganded hybrids contain two ligands either at the α or β subunits, or one type of ligand at the α subunit and another type at β subunit. Hybrids with different prosthetic groups may contain meso- or deuteroporphyrin in the α subunit and, protoporphyrin in the β subunit or vice versa. Mixed hybrids are pairs of the type HbA–HbS, HbA–HbC, and so on.

III. PREPARATION AND CHARACTERIZATION

Two symmetric and six asymmetric intermediate species apart from the two extreme R and T states, as shown in Fig. 2, can be used to study the mechanism behind allosteric transition. The symmetric as well as asymmetric hybrid species can be written as follows:

$$\alpha_1^*\alpha_2^*\beta_1\beta_2,\ \alpha_1\alpha_2\beta_1^*\beta_2^* \qquad \text{(Symmetric)}$$

$$\alpha_1^*\alpha_2\beta_1\beta_2,\ \alpha_1\alpha_2\beta_1^*\beta_2,\ \alpha_1^*\alpha_2\beta_1^*\beta_2,\ \alpha_1^*\alpha_2\beta_1\beta_2^*,\ \alpha_1^*\alpha_2^*\beta_1\beta_2^*,$$

$$\alpha_1^*\alpha_2^*\beta_1^*\beta_2 \qquad \text{(Asymmetric)}$$

TABLE I
Hybrids and Their Mode of Studies

Hybrids	Mode of Studies or Physiochemical Properties	References
	I. Metal Reconstituted Hybrids	
$[\alpha_2(Fe)\beta_2(Mn)]$	B, C, F	166, 302, 304, 305, 319, 320
$[\alpha_2(Mn)\beta_2(Fe)]$	B, C, F	166, 302, 304, 305, 319, 320
$[\alpha_2(Fe-CO)\beta_2(Mn)]$	C, F	304, 319, 320
$[\alpha_2(Mn)\beta_2(Fe-CO)]$	C, F	304, 319, 320
$[\alpha_2(Fe-PMe)\beta_2(Mn)]$	B	166, 307
$[\alpha_2(Mn)\beta_2(Fe)]$	K	171
$[\alpha_2(Mn)\beta_2(Fe-CO)]$	F, K	171, 326
$[\alpha_2(Fe-CO)\beta_2(Mn)]$	A	78
$[\alpha_2(Fe)\beta_2(Mn)]$	F, K	117, 326
$[\alpha(Fe)\beta(Fe)][\alpha(Ni)\beta(Ni)]XL$	D	257, 258
$[\alpha(Fe)\beta(Ni)][\alpha(Ni)\beta(Fe)]XL$	D	258
$[\alpha_2(Fe)\beta_2(Ni)]$	B, D, G	133, 164, 257, 258
$[\alpha_2(Ni)\beta_2(Fe)]$	B, D, G	133, 164, 257, 258
$[\alpha_2(Ni)\beta_2(Fe-CO)]$	B, A, K	80, 164
$[\alpha_2(Fe-CO)\beta_2(Ni)]$	B, G, K	164
$[\alpha_2(Fe-O_2)\beta_2(Ni)]$	G, A	79, 135
[des-Arg(α141)$\alpha_2(Ni)\beta_2(Fe)$]	G	135
[NES-Cys(93)-des-Arg(α141) $\alpha_2(Ni)\beta_2(Fe)$]	G	135
[NES-Cys(93)$\alpha_2(Ni)\beta_2(Fe)$]	G	135
[des-Arg(α141)$\alpha_2(Fe)\beta_2(Ni)$]	D, G	135, 252
[NES-Cys(93)-des-Arg(α141) $\alpha_2(Fe)\beta_2(Ni)$]	D, G	135, 252
[NES-Cys(93)$\alpha_2(Fe)\beta_2(Ni)$]	D, G	135
$[\alpha_2(Zn)\beta_2(Fe-CO)]$	B, C, M	120, 207

TABLE I (*Continued*)

Hybrids	Mode of Studies or Physiochemical Properties	References
	I. Metal Reconstituted Hybrids (Continued)	
$[\alpha_2(Fe-CO)\beta_2(Zn)]$	B, C, M	120, 207
$[\alpha_2(Zn)\beta_2(Fe-CO)]$	B, C, K	120
$[\alpha_2(Fe-CO)\beta_2(Zn)]$	B, C, K	120
$[\alpha_2(Fe-CN)\beta_2(Fe-CO)]$	B, C, K	120
$[\alpha_2(Zn)\beta_2(Fe-CN)]$	B, C, H, K	120, 216, 218, 224
$[\alpha_2(Fe-CN)\beta_2(Zn)]$	B, C, H, K	120, 216, 218, 224
$[\alpha_2(Fe-H_2O)\beta_2(Zn)]$	H, K	120, 222–225
$[\alpha_2(Zn)\beta_2(Fe-H_2O)]$	H, K	120, 222–225
$[\alpha_2(Fe)\beta_2(Zn)]$	C	302, 306
$[\alpha_2(Zn)\beta_2(Fe)]$	C	302, 306
$[\alpha_2(Zn)\beta_2(Fe-X)]$	H	219, 220, 224, 225, 227, 228
$[\alpha_2(Fe-X)\beta_2(Zn)]$	H	219, 220, 224, 225, 227, 228
$[\alpha_2(Mg)\beta_2(Fe)]$	C, A	83, 303
$[\alpha_2(Fe)\beta_2(Mg)]$	C, A	83, 303
$[\alpha_2(Mg)\beta_2(Fe-X)]$	H	219, 220, 224, 225, 227, 228
$[\alpha_2(Fe-X)\beta_2(Mg)]$ (X = F, N_3, Im)	H	219, 220, 224, 225, 227, 228
$[\alpha_2(Zn)\beta_2(Fe-CN)/Cyt.b_5]$	H	221
$[\alpha_2(Fe-CO)\beta_2(Co-O_2)]$	G, I	91, 130
$[\alpha_2(Co-O_2)\beta_2(Fe-O_2)]$	G, I	91, 130
[des-Arg(α141)$\alpha_2(Fe)\beta_2(Co)$]	G, I	91, 92, 130
[des-Arg(α141)$\alpha_2(Co)\beta_2(Fe)$]	G, I	91, 92, 130
$[\alpha_2(Co)\beta_2(Fe)]$	B, C, D, I, G	89, 91, 92, 94, 117, 132, 133, 162, 163, 254, 295, 297–300, 302
$[\alpha_2(Fe)\beta_2(Co)]$	B, C, D, I, G	89, 91, 92, 94, 117, 132, 133 162, 163, 254, 295, 297–300, 302
$[\alpha_2(Fe-O_2)\beta_2(Co)]$	C, I	123, 295, 296

$[\alpha_2(Co)\beta_2(Fe-O_2]$	C, I	123, 295, 296
$[\alpha_2(Co)\beta_2(Fe)]$	C	299
$[\alpha_2(Fe)\beta_2(Co)]$	C	299
$[\alpha_2(Co)\beta_2(Fe-CO)]$	B, C, F, G, I, K	89, 93, 129, 162, 163, 318
$[\alpha_2(Fe-CO)\beta_2(Co)]$	B, C, F, G, I, K	89, 93, 129, 162, 163, 318
$[\alpha(Fe)\beta(Co)]_A[\alpha(Co)\beta(Co)]_C XL$	B, D, G, I	96, 137, 138, 167, 260
$[\alpha(Co)\beta(Fe)]_A[\alpha(Co)\beta(Co)]_C XL$	B, D, G, I	96, 137, 138, 167, 260
$[\alpha(Fe)\beta(Co)]_A[\alpha(Fe)\beta(Co)]_C XL$	B, D, G	96, 137, 138, 167, 260
$[\alpha(Fe)\beta(Co)]_A[\alpha(Fe)\beta(Fe)]_C XL$	B, D, G	96, 137, 138, 167, 260
$[\alpha(Co)\beta(Fe)]_A[\alpha(Co)\beta(Fe)]_C XL$	B, D, G	96, 137, 138, 167, 260
$[\alpha(Co)\beta(Fe)]_A[\alpha(Fe)\beta(Fe)]_C XL$	B, D	167, 260
$[\alpha(Co)\beta(Co)]_A[\alpha(Fe)\beta(Fe)]_C XL$	B, D	167, 260
$[\alpha(Fe)\beta(Co)]_A[\alpha(Fe)\beta(Co)]_C XL$	B, D	167, 260
$[\alpha(Fe)\beta(Fe)]_A[\alpha(Co)\beta(Fe)]_C XI$	B, D	167, 260
$[\alpha(Fe-O_2)\beta(Co-O_2)]_A$ $[\alpha(Co-O_2)\beta(Co-O_2)]_C XL$	B	167
$[\alpha(Fe-CO)\beta(Co)]_A[\alpha(Co)\beta(Co)]_C XL$	B	167
$[\alpha(Co-O_2)\beta(Fe-O_2)]_A$ $[\alpha(Co-O_2)\beta(Co-O_2)]_C XL$	B	167
$[\alpha(Co)\beta(Fe-CO)]_A[\alpha(Co)\beta(Co)]_C XL$	B	167
$[\alpha(Co)\beta(Fe)]_A[\alpha(Fe)\beta(Fe)]_S XL$	I	95
$[\alpha(Fe)\beta(Co)]_A[\alpha(Fe)\beta(Fe)]_S XL$	I	95
$[\alpha_2(Co)\beta_2(Fe)]XL$	I	95
$[\alpha(Co)\beta(Co)]_A[\alpha(Fe)\beta(Fe)]_S XL$	I	95
$[\alpha_2(Fe)\beta_2(Co)]XL$	I	95
$[\alpha(Co)\beta(Fe)]_A[\alpha(Co)\beta(Co)]_S XL$	I	95
$[\alpha(Fe)\beta(Co)]_A[\alpha(Co)\beta(Co)]_S XL$	I	95
$[\alpha_2(Fe)\beta_2(Ru-CO)]$	B, G, K	136, 163
$[\alpha_2(Ru-CO)\beta_2(Fe)]$	B, G, K	136, 163
$[\alpha_2(Fe)\beta_2(Cu)]$	D, I	90
$[\alpha_2(Cu)\beta_2(Fe)]$	D, I	90
$[\alpha_2(Fe-CO)\beta_2(Cu)]$	D, I	90
$[\alpha_2(Cu)\beta_2(Fe-CO)]$	D, I	90

TABLE I (*Continued*)

Hybrids	Mode of Studies or Physiochemical Properties	References
	II. Valency Hybrids	
$[\alpha_2^{+X}\beta_2]$	I	113
$[\alpha_2\beta_2^{+X}]$ ($x = H_2O$)	I	113
$[\alpha_2^{+CN}\beta_2]$	B, C, D, F, G, I, K	113, 126, 128, 143, 152, 183, 242, 244, 245, 274, 309
$[\alpha_2\beta_2^{+CN}]$	B, C, D, F, G, I, K	113, 126, 128, 143, 152, 183, 242, 244, 245, 274, 309
$[\alpha_2^{+X}\beta_2^{+CO}]$	B, I, K	114, 157, 181
$[\alpha_2^{+CO}\beta_2^{+X}]$	B, I, K	114, 157, 181
$[\alpha_2\beta_2^{+}]$	C, D, I, K, L	39–41, 38, 195, 230, 243, 284, 31, 32, 112
$[\alpha_2^{+}\beta_2]$	C, D, I, K, L	38, 39–41, 195, 230, 243, 284, 31, 32, 112
$[\alpha_2^{oxy}\beta_2^{+}]$	B, F, I	114, 150, 317
$[\alpha_2^{+}\beta_2^{oxy}]$	B, F, I	114, 150, 317
$[(\alpha^{+}\beta^{+})(\alpha^{CO}\beta^{CO})]$	I, J	59, 114
$[\alpha^{+}\beta^{+})(\alpha^{oxy}\beta^{oxy})]$	I	114
$[(\alpha^{CO}\beta^{+})(\alpha^{CO}\beta^{+})]$	I, J	59, 114
$[\alpha_2^{CO}\beta_2^{+CN}]$	G, I, J, K	59, 63, 114, 126, 181
$[\alpha_2^{+CN}\beta_2^{CO}]$	G, I, J, K	59, 63, 114, 126, 181
$[\alpha_2^{+CN}\beta_2^{NO}]$	K, I	108, 177
$[\alpha_2^{NO}\beta_2^{+CN}]$	K, I	108, 177
$[\alpha_2^{P}\beta_2^{M(+CN)}]$	G	127
$[\alpha_2^{M(+CN)}\beta_2^{P}]$	G	127
$[\alpha_2^{NO}\beta_2^{a}]$ ($a = N_3^{-}$)	I	108

$[\alpha_2^{a}\beta_2^{NO}]$	I	108
$[\alpha_2^{+X}\beta_2^{NO}]$	I	108
$[\alpha_2^{NO}\beta_2^{+X}]$	I	108
$[\alpha_2^{+F}\beta_2]$	I	113
$[\alpha_2^{a}\beta_2]$	I	113
$[\alpha_2\beta_2^{a}]$	I	113
$[\alpha_2\beta_2^{+F}]$	I	113
$[(\alpha\beta^{+CN})_A(\alpha\beta)_C]XL$	B, D	154, 247, 259
$[(\alpha^{+CN}\beta)_A(\alpha\beta)_C]XL$	B, D	154, 247, 259
$[(\alpha^{+CN}\beta^{+CN})_A(\alpha\beta)_C]XL$	B, D	154, 259
$[(\alpha^{+CN}\beta)_A(\alpha^{+CN}\beta)_C]XL$	B	154
$[(\alpha\beta^{+CN})_A(\alpha^{+CN}\beta)_C]XL$	B	154
$[\alpha_1^{+}\alpha_2^{+}\beta_1^{+}\beta_2]_A XL$	J	299
$[\alpha_1^{+}\alpha_2^{+}\beta_1^{+}\beta_2^{o}]_C$ or $[\alpha_1^{o}\alpha_2^{+}\beta_1^{+}\beta_2^{+}]_C$	C	289
$[\alpha_1^{+}\alpha_2^{o}\beta_1^{o}\beta_2^{o}]_C$ or $[\alpha_1^{o}\alpha_2^{o}\beta_1^{+}\beta_2^{o}]_C$	C	288, 289
(+ = Fe^{III} with H_2O or CN^- and o = Fe^{II} with CO or deoxy)		
$[(1,3\text{-methyl-}^2H)\text{hemin-}\alpha^{+CN}]_2\,[\beta^{+CN}]_2$	B	161
$[\alpha^{+CN}]_2[(1,3\text{-methyl-}^2H)\text{hemin-}\beta^{+CN}]_2$	B	161
$[(2,4\text{-Vinyl-}\alpha\text{-}^2H)\text{hemin-}\alpha^{+CN}]_2\,[\beta^{+CN}]_2$	B	161
$[\alpha^{+CN}]_2\,[(2,4\text{-Vinyl-}\alpha\text{-}^2H)\text{hemin-}\beta^{+CN}]_2$	B	161
$[(\alpha^{+CN}\beta)_A(\alpha\beta)_C]XL$	B	154
$[(\alpha^{CO}\beta^{+CN})_A(\alpha^{+CN}\beta^{CO})_C]XL$	B	154
$[(\alpha^{+CN}\beta^{+CN})_A(\alpha^{CO}\beta^{CO})_C]XL$	B	154
$[(\alpha^{CO}\beta^{+CN})_A(\alpha^{CO}\beta^{CO})_C]XL$	B	154
$[(\alpha^{+CN}\beta^{CO})_A(\alpha^{CO}\beta^{CO})_C XL$	B	154
$[\alpha_2^{+CN}\beta_2^{NO}]$	I	108
$[\alpha_2^{NO}\beta_2^{+CN}]$	I	108
$[\alpha_2^{NO}\beta_2^{a}]$	I	108
$[\alpha_2^{a}\beta_2^{NO}]$	I	108

TABLE I (*Continued*)

Hybrids	Mode of Studies or Physiochemical Properties	References
	II. Valency Hybrids (Continued)	
HbM	K	200
HbM(Boston)[α(E7)(His → Tyr)]	K, G, I	119, 126, 127, 199–201
HbM(Iwate)[α(F8)(His → Tyr)]	A, B, K	71, 146, 199–201
HbM(Sankatoon)[β(E7)(His → Tyr)]	K, I, G	119, 126, 127 199–201
HbM(Hyde Park)[β(F8)(His → Tyr)]	A, K	71, 123, 199–20i
HbM(Milwaukee)[β(E11)(Val → Glu)]	G, K	125, 126, 127, 199–201
	III. Partially Ligated Hybrids	
$[\alpha_2^{CO}\beta_2]$	C	263, 264, 266, 267, 268
$[\alpha_2\beta_2^{CO}$	C	263, 264, 266, 267, 268
$[\alpha_2^{NO}\beta_2]$	C, F, I, K	109, 179, 277, 311
$[\alpha_2\beta_2^{NO}]$	C, F, I, K	109, 179, 277
$[\alpha^{NO}\beta^{NO}\alpha^{deoxy}\beta^{deoxy}]$	I,K	111, 177
$[\alpha_2^{NO}\beta_2^{CO}]$	I, K	108, 177
$[\alpha_2^{CO}\beta_2^{NO}]$	I, K	108, 177
$[(\alpha_1^{CO}\beta_1^{CO})(\alpha_2\beta_2)]$	C, I	114, 255, 271, 276
$[(\alpha_1\beta_1^{CO})(\alpha_2^{CO}\beta_2)]$	C, I	114, 255, 271, 276
$[(\alpha_1^{CO}\beta_1)(\alpha_2\beta_2^{CO})]$	C, I	114, 255, 271, 276
$[(\alpha_1^{CO}\beta_1)(\alpha_2\beta_2)]$	C, I	114, 265, 275, 276, 281
$[(\alpha_1\beta_1^{CO})(\alpha_2\beta_2)]$	C, I	114, 265, 275, 276, 281
$[\alpha_2^{NO}\beta_2^{oxy}]$	I	108
$[\alpha_2^{oxy}\beta_2^{NO}]$	I	108
[(des-Arg$\alpha^{NO}\beta^{NO}$)($\alpha^{deoxy}\beta^{deoxy}$)]	I	108
[(des-Arg-Tyr$\alpha^{NO}\beta^{NO}$)($\alpha^{deoxy}\beta^{deoxy}$)]	I	111

$[(\alpha^{NO}\text{des-His}\beta^{NO})(\alpha^{deoxy}\beta^{deoxy})]$	I	111
$[(\alpha^{NO}\text{des-His-Tyr}\beta^{NO})(\alpha^{deoxy}\beta^{deoxy})]$	I	111
$[(\alpha^{NO}\beta^{NO})(\text{des-Arg } \alpha^{deoxy}\beta^{deoxy})]$	I	111
$[(\alpha^{NO}\beta^{NO})(\text{des-Arg-Tyr}\alpha^{deoxy}\beta^{deoxy})]$	I	111
$[(\alpha^{NO}\beta^{NO})(\alpha^{deoxy}\text{des-His}\beta^{deoxy})]$	I	111
$[(\alpha^{NO}\beta^{NO})(\alpha^{deoxy}\text{des-His-Tyr}\beta^{deoxy})]$	I	111
$[\alpha_1^{NO}\alpha_2^{NO}\beta_1^{NO}\beta_2^{CO}]$ (or) $[\alpha_1^{CO}\alpha_2^{NO}\beta_1^{NO}\beta_2^{NO}]$	C	279
$[\alpha_1^{NO}\alpha_2^{NO}\beta_1^{CO}\beta_2^{CO}]$	C	279
$[\alpha_1^{X}\alpha_2^{X}\beta_1^{X}\beta_2]$XL (or) $[\alpha_1\alpha_2^{X}\beta_1^{X}\beta_2^{X}]$XL ($X = O_2$ or CO)	C	282, 289
$[\alpha_2(Fe{-}O_2)\beta_2(Fe)]$	A	67, 69, 72
IV. Hybrid Hemoglobin with Different Prosthetic Groups		
$[\alpha_2^{P}\beta_2^{M}]$ (P = proto, M = meso and	D, B, K, G	129, 158, 172, 173, 193, 291
$[\alpha_2^{M}\beta_2^{P}]$ D = deutero porphyrins)	B, C, D, G, K	129, 158, 172, 173, 193, 291
$[\alpha_2^{P}\beta_2^{D}]$	C, K	172, 291
$[\alpha_2^{D}\beta_2^{P}]$	C, K	159
$[\alpha_2^{Chl}\beta_2^{+CN}]$	B	159
$[\alpha_2^{+CN}\beta_2^{Chl}]$	B	159
$[\alpha_2^{Chl}\beta_2^{deoxy}]$	B	159
$[\alpha_2^{deoxy}\beta_2^{Chl}]$	B	159
$[(\alpha_2^{P}\beta_2^{M(+CN)})]$	C, G	129, 290
$[(\alpha_2^{M}\beta_2^{P(+CN)})]$	C, G	129, 290
V. Spin Labeled Hybrids		
$[\alpha_2^{SL}\beta_2]$	I	100, 107, 110
$[\alpha_2\beta_2^{SL}]$	I	100, 107, 110
$[\text{met}\alpha_2^{SL}\beta_2]$	I	107
$[\text{met}\alpha_2\beta_2^{SL}]$	I	107

TABLE I (*Continued*)

Hybrids	Mode of Studies or Physiochemical Properties	References
	V. Spin Labeled Hybrids (Continued)	
$[\alpha_2\beta_2^{+SL}]$	I	107
$[\alpha_2^{+SL}\beta_2]$	I	107
	VI. Semihybrids	
$[\alpha_2(+)\beta_2(PP)]$	K	192, 198
$[\alpha_2(PP)\beta_2(+)]$	K	114, 119
$[\alpha_2(Zn)\beta_2(+)]$	K	191
$[\alpha_2(+)\beta_2(Zn)]$	K	191
$[\alpha_2(Zn)\beta_2(PP)]$	K	191
$[\alpha_2(PP)\beta_2(Zn)]$	K	191
$[\alpha_2(PP)\beta_2(Fe)]$	B, K	160, 174, 192, 198
$[\alpha_2(Fe)\beta_2(PP)]$	B, K	160, 174, 192, 198
$[\alpha_2(PP)\beta_2(Fe-CO)]$	B	160
$[\alpha_2(Fe-CO)\beta_2(PP)]$	B	160
	VII. Mixed Hybrids	
HbAC	L	46
$[\alpha_2\beta^S\beta^A]$ HbAS	E, J, L	48, 57
$[\alpha_2\beta^S\gamma]$ HbFS	E, J, L	57
$[\alpha_2\beta\gamma]$XL	E	331
$[\alpha_2\beta^S\beta^Y]$ HbYS		
HbYork (β146His → Pro)	D, L	50, 255
(* and their cross-linked Hb)		
$[(\alpha\beta)_A(\alpha\beta)_C]$XL	B, D, G	137, 155, 258
$[\alpha(\text{des-Arg})\beta]_A[\alpha\beta]_C$XL	B	155

$[\alpha(\text{des-Arg-Tyr})\beta]_A[\alpha\beta]_C$XL	B	155
$[\alpha(\text{des-Arg})\beta(\text{NES})]_A[\alpha\beta]_C$XL	B	155
$[\alpha(\text{des-Arg})\beta]_A[\alpha\beta(\text{NES})]_C$XL	B	155
$[\alpha_2^{\text{Can}}\beta_2^{\text{A}}]$	C, D	233, 294
$[\alpha_2^{\text{A}}\beta_2^{\text{Can}}]$	C, D	233, 294
S/Kariya [α40Lys → Glu]	F, J	311–315
S/Chesapeak [α92 Arg → Leu]	F, J	311–315
S/Dallas [α97Asn → Lys]	F, J	311–315
S/Tarrant [α126Asp → Asn]	F, J	311–315
S/St. Claude [α127Lys → Thr]	F, J	311–315
S/des-Arg	F, J	311–315
S/Austin [β40Arg → Lys]	F, J	311–315
S/A Thens G.A [β40Asp → Lys]	F, J	311–315
S/Zurich [β93Cys]	F, J	311–315
S/Thiomethylated [β93Cys]	F, J	311–315
S/NES [β93Cys]	F, J	311–315
S/Hotel Dieu [β99Asp → Gly]	F, J	311–315
S/Rad diffe [β99Asp → Ala]	F, J	311–315
S/Kempsey [β99Asp → Asn]	F, J	311–315
S/Yakima [β99Asp → His]	F, J	311–315
S/Ypsilant [β99Asp → Tyr]	F, J	311–315
S/St. Mande [β102Asn → Tyr]	F, J	311–315
S/TyGard [β124 Pro → Gly]	F, J	311–315
S/Abruzzo [β143His → Arg]	F, J	311–315
Hb Muel(horse–donkey)	F, J	311–315
[α_2 (human)COβ_2(carp)CO]	C, D, L	30, 238, 292, 293
[α_2(human)CN: β_2(carp)CO]	C	292, 293
[α_2(human)CO: β_2(carp)CN]		292, 293
[α_2(human): β_2(carp)]	D	238, 292, 293
$[(\alpha_2^{\text{Gphilla}}\beta_2^{\text{S}})]$	D, E	237
$[(\alpha_2^{\text{Cat}}\beta_2^{\text{A}})]$	C, D, I	105, 234
$[(\alpha_2^{\text{A}}\beta_2^{\text{Cat}})]$	C, D, I	105, 234

TABLE I (*Continued*)

Hybrids	Mode of Studies or Physiochemical Properties	References
	VII. Mixed Hybrids (Continued)	
$[\alpha_2^{donkey}\beta_2^{monkey})]$	D	232, 235
$[\alpha_2^{monkey}\beta_2^{donkey}]$	D	232, 235
$[\alpha_2^{\chi}\beta_2^{A}]$	D	239
$[\alpha_2^{A}\beta_2^{\chi}]$ (χ = xenopuslaevis Hb) [β93cys → Ser]	D	239
$[\alpha_2^{mouse}\beta_2^{human}]$	D	236
$[\alpha_2^{human}\beta_2^{mouse}]$	D	236
	VIII. Chemically Modified Hybrids	
$[\alpha_2^{CM}\beta_2]$ (CM-carboxy-methylated chain)	D	248, 249
$[\alpha_2\beta_2^{CM}]$	D	248, 249
$[\alpha_2\beta_2^{DHP}]$ (DHP-glycated chain)	D	250
$[\alpha_2^{DHP}\beta]$	D	250
$[\alpha_2\beta_2^{Ac}]$ (Ac = acetylated chain)	D	251
des-Arg(141α)HbA	A, B, K, D	82, 155, 196
des-(Arg, Tyr) HbA	B, K	155, 196

[a]A = XRD/Structural characterization; B = $^{1}H/^{13}C/^{31}P$ NMR studies; C = kinetic study (flash photolysis/stop flow, etc.); D = oxygen equilibrium studies; E = polymerization studies; F = thermodynamic studies; G = resonance Raman studies; H = electron-transfer studies; I = EPR/electronic properties; J = electrophoresis/IEF intermediate isolation and characterization; K = UV/fluorescence/IR/CD studies; L = preparation/physical aspect study; M = EXAFS studies.

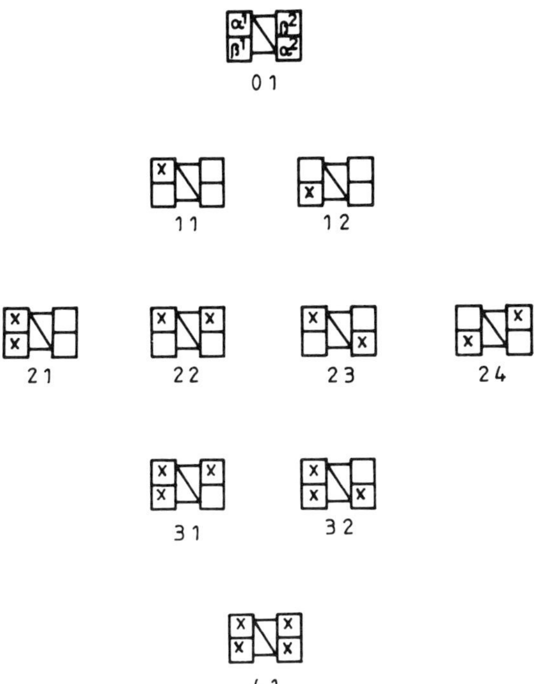

Figure 2. Representation of the 10 ligation states of hemoglobin tetramer: The X denotes ligation of a given chain. The 01, 11, 12, 21, 22, 23, 24, 31, 32, 41 represent appropriate species. [Reproduced with permission from M. A. Daugherty, M.A. Shea, and G. K. Akers, *Biochemistry*, *33*, 10345 (1994). Copyright © 1994 American Chemical Society.]

where the asterisk (*) represents the metal reconstituted subunit, oxidized subunit, or ligated subunit. Symmetric hybrids are more stable compared to the asymmetric hybrids, since dissociation and reassociation of the latter will always lead to establishment of more complex hybrid equilibria. One way to overcome this difficulty is to prepare cross-linked hybrid hemoglobin.

Preparation of hybrid hemoglobin can be done in four different methods (1) by reconstitution procedure, (2) by mixing two different hemoglobins, (3) by partial oxidation or reduction, and (4) by partial ligation.

The reconstitution method involving many steps can be adopted for the preparation of symmetric as well as asymmetric hemoglobins. The first step is to separate the α and β chains (27–29). The second step is to prepare reconstituted protoporphyrin or other porphyrins such as meso- and dueteroporphyrin with transition metal ions other than iron (e.g., Cu, Zn, Mn, Ni, Co, VO, Cr, or Ru). The third step involves mixing of the separated globin part with the reconstituted porphyrin part, followed by chromatographic purification procedures, the details of which are given in the form of a chart in Figs. 3 and 4. Asymmetric hemoglobin hybrids can be prepared from HbA or from two different hemoglobins say HbA–HbC, HbA–HbS, and so on. Both symmetric as well as asymmetric hybrid hemoglobins can be prepared by using cross-linkers such as bis(3,5-dibromo salicyl)fumarate, which binds Lysin 82 residues of the two β chain in oxyhemoglobin and under deoxy condition Lysin 99 residues of two α chains. The former is called $\beta\beta$ cross-linked hybrids and the later is called $\alpha\alpha$ cross-linked hybrid.

The second method of hybrid preparation involved mixing of both the native hemoglobins, followed by dilution and concentration. Hybrids are formed by this method due to the inherent property of hemoglobin, that is, dissociation of hemoglobin tetramers to dimers followed by reassociation to form tetramers as shown in Fig. 5. The resulting hemoglobin hybrids can be separated using chromatographic procedures (30–33). The distribution of the three components follows the binomial expansion $a^2 + 2ab + b^2 = 1$, where a and b are the initial fractions of parent hemoglobins as shown in the Fig. 5. For example, the final solution, obtained by mixing equal amounts of native hemoglobin and oxidized methemoglobin, was found to be composed of 13.2% oxyhemoglobin, 18.0% $(\alpha_2\beta_2^+)$, 37.5% $(\alpha_2^+\beta_2)$, and 31.3% methemoglobin.

The third method of partial oxidation of the hemoglobin is carried out by using reagents such as ferricyanide (34–36), nitrates (37), or by reduction of oxidized hemoglobin with reagents such as dithionite, ferredoxin, ferredoxin–NADP (38), ascorbic acid (39, 40), followed by separation of the hybrids by chromatography (41, 42). The chromatographic materials generally used are cation or anion exchange resins, gel permeation materials, and so on. The details for the preparation of symmetric and asymmetric hybrids by this method are given in the form of a chart in Fig. 6.

The fourth method involves partially ligated hybrid preparation, where α or β chains are partially ligated by ligands such as CO or NO. These hybrids are summarized in Fig. 7. Gaseous ligand in different ratios can be passed into a sealed tube containing deoxyhemoglobin, followed by separation and characterization. Alternatively, gaseous ligands can react with valency hybrids. By a double-mixing technique that is given in the form of a chart in Fig. 7, it is possible to prepare mono- and triliganded hybrids.

Figure 3. Flow chart for the preparation of reconstituted hemoglobin, symmetric mixed-metal hybrid hemoglobins, and symmetric cross-linked hybrid hemoglobins. Each box represents either an α or a β chain with protoheme; M = metalloporphyrin; PMB = p-hydroxymercuribenzoate; DTT = dithiothreitol; XL = cross-linkers such as (a) bis(3,4-dibromo salicyl) fumarate, which binds in the DPG site linking Lys 82(β_1)-Lys82(β_2) in oxyhemoglobin and Lys99(α_1)-Lys99(α_2) in deoxyhemoglobin; (b) 2-nor-2-formyl pyridoxal 5′-phosphate linking Lys82(β_1)-Val1(β_2) in deoxyhemoglobin.

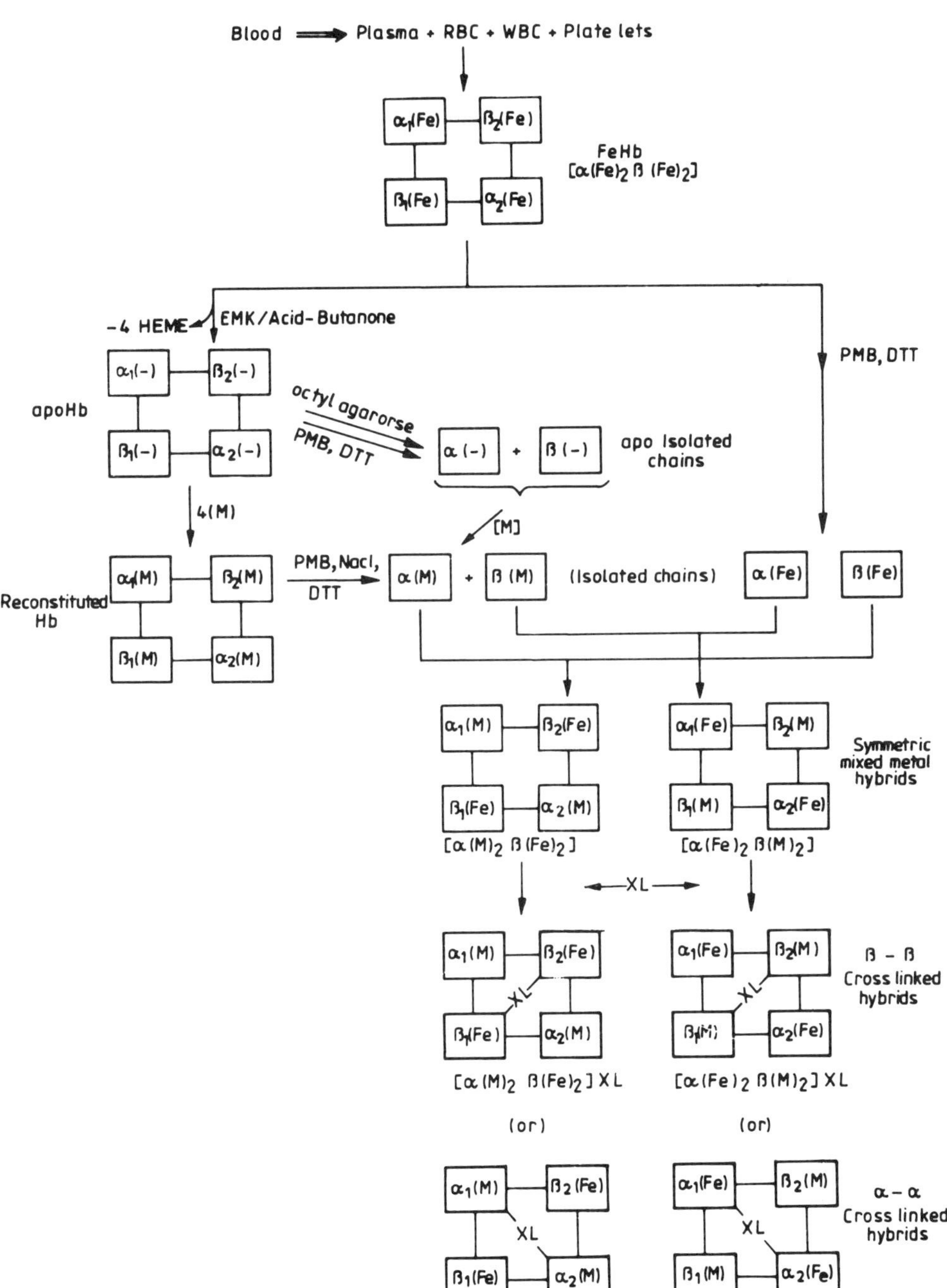
Blood
Plasma + RBC + WBC + Plate lets
α1(Fe)
β2(Fe)
β1(Fe)
α2(Fe)
FeHb
[α(Fe)2 β(Fe)2]
-4 HEME
EMK/Acid-Butanone
PMB, DTT
α1(-)
β2(-)
β1(-)
α2(-)
apoHb
octyl agarorse
PMB, DTT
α(-)
β(-)
apo Isolated chains
4(M)
[M]
α1(M)
β2(M)
β1(M)
α2(M)
Reconstituted Hb
PMB, Nacl, DTT
α(M)
β(M)
(Isolated chains)
α(Fe)
β(Fe)
α1(M)
β2(Fe)
β1(Fe)
α2(M)
[α(M)2 β(Fe)2]
α1(Fe)
β2(M)
β1(M)
α2(Fe)
[α(Fe)2 β(M)2]
Symmetric mixed metal hybrids
XL
α1(M)
β2(Fe)
XL
β1(Fe)
α2(M)
[α(M)2 β(Fe)2] XL
α1(Fe)
β2(M)
XL
β1(M)
α2(Fe)
[α(Fe)2 β(M)2] XL
β - β
Cross linked hybrids
(or)
(or)
α1(M)
β2(Fe)
XL
β1(Fe)
α2(M)
α1(Fe)
β2(M)
XL
β1(M)
α2(Fe)
α - α
Cross linked hybrids

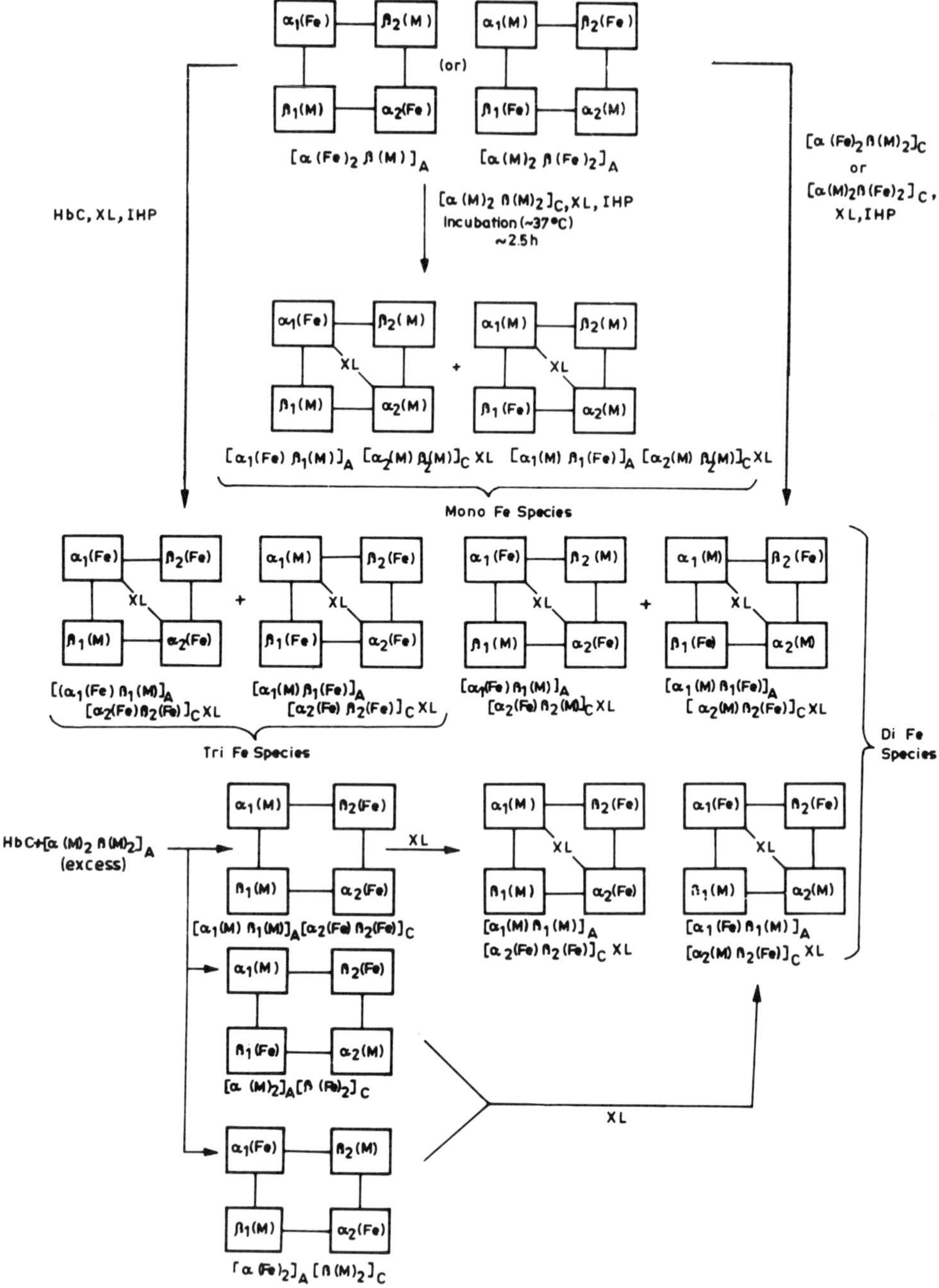

Figure 4. Flow chart for the preparation of asymmetric mixed-metal hybrids, namely, mono-, di- and tri-iron containing species. Here subscripts A and C in each square bracket represents chains belonging to HbA and HbC, respectively. Inisotol hexaphosphate = IHP; cross-linkers = XL. In this case, $\alpha\alpha$ cross-linked hybrids are shown; this is also applicable to $\beta\beta$ hybrids. During incubation, along with the desired hybrids, the starting materials and their corresponding cross-linked, products are obtained that are not shown in the diagram.

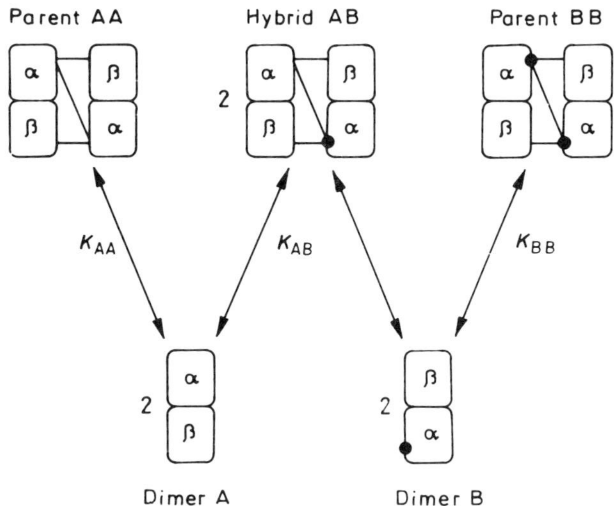

Figure 5. Schematic representation of dissociation of tetramer to dimer followed by reassociation of dimer from one hemoglobin with dimer from another to form hybrid hemoglobin. Here AA represents HbA and BB represents mutant hemoglobin, for example, HbS. Here K_{AA}, K_{BB} are dissociation constants and K_{AB} is association constant. [Reproduced with permission from V. J. LiCata, P. C. Sperose, E. Rovida, and G. K. Ackers, *Biochemistry*, *29*, 9771 (1990). Copyright © 1990 American Chemical Society.]

Chemically modified hemoglobin hybrids are also prepared by similar procedures. The chemical modification can be of two types: the first involves *N*-carboxymethylation, acetylation, glycation, or n-terminal α-chain cross-linking (43), and the second involves removal of certain amino acids by using enzymes such as carboxypeptidase A or carboxypeptidase B (44). This chemical modification can be done in the α or β chain in order to chemically modify hybrid hemoglobins. Modification of hemoglobin is done to identify amino acid chains that participate in a particular function.

By using the above mentioned procedures, it is possible to prepare both symmetric as well as asymmetric hybrids (45–47). The existence of hemoglobin hybrids has been demonstrated and characterized by the following techniques: (1) using cross-linking agents (48–50), (2) using cellulose acetate and polyacrylamide electrophoresis (51–55), (3) using polyacrylamide isoelectric focusing electrophoresis (56, 57), (4) using low-temperature quantitative cryogenic–isoelectric focusing technique (QC–IEF) (58, 59), and (5) using high-performance liquid chromatography (HPLC).

The most sensitive of these methods QC–IEF, which was developed by Perrella, where the hybirds are characterized as such in the gels. Presence of all nine microstates is confirmed by densitometer scans of gel tubes used for the separation by isoelectric focusing at $-25\,^{\circ}\mathrm{C}$ of the products, namely, partially

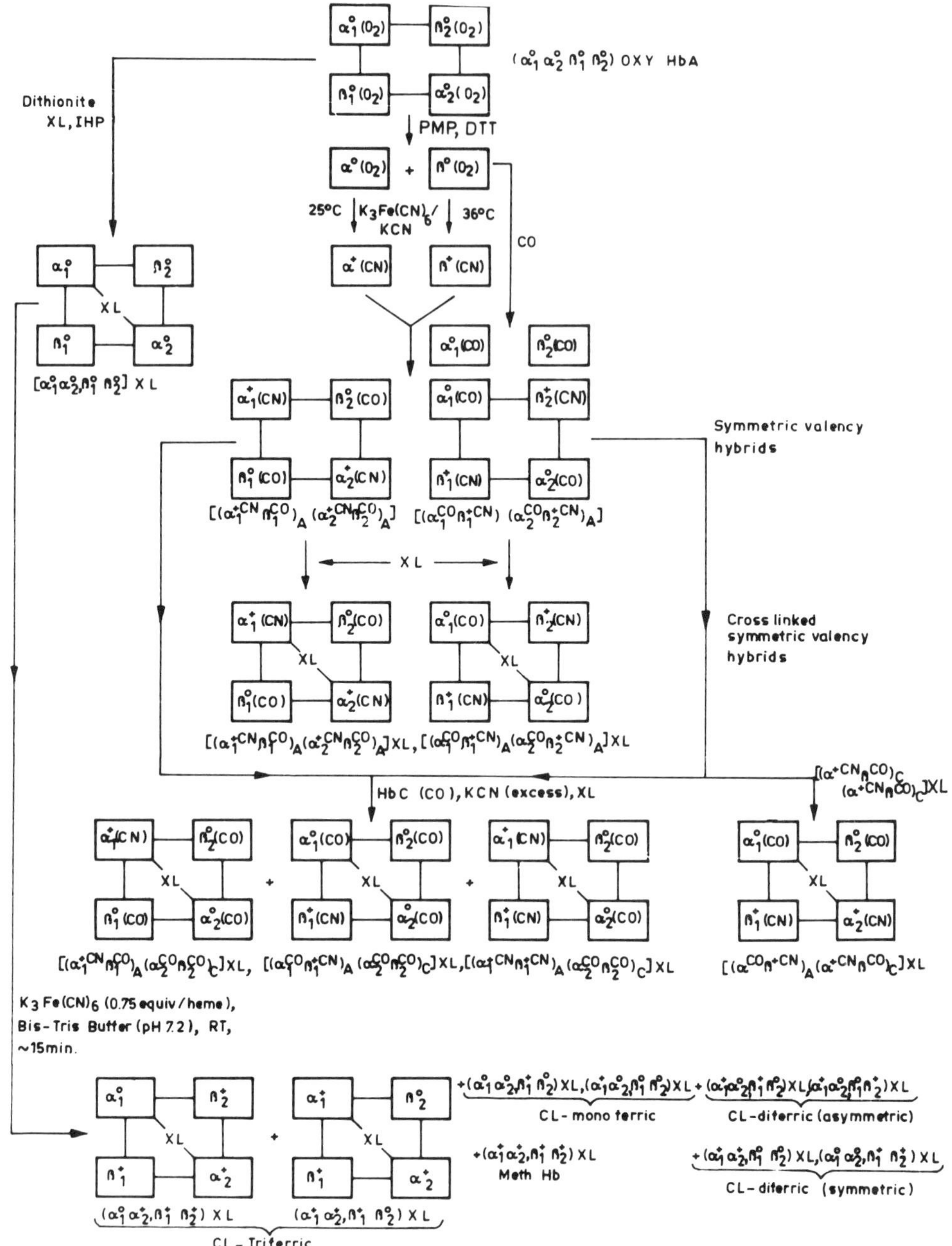

Figure 6. Flow chart for the preparation of symmetric and asymmetric valency hybrid hemoglobins. Since all four subunits contain iron in the ferrous or ferric state, ferrous is represented as α_1^0 or β_1^0, ferric is represented as α_1^+ or β_1^+, and the corresponding ligand attached to the chain is written in the bracket. Here the subscripts A and C in each square brackets represent a chain belonging to HbA and HbC, respectively.

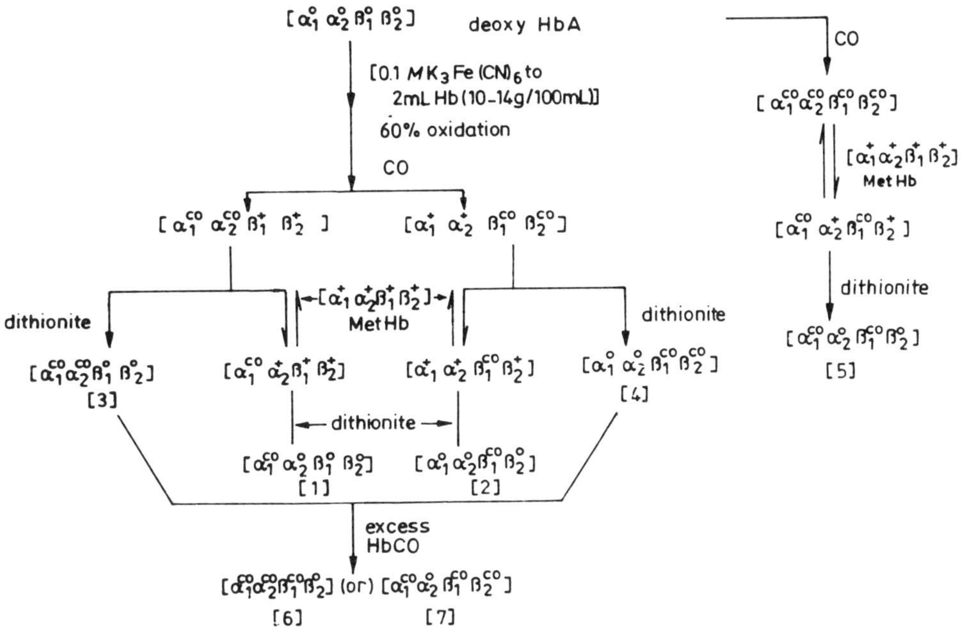

Figure 7. Flow chart for the preparation of partially ligated hybrid using double-mixing technique. Here [1] and [2] are monoligated hybrids; [3], [4], [5] are diligated hybrid; [6], [7], [8] are triligated hybrids.

oxidized HbCO (60–62). Partially oxidized hybrids (60–62), asymmetric hybrids (30), and partially liganded hybrids (63) have been studied by this technique. A number of asymmetric mutant hybrid hemoglobins and intermediates of hemoglobin (between the extreme T and R structures) have been studied with the help of these techniques (64–66).

IV. CRYSTAL STRUCTURE ANALYSIS BY X-RAY DIFFRACTION METHODS

The X-ray crystal structures of hemoglobins have been refined down to 1.5 Å (67) resolution which have become the basis for a great deal of the work on hemoglobin, Perutz's stereochemical mechanism, which provides a structural basis for cooperative oxygenation, was for the most part based on the X-ray structures of oxyhemoglobin and deoxyhemoglobin. There are, however, several limitations of the X-ray structures that affect the development of a comprehensive model for hemoglobin oxygenation.

X-ray diffraction provides data in the solid state that must be related to the functional solution state of hemoglobin. There have been several attempts to directly relate the solution state to the solid state, however, it is the ability to combine solution spectroscopic data with X-ray data that will provide the essential solution structural details.

X-ray structures provide a static picture of the hemoglobin molecule, but in solution, proteins are not static and their fluctuations are essential to explain their structure–function relationships. There are solution measurements such as hydrogen exchange that directly monitor dynamics; however, the major input to this area at present involves molecular dynamics, which simulate the protein fluctuations occurring in solution.

Perhaps the most serious limitation of the initial X-ray structures is that they were limited to the structures of fully liganded R-state oxyhemoglobin and carboxyhemoglobin as well as fully unligated T-state deoxyhemoglobin. Because of this limitation, attempts were made to explain oxygenation on the basis of these results to structures neglecting possible distinct intermediate conformations for partially ligated hemoglobins. This problem is now beginning to be addressed.

Quite recently it was found that chemical cross-linkers such as 3,3-stilbene dicarboxylic acid and trimesic acid seemed to help the formation of intermediates. Structural characterization has been done on these species, cross-linking the amino group of residues val-1(β) and Lys-82(β). Such a cross-linked hemoglobin has a structure similar to that of R state (68). These findings directly indicate the presence of the intermediate in going from the T to the R structure. There have been several X-ray structures showing the effect of binding ligands to hemoglobin in the T state. Recently, there have been attempts to obtain X-ray structures of some partially ligated hybrid hemoglobins.

The complete heme stereochemistry in hemoglobin for oxy-, deoxy-, and semioxyhemoglobins along with the small amount of information available for only three hybrids is given in Table II. Various means have been used to get crystallographic information on hemoglobin intermediate states. These include (a) chemical modifiers to stabilize the R state (69, 70), (b) mutant hybrids where the T state has been stablized (71), (c) crystal lattice constraints to prevent R to T transitions in the quaternary state (72–74), and (d) carrying out unique crystallization of partially ligated hemoglobin in T state (67). Another important method of generating stable half-ligated hemoglobin molecules is through mixed-metal hybrid hemoglobin (20). Below, we compare some important structural features of these hybrids with those of hemoglobin during the ligation process.

The hybrids studied so far by crystallography and their corresponding cell parameters are listed in Table III. Note that all hybrids are in the tensed (T) state. The picture clearly shows that the cell parameters of the identified hybrids

TABLE II

Comparison of Structural Aspects Obtained from the Crystal Structures of Oxy, Deoxy States with One Hybrid

Species	Subunit	Bond Length (Å)[a]								Angle	Reference
		$M-N_{porp}$	$M-N_{\epsilon}$	$M-P_{heme}$	$N_{\epsilon}-P_{heme}$	P_N-P_C	M−O(C)	$N_{\epsilon}-O_2(O)$	$N_{\epsilon}-O_1(C)$	M−O(C)−O	
Hb(deoxy)	α mean	2.08	2.16	0.58	2.72	0.16					76
$[(NH_4)_2SO_4]$	β mean	2.05	2.09	0.50	2.58	0.10					
Hb(deoxy)	α mean	2.03	2.16	0.56	2.72	0.21					73
(PEG)[b]	β mean	2.03	2.21	0.48	2.69	0.05					
HbO_2 (oxy)	α mean	1.99	1.94	0.16	2.1	0.04	1.66	2.6	3.0	153	77
$[NaKHPO_4]$	β mean	1.96	2.07	0.08	2.1	0.06	1.87	3.5	3.2	159	
HbO_2 (T state)	α_1 mean		2.26	0.50			1.61	2.84	3.00	153	84
	α_2 mean		2.35	0.54			1.85	2.47	3.06	166	
	β_1 mean		2.18	0.38			1.72	2.53	2.73	151	
	β_2 mean		2.22	0.50			1.77	2.76	2.98	159	
Semioxygenated Hb	α_1 oxy		2.18	0.20			1.78	2.63	2.86	149	67
	α_2 oxy		2.32	0.20			1.86	2.89	2.88	156	
	β_1 deoxy		2.24	0.30							
	β_2 deoxy		2.11	0.30							
COHbA	α mean	2.00	2.09	0.07	2.15	0.04	1.81	4.02	4.11	179	82c
	β mean	2.00	2.08	0.00	2.07	0.01	1.81	3.42	3.56	181	
$[\alpha_2(Ni)\beta_2(Fe-CO)]$	α mean	1.92	3.23	0.19	3.36	0.05					80
	β mean	1.96	2.23	0.32	2.51	0.10	1.80	2.87	3.31	155	
$[\alpha_2(Mg)\beta_2(Fe)]$	α mean	2.05	2.28	0.71	2.86	0.15					83
	β mean	2.04	2.18	0.58	2.75	0.12					
$[\alpha_2(Mg)\beta_2(Fe-CO)]$	α mean	2.05	2.39	0.71	3.07	0.17					83
	β mean	2.01	2.30	0.14	2.43	0.11	1.82	3.14	3.34	147	

[a] $M-N_{porp}$ = Distance between M and the porphyrin nitrogens; $M-N_{\epsilon}$ = distance between M and N_{ϵ} of F8 histidine; $M-P_{heme}$ = displacement of M from mean plane of prophyrin N and C atoms including the first atom of each side chain; $N_{\epsilon}-P_{heme}$ = displacement of histidine F8 nitrogen (N_{ϵ}) from the mean plane of prophyrin N and C atoms including the first atom of each side chain; P_N-P_C = displacement of plane of prophyrin N atoms from plane of porphyrin C atoms (doming parameter); M−O(C) = distance between M and first oxygen or carbon; M−O(C)−O = angle between M and oxygen molecule or carbon monoxide molecule; $N_{\epsilon}-O_1(C)$, $N_{\epsilon}-O_2(O)$ = distance between E7 His N_{ϵ} and O_1 or carbon, O_2, respectively.

TABLE III
Available Crystal Structural Unit Cell Parameters for Normal and Hybrid Hemoglobins

Species or Hybrids	pH	Space Group	Resolution (Å)	Quaternary Structure	Unit Cell Constants a (Å)	b (Å)	c (Å)	β(deg)	References
Human Hb(deoxy) (2.3*M* $(NH_4)_2SO_4$/0.3 *M*$(NH_4)_2PO_4$)	6.8	$P2_1$	1.74	T	63.2	83.5	53.8	99.3	76
Human HbO_2 (oxy) ($NaKHPO_4$)	6.7	$P4_1 2_1 2_1$	2.1	R	53.7	53.7	193.8		77
Human Hb(deoxy) (8% PEG, 100 m*M* KCl, 10 m*M* K_2PO_4)	7.0	$P2_1 2_1 2_1$	1.9	T	96.9	98.8	65.9		
Semioxygenated		$P2_1 2_1 2_1$	1.5	T	95.6	97.8	65.5		67
HbO_2(T state)	7.2	$P2_1 2_1 2$	2.1	T	96.7	98.4	66.1		84
Human CO—Hb (16% PEG, 100 m*M* sodium caodylate, 75 m*M* Cl^-)	5.8	$P2_1 2_1 2_1$	1.7	R	97.5	101.7	61.1		85
$[\alpha_2\beta^1S^{82}\beta]$XL		$P4_1 2_1 2_1$	2.4	R	54.13	54.13	196.23		68
$[\alpha_2\beta^1Tm^{82}\beta]$XL		*C2*	1.8	R	104.43	72.16	88.03	108.25	68
$[\alpha_2\beta^{1,82}Tm^{82}\beta]$XL XL		*C2*	1.8	R	104.43	72.16	88.03	108.25	68
$[\alpha_2(Ni)\beta_2(Fe-CO)]$		$P2_1$	2.6	T	63.18	82.26	55.06	98.42	80
$[\alpha_2(Fe-CO)\beta_2(Co)]$		$P2_1$	2.9	T	63.15	83.59	53.80	99.63	79
$[\alpha(Fe-O_2)\beta_2(Ni)]$		$P2_1$	3.5	T	63.2	83.4	53.8	99.9	79
$[\alpha_2(Fe-CO)\beta_2(Mn)]$		$P2_1$	3.0	T	63.2	83.5	53.8	99.8	78
$[\alpha_2(Mg)\beta_2(Fe-CO)]$		$P2_1 2_1 2$	1.9	T	99.35	101.0	166.80		83
Hb—M(IWate)(deoxy)		$P2_1 2_1 2_1$	5.0	T	63.1	84.2	109.3	90.0	68
(87α His → Tyr)									147

in this state are isomorphous with deoxyhemoglobin (75, 76) and different from those of the R state (77).

X-ray analysis of the [α_2(Fe—CO)β_2(Mn)] (78) provides another approach for characterizing the ligand-induced changes to the tertiary structure of an α subunit constrained to the T-state quaternary structure. Carbon monoxide binding to the α heme group pulls the iron atom toward the heme plane and this in turn pulls the last turn of the F helix (residues 85–89) approximately 1 Å closer to the heme group. The largest shift (0.2–0.4 Å) occurs in residue His-87 and Ala-88. The direction of this small movement is perpendicular to the axis of F helices.

In [α_2(Fe—CO)β_2(Co)] (79), the structural difference between this hybrid and native deoxyhemoglobin is small and most of the changes occur in the environment of the "allosteric core." Here small changes do occur at the $\alpha_2\beta_1$ interface, although no detectable changes are observed at the $\alpha_1\beta_2$ interface due to crystal packing. Here the estimated shift of Fe in deoxyhemoglobin is 0.35(+0.15) Å compared to 0.42(+0.09) Å in oxyhemoglobin in the α_1 and α_2 subunits. The mean distance between the Co atom and the heme planes for β_1 and β_2 is 0.43(+0.07) Å. The observed largest structural changes are in the region of the F helix and the FG corner. In the F helix reference frame, the displacement of bridging methylene of the porphyrin (between FG3 and FG5) are 1.2 and 1.0 Å, respectively, for the pairs, namely, HbO_2 and [α_2(Fe—CO)β_2(Co)]. Note that ligand binding in the R state does not significantly change the conformation of FG corner. Another important finding is that the Co–heme plane distance in [α_2(Fe—CO)β_2(Co)] is shorter than the Fe–plane distance in the β subunits of deoxyhemoglobin. The features of the [α_2(Fe—CO)β_2(Ni)] are similar to those seen in [α_2(Fe—CO)β_2(Co)], including the asymmetry of structural changes. Comparison of the structural features of these two hybrids were made with native deoxyhemoglobin by difference Fourier synthesis at 2.8, 2.9, and 3.5-Å resolution, respectively.

In order to account for the low ligand affinity of the β heme of T-state hemoglobin, the structure of [α_2(Ni)β_2(Fe—CO)] (80a) was determined. Here the structure behaves as if the metal atoms were displaced from the porphyrin plane by over 1 Å instead of the normal 0.55 Å, and consequently, the proximal histidine gets pushed further away. The coordinating N_ε of F8 His and the Ni atom are separated by 3.2 Å and, hence, the bond between Ni and proximal histidine (His F8) is broken. This result is in agreement with spectroscopic findings (81), which then introduces small structural perturbation in the α-heme pocket. These structures confirm that β chains do bind ligand in the T state. Ligand binding at the T-state β heme results in a shift of the Fe atom from its position on the proximal state of heme by 0.21 Å, that is, about half-way to its position found in the R-state carbonmonoxyhemoglobin. The hydrogen-binding network is not affected by the movement of Fe atom. Through this structure the important role of E11 Val in the regulation of the T-state ligand affinity of the β heme is confirmed. In the reference frame, $\alpha_1\beta/\beta_1$ interface FG and CD corners undergo little structural changes in the β ligated T state.

In $[\alpha_2(Fe^{2+}-O_2)\beta_2(Fe^{2+})]$ (67) a semioxy species, the occupancy of the ligand atoms and the anisotropic behavior of the iron atoms have been refined at 1.5-Å resolution demonstrating that while the α heme groups are only partially ligated there is no ligation of the β heme. Here the α-heme iron atom has not moved into the plane of the heme groups. The lack of movement of the iron atom leads to longer apparent iron–oxygen distances than expected. The β-subunit heme group appears to be distorted in such a way that they approach a stable five coordinate geometry for iron atom (i.e., α irons are about 0.2 Å out of heme plane and β irons are about 0.3 Å out of the heme plane).

These observations suggest that there may be communication between the subunits across the interface within the T structure. The strain energy of ligand binding must be "distributed" to some extent among those other contacts as well as to the FG contacts. In the T state, the heme plane is parallel to the F-helix axis, whereas in the R state it is tilted by roughly 10 Å. Thus the binding of ligand to a T-state tetramer should generate steric strain that can be relieved only by a shift in quaternary structure to the R state.

Due to the important salt bridges between Arg-141 of one α chain and Asp-126 and Lys-127 of the other α chain, the C-terminal arginine has been implicated in the Bohr effect. Hence, by comparing the crystal structure of chemically modified Hb, namely, des-Arg-141(α) Hb (82) with native deoxyHb has shed some light on the structure and functional properties of hemoglobin. The important changes are (a) the new α-subunit COOH terminus and the NH_2 terminus of the opposite α subunit are about 1 Å closer together, (b) the Val-1(α) NH_2 terminus and Tyr-104(α) COOH terminus, however, remains separated by about 5.5 Å and do not interact to form a strong salt bridge, as suggested by Perutz and Ten Eych (82b).

It is known that MgPPIX (83) stabilizes the T quaternary structure, for example, NiPPIX. Also [Mg—Fe] hybrid showed lower oxygen affinity than [Ni—Fe] hybrid. These results indicate that Mg stabilizes the T quaternary structure even stronger than Fe. The X-ray crystal structure of deoxy-$[\alpha_2(Mg)\beta_2(Fe)]$ and CO-liganded $[\alpha_2(Mg)\beta_2(Fe-CO)]$ were analyzed to see if any changes had taken place at the $\alpha_1\beta_1$ interface in these hybrids in comparison to the native deoxy, COHb, and [Ni—Fe] hybrid. In the case of $[\alpha_2(Mg)\beta_2(Fe)]$, the stereochemical relationship, namely, the metal ion and the N_ϵ of proximal histidine, were identical (as shown in Table II) with that of deoxyhemoglobin. In the case of $[\alpha_2(Mg)\beta_2(Fe-CO)]$, no salt bridges specific to deoxyHb was found to be broken and, hence, assumes a T quarternary structure. Also, the position of N_ϵ of the proximal histidine relative to heme in this hybrid can be considered to be intermediate between those of deoxyHbA and COHbA for the β subunit.

When the oxygen molecule enters the deoxyhemoglobin (T state), due to interaction between oxygen and Fe in the heme center, the T state is destabilized and moves to the favorable R state. But the crystal structure of the T state hemoglobin with oxygen bound to all four hemes reported quite recently (84) shows significant changes in the α and β heme pocket as well as in the $\alpha_1\beta_2$ interface in the direction of the R quaternary structure. Most of the shifts and deviations are similar to the deoxy T-state HbA but larger than the T-state metHb and partially ligated T-state Hb (73).

The crystals obtained by using different crystallization procedures for ligated HbA were found to have differing quaternary structures compared to the known ones, namely, R and T. The new quaternary structure containing different species was named as R_2 (85, 86) and it has been proposed that the R_2 state may function as a stable intermediate along an R–R_2–T pathway. Recently, Smith and Simmons (86) demonstrated a third quaternary structure (quaternary Y), which is accessible to fully ligated Hb.

So far, the crystal structure of hybrids studied are either in doubly ligated or unligated states. We do not know a crystal structure of even a single monoligated or a triligated hybrid crystal. The difficulty in preparing crystals of such hybrids may be due to the highly cooperative oxygen uptake capacity of hemoglobin. In the future, if the crystal structures of mono- and triligated hybrids are known, along with the available data for the hybrid and complementary information from other type of experiments, it may be possible to understand the ligand-binding process.

V. SPECTROSCOPIC STUDIES

Direct evidence for alteration in heme structure and environment due to protein configuration has been obtained from spectroscopic studies of chemically modified hemoglobins and hybrid hemoglobins. Apart from XRD, some of the observed spectral changes include changes in the spin equilibria at the iron center in methemoglobin derivatives and hybrid hemoglobins having other metal ions along with iron, exchangeable NMR resonances involving proximal histidines, alteration in visible and near-IR spectra, and changes in Fe-proximal histidine stretching frequency as observed by resonance Raman spectroscopy. This section reviews the results obtained from different spectroscopic methods for studying hybrid hemoglobins, which help to deduce their structural features.

A. EPR, ENDOR, and ESEEM Spectroscopy

An electron paramagnetic resonance (EPR) spectroscopic study of hemoglobin gives us an idea about the specific environment of the metal ion, provided that the metal ion has a paramagnetic center or the ligand attached to the metal ion has an unpaired electron. Four classes of hemoglobin hybrids are generally studied by EPR spectroscopy. These classes may be classified as (a) valency hybrid hemoglobin; (b) mixed-metal hybrid hemoglobin; (c) NO ligated hybrid hemoglobin; and (d) hybrid hemoglobin modified by the attachment of a paramagnetic label. The useful parameters obtained from EPR measurements are g, A, and D tensors from the metal centers and sometimes the A tensors from ligands; the anisotropy of these values will reflect the surrounding of the iron atom and can be used to probe the symmetry.

The earliest EPR studies were performed on methemoglobin, nitric oxide ligated hemoglobins, and spin-labeled hemoglobins. With the advent of reconstitution procedures, EPR study on reconstituted hemoglobins were carried out elaborately in order to determine the relationship between the electronic state of the prosthetic group and the quaternary state of metal substituted hemoglobin. The fruitful results revealed by these studies include the presence of two distinct metal ion geometries in Cu (and Ni) reconstituted hemoglobins (81, 87, 88), which were later confirmed by XRD.

In recent days, hybrid hemoglobins, which are considered to represent the intermediate state during R to T transition, have been studied using EPR. In this context, Co—Fe hybrids were studied more extensively than any other hybrids for two reasons the first is that both deoxy and oxy Co have clearly discernible EPR spectra. Therefore, replacement of neutral iron porphyrin with artificial cobalt porphyrin can be used to probe the intrinsic role of the central metal ion for the R to T transition, the second, is that Co can bind oxygen but not CO and, hence, it is possible to monitor both the ligated and unligated chains in particular hybrid hemoglobins by EPR.

Initial studies by Ikeda-Saito et al. (89) on the EPR of the Co—Fe hybrid reveals the following: (a) EPR spectra of oxygenated [α_2(Co)β_2(Fe)] and [α_2(Fe)β_2(Co)] are quite similar to each other and the "deoxy EPR" spectra of [α_2(Co)β_2(Fe—CO)] and [α_2(Fe—CO)β_2(Co)] are similar to each other in spite of the changes in the metals in the α and β subunits; in addition, the "oxy" EPR spectra of the two hybrids are totally different from the rest; (b) "deoxy EPR spectra of [α_2(Fe)β_2(Co)] is similar to those of [Co—β^{-SH}]; (c) deoxy EPR spectra of [α_2(Co)β_2(Fe)] are different from those of [Co—α^{-SH}]; (d) deoxy[α_2(Co)β_2(Fe)] shows the existence of two different types of cobalt sites while all other hybrids show the existence of a single site. These are shown in Fig. 8.

All these observations can be rationalized as follows: (a) a difference in the distance of proximal histidine relative to the heme plane (heme iron and the N_ε of the proximal histidine) are 2.0 and 2.2 Å, respectively, in deoxy-Fe—Hb and (b) distortion of the porphyrin plane caused by constraints from the heme–globin contacts in a deoxy α subunit in Co—Hb. An investigation of the pH-dependent conformational changes indicated that oxy[α_2(Co)β_2(Fe)] undergoes

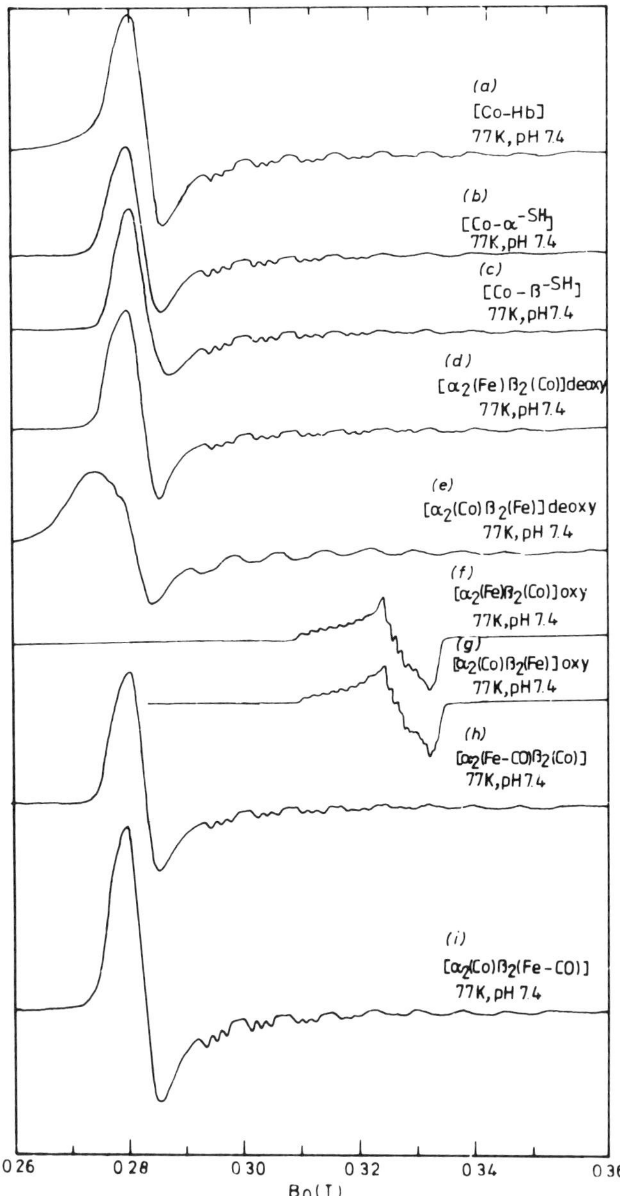

Figure 8. X-band EPR spectra with 100-KHz field modulation at 77 K of the different species: (*a*) Co-Hb tetramer, (*b*) deoxyCoHb α^{-SH} chain, (*c*) deoxyCoHbβ^{-SH}chain, (*d*) deoxy-$[\alpha_2(Fe)\beta_2(Co)]$, (*e*) deoxy$[\alpha_2(Co)\beta_2(Fe)]$, (*f*) oxy$[\alpha_2(Fe)\beta_2(Co)]$, (*g*) oxy$[\alpha_2(Co)\beta_2(Fe)]$, (*h*) $[\alpha_2(Fe{-}CO)\beta_2(Co)]$, (*i*) $[\alpha_2(Co)\beta_2(Fe{-}CO)]$ in 0.05 *M* bis-tris, with pH 7.4 [Reproduced with permission from (89).]

pH-dependent change similar to that of Co—Hb, while oxy[α_2(Fe)β_2(Co)], [α_2(Co)β_2(Fe—CO)], and [α_2(Fe—CO)β_2(Co)] spectra were independent of pH. These observations suggest that part of the alkaline Bohr effect is caused by changes in the electronic structure of α subunits in such a way that Bohr protons are released in fully deoxyHb.

A comparative chart of the EPR parameters obtained for various Co hybrid hemoglobins given in Table IV shows that $g_{\parallel}$ is unaltered throughout but there are slight change in $g_{\perp}$ and $A_{\parallel}$ components. Based on these results it was suggested that the $g_{\perp}$ value of 2.38 corresponds to the low-affinity T conformation, while 2.33 refers to high-affinity R state. Consequently, only [α_2(Co)β_2(Fe)] and not the complementary hybrid has two different types of ligand environments. In all other deoxy or Fe—CO containing hybrids, a single but an identically similar environment is observed while for the oxy hybrids, an environment(s) totally different from that of "deoxy" ones seems to be common both for [α_2(Co)β_2(Fe)] and [α_2(Fe)β_2(Co)]. Somehow, Co—Fe hybrids with or without ligated Fe sites and also fully oxygenated samples do not show any kind of systematic change; the reason may be the poor cooperative behavior of Co-containing Hb.

On the other hand, Cu—Hb as well as Cu—Fe (90) hybrids seem to show a certain trend, as shown in Fig. 9. Pure Cu—Hb easily resolves into two different environments (81); however, the Cu—Fe hybrids and ligated Fe sites show

TABLE IV
EPR Parameters for Hybrid Hemoglobins

Hybrid or Species	pH	$g_{\parallel}$	$g_{\perp}$	$A_{\parallel}$	References
deoxy[α_2(Co)β_2(Fe)]	7.0	2.03	2.38	8.0	89
deoxy[α_2(Fe)β_2(Co)]	7.0	2.03	2.34	7.7	89
[α_2(Co)β_2(Fe—CO)]	7.0	2.03	2.33	7.7	89
[α_2(Fe—CO)β_2(Co)]	7.0	2.03	2.33	7.6	89
des-Arg [α_2(Co)β_2(Fe)]	7.0		2.33		91, 92
des-Arg [α_2(Co)β_2(Fe)] + IHP	7.0		2.39		91, 92
[α_2(VO)β_2(Fe)] + IHP	6.5	1.965	1.994	19.01	Morimoto et al. (unreported)
[α_2(VO)β_2(Fe—CO)]	7.4	1.965	1.994	16.71	Morimoto et al. (unreported)
[α_2(Cu)β_2(Fe)]	7.4	2.197	2.02	19.5	90
[α_2(Fe)β_2(Cu)]	7.4	2.216	2.026	17.6	90
[α_2(Cu)β_2(Fe—CO)]	6.6	2.197	2.02	19.5	90
	8.4	2.213	2.016	17.9	90
[α_2(Fe—CO)β_2(Cu)]	7.4	2.216	2.026	17.6	90
CuHb					
Site 1		2.217	2.054	14.6	81
Site 2		2.178	2.042	14.3	81

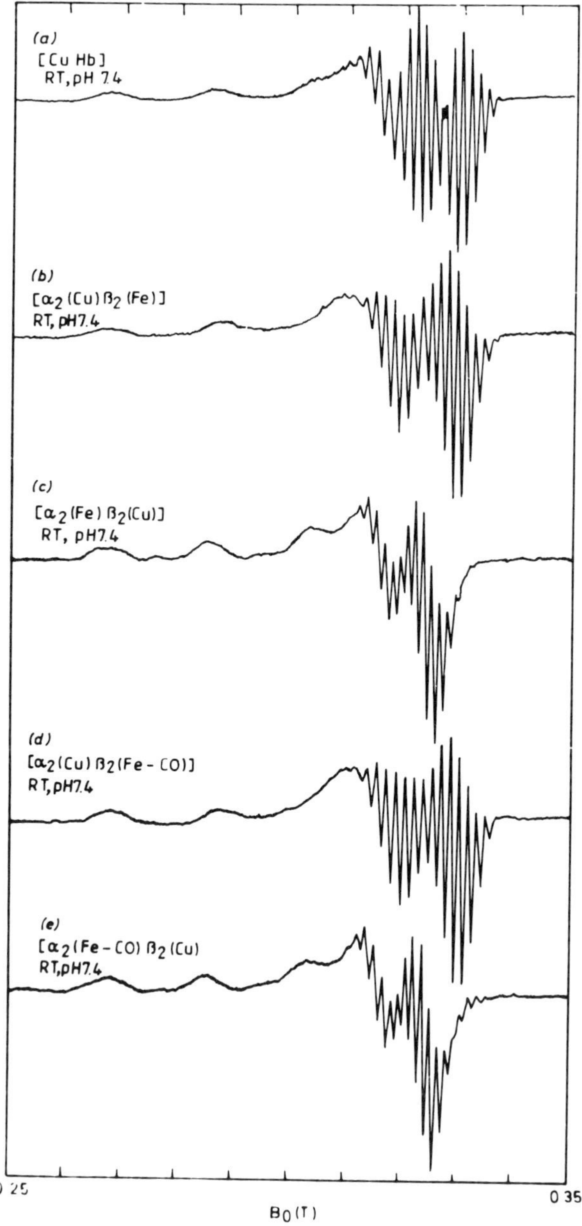

Figure 9. X-band EPR spectra of Cu—Hb, Cu—Fe hybrid hemoglobins recorded at room temperature with pH 7.4. (*a*) Cu-Hb, (*b*) $[\alpha_2(Cu)\beta_2(Fe)]$, (*c*) $[\alpha_2(Fe)\beta_2(Cu)]$, (*d*) $[\alpha_2(Cu)\beta_2(Fe{-}CO)]$, (*e*) $[\alpha_2(Fe{-}CO)\beta_2(Cu)]$. [Reproduced with permission from (90)]

systematic changes in the nature of EPR spectra and are expected to represent a systematic modulation of the heme environment. Unreported results on the iron–vanadyl hybrid are available from the laboratyory of Morimoto and will be useful in this regard, as shown in Fig. 10.

In order to find the effect of a salt bridge on the R to T conformational change (91, 92) Arg-141(α) was removed from the Co hybrids by using carboxypeptidase B, consequently the ionic interaction of Arg-141(α_1) with the α-amino group of Val-1(α_2) and ε-amino groups of Lys-1276(α_2) and the guanidium group of Arg-141(α_1) with the carboxy group of Asp-126(α_2) are removed. In

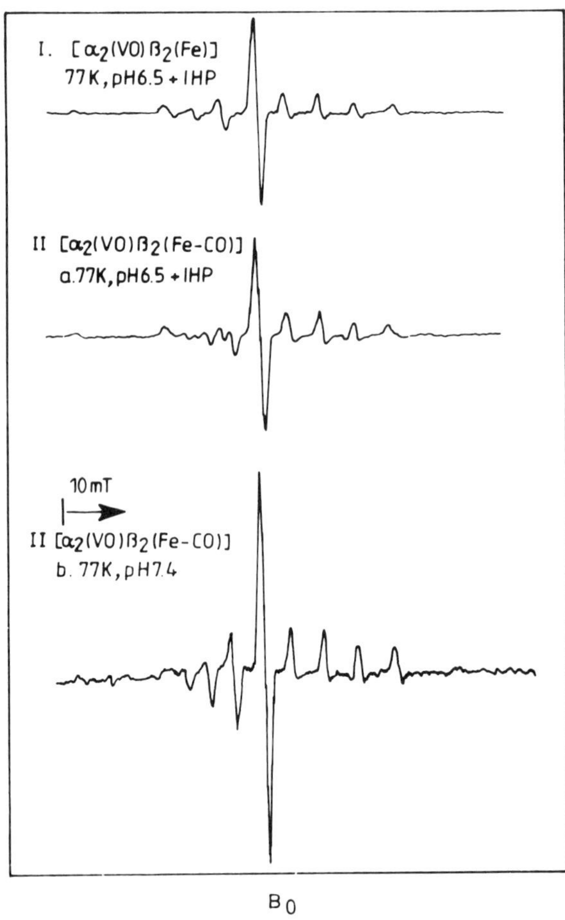

Figure 10. X-band EPR spectra of [Fe—VO] hybrid hemoglobins recorded at 77 K (**I**) [α_2(VO)β_2(Fe)] and (**II**) [α_2(VO)β_2(Fe—CO)]. [Unreported results from Morimoto and Hori.]

the absence of IHP, des-Arg[α_2(Co)β_2(Fe)]deoxy had a $g_\perp$ value of 2.33, which shows the predominancy of the R structure even in the absence of oxygen because of a shift in the allosteric equilibrium; in the presence of IHP, however, there is a small shoulder at $g = 2.39$, which shows that des-Arg[α_2(Co)β_2(Fe)] is in the T structure. This change in g value shows that IHP stabilizes the T structure. Also it is noted that in this hybrid the electronic structure of Co–porphyrin within the hybrid is similar to that of [α_2(Co)β_2(Fe)] in the R structure.

Reexamination of the deoxy[α_2(Co)β_2(Fe)] hybrid at two different microwave frequencies reveals two sets of axially symmetric EPR signals in the perpendicular g region at 2.371 and 2.311 (at Q-band) and at $g = 2.33$ and 2.39 (at X band). This result is considered different from the earlier report of a rhombic symmetry for the $g_\perp$ signal (93). The broad component is associated with the electronic state of metal ion in the T-state Hb and the sharp component with that of the R state. This phenomenon was interpreted in terms of the weakening of the bond between the porphyrin metal and the proximal histidine. Deoxy[α_2(Co)β_2(Fe)] in alkaline pH ($g_\perp = 2.33$) is different from that of neutral pH ($g_\perp = 2.38$) in the presence of IHP, which can be visualized in Fig. 11. This experimental observation may be due to the difference in the distribution of the two species.

With these preliminary ideas obtained from the solution of EPR of various Fe—Co hybrids, it will be interesting to compare them with those from single-

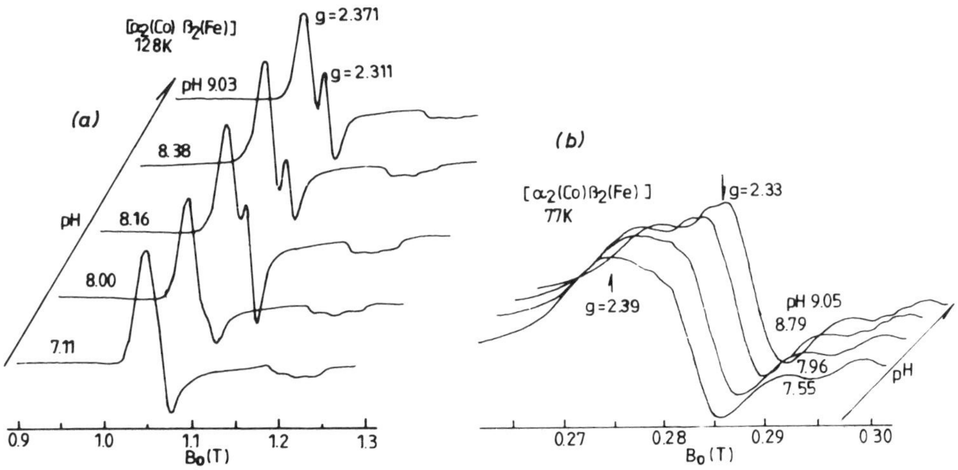

Figure 11. The pH-dependent EPR spectra of the deoxy[α_2(Co)β_2(Fe)] hybrid hemoglobin (*a*) *X* band at 77 K and (*b*) *Q* band at 128 K. [Reproduced with permission from T. Inubushi and T. Yonetani, *Biochemistry*, *22*, 1894 (1983). Copyright © 1983 American Chemical Society.]

crystal EPR studies of oxy- and deoxy[α_2(Co)β_2(Fe)] hybrids (94). Two sets of axially symmetric EPR signals for α(Co), namely, α_1(Co) and α_2(Co), were observed from the EPR experiments within the deoxy[α_2(Co)β_2(Fe)] tetramer, whereas only one set was observed in the β(Co) subunits, which indicates electronic inequivalence between the two α(Co) subunits in [α_2(Co)β_2(Fe)] tetramer.

Upon the addition of a small amount of CO to these hybrids in order to prepare partially ligated species, the $g_{\parallel}$ and $g_{\perp}$ signals corresponding to α_2(Co) become narrow in similarity to the observations seen in powder EPR experiments. These results suggest the presence of close contacts between α and β subunits as noticed in single-crystal XRD studies; carbon monoxide stabilzies the deoxy structure after partial ligation in the crystalline state.

With the advent of a preparation of cross-linked hemoglobin and the ability to prepare asymmetric cross-linked hybrids, the same technique was applied for Fe—Co hybrids, with one, two, and three Co metal centers. In this context, mono- cobalt cross-linked hybrids, namely, [α(Co)β(Fe)]$_A$[α(Fe)β(Fe)]$_S$XL, [α(Fe)β(Co)]$_A$(α(Fe)β(Fe)]$_S$XL, dicobalt species [α((Co)β(Co)]$_A$[α(Fe)-β(Fe)]$_S$XL and tricobalt cross-linked hybrids [α(Co)β(Fe)]$_A$[α(Co)β(Co)]$_S$XL, [α(Fe)β(Co)]$_A$[α(Co)β(Co)]$_S$XL, were prepared both in the fully deoxy and carbonmonoxy(iron) conditions and subjected to EPR measurements (95, 96). A comparative EPR chart for mono-Co, tri-Co cross-linked hybrids is given at various pH values both for deoxy and CO ligated states in Fig. 12. From these

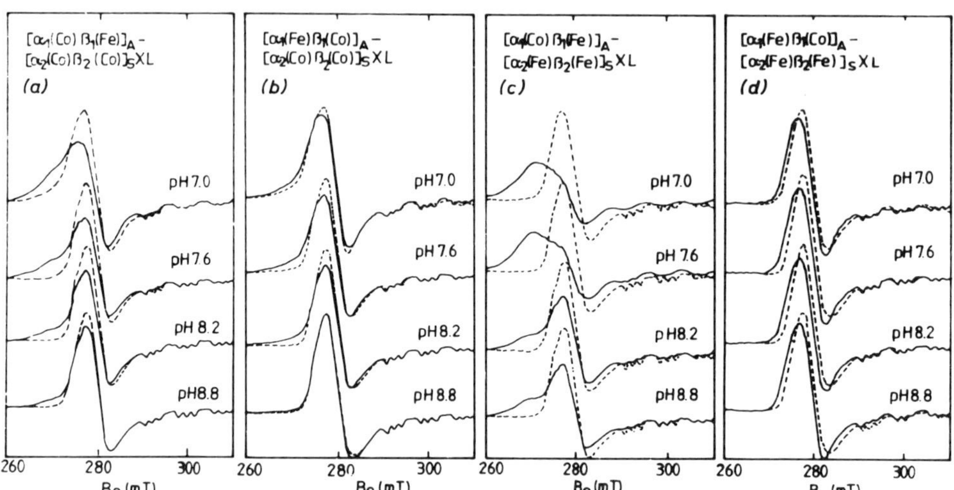

Figure 12. *X* band EPR spectra of the cross-linked hybrids in the pH range between 7.0 and 8.9. Solid and dashed lines represent fully deoxy and carbonmonoxy states, respectively, (*a*) [α(Co)β(Fe)]$_A$[α(Co)β(Co)]$_s$XL, (*b*) [α(Fe)β(Co)]$_A$[α(Co)β(Co)]$_s$XL, (*c*) [α(Co)β(Fe)]$_A$-[α(Fe)β(Fe)]$_s$XL, (*d*)[α(Fe)β(Co)]$_A$[α(Fe)β(Fe)]$_s$XL. [Reproduced with permission from (95).]

results, it is noted that symmetric cross-linked Fe—Co hybrids are indistinguishable from noncross-linked hybrids in both deoxy and carbonomonoxy conditions and, hence, cross-linking does not seem to affect the EPR spectra of the Fe—Co hybrid.

A comparison of noncross-linked Fe—CO hybrids with cross-linked hybrids with one cobaltous ion shows that deoxy$[\alpha(Co)\beta(Fe)]_A[\alpha(Fe)\beta(Fe)]_S$XL is similar to deoxy$[\alpha_2(Co)\beta_2(Fe)]$ and deoxy$[\alpha(Fe)\beta(Co)]_A[\alpha(Fe)\beta(Fe)]_S$XL is similar to the deoxy$[\alpha_2(Fe)\beta_2(Co)]$. Upon addition of CO to deoxy-$[\alpha(Co)\beta(Fe)]_A[\alpha(Fe)\beta(Fe)]_S$XL there is a drastic change in the $g_\perp$ value from 2.39 to 2.34. From these careful experiments, it is concluded that the electronic state of Co in the α subunit appears to reflect the allosteric properties.

The effect of adding IHP to the single cobaltous ion species has also been examined in the presence of IHP. When CO was bound to the β(Fe) subunit, the intensity of the broad signal ($g_\perp = 2.39$) decreased and the intensity of the sharp signal ($g_\perp = 2.34$) increased, but very small changes were observed when CO was added to the α(Fe) subunit. The effect of IHP on the binding of CO to β(Fe) is greater than to α(Fe). This observation may be consistent with the fact that IHP binds between the β subunits.

Another approach to investigate the molecular structure of hemoglobin through EPR is by using a spin label (reporter group). Two techniques have been employed. Technique A: A spin label has been attached to the sulfhydryl groups of the β93 cystein residue (97–99). Technique B: The spin label 2,2,5,5-tetramethyl-3-aminopyrrolidine 1-oxyl (100) has been attached to one of the two propionic groups of hemes. Of the two methods, Technique B seems to be advantageous since propionic acid spin labeling does not seem to affect the functional properties of hemoglobin as evidenced by oxygen-binding studies. Furthermore, the label can be attached to the heme group of either α or β chains; this enables one to monitor conformational changes in the region of the heme.

Technique A is followed because in deoxyHb HC2 Tyr-145(β) occupies a pocket between the F and H helices while in oxyHb this Tyr is expelled and F9 Cys-93(β) occupies this pocket; this residue has thus been identified as a sensitive monitor of the crucial conformational changes occurring in this region of the molecule (101). Earlier spin-label studies on Hb have been interpreted in terms of two orientations, namely, one strongly and the other weakly immobilized positions denoted by A and B in solution, A′ and B′ in polycrystalline sample, and A″ and B″ in single crystals (102). By comparing the EPR spectra of spin labels of different lengths attached to metHb, oxyHb, deoxyHb, and NiHb three different orientations of which two belonged to strongly immobilized orientations denoted by A_1 and A_2 and one weakly immobilized orientation B have been identified. These two immobilized orientations A_1 and A_2 could be related to the two possible stereoisomers that can be formed by the reaction across the C_1—C_2 double bond of N-substituted maleimides with F9

Cys-93(β) (103). This experiment also reveals that the relative amounts of weakly immobilized component increases in the order metHb ~ oxyHB < deoxyHb < NiHb.

The interaction of Cu^{II} with spin labeled hemoglobin has been used to identify four binding sites in human and horse hemoglobin. The reported distances between the Cu^{II} and spin label for the four sites are site 1 (9–13 Å), site 2 (<10 Å) from heme, and site 3(~17 Å), site 4(~7 Å) (104). The effect of adding Cu^{II} ions to $\alpha_2^{cat}\beta_2^{human}$], [$\alpha_2^{human}\beta_2^{cat}$] hybrid Hb has confirmed the earlier experiments on horse and human hemoglobin regarding the Cu^{II} oxidation of the β chains. Thus 2 mol of Cu^{II} ions are required for the rapid oxidation of the catHb and [$\alpha_2^{human}\beta_2^{cat}$], whereas 4 mol of Cu^{II} ions are required for the oxidation of humanHb β chain and [$\alpha_2^{cat}\beta_2^{human}$], which showed that the human β chain contains tight binding sites for Cu^{II} ions (105), which do not lead to oxidation. When Technique B was followed to prepare [$\alpha_2^{SL}(Fe^{3+})\beta_2(Fe{-}O_2)$] and [$\alpha_2(Fe{-}O_2)\beta_2^{SL}(Fe^{3+})$], EPR spectra were found to be different for these two hybrids indicating the nonequivalent conformational properties in Hb (106, 107). The first hybrid altered the line shape and resonance amplitude while the latter altered only the line shape. The difference in amplitude is attributed to changes in the paramagnetism of the heme iron in going from diamagnetic oxyheme to high-spin deoxyheme, since the magnetic dipolar interaction between two paramagnetic components decreases relative to the square of the distance. This experiment was repeated with [$\alpha_2^{SL}\beta_2$][$\alpha_2\beta_2^{SL}$] and a similar trend was observed. The EPR spectra of the effect of adding IHP to these hybrids showed small changes in the central resonance. There is also evidence for the binding of oxygen to the α chain prior to the β chain.

Electron paramagnetic resonance studies on nitric oxide Hb have given valuable information on the structural change in and around the porphyrin moiety (108). The effect of IHP on HbNO showed that nitrosyl Hb in the presence of IHP has two distinct peaks, one of which is characteristic of a five-coordinated NO–heme complex and the other of a six coordinated NO–heme complex, while in the absence of IHP only the latter was observed (109, 110). The same experiments performed on various symmetric hybrids and valency hybrids containing NO on either α or β subunits in the presence and absence of IHP showed that the nitrosyl heme of the α subunit is five coordinated in the T state and six coordinated in the R state. Also, the nitrosyl heme of the β subunit is mainly in the six-coordinated form even in the T state. Further studies have been performed on nitrosyl asymmetric hybrids (111) and chemically modified nitrosyl asymmetric hybrids by removing one or two carboxy terminal residue using carboxy peptidase digestion. A comparative chart of EPR spectra of all nitrosyl symmetric hybrids, valency nitrosyl hybrids (112), and chemically modified asymmetric hybrids is given in Fig. 13. A comparison of their EPR spectra shows there is a reduced relative intensity of the three lines due to NO in the

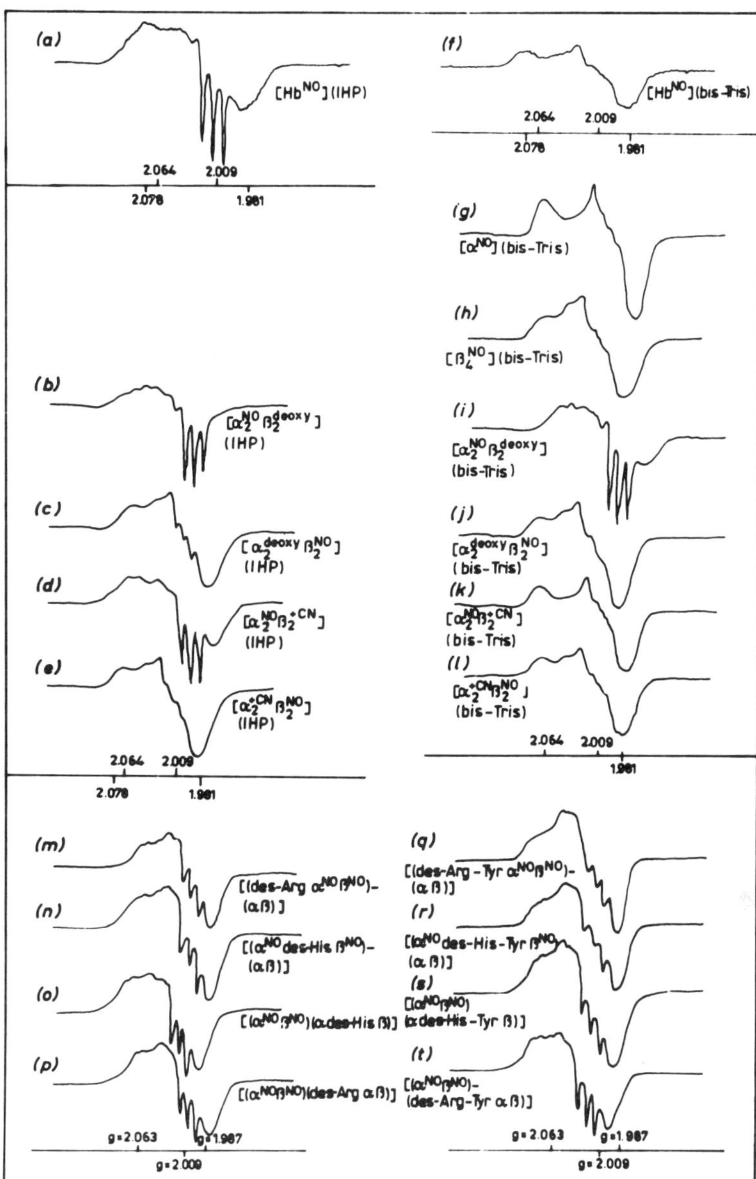

Figure 13. A comparative EPR spectra of nitrosyl hemoglobins: (*a*) Hb^{NO}, (*b*) $[\alpha_2^{NO}\beta_2^{deoxy}]$, (*c*) $[\alpha_2^{deoxy}\beta_2^{NO}]$, (*d*) $[\alpha_2^{NO}\beta_2^{+CN}]$, (*e*) $[\alpha_2^{+CN}\beta_2^{NO}]$ at 4.2 K in the presence of 2 m*M* IHP, Hb concentration 1 m*M* in 0.1 *M* bis–tris pH 6.5 spectra (*f*) and (*i*)–(*l*) are the same Nitrosyl-hemoglobin as mentioned above in the absence of IHP. Spectra (*g*) and (*h*) corresponds to Nitrosyl α and β chains. The other set of spectra includes (*m*) [(des-Arg $\alpha^{NO}\beta^{NO}$)($\alpha\beta$)], (*n*) [(α^{NO}des-Hisβ^{NO})($\alpha\beta$)], (*o*) [($\alpha^{NO}\beta^{NO}$)(αdes-Hisβ)], (*p*) [($\alpha^{NO}\beta^{NO}$)(des-Arg $\alpha\beta$)], (*q*) [des-Arg-Tyr $\alpha^{NO}\beta^{NO}$)($\alpha\beta$)], (*r*) [(α^{NO}des-His-Tyr β^{NO})($\alpha\beta$)], (*s*) [($\alpha^{NO}\beta^{NO}$)(αdes-His-Tyrβ)], (*t*) [($\alpha^{NO}\beta^{NO}$)(des-Arg-Tyr$\alpha\beta$)]. [Reproduced with permission from (111).]

fifth position in the absence of IHP. Furthermore, these lines have decreased intensities for the modified hemoglobins indicating that the noncovalent bonds involving carboxy terminal residue stabilizes the deoxyHb structure. In addition, the relative intensity of the triplet on modifying one specific carboxy terminus of the NO heme in the α_1 subunit is not symmetric, possibly due to nonequivalence of intersubunit interaction.

DeoxyHb is a high-spin paramagnetic system but EPR inactive due to fast electron relaxation. OxyHb is a low-spin diamagnetic system and, hence, EPR inactive. But when Fe^{II} is oxidized to Fe^{III}, which is EPR active (113), there can be two possibilities of Fe^{III} being low- or high-spin state. Oxidized Fe^{III} hemoglobin ligands bound in this ligand pocket includes water, hydroxide, fluoride, azide, thiocyanate, imidazole, and cyanide. The spin state for these complexes depends on the strength of the axial ligand and can be high spin (e.g., fluoride) or low spin (e.g., hydroxide). In some cases, low-spin Fe^{III}Hb complexes also exist when the exogenous ligand is replaced by endogenous amino acid side chains. A detailed review of these hemicromes had been dealt with earlier (114, 115). High-spin Fe^{III} hemoglobin can have two types of geometries whose EPR parameters are given in brackets, namely, tetragonal ($g_{\perp} = 6.0$, $g_{\parallel} = 2.0$) or rhombic complex ($g = 4.3$). Low-spin Fe^{III}Hb is formed in three different ways, namely, (a) with hydroxide at high pH ($g_1 = 2.59$, $g_2 = 2.17$, $g_3 = 1.83$); (b) with metHb frozen by freeze quenching thought to have both the distal histidine and a water associated with the heme ($g_1 = 2.83$, $g_2 = 2.26$, $g_3 = 1.63$); (c) with bis-histidine, that is, hemichrome producing during incubation at 210–250 K ($g_1 = 2.9814$, $g_2 = 2.28$, and $g_3 < 1.6$). With these basic ideas about ferric derivatives of Hb, attempts were made to study various valency hybrids (116) and natural valency hybrids (HbM) (117, 118), whose EPR spectra can be used to find the amount of low- and high-spin species in the solution and the nature of hemichrome formed. It is also found that these distal interactions perturb the $\alpha\beta$ dimer interface producing some changes in quaternary structure. By using EPR, it was found that 50–76% of mutant subunits of HbM (Boston) and HbM (Saskatoon) remained reduced in fresh blood (119).

The EPR studies on Hb, if the species gives a signal or not, play a vital role in understanding not only the R $\rightarrow$ T conformational changes but also in identifying the immediate environment of the metal ion. Electron nuclear double resonance (ENDOR) spectroscopy and electron spin echo envelop modulation (ESEEM) spectroscopy both provide uniquely sensitive probes of structure in the vicinity of the paramagnetic center including distance information. These techniques can pick up minute hyperfine splittings that could not be obtained directly from EPR. Although the ENDOR technique has been applied to a large number of biomolecules, not much work has been done in the area of hybrid hemoglobins. Simolo et al. (120) measured the ^{1}H, ^{13}C ENDOR spectra of

$[\alpha_2(Zn)\beta_2(Fe^{3+}CN)]$. It is interesting to note that there are no differences between the ^{13}C ENDOR spectra of $[\alpha_2(Zn)\beta_2(Fe^{3+}CN)]$ and $[\alpha_2(Fe^{3+}CN)$-$\beta_2(Fe{-}CO)]$. The ^{1}H ENDOR peaks have been assigned both from D_2O substitution results and by comparison with previous measurements (121).

Although the ^{1}H ENDOR has reported poor results, Mulks et al. (121) concluded that these results were similar to those obtained from ^{13}C ENDOR, suggesting that the structure around the metal center of the ligated Hb subunit is independent of the other subunits being ligated or unligated. These ENDOR results seem to be consistent with EXAFS measurement (see below). The coincidence of inferences from ENDOR and EXAFS results suggests that any displacement of metal–ligand bond lengths or angles must be quite small in going from strongly CO binding HbCO to weakly CO binding $[\alpha_2(Zn)\beta_2(Fe{-}CO)]$.

Peisach's group (122) had first applied ESEEM spectroscopy to cobalt substituted oxyHb. Subsequently, they studied $[\alpha_2(Co{-}O_2)\beta_2(Fe{-}O_2)]$ and $[\alpha_2(Fe{-}O_2)\beta_2(Co{-}O_2)]$ by ESEEM spectroscopy to measure electron nuclear hyperfine and nuclear quadrupole coupling constant from the N_ε of proximal histidine and also nuclear hyperfine coupling constants of the exchangeable ^{2}H in oxyCo subunits (123). The ESEEM spectra of $[\alpha_2(Co{-}O_2)\beta_2(Fe{-}O_2)]$ and $[\alpha_2(Fe{-}O_2)\beta_2(Co{-}O_2)]$ are shown in Fig. 14. The ^{14}N hyperfine couplings in oxyCo α subunits is smaller than in oxy Co β subunits. This result suggests that the α-subunit Co—O bond is shorter and more ionic, which correlates with the higher oxygen affinity for $[\alpha_2(Co)\beta_2(Fe{-}O_2)]$ than for $[\alpha_2(Fe{-}O_2)\beta_2(Co)]$. Similarly, the quadrupole coupling constant information reveals a shorter $Co{-}N_\epsilon$ bond in the α subunit. However, the combined results from g and quadrupole tensor orientations reveal a similarity in the Co—O—O bond angle in both oxyCo subunits in concurrence with earlier X-ray crystallographic studies on oxyHb.

B. Resonance Raman and IR Spectroscopy

Resonance Raman (RR) spectroscopy is similar to Raman spectroscopy resulting from inelastic scattering of light $h(\nu_0 + \nu_\nu)$ from the molecule when an incident photon $h\nu_0$ is supplied to the molecule. In RR spectroscopy, the excitation frequency corresponds to that of an electronic transition; this increases the Raman scattered intensity by about 10^6 times. The details of the principles, instrumentation, theory, and experimental approach are dealt with elsewhere (124). With the help of this technique, molecular bonding in and around the heme and heme–globin interactions are studied in detail. The initial studies used the frequency corresponding to the soret band. However, with the advent of advanced instrumentation and the tunable dye laser, any particular frequency in the visible or UV region can be selectively used for excitation to get information

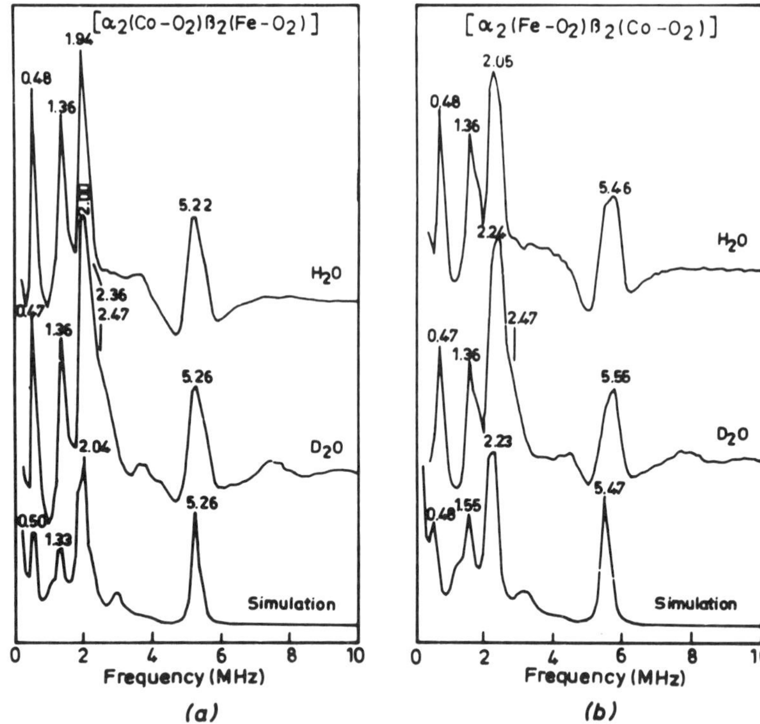

Figure 14. (*a*) ESEEM spectrum of hybrid hemoglobin [α_2(Co—O_2)β_2(Fe—O_2)] (top) in 0.1 *M* Hepes, pH 6.78 and H_2O, collected at 9.9626 GHz, 3506 G (g = 2.03), τ = 201 ns, and temperature = 4.2 K and (middle) in 0.1 *M* Hepes, pH 7.06 and D_2O, collected at 9.9145 GHz, 3490 G (g = 2.03), τ = 202 ns, and temperature = 4.2 K. Simulated (bottom) H_2O spectrum. (*b*) the ESEEM spectrum of hybrid hemoglobin [α_2(Fe—O_2)β_2(Co—O_2)]. Experimental parameters for this hybrid are almost the same as for the complementary hybrid mentioned before. [Reproduced with permission from H. C. Lee, J. Peisach, A. Tsuneshige, and T. Yonetani, *Biochemistry*, *34*, 6883 (1995). Copyright © 1995 American Chemical Society.]

on specific normal mode(s) and thereby get an excitation profile. Hence, a particular normal mode corresponding to the heme or any UV absorbing residue can be monitored. In this way, information can be obtained regarding changes at these centers with respect to pH, ligand binding, and so on. Also the sensitivity of this method has been further enhanced by the possibility of Raman difference technique, which permits reliable detection of frequency difference of less than 0.1 cm^{-1}. Small changes in spectrum using this technique can therefore be correlated to the structure and function of hemoglobin.

The first report of RR spectra on Hb was made by Spiro and co-workers (125), in 1972; subsequently, a large number of experiments were performed on Hb and its derivatives to understand the heme–protein structure and function. Attempts were then made to characterize both symmetric and asymmetric hybrids and to obtain information regarding possible intermediate conformations. Emphasis has been given to studies involving Fe—CO stretching (ν_{Fe-CO}), Fe—C—O bending (δ_{Fe-C-O}), O—O stretching (ν_{O-O}), and Fe—His(F8) stretching (ν_{Fe-His}).

Quite recently, ν_{C-O}, ν_{Fe-CO}, and δ_{Fe-C-O} frequencies of CO—HbM (Iwate) (αF8-His → Tyr), CO—HbM (Hyde Park) (βF8-His → Tyr) obtained from the RR measurement were found to be nearly identical to those of CO—HbA. In contrast, RR spectra of CO-HbM (Boston) (αE7-His → Tyr) and CO—HbM (Saskatoon) (βE7-His → Tyr) showed two new Raman bands, ν_{Fe-CO} at 490 cm^{-1} and ν_{CO} at 1972 cm^{-1}, from the abnormal subunits in addition to 505 cm^{-1} (ν_{Fe-CO}) and 1952 cm^{-1} (ν_{CO}) bands of normal subunits. Based on these results, they predicted that the orientation of CO in the abnormal subunits of HbM (Boston) is linear while in HbM (Iwate) is bent (126).

Valuable informations on the R → T linked change of the ν_{Fe-His} bond in α^{deoxy} and β^{deoxy} subunits have been obtained by RR studies on valency hybrids $[\alpha_2\beta_2^{+CN}]$, $[\alpha_2^{+CN}\beta_2]$, natural hybrids HbM (Milwaukee), HbM (Boston), and isolated chains, namely, α^{deoxy} and β^{deoxy} (127) for which the results are shown in Fig. 15. It is noted that valency hybrids are in the R state in the absence of IHP and in the T state in the presence of IHP. Also, ν_{Fe-His} frequency gives information on the strain imposed on the Fe—His bond; if the strain is more, the Fe—His bond becomes weaker and the ν_{Fe-His} band is shifted toward lower frequencies. It was found that ν_{Fe-His} for $[\alpha_2\beta_2^{+CN}]$ was found at 222 cm^{-1} in the absence of IHP and 207 cm^{-1} in the presence of IHP. The ν_{Fe-His} for $[\alpha_2^{+CN}\beta_2]$ was found at 224 cm^{-1} in the absence of IHP and 220 cm^{-1} in the presence of IHP. From these experiments, it is seen that the shift of ν_{Fe-His} during R → T transition was three times larger in α ($\Delta\nu = 20$ cm^{-1}) than in β ($\Delta\nu = 7$ cm^{-1}). The change in bond length in the α and β chains are $\Delta r_\alpha =$ 0.02 Å and $\Delta r_\beta = 0.008$ Å. By substituting these values in the following equation:

$$V(r) = D_e\{1 - \exp[-a(r - r_e)]\}^2$$

and with $D_e = 10$ kcal mol^{-1} we get $\Delta V_\alpha = 31$ and $\Delta V_\beta = 4$ cal mol^{-1}. So the strain energy developed is eight times larger in the α than in the β subunit. These results indicate that the tertiary structure of the α subunit is much more responsive to the quaternary structural transition in the deoxy form of the protein.

Recently, laser excitation at 230 nm has been employed to enhance the vibrational band of Tyr(Y) and Trp(W) residues (128). It was found that signals from these residues do not respond to tertiary changes associated with ligand or effector binding but only to quaternary arrangement. Either studies indicate that IHP was found to stabilize the T state of the valency hybrid only to a modest extent. In order to confirm this, tryptophan resonance studies were per-

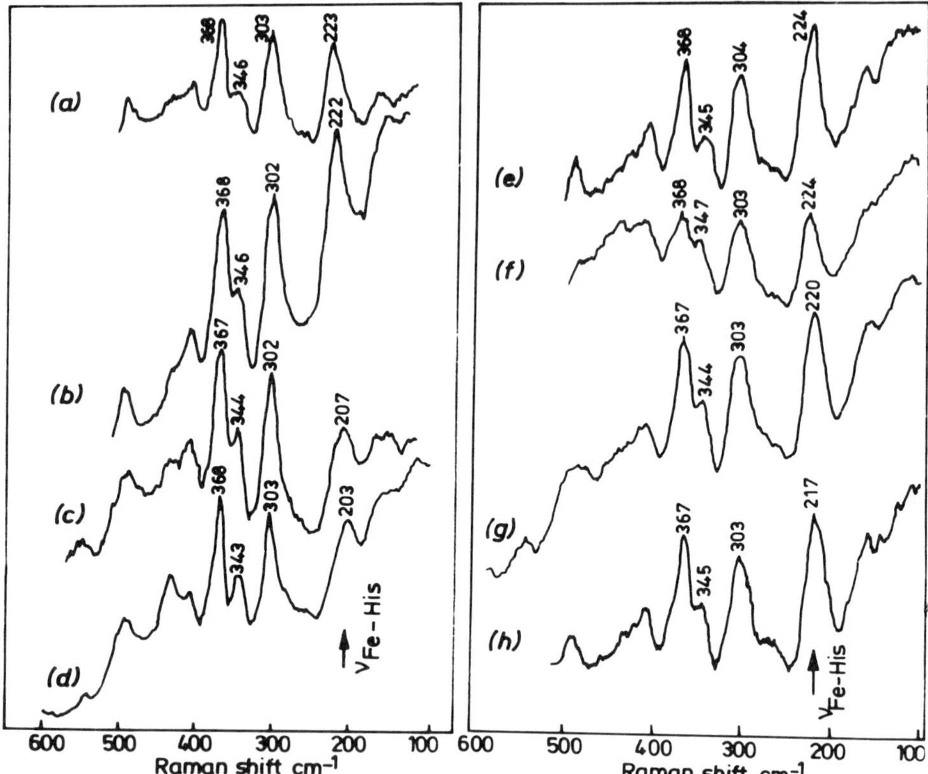

Figure 15. Resonance Raman spectra of isolated chains and valency hybrid hemoglobins excited at 441.6 nm. (*a*) Isolated α^{-SH} (deoxy) (pH 6.5), (*b*) $[\alpha_2^{deoxy}\beta_2^{+CN}]$ (pH 9.0), (*c*) $[\alpha_2^{deoxy}\beta^{+CN}]$ (pH 6.5 + IHP), (*d*) HbM (Milwaukee) (pH 6.5), (*e*) Isolatedβ^{-SH}(deoxy) (pH 6.5), (*f*) $[\alpha_2^{+CN}\beta_2^{deoxy}]$ (pH 9.0), (*g*) $[\alpha_2^{+CN}[\beta_2^{deoxy}]$ (pH 6.5 + IHP (*h*) HbM Boston (pH 6.5). [Reproduced with permission from (127).]

formed. The percentage of T state molecule after the addition of IHP was found to increase from 30 to 40% in $(\alpha\beta^{+CN})_2$ hybrids and 43–53% in $(\alpha^{+CN}\beta)_2$ hybrids from the difference spectrum between the valency hybrid and their corresponding CO adduct. This spectra shows that R and T state molecules are in equilibrium.

The effect of change of the porphyrin moiety (i.e., meso for proto) on ν_{Fe-His} stretching frequency by RR has been studied using various hybrids containing meso heme (129), reconstituted hemoglobin derivatives as deoxy tetramers, and cyanomet hybrids. These studies indicate that ν_{Fe-His} are found not to be substantially different from those of the native system (see comparative Chart, Table V).

TABLE V

Fe—His N_ϵ Stretching Frequency (in cm^{-1}) from RR Spectroscopy

Hybrid or Subunit	α	Hybrid or Subunit	β	Reference
α_P	223	β_P	224	129
α_M	223	β_M	225	129
$[\alpha_P\beta_P^{+CN}]_2$	223	$[\alpha_P^{+CN}\beta_P]$	223	129
$[\alpha_P\beta_P^{+CN}]_2$ + IHP	201/202	$[\alpha_P^{+CN}\beta_P]$ + IHP	219	129
$[\alpha_M\beta_M^{+CN}]_2$	223	$[\alpha_M^{+CN}\beta_M]$	222	129
$[\alpha_M\beta_M^{+CN}]_2$ + IHP	201/203	$[\alpha_M^{+CN}\beta_M]$ + IHP	218	129
$[\alpha_P\beta_M]_2$	206	$[\alpha_M\beta_P]_2$	216	129
$[\alpha_M\beta_P]_2$	206	$[\alpha_P\beta_M]_2$	219	129
$[\alpha_2(Fe)\beta_2(Ru-CO)]$		$[\alpha_2(Ru-CO)\beta_2(Fe)]$		
pH 6.4	206		223	136
pH 7.4	206		223	136
pH 7.4 + IHP	206		219	136
pH 8.4	217		223	136
$[\alpha_2(Fe)\beta_2(Ni)]$	201	$[\alpha_2(Ni)\beta_2(Fe)]$	216	135
des-Arg $[\alpha_2(Fe)\beta_2(Ni)]$	222	des-Arg$[\alpha_2(Ni)\beta_2(Fe)]$	218	135
$[\alpha_2(Fe)\beta_2(Co)]$		$[\alpha_2(Co)\beta_2(Fe)]$		
pH 6.6	201/212		218	135
pH 9.2	201/212		220	135
Freezing	216		222	135
des-Arg $[\alpha_2(Fe)\beta_2(CO)]$	220	des-Arg$[\alpha_2(Co)\beta_2(Fe)]$	222	130
$[\alpha_2(Fe-CO)\beta_2(Co-O_2)]$	223	$[\alpha_2(Co-O_2)\beta_2(Fe-CO)]$	222	130
HbM Milwaukee	203	HbM(Boston)	217	126
$[\alpha_2\beta_2^{+CN}]$ −IHP	210	$[\alpha_2^{+CN}\beta_2]$ −IHP	212	128
+IHP	203	+IHP	212	128
$[\alpha_2^{Oxy}\beta_2^{+CN}]$ −IHP	223	$[\alpha_2^{+CN}\beta_2^{oxy}]$ −IHP	222	128
+IHP	219	+IHP	222	128

In order to reexamine the trend revealed by previous experiments, RR studies were conducted on Fe—Co hybrid hemoglobins (130). From these experiments it was found that ν_{Fe-His} is located for HbA at 216 cm^{-1}, [α_2(Fe)β_2(Co)] at 201–212 cm^{-1}, [α_2(Co)β_2(Fe)] at 218 cm^{-1}, des-Arg[α_2(Fe)β_2(Co)] at 220 cm^{-1}, and des-Arg[α_2(Co)β_2(Fe)] at 222 cm^{-1}. Two peaks for [α_2(Fe)β_2(Co)] are attributed to conformational heterogeneity of the α subunit within the T structure. By photolysis of bound CO in the ferrous subunit, we can generate half-ligated hybrids such as [α_2(Fe)β_2(Co—O_2)] and [α_2(Co—O_2)β_2(Fe)], respectively, from [α_2(Fe—CO)β_2(Co—O_2)] and [α_2(Co—O_2)β_2(Fe—CO)], which are in the R quaternary structure similar to des-Arg[Fe—Co] hybrids in conformity with the EPR results. Transient Raman studies on these hybrids with respect to pH changes show that at high pH these two hybrids are in the R conformation; upon lowering the pH the contribution to the R state, stability for ligand binding in the α subunit is less compared to ligand binding in the β subunit (131) as shown in Fig. 16.

Presence of hygroden bonding between dioxygen and distal histidine (132) has been revealed by comparing ν_{O-O} stretching band for Co—Fe hybrids in nondeuterated and deuterated water using difference spectra.

Recently, 230-nm laser excitation was used to study the Fe—Co hybrid; the difference spectrum reveals a displacement of the E-helix toward the unligated heme as a result of the weakening of a hydrogen bond between the A and E helices Trp-14(α) and Trp-15(β) to the OH group of Thr-67(β) or Ser-72(α); so this destabilization is inferred to be a source of energy for the molecule to change from the R to the T state (133). The corresponding RR spectral details are found in Fig. 17. The T/R difference spectra were obtained with relative intensities by subtracting the spectra of cyanometHb from those of mono-, di- and triligated cyanomet hybrids Hb. By monitoring the hydrogen bond between A and E helices, UV RR spectra suggests a physical mechanism. When the first ligand enters the T state, the contacts are strong enough to withstand the perturbation. Upon entry of the second ligand, if it binds to the second subunit of the same dimer, the contact between $\alpha_1\beta_2$ are weakened; on the other hand, if it binds to the other side of the interface then the $\alpha_1\beta_2$ contact breaks resulting in rearrangement to the R state (if there is no equilibrium between T and R states). Upon entry of the third and fourth ligand the molecules are in the R state (134).

Later, RR studies have been performed on Fe—Ni (135) and Fe—Ru (136) hybrids. In the case of the Fe—Ni hybrid [α_2(Ni)β_2(Fe)], ν_{Fe-His} was observed at 216–218 cm^{-1} and 214–216 cm^{-1} in the presence and absence of IHP, whereas for [α_2(Fe)β_2(Ni)], ν_{Fe-His} was observed at 201–203 cm^{-1} (deoxy-like structure) and 220–221 cm^{-1} (oxy-like structure). The transition between these two stages occur around K_1 = 3–6 mmHg during oxygen binding. But when RR studies on chemically modified Ni—Fe hybrids were made, namely, des-

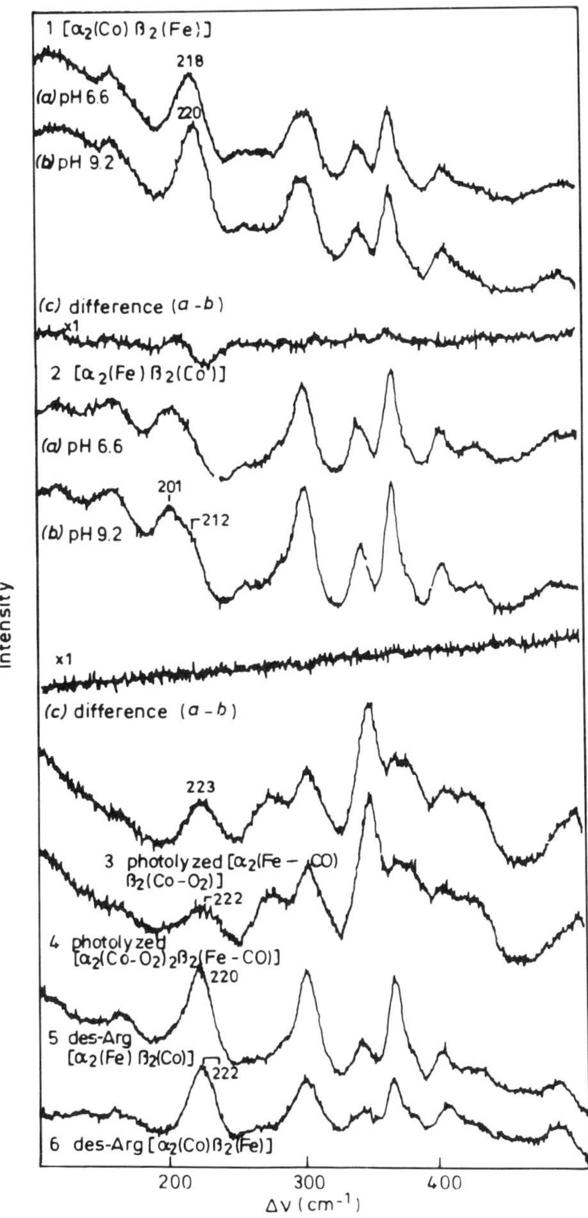

Figure 16. Comparative chart of RR spectra obtained for iron–cobalt hybrid hemoglobins with 441.6-nm excitation. (1*a*) [α_2(Co)β_2(Fe)] at pH 6.6; (1*b*) [α_2(Co)β_2(Fe)] at pH 9.2; (1*c*) difference between 1*a* and 1*b*; (2*a*) [α_2(Fe)β_2(Co)] at pH 6.6, (2*b*) [α_2(Fe)β_2(Co)] at pH 9.2 (2*c*) difference between 2*a* and 2*b*, (3) photolyzed [α_2(Fe—CO)β_2(Co—O_2)]; (4) photolyzed [α_2(Co)β_2(Fe—CO)]; (5) des-Arg [α_2(Fe)β_2(Co)], (6) des-Arg[α_2(Co)β_2(Fe)]. [Reproduced with permission from (130).]

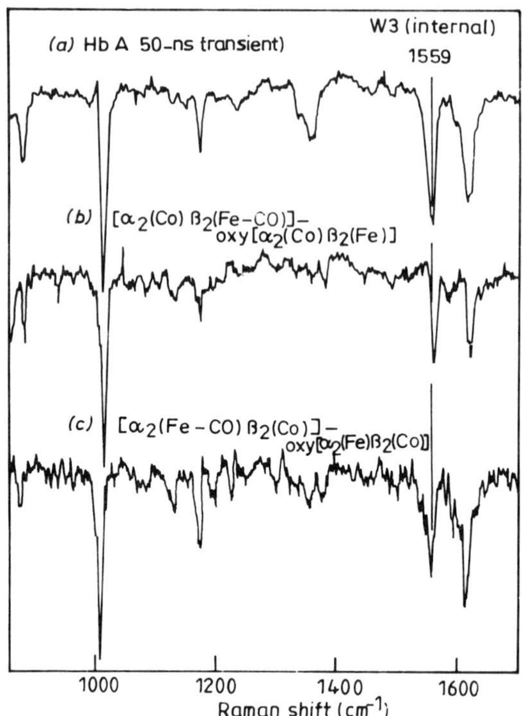

Figure 17. Comparison of the 50-ns transient difference spectrum with static difference spectra of the doubly ligated CO adduct minus the fully ligated O_2 adduct of the [Co—Fe] hybrid Hb with Co substituted for Fe in the α chain (*c*) or in the β chain (*b*). [Reprinted with permission from K. R. Rodgers and T. G. Spiro, *Science, 265,* 1697 (1994). Copyright © 1994 American Association for the Advancement of Science.]

Arg[α_2(Fe)β_2(Ni)], des-Arg[α_2(Ni)β_2(Fe)], ν_{Fe-His} are, respectively, at 222 and 218 cm^{-1}, which shows these two hybrids to be in the R conformation. For the Ru—Fe hybrid, [α_2(Ru—CO)β_2(Fe)], at pH 7.4 with IHP, ν_{Fe-His} was observed at 219 cm^{-1}. This result was characteristic of the T state, but at the increased pH of 8.4, the frequency shifted to 223 cm^{-1}. This result was characteristic of the R structure. For the complementary hybrid, namely, [α_2(Fe)β_2(Ru—CO)], at pH 7.4 with IHP, ν_{Fe-His} was observed at 206 cm^{-1}. As the pH increased to 8.4 the frequency shifted to 217 cm^{-1}, which was similar to the Ni—Fe hybrid.

With the development of methods to produce asymmetric cross-linked Fe—Co hybrids, RR spectral studies were performed, initially with $\beta\beta$ cross-linked hybrids (137) and later with $\alpha\alpha$ cross-linked hybrids (138). It has been confirmed that the quaternary structure including its dynamics are unaltered by cross-linking ν_{Fe-His}, ν_{O-O}, ν_{Fe-CO} stretching frequencies for 1, 2, and 3 iron containing tetramers are shown in Table VI, for $\alpha\alpha$ and $\beta\beta$ cross-linked symmetric and asymmetric hybrids. When comparing the ν_{Fe-His} frequencies of $\alpha\alpha$ and $\beta\beta$ cross-linked hybrids, the Fe—His bond of the α subunit received slightly larger strain in $\alpha\alpha$ than in $\beta\beta$ cross-linked hybrid, which in turn indicates that $\alpha\alpha$ cross-linking stabilizes the T structure slightly more than $\beta\beta$ cross linking. Again $\alpha\alpha$ cross-linked Fe—Co hybrid seems to be much closer to HbA than symmetric or $\beta\beta$ cross-linked Co—Fe hybrid. While comparing ν_{Fe-His} stretching and δ_{Fe-C-O} bending in $\alpha\alpha$ and $\beta\beta$ cross-linked hybrids, Fe—CO stretching is identical in both cases but Fe—C—O bending in $\alpha\alpha$ cross-linked hybrid gives extra peaks, which are assigned as a combination band. Again ν_{O-O} stretching in $\alpha\alpha$ and $\beta\beta$ cross-linked hybrids are identical. Quite recently, with the use of time-resolved RR spectroscopy in the nanosecond to microsecond interval, HbCO photolysis showed the presence of intermediates with different tertiary and quaternary structures (139).

Infrared spectroscopy, which is complementary to Raman spectroscopy, has been used to study the vibration of IR active ligands such as O_2, NO, CO, CN^-, N_3^- when they bind to iron in hemoglobin (140). Also, with the help of these spectroscopic techniques the change in protein structure with respect to pH, effector molecule, and temperature can be monitored. Details of various experimental approaches for the use of IR spectroscopy to study Hb reactions is given elsewhere (140).

For native HbA and Co—HbA, it was observed that band III in the near-IR region corresponds to a charge-transfer transition between the porphyrin π system and the iron π system (i.e., $a_{2u}(\pi) \rightarrow d_{zy}$). Band III was sensitive to (a) ligation and oxidation; (b) protein conformational changes with respect to pH, temperature, and effector; and (c) time-dependent protein dynamics. Hence, the same experiments were done for Fe—Mn hybrids, which are assumed to be intermediate between the R to T conformation (141). By using band III, alterations in the local heme pocket geometry is monitored because band III is sensitive to heme–histidine interactions. Tilting of the Fe—Hist bond would destabilize the Fe d_π orbital relative to porphyrin a_{1u} and a_{2u} orbitals producing a blue shift in the band. With this basic idea, the temperature effect was followed, which reflects contraction of the heme pocket at cryogenic temperature. Interpretations are based on peak position and full width at half-maxima of the peak. Variation in peak position with respect to pH is correlated to ν_{Fe-His} stretching frequencies in Fe—Co hybrids, obtained through RR techniques.

C. NMR Spectroscopy

Proton NMR (^{1}H NMR) is one of the most powerful techniques by means of which one can understand the dynamics, structure, and functional aspect of

TABLE VI

Cross-Linked Hybrid	$\beta\beta$ Cross-Linked Hybrid[a] A	B	C	$\alpha\alpha$ Cross-Linked Hybrid[a] A	B	C	References
I. The $\nu_{Fe-HisN\epsilon}$ Stretching Frequencies from RR Spectroscopy							
[α(Fe)β(Co)][α(Fe)β(Fe)]XL	209	220	220	213	218	218	137, 138
[α(Co)β(Fe)][α(Fe)β(Fe)]XL	216	223	223	214	220	221	137, 138
[α(Fe)β(Co)][α(Fe)β(Co)]XL	214/204	220	220	211/200	212/201	214/202	137, 138
[α(Co)β(Fe)][α(Co)β(Fe)]XL	216	223	223	216	219	221	137, 138
[α(Fe)β(Co)][α(Co)β(Co)]XL	215/206	216/206	219	211/200	212/201	214/202	137, 138
[α(Co)β(Fe)][α(Co)β(Co)]XL	216	220	222	216	218	221	137, 138

Cross-Linked Hybrid	$\beta\beta$ Cross-Linked Hybrid ν_{Fe-CO}	δ_{Fe-C-O}	$\alpha\alpha$ Cross Linked Hybrid ν_{Fe-CO}	δ_{Fe-C-O}	References
II. The ν_{Fe-CO} Stretching and δ_{Fe-C-O} Bending Frequencies					
[α(Fe)β(Co)][α(Fe)β(Fe)]XL	507	581	505	583/552	137, 138
[α(Co)β(Fe)][α(Fe)β(Fe)]XL	506	580	505	585/551	137, 138
[α(Fe)β(Co)][α(Fe)β(Co)]XL	506	581	508	581/550	137, 138
[α(Co)β(Fe)][α(Co)β(Fe)]XL	506	584	504	583	137, 138
[α(Fe)β(Co)][α(Co)β(Co)]XL	506	583	506	583	137, 138
[α(Co)β(Fe)][α(Co)β(Co)]XL	504	585	504	586	137, 138

Cross-Linked Hybrid	$\beta\beta$ Cross-Linked Hybrid ν_{O-O} 1	2	3	$\alpha\alpha$ Cross-Linked Hybrid ν_{O-O} 1	2	3	References
III. The ν_{O-O} Stretching Frequency							
[α(Fe)β(Co)][α(Fe)β(Fe)]XL	1226	1176	1135	1225	1175	1133	137, 138
[α(Co)β(Fe)][α(Fe)β(Fe)]XL	1226	1175	1135	1225	1174	1133	137, 138
[α(Fe)β(Co)][α(Fe)β(Co)]XL	1227	1176	1134	1226	1176	1134	137, 138
[α(Co)β(Fe)][α(Co)β(Fe)]XL	1227	1176	1136	1127	1173	1134	137, 138
[α(Fe)β(Co)][α(Co)β(Co)]XL	1225	1173	1134	1227	1172	1133	137, 138
[α(Co)β(Fe)][α(Co)β(Co)]XL	1227	1174	1136	1227	1174	1134	137, 138

[a]Here A, B, C represent stretching frequencies of deoxy[α_2(Fe)β_2(Co)], CO-photodissociated transient [α_2(Fe$-$CO)β_2(Co)] and CO-photodissociated transient [α_2(Fe$-$CO)β_2(Co$-O_2$)] states, respectively].

hemoglobin in solution from parameters such as chemical shift (δ), spin–spin coupling constant (J), spin–lattice relaxation time (T_1), and spin–spin relaxation time (T_2). The proton resonances are assigned to specific amino acid residues in the Hb molecule by comparing native HbA with various mutant hemoglobins (142, 143). The ^1H NMR spectra was initially used to study structural changes associated with the cooperative oxygenation of Hb. Later, emphasis was put on hybrid hemoglobins, which can provide information about possible intermediates in R $\rightarrow$ T conformation. In this respect, valency hybrids and metal reconstituted hybrids have been studied extensively. Apart from ^1H NMR, ^{31}P and ^{13}C NMR have also been used to investigate the properties of various hybrid derivatives of Hb. The NMR results obtained for various experimental approaches such as one dimensional (1D), two dimensional (2D), proton NMR (^1H NMR), correlated spectroscopy (COSY), and nuclear Overhauser and exchange spectroscopy (NOESY) give an excellent correlation with results obtained from other studies such as oxygen binding and thermodynamics (144–147).

Natural valency hybrids (HbMs), symmetric valency hybrids, and asymmetric valency hybrids have been subjected to considerable NMR investigations. In HbM, only two Fe^{II} ions are available to take up oxygen. The ^1H NMR of metHbA is observed in the range +10 to +90 ppm, whereas that of deoxyHbA is observed in the range from +6 to +20 ppm, both from the HDO position. Therefore, during ligation at the Fe^{II} subunit, the changes occurring in the Fe^{III} subunit can be monitored. Experiments have been done in such a way that initially the ^1H NMR of deoxyHbM (Milwaukee) (148, 149) was followed by stepwise addition of oxygen at the α subunit to monitor the changes in the β subunit. The oxy spectrum at about +48 ppm resonance is shifted to about +45 ppm resonance on deoxygenation, as shown in the Fig. 18. In order to find if the partially ligated valency hybrid is in the T or R state, the ^1H NMR of the deoxy valency hybrid was measured; two characteristic peaks at +9.4 and +6.4 ppm from H_2O reveal that this hybrid is in the T structure, similar to the "marker" peak of deoxyHbA.

However, the absence of these two peaks in the ^1H NMR of partially ligated hybrids indicates that these hybrids are in the R structure. When various concentrations of IHP were added to this ligated hybrid, the two characteristic peaks appeared back at the hybrid. The IHP concentration ratio of 1:0.25 indicates that the R state has been converted to the T state at this concentration. On further addition of IHP, the intensity of these two peaks increases as shown in Fig. 18. Similar ^1H NMR work has also been reported for HbM (Iwate) (α87 His $\rightarrow$ Tyr). However, in the ^1H NMR work of Shulman and co-workers (150–153) cyanomet symmetric valency hybrids, namely, $[(\alpha^{+CN}\beta)_2(\alpha^{+CN}\beta)_2]$, the hyperfine shifted proton resonances associated with high-spin deoxyheme and low-spin cyanomethemes overlap with each other in the spectral region (+10

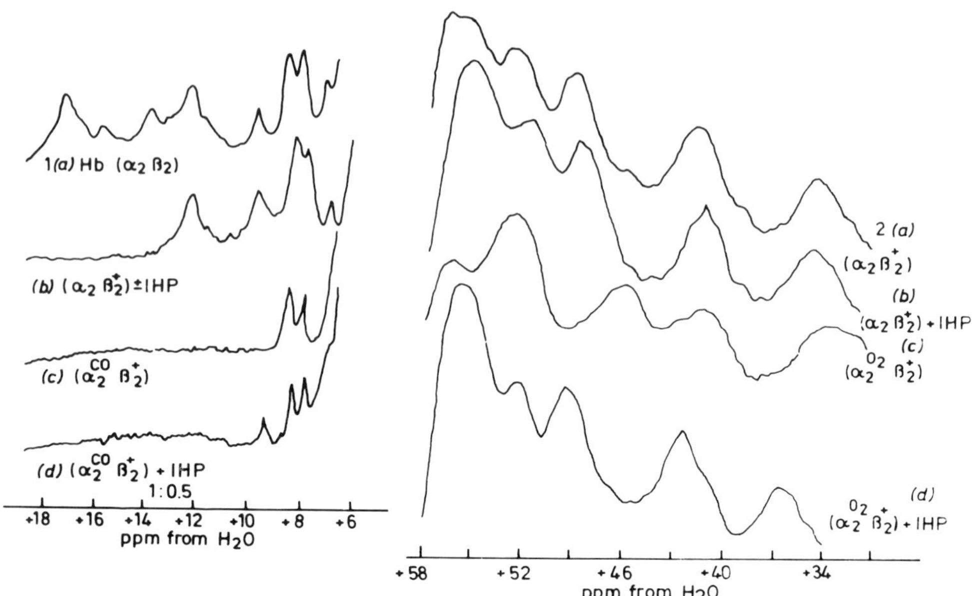

Figure 18. Comparative NMR spectra of HbM Milwaukee of (1*a*) Hb($\alpha_2\beta_2$); (1*b*) ($\alpha_2\beta_2^+$) ± IHP; (1*c*) ($\alpha_2^{CO}\beta_2^+$); (1*d*) ($\alpha_2^{CO}\beta_2^+$) + IHP; (2*a*) Hb($\alpha_2\beta_2^+$); (2*b*) ($\alpha_2\beta_2^+$) + IHP; (2*c*) ($\alpha_2^{O_2}\beta_2^+$); (2*d*) ($\alpha_2^{O_2}\beta_2^+$) + IHP. [Reproduced with permission from L. W., M. Fung, A. P. Minton, T. R. Lindstrom, A. V. Pisciotta, C-Ho *Biochemistry, 16,* 1452 (1977). Copyright © American Chemical Society.]

to +25 ppm from DSS) or (+5 to +20 ppm from HDO). Consequently, this hybrid did not provide sufficient information regarding quaternary structural changes during oxygenation of hemoglobin. A comparative ^{1}H NMR chart of symmetric valency hybrids along with cyanomet subunits α and β is shown in Fig. 19.

Cross-linked asymmetric hybrid Hb molecules containing the cyanomet center in one or two chains, namely, $[(\alpha^{+CN}\beta)_A(\alpha\beta)_C]$XL, $[(\alpha\beta^{+CN})_2(\alpha\beta)_2]$XL, $[(\alpha^{+CN}\beta^{+CN})_2(\alpha\beta)_2]$XL, and $[(\alpha\beta^{+CN})_2(\alpha^{+CN}\beta)_2]$XL (154), were (A and C denotes $\alpha\beta$ dimers from HbA and HbC, respectively, and XL denotes cross-linked Hb) showed hyperfine shifted proton resonances from the heme group and/or protons from the amino acids near the heme center. The chemical shift positions of these protons can be compared to the usual diamagnetic position due to hyperfine interaction between high-spin ferrous heme and low-spin ferric heme as discussed above. Signals from high-spin ferrous heme are broader than those from ferric heme; this is due to a hyperfine interaction between the unpaired electron from the Fe^{III} center and a nearby proton. A comparison of 600 MHz ^{1}H NMR spectra of all the four unsymmetric valency hybrids and their CO

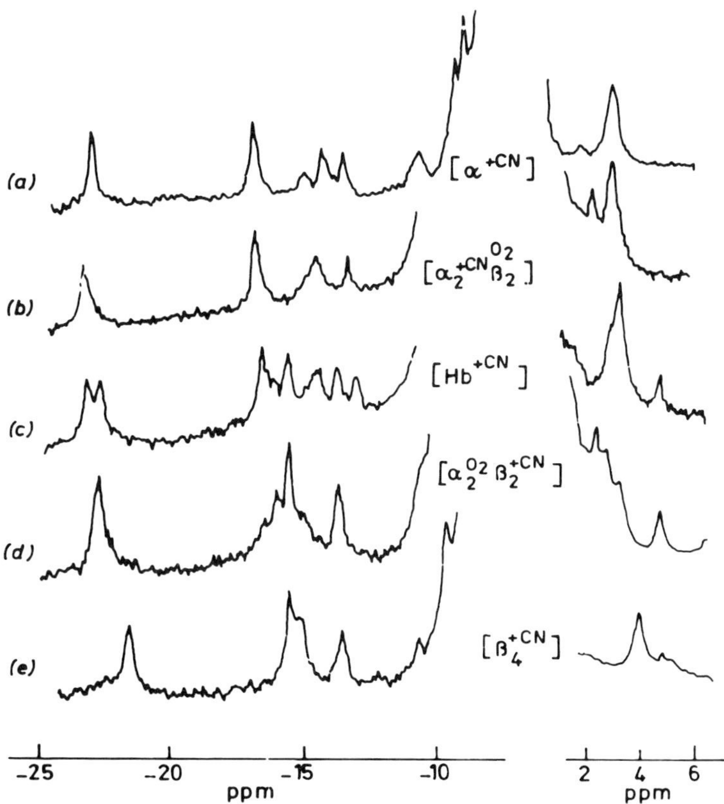

Figure 19. Comparative NMR spectra of oxy derivative of valency hybrids with that of oxidized Hb and oxidized subunit (*a*) $[\alpha^{+CN}]$, (*b*) $[\alpha_2^{+CN}\beta_2O_2]$, (*c*) $[Hb^{+CN}]$, (*d*) $[\alpha_2^{O_2}\beta_2^{+CN}]$, and (*e*) $[\beta^{+CN}]$. [Reproduced with permission from (151).]

derivatives are given in Fig. 20. From the spectral data, resonances at +18.1, +11.8, +9.7, and +7.3 ppm from the HDO position were assigned to the cyanomet α chain and those at +17.7, +10.8, and +8.7 ppm were assigned to the cyanomet β chain. The resonance at +9.2 ppm down field from H_2O was fixed as a marker for the T state of Hb, which arises from the intersubunit hydrogen bond between Asp-99(β) and Tyr-42(α). This resonance was observed for hybrids with one cyanomet chain but absent for hybrids with two cyanomet hybrids, this shows that the former are in the T state, whereas the latter are in the R state. A comparison of the relative signal intensities of the +9.2 ppm peak reveals that the hybrid with one cyanomet center has a reduced intensity in relation to deoxy Hb showing that the $\alpha_1\beta_2$ interfaces in these hybrids are altered.

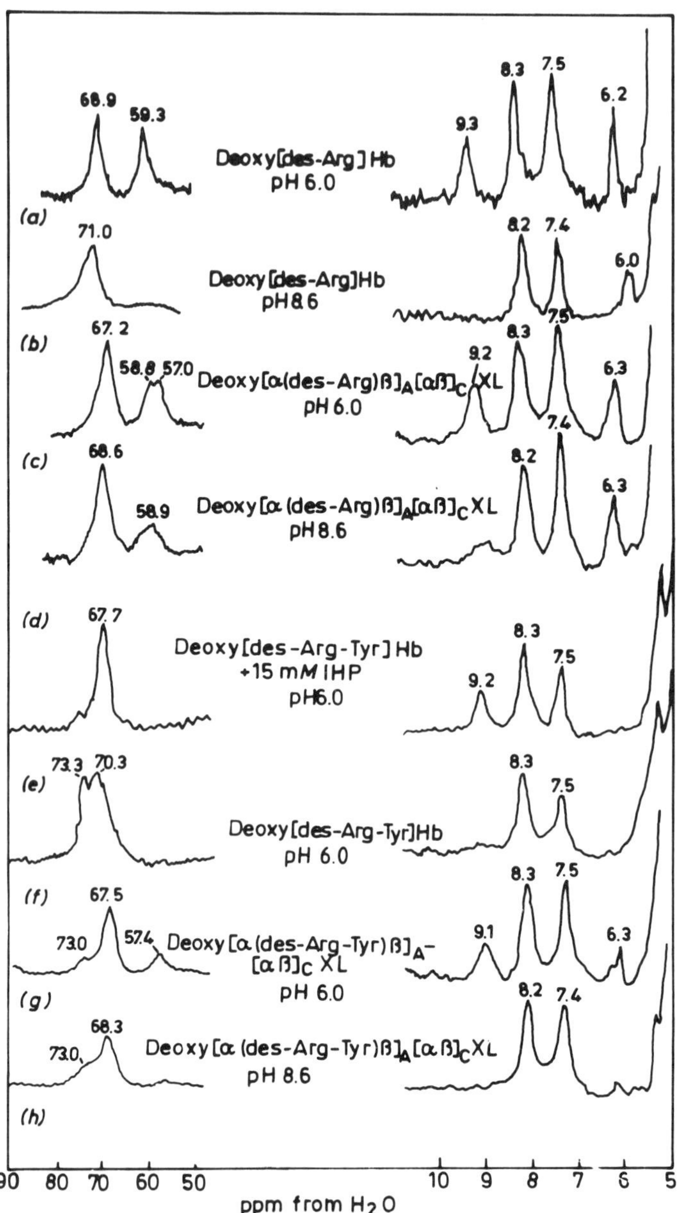

Figure 20. Comparative NMR spectra of cross-linked asymmetric valency hybrids at two different temperatures in both D_2O and H_2O. (*a*) $[(\alpha^{+CN}\beta)_A(\alpha\beta)_C]$XL, (*b*) $[(\alpha^{+CN}\beta^{CO})_A(\alpha^{CO}\beta^{CO})_C]$XL, (*c*) $[(\alpha\beta^{+CN})_A(\alpha\beta)_C]$XL, (*d*) $[(\alpha^{CO}\beta^{+CN})_A(\alpha^{CO}\beta^{CO})_C]$XL, (*e*) $[(\alpha\beta^{+CN})_A(\alpha^{+CN}\beta)_C]$XL, (*f*) $[(\alpha^{CO}\beta^{+CN})_A(\alpha^{+CN}\beta^{CO})_C]$XL, (*g*) $[(\alpha^{+CN}\beta^{+CN})_A(\alpha\beta)_C]$XL, and (*h*) $[(\alpha^{+CN}\beta^{+CN})_A(\alpha^{CO}\beta^{CO})_C]$XL. [Reproduced with permission from S. Miura and C. Ho, *Biochemistry, 21,* 6280 (1982). Copyright © 1982 American Chemical Society.]

In order to see the effect of salt bridges on the tertiary and quaternary structure of hemoglobin, the ^{1}H NMR studies of [(des-Arg $\alpha\beta)_A(\alpha\beta)_C$]XL, [(des-Arg-Tyr $\alpha\beta)_A(\alpha\beta)_C$]XL have been done (155). The ^{1}H NMR spectrum of this hybrid has been compared with those of deoxy[des-Arg]Hb and deoxy[des-Arg-Tyr]Hb at various pH values, as shown in Fig. 21. Exchangeable proton resonances at +58.5 and +71.0 ppm originate from the NH proton of proximal histidines of α and β chains, respectively. It is also possible to use the marker proton signals around 9.2 ppm to understand the T or R structure of cross-linked hemoglobin. In [des-Arg]Hb, at low pH, exchangeable proton resonance occurs at +9.3 ppm showing that it is in the T state and with a raise in pH the disappearance of the resonance peak indicates its transformation to the R state. Also, for this same [des-Arg]Hb there is a proximal histidyl NH exchangeable proton resonance at +59 ppm shifted downfield by 12 ppm on raising the pH. The NMR spectrum of deoxy[α(des-Arg)$\beta]_A[\alpha\beta]_C$ XL at pH 6.0 is identical to that of deoxy[des-Arg]Hb (pH 6.0) except for the appearance of a doublet in the exchangeable NH proton resonance of proximal histidine of the α chain at +58.8 ppm. These peaks correspond to the α chain of the $\alpha\beta$ dimer of HbC, whereas resonance at 57.0 ppm corresponds to the α chain in the [α(des-Arg)β] dimer of HbA. Upon increase in pH, the intensity of exchangeable NH proton resonance of proximal histidine of the α chain is lowered and a shift of +10 ppm is noticed, indicating that the $\alpha_1\beta_2$ or the $\alpha_2\beta_1$ subunit interface is altered. Absence of a peak at +9.4 ppm at a higher pH indicates that this hybrid exists in the R state.

Similar experiments were done on deoxy[des-Arg-Tyr]Hb in the presence and absence of IHP. It was found that this Hb exists in the R state in the absence of the IHP but upon addition of IHP it is converted to the T state, as found, respectively, by the disappearance and appearance of the resonance at +9.2 ppm. Similarly, deoxy[α(des-Arg-Tyr)$\beta]_A[\alpha\beta]_C$ XL at pH 6.0 is in the T state; on increasing the pH from 6.0 to 8.6 it changes to the R state. An increase of pH shifts the α subunit exchangeable NH proton resonance occurring at 57 ppm (from H_2O) to 14 ppm downfield as shown in Fig. 21. This figure shows that intersubunit interfaces are different in these modified hybrid hemoglobins. These valency hybrids were also studied by monitoring ^{19}F NMR, after chemical modification (i.e., attaching trifluoroacetonylate at Cys-93(β) residue). At near 50% ligand saturation the presence of two main intermediates were observed accounting for about 12% of Hb in solution, which were identified as $[\alpha_2\beta_2^{CO}]$ (4%) and $[\alpha_2\beta\beta^{CO}]$ (8%). Similarly, ^{31}P resonances were followed by monitoring the DPG residue as a function of ligand binding. The results permitted the proposal of a concerted transition model (156). In some cases, ^{13}C NMR studies monitored the ^{13}C labeled CO of the hybrid ($\alpha_2^{CO}\beta_2^+$). When H_2O on the ferric β chain was modified to CN, the observed shift of 0.16 ppm was attributed to a change in magnetic susceptibility of the environment and a change in the electronic charge of ^{13}C due to metal–ligand charge transfer (157).

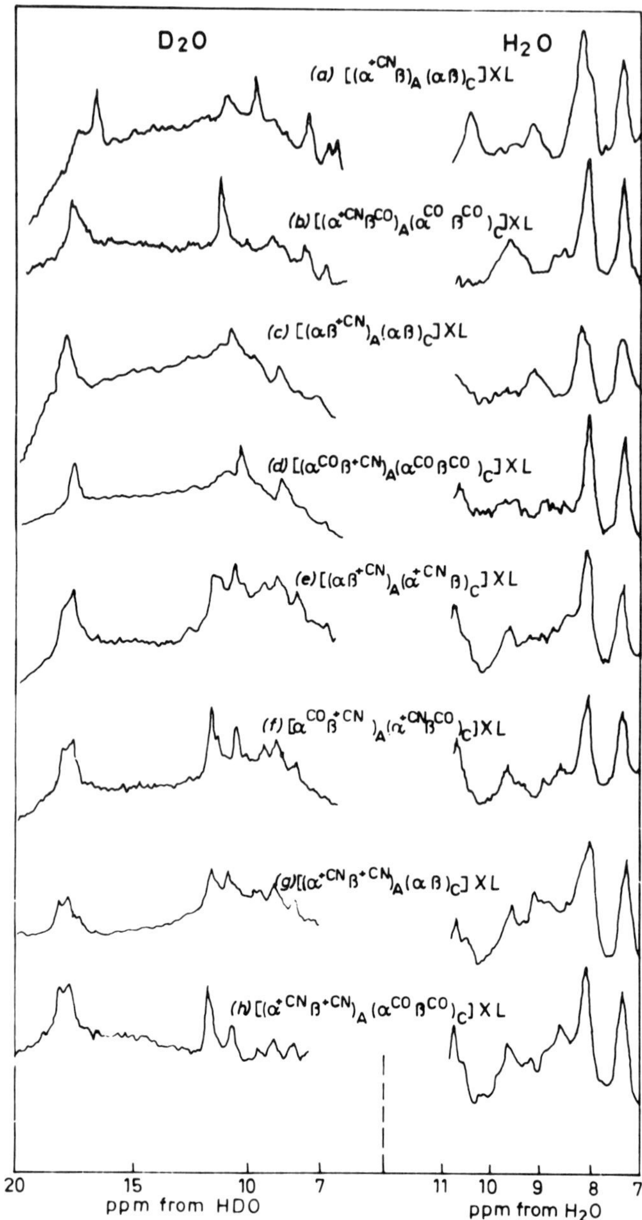

Figure 21. Comparative NMR specta of chemically modified cross-linked hybrid Hb with chemically modified native Hb in the deoxy state at different pH values. (*a*) [des-Arg]Hb at pH 6.0; (*b*) [des-Arg]Hb at pH 8.6; (*c*) $[\alpha(\text{des-Arg})\beta]_A[\alpha\beta]_C$XL at pH 6.0; (*d*) $[\alpha(\text{des-Arg})\beta]_A[\alpha\beta]_C$XL at pH 8.6; (*e*) [des-Arg-Tyr]Hb + 15 m*M* IHP at pH 6.0; (*f*) [des-Arg-Tyr]Hb at pH 6.0; (*g*) $[\alpha(\text{des-Arg-Tyr})\beta]_A[\alpha\beta]_C$XL at pH 6.0; and (*g*) $[\alpha(\text{des-Arg-Tyr})\beta]_A[\alpha\beta]_C$XL at pH 8.6. [Reproduced with permission from S. Miura and C. Ho, *Biochemistry,* 23, 2492 (1984). Copyright © 1984 American Chemical Society.]

Other NMR studies involved (a) hybrids prepared by replacing two of the protoporphyrin IX with other porphyrins such as meso and deutero (157–159); (b) hybrids prepared by removing metals from two out of four of the heme centers (semiHb hybrid) (160); and (c) deuterating one of the prosthetic groups on the porphyrin in Hb (161).

In semiHb [α_2(PP)β_2(Fe)] and [α_2(Fe)β_2(PP)] there is no axial bonding between proximal histidine and protoporphyrin in the α and β chains, respectively, and the strength of metalloporphyrin and proximal histidine in the other center becomes weak. Earlier NMR studies revealed a resonance at +14 ppm to be attributed to the intersubunit hydrogen bond between Tyr-42(α_1) and Asp-99(β_2) and resonance at 11 ppm to intersubunit hydrogen bonding between Asp-94(α_1) and Trp-37(β_2). Monitoring by these markers shows that deoxy-Fe—PP hybrids are in the deoxy quaternary state (T state). Addition of CO to [α_2(PP)β_2(Fe)] yields [α_2(PP)β_2(Fe—CO)] for which NMR markers are the same as [α_2(PP)β_2(Fe)], indicating that after ligation this hybrid is still in the T state. The pH variation and effector concentration on these hybrids shows that NMR markers in [α_2(PP)β_2(Fe)] are less influenced by pH or IHP compared to HbA, whereas those in [α_2(Fe)β_2(PP)] are sensitive to pH and IHP similar to those of deoxyHbA. From these studies, it was proposed that HbA preferentially assumes the deoxy quaternary structure in which there is weakening of Fe—His bonds in the α subunit.

In another set of NMR experiments containing mesoproto- and deuteroprotoporphyrin centers, it has been shown that heme peripheral modification results in a preferential downfield shift of proximal histidine NH signal for the β subunit indicating nonequivalent structural changes induced by heme modification in α and β subunits. In a similar way, NMR studies on chlorophyllide substituted HbA tetramer hybrids, $[\alpha^{chl}\beta]_2[\alpha\beta^{chl}]_2$ (where the chl superscript indicates the center with chlorophyllide–Fe modification) and its cyanomet derivatives have been studied.

The NMR studies related to changes in quaternary structure during ligand binding were made with the help of metal reconstituted hybrid hemoglobins, namely, [Co—Fe] (162, 163) [Ni—Fe] (164), [Ru—Fe] (165), [Mn—Fe] (166) [Zn—Fe] (120) hybrid hemoglobins. Note that metals such as Co, Ni, Mn, and Zn do not bind CO whereas the counterpart, namely, the ferrous iron, can bind CO. This permits structural changes to be observed in half-ligated Hb hybrids. The NMR spectra of these hybrids can then be compared to natural HbA. The NH exchangeable proton resonances for various hybrids are given in Fig. 22. A comparison of [α_2(Co)β_2(Fe)] and [α_2(Fe)β_2(Co)] before and after ligation of the Fe with Co reveals that there is an 8-ppm downfield shift of the proximal NH signal in the α(Co) subunit but only a 1-ppm shift in β(Co). These results show that metal–histidine bonds in deoxyCo subunit are strengthened after ligation of CO in the counterpart ferrous subunits.

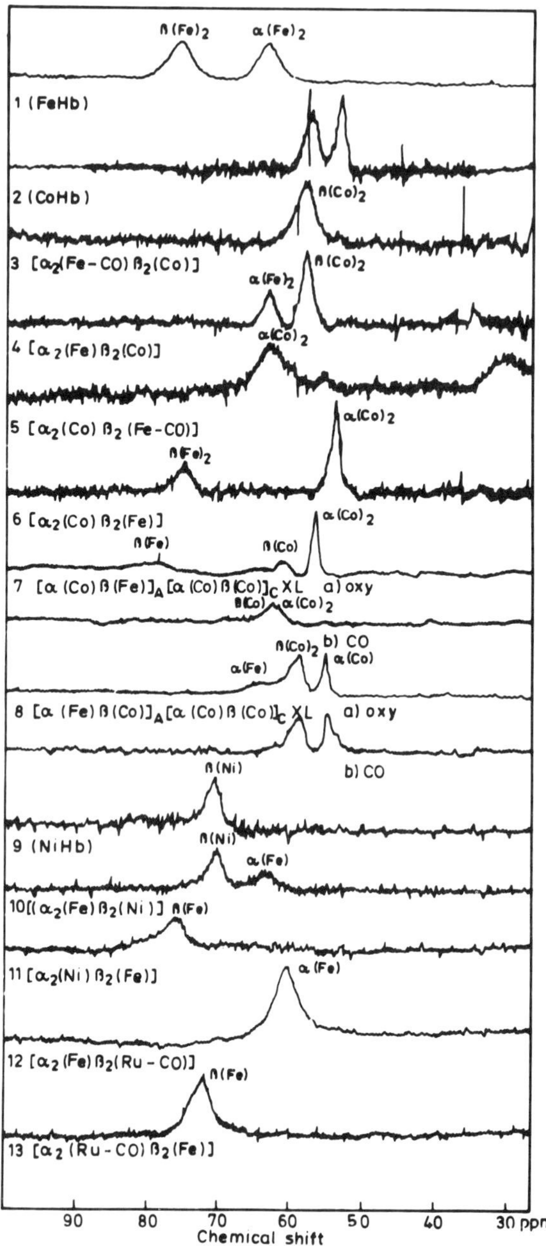

Figure 22. Comparative NMR spectra of hyperfine-shifted proximal histidyl $N_{\delta}H$ proton resonances for various hybrids and native Co—Hb and Ni—Hb compounds. (1) Native Hb; (2) Co—Hb; (3) $[\alpha_2(Fe-CO)\beta_2(Co)]$; (4) $[\alpha_2(Fe)\beta_2(Co)]$; (5) $[\alpha_2(Fe)\beta_2(Fe-CO)]$; (6) $[\alpha_2(Co)\beta_2(Fe)]$; (7*a*) $[\alpha(Co)\beta(Fe)]_A[\alpha(Co)\beta(Co)]_C$XL oxy derivative, (7*b*) carbonmonoxy derivative of the above; (8*a*)

In order to study monoligated Hb, cross-linked asymmetric Fe—Co hybrid containing only one Fe was prepared. The ^{1}H NMR measured on the deoxy, oxy, and Co, (96, 167) derivative of these hybrids shows that deoxy and CO derivatives are in the T state. Also, $[\alpha(Co)\beta(Fe-CO)]_A[\alpha(Co)\beta(Co)]_C$XL has more T-state character than $[\alpha(Fe-CO)\beta(Co)]_A[\alpha(Co)\beta(Co)]_C$XL (167). All these results collectively indicate that the first O_2 enters the α subunit, which induces less change in quaternary structure. This inference is in accordance with the crystallographic results on partially ligated hemoglobin. Similar information and inferences can be obtained from other hybrids for which the ^{1}H NMR spectra are shown in Fig. 22. A comparison of all the proton resonances for various hybrids observed under different conditions is given in the Table VII.

D. Electronic Spectroscopy

1. *UV-Visible*

Absorption spectral studies of hemoglobin in both the visible and UV region help us to monitor the changes in the heme group, which in turn helps in understanding the assembly of the tetramer and ligand-linked quaternary conformational changes in unligated and ligated hemoglobins. The assignments of the spectra have been based on extensive experimental investigation using polarized radiation and theoretical calculations (168). Optical measurements have been made in two ways, namely, static mode and a dynamic mode (169). The static mode of absorption study involved studying Hb in the ligated and unligated state. The dynamic mode of absorption study involves photolysis of the bound ligand by a very short laser pulse (i.e., picoseconds to nanosecond in duration), subsequently studying the ligand reassociation rate kinetically (by stop-flow technique). This technique allows us to understand the kinetics for quaternary transitions in Hb.

The T − R difference spectrum of HbA by laser photolysis experiments was first reported by Gibson (170). This technique has been elaborately applied to hybrid Hb molecules. Apart from normal absorption studies, UV-vis spectroscopy is used for obtaining the oxygen-binding curves that are dealt with separately in the later part of this chapter.

$[\alpha(Fe)\beta(Co)]_A$ $[\alpha(Co)\beta(Co)_C$XL oxy derivative; (8*b*) carbonmonoxy derivative of the above; (9) Ni—Hb; (10) $[\alpha_2(Fe)\beta_2(Ni)]$; (11) $[\alpha_2(Ni)\beta_2(Fe)]$; (12) $[\alpha_2(Fe)\beta_2(Ru-CO)]$; (13) $[\alpha_2(Ru-CO)\beta_2(Fe)]$. [Reproduced with permission from T. Inubushi, M. Ikeda-Saito, and T. Yonetani, *Biochemistry, 22,* 2904 (1983). Copyright © 1983 American Chemical Society. N. Shebayama, T. Inubushi, H. Morimoto, and T. Yonetani, *Biochemistry, 26,* 2194 (1987). Copyright © 1987 American Chemical Society. K. Ishimori and I. Morishima, *Biochemistry, 27,* 4060 (1988). Copyright © American Chemical Society. T. Inubushi, C. D'Ambrosio, M. Ikeda-Saito, and T. Yonetani, *J. Am. Chem. Soc., 108,* 3799 (1986). Copyright © 1986 American Chemical Society.]

TABLE VII

NMR Marker Band for Native and Hybrid Hemoglobins

Native/Hybrid Hb	Condition pH	Fe-His NH α	Fe-His NH β	α 42 Tyr-β 99 Asp	α 126 Asp-β 35 Tyr Deoxy	α 126 Asp-β 35 Tyr Ligated	β 98 Val-β 145 Tyr	α 94 Asp-β 102 Asp	References
Native Hb	6.5	59.5	72.1	9.4	8.3	8.2	6.4	5.8	158
Cross-linked Hb	6.5			9.3	8.2	8.2	6.4	5.8	154
MesoHb	6.5	61.3	76.3	9.6	8.2	8.1	6.3	5.7	158
DeuteroHb	6.5	60.0	73.2	9.6	8.2	8.1	6.6, 6.3	5.8	158
$[\alpha^{M}\beta^{P}]_2$	6.5	61.5	72.0	9.3	8.3	8.1	6.2	5.6	158
$[\alpha^{P}\beta^{M}]_2$	6.5	59.6	76.2	9.6	8.1	8.1	6.3	5.7	158
$[\alpha^{D}\beta^{P}]_2$	6.5	59.8	71.5	9.5	8.3	8.2	6.4, 6.1	5.7	158
$[\alpha^{P}\beta^{D}]_2$	6.5	59.2	73.0	9.6	8.2	8.3	6.6, 6.4	5.8	158
Deoxy[des-Arg]Hb	6.0	59.3	68.9	9.3	8.3		6.2		155
Deoxy[des-Arg]Hb	8.6		71.0		8.2		6.0		155
Deoxy$[\alpha$(des-Arg)$\beta]_A[\alpha\beta]_C$XL	6.0	58.8, 67.2	67.2	9.2	8.3		6.3		155
Deoxy$[\alpha$(des-Arg)$\beta]_A[\alpha\beta]_C$XL	8.6	58.9	68.6		8.2		6.3		155
Deoxy[des-Arg-Tyr]Hb + IHP	6.0		67.7	9.2	8.3				155
Deoxy[des-Arg-Tyr]Hb	8.6		73.3, 70.3		8.3				155
Deoxy$[\alpha$(des-arg-Tyr)$\beta]_A$ $[\alpha\beta_C$XL	6.0	57.4	73.0, 67.5	9.1	8.3		6.3		155
Deoxy$[\alpha$(des-Arg-Tyr)$\beta]_A$ $[\alpha\beta]_C$XL	8.6		68.3, 73.0		8.3				155
Deoxy NES HbA	6.0	57.6	58.6	9.3	8.2		6.4		155
Deoxy NES des-ArgHbA + IHP	6.0	59.5	68.2	9.2	8.3				155
Deoxy NES des-Arg HbA	6.0		71.2, 69.0		8.1				155
Deoxy$[\alpha$(des-Arg)$\beta]_A$ $[\alpha\beta$(NES)$]_C$XL	6.0	59.5	68.6		8.2				155
	8.6		70.2		8.1				155

Deoxy[α(des-Arg)β(NES)]$_A$									
[$\alpha\beta$]$_C$XL	6.0	59.7	69.2		8.2				155
	8.6		68.5, 72.0		8.2				155
[($\alpha^{+CN}\beta$)$_A$($\alpha\beta$)$_C$]XL	6.8			9.2	8.2	8.1			154
[($\alpha\beta^{+CN}$)$_A$($\alpha\beta$)$_C$]XL	6.8			9.1	8.2	8.0			154
[($\alpha^{+CN}\beta^{+CN}$)$_A$($\alpha\beta$)$_C$]XL	6.8			9.2, 9.6	8.1	8.1, 8.6			154
[($\alpha\beta^{+CN}$)$_A$($\alpha^{+CN}\beta$)$_C$]XL	6.8			9.6	8.1, 8.4	8.1, 8.6			154
HbM(Milwaukee)		56.0, 55.9		9.2	8.2	8.1			148
[α_2(Ru—CO)β_2(Ru—CO)]						8.2		5.9	165
[α_2(Ru—CO)β_2(Fe—O_2)]						8.1		5.9	165
[α_2(Fe—O_2)β_2(Ru—CO)]						8.2		5.9	165
[α_2(Ru—CO)β_2(Fe)]	8.7		72.9		8.2				165
	7.6		72.2		8.2				165
	6.5		71.7	9.0	8.2		6.1		165
	7.2		71.3	9.0	8.2		6.1		165
[α_2(Fe)β_2(Ru—CO)]	8.5	58.7		9.5	8.2		6.4		165
	6.9	58.7		9.6	8.1		6.4		165
	6.5	58.4		9.7	8.1		6.4		165
	7.2	58.0		9.8	8.1		6.4		165
CoHb		53.8(Co)	58.4(Co)	14.0			10.7, 10.3		162, 163
[α_2(Co)β_2(Fe)]		53.3(Co)	75.9(Fe)						162, 163
[α_2(Fe)β_2(Co)]		63.2(Fe)	57.9(Co)						162, 163
[α_2(Co)β_2(Fe—CO)]		61.4(Co)							162, 163
[α_2(Fe—CO)β_2(Co)]			58.7(Co)						162, 163
[α(Fe)β(Co)]$_A$									
[α(Co)β(Co)]$_C$XL									
Deoxy derivative		56.1(1Co)	61.0(2Co)	14.1					162, 163
		61.7(1Fe)							
CO derivative		56.1(1Co)	61.0(2Co)	14.1			10.8		167
[α(Co)β(Fe)]$_A$[α(Co)β(Co)]$_C$ XL									
Deoxy derivative		54.9(2Co)	78.7(1Fe)	13.9, 13.6			10.8		167
CO Derivative		62.5	62.5	9.4			10.8	10.5	167

TABLE VII (*Continued*)

Native/Hybrid Hb	Condition	Fe-His NH		α 42 Tyr-	α 126 Asp-β 35 Tyr		β 98 Val-	α 94 Asp-	References
	pH	α	β	β 99 Asp	Deoxy	Ligated	β 145 Tyr	β 102 Asp	
NiHb	7.4		70.8(Ni)				8.2	6.2	164
$[\alpha_2(Fe)\beta_2(Ni)]$	7.4	63.6(Fe)	70.5(Ni)						164
$[\alpha_2(Ni)\beta_2(Fe)]$	7.4		76.1(Fe)						164
$[\alpha_2(Fe-CO)\beta_2(Ni)]$	6.5		73.0(Ni)						164
	7.4		75.5(Ni)						164
	8.4		75.4(Ni)						164
(+IHP)	7.4		70.5(Ni)						164
$[\alpha_2(Ni)\beta_2(Fe-CO)]$	7.4		77.0(Ni)						164
	8.4		77.3(Ni)						164
ZnHb				9.4	8.0			5.6	120
$[\alpha_2(Fe-CO)\beta_2(Zn)]$						8.1		5.8	120
$[\alpha_2(Zn)\beta_2(Fe-CO)]$						8.2		5.8	120

The absorption positions from the spectral data obtained by the static mode for various hybrid hemoglobins–metal reconstituted hybrids (171), porphyrin substitute hybrids (172, 173), and semihemoglobin (174, 175) hybrids both in ligated and unligated states are given in Table VIII.

It is known that the addition of IHP to HbNO drastically changes the visible absorption spectrum. In order to study and elucidate this drastic change, absorption studies were carried out on the effect of IHP on a symmetric nitrosyl hybrid, namely, $[\alpha_2^{CO}\beta_2^{NO}]$ and $[\alpha_2^{NO}\beta_2^{CO}]$ (176–178). For these hybrids, the different absorption spectra have been measured in the presence and absence of IHP, which are shown in Fig. 23. It is seen that the difference spectrum of

TABLE VIII
UV–Visible Spectral Data for Some of the Hybrid Hb

Native or Hybrid Hb	Soret Region	*Q*-Band
Native Hb		
Deoxy	430(133)	555(13.1)
Oxy	415(131)	542(14.4), 577(15.3)
Carbonmonoxy	419(194)	569(14.3), 539(14.2)
Cyanometh	420(116)	543(11.1)
$[\alpha_2(\text{meso})\beta_2(\text{proto})]$		
Deoxy	423	540
Oxy	407	537, 569
Carbonmonoxy	409, 419	534, 561
Cyanometh		495, 625
$[\alpha_2(\text{proto})\beta_2(\text{Meso})]$		
Deoxy	424	550
Oxy	408	533, 571
Carbonmonoxy	410, 416	533, 562
Cyanometh		500, 623
RuHbCO	395(888)	517(54.4), 548(58.0)
$[\alpha_2(\text{Ru}-\text{CO})\beta_2(\text{Fe}-\text{O}_2)]$	395(536), 415(301)	519(39.4), 546(52.2), 576
$[\alpha_2(\text{Ru}-\text{CO})\beta_2(\text{Fe})]$	395(499), 427(253)	518(41.5), 550(51.8)
$[\alpha_2(\text{Fe}-\text{O}_2)\beta_2(\text{Ru}-\text{CO})]$	395(545), 415(310)	520(38.1), 545(50.0), 575
$[\alpha_2(\text{Fe})\beta_2(\text{Ru}-\text{CO})]$	396(498), 428(256)	517(39.4), 548(48.5)
NiHb	398(539), 420(438)	558(98.0)
$[\alpha_2(\text{Ni})\beta_2(\text{Fe})]$	398(528), 420(280)	558(112)
$[\alpha_2(\text{Fe})\beta_2(\text{Ni})]$	420(550)	540(54.1), 574(61.8)
$[\alpha_2(\text{Ni})\beta_2(\text{Fe}-\text{CO})]$	398(539), 420(521)	523(43.5), 558(98.2)
+IHP	398(508), 420(481)	523(45.1), 558(107)
$[\alpha_2(\text{Fe}-\text{CO})\beta_2(\text{Ni})]$	420(718)	539(57.9), 572(60.4)
+IHP	420(736)	539(57.9), 572(63.9)
$[\alpha_2(-)\beta_2(\text{Fe})]$		
Deoxy	430	554
Oxy	415	541, 577
Carbonmonoxy	419.5	538, 568
$[\alpha_2(-)\beta_2(\text{Fe}^{3+})]$	405.5	498, 630

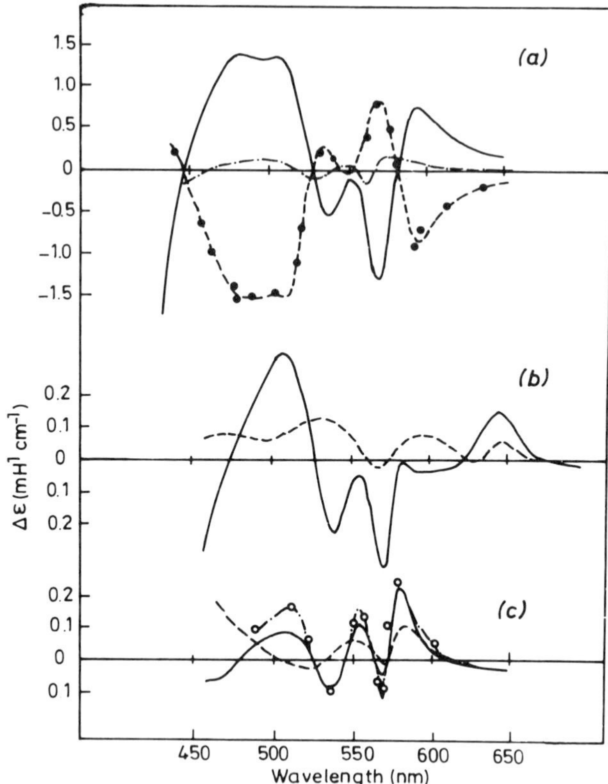

Figure 23. Difference spectra obtained for hybrids in the presence of IHP minus the absence of IHP. (*a*) Effect of IHP on $[\alpha_2^{NO}\beta_2^{CO}]$(—), $[\alpha_2^{CO}\beta_2^{NO}]$ (– – · – – · – – · –), and binding of NO to $\alpha_2^{NO}\beta_2^{deoxy}]$ (· · ·) with hybrid concentration of 30×10^{-6} *M* and IHP 60×10^{-6} *M*. (*b*) Effect of IHP on $[\alpha_2^{CO}\beta_2^{+H2O}]$ (—), $[\alpha_2^{+H2O}\beta_2^{CO}]$ (-----) with the same hybrid concentration of 30 m*M* and IHP 60 m*M* in 0.13 *M* bis–tris buffer, pH 6.6 at 4°C). (*c*) Effect of IHP on $[\alpha_2^{CO}\beta_2^{+CN}[$ (—), $[\alpha_2^{+CN}\beta_2^{CO}]$(-----). The open circles represent the difference spectra for HbCO with the same hybrid concentration, IHP, as above. [Reproduced with permission from (181).]

$[\alpha_2^{NO}\beta_2^{CO}]$ is completely different from that of $[\alpha_2^{CO}\beta_2^{NO}]$. This spectra shows that IHP induces a new quaternary conformation in HbNO with deoxy character. Later, attempts were made to study the asymmetric hybrids $[\alpha^{NO}\beta^{NO}\alpha^{oxy}\beta^{oxy}]$ and $[\alpha^{NO}\beta^{NO}\alpha^{deoxy}\beta^{deoxy}]$ (179). These hybrids were prepared by mixing HbNO and HbO_2, followed by deoxygenation. Here difference spectra for asymmetric hybrids were obtained from two different spectra of HbNO + deoxyHb against HbNO and deoxyHb. The difference spectra due to an asymmetric hybrid $[\alpha^{NO}\beta^{NO}\alpha^{deoxy}\beta^{deoxy}]$ and a symmetric hybrid $[\alpha_2^{NO}\beta_2^{CO}]$ are different, revealing their distinct properties. Carbon monoxide kinetics were measured for both asymmetric and symmetric deoxy hybrids. The results show that the asymmetric hybrid reacts more slowly with CO than the symmetric hybrid. This result in turn reveals that asymmetric hybrids are in a deoxy T state contrary to symmetric hybrids, which are in the ligated R state.

The existence of valency hybrids $[\alpha_2^+\beta_2]$, $[\alpha_2\beta_2^+]$, $[\alpha_2^+\beta_2^{CO}]$, $[\alpha_2^{CO}\beta_2^+]$ in solution during partial reduction of methemoglobin was confirmed by Perutz et al. (180) who prepared samples in such a way that 0.1, 0.2, 0.4, 0.6, 0.8, and 1.2 μM of sodium dithionite was added to corresponding portions of 38 μM metheme and by recording the difference spectra for deoxy and CO derivatives. The difference spectra of hybrid mixtures showed a maxima at 630 nm, corresponding to metHb and a maxima at 645–650 nm, attributed to changes in quaternary structure. The quaternary structure was assumed to alter the tension between heme-linked histidine and the porphyrin ring. This then changes the bond lengths and affects the spin state of the heme perturbing the spectra.

The effect of IHP on the visible absorption spectra of ligated valency hybrids $[\alpha_2^{CO}\beta_2^{+H_2O}]$, $[\alpha_2^{+H_2O}\beta_2^{CO}]$, $[\alpha_2^{+CN}\beta_2^{CO}]$ and $[\alpha_2^{CO}\beta_2^{+CN}]$ (181, 182) have been carried out. The difference spectra obtained for these hybrids in the presence and absence of IHP are shown in the Fig. 23. It is seen from the spectrum that the β^{+H^2O} subunit contributes more to the absorption change rather than α^{+H_2O}; in other words, α^{CO} chains are more sensitive to IHP than β^{CO}. For the same set of hybrids, femtosecond pulse experiments have produced transient absorption difference spectra.

The UV absorption difference spectra between the deoxy and oxy valency hybrids $[\alpha_2^+\beta_2]$, $[\alpha_2\beta_2^+]$ with different ligands bound to the oxidized chain such as F^-, H_2O, N_3^-, and CN^- have been reported (183). These spectra are shown in Fig. 24. From these results, it is seen that the spectral characteristics of high-spin hybrids (with ligands, H_2O) are considered to be similar to those of low-spin hybrids (with ligands, CN^-); but changes in difference spectra are due to changes at the $\alpha_1\beta_2$ interface, which alters the iron proximal His(F8) bond of the parent ferrous subunit.

Transient optical absorption difference spectra from deoxy minus carbonmonoxy were obtained at room temperature in the nanosecond time scale for the [Fe—Co] hybrids [α_2(Co)β_2(Fe—Co)] and [α_2(Fe—CO)β_2(Co)] (184, 185) as shown in Fig. 25. This difference spectra does not show any features in the Co porphyrin absorption region. In order to find the relative contribution of α and β subunits to the germinate recombination and conformation change within the tetramer, ligand rebinding and spectral changes associated with photodissociation for hybrids were followed. Five different relaxation's were observed, the spectral changes associated with each relaxation are shown in Fig. 25. Comparison of the spectral changes accompanying each relaxation after photolysis of [α_2(Co)β_2(Fe—CO)] with HbCO suggests that the α and β hemes also make comparable contribution to the spectral changes in the R-state Hb. Analysis of

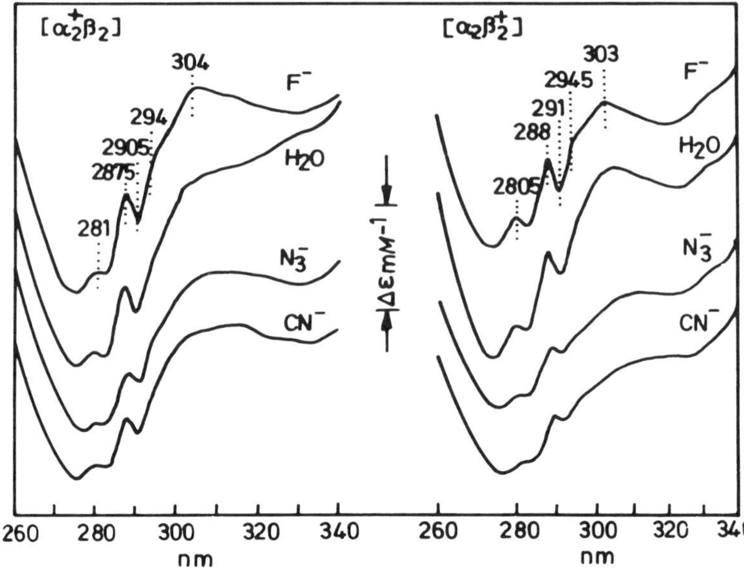

Figure 24. Difference spectra of the hybrids $[\alpha_2^+\beta_2]$ (left) and $[\alpha_2\beta_2^+]$ (right). Here the ferric ligand concentration 0.1 *M* NaF for F^-; 1 m*M* NaN_3 for $N^-{}_3$, 1 m*M* NaCN for CN^- with heme concentration of 60 m*M* in 50 m*M* bis-tris buffer, pH 7.0; the same with 0.1 *M* NaCl at 25°C. [Reprinted from K. Mawatari, S. Matsukawa, Y. Yoneyama, and Y. Takeda, 313, Biochem. Biophys. Acta, 913/313, (1987) Copyright © 1987 with kind permission of Elsevier Science NL, Sara Burgerhartstraat 25, 1055 KV Amsterdam, The Netherlands.]

the spectra for the $[\alpha_2(Fe{-}CO)\beta_2(Co)]$ hybrid are consistent with the above conclusion even though the analysis complicated by the presence of R and T species.

Similar to the above experiments, nanosecond absorption spectra were measured in the soret and near-UV spectral region of HbCO after photolysis in order

---→

Figure 25. (1*a*) Static absorption spectra of $[\alpha_2(Fe{-}CO)\beta_2(Co)]$ and $[\alpha_2(Fe)\beta_2(Co)]$, where the concentration of the hybrid molecule is 171$-\mu M$ heme. (2*a*) Static absorption spectra of $[\alpha_2(Co)\beta_2(Fe\text{-}CO)]$ and $[\alpha_2(Co)\beta_2(Fe)]$, where the concentration of the hybrid molecule is 128-μM heme. (1*b*), and (2*b*) represents unligated minus ligated difference spectra. (1*c*) and 2(*c*) represent ligand rebinding curve for the corresponding hybrids. Here the solid line is calculated from the results of the simultaneous least-squares fits to the first three column of V matrix (107). The dashed line is the corrected ligand-rebinding curve calculated from the sum of five exponential with relaxation time obtained from the fits of the first column of the V matrix. (1*d*–*e*), (2*d*–*e*) represents the deoxyheme spectral change associated with each relaxation. The dashed lines are the normalized difference spectra and the solid lines are the renormalized spectra. [Reproduced with permission from J. Hofrichter, E. R. Henry, J. H. Sommer, R. L. Deutsch, M. Ikeda-Saito, T. Yonetani, and W. A. Eaton, *Biochemistry*, 24, 2667 (1985). Copyright © 1985 American Chemical Society.]

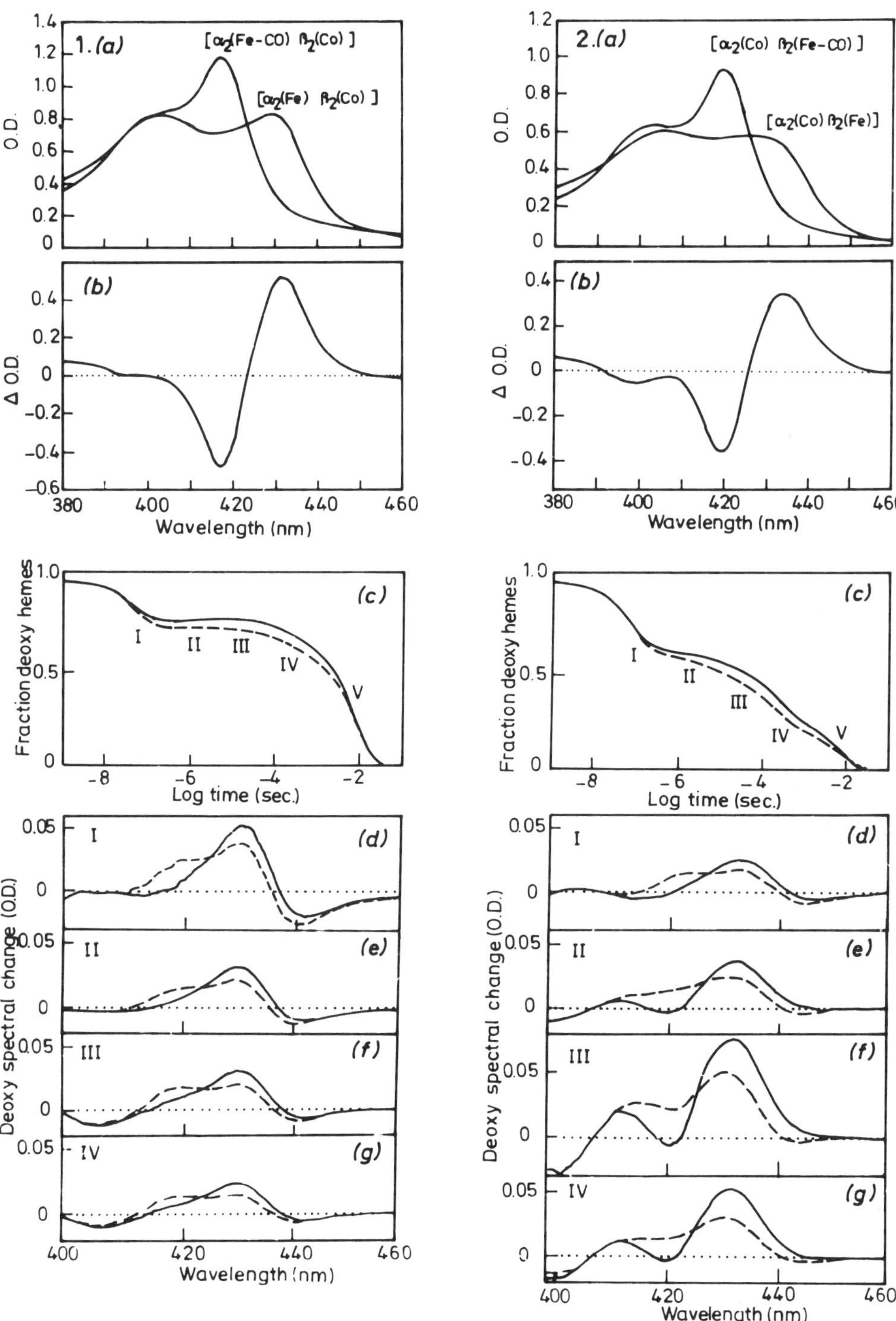
1.(a)
[α2(Fe-CO) β2(Co)]
[α2(Fe) β2(Co)]
O.D.
(b)
Δ O.D.
Wavelength (nm)
(c)
Fraction deoxy hemes
I
II
III
IV
V
Log time (sec.)
(d)
(e)
(f)
(g)
Deoxy spectral change (O.D.)
Wavelength (nm)
2.(a)
[α2(Co) β2(Fe-CO)]
[α2(Co) β2(Fe)]
O.D.
(b)
Δ O.D.
Wavelength (nm)
(c)
Fraction deoxy hemes
Log time (sec.)
(d)
(e)
(f)
(g)
Deoxy spectral change (O.D.)
Wavelength (nm)

to probe the allosteric intermediates in hemoglobin (185a). Intermediates were distinguished on the basis of extent of heme ligation and intraprotein relaxation.

2. *Fluorescence*

Fluorescence is observed when there is a deexcitation from the first or second excited state to the ground state of the same spin multiplicity after absorption of energy by the system. The principles and application of front-face fluorescence and time-resolved fluorescence in the context of Hb is dealt with in detail elsewhere (186). The primary condition in order to study the fluorescence of the heme in Hb is that there should be either free protoporphyrin IX or zinc protoporphyrin. In normal Hb, this condition is not met but we still observe fluorescence, which is due to aromatic amino acid residues such as Trp, Tyr. In going from the R → T state, there is an 18.25% increase in fluorescence emission as a result of quaternary conformational transition and consequent changes in the environment of Tyr and Trp residues (187).

In yet another mode of studying fluorescence, an extrinsic fluorescent probe like 4(5)-(*N*-maleinisoimido)rhodamine B (188) or 5-iodoacetamido fluorescein (189, 190) can be covalently attached to the Cys-93(β) residue, which serves as a reporter group. This method is similar to attaching a spin label to a particular residue or functional group and studying it by EPR. This method can give information on the structure–function relationship and inter- and intramolecular distances.

In order to study the dynamics of protein structure for the intermediate in going from the R → T state by fluorescence, symmetric hybrids of the type [α_2(Fe)β_2(PP)], [α_2(PP)β_2(Fe)], [α_2(Zn)β_2(Fe)], [α_2(Fe)β_2(Zn)], and [α_2(Zn)-β_2(PP)], have been used. These states contain either protoporphyrin (PP) or ZnPP in α or β chains (191).

Fluorescence decay times for these hybrids are given in the Table IX. This

TABLE IX
Fluorescence Lifetime (191)

	τ_f	
Hybrid or Species	Oxy	Deoxy
PP in buffer		14.1
PPHb		18.1
[α_2(Fe)β_2(PP)]	17.8	17.2
[α_2(PP)β_2(Fe)]	13.4	7.1
[α_2(Zn)α_2(Fe)]	2.8	2.7
[α_2(Fe)β_2(Zn)]	2.5	2.6
[α_2(Zn)β_2(Zn)]	(593 emi)	3.5
	(625 emi)	18.5

time is a direct measure of the energy transfer between the two fluorescent porphyrin molecules in the same tetramer. The [α_2(Fe)β_2(PP)] complex has a very short fluorescence lifetime and, hence, has very low energy-tansfer properties, whereas [α_2(PP)β_2(Fe)] has an even shorter fluorescence lifetime and, hence, has much less energy transfer compared to PPHb. This change in decay time reflects the porphyrin orientational change and provides results consistent with NMR and EPR experiments (191).

In order to further study the porphyrin–protein interaction, fluorescence line narrowing spectroscopy (which is a high-resolution fluorescence technique that involves laser excitation producing vibrationally resolved emission spectra) was used on the hybrids, namely, [α_2(Fe)β_2(PP)] and [α_2(PP)β_2(Fe)] in the temperature region 5–50 K (192). The observed spectral change is correlated to the motion of the polypeptide chain and twist or bend in the porphyrin macrocycle. The population distribution function of [α_2(Fe)β_2(PP)], [α_2(PP)β_2(Fe)], and [α_2(PP)β_2(PP)], is given in the Fig. 26. The results from these experiment show that the configuration of porphyrin in the two hybrids is nearly the same.

3. *Circular Dichroism*

Theory of CD has been discussed by Geraci and Parkhurst (193) elsewhere, while its application has been such as recently discussed by Zentz et al. (194).

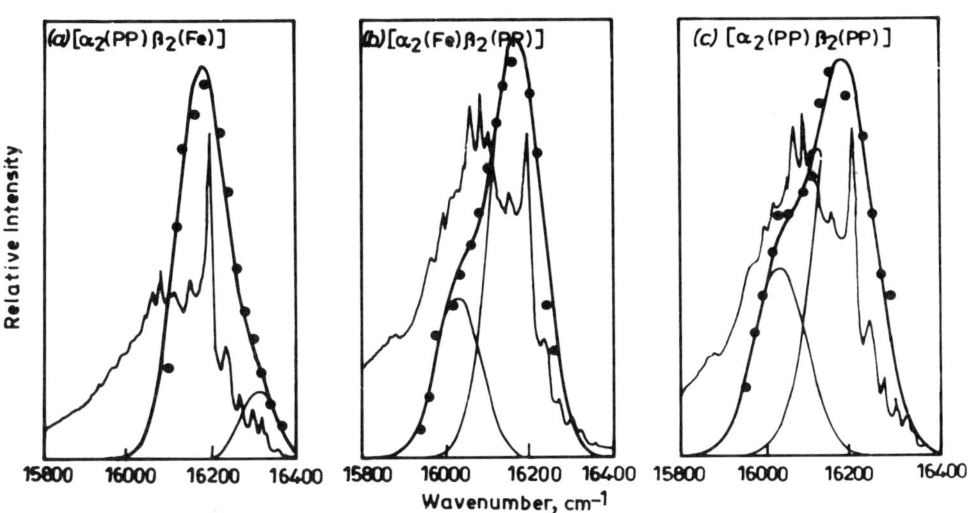

Figure 26. Fluorescence spectra and population distribution function of (*a*) [α_2(Fe)β_2(PP)]; (*b*) [α_2(PP)β_2(PP)]; and (*c*) [α_2(PP)β_2(Fe)]. The bold face line shows the fitted bimodal Gaussian distribution function with individual Gaussian components indicated with narrow lines. [Reproduced with permission from (192).]

Along with other investigation techniques such as XRD, CD measurement is used to study the conformational changes in proteins. Circular dichroism involves studies mainly in four wavelength regions. (a) The far-UV region (below 250 nm) gives information about peptide backbone; (b) the near-UV region 250–300 nm gives information on local environment of Trp, Tyr, and Phe residues, S—S bridge, and the heme group; (c) the soret region (300–480 nm) gives information about interaction of the heme with protein matrix, and (d) the visible region (480–600 nm) provides information on the charge-transfer band between the Fe and axial ligands.

Hybrids such as valency, chemically modified, metal reconstituted, semihemoglobin, and naturally occurring hybrids have been studied by this technique. The soret region CD spectra of valency hybrids, namely, $[\alpha_2^+\beta_2]$ and $[\alpha_2\beta_2^+]$ (195), are considerably different, showing variations in their heme environments indicating the Fe—His bond of both ferric and ferrous ion and orientation of aromatic residues in the vicinity of heme. The soret region CD results on chemically modified hybrids (196) with one or two residue deletion in the α or β chain is shown in Table X indicates the positions of peak maxima at different pH values in the presence and absence of IHP. Here a band at 433 nm corresponds to the T state while the one at 437 nm corresponds to R state, in other words, they can be treated as marker bands for quaternary conformational changes. Such changes are not noticed in absorption spectroscopy (197). The shift in wavelength is attributed to differences in $\alpha_1\beta_1$ subunit contacts in the T and R states. The CD spectra of semiHb (198) were found to be intermediate between those of natural Hb and their corresponding isolated chains.

Phenomenal changes were noticed in the naturally occurring hybrids, namely, HbM(Boston) [α(E7) His → Tyr], HbM(Iwate) [α(F8) His → Tyr], HbM-(Saskatoon) [β(E7) His → Tyr], and HbM(Hyde Park) [β(F8)His → Tyr]

TABLE X
CD Measurement Wavelength Maximas for Chemically Modified Hybrid Hb[a]

Subunit		pH 6		
α-Chain	β-Chain	+IHP	−IHP	pH 9
Normal	Normal	433	433	433
Normal	des-His	433	433	433
Normal	des-His-Tyr	433	433	433
des-Arg	Normal	433	433	437
des-Arg	des-His	433	437	437
des-Arg	des-His-Tyr	433	437	437
des-Arg-Tyr	Normal	433	437	437
des-Arg-Tyr	des-His	433	437	437
des-Arg-Tyr	des-His-Tyr	437	437	437

[a]Taken from (196).

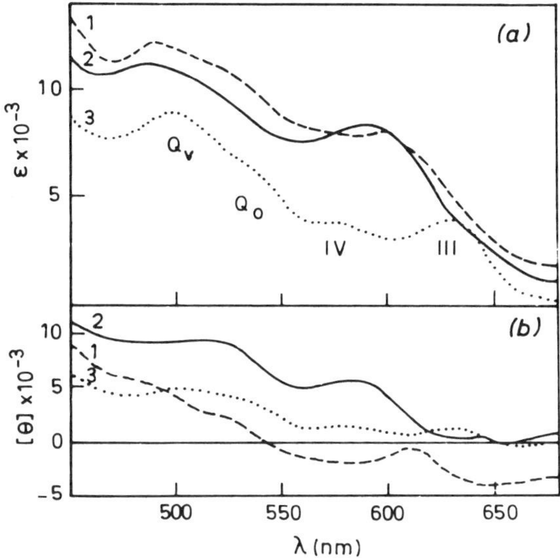

Figure 27. Comparison of (*a*) visible absorption and (*b*) CD spectra for HbM. Here (*a*) HbM (Boston); (*b*) HbM (Iwate); (*c*) metHbA. Concentration of hemoglobin is 100 μM in 0.05 M bis–tris buffer, containing 0.1 M NaCl at pH 7.0. In the absorption spectra Q_v, Q_o represents a $\pi \rightarrow \pi^*$ transition where as III and IV represents a charge-transfer band. [Reproduced with permission from (199).]

(199–201), when studied by CD in the visible region. When proximal (F8) His was replaced by some other residue, a positive CD was observed, whereas a negative CD was observed when (F7) His was replaced by the same such residue. Such changes are not observable in visible absorption spectroscopy, as seen in Fig. 27.

For the metal reconstituted [Fe—Zn] hybrids (202), [α_2(Zn)β_2(Fe—CO)], [α_2(Fe—CO)β_2(Zn)], the CD spectra in the near-UV region showed a band at 280 nm with negative ellipticity, which is attributed to T-state Hb. This result seems to contradict those obtained from other spectroscopic techniques, such as NMR and ENDOR.

Nanosecond time-resolved circular dichroism (TRCD) has been recently used to probe allosteric intermediates by photolyzing HbCO (202a) both in the near-UV (250–400 nm) and soret (400–460 nm) regions. Distinct changes in the CD spectra are (a) in going from 50 and 500 μs an increase in the $+ve$ CD band and a growing of negative ellipticity around 350 nm were noticed; (b) large increase in CD in the region near 260 nm accompanying ligation is attributed to structural changes at the heme site during photolysis; (c) a red shift of the CD maximum and slightly negative ellipticity around 400 nm implying increasing distance of the iron out of the heme plane, and (d) the prompt appearance of a negative ellipticity in the aromatic bands near 285 nm may relate to the R → T transition. Changes at the $\alpha_1\beta_2$ interface probably corresponds to breaking and re-forming of hydrogen and nonbonded interactions pertaining to aromatic amino acid residue Typ-37(β), Tyr-42(α). These results are in accordance with the UV time-resolved RR results (139). Further TRCD results indicate that the *−ve* CD feature is first acquired through conformational changes at the interface during the R′ — R step along the kinetic pathway to the T state (i.e., R′ → R → T). These results are again in accordance with recent X-ray results (85, 86).

VI. EXTENDED X-RAY ABSORPTION FINE STRUCTURE STUDIES

The EXAFS studies give information about the geometric and electronic state of iron in hemoglobin. The first EXAFS of HbA(deoxy) was measured in 1975 by Kincaid et al. (203). Details on experimental and other aspects of X-ray absorption spectroscopy of Hb are dealt with elsewhere (204). This technique provides distance information between the metal and different coordination shells. In the case of Hb molecules, distances between Fe and the nearby proximal histidine and the pyrrole nitrogen can be obtained in the solution state. This method has an advantage over XRD techniques, which require crystallization. In the crystallized state, stereochemically unusual species that are not normally highly populated in solution may be formed. Earlier work was restricted to Hb, HbO_2, HbCO, and other derivatives of Hb (205).

Hybrid Hb EXAFS studies were initiated with the [Zn—Fe] (206, 207) hybrid. This study was possible as a model since the *K*-absorption edge energies for Zn and Fe are very different and, hence, each center can be monitored through EXAFS separately. The Zn and Fe-filtered first-shell EXAFS of these hybrids are shown in Fig. 28. The results obtained for [α_2(Fe—CO)β_2(Zn)] and [α_2(Zn)β_2(Fe—CO)] hybrids are given in the Table XI. These results lead to the inference that the Fe environment in these hybrids are stereochemically similar to native carbonmonoxy Hb, that is, iron–ligand bond lengths in low-affinity half-ligated hybrids are closely similar to those in fully ligated [α_2(Fe—CO)β_2(Fe—CO)].

The EXAFS studies also reveal that the five-coordinate Zn porphyrin is isostructural with five-coordinate deoxyFe porphyrin. There seems to be a structural analogy between the EXAFS results of [α_2(Fe—CO)β_2(Zn)] species and [α_2(Fe—O_2)β_2(Zn)] species studied crystallographically (73), where iron-proximal His bond length in the two crystallographically distinct oxygenated α-heme are found to be 2.10 and 2.39 Å.

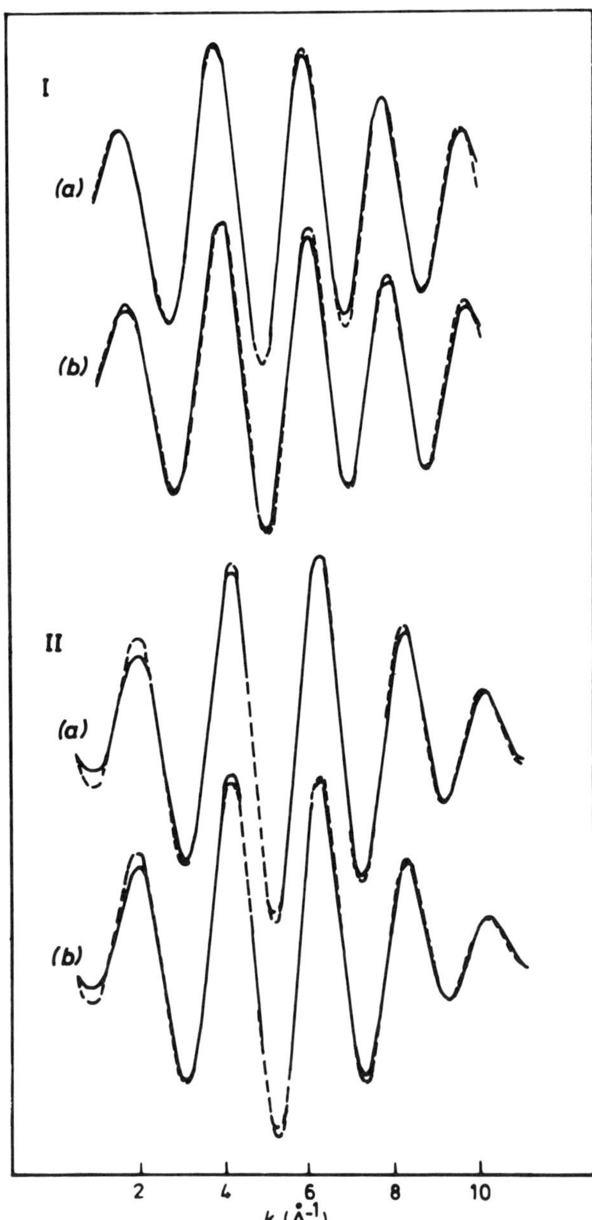

Figure 28. [I] gives comparison of Fe-filtered first-shell EXAFS contributions and [II] gives a comparison of Zn-filtered first-shell EXAFS contributions. Both IA and IIA are [α_2(Fe—CO)β_2(Zn)], IB and IIB are [α_2(Zn)β_2(Fe—CO)]. [Reproduced with permission from K. Simolo, Z. R. Korszun, G. Stuky, K. Moffat, G. McLendon, and G. Bunker, *Biochemistry*, *25*, 3773 (1986). Copyright © American Chemical Society.]

TABLE XI

EXAFS Parameters (206)

Species or Hybrids	Nature of EXAFS	ΔR	ΔA	$\Delta\sigma^2$	ΔE(eV)
$[\alpha_2(Fe-CO)\beta_2(Zn)]$	Fe-EXAFS[a]	-0.003 ± 0.015	-0.02 ± 0.10	$3.8 \times 10^{-5} \pm 4.0 \times 10^{-5}$	0.99
	Zn-EXAFS[b]	-0.010 ± 0.10	0.05 ± 0.10	$-1.7 \times 10^{-3} \pm 9.0 \times 10^{-4}$	−0.10
$[\alpha_2(Zn)\beta_2(Fe-CO)]$	Fe-EXAFS[a]	-0.017 ± 0.015	0.07 ± 0.015	$-7.7 \times 10^{-5} \pm 8.0 \times 10^{-5}$	−1.31
	Zn-EXAFS[b]	-0.020 ± 0.015	-0.05 ± 0.10	$-1.1 \times 10^{-3} \pm 8.0 \times 10^{-4}$	−0.81

[a]The values presented are differences in the least-square refinement parameters between the modified hemoglobin molecules and native HbCO. The parameter ΔR is the difference between average first shell nearest-neighbors distances; ΔA is the difference in backscattering amplitude for the first shell peaks; $\Delta\sigma^2$ is the difference in their Debye–Waller factors; ΔE is the difference in their X-ray absorption thresholds.

[b]The values presented are difference in least-square refinement parameters of the partially substituted molecule and (py)Zn(TpyP), using the fully substituted ZnHb molecule as model.

VII. ELECTRON-TRANSFER STUDIES

Preliminary investigations by electron-transfer studies on synthetic ligand chains, where one heme acts as a donor and the other heme as an acceptor, corroborates the idea of coupling between two or more heme iron units. This allows us to monitor their electron-transfer (ET) properties. The same idea was extended to proteins, especially hemoglobin (where the chromophores could be maintained at the same distance ~25 Å). In order to understand and ultimately mimic biological ET studies, ''intramolecular'' long-distance ET studies on hybrid hemoglobins have been and are being studied in many laboratories. Many general reviews, which include various experimental techniques (208–211), are available. The recent Nobel lecture by Marcus gave the basic theory of ET and a summary of the work done in this area to date (212).

Since the high-resolution X-ray structures are known for some hemoglobins, it facilitates structure–function correlation. To initiate intramolecular ET, photoinitiation is done through pulse radiolysis. In order to utilize photochemial ET, we have to replace the iron containing porphyrin active site in hemoglobin with equivalent photoactive metalloporphyrins, namely, Zn or Mg protoporphyrin (213–215). In general, the ET rate depends on two factors, namely, distance between the chromophores and the medium in which the ET occurs.

Hoffmann and co-workers gave the first set of examples for mixed-metal hybrid hemoglobins (216, 217), namely, $[\alpha_2(Fe^{3+}{-}H_2O)\beta_2(Zn)]$ or $[\alpha_2(Zn)\beta(Fe^{3+}{-}H_2O)]$ undergoing surprisingly efficient net photoreduction $Fe^{III} \rightarrow Fe^{II}$. The hybrids were in fully T state in the presence of IHP and were partially converted to R state in its absence. This result is surprising because in the absence of a sacrificial electron donor, no net photochemistry should occur. So it was concluded that ET takes place from the photoexcited zinc triplet state in one subunit to Fe^{III} in another subunit. Also, ET takes place between the subunits $\alpha_1\beta_2$ and $\alpha_2\beta_1$, where heme separations are 25 Å, and not between the subunits, $\alpha_1\beta_1$ and $\alpha_2\beta_2$, where heme separations are about 35 Å. In the latter, ET would be reduced by several orders of magnitude. As a whole, the hemoglobin molecule can be considered as two independent ET complexes. The ET results have been interpreted by the following scheme, also depicted in Fig. 29.

$$\text{I}\ [\alpha\text{ZnP}/\beta\text{Fe}^{III}\text{P}] \xrightarrow[\text{ISC}]{h\nu} [^3(\alpha\text{ZnP})^*/\beta\text{Fe}^{III}\text{P}]$$

$$\text{II}\ [^3(\alpha\text{ZnP})^*/\beta\text{Fe}^{III}\text{P}] \xrightarrow{k_t} [\alpha\text{Zn}^+\text{P}/\beta\text{Fe}^{II}\text{P}]$$

$$\text{III}\ [^3(\alpha\text{ZnP})^*/\beta\text{Fe}^{III}\text{P}] \xrightarrow{K_D} [\alpha ZnP/\beta\text{Fe}^{III}\text{P}] + \text{heat}$$

$$\text{IV}\ [\alpha\text{ZnP}^+/\beta\text{Fe}^{II}\text{P}] \xrightarrow{K_m} [\alpha\text{ZnP}/\beta\text{Fe}^{II}\text{P}] + \text{oxidized protein residue}$$

$$\text{V}\ [\alpha\text{ZnP}^+/\beta\text{Fe}^{II}\text{P}] \xrightarrow{K_t} [\alpha\text{ZnP}/\beta\text{Fe}^{III}\text{P}]$$

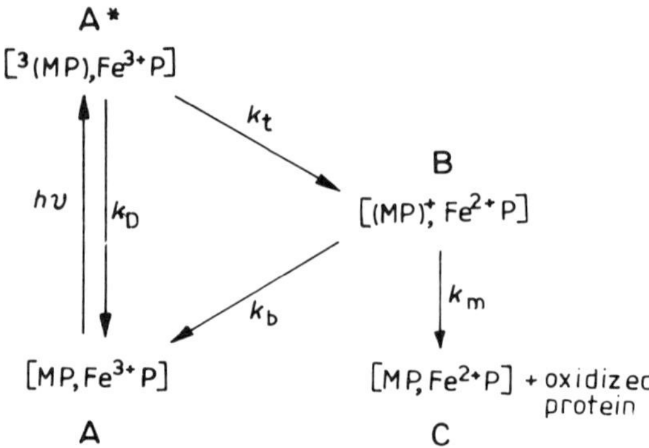

Figure 29. The kinetic scheme for ET following flash photolysis excitation. [MP, Fe^{3+}P] chains (α or β) substituted with metal ions (M) such as Zn, Mg protoporphyrin IX, and chains of the opposite type are oxidized to ferric heme (Fe^{3+} P) state: $h\nu$ is the photolysis energy required to excite A to A*; K_D is the triplet decay constant; K_t is the photoinitiated ET rate constant (or) quenching rate constant; K_m is the rate constant for the process $(MP)^+$ and is reduced by solution impurities (or) unidentified protein residue; K_b is the electron-back-transfer rate as in the figure. [Reproduced with permission from M. J. Natan and B. M. Hoffman, *J. Am. Chem. Soc.*,, *111*, 6468 (1989). Copyright © American Chemical Society.]

Here, the triplet state (^{3}ZnP) is a good reductant ($E_0 \sim -1.1$ V) and the π cation radical product of electron transfer $(ZnP)^+$ is a strong oxidant ($E_0 \sim +0.74$). As a result of ET, a large internal reorganization occurs in the transition from six-coordinate high-spin $Fe^{III}Hb{-}H_2O$ to five-coordinate high-spin Fe^{II}Hb. Interpretation is given in such a way that an electron might hop indirectly from ^{3}(MP) to (Fe^{3+}P) via oxidative triplet quenching by an amino acid with subsequent reduction of Fe^{3+}P by amino acid anion. The ET from ^{3}ZnP to a high-spin ferric heme occurs at a rate of 100 s^{-1} at room temperature. The rate is effectively invariant and the reorganization energy involved is very small. It is also found that triplet decay is enhanced only when the subunit contains $Fe^{III}H_2O$, as shown in Table XII.

When zinc was replaced by magnesium (218), the photoinitiated ET rate constant at room temperature was found to be 35(4) s^{-1}. In general, the rate of ET for Zn hybrids is greater than that for Mg hybrids (219). The explanation given for such a difference was that the square of the ET matrix element describing the electronic coupling between the donor and acceptor metalloporphyrin for Zn may be roughly double that of Mg (220).

Note that the Cytochrome b_5 function in the erythrocyte is to reduce methemglobin back to the functional deoxy form. To study these biological reactions, the photoinitiation study of ZnHb/Fe^{III}Cyt b_5 system was conducted.

$$[^3(ZnP^*)Hb/Cytb_5Fe^{III}] \xrightarrow{K_t} [(ZnP)^+Hb/Cytb_5Fe^{II}]$$

TABLE XII
Electron-Transfer Parameters for Various Hybrids

Donor or Acceptor	K_t (s^{-1})	K_b (s^{-1})	R (Å)	References
[^{3}ZnP/Fe^{3+}(H_2O)P]Hb	100(10)	325(30)	25	216, 217 219, 220, 222
[^{3}ZnP/Fe^{3}Hb/Cyt b_5Fe^{3+}]	8(2) × 10^3		10	221
[^{3}ZnP/Fe^{3+}(imidazole)P]Hb	175(15)	300(35)	20	219
[^{3}ZnP/Fe^{3+}(CN^-)P]Hb	20(5)	243(25)	20	219
[^{3}ZnP/Fe^{3+}(F^-)P]Hb	20(5)	200(35)	20	219
[^{3}ZnP/Fe^{3+}(N_3^-)P]Hb	20(5)	230(40)	20	219
[^{3}MgP/Fe^{3+}(H_2O)P]Hb	35(4)	155(20)	20	217, 219, 220
[^{3}MgP/Fe^{3+}(imidazole)P]Hb	75(25)	150(25)	20	219
[^{3}MgP/Fe^{3+}(CN^-)P]Hb	6(3)	135(10)	20	219
[^{3}MgP/Fe^{3+}(F^-)P]Hb	6(4)	105(10)	20	219
[^{3}MgP/Fe^{3+}(N_3)P]Hb	6(4)	115(15)	20	219
[^{3}ZnP/Fe^{2+}P]Hb	0	0	25	

Here the photoproduction of ^{3}ZnP* is followed by long-range ^{3}ZnP*—Fe^{3+} electron transfer in the Hb/Cyt b_5 complex at a rate of 8000 s^{-1}. This rate is slow for transfer that is believed to involve a relatively short (~10 Å) donor–acceptor distance. With these results in the background, the effect of heme ligation, porphyrin substitutents, and protein matrix on the Fe^{2+}P—$(MP)^+$ long-range ET were investigated (221). Among the porphyrin substituents, the vinyl group of ZnP does not conjugate with the ring sufficiently to serve as an ET antenna (222).

The triplet quenching rates and the corresponding rise and fall in rates also depend on the ligand on the ferric heme. It has been found that hemes with neutral ligands (H_2O or methyl imidazole) quench the triplet states faster than the heme with anionic ligands (OH^-, CN^-, F^-, N_3^-). It was found that K_t decreases in the order imidazole > H_2O > F^- > CN^- ~ N_3^- (223, 224). Anion-binding modulates the rate constant K_t and K_b through a major change in energies. Recently, light modulated EPR has been used to study the intermediates that occur during ET in hybrids (225).

Temperature-dependent ET studies (226) reveal that K_t falls smoothly from the room temperature value to a nonzero value of $K_t = 9 \pm 4$ s^{-1}, which is effectively invariant from 170 down to 77 K as shown in Fig. 30. Temperature dependence indicates coupling of the ET process to thermal vibration and fluctuation (227, 228). The simple expression of Hoppfield's semiclassical model derived from the Forster–Dexter theory of energy transfer relating the rate constant K_t and temperatue in terms of three parameters [α, $T_C(\omega)$ and β] can be depicted as follows:

$$K_t = \alpha[\tanh (T_C/T)]^{1/2} \exp [-\beta \tanh (T_C/T)]$$

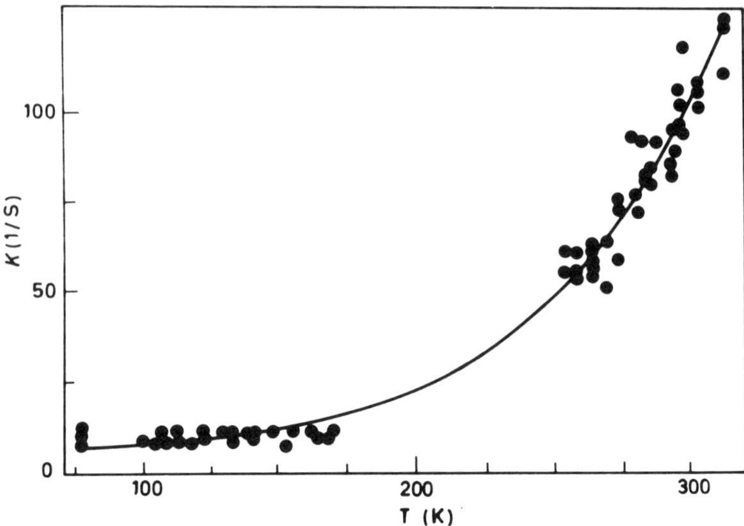

Figure 30. Temperature dependence of the ET rate K_t in $[\alpha_2\ (Fe^{3-}-H_2O)\beta_2\ (Zn)]$ hybrid hemoglobin as probed by triplet-state quenching. [Reproduced with permission from D. Kuila, W. W. Baxter, M. J. Natan, and B. M. Hoffman, *J. Phys. Chem.*, *95*, 1(1991). Copyright © 1991 American Chemical Society.]

Here, $T_C \equiv \hbar\omega/2k_B$, and the other parameters are interpreted as follows: $\beta \equiv E^{\#}/k_B T_C$, where the activation energy in the high-temperature, classical limit is $E^{\#} = (\Delta E_0' - \lambda)^2/4\lambda$, where $\Delta E_0'$ is the redox potential difference with in the hybrid and λ is the reorganization energy; $\alpha = (2\pi/\hbar)|H|^2/(2\pi\hbar\omega\lambda)^{1/2}$, where H is the electron tunneling matrix element.

The exact quantum mechanical description of the ET rate constant in the case of single mode-vibronic coupling is also available and can be formulated in terms of three parameters (α_Q, T_C, S):

$$K_t = \alpha_Q I_p[S/\sinh(T_C/T)] \exp[(pT_C/T) - S\coth(T_C/T)]$$

where I_P is the modified Bessel function and p is the ratio of the exoergicity $(\Delta E_0')$ to the vibrational spacing $(\hbar\omega)$; $\alpha_Q = 2\pi|H|^2/\hbar^2\omega$; $T_C = \hbar\omega/2k_B$ and $S = \lambda/\hbar\omega$. From quantum mechanical and semiclassical theories, the parameters are the result of fitting procedures. It has been found that the parameters are different for $[\alpha_2(Zn)\beta_2(Fe^{3+}-H_2O)]$ and $[\alpha_2(Fe^{3+}-H_2O)\beta_2(Zn)]$.

By using saturation-transfer experiments (based on ^1H NMR) intra- and intermolecular ET have been followed for [Fe, Mn] hybrid hemoglobin, namely, $[\alpha_2(Fe^{3+}-PMe_3)\beta_2(Mn)]$ and $[\alpha_2(Mn)\beta_2(Fe^{3+}-PMe_3)]$ in an attempt to investigate the mechanism of ET in trimethylphosphine complexed Hb (229). For these experiments, an appropriate equivalent of sodium dithionite was added to met-hybrid hemoglobin in D_2O to make a mixture of metHb/Hb. The dependence of the lifetime of the PMe_3 resonance of both α and β chains as a function of concentration of exchanging species was first investigated. The exchange process can be described as follows:

$$[\alpha^{3+}PMe_3] + [\alpha^{2+}PMe_3] \underset{k'_{11}}{\overset{k_{11}}{\rightleftharpoons}} [\alpha^{2+}PMe_3] + [\alpha^{3+}PMe_3]$$

$$[\beta^{3+}PMe_3] + [\beta^{2+}PMe_3] \underset{k'_{22}}{\overset{k_{22}}{\rightleftharpoons}} [\beta^{2+}PMe_3] + [\beta^{3+}PMe_3]$$

$$[\alpha^{2+}PMe_3] + [\beta^{3+}PMe_3] \underset{k_{21}}{\overset{k_{12}}{\rightleftharpoons}} [\alpha^{3+}PMe_3] + [\beta^{2+}PMe_3]$$

The rate constant for these processes were found to be $k_{11} = 3200\ M^{-1}s^{-1}$, $k_{22} = 2090\ M^{-1}s^{-1}$, $k_{12} = 1020\ M^{-1}s^{-1}$, and $k_{21} = 430\ M^{-1}s^{-1}$.

Note that the photosynthetic reaction is much more rapid than ET from [^{3}ZnP] to [Fe^{3+}—(H_2O)P] in the α_1–β_2 complex, that is, the ET rate in the former is 10^4 times larger at ambient temperature. Such studies of ET in proteins provide some surprising correlation of ET rates with protein structures.

VIII. OXYGEN-BINDING STUDIES

Functional reduced hemoglobin binds a number of gaseous ligands such as O_2, CO, and NO. The affinity for CO is 200 times greater than that for oxygen, and the affinity for NO is approximately 1500 times greater than that for CO. While many studies have dealt with hybrids with CO or NO bound to certain chains, the studies on equilibrium binding are limited to oxygen, because of the difficulties in measuring equilibrium data for the other ligands. In addition, as mentioned above, the major impetus of much of the recent hybrid studies have been directed at understanding the cooperative nature of oxygen binding. For this purpose, the use of hybrids to provide potentially unambiguous data on the changes in oxygen affinity for different partially ligated intermediates is of utmost importance.

All the methods used for studying oxygen equilibrium with hybrids are based on the spectrophotometric changes that occur when oxygen is bound to hemoglobin (Fig. 31). Methods include the use of tonometers, for which data on a series of points equilibrated with different oxygen partial pressures are used, and the automated continuous methods that provide much more data. If proper precautions are taken, it can be assumed that valid equilibrium data is being obtained. While the validity of this spectrophotometric data has been questioned

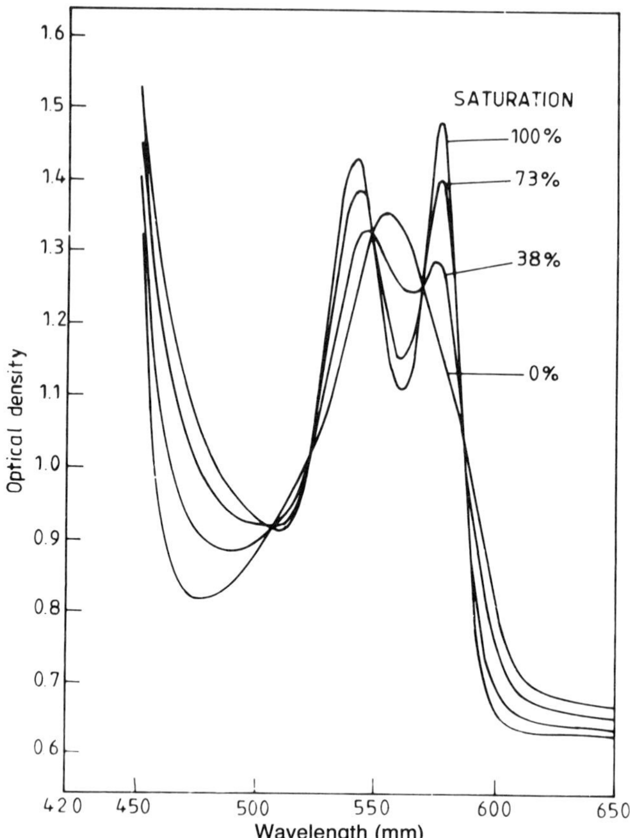

Figure 31. Visible-spectral absorption changes of Hb—A sample at pH 7.02, 0.2 *M* NaCl, and 0.2 *M* bis–tris buffer taken at various percentages of oxygen. [Reproduced with permission from (230).]

(230), it is expected that the relative changes in oxygen-binding data for different hybrids may still be valid.

The oxygen equilibrium curves were expressed as conventional *Y* versus log *P* plots (*Y* is the percentage of oxygen saturation, P is the oxygen pressure) and further analyzed by the Hill plot log $[Y(1 - Y)]$ versus log *P* plot (*Y* is the fractional oxygen saturation) as shown in Fig. 32. The oxyen affinity and the extent of cooperativity in oxygen binding are expressed by the oxygen pressure at $y = 50$, P_{50}, and the slope of the Hill plots *n*, respectively (231).

Finding the oxygen equilibrium parameters for hybrids provides a means to investigate the relative role of α and β chains in the functional properties of hemoglobin. This work was initiated by Riggs and Herner (232) with the hy-

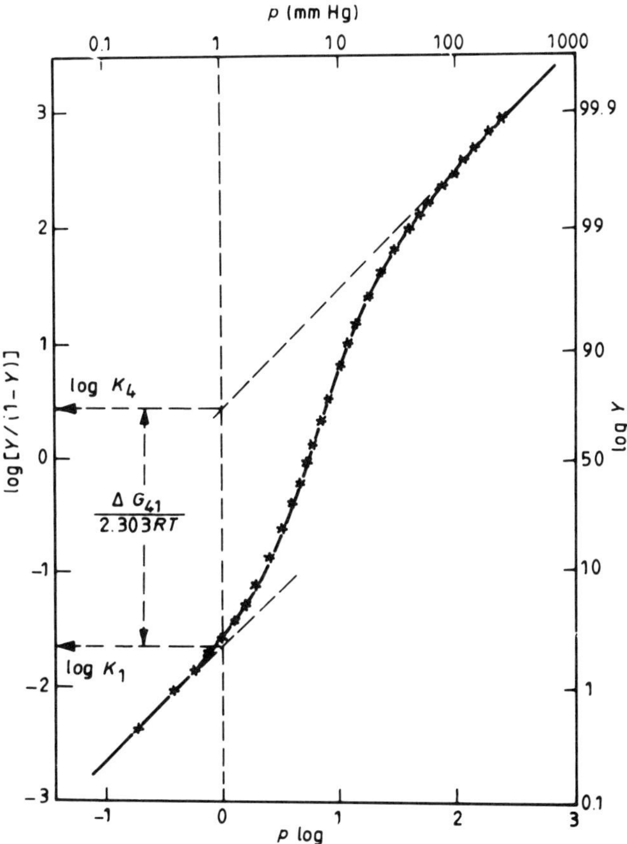

Figure 32. Hill plot of oxygen binding of Hb-KA, which is sigmoidal in shape. Here the plot is between log P versus log $[Y/(1 - Y)]$, where P is the partial pressure of oxygen and Y is the fractional saturation with oxygen. [Reproduced with permission from (230).]

brids of mouse and donkey, since mouse Hb exhibited relatively lower oxygen affinity and higher Bohr effect than those of donkey. Their investigation concluded that oxygen affinity were controlled by β chain. Subsequently, Antonini et al (233) reported that the oxygen equilibrium of human–dog hybrids differed considerably from one another even though the parent molecules were practically identical and they concluded that functional characteristics of hemoglobin could not depend only on β chains. In a similar fashion, the oxygen equilibrium studies were done on human hybrids by Taketa et al (234). A comparison with the HbA parent shows that the n_{max} value of both hybrids are somewhat lower whereas the oxygen affinity of one hybrid ($\alpha_2^A\beta_2^{cat}$) is slightly lower than that of the other ($\alpha_2^{cat}\beta_2^A$). They concluded that functional properties are determined by both the structures and specific interaction between α and β chains. Recently, Condo et al. (235) studied the oxygen equilibrium and the effect of Cl^-, DPG, and IHP on the hybrid Hb present in vivo in erythrocytes from mule and on its parent Hb molecules from horse and donkey. Recent studies on hybrid composed of mouse and human showed abnormal oxygen affinity (236).

Oxygen equilibrium studies by Parkhurst and co-workers (238) on carp–human Hb (237) hybrids showed that Hb (α carp : β human) and Hb (α human : β carp) have P_{50} values resemblance closer to carp Hb(β43 Glu → Ala) than to human Hb. In the pH region 7 for both the hybrid Hb molecules, the Hill number is much reduced from that observed for the two parent Hb molecules. Based on these results, they suggested that the F9 Ser in the carp β chain as well as $\alpha_1\beta_1$ interaction are important in controlling the allosteric transitions in these hybrids. Parkhurst and Goss (239) evidenced the presence of an intermediate state denoted S in which the α chain (human) has R-state properties and the β chain (carp) has T-state propertis. In a similar fashion, *Xenopus Laevis*–human hybrid Hb molecules were prepared by Condo et al. (240) and their oxygen equilibrium studies showed P_{50} to have properties in between the respective hemoglobins and *n* value almost similar to *X laevis* hemoglobin. They proposed that novel $\alpha\beta$ contacts allow competent fit of the surfaces and are designed to keep the deoxygenated Hb in a low-affinity conformation, thus leading to a quaternary oxygen-linked allosteric transition. Oxygen saturation studies on [α_2(meso)β_2(proto)] and [α_2(proto)β_2(meso)] by Makino and Sugita (241) concluded that the difference between the chains will not seriously affect the validity of cooperativity models, which is widely accepted.

Much data on oxygen equilibrium properties of valency hybrid Hb molecules have been documented. Here artificially some of the subunits (either α chain or β chain) are in an inactive state, leaving their partner chains alone to undertake ligand binding (242–244). By measuring the oxygen saturation parameters, $\alpha\alpha$ and $\beta\beta$ interactions are studied. For [$\alpha_2^{+CN}\beta_2$], Haber and Koshland (245) found PO_2 to be 4.2 and 2.4 in the presence and absence of DPG, respectively, compared to the value of 11.0 for the parent in the presence of DPG and 6.5 in the absence of DPG. This made them conclude that the binding of DPG stabilized the Hb molecule in the deoxy conformation and DPG did not alter the interaction of the two β subunits. Maeda et al. (246), who studied the effect of DPG on the half-cyanomet hybrid Hb molecules, found that although the phosphate decreased the oxygen affinity, the effects were several times larger for [α_2^+(CN)β_2(O_2)] than for [α_2(O_2)β_2^+(CN)]. This result suggests that β subunits change their conformation when the partner α subunit bind the oxygen molecule. The present results do not follow original simple allosteric models of MCW and KNF, which assumes progressive or sequential changes in the subunit conformation. But introducing of nonequivalence of α and β subunits in both the models permits the oxygen equilibrium curves of the hybrids to be affected by DPG in different manner.

Recently, oxygen-binding results were obtained from a combination of direct and indirect method by Ackers and co-worker (247). They demonstrated the same combinatorial aspect to cooperativity that is predicted by symmetry rule. Here, 6 of the 10 ligation microstates cannot be studied in isolated form due to $\alpha\beta$ dimer rearrangement reaction, but they are studied as hybrid mixtures in the presence of two parent species. By a combination of different techniques Ackers and co-worker (247) found the oxygen-binding parameters, namely, P_{50} and n_{max}, for all eight intermediate species as seen in Table XIV.

Oxygen-binding properties on selective or specific modified hemoglobin hybrids were also carried out in order to find the importance of these sites in governing the oxygen affinity of hemoglobin and to estimate the contribution of some of the chloride-binding sites to the alkaline Bohr effect. In this respect, the hybrids such as $[\alpha_2^{CM}\beta_2]$, $[\alpha_2\beta_2^{CM}]$ (248, 249) (CM is carboxymethylated), and $[\alpha_2^{DHP}\beta_2]$, $[\alpha_2\beta_2^{DHP}]$ (250) (DHP–glycation), $[\alpha_2\beta_2^{AC}]$ (251) (Ac is acetylation) have been prepared and subjected to oxygen-binding studies. In a similar approach, oxygen-binding studies on hemoglobin hybrids $[\alpha_2\beta^S\beta^Y]$, which are formed as abnormal hemoglobins such as HbY, HbM, and HbS, along with HbA give us an idea on the relative roles of various chloride-binding sites that determine the oxygen affinity of hemoglobin.

Strong cooperativity of hemoglobin oxygenation makes it difficult to characterize the oxygenation intermediates. The properties of metal reconstituted hemoglobin hybrids depend on the electronic configuration of the substituted metal ion. Some metalloporphyrins do not bind oxygen. Therefore metal hybrids may behave as good models for oxygenated intermediates of parent hemoglobin. Shibayama et al. (252) reported that the oxygen affinity of Ni—Fe hybrid Hb moleculues was very low compared to the parent HbA; also in $[\alpha_2(Fe)\beta_2(Ni)]$, n was unity at pH 6.5 and increased as the pH was raised, while $[\alpha_2(Ni)\beta_2(Fe)]$ takes up oxygen noncooperatively under all conditions as shown in Table XIII. They concluded that for $[\alpha_2(Ni)\beta_2(Fe)]$, the structural change of the β subunit induced by binding oxygen is transmitted to the α subunit and that the α subunit changes its structure by changing the coordination state of NiPPIX. The $[Cu^{II}—Fe^{II}]$ hybrid Hb molecules behave just like the Ni—Fe hybrid Hb, while other metal-substituted hybrid Hb moleculues gave a Hill coefficient significantly higher than unity.

It was found in earlier studies of Fe—Hb that the β subunits had a higher oxygen affinity than the α subunits. The difference becomes smaller as the pH decreases but in Co—Hb the affinity is slightly in the reverse order (Co binds oxygen but not CO). In order to compare the affinity difference between Fe—Hb and Co—Hb, oxygen-binding studies for the Fe—Co hybrid (253) were done by Imai et al (254a). These studies reveal that the results obtained are not con-

TABLE XIII
Oxygen-Binding Parameters for [Fe—Ni] Hybrid

			$[\alpha_2(Fe)\beta_2(Ni)]$		$[\alpha_2(Fe)\beta_2(Ni)]$	
Hybrid or Species	pH	IHP	P_{50} (mmHg)	n	P (mmHg)	n
Unmodified	6.5	−	73.00	1.0	110.00	1.0
		+	170.00	1.0	200.00	1.0
	7.5	−	27.00	1.1	44.00	1.0
		+	170.00	1.0	220.00	1.0
	8.5	−	5.80	1.3	16.00	1.0
		+	36.00	1.2	110.00	1.0
des-Cys(β93)	6.5	−	3.40	1.3	30.00	1.1
		+	120.00	1.0	—	—
	7.5	−	2.80	1.3	18.00	1.1
		+	51.00	1.0	150.00	1.0
	8.5	−	1.50	1.3	11.00	1.1
		+	4.10	1.3	60.00	1.0
des-Arg(α141)	6.5	−	1.70	1.4	5.70	1.0
		+	16.00	1.0	11.00	1.1
	7.5	−	0.90	1.2	3.20	1.0
		+	13.00	1.1	7.50	1.0
	8.5	−	0.57	1.1	1.00	1.1
		+	0.72	1.3	3.10	1.0
Des-Cys(β93) and Arg(α141)	6.5	−	0.74	1.1	—	—
		+	3.30	1.4	—	—
	7.5	−	0.83	1.1	1.70	1.3
		+	1.60	1.1	10.00	1.0
	8.5	−	0.62	1.0	1.60	1.1
		+	0.65	1.1	1.80	1.0
Des-His(β146) and Tyr(β145)	6.5	−	—	—	0.78	1.0
		+	—	—	11.80	1.0
	7.5	−	—	—	0.68	1.0
		+	—	—	5.00	1.0
	8.5	−	—	—	0.81	1.0
		+	—	—	0.85	1.0

[a]Condition 0.1 *M* Cl, 50 m*M* Tris/HCl, 25°C.

sistent. Oxygen equilibrium properties of $[Cr^{III}—Fe^{II}]$ hybrid Hb molecules reported recently almost agreed with those of the corresponding cyanomet valency hybrid hemoglobins (254b). Since cyanomet valency hybrids are unstable and undergo autooxidation, they result in asymmetric oxygen equilibrium curves, while such a possibility is ruled out for $[Cr^{III}—Fe^{II}]$ hybrid Hb molecules. Hence, the [Cr—Fe] hybrid can be taken as a reasonable substitute for the cyanomet valency hybrid.

From the available oxygen-binding parameters, the oxygen affinity of $M-Fe^{II}$ hybrids, containing the first transition metal ions, are in the following order: $[Zn^{II}-Fe^{II}](d^{10}; d\gamma^4) < [Cu^{II}-Fe^{II}](d^9; d\gamma^3) \simeq [Ni^{II}-Fe^{II}](d^8; d\gamma^2) < \cdots [Fe^{III}F-Fe^{II}](d^5; d\gamma^2) \simeq [Mn^{III}H_2O-Fe^{II}](d^4; d\gamma^1) \simeq [V{=}O^{II}-Fe^{II}](d^1; d\gamma^0) < [Co^{III}CN^{-}-Fe^{II}](d^6; d\gamma^0) \simeq [Fe^{III}CN^{-}-Fe^{II}](d^5; d\gamma^0) \simeq [Cr^{III}H_2O-Fe^{II}](d^3; d\gamma^0)$. Thus by comparing the oxygen affinity of various hybrids it becomes evident that the number of $d\gamma$ electrons of the first metal in the center is important for determining the structure of the globin moiety.

Recently, symmetric and asymmetric hybrids have been studied in their cross-linked (255) form. As noted before, this study was required because without cross-linking the asymmetric hybrid Hb rearranged due to dimer dissociation. Compared to HbA, the cross-linking between two $\alpha\beta$ dimers appears to cause a decrease in cooperativity and an increase in oxygen affinity as measured by the Hill coefficient (i.e., the Hill coefficient values decrease as we go from HbA to cross-linked Hb) (see Table XIV). This table gives a clear picture of the structure–function relationship of the hybrids, which are supposed to form in the intermediate stage of the oxygen-binding process.

So far, the oxygen-binding studies on asymmetric hybrids have been performed on the following systems [Ni—Fe]XL (256), [Co—Fe]XL, $[(\alpha\beta)_A(\alpha\beta)_C]$XL, $[(\alpha^{+CN}\beta)_A(\alpha\beta)_C]$XL, $[(\alpha\beta^{+CN})_A(\alpha\beta)_C]$XL, $[(\alpha^{+CN}\beta^{+CN})_A(\alpha\beta)_C]$XL, and $[(\alpha\beta^{+CN})_A(\alpha^{+CN}\beta)_C]$XL, where A = HbA and C = HbC (β6Glu → Lys). For [α(Fe)β(Fe)] [α(Ni) β(Ni)]XL (257), the Hill coefficient increased from 1.41 to 1.53 upon a pH change from 7.4 to 8.4. These results made them propose that the $\alpha_1\beta_1$ interactions in the initial stage of oxygenation of Hb were neither inert as predicted by crystallographic studies nor strong as in the case of the cyanomet ligation system. But, the Hill coefficient for [α(Fe)β(Ni)] [α(Ni)β(Fe)]XL (258) increased from 1.64 to 1.73 upon changing the pH from 7.4 to 8.4; this shows that $\alpha_1\beta_2$ oxygenation is always more cooperative than $\alpha_1\beta_1$ oxygention at all pH values. By measuring P_{50} on all five asymmetric cyanomethybrids, Miura et al. (259) found that they can be arranged in the following decreasing order of oxygen binding: $[(\alpha\beta)_A(\alpha\beta)_C]$XL > $[(\alpha^{+CN}\beta)_A(\alpha\beta)_C]$XL > $[(\alpha\beta^{+CN})_A(\alpha\beta)_C]$XL > $[(\alpha^{+CN}\beta^{+CN})_A(\alpha\beta)_C]$XL > $[(\alpha\beta^{+CN})_A(\alpha^{+CN}\beta)_C]$XL. For the asymmetric hybrid hemoglobin with one cyanomet heme, namely, $[(\alpha^{+CN}\beta)_A(\alpha\beta)_C]$XL > $[(\alpha\beta^{+CN})_A(\alpha\beta)_C]$XL. Each oxygen equilibrium constant increases on raising the pH and consequently the Hill coefficient remains constant over the pH range. The plot of P_{50} of both hybrids gave a slope of 0.6, which is steeper than that for $[(\alpha\beta)_A(\alpha\beta)_C]$XL, indicating that about 0.6 protons per ferrous heme and 1.8 protons per tetramer, $[(\alpha^{+CN}\beta)_A(\alpha\beta)_C]$XL, were released on oxygenation. This number was comparable to $[(\alpha\beta)_A(\alpha\beta)_C]$XL. Based on this result, it was proposed that the ox-

TABLE XIV
Oxygen-Binding Parameters for Various Hybrid Hemoglobins

Hybrids or species	Condition	pH	P_{50}	n (torr)	References
Human hemoglobin	0.1 *M* Phosphate	7.0	7.4	2.8	50
	0.1*M* + 2 m*M* DPG	7.0	11.7	2.7	50
	0.1*M* + 2 m*M* IHP	7.0	38.0	2.7	50
	0.1 *M* NaCl, 0.05 *M* Bis–tris	7.0	4.0	2.8	50
	0.1 *M* + 2 m*M* DPG	7.0	11.8	2.7	50
	0.1 *M* + 2 m*M* IHP	7.0	38.5	2.7	50
HbA-XL	0.1 *M* phosphate	7.0	3.2	2.4	50
	0.1 *M* + 2 m*M* DPG	7.0	3.3	2.4	50
	0.1 *M* + 2 m*M* IHP	7.0	3.3	2.4	50
	0.1 *M* NaCl, 0.05 *M* Bis–tris	7.0	2.8	2.3	50
	0.1 *M* + 2 m*M* DPG	7.0	2.8	2.3	50
	0.1 *M* + 2 m*M* IHP	7.0	2.8	2.2	50
Cat hemoglobin		6.87	20.1	1.80	234
		7.71	14.2	2.00	234
$[\alpha_2^{cat}\beta_2^{A}]$		6.81	7.7	1.90	234
		7.31	4.3	2.20	234
$[\alpha_2^{A}\beta_2^{cat}]$		6.83	11.4	2.00	234
		7.23	7.3	1.90	234
Hb Donkey	0.1 *M* phosphate	7.1	12.3	2.4	228
Hb Mouse	0.1 *M* Phosphate		4.7	2.3	228
$[\alpha_2^{mouse}\beta_2^{donkey}]$	0.1 *M* Phosphate		14.4	2.0	228
$[\alpha_2^{donkey}\beta_2^{mouse}]$	0.1 *M* Phosphate	7.1	4.7	2.0	228
Carp hemoglobin	0.1 *M* Phosphate	9.0	1.2 + 0.1	1.10 + 0.1	238
	0.1 *M* Phosphate	7.1	24.0 + 1.0	2.40 + 0.09	238
	IHP	6.0	94.0 + 4.0	0.90 + 0.1	238
Hb[αcarp:βhuman]	IHP	9.0	6.1 + 0.5	1.00 + 0.1	238
	IHP	7.0	32.4 + 0.9	1.40 + 0.1	238
	IHP	6.0	42.0 + 2.0	1.10 + 0.1	238

Hb[αhuman:βcarp]	IHP	9.0	5.3 + 0.5	1.20 + 0.1	238
	IHP	6.9	46.5 + 0.9	1.45 + 0.09	238
	IHP	6.0	59.0 + 3.0	1.08 + 0.09	238
χ-Laevis Hb	0.1 *M* NaCl, 0.1 *M* Bis–tris	6.0	152.0	1.00	240
	0.1 *M* + 3 m*M* IHP	7.0	89.0	1.7	240
	0.1 *M* + 3 m*M* IHp	8.5	8.0	2.5	240
[$\alpha_2^{\chi}\beta_2$]	0.1 *M* + 3 m*M* IHP	6.0	20.0	1.5	240
	0.1 *M* + 3 m*M* IHP	7.0	17.5	1.5	240
	0.1 *M* + 3 m*M* IHP	8.5	4.0	1.6	240
[$\alpha_2\beta_2^{\chi}$]	0.1 *M* + 3 m*M* IHP	6.0	41.5	1.0	240
	0.1 *M* + 3 m*M* IHP	7.0	28.0	1.2	240
	0.1 *M* + 3 m*M* IHP	8.5	6.3	1.6	240
[$\alpha_2^{A}\beta_2^{Ca}$](Ca-Canine Hb)	0.2 *M* Phosphate	6.10	1.10	1.7	233
	0.2 *M* Phosphate	7.01	0.99	1.6	233
	0.05 *M* Boarate	8.73	0.03	2.5	233
[$\alpha_2^{Ca}\beta_2^{A}$]	0.2 *M* Phosphate	6.00	0.75	2.7	233
	0.2 *M* Phosphate	7.01	0.68	2.1	233
	0.05 *M* Boarate	8.85	0.92	2.4	233
[HbS]	0.05 *M* Bis-tris(0.1 *M* NaCl)	7.0	2.50(6.4)	2.8(2.8)	255
	0.05 *M* (+DPG)(0.1 *M* NaCl)	7.0	5.40(15.9)	2.8(2.8)	255
	0.05 *M* (+IHP)(0.1 *M* NaCl)	7.0	5.50(57.0)	2.7(2.7)	255
[HbY]	0.05 *M* Bis-tris(0.1 *M* NaCl)	7.0	0.80(0.95)	1.4(1.4)	255
	0.05 *M* (+DPG)(0.1 *M* NaCl)	7.0	1.25(1.91)	1.4(1.4)	255
	0.05 *M* (+IHP)(0.1 *M* NaCl)	7.0	7.45(7.20)	1.8(1.8)	255
[HbSY]($\alpha_2\beta^S\beta^Y$)	0.05 *M* Bis–tris(0.1 *M* NaCl)	7.0	0.93(1.35)	1.25(1.25)	255
	0.05 *M* (+DPG)(0.1 *M* NaCl)	7.0	1.15(1.52)	1.25(1.25)	255
	0.05 *M* (+IHP)(0.1 *M* NaCl)	7.0	1.15(1.52)	1.25(1.20)	255
HbAs-XL	0.1 *M* Phosphate	7.0	3.4	2.6	48
	0.01 *M* + 2 m*M* DPG	7.0	3.7	2.4	48
	0.01 *M* + 2 m*M* IHP	7.0	3.8	2.4	48
	0.1 *M* NaCl, 0.05 *M* Bis–tris	7.0	3.0	2.5	48
	0.01 *M* + 2 m*M* DPG	7.0	3.6	2.3	48
	0.01 *m* + 2 m*M* IHP	7.0	3.6	2.2	48

TABLE XIV (*Continued*)

Hybrids or species	Condition	pH	P_{50}	n (torr)	References
HbF	0.05 *M* Bis–tris, 0.1 M NaCl	7.26	5.7	3.0	332
HbS−XL	0.05 *M* Bis–tris, 0.1*M* NaCl	7.26	3.4	2.2	332
HbSF−XL	0.05 *M* Bis–tris, 0.1 *M* NaCl	7.26	3.5	2.2	332
HbYS−XL	0.1 *M* Phosphate	7.0	1.1	1.3	332
	0.1 *M* + 2 m*M* DPG	7.0	1.2	1.3	332
	0.1 *M* + 2 m*M* IHP	7.0	1.4	1.2	332
	0.1 *M* NaCl, 0.05 *M* Bis–tris	7.0	0.9	1.4	332
	0.1 *M* + 2 m*M* DPG	7.0	1.0	1.3	332
	0.1 *M* + 2 m*M* IHP	7.0	1.0	1.3	332
[$\alpha_2\beta_2^{CM}$] Carboxymethylated (CM)		7.0	12.0	2.4	247–249
[$\alpha_2^{CM}\beta_2$] Val1(α)		7.0	17.0	2.4	247–249
[$\alpha_2\beta_2^{AC}$] Acetylated Hb [Val1(β), Lys82(β)]		7.0	28.0	2.3	247–249
[$\alpha_2\beta_2^{DHP}$] Glycated Hb		7.0	5.0	2.5	247–249
[Val-1(α) + Lys-16(α), 59(β), 82(β), 120(β)]					
[α(Fe)β(Fe)]					
[α(Ni)β(Ni)]XL	0.05 *M* Tris, 0.1 *M* Cl^-	6.4	86.2	1.03	257
	0.05 *M* Tris, 0.1 *M* Cl^-	7.4	24.9	1.41	257
	0.05 *M* Tris + IHP	7.4	25.4	1.39	257
[α(Fe)β(Ni)]	0.05 *M* Tris + IHP	8.4	7.7	1.33	257
[α(Ni)β(Fe)]XL	0.05 *M* Tris, 0.1 *M* Cl^-	6.4	101.0	1.33	258
	0.05 *M* Tris, 0.1 *M* Cl^-	7.4	29.4	1.64	258
	0.05 *M* Tris + IHP	7.4	30.7	1.60	258
	0.05 *M* Tris + IHP	8.4	10.1	1.73	258
[α_2^D(Ru−CO)β_2^D(Fe)]	−IHP	7.4	0.85	1.31	Ishimo et al.
[α_2^D(Fe)β_2^D(Ru−CO)]	−IHP	7.4	0.6	1.21	(unreported)
[α_2(Ru−CO)β_2(Fe)]	0.05 *M* Bis–tris, 0.1 *M* NaCl	6.4	1.5	1.32	136
		6.9	1.0	1.22	136
		7.4	0.6	1.22	136

	0.05 *M* Tris, 0.1 *M* NaCl	7.9	0.46	1.16	136
$[\alpha_2(Fe)\beta_2(Ru-CO)]$	0.05 *M* Bis–tris, 0.1 *M* NaCl	6.4	3.3	1.51	136
		6.9	1.5	1.33	136
		7.4	0.85	1.31	136
	0.05 *M* Tris, 0.1 *M* NaCl	7.9	0.46	1.22	136
des-Arg FeHbA	0.1 *M* Phosphate	7.0	0.91	1.7	92
des-Arg CoHb	0.1 *M* Phosphate	7.0	20.1	1.1	92
des-Arg$[\alpha_2(Fe)\beta_2(Co)]$	0.05 *M* Bis–tris				
	0.1 *M* NaCl	7.5	56.20	1.0	92
	0.1 *M* NaCl + 2 m*M* IHP	7.5	60.70	1.0	92
des-Arg$[\alpha_2(Co)\beta_2(Fe)]$	0.05 *M* Bis–tris				
	0.1 *M* NaCl	7.5	0.77	1.0	92
	0.1 *M* NaCl + 2 m*M* IHP	7.5	3.50	1.2	92
$[\alpha_2(Co^{+CN})\beta_2(Fe)]$	100 m*M* Cl	7.5	0.5	1.21	Mochizuki et al.
	100 m*M* Cl + IHP	7.4	9.2	1.61	(unreported)
$[\alpha_2(Fe)\beta_2(Co^{+CN})]$	100 m*M* Cl	7.4	0.4	1.41	(unreported)
	100 m*M* Cl + IHP	7.4	2.6	1.31	(unreported)
$[\alpha_2(Zn)\beta_2(Fe)]$	100 m*M* Cl	7.4	54.0	1.01	Mizazaki et al.
	100 m*M* Cl + IHP		166.0	1.01	(unreported)
$[\alpha_2(Fe)\beta_2(Zn)]$	100 m*M* Cl	7.4	77.0	1.01	(unreported)
	100 m*M* Cl + IHP	7.4	164.0	1.01	(unreported)
$[\alpha_2(Cu)\beta_2(Fe)]$	100 m*M* Cl	7.4	38.0	1.01	90
	100 m*M* Cl + IHP	7.4	113.0	1.01	90
$[\alpha_2(Fe)\beta_2(Cu)]$	100 m*M* Cl	7.4	26.0	1.11	90
	100 m*M* Cl + IHP	7.4	150.0	1.11	90
$[\alpha_2(Mn)\beta_2(Fe)]$	100 m*M* Cl	7.4	3.2	1.41	Masuda et al.
	100 m*M* Cl + IHP	7.4	150.0	1.01	(unreported)
$[\alpha_2(Fe)\beta_2(Mn)]$	100 m*M* Cl	7.4	1.3	1.31	(unreported)
	100 m*M* Cl + IHP	7.4	15.0	1.31	(unreported)
$[\alpha_2(Cr)\beta_2(Fe)]$	100 m*M* Cl	7.4	0.4	1.31	Unzai et al.
	100 m*M* Cl + IHP	7.4	6.1	1.41	(unreported)
$[\alpha_2(Fe)\beta_2(Cr)]$	100 m*M* Cl + IHP	7.4	0.6	1.21	(unreported)
	100 m*M* Cl + IHP	7.4	17.0	1.61	(unreported)

TABLE XIV (*Continued*)

Hybrids or species	Condition	pH	P_{50}	*n* (torr)	References
$[\alpha_2(V{=}O(\beta_2(Fe)]$	100 m*M* Cl	7.4	2.5	1.41	Nasu et al.
	100 m*M* Cl + IHP	7.4	189.0	1.21	(unreported)
$[\alpha_2(Fe)\beta_2(V{=}O)]$	100 m*M* Cl	7.4	0.7	1.51	(unreported)
	100 m*M* Cl + IHP	7.4	22.0	1.31	(unreported)
$[\alpha_2^M(V{=}O)\beta_2(Fe)]$ (M = mesoporphyrin)	100 m*M* Cl	7.4	0.8	1.351	(unreported)
	100 m*M* Cl + IHP	7.4	4.0	1.31	(unreported)
$[\alpha_2(Fe)\beta_2^M(V{=}O)]$	100 m*M* Cl	7.4	22.0	1.251	(unreported)
	100 m*M* Cl + IHP	7.4	143.0	1.051	(unreported)
$[\alpha_2(Mg)\beta_2(Fe)]$	100 m*M* Cl	7.4	86.0	1.01	Matsubara et al.
	100 m*M* Cl + IHP	7.4	171.0	1.01	(unreported)
$[\alpha_2(Fe)\beta_2(Mg)]$	100 m*M* Cl	7.4	63.0	1.21	(unreported)
	100 m*M* Cl + IHP	7.4	127.0	1.21	(unreported)
$\alpha_2^M(Ti{=}O)\beta_2(Fe)]$	100 m*M* Cl	7.4	0.6	1.2	Minagawa et al.
	100 m*M* Cl + IHP	7.4	17.0	1.41	(unreported)
$[\alpha_2(Fe)\beta_2^M(Ti{=}O)]$	100 m*M* Cl	7.4	1.2	1.41	(unreported)
	100 m*M* Cl + IHP	7.4	76.0	1.151	(unreported)
$[\alpha_2\beta_2^{+CN}]$	NaCl Free(stripp)	7.4	0.30	1.08	246
	NaCl Free (2 m*M* DPG)	7.4	1.41	1.41	246
	0.1 *M* NaCl(Stripped)	7.4	0.40	1.17	246
	0.1 *M* NaCl(2m*M* DPG)	7.4	1.38	1.52	246
	0.1 *M* Cl (+IHP)	7.4	2.8	1.5	246
$[\alpha_2^{+CN}\beta_2]$	NaCl Free(stripped)	7.4	0.40	1.17	246
	NaCl Free (2 m*M* DPG)	7.4	0.83	1.13	246
	0.1 *M* NaCl(stripped)	7.4	0.47	1.10	246
	0.1 *M* NaCl(2 m*M* DPG)	7.4	0.79	1.14	246
	0.1 *M* Cl (+IHP)	7.4	21.0	1.5	246
$[(\alpha\beta)_A(\alpha\beta)_C]XL$	0.1 *M* Phosphate buffer	7.41		2.30	259
$[(\alpha\beta^{+CN})_A(\alpha^{+CN})\beta)_C]XL$	0.1 *M* DPG	7.1	0.81	1.2	259

$[(\alpha^{+CN}\beta^{+CN})_A(\alpha\beta)_C]XL$	0.1 *M* DPG	7.1	1.04	1.1	259
$[\alpha_2^{+F}\beta_2]$	100 m*M* Cl	7.4	1.8	1.41	Nagai et al.
	100 m*M* Cl + IHP	7.4	47.0	1.41	(unreported)
$[\alpha_2\beta_2^{+F}]$	100 m*M* Cl	7.4	1.9	1.21	(unreported)
	100 m*M* Cl + IHP	7.4	3.0	1.71	(unreported)
$[\alpha_2(PP)\beta_2(Fe)]$	100 m*M* Cl	6.5	128.0	1.05	Fuji et al.
	100 m*M* Cl + IHP	6.5	197.0	1.07	(unreported)
	100 m*M* Cl	7.4	89.3	1.1	(unreported)
	100 m*M* Cl + IHP	7.4	190.0	1.05	(unreported)
$[\alpha_2(Fe)\beta_2(PP)]$	100 m*M* Cl	7.4	29.0	1.3	(unreported)
	100 m*M* Cl	8.5	5.2	1.34	(unreported)
	100 m*M* Cl + IHP	7.4	182.0	1.1	(unreported)
$[Fe^{II}/Fe^{III}CN]$					
Species 01			5.3	3.3	247
11			2.3	2.0	247
12			2.3	2.0	247
21			5.1	1.9	247
22			0.4	1.0	247
23			0.4	1.2	247
24			0.5	1.2	247
31			0.2	1.0	247
32			0.2	1.0	247
41			n.a[a]	n.a.[a]	247
$[\alpha_2^{SL}\beta_2]$	0.1 *M* Phosphate	7.0	8.5	2.62	
$[\alpha_2\beta_2^{SL}]$	0.1 *M* Phosphate	7.0	8.4	2.72	

[a]Not available = n.a.

ygenation of monocyanomet hybrid Hb gave the same structural changes as that of $[(\alpha\beta)_A(\alpha\beta)_C]$XL. For asymmetric hybrid Hb molecules with two cyanomet hemes, namely, $[(\alpha^{+CN}\beta^{+CN})_A(\alpha\beta)_C]$XL and $[(\alpha\beta^{+CN})_A(\alpha^{+CN}\beta)_C]$XL, the oxygen affinities were almost similar. Thus the oxygen-binding property depends on both the number of cyanomet hemes and on the distribution of cyanomet hemes among the four subunits.

Tsuneshige et al. (260) reported the oxygen-binding studies on both symmetric and asymmetric $\alpha\alpha$ cross-linked [Fe—Co] hybrids and compared their results with those of $\beta\beta$ cross-linked cyanomet valency hybrid Hb molecules by Miura et al. (259). These two studies show close resemblance to the variation of pH with respect to both n_{max} and P_{50}, values, despite the fact that (a) the cyanomet hybrid has cyanide bound to the ferric heme and the Fe—Co hybrid has CO bound to ferrous iron, (b) the heme metals involved in oxygenation process is Fe^{II} in cyanomet hybrid and Co^{II} in Co—Fe Hybrid, and (c) the $\beta\beta$ and $\alpha\alpha$ cross-linking shift the equilibrium toward high- and low-affinity states, respectively. Hence, they proposed that these hybrids are functionally and energetically important intermediate structures of HbA in going from the deoxy to oxy state.

For many hybrid species, the oxygen-binding studies are followed by other kinetic and spectroscopic measurements, which help provide the rationale for a decrease or increase in oxygen binding. In spite of the fact that a large amount of work on oxygen-binding studies on Hb have been carried out, more work is needed to clarify various published data on the oxygenation of hybrid Hb and to understand the structure–function relationship of intermediate ligated species found during oxygenation of hemoglobin.

IX. KINETIC STUDIES

A detailed account of experimental techniques such as stop-flow, flash photolysis, and temperature jump methods by which the kinetics of the hybrids are generally followed is given elsewhere (11). A review by Parkhurst and coworker [11, 261] give a detailed picture of kinetic studies on diverse hemoglobins. The kinetics of (a) partially ligated hybrids, (b) valency hybrids, (c) metal reconstituted hybrids, and (d) hybrids with different hemes have been studied in depth. The kinetic studies of ligand interaction in hybrid hemoglobins have provided important information on the magnitude of association and dissociation rate constants of ligands such as O_2, CO, and NO, in the T and R quaternary states (262). A scheme representing the 16 microscopic intrinsic equilibrium constants for the individual ligand-binding steps of tetrameric hemoglobin is shown in Fig. 33. Apart from the normal rate constant measurements, the

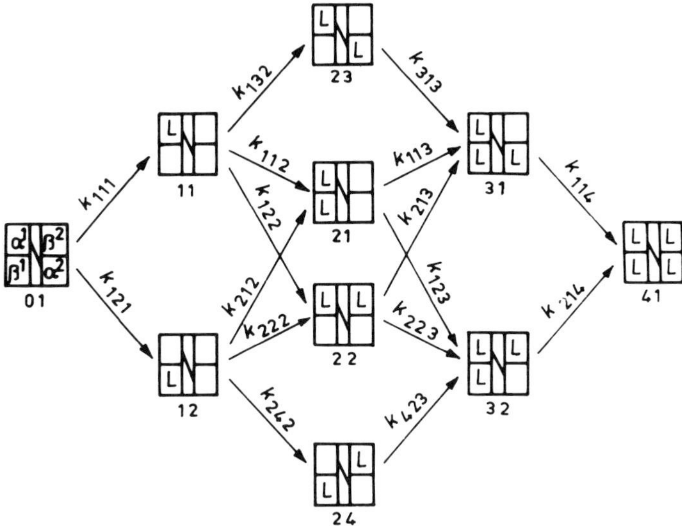

Figure 33. Scheme to represent microscopic intrinsic equilibrium constant for individual ligand-binding step of tetrameric hemoglobin. The 10 *ij* species are identified by the state of ligation $i(i = 1–4)$ and $j(j = 1–4)$. the configuration of α and β subunits in a tetramer are identified in species (01), the line joining them is the intersubunit contact. The indexes that identify a rate constant K_{kji} refer to the reactant subunit, product subunit and product ligation state, respectively. [Reproduced with permission from (264).]

effect of temperature, pH, and effector concentration such as DPG and IHP on rate constants are also generally followed.

Double mixing is one of the methods used for kinetic studies of ligand binding in partially ligated intermediates (i.e., isomers of partially ligated species). The partially ligated hybrid hemoglobins are generated during the experiment. The flow chart for the preparation of partially ligated hybrids with monoligated $[\alpha_1^{CO}\alpha_2\beta_1\beta_2]$, $[\alpha_1\alpha_2\beta_1^{CO}\beta_2]$ (263–269); diligated $[\alpha_1^{CO}\alpha_2^{CO}\beta_1\beta_2]$, $[\alpha_1\alpha_2\beta_1^{CO}\beta_2^{CO}]$, $[\alpha_1^{CO}\alpha_2\beta_1^{CO}\beta_2]$ (269–271); and triligated $[\alpha_1^{CO}\alpha_2^{CO}\beta_1^{CO}\beta_2]$, $[\alpha_1^{CO}\alpha_2\beta_1^{CO}\beta_2^{CO}]$ (167a–c) species using the double-mixing technique is shown in Fig. 6. Since there are four centers in Hb for ligand binding, the stepwise ligand combination and dissociation rate constants can be given as follows:

$$\mathrm{Hb} \underset{\ell_1}{\overset{\ell'_1}{\rightleftharpoons}} \mathrm{HbL_1} \underset{\ell_2}{\overset{\ell'_2}{\rightleftharpoons}} \mathrm{HbL_2} \underset{\ell_3}{\overset{\ell'_3}{\rightleftharpoons}} \mathrm{HbL_3} \underset{\ell_4}{\overset{\ell'_4}{\rightleftharpoons}} \mathrm{HbL_4}$$

The details of the experimental approach is discussed elsewhere. For finding the combination rate constant, any one of the mono- or di- or triligated hybrids is prepared in the first step of a double-mixing experiment, the saturated CO solution (929.64 μM at 1 atm and 25°C) is added in the second step, and the kinetics is then followed. By using this technique, the four CO combination rate constants were determined as $\ell_1' = 0.1 \times 10^6\ M^{-1}s^{-1}$, $\ell_2' = 0.7 \times 10^6\ M^{-1}s^{-1}$, $\ell_3' = 0.2 \times 10^6\ M^{-1}s^{-1}$, and $\ell_4' = 4.8 \times 10^6\ M^{-1}s^{-1}$. Similarly, the stepwise CO dissociation rate constants were obtained first by taking any one of the mono-, di- or triligated hybrids, then adding microperoxidase in the second step, and finally observing the kinetics.

The results obtained through the double-mixing technique are in agreement with those from other kinetic methods such as flash photolysis (272–274) and single-mixing studies with cyanometHb symmetric diligated hybrids. They are also consistent with the results of Perrella et al. (264, 275), who separated partially ligated isomers at -30°C and obtained relative values for the stepwise rate constants using ℓ_4' as the reference rate constant. The details of these low-temperature quenching experiment are discussed elsewhere (268). This experimental approach, unlike the stop-flow double-mixing experiment, does not provide a direct measurement of the rates of reaction, but gives us an idea about the ratios of the rates along a reaction pathway. The ratio of the rate constant for association of CO with Hb was found to be $\ell_2/\ell_1 = 1.7$, $\ell_3/\ell_2 = 1.7$, $\ell_4/\ell_3 = 3.7$, and $\ell_4/\ell_1 = 14$. Similar experiments have been done to measure ℓ_1 and ℓ_2 in the presence and absence of IHP at various pH values (276).

Apart from kinetic studies of CO dissociation and the association of partially ligated CO hybrid intermediates, the kinetics of CO ligand association was found earlier for partially NO ligated hybrids (277). The relative order of rates for CO combination were $[\alpha_2\beta_2^{NO}] < [\alpha_2^{NO}\beta_2] < [\alpha_2\beta_2]$. Also, the rate constant was found for the CO $\rightarrow$ NO replacement reaction in the chemically modified hybrid NES des-Arg Hb and HbA (278). The results obtained are HbA (0.007), NES des-Arg (0.009), and HbA + IHP (0.029). These results show that in the presence of IHP, the replacement of CO by NO senses a switch in protein structure from the high- to the low-affinity state.

In order to find the rate for the combination of CO with triligated Hb, germinate CO recombination kinetics using flash photolysis have been measured for Hb with 90% of the ligand-binding sites occupied by NO. Also, CO bimolecular recombination kinetics have been done on symmetric hybrids $[\alpha_2^{NO}\beta_2^{CO}]$ or $[\alpha_2^{CO}\beta_2^{NO}]$ and asymmetric hybrid $[\alpha^{NO}\beta^{NO}/\alpha^{CO}\beta^{NO}]$ (279). For symmetric hybrids, the recombination rate was $7 \times 10^4\ M^{-1}s^{-1}$ or $20 \times 10^4\ M^{-1}s^{-1}$ respectively, for asymmetric hybrids the biphasic kinetics was obtained with rates of $20 \times 10^4\ M^{-1}s^{-1}$ and $7 \times 10^4\ M^{-1}s^{-1}$; for the hybrid $[Hb(NO)_3(CO)]$ the rate was $17 \times 10^4\ M^{-1}s^{-1}$. For the above species, when photolysis was done in the presence of oxygen, $[Hb(NO)_3(O_2)]$ was found to be formed. These results give evidence for the existence of T-state species with four ligands bound, contradicting the two-state model. Based on this result it was concluded that there is no unique R and T state, but that a number of R and T like states are present. Also, the number of bound ligands distributed at various sites seems to be critical for deciding the nature of the quaternary state.

Another branch of kinetic perturbation method is the modulated excitation technique. In this technique, a weak oscillatory light is used to excite the molecule during which process one to three ligands can be knocked off. By adopting suitable strategies, intermediates formed during the experiment according to binomial distribution can be studied. The details of the experimental procedures are described elsewhere (280). This technique was applied on single ligand bound Hb species to find the conformational change of Hb at this state and to study the ligand-binding kinetics. The results from these experiments reveal that there is no significant mixing of the quaternary state at the first ligation step. In another experiment, the modulation excitation was done on triply oxidized $[\alpha_1^+\alpha_2^+\beta_2^+\beta_2]$XL and triply ligated $[\alpha_1^x\alpha_2^x\beta_1^x\beta_2]$XL (281, 282) (x = CO or O_2) hemoglobin, and their equilibrium constants were compared. For triply ligated hybrid R and T structures, the equilibrium constant is of the order of unity for O_2 or CO (1.1–1.5 for $3O_2$ and 0.7 for 3CO), whereas with three aquamet subunits, it is greater than 23. Also, there is a decrease in affinity due to cross-linking in agreement with the oxygen-binding studies. The rate constant of association and dissociation for $[\alpha_1^x\alpha_2^x\beta_1^x\beta_2]$XL (x = O_2) in both the phosphate and bis–tris buffers was found to be $8.0 \times 10^6\ M^{-1}s^{-1}$ and $41 \times 10\ M^{-1}s^{-1}$, respectively.

The kinetic properties of valency hybrids have been reviewed earlier by Baldwin (262) and Parkhurst and co-worker (11, 261); they reported that the ferrous heme was binding slower with oxygen in the hybrid containing ferric heme with CN^- (low spin) than with the one containing ferric heme in the high-spin met state. Cassoly and Gibson (274), who carried out a kinetic study on CN^- valency hybrids, found that the rate constants for association with ligand for valency hybrids were higher than the same study for unligated normal hemoglobin. In addition, the reaction rate is higher when the β chain of the hybrid is free to react, except in the case of CO, where the α chain reacts more rapidly. Brunori et al. (283a) studied the kinetics of CO in partially oxidized trout Hb. The flash photolysis results showed an increasingly faster rate as the met fraction increased in the hybrid. The rate increases by 30–60% when CN^- is bound to the ferric site. Brunori et al. (283) reported CO binding rates for the valency hybrids $[\alpha_2\beta_2^{+CN}]$ and $[\alpha_2^{+CN}\beta_2]$. The rates for the slow components in CO binding were similar, but about twice as fast as for Hb. Also, the rate for the ferrous α chain was slightly higher than that for the β chain in the other hybrid. Further, the kinetics of CO binding were elaborated on symmetric valency hybrids $[\alpha_2(Fe^{3+}{-}X)\beta_2(Fe^{2+})]$ and $[\alpha_2(Fe^{2+})\beta_2(Fe^{3+}{-}X)]$, (where X = F, aqua, SCN, azide, or CN) (283b).

The kinetics of methemoglobin reduction by enzyme (NADH or Cytβ_5 reductase) (284) showed the presence of valency hybrid as intermediates. Such a reduction involves the following pathway:

$$\text{metHb} \xrightarrow{K_1} [\alpha_2^+\beta_2] \xrightarrow{K_2} \text{oxyHb}$$

$$\text{metHb} \xrightarrow{K_3} [\alpha_2\beta_2^+] \xrightarrow{K_4} \text{oxyHb}$$

From the rate constants K_1 and K_3, the percentage of hybrids formed in the absence of IHP were estimated as $[\alpha_2^+\beta_2]$ (65%) and $[\alpha_2\beta_2^+]$ (35%), whereas in the presence of IHP they were $[\alpha_2^+\beta_2]$ (80%) and $[\alpha_2\beta_2^+]$ (20%). The oxygen-binding features for natural hybrids such as HbM(Saskatoon) (β63 His → Tyr) (285), HbM(Milwaukee) (286), and artificial valency hybrid $[\alpha_2\beta_2^{+\text{CN}}]$, $[\alpha_2^{+\text{CN}}\beta_2]$ (287) are also known.

The cyanomet, singly oxidized (288) aquamet and triply oxidized (288, 289) cross-linked carpHb have been isolated and the rate of CO binding followed at various pH values in the presence and absence of IHP. The values of rate constants are given in Table XV. The experimental data from these experiments show that at pH 6.1 and 6.4 the cyanide is more effective than water in altering the T → R equilibrium and, hence, the T-state properties seem to be sensitive to pH. The kinetics of CO binding on hemoglobin with differing heme centers had also been reported. In this respect, the hybrids, $[\alpha_2(\text{proto})\beta_2(\text{meso})]$, $[\alpha_2(\text{proto})\beta_2(\text{deutero})]$, $[\alpha_2(\text{meso})\beta_2(\text{proto})]$, and $[\alpha_2(\text{deutero})\beta_2(\text{proto})]$ (290, 291) have been prepared and their kinetics of CO binding followed. In yet another set of studies on ligand binding, the hybrid hemoglobins $[\alpha_2(\text{human})\beta_2(\text{carp})]$, $[\alpha_2(\text{carp})\beta_2(\text{human})]$ (292, 293), $[\alpha_2(\text{canine})\beta_2(\text{human})]$, and $[\alpha_2(\text{human})\beta_2(\text{canine})]$ (294) have been employed.

With the advent of metal reconstituted hybrids, the kinetic studies on CO binding to symmetric hybrids containing Fe in one center and Co, Mn, and Zn in the other center have been made. These hybrids act as models for the investigation of individual specific kinetic properties of α and β subunits in the Hb tetramer because none of the metals except Fe can bind CO. The rate constants of CO binding to the ferrous subunit in all metal-reconstituted hybrid hemoglobin are summarized in Table XV. Among all the hybrids, the kinetics of [Fe—Co] hybrids have been studied in detail. The studies include germinate combination of CO (294–296), O_2 (297), and NO (298) to the Fe center of the hybrid using a nanosecond pulse. The results from these experiments show that there is a marked difference between the subunits. In one set of experiments, the kinetics of replacement of oxygen with CO on the hybrid $[\alpha_2^{\text{P}}(\text{Fe})\beta_2^{\text{M}}(\text{Co})]$, $[\alpha_2^{\text{M}}(\text{Fe})\beta_2^{\text{P}}(\text{Co})]$ (299) (P = proto, M = meso) have been followed. The results from these experiments show that the presence of Co–meso

TABLE XV
Kinetics Parameters for Various Hybrids

Hybrids	Condition	pH	Rate Constant (M^{-1} s^{-1})				Reference
			Rapid Process		Slow Process		
			−IHP	+IHP	−IHP	+IHP	
$[\alpha_2(Fe)\beta_2(Co)]$	0.05 *M* Bis–tris	7.0	1.1×10^5	6.2×10^4			295
	0.01 *M* Dis–tris	6.6			0.09×10^6	0.09×10^6	295
	0.1 *M* Phosphate	6.6	6.4×10^4				295
	0.1 *M* Phosphate	7.9	7.3×10^4				295
$[\alpha_2(Co)\beta_2(Fe)]$	0.05 *M* Bis–tris	7.0	1.0×10^5	3.9×10^4			295
	0.05 *M* Bis–tris	6.6			0.10×10^6	0.05×10^6	295
	0.1 *M* Phosphate	6.6	1.0×10^5				295
	0.1 *M* Phosphate	7.9	1.4×10^5				295
$[\alpha_2(Fe)\beta_2(Mn)]$	0.01 *M* Bis–tris–HCl	6.6			0.15×10^6	0.11×10^6	302
$[\alpha_2(Mn)\beta_2(Fe)]$	0.01 *M* Bis–tris–HCl	6.6			0.14×10^6	0.05×10^6	302
$[\alpha_2(Fe)\beta_2(Zn)]$	0.01 *M* Bis–tris–HCl	6.6			0.10×10^6	0.10×10^6	302
$[\alpha_2(Zn)\beta_2(Fe)]$	0.01 *M* Bis–tris–HCl	6.6			0.10×10^6	0.05×10^6	302
$[\alpha_2(Mg)\beta_2(Fe)$	0.05 *M* Bis–tris	6.6	3.6×10^6		0.20×10^6	0.07×10^6	274
$[\alpha_2\beta_2^{+CN}]$	0.05 *M* Bis–tris	6.6	4.8×10^6		0.27×10^6	0.18×10^6	274
$[\alpha_1^{+X}\alpha_2\beta_1\beta_2]/[\alpha_1\alpha_2\beta_1^{+X}\beta_2]$	x = H_2O	6.09–6.20	0.18×10^6	0.16×10^6		0.05×10^6	289
	CN^-	6.09–6.20	0.30×10^6	0.17×10^6		0.05×10^6	289
	H_2O	6.36–6.46	0.28×10^6	0.18×10^6		0.05×10^6	289
	CN^-	6.36–6.46	0.43×10^6	0.17×10^6	0.14×10^6	0.05×10^6	289
	H_2O	6.93–7.02	0.87×10^6	0.16×10^6		0.07×10^6	289
	CN^-	6.93–7.02	1.00×10^6	0.15×10^6		0.07×10^6	289
	H_2O	8.23–8.25	1.40×10^6		0.50×10^6		289
	CN^-	8.23–8.25	1.20×10^6				289
$[\alpha_1^{+X}\alpha_2^{+X}\beta_1^{+X}\beta_2]/[\alpha_1^{+X}\alpha_2^{+X}\beta_1\beta_2^{+X}]$	x = H_2O	6.08–6.18	1.60×10^6	0.23×10^6	0.39×10^6	0.06×10^6	289
	CN^-	6.08–6.18	2.40×10^6	0.37×10^6	0.80×10^6	0.15×10^6	289
	H_2O	6.53–6.61	2.10×10^6	0.22×10^6	0.55×10^6	0.07×10^6	289
	CN^-	6.53–6.61	2.50×10^6	0.36×10^6	0.87×10^6	0.13×10^6	289
	H_2O	6.87–6.92	2.70×10^6	0.23×10^6	0.70×10^6	0.07×10^6	289
	CN^-	6.87–6.92	2.70×10^6	0.66×10^6	1.00×10^6	0.17×10^6	289
	H_2	7.03–7.08	2.80×10^6	0.32×10^6	0.75×10^6	0.12×10^6	289
	CN^-	7.03–7.08	3.00×10^6	2.50×10^6	1.20×10^6	0.55×10^6	289

heme on the β subunit alters the ligand association properties of the α subunit. This change has been attributed to the interaction of π electrons of porphyrins with the protein moiety. In particular, the vinyl group of porphyrin interacts with Phe CD1. In the case of $[\alpha_2^P(Fe)\beta_2^P(Co)]$ the contact is closer but the contact becomes weaker in $\alpha_2^P(Co)\beta_2^P(Fe)]$. On the other hand, this contact is almost absent in $[\alpha_2^P(Fe)\beta_2^M(Co)]$, since in meso porphyrin, the vinyl groups of normal heme are replaced by ethyl groups.

For [Fe—Co] hybrids, the oxygen dissociation rate constants only from Fe subunits, that is, $[\alpha_2(Fe{-}O_2)\beta_2(Co)]$, $[\alpha_2(Co)\beta_2(Fe{-}O_2)]$ (300), are considered because the oxygen dissociation rate from the Co subunits are about 150 times faster than those of the Fe subunit (i.e., too fast to be observed by stop-flow technique). The scheme of oxygen dissociation in [Fe, Co] hybrids both in the T and R states can be represented as follows:

$$\begin{array}{ccccc} [\alpha_2(Fe{-}O_2)\beta_2(Co)] & \xrightarrow{K_R} & [\alpha(Fe{-}O_2)\alpha(Fe)\beta_2(Co)] & \xrightarrow{K_R} & [\alpha_2(Fe)\beta_2(Co)]\ (\text{R state}) \\ \upharpoonleft\downharpoonright L_2 & & \upharpoonleft\downharpoonright L_1 & & \upharpoonleft\downharpoonright L_0 \\ [\alpha_2(Fe{-}O_2)\beta_2(Co)] & \xrightarrow{K_T} & [\alpha(Fe{-}O_2)\alpha(Fe)\beta_2(Co)] & \xrightarrow{K_T} & [\alpha_2(Fe)\beta_2(Co)]\ (\text{T state}) \end{array}$$

The value of L_1 and L_2 can be represented by L_0, which is the ratio of the population of the T state to that of the R state in the absence of oxygen:

$$L_1 = L_0 c \qquad L_2 = L_0 c^2 \qquad c = K_R/K_T$$

where c is the ratio of the oxygen dissociation constant for the R state to that for T state.

The oxygen dissociation rates for $[\alpha_2(Fe{-}O_2)\beta_2(Co)]$ were estimated to be greater than 1300 s^{-1} (T state) and less than 13 s^{-1} (R state); for $[\alpha_2(Co)\beta_2(Fe{-}O_2)]$ the oxygen dissociation rates are estimated as greater than 180 s^{-1} (T state) and less than 5 s^{-1} (R state). The pH-dependent studies of oxygen dissociation showed that these hybrids exist in the R state at pH 8.8 and in the T state at pH 6.6.

The stop-flow measurements indicate that the rate of CO binding to the Fe^{II} chain of unligated [Fe—Mn] (301–305) and [Fe—Zn] (302, 306) hybrids are found to be homogeneous and first order in the presence and absence of IHP compared to the [Fe, Co] hybrid, which is not homogeneous. The difference in rate constants in the presence and absence of IHP is attributed to the T form of the protein being fully tetrameric but the R form being in equilibrium with the symmetric dimer; the addition of IHP stabilizes the T state relative to R and reduces the dissociation of R into dimers. It has been concluded from these measurements that the CO binding rate of an Fe chain in the Hb molecule are not influenced by the metal occupying the complementary chain. The pH-dependent rate constants for these hybrids have also been reported.

The kinetics of ligand binding such as Co, O_2, and NO to [α_2(Ni)β_2(Fe)] and [α_2(Fe)β_2(Ni)] hybrid hemoglobins has been carried out using a flash photolysis technique (307). For reactions with oxygen, the on rates are 4.8×10^{-6} $M^{-1}s^{-1}$ for [α_2(Fe)β_2(Ni)] and 7.5×10^{-6} $M^{-1}s^{-1}$ for [α_2(Ni)β_2(Fe)], while the off rate is about 2×10^{-3} $M^{-1}s^{-1}$ for both hybrids. For reactions with NO (nanosecond germinate recombination), [α_2(Fe)β_2(Ni)] recombines at 30 ns^{-1} while it is 0.3 ns^{-1} for [α_2(Ni)β_2(Fe)]. These results reveal the earlier and regular departure of the ligand from the βFe atom but the ligand seems to remain near the αFe atom. The low rate of germinate recombination by βFe has been explained by using molecular dynamics simulation as follows: in [α_2(Ni)β_2(Fe)] the ligand molecule moves rapidly away from the iron leaving no open return pathway behind it, which is quite an unusual behavior.

X. ENERGETICS OF INTERMEDIATES

Based on recent ^{1}H NMR spectra and by using the inverse recovery method, the rate constant for ET was found for the [Fe, Mn] hybrid (308). While evidence for intermediate conformations has been seen by many investigators the relative concentrations of these intermediates and the functional relevance depend on energetics.

The first evaluation of Gibb's free energies for all intermediate ligation states of hemoglobin, through experimental techniques, namely, kinetics and equilibrium studies, was achieved, by Smith and Ackers in 1985 (309). The number of distinguishable arrangements from fully unligated to complete ligated conformations will be made up of 10 ligation states (310), as shown in Fig. 2. The isolation and study of each partially oxygenated tetramer is not possible because oxygen moves from one site to another rapidly, as a result of which the hybrid pattern mixtures had to be studied by a combination of different experimental methods such as haptoglobin kinetics, quantitative cryogenic isoelectric focusing (311–315), and analytical gel permeation chromatography (316).

It has been shown that the Adair ligand-binding constant (A_{4i}) and the intrinsic binding constants A_x from the expression, $\Delta G_x = -RT \ln A_x$ are related by the equation

$$A_{4i} = A_x^i \Sigma g_{ij} e^{-\Delta G_c/RT}$$

where g represents the static degeneracy of the species ij (when $i = 0, 1, 2, 3, 4$, $\Sigma g_{ij} = 1, 4, 6, 4, 1$), and ΔG_C is the cooperative free energy of ligand binding, which can be determined by measuring the dimer–tetramer dissociation equilibrium constant at different levels of ligand saturation.

The scheme for measurement of a cooperative free energy ΔG_C can be obtained as shown in Fig. 34. The free energy of association in forming unligated tetramer from two unligated dimers is given by ${}^{0}\Delta G_{21}$. Also the ${}^{1}\Delta G_{21}$ is the free energy association in forming singly ligated tetramer from one unligated dimer and one monoligated dimer. The ΔG_x parameter is the free energy difference between unligated and singly ligated dimer. Also, $(\Delta G_x + \Delta G_C)$ is the free energy difference between singly ligated tetramer and unligated dimer. Since Gibbs free energy is a state function, the free energies around the cycle must be zero, which results in the following equation:

$$\Delta G_C = {}^{1}\Delta G_{21} - {}^{0}\Delta G_{21}$$

Hence, by measuring ${}^{1}\Delta G_{21}$ and ${}^{0}\Delta G_{21}$, ΔG_C can be calculated.

The same principle can be applied to calculate cooperative free energy for a hemoglobin tetramer where four ligands are attached one at a time, reflecting ligand-induced conformational work performed against a quaternary T interface, which serves as a mechanical constraint.

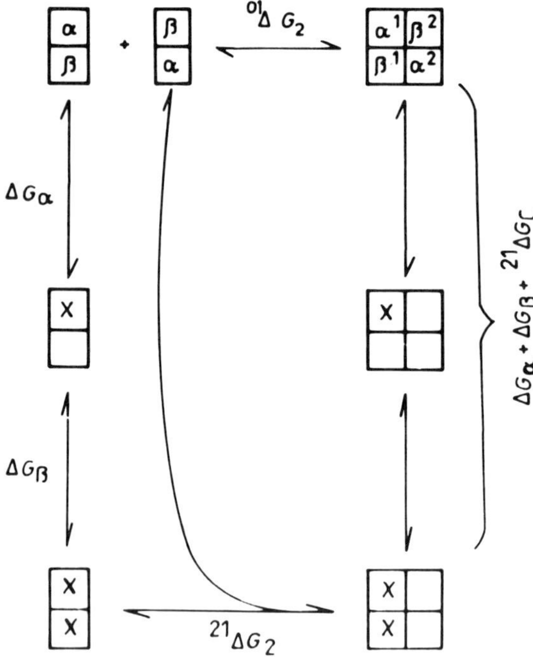

Figure 34. Schematic representation showing the measurement of cooperative free energy ΔG_C from the dimer–tetramer assembly free energies for binding of ligand X to the α_1 chain of hemoglobin. [Reprinted by permission of John Wiley & Sons, Inc. V. J. LiCata, P. M. Dalessio, and G. K. Ackers, *Proteins Structure Function and Genetics*, *17*, 279 (1993). Copyright © 1993.]

$$\Delta G_C = {}^{i}G_{2j} - {}^{0}\Delta G_{21}$$

where ${}^{0}\Delta G_{21}$ and ${}^{i}\Delta G_{2j}$ are the free energies of dimer–tetramer assembly of deoxyhemoglobin and the species with i ligands, respectively. The fractional population $[f_{4i}]$ and binding isotherm (fraction of sight occupied by ligand) $[\overline{\mathrm{Y}}]$ of the i-ligated intermediate can be calculated from the equation.

$$f_{4i} = A_{4i}\,[\mathrm{X}]^i \Big/ \sum_{i=0}^{4} A_{4i}\,[\mathrm{X}]^i \qquad [\overline{\mathrm{Y}}] = \sum_{i=0}^{4} A_{4i}\,[\mathrm{X}]^i \Big/ 4 \sum_{i=0}^{4} A_{4i}\,[\mathrm{X}]^i$$

where [X] is the ligand concentration. In practice ${}^{i}\Delta G_{2j}$ is obtained by measuring the rate constant for the forward and reverse reactions; this measurement is possible only for four specis (01, 23, 24, 41). The assembly free energies for other species are measured from hybrid mixtures.

Initially, this principle has been adopted to determine the cooperative free energies of a tetramer in all 10 ligation states for deoxy–cyanometHb systems. This procedure reveals that there are three distinct free energies among the 10 ligation states, which are depicted in Fig. 35. The lowest free energy (-14.3 kcal mol^{-1}) was observed for species 01 and the highest free energy (-8.5 kcal mol^{-1}) was observed for five species, namely, 31, 32, 23, 24, 41. In between these two free energy values a third free energy (-11.4 kcal mol^{-1}) was ob-

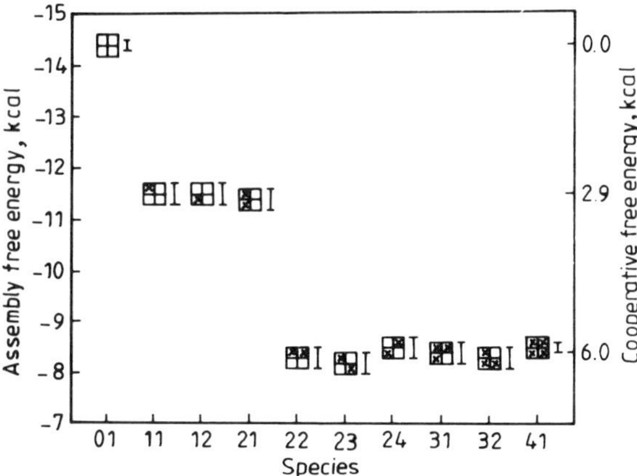

Figure 35. Tetrameric free energy levels versus ligation species in general for [Co^{II}/Fe^{II}—CO], [Fe^{II}/Fe^{III}—CN], [Fe $^{II}/Mn^{III}$] and [Mn^{II}/Fe^{II}—CO] systems, the 10 species of this system define descrete cooperative free energy levels. The spacing between the cooperative free energy levels can vary dramatically with pH. [Reprinted with permission of John Wiley & Sons, Inc. V. J. LiCata, P. M. Dalessio, and G. K. Ackers, *Proteins Structure Function and Genetics*, *17*, 279 (1993). Copyright © 1993.]

served for species 11, 12, 21, and 22. Hence, during the course of ligation the total cooperative free energy of 6 kcal mol^{-1} is spent in two transitions (i.e., 3 kcal mol^{-1} for each transition).

The analysis of species of hybrid mixtures 11, 21, and 32 were studied as the hybrid AB by mixing the two parent tetramers AA and BB. By knowing f_{AA}, F_{BB}, and f_{AB}, the fractional populations of tetramers AA, BB, and AB, respectively, the deviation free energy δ can be calculated by the equation.

$$\delta = -RT \ln f_{AB}/2\sqrt{f_{AA} f_{BB}}$$

By finding the parent assembly free energy by kinetic methods, the assembly free energy of the hybrid AB is calculated.

$$\Delta G_{AB} = \delta + \tfrac{1}{2}(\Delta G_{AA} + \Delta G_{BB})$$

With the help of the above mentioned principle and experimental background, initial studies were carried out on deoxy–cyanomet intermediates. In order to find out if ligand-linked transitions depend on change in both the number and specific configuration of bound ligand, intermediate hybrids (317) with CO and NO were studied in a similar way. It was found in these cases that the cooperative free energy of $\Delta G_C = 3.3$ kcal mol^{-1} for singly ligated intermediate, one-half of the total energy is approximately the same as that for the earlier studied systems. Later on, hybrid hemoglobins were used to check if cooperative switching was a special feature of ligated hemoglobins. In these cases, the heme iron was replaced by other metals such as Co (318, 319) and Mn (320) at two centers so that the ligation state of the heme can be stabilized. The [Fe^{II}/Mn^{III}], [Mn^{II}/Fe^{II}—CO], and [Co^{II}/Fe^{II}—CO] systems also exhibited patterns similar to that of cyanomet hemoglobin, but with variations in spacings of the energy level, as represented in Table XVI.A. These results show that the energetics of cooperative interaction depends not only on the number of ligands bound but also on the specific configuration of ligated subunits, a characteristic property of hemoglobin.

In order to understand the structure–function characteristics of hemoglobin and its intermediate allosteric states from thermodynamic parameters, studies involving the effect of pH, temperature, and single site mutation on the free energy of quaternary assembly were carried out on 20 Hb mutants (311–315). The hybrid mixture comprising of normal hemoglobin and mutant hemoglobin was studied using QC–IEF technique (312). The results from these experiments indicate that there are both independent and nonindependent elements to the energetic communication during cooperative ligation in Hb. Also, these results again confirm the presence of intermediate states. Furthermore, the intermediate allosteric tetramer has a structure similar to that of the deoxy(T) quaternary state

TABLE XVI.A

Assembly Free Energies for Various Ligation Systems (in kcals)[a]

Species or Hybrids	pH	01	11	12	21	22	23	24	31	32	41	References
$[Fe^{II}/Fe^{III}CN]$	7.0	−14.77	−10.90	−10.97	−11.11	−7.28	−7.26	−7.41	−7.32	−7.29	−7.66	309, 328
	7.4	−14.35	−11.15	−10.89	−11.14	−8.09	−8.38	−7.92	−7.97	−8.24	−8.34	
	8.0	−13.99	−11.14	−11.02	−10.84	−8.85	−9.10	−8.74	−8.81	−8.99	−9.10	
	8.5	−13.27	−10.94	−10.69	−10.41	−9.30	−9.60	−9.11	−9.27	−9.55	−9.60	
	8.8	−12.74	−10.00	−10.00	−09.74	−8.89	−9.10	−8.88	−8.94	−9.07	−9.22	
	9.0	−12.35			−09.30						−8.99	
	9.5	−11.69	−09.06	−09.09	−08.98	−8.26	−8.60	−8.14	−8.39	−8.62	−8.82	
$[Fe^{II}/Mn^{III}]$	7.4	−14.4	−11.5	−10.7	−11.0	−7.8	−7.6	−8.2	−7.9	−7.9	−7.5	319, 326
$[Mn^{II}/Fe^{II}—CO]$	7.4	−15.6			−13.1		−7.8	−8.3			−8.0	319, 326
$[Co^{II}/Fe^{II}—CO]$	7.4	−10.6	−09.1	−08.6	−08.5	−7.6	−7.5	−7.4	−7.6	−7.5	−8.0	318
$[Co^{II}/Fe^{III}CN]$	7.4	−10.6	−08.9	−08.5	−09.0	−7.5	−7.7	−7.5	−7.7	−7.9	−8.3	330a
$[Fe^{II}/Fe^{II}—O_2]$	7.4	−14.4	−11.5	−11.5	−09.2	−7.2	−7.2	−7.2	−7.2	−7.2	−8.0	321, 322
	8.9	−12.4	−10.1	−10.1	−08.8	−8.1	−8.1	−8.1	−8.1	−8.1	−9.1	317
$[Fe^{II}—CO/Fe^{III}]$	7.0	−14.8	−11.3	−11.3	−08.3	−7.8	−7.8	−7.8	−7.8	−7.8	−7.8	
$[Zn/FeO_2]$	7.4	−14.4	−11.6	−11.6	−09.4	−7.7	−7.6	−7.8	−7.5	−7.5	−8.1	330b

[a]Under the experimental condition 0.1 m tris–HCl, 0.1 *M* NaCl, pH 7.4, *T* 21.5°C. These assembly energies are calculated by a combination of different methods (01, 11, 22) dissociation kinetics, (12, 21, 31, 32) by cryogenic isoelectric focusing, (23, 24, 41) by gel permeation technique. (Uncertainty in these measurements ranges from ± 0.1 to 0.3.)

(321). Recent measurements of the temperature dependence of the equilibrium constants for species 21 and 41 are found to exhibit a linear van't Hoff relationship from which other thermodynamic parameters such as change in enthalpies (ΔH) and change in entropies (ΔS) have been estimated (322).

Thermodynamic parameters have also been obtained from the oxygen-binding isotherms measured with a continuous-flow spectrometer for normal hemoglobin (323–326) and mutant hemoglobin (318, 327) (prepared by protein engineering). The Hill coefficient, n_H of 1.7 for $[\alpha_2\beta_2^{+CN}]$, as mentioned before, indicated a high degree of positive cooperativity in partially ligated tetramers. From this Hill coefficient, it is possible to calculate the free energy increment between the third and fourth binding steps.

$$\Delta G = -RT \ln n_H^2/(2 - n_H)$$

which gives $\Delta G_{3,4}$ as -2.03 kcal mol. Since the uncertainty in measuring the Hill coefficient is $+0.2$, the free energy obtained by the above mentioned procedure is identical to that of cyanomet ligated hybrid, which is calculated from assembly free energies.

The free energies for quaternary assembly have been recently determined for all the 10 ligation species of $[Fe^{II}/Fe^{III}]$ in the pH region 7–9, as shown in Table XVI.A (328). From these free energies, the Bohr free energy of tetramers relative to dimers can be evaluated at each pH using the following equation (the complete scheme for the mode of calculation is given in Fig. 36).

$$
\begin{aligned}
{}^{ij}\Delta G_{\text{Bohr}}^{\text{Ph}'} &= {}^{ij}\Delta G_{\text{H}^+} - {}^{01}\Delta G_{\text{H}^+} \\
{}^{ij}\Delta G_{\text{Bohr}}^{\text{pH}'} &= {}^{ij}\Delta G_{\text{x}}^{\text{pH}'} - {}^{ij}\Delta G_{\text{x}}^{\text{ref}} \\
{}^{ij}\Delta G_{\text{x}}^{\text{pH}} &= {}^{ij}\Delta G_{2}^{\text{pH}} - {}^{01}\Delta G_{2}^{\text{pH}} + 2\overline{\Delta G}_{\text{x}}^{\text{pH}}
\end{aligned}
$$

Writing one such relationship for pH′ and a second one for "ref", we get

$${}^{ij}\Delta G_{\text{Bohr}}^{\text{pH}'} = ({}^{ij}\Delta G_2^{\text{pH}'} - {}^{ij}\Delta G_2^{\text{ref}}) - ({}^{01}\Delta G_2^{\text{pH}'} - {}^{01}\Delta G_2^{\text{ref}}) + 2(\overline{\Delta G}_{\text{x}}^{\text{pH}'} - \overline{\Delta G}_{\text{x}}^{\text{ref}})$$

Here the free energy value at pH 9.5 data is used as "reference" since the tetramer Bohr effect vanishes at this pH.

The results obtained from such evaluations are given in Table XVI.B. At low pH, {11}, {12}, and {21} have common proton-linked free energy, which is different in magnitude from that for the remaining six species {22}, {23}, {24}, {31}, {32}, and {41}. At high pH (i.e., 8.8) the free energy of the species {11}, {12}, {21} are similar to those of {41}. These results are in agreement with those of Perrella et al. (329a) who evaluated the Bohr proton release using oxygen binding to the vacant sites of the $[Fe^{II}/Fe^{III}CN]$ species.

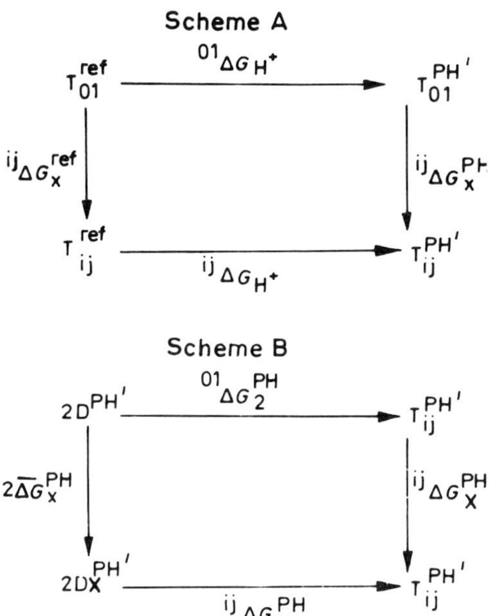

Figure 36. Scheme to obtain Gibbs free energy of each Bohr effect. Scheme A: Here T_{ij} is any one of the nine ligated tetramers and T_{01} is the unligated tetramer (01); subscripts X and H^+ represents heme site ligands and protons, respectively. The superscript "ref" and pH′ indicates pH at 9.5 and other pH of interest, respectively. Scheme B: (01) and (*ij*) represent assembly of tetramers from their constituent dimers, respectively; $DX_{(i)}^{pH}$ denotes the set of constituent dimers similar to T_{ij}. The parameter ΔG_X^{pH} is the mean ligation free energy for the two dimers. [Reproduced with permission from M. A. Daugherty, M. A. Shea, and G. K. Ackers, *Biochemistry*, *33*, 10345 (1994). Copyright © 1994 American Chemical Society.

TABLE XVI.B
Bohr Proton Release for Different Species in Hemoglobin (328)

Tetrameric States	pH 7.0	pH 7.4	pH 8.0	pH 8.5	pH 8.8
[11]	1.24 ± 0.3	0.57 ± 0.2	0.22 ± 0.3	−0.30 ± 0.3	0.10 ± 0.3
[12]	1.20 ± 0.3	0.86 ± 0.2	0.37 ± 0.2	−0.02 ± 0.3	0.14 ± 0.2
[21]	0.95 ± 0.4	0.50 ± 0.3	0.44 ± 0.3	0.15 ± 0.4	0.29 ± 0.4
[22]	4.06 ± 0.3	2.89 ± 0.3	1.71 ± 0.3	0.54 ± 0.3	0.42 ± 0.3
[23]	4.42 ± 0.2	2.88 ± 0.2	1.80 ± 0.3	0.58 ± 0.2	0.55 ± 0.2
[24]	3.81 ± 0.2	2.88 ± 0.2	1.70 ± 0.2	0.61 ± 0.2	0.31 ± 0.2
[31]	4.15 ± 0.3	3.08 ± 0.2	1.88 ± 0.3	0.70 ± 0.2	0.50 ± 0.2
[32]	4.41 ± 0.3	3.04 ± 0.3	1.93 ± 0.4	0.65 ± 0.2	0.60 ± 0.3
[41]	4.24 ± 0.2	3.14 ± 0.2	2.02 ± 0.4	0.80 ± 0.3	0.65 ± 0.2

These results show that along with quaternary switching, Bohr proton release also takes place in accordance with the symmetry rule behavior. Also, switching from the T to the R conformation takes place after species {21} is formed, in agreement with the oxygen-binding parameters for these species. For these same species, the effect of chloride in the presence of sucrose was done and the quaternary assembly free energies for the 10 ligation states was also analyzed (329b).

A more recent work of Huang and Ackers (330a) talks about the development of a new strategy for quantitatively translating the distribution of cooperative free energy between different oxygenation analogues of hemoglobin.

In the case of mixed-metal hybrid (M/Fe) the cooperative free energy is obtained by using the following equation:

$$^{ij}\Delta G_{\mathrm{C}}^{\mathrm{M/FeX}} = {}^{ij}\Delta G_{2}^{\mathrm{M/FeX}} - {}^{ij}\Delta G_{2}^{\mathrm{M/Fe}}$$

The perturbation free energy, which is a measure of free energy of structural perturbation induced by metal substitution with in the tetramer relative to that of the dimers with similar metal substitutions, is given by

$$^{ij}\Delta G_{\mathrm{P}}^{\mathrm{M/Fe}} = {}^{ij}\Delta G_{2}^{\mathrm{M/Fe}} - {}^{01}\Delta G_{2}^{\mathrm{Fe/Fe}}$$

Similarly, for mixed-metal hybrid species with bound ligand (M/FeX) the perturbation free energy is determined by

$$^{ij}\Delta G_{\mathrm{P}}^{\mathrm{M/FeX}} = {}^{ij}\Delta G_{2}^{\mathrm{M/FeX}} - {}^{ij}\Delta G_{2}^{\mathrm{Fe/FeX}}$$

By using the above parameters from metal-mixed system, apparent cooperative free energies of native FeHb is derived as follows:

$$^{ij}\Delta G_{\mathrm{C}}^{\mathrm{Fe/FeX}} = {}^{ij}\Delta G_{\mathrm{C}}^{\mathrm{M/FeX}} + ({}^{ij}\Delta G_{\mathrm{P}}^{\mathrm{M/Fe}} - {}^{ij}\Delta G_{\mathrm{P}}^{\mathrm{M/FeX}})$$

This strategy has been used to resolve the cooperative free energies of all eight carbon monoxide binding intermediates. By using hybridized combinations of normal and cobalt substituted Hb, ligation analogue systems Co/FeX (X = CO or CN) were constructed and experimentally quantified. Energetics of cobalt-induced structural perturbation were determined for all species of both the mixed-metal Co/Fe and the ligation Co/FeCN system. The results from these experiments show that major energetic perturbation of the Co/Fe hybrid species originate from pure cobalt substitution effects on the α subunit. These perturbations are then transduced to the β subunit within the same dimer of the tetramer. A similar conclusion was arrived at when the above experiments were extended to the Zn/FeO_2 system (330b).

Recently, by using hybridization experiments, assembly of the free energies of heterometallic hybrid hemoglobins versus homometallic hemoglobins were compared in order to find if there was any change in tetramer behavior due to perturbation at the heme sites (331). The tetrameric hybrids were asymmetric in nature having the configuration $[\alpha_1(M)\beta_1(M)\alpha_2(Fe{-}X)\beta_2(Fe{-}X)]$, where $M = Fe^{2+}$, Co^{2+}, Mg^{2+}, Mn^{2+}, Mn^{3+}, Ni^{2+}, or Zn^{2+} and $X = CN^-$, O_2, CO, or no ligand. The results of these experiments reveal that Mg^{2+}Hb, Mn^{2+}Hb, Ni^{2+}Hb, and Zn^{2+}Hb did not form detectable amount of heterometallic hybrids with ligated and unligated Fe^{2+} even after prolonged incubation. However, the Mn^{3+} and Co^{2+} Hb molecules readily exhibited heterometallic hybrid hemoglobins with assembly free energies -10.8 kcal and -12.4 kcal, respectively, even at normal incubation conditions. The probable reasons for destabilization of hybrids in the previous case may be (a) two dimeric half-tetramers are mismatched structurally with respect to their heme site and (b) thermodynamic stability and not the result of slow kinetics.

The results on the systems studied to date give us a gross idea that allosteric transitions occur in hemoglobin with a common set of rules. Of course, these rules can be elucidated only by studying a variety of systems. These studies have provided evidences for ET, which could be explained by a combination of indirect long-range pathways (energetics communication between two sites from different parts of the same molecule) and direct long-range pathways (energetics communication between two residue sites within the molecule).

XI. SUMMARY

As indicated above, there have been tremendous advances in the study of hybrid hemoglobins. Although a thorough understanding of the intermediates formed during oxygenation should provide a detailed understanding of hemoglobin oxygenation, when we evaluate the literature it becomes necessary to consider to what extent the modified subunits properly mimic partial oxygenation.

Changing iron for another metal center is clearly going to affect the electronic properties of the metal and therefore the ligand-binding properties. This result is clearly evident in the comparison of CoHb with normal hemoglobin. Perhaps the more important question is to what extent changing the metal center influences the transmission of conformational effects responsible for cooperative oxygenation. In using valency hybrid, the assumption is that Fe^{III} mimics an oxygenated subunit since the X-ray structures of methemoglobin are isomorphous with oxyhemoglobin. Metal centers that do not bind ligands are assumed to mimic the deoxygenated subunits. While there are clearly conformational similarities, differences do exist. The partially ligated Fe^{II} hybrids with

NO bound to certain subunits and double-mixing kinetic hybrids with CO bound to certain hybrids are much less stable, which limit the types of experiments that can be performed. Although these hybrids have Fe^{II} on all the chains, there is clear evidence that replacing oxygen with NO and even CO produces significant structural perturbations.

It is surprising that not many Mössbauer studies on these hybrids with Fe centers have been performed thus far and, hence, these studies could be a potential area of research. Studies of hybrids using this technique are likely to yield more information about the dynamics involved in and around the iron center. Similarly, many other important spectroscopic investigations, such as the use of NMR and EPR directly on the hybrid and their ligated species, deserve careful attention. These investigations are definitely possible provided one can make electrophoretically pure protein.

While it is necessary to be fully cognisant of these reservations in interpreting results on hybrid hemoglobins, it is clear that these studies can provide important insights into understanding hemoglobin function, which cannot otherwise be obtained. A valid approach to deal with these questions is to compare data on a number of different types of hybrids and then to test whether the insights gained from these studies can help explain the ambiguities still found in other hemoglobin studies.

ACKNOWLEDGMENTS

This work was taken up as part of the project supported by the US–India Rupee Fund awarded to PTM and JMR. We thank Professor H. Morimoto, Dr. N. Shibayama, and Dr. H. Hori of Osaka University for having given us some unreported results in advance. Also BV thanks the CSIR, New Delhi for a fellowship and PTM thanks the CSIR, New Delhi for an Emeritus scientistship. We like to thank all the authors of the various papers cited in this chapter for permitting us to use their figures. We also thank Mr. S. Ramasamy and Mr. K. Suryaprakasam for help in drafting the manuscript.

ABBREVIATIONS

AC	Acetylation
bis–Tris	[Bis(2-hydroxyethyl)imino-tris (hydroxymethyl)-methane)]
COSY	Correlated spectroscopy
CD	Circular dichroism
CM	Carboxymethylation
DPG	Diphosphoglycerate
DHP	Glycation
DTT	Dithiothreitol
des	Deletion of amino acid residue

1D	One dimensional
2D	Two dimensional
ET	Electron transfer
EPR	Electron paramagnetic resonance
ESEEM	Electron spin echo envelop modulation
ENDOR	Electron nucler double resonance
EXAFS	Extended X-ray absorption fine structures
HbA	Hemoglobin (normal)
HbC	Carp hemoglobin
HbS	Sickle cell hemoglobin
IHP	Inisitol hexaphosphate
HPLC	High-performance liquid chromatography
^{1}H NMR	Proton nuclear magnetic resonance
IR	Infrared
KNF	Koshland, Nemethy, Filmer
MCW	Monod, Wyman, Changex
NADP	Nicotinamide adinine dinucleotide phosphate
NOESY	Nuclear Overhauser and exchange spectroscopy
NES	*N*-Ethyl succinimide
NMR	Nuclear magnetic resonance
PP	Protoporphyrin
PEG	Polyethyleneglycol
PMB	*p*-Hydroxymercuribenzoate
py	Pyridine
QIEF	Quantitative isoelectric focusing
QC–IEF	Quantitative cryogenic isoelectric focusing
R	Relaxed state
RR	Resonance Raman
SL	Spin labeled
TRCD	Time-resolved circular dichroism
T	Tensed state
Tris	Tris(hydroxymethyl)amino methane
TFA	Trifluoro acetonylated
UV–vis	Ultraviolet–visible
XL	Cross-linked ($\alpha\alpha$ or $\beta\beta$)
XRD	X-ray diffraction

REFERENCES

1. M. F. Perutz, *Ann. Rev. Biochem.*, *48*, 327 (1979).
2. R. E. Dickerson and I. Geis, *Hemoglobin Structure Function Evolution and Pathology*, Benjamin Cummings, Menlo Park, CA 1983.

3. E. Antonini and M. Brunori, *Hemoglobin and Myoglobins in their Reactions with Ligands*, North-Holland, Amsterdam, London, 1971.
4. C. R. Cantor and P. P. Schimmel, Eds. in *The Conformation of Biological Macromolecules*, Freeman, San Fransisco, 1980.
5. J. Baldwin and C. Chothia, *J. Mol. Biol.*, *129*, 175 (1979).
6. J. M. Rifkind, *Advances in Inorganic Biochemistry*, Elsevier North-Holland, New York, 1988, Vol. 7, p. 155
7. J. M. Rifkind, *Inorganic Biochemistry*, G L. Eichborn, Ed., Elsevier, Amsterdam, The Netherlands, 1973, p. 832.
8. M. F. Perutz, *Nature (London)*, *228*, 726 (1970).
9. R. Hill and F. Holden, *Biochem. J.*, *20*, 1326 (1926).
10. B. M. Hoffman, *The Porphyrins*, *7*, 403 (1979).
11. L. Parkhurst, *J. Ann. Rev. Phy. Chem.*, *30*, 503 (1979).
12. D. M. Scholler, M. Y. R. Wang, and B. M. Hoffman, *Methods in Enzymology*, Academic, San Diego, CA, 1978, Vol. 52, p. 487.
13. R. Cassoly, *Methods in Enzymology*, Academic, San Diego, CA, 1981, Vol. 76, p. 106.
14. R. Cassoly, *Methods in Enzymology*, Academic, San Diego, CA, 1981, Vol. 76, p. 121.
15. M. Ikeda-Saito, T. Inubushi, and T. Yonetani, *Methods in Enzymology*, Academic, San Diego, CA, 1981, Vol. 76, p. 113.
16. H. F. Bunn, *Methods in Enzymology*, Academic, San Diego, CA, 1981, Vol. 46, p. 126.
17. F. Ascoil, M. R. R. Fanelli, and E. Antonini, *Methods in Enzymology*, Academic, San Diego, CA, 1981, Vol. 76, p. 72.
18. M. Perrella and L. Rossi-Bernardi, *Methods in Enzymology*, Academic, San Diego, CA, 1981, Vol. 76, p. 133.
19. (a) M. Perrella and L. Rossi-Bernardi, *Methods in Enzymology*, Academic, San Diego, CA, 1994, Vol. 232, p. 445. (b) M. Perrella, M. Samaja, and L. Rossi-Bernardi, *J. Biol. Chem.*, *254*, 8748 (1979).
20. M. F. Perutz, G. Fermi, B. Luisi, B. Shaanan, and R. C. Liddington, *Acc. Chem. Res.*, *20*, 309 (1987).
21. G. S. Adair, *J. Biol. Chem.*, *63*, 529 (1925).
22. J. Monod, J. Wyman, and J. P. Changex, *J. Mol. Biol.*, *12*, 88 (1965).
23. D. E. Koshland, G. Jr. Nemethy, and D. Filmer, *Biochemistry*, *5*, 365 (1966).
24. M. F. Perutz, *Br. Med. Bull.*, *32*, 195 (1976).
25. B. R. Gelin and M. Karplus, *Proc. Natl. Acad. Sci. USA*, *74*, 801 (1977).
26. A. Warshel, *Proc. Natl. Acad. Sci. USA*, *74*, 1789 (1977).
27. F. K. Friedman, K. Alston, and A. N. Schechter, *Anal. Biochem.*, *117*, 103 (1981).
28. K. M. Parkhurst and L. J. Parkhurst, *Int. J. Biochem.*, *24*, 993 (1992).

29. A. Tsuneshige and T. Yonetani, *Methods in Enzymology*, Academic, San Diego, CA, 1994, Vol. 231, p. 215.
30. M. Perrella, M. Samaja, and L. Rossi-Bernardi, *J. Biol. Chem.*, *254*, 8748 (1979).
31. S. C. Bernstein and J. E. Bowman, *Biochem. Biophys. Acta*, *427*, 512 (1976).
32. H. F. Bunn, Hemoglobin, *Red Cell Structure and Function*, Hrewer, 4th ed., 1972, p. 41.
33. H. F. Bunn, *Ann. N. Y. Acad. Sci.*, *209*, 345 (1973).
34. H. F. Bunn and J. W. Drysdale, *Biochem. Biophys. Acta*, *229*, 51 (1971).
35. A. Tomoda and Y. Yoneyama, *Biochem. Biophys. Acta*, *581*, 128 (1979).
36. W. H. Ford and S. Ainsworth, *Biochem. Biophys. Acta*, *160*, 1 (1968).
37. T. H. J. Huisman, *Arch. Biochem. Biophys.*, *133*, 427 (1966).
38. M. Nagai, A. Tomada, and Y. Yoneyama, *J. Biol. Chem.*, *256*, 9195 (1981).
39. A. Tomoda, A. Tsuji, S. Matsukawa, M. Takeshita, and Y. Yoneyama, *J. Biol. Chem.*, *253*, 7420 (1978).
40. A. Tomoda, T. Masazumi, and Y. Yoneyama, *J. Biol. Chem.*, *253*, 7415 (1978).
41. A. Tomoda and Y. Yoneyama, *Anal. Biochem.*, *110*, 431 (1981).
42. A. Tomoda, Y. Yoneyama, and A. Tsuji, *Biochem. J.*, *195*, 485 (1981).
43. J. M. Manning, *Methods in Enzymology*, Academic, San Diego, CA, 1994, Vol. 231, p. 225.
44. J. M. Manning, *Methods in Enzymology*, Academic, San Diego, CA, 1981, Vol. 76, p. 159.
45. M. Anbari, K. Adachi, C. Y. Ip, and T. Asakura, *J. Biol. Chem.*, *260*, 15522 (1985).
46. C. Y. Ip and T. Asakura, *Anal. Biochem.*, *139*, 427 (1984).
47. R. C. Williams, Jr., and H. Kim, *Arch. Biochem. Biophys.*, *170*, 368 (1975).
48. R. M. Macleod and R. Hill, *J. Biol. Chem.*, *248*, 100 (1973).
49. J. A. Walder, R. Y. Walder, and A. Arnone, *J. Mol. Biol.*, *141*, 195 (1980).
50. K. Kikugawa, K. Adachi, H. Kosugi, and T. Asakura, *Hemoglobin*, *7*, 533 (1983).
51. M. K. McCormack, T. G. Westbrook, R. M. Paull, and S. Berman, *Hemoglobin*, *5*, 251 (1981).
52. H. Lehmann and R. G. Huntsman, in *Mans Hemoglobins*, 2 ed. J. B. Lippincott, Ed. Philadelphia, 1974, pp. 144 and 328.
53. H. Harris, *The Principles of Human Biochemical Genetics*, 2 ed., North-Holland, Amsterdam, The Netherlands, 1975, p. 30.
54. C. Y. Ip and T. Asakura, *Anal. Biochem.*, *156*, 348 (1980).
55. S. C. Bernstein and J. E. Bowman, *Biochem. Biophys. Acta*, *427*, 512 (1976).
56. C. M. Park, *Ann. N. Y. Acad. Sci.*, *209*, 237 (1973).
57. H. F. Bunn and M. McDonough, *Biochemistry*, *13*, 988 (1974).
58. M. Perrella, A. Heyda, A. Mosca, and L. Rossi-Bernardi, *Anal. Biochem.*, *88*, 212 (1978).

59. M. Perrella, L. Cremonesi, L. Benazzi, and L. Rossi-Bernardi, *J. Biol. Chem.*, *256*, 11098 (1981).
60. M. Perrella, L. Sabbioneda, M. Samaja, and L. Rossi-Bernardi, *J. Biol. Chem.*, *261*, 8391 (1986).
61. M. Perrella, L. Benazzi, L. Cremonesi, S. Vesely, G. Viggiano, and R. Berger, *J. Biochem. Biophys. Methods*, *7*, 187 (1983).
62. M. Perrella, L. Benazzi, L. Cremonesi, S. Vesely, G. Viggiano, and L. Rossi-Bernardi, *J. Biol. Chem.*, *258*, 4511 (1983).
63. M. Perrella, A. Heyda, A. Mosca, and L. Rossi-Bernardi, *Anal. Biochem.*, *88*, 212 (1978).
64. A. Mosca, L. Rossi-Bernardi, and M. Niggeler, *Hemoglobin*, *9*, 495 (1985).
65. T. B. Bradley and R. C. Wohl, *Science*, *157*, 1581 (1967).
66. R. Benesch, R. E. Benesch, and I. Tyuma, *Proc. Natl. Acad. Sci. USA*, *56*, 1268 (1966).
67. D. A. Waller and R. C. Liddington, *Acta Crystallogr.*, *B46*, 409 (1990).
68. M. A. Schumacher, M. M. Dixon, R. Kluger, R. T. Jones, and R. G. Brennan, *Nature (London)*, *375*, 84 (1995).
69. R. Liddington, Z. Derewenda, E. Dodson, R. Hubbard, and G. Dodson, *J. Mol. Biol.*, *228*, 551 (1992).
70. I. Lalezari, P. Lalezari, C. Poyart, M. Marden, J. Kister, B. Bohn, G. Fermi, and M. F. Perutz, *Biochemistry*, *29*, 1515 (1990).
71. J. Greer, *J. Mol. Biol.*, *59*, 107 (1971).
72. D. J. Abraham, R. A. Peascoe, R. S. Randad, and J. Panikker, *J. Mol. Biol.*, *227*, 480 (1992).
73. A. Brzozowski, Z. Derewenda, E. Dodson, G. Dodson, M. Grabowski, R. Liddington, T. Skarzynski, and D. Vallely, *Nature (London)*, *307*, 74 (1984).
74. J. S. Kavanaugh, P. H. Rodgers, D. A. Case, and A. Arnone, *Biochemistry*, *31*, 4111 (1992).
75. L. F. TenEyck and A. Arnone, *J. Mol. Biol.*, *100*, 3 (1976).
76. G. Fermi, M. F. Perutz, B. Shaanan, and R. Fourme, *J. Mol. Biol.*, *175*, 159 (1984).
77. B. Shaanan, *J. Mol. Biol.*, *171*, 31 (1983).
78. A. Arnone, P. Rogers, N. V. Blough, J. L. McGourty, and B. M. Hoffman, *J. Mol. Biol.*, *188*, 693 (1986).
79. B. Luisi and N. Shibayama, *J. Mol. Biol.*, *206*, 723 (1989).
80. B. Luisi, B. Liddington, G. Fermi, and N. Shibayama, *J. Mol. Biol.*, *214*, 7 (1990).
81. P. T. Manoharan, K. Alston, and J. M. Rifkind, *J. Am. Chem. Soc.*, *108*, 7095 (1986).
82. (a) J. S. Kavanaugh, D. R. Chafin, A. Arnone, A. Mozzarelli, C. Rivetti, G. L. Rossi, L. D. Kwiatkowski, and R. W. Noble, *J. Mol. Biol.*, *248*, 136 (1995). (b)

M. F. Perutz and L. F. Ten Eych, *Cold Spring Harbor Symp. Q. Biol.*, *36*, 295 (1972). (c) J. M. Baldwin, *J. Mol. Biol.*, *130*, 103 (1980).

83. S. Y. Park, A. Nakagava, and H. Morimoto, *J. Mol. Biol.*, *255*, 726 (1996).
84. M. Poali, R. Liddington, J. Tame, A. Wilkinson, and G. Dodson, *J. Mol. Biol.*, *256*, 775 (1996).
85. M. M. Silva, P. H. Rogers, and A. Arnone, *J. Biol. Chem.*, *267*, 17248 (1992).
86. F. R. Smith and K. C. Simmons, *Proteins*, *18*, 295 (1994).
87. P. T. Manoharan, K. Alston, and J. M. Rifkind, *Biochemistry*, *28*, 7148 (1989).
88. P. T. Manoharan, *Proc. Ind. Acad. Sci.*, 337 (1990).
89. M. Ikeda-Saito, H. Yamamoto, and T. Yonetani, *J. Biol. Chem.*, *252*, 8639 (1977).
90. N. Shibayama, M. Ikeda-Saito, H. Hori, K. Itaroku, H. Morimoto, and S. Saigo, *FEBS Lett.*, *372*, 126 (1995).
91. M. Tsubaki and K. J. Nagai, *Biochemistry*, *86*, 1029 (1979).
92. M. Ikeda-Saito, *J. Biol. Chem.*, *255*, 8497 (1980).
93. T. Inubushi and T. Yonetani, *Biochemistry*, *22*, 1894 (1983).
94. H. Hori and T. Yonetani, *J. Biol. Chem.*, *261*, 13693 (1986).
95. K. Kitagishi, C. D'Ambrosio, and T. Yonetani, *Arch. Biochem. Biophy.*, *264*, 176 (1988).
96. Z. Yu-Xiang, F. Yu-Ping, and T. Yonetani, *Sci. China(B)*, *34*, 850 (1991).
97. H. M. McConnell and B. G. Mc Farland, *Quant. Rev. Biophy.*, *3*, 91 (1970).
98. S. Ohnish, J. C. A. Boeyens, and H. M. McConnell, *Proc. Natl. Acad. Sci. USA*, *56*, 809 (1966).
99. S. Ogawa and H. M. McConnell, *Proc. Natl. Acad. Sci. USA*, *58*, 19 (1967).
100. T. Asakura and H. R. Drott, *Biochem. Biophys. Res. Commun.*, *44*, 1199 (1971).
101. J. K. Moffat, *J. Mol. Biol.*, *55*, 135 (1971).
102. J. C. W. Chien, *J. Mol. Biol.*, *133*, 385 (1979).
103. P. T. Manoharan, J. T. Wang, K. Alston, and J. M. Rifkind, *Hemoglobin*, *14*, 41 (1990).
104. W. E. Antholine, F. Taketa, J. T. Wang, P. T. Manoharan, and J. M. Rifkind, *J. Inorg. Biochem.*, *25*, 95 (1985).
105. F. Taketa and W. E. Antholine, *J. Inorg. Biochem.*, *17*, 109 (1982).
106. T. Asakura, *J. Biol. Chem.*, *249*, 4495 (1974).
107. P-W. Lau and T. Asakura, *J. Biol. Chem.*, *254*, 2595 (1979).
108. Y. Henry and R. Banerjee, *J. Mol. Biol.*, *73*, 469 (1973).
109. K. Nagai, H. Hori, S. Yoshida, H. Sakamoto, and H. Morimoto, *Biochem. Biophys. Acta*, *32*, 17 (1978).
110. T. Asakura and P-W. Lau, *Proc. Natl. Acad. Sci. USA*, *75*, 5462 (1978).
111. S. Miura and H. Moriomoto, *J. Mol. Biol.*, *143*, 213 (1980).

112. R. Kruszyna, H. Kruszyna, R. P. Smith, C. D. Thron, and D. E. Wilcox, *Pharm. Exptal. Therap.*, *241*, 307 (1987).

113. R. Banerjee, F. Stetzkowski, and Y. Henry, *J. Mol. Biol.*, *73*, 455 (1973).

114. J. M. Rifkind, O. Abugo, A. Levy, and J. Heim, *Methods in Enzymology*, Academic, San Diego, CA, 1994, Vol. 231, p. 449.

115. A. Levy, V. S. Sharma, L. Zhang, and J. M. Rifkind, *Biophys. J.*, *61*, 750 (1992).

116. S. Ogawa and R. G. Shulman, *J. Mol. Biol.*, *70*, 315 (1972).

117. A. Hayashi, T. Suzuki, A. Shimizu, H. Morimoto, and H. Watari, *Biochem. Biophys. Acta*, *147*, 407 (1967).

118. L. W. M. Fung, A. P. Minton, and C. Ho, *Proc. Natl. Acad. Sci. USA*, *73*, 1581 (1976).

119. M. Nagai, K. Mawatari, Y. Nagai, S. Horita, Y. Yoneyama, and H. Hori, *Biochem. Biophys. Res. Commun.*, *210*, 483 (1995).

120. K. Simolo, G. Stucky, S. Chen, M. Bailey, C. Scholes, and G. McLendon, *J. Am. Chem. Soc.*, *107*, 2865 (1985).

121. C. F. Mulks, C. P. Scholes, L. C. Dickinson, and A. Lapidot, *J. Am. Chem. Soc.*, *101*, 1645 (1979).

122. H. C. Lee, J. Peisach, Y. Dou, and M. Ikeda-Saito, *Biochemistry*, *33*, 7609 (1994).

123. H. C. Lee, J. Peisach, A. Tsuneshige, and T. Yonetani, *Biochemistry*, *34*, 6883 (1995).

124. S. A. Asher, *Methods in Enzymology*, Academic, San Diego, CA, 1981, Vol. 76, p. 371.

125. T. C. Strekas and T. G. Spiro, *Biochem. Biophys. Acta*, *263*, 830 (1972).

126. M. Nagai, Y. Yoneyama, and T. Kitagawa, *Biochemistry*, *30*, 6495 (1991).

127. K. Nagai and T. Kitagawa, *Proc. Natl. Acad. Sci. USA*, *77*, 2033 (1980).

128. I. Mukerji and T. G. Spiro, *Biochemistry*, *33*, 13132 (1994).

129. S. Jeyarajah and J. R. Kincaid, *Biochemistry*, *29*, 5087 (1990).

130. M. R. Ondrias, D. L. Rousseau, T. Kitagawa, M. Ikeda-Saito, T. Inubushi, and T. Yonetani, *J. Biol. Chem.*, *257*, 8766 (1982).

131. T. W. Scott, J. M. Friedman,M. Ikeda-Saito, and T. Yonetani, *FEBS*, 158, 68 (1983).

132. T. Kitagawa, M. R. Ondrias, D. L. Rousseau, M. Ikeda-Saito, and T. Yonetani, *Nature* (*London*), *298*, 869 (1982).

133. K. R. Rodgers and T. G. Spiro, *Science*, *265*, 1697 (1994).

134. V. Jayaraman and T. G. Spiro, *Biochemistry*, *34*, 4511 (1995).

135. N. Shibayama, H. Morimoto, and T. Kitagawa, *J. Mol. Biol.*, *192*, 331 (1986).

136. K. Ishimori, A. Tsuneshige, K. Imai, and I. Morishima, *Biochemistry*, *28*, 8603 (1989).

137. S. Kaminaka, T. Ogura, K. Kitagishi, T. Yonetani, and T. Kitagawa, *J. Am. Chem. Soc.*, *111*, 3787 (1989).

138. S. Kaninaka, Y. X. Zhou, A. Tsuneshige, T. Yonetani, and T. Kitagawa, *J. Am. Chem. Soc.*, *116*, 1683 (1994).
139. V. Jayaraman, K. R. Rodgers, I. Mukerji, and T. G. Spiro, *Science*, *269*, 1843 (1995).
140. A. Dong and W. S. Chaughey, *Methods in Enzymology*, Academic, San Diego, CA, 1994, Vol. 232C, p. 139.
141. M. D. Chavez, S. H. Courtney, M. R. Chance, D. Kiula, K. Nocek, B. M. Hoffman, J. M. Friedman, and M. R. Ondrias, *Biochemistry*, *29*, 4844 (1990).
142. C. Ho and I. M. Russ, *Methods in Enzymology*, Academic, San Diego, CA, 1981, Vol. 76, p. 275.
143. C. Ho and J. R. Perussi, *Methods in Enzymology*, Academic, San Diego, CA, 1994, Vol. 232C, p. 97.
144. C. Ho, *Adv. Protein. Chem.*, *43*, 153 (1992).
145. A. V. Pisciotta, S. N. Ebbe, and J. E. Hinz, *J. Lab. Clin. Med.*, *54*, 73 (1959).
146. O. Mayer, S. Ogawa, and K. Gersonde, *J. Mol. Biol.*, *81*, 187 (1973).
147. M. F. Perutz, P. D. Pulsinelli, and H. M. Ranney, *Nature* (*London*), *237*, 259 (1972).
148. L. W.-M. Fung, A. P. Minton, T. R. Lindstrom, A. V. Pisciotta, and C. Ho *Biochemistry*, *16*, 1452 (1977).
149. R. G. Shulman, S. Ogawa, K. Wuthrich, J. Yamone, and J. Peisach, *Science*, *165*, 215 (1969).
150. S. Ogawa and R. G. Shulman, *J. Mol. Biol.*, *70*, 315 (1972).
151. S. Ogawa, R. G. Shulman, M. Fujiwara, and T. Yamane, *J. Mol. Biol.*, *70*, 301 (1972).
152. R. Cassoly, Q. H. Gibson, S. Ogawa, and R. G. Shulman, *Biochemistry*, *44*, 1015 (1971).
153. S. Ogawa and R. G. Shulman, *Biochem. Biophys. Res. Commun.*, *42*, 9 (1971).
154. S. Miura and C. Ho, *Biochemistry*, *21*, 6280 (1982).
155. S. Miura and C. Ho, *Biochemistry*, *23*, 2492 (1984).
156. W. H. Huestis and M. A. Raftery, *Biochemistry*, *14*, 1886 (1975).
157. R. Banerjee, F. Stetzkowski, and J. M. Lhoste, *FEBS Lett.*, *70*, 171 (1976).
158. K. Ishimori and I. Morishima, *Biochemistry*, *25*, 4892 (1986).
159. A. Kuki and S. G. Boxer, *Biochemistry*, *22*, 2923 (1983).
160. M. Fujii, H. Hori, G. Miyazaki, H. Morimoto, and T. Yonetani, *J. Biol. Chem.*, *268*, 15386 (1993).
161. G. N. La mar, T. Jue, K. Nagai, K. M. Smith, Y. Yamamot, R. J. Kauten, V. Thanabal, K. C. Langry, R. K. Pandey, and H-K. Leung, *Biochem. Biophys. Acta*, *952*, 131 (1988).
162. T. Inubushi, M. Ikeda-Saito, and T. Yonetani, *Biochemistry*, *22*, 2904 (1983).
163. T. Inubushi, M. Ikeda-Saito, and T. Yonetani, *Biophys. J.*, *47*, 74a(Abst) (1985).

164. N. Shibayama, T. Inubushi, H. Morimoto, and T. Yonetani, *Biochemistry*, *26*, 2194 (1987).

165. K. Ishimori and I. Morishima, *Biochemistry*, *27*, 4060 (1988).

166. A. Bondon and G. Simonneaux, *Biophys. Chem.*, *37*, 407 (1990).

167. T. Inubushi, C. D'Ambrosio, M. Ikeda-Saito, and T. Yonetani, *J. Am. Chem. Soc.*, *108*, 3799 (1986).

168. W. A. Eaton and J. Hofrichter, *Methods in Enzymology*, Academic, San Diego, CA, 1981, Vol. 76, p. 175.

169. A. Bellelli and M. Brunori, *Methods in Enzymology*, Academic, San Diego, CA, 1994, Vol. 232, p. 56.

170. Q. H. Gibson, *Biochem. J.*, *71*, 293 (1959).

171. M. R. Waterman and T. Yonetani, *J. Biol. Chem.*, *245*, 5847 (1970).

172. T. Nakamura, Y. Sugita, and S. Bannai, *J. Biol. Chem.*, *248*, 4119 (1973).

173. H. Yamamoto and T. Yonetani, *J. Biol. Chem.*, *249*, 7964 (1974).

174. M. R. Waterman, R. Gondko, and T. Yonetani, *Arch. Biochem. Biophys.*, *145*, 448 (1971).

175. R. Gondko, M. J. Obrebska, and M. R. Waterman, *Biochem. Biophys. Res. Commun.*, *56*, 444 (1974).

176. R. Cassoly, *C. R. H. Acad. Sci. Paris*, *278(Series D.)*, 1417 (1974).

177. R. Cassoly, *J. Mol. Biol.*, *98*, 581 (1975).

178. Y. Sugita, *J. Biol. Chem.*, *250*, 1251 (1975).

179. R. Cassoly, *J. Biol. Chem.*, *253*, 3602 (1978).

180. M. F. Perutz, *Nature (London)*, *237*, 495 (1972).

181. R. Cassoly, *Eur. J. Biochem.*, *65*, 461 (1976).

182. J. W. Petrich, C. Poyart, and J. L. Martin, *Biochemistry*, *27*, 4049 (1988).

183. K. Mawatari, S. Matsukawa, Y. Yoneyama, and Y. Takeda, *Biochem. Biophys. Acta*, *913*, 313 (1987).

184. J. Hofrichter, E. R. Henry, J. H. Sommer, R. L. Deutsch, M. Ikeda-Saito, T. Yonetani, and W. A. Eaton, *Biochemistry*, *24*, 2667 (1985).

185. L. P. Murray, J. Hofrichter, E. R. Henry, and W. A. Eaton, *Biophys. Chem.*, *29*, 63 (1988). (a) R. A. Goldbeck, S. J. Paquette, S. C. Bjorling, and D. S. Klige, *Biochemistry*, *35*, 8628 (1996).

186. R. E. Hirsch, *Methods in Enzymology*, Academic, San Diego, CA, 1994, Vol. 232, p. 231.

187. R. E. Hirsch and R. L. Nagel, *J. Biol. Chem.*, *256*, 1080 (1981).

188. E. Ya. Alfimova and G. I. Likhtenshtein, *Biofizika*, *17*, 49 (1972).

189. R. E. Hirsch, R. S. Zukin, and R. L. Nagel, *Biochem. Biophys. Res. Commun.*, *138*, 489 (1986).

190. R. E. Hirsch and R. L. Nagel, *Anal. Biochem.*, *176*, 19 (1989).

191. J. J. Leonard, T. Yonetani, and J. B. Callis, *Biochemistry*, *13*, 1460 (1989).

192. K. Sudhakar, S. Loe, T. Yonetani, and J. M. Vanderkooi, *J. Biol. Chem*, *269*, 23095 (1994).

193. G. Geraci and L. J. Parkhurst, *Methods in Enzymology*, Academic, San Diego, CA, 1981, Vol. 76, p. 262.

194. C. Zentz, S. Pin, and B. Alpert, *Methods in Enzymology*, Academic, San Diego, CA, 1994, Vol. 232, p. 247.

195. K. Mawatari, S. Matsukawa, and Y. Yoneyama, *Biochem. Biophys. Acta*, *748*, 381 (1983).

196. Y. Kawamura-Konishi and H. Suzuki, *Biochem. Biophys. Res. Commun.*, *156*, 348 (1988) (and references cited therein).

197. J. V. Kilmartin and J. A. Hewitt, *Cold Spring Harbor Symp. Quant. Biol.*, *36*, 311 (1971).

198. R. Cassoly, R. Banerjee, *Eur. J. Biochem.*, *19*, 514 (1971).

199. M. Nagai, S. Takama, and Y. Yoneyama, *Biochem. Biophy. Res. Commun.*, *128*, 689 (1985).

200. M. Nagai, S. Takama, and Y. Yoneyama, *Stud. Biophys.*, *116*, 135 (1986).

201. M. Nagai, S. Takama, and Y. Yoneyama, *Acta Haematol.*, *78*, 95 (1987).

202. H. D. Fiechtner, G. McLendon, and M. W. Bailey, *Biochem. Biophy. Res. Commun.*, *96*, 618 (1980). (a) S. C. Bjorling, R. A. Goldbeck, S. J. Paquette, S. J. Milder, and D. S. Kliger, *Biochemistry*, *35*, 8619 (1996).

203. B. M. Kincaid, P. Fisenberger, K. O. Hodgson, and S. Doniach, *Proc. Natl. Acad. Sci. USA*, *72*, 2340 (1975).

204. S. Pin, B. Alpert, A. Congiu-Castellano, S. D. Longa, and A. Bianconi, *Methods in Enzymology*, Academic, San Diego, CA, 1994, Vol. 232, p. 266.

205. P. Eisenberger, R. G. Shulman, B. M. Kincaid, G. S. Brown, and S. Ogawa, *Nature* (*London*), *274*, 30 (1978).

206. W. R. Scheidt, *Acc. Chem. Res.*, *10*, 339 (1977).

207. K. Simolo, Z. R. Korszun, G. Stuky, K. Moffat, G. McLendon, and G. Bunker, *Biochemistry*, *25*, 3773 (1986).

208. T. M. Bednarski, and J. Jordan, *J. Am. Chem. Soc.*, *89*, 1552 (1967).

209. G. McLendon, *Acc. Chem. Res.*, *21*, 160 (1988).

210. S. L. Mayo, W. R. Ellis, Jr., R. J. Crutchley, and H. B. Gray, *Science*, *233*, 948 (1986).

211. G. McLendon and J. Feitelson, *Methods in Enzymol.*, Academic, San Diego, CA, 1994, Vol. 232, p. 86.

212. R. A. Marcus, *Ang. Chem.*, *32*, 1111 (1993).

213. G. McLendon, T. Guarr, M. McGuire, K. Simolo, S. Strauch, and K. Taylor, *Coord. Chem. Rev.*, *64*, 113 (1985).

214. S. E. Peterson-Kennedy, T. L. McGourty, P. S. Ho, C. J. Sutoris, N. Liang, H. Zemel, N. V. Blough, E. Margoliash, and B. M. Hoffman, *Coord. Chem. Rev.*, *64*, 125 (1985).

215. H. Zemel and B. M. Hoffman, *J. Am. Chem. Soc.*, *103*, 1192 (1981).

216. J. L. McGourty, N. V. Blough, and B. M. Hoffman, *J. Am. Chem. Soc.*, *105*, 4470 (1983).

217. B. M. Hoffman and M. A. Ratner, *J. Am. Chem. Soc.*, *109*, 6237 (1987).

218. M. J. Natan and B. M. Hoffman, *J. Am. Chem. Soc.*, *111*, 6468 (1989).

219. M. J. Natan, D. Kuila, W. W. Baxter, B. C. King, F. M. Hawkridge, and B. M. Hoffman, *J. Am. Chem. Soc.*, *112*, 4081 (1990).

220. D. Kuila, M. J. Natan, P. Rogers, D. J. Gingrich, W. W. Baxter, A. Arnone, and B. M. Hoffman, *J. Am. Chem. Soc.*, *113*, 6520 (1991).

221. S. E. Peterson-Kennedy, J. L. McGourty, J. A. Kalweit, and B. M. Hoffman, *J. Am. Chem. Soc.*, *108*, 1739 (1986).

222. D. J. Gingrich, J. M. Nocek, M. J. Natan, and B. M. Hoffman, *J. Am. Chem. Soc.*, *109*, 7533 (1987).

223. J. L. McGourty, S. E. Peterson-Kennedy, W. Y. Ruo, and B. M. Hoffman, *Biochemistry*, *26*, 8302 (1987).

224. K. P. Simolo, G. L. McLendon, M. R. Mauk, and A. G. Mauk, *J. Am. Chem. Soc.*, *106*, 5012 (1984).

225. W. W. Baxter, K. E. Hines, B. M. Hoffman, S. L. J. Nebolsky, J. M. Nocek, D. L. Overdeck, M. Thurnauer, and Y. Zhang, *J. Inorg. Bio. Chem.*, 171 (Abs) (1993).

226. S. E. Peterson-Kennedy, J. L. McGourty, and B. M. Hoffman, *J. Am. Chem. Soc.*, *106*, 5010 (1984).

227. D. Kuila, W. W. Baxter, M. J. Natan, and B. M. Hoffman, *J. Phys. Chem.*, *95*, 1 (1991).

228. L. A. Dick, I. Malfant, D. Kuila, and B. M. Hoffman, *J. Inorg. Biochem.*, 256, (Abs) (1993).

229. C. Brunel, A. Bondon, and G. Simonneaux, *J. Am. Chem. Soc.*, *116*, 11827 (1994).

230. K. Imai, *Methods in Enzymology*, Academic, San Diego, CA, 1981, Vol. 72, p. 438.

231. K. Nagai, *J. Mol. Biol.*, *111*, 41 (1977).

232. A. Riggis and A. E. Herner, *Proc. Natl. Acad. Sci. USA*, *48*, 1664 (1962).

233. E. Antonini, J. Wyman, E. Bucci, C. Fronticelli, M. Brunori, M. Reichlin, and A. R. Fanelli, *Biochem. Biophys. Acta.*, *104*, 160 (1965).

234. F. Taketa, M. R. Smits, F. J. Dibona, and J. L. Lessard, *Biochemistry*, *6*, 3809 (1967).

235. S. G. Condo, M. Coletta, R. Cicchetti, G. Argentine, P. Guerrieri, S. Marini, S. El-Sherbini, and B. Giardina, *Biochem. J.*, *282*, 595 (1992).

236. R. P. Roy, *J. Prot. Chem.*, *14*, 81 (1995).

237. E. Antonini, J. Wyman, E. Bucci, C. Fronticelli, M. Brunori, M. Reichlin, and A. R. Fanelli, *Biochem. Biophys. Acta*, *104*, 160 (1965).

238. T. Causgrove, D. J. Goss, and L. J. Parkhurst, *Biochemistry*, *23*, 2168 (1984).
239. L. J. Parkhurst and D. T. Goss, in: *Hemoglobin and Oxygen Binding*, H. Chien, Ed., Elsevier/North-Holland, New York, 1982.
240. S. G. Condo, B. Giardina, A. Bellelli, and M. Brunori, *Biochemistry*, *26*, 6718 (1987).
241. N. Makino and Y. Sugita, *J. Biol. Chem.*, *253*, 1174 (1978).
242. J. E. Haber and D. E. Koshland, *Biochem. Biophys. Acta*, *194*, 339 (1969).
243. R. Banerjee and R. Cassoly, *J. Mol. Biol.*, *42*, 351 (1969).
244. M. Brunori, G. Amiconi, E. Antonini, J. Wyman, and K. H. Winterhalter, *J. Mol. Biol.*, *49*, 461 (1970).
245. J. E. Haber and D. E. Koshland, *J. Biol. Chem.*, *246*, 7790 (1971).
246. R. Maeda, K. Imai, and I. Tyuma, *Biochemistry*, *11*, 3685 (1972).
247. M. L. Doyle and G. K. Ackers, *Biochemistry*, *31*, 11182 (1992).
248. A. DiDonato, W. J. Fantl, A. S. Acharya, and J. M. Manning, *J. Biol. Chem.*, *258*, 11890 (1983).
249. W. J. Fantl, A. DiDonato, J. M. Manning, P. H. Rodgers, and A. Arnone, *J. Biol. Chem.*, *262*, 12700 (1987).
250. Y. Bai, H. Ueno, and J. M. Manning, *J. Prot. Chem.*, *8*, 299 (1989).
251. H. Ueno and J. M. Manning, *J. Protein Chem.*, *11*, 77 (1992).
252. N. Shibayama, H. Morimoto, and G. Miyazaki, *J. Mol. Biol.*, *192*, 323 (1986).
253. H. Yamamoto, M. Ikeda-Saito, and T. Yonetani, *Fed. Proc.*, *35*, 1392 (1975).
254. (a) K. Imai, M. Ikeda-Saito, H. Yamamoto, and T. Yonetani, *J. Mol. Biol.*, *138*, 635 (1980). (b) S. Unzai, H. Hori, G. Miyazaki, N. Shibayama, and H. Morimoto, *J. Biol. Chem.*, *271*, 12451 (1996).
255. H. Adachi, T. Asakura, and K. Adachi, *J. Biol. Chem.*, *258*, 13422 (1983).
256. N. Shibayama, K. Imai, H. Hirata, H. Hiraiwa, H. Morimoto, and S. Saigo, *Biochemistry*, *30*, 8158 (1991).
257. N. Shibayama, K. Imai H. Morimoto, and S. Saigo, *Biochemistry*, *32*, 8792 (1993).
258. N. Shibayama, K. Imai, H. Morimoto, and S. Saigo, *Biochemistry*, *34*, 4773 (1995).
259. S. Miura, M. Ikeda-Saito, T. Yonetani, and C. Ho, *Biochemistry*, *26*, 2149 (1987).
260. A. Tsuneshige, Y. X. Zhou, and T. Yonetani, *J. Biol. Chem.*, *268*, 23031 (1993).
261. L. J. Parkhurst, G. Geraci, and Q. H. Gibson, *J. Biol. Chem.*, *245*, 4131 (1970).
262. J. M. Baldwin, *Prog. Biophys. Mol. Biol.*, *29*, 225 (1975).
263. V. S. Sharma, *J. Mol. Biol.*, *166*, 677 (1983).
264. M. Perrella, N. Davids, and L. Rossi-Bernardi, *J. Biol. Chem.*, *267*, 8744 (1992).
265. V. S. Sharma, D. Bandyopadhyay, M. Berjis, J. M. Rifkind, and G. R. Boss, *J. Biol. Chem.*, *266*, 24491 (1991).
266. V. S. Sharma, *J. Biol. Chem.*, *263*, 2292 (1988).

267. M. Berjis, D. Bandyopadhyay, and V. S. Sharma, *Biochemistry*, *29*, 10106 (1990).
268. D. Bandyopadhyay, M. Magde, T. G. Traylor, and V. S. Sharma, *Biophys. J.*, *63*, 673 (1992).
269. V. S. Sharma and H. M. Ranney, *J. Mol. Biol.*, *158*, 551 (1982).
270. V. S. Sharma, *Methods in Enzymology*, Academic, San Diego, CA, 1994, Vol. 232, p. 430.
271. V. S. Sharma, *J. Biol. Chem.*, *264*, 10582 (1989).
272. M. A. Khaleque and C. A. Sawicki, *Photobiochem. Photobiophys.*, *13*, 155 (1986).
273. I. A. Zahroon and C. A. Sawicki, *Biophys. J.*, *56*, 947 (1989).
274. R. Cassoly and Q. H. Gibson, *J. Biol. Chem.*, *247*, 7332 (1972).
275. M. Perrella, L. Subbioneda, M. Samaja, and L. Rossi-Bernardi, *J. Biol. Chem.*, *261*, 8391 (1986).
276. M. Samaja, E. Rovida, M. Niggeler, M. Perrella, and L. Rossi-Bernardi, *J. Biol. Chem.*, *262*, 4528 (1987).
277. E. Antonini, M. Brunori, J. Wyman, and R. W. Noble, *J. Biol. Chem.*, *241*, 3236 (1966).
278. J. M. Salhany, S. Ogawa, and R. G. Shulman, *Proc. Natl. Acad. Sci. USA*, *71*, 3359 (1974).
279. L. Kiger, C. Poyart, and M. C. Marden, *Biophys. J.*, *65*, 1050 (1993).
280. F. A. Ferrone, *Methods in Enzymology*, Academic, San Diego, CA, 1994, Vol. 232, p. 292.
281. D. Liao, J. Jiang, M. Zhao, and F. A. Ferrone, *Biophys. J.*, *65*, 2059 (1993).
282. M. Zhao, J. Jiang, M. Greene, M. E. Andracki, S. A. Fowler, J. A. Walder, and F. A. Ferrone, *Biophys. J.*, *64*, 1520 (1993).
283. (a) M. Brunori, B. Giardina, and E. E. DiIorio, *FEBS Lett.*, *46*, 312 (1974). (b) J. S. Philo, U. Dreyer, and J. W. Lary, *Biophysical J.*, *70*, 1949 (1996).
284. A. Tomoda, T. Yubisui, A. Tsuji, and Y. Yoneyama, *J. Biol. Chem.*, *254*, 3119 (1979).
285. N. Makino, Y. Sugita, and T. Nakamura, *J. Biol. Chem.*, *254*, 10862 (1979).
286. N. Makino, Y. Sugita, and T. Nakamura, *J. Biol. Chem.*, *254*, 2353 (1979).
287. M. C. Marden, L. Kiger, J. Kister, and C. Poyart, *Biophys. J.*, *63*, 1681 (1992).
288. S. A. Fowler, J. Walder, A. DeYoung, L. D. Wiatkowski, and R. W. Noble, *Biochemistry*, *31*, 717 (1992).
289. L. D. Kwiatkowski, A. DeYoung, and R. W. Noble, *Biochemistry*, *33*, 5884 (1994).
290. L. J. Parkhurst, G. Geraci, and Q. H. Gibson, *J. Biol. Chem.*, *245*, 4131 (1970).
291. Y. Sugita, S. Bannai, Y. Yoneyama, and T. Nakamura, *J. Biol. Chem.*, *247*, 6092 (1972).
292. D. J. Goss and L. J. Parkhurst, *Biochemistry*, *23*, 2174 (1984).
293. L. J. Parkhurst and D. J. Goss, *Biochemistry*, *23*, 2180 (1984).

294. E. R. Huehns, E. M. Shooter, and G. H. Beaven, *Biochem. J.*, *91*, 331 (1964).
295. M. Ikeda-Saito and T. Yonetani, *J. Mol. Biol.*, *138*, 845 (1980).
296. L. P. Murray, J. Hofrichter, E. R. Henry, M. Ikeda-Saito, K. Kitagishi, T. Yonetani, and W. A. Eaton, *Proc. Natl. Acad. Sci. USA*, *85*, 2151 (1988).
297. R. J. Morris, Q. H. Gibson, M. Ikeda-Saito, and T. Yonetani, *J. Biol. Chem.*, *259*, 6701 (1984).
298. Q. H. Gibson, M. Ikeda-Saito, and T. Yonetani, *J. Biol. Chem.*, *260*, 14126 (1985).
299. M. Oertle, K. H. Winterhalter, and E. E. Di-Iorio, *FEBS Lett.*, *153*, 213 (1983).
300. K. Kitagishi, M. Ikeda-Saito, and T. Yonetani, *J. Mol. Biol.*, *203*, 1119 (1988).
301. B. M. Hoffman, Q. H. Gibson, C. Bull, R. H. Crepeau, S. J. Edelstein, R. G. Fisher, and M. J. McDonald, *Ann. N. Y. Acad. Sci.*, *244*, 174 (1975).
302. N. V. Blough, H. Zemel, B. M. Hoffman, T. C. K. Lee, and Q. H. Gibson, *J. Am. Chem. Soc.*, *102*, 5683 (1980).
303. N. V. Blough and B. M. Hoffman, *J. Am. Chem. Soc.*, *104*, 4247 (1982).
304. N. V. Blough and B. M. Hoffman, *Biochemistry*, *23*, 2875 (1984).
305. N. V. Blough, H. Zemel, and B. M. Hoffman, *Biochemistry*, *23*, 2883 (1984).
306. M. D. Fiechtner, G. McLendon, and M. W. Bailey, *Biochem. Biophys. Res. Commun.*, *96*, 618 (1980).
307. N. Shibayama, T. Yonetani, R. M. Regan, and Q. H. Gibson, *Biochemistry*, *34*, 14658 (1995).
308. C. Brunel, A. Bondon, and G. Simonneaux, *J. Am. Chem. Soc.*, *116*, 11827 (1994).
309. F. R. Smith and G. K. Ackers, *Proc. Natl. Acad. Sci. USA*, *82*, 5347 (1985).
310. G. K. Ackers and F. R. Smith, *Ann. Rev. Biophys. Biophys. Chem.*, *16*, 583 (1987).
311. V. J. LiCata, P. C. Sperose, E. Rovida, and G. K. Ackers, *Biochemistry*, *29*, 9771 (1990).
312. M. Perrella, L. Benazzi, M. Shea, and G. K. Ackers, *Biophys. Chem.*, *35*, 97 (1990).
313. V. J. LiCata, P. M. Dalessio, and G. K. Ackers, *Proteins Structure Function Genet.*, *17*, 279 (1993).
314. G. K. Ackers and J. H. Hazzard, *Trends Biochem. Sci.*, *18*, 385 (1993).
315. J. M. Holt and G. K. Ackers, *FASEB*, *9*, 210 (1995).
316. R. Valdes, Jr. and G. K. Ackers, *Methods in Enzymology*, Academic, San Diego, CA, Vol. 61, p. 125.
317. M. Perrella, A. Colosimo, L. Benazzi, M. Ripamonti, and L. Rossi-Bernardi, *Biophys. Chem.*, *37*, 211 (1990).
318. P. C. Speros, V. J. LiCata, T. Yonetani, and G. K. Ackers, *Biochemistry*, *30*, 7254 (1991).

319. M. L. Doyle, P. C. Speros, V. J. LiCata, D. Gingrich, B. M. Hoffman, and G. K. Ackers, *Biochemistry*, *30*, 7263 (1991).

320. F. R. Smith, D. Gingrich, B. M. Hoffman, and G. K. Ackers, *Proc. Natl. Acad. Sci. USA*, *84*, 7089 (1987).

321. G. K. Ackers, M. L. Doyle, D. Myers, and M. A. Daugherty, *Science*, *255*, 54 (1992).

322. Y. Huang and G. K. Ackers, *Biochemistry*, *34*, 6316 (1995).

323. M. A. Daugherty, M. A. Shea, J. A. Johnson, V. J. Li Cata, G. J. Turner, and G. K. Acker, *Proc. Natl. Acad. Sci. USA*, *88*, 1110 (1991).

324. M. Straume and M. L. Johnson, *Biophys. J.*, *56*, 15 (1989).

325. G. K. Ackers and M. L. Johnson, *Biophys. Chem.*, *37*, 265 (1990).

326. G. K. Ackers, *Biophys. Chem.*, *37*, 371 (1990).

327. M. L. Doyle, G. Lew, A. D. Young, A. K. Wierzba, R. W. Noble, and G. K. Ackers, *Biochemistry*, *31*, 8629 (1992).

328. M. A. Daugherty, M. A. Shea, and G. K. Ackers, *Biochemistry*, *33*, 10345 (1994).

329. (a) M. Perrella, L. Benazzi, M. Ripamonti, and L. Rossi-Bernardi, *Biochemistry*, *33*, 10358 (1994). (b) Y. Huang, M. L. Koestner, and G. K. Ackers, *Biophys. J.*, *71*, 2106 (1996).

330. (a) Y. Huang and G. K. Ackers, *Biochemistry*, *35*, 704 (1996). (b) Y. Huang, M. I. Doyle, and G. K. Ackers, *Biophys. J.*, *71*, 2094 (1996).

331. Y. Huang, T. Yonetani, A. Tsuneshige, B. M. Hoffman, and G. K. Ackers, *Proc. Natl. Acad. Sci. USA*, *93*, 4425 (1996).

332. K. Nibu and K. Adachi, *Biochem. Biophys. Acta*, *829*, 97 (1985).

Three-Coordinate Complexes of "Hard" Ligands: Advances in Synthesis, Structure and Reactivity

CHRISTOPHER C. CUMMINS

Massachusetts Institute of Technology
Department of Chemistry
Cambridge, MA

CONTENTS

Progress in Inorganic Chemistry, Vol. 47, Edited by Kenneth D. Karlin.
ISBN 0-471-24039-7 © 1998 John Wiley & Sons, Inc.

I. INTRODUCTION

This chapter represents an attempt to review the chemistry of three-coordinate complexes of the type pioneered by Bürger and Wannagat, Bradley, Lappert, Andersen, and their co-workers. The term "three-coordinate complex" is used here to describe *mononuclear* entities having the following properties: (a) the central metal ion possesses a d^n $(0 < n < 10)$ or f^n $(0 < n < 14)$ formal electron configuration, and (b) the central metal is ligated by exactly three monodentate ligands or the equivalent thereof, not counting agostic or other low-energy interactions, and (c) each of the ligands is either formally monoanionic (e.g., alkoxide) dianionic (e.g., imido), or trianionic (e.g., alkylidyne), but not neutral or cationic. A classic example of such a three-coordinate complex is $Cr(N{-}i\text{-}Pr_2)_3$. The state of the art in this research area is evident in the material covered in this chapter. While many three-coordinate complexes have been described since the early 1960s, this chapter makes it obvious that much work remains to be done. Previous reviews describing three-coordinate complexes are out of date (1–4). This chapter covers in large part the literature since 1977. Older work is included in some cases, in order to draw attention to known but largely forgotten substances that might be useful as reagents in modern applications.

Peripherally related reviews have covered three coordination in metal nitrides (5), the chemistry of metal amides including three-coordinate ones (6, 7), and the chemistry of open-shell two-coordinate complexes (8, 9).

No search of the literature is perfect. In the present case, data bases including the Cambridge Structural Database, the Chemical Abstracts, and the ISI Chemistry Citation Index have been employed to harvest relevant publications. I apologize in advance for omission of any compounds falling within the intended scope of this chapter, as enunciated in Section II.

A final introductory comment is the caveat that most of the complexes described here, unless stated otherwise, are oxygen and/or water sensitive and have been prepared and studied using standard techniques for the manipulation of air-sensitive substances.

II. SCOPE

This chapter aims to stimulate new explorations involving three-coordinate complexes. The somewhat nebulous term "three coordinate" is used herein with reference to the number of atoms in the first coordination sphere of the metal in question. Relatively weak interactions such as agostic interactions involving C—H or C—F bonds, or reversible intramolecular π complexation of a ligand aryl substituent, are not counted, for our purposes, toward the overall coordination number.

As mentioned in Section I, prototypical of the type of compound included in this chapter are the classic neutral molecules $Cr(N{-}i\text{-}Pr_2)_3$, $U[N(SiMe_3)_2]_3$, and $Os(N{-}2,6{-}C_6H_3{-}i\text{-}Pr_2)_3$. Mixed-ligand (heteroleptic) species such as the beautiful compound $W(OSi{-}t\text{-}Bu_3)_2(N{-}t\text{-}Bu)$ are also included. It is noteworthy that essentially all of the three-coordinate complexes I uncovered in the preparation of this chapter involve monodentate ligands exclusively. Obviously, much work remains to be done.

Targeted here are three-coordinate complexes of (formally) negatively charged ligands (e.g., alkyl, aryl, amide, imide, and alkoxide), which do not readily dissociate in low dielectric media. For this reason, complexes containing potentially labile neutral donors, for example, $Co[N(SiMe_3)_2]_2(OPPh_3)$ or $Fe(Mes)_2(col)$ (10) are not included; such complexes may be considered "masked" two-coordinate species. Some masked three-coordinate species *are* included, for example, $(thf)V(Mes)_3$, inasmuch as they *may* give rise to three-coordinate complexes via dissociation of the neutral ligand and thereby serve to transfer the tricoordinate fragment to interesting substrates (e.g., N_2). Related trigonal monopyramidal complexes such as $V(Me_3SiNCH_2CH_2)_3N$ are not considered here because they are effectively constrained by chelation to be four coordinate; the chemistry of triamidoamine complexes was reviewed recently (11).

Bi- and polymetallic or oligomeric complexes are excluded from coverage in order to preserve the focus on mononuclear entities. Ionic species with separate anions and cations such as $[Mn(Mes)_3]$ $[Li(thf)_4]$ (12) are therefore included, while intimately associated "ate" complexes, of which $[(Me_3Si)_2N]Mn[\mu{-}N(SiMe_3)_2]_2Li(thf)$ (13) serves as a representative example, are not.

Only those metal complexes are included that may potentially act as reducing agents by virtue of the presence of *d* electrons, or chemically accessible *f* electrons. This qualification focuses attention *away* from the Lewis acid or σ-bond metathesis (including protolytic) chemistry of three-coordinate metal complexes exemplified by $M[N(SiMe_3)_2]_3$, where M = Al, Sc, and Y, or a lanthanide element (14). The purpose of this qualification is to highlight recent examples of dramatic small-molecule [e.g., CO (15), N_2 (16)] reductive cleavage reactions effected by three-coordinate complexes. Such reactions are to be anticipated mainly for d^n $(0 < n < 10)$ early transition metal derivatives, or for actinide elements (especially uranium).

Not included are ubiquitous three-coordinate complexes of d^{10} metal ions including Ni^0, Pd^0, Pt^0, Cu^{1+}, Ag^{1+}, Au^{1+}, Zn^{2+}, Cd^{2+}, and Hg^{2+} (2). While mononuclear three-coordinate complexes of the latter ions atoms may indeed be redox active, they are usually supported by readily dissociable neutral ligands (e.g., triorganophosphine or organonitrile ligands), on which basis they are excluded from this chapter.

III. ELECTRONIC STRUCTURE OF THREE-COORDINATE COMPLEXES

Several lucid discussions of the orbital structure found for coordination number 3 have been given (15, 17–19), such that a brief introduction will suffice here.

An illustrative example is hypothetical D_{3h} $Mo(NH_2)_3$, as considered in Figs. 1 and 2; these figures derive from an extended Hückel (20) calculation carried out using the CACAO program suite (21). The frontier orbitals ($3a_1'$ and $2e''$) are those derived largely from d_z^2, d_{xz}, and d_{yz}; $3a_1'$ (d_z^2) evinces a small degree of antibonding character with respect to the Mo—N σ bonds, while the $2e''$ (d_{xz}, d_{yz}) pair is essentially nonbonding. The two molecular orbitals derived largely from $d_{x^2-y^2}$ and d_{xy} ($4e'$) are strongly Mo—N σ and π antibonding in character.

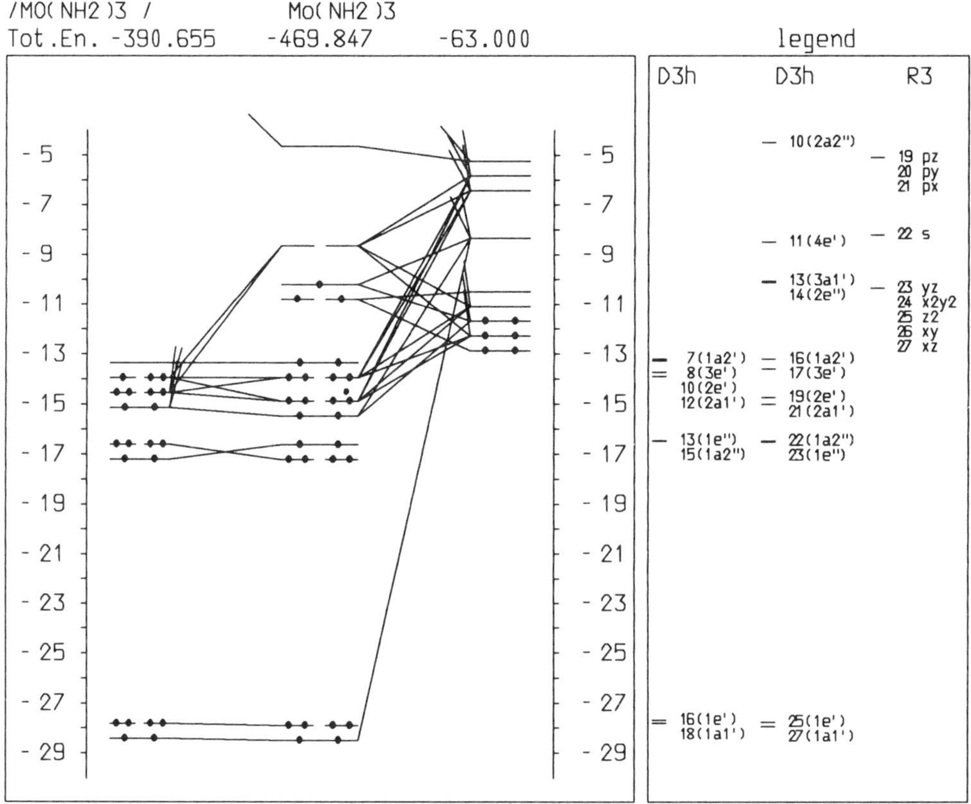

Figure 1. Energy level diagram for D_{3h} $Mo(NH_2)_3$.

Next lowest lying is the highly energetic $2a_2''$ orbital, which is essentially a nonbonding molybdenum $4p_z$ orbital.

The above features neatly explain the stability of high-spin d^3 complexes like $Cr(N-i\text{-}Pr_2)_3$ and $Mo[N(R)Ar]_3$ [R = $C(CD_3)_2Me$, Ar = 3,5-$C_6H_3Me_2$]. In fact, the splitting of the *d*-orbitals in a two-above-three pattern is reminiscent of the familiar situation for octahedral complexes, where the high-spin d^3 case is similarly ubiquitous. The criterion for stability is seen to be fulfilled as follows: no low-lying levels are empty, and no orbitals having significant antibonding character are populated. For molecules that on casual inspection appear to defy the foregoing prescription, for example, $Ti[N(R)Ar]_3$ with its trigonal planar TiN_3 core and d^1 electron count, it is frequently the case that weak interactions take place between the low-lying empty orbitals based on the metal and some supplemental source of electron density provided by the ligand. In

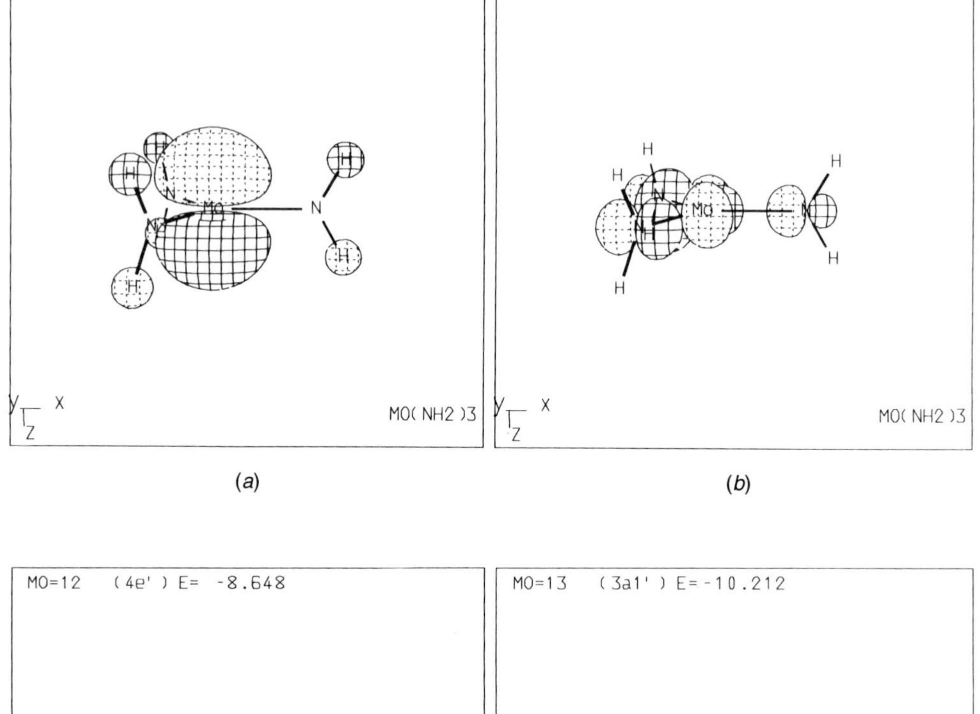

(a) (b)

(c) (d)

Figure 2. Selected molecular orbitals of $Mo(NH_2)_3$.

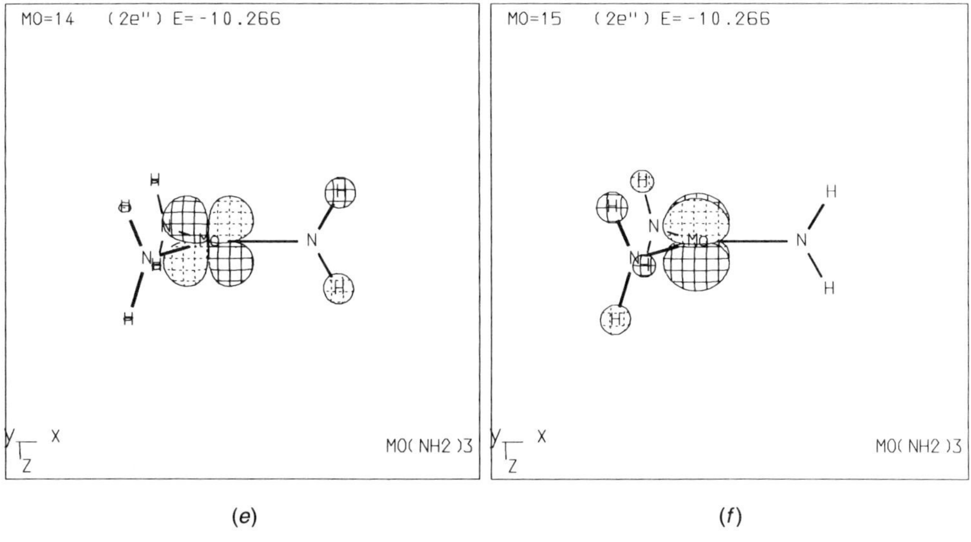

Figure 2. (*Continued.*)

the particular case of Ti[N(R)Ar]$_3$, the titanium center binds two of the aryl rings as revealed by X-ray crystallography. This result is in stark contrast to the threefold symmetric structures of d^3 M[N(R)Ar]$_3$ (M = Cr or Mo). Ligand → metal π bonding is also occasionally invoked as a supplemental interaction, as in the case of low-spin d^2 Ta(OSi—*t*-Bu$_3$)$_3$. Here, the $2e''$ (d_{xz}, d_{yz}) pair acquires antibonding character with respect to dative O → Ta π bonds, making $3a_1'$ (d_z^2) the lowest lying of the five *d*-derived molecular orbitals.

From a reactivity standpoint, it is instructive to examine isolobal relationships (22). High-spin trigonal d^3 complexes such as Cr(N-*i*-Pr$_2$)$_3$ are isolobal with a nitrogen atom, in accord with the finding that N≡Cr(N—*i*-Pr$_2$)$_3$ (isolobal with N≡N) is a robust diamagnetic pseudotetrahedral entity. Furthermore, low-spin d^2 Ta(OSi—*t*-Bu$_3$)$_3$, with its tantalum-based "lone-pair" $3a_1'(d_z^2)$, is isolobal with a phosphite, P(OR)$_3$; the tantalum species, however, is a far more potent reductant and oxophile! The d^1 titanium complex Ti[N(R)Ar]$_3$ is analogous to atomic lithium, making ClTi(N[R]Ar)$_3$, with its two symmetry–allowed Cl → Ti dative π interactions, analogous to *molecular* lithium chloride, which would have a formal Li—Cl bond order of 3. Thus it can be seen that these early metal-based three-coordinate complexes are poised to behave as atom acceptors in single- or multielectron redox reactions. Many such reactions have been realized, and are detailed in the body of this chapter.

Because the field imposed on a metal ion by three monodentate uninegative ligands is relatively weak, most three-coordinate complexes are high spin. Complexes including high-spin d^5 $Fe(N[SiMe_3]_2)_3$ therefore possess electrons in the $d_{x^2-y^2}$ and d_{xy} ($4e'$) levels, which are strongly M—N σ antibonding in character. Consequently, relatively high lability is anticipated for such species. Extremely interesting and unusual are the pyramidal low-spin d^6 complexes $M(Mes)_3$ (see below; M = Rh, Ir; Mes = 2,4,6-$C_6H_2Me_3$) for which Figs. 1 and 2 do not apply. In addition to the three metal–aryl σ bonds, these complexes possess three intramolecular M · · · HC agostic interactions that stabilize a pseudooctrahedral structure (see details below). Landis et al. (23) give an elegant explanation for the pyramidal geometry of $Rh(Mes)_3$ and related compounds.

IV. GROUP 4 (IVB) COMPLEXES

A. $Ti[CH(SiMe_3)_2]_3$

1. *Synthesis and Characterization*

This intriguing homoleptic titanium(III) alkyl (24, 25) was mentioned briefly in reports also describing related vanadium(III) and chromium(III) homoleptic alkyls. The procedure given involves addition of $Li[CH(SiMe_3)_2]$ (3 equiv) to cold ethereal $TiCl_3(NMe_3)_2$ (3.26 mmol) (Scheme 1). A blue-black mixture formed, from which a blue-green oil was obtained by extraction with hexane subsequent to removal of volatile matter. The $Ti[CH(SiMe_3)_2]_3$ complex was subsequently obtained as extremely sensitive blue-green crystals from cold hexane in 6% yield. Due to the extreme sensitivity of the complex a sample for elemental analysis was not obtained; however, hydrolysis of the compound in an NMR tube gave the expected quantity of $CH_2(SiMe_3)_2$ as measured by integration against an internal standard. A solution of $Ti[CH(SiMe_3)_2]_3$ in toluene (20°C) gave a signal (g = 1.968) exhibiting hyperfine coupling (quartet) attributed to the three equivalent α protons.

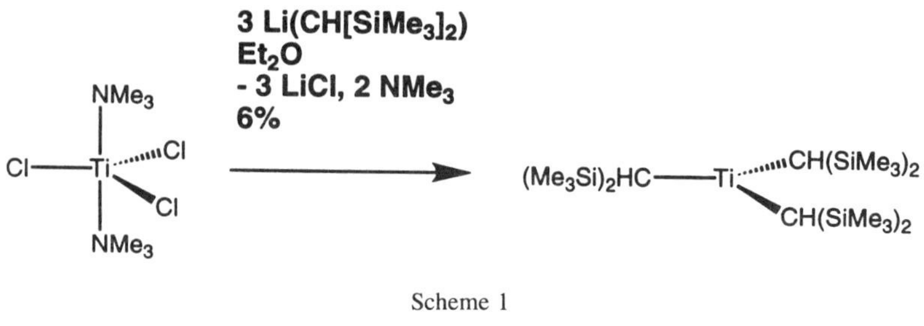

$$TiCl_3(NMe_3)_2 \xrightarrow[\text{6\%}]{\text{3 Li(CH[SiMe}_3]_2\text{), Et}_2\text{O, - 3 LiCl, 2 NMe}_3} Ti[CH(SiMe_3)_2]_3$$

Scheme 1

$$TiCl_3 \xrightarrow[-\,3\,LiCl]{3\,Li(N\text{-}i\text{-}Pr_2)\ \text{ethereal solvent}} i\text{-}Pr_2\text{-}N\text{—}Ti(N\text{-}i\text{-}Pr_2)_2$$

Scheme 2

B. $Ti(N{-}i\text{-}Pr_2)_3$

1. *Synthesis and Characterization*

Tris(diisopropylamido)titanium(III) was considered in a paper devoted to the EPR spectra of three-coordinate compounds of titanium and chromium (26). With regard to its synthesis, it was remarked that the compound was prepared from titanium(III) chloride and diisopropylamine according to the method of Bradley and Thomas (27) (Scheme 2). The cited paper is concerned only with the preparation of metal(IV) tetrakis(dialkylamido) complexes, and makes no mention of $Ti(N{-}i\text{-}Pr_2)_3$. A fair assumption is that Chien and Kruse (26) prepared $Ti(N{-}i\text{-}Pr_2)_3$ by reaction of 3 equiv of $LiN{-}i\text{-}Pr_2$ with $TiCl_3$, $TiCl_3(thf)_3$, or $TiCl_3(NMe_3)_2$. However, Bradley and co-workers in 1969 (28) stated that "complete replacement of all the chloride ligands of MCl_3 by $N{-}i\text{-}Pr_2$ (from $LiN{-}i\text{-}Pr_2$) was not achieved for M = Ti or V, in contrast to M = Cr; this result may be related to the relative gain ($d^3 >> d^1$ or d^2) in crystal field stabilization energy (CFSE) in forming trigonal $M(N{-}i\text{-}Pr_2)_3$ from tetrahedral $[ClM(N{-}i\text{-}Pr_2)_2]_2$" (29). The latter dimeric species do not appear to have been characterized.

The EPR spectra of $Ti(N{-}i\text{-}Pr_2)_3$ in toluene, heptane, or THF are reported to consist of a single, symmetric signal (g = 1.987, line width 20.9 G) at temperatures between −85 and 25°C (26). The foregoing spectral parameters were insensitive to temperature in the stated range. The powder spectrum recorded at −145°C consisted of an asymmetric signal having a line width of 20 G indicating that $g(\|) \approx g(\perp)$. No hyperfine interactions with ^{47}Ti, ^{49}Ti, or ^{14}N were resolved. The EPR results indicate the ground state of $Ti(N{-}i\text{-}Pr_2)_3$ to be $^2A_1$, according to the theory of titanium(III) in a trigonal field (30). The excited ^2E state is indicated to lie about 3000 cm^{-1} higher in energy.

C. $Ti(NPh_2)_3$

1. *Synthesis and Characterization*

Very little appears to have been written about this compound. Its EPR spectrum has been studied, and was found to have properties very similar to $Ti(N{-}i\text{-}Pr_2)_3$ in that regard (see above) (26).

D. Ti[N(R)Ar]$_3$ [R = C(CD$_3$)$_2$Me, Ar = 3,5-C$_6$H$_3$Me$_2$]

1. Synthesis and Characterization

One preparative procedure for Ti[N(R)Ar]$_3$ (31–33) involves reduction of ClTi[N(R)Ar]$_3$ with 0.5% Na/Hg (Scheme 3). In such an experiment, carried out on a scale of about 1 mmol, forest green Ti[N(R)Ar]$_3$ was obtained as a crystalline solid in 73% yield. The sodium amalgam reduction protocol mirrors that used in the preparation of Ti(OSi—*t*-Bu$_3$)$_3$ (see below). An alternative procedure for preparing Ti[N(R)Ar]$_3$ calls for mixing blue TiCl$_3$(thf)$_3$ with 3 equiv of the amido transfer reagent Li[N(R)Ar](OEt$_2$) (34) in an ether/tmeda (tmeda = Me$_2$NCH$_2$CH$_2$NMe$_2$) mixture in a carefully prescribed manner. First, 2 equiv of Li[N(R)Ar](OEt$_2$) are added, the third equivalent being added when the mixture undergoes a sudden color change to red-brown. An experiment of the latter variety provided large green crystals of Ti[N(R)Ar]$_3$ in 83% yield subsequent to removal of tmeda and salt, and recrystallization from pentane at −35°C, on a scale of about 2 mmol. Inspiration for the latter synthetic protocol came from the work of Gambarotta and co-workers (35) in which "ate" complexes such as (Cy$_2$N)$_2$Ti(μ-Cl)$_2$Li(tmeda) were prepared using 2 equiv of lithium amide in conjunction with TiCl$_3$(thf)$_3$ and tmeda, under similar conditions. The Ti[N(R)Ar]$_3$ complex displays a single signal (δ 4.64, $\Delta\nu_{1/2}$ = 46 Hz) in its ^{2}H NMR spectrum as recorded in C$_6$D$_6$. Partially deuterated *tert*-butyl groups were employed in order to facilitate NMR analyses involving this paramagnetic entity (36–40). The EPR data (toluene, 107 K: g_1 = 1.995; g_2 = 1.964; g_3 = 1.949) and magnetic susceptibility data (μ_{eff} = 2.2 μ_B) are in agreement with formulation of the compound as a d^1 monomeric complex. Mass spectral and analytical data were as expected for the formula as given.

An X-ray structure determination was carried out for Ti[N(R)Ar]$_3$, revealing the trigonal planar nature of the TiN$_3$ moiety (Fig. 3). The three N—Ti—N angles are 116.8(1), 125.7(1), and 117.5(1)°, summing to 360° while revealing some molecular asymmetry. The molecule belongs to the point group C_1. A three-coordinate titanium(III) complex would be expected to exhibit a significant degree of Lewis acidity, or electrophilic character, due to the presence of two empty, low-lying orbitals rich in *d* character (see Section III). Taking the TiN$_3$ plane as the *xy* plane, the metal *d* orbitals in question are d_{z^2}, d_{xz}, and d_{yz}. A molecular orbital having a substantial percentage of one of these three would be singly occupied, while two molecular orbitals rich in the remaining two would be empty. This unfavorable situation can be alleviated through addition of a Lewis base to the metal center, or through the formation of relatively weak agostic or related interactions. In the present case, the structure of Ti[N(R)Ar]$_3$ does display interactions between two axially disposed aryl residues and the metal center [Ti—C(ipso) distances: 2.495(3), 2.537(4); Ti—C(ortho) dis-

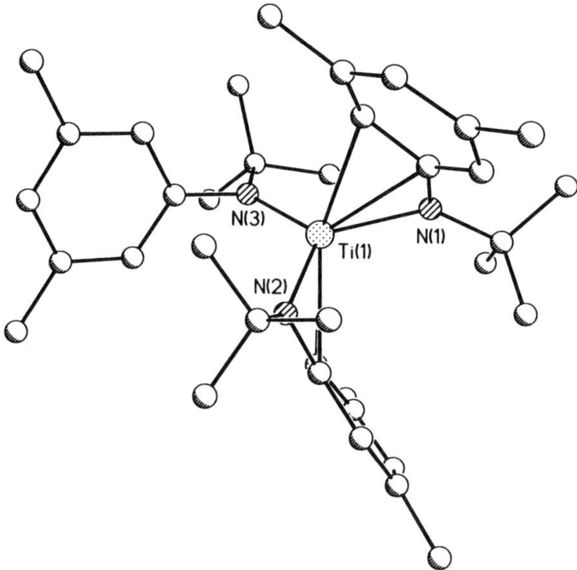

Figure 3. Structural drawing of $Ti[N(R)Ar]_3$; R = $C(CD_3)_2Me_3$, Ar = 3,5-$C_6H_3Me_2$.

tances: 2.523(4), 2.510(4) Å]. The third aryl residue is distant from the metal center (no Ti—C distances < 3 Å) and does not interact. A slight degree of pyramidalization is noted for the two nitrogens bearing the interacting aryls. The mode of arylamide–titanium interaction for the two "unusual" ligands is reminiscent of the η^3 bonding mode sometimes observed for benzyl ligands (41). Likewise, bond length alternation about the interacting aryl rings is evident. The related d^2 complex $V[N(Ad)Ar]_3$ (Ad = 1-adamantyl, see below) exhibits one interacting aryl residue, while the structures of d^3 $Mo[N(R)Ar]_3$ and $Cr[N(R)Ar]_3$ display no such interactions.

2. *Reaction with Methyl Iodide*

This facile reaction (Scheme 3) is useful as a test for the presence of $Ti[N(R)Ar]_3$ because it is complete essentially upon mixing, according to the rapid color change from green to red-orange, in solvents such as pentane even at low temperatures (ca. −78°C). The reaction gives a ^{1}H NMR spectroscopically pure 1:1 mixture of $ITi[N(R)AR]_3$ and $MeTi[N(R)Ar]_3$. Both compounds have been prepared independently and have been characterized by elemental analysis, and standard spectroscopies. The reaction of $Ti[N(R)Ar]_3$ with methyl iodide represents a standard mode of dinuclear oxidative addition (42).

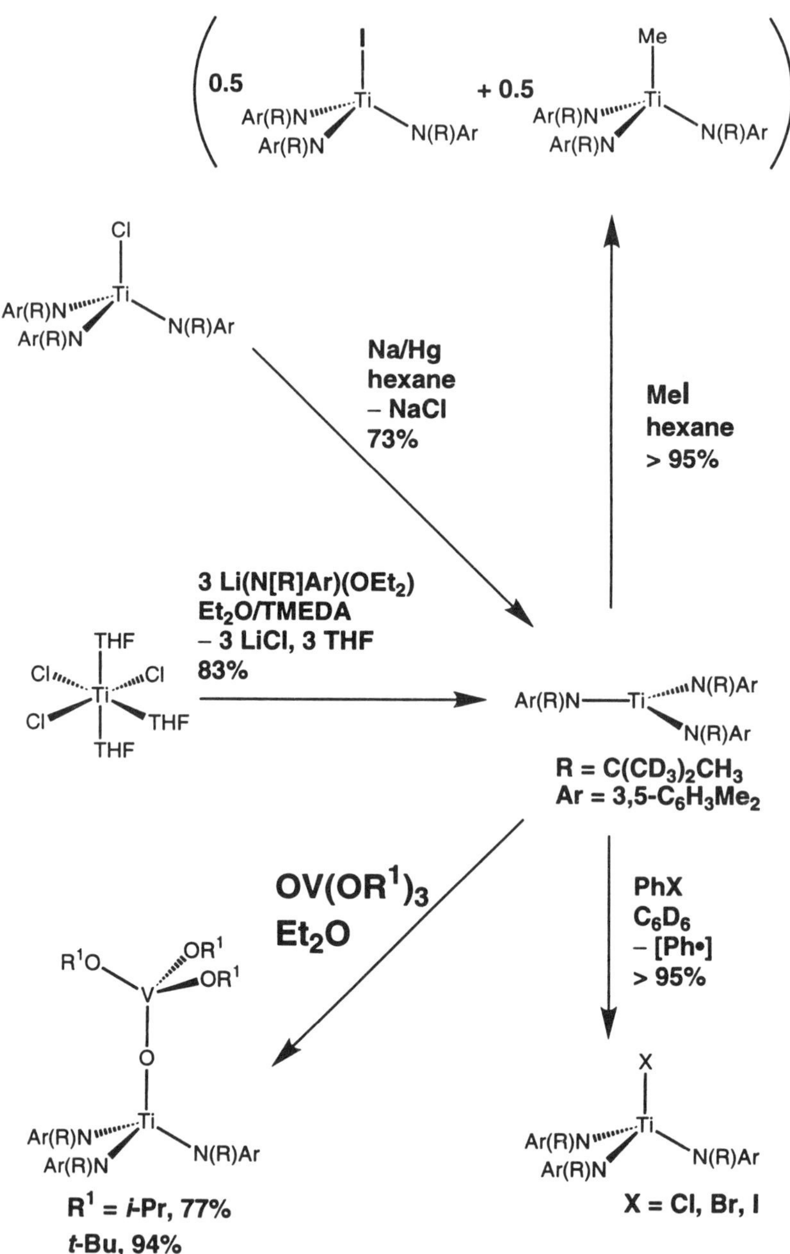

0.5
I
Ar(R)N
Ti
N(R)Ar
Ar(R)N
+ 0.5
Me
Cl
Na/Hg
hexane
– NaCl
73%
MeI
hexane
> 95%
3 Li(N[R]Ar)(OEt2)
Et2O/TMEDA
– 3 LiCl, 3 THF
83%
THF
Cl
THF
R = C(CD3)2CH3
Ar = 3,5-C6H3Me2
OV(OR1)3
Et2O
PhX
C6D6
– [Ph•]
> 95%
R1O
OR1
V
O
X
R1 = i-Pr, 77%
t-Bu, 94%
X = Cl, Br, I

Scheme 3

3. *Reaction with Aryl Halides*

The $Ti[N(R)Ar]_3$ complex reacts with iodobenzene giving $ITi[N(R)Ar]_3$ in essentially quantitative fashion (Scheme 3); no evidence was obtained for formation of $PhTi[N(R)Ar]_3$ (43). Bromobenzene and chlorobenzene react similarly, giving $BrTi[N(R)Ar]_3$ and $ClTi[N(R)Ar]_3$, respectively; qualitatively the rates for these reactions go in the order IPh > BrPh >> ClPh. No reaction was observed between FPh and $Ti[N(R)Ar]_3$ when these were mixed in a 1 : 1 ratio (benzene, 25°C, 36 h).

4. *Reaction with OV(O—i-Pr)₃*

Incomplete oxo transfer involves μ-oxo bridge formation with concomitant electron transfer across the bridge (44–46). Such a reaction has been observed on addition of $Ti[N(R)Ar]_3$ (~0.7 mmol) to $OV(O{-}i\text{-}Pr)_3$ (1 equiv) (20, 21) in ether (Scheme 3); the μ-oxo complex $(i\text{-}PrO)_3VOTi[N(R)Ar]_3$ was thereby obtained in 77% yield subsequent to multiple recrystallizations (pentane, −35°C) (31, 32). That this μ-oxo complex can be thought of as a metallo-analogue of the titanium ketyl complexes described below for $Ti(OSi{-}t\text{-}Bu)_3$, is indicated by EPR and NMR studies consistent with the oxidation state assignment titanium(IV)/vanadium(IV). The EPR spectrum of $(i\text{-}PrO)_3$-$VOTi[N(R)Ar]_3$ (240 K, toluene, $g_{iso} = 1.957$, $a_{iso}(^{51}V) = 63.9 \times 10^{-4}\ cm^{-1}$) is quite similar to that reported for $V(O{-}t\text{-}Bu)_4$ (47).

5. *Reaction with OV(O—t-Bu)₃*

An adduct, $(t\text{-}BuO)_3VOTi[N(R)Ar]_3$, was obtained in 94% yield as orange-brown needles on treatment of $Ti[N(R)Ar]_3$ with 1 equiv of $OV(O{-}t\text{-}Bu)_3$ (32) in cold ether. The thermal stability of this compound stands in stark contrast to related adducts formed by addition of $Ti[N(R)Ar]_3$ to $NMo(O{-}t\text{-}Bu)_3$ (see below).

6. *Reaction with NMo(O—t-Bu)₃*

The reaction between 1 equiv of $Ti[N(R)Ar]_3$ and $NMo(O{-}t\text{-}Bu)_3$ (48), when carried out in benzene (~28°C, 8 h) on a scale of about 1.5 mmol, led to smooth formation of $(t\text{-}BuO)_2(N)MoOTi[N(R)Ar]_3$, a titanium(IV)/molybdenum(VI) compound (32) (Scheme 4). The latter diamagnetic species was isolated in 83.3% yield as an orange powder. When prepared with ^{15}N-enriched $NMo(O{-}t\text{-}Bu)_3$ (49), $(t\text{-}BuO)_2(^{15}N)MoOTi[N(R)Ar]_3$ was obtained similarly; this species exhibited a characteristic ^{15}N NMR (25°C, C_6D_6) signal at 827-ppm downfield of liquid NH_3 for its terminal nitrido (50) functionality. The

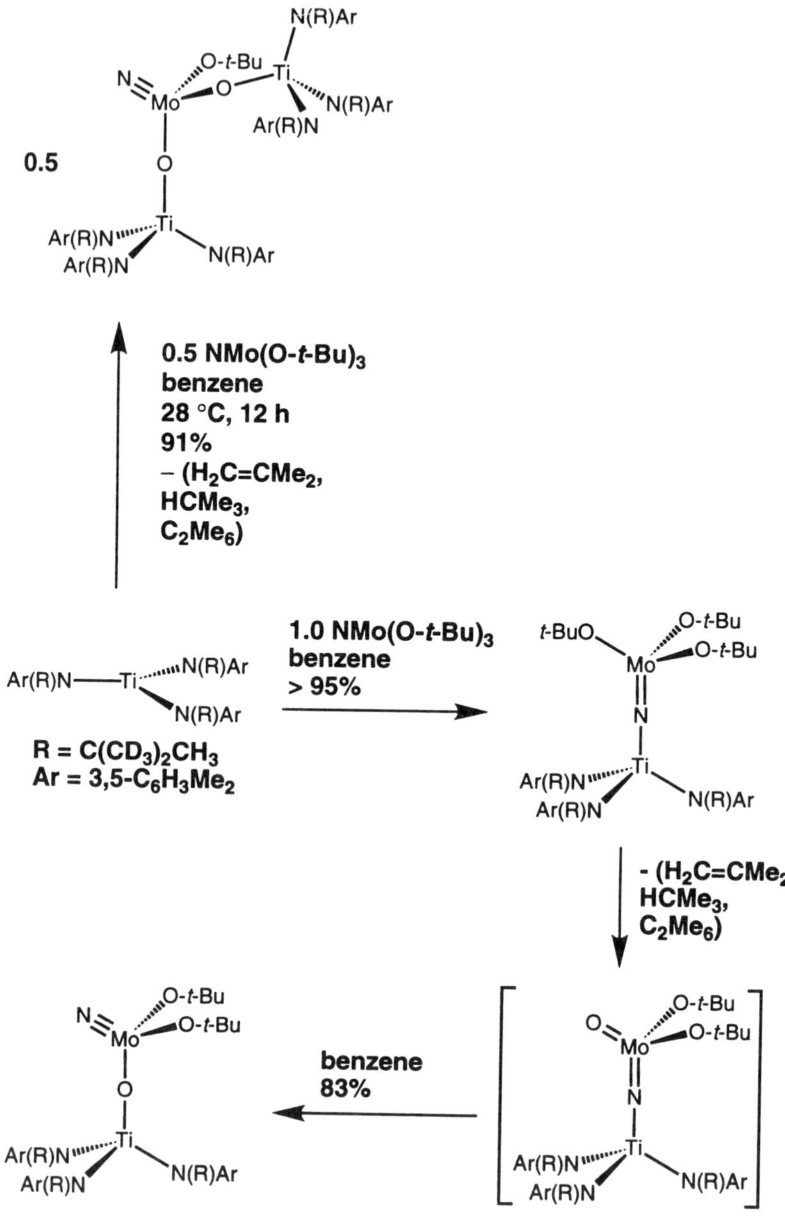
N(R)Ar
O-t-Bu
N
Mo
O
Ti
N(R)Ar
Ar(R)N
0.5
O
Ar(R)N
Ti
Ar(R)N
N(R)Ar
0.5 NMo(O-t-Bu)3
benzene
28 °C, 12 h
91%
– (H2C=CMe2,
HCMe3,
C2Me6)
1.0 NMo(O-t-Bu)3
benzene
> 95%
Ar(R)N
Ti
N(R)Ar
N(R)Ar
R = C(CD3)2CH3
Ar = 3,5-C6H3Me2
t-BuO
O-t-Bu
Mo
O-t-Bu
N
Ti
Ar(R)N
Ar(R)N
N(R)Ar
- (H2C=CMe2
HCMe3,
C2Me6)
O-t-Bu
N
Mo
O-t-Bu
O
Ti
Ar(R)N
Ar(R)N
N(R)Ar
benzene
83%
O
Mo
O-t-Bu
O-t-Bu
N
Ti
Ar(R)N
Ar(R)N
N(R)Ar

Scheme 4

structure of the bridging oxo species $(t\text{-BuO})_2(N)MoOTi[N(R)Ar]_3$ was determined by X-ray diffraction; both metals exhibit pseudotetrahedral coordination geometries and the oxo bridge is linear.

Formation of $(t\text{-BuO})_2(N)MoOTi[N(R)Ar]_3$ occurs via an observable intermediate. All characterization data available for this intermediate (EPR, NMR, and chemical reactivity) are consistent with its formulation as $(t\text{-BuO})_3$-$MoNTi[N(R)Ar]_3$, a nitrido-bridge adduct reminiscent of oxo-bridged adducts $(t\text{-BuO})_3VOTi[N(R)Ar]_3$ and $(i\text{-PrO})_3VOTi[N(R)Ar]_3$ (see above). The EPR data obtained for $(t\text{-BuO})_3MoNTi[N(R)Ar]_3$ are consistent with the formulation molybdenum(V)/titanium(IV). This formulation makes sense in view of the fact that this thermally unstable species fragments in a first-order fashion, giving rise to $(t\text{-BuO})_2(N)MoOTi[N(R)Ar]_3$ (after migration of Ti from N to O) and products of *tert*-butyl radical disproportionation (isobutylene and isobutane) and conproportionation (hexamethylethane). The latter products were collected and identified by comparison of their ^1H NMR spectra with those for authentic samples.

Reaction of $NMo(O{-}t\text{-Bu})_3$ with 2 equiv of $Ti[N(R)Ar]_3$ gave rise to formal displacement of two *tert*-butyl radicals. In such a reaction carried out on an approximate 0.7-mmol scale (benzene, ~28°C, 12 h) the product, orange diamagnetic $(t\text{-BuO})(N)Mo\{OTi[N(R)Ar]_3\}_2$, was isolated in 91% yield. Formation of this molybdenum(VI)/titanum(IV) product does occur by way of a second observable intermediate thought to be $(t\text{-BuO})_2\{[Ar(R)N]_3TiO\}$-$MoNTi[N(R)Ar]_3$ in which the *tert*-butyl radical has not yet been lost. The ^{15}N-enriched $(t\text{-BuO})(N)Mo\{OTi[N(R)Ar]_3\}_2$ exhibited a characteristic ^{15}N NMR (25°C, C_6D_6) signal at 845-ppm downfield of liquid NH_3 for its terminal nitrido (50) functionality.

E. Ti[N(*t*-Bu)Ph]$_3$

1. *Synthesis and Characterization*

The compound $Ti[N(t\text{-Bu})Ph]_3$ was obtained in 75% yield on a scale of about 9 mmol upon treatment of $TiCl_3(thf)_3$ with 3 equiv of $Li[N(t\text{-Bu})Ph](Et_2O)$ in a carefully prescribed manner (Scheme 5). The $Ti[N(t\text{-Bu})Ph]_3$ complex is recovered as green crystals. Analytical data support the proposed formulation. The ^1H NMR data pertaining to $Ti[N(t\text{-Bu})Ph]_3$ reveal two broad resonances, located at δ 4.8 ($\Delta\nu_{1/2}$ = 540 Hz) and δ 9.15 ($\Delta\nu_{1/2}$ = 210 Hz).

2. *Reaction with* $O_2Mo(O{-}t\text{-Bu})_2$

The dioxomolybdenum(VI) butoxide $O_2Mo(O{-}t\text{-Bu})_2$ of Chisholm et al. (51, 52), obtained from reaction of $Mo_2(O{-}t\text{-Bu})_6$ with O_2, was treated with

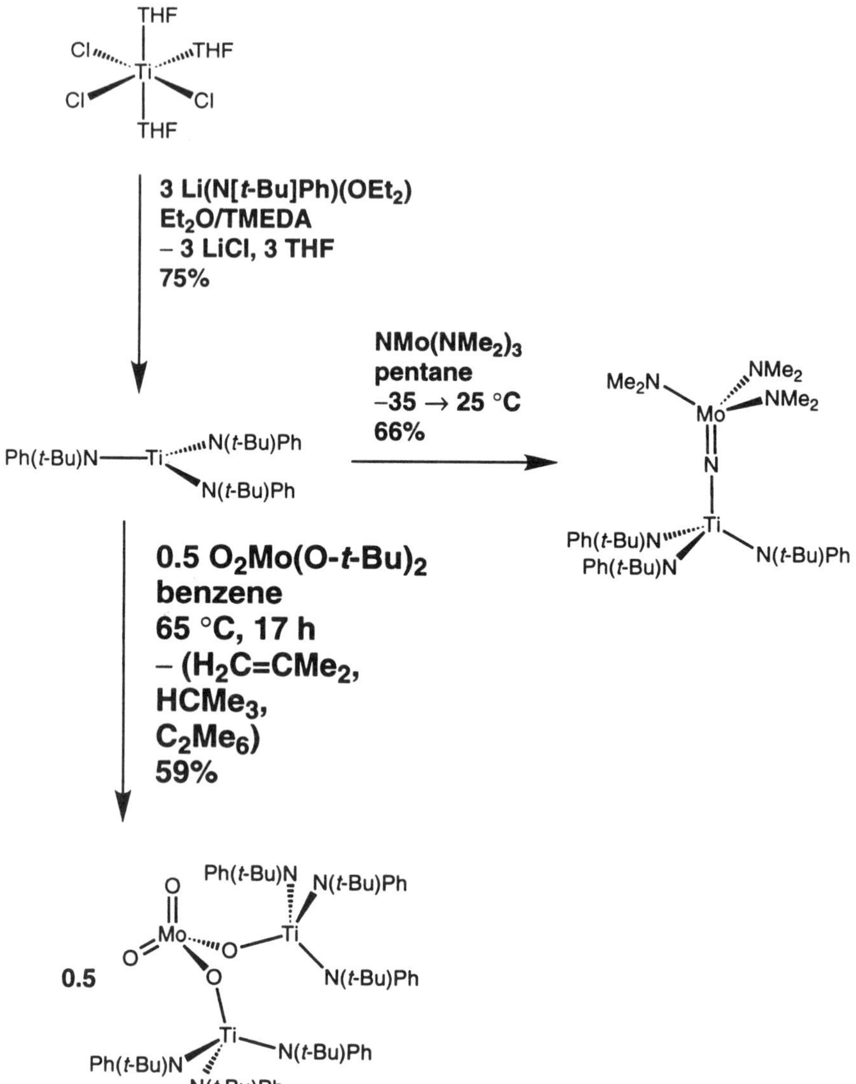

THF
Cl, THF, Ti, Cl, Cl, THF

3 Li(N[*t*-Bu]Ph)(OEt_2)
Et_2O/TMEDA
– 3 LiCl, 3 THF
75%

Ph(*t*-Bu)N—Ti N(*t*-Bu)Ph N(*t*-Bu)Ph

$NMo(NMe_2)_3$
pentane
$-35 \rightarrow 25$ °C
66%

Me_2N, NMe_2, NMe_2, Mo, N, Ti, Ph(*t*-Bu)N, Ph(*t*-Bu)N, N(*t*-Bu)Ph

0.5 $O_2Mo(O\text{-}t\text{-}Bu)_2$
benzene
65 °C, 17 h
– ($H_2C{=}CMe_2$,
$HCMe_3$,
C_2Me_6)
59%

0.5 O, Mo, O, O, O, Ti, Ph(*t*-Bu)N, N(*t*-Bu)Ph, N(*t*-Bu)Ph, Ti, Ph(*t*-Bu)N, N(*t*-Bu)Ph, N(*t*-Bu)Ph

Scheme 5

Ti[N(*t*-Bu)Ph]$_3$ in a reaction leading to net displacement of two *tert*-butyl radicals by incoming titanium radicals (Scheme 5). Accordingly, in a procedure carried out on a scale of about 0.7 mmol, a solution of Ti[N(*t*-Bu)Ph]$_3$ (2 equiv) in benzene was added to a stirring yellow solution containing $O_2Mo(O{-}t\text{-}Bu)_2$, freshly prepared *in situ*. The mixture was stirred subsequently at 65°C for 17 h in a closed vessel. The bright orange powder that was subsequently obtained in 59% isolated yield exhibited low solubility in pentane when pure. This orange diamagnetic product exhibited ^{1}H and ^{13}C NMR data consistent with its formulation as $O_2Mo\{OTi[N(t\text{-Bu})Ph]_3\}_2$. The latter compound was subjected to characterization via a single-crystal X-ray diffraction study, confirming the proposed connectivity, and displaying the three pseudotetrahedral metal centers. Several reactions carried out to probe the chemistry of $O_2Mo\{OTi[N(t\text{-Bu})Ph]_3\}_2$ did indicate, when considered collectively, that the species is best thought of as containing a central MoO_4^{2-} ion encapsulated by two $\{Ti[N(t\text{-Bu})Ph]_3\}^+$ cations.

3. *Reaction with* $NMo(NMe_2)_3$

In an experiment carried out on a scale of about 0.5 mmol, solid $Ti(N[t\text{-Bu}]Ph)_3$ was added to a stirring suspension of $NMo(NMe_2)_3$ (53) at −35°C (Scheme 5). A homogeneous forest-green mixture was soon in evidence (32). After stirring 1 h at 25°C, concentration and cooling produced a batch of green crystals (~66% isolated yield). The thermal stability of the product, formulated as $(Me_2N)_3MoNTi[N(t\text{-Bu})Ph]_3$ was assessed by heating a benzene-d_6 a solution of the compound to 70°C for 13 h; under such conditions no appreciable decay was detected by ^{1}H NMR monitoring. The NMR and EPR data were in accord with formulation of paramagnetic $(Me_2N)_3MoNTi[N(t\text{-Bu})Ph]_3$ as a threefold symmetric nitrido-bridged species, reminiscent of the intermediate complexes produced initially upon attack on $(t\text{-BuO})_3MoN$ by $Ti[N(R)Ar_3]$ (see above).

F. $Ti[N(R)Ar]_2[CH(SiMe_3)_2]$ [$R = C(CD_3)_2Me$, $Ar = 3,5\text{-}C_6H_3Me_2$]

1. *Synthesis and Characterization*

This heteroleptic bis(amido) alkyl titanium(III) species was obtained recently (54) in two steps from $TiCl_3(thf)_3$ (Scheme 6) (55,56). First, a titanium/lithium species proposed to be $[Ar(R)N]_2Ti(\mu\text{-Cl})_2Li(tmeda)$ by analogy with known, structurally characterized $(Cy_2N)_2Ti(\mu\text{-Cl})_2Li(tmeda)$ (57), was prepared by treatment of $TiCl_3(thf)_3$ (2.7 mmol) with 2 equiv of $Li[N(R)Ar](OEt_2)$ (34) in a tmeda–thf mixture. The $[Ar(R)N]_2Ti(\mu\text{-Cl})_2Li(tmeda)$ complex was isolated in 57% yield as green needles; characterization data including ^{2}H NMR, elemental analysis, UV–vis, and EPR data are consistent with the given formulation. Treatment of $[Ar(R)N]_2Ti(\mu\text{-Cl})_2Li(tmeda)$ (~2 mmol) with $LiCH(SiMe_3)$ (58) (1 equiv) in pentane led to a dark green mixture. Subsequent to salt removal and crystallization (pentane, −35°C), $Ti[N(R)Ar]_2[CH(SiMe_3)_2]$ was obtained in 72% yield as dark green blocks.

In benzene the heteroleptic three-coordinate complex exhibits a single ^{2}H

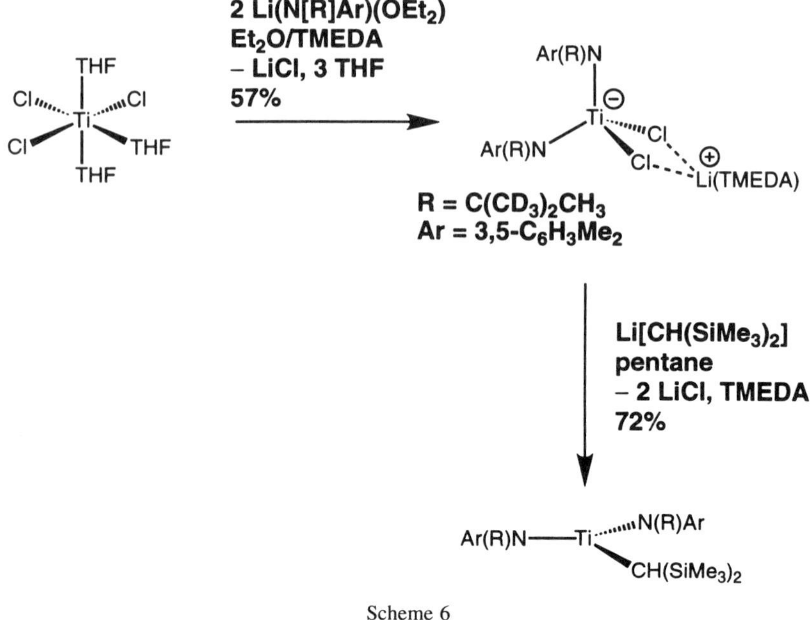

Scheme 6

NMR signal (δ 3.97 ppm, $\Delta\nu_{1/2}$ 17.9 Hz), for the deuterated amino *tert*-butyl residues. An EPR spectrum obtained in frozen toluene revealed g values consistent with unsymmetrical titanium(III) ($g_1 = 1.997$; $g_2 = 1.962$, $g_3 = 1.918$); the isotropic signal observed in toluene at 298 K appeared at $g = 1.95$. Magnetic susceptibility data obtained by SQUID magnetometry (59) revealed $\mu = 1.66\ \mu_B$; the complex exhibited clean Curie–Weiss behavior over the temperature range 5–300 K.

The structure of $Ti[N(R)Ar]_2[CH(SiMe_3)_2]$ was determined by X-ray diffraction. In this manner it was observed that one of the N(R)Ar ligands bonds in the *N*-only fashion, while the other exhibits the η^3-*N*,*C*(ipso),*C*(ortho) bonding mode as has been seen for related unsaturated complexes including $Ti[N(R)Ar]_3$ and $V(N[Ad]Ar)_3$ (see above and below, respectively). The Ti—N bond distances are 1.917(5) Å (to the *N*-only amide) and 1.962(5) Å (to the η^3 amide), while the Ti—C distance to the $CH(SiMe_3)_2$ ligand is 2.137(7) Å. Substantially longer distances to the ipso and ortho carbons of the η^3 amide are seen, 2.407(7) and 2.440(8) Å. That the structure is based on a distorted trigonal titanium(III) center is evidenced by the sum of the three angles about titanium, 355°. The unsaturated complex acquires additional stabilization, presumably, by the weak interactions with the ipso and ortho carbons of the η^3

amide. It is of interest that the η^3-*N*,*C*(ipso),*C*(ortho) bonding mode is not observed for high-spin d^3 complexes such as $M[N(R)Ar]_3$ (M = Cr or Mo) in which there are, evidently, no low-lying empty orbitals.

Note that the synthetic strategy described here, while successful in the special case of the hindered alkyl $CH(SiMe_3)_2$, failed for other alkyls (e.g., R′ = neopentyl or CH_2SiMe_3), disproportionation to $R'_2Ti[N(R)Ar]_2$ species being evident under such circumstances (57). In order to obtain discrete, robust Ti(alkyl)(amido)$_2$ entities, improved amido ligands clearly are required.

2. *Reaction with Benzonitrile*

Ethereal $Ti[N(R)Ar]_2[CH(SiMe_3)_2]$, upon treatment at −35°C with benzonitrile (1 equiv), gave a rapid color change to dark blue, presumably indicative of the adduct $(PhCN)Ti[N(R)Ar]_2[CH(SiMe_3)_2]$ (Scheme 7). On warming to about 25°C the mixture attained an orange color. The orange diamagnetic product (82% isolated yield) exhibited low solubility in pentane and ether, but could be recrystallized from hot THF. The orange product was formulated as the benzonitrile-coupled species (μ-NCPhCPhN) $\{Ti[N(R)Ar]_2[CH(SiMe_3)_2]\}_2$, a formulation consistent with ^1H and ^{13}C NMR data and analytical data obtained for the complex. Similar nitrile coupling reactions were noted previously for titanium(III) (60).

3. *Reaction with Trimethylsilylazide/Pivalonitrile*

It was noted that $Ti[N(R)Ar]_2[CH(SiMe_3)_2]$ reacts with pivalonitrile, giving a light green adduct that does not appear to couple readily, in contrast to the benzonitrile reaction noted above (Scheme 7). Accordingly, treatment of the heteroleptic three-coordinate complex with 4 equiv of pivalonitrile followed by 1.4 equiv of Me_3SiN_3 led over 2 h to a yellow-orange color. A waxy residue, obtained on removal of volatile matter, was recrystallized (pentane, −35°C) to provide diamagnetic $(N_3)Ti[N(R)Ar]_2[CH(SiMe_3)_2]$ in 37% yield as a light orange powder. The low yield was attributed to inefficient isolation, as NMR examination of the crude reaction mixture indicated clean conversion. The IR data confirmed the presence of the azido functionality [$\nu(N_3)$, 2107 cm^{-1}], while ^1H and ^{13}C NMR spectra, and microanalytical data, were consistent with the formulation as given. The use of pivalonitrile served to temper the reactivity of $Ti[N(R)Ar]_2[CH(SiMe_3)_2]$; a similar experiment carried out in the absence of this additive led to an intractable product mixture. The reaction represents formal abstraction of the neutral N_3 radical by the titanium(III) center. Related reactions of organoazides have been observed for the vandium(II)/(III) couple (61).

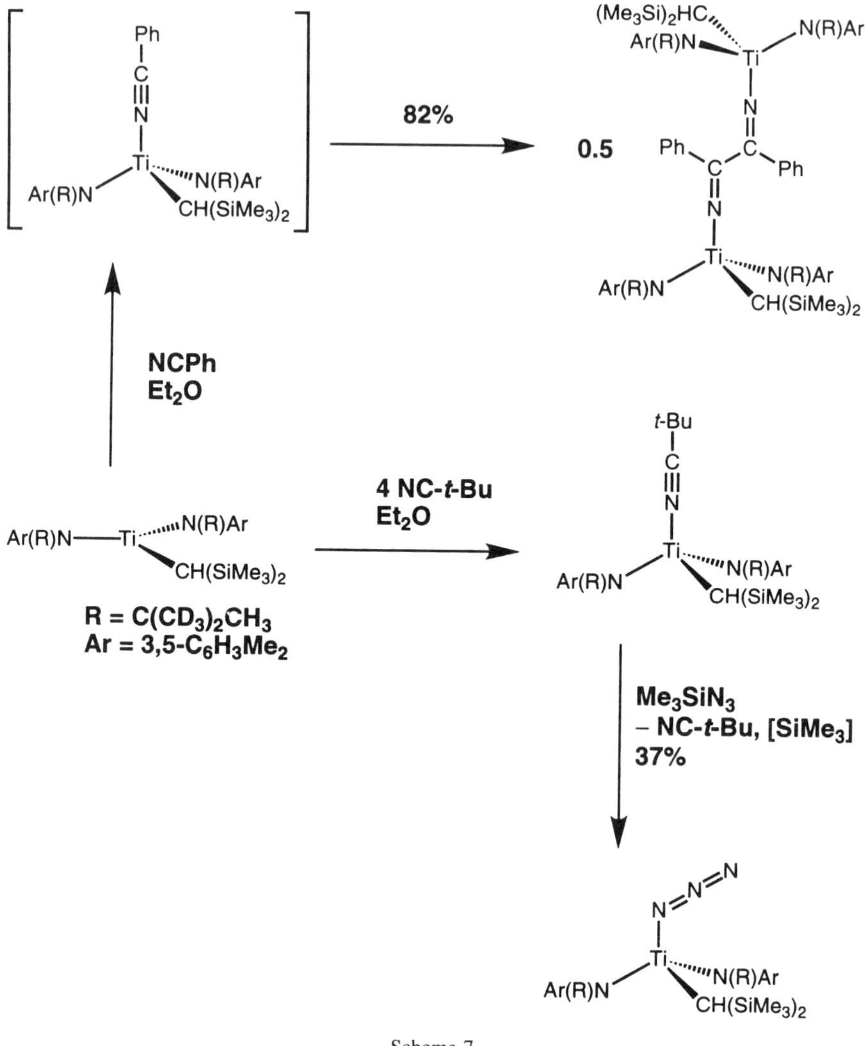

Scheme 7

G. $Ti(OSi{-}t\text{-}Bu_3)_3$

1. *Synthesis and Characterization*

Three-coordinate $Ti(OSi{-}t\text{-}Bu_3)_3$ was obtained via sodium amalgam reduction of the chloro precursor $ClTi(OSi{-}t\text{-}Bu_3)_3$, itself prepared in 85% yield on an approximate 11 mmol scale from $Na(OSi{-}t\text{-}Bu_3)$ (3 equiv) and $TiCl_4(thf)_2$

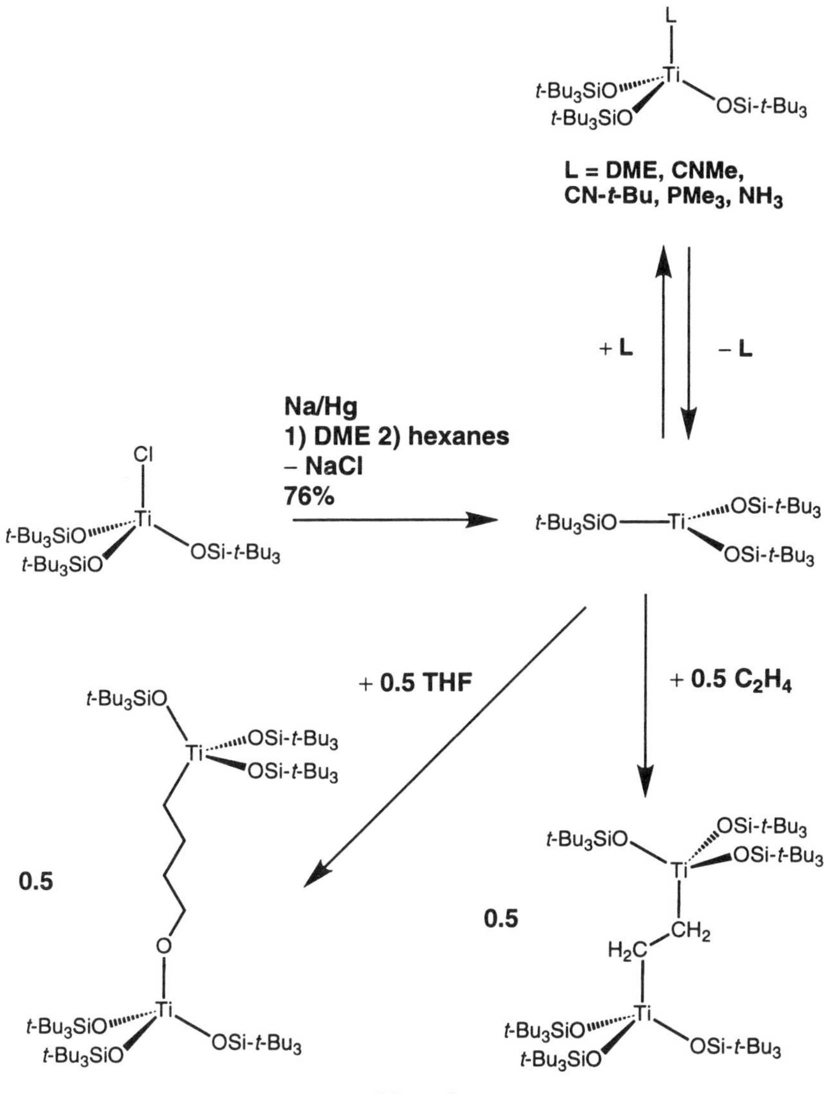

Scheme 8

(Scheme 8). Accordingly, $ClTi(OSi{-}t\text{-}Bu_3)_3$ (~4 mmol) was treated with sodium amalgam in 1,2-dimethoxyethane (dme). Stirring was effected with a glass-encased stirbar. The mixture adopted a green color, attributed to the formation of $(dme)Ti(OSi{-}t\text{-}Bu_3)_3$. The crude material obtained on removal of volatile matter was subjected to repeated trituration with toluene. Subsequent to salt removal and crystallization from hexanes, bright orange $Ti(OSi{-}t\text{-}Bu_3)_3$ was obtained in 76% yield (19, 62).

The $Ti(OSi{-}t\text{-}Bu_3)_3$ complex exhibits a broad singlet in its ^1H NMR spectrum (δ 1.5, $\Delta\nu_{1/2}$ = 110 Hz) for its 81 equivalent hydrogens. Analytical data were consistent with the formula given. A detailed description of the EPR and UV–vis properties of $Ti(OSi{-}t\text{-}Bu_3)_3$ was given, and a comparison was made to two other TiX_3 complexes. In brief, the UV–vis spectrum of $Ti(OSi{-}t\text{-}Bu_3)_3$ exhibited a band at 17,400 cm^{-1} attributed to a ($^2A_1' \rightarrow {}^2E'$) transition. The anticipated lower energy ($^2A_1' \rightarrow {}^2E''$) transition was not observed despite exhaustive attempts to locate it. The EPR spectra of $Ti(OSi{-}t\text{-}Bu_3)_3$ are characteristic of an axially symmetric d^1 species, with g_{iso} = 1.9554, a_{iso} = 155 MHz (~56.7 G), $g_\perp$ = 1.9323, $g_\parallel$ = 1.9997.

2. *Adducts of* $Ti(OSi{-}t\text{-}Bu_3)_3$

The 1 : 1 adducts were observed between $Ti(OSi{-}t\text{-}Bu_3)_3$ and the Lewis bases DME, CNMe, CN—*t*-Bu, NC—*t*-Bu, PMe_3, and NH_3 (Scheme 8). The adducts were pale green, and in most cases (e.g., DME, PMe_3, and NH_3) the Lewis base appeared weakly bound and was readily lost in vacuo. In addition, the ^1H NMR spectra recorded in C_6D_6 indicated a significant degree of dissociation. Presumed adducts with THF, ether, and water were also light green but eluded characterization due to their instability. In the case of THF, ring-opening occurred giving a diamagnetic species formulated as $(t\text{-}Bu_3SiO)TiO(CH_2)_4Ti(OSi{-}t\text{-}Bu_3)_3$. Such reactivity is in contrast to that observed for amido complexes such as $Ti(N[R]Ar)_3$, which are soluble and relatively robust in neat THF at about 25°C (see above).

3. *Ketyl Complexes of* $Ti(OSi{-}t\text{-}Bu_3)_3$

Carbonyl compounds are readily scavenged by $Ti(OSi{-}t\text{-}Bu_3)_3$, in some cases to give observable ketyl complexes (Scheme 9). The latter are formulated with a carbon-centered radical and titanium(IV), whereas an alternative resonance structure would place the radical on Ti for titanium(III) while maintaining an intact C—O π-bond.

In the case of acetone, the products observed were those expected for disproportionation of the intermediate ketyl radical complex, namely the enolate complex $[Me(H_2C)CO]Ti(OSi{-}t\text{-}Bu_3)_3$ and the isopropoxide species $(Me_2HCO)Ti(OSi{-}t\text{-}Bu_3)_3$, which were formed in a 1 : 1 ratio. Characteristic ^1H NMR singlets at 4.16 and 4.62 ppm were observed for the vinylic hydrogens of the O-bound enolate. It is noteworthy that for the substrates acetone and acetaldehyde, no evidence for radical combination was obtained; only disproportionation products were observed.

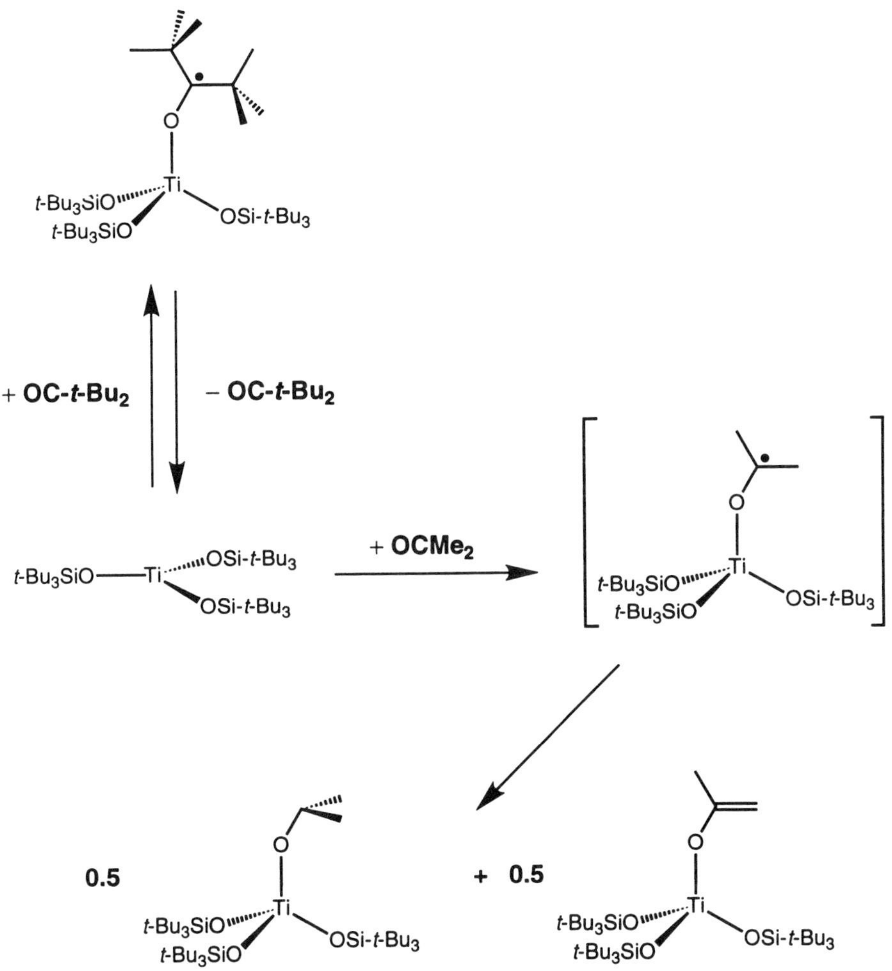

$+ OC\text{-}t\text{-}Bu_2$ $- OC\text{-}t\text{-}Bu_2$

$t\text{-}Bu_3SiO{-}Ti(OSi\text{-}t\text{-}Bu_3)_2$ $\xrightarrow{+ OCMe_2}$ $[(Me_2\dot{C}O)Ti(OSi\text{-}t\text{-}Bu_3)_3]$ → 0.5 $(Me_2HCO)Ti(OSi\text{-}t\text{-}Bu_3)_3$ + 0.5 $(CH_2{=}C(Me)O)Ti(OSi\text{-}t\text{-}Bu_3)_3$

Scheme 9

Di-*tert*-butyl ketone was selected as a substrate because it was anticipated that the ketyl complex derived from its addition to $Ti(OSi{-}t\text{-}Bu_3)_3$ would not be subject to disproportionation or combination. Accordingly, reversible binding of di-*tert*-butyl ketone by $Ti(OSi{-}t\text{-}Bu_3)_3$ gave ink-blue solutions containing $(t\text{-}Bu_2CO)Ti(OSi{-}t\text{-}Bu_3)_3$. A new EPR signal attributed to the ketyl complex was observed at $g = 1.9985$. Titanium hyperfine coupling ($a_{iso} = 4$ G) was observed, as was a coupling to ^{13}C ($a_c \sim 26$ G, natural abundance).

The pivaldehyde-derived ketyl complex $[t\text{-}Bu(H)CO]Ti(OSi{-}t\text{-}Bu_3)_3$ was studied at 203 K; this species degraded to a mixture of unidentified products on warming to about 25°C.

3,3,5,5-Tetramethylcyclohexanone binds at 218 K to $Ti(OSi{-}t\text{-}Bu_3)_3$, giving a ketyl complex that was studied by EPR spectroscopy at that temperature. This ketone appeared to dissociate reversibly at temperatures above 218 K.

Acetophenone reacted with $Ti(OSi{-}t\text{-}Bu_3)_3$ giving products indicative of disproportionation of the corresponding ketyl radical, although these complexes, the acetophenone enolate and the α-phenylethoxide complex, were not extensively characterized.

The benzophenone adduct obtained was reminiscent of Gomberg's dimer in being a head-to-tail radical coupling product (Scheme 10); it was procured as a

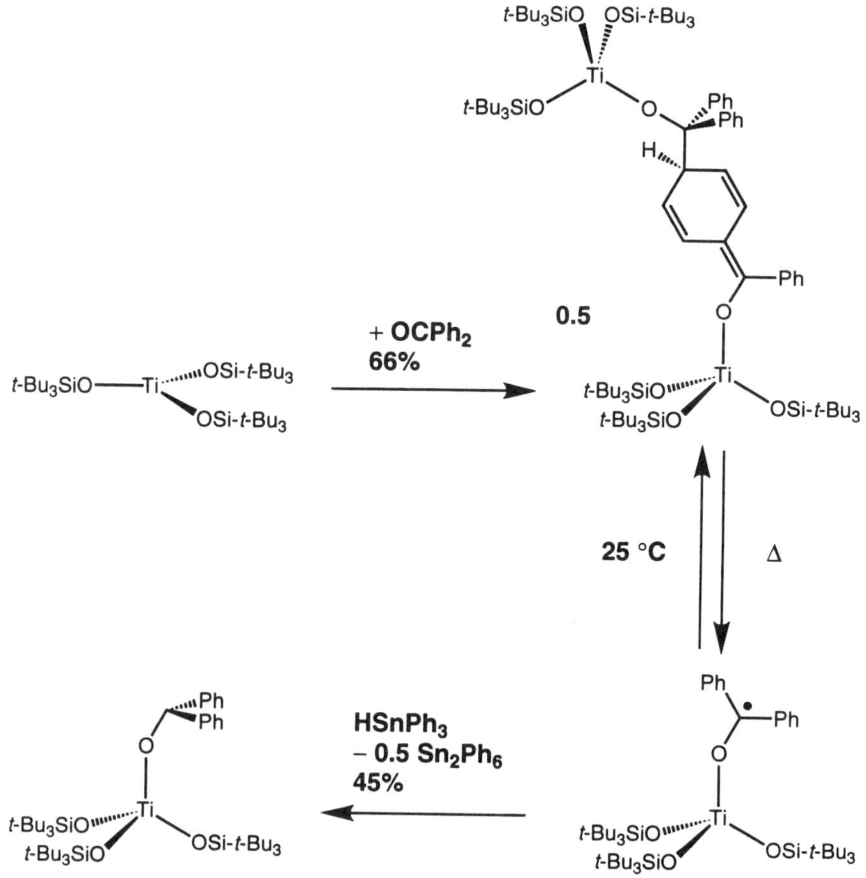

Scheme 10

yellow powder on a scale of about 1 mmol, in 66% isolated yield. At elevated temperatures in toluene, C—C fragmentation to give significant concentrations of the benzophenone ketyl complex $(Ph_2CO)Ti(OSi{-}t\text{-}Bu_3)_3$ took place. The latter could be trapped efficiently by hydrogen-atom donors such as $HSnPh_3$, giving diphenylmethoxide, $(Ph_2HCO)Ti(OSi{-}t\text{-}Bu_3)_3$. The latter titanium(IV) species was obtained in 45% yield as yellow microcrystals from such an experiment carried out on a scale of about 0.2 mmol; characterization data were fully consistent with its formulation. Equilibrium studies of the reversible dimerization of $(Ph_2CO)Ti(OSi{-}t\text{-}Bu_3)_3$ were conducted by UV–vis spectroscopy (λ_{max} = ~646 nm) as a function of temperature, giving a dissociation enthalpy for the C—C bond in question of 18 kcal mol^{-1}. This value is to be compared with that for trityl radical of 12 kcal mol^{-1}. The equilibrium constant for dissociation of $[(Ph_2CO)Ti(OSi{-}t\text{-}Bu_3)_3]_2$ ($K_{eq} = 7.5 \times 10^{-7}$) is such at 22°C that about 1.3% dissociation obtains under these conditions.

Di-*p*-tolyl ketone gave a blue ketyl complex, $[(p\text{-}MeC_6H_4)_2CO]Ti(OSi{-}t\text{-}Bu_3)_3$, which was studied at 218 K by EPR spectroscopy. The complex decomposed to unidentified products upon warming.

4. *Heterocycle binding by Ti(OSi—t-Bu₃)₃*

Reactions of pyridine, 2-picoline, 4-picoline, and 2,6-lutidine with $Ti(OSi{-}t\text{-}Bu_3)_3$ were investigated (Scheme 11). It was shown that pyridine binds in a straightforward η^1 fashion, although its IR spectrum differs from the related adducts $(\eta^1\text{-}py(M(OSi{-}t\text{-}Bu_3)_3$ (M = Sc, V) in which, it was asserted, the metals behave as simple Lewis acids. The titanium(III) center, on the other hand, being a relatively strong $1e^-$ reductant, gives rise to partial reduction of

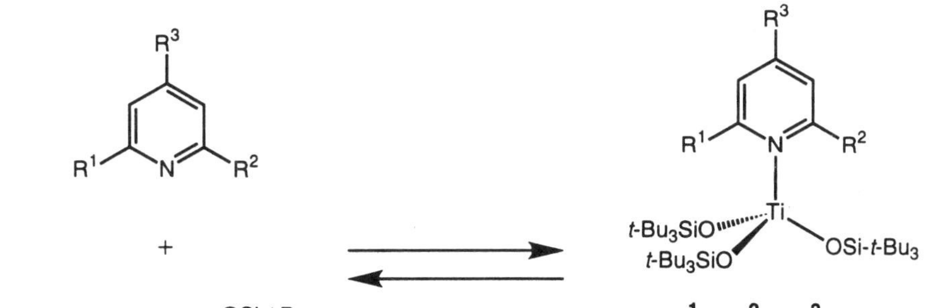

1: R¹ = R² = R³ = H
2: R¹ = Me, R² = R³ = H
3: R³ = Me, R¹ = R² = H
4: R¹ = R² = Me, R³ = H

Scheme 11

the bound pyridine in the adduct (η^1-py)Ti(OSi$-$$t$-$Bu_3$)$_3$. The equilibrium constant for 2-picoline binding to Ti(OSi$-$$t$-$Bu_3$)$_3$ was estimated to be about 18 M^{-1} at about 20°C. 2,6-Lutidine exhibited no affinity for Ti(OSi$-$$t$-$Bu_3$)$_3$, in contrast to the chemistry observed between Ta(OSi$-$$t$-$Bu_3$)$_3$ and the same substrate.

5. *Reaction With Ethylene*

Addition of 0.5 equiv of ethylene to Ti(OSi$-$$t$-$Bu_3$)$_3$ ($\sim$0.5 mmol), as a solution in hexanes, led to production of a yellow precipitate over a 5–10 min period. A material formulated as (μ-C_2H_4) [Ti(OSi$-$$t$-$Bu_3$)$_3$]$_2$ was subsequently isolated in 82% yield (Scheme 8). The compound exhibited very low solubility in hexanes, benzene, toluene, ether, and THF. Analytical data were consistent with the compound's proposed formulation.

H. Ti(O$-$2,6-$C_6H_3$$-$$t$-$Bu_2$)$_3$

1. *Synthesis and Characterization*

Subsequent to unsuccessful attempts to prepare this compound from $TiCl_3$(thf)$_3$ and Li(O$-$2,6-$C_6H_3$$-$$t$-$Bu_2$), it was discovered that blue Ti(O$-$2,6-$C_6H_3$$-$$t$-$Bu_2$)$_3$ could be obtained in good yield (yield and reaction scale unspecified) upon treatment of $TiCl_3$($HNMe_2$)$_2$ with an excess of the lithium aryloxide in benzene (Scheme 12) (18). Although characterization of the presumed three-coordinate species by X-ray crystallography was not effected, the suggestion was posited that it is probably related structurally to the scandium complex Sc(O$-$2,6-t-$Bu_2$$-$4-Me-$C_6H_2$)$_3$ (14), of known structure. Rothwell and co-workers (18) present a lucid discussion of the electronic structure of Ti(O-2,6$-$$C_6H_3$$-$$t$-$Bu_2$)$_3$, drawing on related analyses authored by Bradley and co-workers (63, 64). The compound is characterized by EPR ($g_{\parallel}$ = 1.998, $g_{\perp}$ = 1.926; frozen toluene) and UV–vis (λ_{max} = 582 nm) parameters consistent with its formulation as a monomeric, three-coordinate species. That a low-energy absorbtion band in the electronic spectrum of Ti(O$-$2,6-$C_6H_3$$-$$t$-$Bu_2$)$_3$ could not be observed prevented the assignment of a spin–orbit coupling constant for the complex; with reference to the work of Bradley and co-workers (63, 64) this difficulty was rationalized on the basis of the presumed relative π-donating ability of the ancillary ligands: N($SiMe_3$)$_2$ >> O-2,6$-$$C_6H_3$$-$ t-Bu_2.

2. *Reaction with CX_4 (X = Cl, Br)*

The ability of Ti(O-2,6-$C_6H_3$$-$$t$-$Bu_2$)$_3$ to serve as a halogen-atom acceptor is manifested in its reactions with CCl_4 and CBr_4. Condensation of about 4

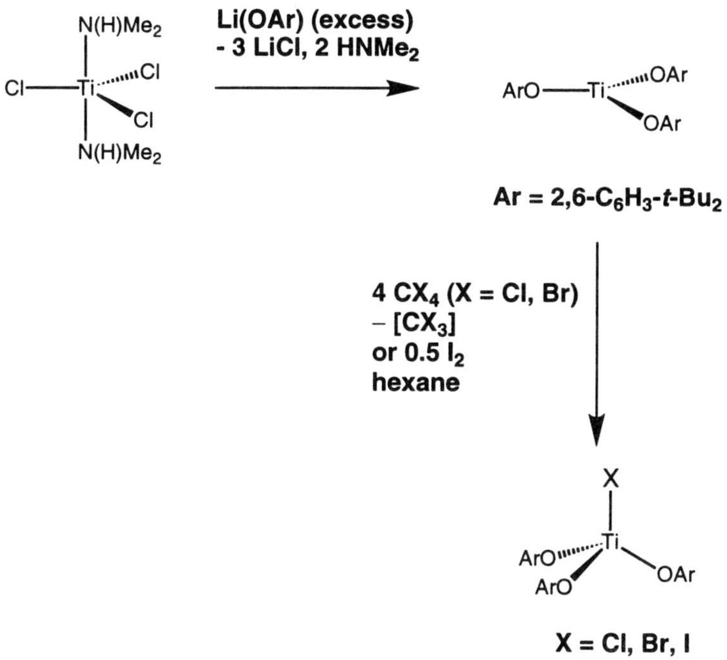

Scheme 12

equiv of either tetrahalide from a calibrated gas manifold into a hexane solution of Ti(O$-$2,6-$C_6H_3$$-$$t$-$Bu_2$)$_3$ led to reddish-orange solutions over a period of a few hours. Crystal of XTi(O$-$2,6-$C_6H_3$$-$$t$-$Bu_2$)$_3$ (X = Br, Cl) were subsequently obtained by cooling to -15°C. The diamagnetic XTi(O$-$2,6-$C_6H_3$$-$$t$-$Bu_2$)$_3$ compounds were characterized by elemental analysis and by NMR spectroscopic measurements. The latter data were indicative of a fluxional process involving exchange of proximal and distal (with respect to the X substituent in the C_3 symmetric complexes) t-Bu groups. Coalescence temperatures of 265 and 268 K were noted for X = Cl and Br, respectively, giving activation energies of 12.3 and 12.4 kcal mol^{-1} upon consideration of the relevant chemical shifts in the low-temperature spectrum.

3. *Reaction with Iodine*

The iodo complex ITi(O$-$2,6-$C_6H_3$$-$$t$-$Bu_2$)$_3$ was obtained by addition of iodine crystals to a blue hexane solution of Ti(O$-$2,6-$C_6H_3$$-$$t$-$Bu_2$)$_3$. Crystals were obtained by cooling the reaction mixture to an unspecified temperature. An X-ray crystal structure determination carried out for ITi(O$-$2,6-$C_6H_3$$-$$t$-$Bu_2$)$_3$ highlighted the expected C_3 molecular symmetry, which was not imposed

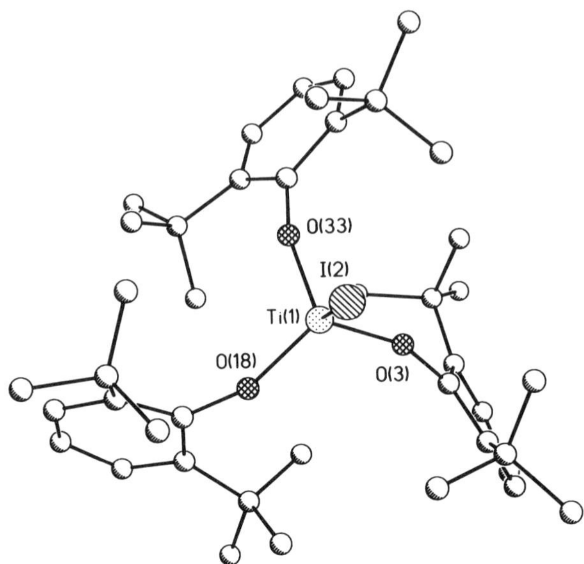

Figure 4. Structural drawing of $ITi(O-2,6-C_6H_3-t-Bu_2)_3$.

crystallographically (Fig. 4). The Ti—I bond length [d(Ti—I), 2.634(8) Å] is as expected for a four-coordinate titanium(IV) iodide. As for the related chloro and bromo species (see above), a fluxional process was found (^{1}H NMR) to exchange proximal and distal t-Bu groups. An activation energy (12.3 kcal mol^{-1}) identical to that for the chloro compound was extracted from data obtained as a function of temperature.

V. GROUP 5 (VB) COMPLEXES

A. (thf)V(Mes)$_3$

1. *Synthesis and Characterization*

The (thf)V(Mes)$_3$ complex was evidently first prepared in 1974, (65, 66) but other descriptions of its synthesis have subsequently appeared. A 1977 procedure (65) describes the reaction of $VCl_3(thf)_3$ with mesityl Grignard, prepared *in situ* on a large scale (Scheme 13). This procedure gave the desired (thf)V(Mes)$_3$ as a blue solid in 64% yield (28–32 g). A slightly modified procedure (67) gave (thf)V(Mes)$_3$ in 60% yield. The 1977 paper contains a de-

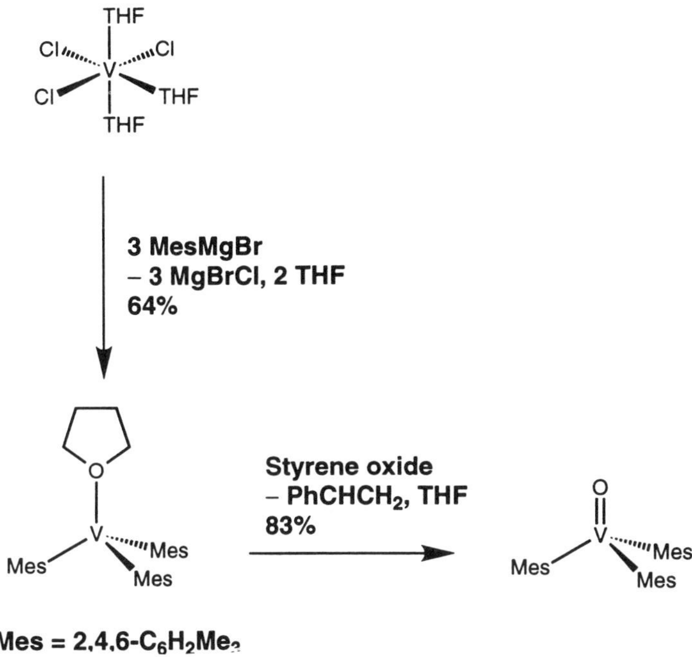

Scheme 13

scription of the electronic spectrum of the compound, and gave magnetic susceptibility data as follows: $\mu_{eff} = 2.88\ \mu_B$.

The X-ray structure of (thf)V(Mes)$_3$ has been described (Fig. 5) (67). The coordination polyhedron was described as a trigonal pyramid, a geometry that has been documented for vanadium(III) complexes of triamidoamine ligands (68). That the thf ligand is located in the apical site is suggestive of this ligand's lability, in contrast with the relatively short V—O distance of 2.067(4) Å. It was stated that "The thf is tightly bound to the metal and is not lost when (thf)V(Mes)$_3$ is recrystallized from toluene or benzene." Notwithstanding the latter comment, most or possibly all of the compound's reactions result in loss of the thf ligand. Analogy with base-free compounds such as V[N(SiMe$_3$)$_2$]$_3$ and V[CH(SiMe$_3$)$_2$]$_3$ renders dissociation of thf from (thf)V(Mes)$_3$ to give "naked" V(Mes)$_3$ plausible.

2. *Epoxide Deoxygenation*

Styrene oxide, upon treatment with (thf)V(Mes)$_3$, produced styrene and the oxovanadium(V) complex OV(Mes)$_3$ (69, 70) (Scheme 13). The latter was iso-

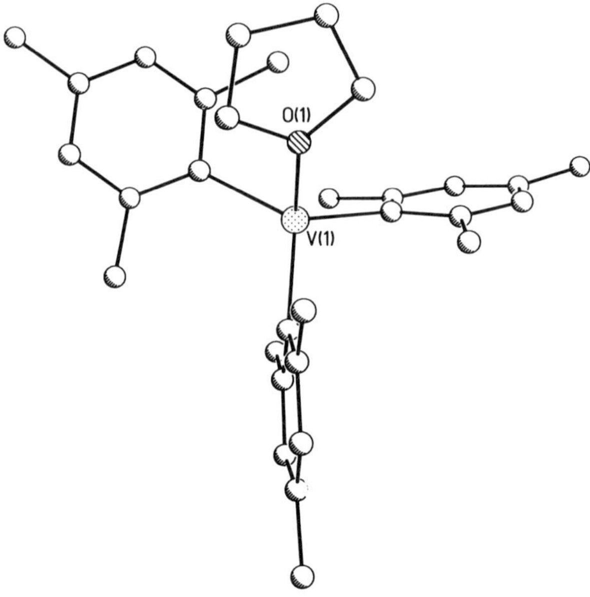

Figure 5. Structural drawing of (thf)V(Mes)$_3$.

lated in 83% yield on an approximate 5 mmol scale as an orange solid, subsequent to crystallization from hexane. X-ray crystallography was used to verify the monomeric nature of pseudotetrahedral OV(Mes)$_3$ (Fig. 6); the compound is reminiscent both structurally and from a synthetic viewpoint to OIr(Mes)$_3$ (described below).

Speculation on the mechanism of styrene oxide deoxygenation by (thf)V(Mes)$_3$ was posited by Floriani and co-workers (70). An oxametallacycle intermediate derived from formal insertion of the d^2 metal center into an epoxide O—C bond was postulated without evidence, a typical occurrence for that particular class of intermediate as evinced in a recent review article (71). Related vanadium(III)-mediated epoxide deoxygenations of *cis*- and *trans*-2-butene oxides are more consistent with a radical ring-opening process (72).

3. *Deoxygenation of Coordinated Nitric Oxide*

The reactions described here represent the confluence of three-coordinate chemistry of Groups 5 (VB) and 6 (VIB). See below for the preparation of chromium nitrosyl complexes such as (ON)Cr(N—*i*-Pr$_2$)$_3$ by direct addition of gaseous NO to three-coordinate precursors. It was discovered that nitric oxide reductive cleavage could be effected by treatment of (ON)Cr(N—*i*-Pr$_2$)$_3$ with

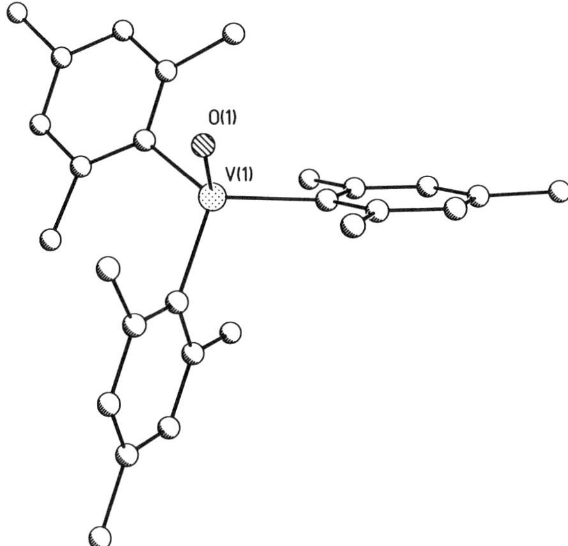

Figure 6. Structural drawing of $OV(Mes)_3$.

$(thf)V(Mes)_3$ in toluene, giving the first example of a chromium(VI) nitrido complex, $NCr(N{-}i\text{-}Pr_2)_3$, in 67% yield as a beet-colored crystalline solid (Scheme 14) (73). Two equiv of $(thf)V(Mes)_3$ were employed so that the initially formed $OV(Mes)_3$ would be trapped by the additional $(thf)V(Mes)_3$ to form the oxo-bridged dinuclear species $(\mu\text{-}O)V_2(Mes)_6$ (69), the low solubility of which facilitates separation from the desired nitrido complex. Preparation of the labeled complex $^{15}NCr(N{-}i\text{-}Pr_2)_3$ permitted the nitrido nitrogen chemical shift of $\delta = 979$ ppm (relative to liquid NH_3 taken as 0 ppm) to be assigned, and the chromium–nitrogen stretching frequency for $^{15}NCr(N{-}i\text{-}Pr_2)_3$ (1029 cm^{-1}) to be compared with that for $NCr(N{-}i\text{-}Pr_2)_3$ (1054 cm^{-1}).

X-ray crystallography demonstrated $NCr(N{-}i\text{-}Pr_2)_3$ to be a pseudotetrahedral monomer as anticipated by analogy with certain molybdenum and tungsten analogues (48, 50, 74–76). The chromium–nitrogen triple bond in $NCr(N{-}i\text{-}Pr_2)_3$ is exceedingly short, 1.544(3) Å. Other methodology has since been utilized in the preparation of the related nitrido complex $NCr(O{-}t\text{-}Bu)_3$, which was also characterized by single-crystal X-ray crystallography and shown to be monomeric (77).

Other chromium nitrosyl complexes that underwent deoxygenation smoothly upon treatment with $(thf)V(Mes)_3$ include $(ON)Cr[N(R)Ar_F]_3$ [R = $C(CD_3)_2Me$, Ar = 2,5-C_6H_3FMe], $(O^{15}N)Cr[N(R)Ar_F]_3$, and $(ON)Cr[N(R)Ar]_3$ [R = $C(CD_3)_2Me$, Ar = 3,5-$C_6H_3Me_2$] (45).

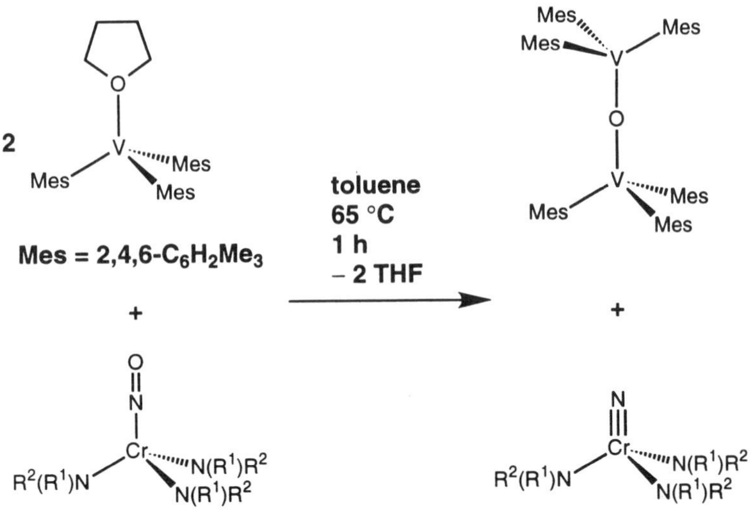

Mes = 2,4,6-$C_6H_2Me_3$

toluene
65 °C
1 h
– 2 THF

1: $R^1 = R^2 =$ *i*-Pr; 67%
2: $R^1 = C(CD_3)_2CH_3$, $R^2 =$ 2,5-C_6H_3FMe; 91%
3: $R^1 = C(CD_3)_2CH_3$, $R^2 =$ 3,5-$C_6H_3Me_2$; 73%

Scheme 14

From an electron-counting viewpoint, it is gratifying that d^2 vanadium(III) acts in concert with d^3 chromium(III) to effect a five-electron reductive cleavage of NO, resulting in terminal d^0 oxo and nitrido complexes of vanadium and chromium, respectively [ignoring the synthetic trick utilized to scavenge the $OV(Mes)_3$]. In this regard, the nitric oxide reductive cleavage reaction is closely related to the reductive scission of dinitrogen to two terminal nitrido moieties as effected by two d^3 molybdenum centers acting cooperatively (see the section below devoted to chemistry of $Mo[N(R)Ar]_3$). The reaction is also indicative that (thf)$V(Mes)_3$ should be regarded as a powerful oxygen-atom acceptor, in that other potential deoxygenating agents (e.g., neat PEt_3 at $\sim 80°C$) failed to react with the chromium nitrosyl complex $(ON)Cr[N(R)Ar_F]_3$. These reactivity considerations underscore the isolobal relationships developed in Section III.

4. *Alkali Metal Reduction: Reactions with Dinitrogen*

Floriani and co-workers in 1993 (78) produced an interesting set of results based on sodium or potassium metal reductions of (thf)$V(Mes)_3$ in diglyme. The work was motivated by a desire to shed light on vanadium–dinitrogen interactions that may augment our understanding of vanadium-containing nitrogenases.

Sodium reduction of (thf)V(Mes)$_3$ in diglyme under Ar produced a ligand-exchange product, [Na(diglyme)$_2$][V(Mes)$_4$], as the major characterizable species (Scheme 15). In contrast, sodium–diglyme reduction of (thf)V(Mes)$_3$ in the presence of a dinitrogen atmosphere led to the near-quantitative formation of the d^3–d^3 μ-N$_2$ complex [Na(diglyme)$_2$][Na(μ-Mes)$_2$(μ-N$_2$)V$_2$(Mes)$_4$]. Conjecture that the latter d^3–d^3 bridging N$_2$ complex forms via a d^3–d^2 bridging N$_2$ species was supported by the results of potassium metal reduction of (thf)V(Mes)$_3$ in diglyme. Under the latter conditions, the ion pair [K(diglyme)$_2$][(μ-N$_2$)V$_2$(Mes)$_6$] containing the pertinent d^3–d^2 anion was obtained. Both of the bridging N$_2$ complexes display essentially linear VNNV

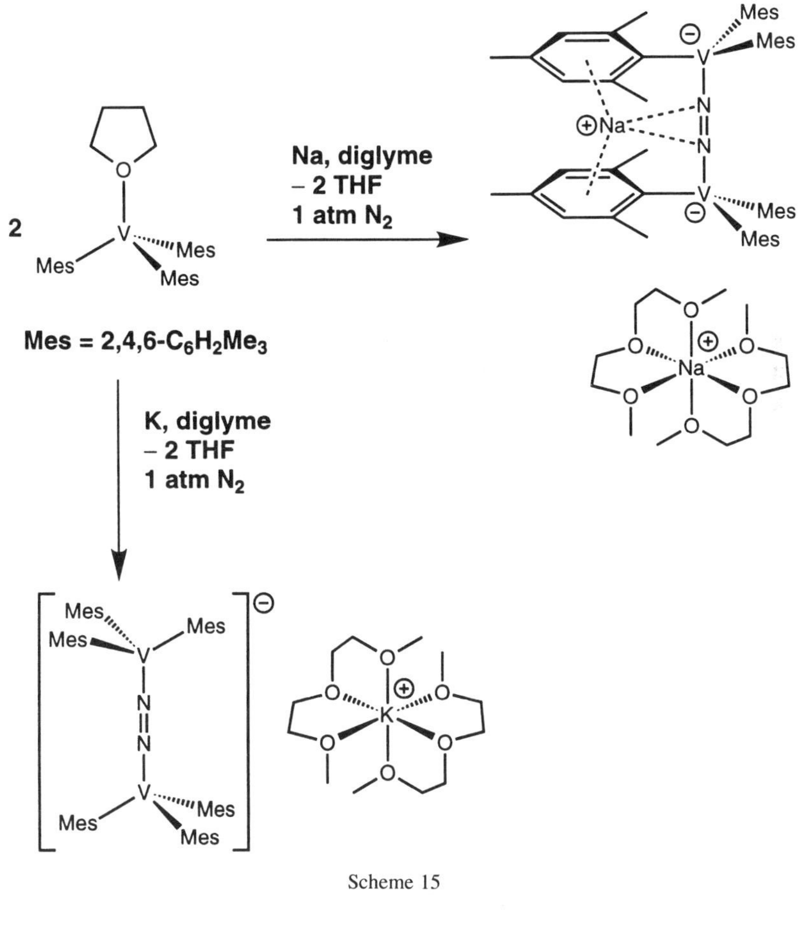

Scheme 15

cores, and in both electronic communication between the metal centers via the N_2 ligand is indicated by magnetic susceptibility measurements. The d^3–d^2 species exhibits $\mu_{eff} = 1.83\ \mu_B$, indicative of a single unpaired electron, while the d^3–d^3 dinuclear complex exhibits $\mu_{eff} = 3.38\ \mu_B$, which is nominally consistent with two unpaired electrons.

B. V[CH(SiMe$_3$)$_2$]$_3$

1. *Synthesis and Characterization*

This homoleptic alkyl was obtained in a manner similar to the analogous titanium(III) and chromium(III) species (Scheme 16). Accordingly, ethereal $VCl_3(NMe_3)_2$ (2.4 mmol) was treated with Li[CH(SiMe$_3$)$_2$] (3 equiv) giving a green solution. Removal of volatile matter and extraction with hexane provided a green extract, concentration and cooling of which gave V[CH(SiMe$_3$)$_2$]$_3$ (4, 24, 25) as blue-green crystals (yield 14%). It was observed that the substance was quite sensitive to handling and readily became contaminated with a black tar.

C. V(N—*i*-Pr$_2$)$_3$

1. *Synthesis and Characterization*

As in the case of the titanium derivative, little is known about the homoleptic amide V(N—*i*-Pr$_2$)$_3$. Bradley and co-workers (29) investigated the reaction of LiN—*i*-Pr$_2$ with an unspecified form of VCl_3, but indicated that incomplete replacement of chloride hampered identification or isolation of V(N—*i*-Pr$_2$)$_3$. Recently, Gambarotta and co-workers (79) investigated the reaction of VCl_3(thf)$_3$ with 3 equiv of LiN—*i*-Pr$_2$ in THF at −40°C, which gave rise to a green solution. It was stated that the low solubility of the presumed product, (thf)V(N—*i*-Pr$_2$)$_3$, precluded its isolation.

3 Li(CH[SiMe$_3$]$_2$)
Et$_2$O
− 3 LiCl, 2 NMe$_3$
14%

Scheme 16

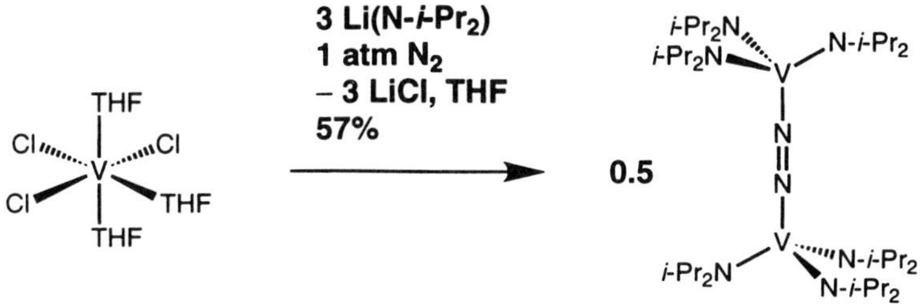

$VCl_3(THF)_3$ → [3 Li(N-*i*-Pr_2), 1 atm N_2, – 3 LiCl, THF, 57%] → 0.5 $(i\text{-}Pr_2N)_3V{-}N{\equiv}N{-}V(N\text{-}i\text{-}Pr_2)_3$

Scheme 17

2. *Reaction with Dinitrogen*

Despite the lack of available information with respect to the properties of putative $V(N{-}i\text{-}Pr_2)_3$, evidence from reaction chemistry shows that the fragment can, in any event, be transferred to N_2 in good yield (Scheme 17) (79). A THF suspension of $VCl_3(thf)_3$ was treated with $LiN{-}i\text{-}Pr_2$ (3 equiv) at −40°C. The deep-green solution was allowed to warm to room temperature, after which time the solvent was removed. The residual solid was dissolved in hexane and the resulting solution was allowed to stand for 5 days at 4°C, whereupon deep orange crystals of the dinuclear linear bridging N_2 complex $(\mu\text{-}N_2)[V(N{-}i\text{-}Pr_2)_3]_2$ separated in 57% yield. The $(\mu\text{-}N_2)[V(N{-}i\text{-}Pr_2)_3]_2$ complex was characterized by X-ray crystallography (Fig. 7). It was asserted that

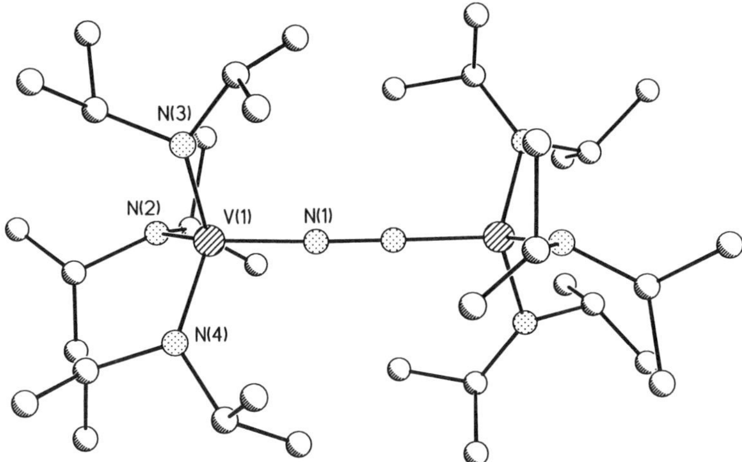

Figure 7. Structural drawing of $(\mu\text{-}N_2)[V(N{-}i\text{-}Pr_2)_3$.

(μ-N_2)[V(N—*i*-Pr_2)$_3$]$_2$ possesses a robust V—N_2—V frame that is unreactive toward (unspecified) azide and diazo reagents. It was further stated that attempts to replace the bridging N_2 ligand in (μ-N_2)[V(N—*i*-Pr_2)$_3$]$_2$ with other ligands (PR_3, CO, RNC) were unsuccessful. Decomposition of (μ-N_2)[V(N—*i*-Pr_2)$_3$]$_2$ with HCl and MeI was said to release N_2 without concomitant production of detectable amounts of the corresponding hydrazine.

D. V[N($SiMe_3$)$_2$]$_3$

1. *Synthesis and Characterization*

This compound is the premier example of three-coordinate vanadium(III). Originally synthesized by Bradley and co-workers (80), descriptions of this compound are in the older literature. See Section I above for references. The X-ray structure of this compound showed it to be isostructural (D_3 molecular symmetry) with the other transition metal M(N[$SiMe_3$]$_2$)$_3$ complexes; see below for the structure of Mn[N($SiMe_3$)$_2$]$_3$. Very little appears to have been done with V[N($SiMe_3$)$_2$]$_3$ recently.

E. {V[N($SiMe_3$)$_2$]$_3$}$^+$

1. *Synthesis and Characterization*

Highly electrophilic, cationic Group 4 (IV.B) metal complexes have emerged as useful polymerization catalysts, a development that has relied heavily on weakly coordinating anions such as methyl alumoxane derived species or fluorinated tetraarylborates. The cation of interest in this section, [V{N($SiMe_3$)$_2$}$_3$]$^+$, evidently is not electrophilic since Gambarotta and co-worker (81) reported its crystallization as a cyanide salt. Accordingly, [V{N($SiMe_3$)$_2$}$_3$] [CN] is said to have resulted from the reaction in hexane of a dimeric vanadium(III) species, {V(μ-η^2-$CH_2SiMe_2NSiMe_3$) [N($SiMe_3$)$_2$]}$_2$, with *tert*-butyl isocyanide (Scheme 18). The cyanide salt, [V{N($SiMe_3$)$_2$}$_3$]

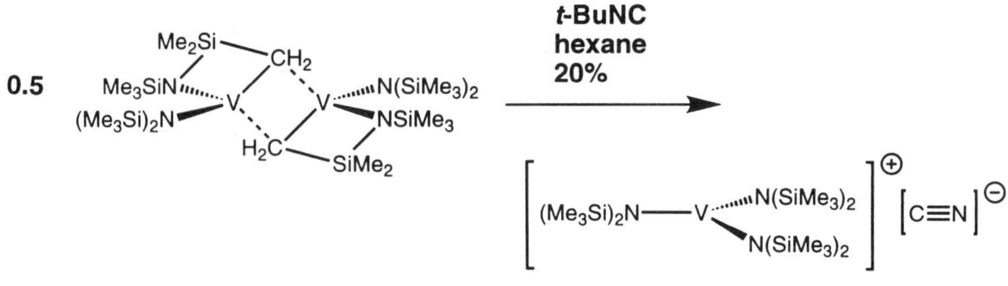

Scheme 18

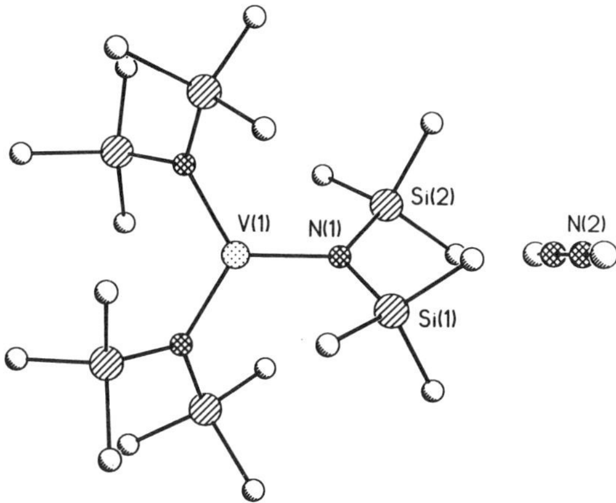

Figure 8. Structural drawing of $\{V[N(SiMe_3)_2]_3\}(CN)$.

[CN], was isolated in 20% yield as a brown-black crystalline solid exhibiting a strong, sharp band in its IR spectrum [ν(CN) = 2159 cm^{-1}]. The observed magnetic moment (μ_{eff} = 2.37 μ_B) is significantly higher than expected for a d^1 species, although the sample manifested clean Curie behavior in the 6–307 K temperature range.

As Berno and Gambarotta (81) state, "the ionic structure of $[V\{N(SiMe_3)_2\}_3]$ [CN] is somewhat surprising." It was, however, verified by a single-crystal X-ray diffraction study (Fig. 8), which appeared to show that the system contains two separate ionic fragments. The salt crystallized in the trigonal space group, $P\bar{3}$, with the vanadium atom located on a threefold axis, relegating the cationic fragment to the point group D_3. In this respect, and also with respect to the observed vanadium–nitrogen bond length [d(VN) = 1.91(1) Å], the structure is exceedingly reminiscent of neutral $V[N(SiMe_3)_2]_3$. The free cyanide bond length is given as 1.21(2) Å. Berno and Gambarotta invoke (81) the d^1 electronic configuration of the vandium atom to explain the observation of separate anions and cations, since steric factors alone cannot justify the ionic structure.

F. $V[N(Ad)Ar]_3$ (Ad = 1-Adamantyl, Ar = 3,5-$C_6H_3Me_2$)

1. Synthesis and Characterization

Citing especially interest in dinitrogen and organometallic chemistries, Gambarotta and co-workers (130) recently prepared the new homoleptic vanadium(III) complex $V[N(Ad)Ar]_3$ and initiated exploration of its utility as a chalcogen atom acceptor (82). The compound was prepared by reaction of $VCl_3(thf)_3$ (56) with 3 equiv of $Li[N(Ad)Ar](OEt_2)$ in THF; the final color of the reaction

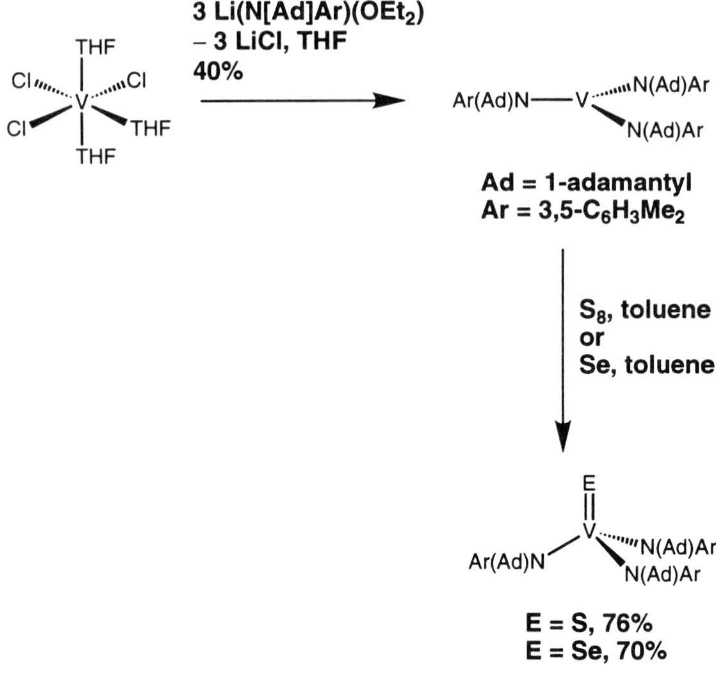

Scheme 19

mixture prior to work up was green-brown (Scheme 19). Dark green crystals of V[N(Ad)Ar]$_3$ were isolated in 40% yield on a 2.0-mmol scale. Analytical and magnetic susceptibility data ($\mu_{eff} = 2.72\ \mu_B$) were in agreement with the compound as formulated. The V[N(Ad)Ar]$_3$ complex was found to be EPR silent (conditions not stated).

X-ray crystallography showed V[N(Ad)Ar]$_3$ to possess a trigonal planar VN_3 core supplemented by interaction of the vanadium center with an ipso and ortho carbon of one of the aryl rings (Fig. 9). Such interactions were observed earlier for the complex Ti[N(R)Ar]$_3$ [R = C(CD_3)$_2$Me; Ar = 3,5-$C_6H_3Me_2$] and likened to the η^3-bonding mode sometimes observed for benzyl ligands.

2. *Reaction with Sulfur*

Red-violet solutions were produced on addition of S_8 to a toluene solution of V[N(Ad)Ar]$_3$ (Scheme 19). Isolated as a crystalline solid in 76% yield, sulfido SV[N(Ad)Ar]$_3$ is diamagnetic, and was said to react readily with acids to release H_2S. Two-dimensional NMR techniques were utilized to analyze the resonances associated with the 1-adamantyl substituent, and a C_3 geometry static

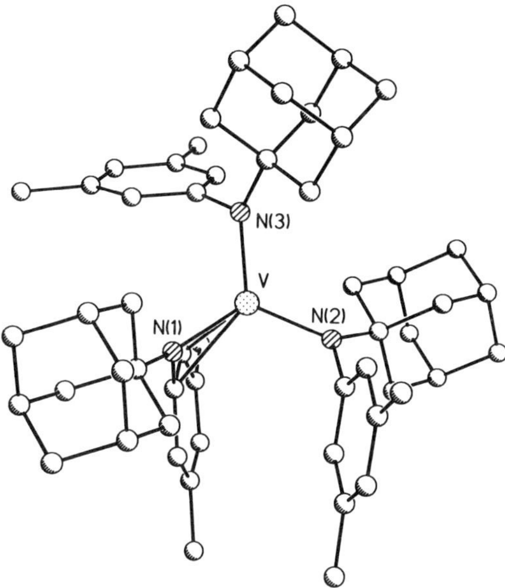

Figure 9. Structural drawing of V[N(Ad)Ar]$_3$; Ad = 1-adamantyl, Ar = 3,5-$C_6H_3Me_2$.

on the NMR time scale is indicated for the complex by the observation of three distinct aryl proton signals. The structure of SV[N(Ad)Ar]$_3$ was determined by X-ray crystallography to feature a pseudotetrahedral vanadium center and typical dimensions for the thiovanadyl function [2.045(3) Å].

3. *Reaction with Selenium*

Elemental selenium reacted with V[N(Ad)Ar]$_3$ to give SeV[N(Ad)Ar]$_3$ in 70% isolated yield, in a manner similar to the sulfur reaction described above. The NMR data for SeV[N(Ad)Ar]$_3$ were similar to those for the sulfido analogue; no signal could be detected in the ^{77}Se NMR spectrum of the complex, presumably due to coupling to the ^{51}V quadrupole. Such coupling has been invoked previously to explain difficulties encountered in detecting ^{77}Se NMR signals associated with vanadium(V) selenide complexes (72).

G. V[SeSi(SiMe$_3$)$_3$][N(SiMe$_3$)$_2$]$_2$

1. *Synthesis and Characterization*

Treatment of (thf)V(Br)[N(SiMe$_3$)$_2$]$_2$ with 1 equiv of Li(thf)$_2$[SeSi(SiMe$_3$)$_3$] in hexanes led to the expulsion of LiBr and THF, and to the formation of forest

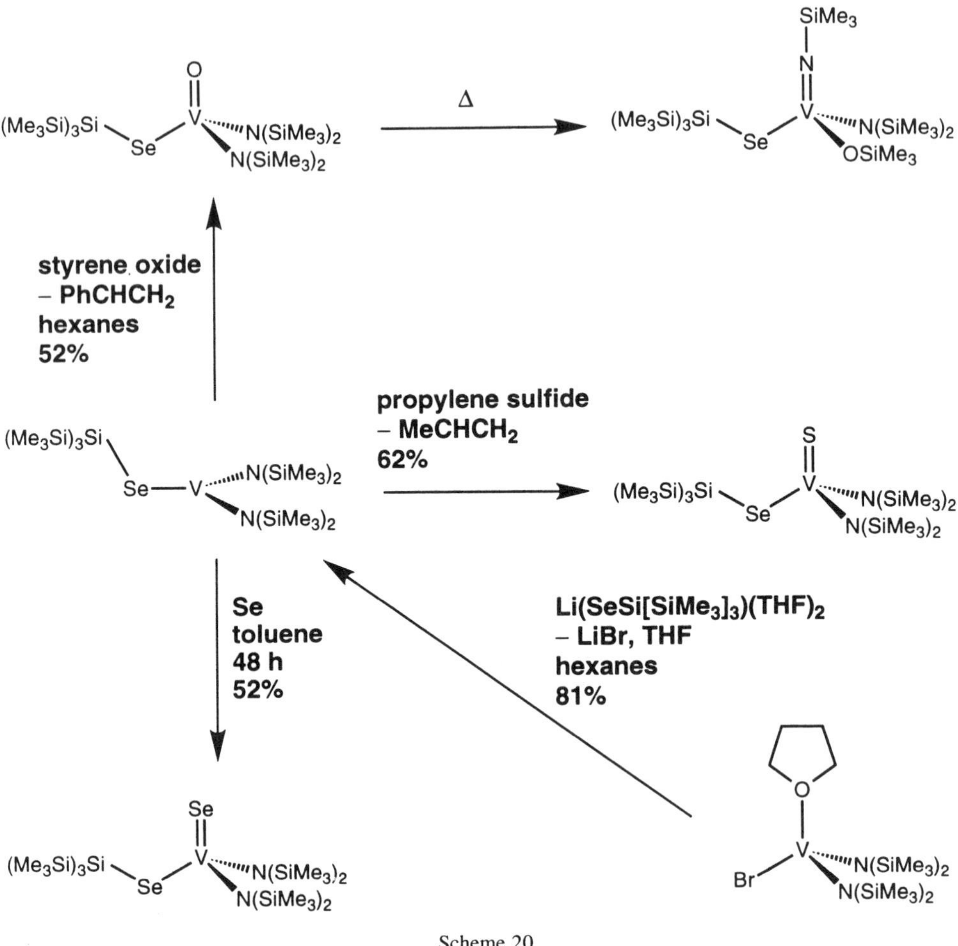

Scheme 20

green $V[SeSi(SiMe_3)_3][N(SiMe_3)_2]_2$ (83), which was isolated as crystals in 81% yield (Scheme 20).

The structure of $V[SeSi(SiMe_3)_3][N(SiMe_3)_2]_2$ (34) was determined by X-ray crystallography (Fig. 10). The coordination about the vanadium center is trigonal planar (VN_2Se). The expected bent geometry at selenium was observed: V—Se—Si, 124.07(4)°, and the vanadium–selenium bond length of 2.451(1) Å appeared normal. Two methyl groups from the $N(SiMe_3)_2$ ligands were discerned to engage in agostic interactions with the vanadium center, bringing the nominal coordination number for the complex to 5 and giving it a pseudotri-

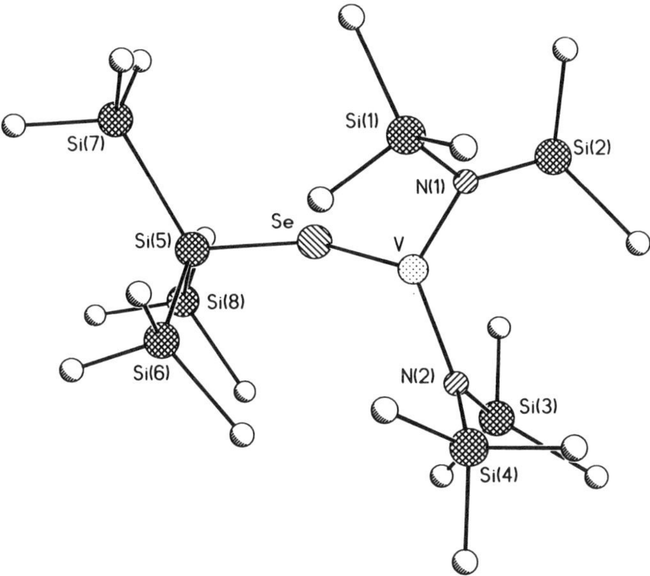

Figure 10. Structural drawing of $V[SeSi(SiMe_3)_3][N(SiMe_3)_2]_2$.

gonal bipyramidal geometry. The V · · · C distances to the two carbon atoms in question are 2.504(4) and 2.624(4) Å.

2. *Reaction with Styrene Oxide*

Addition of styrene oxide to a hexane solution of $V[SeSi(SiMe_3)_3][N(SiMe_3)_2]_2$ led rapidly to a wine red reaction mixture. After crystallization from $O(SiMe_3)_2$ the oxovanadium(V) compound $OV[SeSi(SiMe_3)_3][N(SiMe_3)_2]_2$ was obtained in 52% yield. The compound exhibits a signal in its ^{51}V NMR spectrum at δ 245 ($\Delta\nu_{1/2}$ ~275 Hz).

It is noteworthy that upon thermolysis (toluene, reflux, 16 h) the oxo compound $OV[SeSi(SiMe_3)_3][N(SiMe_3)_2]_2$ undergoes rearrangement to $(Me_3SiN)V[SeSi(SiMe_3)_3][N(SiMe_3)_2](OSiMe_3)$, with migration of an amido trimethylsilyl substituent to the oxo moiety, resulting in imido and siloxide functionalities.

3. *Reaction with Propylene Sulfide*

The maroon sulfido complex $SV[SeSi(SiMe_3)_3][N(SiMe_3)_2]_2$ was obtained in 62% yield subsequent to treatment of $V[SeSi(SiMe_3)_3][N(SiMe_3)_2]_2$ with propylene sulfide. A characteristic ^{51}V NMR signal (δ 1102, $\Delta\nu_{1/2}$ ~ 300 Hz) was noted for this species

4. *Reaction with Selenium*

Upon stirring $V[SeSi(SiMe_3)_3][N(SiMe_3)_2]_2$ together with elemental selenium in toluene, conversion to a maroon reaction mixture was noted over 48 h. The selenidovanadium(V) complex $SeV[SeSi(SiMe_3)_3][N(SiMe_3)_2]_2$ was subsequently isolated in 52% yield by crystallization from $O(SiMe_3)_2$. In its ^{51}V NMR spectrum, the terminal selenido complex exhibits a characteristic signal (δ 1444, $\Delta\nu_{1/2}$ ~ 270 Hz).

H. $V[TeSi(SiMe_3)_3][N(SiMe_3)_2]_2$

1. *Synthesis and Characterization*

Treatment of $(thf)V(Br)[N(SiMe_3)_2]_2$ with 1 equiv of $Li(thf)_2[TeSi(SiMe_3)_3]$ in hexanes led to the expulsion of LiBr and THF (83), and to the formation of brown $V[TeSi(SiMe_3)_3][N(SiMe_3)_2]_2$, which was isolated as crystals in 61% yield (Scheme 21).

As for the corresponding selenolate species (see above), the complex $V[TeSi(SiMe_3)_3][N(SiMe_3)_2]_2$ was characterized by X-ray crystallography (Fig. 11). Except for the V—Te bond length of 2.6758(9) Å, and the V—Te—Si angle of 117.70(4)°, the three-coordinate silylselenolate and silyltellurolate complexes display quite similar metrical parameters. Two methyl groups based on the $N(SiMe_3)_2$ ligands appear to engage in agostic interactions with the vanadium center for $V[TeSi(SiMe_3)_3][N(SiMe_3)_2]_2$ as was the case for the selenium analogue.

2. *Reaction with Styrene Oxide*

On a scale of just under 1 mmol, reaction of styrene oxide with $V[TeSi(SiMe_3)_3][N(SiMe_3)_2]_2$ led to production of the oxovanadium(V) species

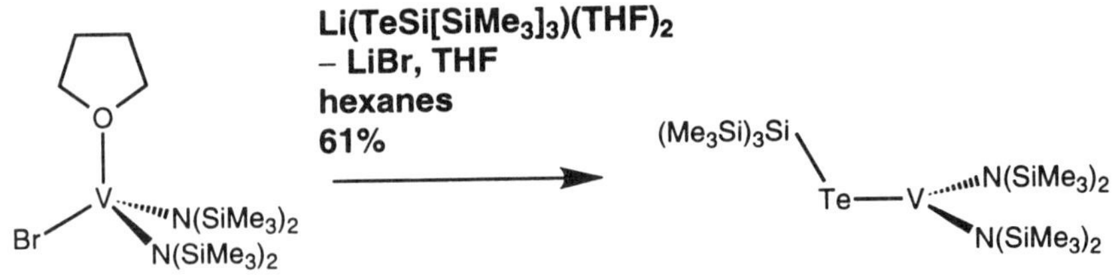

$Li(TeSi[SiMe_3]_3)(THF)_2$
– LiBr, THF
hexanes
61%

Scheme 21

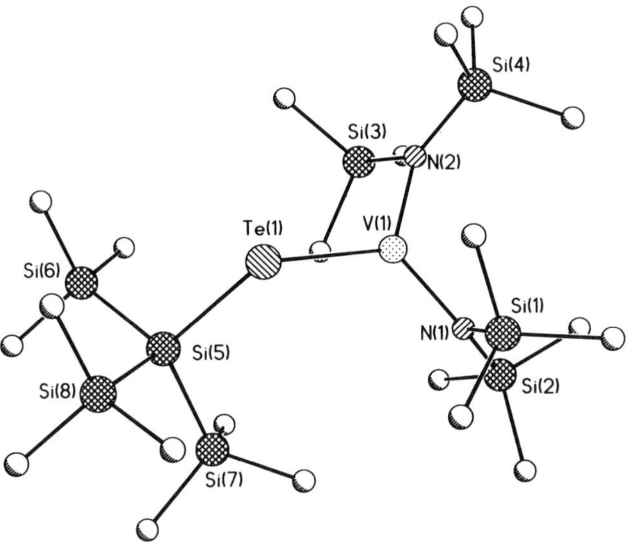

Figure 11. Structural drawing of $V[TeSi(SiMe_3)_3][N(SiMe_3)_2]_2$.

$OV(TeSi[SiMe_3]_3)[N(SiMe_3)_2]_2$ (Scheme 22). Blue-black crystals of the complex were isolated in 47% yield; these dissolve to give deep aqua solutions. Deemed photosensitive, $OV[TeSi(SiMe_3)_3][N(SiMe_3)_2]_2$ was not isolated in analytically pure form due to contamination with ditelluride $[TeSi(SiMe_3)_3]_2$. A characteristic ^{51}V NMR signal (δ 504, $\Delta\nu_{1/2}$ ~ 400 Hz) was noted for the oxovanadium(V) tellurolato complex.

3. *Reaction with Propylene Sulfide*

Sulfurization of three-coordinate $V[TeSi(SiMe_3)_3][N(SiMe_3)_2]_2$ was effected by treatment of a solution of the complex with propylene sulfide in an NMR tube scale reaction performed in C_6D_6. Analysis of the resulting black mixture by NMR spectroscopy revealed a ^{51}V NMR signal (δ 1410, $\Delta\nu_{1/2}$ ~ 3.75 Hz) indicative of $SV[TeSi(SiMe_3)_3][N(SiMe_3)_2]_2$. Under these conditions, the latter species exhibited a half-life of about 10 h, and was presumed to be decomposing to $[TeSi(SiMe_3)_3]_2$ and $\{(\mu\text{-}S)V[N(SiMe_3)_2]_2\}_2$, the selenido analogue of which was characterized structurally in the same study.

4. *Reaction with Triphenylphosphine Selenide*

A dark maroon mixture resulted upon treating $V[TeSi(SiMe_3)_3][N(SiMe_3)_2]_2$ with Ph_3PSe in C_6D_6. Formation of diamagnetic SeV[TeSi-

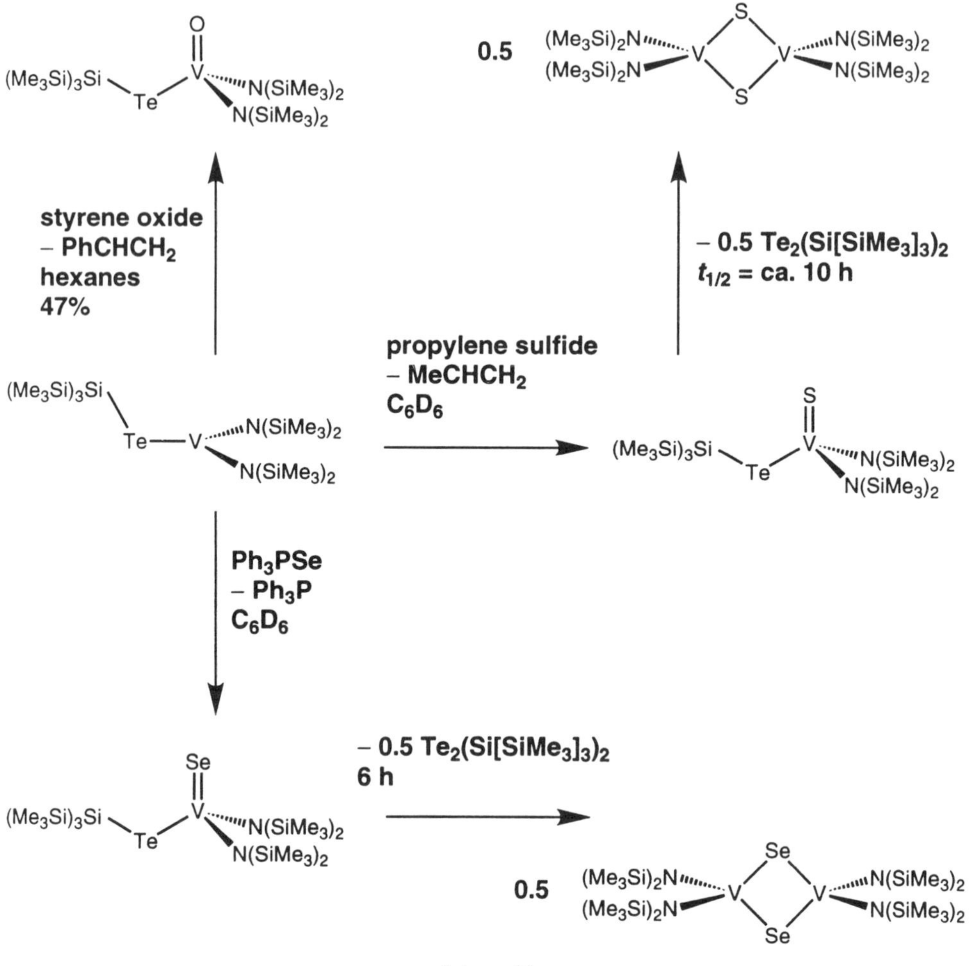

Scheme 22

$(SiMe_3)_3][N(SiMe_3)_2]_2$ was indicated by ^{51}V NMR data (δ 1755, $\Delta\nu_{1/2}$ ~ 410 Hz), while formation of Ph_3P was confirmed by ^{31}P NMR spectroscopy. The maroon mixture turned emerald green over 6 h at about 25°C under ambient lighting. Subsequent investigation revealed complete and essentially quantitative formation of $\{(\mu\text{-Se})V[N(SiMe_3)_2]_2\}_2$, a nondiamagnetic ($\mu = 0.41\ \mu_B$ per V) compound. The latter compound was synthesized alternatively by treatment of $V[TeSi(SiMe_3)_3][N(SiMe_3)_2]_2$ with elemental selenium, as described below, and was characterized in an X-ray diffraction study. The related paramagnetic oxo compound $\{(\mu\text{-O})V[N(SiMe_3)_2]_2\}_2$ was prepared recently via a completely unrelated method (84).

5. *Reaction with Selenium*

The $V[TeSi(SiMe_3)_3][N(SiMe_3)_2]_2$ complex was treated with selenium (1 equiv) in toluene on a scale of about 1 mmol, and the mixture was stirred for 2 days. Dark green crystals of $\{(\mu\text{-Se})V[N(SiMe_3)_2]_2\}_2$ were subsequently obtained in 89% yield. An unidentified impurity was observed by ^1H NMR spectroscopy; the unknown byproduct could be removed by recrystallization from hexanes.

The X-ray diffraction study carried out for $\{(\mu\text{-Se})V[N(SiMe_3)_2]_2\}_2$ showed that the compound crystallizes in a space group ($P2_1/c$) different from that found ($Pbcn$) for the related oxo derivative $\{(\mu\text{-O})V[N(SiMe_3)_2]_2\}_2$ (131), although the two compounds display similar metrical parameters for the $V_2[N(SiMe_3)_2]_4$ moieties. Both compounds exhibit nominal D_{2h} point symmetry. In this they resemble $\{(\mu\text{-N})Cr[N{-}i\text{-}Pr_2]_2\}_2$ (85) and $[\{(\mu\text{-N})V[N(SiMe_3)_2]_2\}_2]^-$ (86).

I. $V(SeSiPh_3)[N(SiMe_3)_2]_2$

1. *Synthesis and Characterization*

Addition of toluene to a mixture of $(thf)V(Br)[N(SiMe_3)_2]_2$ and $Li(SeSiPh_3)(thf)_2$ gave a green mixture that yielded a purple solid upon removal of volatile matter. Separation from salt and recrystallization gave $V(SeSiPh_3)[N(SiMe_3)_2]_2$ as purple crystals in 67% yield (83) (Scheme 23). Magnetic susceptibility data ($\mu_{eff} = 2.45\ \mu_B$) are in accord with the compound as formulated, as are the analytical data. Although this compound was not characterized structurally by X-ray diffraction, the corresponding triphenylsilyltellurolate compound was verified in that way to possess a three-coordinate vanadium center (see below).

2. *Reaction with Styrene Oxide*

The reaction of $V(SeSiPh_3)[N(SiMe_3)_2]_2$ with styrene oxide in C_6D_6 gave a dark orange solution. Formation of diamagnetic $OV(SeSiPh_3)[N(SiMe_3)_2]_2$ was indicated by ^{51}V NMR data (δ 155, $\Delta\nu_{1/2} \sim 430$ Hz) and by ^1H NMR data; the latter confirmed also the formation of styrene.

3. *Reaction with Propylene Sulfide*

The reaction of $V(SeSiPh_3)[N(SiMe_3)_2]_2$ with propylene sulfide in C_6D_6 gave a dark orange-red mixture. Formation of diamagnetic $SV(SeSiPh_3)[N(SiMe_3)_2]_2$ was indicated by ^{51}V NMR data (δ 1012, $\Delta\nu_{1/2} \sim 450$ Hz). The ^1H NMR data

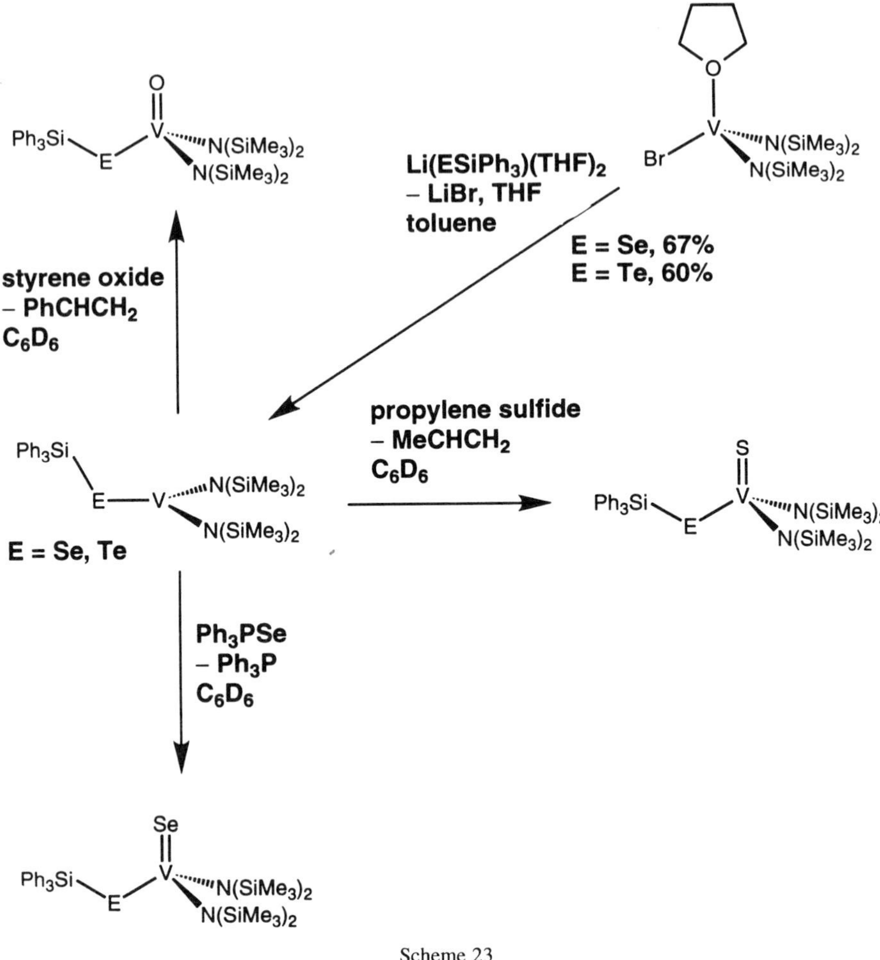

Scheme 23

were in agreement with this interpretation and, furthermore, the expected resonances for propylene were observed.

4. *Reaction with Triphenylphosphine Selenide*

Mixing $V(SeSiPh_3)[N(SiMe_3)_2]_2$, Ph_3PSe, and C_6D_6 gave a dark orange-red mixture with ^{51}V NMR data (δ 1369, $\Delta\nu_{1/2}$ ~ 460 Hz) in accord with formation of diamagnetic $SeV(SeSiPh_3)[N(SiMe_3)_2]_2$.

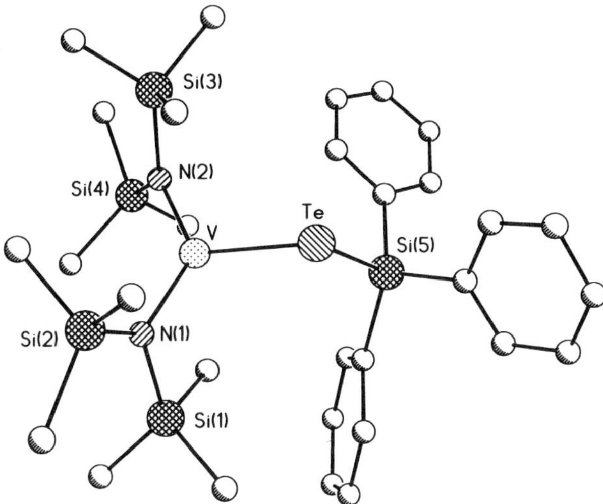

Figure 12. Structural drawing of $V(TeSiPh_3)[N(SiMe_3)_2]_2$.

J. $V(TeSiPh_3)[N(SiMe_3)_2]_2$

1. Synthesis and Characterization

This brown crystalline triphenylsilyltellurolate was obtained in 60% yield subsequent to treatment of $(thf)V(Br)[N(SiMe_3)_2]_2$ with $Li(TeSiPh_3)(thf)_2$ in toluene (Scheme 23). In addition to characterization by magnetic susceptibility measurements ($\mu_{eff} = 2.52\ \mu_B$) and the usual analytical methods, $V(TeSiPh_3)[N(SiMe_3)_2]_2$ was additionally characterized by X-ray crystallography (Fig. 12). Metrical parameters for the complex were similar to the analogue described above, $V[TeSi(SiMe_3)_3][N(SiMe_3)_2]_2$. Close contacts between two axially disposed silylamido methyl groups and the vanadium center were again observed. The postulate was posited that the tellurolate ligand, being predominantly a σ donor, renders the metal rather electrophilic thus favoring the observed agostic interactions. The following intriguing contrast remains: compounds such as $(thf)V(OSi{-}t\text{-}Bu_3)_3$ fail to lose thf even under forcing conditions, but $V(TeSiPh_3)[N(SiMe_3)_2]_2$, which is prepared in the presence of thf, evidently loses it in favor of the observed agostic interactions. Solvent-free arylamido vanadium and titanium compounds display M–aryl interactions that take the place of solvent binding or agostic interactions, as described above for $Ti[N(R)Ar]_3$ and $V[N(Ad)Ar]_3$.

2. *Reaction with Styrene Oxide*

Styrene oxide reacted with $V(TeSiPh_3)[N(SiMe_3)_2]_2$ in C_6D_6, giving an ink purple solution, evidently containing $OV(TeSiPh_3)[N(SiMe_3)_2]_2$ (^{51}V NMR data: δ 387, $\Delta\nu_{1/2} \sim$ 550 Hz).

3. *Reaction with Propylene Sulfide*

A dark brown-red mixture was produced upon combining $V(TeSiPh_3)[N(SiMe_3)_2]_2$ with propylene sulfide in C_6D_6. The NMR (^1H and ^{51}V) analysis confirmed formation of propylene and $SV(TeSiPh_3)[N(SiMe_3)_2]_2$ (^{51}V NMR data: δ 1287, $\Delta\nu_{1/2} \sim$ 430 Hz).

4. *Reaction with Triphenylphosphine Selenide*

A dark orange-red mixture was produced upon combining $V(TeSiPh_3)[N(SiMe_3)_2]_2$ with triphenylphosphine selenide in C_6D_6. The NMR (^1H and ^{51}V) analysis confirmed formation of Ph_3P and $SeV(TeSiPh_3)[N(SiMe_3)_2]_2$ (^{51}V NMR data: δ 1648, $\Delta\nu_{1/2} \sim$ 400 Hz).

K. $V(I)[N(R)Ar_F]_2$ [$R = C(CD_3)_2Me$, Ar_F = 2,5-C_6H_3FMe]

1. *Synthesis and Characterization*

If this compound (87) is a trigonal planar monomeric entity, it will be the only such species [other than anionic $Fe(Cl)(Si[SiMe_3]_3)_2^-$, perhaps: see below] to have such credentials with a relatively nonsterically demanding *halide* ligand in the coordination sphere. However, it has been noted that arylamido ligands bearing ortho-fluorine substitution occasionally revert to coordination (via chelation) of the ortho fluorine to the metal in cases where the metal would otherwise be low coordinate (88). This phenomenon would be expected to favor a monomeric formulation for $V(I)[N(R)Ar_F]_2$.

The synthesis of $V(I)[N(R)Ar_F]_2$ was achieved in three steps from $VCl_3(thf)_3$ (87) (Scheme 24). Treatment of $VCl_3(thf)_3$ with 2 equiv of $Li[N(R)Ar_F](OEt_2)$ in THF on a scale of about 4 mmol led to the isolation of $(thf)V(Cl)[N(R)Ar_F]_2$ in 65% yield. This species was relieved of THF by toluene trituration, providing forest green $V(Cl)[N(R)Ar_F]_2$ in 92% yield; the latter species is probably not monomeric in view of its μ_{eff} of 2.60 μ_B in benzene-d_6 at about 25°C. For this reason, and because of the relatively low solubility of $V(Cl)[N(R)Ar_F]_2$, a chloride-for-iodide exchange was effected by reaction with $ISiMe_3$. The $V(I)[N(R)Ar_F]_2$ was thereby obtained in 88% isolated yield subsequent to an experiment performed in ether on a scale of about 3 mmol. Magnetic suscep-

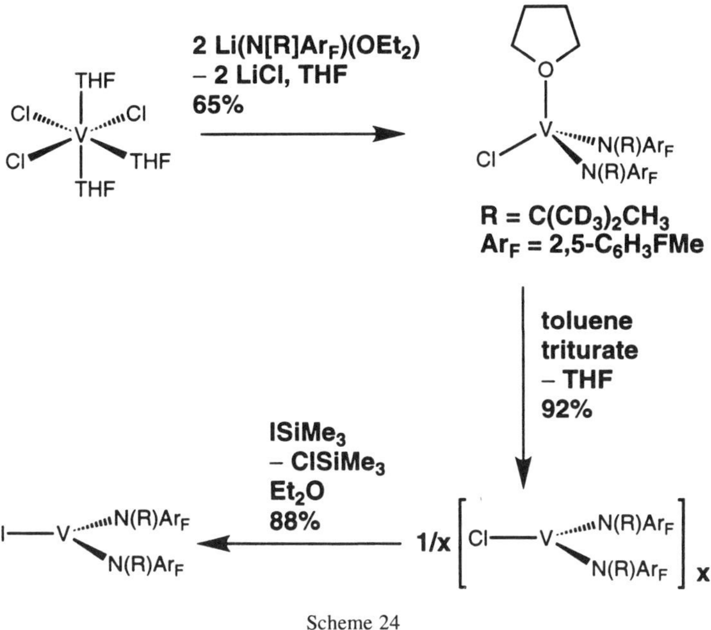

Scheme 24

tibility data ($\mu_{eff} = 3.01\ \mu_B$) for $V(I)[N(R)Ar_F]_2$ are indicative of monomeric, high-spin vanadium(III). In addition, the iodide complex is much more soluble (qualitatively speaking) in hydrocarbon solvents than is the corresponding chloro species. Thus far, $V(I)[N(R)Ar_F]_2$ has eluded structural characterization. The $V(I)[N(R)Ar_F]_2$ complex is thermochromic (green at $-100\,°C$ in ether, brown at $25\,°C$) for reasons that remain mysterious.

2. *Reaction with Mesityl Azide*

The key feature of this reaction is that, when carried out at the temperature of melting ether (1 atm), mesityl azide reacts with $V(I)[N(R)Ar_F]_2$ to give, in 79% isolated yield (~2 mmol), the mesitylazide complex $(MesN_3)V(I)$-$[N(R)Ar_F]_2$ (Scheme 25). The latter species is relatively long lived at $25\,°C$. Diamagnetic $(MesN_3)V(I)[N(R)Ar_F]_2$ was characterized by ^{1}H, ^{13}C, ^{19}F, and ^{51}V NMR spectroscopies (δ ^{51}V = 140.6 ppm), and by single-crystal X-ray crystallography (Fig. 13). The complexed organoazide moiety can well be described as a "diazenylimido" ligand in that the observed d(VN) of 1.662(4) Å is quite similar to that [1.645(7) Å] for the corresponding mesitylimido complex $(MesN)V(I)[N(R)Ar_F]_2$. The latter complex is in fact obtained upon thermolysis of $(MesN_3)V(I)[N(R)Ar_F]_2$. The mesityl azide complex

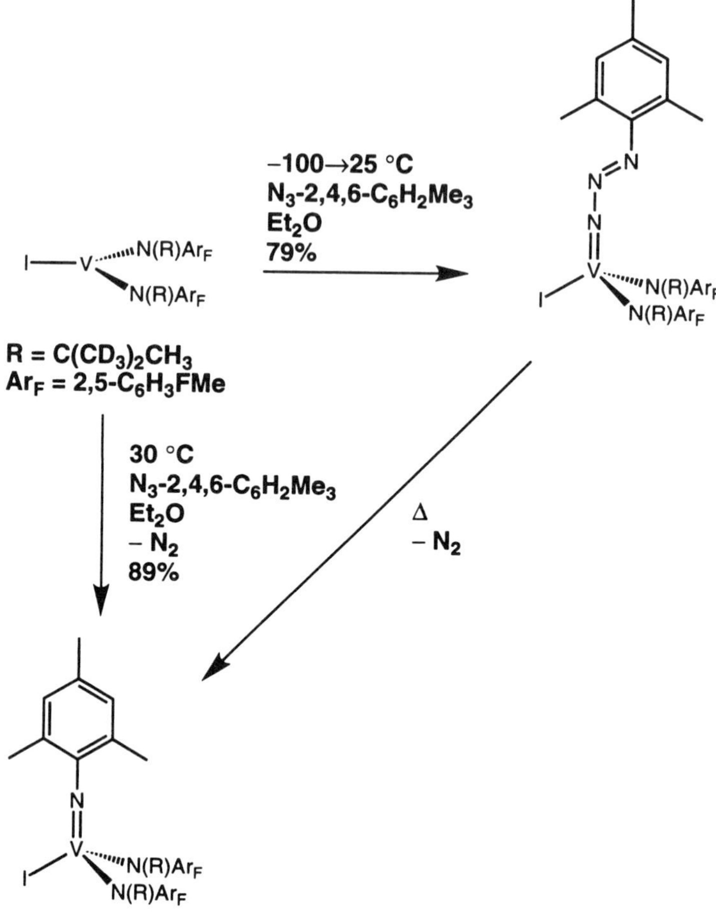

Scheme 25

$(MesN_3)V(I)[N(R)Ar_F]_2$ is formally reminiscent of Staudinger phosphazides, species in which a triorganophosphine has added to the terminal nitrogen of an organoazide, which have been discussed as intermediates in the formation of phosphinimines (89). Another transition metal complex in which an organic azide is affixed via its terminal nitrogen, $Cp_2Ta(Me)(N_3Ph)$, has been synthesized and structurally characterized (90).

Carrying out the reaction between $V(I)[N(R)Ar_F]_2$ and mesityl azide at 30°C led directly to the isolation (0.5 mmol, 89% yield) of the dark red imido complex $(MesN)V(I)[N(R)Ar_F]_2$, despite the fact that pure $(MesN_3)V(I)[N(R)Ar_F]_2$ is relatively robust at that temperature.

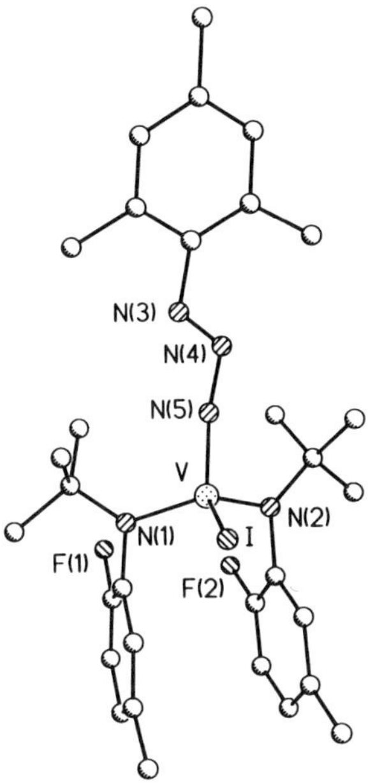

Figure 13. Structural drawing of $(MesN_3)V(I)[N(R)Ar_F]_2$; R = $C(CD_3)_2Me$, Ar_F = 2,5-C_6H_3FMe.

Attempts to generate the putative nitrous oxide complex (ONN)-$V(I)[N(R)Ar_F]_2$ by addition of nitrous oxide to $V(I)[N(R)Ar_F]_2$ led only to the isolation of the oxo complex $OV(I)[N(R)Ar_F]_2$, which was found by X-ray diffraction to have a structure reminiscent of the mesityl azide and mesitylimido complexes described above. Nitrous oxide is expected to exhibit a preference for binding to transition metals via its terminal nitrogen atom (91).

L. $V(PCy_2)_3$

1. *Synthesis and Characterization*

A 1964 report from Issleib and Wenschuh describes the preparation and some properties of $V(PCy_2)_3$ (92). To a suspension of $VCl_3(thf)_3$ in benzene was added a THF solution of $Li(PCy_2)$ giving a dark green solution. After removal of LiCl, $V(PCy_2)_3$ (mp 215°C) was isolated in 61% yield (Scheme 26). Phosphorus and vanadium analytical data were in agreement with the proposed composition, and a cryoscopic molecular weight determination carried out in benzene was consistent with a monomeric formulation (612.0 found, 642.8 calcd for $C_{36}H_{66}P_3V$). The $V(PCy_2)_3$ complex does not appear to have been characterized structurally. Magnetic susceptibility measurements provided an unusually low value of $\mu_{eff} = 1.0\ \mu_B$ for $V(PCy_2)_3$, a value difficult to reconcile

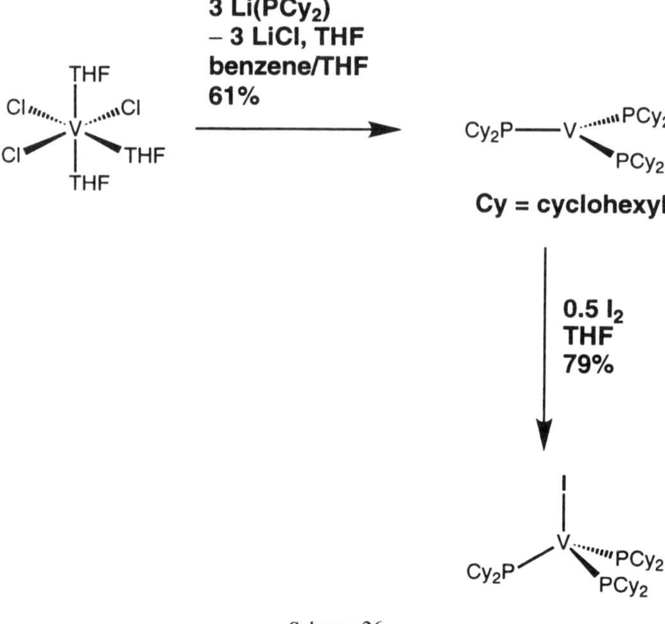

Scheme 26

with formulation of this compound as a monomer; however, it was not clear whether the magnetic measurements were carried out in solution or in the solid state. The $V(PCy_2)_3$ complex could conceivably exist as a dimer or higher oligomer in the solid state, and yet be largely monomeric in benzene solution as indicated by the cryoscopic data.

2. *Reaction with Iodine*

A color change to dark brown was observed upon treatment of $V(PCy_2)_3$ with iodine in THF. The vanadium(IV) iodide species $IV(PCy_2)_3$ subsequently was isolated in 79% yield.

M. $Ta(OSi{-}t\text{-}Bu_3)_3$

1. *Synthesis and Characterization*

In the initial report describing $Ta(OSi{-}t\text{-}Bu_3)_3$ the compound was prepared via sodium amalgam reduction of tantalum(V) precursor $Cl_2Ta(OSi{-}t\text{-}Bu_3)_3$, in THF solvent. Pale blue $Ta(OSi{-}t\text{-}Bu_3)_3$ was purified by crystallization from

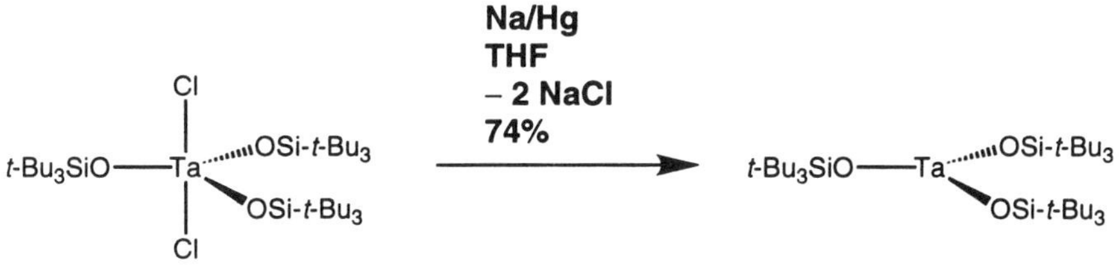

$$t\text{-Bu}_3\text{SiO—TaCl}_2(\text{OSi-}t\text{-Bu}_3)_2 \xrightarrow[\text{74\%}]{\text{Na/Hg, THF, } -2\text{ NaCl}} t\text{-Bu}_3\text{SiO—Ta}(\text{OSi-}t\text{-Bu}_3)_2$$

Scheme 27

hexane and isolated in 60% yield (93) (Scheme 27). A similar procedure carried out on an approximate 5 g scale provided $Ta(OSi{-}t\text{-}Bu_3)_3$ in 74% yield (15). Benzene cryoscopy established a monomeric formulation for the complex in that solvent. Magnetic susceptibility measurements and sharp NMR spectra indicate that $Ta(OSi{-}t\text{-}Bu_3)_3$ is diamagnetic, a finding accounted for on the basis of an $(a_1', d_{z^2})^2$ configuration and $^1A_1'$ ground state for the species, thought to possess local D_{3h} symmetry for the $Ta(OSi)_3$ core. Extended Hückel calculations pointed to a preference for a trigonal planar geometry over alternative T shaped or trigonal pyramidal structures.

2. *Thermal Cyclometalation*

Intramolecular C—H oxidative addition constitutes a relatively slow (benzene, $t_{1/2} \sim 90$ h) decomposition mode for $Ta(OSi{-}t\text{-}Bu_3)_3$ at 25°C (93) (Scheme 28). Cyclometalation to the colorless tantalum(V) hydrido species also proceeds slowly at 25°C in the solid state, the product being isolable in about 84% yield (19). A characteristic downfield ^1H NMR signal (δ 21.97) and IR band (1770 cm^{-1}) were noted for the terminal hydride ligand.

3. *Olefin Binding*

Ethylene and propylene were found to form isolable adducts with $Ta(OSi{-}t\text{-}Bu_3)_3$; $(C_2H_4)Ta(OSi{-}t\text{-}Bu_3)_3$ and $(MeCHCH_2)Ta(OSi{-}t\text{-}Bu_3)_3$ were isolated in yields of 63 and 56% as yellow-orange and orange-red species, respectively (93). The NMR monitoring of reactions leading to these adducts indicated quantitative formation. The olefin complexes may be regarded as metallacyclopropanes in view of the strong π-donor nature of the tantalum center (19).

Interestingly, while *cis*-2-butene reacted slowly (1 : 1 stoichiometry, 25°C), over 3 days, with $Ta(OSi{-}t\text{-}Bu_3)_3$ to produce a corresponding adduct, no reaction was observed with *trans*-2-butene under similar conditions. The selectivity was ascribed to steric factors, consistent also with the failure of $Ta(OSi{-}t\text{-}Bu_3)_3$ to bind *tert*-butylethylene or tetramethylethylene (19).

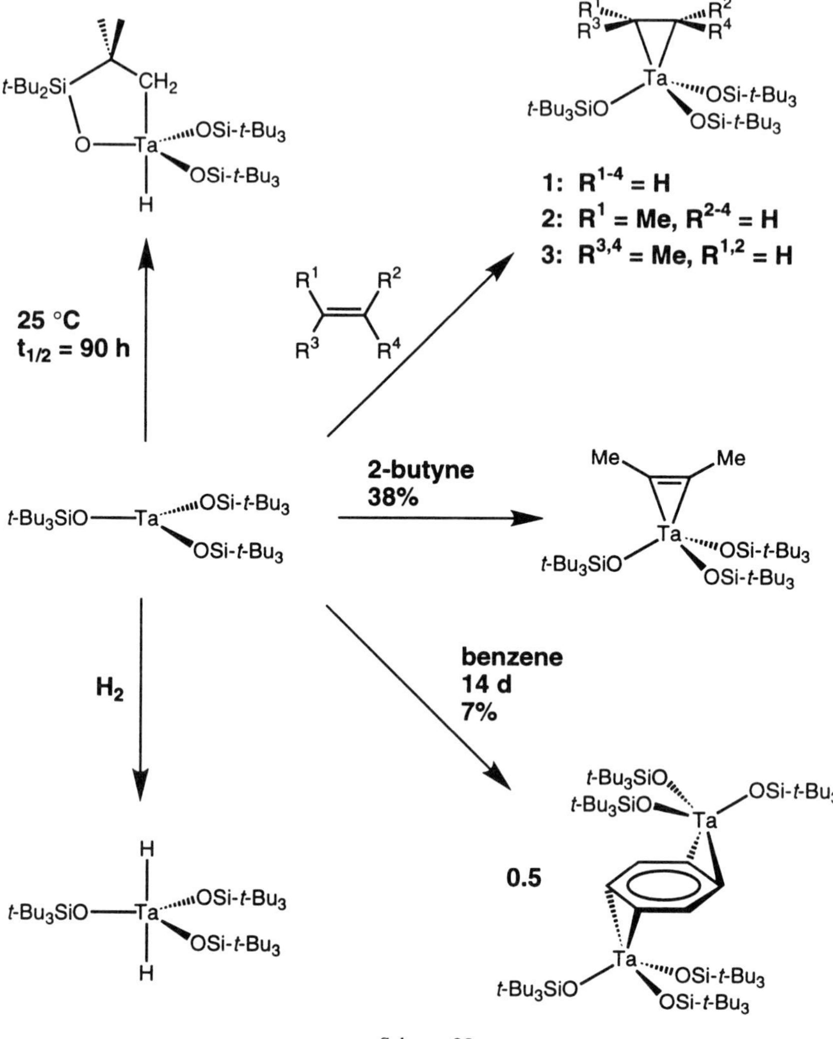

Scheme 28

Solution ^{1}H NMR studies of the isolated (olefin)Ta(OSi—*t*-Bu$_3$)$_3$ complexes revealed equivalent siloxide moieties, consistent with facile rotation of the olefin π face with respect to the axially symmetric, pyramidal Ta(OSi—*t*-Bu$_3$)$_3$ fragment. Bonding, rotational barriers, and conformational preferences in transition metal olefin complexes have been subjected to detailed theoretical treatments (94).

4. *Alkyne Binding*

Treatment of $Ta(OSi{-}t\text{-}Bu_3)_3$ with 2-butyne led to gradual ($t_{1/2} \sim 90$ h) formation of the colorless adduct $(C_2Me_2)Ta(OSi{-}t\text{-}Bu_3)_3$, isolated in 38% yield.

5. *Benzene Complexation*

On standing at room temperature for 2 weeks, a blue benzene solution of $Ta(OSi{-}t\text{-}Bu_3)_3$ deposited brown crystals of $\{\mu\text{-}\eta^2(1,2):\eta^2(4,5)\text{-}C_6H_6\}[Ta(OSi{-}t\text{-}Bu_3)_3]_2$ in 7% isolated yield (19, 95). Competitive cyclometalation (see above) occurred during the time period required for deposition of the 2 : 1 tantalum–benzene complex. An unfortunate physical property of the benzene adduct, from the standpoint of characterization, was its low solubility in hydrocarbon and ethereal solvents; however, its formulation was confirmed by elemental analysis and X-ray crystallography (Fig. 14). The latter investigation indicated a roughly C_i symmetric core structure, with η^2-binding of the two trans Ta atoms in the 1,2–4,5-positions. Disorder in the central benzene fragment, however, prevented an alternative η^3, η^3 bis(allyl) formulation from being ruled out conclusively. The benzene adduct is diamagnetic (confirmed by

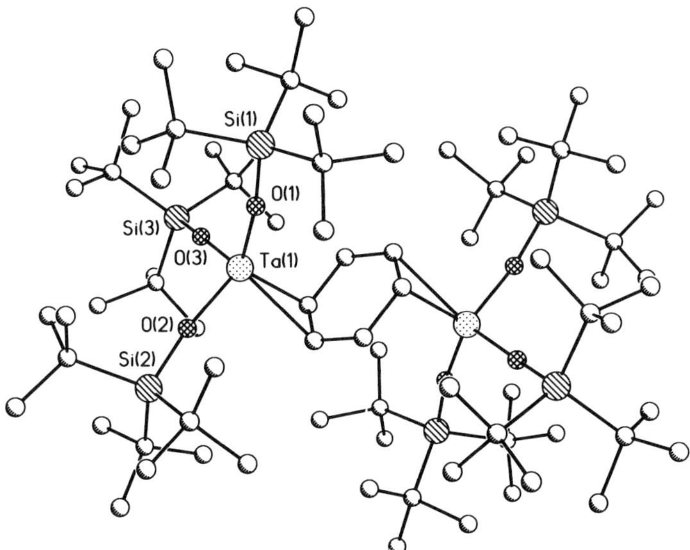

Figure 14. Structural drawing of $(\mu\text{-}C_6H_6)[Ta(OSi{-}t\text{-}Bu_3)_3]_2$.

magnetic measurements), an observation easy to rationalize for this low-symmetry, even-electron compound.

6. *Dihydrogen Oxidative Addition*

A potent reductant, d^2 $Ta(OSi{-}t\text{-}Bu_3)_3$ is converted readily to dihydride $H_2Ta(OSi{-}t\text{-}Bu_3)_3$ upon exposure to dihydrogen (93). The tantalum(V) dihydride had been prepared previously by reduction of dichloride $Cl_2Ta(OSi{-}t\text{-}Bu_3)_3$ under a dihydrogen atmosphere (96).

7. *Reaction with Arylamines: C—N Bond Oxidative Addition*

Reactions of $Ta(OSi{-}t\text{-}Bu_3)_3$ with a variety of substituted anilines were recently described (97) (Scheme 29). A key observation is that of competitive C—N and N—H oxidative additions to the d^2 tantalum center. The product of one such C—N oxidative addition, $Ta(NH_2)(4\text{-}C_6H_4CF_3)(OSi{-}t\text{-}Bu_3)_3$, was the subject of an accompanying X-ray diffraction study; the compound displays a slightly distorted square pyramidal geometry at Ta, with one of the bulky siloxide ligands in the apical position.

Substituent effects on the N—H versus C—N oxidative additions illuminated mechanistic features of these energetically competitive processes. Electron-donating substituents, which increase the basicity of the aniline nitrogen atom, led to a preference for N—H oxidative addition over C—N bond cleavage. Thus it was inferred that binding of the nitrogen lone pair to an empty metal-based orbital precedes N—H bond cleavage. On the other hand, anilines substituted with electron-withdrawing groups led to preferential C—N oxidative addition, a result interpreted as meaning that attack of the filled d_{z^2} orbital of $Ta(OSi{-}t\text{-}Bu_3)_3$ on the aniline ipso carbon is a governing interaction in this case. The C—N bond oxidative addition in this instance therefore is reminiscent of nucleophilic aromatic substitution.

8. *Imido, Phosphinidene, and Arsinidene Complexes*

In a contribution to the chemistry of metal–element multiply bonded comounds, Wolczanski and co-workers (98) reported the formation of tantalum-E (E = N, P, and As) multiple bonds by dihydrogen elimination upon reaction of H_2EPh with $Ta(OSi{-}t\text{-}Bu_3)_3$. The reactions were found to proceed by way of intermediates representing H-EHPh oxidative addition products. For example, treatment of $Ta(OSi{-}t\text{-}Bu_3)_3$ with phenylphosphine at 25°C led to formation of $HTa(P[H]Ph)(OSi{-}t\text{-}Bu_3)_3$, a compound exhibiting a characteristic

x $t\text{-}Bu_3SiO\text{—}Ta(Ar)(NH_2)(OSi\text{-}t\text{-}Bu_3)_2$

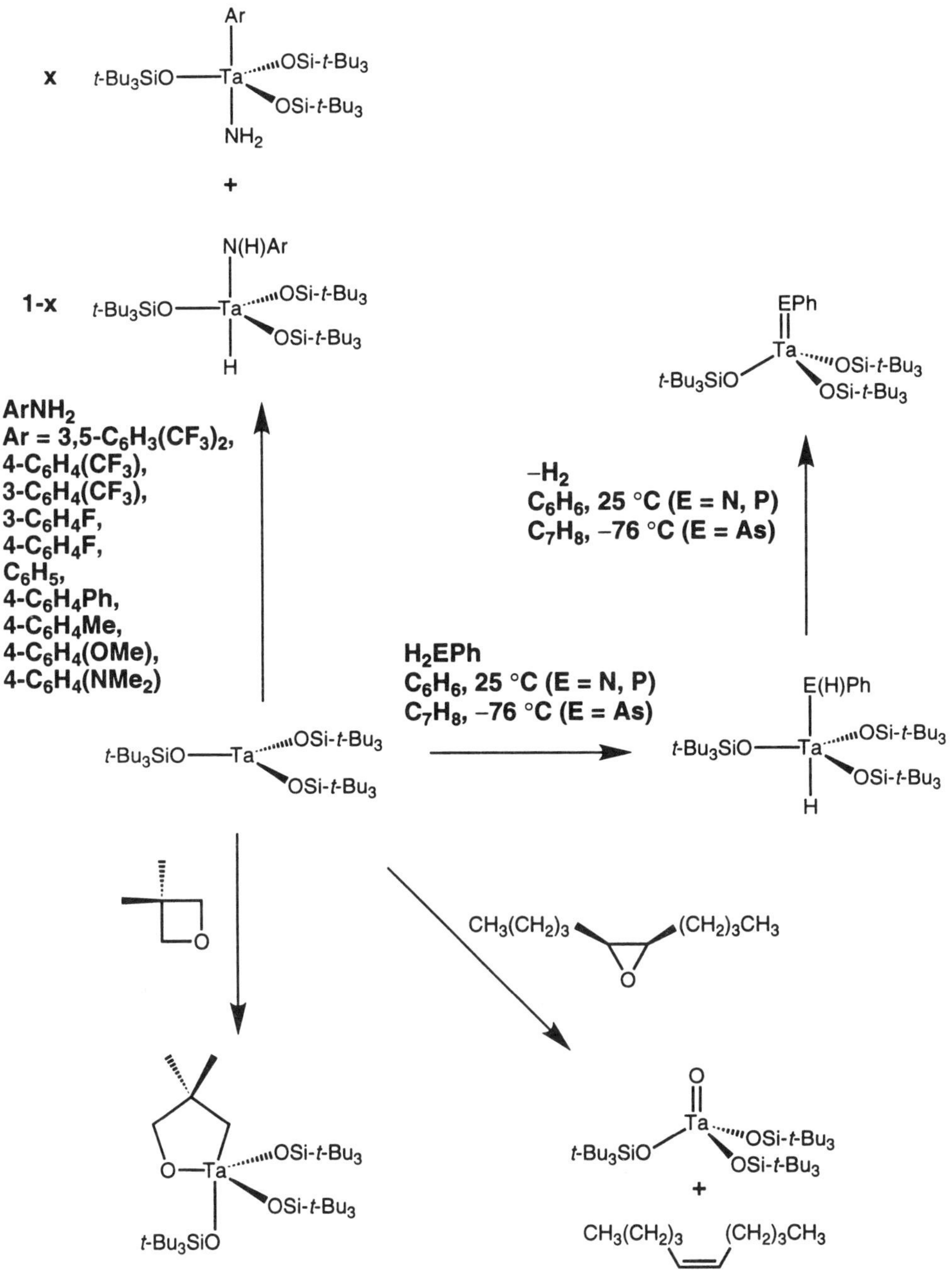

+

1-x $t\text{-}Bu_3SiO\text{—}Ta(N(H)Ar)(H)(OSi\text{-}t\text{-}Bu_3)_2$

$ArNH_2$
Ar = 3,5-$C_6H_3(CF_3)_2$, 4-$C_6H_4(CF_3)$, 3-$C_6H_4(CF_3)$, 3-C_6H_4F, 4-C_6H_4F, C_6H_5, 4-C_6H_4Ph, 4-C_6H_4Me, 4-$C_6H_4(OMe)$, 4-$C_6H_4(NMe_2)$

$t\text{-}Bu_3SiO\text{—}Ta(OSi\text{-}t\text{-}Bu_3)_2$

H_2EPh
C_6H_6, 25 °C (E = N, P)
C_7H_8, –76 °C (E = As)

$t\text{-}Bu_3SiO\text{—}Ta(E(H)Ph)(H)(OSi\text{-}t\text{-}Bu_3)_2$

$-H_2$
C_6H_6, 25 °C (E = N, P)
C_7H_8, –76 °C (E = As)

$t\text{-}Bu_3SiO\text{—}Ta(=EPh)(OSi\text{-}t\text{-}Bu_3)_2$

$CH_3(CH_2)_3$ — epoxide (O) — $(CH_2)_3CH_3$

$t\text{-}Bu_3SiO\text{—}Ta(=O)(OSi\text{-}t\text{-}Bu_3)_2$

+

$CH_3(CH_2)_3$ $(CH_2)_3CH_3$ (cis-alkene)

O (3,3-dimethyloxetane)

O—Ta metallacycle with $OSi\text{-}t\text{-}Bu_3$, $OSi\text{-}t\text{-}Bu_3$, $t\text{-}Bu_3SiO$

Scheme 29

^{1}H NMR signal (δ 21.61, $^2J_{PH}$ = 56, $^3J_{HH}$ = 9 Hz) ascribed to the terminal hydrido functionality. None of the HTa(E[H]Ph)(OSi—t-Bu_3)$_3$ (E = N, P, As) complexes was thermally stable, with the arsenic derivative exhibiting the greatest thermal lability. The first-order loss of dihydrogen from the pnictide hydride species to give corresponding Ta(EPh)(OSi—t-Bu_3)$_3$ complexes occurred with the following relative rates: N(15), P(1), and As(2.6 $\times$ 10^7).

X-ray diffraction studies revealed an acute angle [110.2(4)°] at phosphorus for the phosphinidene complex Ta(PPh)(OSi—t-Bu_3)$_3$ and a corresponding bend at arsenic [107.2(4)°] in the arsinidene derivative (Fig. 15). In both complexes, tantalum exhibits a pseudo-T_d geometry. Both the propensity of P and As to form bonds rich in p character and tantalum–oxygen π bonding were invoked to explain the acute angles observed for the phosphinidene and arsinidene linkages. The result is effectively a tantalum–E double bond, with the "lone pair" at E being rich in s character. Tantalum–element bond lengths of 2.317(4) and 2.428(2) Å were observed to P and As, respectively. A ^{31}P NMR signal (δ 334.6) was located for the red phosphinidene complex Ta(PPh)(OSi—t-Bu_3)$_3$, the upfield nature of which was argued to be an indicator of a relatively strong interaction by comparison with values for other bent phosphinidene complexes (99–102). Known linear phosphinidene complexes (103–105) may manifest, effectively, a metal–phosphorus triple bond.

Dihydrogen elimination leading to metal–carbon triple bonds has been ob-

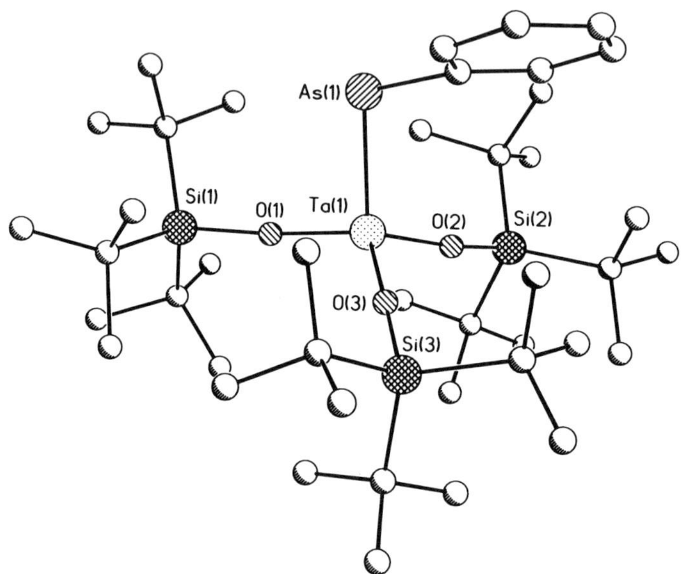

Figure 15. Structural drawing of (PhAs)Ta(OSi—t-Bu_3)$_3$.

served recently (106), a reaction exhibiting formal parallels to the imide, phosphinidene, and arsinidene complex syntheses synopsized here.

9. *Reaction with Epoxides*

Cis- and *trans*-decene oxides were deoxygenated by $Ta(OSi{-}t\text{-}Bu_3)_3$ to give $OTa(OSi{-}t\text{-}Bu_3)_3$ and, respectively, the corresponding (Z) and (E) olefins (15, 97). The stereospecific nature of these deoxygenations is consistent with a concerted mechanism. Several related epoxide deoxygenations by reducing metal complexes have been reported (107, 108).

10. *Reaction with 3,3-Dimethyloxetane*

Treatment of $Ta(OSi{-}t\text{-}Bu_3)_3$ with 3,3-dimethyloxetane resulted in rapid (<5 min, 25°C, hexane) oxidative addition of a carbon–oxygen bond, giving a five-membered oxatantalacycle (97). No extrusion of $OTa(OSi{-}t\text{-}Bu_3)_3$ was observed upon heating the 3,3-dimethyloxetane oxidative addition product (100°C, weeks).

11. *Carbon Monoxide Cleavage*

The original communication (93) describing CO cleavage by $Ta(OSi{-}t\text{-}Bu_3)_3$ reported as cleavage products the mononuclear oxo species $OTa(OSi{-}t\text{-}Bu_3)_3$ and the paramagnetic ($\mu = 3.0\ \mu_B$) dinuclear carbide complex $(\mu\text{-}C_2)[Ta(OSi{-}t\text{-}Bu_3)_3]_2$ (109) (Scheme 30). The latter was characterized structurally by X-ray diffraction (Fig. 16), from which study metrical parameters in the essentially linear TaCCTa moiety were consistent with a valence bond description involving tantalum–carbon and carbon–carbon double bonds. Accordingly, treatment of a benzene solution of $Ta(OSi{-}t\text{-}Bu_3)_3$ with 1.00 equiv of CO at 25°C resulted in the uptake of 0.47 CO (Toepler pump). Oxo $OTa(OSi{-}t\text{-}Bu_3)_3$ was isolated in 47% yield (based on tantalum), while the relatively insoluble dicarbide $(\mu\text{-}C_2)[Ta(OSi{-}t\text{-}Bu_3)_2]_2$ was isolated in 46% yield (again based on Ta). The identity of a 709 cm^{-1} IR band was probed though experiments involving treatment of $Ta(OSi{-}t\text{-}Bu_3)_3$ with a $^{12}CO/^{13}CO$ mixture; the dicarbide sample thereby obtained exhibited IR bands at 709, 695, and 682 cm^{-1} in an approximate 1 : 2 : 1 ratio.

In a 1989 full paper (15), it was shown that the CO cleavage products are different when $Ta(OSi{-}t\text{-}Bu_3)_3$ is treated with CO at low temperature (10 equiv CO, toluene, $-78 \rightarrow 25$°C). Under these conditions oxo $OTa(OSi{-}t\text{-}Bu_3)_3$ is produced along with an equimolar quantity of the ketenylidene complex $(OCC)Ta(OSi{-}t\text{-}Bu_3)_3$. It was noted that "color changes and the formation and dissolution of precipitates observed during the course of ketenylidene formation

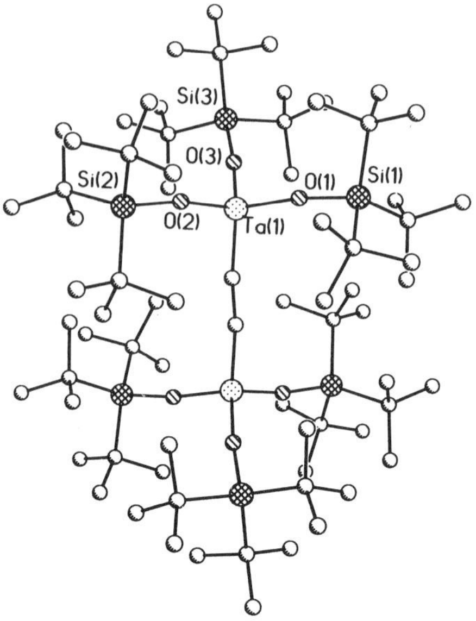

Figure 16. Structural drawing of (μ-C_2)-[Ta(OSi—t-Bu_3)$_3$]$_2$.

were indicative of the complexity of this reaction'' (15). Characterization of ketenylidene (OCC)Ta(OSi—t-Bu_3)$_3$ included ^{13}C NMR experiments, which showed two intense doublets at δ 135.96 and 142.52 ($^1J_{CC}$ = 100 Hz) for ($O^{13}C^{13}C$)Ta(OSi—t-Bu_3)$_3$, prepared using ^{13}CO. A strong IR band at 2076 cm^{-1} was attributed to the CO stretching mode in the ketenylidene complex.

Extensive mechanistic studies were carried out pertaining to CO cleavage by Ta(OSi—t-Bu_3)$_3$; just a few of the salient results are abstracted here. For more in-depth information the reader is directed to the full paper (15). Carbonylation of Ta(OSi—t-Bu_3)$_3$ at low temperature (−78°C) in THF allowed observation by low-temperature NMR spectroscopy of an adduct formulated as paramagnetic (OC)(thf)Ta(OSi—t-Bu_3)$_3$. Ketenylide(OCC)Ta(OSi—t-Bu_3)$_3$ is thought to be an intermediate in the formation of dicarbide (μ-C_2)[Ta(OSi—t-Bu_3)$_3$]$_2$ in the following way: deoxygenation of the ketenylidene complex by Ta(OSi—t-Bu_3)$_3$ leads to OTa(OSi—t-Bu_3)$_3$ along with a transient linear vinylidene complex [(CCTa(OSi—t-Bu_3)$_3$], which electronically resembles CO and trapping of which by Ta(OSi—t-Bu_3)$_3$ produces (μ-C_2)[Ta(OSi—t-Bu_3)$_3$]$_2$. The most plausible manner in which to account for formation of the ketenylidene complex itself is via an unsymmetrical dimerization of the monocarbonyl species [(OC)Ta(OSi—t-Bu_3)$_3$]. The proposed unsymmetrical dimer features a TaOCC four-membered ring that can break up in a productive metathetical process leading directly to the oxo and ketenylidene complexes. A red precipitate observed

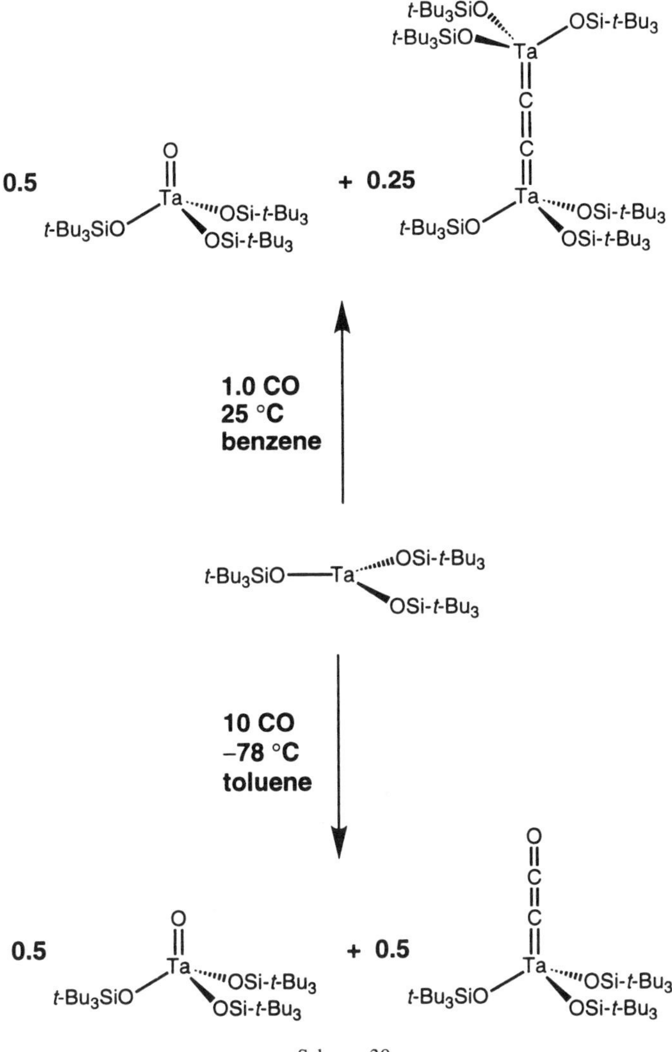

Scheme 30

during the low-temperature carbonylation of $Ta(OSi{-}t\text{-}Bu_3)_3$ carried out in toluene is thought to be $[(OC)Ta(OSi{-}t\text{-}Bu_3)_3]_n$.

12. Reaction with Dioxygen

Oxygenation of $Ta(OSi{-}t\text{-}Bu_3)_3$ (~0.4 mmol) as a benzene solution at 25°C with O_2 (excess, 1 atm) gave a color change from pale blue to pale orange.

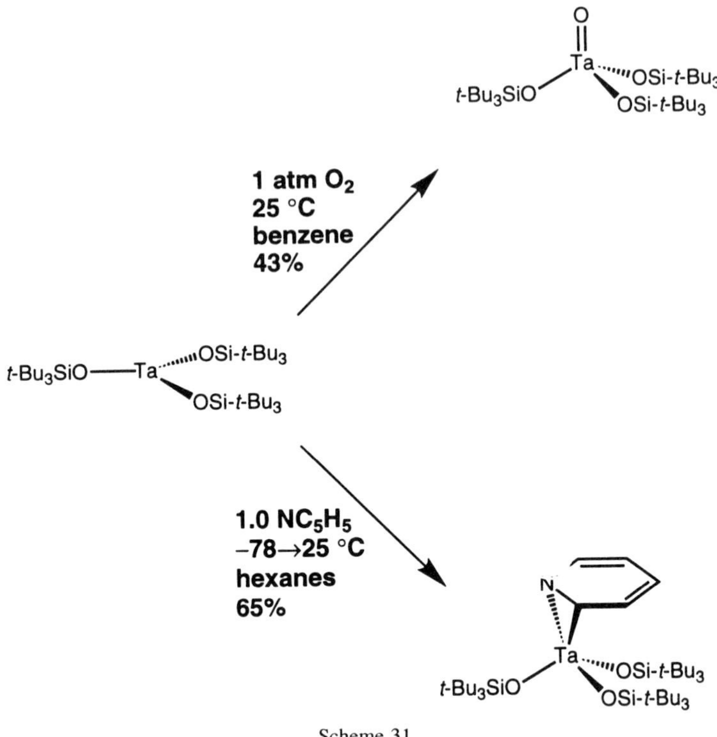

Scheme 31

Isolated in 43% yield by crystallization from Et_2O was the diamagnetic off-white oxo complex $OTa(OSi{-}t\text{-}Bu_3)_3$ (Scheme 31) (15, 93).

N. (thf)$V(OSi{-}t\text{-}Bu_3)_3$

1. *Synthesis and Characterization*

Blue crystalline (thf)$V(OSi{-}t\text{-}Bu_3)_3$ was obtained in 51% yield subsequent to treatment of vanadium trichloride with $Na(OSi{-}t\text{-}Bu_3)$ in THF solvent (Scheme 32). Procedures including (a) recrystallization from hot toluene and (b) heating in vacuo (150°C, 48 h) did not appear to degrade (thf)$V(OSi{-}t\text{-}Bu_3)_3$ or result in loss of the thf ligand (19, 110).

2. *Reaction with* $OCCTa(OSi{-}t\text{-}Bu_3)_3$

Treatment of (thf)$V(OSi{-}t\text{-}Bu_3)_3$ with a mixture of $OCCTa(OSi{-}t\text{-}Bu_3)_3$ and $OTa(OSi{-}t\text{-}Bu_3)_3$ (see above) in hexanes resulted in precipitation of the

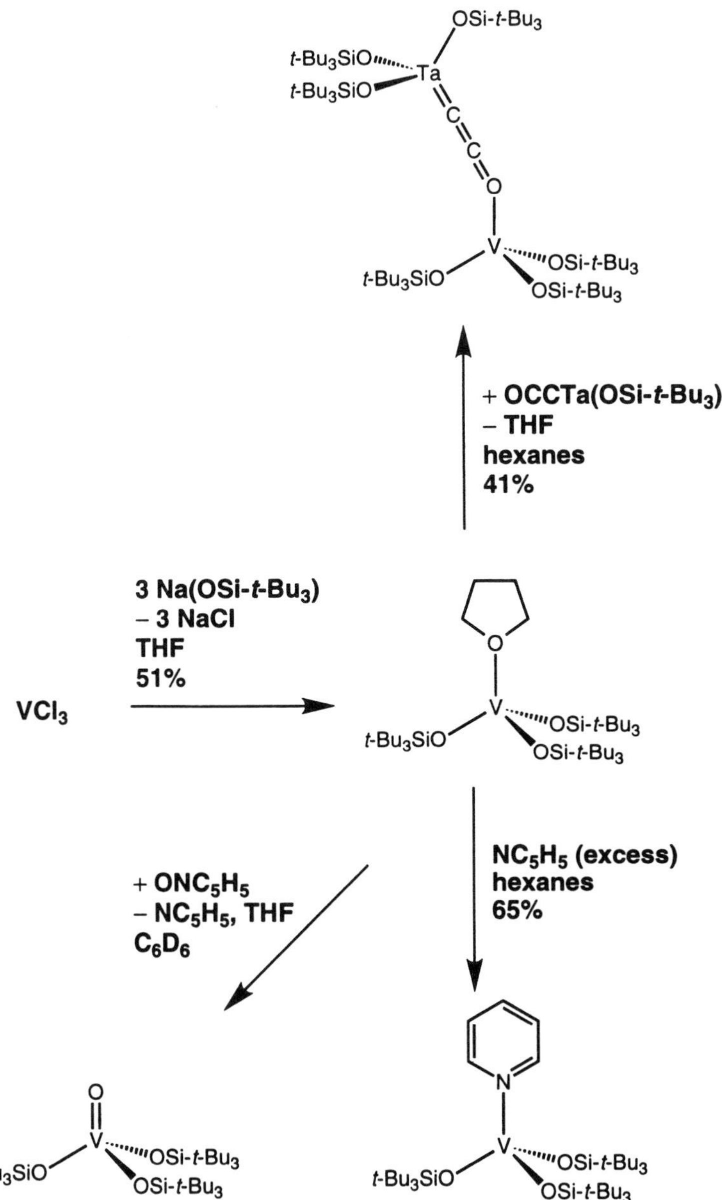
OSi-t-Bu3
t-Bu3SiO
t-Bu3SiO
Ta
C
C
O
V
t-Bu3SiO
OSi-t-Bu3
OSi-t-Bu3
+ OCCTa(OSi-t-Bu3)
− THF
hexanes
41%
3 Na(OSi-t-Bu3)
− 3 NaCl
THF
51%
VCl3
O
V
t-Bu3SiO
OSi-t-Bu3
OSi-t-Bu3
+ ONC5H5
− NC5H5, THF
C6D6
NC5H5 (excess)
hexanes
65%
O
V
t-Bu3SiO
OSi-t-Bu3
OSi-t-Bu3
N
V
t-Bu3SiO
OSi-t-Bu3
OSi-t-Bu3

Scheme 32

adduct $(t\text{-Bu}_3SiO)_3VOCCTa(OSi{-}t\text{-Bu}_3)_3$, which was subsequently isolated in 41% yield as emerald green crystals. The low solubility of the dinuclear Ta/V complex precluded measurement of its NMR spectra; this complex was characterized by elemental analysis, IR spectroscopy [ν(CCO) = 2059 cm^{-1}], and X-ray crystallography.

3. Reaction with Pyridine-N-oxide

The experiment was carried out in an NMR tube in C_6D_6 solvent. The blue color due to $(thf)V(OSi{-}t\text{-Bu}_3)_3$ gave way to green rapidly upon mixing with ONC_5H_5 at 25°C. One equivalent of thf was liberated, and the vanadium product deemed to be threefold symmetric $OV(OSi{-}t\text{-Bu}_3)_3$ on the basis of its ^1H, ^{13}C, and ^{51}V NMR spectra (δ ^{51}V = −733.7 ppm).

4. Reaction with Pyridine

The complex $(thf)V(OSi{-}t\text{-Bu}_3)_3$ was treated with pyridine (excess) in hexanes at −78°C. The solution adopted a green color upon warming to 25°C. The complex $(C_5H_5N)V(Osi{-}t\text{-Bu}_3)_3$ was isolated subsequently in 65% yield as green crystals (1-mmol scale). A broad *t*-Bu signal was observed in the ^1H NMR spectrum of the paramagnetic complex; resonances due to the pyridine ligand were absent from the spectrum. The substance was characterized additionally by elemental analysis and UV–vis spectroscopy.

O. $V(CH_2{-}t\text{-Bu}_3)_3$

1. Reaction of $VCl_3(thf)_3$ with Neopentyllithium

The $VCl_3(thf)_3$ complex (111) was treated with 3 equiv $LiCH_2{-}t\text{-Bu}$ in diethyl ether. Subsequent to the reaction, $(\mu\text{-}N_2)[V(CH_2{-}t\text{-Bu})_3]_2$ was obtained in 45% yield as red-brown crystalline material (112) (Scheme 33). The complex was characterized structurally in an X-ray diffraction study [*d*(NN) = 1.250(3) Å], revealing a linear VNNV core. That $(\mu\text{-}N_2)[V(CH_2{-}t\text{-Bu})_3]_2$ can be considered a source of $V(CH_2{-}t\text{-Bu})_3$ is indicated by several reactions of the dinuclear species in which dinitrogen is liberated.

2. Reaction of $(\mu\text{-}N_2)[V(CH_2{-}t\text{-Bu})_3]_2$ with Lewis Bases

Evidence for quantitative N_2 dissociation upon dissolution of $(\mu\text{-}N_2)[V(CH_2{-}t\text{-Bu})_3]_2$ in THF was obtained via Toepler pump. The presumed vanadium product is $(thf)V(CH_2{-}t\text{-Bu})_3$, although this species appears not yet to have been characterized fully. Other adducts $(L)V(CH_2{-}t\text{-Bu})_3$ (L = PMe_3, py, *t*-BuCN) were prepared similarly.

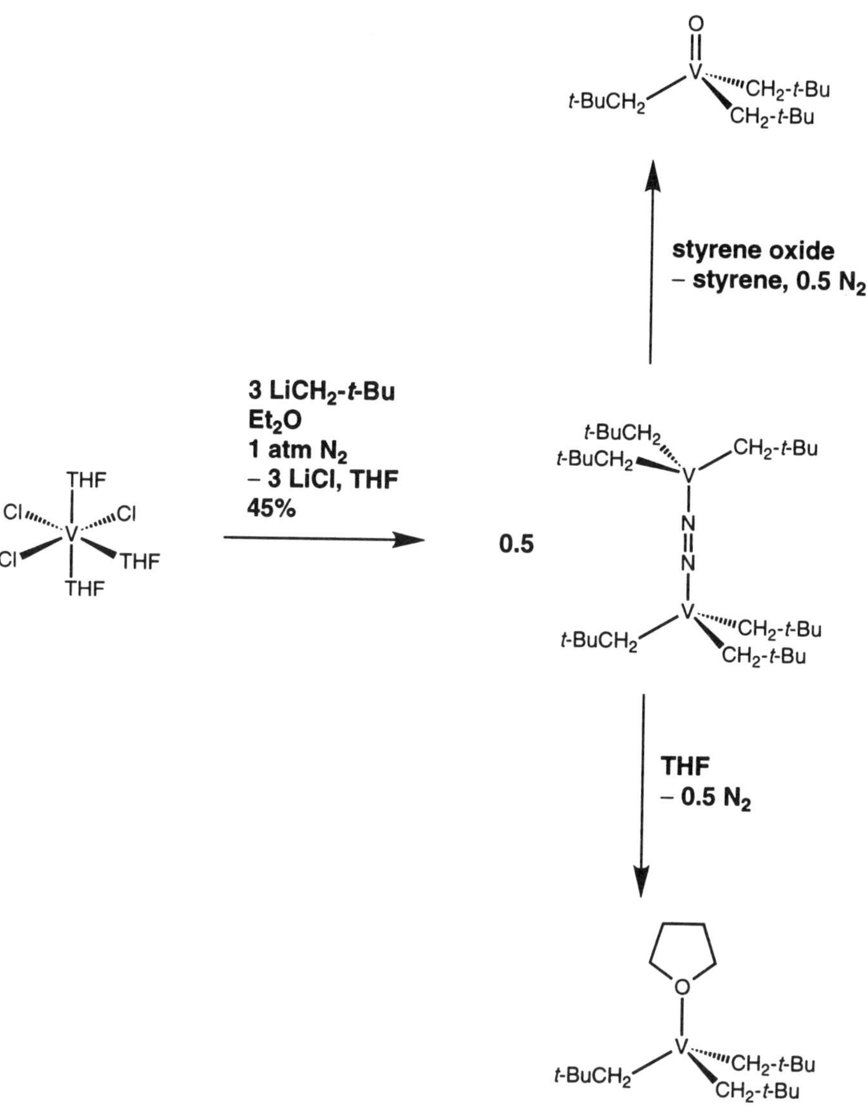

Scheme 33

3. *Reaction of* $(\mu\text{-}N_2)[V(CH_2{-}t\text{-}Bu)_3]_2$ *with Styrene Oxide*

Deoxygenation of styrene oxide to produce $OV(CH_2{-}t\text{-}Bu)_3$ (δ ^{51}V = 1212 ppm) and styrene was observed upon treatment of 0.5 equiv $(\mu\text{-}N_2)[V(CH_2{-}t\text{-}Bu)_3]_2$ with styrene oxide. The low-field ^{51}V NMR signal for $OV(CH_2{-}t\text{-}Bu)_3$ is deemed indicative of electronic unsaturation at the vanadium center (113).

VI. GROUP 6 (VIB) COMPLEXES

A. $Cr[CH(SiMe_3)_2]_3$

1. Synthesis and Characterization

Lappert and co-workers (24, 25) described homoleptic MR_3 (M = Ti, V, Cr) complexes. Bright green $Cr[CH(SiMe_3)_2]_3$ was obtained in 71% yield via treatment at 0°C of $CrCl_3$ (1.5 mmol) in ether with 3 equiv of the lithium alkyl, $LiCH(SiMe_3)_2$ (Scheme 34). Benzene cryoscopy showed $Cr[CH(SiMe_3)_2]_3$ to be monomeric in that solvent. Infrared bands assigned to chromium–carbon stretches were observed at 403 and 449 cm^{-1}. The ESR spectroscopic measurements yielded data consistent with trigonal planar chromium(III), namely, g values at about 2 and about 4 in frozen toluene. Analytical data also were consistent with the formulation given for the complex.

An X-ray diffraction study revealed a slightly pyramidal C_3-symmetric struc-

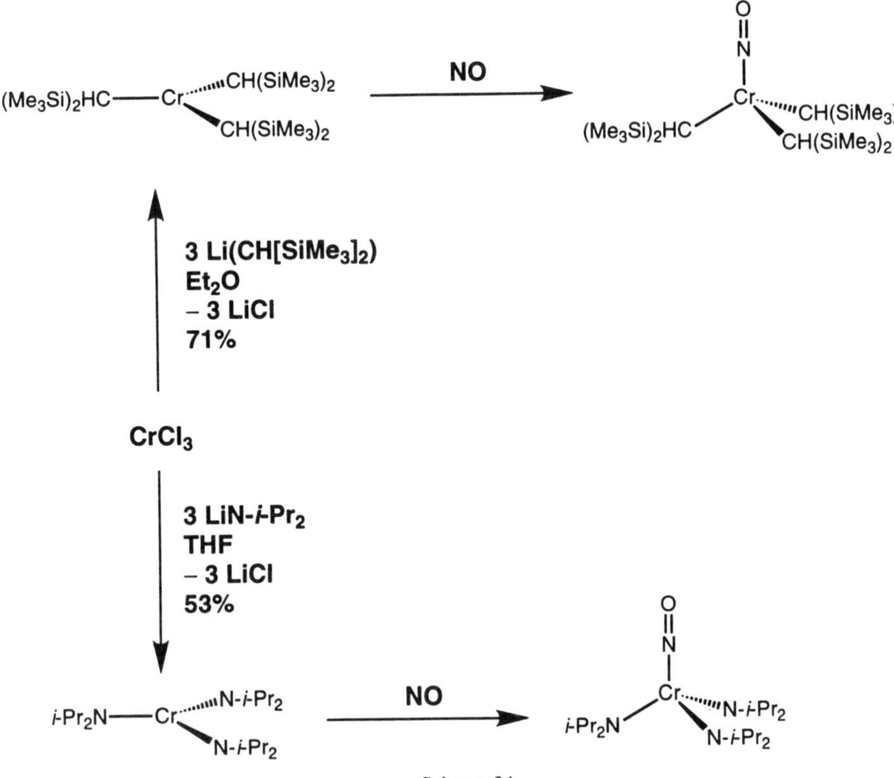

Scheme 34

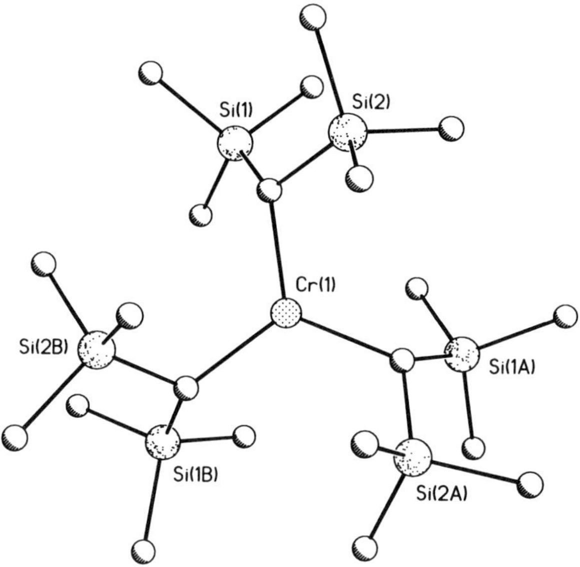

Figure 17. Structural drawing of $Cr[CH(SiMe_3)_2]_3$.

ture for $Cr[CH(SiMe_3)_2]_3$ (Fig. 17); the chromium atom displacement from the plane defined by the three ligating carbon atoms is 0.32 Å (25). The pyramidal nature of $Cr[CH(SiMe_3)_2]_3$ may allow for agostic interactions as observed for $U[CH(SiMe_3)_2]_3$ (see below). The Cr—C bond lengths are 2.07(1) Å.

2. *General Comments on Reactivity*

It was noted that the hindering alkyl ligands in $Cr[CH(SiMe_3)_2]_3$ appear to diminish the reactivity of this species beyond expectations. No reaction was observed with pyridine, hexamethyldisilazane, or carbon monoxide. In fact, such behavior is typical of known three-coordinate chromium(III) complexes. That access to Cr is not completely restricted in $Cr[CH(SiMe_3)_2]_3$ is evidenced by this complex's reaction with nitric oxide leading to diamagnetic $(ON)Cr[CH(SiMe_3)_2]_3$ (ν(NO) = 1672 cm^{-1}], a result presaged by similar experiments performed with homoleptic chromium(III) amides.

B. $Cr(N-i\text{-}Pr_2)_3$

1. *Synthesis and Characterization*

In a classic report (28) on three- and four-coordinate complexes of chromium(III), one finds instructions for the synthesis of the remarkable tris(diisopropylamide) $Cr(N-i\text{-}Pr_2)_3$ (Scheme 34). The compound was produced upon treatment of (anhydrous) chromium trichloride with lithium diisopropylamide in THF solvent. The $Cr(N-i\text{-}Pr)_3$ complex was obtained as a red-brown solid. An isolated yield of 53% has been reported for the preparation of $Cr(N-i\text{-}Pr)_3$ by this procedure (73).

In the visible region, $Cr(N-i\text{-}Pr_2)_3$ exhibits two absorption bands, at 6900 (br, $\epsilon_{max} \sim 170$) and 23,200 cm^{-1} (sh, $\epsilon_{max} \sim 2800$), values incompatible with tetrahedral or octahedral chromium(III). Cryoscopic molecular weight determinations confirmed the monomeric formulation for the complex, and solid-state magnetic susceptibility measurements gave $\mu_{eff} = 3.80\ \mu_B$. It was remarked that, in addition to the steric protection afforded by the bulky diisopropylamide substituents, the ligand-field stabilization energy of the d^3 configuration in a trigonal field may well be a significant factor contributing to the stability of $Cr(N-i\text{-}Pr_2)_3$ (28, 114).

The X-ray structure of $Cr(N-i\text{-}Pr_2)_3$ was reported by Bradley and co-workers (114) in 1971. Of interest to these workers was the observation of planarity in this compound at nitrogen because of the possible implications with regard to $Cr-N$ π bonding. Noteworthy in this respect is the subsequent observation that the *Angewandte Chemie* cover molecule $N-i\text{-}Pr_3$ is planar at nitrogen (115). The CrN_3 unit in $Cr(N-i\text{-}Pr_2)_3$ was shown to be planar within the limits of experimental error.

Bradley and co-workers (114) found that while $Cr(N-i\text{-}Pr_2)_3$ does not exhibit an EPR signal at room temperature, the compound does so in frozen solutions at low temperature with $g(\|) = 2.0$ and $g(\perp) = 4.0$. A more detailed analysis of the EPR spectrum of $Cr(N-i\text{-}Pr_2)_3$ was given by Chien and Kruse (26) in 1970. Contradicting the report by Bradley and co-workers (114), these authors find an EPR signal ($g = 1.975$, line width 10.8 G) for $Cr(N-i\text{-}Pr_2)_3$ in solution at 25°C. The principal values were found to be $g(\perp) = 1.982$ and $g(\|) = 1.966$ at -195°C. The parallel and perpendicular transitions were assigned as $(-\frac{1}{2} \leftrightarrow -\frac{3}{2})$ and $(\frac{1}{2} \leftrightarrow \frac{3}{2})$ transitions, respectively. A ligand-field approach has been used to predict EPR properties of various three-coordinate first-row transition metal complexes (17).

2. *Reaction with Nitric Oxide*

In a classic 1970 paper, Bradley and Newing (116) described the preparation of diamagnetic, red-brown $Cr(NO)(N-i\text{-}Pr_2)_3$ by direct addition of gaseous nitric oxide to three-coordinate $Cr(N-i\text{-}Pr_2)_3$ (Scheme 34). The conversion of $Cr[N(SiMe_3)_2]_3$ to $Cr(NO)[N(SiMe_3)_2]_3$ was reported in the same study, which includes an X-ray study of the latter complex (116).

From the viewpoint of electronic structure $Cr(NO)(N-i\text{-}Pr_2)_3$, with its presumed linear CrNO moiety, may be thought of (see Section III above) as containing NO^+ coordinated to chromium(II) configured as follows: $(d_{z^2}, a, \sigma^*)^0$, $([d_{x^2-y^2}, d_{xy}], e, \sigma^*)^0$, $([d_{xz}, d_{yz}], e, \pi)^4$ (116). The π bonding in the CrNO moiety then comprises $8e^-$ in two doubly degenerate levels, making the fragment isolobal (22) with nitrous oxide. This bonding picture is consistent with the observation of a low NO stretching frequency for $Cr(NO)(N-i\text{-}Pr_2)_3$: 1641 cm^{-1}.

Chemical characterization of $Cr(NO)(N-i\text{-}Pr_2)_3$ was provided through treatment with alcohols *t*-BuOH and *i*-PrOH. The CrNO moiety was found to remain intact under such conditions, with the result that new $Cr(NO)L_3$ complexes, where L = *t*-BuO or *i*-PrO, were obtained smoothly as a consequence of diisopropylamine elimination (116). The mixed-ligand complex $(ON)Cr(N-i\text{-}Pr_2)(O-t\text{-}Bu)_2$ was also obtained via dialkylamide protonolysis (116).

3. *Reaction with Dioxygen*

In the same year as the nitric oxide chemistry described above was pursued, Bradley and co-workers (117) also investigated reactions of $Cr(N-i\text{-}Pr_2)_3$ with O_2. The products obtained seemed a function of the conditions employed. Addition of small amounts of oxygen at about −90°C to dilute heptane or toluene solutions of $Cr(N-i\text{-}Pr_2)_3$ elicited disappearance of the g = 1.975 ESR signal (A). A new signal (B) appeared concomitantly, only to give way to a third signal (C). Species B exhibited coupling to one nitrogen nucleus and two equivalent protons; species B was tentatively assigned as the chromium(IV) nitroxide complex $Cr(O)(N-i\text{-}Pr_2)_2(ON-i\text{-}Pr_2)$. Species C gave a clearly resolved ESR spectrum indicative of free diisopropylnitroxide (g = 2.0026, a_N = 15.0, a_H = 3.9 G); the ESR data for diisopropylnitroxide have also been given as g = 2.0055, a_N = 17.0, a_H = 4.75 G, representing fairly good agreement (118). Whereas species B was unstable even at −90°C, species C was reasonably long lived even at room temperature. A further observation is that species C was obtained directly upon addition of an excess of oxygen directly to $Cr(N-i\text{-}Pr_2)_3$, presumably at low temperature. Conversely, addition of excess oxygen to dilute solutions of $Cr(N-i\text{-}Pr_2)_3$ at room temperature caused disappearance of the complex without the appearance of B or C.

The following surprising observation was also made: When oxygen was added to more concentrated solutions (~ 1 *M*) of $Cr(N-i\text{-}Pr_2)_3$ in pentane or toluene at 0°C, the ESR signal due to $Cr(N-i\text{-}Pr_2)_3$ (g = 1.977) did not disappear but was considerably enhanced in intensity without changing its g value. Furthermore, it was possible upon treatment of pentane solutions of $Cr(N-i\text{-}Pr_2)_3$ with oxygen at −10 to +5°C to isolate a very volatile, unstable, blue crystalline compound analyzing satisfactorily for $Cr(O_2)(N-i\text{-}Pr_2)_3$ or $Cr(O)(N-i\text{-}Pr_2)_2(ON-i\text{-}Pr_2)$ (analytical data not provided). The unknown blue species exhibited a temperature and magnetic field dependent magnetic moment, with values of μ_{eff} = 2.03 and 1.58 μ_B observed at 0 and −175°C, respectively. A strong IR band at 980 cm^{-1} was assigned tentatively to ν(CrO) for the unknown substance.

The rich chemistry described in this section warrants more attention. Other chromium(III) three-coordinate amido complexes are known to provide diamagnetic $Cr(O)_2(NR_2)_2$ species in high yield upon treatment with O_2 (see below). Such reactions are reminiscent of the reaction of $Mo_2(O—t\text{-}Bu)_6$ with O_2, which gives $Mo(O)_2(O—t\text{-}Bu)_2$ (51, 52).

C. $Cr(N[SiMe_3]_2)_3$

1. *Synthesis and Characterization*

This compound (119) has been known for over 20 years, having been synthesized originally by Bürger and Wannagat (120). For older literature pertaining to this species, the reader is referred to the review articles mentioned in Section 1. The $Cr[N(SiMe_3)_2]_3$ complex has been characterized structurally by X-ray diffraction; reference to its structure is made in a recent article that reviews the structures of various $M[N(SiMe_3)_2]_3$ species.

A recent paper gave a new synthesis of $Cr[N(SiMe_3)_2]_3$ and an improved X-ray structure determination (119). The complex of chromium trichloride with *N,N′,N″*-trimethyl-1,3,5-triazacyclohexane, abbreviated Me_3tac, was treated in petroleum ether with 2.8 equiv of $LiN(SiMe_3)_2$, giving a deep green solution and a nearly colorless precipitate. Subsequent to removal of salt and volatile matter, crude $Cr(N[SiMe_3]_2)_3$ was obtained in 97% yield and determined to be of high purity (Scheme 35). The Me_3tac was intact among the volatiles. An interesting point made by Köhn et al. (47) is that the reaction proceeds to completion in the presence of unreacted $(Me_3tac)CrCl_3$, due to the low solubility of the latter complex.

In the new X-ray structure determination carried out for $Cr[N(SiMe_3)_2]_3$, a key feature was cocrystallization of the compound with Si_2Me_6, the crystalli-

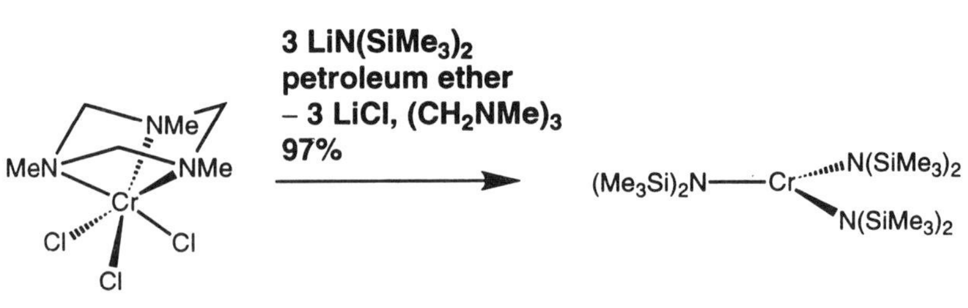

Scheme 35

zation solvent. It was asserted that D_3-symmetric $M[N(SiMe_3)_2]_3$ molecules tend to crystallize with severely disordered solvent molecules present. In the present study, the non-hydrogen atoms of the included Si_2Me_6 molecules were refined anisotropically, although two positions for Si_2Me_6 were observed in the hexagonal channels, both assigned half-occupancy.

The air-stable nature of $(Me_3tac)CrCl_3$ and ready displacement of the volatile Me_3tac moiety upon introduction of bulky hexamethyldisilazane ligands is taken as an indication that Me_3tac may prove useful in the preparation of presently unknown three-coordinate complexes.

D. $Cr[N(R)Ar]_3$ [$R = C(CD_3)_2Me$, $Ar = 3,5\text{-}C_6H_3Me_2$]

1. *Synthesis and Characterization*

Anhydrous $CrCl_3$ (~4 mmol) and solid $Li[N(R)Ar](OEt_2)$ (34) (~3 equiv) were added to thawing ether. The reaction mixture was warmed to 28°C with stirring, and it attained a brown color. After about 12 h of stirring at 28°C, work up gave a pentane-soluble black solid, which was purified by crystallization from a minimum of pentane at −35°C (76% yield) (Scheme 36). A characteristic ^2H NMR signal for the complex appears at δ 42.3 ppm in ethereal solvent. That $Cr(N[R]Ar)_3$ is a high-spin chromium(III) complex is evidenced by its effective magnetic moment (benzene-d_6, 25°C) of 3.87 μ_B. Other characterization data also were in agreement with formulation of this complex as a monomeric complex of high-spin chromium(III) (73).

The single-crystal X-ray structure determination carried out for $Cr[N(R)Ar]_3$ revealed (Fig. 18) that the complex is isostructural with $Mo[N(R)Ar]_3$ (see below). Both complexes crystallize in the space group $P\bar{1}$, with similar cell constants. The chromium–nitrogen bond lengths of 1.854(7), 1.875(7), and 1.864(7) Å are uniformly about 0.1 Å shorter than the molybdenum–nitrogen bond lengths in $Mo[N(R)Ar]_3$. In both of the crystallographically independent but chemically identical molecules of $Cr[N(R)Ar]_3$ in the unit cell, the sum of N—Cr—N bond angles is 359.4°, reflecting the regular trigonal planar geometry at chromium. The NC_2 amido planes are oriented nearly perpendicular to the CrN_3 plane, such that any metal–nitrogen π bonding is relegated largely to the xy plane, taking the z axis as normal to the CrN_3 plane. The same considerations apply to $Mo[N(R)Ar]_3$. Both complexes orient with the three aryl groups packed together on one side of the molecule, the *tert*-butyl groups thereby being relegated to the opposing hemisphere. For a fuller discussion of this phenomenon, see the section below devoted to $Mo[N(R)Ar]_3$. While it is noted that the observed structure is consistent with the solution NMR spectra for the complex, the available data do not rule out an unsymmetrical fluxional structure for $Cr[N(R)Ar]_3$ in solution.

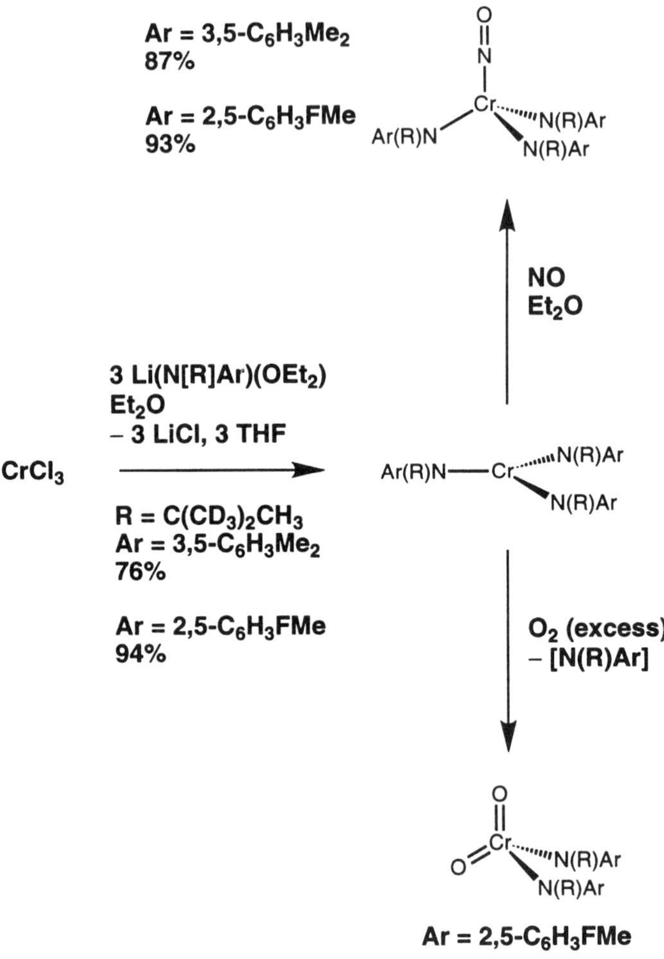

Scheme 36

The $Cr[N(R)Ar]_3$ complex has been found to be unreactive toward CO and N_2O under mild conditions (benzene, 1 atm CO or N_2O, 25°C).

2. *Reaction with Nitric Oxide*

An ethereal solution containing $Cr[N(R)Ar]_3$ (~2 mmol) was treated with gaseous nitric oxide, added via syringe. A rapid color change from dark brown to orange-red was noted. The $(ON)Cr[N(R)Ar]_3$ complex was subsequently obtained in 87.3% yield as orange-red crystals. A strong IR band at 1662 cm^{-1}

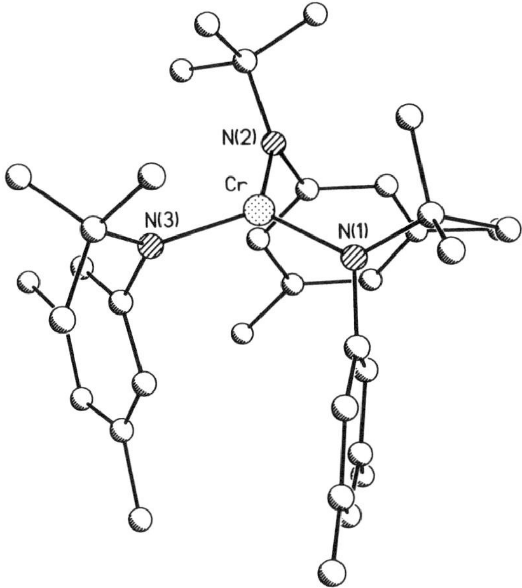

Figure 18. Structural drawing of $Cr[N(R)Ar]_3$; R = $C(CD_3)_2Me$, Ar = 3,5-$C_6H_3Me_2$.

was attributed to ν(NO). This value can be compared with that for $(ON)Mo[N(R)Ar]_3$: 1604 cm^{-1}. Diamagnetic $(ON)Cr[N(R)Ar]_3$ exhibited ^1H and ^{13}C NMR signals consistent with a threefold symmetric species, and other characterization data were consistent with the given formulation. Deoxygenation of $(ON)Cr[N(R)Ar]_3$ to give $NCr[N(R)Ar]_3$ is described in Section V.A devoted to (thf)$V(Mes)_3$.

E. $Cr[N(R)Ar_F]_3$ [R = $C(CD_3)_2Me$, Ar_F = 2,5-C_6H_3FMe]

1. Synthesis and Characterization

Anhydrous $CrCl_3$ (27.2 mmol) and solid $Li[N(R)Ar_F](OEt_2)$ (88) (~3 equiv) were added to thawing ether. The reaction mixture was warmed to 28°C with stirring, and it attained a brown color (Scheme 36). After about 12 h of stirring at 28°C, work-up gave a pentane-soluble black solid, which was purified by crystallization from a minimum of pentane at −35°C (94% yield) (73). A characteristic ^2H NMR signal for the complex appears at δ 44.2 ppm in ethereal solvent. That $Cr[N(R)Ar_F]_3$ is a high-spin chromium(III) complex is evidenced by its effective magnetic moment (benzene-d_6, 25°C) of 3.83 μ_B. Other characterization data also were in agreement with formulation of this complex as a monomeric complex of high-spin chromium(III). Although $Cr[N(R)Ar_F]_3$ gives large faceted black crystals, which are attractive to the eye, several X-ray diffraction experiments provided data sets from which a successful structure solution was not obtained. The same was true for the corresponding nitrido complex $NCr[N(R)Ar_F]_3$ (see below).

It has been determined that $Cr[N(R)Ar_F]_3$ is unreactive toward CO and N_2O under mild conditions (benzene, 1 atm CO or N_2O, 25°C).

2. *Reaction with Nitric Oxide*

An ethereal solution containing $Cr[N(R)Ar_F]_3$ (~0.5 mmol) was treated with gaseous nitric oxide, added via syringe. A rapid color change from dark brown to orange-red was noted. The $(ON)Cr[N(R)Ar_F]_3$ complex was subsequently obtained in 92.5% yield as orange-red crystals. A strong IR band at 1673 cm^{-1} was attributed to ν(NO). The $(O^{15}N)Cr[N(R)Ar_F]_3$ complex was prepared similarly using ^{15}NO and exhibited $\nu(^{15}NO)$ = 1637 cm^{-1}. Diamagnetic (ON)Cr-$[N(R)Ar_F]_3$ exhibited ^1H and ^{13}C NMR signals consistent with a threefold symmetric species, and other characterization data were consistent with the given formulation. Deoxygenation of $(ON)Cr[N(R)Ar_F]_3$ to give $NCr[N(R)Ar_F]_3$, and $(O^{15}N)Cr[N(R)Ar_F]_3$ to give $^{15}NCr[N(R)Ar_F]_3$, is described in Section V.A, which is devoted to $(thf)V(Mes)_3$.

3. *Reaction with Dioxygen*

Bradley and co-workers (117) noted that $Cr(N-i\text{-}Pr_2)_3$ reacts with O_2 (see above). Following up on their initial efforts, the reaction of $Cr[N(R)Ar_F]_3$ with O_2 was found to constitute an effective means for synthesizing $O_2Cr[N(R)Ar_F]_2$ in high yield (121). This is done by treating an ethereal solution of $Cr[N(R)Ar_F]_3$ with an excess of dry oxygen at about 25°C, the reaction being accompanied by a color change from dark brown to brilliant purple. The amido ligand lost in this reaction is presumably extruded as the aminyl radical $[N(R)Ar_F]$ at some point along the reaction's path; the reaction is likely to be mechanistically complex based on the early observations of Bradley and co-workers (see above) (117).

The structure of $O_2Cr[N(R)Ar_F]_2$ has been studied by X-ray diffraction (Fig. 19). The pseudo-C_2 molecular symmetry is broken by the orientation of the ortho fluorine substituents, one of which is proximal to the dioxo functionality and the other of which is distal thereto. A closely related compound, $O_2Cr[N(SiMe_3)_2]_2$, was prepared using an entirely different synthetic regimen (122).

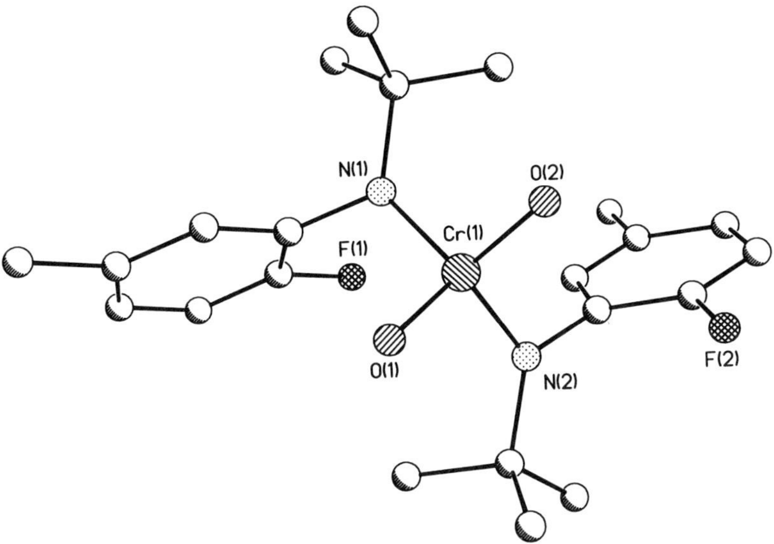

Figure 19. Structural drawing of $O_2Cr[N(R)Ar_F]_2$; R = $C(CD_3)_2Me$, Ar_F = 2,5-C_6H_3FMe.

F. $Cr\{OSiMe_2[C(SiMe_3)_3]\}_3$

1. *Synthesis and Characterization*

A 1988 report describes the use of a sterically demanding siloxide ligand referred to as "trisilox," $OSiMe_2[C(SiMe_3)_3]$, for the synthesis of a three-coordinate chromium(III) derivative (123). A THF solvate of the lithium salt of this alkoxide had been reported previously (124). Hitchcock et al. (124) attempted the preparation of putative $Cr(Cl)\{OSiMe_2[C(SiMe_3)_3]\}_2$ by treatment of $CrCl_3$ in THF with 2 equiv of $Li\{OSiMe_2[C(SiMe_3)_3]\}$ as a THF solution. The reaction produced a royal blue solution, concentration of which provided a green solid. The green solid was extracted into pentane, and the pentane extracts were filtered and concentrated to provide a pale blue solid. Recrystallization from concentrated THF solutions provided $Cr\{OSiMe_2$-$[C(SiMe_3)_3]\}_3$ as turquoise needles in 20% yield based on chromium (Scheme 37).

Pointing to a monomeric, three-coordinate formulation for $Cr\{OSiMe_2$-$[C(SiMe_3)_3]\}_3$ are an observed $\mu_{eff} = 3.6\ \mu_B$, EPR data as follows (frozen THF): $g(\perp) \sim 3.9$ and $g(\|) \sim 2.1$, and a preliminary X-ray structure determination.

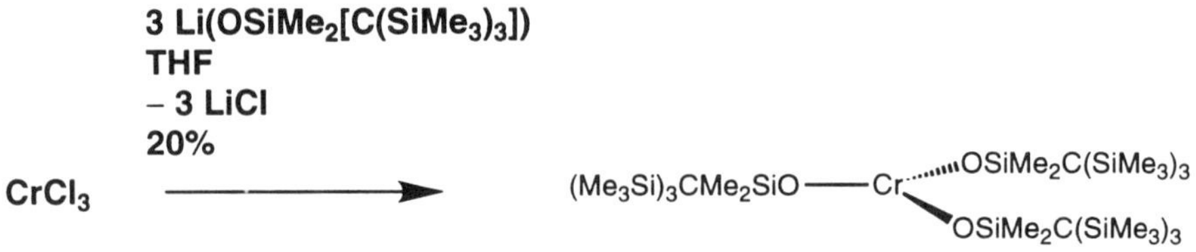

3 $Li(OSiMe_2[C(SiMe_3)_3])$
THF
− 3 LiCl
20%

$CrCl_3$ ⟶ $(Me_3Si)_3CMe_2SiO—Cr(OSiMe_2C(SiMe_3)_3)_2$

Scheme 37

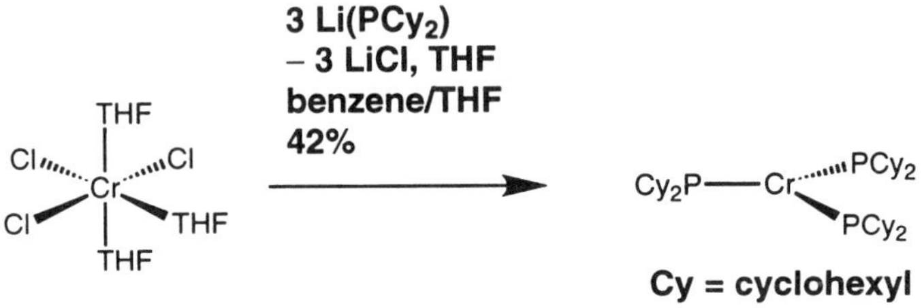

$CrCl_3(THF)_3$ —[3 $Li(PCy_2)$, − 3 LiCl, THF, benzene/THF, 42%]⟶ $Cy_2P—Cr(PCy_2)_2$

Cy = cyclohexyl

Scheme 38

G. $Cr(PCy_2)_3$

1. Synthesis and Characterization

The $Cr(PCy_2)_3$ (mp 160°C) complex was prepared in 42% yield from $CrCl_3(thf)_3$ and $Li(PCy_2)$ in a fashion similar to the vanadium analogue described above (Scheme 38) (92). Chromium and phosphorus analytical data were consistent with the proposed composition, and a cryoscopic molecular weight determination (benzene; found 634.0, calcd 643.8) was consistent with formulation of $Cr(PCy_2)_3$ as a monomer in solution. As was the case for $V(PCy_2)_3$, magnetic susceptibility measurements performed on $Cr(PCy_2)_3$ yielded an unexpectedly low moment: $\mu_{eff} = 2.93\ \mu_B$.

H. $(thf)Cr(Mes)_3$

1. Synthesis and Characterization

Chromium(III) trimesityl is not an extremely well-known compound, and very little use seems to have been made of it in recent times. As reported, its synthesis is straightforward, and results in a thf adduct, $(thf)Cr(Mes)_3$ (Scheme 39) (125). The thf ligand is probably labile (cf. $(thf)V(Mes)_3$, see above). The intriguing feature of chromium(III) trimesityl, to the author responsible for its

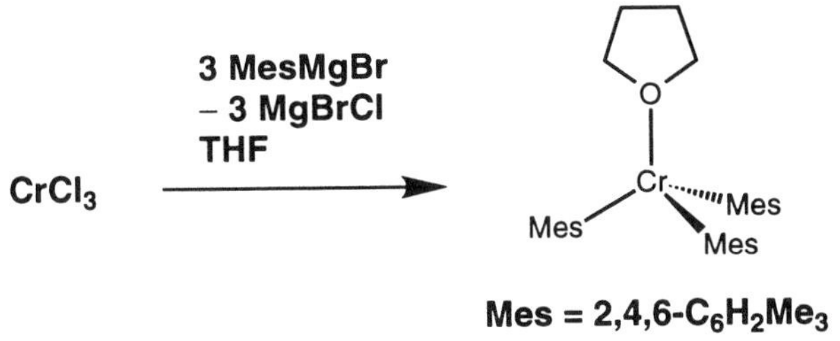

$$CrCl_3 \xrightarrow[\text{THF}]{\text{3 MesMgBr} \\ -\text{3 MgBrCl}} (\text{thf})Cr(\text{Mes})_3$$

Mes = 2,4,6-$C_6H_2Me_3$

Scheme 39

preparation, was its failure to rearrange into π aryl species, a typical reaction for other chromium aryl derivatives.

Mesityl Grignard was prepared first from Mg (~0.33 mol) and 1 equiv of mesityl bromide, in THF (500 mL). Subsequent to filtration of the Grignard solution, chromium(III) chloride (~0.11 mol) was added to the chilled mixture (−20°C). The light blue solution was stirred subsequently at room temperature for 1–2 h, at which point dioxane (100 mL) was added. Removal of magnesium halide and crystallization provided blue (thf)Cr(Mes)$_3$. Magnetic susceptibility measurements were made for (thf)Cr(Mes)$_3$ (μ_{eff} = 3.74 μ_B; 291.1°C), indicative of high-spin chromium(III). Dioxane and benzene cryoscopy gave molecular weights (respectively, 231.8 and 257.5) consistent with a monomeric formulation.

I. Mo(N—*i*-Pr$_2$)$_3$

1. Synthesis and Characterization

In a review on dialkylamides and alkoxides of molybdenum and tungsten, Chisholm et al. (126) state: "Our characterization of Mo(N—*i*-Pr$_2$)$_3$ is incomplete, but it does appear that this compound is quite different from the dimethylamide."

J. Mo[N(SiMe$_3$)$_2$]$_3$

1. Synthesis and Characterization

In a review on transition metal dialkylamides and disilylamides, Bradley and Chisholm (6) provide the following comment: "The characterization of three-coordinated Cr^{III} leads us to speculate on whether the use of appropriately bulky NR_2^- ligands will allow the isolation of three-coordinated Mo^{III} and W^{III} compounds. This possibility has not yet been fully investigated, although from a small amount of sublimate obtained from the reaction between $MoCl_3$ and $LiNSi_2Me_6$, a base peak MoL_3^+ together with cluster species $M_2L_3^+$ and $M_3L_3^+$ (L = NSi_2Me_6) was observed in the mass spectrum.'' The reference given was to unpublished results (Bradley and Smallwood).

K. Mo[N(R)Ar]$_3$ [R = $C(CD_3)_2Me$, Ar = 3,5-$C_6H_3Me_2$]

1. Synthesis and Characterization

This compound was reported initially in a 1995 communication (127). Details concerning its preparation in 70% yield as orange-red crystals were elaborated upon in a subsequent full paper (16). The synthesis of Mo[N(R)Ar]$_3$ involves simple treatment of $MoCl_3(thf)_3$ (128–134) with 2 equiv of the amido delivery agent Li[N(R)Ar](OEt_2) (Scheme 40) (34). Purification of the compound involves recrystallization from Et_2O at −35°C under Ar or in a partially evacuated system. Reaction conditions leading to the optimum yield of Mo[N(R)Ar]$_3$ were determined by varying the parameters solvent, stoichiometry, and initial temperature, monitoring the reaction's progress in each case as a function of time by deuterium NMR spectroscopy; this approach being facil-

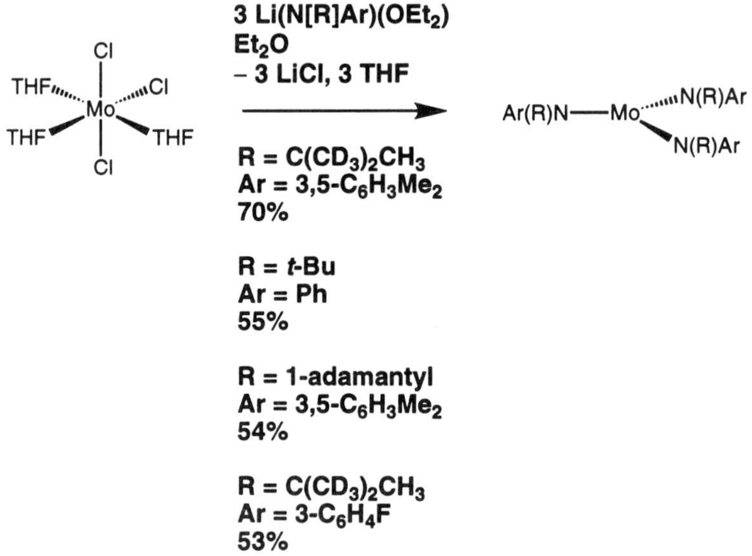

$MoCl_3(THF)_3$ → Ar(R)N—Mo(N(R)Ar)(N(R)Ar)

3 Li(N[R]Ar)(OEt_2)
Et_2O
− 3 LiCl, 3 THF

R = $C(CD_3)_2CH_3$
Ar = 3,5-$C_6H_3Me_2$
70%

R = *t*-Bu
Ar = Ph
55%

R = 1-adamantyl
Ar = 3,5-$C_6H_3Me_2$
54%

R = $C(CD_3)_2CH_3$
Ar = 3-C_6H_4F
53%

Scheme 40

itated by a characteristic, relatively sharp ^2H NMR signal for $Mo[N(R)Ar]_3$ at δ 64 ppm. That 2 rather than 3 equiv of $Li[N(R)Ar](OEt_2)$ were found to be optimal is likely the result of the interplay of several variables including the relatively low solubility of $MoCl_3(thf)_3$ in Et_2O. A control experiment indicated that $MoCl_3(thf)_3$ reacts relatively slowly with $Mo[N(R)Ar]_3$ to form as yet unidentified products. When $MoCl_3(thf)_3$ is treated with 3 equiv of $Li[N(R)Ar]$-(OEt_2) in Et_2O, significant side reactions set in (according to ^2H NMR) prior to complete consumption of the lithium amide. It is possible that the reaction between $MoCl_3(thf)_3$ and $Li[N(R)Ar](OEt_2)$ could be optimized for the production of species other than $Mo[N(R)Ar]_3$, although no such compounds have been characterized thus far.

A monomeric formulation for $Mo[N(R)Ar]_3$ was indicated initially by solution magnetic susceptibility measurements ($\mu_{eff} = 3.56\ \mu B$) consistent with high-spin molybdenum(III), in stark contrast with the diamagnetism evinced by the dimeric molybdenum(III) amide $Mo_2(NMe_2)_6$ (135, 136). Solid-state magnetic susceptibility measurements also were consistent with monomeric high-spin molybdenum(III) in $Mo[N(R)Ar]_3$; the compound exhibited Curie–Weiss behavior between 5 and 300 K ($\mu_{eff} = 3.82\ \mu B$). Deuterium NMR and EXAFS data are consistent with a symmetrical structure for $Mo(N[R]Ar)_3$ in solution (16).

X-ray crystallography confirmed the monomeric nature of $Mo[N(R)Ar]_3$ and the molybdenum atom's trigonal planar coordination geometry (16, 127). Each N(R)Ar ligand was observed to adopt the standard conformation in which the two planes defined by (a) the trigonal planar nitrogen atom, and (b) the aromatic residue, are mutually perpendicular (to within 10°). Furthermore, the overall shape of the molecule approximates C_3 point symmetry, such that the *tert*-butyl moieties adorn one coordination hemisphere and the aromatic residues the other. Orange-red $Mo[N(R)Ar]_3$ is isostructural with chocolate brown $Cr[N(R)Ar]_3$ (73), the structure of which has not yet been reported in detail.

2. *Reaction with Nitrous Oxide*

It was thought that $Mo[N(R)Ar]_3$ should behave as a nitrogen-atom acceptor or "azophile," due to the existence of robust terminal molybdenum(VI) nitrido compounds (48, 50, 74, 76, 137) as exemplified by $NMo(NPh_2)_3$ (75). Consideration of possible nitrogen-atom donors led to the selection of nitrous oxide as a candidate, since nitric oxide is a potentially viable leaving group, and because N_2O could lose a nitrogen atom with a minimum of reorganization. On the other hand, thermochemical considerations favor nitrous oxide NO bond cleavage over NN bond cleavage by about 75 kcal mol^{-1} (138), consistent with perception and implementation of nitrous oxide as a clean oxidant (139–144).

Treatment of $Mo[N(R)Ar]_3$ in Et_2O with an excess of gaseous nitrous oxide (3–4 equiv) resulted in a color change from red-orange to amber over 5–10 min. Removal of volatile material followed by ^1H and ^2H NMR analysis of the crystalline residue indicated formation of a 1:1 mixture of $NMo[N(R)Ar]_3$ and $(ON)Mo[N(R)Ar]_3$. Although the diamagnetic nitrido and nitrosyl complexes were not separated in pure form from the mixture, they were synthesized in pure form by reactions of $Mo[N(R)Ar]_3$ with N_3Mes and NO, respectively (see below), permitting their characterization and determination of their spectroscopic properties. The paramagnetic oxo compound $OMo[N(R)Ar]_3$, which was *not* produced in the reaction of $Mo[N(R)Ar]_3$ with N_2O, has been synthesized independently (see below) for determination of its properties and spectroscopic characteristics.

Possible mechanisms for N_2O dissociation mediated by $Mo[N(R)Ar]_3$ include (a) a monometallic mechanism involving an η^1-*N*-bonded nitrous oxide intermediate that fragments to $NMo[N(R)Ar]_3$ and free nitric oxide, followed by fast trapping of NO by a second equivalent of $Mo[N(R)Ar]_3$, and (b) a bimetallic mechanism involving an N_2O-bridged species that fragments directly to $NMo[N(R)Ar]_3$ and $(ON)Mo[N(R)Ar]_3$. Both mechanisms are plausible based on ethalpic considerations.

3. *Reaction with Nitric Oxide*

Lemon yellow crystals of diamagnetic $(ON)Mo[N(R)Ar]_3$ were obtained in essentially quantitative yield upon removal of solvent subsequent to treatment of ethereal $Mo[N(R)Ar]_3$ with a slight excess (1.5 equiv) of gaseous NO (Scheme 41) (127). This result is related to the classic syntheses of $(ON)Cr(N{-}i\text{-}Pr_2)_3$ and $(ON)Cr[N(SiMe_3)_2]_3$ from corresponding three-coordinate precursors. The $(ON)Mo[N(R)Ar]_3$ complex exhibits an intense low-energy IR band (1604 cm^{-1}) attributable to ν(NO). By analogy with structurally characterized $(ON)Cr[N(SiMe_3)_2]_3$ (116), it was reasoned that $(ON)Mo[N(R)Ar]_3$ contains a linear metal–nitrosyl moiety. Although X-ray diffraction data for $(ON)Mo[N(R)Ar]_3$ eluded acquisition, the related derivative $(ON)Mo[N(Ad)Ar]_3$ [prepared similarly (145)] was amenable to structural characterization by X-ray diffraction methods, whereupon it was found to exhibit the expected linear nitrosyl function [179.7(4)°]. The short molybdenum–nitrosyl nitrogen bond in $(ON)Mo[N(Ad)Ar]_3$ (1.724(4) Å] is indicative of substantial π back-bonding from the strong d^4 π-donor fragment $\{Mo[N(Ad)Ar]_3\}^-$ to the strong π-acceptor ligand $(NO)^+$. An alternative description of the $(ON)Mo[N(R)Ar]_3$ electronic structure recognizes that $Mo[N(R)Ar]_3$ is isolobal with a free nitrogen atom (22), and that $(ON)Mo[N(R)Ar]_3$ is therefore electronically analogous to N_2O in terms of the bonding in the π system of its linear MoNO moiety: $(1e,\pi)^4(2e,\pi)^4$.

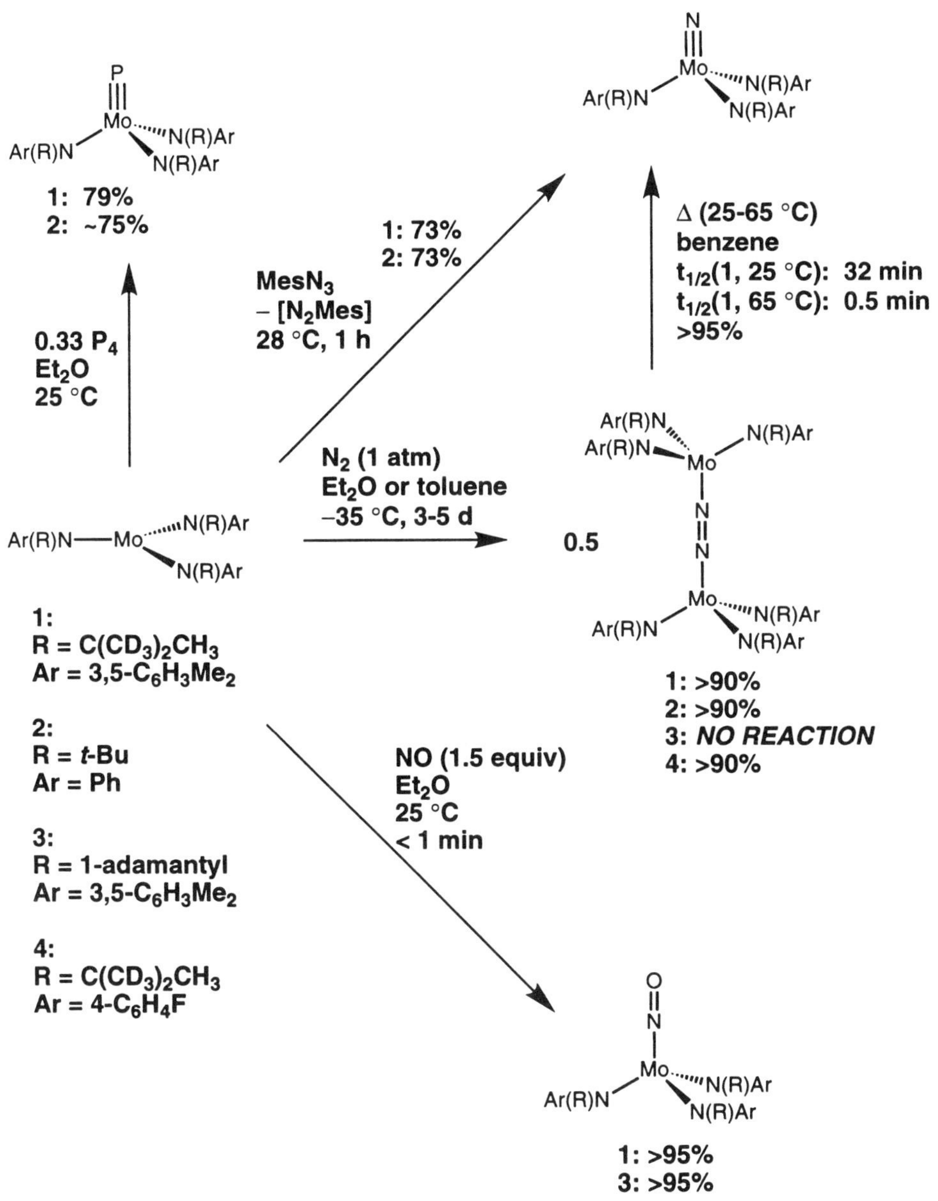
P
Mo
Ar(R)N
N(R)Ar
N(R)Ar
1: 79%
2: ~75%
0.33 P4
Et2O
25 °C
MesN3
– [N2Mes]
28 °C, 1 h
1: 73%
2: 73%
N
Mo
Ar(R)N
N(R)Ar
N(R)Ar
Δ (25-65 °C)
benzene
t1/2(1, 25 °C): 32 min
t1/2(1, 65 °C): 0.5 min
>95%
Ar(R)N—Mo
N(R)Ar
N(R)Ar
N2 (1 atm)
Et2O or toluene
–35 °C, 3-5 d
0.5
Ar(R)N
Ar(R)N
Mo
N(R)Ar
N
N
Mo
Ar(R)N
N(R)Ar
N(R)Ar
1: >90%
2: >90%
3: NO REACTION
4: >90%
1:
R = C(CD3)2CH3
Ar = 3,5-C6H3Me2
2:
R = t-Bu
Ar = Ph
3:
R = 1-adamantyl
Ar = 3,5-C6H3Me2
4:
R = C(CD3)2CH3
Ar = 4-C6H4F
NO (1.5 equiv)
Et2O
25 °C
< 1 min
O
N
Mo
Ar(R)N
N(R)Ar
N(R)Ar
1: >95%
3: >95%

Scheme 41

4. *Reaction with Mesityl Azide*

This reaction, (127) like the reaction of $Mo[N(R)Ar]_3$ with N_2O, runs counter to usual expectations for the substrate in question. Specifically, organic azides N_3R' have been employed as imido or nitrene delivery agents that transfer [NR′] to a metal center, oxidizing it by two units in a reaction accompanied by expulsion of dinitrogen (87, 90, 146). The $Mo[N(R)Ar]_3$ complex, on the other hand, undergoes a three-electron oxidation upon reaction with N_3Mes, yielding $NMo[N(R)Ar]_3$ (quantitative according to ^1H NMR, 73% isolated) and *formally* expelling the diazenyl radical (NNMes).

The diazenyl radical (NNMes) has been the study of low-temperature EPR studies showing that it decomposes to N_2 and mesityl radical-derived products with a half-life of 2.4×10^{-3} at $-73°C$ in cyclopropane solution (147, 148). The relative instability of (NNMes) may account for the lack of formation of the aryldiazonium complex $(MesN_2)Mo[N(R)Ar]_3$: that is, (NNMes) decomposes too rapidly for efficient trapping by $Mo[N(R)Ar]_3$. On the other hand, certain metal complexes have been known to abstract the azide radical, neutral N_3, from aryl azides, with concomitant $1e^-$ oxidation of the metal center (54, 61). If this pathway is operative for $Mo[N(R)Ar]_3 + N_3Mes$, then the presumed intermediate molybdenum(IV) complex $(N_3)Mo[N(R)Ar]_3$ evidently loses N_2 in a fast step to generate $NMo[N(R)Ar]_3$. Decomposition of intermediate molybdenum(IV) azido complexes to produce molybdenum(VI) nitrido species represents a known and valuable synthetic methodology (48, 149). While the terminal nitrido complex $NMo[N(R)Ar]_3$ has eluded structural characterization by X-ray diffraction, the related derivative $NMo[N(t\text{-}Bu)Ph]_3$ has been characterized structurally [$d(Mo \equiv N) = 1.658(5)$ Å; see below].

Terminal nitrogen-atom abstraction from the organic azide was verified by reaction of selectively labeled $^{15}N(N_2{-}4\text{-}C_6H_4Me)$ (150) with $Mo[N(R)Ar]_3$, giving $^{15}NMo[N(R)Ar]_3$ according to IR [$\nu(MoN) = 1014\ cm^{-1}$ versus 1042 cm^{-1} for the unlabeled complex], and ^{15}N NMR (δ = 840-ppm downfield of the signal for liquid ammonia).

5. *Reaction with Dinitrogen*

Unlike the reactions described above with N_2O, NO, and N_3Mes, the reaction of $Mo[N(R)Ar]_3$ with N_2 is relatively slow, requiring a minimum of several days for complete consumption of the starting molybdenum(III) complex when conducted at 1-atm N_2 (16, 151). Red-orange solutions of $Mo[N(R)Ar]_3$ ($-35°C \sim 0.05$ *M* in Et_2O or toluene, 1-atm N_2) adopt the intense purple color of $(\mu\text{-}N_2)\{Mo[N(R)Ar]_3\}_2$ (λ_{max} = 547 nm) in less than 3 h; the reaction's progress can be monitored by ^2H NMR spectroscopy (appearance of a single new peak at ~ 14 ppm) of aliquots. Under such conditions, conversion of

Mo[N(R)Ar]$_3$ to (μ-N_2){Mo[N(R)Ar]$_3$}$_2$ requires approximately 76 h to reach completion (Scheme 41). Characterization data for thermally unstable, paramagnetic (μ-N_2){Mo[N(R)Ar]$_3$}$_2$ include resonance Raman spectra [ν(NN) = 1630 cm^{-1} compared with 1577 cm^{-1} for (μ-$^{15}N_2$)[Mo[N(R)Ar]$_3$]$_2$, and an EXAFS structural study. The latter indicated a linear MoNNMo moiety in the complex with d(NN) = 1.19(2) Å, rendering the complex's core geometrically similar to that of the dinuclear molybdenum(III) dinitrogen complex (μ-N_2)[Mo(t-BuMe$_2$SiNCH$_2$CH$_2$)$_3$N]$_2$, which was the subject of a recent single-crystal X-ray diffraction study (152). Magnetic measurements performed on the related, more crystalline, derivative (μ-N_2){Mo[N(t-Bu)Ph]$_3$}$_2$ indicated two unpaired electrons for the complex (μ_{eff} = 2.85 μ_B). Also, the electronic structure and thermal chemistry of the hypothetical bridging N_2 complex (μ-N_2)[Mo(NH$_2$)$_3$]$_2$ were the subject of a recent theoretical study (153).

Although linear dinuclear bridging N_2 complexes are quite numerous and have long been known (154), (μ-N_2){Mo[N(R)Ar]$_3$}$_2$ is the first such species to undergo spontaneous NN bond cleavage: (μ-N_2){Mo[N(R)Ar]$_3$}$_2$ $\rightarrow$ 2NMo[N(R)Ar]$_3$ (16, 151, 155, 156). The overall reaction, production of two terminal nitrido molybdenum(VI) complexes from each molecule of Mo[N(R)Ar]$_3$ in the presence of 1-atm N_2, represents a homogeneous example of dissociative chemisorption of N_2, the heterogeneous version of which constitutes the rate-determining step in the Haber–Bosch ammonia synthesis. A reaction exhibiting a first-order kinetic profile, dissociation of (μ-N_2)-{Mo[N(R)Ar]$_3$}$_2$ is characterized by activation parameters (ΔH‡ = 23.3 ± 0.3 kcal mol^{-1} and ΔS‡ = 2.9 ± 0.8 cal mol^{-1} K^{-1}) and a primary kinetic isotope effect (~ 1.1 for $^{14}N_2/^{15}N_2$ over the temperature range 25–65°C) consistent with rate-determining NN bond cleavage (16). Based on the above activation parameters the half-life of purple (μ-N_2){Mo[N(R)Ar]$_3$}$_2$ is about 2.7 years at −35°C, whereas at 25°C $t_{1/2}$ rises to approximately 1 h. The NMo[N(R)Ar]$_3$ complex is the first terminal nitrido complex to be synthesized directly from N_2, having been isolated in 76% recrystallized yield from the reaction. Samples of NMo[N(R)Ar]$_3$ and ^{15}NMo[N(R)Ar]$_3$ prepared from treatment of Mo[N(R)Ar]$_3$ with N_2 and $^{15}N_2$, respectively, exhibited spectroscopic properties essentially identical with those for samples prepared independently through treatment of Mo[N(R)Ar]$_3$ with N$_3$Mes and ^{15}N(N$_2$—4-C$_6$H$_4$Me), respectively.

A zigzag conformation of the MoNNMo moiety (approximate C_{2h} point symmetry) has been proposed/calculated for the transition state of the following hypothetical fragmentation reaction: (μ-N_2)[Mo(NH$_2$)$_3$]$_2$ $\rightarrow$ 2 NMo(NH$_2$)$_3$ (153). Such a transition structure is deemed sterically plausible for the corresponding fragmentation of (μ-N_2){Mo[N(R)Ar]$_3$}$_2$, and it has been pointed out that an alternative linear transition structure is "forbidden" by orbital symmetry restrictions (16). A qualitative orbital correlation diagram for N_2 cleavage via a C_{2h}-symmetric transition state has been given (16).

6. *Reaction with White Phosphorus: Terminal Phosphido (M≡P) Complexes*

Various interesting attempts to generate the terminal M≡P functionality, the phosphorus analogue of the nitrido linkage, have been documented (157). In such attempts, white phosphorus or phosphaalkynes (RCP) have typically served as the source of phosphorus. In 1995, it was reported that $Mo[N(R)Ar]_3$ reacts smoothly with a 40% excess of P_4 to give an isolable, robust terminal phosphido complex, $PMo[N(R)Ar]_3$, in about 79% isolated yield on a scale of about 3 mmol (158). Reported simultaneously were syntheses of the Group 6 (VIB) terminal phosphido complexes $PM[(Me_3SiNCH_2CH_2)_3N]$ (M = Mo, W), which were prepared (on scales of ~ 0.1 mmol) via the breakdown of M^{IV} phenylphosphido[P(H)Ph] precursors (159).

An unusually large downfield ^{31}P NMR shift (δ 1216 ppm) was noted for terminal phosphido complex $PMo[N(R)Ar]_3$; spectroscopic and electronic structure characteristics of this and other terminal phosphido complexes provided the basis for a combined solid-state ^{31}P NMR/theoretical study (160).

The $PMo[N(R)Ar]_3$ complex was characterized structurally via an X-ray diffraction study (158). Consistent with the formulation featuring a triple bond is the short *d*(Mo≡P) of 2.119(4) Å. The overall conformation of the molecule, the termal P shielded by the three R groups, is exactly reminiscent of the structures of $NMo[N(t\text{-Bu})Ph]_3$ and $PMo[N(t\text{-Bu})Ph]_3$ (see below for a description of the latter), both of which display C_3 point symmetry.

Preliminary reactivity studies involving $PMo[N(R)Ar]_3$ (158) are beyond the scope of this chapter.

7. *Reaction with Terminal Nitrido Complex NMo(O—t-Bu)₃*

Intermetal nitrogen-atom transfer reactions are of considerable intrinsic interest because documented examples involve the concurrent breakage and formation of an exceedingly strong linkage, a metal–nitrogen triple bond, and yet occur rapidly under mild conditions (45, 161–163). From a synthetic point of view nitrogen-atom abstraction is unique in that it simultaneously (a) makes available an open coordination site, and (b) effects a three-electron reduction. With these issues in mind the reaction between $Mo[N(R)Ar]_3$ and Chisholm and co-workers $NMo(O\text{—}t\text{-Bu})_3$ was studied under a variety of conditions (49).

Upon treatment with $NMo(O\text{—}t\text{-Bu})_3$ (48) (degassed benzene, 28°C), $Mo[N(R)Ar]_3$ was completely converted to $NMo[N(R)Ar]_3$ over the course of approximately 12 h (Scheme 42). Thereby stripped of its terminal nitrogen atom, $NMo(O\text{—}t\text{-Bu})_3$ was concomitantly converted to the well-known dimer $Mo_2(O\text{—}t\text{-Bu})_6$(M≡M), presumably via dimerization of the reactive unobserved fragment $[Mo(O\text{—}t\text{-Bu})_3]$. Nitrido $NMo[N(R)Ar]_3$ and dimer $Mo_2(O\text{-}t\text{-Bu})_6$ were separated on the basis of solubility differences and spectroscopically characterized.

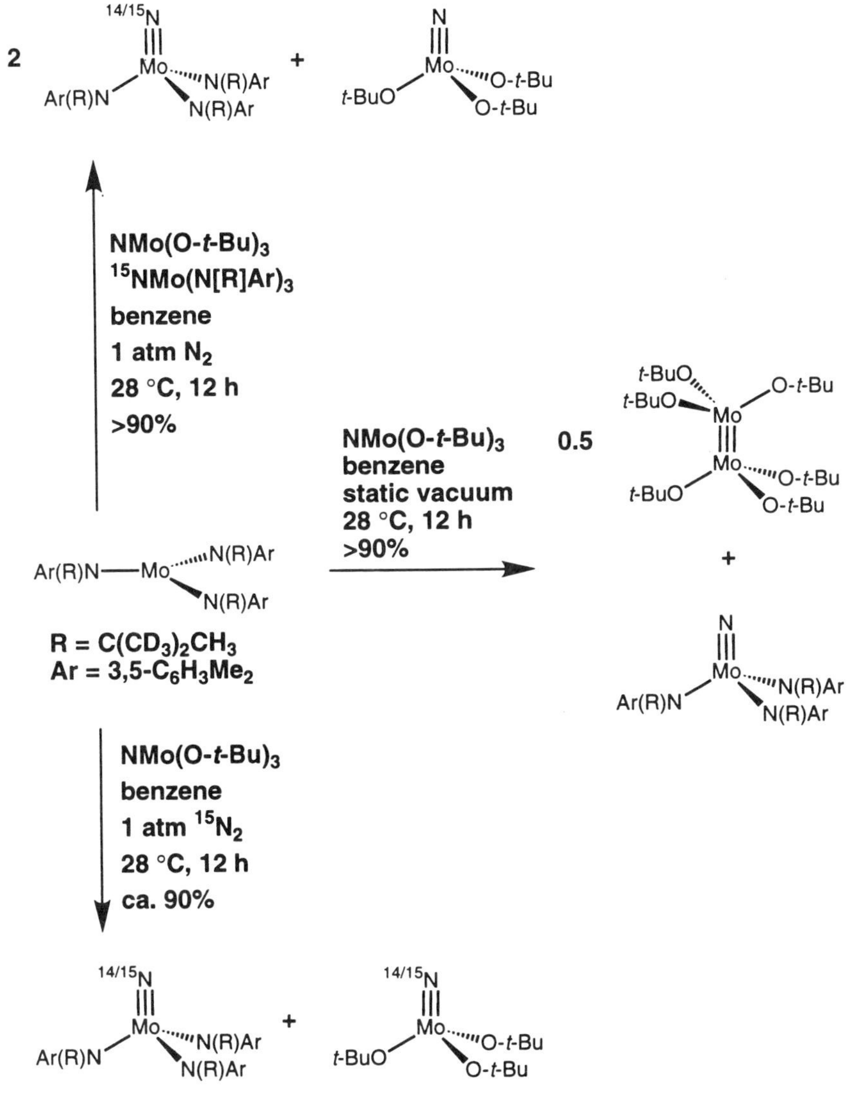

2 $^{14/15}N\equiv Mo[N(R)Ar]_3$ + $N\equiv Mo(O\text{-}t\text{-}Bu)_3$

$NMo(O\text{-}t\text{-}Bu)_3$
$^{15}NMo(N[R]Ar)_3$
benzene
1 atm N_2
28 °C, 12 h
>90%

$Mo[N(R)Ar]_3$

R = $C(CD_3)_2CH_3$
Ar = 3,5-$C_6H_3Me_2$

$NMo(O\text{-}t\text{-}Bu)_3$
benzene
static vacuum
28 °C, 12 h
>90%

0.5 $(t\text{-}BuO)_3Mo\equiv Mo(O\text{-}t\text{-}Bu)_3$ + $N\equiv Mo[N(R)Ar]_3$

$NMo(O\text{-}t\text{-}Bu)_3$
benzene
1 atm $^{15}N_2$
28 °C, 12 h
ca. 90%

$^{14/15}N\equiv Mo[N(R)Ar]_3$ + $^{14/15}N\equiv Mo(O\text{-}t\text{-}Bu)_3$

Scheme 42

Noting that $NMo(N[R]Ar)_3$ reacts much more rapidly with $NMo(O{-}t\text{-}Bu)_3$ than with N_2 (1 atm, benzene, 28°C), it was of interest to determine whether putative transient $[Mo(O{-}t\text{-}Bu)_3]$ could be diverted from dimerization through interaction with N_2. Accordingly, $Mo[N(R)Ar]_3$ was subjected to treatment with $NMo(O{-}t\text{-}Bu)_3$ under N_2 (1 atm, benzene, 28°C), and the reaction was allowed to proceed to completion. Spectroscopic interrogation of the product mixture revealed mainly $NMo[N(R)Ar]_3$ and $NMo(O{-}t\text{-}Bu)_3$, along with a small amount (~ 7%) of dimer $Mo_2(O{-}t\text{-}Bu)_6$, results implicating $[Mo(O{-}t\text{-}Bu)_3]$ in a N_2 cleavage process. The latter inference was confirmed upon carrying out the reaction under similar conditions, substituting $^{15}N_2$ for unlabeled N_2. The products thus obtained were $^{14/15}NMo[N(R)Ar]_3$, $^{14/15}NMo(O{-}t\text{-}Bu)_3$, and about 7% of $Mo_2(O{-}t\text{-}Bu)_6$. Both nitrido products, which were physically separated and analyzed, exhibited ^{15}N-enrichment in the nitrido position to the extent of about 45%. A control experiment established that $^{15}NMo[N(R)Ar]_3$ (prepared independently) does not exchange its nitrido nitrogen atom with $NMo(O{-}t\text{-}Bu)_3$, implying that ^{15}N-enrichment in $^{14/15}NMo(O{-}t\text{-}Bu)_3$ derives in some direct fashion from reaction of $[Mo(O{-}t\text{-}Bu)_3]$ with N_2, or N_2 complexed to $Mo[N(R)Ar]_3$.

The above experiments demonstrate accelerated N_2 consumption by $Mo[N(R)Ar]_3$ in the presence of $NMo(O{-}t\text{-}Bu)_3$. The nature of the acceleration/catalysis is clear: nitrogen-atom abstraction coupled with N_2 cleavage (164, 165).

It was postulated that the nitrogen-atom transfer reaction involves an observable intermediate, $(t\text{-}BuO)_3Mo(\mu\text{-}N)Mo[N(R)Ar]_3$, responsible for the blue color observed at intermediate reaction times. Related nitrogen- and phosphorus-atom bridged complexes have since been isolated and structurally characterized (see below).

Although $Mo[N(R)Ar]_3$ is indisputably a potent nitrogen-atom abstraction reagent, the fact that it fails to react with $NW(O{-}t\text{-}Bu)_3$ under conditions similar to those utilized for the $NMo(O{-}t\text{-}Bu)_3$ experiments is noteworthy (166).

8. *Reaction with Pyridine-N-Oxide*

Treatment of $Mo[N(R)Ar]_3$ with ONC_5H_5 led to essentially quantitative formation of paramagnetic $OMo[N(R)Ar]_3$, which was isolated in 80% yield as brown crystals (167) (Scheme 43). The superconductivity quantum interference device (SQUID) magnetic susceptibility measurements showed $OMo[N(R)Ar]_3$ to be a Curie paramagnet from 5–300 K. An IR band at 894 cm^{-1} tentatively was ascribed to ν(MoO). Taking into account literature values of the N—O bond dissociation enthalpy (BDE) for pyridine-*N*-oxide, a calorimetric study of the $Mo[N(R)Ar]_3/ONC_5H_5$ reaction yielded an estimate of the MoO BDE: 155.6 $\pm$ 1.6 kcal mol^{-1}. This surprisingly high value (168) is all the more remarkable when it is recalled that $OMo[N(R)Ar]_3$ evidently does not form in the reaction of $Mo[N(R)Ar]_3$ with N_2O [$d(N_2{-}O)$ = ~ 40 kcal mol^{-1}] (127).

The structure of $OMo[N(R)Ar]_3$ was determined by X-ray crystallography (Fig. 20). As might be expected for a nonlinear species isolobal with nitric oxide $[(e\pi^4(1e\pi^*)^1]$, obvious distortions from ideal C_3 point symmetry are evident for the compound. The relationship carries further in that $OMo[N(R)Ar]_3$,

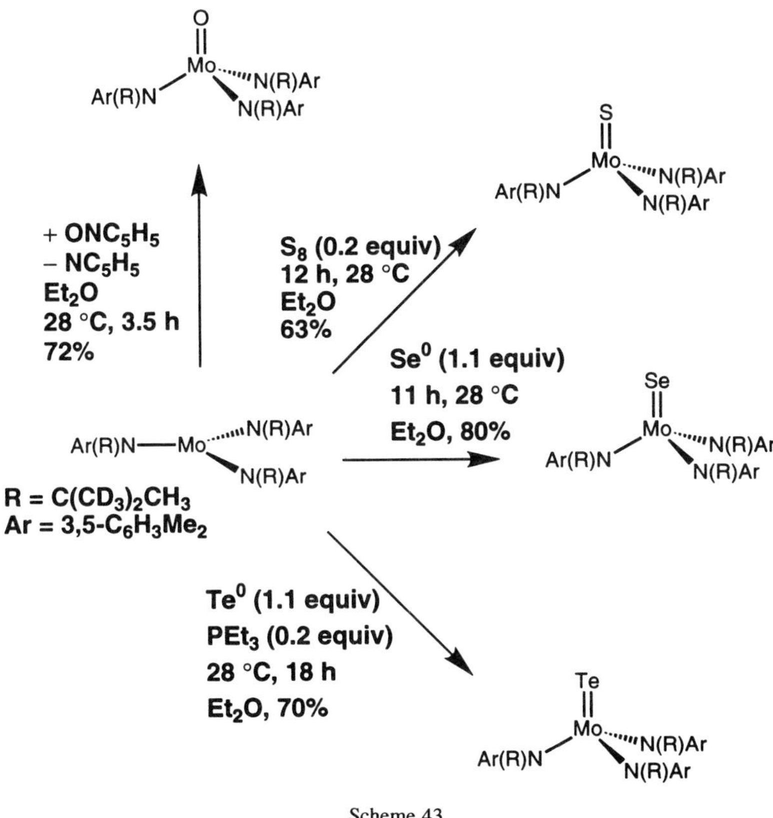

Scheme 43

like NO, is readily oxidized to a stable cationic species: $\{OMo[N(R)Ar]_3\}^+$, isolable as iodide or triflate salts. Oxo-cation$\{OMo[N(R)Ar]_3\}^+$ is, of course, isoelectronic to the nitrido complex $NMo[N(R)Ar]_3$ that results from reactions of $Mo[N(R)Ar]_3$ with N_2, N_2O, N_3Mes, and $NMo(O{-}t\text{-}Bu)_3$, as described above. The molybdenum–oxygen distance in $OMo[N(R)Ar]_3$ [1.706(2) Å] is toward the high end of the range for structurally characterized molybdenum oxo species (169).

9. *Reactions with Chalcogen Atom (E = S, Se, Te) Sources*

The isostructural molybdenum(V) chalcogenido complexes $EMo[N(R)Ar]_3$ (E = S, Se, Te) were obtained in good yield following treatment of $Mo[N(R)Ar]_3$ with sulfur, selenium, and tellurium (in the presence of 0.1 equiv PEt_3), respectively (Scheme 43). The $SMo[N(R)Ar]_3$ complex was obtained in

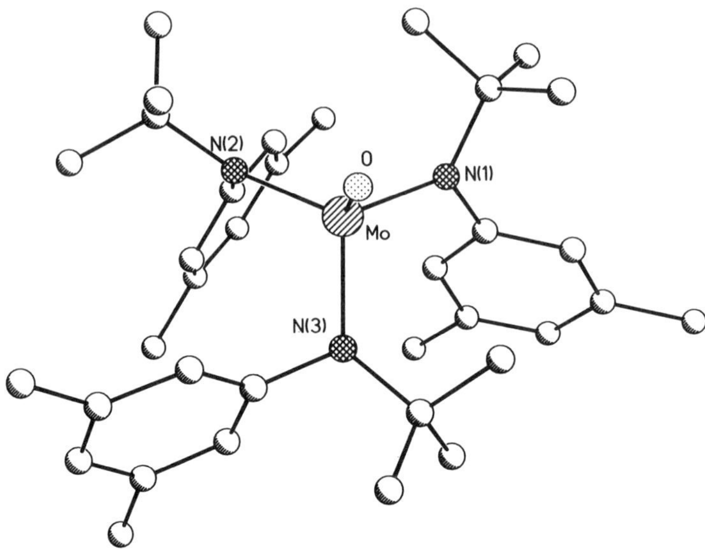

Figure 20. Structural drawing of $OMo[N(R)Ar]_3$; R = $C(CD_3)_2Me$, Ar = 3,5-$C_6H_3Me_2$.

63% recrystallized yield on an approximately 0.7-mmol scale, $SeMo[N(R)Ar]_3$ was obtained in 80% recrystallized yield on an approximately 0.8-mmol scale, while $TeMo[N(R)Ar]_3$ was obtained in 73% recrystallized yield on an approximately 0.8-mmol scale. These $EMo[N(R)Ar]_3$ complexes exhibit pseudo-C_s symmetry in the solid state (Fig. 21), a result attributable to a Jahn–Teller distortion from threefold symmetry, assuming the odd electron to reside in a molecular orbital rich in MoE π^* character. The MoE bond lengths in the complexes are as follows: *d*(MoS), 2.1677(12) Å; *d*(MoSe), 2.3115(6) Å, and *d*(MoTe), 2.5353(6) Å.

L. Mo[N(*t*-Bu)Ph]$_3$

1. General Comments

This nondeuterated analogue of $Mo[N(R)Ar]_3$ (see above) can be considered the "parent" three-coordinate molybdenum(III) complex based on *N-tert*-hydrocarbylarylamido ligation (16). Although not currently commercially available, *N-tert*-butylaniline can be prepared rapidly and in large quantities (~100 mL per batch) by the literature method involving benzyne addition to *tert*-butylamine. Deprotonation with butyllithium, and crystallization of the resulting salt in the presence of ether gives Li[N(*t*-Bu)Ph](OEt_2) as a convenient amido-transfer agent.

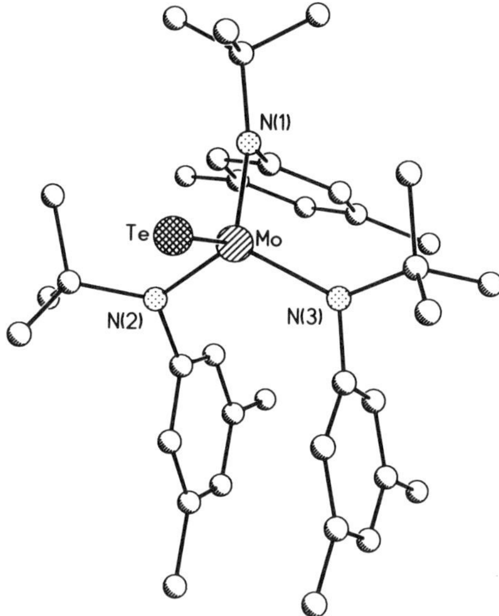

Figure 21. Structural drawing of TeMo[N(R)Ar]$_3$; R = C(CD$_3$)$_2$Me, Ar = 3,5-$C_6H_3Me_2$.

The reactivity of Mo[N(*t*-Bu)Ph]$_3$ with reagents such as dinitrogen, organic azides, white phosphorus, NMo(O—*t*-Bu)$_3$, and chalcogen atom sources, is qualitatively identical to the same reactions involving Mo[N(R)Ar]$_3$ (see above).

2. *Synthesis and Characterization*

The preparation and characterization data pertaining to Mo[N(*t*-Bu)Ph]$_3$ were given in a recent full paper (16). The compound is obtained as burgundy crystals in 55.4% yield (scale ~9 mmol) in a procedure conducted in the absence of N_2 due to the compound's reactivity toward same. The procedure involves treatment of $MoCl_3$(thf)$_3$ with 3 equiv of Li[N(*t*-Bu)Ph](OEt_2) in Ar-sparged Et_2O, salt removal, and recrystallization. Four isotropically shifted and broadened signals for Mo[N(*t*-Bu)Ph]$_3$ were found in its ^{1}H NMR spectrum as expected, at 66.5, 22.4, −23, and −50 ppm, assigned to the *tert*-butyl, meta or ortho, meta or ortho, and para protons, respectively. Indicative of high-spin molybdenum(III) is the compound's effective magnetic moment (3.41 μ_B) as obtained by the method of Evans (170, 171), in toluene-d_8 at 23°C.

That Mo[N(*t*-Bu)Ph]$_3$ possesses a pseudo-C_3 symmetric molecular structure, consistent with its solution ^{1}H NMR spectrum, was confirmed by an X-ray diffraction study. The latter study has not yet been submitted for publication, and so the structure of Mo[N(*t*-Bu)Ph]$_3$ will not be described in detail here. The structure of Mo[N(t-Bu)Ph]$_3$ is, however, quite similar to that reported for Mo[N(R)Ar]$_3$ (see above), with a trigonal planar coordination geometry at molybdenum.

3. *Reaction with Dinitrogen*

As has been found for Mo[N(R)Ar]$_3$ (see above), Mo[N(t-Bu)Ph]$_3$ reacts with N_2 (1 atm, pentane, 2 days, $-35\,°C$) to give a purple dinuclear N_2 complex (μ-N_2){Mo[N(t-Bu)Ph]$_3$}$_2$, which presumably possesses a linear MoNNMo moiety. The (μ-N_2){Mo[N(t-Bu)Ph]$_3$}$_2$ complex exhibits relatively low pentane solubility, and was obtained as a purple-black microcrystalline precipitate. The dinuclear N_2 complex is paramagnetic, exhibiting isotropically shifted and broadened peaks in its ^1H NMR spectrum; that assigned to the six *tert*-butyl residues appears at δ 12.6 ppm ($\Delta\nu_{1/2}$ = 33 Hz). Temperature-dependent magnetic susceptibility measurements obtained on a powder sample gave good agreement with Curie–Weiss behavior in the temperature range 29–300 K, with the observed magnetic moment (2.85 μ_B) indicative of a triplet ground state for the complex as expected assuming a linear MoNNMo core.

Like (μ-N_2){Mo[N(R)Ar]$_3$}$_2$, (μ-N_2){Mo[N(t-Bu)Ph]$_3$}$_2$ fragments cleanly to terminal molybdenum nitrido species on warming to greater than or equal to 25 °C, although detailed kinetic studies have not been carried out for this derivative. Here, diamagnetic NMo[N(t-Bu)Ph]$_3$ is produced. This nitrido complex was characterized structurally by X-ray diffraction (Fig. 22). The Mo–nitrido

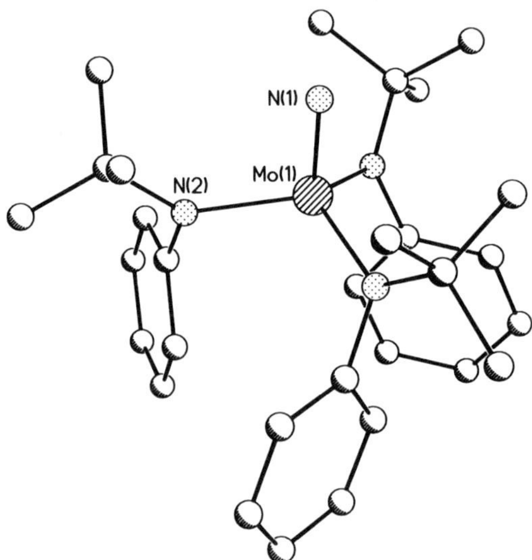

Figure 22. Structural drawing of NMo[N(t-Bu)Ph]$_3$.

nitrogen bond length of 1.658(5) Å is characteristic of this functional group. In the crystal, $NMo[N(t\text{-}Bu)Ph]_3$ was found to reside with its Mo–nitrido vector on a crystallographic C_3 axis, the nitrido nitrogen being encapsulated within a sheath comprised of the three *tert*-butyl residues, the phenyl groups packing tightly in a triangular edge-to-face manner on the reverse side of the molecule.

4. *Reaction with Mesityl Azide*

A convenient and rapid synthesis of $NMo[N(t\text{-}Bu)Ph]_3$ involves the addition of mesityl azide to a solution of $Mo[N(t\text{-}Bu)Ph]_3$ in ether. After fading of the initial dark purple color (~ 3 h), recrystallization of the crude material provided nitrido $NMo[N(t\text{-}Bu)Ph]_3$ in 73% yield (scale ~1 mmol).

5. *Reaction with $NMo(NMe_2)_3$: Synthesis of $(Me_2N)_3MoNMo(NMe_2)_3$*

A dark orange-brown solution of $Mo[N(t\text{-}Bu)Ph]_3$ (0.19 mmol) in ether was added to a yellow solution of $NMo(NMe_2)_3$ (2 equiv) in ether at 28°C (Scheme 44) (53). The resulting dark teal mixture was stirred rapidly and sparged with Ar for 30 s before the vial was capped tightly, taped, and placed in a −35°C freezer. Large black blocks formed at the bottom of the vial over several hours; an orange plate of $NMo[N(t\text{-}Bu)Ph]_3$ formed on the side of the vial above the surface of the dark teal mother liquor. The solution was decanted and the crystals were dried under vacuum. The single crystal of $NMo[N(t\text{-}Bu)Ph]_3$ remained apart from the dark blocks, which are presumed to consist largely of $(Me_2N)_3MoNMo(NMe_2)_3$ (57.1% yield); neither $Mo[N(t\text{-}Bu)Ph]_3$ nor $NMo[N(t\text{-}Bu)Ph]_3$ was observed in the ^1H NMR spectrum of the isolated product. The ^1H NMR spectrum of $(Me_2N)_3MoNMo(NMe_2)_3$ (500 MHz, C_7D_8, 22°C) is as follows: δ 7.13 (s, $\Delta\nu_{1/2}$ = 74 Hz). The thermally unstable compound was characterized chemically in the following way: dark shiny crystals of $(Me_2N)_3MoNMo(NMe_2)_3$ (~0.1 mmol) were placed in a small glass vessel to which just-melted ether was added. The vessel was sealed under nitrogen and allowed to warm to 28°C while stirring. The initially dark teal solution acquired a green hue over approximately 5 min, and became brown within 90 min, at which point volatile matter was removed. The ^1H and $^{13}C\{^1H\}$ NMR spectra indicated the presence of $Mo_2(NMe_2)_6$ and $NMo(NMe_2)_3$ in a 1:4 mole ratio. No signals attributable to $[N(t\text{-}Bu)Ph]$-containing compounds were observed.

The structure of $(Me_2N)_3MoNMo(NMe_2)_3$ was interrogated by X-ray diffraction, the specimen used being a serendipitous 1:1 cocrystal of the compound of interest with $Mo_2(NMe_2)_6$ (Fig. 23). Accordingly, a single crystal obtained by cooling a solution in which equimolar quantities of $NMo(NMe_2)_3$ and $Mo(N[t\text{-}Bu]Ph)_3]$ had been combined at 28°C was found to contain a 1:1

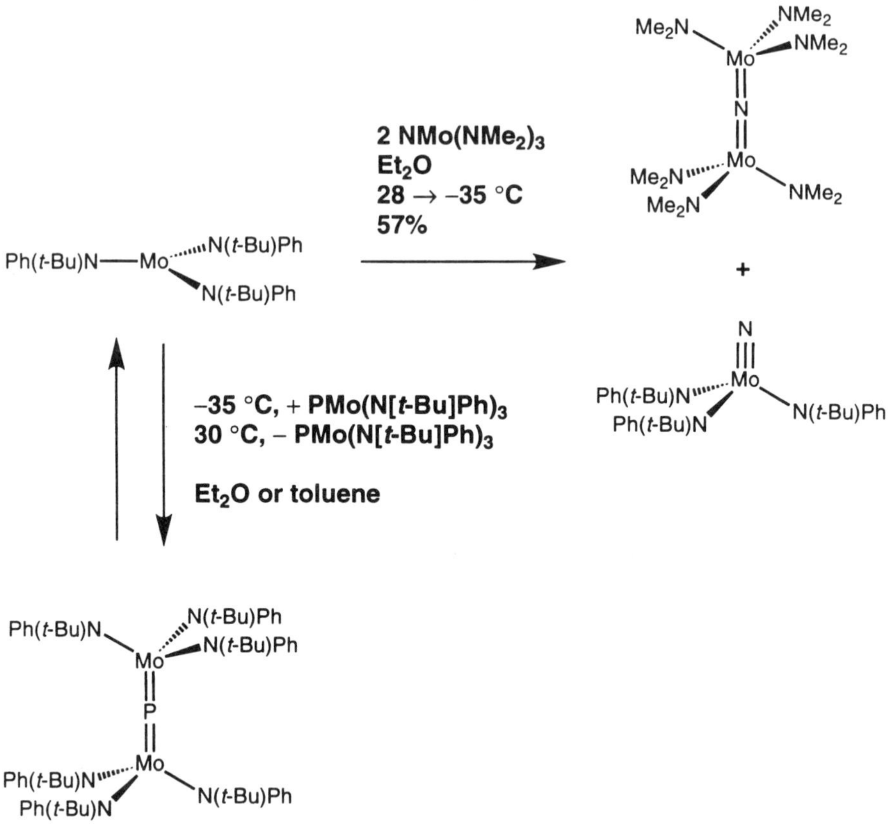

Ph(*t*-Bu)N—Mo(N(*t*-Bu)Ph)(N(*t*-Bu)Ph)

2 $NMo(NMe_2)_3$
Et_2O
28 → −35 °C
57%

Me_2N—Mo(NMe_2)(NMe_2)=N=Mo(NMe_2)(NMe_2)(NMe_2)

+

N≡Mo(N(*t*-Bu)Ph)(N(*t*-Bu)Ph)(N(*t*-Bu)Ph)

−35 °C, + PMo(N[*t*-Bu]Ph)$_3$
30 °C, − PMo(N[*t*-Bu]Ph)$_3$

Et_2O or toluene

Ph(*t*-Bu)N—Mo(N(*t*-Bu)Ph)(N(*t*-Bu)Ph)=P=Mo(N(*t*-Bu)Ph)(N(*t*-Bu)Ph)(N(*t*-Bu)Ph)

Scheme 44

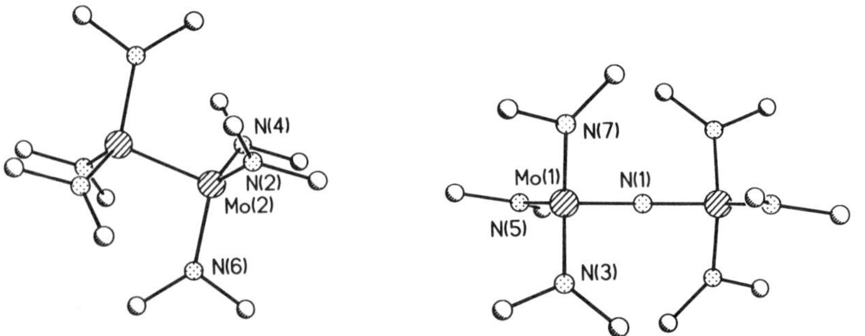

Figure 23. Structural drawing of (μ-N)[$Mo(NMe_2)_3$]$_2$ · $Mo_2(NMe_2)_6$.

mixture of known dimer $Mo_2(NMe_2)_6$ and symmetrical μ-nitrido $(Me_2N)_3$-$MoNMo(NMe_2)_3$. The structure of dimer $Mo_2(NMe_2)_6$ is as reported previously by Chisholm, Cotton, and coworkers (136). The structure of μ-nitrido $(Me_2N)_3MoNMo(NMe_2)_3$ suggests that the complex may be considered the product of reaction between putative transient $[Mo(NMe_2)_3]$ and terminal nitrido $NMo(NMe_2)_3$. The μ-nitrido nitrogen atom resides at a crystallographic inversion center, which when combined with the absence of a crystallographic threefold axis along the linear Mo(μ-N)Mo vector leads to the point group assignment C_i for the molecule. The unique Mo—μ-N distance corresponds approximately to bond order 2. The geometry at molybdenum is pseudotetrahedral. From a chemical point of view, the molecule $(Me_2N)_3MoNMo(NMe_2)_3$ is probably best regarded as a source of putative transient $[Mo(NMe_2)_3]$ along with 1 equiv of $NMo(NMe_2)_3$, since disintegration of either Mo—μ-N linkage would produce $[Mo(NMe_2)_3]$ along with terminal nitrido $NMo(NMe_2)_3$. Dimerization of three-coordinate $[Mo(NMe_2)_3]$ would produce $Mo_2(NMe_2)_6$. On standing in solution at 25°C, $(Me_2N)_3MoNMo(NMe_2)_3$ is in fact seen to transform irreversibly to a mixture of $Mo_2(NMe_2)_6$ and $NMo(NMe_2)_3$, as described in the preceding paragraph.

6. *Reaction with White Phosphorus*

Like $Mo(N[R]Ar)_3$, $Mo[N(t\text{-Bu})Ph]_3$ reacts with P_4 in ether to give a terminal phosphido complex in high yield (Scheme 44). The $PMo[N(t\text{-Bu})Ph]_3$ complex exhibits a ^{31}P NMR signal at 1226 ppm in solution at 25°C. The solid-state ^{31}P NMR spectrum of $PMo[N(t\text{-Bu})Ph]_3$ and a theoretical interpretation thereof has been published (160). The solid-state structure of $PMo[N(t\text{-Bu})Ph]_3$ is known from a single-crystal X-ray diffraction study Fig. 24; the phosphorus atom is one-coordinate and the observed d(MoP) is 2.111 Å, consistent with an MoP triple bond as for $PMo[N(R)Ar]_3$.

7. *Reaction with $PMo[N(t\text{-Bu})Ph]_3$*

By way of introduction, it was discovered that mixing equimolar quantities of $PMo[N(R)Ar]_3$ and $Mo[N(t\text{-Bu})Ph]_3$ in toluene leads to reversible phosphorus-atom transfer at 25°C, with ^{31}P NMR signals observed at δ 1216 and 1226 ppm for terminal phosphido complexes $PMo[N(R)Ar]_3$ and $PMo[N(t\text{-Bu})Ph]_3$, respectively (53). Cooling (to −35°C) an equimolar toluene solution of three-coordinate $Mo[N(t\text{-Bu})Ph]_3$ and terminal phosphido $PMo[N(t\text{-Bu})Ph]_3$ elicits a color change from red-brown to purple that is reversible upon repeated warming and cooling cycles. The purple color is attributed to the symmetrical bridging phosphido complex $[Ph(t\text{-Bu})N]_3MoPMo[N(t\text{-Bu})Ph]_3$, which was isolated and characterized structurally by X-ray crystallography (Fig. 25).

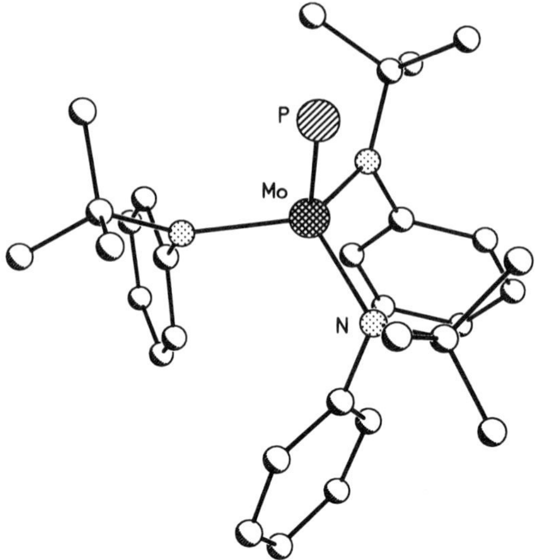

Figure 24. Structural drawing of PMo[N(*t*-Bu)Ph]$_3$.

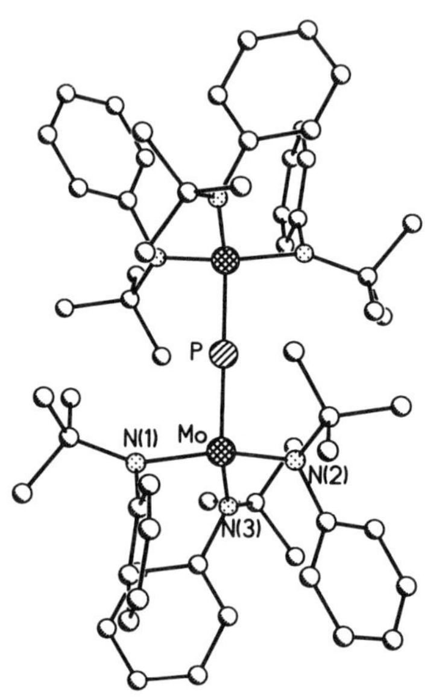

Figure 25. Structural drawing of (μ-P){Mo[N(*t*-Bu)-Ph]$_3$}$_2$.

Molecular μ-phosphido $[Ph(t\text{-}Bu)N]_3MoPMo[N(t\text{-}Bu)Ph]_3$ possesses C_i symmetry in the crystal, reminiscent of the structure of μ-nitrido $(Me_2N)_3$-$MoNMo(NMe_2)_3$, as described above. Two molybdenum atoms at 2.2430(6) Å flank the symmetrically bridging phosphorus atom in $[Ph(t\text{-}Bu)N]_3MoPMo[N(t\text{-}Bu)Ph]_3$, a distance to be compared with *d*(MoP) values of 2.119(4) and 2.111(2) Å for the terminal phosphido complexes $PMo[N(R)Ar]_3$ and $PMo[N(t\text{-}Bu)Ph]_3$, respectively. Based on the observed *d*(MoP) in $[Ph(t\text{-}Bu)N]_3MoPMo[N(t\text{-}Bu)Ph]_3$ and on the probable electronic configuration of its $Mo(\mu\text{-}P)Mo$ π-system, specifically $(1\pi u)^4(1\pi_g)^3$ assuming a class III type mixed-valence complex, a MoP bond order of 2 follows for the complex. Metrical parameters for $[Ph(t\text{-}Bu)N]_3MoPMo[N(t\text{-}Bu)Ph]_3$, apart from those associated with the phosphorus atom, indicate that little reorganization takes place during its formation.

M. $Mo[N(Ad)Ar]_3$ (Ad = 1-adamantyl, Ar = 3,5-$C_6H_3Me_2$)

1. *Synthesis and Characterization*

The procedure for preparing $Mo[N(Ad)Ar]_3$ is quite reminiscent of that followed in the preparation of $Mo[N(R)Ar]_3$ [see above, R = $C(CD_3)_2Me$]. Accordingly, treatment of $MoCl_3(thf)_3$ (~5 mmol) with $Li(N[Ad]Ar)(OEt_2)$ (2 equiv) in ether at 25°C led to the formation of $Mo[N(Ad)Ar]_3$, which was isolated in 54% yield as light orange crystals (145). Unlike $Mo[N(R)Ar]_3$, $Mo[N(Ad)Ar]_3$ does not *appear* to react with N_2, even in solution under 1-atm N_2 at −35°C. Therefore, some of the special handling considerations that apply to $Mo[N(R)Ar]_3$ do not apply to $Mo[N(Ad)Ar]_3$. The apparent lack of reaction between $Mo[N(Ad)Ar]_3$ and N_2 is attributed to the increased steric demands introduced by the substitution of 1-adamantyl for *tert*-butyl, since the 1-adamantyl substituent extends outward radially from the metal farther by about 1.5 Å than the *tert*-butyl substituent, all other things being equal. Thus access to a dinuclear linear μ-N_2 complex, which requires an intermolybdenum separation of about 5.0 Å, is restricted. The structure of $Mo[N(Ad)Ar]_3$ was determined by X-ray diffraction in a study showing that the compound is, from a structural point of view, directly analogous to $Mo[N(R)Ar]_3$, the only significant difference being the presence of 1-adamantyl substituents in place of *tert*-butyl groups Fig. 26.

2. *Reaction with Nitric Oxide*

Like $Mo[N(R)Ar]_3$, $Mo[N(Ad)Ar]_3$ readily absorbs 1 equiv of NO giving a diamagnetic adduct, here $(ON)Mo[N(Ad)Ar]_3$. The latter nitrosyl complex was

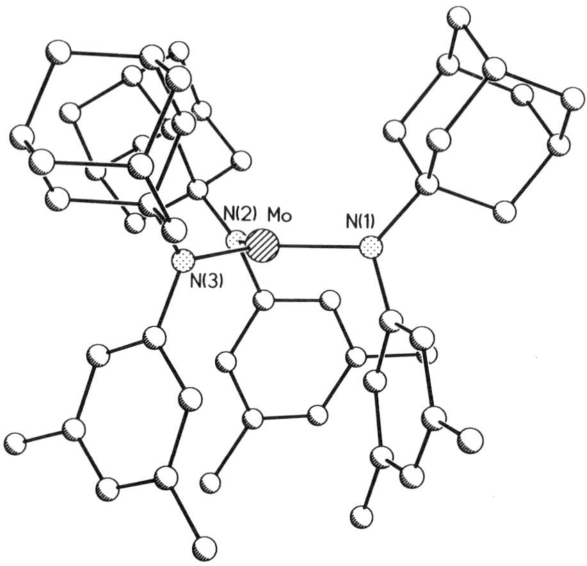

Figure 26. Structural drawing of $Mo[N(Ad)Ar]_3$; Ad = 1-adamantyl, Ar = 3,5-$C_6H_3Me_2$.

characterized structurally by X-ray diffraction, in a study revealing that NO indeed binds in the deep pocket presented by the three 1-adamantyl substituents (Fig. 27). Other than NO binding, the only significant structural reorganization vis-a-vis $Mo[N(Ad)Ar]_3$ is a pyramidalization at molybdenum to accommodate a pseudotetrahedral geometry in the four-coordinate complex.

N. $Mo[N(R)Ar_{4F}]_3$ [R = $C(CD_3)_2Me$, Ar_{4F} = 4-C_6H_4F]

1. Synthesis and Characterization

White $Li[N(R)Ar_{4F}](OEt_2)$ was prepared according to the published procedure for $Li[N(R)Ar](OEt_2)$ (34), which was substituting 4-fluoroaniline for 3,5-dimethylaniline. A solution of $Li[N(R)Ar_{4F}](OEt_2)$ (20.1 mmol) in ether was chilled until frozen, and then allowed to thaw. To the thawing mixture was added $MoCl_3(thf)_3$ (10.0 mmol). The resulting orange suspension was stirred and, after 30 min, rapidly acquired a brown color. Monitoring by ^2H NMR spectroscopy indicated an extent of reaction of 70% after 90 min, and greater than 80% after 2 h. The mixture was filtered through Celite on a frit after 3 h, and all volatile matter was removed from the filtrate. The crude dark brown powder was dissolved in ether, and the solution was sparged with Ar prior to storage in a tightly capped vessel at −35°C overnight. Green-tinted yellow-

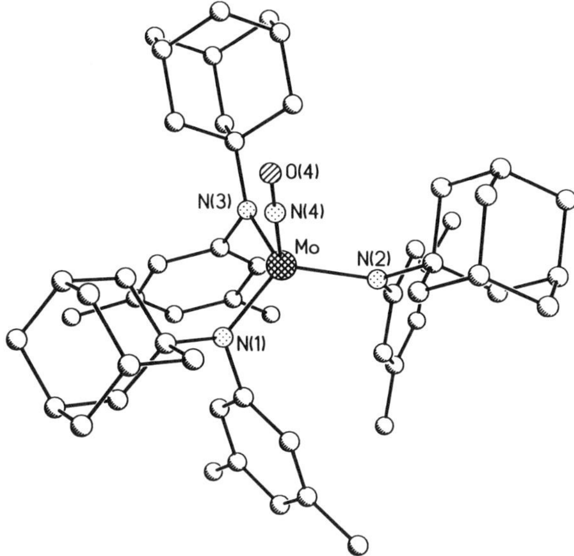

Figure 27. Structural drawing of $(ON)Mo[N(Ad)Ar]_3$; Ad = 1-adamantyl, Ar = 3,5-$C_6H_3Me_2$.

brown powder was isolated by filtration on a sintered glass frit (52.5% yield) (53). The characteristic ^2H NMR signal for $Mo[N(R)Ar_{4F}]_3$ appears at δ 69.37 ppm ($\Delta\nu_{1/2}$ = 36 Hz). Magnetic susceptibility data [$\mu_{eff}(C_6D_6$, 21.0°C) = 3.74 μ_B] indicate a configuration with three unpaired electrons for the complex, as for the other known three-coordinate molybdenum(III) amido complexes. Like $Mo[N(R)Ar]_3$, $Mo[N(R)Ar_{4F}]_3$ is reactive toward N_2 such that it is best purified and manipulated in the absence thereof.

2. *Reaction with $NMo(NMe_2)_3$*

This reaction was carried out in the course of studying complete intermetal nitrogen-atom transfer reactions (53). The complexes $NMo(NMe_2)_3$ (~0.5 mmol) and $Mo[N(R)Ar_{4F}]_3$ (1 equiv) were mixed as solids in a scintillation vial. Ether was added to the mixture at 28°C, while stirring vigorously. After about 10 s, the resulting dark blue-green homogeneous solution was placed in a freezer (−35°C). Over a period of 4 h, dark green crystals and powder separated from the solution (Scheme 45). The solid product, formulated as $(Me_2N)_3MoNMo$-$[N(R)Ar_{4F}]_3$ was isolated in 74% yield and exhibited a characteristic ^2H NMR signal (46 MHz, C_7H_8, 22°C) at δ 5.49 ppm ($\Delta\nu_{1/2}$ = 14 Hz). Fluorine-19 NMR, analytical, and SQUID magnetic susceptibility data [μ_{eff}(5–300 K) =

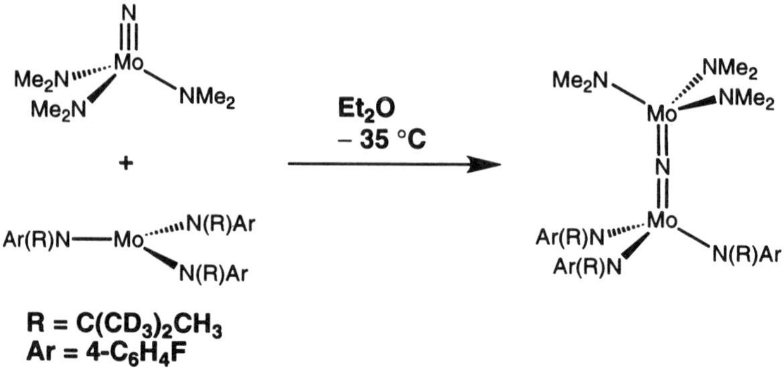

$N \equiv Mo(NMe_2)_3$ + $Mo[N(R)Ar]_3$ $\xrightarrow[-35\ ^\circ C]{Et_2O}$ $(Me_2N)_3Mo{=}N{=}Mo[N(R)Ar]_3$

$R = C(CD_3)_2CH_3$
$Ar = 4\text{-}C_6H_4F$

Scheme 45

1.56 μ_B] were consistent with formulation of the complex as a linear μ-nitrido dimolybdenum complex containing one unpaired electron.

Salient features of the structure of $(Me_2N)_3MoNMo[N(R)Ar_{4F}]_3$, as determined by X-ray crystallography (Fig. 28), are a linear μ-nitrido bridge and crystallographic threefold symmetry about the Mo(μ-N)Mo axis. Disorder about the threefold axis, evident in the $(Me_2N)_3Mo$ moiety, was accounted for by a model in which one dimethylamido group is oriented with its NC_2 plane per-

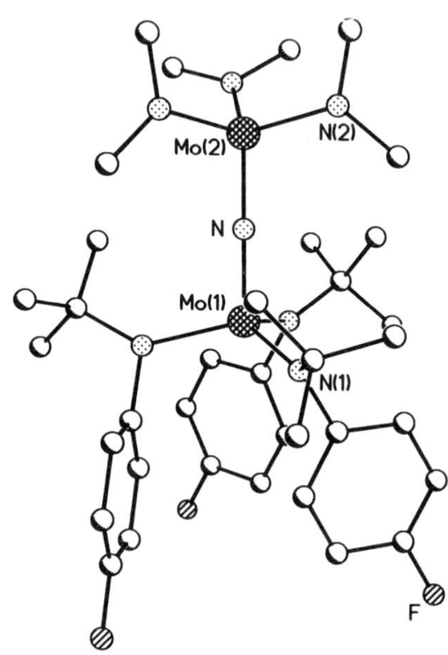

Figure 28. Structural drawing of $(Me_2N)_3Mo(\mu\text{-}N)Mo(N[R]Ar_F)_3$; R = $C(CD_3)_2Me$, Ar_F = $4\text{-}C_6H_4F$.

pendicular to the molecular threefold axis, and the other two NMe_2 groups adopt a parallel orientation thereto. In this model, although each molecule of $(Me_2N')_3MoNMo[N(R)Ar_{4F}]_3$ conforms to C_1 point symmetry, three equally probable orientations of the molecule in the unit cell result in crystallographic threefold symmetry and the observed disorder. Observation of the C_1 symmetric structure of $(Me_2N)_3MoNMo[N(R)Ar_{4F}]_3$ as opposed to an alternative C_3 symmetric structure can be attributed to a Jahn–Teller distortion for this $(1\pi_u)^4(1\pi_g)^3$ system. Although the μ-nitrido nitrogen atom appears to be symmetrically disposed between the inequivalent Mo centers, relatively high estimated standard deviations associated with this atom prohibit a firm conclusion in this regard. The thermal stability of $(Me_2N)_3MoNMo(N[R]Ar_{4F})_3$ is probably not great [cf. the reaction of $Mo[N(t\text{-}Bu)Ph]_3$ with $NMo(NMe_2)_3$ described above].

O. W(OSi—*t*-Bu$_3$)$_2$(N—*t*-Bu)

1. *Synthesis and Characterization*

The four-coordinate tungsten(VI) bis(imido)bis(amido) complex $W(N{-}t\text{-}Bu)_2(NH{-}t\text{-}Bu)_2$ served as the tungsten starting material in the preparation of three-coordinate $W(OSi{-}t\text{-}Bu_3)_2(N{-}t\text{-}Bu)$ (172, 173) (Scheme 46). Heating $W(N{-}t\text{-}Bu)_2(NH{-}t\text{-}Bu)_2$ together with 2 equiv of $HOSi{-}t\text{-}Bu_3$ in benzene led to expulsion of 2 equiv of *tert*-butylamine and to formation of colorless $W(N{-}t\text{-}Bu)_2(OSi{-}t\text{-}Bu_3)_2$, which was isolated in 81% yield. Treatment in benzene of the latter complex with 3 equiv of dry HCl led to loss of *tert*-butylamine hydrochloride, and to formation of light-yellow $(Cl)_2W(OSi{-}t\text{-}Bu_3)_2(N{-}t\text{-}Bu)$, isolated in 88% yield subsequent to crystallization from hexane. Conversion to three-coordinate emerald green $W(OSi{-}t\text{-}Bu_3)_2(N{-}t\text{-}Bu)$ was accomplished in 91% isolated yield by treatment with magnesium dust in ether, the byproduct being $MgCl_2$. The $W(OSi{-}t\text{-}Bu_3)_2(N{-}t\text{-}Bu)$ complex was found to have limited thermal stability in hydrocarbon solvents ($t_{1/2}$ = ~1 h), although the compound is stable in the solid form. The ^1H NMR and IR spectra of $W(OSi{-}t\text{-}Bu_3)_2(N{-}t\text{-}Bu)$ were indicative of a monomeric, diamagnetic species.

An X-ray structural study carried out for $W(OSi{-}t\text{-}Bu_3)_2(N{-}t\text{-}Bu)$ verified the proposed monomeric formulaton (Fig. 29). A trigonal planar coordination geometry at tungsten is indicated by the sum of bond angles (359.9°) about that element. Relatively short tungsten–oxygen bonds were observed (~1.82 Å), and the O—W—O angle [127.4(6)°] is larger than the two N—W—O angles (~116°). The tungsten–nitrogen bond length [1.658(17) Å] is consistent with a pseudotriple bond.

Given its structure, diamagnetism, and electron count, $W(OSi{-}t\text{-}Bu_3)_2(N{-}t\text{-}Bu)$ is isolobal (22) with $Os(N{-}2,6\text{-}C_6H_3{-}t\text{-}Pr_2)_3$ and $Ta(OSi{-}t\text{-}Bu_3)_3$ (see below and above, respectively), explaining why the complex does not read-

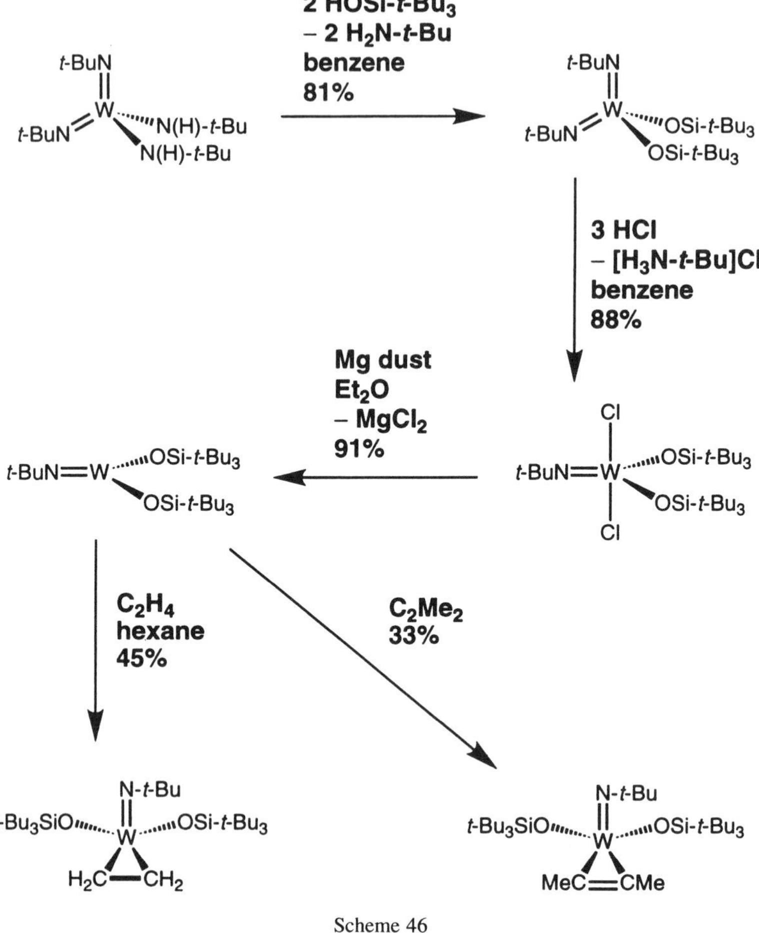

Scheme 46

ily bind simple σ donors (e.g., thf). The tungsten-based ''lone pair'' of electrons is thought to reside in a molecular orbital largely of d_{z^2} character, taking the NO_2 plane as the *xy* plane.

2. *Reaction with Ethylene*

Treatment of a hexane solution of W(OSi—*t*-Bu$_3$)$_2$(N—*t*-Bu) with ethylene generated orange diamagnetic (H$_4$C$_2$)W(OSi—*t*-Bu$_3$)$_2$(N—*t*-Bu) which, according to its ^{1}H and ^{13}C NMR spectra is C_s symmetric with the ethylene C—C vector parallel with the two oxygens atoms. Such an experiment carried out on

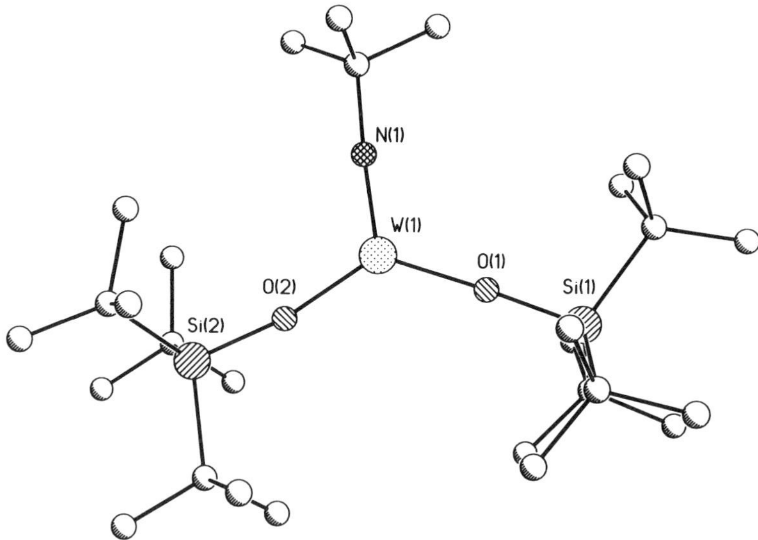

Figure 29. Structural drawing of $W(N{-}t\text{-}Bu)(OSi{-}t\text{-}Bu_3)_2$.

a scale of about 0.7 mmol led to the isolation of the ethylene complex in 45% yield. Variable-temperature NMR studies indicated a barrier to rotation about the tungsten–ethylene linkage of 15.3 kcal mol^{-1} (T_{coal} = 69°C).

3. *Reaction with 2-Butyne*

A reaction of $W(OSi{-}t\text{-}Bu_3)_2(N{-}t\text{-}Bu)$, carried out on an approximate 0.9-mmol scale, with 2-butyne resulted in clean formation of colorless diamagnetic $(Me_2C_2)W(OSi{-}t\text{-}Bu_3)_2(N{-}t\text{-}Bu)$, which was isolated subsequently in 33% yield. The ^1H NMR spectra of the complex are indicative of a C_s symmetric species.

VII. GROUP 7 (VII B) COMPLEXES

A. $Mn[N(SiMe_3)_2]_3$

1. *Synthesis and Characterization*

This tris(disilylamide) of manganese(III) was sought unsuccessfully (174) until, in 1989, Power and co-workers (175) described a successful synthesis

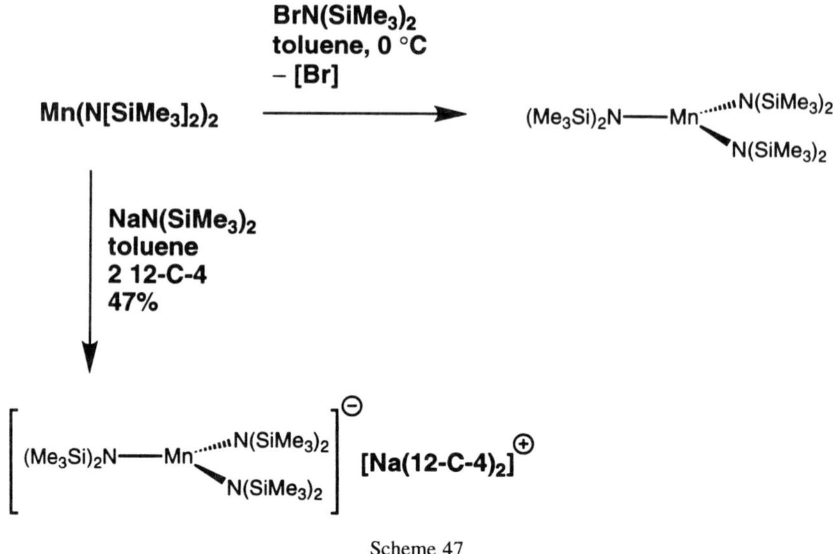

Scheme 47

employing the unorthodox reagent $BrN(SiMe_3)_2$. In this approach, the readily accessible manganese(II) bis(disilylamide) $Mn[N(SiMe_3)_2]_2$ was treated with $BrN(SiMe_3)_2$ in toluene at about 0°C (Scheme 47). The $Mn[N(SiMe_3)_2]_3$ complex was isolated as violet rod-like crystals (mp 108–110°C) in unspecified yield. The method was also applicable to preparation of $Co[N(SiMe_3)_2]_3$ and $Fe[N(SiMe_3)_2]_3$ (see below) from their respective metal(II) bis(disilylamide) precursors. By way of explaining the success of this synthetic method, it was pointed out that among the trivalent first-row transition metal trihalides the manganese and cobalt variants are the most strongly oxidizing. Accordingly, reactions between, for example, $LiN(SiMe_3)_2$ and $MnCl_3$, presumably proceed with reduction of the metal center. In commendable fashion, Power and co-workers (175) did not speculate as to the mechanism of formation of $Mn[N(SiMe_3)_2]_3$ ensuing from treatment of $Mn[N(SiMe_3)_2]_2$ with $BrN(SiMe_3)_2$; neither was the fate of the bromine atom commented upon.

Transitions in the electronic spectrum of high-spin d^4 $Mn[N(SiMe_3)_2]_3$, assigned a ground-state ^5D term, were observed at 568 and 470 nm. These were tenatively assigned as *d–d* transitions (cf. the cobalt case).

X-ray diffraction revealed $Mn[N(SiMe_3)_2]_3$ to be isomorphous with the other structurally characterized first-row transition metal $Mn[N(SiMe_3)_2]_3$ complexes (M = Ti, V, Cr, Fe, Co) (Fig. 30). The molecules belong to the D_3 point group.

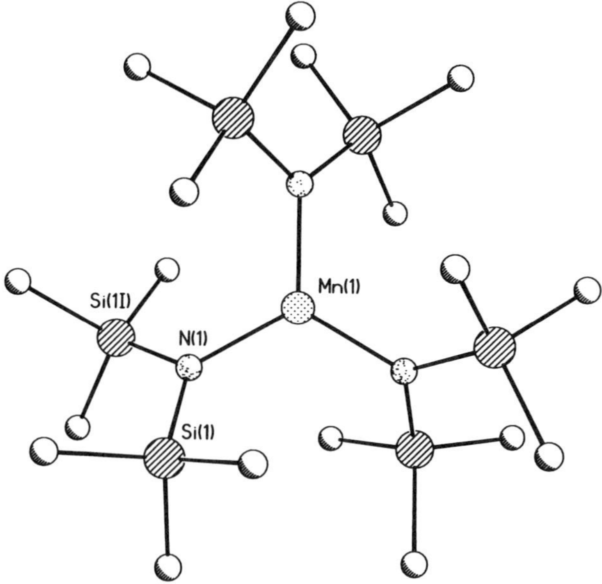

Figure 30. Structural drawing of $Mn(N[SiMe_3]_2)_3$.

B. $\{Mn[N(SiMe_3)_2]_3\}^-$

1. *Synthesis and Characterization*

Isoelectronic with neutral iron tris(hexamethyldisilazide) is the anion $\{Mn[N(SiMe_3)_2]_3\}^-$, which has recently been isolated as its $[Na(12\text{-crown-}4)_2]^+$ salt (176). This was accomplished via treatment of the manganese(II) bis(hexamethyldisilazide) complex $Mn(N[SiMe_3]_2)_2$, itself known to be dimeric in the solid state, with 1 equiv of $NaN[SiMe_3]_2$ in toluene in the presence of 2 equiv of 12-crown-4. The dark brown precipitate was isolated by filtration, washed with hexane, and dried under vacuum to provide $\{Mn[N(SiMe_3)_2]_3\}$ $[Na(12\text{-crown-}4)_2]$ in 47% yield (scale ~20 mmol).

X-ray crystallography showed the $\{Mn[N(SiMe_3)_2]_3\}^-$ ion to exhibit D_3 point symmetry (Fig. 31), much as was found (see below) for neutral $Mn[N(SiMe_3)_2]_3$. The propellor dihedrals in the complex are between 44 and 54°; these angles describe the relationship between the interpenetrating MnN_3 and $NMnSi_2$ trigonal planes. The manganese–nitrogen bond lengths of about 2.075(5) Å for the anion $\{Mn[N(SiMe_3)_2]_3\}^-$ are longer by about 0.018 Å than for neutral $Mn[N(SiMe_3)_2]_3$, indicative of high-spin manganese(II).

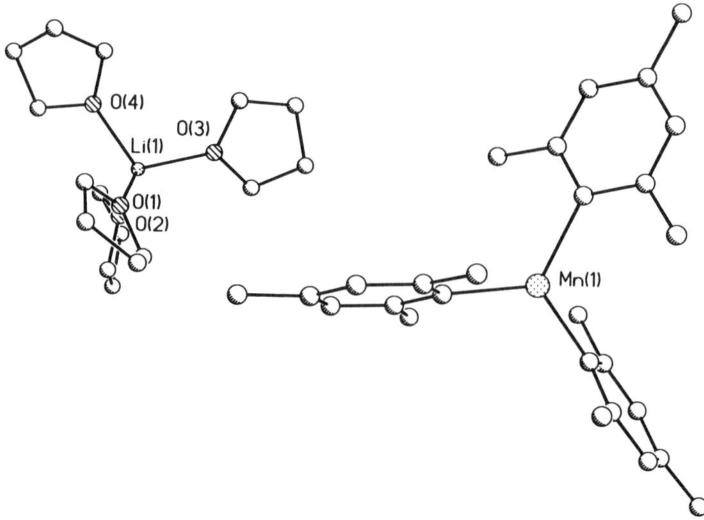

Figure 31. Structural drawing of $[Mn(Mes)_3][Li(thf)_4]$.

C. $[Mn(Mes)_3]^-$

1. Synthesis and Characterization

Power and co-workers (12) prepared this anionic trigonal manganese(II) complex as its $[Li(thf)_4]^+$ salt (12, 177). Since several two-coordinate manganese(II) complexes have been authenticated, it is reasonable to view $[Mn(Mes)_3][Li(thf)_4]$ as a functional equivalent of $Mn(Mes)_2$ and mesityllithium. The salt, $[Mn(Mes)_3][Li(thf)_4]$, was prepared by addition of 3 equiv of mesityllithium to MnI_2 in a THF/Et_2O/dioxane solvent mixture (Scheme 48). The $[Mn(Mes)_3][Li(thf)_4]$ complex was isolated in 31% yield as pale brown crystals.

X-ray crystallography was used to characterize $[Mn(Mes)_3][Li(thf)_4]$, which crystallizes with separate cations and anions (Fig. 31). The anion $[Mn(Mes)_3]^-$

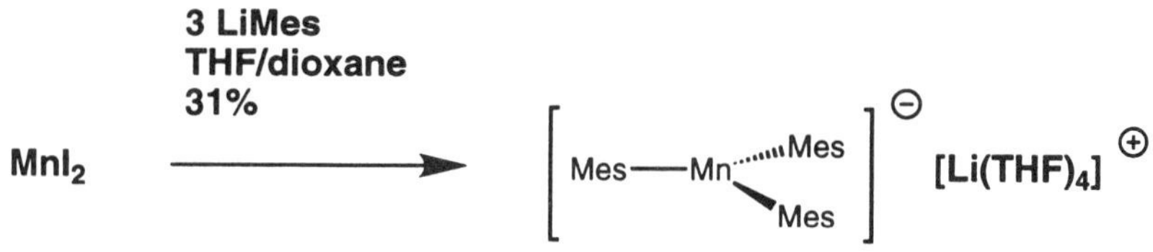

MnI_2 —(3 LiMes, THF/dioxane, 31%)→ $[Mes{-}Mn(Mes)_2]^{\ominus}$ $[Li(THF)_4]^{\oplus}$

Scheme 48

exhibits planarity at manganese, with irregular C—Mn—C angles. Power and co-workers (12) comment that the mononuclear nature of $[Mn(Mes)_3]^-$ can be attributed to the steric requirements of the mesityl groups, given that related anions prepared with, for example, phenyllithium exist as oligomers associated via bridging phenyl groups (12).

D. $Mn[3\text{-}Me\text{-}1,5\text{-}(Me_3Si)_2C_5H_4]_3K$

1. Synthesis and Characterization

Ernst and co-workers (63) discovered that treatment of $K[3\text{-}Me\text{-}1,5\text{-}(Me_3Si)_2C_5H_4]_3$ with $MnCl_2$ in THF led to production of a red-brown color. Blood red crystals of the highly lipophilic manganese–potassium complex $Mn[3\text{-}Me\text{-}1,5\text{-}(Me_3Si)_2C_5H_4]_3K$ were isolated in good yield from hexanes at −90°C (several crops) (Scheme 49). Magnetic susceptibility measurements obtained in THF revealed a high-spin configuration for the manganese center, with μ = 5.91 μ_B (178).

X-ray crystallography revealed an intriguing threefold symmetric structure for $Mn[3\text{-}Me\text{-}1,5\text{-}(Me_3Si)_2C_5H_4]_3K$ (Fig. 32) (178). As necessitated by symmetry, both metal centers lie on the crystallographic threefold axis, while the pentadienyl units adopt their fully extended conformation. The potassium ion makes close contacts with nine dienyl carbon atoms, three from each chain. The three-coordinate manganese center lies at the dienyl terminus, distal from potassium. The structure is intriguing in that the aggregate of three extended $[3\text{-}Me\text{-}1,5\text{-}(Me_3Si)_2C_5H_4]_3$ anions proffers two mutually remote metal ion binding sites. Thus the structure can be seen as comprised of a three-coordinate manganese(II) anion, similar to $[Mn(Mes)_3]^-$, conjoined with a potassium ion binding site. Zwitterionic in this sense, the compound $Mn[3\text{-}Me\text{-}1,5\text{-}(Me_3Si)_2C_5H_4]_3K$ retains the advantages of high solubility in hydrocarbon solvents and a definite stoichiometry.

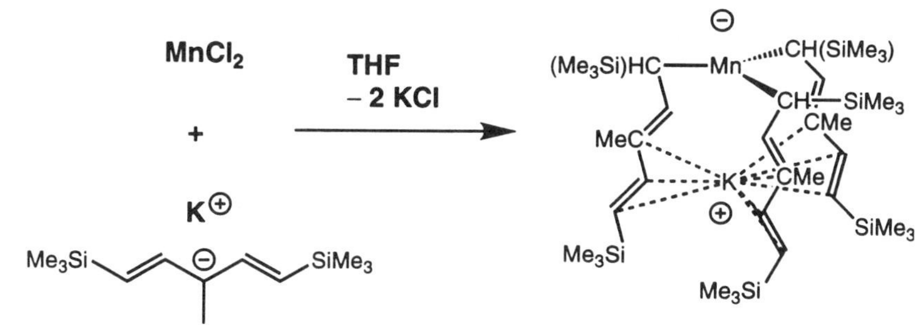

Scheme 49

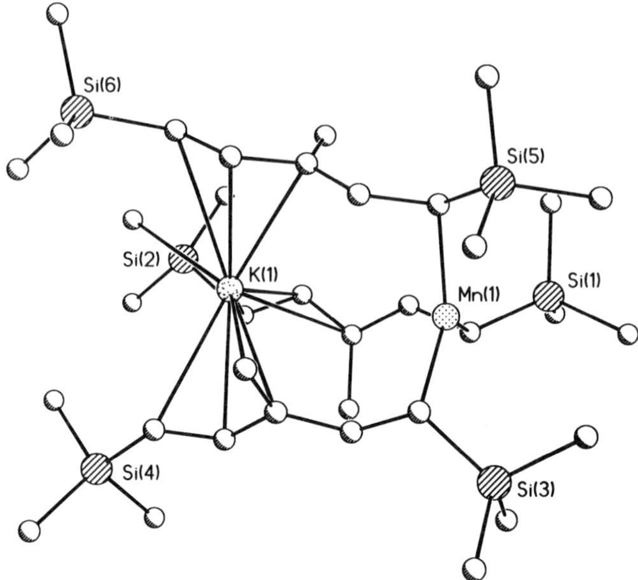

Figure 32. Structural drawing of $Mn[3\text{-}Me\text{-}1,5\text{-}(Me_3Si)_2C_5H_4]_3K$.

E. $[Re(N{-}2,6\text{-}C_6H_3{-}i\text{-}Pr_2)_3]^-$

1. Synthesis and Characterization

The isolation of this trigonal anion (179, 180) (as various salts, see below) came on the heels of the first description of the analogous neutral osmium tris(imido) complex $Os(NAr')_3$ ($Ar' = 2,6\text{-}C_6H_3{-}i\text{-}Pr_2$). The latter is described below. Entry into the rhenium system involves reduction of the dark red crystalline chloro precursor $ClRe(NAr')_3$, which itself is obtained in 99% yield on a scale of 25 mmol from treatment of Re_2O_7 with H_2NAr', triethylamine, and chlorotrimethylsilane in dichloromethane.

Reduction of $ClRe(NAr')_3$ with 0.5% sodium/mercury amalgam in THF, on a scale of about 0.7 mmol, produced the salt $[Re(NAr')_3][Na(thf)_2]$ as brown microcrystals in 97% yield. This particular salt of the $[Re(NAr')_3]^-$ anion is unlikely to contain well-separated anions and cations. It was proposed (179, 180) that sodium coordination to the imido nitrogens is probably operative. However, this salt is a useful synthon giving chemistry representative of $[Re(NAr')_3]^-$.

Reduction of $ClRe(NAr')_3$ with 0.5% sodium/mercury amalgam in THF, in the presence of $[NEt_4]Cl$, afforded the salt $[Re(NAr')_3][NEt_4]$ as a tan powder in 88% yield on an approximate 7-mmol scale (Scheme 50). Recrystallization of the salt from THF/Et_2O yielded red needles. The salt dissolves in PhBr, in

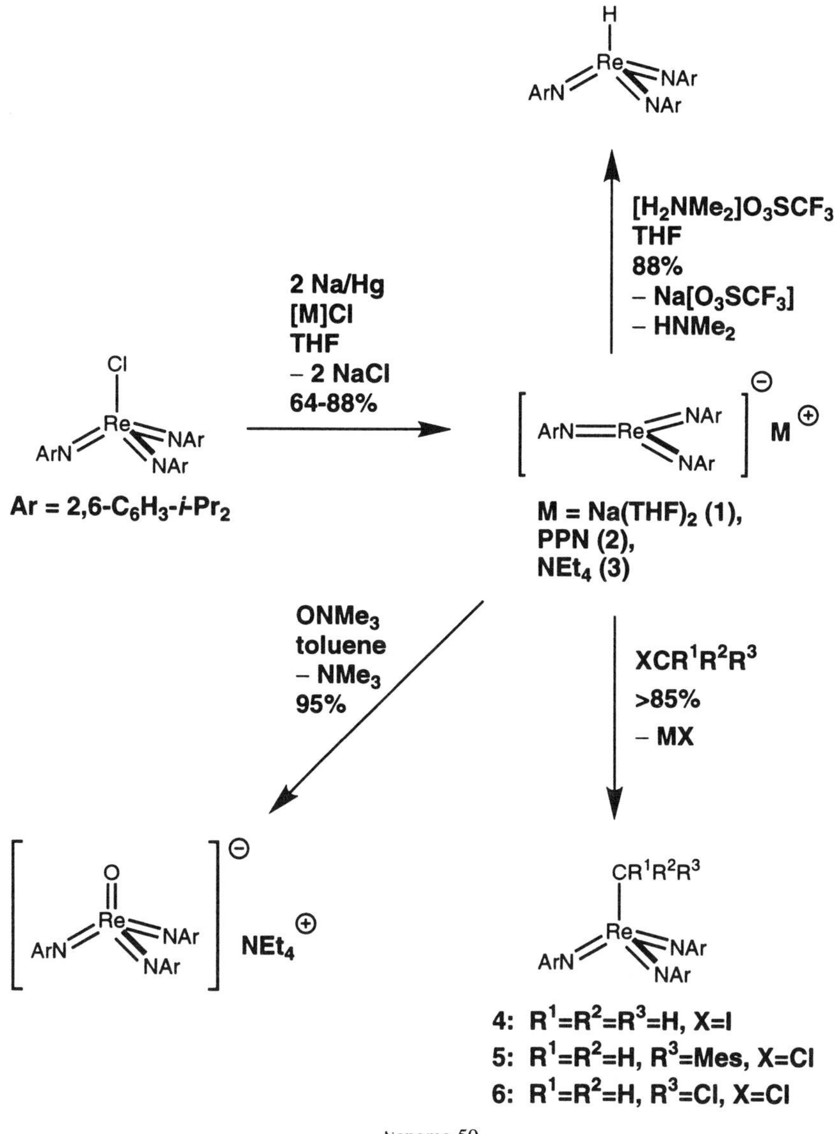

Scheme 50

which solvent it gives ^{1}H NMR data consistent with the proposed trigonal planar structure of the diamagnetic anion.

Reduction of $ClRe(NAr')_3$ with 0.5% sodium/mercury amalgam in THF, in the presence of $[N(PPh_3)_2]Cl$, afforded the salt $[Re(NAr')_3][N(PPh_3)_2]$ as ruby red crystals in 64% yield on an approximate 1-mmol scale. The ^{1}H NMR data

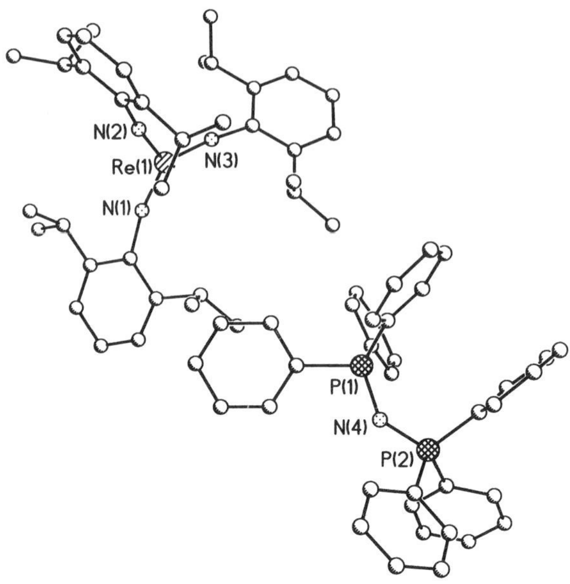

Figure 33. Structural drawing of $[Re(N{-}2,6\text{-}C_6H_3{-}i\text{-}Pr_2)_3][N(PPh_3)_2]$.

were difficult to obtain for the compound due to its low solubility; analytical data and an X-ray diffraction study were consistent with the proposed formulation (Fig. 33).

The $[Re(NAr')_3][N(PPh_3)_2]$ complex crystallized in the space group $P\bar{1}$; discrete, well-separated anions and cations are evident in the structure. The rhenium center is trigonal planar, as had been found previously for neutral, isoelectronic, $Os(NAr')_3$. In the structure of $[Re(NAr')_3]^-$, each imido ligand was found to be crystallographically distinct, with the three N—Re—N angles [116.1(4), 128.0(4), and 115.8(4)°] reflective of lower than ideal threefold symmetry. The asymmetry also is maintained in the three dihedral angles (73, 95, and 68°) relating the aryl planes with the N_3 plane.

It was stated that $[Re(NAr')_3][NEt_4]$ is unreactive toward ethylene, acetylene, and acetone, in contrast to the behavior of the osmium analogue $Os(NAr')_3$. Some reactions of $[Re(NAr')_3]^-$, in its various salts, are synopsized below.

2. *Reaction with a Proton Source*

Dimethylammonium triflate underwent a reaction with $[Re(NAr')_3][Na(thf)_2]$ in THF, on a scale of 3 mmol, to provide a magenta solution from which $HRe(NAr')_3$ was isolated as a magenta powder in 88% yield. The neutral diamagnetic hydrido complex exhibited a ^1H NMR signal (δ 7.24) attributed to the hydrido function. Also, treatment of $HRe(NAr')_3$ with CCl_4 appeared, according to a small-scale ^1H NMR experiment, to return $ClRe(NAr')_3$ in quantiative fashion, as might be expected. Other characterization data were likewise in accord with the given formulation of $HRe(NAr')_3$.

3. Reaction with Alkyl Halides

The nucleophilic character of $[Re(NAr')_3]^-$ is evident in its reactions with alkyl halides. Along these lines, methyl iodide was found to convert $[Re(NAr')_3][Na(thf)_2]$ to the methyl complex $MeRe(NAr')_3$, which was isolated as red crystals in 94% yield, in an experiment carried out on an approximate 0.6-mmol scale.

Also, the benzylic chloride $ClCH_2Mes$ (Mes = 2,4,6-$C_6H_2Me_3$) converted $[Re(NAr')_3][Na(thf)_2]$ to $MesCH_2Re(NAr')_3$, a red crystalline compound isolated in 98% yield from a reaction carried out on a scale of about 0.6 mmol. Evidently purification of $(MesCH_2)Re(NAr')_3$ was hampered due to the presence of small amounts of 1,2-dimesitylethane.

In an unusual display of nucleophilicity, the anionic component of $[Re(NAr')_3][NEt_4]$ was converted to $ClCH_2Re(NAr')_3$ upon treatment with dichloromethane. Purple $ClCH_2Re(NAr')_3$ was obtained in 85% yield from an experiment carried out on a scale of about 0.7 mmol. A ^1H NMR signal (δ 5.67) integrating correctly was assigned to the chloromethyl substituent.

4. Reaction with Trimethylamine-N-Oxide

The anionic oxo complex $[ORe(NAr')_3]^-$ was obtained in 95% yield as an orange powder, in an experiment carried out on a scale of about 0.6 mmol, upon treatment of $[Re(NAr')_3][NEt_4]$ with Me_3NO in toluene.

VIII. GROUP 8 (VIII) COMPLEXES

A. $[Fe(Mes)_3]^-$

1. Synthesis and Characterization

Nearly colorless trimesitylferrate, as its $[Li(dioxane)_4]^+$ salt, was obtained in very good (89%) yield according to the following procedure (Scheme 51) (181). To mesityllithium suspended in Et_2O was added $\frac{1}{3}$ equiv ferrous bromide in portions. Crystals of $[Fe(Mes)_3][Li(dioxane)_4]$ were obtained subsequent to addition of dioxane and removal of LiBr. Magnetic measurements performed

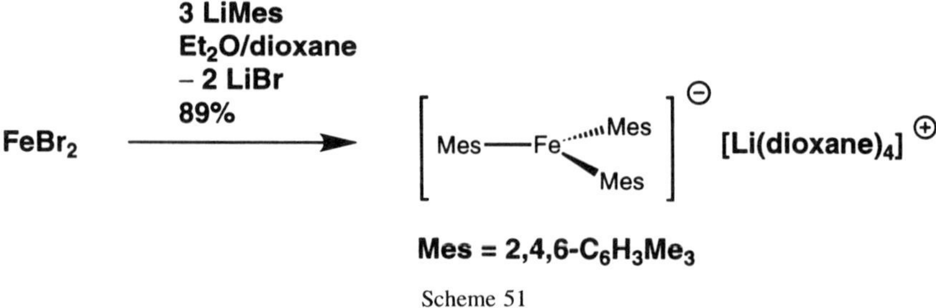

Mes = 2,4,6-$C_6H_3Me_3$

Scheme 51

on $[Fe(Mes)_3][Li(dioxane)_4]$ are consistent with high-spin iron(II) ($\mu_{eff} = 5.56$ μ_B). Although the nuclearity of trimesitylferrate has not been determined beyond a shadow of a doubt, it is indeed likely that $[Fe(Mes)_3]^-$ can exist as a discrete anion if paired with an appropriate cation. This compound, then, is a member of the expanding series of compounds $[M(Mes)_3]^{n-}$ ($n = 1$, M = Mn, Fe, Co, Ni; $n = 0$, M = V, Cr, Rh, Ir).

B. $\{Fe[N(SiMe_3)_2]_3\}$

1. Synthesis and Characterization

While the neutral iron tris(hexamethyldisilazide) is one of the earliest known three-coordinate transition metal complexes, the anionic version was first described in 1996 (Scheme 52). It was obtained from the reaction of neutral $Fe[N(SiMe_3)_2]_3$ with an equivalent of $NaN(SiMe_3)_2$ in the presence of 2 equiv of 12-crown-4 (176). In using sodium hexamethyldisilazide as the sacrificial reductant, 0.5 equiv of $(Me_3Si)_2NN(SiMe_3)_2$ was the proposed byproduct. From such an experiment, carried out on an approximate 2.0-mmol scale in toluene,

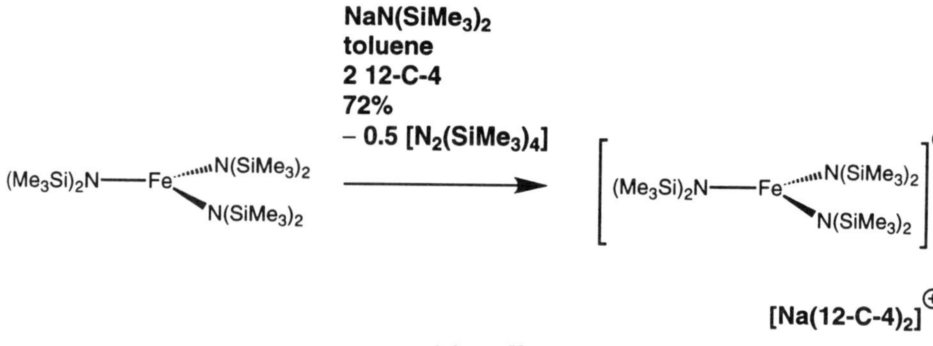

$NaN(SiMe_3)_2$
toluene
2 12-C-4
72%
– 0.5 $[N_2(SiMe_3)_4]$

$(Me_3Si)_2N$—Fe$N(SiMe_3)_2$ $N(SiMe_3)_2$ → $[(Me_3Si)_2N$—Fe$N(SiMe_3)_2$ $N(SiMe_3)_2]^-$ $[Na(12\text{-}C\text{-}4)_2]^+$

Scheme 52

the salt $\{Fe[N(SiMe_3)_2]_3\}[Na(12\text{-crown-}4)_2]$ was isolated as a dark green precipitate in 72% yield.

Characterization of $\{Fe[N(SiMe_3)_2]_3\}[Na(12\text{-crown-}4)_2]$ by X-ray diffraction showed, for the anion, pseudo-D_3 symmetry quite reminiscent of the neutral tris(hexamethyldisilazide). The Fe—N bond lengths for the anion are about 1.983(5) Å, approximately 0.07 Å longer than is the case for the neutral species.

Reported in the same paper are syntheses and structural determinations for anionic manganese(II) and cobalt(II) tris(hexamethyldisilazide) complexes, again as salts with the $[Na(12\text{-crown-}4)_2]$ sandwich cation.

C. $Fe[N(R)Ar]_3$ [$R = C(CD_3)_2Me_3$, $Ar = 3,5\text{-}C_6H_3Me_2$]

1. *Synthesis and Characterization*

Whereas ferric chloride appeared to undergo reduction to iron(II) species upon treatment with the amido transfer agent $Li[N(R)Ar](OEt_2)$ (34), smooth substitution chemistry occurred upon addition of ferrous chloride to a benzene solution containing 3 equiv of the same amido transfer agent (Scheme 53). In such an experiment, carried out on a scale of about 8 mmol, the "ate" complex $[Ar(R)N]Fe[\mu\text{-}N(R)Ar]_2Li(OEt_2)$ was obtained in 86% recrystallized yield as a yellow-green solid (88). The structural assignment with two bridging and one terminal amide is consistent with ^2H NMR data for the complex (two peaks in a 2:1 ratio at δ −2.35 and 44.2 ppm, respectively). Oxidation of $[Ar(R)N]Fe[\mu\text{-}N(R)Ar]_2Li(OEt_2)$ with ferrocenium triflate (182) led to clean formation of the desired three-coordinate complex $Fe[N(R)Ar]_3$, which was easily separated from the byproducts ferrocene and lithium triflate. Dark blue $Fe[N(R)Ar]_3$, exhibiting a single peak at δ 54.6 ppm ($\Delta\nu_{1/2}$, 93 Hz) in its ^2H NMR spectrum, was isolated in 85% yield on an approximate 2-mmol scale from such a sequence. Analytical data and an Evans' method magnetic susceptibility measurement (25°C, $\mu_{eff} = 6.22\ \mu_B$) were consistent with the formulation of $Fe[N(R)Ar]_3$ as a homoleptic amide of high-spin iron(III). Chemical oxidation of so-called "ate" complexes has been used previously to synthesize neutral amides of iron(IV) (183) and manganese(III) (68).

D. $Fe[N_2Mo(N_3N)]_3$ [$N_3N = (Me_3SiNCH_2CH_2)_3N$]

1. *Synthesis and Characterization*

Plum colored, paramagnetic $Fe[N_2Mo(N_3N)]_3$ was obtained from reaction of the magnesium salt $(thf)_2Mg[N_2Mo(N_3N)]_2$, itself obtained by magnesium reduction of $ClMo(N_3N)$ in the presence of N_2, with ferrous chloride (Scheme

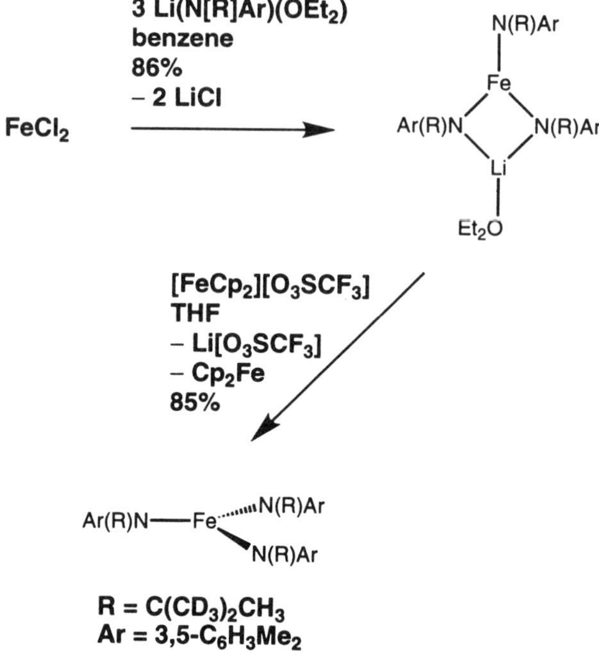

Scheme 53

54). Accordingly, treatment of $FeCl_2$ with 1 equiv of $(thf)_2Mg[N_2Mo(N_3N)]_2$ on a scale of about 0.5 mmol led to the isolation of $Fe[N_2Mo(N_3N)]_3$ as a pentane-soluble black/purple solid in 63% crude yield (184). A black magnetic solid presumed to be elemental iron was formed as a byproduct in the reaction, allowing a stoichiometric equation to be proposed for the process in which two-thirds of the iron goes to $Fe[N_2Mo(N_3N)]_3$, while the other one-third goes to iron metal.

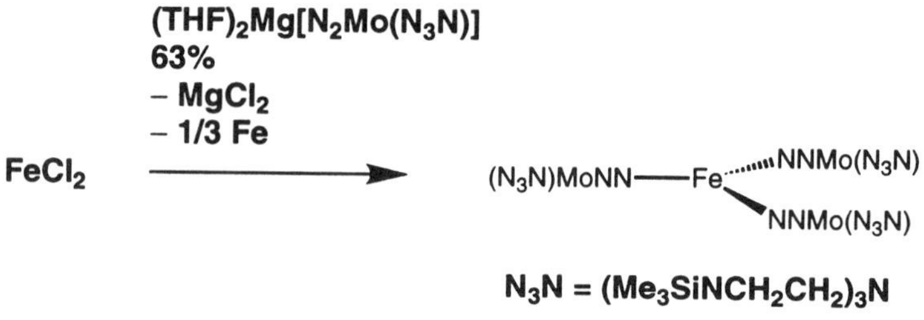

(THF)$_2$Mg[N$_2$Mo(N$_3$N)]
63%
– MgCl$_2$
– 1/3 Fe
FeCl$_2$ ⟶ (N$_3$N)MoNN—Fe(NNMo(N$_3$N))(NNMo(N$_3$N))

N$_3$N = (Me$_3$SiNCH$_2$CH$_2$)$_3$N

Scheme 54

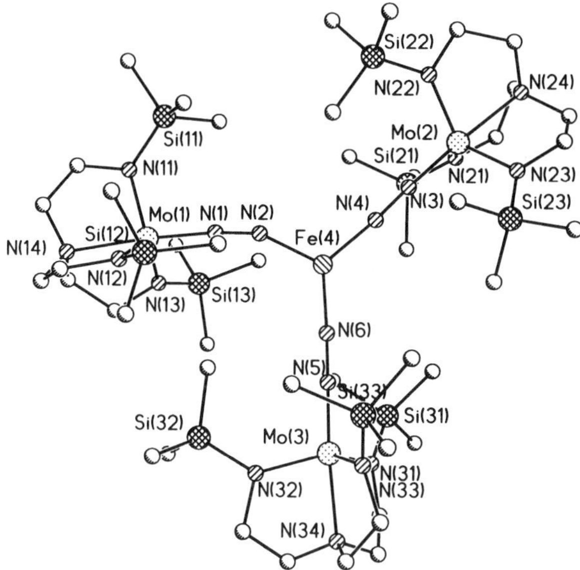

Figure 34. Structural drawing of $Fe[N_2Mo(N_3N)]_3$; $N_3N = (Me_3SiNCH_2CH_2)_3N$.

An X-ray structural study verified the trigonal planar geometry at iron for $Fe[N_2Mo(N_3N)]_3$ (Fig. 34). An intriguing facet of the structure is that, while two of the $N_2Mo(N_3N)$ ligands bind to iron with essentially linear $Fe(\mu\text{-}N_2)Mo$ segments, the other exhibits a distinct bend [156(2)°] at the iron-bound nitrogen atom.

Not surprisingly, given the complex makeup of the ligand structure about iron, the assignment of formal oxidation states to the metals in the tetranuclear complex represents a formidable challenge. Temperature-dependent solid-state magnetic susceptibility studies revealed $\mu = 6.03(3)\ \mu_B$ for $Fe(N_2Mo[N_3N])_3$, consistent with the presence of a sextet ground state for the complex. This result can be rationalized in two complementary simplistic ways. First off, if the central iron is assigned as high-spin iron(III), then the prediction of five unpaired electrons follows. Second, if the assumption is made that $Fe[N_2Mo(N_3N)]_3$ actually contains two different types of $N_2Mo(N_3N)$ ligand, two anionic and one neutral, then the central iron is high-spin iron(II) and would contribute four unpaired electrons to the total spin. The neutral $N_2Mo(N_3N)$ ligand, which one would be tempted to assign as the one exhibiting the bent FeNN function, represents an instance of low-spin molybdenum(III) and as such would contribute one unpaired electron to the overall spin. The observation of a bent FeNN moiety is consistent with a Jahn–Teller distortion brought about by the presence of three electrons in the doubly degenerate HOMO of threefold symmetric

$N_2Mo[N_3N]$. A complete description of the electronic structure of $Fe[N_2Mo(N_3N)]_3$ could conceivably involve some admixture of the two states alluded to here.

Synthetic, structural, and other characterization data pertaining to the neutral molecule $N_2Mo(N_3N)$ were in fact also given in the paper reporting on $Fe[N_2Mo(N_3N)]_3$ (184). It was remarked that $Fe[N_2Mo(N_3N)]_3$ is significant in being perhaps the only known iron/molybdenum bridging N_2 complex, in containing a trigonal planar iron atom, and in containing a new metalloligand, $N_2Mo(N_3N)$, which can be obtained in both anionic and neutral forms. It is also significant and instructive that whereas three-coordinate $Mo[N(R)Ar]_3$ appears to have a small equilibrium constant for N_2 binding (see above), the neutral complex $N_2Mo(N_3N)$ does not lose N_2 readily. The contrast is explicable in terms of stabilization of the low-spin form of hypothetical $Mo(N_3N)$ via binding of the axial tertiary amine.

E. $[Fe(SAr)_3]^-$ (Ar = 2,4,6-C_6H_2—*t*-Bu_3)

1. *Synthesis and Characterization*

In the wake of the discovery by X-ray crystallography of trigonal iron atoms in the FeMo-co of nitrogenase, an increased interest in unusual coordination environments for iron has developed. The anion described here is a manifestation of this increased interest (Scheme 55). A tetraphenylphosphonium salt of the anionic trigonal iron arylthiolate $[Fe(SAr)_3]^-$ (Ar = 2,4,6-C_6H_2—*t*-Bu_3) was prepared as follows (185). A pale yellow solution of Li(SAr) generated from HSAr and BuLi in toluene was added dropwise to a toluene solution of $[Fe(SAr)_2]_2$ (186). The mixture was stirred for 2 h, and then was added to a

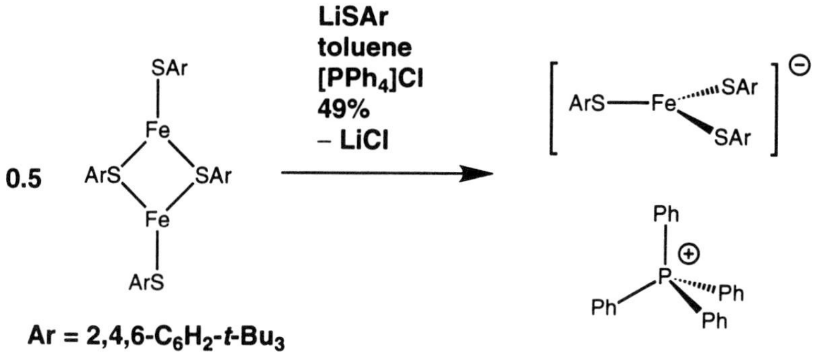

0.5 $[Fe(SAr)_2]_2$ —(LiSAr, toluene, $[PPh_4]Cl$, 49%, – LiCl)→ $[Fe(SAr)_3]^-$ $[PPh_4]^+$

Ar = 2,4,6-C_6H_2-*t*-Bu_3

Scheme 55

toluene suspension of Ph_4PCl. Ultimately a dark yellow precipitate formed, which was almost completely redissolved by the addition of a small volume of acetonitrile. Filtration, concentration, and storage of the solution at $-20°C$ provided yellow crystals of $[Fe(SAr)_3][PPh_4]$ in 49% yield as a toluene and acetonitrile solvate. Characterization data are as follows: $\mu_{eff} = 5.05\ \mu_B$ (298 K); mp > 95°C (dec). Zero-field Mössbauer spectra of $[Fe(SAr)_3][PPh_4]$ were obtained at 4.2 and 100 K. At these temperatures, respectively, δ 0.57 and 0.53 and $\Delta Eq = 0.81$ and 0.81 mm s^{-1}; these values serve to distinguish the trigonal monomeric species from related binuclear or tetrahedral iron thiolates. It was asserted that the small quadrupole splitting observed for $[Fe(SAr)_3]^-$ is consistent with its probable ground-state electron configuration, $(a')^2(e'')^2(e')^2$, in idealized C_{3h} symmetry.

The structure of $[Fe(SAr)_3][PPh_4]\cdot(\text{toluene})\cdot(\text{acetonitrile})_2$ was determined by X-ray crystallography (185) revealing pseudo-C_{3h} symmetry for the trigonal planar anion (Fig. 35). The structure of $[Fe(SAr)_3]^-$ was scrutinized carefully for agostic interactions with the result that the shortest Fe· · ·H contact, 2.74 Å, is greatly in excess of the sum (~ 1.7 Å) of iron and hydrogen covalent radii. In contrast, marginal agostic interactions are indicated for $[Co_2(SAr)_4]$, with d(Co· · ·H) = ~1.8 Å (186). It was concluded that no M· · ·H or M· · ·C interaction in $[Fe(SAr)_3]^-$ could be construed as coordinative.

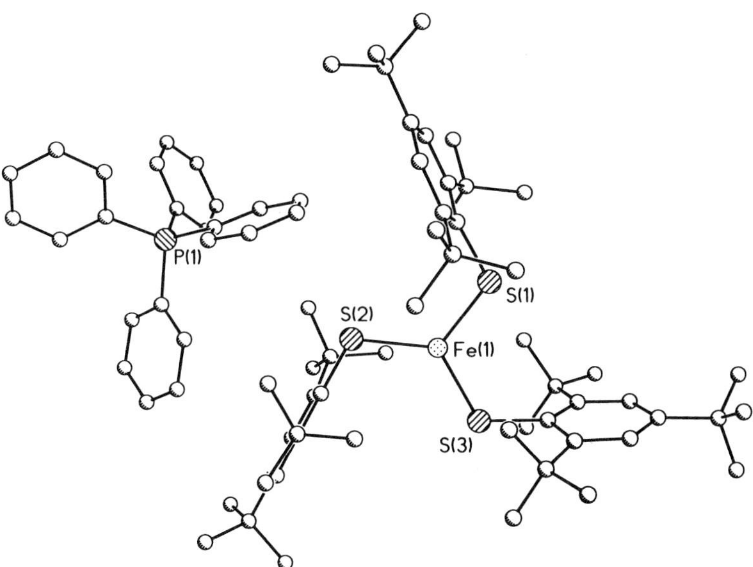

Figure 35. Structural drawing of $[Fe(SAr)_3][PPh_4]$; Ar = 2,4,6-C_6H_2—t-Bu_3.

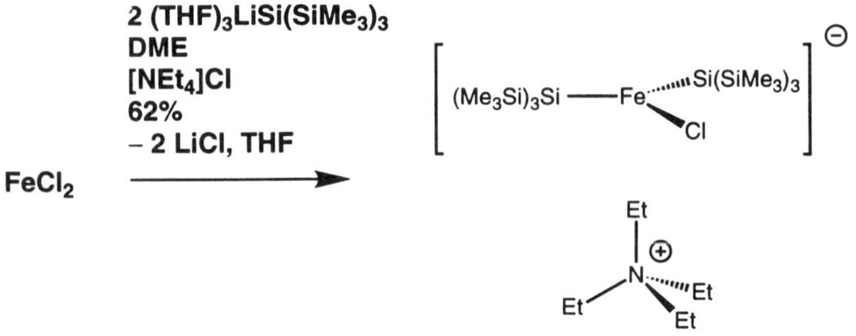

$$\mathrm{FeCl_2} \xrightarrow[\text{62\%}\ -2\ \text{LiCl, THF}]{2\ (\text{THF})_3\text{LiSi(SiMe}_3)_3,\ \text{DME},\ [\text{NEt}_4]\text{Cl}} [(\mathrm{Me_3Si})_3\mathrm{Si{-}Fe(Cl){-}Si(SiMe_3)_3}]^{\ominus}\ [\mathrm{NEt_4}]^{\oplus}$$

Scheme 56

F. $\{Fe(Cl)[Si(SiMe_3)_3]_2\}^-$

1. *Synthesis and Characterization*

In conjunction with their studies on the chemistry of compounds containing a metal–silicon bond, Tilley and co-workers (187) reported some three-coordinate derivatives of the hindered $Si(SiMe_3)_3$ ligand. For example, treatment of $FeCl_2$ in DME with 2 equiv of $(thf)_3Li[Si(SiMe_3)_3]$ produced purple $[Li(dme)_2]$-$\{Fe(Cl)[Si(SiMe_3)_3]_2\}$ ($\mu_{eff} = 6.3\ \mu_B$) in 76% yield (Scheme 56). The latter salt reacts in DME with NEt_4Cl to give $[NEt_4]\{Fe(Cl)[Si(SiMe_3)_3]_2\}$ ($\mu_{eff} = 5.8\ \mu_B$) in 62% yield. Structural characterization of the tetraethylammonium salt was achieved (X-ray diffraction), revealing unassociated anions and cations and a trigonal planar geometry at iron (Fig. 36). The large Si—Fe—Si angle [136.9(2)°] was attributed to steric demands of the gargantuan $Si(SiMe_3)_3$ ligands. The Fe—Si bond distances [av 2.490(6) Å] were thought to be the longest known at the time. The related anions $\{M(Cl)[Si(SiMe_3)_3]_2\}^-$ (M = Cr, Mn) were also described.

G. $Os(N{-}2,6\text{-}C_6H_3{-}i\text{-}Pr_2)_3$

1. *General*

This trigonal planar molecule (188, 189) has attracted attention because it presents an element of simplicity and elegance, stemming from the low coordination number, which has little precedent among metal–ligand multiply bonded systems. Its motif was one of several considered in a recent review on group theoretical analyses of metal–ligand multiply bonded transition metal systems (190).

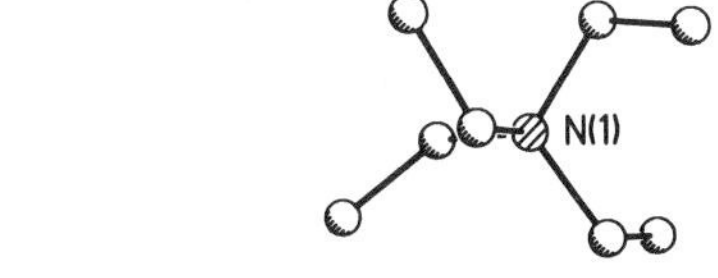

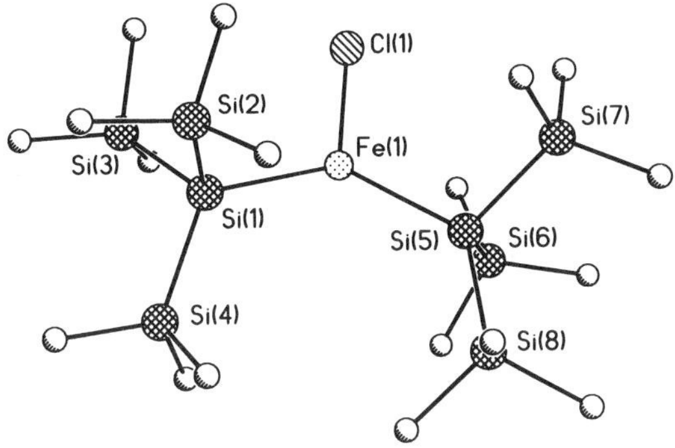

Figure 36. Structural drawing of $\{Fe(Cl)[Si(SiMe_3)_3]_2\}[NEt_4]$.

2. *Synthesis and Characterization*

$Os(NAr')_3$ ($Ar' = 2,6\text{-}C_6H_3\text{—}i\text{-}Pr_2$) has been obtained in 50% yield as red-brown prisms from the reaction of osmium tetroxide with 2,6-diisopropylphenyl isocyanate (3 equiv) in refluxing heptane, carried out on an approximate 8-mmol scale (Scheme 57). Introduction of the imido ligand by reaction of a metal oxo with an organic isocyanate, proceeding with elimination of carbon dioxide, is a standard synthetic strategy (146). In the present manifestation, however, the fate of the fourth oxo ligand was not investigated in detail. The three Ar′ residues in diamagnetic $Os(NAr')_3$ are equivalent on the ^1H and ^{13}C NMR time scales, and analytical and mass spectroscopic data agree with the formula of the compound as given.

An X-ray diffraction study was carried out for $Os(NAr')_3$ (Fig. 37), the structure of which was found to be quite similar to that of the isoelectronic rhenium analogue $[Re(NAr')_3]^-$ described above. The three N—Os—N angles were found to be 119.8(3), 120.1(2), and 120.1(2)°, reflecting a regular trigonal planar geometry at osmium; the molecule happens to display crystallographic twofold symmetry about one of the osmium–nitrogen bonds. Two of the Ar′ residues are oriented "up", with their best planes nearly perpendicular to the

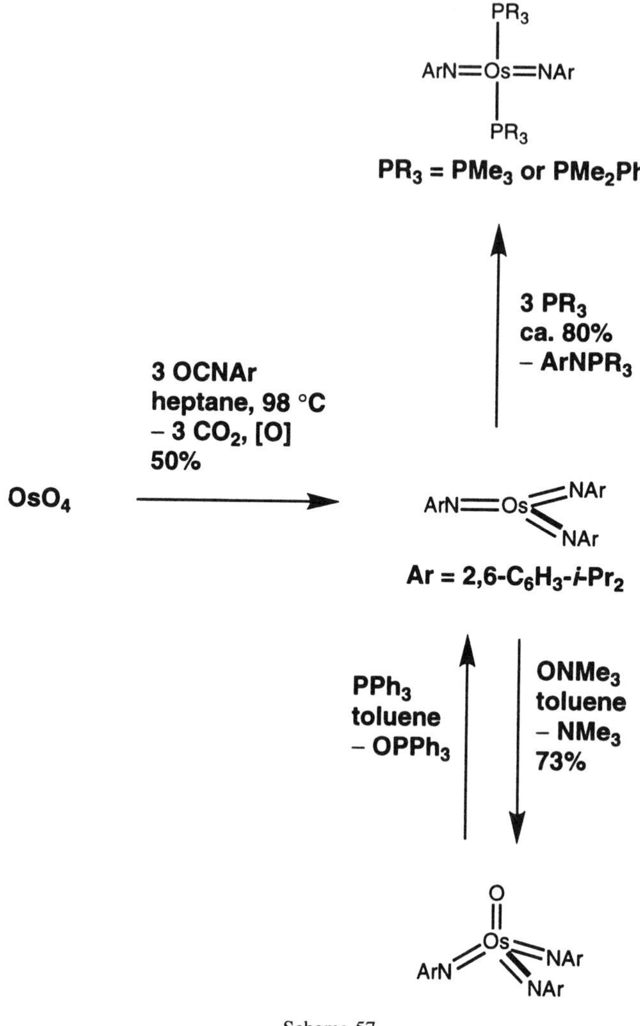

Scheme 57

N_3 plane, while the third Ar′ residue, the one lying on the twofold axis, orients its best plane nearly coplanar with the N_3 plane.

The electronic structure of $Os(NAr')_3$ has been discussed in terms of group theoretical analyses (188, 190) and it has also been the subject of an SCF–Xα–SW theoretical study (189). The X-ray structure showed $Os(NAr')_3$ to possess three linear imido linkages, that is, the three M—N—C(ipso) angles are all close to 180°. The group theoretical arguments and the MO calculations

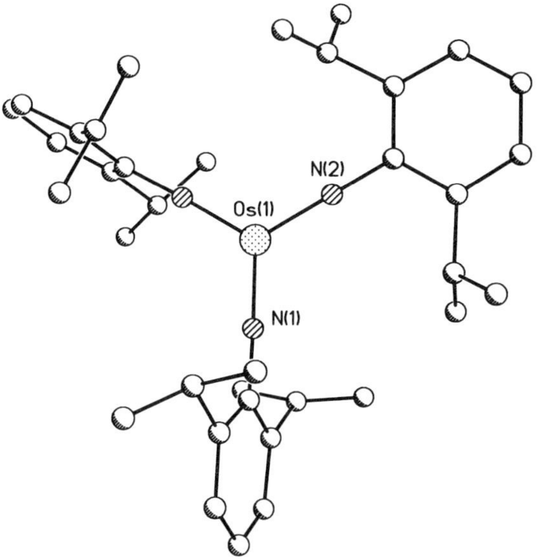

Figure 37. Structural drawing of $Os(N{-}2,6\text{-}C_6H_3{-}i\text{-}Pr_2)_3$.

both predict that, given idealized D_{3h} point symmetry, there is one nitrogen lone-pair symmetry adapted linear combination transforming as a_2' that finds no match on the metal and so must remain ligand localized. Were this not the case, the formal electron count would be 20 for the complex, since linear imido linkages are typically represented as dianions capable of donating $6e^-$ to a metal center. Other examples of linear imido linkages wherein full $6e^-$ donation by the (formally) dianionic imido would result in an unacceptably high formal electron count are known (191). Furthermore, Rothwell and co-workers (192) have shown that metal–oxygen distances do not correlate with M—O—C (ipso) angles in early metal aryloxide complexes.

3. *General Comments on Reactivity*

It was stated that $Os(NAr')_3$ does not react with Lewis bases including PPh_3, $AsPh_3$, quinuclidine, py, or ethyldimethylamine. Relatively small phosphines do react, as described below. The $Os(NAr')_3$ complex is stable to moist air in both the solid state and solution at 25°C, and the compound is both insoluble in and stable to water. Other reagents that, under various conditions, failed to react with $Os(NAr')_3$ include gaseous HCl in ether, $OPPh_3$, propylene oxide, norbornene, cyclopentene, and styrene. This level of inertness is somewhat surprising given the isolobal relationship of the compound with highly reactive

$Ta(OSi{-}t\text{-}Bu_3)_3$ (see above), but is readily rationalized in terms of oxidation state preferences for the two metals in question.

4. *Reaction with Phosphines*

Treatment of $Os(NAr')_3$ with 3 equiv of phosphines, such as PMe_3 or PMe_2Ph, in pentane solution at room temperature, leads to elimination of 1 equiv of phosphinimine and production of square planar bis(imido)-bis(phosphine) osmium complexes. In such an experiment, carried out on a scale of about 2 mmol, $Os(NAr')_2(PMe_2Ph)_2$ was obtained in 80% yield as red-purple crystals. The trans-$Os(NAr')_2(PMe_2Ph)_2$ complex was characterized structurally by X-ray diffraction.

5. *Reaction with Trimethylamine-N-Oxide*

Red-black crystalline $OOs(NAr')_3$ was obtained in 73% yield subsequent to treatment of $Os(NAr')_3$ with $ONMe_3$ in toluene at about 25°C, in an experiment carried out on a scale of about 0.3 mmol. The $OOs(NAr')_3$ complex was characterized by conventional spectroscopic techniques, and elemental analysis, and its reaction chemistry with olefins was also investigated. It was found that PPh_3 reacts with $OOs(NAr')_3$ regenerating $Os(NAr')_3$ and the corresponding phosphine oxide. The $OOs(NAr')_3$ complex is an analogue of the known species $OOs(N{-}t\text{-}Bu)_3$ (193).

IX. GROUP 9 (VIII) COMPLEXES

A. $Co[N(SiMe_3)_2]_3$

1. *Synthesis and Characterization*

This tris(disilylamide) of cobalt(III) was sought unsuccessfully (174) until, in 1989, Power and co-workers (175) described a successful synthesis employing the unorthodox reagent $BrN(SiMe_3)_2$ (cf. $Mn[N(SiMe_3)_2]_3$, described above). The $Co[N(SiMe_3)_2]_3$ complex was obtained upon treatment of $Co[N(SiMe_3)_2]_2$ with $BrN(SiMe_3)_2$ in toluene at about 0°C. The $Co[N(SiMe_3)_2]_3$ complex was isolated as dark olive-green crystals (mp 86–88°C) in unspecified yield.

Transitions in the electronic spectrum of high-spin d^6 $Co[N(SiMe_3)_2]_3$, assigned a ground-state ^5D term, were observed at 600 and 454 nm. These were tentatively assigned as *d–d* transitions (cf. the manganese case).

X-ray diffraction revealed $Co[N(SiMe_3)_2]_3$ to be isomorphous with the other structurally characterized first-row transition metal $M[N(SiMe_3)_2]_3$ complexes (M = Ti, V, Cr, Fe, Co). The molecules belong to the D_3 point group.

B. $\{Co[N(SiMe_3)_2]_3\}^-$

1. *Synthesis and Characterization*

Isoelectronic with the unknown neutral nickel tris(hexamethyldisilazide) is the anion $\{Co(N[SiMe_3]_2)_3\}^-$, which has recently been isolated as its $[Na(12\text{-crown-}4)_2]^+$ salt (176). This reaction was accomplished via treatment of the cobalt(II) bis(hexamethyldisilazide) complex $Co[N(SiMe_3)_2]_2$, itself known to be dimeric in the solid state, with 1 equiv of $NaN(SiMe_3)_2$ in toluene in the presence of 2 equiv of 12-crown-4 (Scheme 58). The dark green precipitate was isolated by filtration, washed with hexane, and dried under vacuum to provide $\{Co[N(SiMe_3)_2]_3\}[Na(12\text{-crown-}4)_2]$ in 67% yield (scale ~6 mmol).

X-ray crystallography showed the $\{Co[N(SiMe_3)_2]_3\}^-$ ion to exhibit D_3 point symmetry, much as was found (see below) for neutral $Co[N(SiMe_3)_2]_3$. The propellor dihedrals in the complex are between 51 and 55°; these angles describe the relationship between the interpenetrating CoN_3 and $NCoSi_2$ trigonal planes. The cobalt–nitrogen bond lengths of about 1.975(2) Å for the anion $\{Co[N(SiMe_3)_2]_3\}^-$ are longer by about 0.011 Å than for neutral $Co[N(SiMe_3)_2]_3$, indicative of high-spin cobalt(II).

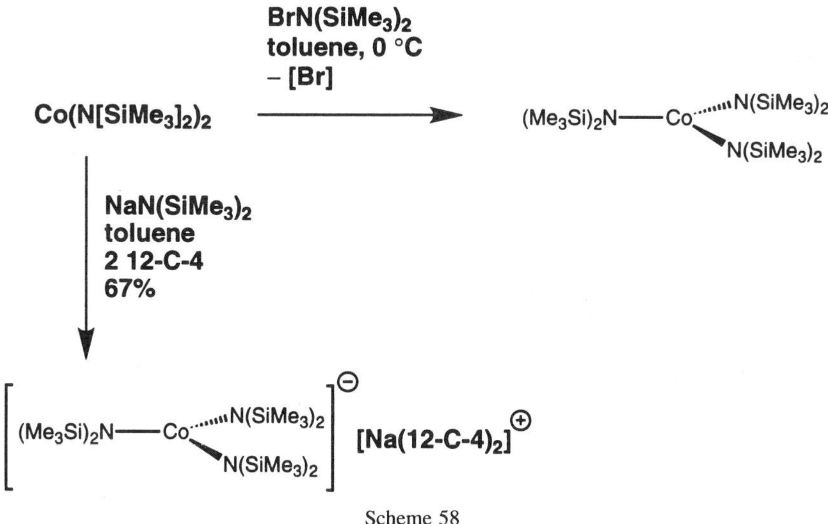

Scheme 58

C. $\{Co[N(SiMe_3)_2](OC{-}t\text{-}Bu_3)_2\}$

1. *Synthesis and Characterization*

A discrete three-coordinate cobalt(II) anion, displaying mixed-alkoxide/silylamide ligation, has been isolated and characterized as its $[Li(thf)_4]^+$ salt (194). Alcoholysis of the cobalt(II) silylamide $Co[N(SiMe_3)_2]_2$ with the hindered carbinol $HOC{-}t\text{-}Bu_3$ (2 equiv), followed by addition of $LiN(SiMe_3)_2$, gave $\{Co[N(SiMe_3)_2](OC{-}t\text{-}Bu_3)_2\}[Li(thf)_{4.5}]$ subsequent to filtration and crystallization from a THF/hexane mixture (Scheme 59). The salt was obtained as blue crystals in low (unspecified) yield.

An X-ray structure determination represents the only characterization data given for $\{Co[N(SiMe_3)_2](OC{-}t\text{-}Bu_3)_2\}[Li(thf)_{4.5}]$ (Fig. 38) (194). The anion displays a planar, three-coordinate arrangement at cobalt, with fairly regular angles about the metal atom. The widest angle at the metal [124.9(4)°] is the O—Co—O angle, consistent with a larger cone angle for the sterically isotropic $OC{-}t\text{-}Bu_3$ ligand than for the quasiplanar silylamide. The $\{Co[N(SiMe_3)_2]$-$(OC{-}t\text{-}Bu_3)_2\}^-$ complex displays no short Co···H contacts.

D. $Co(Cl)[CH(SiMe_3)_2]_2\}^-$

1. *Synthesis and Characterization*

Initial attempts to isolate crystalline salts from reactions of $CoCl_2$ with lithium alkyls did not result in tractable products, but the work summarized here illustrates a successful procedure (Scheme 60). Addition of excess (6 equiv) ethereal lithium reagent to a mixture of $CoCl_2$ and tmeda at −78°C was carried out, and the mixture subsequently was allowed to warm to room temperature. Red and then green colors were noted, and a black solid was deposited. After

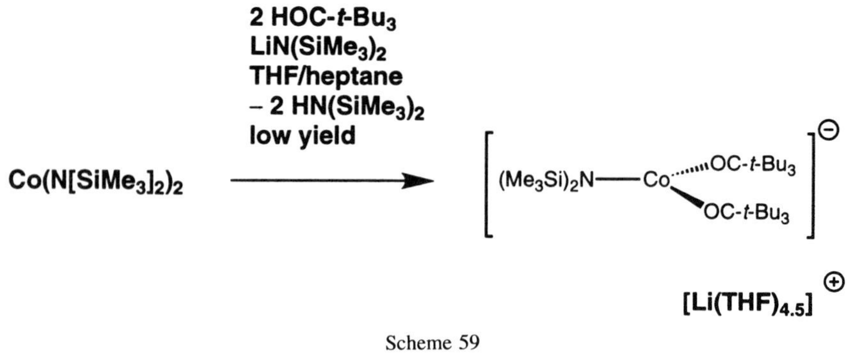

$$Co(N[SiMe_3]_2)_2 \xrightarrow[\text{low yield}]{\substack{2\ HOC\text{-}t\text{-}Bu_3 \\ LiN(SiMe_3)_2 \\ \text{THF/heptane} \\ -\,2\ HN(SiMe_3)_2}} \left[(Me_3Si)_2N{-}Co(OC\text{-}t\text{-}Bu_3)_2\right]^{\ominus}\ [Li(THF)_{4.5}]^{\oplus}$$

Scheme 59

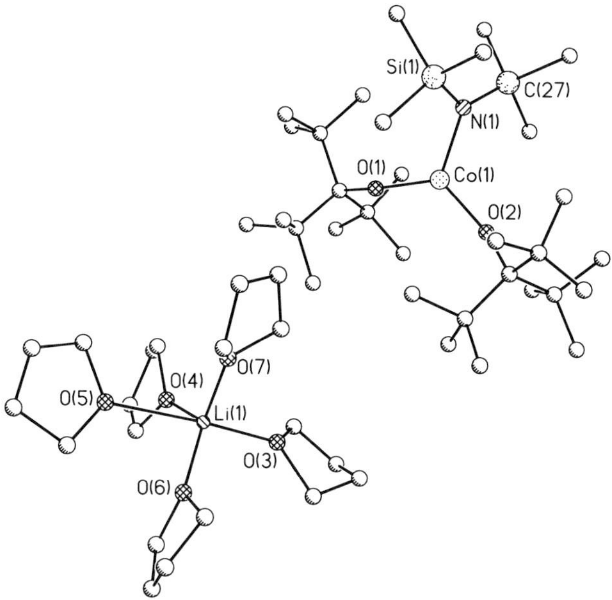

Figure 38. Structural drawing of $\{Co[N(SiMe_3)_2](OC{-}t\text{-}Bu_3)_2\}[Li(thf)_4]$.

removal of volatile matter, extraction with hexane gave a blue solution which, when cooled to −20°C, provided $[Li(tmeda)_2]\{Co(Cl)[CH(SiMe_3)_2]_2\}$. The latter was isolated in 20% yield, and was characterized structurally (X-ray diffraction) (195).

Remarkably, the anion $\{Co(Cl)[CH(SiMe_3)_2]_2\}^-$ exhibits three nearly equal angles at cobalt that sum to 360°. Wilkinson and co-workers (195) suggest that the lack of distortion from a regular trigonal planar geometry in this anion may reflect an increased covalency of the cobalt–ligand bonding vis-à-vis analogous

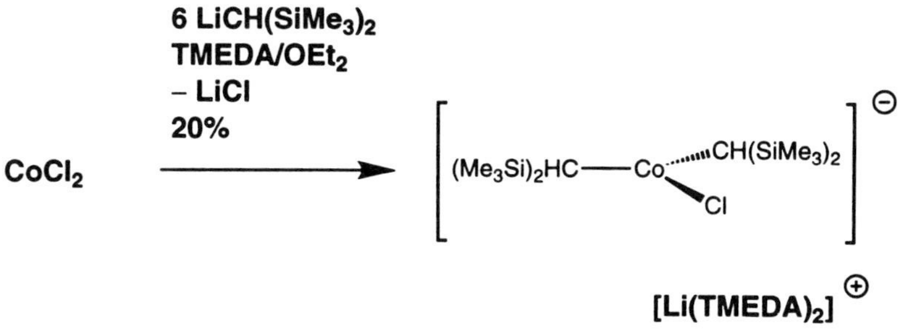

6 $LiCH(SiMe_3)_2$
TMEDA/OEt_2
− LiCl
20%

$CoCl_2$ ⟶ $[(Me_3Si)_2HC{-}Co(CH(SiMe_3)_2)(Cl)]^{\ominus}$

$[Li(TMEDA)_2]^{\oplus}$

Scheme 60

alkoxide or amide complexes. Also studied was the EPR spectrum of $[Li(tmeda)_2]\{Co(Cl)[CH(SiMe_3)_2]_2\}$ in frozen toluene, yielding the parameters $g(p_{\|}) = 2.550$ and $g_{\perp} = 2.065$. The spectrum was analyzed in terms of a low-spin d^7 ion in the point group C_{2v}; magnetic susceptibility data were not reported.

E. $[Co(Mes)_3]^-$

1. *Synthesis and Characterization*

Light green samples of this anion, as its lithium tetrakis(tetrahydrofuranate) salt, were obtained and described in detail in 1989 (196), although another synthesis of the same compound dates to 1977 (Scheme 61) (177). Accordingly, mesityl bromide was converted to mesityllithium through treatment at low temperature (−78°C) with *tert*-butyllithium (2 equiv). Subsequent treatment with $CoCl_2(thf)$ ($\frac{1}{3}$ equiv, ~7 mmol) ultimately provided a light green solution. Work-up provided $[Co(Mes)_3][Li(thf)_4]$ (76% yield) as a green ether-immiscible oil, which crystallized upon cooling to −35°C. The PPN salt of $[Co(Mes)_3]^-$ was obtained by adding $[Co(Mes)_3][Li(thf)_4]$ to a THF suspension of [PPN][Cl]; filtration of the mixture and recrystallization from Et_2O/THF at −35°C provided yellow-green crystals of $[Co(Mes)_3]$[PPN] (yield 78%, scale ~0.3 mmol). The ^1H NMR data were given for both $[Co(Mes)_3][Li(thf)_4]$ and $[Co(Mes)_3]$[PPN]; the spectra indicate a threefold symmetric anion. In addition, the mesityl chemical shifts in THF-d_8 were essentially the same for both the $[Li(thf)_4]$ and the [PPN] salts, indicating merely electrostatic interactions between anion and cation in both cases. Curie behavior in the 4–300 K temperature range was observed for $[Co(Mes)_3][Li(thf)_4]$ ($\mu_{eff} = 3.8\ \mu_B$). The partially deuterated trimesityl species $\{Co[2,4,6\text{-}C_6H_3(CH_2D)_3]\}[Li(thf)_4]$ was prepared

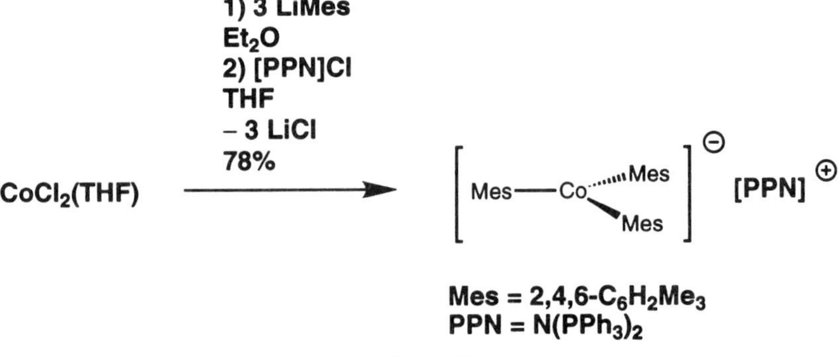

Scheme 61

to facilitate NMR studies of the dimeric complex $[Co(\mu\text{-Mes})(Mes)]_2$, a species obtained in 54% yield (scale, ~1 mmol) subsequent to treatment of $[Co(Mes)_3][Li(thf)_4]$ with HBF_4-etherate (1 equiv).

F. $Rh(Mes)_3$

1. *Synthesis and Characterization*

A THF solution of mesityl Grignard reagent was added to $RhCl_3(tht)_3$ (~1 mmol), as an ether solution kept at −78°C while stirring vigorously. The $Rh(Mes)_3$ complex subsequently was isolated in 35% yield as orange crystals (Scheme 62) (197–199). The ^1H and ^{13}C NMR data for $Rh(Mes)_3$ are as expected for a diamagnetic trimesityl compound. Variable-temperature ^1H NMR spectra showed that the mesityl groups rotate about the Rh—C bonds in a synchronized manner, with $\Delta G^{\ddagger}$ for the fluxional process being about 64 kJ mol^{-1}; the observation of inequivalent ortho methyl groups in the low-temperature limit corresponds with the solid-state structure of the complex as determined by X-ray diffraction. The $Rh(Mes)_3$ complex is strongly pyramidalized at rhodium with C—Rh—C angles of about 104°. One ortho methyl from each mesityl group makes a close (agostic) approach to rhodium ($d(Rh\cdots C_{agostic}$, ~2.2 Å); by way of comparison the $Rh-C_{ipso}$ distances are about 1.96 Å. Thus, the authors suggest that $Rh(Mes)_3$ is best thought of as a distorted low-spin d^6 pseudooctrahedral complex with agostic C—H interactions filling the three available coordination sites trans to the $Rh-C_{ipso}$ bonds. Much the same result was obtained for $Ir(Mes)_3$ (see below).

2. *General Comments*

The $Rh(Mes)_3$ complex was found to be unreactive under ambient conditions toward the following reagents: H_2, O_2, SO_2, ethylene, $HSiPh_3$, and *t*-BuOO—*t*-Bu. Rhodium trimesityl is air stable both in the solid state and in solution. Attempted reduction of $Rh(Mes)_3$ with Na/Hg in THF gave decomposition to black insoluble matter. Given its highly unusual structure, the relatively low reactivity of $Rh(Mes)_3$ is surprising.

3. *Reaction with Carbon Monoxide*

This reaction (−78°C, Et_2O, excess CO) is catastrophic to the trimesitylrhodium core, giving an unsymmetrical dinuclear species with the nominal formula $Rh_2(CO)_4(Mes)_2$, along with dimesityl ketone. In fact, the dirhodium species contains two bridging moieties: a μ^2, η^1, η^6-mesityl group and a μ^2-2,4,6-

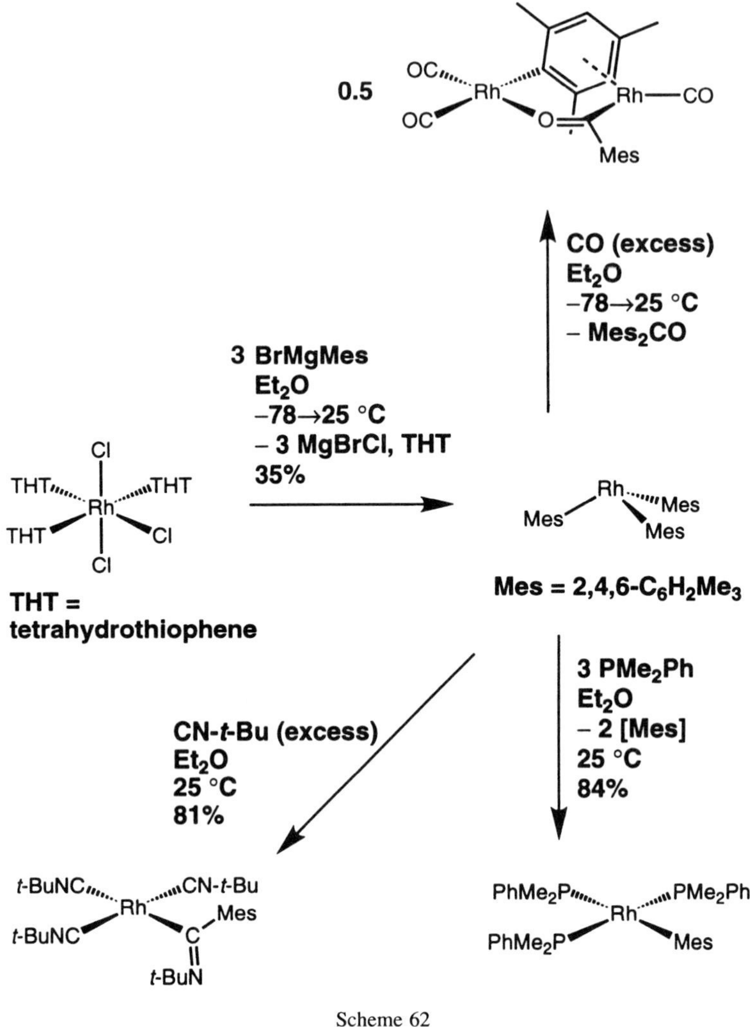

Scheme 62

trimethylphenacyl group. The light yellow crystalline compound was characterized by X-ray crystallography to reveal the nature of the bridging groups.

4. *Reaction with PMe_2Ph*

Reduction to rhodium(I) occurred when $Rh(Mes)_3$ was treated with dimethylphenylphosphine in Et_2O at about 25°C. The product, obtained in 84% yield as yellow crystals, was found to be square planar $Rh(Mes)(PMe_2Ph)_3$, which compound was the subject of a single-crystal X-ray diffraction study.

5. *Reaction with t-BuNC*

Exposure of $Rh(Mes)_3$ to an excess of *tert*-butyl isocyanide in ether at about 25°C gave an orange crystalline rhodium(I) product in 81% yield. This product, $Rh(CN{-}t\text{-}Bu)_3(2,4,6\text{-}Me_3C_6H_2C{=}N{-}t\text{-}Bu)$, is a tris(isocyanide) iminoacyl species. The GCMS analysis of the products revealed only the presence of mesitylene; the hexamethylbiphenyl product that would result from coupling two mesityl groups was not observed.

G. $Ir(Mes)_3$

1. *Synthesis and Characterization*

Iridium trimesityl was described in 1992 by Wilkinson and co-workers (200). The compound was obtained by essentially the same procedure as that used for the rhodium analogue (see above). Mesitylmagnesium bromide was added to a suspension of $IrCl_3(tht)_3$ in Et_2O, and the mixture was refluxed for 4 h (Scheme 63). Following this, the solvent was removed in vacuo to provide a brown-black residue, which was extracted with hexane. Upon concentration and cooling, brown-red crystals of diamagnetic $Ir(Mes)_3$ were isolated in 20.5–26% yield. It was noted that $Ir(Mes)_3$ is exceedingly sensitive to oxygen.

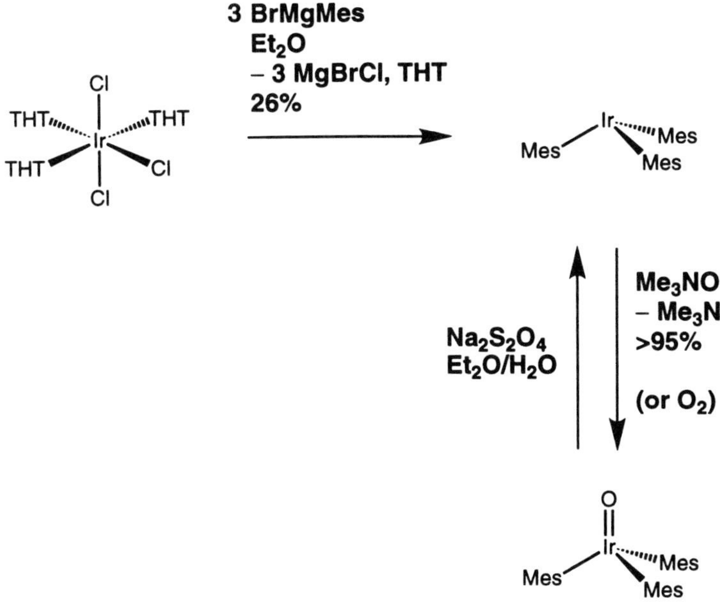

Scheme 63

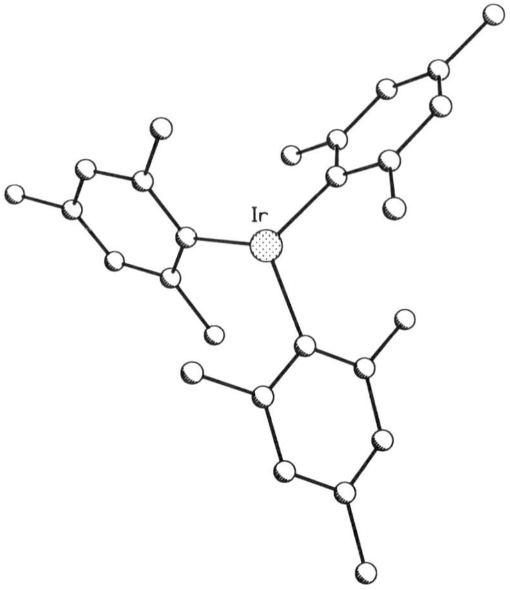

Figure 39. Structural drawing of $Ir(Mes)_3$.

The geometry of $Ir(Mes)_3$ was determined by X-ray crystallography (Fig. 39) to exhibit a moderately pyramidal IrC_3 core [the C—Ir—C bond angles range from 106.1(4)–109.9(4)°]. On the top side of the pyramid, three close (agostic) contacts to mesityl methyl hydrogens were found to complete a quasioctahedral geometry for the molecule. The overall confirmation of the molecule is such that the planar mesityl groups are arranged about the central iridium atom in a propellor-like fashion.

2. *General Comments*

It was reported (200) that the compound does not react with water, tetramethylpiperidine, or THF. Reactions leading to uncharacterized product mixtures occurred with dihydrogen, carbon monoxide, diphenylacetylene, CS_2, *tert*-butylamine, *tert*-butylisocyanide, and $[NO][PF_6]$. No explicit reference was made to the conditions employed in the latter reactivity studies.

3. *Reaction with Trimethylamine Oxide*

A quantitative reaction leading to $Ir(O)(Mes)_3$, the first terminal oxo complex of iridium, was observed between $Ir(Mes)_3$ and trimethylamine oxide. The $Ir(O)(Mes)_3$ complex is colored green in solution, and green-black in the crystal. It was stated that the "greenish-black air-stable crystals of $Ir(O)(Mes)_3$ can be obtained with some difficulty." They were grown from petroleum (201).

The $Ir(Mes)_3$ complex was regenerated from $Ir(O)(Mes)_3$ by treatment of an Et_2O solution with aqueous sodium dithionite, $Na_2S_2O_4$. The $Ir(Mes)_3$ complex was isolated from the ether layer and characterized by ^1H NMR spectroscopy. Survey reactions carried out with $Ir(O)(Mes)_3$ revealed that the compound is unreactive toward CO_2, that clean products could not be isolated upon treatment with HCl in Et_2O, and that attempts to convert the terminal oxo linkage to a terminal imido using organic isocyanates or trimethylsilyl azide gave only $Ir(Mes)_3$ and decomposition products.

As determined by X-ray crystallography (Fig. 40), the geometry of $Ir(O)(Mes)_3$ is distorted tetrahedral. While the Ir—C bond lengths were normal [1.99–2.03(1) Å], the Ir—O distance [1.725(9) Å] seems long, though it must be borne in mind that a suitable reference compound is lacking. The iridium oxo stretch was tentatively assigned as follows: $\nu(\text{IrO}) = 802\ \text{cm}^{-1}$. Although it was stated that the ^1H NMR spectrum of $Ir(O)(Mes)_3$ is consistent with its solid-state structure, the spectrum was not provided.

4. *Reaction with Dioxygen*

An experiment was conducted wherein $Ir(Mes)_3$ in benzene-d_6 was treated with an O_2/N_2 mixture, giving a rapid reaction that was followed by ^1H NMR

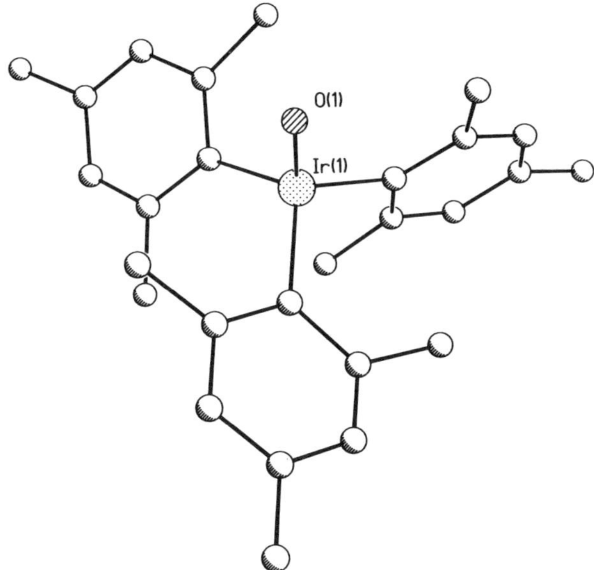

Figure 40. Structural drawing of $OIr(Mes)_3$.

spectroscopy (201). The initial spectrum of $Ir(Mes)_3$ changed to that of an intermediate, labile species, possibly $Ir(\eta^2\text{-}O_2)(Mes)_3$. The final spectrum was that of $Ir(O)(Mes)_3$, which may arise through deoxygenation of putative $Ir(\eta^2\text{-}O_2)(Mes)_3$ by $Ir(Mes)_3$.

X. GROUP 10 (VIII) METAL COMPLEXES

A. $[Ni(NPh_2)_3]^-$

1. Synthesis and Characterization

Low-coordinate complexes of the late transition metals are less ubiquitous than those of the early transition metals. In order to extend studies of low coordinate species to the right-hand side of the periodic table, Power and coworkers (202) employed the nitrogen-donor ligand (diphenylamide) that had given rise to the first transition metal amido complex, $Ti(NPh_2)_4$. Thus, a $-78\,°C$ THF slurry of anhydrous $NiCl_2$ was treated with a THF/hexane solution of $LiNPh_2$ (Scheme 64). A navy blue color developed upon warming to 25 °C. Addition of a 1 : 1 : 1 ether/hexane/toluene mixture at this point, filtration, and cooling gave navy blue crystals of $[Ni(NPh_2)_3][Li(thf)_4] \cdot 0.5PhMe$ (mp 115 °C) in 40% yield. Alternatively, removal of volatiles from the blue reaction mixture, followed by extraction of the residue with hot toluene, filtration, and crystallization, provided crystals of the dark green dimer $[Ni(NPh_2)_2]_2$ (mp 140 °C) in 50% yield. Analogous cobalt complexes were isolated by substituting cobalt for nickel in the above procedures (see above).

A trigonal planar geometry prevails at nickel for $[Ni(NPh_2)_3]^-$, according to a single-crystal X-ray diffraction study carried out for $[Ni(NPh_2)_3][Li(thf)_4] \cdot 0.5PhMe$ (Fig. 41). The approximate local D_{3h} symmetry at nickel leads to the prediction of two unpaired electrons in the e' orbitals, consistent with the

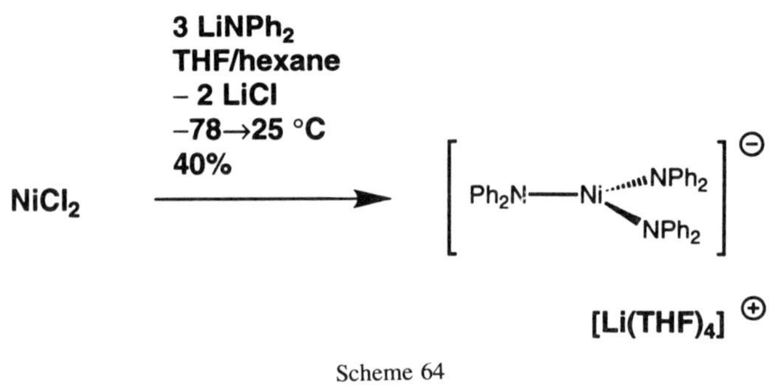

Scheme 64

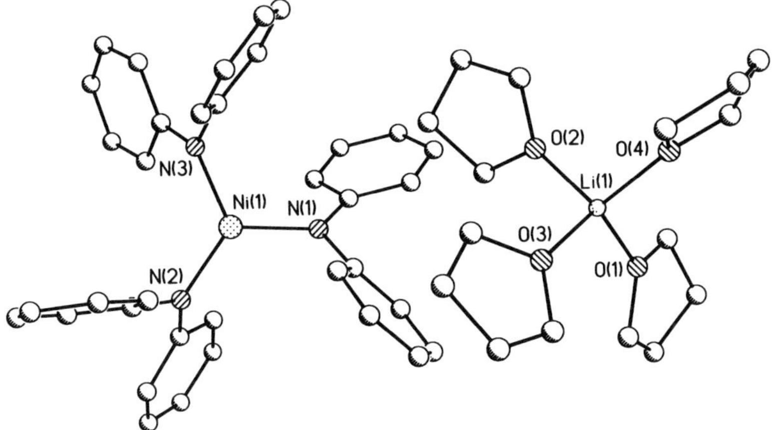

Figure 41. Structural drawing of $[Ni(NPh_2)_3][Li(thf)_4]$.

observed magnetic moment: $\mu_{eff} = 2.6\ \mu_B$ (298 K). The point symmetry of $[Ni(NPh_2)_3]^-$, when the orientation of the phenyl groups is taken into account, approximates D_3, just as for the crystallographically characterized $M[N(SiMe_3)_2]_3$ complexes (M = Ti, V, Cr, Mn, Fe, Co).

B. $[Ni(Mes)_3]^-$

1. *Synthesis and Characterization*

The blue-purple salt $[Li(thf)_4][Ni(Mes)_3]$ was obtained in about 80–85% yield subsequent to treatment of $NiCl_2(dme)$ (0.91 mmol) with 3 equiv of $Li(Mes)(OEt_2)$ in a THF/ether mixture (Scheme 65). It was noted that attempts to obtain reliable microanalytical data were not successful. A broad signal (*g*

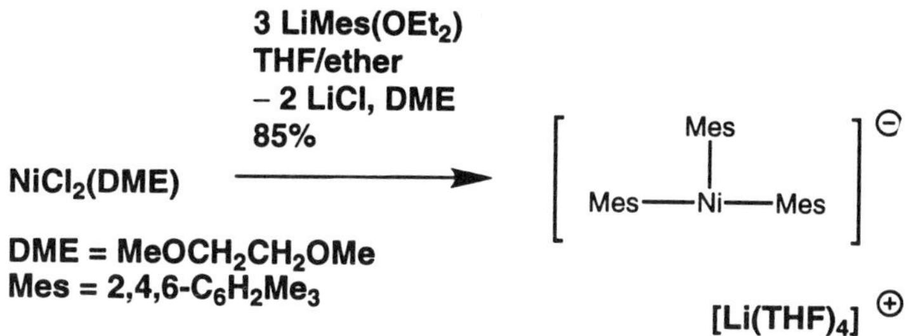

Scheme 65

= 1.999) was noted in the X-band EPR spectrum of $[Li(thf)_4][Ni(Mes)_3]$ taken in THF (203).

X-ray crystallography showed the anion $[Ni(Mes)_3]^-$ to contain an interesting T-shaped NiC_3 core, supplemented by agostic interactions to two mesityl methyl groups in the fourth coordination site of a pseudosquare planar structure. The structure can also be viewed as a distorted trigonal bipyramid, with the two agostic interactions utilizing cis-equatorial sites. The $Ni—C_{ipso}$ distances [1.838(5)–1.932(6) Å] are significantly shorter, of course, than the two agostic $Ni—C_{methyl}$ distances (2.88 and 2.89 Å), and the two hydrogens pertinent to the agostic interactions were located and refined [*d*(U· · ·H), 2.17(1) and 2.13(1) Å].

2. *Reaction with $CN—2,6\text{-}C_6H_3Me_2$*

The only product isolated upon treatment of $[Li(thf)_4][Ni(Mes)_3]$ with this aryl isocyanide was the tetraisocyanide nickel(0) complex $Ni(Cn—2,6\text{-}C_6H_3Me_2)_4$, the identity of which was established by NMR spectroscopy and elemental analysis.

XI. URANIUM COMPLEXES

A. $U[N(SiMe_3)_2]_3$

1. *Synthesis and Characterization*

In an investigation aimed at describing well-behaved complexes of uranium(III), Andersen (204) disclosed a preparation of $U[N(SiMe_3)_2]_3$. Because uranium trichloride was difficult to prepare, Andersen utilized an *in situ* reduction of uranium tetrachloride with sodium naphthalenide to produce a uranium(III) chloride species, which was not characterized. Addition to this "UCl_3" of sodium (hexamethyldisilyl)amide in THF produced red $U[N(SiMe_3)_2]_3$, which was isolated in 76% yield on a multigram scale.

More recently, a Los Alamos team reported that the well-behaved uranium(III) starting material $UI_3(thf)_4$ can be obtained conveniently upon treatment of amalgamated uranium metal turnings with elemental iodine in THF solvent (205). The structure of monomeric $UI_3(thf)_4$ was determined by X-ray crystallography (205). The Los Alamos group showed further that $UI_3(thf)_4$ smoothly provides $U[N(SiMe_3)_2]_3$ upon treatment with $NaN(SiMe_3)_2$ in THF at 25°C; the yield of $U[N(SiMe_3)_2]_3$ obtained as purple needles was 82% on a scale of 33 mmol (Scheme 66) (205). A similar approach was used to prepare analogous Np and Pu silylamides (205).

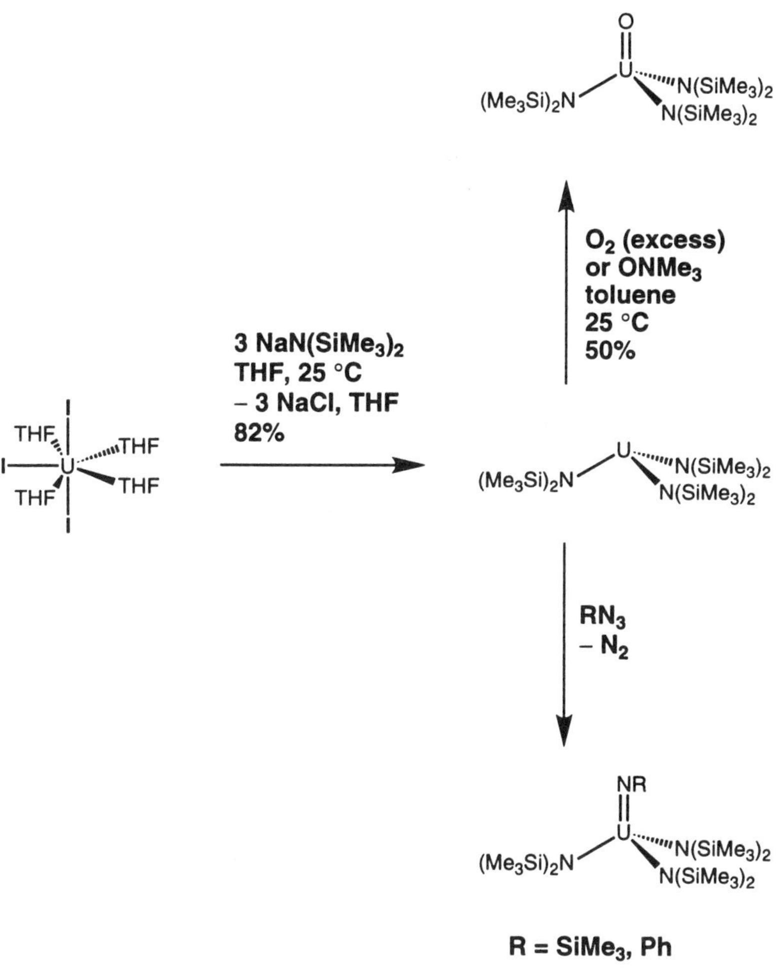

Scheme 66

Infrared spectra of $U[N(SiMe_3)_2]_3$ indicate that the compound is pyramidal, by comparison with spectra for trivalent scandium and 4*f* series (neodymium, europium, and ytterbium) elements. Unfortunately, single crystals suitable for an X-ray analysis of $U[N(SiMe_3)_2]_3$ proved elusive to obtain. The compound exhibits a singlet at -18.1 ppm ($\Delta\nu_{1/2} = 9$ Hz) in its ^1H NMR spectrum. A magnetic susceptibility study showed that $U[N(SiMe_3)_2]_3$ is a Curie–Weiss paramagnet in the temperature range 10–70 K, with $\mu_{eff} = 2.51\ \mu_B$ (204).

2. *General Reactivity Comments*

From a reaction chemistry point of view, it is noteworthy that $U[N(SiMe_3)_2]_3$ was found not to yield isolable complexes with CO triethylphosphine, trimethylphosphine oxide, THF, trimethylamine, py, *tert*-butyl isocyanide, or *tert*-butyl cyanide at room temperature and atmospheric pressure (204). It was noted that the lack of Lewis acidity exhibited by $U[N(SiMe_3)_2]_3$ is best accounted for in terms of steric effects.

3. *Reaction with Dioxygen*

The oxo compound $OU[N(SiMe_3)_2]_3$ was prepared via treatment of $U[N(SiMe_3)_2]_3$ with molecular oxygen (204). The reaction was carried out by passing O_2 through a toluene solution of $U[N(SiMe_3)_2]_3$ for 5 min. The $OU[N(SiMe_3)_2]_3$ complex was isolated as green-yellow prisms, in 50% yield.

The oxouranium(V) tris(silylamide) exhibits $\mu_{eff} = 1.82\ \mu_B$ over the temperature range 10–50 K, in which range it exhibits Curie–Weiss behavior. Its ^1H NMR spectrum consists of a singlet at 0.44 ppm in benzene, and a band at 930 cm^{-1} in its IR spectrum was assigned to $\nu(UO)$.

4. *Reaction with Trimethylamine Oxide*

The $OU[N(SiMe_3)_2]_3$ complex was prepared independently in 51% yield using trimethylamine oxide as the oxo transfer agent (204).

5. *Reaction with Organic Azides*

Treatment of $U[N(SiMe_3)_2]_3$ with trimethylsilyl azide gave $(Me_3SiN)U[N(SiMe_3)_2]_3$, while similar treatment with phenyl azide produced $(PhN)U[N(SiMe_3)_2]_3$ (206–208). The silylimido uranium(V) complex was characterized structurally by X-ray diffraction (Fig. 42) (209).

Further investigations (208) involving the above uranium organoimido complexes produced the finding that $(Me_3SiN)U[N(SiMe_3)_2]_3$ and $(PhN)U[N(SiMe_3)_2]_3$ possess reversible $1e^-$ oxidation waves at -0.41 and -0.42 V, respectively, with reference to a ferrocene internal standard. This result indicated that mild chemical oxidants would suffice for the production of $[(imido)U[N(SiMe_3)_2]_3]^+$ species. The use of silver hexafluorophosphate as oxidant led to the isolation of trigonal bipyramidal $F(RN)U[N(SiMe_3)_2]_3$ complexes. For R = Me_3Si, the color is cherry red, while for R = Ph, the color is red-purple. Both uranium(VI) imido–fluoride complexes were characterized by

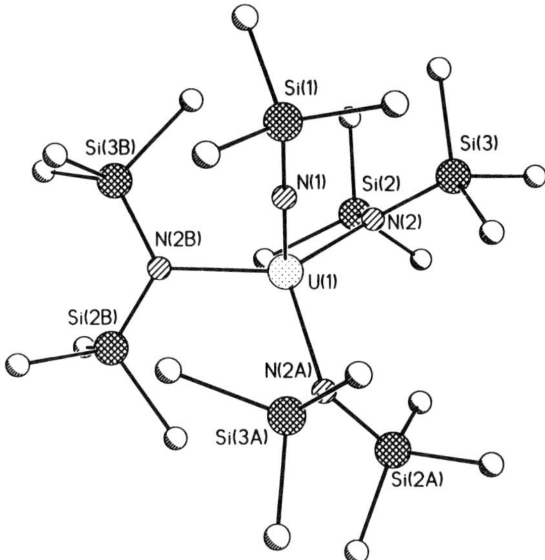

Figure 42. Structural drawing of $(Me_3SiN)U[N(SiMe_3)_2]_3$.

X-ray diffraction (Fig. 43), revealing the anticipated equatorial disposition of the voluminous $N(SiMe_3)_2$ ligands.

B. $U[CH(SiMe_3)_2]_3$

1. *Synthesis and Characterization*

This royal blue homoleptic uranium(III) alkyl complex was obtained in about 40% yield subsequent to treatment of $U(O{-}2,6\text{-}C_6H_3{-}t\text{-}Bu_2)_3$ (see below) in hexane with $LiCH(SiMe_3)_2$ (Scheme 67) (210). Only one ^1H NMR signal (C_6D_6, δ −5.8 ppm) was noted for the methyl groups of $U(CH[SiMe_3]_2)_3$; a resonance assignable to the three α-hydrogen atoms was not found. The $U[CH(SiMe_3)_2]_3$ complex was found to follow clean Curie–Weiss behavior over the temperature range 100–270 K, with $\mu_{eff} = 3.0\ \mu_B$; the latter is significantly lower than the spin-only value 3.87 μ_B for three unpaired electrons, but is typical of uranium(III) complexes including Andersen's $U(N[SiMe_3]_2)_3$ (see above).

Unlike the compounds $M[CH(SiMe_3)_2]_3$ (M = Ti, V, Cr; see above), $U[CH(SiMe_3)_2]_3$ displays a distinctly pyramidal MC_3 core, as determined by a single-crystal X-ray diffraction study (Fig. 44). The uranium atom lies 0.90 Å out of the C_3 plane, commensurate with a unique C—U—C angle of 107.7(4)°.

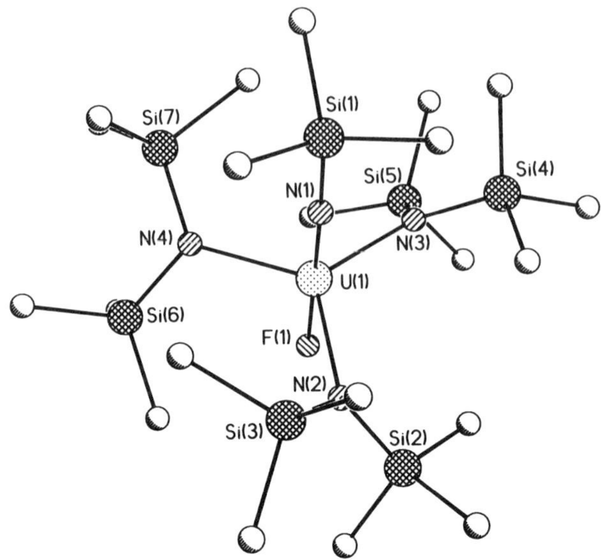

Figure 43. Structural drawing of $(Me_3SiN)U(F)[N(SiMe_3)_2]_3$.

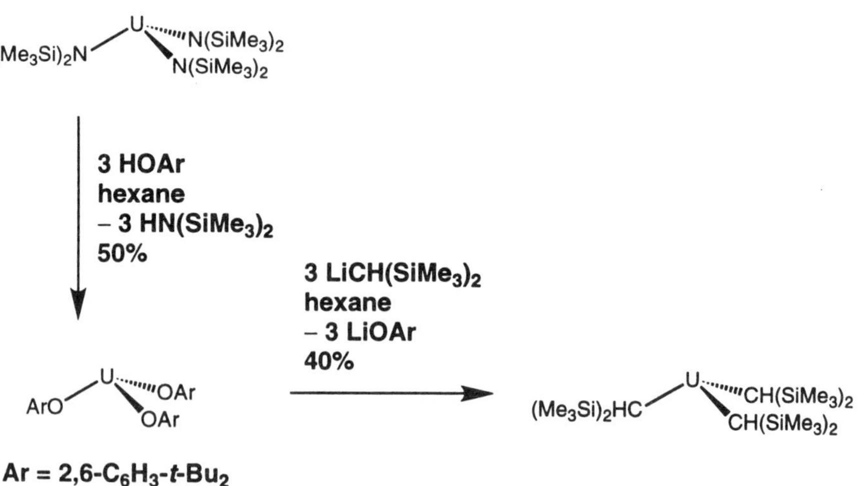

Scheme 67

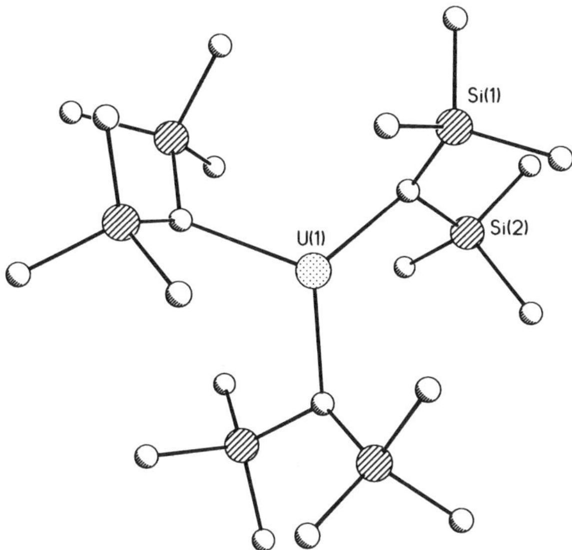

Figure 44. Structural drawing of $U[CH(SiMe_3)_2]_3$.

The pyramidalization permits three agostic $U\cdots\gamma C$ interactions characterized by distances significantly longer [$d(U\cdots\gamma C)$, 3.09(2) Å] than the $U\cdots\alpha C$ bond lengths [$d(U\cdots\alpha C)$, 2.48(2) Å]. The molecular symmetry of $U[CH(SiMe_3)_2]_3$ in the crystal is C_3.

C. $U(O{-}2,6\text{-}C_6H_3{-}i\text{-}Pr_2)_3$

1. Synthesis and Characterization

Homoleptic uranium(III) alkoxide complexes were described in the 1988 communication summarized here (211). The preparation in particular of $U(O{-}2,6\text{-}C_6H_3{-}i\text{-}Pr_2)_3$ was achieved by treating a hexane solution of $U[N(SiMe_3)_2]_3$ with $HO{-}2,6\text{-}C_6H_3{-}i\text{-}Pr_2$ (3 equiv). The isolated yield of dark purple $U(O{-}2,6\text{-}C_6H_3{-}i\text{-}Pr_2)_3$ was about 50%.

The structure of $U(O{-}2,6\text{-}C_6H_3{-}i\text{-}Pr_2)_3$ was shown to be pyramidal, with O—U—O angles of about 105°; this value should be compared with that noted for the C—U—C angles [107.7(4)°] found for $U[CH(SiMe_3)_2]_3$. While in the latter species, agostic interactions with the γ-carbon atoms supplemented the three primary bonding interactions, in the case of $U(O{-}2,6\text{-}C_6H_3{-}i\text{-}Pr_2)_3$ the complex was found to be a dimer in the solid state (Fig. 45). The mode of

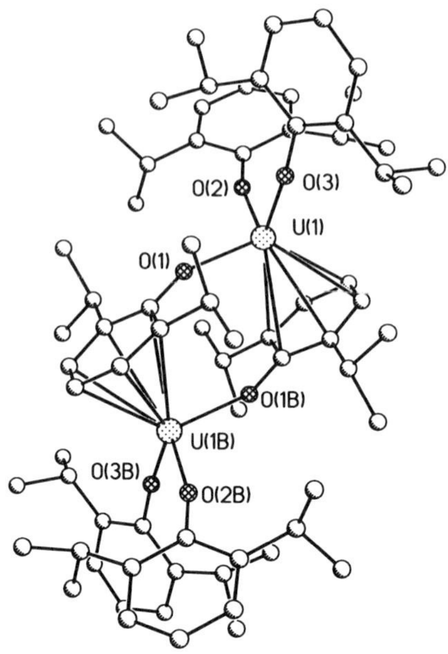

Figure 45. Structural drawing of $[U(O{-}2,6\text{-}C_6H_3{-}i\text{-}Pr_2)_3]_2$.

dimerization was, surprisingly, via arene coordination rather than by phenoxide oxygen bridging. The coordinated arene exhibits relatively long uranium–carbon interactions [$d(U\cdot\cdot\cdot C)$, ~2.92(2) Å] indicative of this moiety's lability. In solution, only one type of phenoxide aryl residue is observed (^1H NMR, C_6D_6), which is consistent with monomeric $U(O{-}2,6\text{-}C_6H_3{-}i\text{-}Pr_2)_3$, a benzene-complexed species, or a fluxional dimer.

It was stated that $U(O{-}2,6\text{-}C_6H_3{-}i\text{-}Pr_2)_3$ forms presumably pseudotetrahedral 1 : 1 adducts with Lewis bases, the citation being to work in progress.

D. $U(O{-}2,6\text{-}C_6H_3{-}t\text{-}Bu_2)_3$

1. Synthesis and Characterization

Dark green $U(O{-}2,6\text{-}C_6H_3{-}t\text{-}Bu_2)_3$ was prepared in a manner similar to the diisopropylphenoxide analogue (see above) by adding $HO{-}2,6\text{-}C_6H_3{-}t\text{-}Bu_2$ (3 equiv) to $U[N(SiMe_3)_2]_3$ in hexane (Scheme 67) (211). The complex was thereby obtained in about 50% yield. Only one type of phenoxide aryl residue was observed (^1H NMR, C_6D_6). Although structural characterization data pertaining to $U(O{-}2,6\text{-}C_6H_3{-}t\text{-}Bu_2)_3$ were not obtained, it was proposed on the basis of IR data that the complex is monomeric in the solid state. It was shown that, in C_6D_6, $U(O{-}2,6\text{-}C_6H_3{-}t\text{-}Bu_2)_3$ takes up 1 equiv of THF forming (thf)U(O$-$2,6-$C_6H_3$$-$*t*-$Bu_2$)$_3$. Also, thf was not lost from this adduct when a sample was placed under high vacuum (10^{-6} torr, ~25°C).

ABBREVIATIONS

Ad	1-Adamantyl
Bu	Butyl
Col	2,4,6-Trimethylpyridine
Cp	η^5-C_5H_5
Cy	Cyclohexyl
CFSE	Crystal field stabilization energy
NMR	Nuclear magnetic resonance
dme	1,2-Dimethoxyethane (ligand)
CSD	Cambridge structural database
HOMO	Highest occupied molecular orbital
MO	Molecular orbital
EPR	Electron paramagnetic resonance
TAC	1,3,5-Triazacyclohexane
Me	Methyl
Ph	Phenyl
Pr	Propyl
THF	Tetrahydrofuran (solvent)
thf	Tetrahydrofuran (ligand)
IR	Infrared
UV–vis	Ultraviolet–visible
SCF	Self-consistent field
BDE	Bond dissociation enthalpy
Me_3TAC	*N,N',N"*-Trimethyl-1,3,5-triazacyclohexane
Mes	Mesityl
tmeda	*N,N,N',N"*-Tetramethylethylenediamine
GCMS	Gas chromatography–mass spectrometry
THT	Tetrahydrothiophene

ACKNOWLEDGMENTS

Aaron L. Odom, a graduate student, is gratefully acknowledged for preparation of the structural figures. For the inspiration to study the chemistry of three-coordinate complexes, I thank former mentors Peter T. Wolczanski and Richard R. Schrock in addition to the pioneers in this area, especially Donald C. Bradley. Any contributions to date from my laboratory to the research area of three-coordinate metal complexes are the direct consequence of the strenuous efforts of my early research team comprised principally of undergraduate students Catalina E. Laplaza and Paulus W. Wanandi, and graduate students Sheree L. Stokes, Michael G. Fickes, Adam R. Johnson, Marc J. A. Johnson, Aaron L. Odom, and Jonas C. Peters. The funding that has permitted me to devote time to this research area has come from the following sources: the MIT Department of Chemistry, the National Science Foundation, the Packard Foundation, DuPont, Union Carbide, 3M, and Monsanto. CCC is a Sloan Foundation Fellow (1997–1998).

REFERENCES

1. D. C. Bradley, *Chem. Br.*, *11*, 393 (1975).
2. P. G. Eller, D. C. Bradley, M. B. Hursthouse, and D. W. Meek, *Coord. Chem. Rev.*, *24*, 1 (1977).
3. D. C. Bradley, *Adv. Chem. Ser.*, *150*, 266 (1976).
4. M. F. Lappert, *Adv. Chem. Ser.*, *150*, 256 (1976).
5. R. B. King, *Can. J. Chem.*, *73*, 963 (1995).
6. D. C. Bradley and M. H. Chisholm, *Acc. Chem. Res.*, *9*, 273 (1976).
7. M. F. Lappert, P. P. Power, A. R. Sanger, and R. C. Srivastava, *Metal and Metalloid Amides*, Ellis Horwood, Chichester, 1980.
8. P. P. Power, *Comments Inorg. Chem.*, *8*, 177 (1989).
9. P. P. Power, *Chemtracts-Inorg. Chem.*, *6*, 181 (1994).
10. H. Müller, W. Seidel, and H. Görls, *Z. Anorg. Allg. Chem.*, *622*, 1269 (1996).
11. R. R. Schrock, *Acc. Chem. Res.*, *30*, 9 (1997).
12. R. A. Bartlett, M. M. Olmstead, P. P. Power, and S. C. Shoner, *Organometallics*, *7*, 1801 (1988).
13. B. D. Murray and P. P. Power, *Inorg. Chem.*, *23*, 4584 (1984).
14. P. B. Hitchcock, M. F. Lappert, and A. Singh, *J. Chem. Soc. Chem. Commun.*, 1499 (1983).
15. D. R. Neithamer, R. E. LaPointe, R. A. Wheeler, D. S. Richeson, G. D. Van Duyne, and P. T. Wolczanski, *J. Am. Chem. Soc.*, *111*, 9056 (1989).
16. C. E. Laplaza, M. J. A. Johnson, J. C. Peters, A. L. Odom, E. Kim, C. C. Cummins, G. N. George, and I. J. Pickering, *J. Am. Chem. Soc.*, *118*, 8623 (1996).
17. N. D. Fenton and M. Gerloch, *J. Chem. Soc. Dalton Trans.*, 2201 (1988).
18. S. L. Latesky, J. Keddington, A. K. McMullen, I. P. Rothwell, and J. C. Huffman, *Inorg. Chem.*, *24*, 995 (1985).
19. K. J. Covert, D. R. Neithamer, M. C. Zonnevylle, R. E. LaPointe, C. P. Schaller, and P. T. Wolczanski, *Inorg. Chem.*, *30*, 2494 (1991).
20. R. Hoffmann, *J. Chem. Phys.*, *39*, 1397 (1963).
21. C. Mealli and D. M. Proserpio, *J. Chem. Educ.*, *67*, 399 (1990).

22. R. Hoffmann, *Angew. Chem. Int. Ed. Engl.*, *21*, 711 (1982).
23. C. R. Landis, T. Cleveland, and T. K. Firman, *J. Am. Chem. Soc.*, *117*, 1859 (1995).
24. G. K. Barker and M. F. Lappert, *J. Organomet. Chem.*, *76*, C45 (1974).
25. G. K. Barker, M. F. Lappert, and J. A. K. Howard, *J. Chem. Soc. Dalton Trans.*, 734 (1978).
26. J. C. W. Chien and W. Kruse, *Inorg. Chem.*, *9*, 2615 (1970).
27. D. C. Bradley and I. M. Thomas, *J. Chem. Soc.*, 3857 (1960).
28. E. C. Alyea, J. S. Basi, D. C. Bradley, and M. H. Chisholm, *J. Chem. Soc. Chem. Commun.*, 495 (1968).
29. E. C. Alyea, D. C. Bradley, M. F. Lappert, and A. R. Sanger, *J. Chem. Soc. Chem. Commun.*, 1064 (1969).
30. H. M. Gladney and J. D. Swalen, *J. Chem. Phys.*, *42*, 1999 (1965).
31. P. W. Wanandi, W. M. Davis, C. C. Cummins, M. A. Russell, and D. E. Wilcox, *J. Am. Chem. Soc.*, *117*, 2110 (1995).
32. J. C. Peters, A. R. Johnson, A. L. Odom, P. W. Wanandi, W. M. Davis, and C. C. Cummins, *J. Am. Chem. Soc.*, *118*, 10175 (1996).
33. A. R. Johnson, P. W. Wanandi, C. C. Cummins, and W. M. Davis, *Organometallics*, *13*, 2907 (1994).
34. C. E. Laplaza, W. M. Davis, and C. C. Cummins, *Organometallics*, *14*, 577 (1995).
35. R. K. Minhas, L. Scoles, S. Wong, and S. Gambarotta, *Organometallics*, *15*, 1113 (1996).
36. G. N. La Mar, W. D. Horrocks Jr., and R. H. Holm, *NMR of Paramagnetic Molecules*, Academic, New York, 1973.
37. A. Johnson and G. W. Everett Jr., *J. Am. Chem. Soc.*, *94*, 1419 (1972).
38. W. D. Wheeler, S. Kaizaki, and J. I. Legg, *Inorg. Chem.*, *21*, 3248 (1982).
39. D. H. Hill and A. Sen, *J. Am. Chem. Soc.*, *110*, 1650 (1988).
40. D. H. Hill, M. A. Parvez, and A. Sen, *J. Am. Chem. Soc.*, *116*, 2889 (1994).
41. F. A. Cotton and M. D. LaPrade, *J. Am. Chem. Soc.*, *90*, 5418 (1968) .
42. G. Trinquier and R. Hoffmann, *Organometallics*, *3*, 370 (1984).
43. P. W. Wanandi and C. C. Cummins, manuscript in preparation.
44. R. H. Holm, *Chem. Rev.*, *87*, 1401 (1987).
45. L. K. Woo, *Chem. Rev.*, *93*, 1125 (1993).
46. B. O. West, *Polyhedron*, *8*, 219 (1989).
47. G. F. Kokoszka, H. C. Allen, Jr., and G. Gordon, *Inorg. Chem.*, *5*, 91 (1966).
48. D. M.-T. Chan, M. H. Chisholm, K. Folting, J. C. Huffman, and N. S. Marchant, *Inorg. Chem.*, *25*, 4170 (1986).
49. C. E. Laplaza, A. R. Johnson, and C. C. Cummins, *J. Am. Chem. Soc.*, *118*, 709 (1996).

50. K. Dehnicke and J. Strähle, *Angew. Chem. Int. Ed. Engl.*, *31*, 955 (1992).
51. M. H. Chisholm, K. Folting, J. C. Huffman, and C. C. Kirkpatrick, *Inorg. Chem.*, *23*, 1021 (1984).
52. M. H. Chisholm, K. Folting, J. C. Huffman, C. C. Kirkpatrick, and A. L. Ratermann, *J. Am. Chem. Soc.*, *103*, 1305 (1981).
53. M. J. A. Johnson, P. M. Lee, A. L. Odom, W. M. Davis, and C. C. Cummins, *Angew. Chem. Int. Ed. Engl.*, *36*, 87 (1997).
54. A. R. Johnson, W. M. Davis, and C. C. Cummins, *Organometallics*, *15*, 3825 (1996).
55. M. Handlovic, D. Miklos, and M. Zikmund, *Acta Crystallogr.*, *B37*, 811 (1981).
56. L. E. Manzer, *Inorg. Synth.*, *21*, 135 (1982).
57. L. Scoles, R. Minhas, R. Duchateau, J. Jubb, and S. Gambarotta, *Organometallics*, *13*, 4978 (1994).
58. P. J. Davidson, D. H. Harris, and M. F. Lappert, *J. Chem. Soc. Dalton Trans.*, 2268 (1976).
59. O. Kahn, *Molecular Magnetism*, VCH, New York, 1993.
60. F. Rehbaum, K.-H. Thiele, and S. I. Trojanov, *J. Organomet. Chem.*, *410*, 327 (1991).
61. J. H. Osborne, A. L. Rheingold, and W. C. Trogler, *J. Am. Chem. Soc.*, *107*, 7945 (1985).
62. K. J. Covert, P. T. Wolczanski, S. A. Hill, and P. J. Krusic, *Inorg. Chem.*, *31*, 66 (1992).
63. E. C. Alyea, D. C. Bradley, R. G. Copperthwaite, and K. D. Sales, *J. Chem. Soc. Dalton Trans.*, 185 (1973).
64. D. C. Bradley, R. G. Copperthwaite, S. A. Cotton, K. D. Sales, and J. F. Gibson, *J. Chem. Soc. Dalton Trans.*, 191 (1973).
65. W. Seidel and G. Kreisel, *Z. Anorg. Allg. Chem.*, *435*, 146 (1977).
66. W. Seidel and G. Kreisel, *Z. Chem.*, *14*, 25 (1974).
67. M. Vivanco, J. Ruiz, C. Floriani, A. Chiesi-Villa, and C. Rizzoli, *Organometallics*, *12*, 1794 (1993).
68. C. C. Cummins, J. Lee, R. R. Schrock, and W. M. Davis, *Angew. Chem. Int. Ed. Engl.*, *31*, 1501 (1992).
69. M. Vivanco, J. Ruiz, C. Floriani, A. Chiesi-Villa, and C. Rizzoli, *Organometallics*, *12*, 1802 (1993).
70. J. Ruiz, M. Vivanco, C. Floriani, A. Chiesi-Villa, and C. Guastini, *J. Chem. Soc. Chem. Commun.*, 762 (1991).
71. K. A. Jørgensen and B. Schiøtt, *Chem. Rev.*, *90*, 1483 (1990).
72. C. C. Cummins, R. R. Schrock, and W. M. Davis, *Inorg. Chem.*, *33*, 1448 (1994).
73. A. L. Odom, C. C. Cummins, and J. D. Protasiewicz, *J. Am. Chem. Soc.*, *117*, 6613 (1995).
74. K. G. Caulton, M. H. Chisholm, S. Doherty, and K. Folting, *Organometallics*, *14*, 2585 (1995).

75. Z. Gebeyehu, F. Weller, B. Neumüller, and K. Dehnicke, *Z. Anorg. Allg. Chem.*, *593*, 99 (1991).

76. W. A. Herrmann, S. Bogdanovic, R. Poli, and T. Priermeier, *J. Am. Chem. Soc.*, *116*, 4989 (1994).

77. H.-T. Chiu, Y.-P. Chen, S.-H. Chuang, J.-S. Jen, G.-H. Lee, and S.-M. Peng, *J. Chem. Soc. Chem. Commun.*, 139 (1996).

78. R. Ferguson, E. Solari, C. Floriani, A. Chiesi-Villa, and C. Rizzoli, *Angew. Chem. Int. Ed. Engl.*, *32*, 396 (1993).

79. J.-I. Song, P. Berno, and S. Gambarotta, *J. Am. Chem. Soc.*, *116*, 6927 (1994).

80. D. C. Bradley and R. G. Copperthwaite, *Inorg. Synth.*, *18*, 112 (1978).

81. P. Berno and S. Gambarotta, *J. Chem. Soc. Chem. Commun.*, 2419 (1994).

82. K. B. P. Ruppa, N. Desmangles, S. Gambarotta, G. Yap, and A. L. Rheingold, *Inorg. Chem.*, *36*, 1194 (1997).

83. C. P. Gerlach and J. Arnold, *Inorg. Chem.*, *35*, 5770 (1996).

84. Z. B. Duan, M. Schmidt, V. G. Young, X. B. Xie, R. E. McCarley, and J. G. Verkade, *J. Am. Chem. Soc.*, *118*, 5302 (1996).

85. A. L. Odom and C. C. Cummins, *Organometallics*, *15*, 898 (1996).

86. P. Berno and S. Gambarotta, *Angew. Chem. Int. Ed. Engl.*, *34*, 822 (1995).

87. M. G. Fickes, W. M. Davis, and C. C. Cummins, *J. Am. Chem. Soc.*, *117*, 6384 (1995).

88. S. L. Stokes, W. M. Davis, A. L. Odom, and C. C. Cummins, *Organometallics*, *15*, 4521 (1996).

89. Y. G. Gololobov and L. F. Kasukhin, *Tetrahedron*, *48*, 1353 (1992).

90. G. Proulx and R. G. Bergman, *J. Am. Chem. Soc.*, *117*, 6382 (1995).

91. D. F. Tuan and R. Hoffmann, *Inorg. Chem.*, *24*, 871 (1985).

92. K. Issleib and E. Wenschuh, *Chem. Ber.*, *97*, 715 (1964).

93. R. E. LaPointe, P. T. Wolczanski, and J. F. Mitchell, *J. Am. Chem. Soc.*, *108*, 6382 (1986).

94. T. A. Albright, R. Hoffmann, J. C. Thibeault, and D. L. Thorn, *J. Am. Chem. Soc.*, *101*, 3801 (1979).

95. D. R. Neithamer, L. Párkányi, J. F. Mitchell, and P. T. Wolczanski, *J. Am. Chem. Soc.*, *110*, 4421 (1988).

96. R. E. LaPointe and P. T. Wolczanski, *J. Am. Chem. Soc.*, *108*, 3535 (1986).

97. J. B. Bonanno, T. P. Henry, D. R. Neithamer, P. T. Wolczanski, and E. B. Lobovsky, *J. Am. Chem. Soc.*, *118*, 5132 (1996).

98. J. B. Bonanno, P. T. Wolczanski, and E. B. Lobkovsky, *J. Am. Chem. Soc.*, *116*, 11159 (1994).

99. P. B. Hitchcock, M. F. Lappert, and W.-P. Leung, *J. Chem. Soc., Chem. Commun.*, 1282 (1987).

100. P. Bohra, P. B. Hitchcock, M. F. Lappert, and W.-P. Leung, *Polyhedron*, *8*, 1884 (1989).

101. Z. Hou and D. W. Stephan, *J. Am. Chem. Soc.*, *114*, 10088 (1992).
102. J. Ho, R. J. Drake, and D. W. Stephan, *J. Am. Chem. Soc.*, *115*, 3792–3793 (1993).
103. A. H. Cowley, B. Pellerin, J. L. Atwood, and S. G. Bott, *J. Am. Chem. Soc.*, *112*, 6734 (1990).
104. A. H. Cowley and A. R. Barron, *Acc. Chem. Res.*, *21*, 81 (1988).
105. C. C. Cummins, R. R. Schrock, and W. M. Davis, *Angew. Chem. Int. Ed. Engl.*, *32*, 756 (1993).
106. K.-Y. Shih, K. Totland, S. Seidel, and R. R. Schrock, *J. Am. Chem. Soc.*, *116*, 12103 (1994).
107. L. M. Atagi, D. E. Over, D. R. McAlister, and J. M. Mayer, *J. Am. Chem. Soc.*, *113*, 870 (1991).
108. C. C. Cummins, R. R. Schrock, and W. M. Davis, *Inorg. Chem.*, *33*, 1448 (1994).
109. P. T. Wolczanski, *Polyhedron*, *14*, 3335 (1995).
110. D. R. Neithamer, Ph. D. Thesis, 1989, ''Preparation of Novel P_i-Complexes and Deoxygenation of Small Molecules and Organic Oxygenates by Tres (tri—*tert*—butylsiloxide) tantalum Including a Detailed Mechanistic Investigation Into the Carbon Monoxide Cleavage Reaction,'' 1989, Cornell University, NY.
111. F. A. Cotton, S. A. Duraj, G. L. Powell, and W. J. Roth, *Inorg. Chim. Acta*, *113*, 81 (1986).
112. J.-K. F. Buijink, A. Meetsma, and J. H. Teuben, *Organometallics*, *12*, 2004 (1993).
113. D. D. Devore, J. D. Lichtenhan, F. Takusagawa, and E. A. Maatta, *J. Am. Chem. Soc.*, *109*, 7408–7416 (1987).
114. D. C. Bradley, M. B. Hursthouse, and C. W. Newing, *J. Chem. Soc. Chem. Commun.*, 411 (1971).
115. H. Bock, I. Goebel, Z. Havlas, S. Liedle, and H. Oberhammer, *Angew. Chem. Int. Ed. Engl.*, *30*, 187 (1991).
116. D. C. Bradley and C. W. Newing, *J. Chem. Soc. Chem. Commun.*, 219 (1970).
117. J. C. W. Chien, W. Kruse, D. C. Bradley, and C. W. Newing, *J. Chem. Soc. Chem. Commun.*, 1177 (1970).
118. L. Grossi, *Tetrahedron Lett.*, *28*, 3387 (1987).
119. R. D. Köhn, G. Kociok-Köhn, and M. Haufe, *Chem. Ber.*, *129*, 25 (1996).
120. H. Bürger and U. Wannagat, *Monatsh.*, *95*, 1099 (1964).
121. A. L. Odom and C. C. Cummins, manuscript in preparation.
122. H. W. Lam, G. Wilkinson, B. Hussainbates, and M. B. Hursthouse, *J. Chem. Soc. Dalton Trans.*, 1477 (1993).
123. H. Blanchard, M. B. Hursthouse, and A. C. Sullivan, *J. Organomet. Chem.*, *341*, 367 (1988).
124. P. B. Hitchcock, N. H. Buttrus, and A. C. Sullivan, *J. Organomet. Chem.*, *303*, 321 (1986).

125. G. Stolze, *J. Organomet. Chem.*, *6*, 383 (1966).
126. M. H. Chisholm, M. Extine, and W. Reichert, *Adv. Chem. Ser.*, *150*, 273 (1976).
127. C. E. Laplaza, A. L. Odom, W. M. Davis, C. C. Cummins, and J. D. Protasiewicz, *J. Am. Chem. Soc.*, *117*, 4999 (1995).
128. J. R. Dilworth and J. Zubieta, *Inorg. Synth.*, *24*, 193 (1986).
129. M. J. Fernandeztrujillo, M. G. Basallote, P. Valerga, M. C. Puerta, and D. L. Hughes, *J. Chem. Soc. Dalton Trans.*, 3149 (1991).
130. P. Hofacker, C. Friebel, K. Dehnicke, P. Bäuml, W. Hiller, and J. Strähle, *Z. Naturforsch.*, *44b*, 1161 (1989).
131. J. H. Matonic, S. J. Chen, L. E. Pence, and K. R. Dunbar, *Polyhedron*, *11*, 541 (1992).
132. R. Poli and H. D. Mui, *J. Am. Chem. Soc.*, *112*, 2446 (1990).
133. R. Poli and J. C. Gordon, *Inorg. Chem.*, *30*, 4550 (1991).
134. D. S. Zeng and M. J. Hampden-Smith, *Polyhedron*, *11*, 2585 (1992).
135. M. H. Chisholm and W. Reichert, *J. Am. Chem. Soc.*, *96*, 1249 (1974).
136. M. H. Chisholm, F. A. Cotton, B. A. Frenz, W. W. Reichert, L. W. Shive, and B. R. Stults, *J. Am. Chem. Soc.*, *98*, 4469 (1976).
137. K. Dehnicke and J. Strahle, *Angew. Chem. Int. Ed. Eng.*, *20*, 413 (1981).
138. P. A. Hintz, M. B. Sowa, S. A. Ruatta, and S. L. Anderson, *J. Chem. Phys.*, *94*, 6446 (1991).
139. J. C. Kramlich and W. P. Linak, *Prog. Energy Combust. Sci.*, *20*, 149 (1994).
140. P. T. Matsunaga, G. L. Hillhouse, and A. L. Rheingold, *J. Am. Chem. Soc.*, *115*, 2075 (1993).
141. G. A. Vaughan, C. D. Sofield, G. L. Hillhouse, and A. L. Rheingold, *J. Am. Chem. Soc.*, *111*, 5491 (1989).
142. F. Bottomley, *Polyhedron*, *11*, 1707 (1992).
143. M. R. Smith, P. T. Matsunaga, and R. A. Andersen, *J. Am. Chem. Soc.*, *115*, 7049 (1993).
144. W. H A. Howard and G. Parkin, *J. Am. Chem. Soc.*, *116*, 606 (1994).
145. J. C. Peters, A. L. Odom, and C. C. Cummins, manuscript in preparation.
146. D. E. Wigley, *Progress in Inorganic Chemistry*, Wiley-Interscience, New York, 1994, Vol. 42, p. 239.
147. T. Suehiro, S. Masuda, R. Nakausa, M. Taguchi, A. Mori, A. Koike, and M. Date, *Bull. Chem. Soc. Jpn.*, *60*, 3321 (1987).
148. T. Suehiro, S. Masuda, T. Tashiro, R. Nakausa, M. Taguchi, A. Koike, and A. Rieker, *Bull. Chem. Soc. Jpn.*, *59*, 1877 (1986).
149. M. Kol, R. R. Schrock, R. Kempe, and W. M. Davis, *J. Am. Chem. Soc.*, *116*, 4382 (1994).
150. G. L. Hillhouse, G. V. Goeden, and B. L. Haymore, *Inorg. Chem.*, *21*, 2064 (1982).

151. C. E. Laplaza and C. C. Cummins, *Science*, *268*, 861 (1995).
152. K.-Y. Shih, R. R. Schrock, and R. Kempe, *J. Am. Chem. Soc.*, *116*, 8804 (1994).
153. Q. Cui, D. G. Musaev, M. Svensson, S. Sieber, and K. Morokuma, *J. Am. Chem. Soc.*, *117*, 12366 (1995).
154. M. Hidai and Y. Mizobe, *Chem. Rev.*, *95*, 1115 (1995).
155. G. J. Leigh, *Science*, *268*, 827 (1995).
156. R. Lipkin, *Sci. News*, *147*, 294 (1995).
157. M. Scheer, *Angew. Chem. Int. Ed. Engl.*, *34*, 1997 (1995).
158. C. E. Laplaza, W. M. Davis, and C. C. Cummins, *Angew. Chem. Int. Ed. Engl.*, *34*, 2042 (1995).
159. N. C. Zanetti, R. R. Schrock, and W. M. Davis, *Angew. Chem. Int. Ed. Engl.*, *34*, 2044 (1995).
160. G. Wu, D. Rovnyak, M. J. A. Johnson, N. C. Zanetti, D. G. Musaev, K. Morokuma, R. R. Schrock, R. G. Griffin, and C. C. Cummins, *J. Am. Chem. Soc.*, *118*, 10654 (1996).
161. L. K. Woo, J. G. Goll, D. J. Czapla, and J. A. Hays, *J. Am. Chem. Soc.*, *113*, 8478 (1991).
162. F. L. Neely and L. A. Bottomley, *Inorg. Chim. Acta*, *192*, 147 (1992).
163. L. A. Bottomley and F. L. Neely, *J. Am. Chem. Soc.*, *111*, 5955 (1989).
164. M. Rouhi, *Chem. Eng. News*, *74*, 7 (1996).
165. T. Hughbanks, *Chem. Ind.*, 222 (1996).
166. F. A. Cotton, C. C. Cummins, and M. J. A. Johnson, unpublished results.
167. A. R. Johnson, W. M. Davis, C. C. Cummins, S. Serron, S. P. Nolan, D. G. Musaev, and K. Morokuma, *J. Am. Chem. Soc.*, *119*, submitted for publication.
168. R. H. Holm and J. P. Donahue, *Polyhedron*, *12*, 571 (1993).
169. W. A. Nugent and J. M. Mayer, *Metal-Ligand Multiple Bonds*, Wiley, New York, 1988.
170. D. F. Evans, *J. Chem. Soc.*, 2003 (1959).
171. S. K. Sur, *J. Magn. Reson.*, *82*, 169 (1989).
172. D. F. Eppley, P. T. Wolczanski, and G. D. Van Duyne, *Angew. Chem. Int. Ed. Engl.*, *30*, 584 (1991).
173. W. A. Nugent and R. L. Harlow, *Inorg. Chem.*, *19*, 777 (1980).
174. E. C. Alyea, D. C. Bradley, and R. G. Copperthwaite, *J. Chem. Soc. Dalton Trans.*, 1580 (1972).
175. J. J. Ellison, P. P. Power, and S. C. Shoner, *J. Am. Chem. Soc.*, *111*, 8044 (1989).
176. M. A. Putzer, B. Neumüller, K. Dehnicke, and J. Magull, *Chem. Ber.*, *129*, 715 (1996).
177. W. Seidel and I. Burger, *Z. Chem.*, *17*, 31 (1977).
178. M. S. Kralik, L. Stahl, A. M. Arif, C. E. Strouse, and R. D. Ernst, *Organometallics*, *11*, 3617 (1992).

179. D. S. Williams, J. T. Anhaus, M. H. Schofield, R. R. Schrock, and W. M. Davis, *J. Am. Chem. Soc.*, *113*, 5480 (1991).
180. D. S. Williams and R. R. Schrock, *Organometallics*, *12*, 1148 (1993).
181. W. Seidel and K.-J. Lattermann, *Z. Anorg. Allg. Chem.*, *488*, 69 (1982).
182. R. R. Schrock, L. G. Sturgeoff, and P. R. Sharp, *Inorg. Chem.*, *22*, 2801 (1983).
183. C. C. Cummins and R. R. Schrock, *Inorg. Chem.*, *33*, 395 (1994).
184. M. B. O'Donoghue, N. C. Zanetti, W. M. Davis, and R. R. Schrock, *J. Am. Chem. Soc.*, *119*, 2753 (1997).
185. F. M. MacDonnell, K. Ruhlandt-Senge, J. J. Ellison, R. H. Holm, and P. P. Power, *Inorg. Chem.*, *34*, 1815 (1995).
186. P. P. Power and S. C. Shoner, *Angew. Chem. Int. Ed. Engl.*, *30*, 330 (1991).
187. D. M. Roddick, T. D. Tilley, A. L. Rheingold, and S. J. Geib, *J. Am. Chem. Soc.*, *109*, 945 (1987).
188. J. T. Anhaus, T. P. Kee, M. Schofield, and R. R. Schrock, *J. Am. Chem. Soc.*, *112*, 1642 (1990).
189. M. H. Schofield, T. P. Kee, J. T. Anhaus, R. R. Schrock, K. H. Johnson, and W. M. Davis, *Inorg. Chem.*, *30*, 3595 (1991).
190. Z. Y. Lin and M. B. Hall, *Coord. Chem. Rev.*, *123*, 149 (1993).
191. P. J. Walsh, F. J. Hollander, and R. G. Bergman, *J. Am. Chem. Soc.*, *110*, 8729 (1988).
192. B. D. Steffey, P. E. Fanwick, and I. P. Rothwell, *Polyhedron*, *9*, 963 (1990).
193. A. O. Chong, K. Oshima, and K. B. Sharpless, *J. Am. Chem. Soc.*, *99*, 3420 (1977).
194. M. M. Olmstead, P. P. Power, and G. Sigel, *Inorg. Chem.*, *25*, 1027 (1986).
195. R. S. Hay-Motherwell, G. Wilkinson, B. Hussain, and M. B. Hursthouse, *Polyhedron*, *9*, 931 (1990).
196. K. H. Theopold, J. Silvestre, E. K. Byrne, and D. S. Richeson, *Organometallics*, *8*, 2001 (1989).
197. R. S. Hay-Motherwell, B. Hussain-Bates, M. B. Hursthouse, and G. Wilkinson, *J. Chem. Soc. Chem. Commun.*, 1242 (1990).
198. R. S. Hay-Motherwell, S. U. Koschmieder, G. Wilkinson, B. Hussain-Bates, and M. B. Hursthouse, *J. Chem. Soc. Dalton Trans.*, 2821 (1991).
199. S. U. Koschmieder and G. Wilkinson, *Polyhedron*, *10*, 135 (1991).
200. R. S. Hay-Motherwell, G. Wilkinson, B. Hussain-Bates, and M. B. Hursthouse, *J. Chem. Soc. Dalton Trans.*, 3477 (1992).
201. R. S. Hay-Motherwell, G. Wilkinson, B. Hussain-Bates, and M. B. Hursthouse, *Polyhedron*, *12*, 2009 (1993).
202. H. Hope, M. M. Olmstead, B. D. Murray, and P. P. Power, *J. Am. Chem. Soc.*, *107*, 712 (1985).
203. R. Hay-Motherwell, G. Wilkinson, T. K. Sweet, and M. B. Hursthouse, *Polyhedron*, *15*, 3163 (1996).

204. R. A. Andersen, *Inorg. Chem.*, *18*, 1507 (1979).
205. L. R. Avens, S. G. Bott, D. L. Clark, A. P. Sattelberger, J. G. Watkin, and B. D. Zwick, *Inorg. Chem.*, *33*, 2248 (1994).
206. J. L. Stewart, Ph.D. Thesis, 1988, University of California-Berkeley, CA.
207. J. G. Brennan, Ph.D. Thesis, 1985, University of California-Berkeley, CA.
208. C. J. Burns, W. H. Smith, J. C. Huffman, and A. P. Sattelberger, *J. Am. Chem. Soc.*, *112*, 3237 (1990).
209. A. Zalkin, J. G. Brennan, and R. A. Andersen, *Acta Crystallogr.*, *C44*, 1553 (1988).
210. W. G. Van Der Sluys, C. J. Burns, and A. P. Sattelberger, *Organometallics*, *8*, 855 (1989).
211. W. G. Van der Sluys, C. J. Burns, J. C. Huffman, and A. P. Sattelberger, *J. Am. Chem. Soc.*, *110*, 5924 (1988).

Metal–Carbohydrate Complexes in Solution

JEAN-FRANÇOIS VERCHÈRE

Université de Rouen
Faculté des Sciences
UMR/CNRS 6522
76821 Mont-Saint-Aignan, France

STELLA CHAPELLE

Université de Rouen
Faculté des Sciences
URA/CNRS 464
Laboratoire de RMN
76821 Mont-Saint-Aignan, France

FEIBO XIN and DEBBIE C. CRANS

Department of Chemistry
Colorado State University
Fort Collins, Colorado

CONTENTS

Progress in Inorganic Chemistry, Vol. 47, Edited by Kenneth D. Karlin.
ISBN 0-471-24039-7 © 1998 John Wiley & Sons, Inc.

I. INTRODUCTION

Carbohydrates are a class of biologically active molecules with many functions. Glycobiology, a subbranch in biology, is devoted to the biologically interesting carbohydrates (1). Metal ions are known to participate in many biochemical reactions and are widely distributed in living organisms. Therefore, metal–carbohydrate interactions may be expected to take place in biological fluids and complexation would certainly affect the biological activity of biomolecules such as glycoproteins or lipopolysaccharides. Although all the functions and mechanism of action of carbohydrates are not fully understood, their various roles will presumably appear more clearly as more molecular information is obtained on the detailed mechanism of immunological responses and glycoprotein actions. Metal–carbohydrate complexes have many applications in chemistry including chromatography, electrophoresis, and various other analytical methods. The number of chemical applications of metal–carbohydrate complexes is rapidly growing, fueled by the discoveries of new reactions like the metal-mediated epimerization of aldoses and the recent interest in using chiral carbohydrate ligands as catalysts or stoichiometric reagents in organic syntheses. In all these systems, the metal–carbohydrate complex acts as a solution species. Despite this fact, most of the previous reviews concerning metal–carbohydrate complexes focused on aspects other than the aqueous solution chemistry of metal–carbohydrate complexes.

The first review describing metal–carbohydrate complexes appeared in 1966 (2). A special volume of the *Advances in Chemistry Series* published in 1973 (3) contained contributions of special interest to carbohydrate complexes in solution. Rendleman (4) discussed the ionization of carbohydrates in the presence of metal hydroxides and oxides. Furthermore, Angyal (5) reviewed the complexes of sugars with cations and Montgomery (6) described the use of potentiometric titration for the determination of the stability constants of carbohydrate complexes. A review by Haines (7) on the relative reactivities of hydroxyl groups in carbohydrates is useful for the rationalization of data on the structures of carbohydrate complexes. Since NMR spectroscopy has become a key technique used for the examination of metal–carbohydrate complexes in solution, two particularly useful reviews describe applications of ^{13}C NMR spectroscopy in studies of mono- (8) and oligosaccharides (9). Relaxation techniques in ^{13}C NMR spectroscopy also proved useful in studying the interactions of metal ions with carbohydrates (10).

Over a period of 20 years, several reviews describing the developments in the area of metal cation interactions with carbohydrates have been published by Angyal (11–14). Together, these reviews show that although the understanding of the reactions of Ca^{II} and lanthanides with carbohydrates has tremendously increased, little progress was made with other metal ions. The use of columns containing various immobilized inorganic ions for the separation of carbohydrates by HPLC was reviewed (15).

The syntheses and characterization of transition metal complexes containing a substituted diamine and a monosaccharide, and the epimerization reactions that occur in such complexes have been reviewed (16). Several recent reviews have focused on the biological importance of sugar–metal complexes (17–19). The structures and function of metal coordination to carbohydrates with a special emphasis on sugar acids (18) or uronic acids (20) have recently been reviewed. One of these articles contains a useful table of stability constants (18). In some cases, the carbohydrate portions of nucleosides and nucleotides play a key role in complexation with metal ions, as shown by Yano and Otsuka (19) for both main group and transition metal ions. Finally, an emerging new area relating to the metal–carbohydrate complexes in organic solvents and their applications in synthesis was recently reviewed by Piarulli and Floriani (21), who suggest that metal–carbohydrate complexes will soon be part of the organic synthetic chemists' arsenal of reagents.

Given the complexity of carbohydrates as ligands, we will briefly summarize the chemistry of carbohydrates in aqueous solution. A more in-depth discussion of this topic can be found in previous reviews by Angyal (22–24). The focus of this chapter is on complexes that form between metal ions and ligands containing a sequence of weakly chelating hydroxyl groups. Thus, sugar derivatives containing other functionalities (amino-, thio-sugars, sugar phosphates) will not be discussed here in detail and the reader is referred elsewhere (17–19). The sugar acids, namely, glycuronic acids $HOOC{-}(CHOH)_{n-2}{-}CHO$, aldonic acids $HOCH_2{-}(CHOH)_{n-2}{-}COOH$ and aldaric acids $HOOC{-}(CHOH)_{n-2}{-}COOH$, are only described here when they form complexes that do not involve the COOH group. It is rarely the case, since in a review of the chemistry, stability constants, and applications of metal complexes of D-gluconic acid by Sawyer (25), all complexes that could be characterized were shown to involve the carboxylate group. Moreover, our discussion will be limited to monosaccharides (aldoses and ketoses) and oligosaccharides, together with their reduced derivatives alditols and cyclitols. Disaccharides in which the ring oxygen or/and the glycosidic oxygen (for disaccharides) are replaced by a methylene group (*C*-saccharides or carbasugars) will also be described here.

Before we embark on the detailed discussion of metal–carbohydrate complexes, we wish to summarize various aspects of the structure of carbohydrates in solution (Section I.A), solid-state structural information (Section I.B), techniques for obtaining structural information in solution (Section I.C), biological importance (Section I.D), and chemical applications of metal–carbohydrate complexes (Section I.E).

A. The Chemistry of Carbohydrates and Ligand Properties in Solution

Carbohydrates are molecules of the general formula $(CH_2O)_n$, which encompass several series of closely related configurational isomers. The most common natural compounds are aldoses $HOCH_2-(CHOH)_{n-2}-CHO$ or 2-ketoses $HOCH_2-(HOCH)_{n-3}-CO-CH_2OH$ with $n = 5$–7. Carbohydrates containing a large number of asymmetric carbon atoms will rapidly generate many isomers, which are cumbersome to name using conventional nomenclature and thus, specific names have been retained for diastereoisomers up to four asymmetric carbons. The correspondence between configurations and names is shown in Table I. The names of higher carbon sugars ($n > 6$) are formed from sequences of shorter chains containing at most four asymmetric carbons (26).

Aldoses and ketoses exist almost exclusively in the form of cyclic hemiacetals resulting from the reaction of the carbonyl group with either HO4 or HO5. These cyclization reactions, respectively, generate five-membered rings named furanoses (f) or six-membered rings named pyranoses (p), which contain one oxygen atom. Moreover, the achiral carbonyl group is transformed into a chiral center bearing a new HO group with two possible configurations. Thus, two isomers, referred to as the α and β anomers, are created in unequal proportions. In the α anomer, the HO1 group points in the same direction as that, prior to cyclization, of the HO group of the last chiral center, that is, HO5 in an aldohexose. The α–β isomerization may be monitored by the variations in optical rotation of the solution and is referred to as mutarotation. The transition between the p and f forms, and the α and β anomers involves the acyclic aldehyde or ketone. Thus, carbohydrates that can open may be easily oxidized and are commonly referred to as reducing sugars. Three reviews (22–24) give a clear overview of the equilibria of reducing sugars in solution, which are illustrated in Schemes 1 (aldose) and 2 (ketose), and the relative proportions of their various forms. Nonreducing sugars are obtained when the anomeric HO group is blocked by esterification or etherification. In addition to these complexities, each carbohydrate form can exist in several conformations of which usually one is predominant. In aqueous solution, aldoses and ketoses exist in very low amounts ($<1\%$) in their open forms represented by the aldehyde and ketone. These acyclic molecules readily undergo hydration of the carbonyl group. The [hydrate]/[carbonyl] ratio is generally close to 10. At equilibrium, the greatest proportion of aldoses and ketoses are present as the cyclic p and f forms.

The p form, existing as a flattened cyclohexane ring, prevails for most carbohydrates with $n \geq 5$ and may adopt many conformations. In most conformations, four atoms in the ring are in one plane. The more stable conformations are two chair (C) conformers given by the notations ${}^1C_4$ and ${}^4C_1$, which indicate the carbon atoms located above (left, i.e., 1 or 4) and below (right, i.e., 4 or

TABLE I
Official Names of Common Carbohydrates (26)

Carbohydrate	n^a	Name	Fischer Projection
Aldose	4	D-Erythrose	$HOCH_2$ — CHO
		D-Threose	$HOCH_2$ — CHO
	5	D-Ribose	$HOCH_2$ — CHO
		D-Lyxose	$HOCH_2$ — CHO
		D-Arabinose	$HOCH_2$ — CHO
		D-Xylose	$HOCH_2$ — CHO
	6	D-Allose	$HOCH_2$ — CHO
		D-Talose	$HOCH_2$ — CHO
		D-Mannose	$HOCH_2$ — CHO
		D-Gulose	$HOCH_2$ — CHO
		D-Galactose	$HOCH_2$ — CHO
		D-Altrose	$HOCH_2$ — CHO
		D-Glucose	$HOCH_2$ — CHO
		D-Idose	$HOCH_2$ — CHO
Ketose	6	D-Psicose	$HOCH_2$ — $CO—CH_2OH$
		D-Tagatose	$HOCH_2$ — $CO—CH_2OH$
		D-Fructose	$HOCH_2$ — $CO—CH_2OH$
		D-Sorbose	$HOCH_2$ — $CO—CH_2OH$
Alditol	4	Erythritol	$HOCH_2$ — CH_2OH
		D-Threitol	$HOCH_2$ — CH_2OH
	5	Ribitol	$HOCH_2$ — CH_2OH
		D-Arabinitol	$HOCH_2$ — CH_2OH
		Xylitol	$HOCH_2$ — CH_2OH
	6	Allitol	$HOCH_2$ — CH_2OH
		D-Altritol	$HOCH_2$ — CH_2OH
		D-Mannitol	$HOCH_2$ — CH_2OH
		D-Iditol	$HOCH_2$ — CH_2OH
		Galactitol	$HOCH_2$ — CH_2OH
		D-Glucitol	$HOCH_2$ — CH_2OH

[a]The total number of carbon atoms is n. Note that Fischer's projections are not written vertically as in the usual convention, but horizontally with the top orienting group C1 on the right-hand side. Thus, in the depicted D series, HO-(n − 1) is oriented downward.

1) the plane defined by O and C2,3,5. In the chair forms, the substituents of the ring may be in axial (a) or equatorial (e) orientation. Most sugars preferentially adopt the $^4C_1$ conformation (Schemes 3 and 4). Other conformations of higher energy frequently encountered are the boat forms (B, cf. $^{2,5}B$ or $B_{2,5}$), in which both C2,5 are above or below the O,C1,3,4 plane, respectively, and the half-chair forms (H, cf. $^3H_4$) in which two adjacent carbons are above (C3) and below (C4) the O,C1,2,5 plane, respectively. The skew (S, cf. $^1S_3$) conformers are distorted boat forms in which, for example, C1 is above and C3 is below the ring plane defined by the three atoms O,C4,5

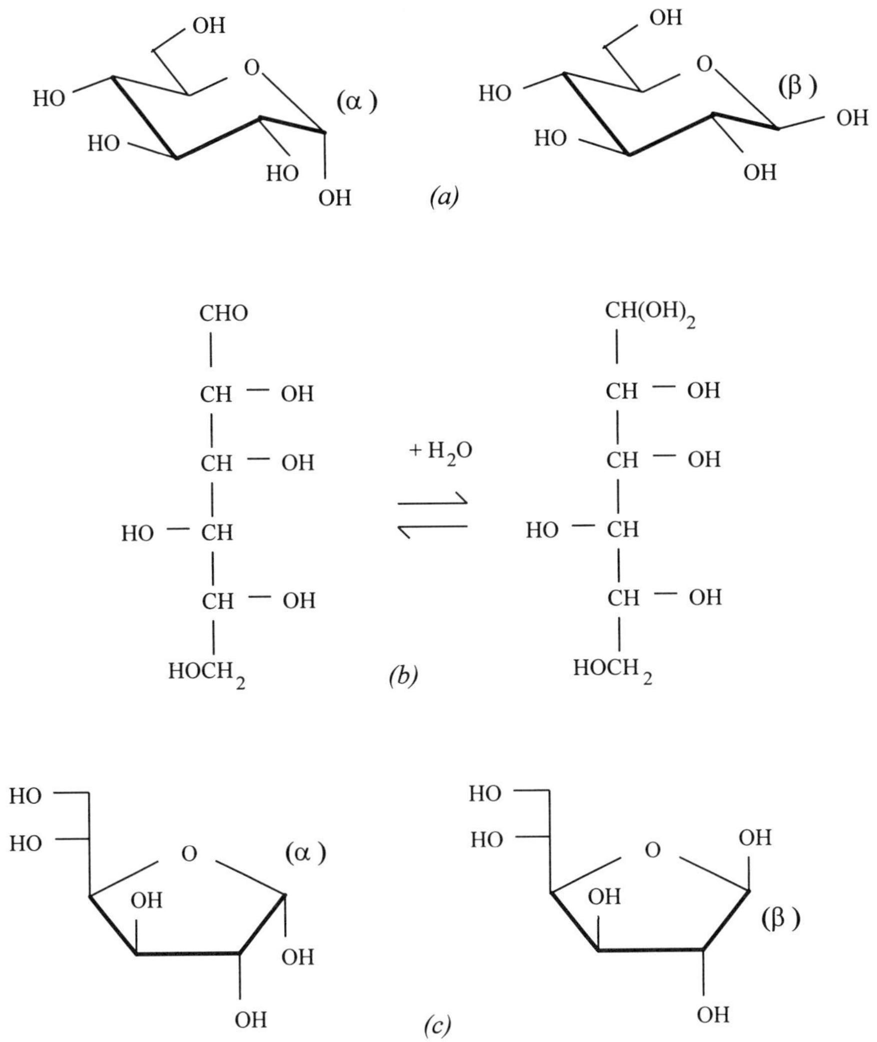

OH
O
HO (α)
HO
HO
OH
(a)
OH
HO O (β)
HO OH
OH

CHO | CH — OH | CH — OH | HO — CH | CH — OH | $HOCH_2$

$+ H_2O$ ⇌

$CH(OH)_2$ | CH — OH | CH — OH | HO — CH | CH — OH | $HOCH_2$

(b)

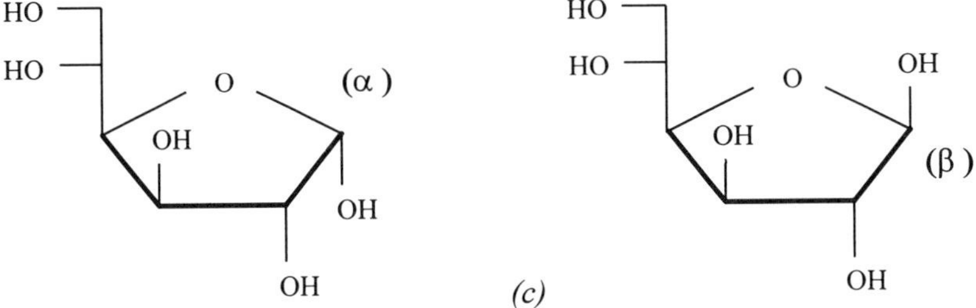

Scheme 1. Forms of D-glucose in aqueous solution. (*a*) Pyranoses in ${}^4C_1$ conformation. (*b*) Aldehyde and hydrate. (*c*) Furanoses. The Fischer projection of the aldehyde form may also be represented as $HOCH_2$—┬—┬—┴—┬—CHO, where C1 is on the right-hand side.

Aldotetroses ($n = 4$) and ketopentoses (pentuloses, $n = 5$) contain too few carbon atoms to form pyranoses and exist as mixtures of furanose and hydrated forms. However, for aldoses with $n > 4$ and ketoses with $n > 5$, the *f* forms are generally much less abundant than the *p* forms. Nevertheless, some ketoses and aldoses of the ribo series afford appreciable proportions of *f* forms as both

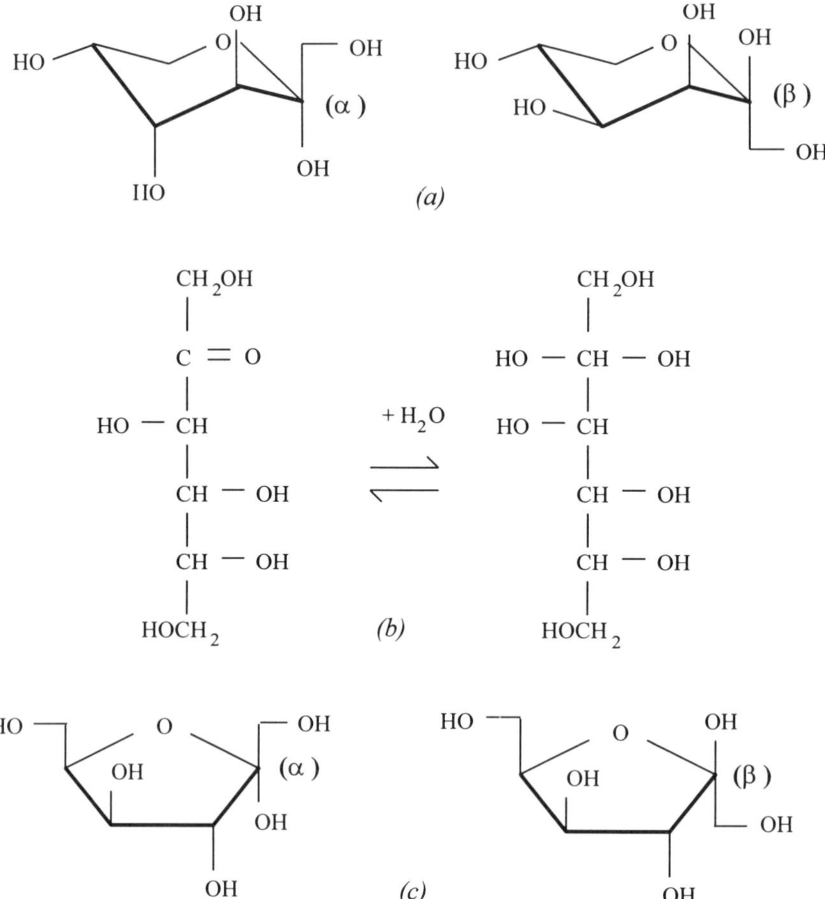

Scheme 2. Forms of D-fructose in aqueous solution. (*a*) Pyranoses in $^4C_1$ conformation. (*b*) Ketone and hydrate. (*c*) Furanoses. The Fischer projection of the ketone form may also be represented as $HOCH_2$ ——┬——┬——┴—— $CO-CH_2OH$, where C1 is on the right-hand side.

α or β anomers. The various conformations of *f* forms, like the envelope (E) or twist (T) forms may be described as quasiplanar rings.

Alditols, $HOCH_2-(CHOH)_{n-2}-CH_2OH$, the reduction products of aldoses and ketoses, are also natural compounds of biological importance. Nomenclature is presented in Table I. The acyclic polyols represent convenient models for the hydrated forms of carbohydrates. Alditols are found in two important conformations. The zigzag form is fully extended, whereas a semicircular form is named sickle. The generation of a given conformation is governed by minimizing unfavorable interactions between hydroxyl groups and the car-

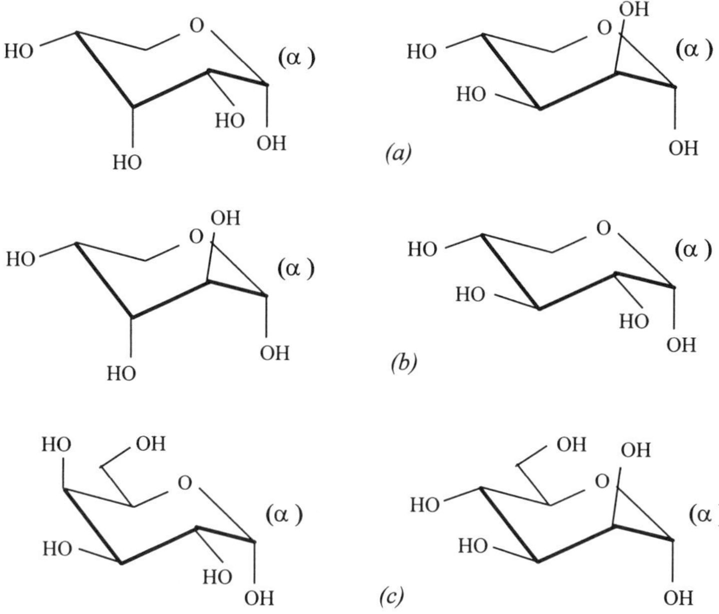

Scheme 3. The α-pyranose forms in ${}^4C_1$ conformation of the pentoaldoses and two common hexoaldoses of the D series. (*a*) Pentoses with erythro HO2,3: ribose (left) and lyxose (right). (*b*) Pentoses with threo HO2,3: arabinose (left) and xylose (right). (*c*) Hexoses: galactose (left) and mannose (right).

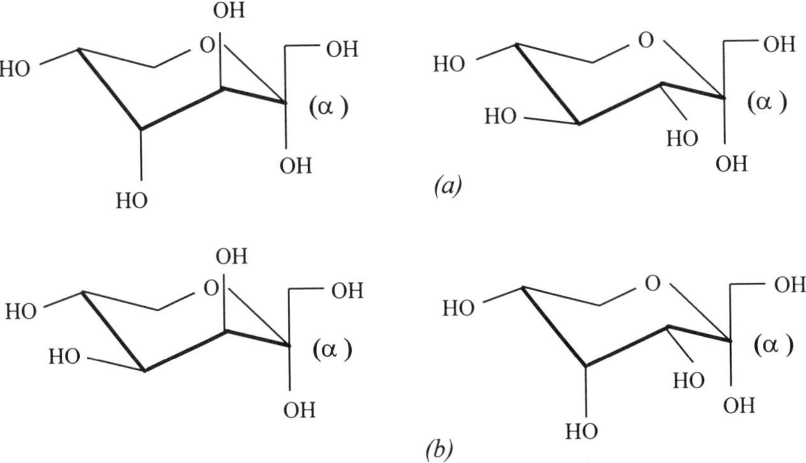

Scheme 4. The four hexoketoses of the D series in the α-*p* form in ${}^4C_1$ conformation. (*a*) Ketoses with threo HO3,4: fructose (left) and sorbose (right). (*b*) Ketoses with erythro HO3,4: tagatose (left) and psicose (right).

bon backbone. Cyclic alditols (cyclitols) are another interesting family of biologically active carbohydrates. The hexahydroxycyclohexanes, widely present in plants, are named inositols. Alditols and cyclitols, in contrast to aldoses and ketoses, exist in a single form and represent simpler systems compared to sugars for studying the ability of carbohydrates to coordinate metal ions.

Since the carbonyl group of sugars is always engaged in hemiacetal formation, the coordinating properties of carbohydrates are solely due to the hydroxyl groups. Some metal complexes of monodentate sugar derivatives have been reported, but usually, more than one hydroxyl group is involved and the resulting chelates are therefore stronger than metal alkoxides of monoalcohols. In complexes formed with dinuclear metal cores, the carbohydrate ligands may be tetra- or pentadentate. The complexes of carbohydrates with a given metal ion can span a wide range of stabilities, depending on the orientation of the hydroxyl groups, that is, the specific configuration of each ligand. These stability differences provide the basis for the separation of carbohydrates after metal ion complexation. The structure of a metal ion–carbohydrate complex depends not only on carbohydrate stereochemistry, but also on the coordination number of the metal ion. Most ions of alkali metals, alkaline earths, transition metals, and lanthanides ions have coordination numbers ranging from 6–12. Therefore, the coordination sphere of a metal ion can include more than one carbohydrate. Alternatively, carbohydrates with six hydroxyl groups or more may chelate two metal ions at two tridentate sites or the carbohydrate may possess pendant hydroxyl groups. It allows the formation of polymeric complexes with a variety of different stoichiometries, which possess a network of bridged metal ions and carbohydrate molecules.

Since carbohydrates chelate metal ions through oxygen atoms only, they are expected to form more stable complexes with "hard" Lewis acid type metal ions (large charge and small radius). On the other hand, "soft" metal ions (small charge and large radius) form weak complexes with carbohydrates. Some soft Lewis acids prefer to bind at carbon atoms, generating organometallic compounds described in Section II.M. Presently, few examples of organometallic species have been described. However, the number of studies in this field is rapidly growing. Several reactions of carbohydrates with organometallics have been described, in which the sugar presumably forms a transient complex with the metal center.

A number of elements, which readily form carbohydrate complexes (typically vanadium, molybdenum, and tungsten) are characterized by the existence of oxo ions in which one or several metal atom(s) is (are) surrounded by oxygen atoms (27). In this respect, they behave like nonmetal elements, like boron and iodine, which are not reviewed here. The chemistry of such oxo ions–carbohydrate complexes present similarities with that of carbohydrate–borate complexes (28), especially for the large differences in complex stability induced by changes in the configuration or conformation of the ligand.

Oxo ion complexes have been invoked as intermediate species in the oxidation of polyols and sugars by periodate, chromate(VI), manganate(VII), and osmate(VIII).

In conclusion, two quite different types of complexes can be distinguished: Those with cations and those with oxo ions. The former are held together by electrostatic forces and hydrogen bonds, and are therefore rather weak; their formation is very rapid. On the other hand, the latter involve covalent bonds; their formation is slower but they are much more stable; in most cases, their existence is pH dependent.

B. Solid-State Information on Metal–Carbohydrate Structure

X-ray crystallography provides information on the molecular structure of complexes that have been crystallized and yield solvable diffraction patterns. Characterization of the solid material is furthermore supported by elemental analysis, magnetic susceptibility, and IR, Mössbauer, and EPR spectroscopies. Structural data are important in documenting and providing evidence for the existence of various types of complexes and their modes of bonding. However, it is often difficult to know whether the compound isolated and structurally characterized was indeed the major species present in solution, especially when mixtures of complexes are obtained. The reason is that the precipitation of a particular species, according to Le Chatelier principle, would shift the equilibrium in the bulk solution. Thus, the isolated compound may have been a minor constituent of the mother solution. There are even examples in which a compound isolated as a solid under certain conditions, does not form in a solution prepared under different experimental conditions. For example, for the vanadium(V) complexes of triethanolamine and tri(2-propanol)amine, the materials isolated and characterized in the solid state are neutral complexes, different from the anionic species present in aqueous solution (29). The nature of the cation used for ensuring precipitation may also favor a minor species that can form a particularly insoluble salt. For example, in the molybdate–oxalate system, a solid with 1 : 2 Mo/oxalate ratio was precipitated as a $[Co(en)_3]^{3+}$ salt (30), whereas the major complexes present in solution are 1 : 1 or 2 : 2 species depending on the pH (31).

It is important to recognize that solid compounds are generally precipitated from concentrated solutions, whereas the physicochemical studies are often performed in dilute solutions as dictated by activity coefficients. Furthermore, metal–carbohydrate complexes show biological and catalytic activity in dilute solutions and accordingly, characterizations under such conditions are of greater biological and chemical interest. There are several important reasons why caution must be exerted when relating results obtained in the solid state with those in solution.

Solid-state characterization of many metal–carbohydrate complexes have recently resulted in an increasing number of compounds containing cores of metal atoms bridged by oxygen atoms. These species are commonly referred to as oxometalates. Most of them are formed with elements such as W, Mo, and V, but examples are also known with Nb and Ta (32). Most metal ions in oxometalates are in octahedral geometry linked by one or several bridging oxygen atoms. The carbohydrate ligand often surrounds the metal oxide core. There is no doubt that new types of oxometalates will be discovered in the future and that new information on their formation and dissociation into smaller components will prove to be important in understanding the catalytic reactivity of oxometalates.

Given the polynuclear nature of most oxometalate cores and the polydentate nature of carbohydrate ligands, the structures of such complexes are difficult to predict both in the solid state and particularly in solution, where various equilibria can complicate matters even further. As we will describe in Section I.C, obtaining structural information on complexes in solution is not a trivial problem, even if solid-state structural data is available.

C. Techniques for Obtaining Structural Information in Solution

The determination of equilibrium constants is essential for the quantitative characterization of the thermodynamic stabilities of complexes. For this purpose, the stoichiometries need to be determined first (by Job's method and other related procedures). Carbohydrate complexes are particularly intractable because of the lack of suitable techniques for monitoring their formation, combined with the fact that many complexes are fairly weak in aqueous solution. For example, photometric methods can seldom yield useful information, since carbohydrate ligands do not possess a chromophoric group. Classical potentiometric methods are also rarely used for the weak complexes of neutral carbohydrates, but these methods are very useful for studies on the stronger metal complexes of sugar derivatives (amines or acids) that possess acid–base properties. Computer software are nowadays available for the determination of the stoichiometries of the complexes and the corresponding equilibrium constants. The use of various programs has been described in several reviews (33–35). In the case of oxo ions, such as vanadate, molybdate, and tungstate, the complex-forming reactions involve uptake or loss of protons, even for neutral sugars and the reaction can be monitored through pH variations. In such cases, the number of protons involved in the overall reaction can be derived from a careful study of the variations of the conditional stability constants with the pH. For the determination of the thermodynamic parameters of complex formation, however, the calorimetric method is superior to techniques in which the variations of log K (the equilibrium constant) are plotted versus the reciprocal of temperature (van't Hoff plots).

Since most carbohydrates are chiral ligands, their complexes also have optical rotations that are often enhanced, due to the limitation of possible conformational variations in the ligand. These phenomena are especially useful for alditols, which generally exhibit very small optical rotations when uncomplexed. For example, measurements of the specific optical rotation of alditols in acidic molybdate solution were used for the identification (36) and even the determination of unknown natural polyols (37). Nowadays, more specific measurements use optical rotatory dispersion (ORD) and circular dichroism (CD) apparatus. These techniques allow the detection of Cotton effects, which are more sensitive to the conformation of the carbohydrate in the metal complex.

Some techniques have been ingeniously designed for the characterization of metal–carbohydrate complexes. For example, the dissolution of poorly soluble calcium salts is enhanced in the presence of ligands that complex the Ca^{2+} cation (38). Thus, the stability constants of various weak calcium–carbohydrate were obtained from the dependence of the conditional solubility product of calcium sulfate on the carbohydrate concentration. More recently, the photometric method has been employed in the indirect mode for the characterization of metal–carbohydrate complexes (39). In this method, the formation of a colorless complex between a carbohydrate and a metal ion was monitored by following the decrease in absorption due to the dissociation of a colored, weak complex initially formed between this metal ion and an auxiliary ligand.

In early studies, information on the order of relative strengths of the complexes was obtained by chromatographic methods. The rate of migration of a carbohydrate on a porous support (paper or TLC plate) impregnated with a solution of a metal salt can be related to the strength of the complex. The same principle was applied to electrophoresis or ionophoresis in aqueous solutions of metal salts.

Structural information on the complexes in solution may be obtained by NMR spectroscopy of the ^{1}H, ^{13}C, and ^{17}O nuclei for the ligand and a suitable isotope of the metal, for example, ^{51}V, ^{95}Mo, ^{119}Sn, ^{183}W, and ^{195}Pt. All the protons and carbon atoms of the organic ligands can be assigned using two-dimensional (2D) modern NMR experiments. Couplings between vicinal protons are detected by 2D homonuclear H—H correlation (COSY) experiments (40, 41), whereas 2D heteronuclear correlations between carbons and protons are used for determining the structure of the site of chelation (42).

Many transition metal ions are paramagnetic and cannot be submitted to conventional NMR studies, since the presence of the unpaired electrons will cause considerable broadening of the NMR signals from the neighboring nuclei. In this case, EPR and ENDOR spectroscopies are available to detect unpaired electrons. However, other effects of the paramagnetism may be exploited in order to overcome this problem, such as the wide window and the possibility of locating by 2D experiments a cross-broadened signal under a diamagnetic envelope. The recent flourish of studies of systems with unpaired electrons using paramagnetic NMR spectroscopy suggests that this method will soon be applied to the study of metal–carbohydrate complexes (43–45). Up to now, a single application can be mentioned (46).

Extended X-ray absorption fine structure spectroscopy has been found to be successful in the determination of coordination numbers and coordinating atoms around a central metal atom, and typically the first-shell bonds distances (47). The advantage of this method is that it can be used both in solid state and in aqueous solution. The local structure environment of a series of Fe^{III}–sugar complexes (48) was determined by EXAFS spectroscopy. A complementary method, XANES spectroscopy, can be used to obtain information on oxidation states of metal ions and electron spin state in the complexes (47). For complexes containing metal ions with Mössbauer-effect isotopes, such as Fe and Sn, Mössbauer spectroscopy is a powerful method for the study of intramolecular interactions in the metal–carbohydrate complexes systems (49).

The kinetic aspects of the reactions between metal ions and carbohydrates have so far received little attention. With most metal ions, complexation is very fast, and equilibrium is reached instantaneously. However, slow complex formation takes place with specific metal ions and the case of d^3 ions, such as Cr^{III}, is well known to every inorganic chemist. Sluggish reactions are also observed with ions subject to hydrolysis or polymerization, such as Al^{III} and W^{VI}, because the kinetically inert polyions must dissociate to form the reactive metal species. In many cases, the rates for complex formation are sufficiently slow, on the NMR scale, so that different signals are generated for the atoms of the uncomplexed and chelated ligand. When the rates for complex formation increase, fast exchange perturbs the NMR spectra (50).

Mixtures of complexes are frequently obtained for a given carbohydrate–metal ion system. Besides, some metal ions prone to oxometalate formation (Al, V, Mo, W, U) yield mixtures of mononuclear and polynuclear species. Sometimes, differences between the complexes only occur in the structure of the ligand (conformation or site of chelation) or in the mode of metal coordination. The distribution of complexes highly depends on the specific pH, temperature, ratio of reagents, nature of solvent, added salts, and other experimental conditions. When several isomeric complexes of identical stoichiometry $(x:y)$ exist, the equilibrium constant K_{xy} determined by macroscopic methods (potentiometry or electrochemistry) encompasses contributions from all of the individual values of the different complexes. However, methods such as NMR spectroscopy, which often allow the simultaneous and separate detection of several complexes, may be employed for the determination of the individual equilibrium constants. Obviously, combination of several methods is extremely powerful and provides information that is particularly well substantiated in this difficult area of solution characterization (51).

D. Biological Importance of Metal–Carbohydrate Complexes

The biological importance of metal–carbohydrate complexes is yet to be documented at the desirable molecular level and in mechanistic detail (17–19). There is, however, no doubt that the interaction of metal ions with isolated simple carbohydrates and carbohydrate markers on biomolecules is and will in the future be recognized as an important mode of recognition in biology. In metabolic processes, the interaction of metal ions with simple carbohydrates may play a role by ensuring the presence of the carbohydrate in particular forms. However, since aldoses and ketoses form weak complexes with most metal ions, carbohydrate derivatives containing acidic groups are the most likely contenders for this type of mechanism (18). The essential role of carbohydrates in the immunological events as well as recognition events, in antibiotics, is well documented. Although much has been learned on the detailed mechanism of such processes, the detailed molecular aspects of synthesis and processing of information are still areas in their infancy. Accordingly, structural information, whether specific metal–carbohydrate complexes play a role in these processes, still remain to be determined in most cases (52–54). For example, the specific role of the divalent metal when various glycosyl transferases transfer a glycosyl group from the donor to the acceptor is not known (52). Nor is the mechanism for cell agglutination mediated by concanavalin A containing one Mn^{II} and one Ca^{II} for each protein monomer near the sugar binding site (53, 54) understood in detail. On the other hand, detailed information concerning the network of protein–carbohydrate–calcium and water interactions in calcium-dependent C-type lectins (55–58) are currently investigated at the molecular level, providing information about both structure and mechanism of action (59–61). The latter studies document one important example of the direct role of metal–carbohydrate complexes in biology.

E. Chemical Applications of Metal–Carbohydrate Complexes

The interest of metal–carbohydrate complexes is by no means restricted to the biological area. Carbohydrates are nowadays recognized as a "chiral pool" of enantiomeric pure compounds that are invaluable synthons for new synthetic methods (62). Thus, the formation of metal–carbohydrate complexes is not only useful in separation and analysis of chiral compounds, but also in the conversions in stoichiometric or catalytic stereoselective syntheses. Early studies of metal–carbohydrate complexes are aimed toward the separation of mixtures of sugars (of natural or synthetic origin) by chromatography or electrophoresis. Various methods used paper chromatography, TLC plates, HPLC, and capillary zone electrophoresis. Since the first separations were achieved with borate buffers (pH 9), solutions of several metal ions, often described as "buffers," were examined as alternatives for use in acidic media. The efficacy of the separation depends on the relative stabilities and mobilities of the complexes.

The formation of polyol complexes may be used for the analysis of the metal ions, especially when the acid–base or redox properties of the metal ions are strongly modified. Processes initially designed with borate were subsequently extended to other inorganic species such as tungstate (63). Inversely, carbohydrates are sometimes determined after reacting with a suitable complex-forming metal ion. For example, alditols are determined by polarimetry after complexation by molybdate (37).

Oxidation of carbohydrates by metal ions could be the topic for a specific review, since an overwhelming amount of work is continuously being generated in this area, given its importance in organic synthesis. In most processes, the reaction involves a transient metal–carbohydrate complex that sometimes is characterized by spectroscopy. Generally, an indirect proof is provided by the observation of saturation kinetics following the Michaelis–Menten equation. However, detailed structural information is seldom available on such short-lived species. In fact, most classical oxidizing agents such as Mn^{VI}, Cr^{VI}, V^{V}, Mn^{III}, Co^{III}, and Ce^{IV} react with alcohols and diols as well as with carbohydrates on a similar time scale. This finding is consistent with the expectation that sugar oxidation may proceed through intermediates similar to simple metal–alcohol adducts (64, 65). The oxidations of sugars by Cu^{II} (66) and ruthenium tetraoxide (67) have previously been reviewed. Oxidizing agents such as periodate salts and lead tetraacetate would be expected to interact with carbohydrates in a specific manner, since they cleave the carbon–carbon bond of 1,2-diols, whereas monofunctional alcohols are not oxidized. Here, the formation of transient chelates in these systems is demonstrated by the fact that threo and cis diols are, respectively, more reactive than their erythro and trans isomers. Complexation by a diol also causes considerable variations of the pH of aqueous periodate solutions. However, the observation that some tertiary trans diols (which are inert to periodate) are oxidized by lead tetraacetate demonstrate that with this oxidant, pathways exist in which such chelates are not necessary intermediates (68).

A section is devoted to the emerging area of the use of metal–carbohydrate complexes in synthetic chemistry. Many inorganic salts catalyze various reactions of carbohydrates but, unfortunately, the intermediate metal complexes have not yet been characterized in detail, with the exception of the well-characterized tin-mediated reactions. Isomerization reactions of glucosamine derivatives mediated by Co^{III} ternary complexes with ethylenediamine (en) were also recently reported (69). Another reaction of synthetic interest is the epimerization of aldoses, in which the C2 configuration is reversed, allowing the easy syntheses of rare sugars from readily available aldoses. This reaction is discussed in the section on molybdenum (Section II.E.3) because it was initially discovered with a Mo^{VI} calayst. However, it is now known to proceed with several other metal ions such as Ni^{II} and Ca^{II} (16, 19).

This chapter specifically focuses on the formation of metal complexes with polyhydroxy compounds in solution. This aspect was somewhat neglected in previous reviews, which developed the study of metal complexes of carbohydrates containing donating atoms other than oxygen, in view of their considerable biological relevance. This chapter documents the fact that, despite the weaker coordinating abilities of simple sugars and polyols than those of carbohydrates containing amino, carboxylic, or phosphate groups, these simple sugars are involved in a very rich coordination chemistry with a large range of main group metals and transition metals.

II. COMPLEXES OF METALS STUDIED AS GROUPS OF THE PERIODIC TABLE

In the following sections, each metal element will be discussed as dictated by group relationship, and/or chemistry with carbohydrates. We will begin describing the alkali metals of Group 1 (Section II.A). Then we will describe the alkaline earth metals. Given the similarities of many of the carbohydrate complexes of divalent metal ions in this group with carbohydrate complexes of other divalent and trivalent metal ions (including lanthanides), we are going to discuss Groups 2 (IIA) and 3 (IIIB) as well as selected lanthanides together in one section (Section II.B). This organization will avoid redundancies in description of this type of metal–carbohydrate complexes since the analogies of these complexes will be reflected in all aspects of complex properties. Then, we will proceed to describe all other groups in the periodic table.

A. Group 1 (IA), Alkali Metal Ions

Metals of this group include Li, Na, K, Rb, and Cs. Structural characterization of alkali metal adducts with carbohydrates documents the interaction between these monovalent metal ions and the un-ionized hydroxyl groups of the carbohydrate (2, 70, 71). In aqueous solution, however, the metal cation is preferentially coordinated by water molecules. In general, these types of metal complexes are not mentioned in the literature, owing to their low stability in aqueous solution. Complex formation is commonly neglected if the proportion of complex at equilibrium is less than 1% (14). In contrast, the complexes formed between carbohydrates and alkali metal ions in alkaline solution (pH > 11) are salts of the deprotonated carbohydrate ligand and have been described previously in great detail (2, 4).

Coordination of carbohydrates (L) with alkali metal ions M^{I} has recently proved useful for the structural analysis of oligosaccharides by tandem mass spectrometry (MS/MS) following fast atom bombardment (FAB) ionization (72–74). Contrary to the behavior of protonated molecular ions, which do not exhibit ring cleavage, specific metal-induced cleavage occurs when an M^{I} ion replaces hydrogen in the cationic precursor. Thus, the formation of the $[L + M]^{+}$ ion results in increased sensitivity for the determination of the carbohydrates or informative MS/MS data for structural analysis. The mechanisms of dissociation of the corresponding lithium complexes of oligosaccharides have been studied (72–75). Large oligosaccharides, which poorly coordinate the small Li^{I} ion, can be analyzed using Ca^{II} as the coordinating metal (76). Coupling of liquid chromatography with MS/MS was achieved with postcolumn addition of metal chlorides in order to improve the analysis of oligosaccharides (77). The more efficient monovalent cations for enhanced sensitivity were Li^{I} and Na^{I}, whereas the addition of $CoCl_2$ proved useful in assisting the elucidation of the carbohydrate structures, by inducing structure-specific fragmentation of the molecules.

In the following sections, the criterion for the formation of weak metal–carbohydrate complexes in aqueous solution will be taken as characterization by suitable physical methods including electrophoresis, chromatography, NMR spectroscopy, potentiometry, calorimetry, or optical rotation measurements.

B. Groups 2 (IIA) and 3 (IIIB) and Other Relevant Divalent and Trivalent Metal Ions, Including Lanthanides

Metals of Group 2 (IIA) include Be, Mg, Ca, Sr, and Ba and those of Group 3 (IIB) include Sc and Y. As mentioned above, the metal–carbohydrate complexes of all these cations are very similar to those of other divalent and trivalent elements, which will also be discussed in this section for the sake of eliminating redundancies. $Calcium^{II}$ is clearly the most studied cation in this group and may be used as a model for the chemistry of other cations as different as Ni^{II}, Cu^{II}, Pb^{II}, and $lanthanides^{III}$. Strong evidence for such analogies will appear throughout the text when considering the chromatographic separations of carbohydrates on cation-exchange resins, NMR spectroscopy, and thermodynamic data, and even the catalytic activity in the epimerization of aldoses, discussed within the molybdenum section (Section II.E.3). Various carbohydrate derivatives have been proposed for application as innocuous calcium-sequestering agents in environmental-friendly detergents, instead of phosphates (78).

1. *Sites of Chelation*

It is now established that sugars in the *p* form chelate Ca^{II} and many other metal cations by a cis,cis-1,2,3-triol system in an *a–e–a* conformation (14). For

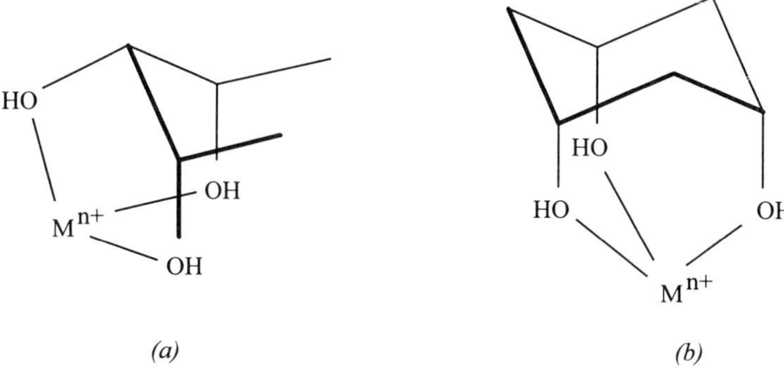

Figure 1. Coordination of a metal cation M^{n+} to (*a*) a carbohydrate containing a cis-HO1,2,3 tridentate site of chelation in *a–e–a* conformation and (*b*) an inositol containing a tridentate cis-HO1,3,5 site of chelation in *a–a–a* conformation.

sugars in the *f* form, a cis-1,2,3-triol system is also the preferred site of chelation. A less common complexing site is the triaxial 1,3,5-triol system of some inositols (cis- and epi-inositol) with the appropriate configuration (Fig. 1). When possible, a carbohydrate may shift its normal conformation to a less favored one, so that a better site of chelation becomes available. The overall stability of a complex depends on the energy cost of this initial conformational change, which reduces its apparent stability. In cases where complex formation do not supply sufficient free energy for the conformational change to occur, no complex will exist.

Molecular mechanics calculations were applied to the complexation of metal cations with inositols (79), allowing the rationalization of previous results showing that the variations of complex stabilities with the size of the metal ions are partly governed by the size of the ring containing the metal ion (11). Accordingly, it was found (79) that five-membered chelate rings favor complex formation with large metal ions having radii close to 1.07 Å, while six-membered chelate rings favor ions having much smaller radii. This finding explains why most metal ions are chelated by carbohydrates using 1,2-diol groups that form five-membered rings.

2. *Solid-State Metal–Carbohydrate Complexes*

Many crystalline adducts of sugars and calcium salts have been isolated and characterized by X-ray crystallography. The structural characteristics of calcium–carbohydrate complexes in the solid state can be summarized from data obtained on a series of crystal structures (80). The calcium ions are generally coordinated to two or more carbohydrate molecules as well as water molecules. However, in aqueous solution, the bonding is likely to be different since both

Ca^{II} and the carbohydrate may bind extra water molecules (14). More recent studies were performed on Pb^{II} complexes of polyols and cyclodextrins (81). These polymeric complexes form ribbons in which the polyol molecules chelate two Pb atoms by a 1,2-diol group and are linked by Pb atoms. The Pb^{II} complex of γ-cyclodextrin (γ-CD) possesses a similar structure in which the ribbon is closed to form the CD ring containing 8 glucose moieties chelating 16 lead atoms (81).

Several studies on the structures of the complexes of a series of metal cations with 1,3,5-triamino-1,3,5-trideoxy-cis-inositol (taci) have brought information on the factors that govern the selection of the site of chelation by a distinct metal ion (82, 83). The ligand taci presents a remarkable versatility for metal binding since, in its two chair conformations, this ligand possesses four possible tridentate sites: a cis-2,4,6-triol system, a cis-1,3,5-triamine system, a 3,4,5-hydroxydiamino system, and a 1,2,6-aminodiol system (Fig. 2). The hard alkaline earth metal cations Mg^{II}, Ca^{II}, Sr^{II}, and Ba^{II} are coordinated to the six hydroxyl groups of two taci molecules, whereas the soft Cd^{II} ion is bound to six nitrogen donors of two taci molecules (82). The Ni^{II}, Cu^{II}, and Zn^{II} ions are exclusively bound by nitrogen atoms (83), showing that the selection of a particular binding site depends of the electron configuration of the metal ion.

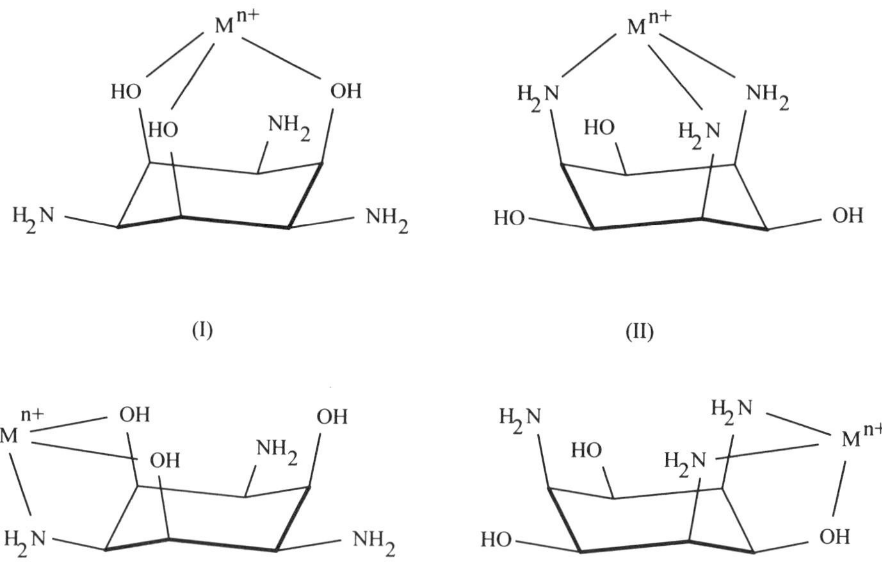

Figure. 2 The two chair conformers of taci, showing the four different types of chelates with a metal cation M^{n+}. In conformer **I**, all HO groups are axial. In conformer **II**, all H_2N groups are axial. [Adapted from (82, 83).]

It was recognized early that the stabilities of the complexes formed between calcium(II) and carbohydrates depended on the configuration and conformation of the sugar. Therefore, progress in the understanding of the formation of complexes between carbohydrates and metal cations was linked to the development of electrophoretic and chromatographic methods for the separation of the mixtures of carbohydrates.

3. *Chromatographic Studies*

In 1962, Mills (84) reported that Ca^{II}, Sr^{II}, and Ba^{II} were the best cations for increasing the electrophoretic mobility of various alditols in aqueous solution. Since cis-inositol showed the greatest mobility with all cations, this cyclitol was chosen as the reference for the measurement of migration ratios. Liquid chromatography on cation-exchange columns was applied to the analysis of mixtures of carbohydrates. The effect of the nature of the cation immobilized on the column was investigated by Goulding (85). A method was described for the separation of sugars on the preparative scale using columns in the Ca^{II} form (86) and the results were discussed in relation to the mechanism of the separation. However, the best results for the chromatographic separation of 14 alditols on a cation-exchange resin were obtained with La^{III} (87), whereas aldoses were better retained on resins loaded with Ba^{II} (88, 89). Rapid high-resolution separation of oligosaccharides obtained by hydrolysis of corn syrups was carried out on cation-exchange resins in the Ag^{I} form (90). Nevertheless, for the separation of monosaccharides, Ca^{II} is more selective than Ag^{I} (90).

Procedures have also been developed for the separation of carbohydrates by TLC, using commercial plates coated with cation-exchange resins loaded with various cations. Besides Cu^{II}, which probably forms dinuclear complexes (see copper Section II.H), the more interesting results were obtained with Ca^{II} and La^{III} (91). Lanthanum(III) is particularly advantageous in the separation of alditols (92). Comparing the complexation of sugars and alditols of various configurations by ion-exchange chromatography, good separations of aldoses and ketoses were achieved only with Ca^{II}, La^{III}, and Pb^{II} (93). The separation of carbohydrates (hexoses, pentoses, and corresponding polyols) by liquid chromatography was achieved using ligand-exchange on a cation-exchange resin column with water as eluent. Seven cations (Ca^{2+}, Sr^{2+}, Ba^{2+}, Pb^{2+}, Y^{3+}, La^{3+}, and Pr^{3+}) were tested (94). The separation is directly reflecting the stability of the complex formed with the cation: The more stable the complex, the longer the molecule will remain immobilized on the column. Similarly, the chromatographic behavior of several methyl aldopyranosides on a cation-exchange resin in the Ca^{II} form was found to be determined by the stability constants of the corresponding calcium complexes (95).

A direct relationship was found between ionic radii and stability constants for the lanthanide–alditol complexes (96). Alditols form stronger complexes with Nd^{III}, Sm^{III}, and Eu^{III} than with La^{III} and the higher lanthanides that have smaller ionic radii. An interesting consequence of the excellent separations obtained with lanthanide cations is that smaller samples can be processed, since the volume of the chromatographic ion-exchange columns can be reduced when using Nd^{III} rather than the usual Ca^{II} columns (96). Galactitol can bind two lanthanide ions, as shown by the crystal structure of a solid formulated as [galactitol$(PrCl_3)_2$]$\cdot 14H_2O$ in which galactitol is in the planar zigzag form with $2Pr^{III}$ cations attached to O1,2,3 and O4,5,6 (93). The ^{1}H NMR data were in agreement with these assignments (78).

4. NMR Studies

In a pioneering work, Angyal and Davies (97) demonstrated that when $CaCl_2$ is added to a solution of *epi*-inositol in D_2O, the proton signals are shifted to lower field. The largest shifts are observed for the protons bound to the C1,2,3 atoms that bear the chelating hydroxyl groups in *a–e–a* arrangement. In the case of sugars which, contrary to inositols, exist as mixtures of interconverting forms, the equilibrium can be shifted by complex formation. For example, uncomplexed D-allose (98) exists mainly in pyranose forms and the α/β ratio is 13.8 : 77.5 at 30°C. In a 0.85 *M* solution of $CaCl_2$, this ratio has been changed to 37.2 : 54.5. Since the signals of the α-*p* form are shifted and enhanced in the presence of Ca^{II}, while those of the β-*p* form are not, these data suggest that the site of chelation is the cis,cis-1,2,3 triol system. Knowing the proportions of all the forms of allose from NMR data, the stability constant of the 1 : 1 Ca^{II}-α-D-allopyranose complex was calculated to be $K = 6$. By using the same analysis, the complexing abilities of various cations were compared: La^{III}, $K = 10$; Sr^{II}, $K = 5$; Ba^{II}, $K = 3$.

The interaction of Ca^{II} and La^{III} ions with D-lyxose and D-ribose was studied by ^{1}H NMR spectroscopy. The data were interpreted as indicating the formation of 1 : 1 complexes of the pyranose conformer, which has an *a–e–a* arrangement of three adjacent hydroxyl groups (99). Thus, D-lyxose is complexed in β-*p* form (O1,2,3 site, $^1C_4$ conformation) and D-ribose in α-*f* (O1,2,3), α-*p* (O1,2,3–$^4C_1$) and β-*p* (O2,3,4–$^1C_4$) forms, in the order of decreasing stability. The origin of the ^{1}H NMR chemical shifts in the Ca^{II} complex of ribose (100) and a series of monosaccharides (101) were discussed, showing that the presence of the HO1,2 system is essential. The Ca^{II} complexes of xylose, galactose, and glucose are weak, whereas D-mannose forms a stronger calcium complex at the O1,2,3 site of the β-*p* and β-*f* forms. The two ketoses examined, fructose and sorbose, also gave weak complexes with Ca^{II}.

Two alditol complexes of Ca^{II}, Y^{III}, and Pr^{III} were compared because the three cations are of similar sizes (78, 102, 103). A ^{1}H NMR study of the Pr^{III}

complexes of 10 alditols showed preferential chelation at tridentate xylo sites (78). Considering the ^{13}C NMR coordination induced shifts or line-broadening patterns, the site of chelation of D-glucitol was assigned as O2,3,4 in the Pr^{III} (78) and Y^{III} (103) complexes. However, in the Ca^{II} complex, the deshielding pattern of glucitol is C3 > C1,4,6 > C2,5 (102). This difference may suggest that the Ca^{II} and Pr^{III} or Y^{III} complexes have different structures in solution. Useful tables containing stability constants, thin-layer, and electrophoretic mobilities for Ca^{II} and La^{III} complexes of carbohydrates have been previously reported (14).

The rate of Ca^{II} binding to D-glucitol was determined by ultrasonic relaxation techniques (102). Since the equilibrium constant found by ^{13}C NMR spectroscopy is close to unity, the rates of formation r_f and dissociation r_d are almost equal. The values of the corresponding rate constants are $k_f = (1\text{–}2) \times 10^8$ 1 mol^{-1} s^{-1} and $k_d = (1\text{–}2) \times 10^8$ s^{-1}.

Interesting examples of M^{II} compounds in which complexation occurs at more than three oxygen atoms are offered by the complexes of di-D-fructose and di-L-sorbose dianhydrides with Ca^{II}, Sr^{II}, and Ba^{II} (104).

Another approach to the study of sugar–lanthanide interactions in aqueous solution is the use of ^{139}La NMR spectroscopy (105). The chemical shift of ^{139}La (referenced to a 0.040 *M* aqueous $LaCl_3$ solution) is not changed by D-arabinose, which is generally considered as a nonchelating sugar. On the other hand, the ^{139}La signal is shifted by 11 ppm and broadened by the addition of D-ribose. The stability constant of the La^{III}–ribose complex, determined from NMR data ($K = 2.8$ M^{-1} at $t = 26°C$) is in agreement with calorimetric data.

5. *Calorimetric Studies*

The calorimetric technique has often been used for the determination of stability constants. In addition, other techniques may be employed; for example, the solubility method in which the enhancement of the solubility of calcium sulfate is measured in the presence of a calcium-complexing alditol (38). It is found that threo alditols form the more stable complexes: glucitol, $K = 1.5 >$ xylitol $\approx$ arabinitol, $K = 1.2 >$ mannitol, $K = 0.9$.

All calorimetric determinations of the enthalpies and entropies of complexation of carbohydrates with various cations establish that the complex stability is mainly governed by the enthalpy component (106–112). The reactions of a series of trivalent lanthanide cations (La to Tb) with ribose (113), xylitol and glucitol (114) were studied, showing that the stability constants of the alditol complexes are the largest with Nd^{3+} and Sm^{3+} (Table II). For xylitol and glucitol, a comparison was made with divalent cations of Mg, Ca, Sr, Ba, and Pb. A detailed study of the thermodynamics of the weak interaction between Sm^{III} and xylitol in water showed the existence of a 1 : 1 complex with O2,3,4 site of

TABLE II
Stability Constants (K) and Thermodynamic Parameters of the Specific Interaction of threo Alditols with Cations (114)

Alditol	Xylitol				Glucitol			
Cation	K^a (l mol^{-1})	$\Delta G°$ (kJ mol^{-1})	$\Delta H°$ (kJ mol^{-1})	$T\Delta S°$ (kJ mol^{-1})	K^a (l mol^{-1})	$\Delta G°$ (kJ mol^{-1})	$\Delta H°$ (kJ mol^{-1})	$T\Delta S°$ (kJ mol^{-1})
Ca^{II}	0.6	1.3	−12.1	−13.4	1.1	−0.2	−9.0	−8.8
Sr^{II}	nd				0.7	0.9	−9.9	−10.8
Ba^{II}	nd				0.9	0.3	−7.2	−7.5
Pb^{II}	1.1	−0.2	−10.1	−9.9	1.3	−0.7	−9.5	−8.8
La^{III}	2.4	−2.2	−10.6	−8.4	2.7	−2.5	−10.2	−7.7
Ce^{III}	3.7	−3.2	−11.7	−8.5	4.7	−3.8	−10.7	−6.9
Pr^{III}	5.2	−4.1	−12.7	−8.6	6.3	−4.6	−11.6	−7.0
Nd^{III}	7.3	−4.9	−12.4	−7.5	8.9	−5.4	−11.6	−6.2
Sm^{III}	8.1	−5.2	−13.7	−8.5	8.7	−5.4	−13.6	−8.2
Eu^{III}	5.5	−4.2	−14.9	−10.7	6.5	−4.6	−14.2	−9.6
Gd^{III}	3.4	−3.0	−15.1	−12.1	4.0	−3.4	−13.8	−12.4
Tb^{III}	1.3	−0.7	−15.5	−14.8	2.1	−1.8	−11.7	−9.9

[a]The parameter K is the stability constant of the ML complex, according to reaction: $M + L \rightleftharpoons ML$, where M is the metal cation and L is the alditol. Not determined = nd.

chelation (115). The alditols form stronger complexes than ribose, but the reactivity of ribose should be corrected for the amount of the complexing isomers really present at equilibrium. The interactions in these complexes are mainly electrostatic, since a linear correlation exists between $\Delta H°$, the standard enthalpies of complexation and the z/r^2 ratio, in which z is the metal charge and r is the ionic radius of the cation. Thus, the HO groups of the binding site of xylitol or ribose can substitute for water molecules in the coordination shell of the cation.

The thermodynamic properties that characterize the interaction between a metal cation and the chelating site in a carbohydrate were obtained by a procedure in which the specific (s) interaction is isolated by determining the nonspecific (ns) interaction of the same cation with a noncomplexing epimer. Noncomplexing carbohydrates were arabinose in the aldose series, and ribitol in the alditol series. Evidence for the weak interactions of arabinose with divalent metal cations in solution is well documented in the cases of Ca, Mg, Zn, Cd, and Hg (116). The term characterizing the s interaction is obtained (Table III) by substracting from the energies of transfer of the cation from water to xylitol or glucitol solutions (ns + s), those calculated for the transfer from water to ribitol solutions (ns), at exactly the same concentration. Comparisons between TLC and calorimetric data were made for the rationalization of the complexation of Ca^{II} and Ln^{III} cations by sugars and alditols in water (117) and solvent effects (118).

TABLE III
Thermodynamic Parameters of the Nonspecific (ns) and Specific (s) Interaction of threo Alditols with Pb^{II} and Nd^{III} (114)

Alditol[a]	Xylitol	Xylitol	Glucitol	Glucitol
Pb^{II}	s	ns + s	s	ns + s
K (L mol^{-1})	1.1	0.7	1.3	0.8
$\Delta G°$ (kJ mol^{-1})	−0.2	0.9	−0.7	0.6
$\Delta H°$ (kJ mol^{-1})	−10.1	−24.1	−9.5	−24.2
$T\Delta S°$ (kJ mol^{-1})	−9.9	−25.0	−8.8	−24.8
Nd^{III}	s	ns + s	s	ns + s
K (L mol^{-1})	7.3	5.7	8.9	6.4
$\Delta G°$ (kJ mol^{-1})	−4.9	−4.3	−5.4	−4.6
$\Delta H°$ (kJ mol^{-1})	−12.4	−15.5	−11.6	−15.4
$T\Delta S°$ (kJ mol^{-1})	−7.5	−11.2	−6.2	−10.8

[a]The parameter K is the stability constant of the ML complex, according to reaction: $M + L \rightleftharpoons ML$, where M is the metal cation and L is the alditol. The specific interaction is obtained by reference to ribitol, assumed to be a nonligating alditol.

6. *The Binding of Carbohydrates by C-Type Lectins*

Lectins are widespread globular proteins that bind carbohydrates with a remarkable configurational specificity. A group of animal lectins, known as C-type lectins, presents a calcium-dependent activity (55–58). These lectins share a highly conserved carbohydrate recognition domain (CRD) of approximately 120 amino acids. An important class of C-type serum lectins is represented by the mannose-binding proteins (MBPs), which are involved in the mechanism of response to infection. Rat MBP contains two or three essential calcium ions. The crystal structure of rat MBP bound to an oligosaccharide has been reported, showing that one calcium ion is directly coordinated to HO3,4 of mannose and to six oxygen atoms from five amino acids (119). The specificity of MBP can be changed from D-mannose to D-galactose by site-directed mutagenesis (120). Molecular mechanics calculations were used to examine the binding specificity of the rat MBP for D-mannose (60, 61). The binding site showing D-mannose bound to Ca^{II} is represented in Fig. 3.

7. *Influence of Complex Formation on the Reactivity of Sugars*

The reactions (isomerization and degradation) of monosaccharides in alkaline medium are accelerated by calcium ions (121). Accordingly, an enhancement of the enolization rate was observed in the cases of D-fructose and D-psicose. This rate enhancement was attributed to the formation of Ca^{II} complexes involving the anomeric hydroxyl group of the sugar.

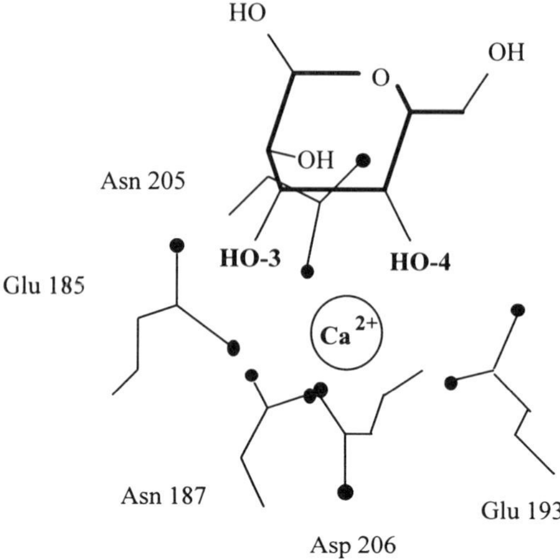

Figure 3. The binding site of rat MBP bound to a mannose ligand, showing bonding of mannose to the calcium cation (119). [Adapted from (60). The gray balls represent the carboxylic oxygen atoms of the amino acids of the protein.]

The complexation of furanoside derivatives has been compared to that of pyranosides. Calorimetric data were reported for the determination of enthalpies of complexation of various metal ions in water (122, 123). The complexation of methyl glycofuranosides by Na, Ca, or La ions has a rate-retarding effect on their solvolysis in acidic methanol (124). Stability constants were determined in methanol by calorimetry and potentiometry, using a calcium-selective electrode (125). The Eu^{III} complexes of some methyl glycosides (metal/ligand ratio, 1 : 1 and 1 : 2) were studied by luminescence excitation spectroscopy (126). The separation of several methyl glycofuranosides was studied by using columns loaded with Ca^{II}, Sr^{II}, Ba^{II}, and Pb^{II}, which showed similar behaviors, whereas La^{III} proved to be less efficient (127). Some aldopentofuranosides such as arabinofuranosides and xylofuranosides, which do not contain three cis oxygen atoms, are thus poor complexing ligands. This expectation is supported by the observed low stabilities of the complexes formed by methyl β-D-ribofuranoside, in which only HO2,3 are cis. Aldopentofuranosides containing three cis oxygen atoms form stronger complexes, in which the sites of chelation were suggested to be O2,3,5 for methyl α-D-lyxofuranoside and O1,2,3 for methyl α-D-ribofuranoside (127). The formation of a Ca^{II} complex with *n*-octyl β-D-mannofuranoside was also reported (128).

A noteworthy difference was reported in the affinity of *C*-sucrose and *O*-sucrose (α-D-glucopyranosyl-β-D-fructofuranoside) toward calcium ions

(129). *C*-Sucrose is an analogue of sucrose in which the glycosidic oxygen is replaced by a methylene group, and the structures of both free disaccharides are very similar (130). *O*-Sucrose does not complex metal cations to any appreciable extent in either aqueous or methanolic solution (14). Nor does *C*-sucrose interact with metal cations in aqueous solution (129). In contrast, in methanol containing 1–6 equiv $CaCl_2$, ^1H NMR chemical shifts, coupling constants, nuclear Overhauser effect (NOE) enhancement and optical rotation data indicate that *C*-sucrose undergoes a conformational change. No complexation is observed with Na^I and Mg^{II}, but Sr^{II} and Ba^{II} show Ca-like chemical shifts. A structure was proposed in which the complexed *C*-sucrose adopts an extended conformation, and the calcium ion is surrounded by an oxygen-rich pocket comprised of the glucopyranosyl and fructofuranosyl ring ether oxygens and HO1′,6′,6. Since the structures of both free disaccharides are similar (130), the lack of affinity of *O*-sucrose toward cations is probably due to the presence of lone pairs of electrons on the glycosidic oxygen. This different complexing behavior may prove to be significant in view of the current interest in the use of *C*-saccharides as surrogates for natural carbohydrates, since it is not commonly recognized that the chemical properties of *C*-saccharides may differ from those of the natural disaccharides.

C. Group 4 (IVB) Metal Ions

Very few carbohydrate complexes of Ti, Zr, Hf are known. In all reported complexes, the monodentate ligand is the protected sugar 1,2:5,6-di-*O*-isopropylidene-α-D-glucofuranose (HL, "diacetoneglucose" = DAG) in which only O3 can bind the metal atom (Fig. 4). Thus, the complexes are merely metal–alcoholate derivatives.

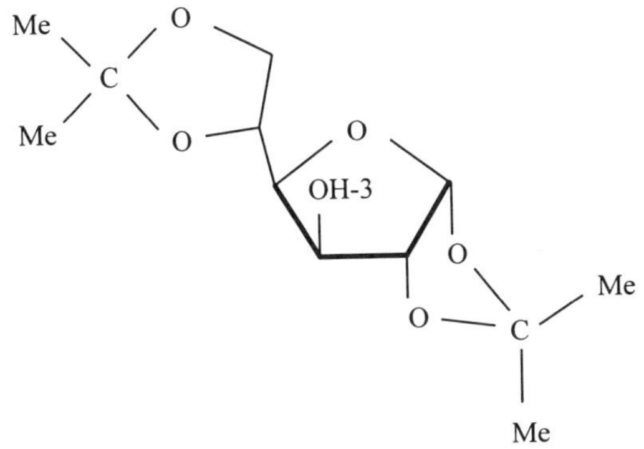

Figure 4. The monodentate ligand DAG showing the free HO3 group.

The synthesis and structure of a titanium–carbohydrate complex of HL, namely, chloro(cyclopentadienyl)bis(1,2 : 5,6-di-*O*-isopropylidene-α-D-glucofuranose-3-*O*-yl)titanate(IV) were reported (131). Complexes of all the Group 4 (IVB) metal ions (M = Ti, Zr, and Hf) with HL have been synthesized (132) from $M(Ph{-}CH_2)_4$ by ligand displacement. They are the "first members of a series of homoleptic DAG complexes" with various metal ions including Al^{III} (132) and Mo^{III} or W^{III} (133). All complexes afforded py adducts and the structure of the zirconium complex $ZrL_4(py)_2$ was determined (132). Zirconium is in a pseudooctahedral geometry involving four sugars and two py groups in a cis arrangement. A 1 : 1 adduct of TiL_4 with 1,10-phenanthroline, TiL_4(phen), was also characterized. In these complexes, the carbohydrate ligand functions as a highly functionalized chiral alkoxide, so that other ligands stereoselectively interact with the metal. Such titanium–carbohydrate complexes were reported to be useful as enantioselective catalysts for allyl and ester additions to aldehydes (131, 134, 135) and aldol condensation (136).

D. Group 5 (VB) Metal Ions

Group 5 (IV) transition metals include V, Nb, and Ta, whereas the related Group 15 (VA) elements are As, Sb, and Bi. There are many similarities between these elements in the same column as well as between columns. By far, most studies of metal ion–carbohydrate complexes have been carried out with vanadium. Since the findings with this transition metal should provide information for the other elements in Group 5 (VB), we will briefly summarize the generalizations that can be made for vanadium–carbohydrate complexes at this time and then processed to discuss each metal ion in greater detail.

1. Vanadium

Vanadium can form complexes with carbohydrate-type ligands in several oxidation states including V, IV, and III. Like other metal cations, vanadium may form stable complexes with polyols containing triol systems in *a–e–a* geometry (137). However, at this time, most studies have been carried out with ligands unable to form a stable conformer with the *a–e–a* triol arrangement. It appears that general rules for the formation of other types of complexes can be made and these will be summarized as follows. Vanadium(V) froms 2 : 2 complexes with ligands containing two adjacent OH groups. The nature of such complexes in aqueous solution has been controversial for the past decade, and only recently is the literature reaching a consensus that this type of complex typically contains five-coordinate vanadium atoms in a distorted trigonal–bipyramidal geometry [Fig. 5(*a*)]. Alternative stoichiometries and structures have been proposed and some of these proposals are supported by related

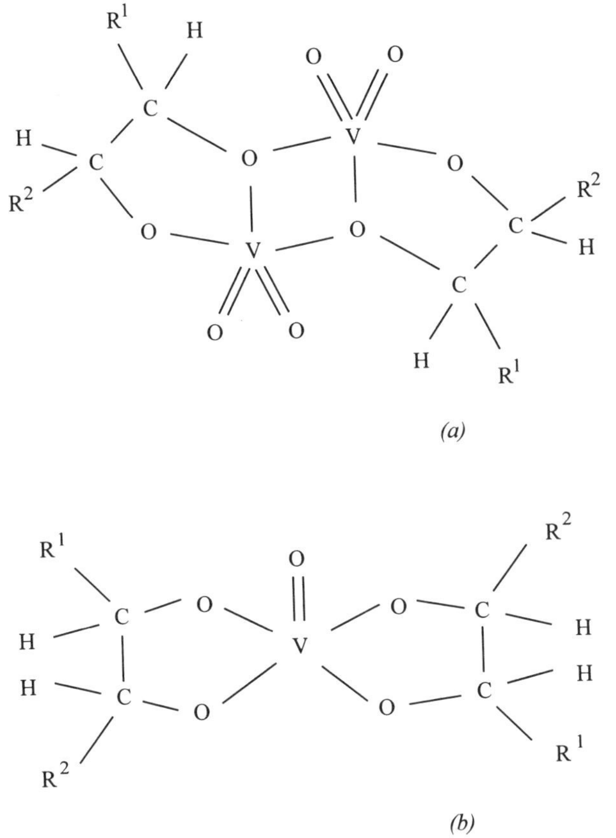

(a)

(b)

Figure 5. Schematic drawing of the structures proposed for the predominant complexes of (*a*) vanadium(V)-vicinal diols and (*b*) vanadium(IV)-vicinal diols in aqueous solution. Only one of the possible isomers is shown.

compounds characterized crystallographically (138–142). Vanadium(IV) forms stable 1 : 2 complexes with ligands containing adjacent OH groups. In these complexes, the vanadium atom coordinates the two bidentate ligands in a square pyramidal geometry shown in Fig. 5(*b*). This structure corresponds to the predominant product formed in most of the vanadyl–sugar systems.

Vanadium in several oxidation states has recently been found to form compounds of greater complexity with multidentate carbohydrates and carbohydrate-like ligands. However, few of these types of complexes have been characterized in sufficient detail, so that a structural motif is not emerging at this time.

a. Solid-State Structures. There are currently no examples of a vanadium complex with a reducing sugar, although significant progress has recently been made with nonreducing sugars. Synthesis and characterization of solid state vanadium(IV, V)–sugar/nucleoside complexes is of great interest, since it may provide details of geometry information. Two approaches have been used to isolate these complexes. A first approach is to crystallize a complex from aqueous solution. A second approach is to synthesize a complex from an organic solvent, because in aqueous solution, the vanadium–sugar/nucleoside system normally contains several species and it is difficult to obtain crystals. However, in nonaqueous solvent, less complexes would normally form, especially in the case of vanadate oligomers. Thus, crystals of the dimeric vanadium(V) complex of a protected carbohydrate, methyl *O*-4,6-benzylidene-α-D-mannopyranoside (H_2L), $[TBA]_2[(VO_2L)_2]$ have been obtained by reaction of $(TBA)H_2VO_4$ with H_2L in acetonitrile [Fig. 6(*a*)] (143). On the other hand,

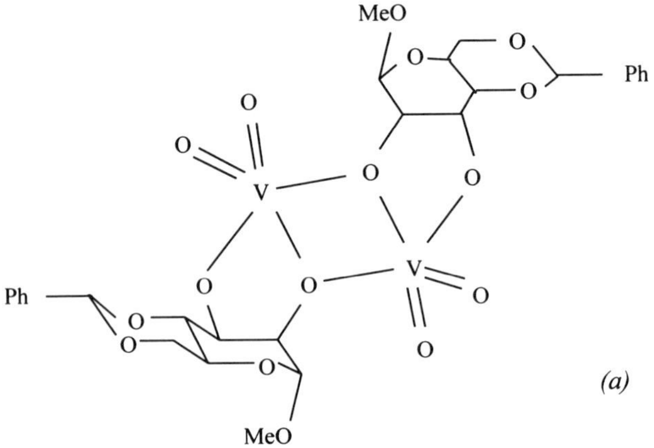

MeO O O Ph O O O V O O V O O O Ph O O O MeO

(*a*)

O O V O B O O O OH HO O O O V O O B

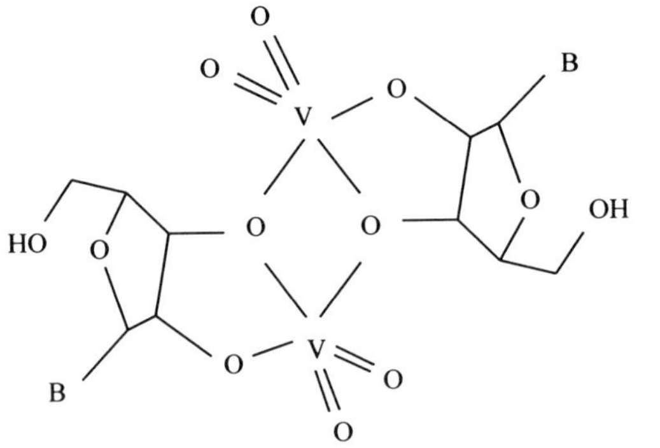

(*b*)

Figure 6. The X-ray structures of (*a*) $[TBA]_2[(VO_2L)_2]$ (143) and (*b*) $[TEA]_2[(VO_2Ad)_2]$ (144). Adenine is HB [Adapted from (143 and 144).]

Tracey and co-workers (144) separated crystals of $[TEA]_2[(VO_2Ad)_2]$ from an aqueous solution containing $(TEA)_2HVO_4$ and adenosine (H_2Ad) [Fig. 6(*b*)]. Crystallographic analysis of both complexes clearly shows that the vanadium(V) is in a distorted trigonal bipyramidal coordination geometry as shown in Fig. 5(*a*). Thus, the V^V complexes with ligands containing adjacent OH groups, and presumably also the V^{IV} complexes, possess the structural features shown in the mannopyranoside and adenosine complexes and can probably be used to project the expected structures of solution complexes (Fig. 7) (145–147).

Apparently, the perturbation from 1,2-diols to ligands containing 1,3-diols or three adjacent coordinating OH groups is not a simple extrapolation in the complex, since the compounds are likely to involve more complex types of structures (Fig. 8) (148–152). Tetranuclear vanadium polyoxo-1,3-diolate com-

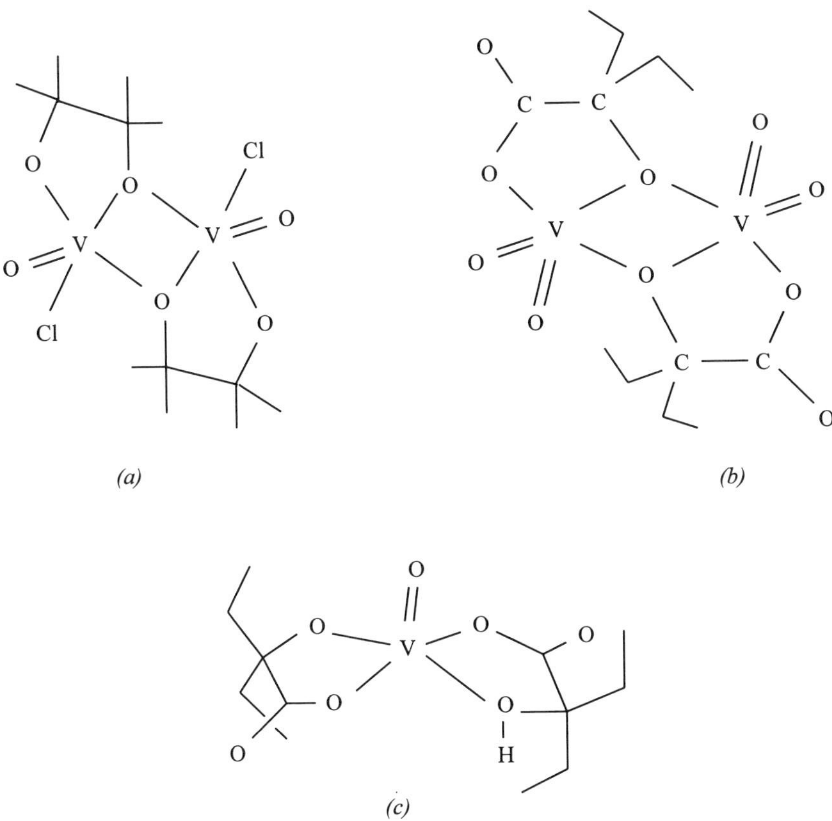

Figure 7. Graphical representation of (*a*) $[VOCl(Me_2C(O){-}(O)CMe_2)]_2$ (145), (*b*) $[VO_2(OCEt_2COO)]_2$ (146), and (*c*) $\{VO[HOC(Et)_2COO][OC(Et)_2COO]\}^-$ (147). [Adapted from (145–147).]

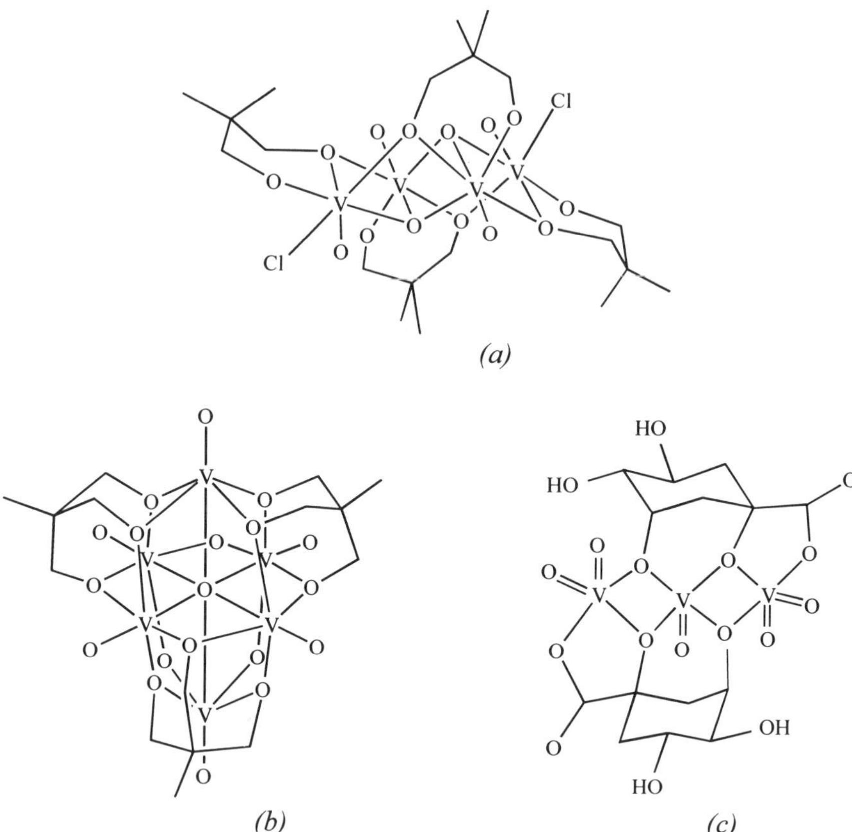

Figure 8. Graphical representation of (*a*) $\{(VOCl)_2[VO(OH)]_2(L)_4\}$ (148), (*b*) $\{V_5^{IV}V^{V}O_7(OH)_3[Me[MeC(CH_2O)_3]_3(CH_2O)_3]_3\}^-$ (150), and (*c*) $[(V^{V}O_2)_2(V^{IV}O)(quinato)_2]^{2-}$ (152). [Adapted from (148, 150, and 152).]

plexes were prepared from nonaqueous solution, and were structurally characterized by X-ray crystallography (148, 149). X-ray structures were obtained on hexavanadium polyoxotriolate complexes, which were prepared through hydrothermal synthesis (150, 151). By using the 2,6-dideoxy-inositol derivative quinic acid [1(*R*),3(*R*),4(*R*),5(*R*)-tetrahydroxycyclohexanecarboxylic acid], furthermore, Lay and co-workers (152) isolated crystals of a linear mixed-valence trinuclear (V^{V}–V^{IV}–V^{V}) complex sandwiched between two tridentate (O1,3 and one carboxylic O) quinato ligands.

b. Solution Structures. Most of the vanadium–carbohydrate complexes that form in aqueous solution are very weak. The available studies concern the determination of the stoichiometry, stability, and pH dependence of the mixture of such complexes. However, given the recent structural characterization of two vanadium(V) carbohydrate derivatives combined with the model systems available, it is now possible to provide a much more detailed structural characterization of the major complexes that form in solution. Accordingly, whenever possible, we have chosen to focus our discussion on the structure of the major vanadium complexes in solution, and to compare their relative stability rather than to describe all the different and weak complexes that also form. Information on the minor complexes can be found elsewhere (139, 153). Careful consideration of the solution structure of vanadium complexes is important since solution and solid-state structures are often different (145, 149, 152, 154, 155). Furthermore, the lability of a few well-characterized complexes of simple vanadium–alkoxide complexes has been demonstrated in solution (149, 154, 156). The following discussion will first describe the V^{V} complexes and then proceed to the V^{IV} and V^{III} complexes with different ligands.

i. ***Vanadium(V)–Carbohydrate Interactions.*** The reaction in aqueous solution of vanadate with any ligand containing hydroxyl groups generates vanadate alkoxides, which are commonly obtained from simple alcohols such as methanol, ethanol, butanol, phenol, and di- and polyfunctional alcohols. Such complexes are very weak ($K_f \sim 0.1$) and can only be observed by ^{51}V NMR spectroscopy and enzymatic techniques (157–160). A series of vanadate alkoxides have been synthesized from organic solvents and characterized by X-ray crystallography and solid-state NMR spectroscopy (161). These types of carbohydrate complexes probably contain four-coordinated vanadate, and possess biological activity as analogues of phosphate esters (161).

The interaction of vanadate with a series of 1,2-diols such as ethyleneglycol (138) and 1,2-cyclohexanediol (162a), as well as a series of carbohydrates (162a,b; 163) has been reported. Depending on the concentrations of ligand and vanadate in the solutions, complexes with stoichiometries 1 : 1, 1 : 2, and 2 : 2 have been observed. In recent and exhaustive studies, the major product in (presumably) all systems has been found to be 2 : 2 complex with a characteristic ^{51}V NMR chemical shift of −520 to −523 ppm (recent studies on selected systems do not confirm some early studies in this area) (163–165). Careful speciation studies are necessary to characterize the multiple complexes forming in even the simple systems, but an exhaustive description would go beyond this chapter.

The solution structure of the 2 : 2 complex has been a matter of controversy for the past decade, resulting in five different structural proposals for these systems [see, e.g., (155)]. Although there is not yet general agreement on this point, most groups currently favor the structure shown in Fig. 5(*a*). This consensus is being reached in part because of the recent characterization in the solid state of the mannopyranoside and adenosine complexes as well as analogous model systems (143–145, 147). In addition, the detailed solution studies now available on simple and carbohydrate-type ligands, using a wide variety of techniques including multinuclear (^{1}H, ^{13}C, ^{17}O, and ^{51}V) NMR, IR, CD, and Raman spectroscopy (155), show that only the solution structure of the type shown in Fig. 5(*a*) is consistent with the spectroscopic characterizations of a variety of complexes. There is, however, no doubt that both solid-state and solution investigations are necessary to determine the geometry of the major complexes that exist in solution; a fact that has been recognized by several workers in this field (145, 149, 152, 154, 155). Unfortunately, the problem is not only the lack of structurally characterized complexes, but also the fact that several oxovanadium(V) alkoxide complexes simply do not remain intact upon dissolution [for review, see (161) and references cited therein]. Furthermore, even complexes that remain intact in aqueous solution undergo ligand-exchange and other dynamic processes that may affect the actual structure of the material in solution (152, 166).

Unlike 1,2-diols, 1,3-diols do not form stable complexes with vanadate in aqueous solution (138). Although such complexes can form from nonaqueous solution, the structural characterization of these systems suggests that the aqueous species will not be as stable as for ligands containing the 1,2-diol system (148, 167, 168).

Carbohydrates containing the free 1,2-diol functionality can form this type of 2:2 complex, which have indeed been observed for furanosides (155) and pyranosides (162) as well as free reducing sugars (163). For example, methyl galactopyranoside, mannopyranoside, and ribofuranoside form complexes with the characteristic ^{51}V NMR chemical shift around −520 ppm. The fact that methyl β-D-glucopyranoside does not form this type of complex is consistent with the possibility that adjacent OH groups in a cis configuration are more favorable to complex formation. A very minor product from vanadate and methyl α-*D*-glucopyranoside, obtained by using very large concentrations of glucoside, gives a signal in the appropriate range of chemical shifts. However, since commercially available methyl α-D-glucopyranoside is 99% pure and contains glucose as an impurity, it is possible that the observed signal is due to a V_2Glc_2 complex. Alternatively, this signal could be due to the vicinal trans-diols of the α-glucoside, since the V_2L_2 complex is observed for trans-cyclohexane-1,2-diol. The much smaller formation constant for methyl α-D-glucopyranoside can be attributed to flattening of the cyclohexane ring when an oxygen atom takes the place of a methylene group (162). However, such effect should be similar for both methyl α- and β-D-glucosides. Since the carbohydrates are asymmetric, there is the possibility for formation of three different 2:2 complexes depending on the ligand chelation to the V_2O_2 unit. Three isomers or more have been observed by multinuclear NMR spectroscopy (155, 169).

The formation constants for the V_2L_2 complexes of various diols, carbohydrates, and nucleosides compared in Table IV vary over five orders of magnitude depending on diol–ligand (entries 1 and 7), and less than one order of magnitude depending on conditions (entries 9 and 11–14). The formation constants for complexes formed from vanadate and vicinal diols on pyranoses (six-membered rings) are in the range 10^3–10^4, two orders of magnitude greater than the complex formed with ethyleneglycol. The most stable complexes are, however, observed with vicinal diols on ribofuranoses (five-membered rings): Their

TABLE IV
Formation Constants K for 2:2 V^V Complexes of Diols, Carbohydrates, and Nucleosides[a]

Entry	Ligand	K (mol^{-3} L^3)
1	Ethyleneglycol	$(4.6 \pm 1.7) \times 10^2$[b]
2	*cis*-Cyclohexanediol	$(1.9 \pm 0.2) \times 10^4$[c]
3	*trans*-Cyclohexanediol	$(2.3 \pm 0.2) \times 10^3$[c]
4	Methyl β-D-galactopyranoside	$(1.0 \pm 0.1) \times 10^3$[d]
5	Methyl α-D-glucopyranoside	8.6[d]
6	Methyl β-D-glucopyranoside	nf[e]
7	Methyl β-D-ribofuranoside	$(2.9 \pm 0.3) \times 10^7$[f]
8	Inosine	$(7.0 \pm 0.5) \times 10^7$[e]
9	Adenosine	$(4.1 \pm 0.2) \times 10^7$[f]; 6.6×10^7[g]
10	AMP	$(5.1 \pm 0.4) \times 10^6$[h]
11	Guanosine	$(3.6 \pm 0.5) \times 10^7$[f]; 2.3×10^7[g]
12	Uridine	$(4.1 \pm 0.3) \times 10^7$[f]; 5.2×10^7[f]
13	Cytidine	$(2.5 \pm 0.3) \times 10^7$[f]; 1.5×10^7[g]
	Reducing sugars	
14	D-Ribose	$(6.8 \pm 3.0) \times 10^6$[i]
15	D-Mannose	1.9×10^5[i]
16	D-Glucose	$(6.1 \pm 1.0) \times 10^4$[i]

[a]The formation constant K is the equilibrium constant of the reaction: $2\ V_1 + 2\ L \rightleftharpoons V_2L_2$, where V_1 is the monomeric form of inorganic vanadate in aqueous solution and L is the ligand. Conditions of the experiments were as follows:

[b]20 mM tris–Cl at pH 7.5 (138).

[c]20 mM HEPES at pH 7.5, 1 M KCl (162).

[d]20 mM HEPES at pH 8.5 (162).

[e]The abbreviation nf = not found by ^{51}V NMR spectroscopy. All formation constants have been obtained under conditions where activity coefficients (not equal to 1) have been incorporated into the formation constant.

[f]30 mM HEPES at pH 7.0, 1 M KCl (153).

[g]100-mM imidazole–Cl at pH 7.0, 0.4 M KCl (141).

[h]20 mM HEPES at pH 7.0, 1.0 M KCL, (188).

[i]No buffer, pH 7.0. The formation constants were calculated from data reported by Geraldes and Castro (163).

formation constants are in the range of 10^7. Those for complexes formed from vanadate and reducing sugars are in the range of 10^4–10^6 depending on the specific aldose.

The stability order of V_2L_2 complexes is as follows: ribofuranosides (10^7) > ribose (10^6) > mannose (10^5) > glucose (10^4) ≈ *cis*-cyclohexane-1,2-diol (10^4) > *trans*-cyclohexane-1,2-diol (10^3) ~ mannopyranoside ~ galactopyranoside (10^3) > ethyleneglycol (10^2) >> methyl α-D-glucopyranoside (10) > methyl β-D-glucopyranoside (no complex observed) (Table IV). The order for glycosides suggests that (a) adjacent OH groups on furanosides form more stable complexes with vanadate than those on pyranosides, (b) pyranosides containing adjacent *cis*-OH form more stable complexes than those containing only *trans*-OH; such an order was previously reported by Angyal (14). The fact that *trans*-cyclohexane-1,2-diol forms a reasonably stable complex with vanadate, whereas glucopyranosides form significantly less stable complexes, is attributed to the ring flattening associated with the replacement of a methylene by an oxygen (162).

The ^{51}V NMR spectra of complexes of all the reducing sugars examined show a broad peak at about δ −523 ppm with overlapping signals. The stability order of vanadium(V) complexes generated from the reducing sugars is as follows: ribose > mannose > glucose (entries 14–16 in Table IV). Mannose and glucose exist in aqueous solution mainly in the pyranose form, while ribose can be found in both the pyranose and the furanose forms (23).

In the *α-p* form, D-ribose contains up to four adjacent *cis*-OH groups, while β-D-mannose contains three *cis*-OH groups, and α-D-glucose only two *cis*-OH groups. Therefore, all these sugars can form a minimum of one stable vicinal cis-diol–vanadate complex, which follows this stability order: vicinal cis-diol on five-membered ring > vicinal cis-diol on six-membered ring > vicinal trans-diol on six-membered ring. Perhaps this observation indicates the generation of complexes with structures related to that shown in Fig. 5(*a*). Alternatively, it is also possible to attribute the order of the complex formation to the chelation of all three hydroxyl functionalities in ribose and mannose, presumably the *a–e–a* hydroxyl groups (163). This type of complex is ubiquitous with divalent and trivalent metal ions (14). We favor the former assignment (bidentate ligands), since the formation constant of the V^V–ribose complex is at the same magnitude as the V^V–ribofuranoside complex, and a larger formation constant would be expected for a tridentate ligand [see (14)]. The formation constants of V_2L_2 complexes of reducing sugars range from 10^4 to 10^6. The formation constant of each V^V-reducing sugar complex presumably reflects the stability of the strongest V^V–furanose complex and the energy required to convert each sugar to the furanose form. Detailed structural characterization of V^V-reducing sugar complexes will require isotopic labeling of the carbohydrate and/or simpler ligands or modified sugars in order to describe and sort out the resulting mixture of several possible V^V complexes. Accordingly, it is not surprising that studies of the reactions of vanadate with reducing sugars reported so far only provided limited structural information.

Vanadate forms very stable 2:2 complexes with ribofuranosides including adenosine and uridine. These complexes are briefly described here in view of their similarity with the simple carbohydrate complexes and the recent structural characterization of the vanadate–adenosine complex (144). The solution containing vanadate and uridine is a potent inhibitor of ribonuclease (170–172). The active species is presumably a weak 1:1 complex in equilibrium with the major 2:2 complex, which exists in three isomeric forms because of the asymmetry of the ligands. Recent studies have shown that the presence of a suitably chosen buffer will generate a major new species (at about δ −510 ppm) containing one molecule of each vanadate, ligand and buffer (141, 169). No information is available on whether such solutions are more potent inhibitors of ribonuclease than solutions containing only the 2:2 complexes.

When vanadate interacts with ligands containing a 1,2,3-triol system, the vanadium can form different 2:2 complexes of the bidentate ligand or chelate all three functionalities simultaneously. Simple triols have been examined and structural information is available for tris(hydroxymethyl)ethane and tris(hydroxymethyl)aminomethane, but these systems result in polynuclear vanadium complexes in the solid state (150, 151, 168), which are not likely to resemble the 1:1, 1:2, and 2:2 complexes observed with carbohydrates in aqueous solution.

Vanadate also forms complexes with nucleotides such as ADP, ATP (166, 173, 174), and NAD (175). These complexes can act as ATP and NADP analogues in biological systems (176, 177). However, little evidence has been reported on the structure of these compounds.

Mechanistic studies on the oxidation of monosaccharides by vanadium(V) in aqueous acidic medium provided evidence for the existence of other types of complexes in these systems [(177–180) and references cited therein]. The focus of these works is the characterization of the kinetics of oxidation of the carbohydrate (180). The kinetic studies suggest the formation of some activated vanadium(V)–monosaccharide complexes, although no information is available on the specific structure of such species. The order of oxidation rates of monosaccharides is as follows: L-sorbose ~ D-fructose > D-ribose ~ D-xylose > L-arabinose > D-galactose > D-mannose > D-glucose. This order is identical to the order of proportions of open-chain forms of the carbohydrates (181). Thus, it was proposed that monosaccharides are oxidized through the vanadium–open-chain complex. In any event, these studies support the existence of intermediate minor vanadium(V)–carbohydrate complexes that are presumably similar with the types of complexes formed between alditols and divalent metal ions.

ii. Vanadium(IV)–Carbohydrate Interactions. The major species formed from solutions of V^{IV} and 1,2-diols are 1:2 complexes, the structure of which is believed to be that shown in Fig. 5(*b*) both in solution and in the solid state (146, 182–186). These weak complexes are much less characterized than those of V^{V} both with respect to solution and solid-state studies.

The reactions between vanadium(IV) and D-glucuronic, D-galacturonic, and lactobionic acids were examined. The combination of EPR, UV–vis, and CD spectroscopy with potentiometric studies shows that complexes with a 1:2 stoichiometry form in the pH range 4–10 (182, 184). At pH > 7.5, V^{IV} is chelated by two adjacent hydroxyl groups to form a 1:2 five-membered cyclic complex [Fig. 5(*b*)]. At lower pH, V^{IV} coordinates to at least one carboxyl group to generate complexes which are not described further in this chapter. At pH 10.8, a species with the stoichiometry 2:2 forms. Based on ENDOR spectra, this complex was assigned to an unusual structure of two nonbridged five-coordinated vanadium(IV) ions sandwiched between two open-chain molecules of D-galacturonic acid (183).

The binding of oxovanadate(IV) with simple sugars was investigated similarly by EPR and UV–vis spectroscopies (187). Three species were observed to form at different pH: VL, VL_2, and an EPR silent V_2L_2. Complexation is favored in basic media. The main product is a 1:2 five-membered cyclic complex with vanadyl coordinated to two adjacent deprotonated hydroxyls of sugar molecules [Fig. 5(*b*)]. The sugars with several adjacent *cis*-OH groups form more stable complexes with vanadyl than sugars with less or no adjacent *cis*-OH do. For example, it is observed that the pyranose forms of galactose, mannose, and glucose give more stable complexes than their pyranoside counterparts (galactoside, mannoside, and glucoside), presumably because the pyranoside contains less adjacent *cis*-OH than the corresponding pyranose. The synthesis of a solid vanadium(IV)–glucose complex from methanol/Na solution was reported (188, 189). The product was characterized by elemental analysis, IR, UV, and EPR, but further analysis to confirm the binding of vanadium with carbohydrate is needed.

iii. Vanadium(III)–Carbohydrate Interactions. Vanadium(III)–alcohol complexes can be prepared from nonaqueous solution. A chiral vanadium(III) complex VL_3, where the monodentate ligand (HL) is DAG, a protected glucofuranose, has been prepared from toluene (190). Structure determination indicated that the vanadium atom binds the three O3 atoms in the equatorial plane of a trigonal bipyramid. The average V—O distance is 1.85 Å. This kind of complex may have synthetic activity in preparation of chiral complexes. No vanadium(III)–sugar complexes have been reported yet in aqueous solution.

2. *Niobium and Tantalum*

In addition to vanadium, Group 5 (VB) transition metals include Nb and Ta. Similarities are expected with vanadium, but no information is currently available on carbohydrate complexes with these elements in any oxidation state. Little information has been reported on complexes of carboxylate ligands and multidentate alkoxide ligands of these metal ions (191–193).

E. Group 6 (VIB) Metal Ions and Uranium

Group 6 (VIB) transition metals include Cr, Mo, and W. The latter elements markedly differ from chromium and present some analogies with vanadium and uranium. In the (VI), (V), and (IV) oxidation states, Cr, Mo, and W readily generate oxometalates, which can form polynuclear complexes. In the (0) oxidation state, Cr, Mo, and W afford organometallic species that are discussed in the corresponding section.

1. *Chromium*

Chromium(VI) compounds, such as chromates and dichromates, are strong oxidizing agents and do not form stable adducts with reducing sugars. The reducing abilities of various carbohydrates toward Cr^{VI} have been compared (194). However, lower valences do form complexes with carbohydrates. A complex of Cr^{III} with taci, a *cis*-triaminotrihydroxycyclohexane, characterized in the solid state, was shown to contain two different isomers in 1 : 2 ratio in the same crystal structure (195). In both isomers, an octahedral Cr^{III} ion is bound to two tridentate taci molecules. In the minor species, both taci molecules chelate Cr^{III} through the O2,4,6 donors, whereas in the major species, one taci is an oxygen donor and the second taci is a nitrogen donor (N1,3,5).

Long-lived glycosyl–Cr^{III}–L complexes have been invoked as products of the first step in the reaction of sugar derivatives with Cr^{II}–L complexes (L = edta or other aminopolycarboxylic acid) (196). The intermediate Cr^{III} complex slowly forms glycals by an E1cB-type elimination of the C2 substituent of the sugar. Stable 1 : 2 complexes of Cr^{V} with 1,2-diols (ethyleneglycol, 1,2-propanediol, and 2,3-butanediol) have been studied by ^{1}H ENDOR spectroscopy in deuterated methanol solution (197, 198). The structures (Fig. 9) are analogous to those of oxovanadium(IV) complexes. In the 1,2-propanediol complex, the presence of the methyl group makes the *a–e* interconversion sufficiently slow for the observation of the magnetically different protons in the puckered five-membered chelated ring (198).

The 1 : 2 Cr^{V}–diol complexes are similar to the Cr^{V} complexes of substituted

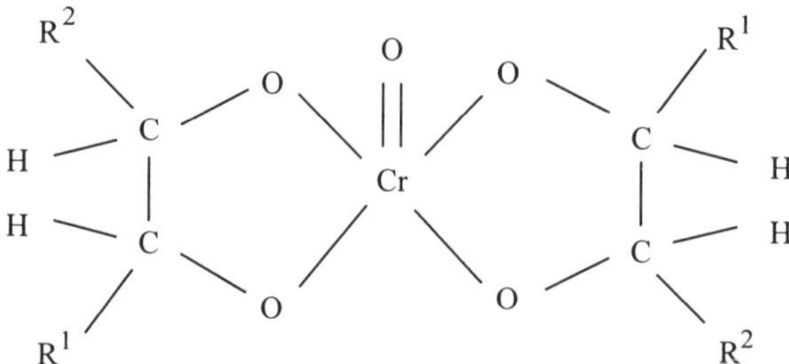

Figure 9. The bis-chelated oxochromium(V) complexes of 1,2-diols ($R^1-CHOH-CHOH-R^2$). Ethyleneglycol, $R^1 = R^2 = H$; 1,2-propanediol, $R^1 = H$, $R^2 = Me$; 2,3-butanediol, $R^1 = R^2 = Me$. [Adapted from (198).]

α-hydroxyacids (199) such as 2-alkyl-2-hydroxybutanoic acids (alkyl = Me or Et). X-ray analysis of these complexes has shown that the geometry at the five-coordinated chromium atom is distorted to an intermediate geometry between the square pyramid and the trigonal bipyramid (200, 201). The two EPR-active species present in equilibrium in the solution (202) were examined by ^1H ENDOR spectroscopy (203) and found to be the cis and trans isomers illustrated in Fig. 10.

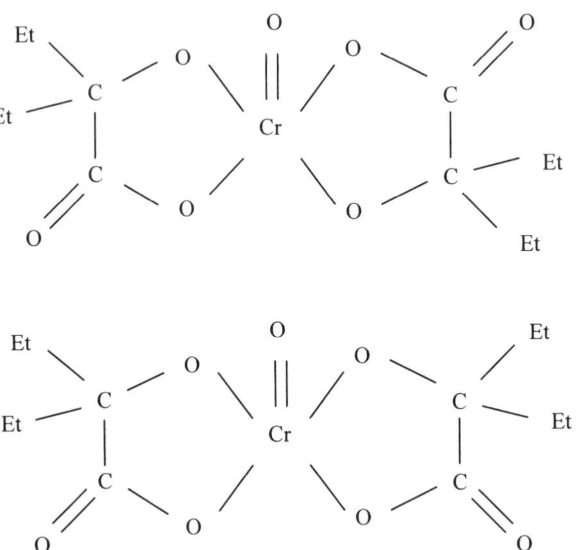

Figure 10. The two isomeric bis[2-ethyl-2-hydroxybutanoato(2^-)]oxochromate(V) complexes. [Adapted from (203).]

The Cr^{V} intermediates generated in the Cr^{VI} oxidations of organic substrates have been implicated in the carcinogenic effects of Cr compounds. By using as models the Cr^{V} complexes of oxalate and 2-ethyl-2-hydroxybutanoate, correlations were developed for distinguishing the coordination number (5 or 6) of the complexes and the donor groups by EPR spectroscopy (204). By this method, the 1:2 Cr^{V} complex formed with quinic acid in methanol was proposed to involve the ligand chelating by the C(OH)—COOH moiety instead of the trihydroxycyclohexane system (204). Bis-chelated Cr^{V} complexes $[Cr(O)L_2]^-$ have been characterized by EPR spectroscopy, in which L is a 1,2-diol or a 1,2,3-triol (205).

2. *Molybdenum and Tungsten*

a. Complexes of Mo^{V}, W^{V}, Mo^{III}, and W^{III}. The Mo^{V} forms diamagnetic complexes with many aldoses and ketoses in aqueous solution of pH 8 (206). These complexes are similar to the dimeric tartrate complexes (207). Their UV spectra show an absorption band near 310 nm attributed to the dimeric $Mo_2^VO_2$ unit. The ligands can be separated into two groups according to the behavior of their Mo^{V} complexes in CD spectroscopy (206). Mannose, lyxose, rhamnose, ribose, fructose, and sorbose give well-defined Cotton effects at 400 nm, whereas galactose, arabinose, glucose, and xylose do not. A possible explanation for these spectroscopic differences is that sugars of the latter group possess HO2,3 in the threo configuration, which are thus in trans orientation and incapable of complex formation when the sugars are in cyclic form.

The irradiation by UV light of solutions of the Mo^{VI} complexes of aldoses or ketoses, in the pH range 2–8, results in the oxidation of the sugar (used in excess) to the aldonic acid and reduction of Mo^{VI} to Mo^{V}. Mannose, glucose, and fructose all yield gluconic acid, as expected if enolization precedes the oxidation step. Then the aldonic acid chelates the resulting Mo^{V}, forming dinuclear complexes (208). Since the CD spectra changed as a function of pH and time, the initially formed 1:1 complexes slowly convert to more stable complexes, the structure of which depend on the chain length of the acid (209). The C_4 and C_5 aldonic acids were reported to form Mo_2L_2 complexes as tridentate ligands, whereas C_6 acids would form bridged Mo_2L complexes. The characteristic patterns of Cotton effects observed for the aldose and ketose complexes were not observed for the complexes with aldonic acids. The 2:2 Mo^{V} complexes of alditols (glucitol, mannitol, and arabinitol) are formed in the pH range 6–11 (209), which is higher than the pH range 2.5–6.5 for the 2:1 Mo^{VI}–alditol complexes.

An ORD study indicated the formation of complexes of W^{V} with aldoses at pH 6–9, alditols at pH 7–10, and aldonic acids at pH 2–6.5 (210). The complexes are presumably dimeric, since they are diamagnetic. The W^{V} complexes show Cotton effects at 270 nm for aldoses, 310 nm for aldonic acids, and 340 nm for alditols.

Both Mo^{III} and W^{III} react with the monodentate protected glucofuranose DAG (HL), to form complexes formulated as M_2L_6, which contain M≡M triple bonds (133). These species were characterized by IR, ^1H and ^{13}C NMR spectroscopies and X-ray structural analysis.

b. Crystal Structures of Mo^{VI} Complexes of Carbohydrates. The structures of the salts of dinuclear Mo^{VI} complexes of several alditols and one aldose, namely, D-lyxose, have been determined (211–216). All these complexes have a 2 : 1 Mo/ligand stoichiometry. The counterions are mostly bulky tetraalkylammonium cations such as TEA or TBA. Both Mo atoms in a cis-dioxo geometry are bridged by μ-oxygen atoms. These structures are retained in aqueous solution, since the results of potentiometric measurements (211) show that the stoichiometries of Mo^{VI}–carbohydrate complexes are (2,1,2) and (2,1,3) (Eqs. 1 and 2), in which the symbol (p, q, r) represents a complex formed from p MoO_4^{2-} units, q alditol, or sugar molecules (L) and r protons.

$$2MoO_4^{2-} + L + 2H^+ \rightleftharpoons (2,1,2)^{2-} + H_2O \quad (1)$$

$$2MoO_4^{2-} + L + 3H^+ \rightleftharpoons (2,1,3)^{-} + H_2O \quad (2)$$

At this time, no Mo^{VI} complex of an alditol with threo configuration has been structurally characterized. However, corresponding complexes have been characterized by NMR spectroscopy for the tetradentate threo polyols (211) in agreement with the structure found by X-ray determination (212) for the related Mo^{VI} complex of 1,4-dithiothreitol (dtt), *cis*-$(TEA)_2[Mo_2O_5(C_4H_6O_2S_2)]$. The ligand is in extended zigzag conformation and binds the Mo atoms at the S1,4;O2,3 site (Table V). Each bridging oxygen atom has one short (2.06 Å) and one long (2.65 Å) Mo—O bond (212).

All Mo^{VI}–alditol complexes isolated so far contain a tetradentate ligand of erythro configuration in sickle conformation (Table V) (213–215). These species possess a bent $[Mo_2O_5]^{2+}$ core made of two cis-dioxo MoO_2 groups linked by an oxygen atom. Two alternate oxygen atoms of the ligand provide additional bridging. The Mo^{VI}–lyxose complex belongs to a similar type (Table V), but the chelating site of the sugar (C1,2,3,5) does not involve the C4 atom (216).

Two Mo^{VI}–erythritol complexes isolated from methanol as solid salts (Table V) (213) correspond to the $(2,1,2)^{2-}$ and $(2,1,3)^{-}$ complexes characterized in solution (211). The complex $(TBA)_2[Mo_2O_5(Ery)]$, where Ery represents $(C_4H_6O_4)^{4-}$, was obtained from $(TBA)_2Mo_2O_7$. Protonation with chloroacetic acid gave $(TBA)[Mo_2O_5(HEry)]$. In both complexes, alternate oxygen

TABLE V
Characterization Data and Structures of Dinuclear Mo^{VI}–Carbohydrate Complexes and Related Species[a,b]

Compound	*cis*-MoO_2 IR	Mo—O—Mo	Mo—O_A	Mo—O_B Mo—O—Mo	Mo—Mo Distance
Mo_2O_5(dtt)[c] TEA salt (212)	1.68–1.72 θ, 105 ν, 882–916	1.89–1.93 θ, 113	2.44–2.46 (S atoms)	2.03 and 2.70 2.10 and 2.67 θ, 84 and 85	3.17–3.20
Mo_2O_5(Ery) TBA salt (213)	1.66–1.69 θ, nd ν, 905–920	1.92–1.95 θ, 104	1.93 (O4) 1.97 (O2)	2.20–2.21 (O3) 2.27–2.29 (O1) θ, 88 (O3) θ, 84 (O1)	3.065
Mo_2O_5(HEry) TBA salt (213)	1.63–1.74 1.70–1.71 θ, nd ν, 905–935	1.92 θ, 109	1.91 (O4) 1.93 (O2)	2.13–2.16(O3) 2.50–2.53 (HO1) θ, 94 (O3) θ, 77 (HO1)	3.14
Mo_2O_5(HMan) NH_4 salt (214)	1.68–1.72 θ, 107	1.94–1.95 θ, 108	1.96 (O4) 2.00 (O2)	2.12–2.17 (O3) 2.47–2.53 (HO1) θ, 94 (O3) θ, 78 (O1)	3.15
Mo_2O_5(HMan) Na salt (215)	1.69–1.73 θ, 107	1.94–1.95 θ, 108	1.93 (O4) 1.93 (O2)	2.11–2.16 (O3) 2.46–2.49 (HO1) θ, 95 (O3) θ, 79 (O1)	3.14
Mo_2O_5(HLyx) NH_4 salt (216)[d]	1.71	1.94	1.93 (O1) 1.94 (O3)	2.13–2.15 (O2) 2.44–2.52 (HO5)	nd
$Mo_2O_5(dbc)_2$ (218)	1.67–1.71 θ, 105	1.91–1.92 θ, 109	1.97–1.98	2.16, 2.40 2.15, 2.37	3.13
$Mo_2O_5(dhn)_2$ NH_4 salt (219)	1.70–1.72 θ, 105 ν, 883–942	1.93–1.94 θ, 109	1.97–1.99	2.17 and 2.45 2.155 and 2.405 θ, 86 and 87	3.15
$Mo_2O_5(pq)_2$ (220)	1.68–1.69 θ, 105	1.89–1.90 θ, 113	2.04	2.24, 2.51 2.24, 2.48	3.16

[a]Bond lengths (Å), angles (θ, °), and IR frequency (ν in cm^{-1}). O = bridging oxygen; O_A = singly bound oxygen; and O_B = bridging oxygen of the ligand.
[b]Ery = erythritolate, Man = mannitolate. dbc = 3,5-di-*tert*-butylcatechol. dhn = 2,3-dihydroxynaphthalene. pq = 9,10-phenanthrenequinone.
[c]1,4-Dithiothreitol, three independent dimers (mean values).
[d]Lyxose = Lyx, angles were not reported. nd = not determined.

donors (O1,3) bridge the Mo atoms. The geometry of the $[Mo_2O_5(Ery)]^{2-}$ ion is roughly symmetrical, whereas that of the $[Mo_2O_5(HEry)]^-$ ion is asymmetric and the bonds between the Mo atoms and HO1 average 2.52 Å versus 2.1–2.3 Å in the unprotonated complex (213).

Crystal structures are available for two mannitol (H_4Man) complexes, $(NH_4)[Mo_2O_5(HMan)]$ (214) and $Na[Mo_2O_5(HMan)]$ (215). In both complexes (Table V), the site of chelation is O1,2,3,4 with O1,3 bridging atoms. The ligand is indeed the $(C_6H_{11}O_6)^{3-}$ hydrogen mannitolate ion ($HMan^{3-}$), in which HO2,3,4 are ionized whereas HO1 bridges the Mo atoms. As in the protonated erythritolate complex, the Mo—OH bonds are longer than the Mo—O bonds.

A salt of the Mo^{VI}–D-lyxose complex was accidentally isolated during an attempt to prepare a Mo^{VI}–D-xylose complex (216), probably because D-xylose was epimerized by molybdate. In this crystal structure, the β-D-lyxofuranose binds two Mo atoms at the O1,2,3,5 site. The bonds between Mo and the bridging oxygens are short for O2 and long for O5, suggesting that O5 is protonated and that the solid salt contains the (2,1,3) complex.

In all the above complexes, the dimolybdate moiety resembles a cofacial octahedron, two distorted MoO_6 octahedra sharing a common face. Such a structure is not specific for Mo^{VI}–carbohydrate complexes, but is also found in a variety of $[Mo_2O_5L_2]^{2-}$ complexes formed with simple bidentate oxygen-donating ligands H_2L. Typical examples are 1,2-dihydroxybenzene or substituted derivatives (catechols) (217, 218), 2,3-naphthalenediol (H_2nhd) (219) and 9,10-phenanthrenequinone (pq) (220). The structural data in Table V show the remarkable analogy of the $[Mo_2O_5]^{2+}$ moiety in all these species. It is unusual (27) for MoO_6 octahedra to share faces (compared with edge or corner sharing). For inorganic compounds, this has been reported only in the structure of $(NH_4)_2H_6(CeMo_{12}O_{42})\cdot 12H_2O$ (221).

Although the triply bridged dimolybdate structure described above is very common for complexes of the Mo_2O_5(O,O,O,O) or $Mo_2O_5(O,O)_2$ types, in which (O,O,O,O) and (O,O) represent tetradentate or bidentate oxygen ligands, different structural types for dinuclear Mo^{VI} complexes of bidentate ligands have been characterized. For example, two MoO_6 octahedra share a corner in the dipotassium salt of a 2:2 Mo^{VI}–oxalate complex containing the $[Mo_2O_5(C_2O_4)_2(H_2O)_2]^{2-}$ anion (222). In this species, the single oxo bridge between the cis-dioxo MoO_2 groups is linear. Each oxalate dianion chelates a single Mo atom and a water molecule occupies the sixth-coordination position about each Mo atom.

c. Complexes of Mo^{VI} and W^{VI} with Alditols in Solution. The chemistry of the Mo^{VI} and W^{VI} complexes of carbohydrates in solution was reviewed in 1963 by Weigel (223) who focused on the electrophoretic separations of mixtures of carbohydrates. The structures of the W^{VI} complexes are generally assumed to be similar (27) to those of Mo^{VI} compounds characterized in the solid state (213–215). Several NMR studies documented this expectation for erythro alditols, but not for threo alditols (224–226). In this chapter, the complexes of alditols are discussed first, since they present less structural variety than the complexes of sugars.

In solution, the Mo^{VI}–alditol complexes were shown to have the 2:1 metal/alditol ratio by polarimetric and potentiometric studies (227–230), in agreement with the solid-state structural determinations. Alditols, which have low specific optical rotations when uncomplexed, show enhanced optical rotation (36, 37) and Cotton effect between 230 and 250 nm (227, 231) when complexed by Mo^{VI}. Measuring the optical rotation in ammonium molybdate solution is a standard procedure for the characterization of unknown alditols (36). D-Mannitol and D-glucitol were determined by polarimetry after addition of molybdate (37). In a polarimetric study of the W^{VI}–D-glucitol complexes, a mononuclear complex, $[WO(OH)(C_6H_{12}O_6)_2]^-$, was found at high pH (up to 11.5) and two dinuclear complexes at pH < 9.5 (232). The species initially present at pH 7, $[W_2O_3(OH)_4(C_6H_{10}O_6)]^{2-}$, was unstable and fully transformed after 72 h at pH 4 into the more stable $[W_2O_3(OH)(C_6H_{11}O_6)_2]^-$ complex. Two solid W^{VI}–D-mannitol complexes formulated as $Na_2[W_2O_3(OH)_4(C_6H_{10}O_6)]$ and $Na[W_2O_3(OH)(C_6H_{11}O_6)_2]$ have been isolated form acidic solutions and characterized by IR spectroscopy. However, the structures proposed from ^{13}C NMR data (233) are different from those characterized in the solid state for Mo^{VI} complexes (214, 215) and in solution for the Mo^{VI} (211) and W^{VI} (234) complexes of mannitol.

Multinuclear ^1H and ^{13}C NMR spectroscopy is a powerful technique for the determination of the structures of Mo^{VI} and W^{VI} complexes in solution. Unidimensional and 2D homo- and heteronuclear experiments are generally necessary for the characterization of the free and chelated ligand (235). The carbon atoms or hydrogen atoms in the chelating sites of each carbohydrate are identified since they shift upon coordination. Such variations of the chemical shifts, $\Delta\delta$, are referred to as ^{13}C coordination induced shifts or CIS and define characteristic CIS patterns for the different types of complexes.

The formation constants of the Mo^{VI} (51, 211, 224, 236, 237) and W^{VI} complexes (63, 225, 229, 238–240) of various alditols have been determined by potentiometry (Tables VI and VII). All alditols complex more strongly with W^{VI} than with Mo^{VI}, since the formation constants K_M (M = Mo or W) are larger for W^{VI} than for Mo^{VI} ($\log K_W - \log K_{Mo} \approx 3$). Consequently, tungstate ions can be determined by acidic titration after complexation by threo alditols such as xylitol and D-glucitol (63), which react faster than erythro alditols such as erythritol, D-mannitol, and galactitol (Table VII). The accuracy of the potentiometric determination of molybdate and tungstate complexed by D-mannitol and D-glucitol can furthermore be improved by the use of Gran's method (236).

The structures of Mo^{VI}–alditol complexes in solution were investigated by ^{95}Mo NMR spectroscopy (211, 241), resulting in limited structural information, due to the broad signals of the spin $\frac{5}{2}$ ^{95}Mo nucleus. Nevertheless, two types of complexes are characterized, since complexes of alditols with a threo diol group display a single Mo signal near 22 ppm, whereas those with an erythro group give a broad Mo signal near 32 ppm (241).

TABLE VI
Formation Constants K_{212} for Mo^{VI} Complexes of Aldoses, Ketoses, Alditols, and Carbohydrate Derivatives[a]

Carbohydrate	Log K_{212}	
D-Arabinose	13.20[b]	
D-Glucose	13.30[b]	
2-Deoxy-D-*arabino*-hexose	13.60[b]	
D-Xylose	13.60[b]	
L-Rhamnose	13.89[c]	
D-Galactose	14.25[b]	
D-Mannose	14.50[c]	
D-Lyxose	14.98[c]	
L-Sorbose	14.15[d]	
D-Fructose	14.45[d]	
D-Psicose	16.30[d]	
D-Tagatose	16.35[d]	
DL-Threitol	14.60	
Erythritol	15.20	
Ribitol	15.55	15.57 (237)
Xylitol	16.25	
D-Arabinitol	16.35	
D-Arabinose diethyl dithioacetal	16.65	
D-Glucitol	16.60	16.90 (236)
D-Mannitol	16.70	16.89 (236)
Galactitol	17.30	17.50 (237)
Perseitol[e]	17.60	
1-Deoxy-1-methylamino-D-glucitol	16.85[b]	

[a]The formation constant K_{212} (mol^{-4} L^4) is the equilibrium constant of the reaction: 2 MoO_4^{2-} + L + 2 H^+ $\rightleftharpoons$ $(2,1,2)^{2-}$ + n H_2O, where L is the carbohydrate. All formation constants have been obtained under conditions where activity coefficients (not equal to 1) have been incorporated into the formation constant. Values determined by potentiometry (0.1 M KCl, t, 25°C). Accuracy ± 0.10 (211).
[b]Values determined by indirect spectrophotometry (39).
[c]Reference (229).
[d]Reference (239).
[e]D-*Glycero*-D-*galacto*-heptitol (224).

The first application of ^{183}W NMR spectroscopy to the study of W^{VI}–carbohydrate complexes in aqueous solution by Chapelle and Verchère (242), showed that this nucleus is more amenable to structural studies. The ^{183}W nucleus generates narrow NMR signals due to its nuclear spin of $\frac{1}{2}$, and given the large spectral window, its chemical shifts are very sensitive to the electronic environment of different types of tungsten atoms, for example, in polytungstates

TABLE VII
Formation Constants[a] and Rates of Formation of W^{VI} Complexes of Alditols and Carbohydrates[a]

Ligand	Log K_{212}		Rate
Alditols			
Erythritol	18.00		Slow
DL-Threitol	16.95		Fast
Ribitol	18.10		Slow
Xylitol	18.50		Fast
D-Arabinitol	18.65		Slow
D-Mannitol	19.65	18.78[b]; 20.24[c]	Slow
D-Glucitol	19.15	19.26[b]	Fast
Galactitol	20.10		Slow
Perseitol	20.35		Slow
Carbohydrates			
D-Lyxose	18.08[d]		nd
D-Ribose	16.90		Slow
D-Galactose	17.00		Slow
D-Mannose	17.50[d]		Slow
L-Rhamnose	17.04[d]		nd
D-Fructose	16.90		Slow
L-Sorbose	16.40		Slow
D-Tagatose	19.10[e]		nd

[a]The formation constant K_{212} (mol^{-4} L^4) is the equilibrium constant of the reaction: $2WO_4^{2-} + L + 2H^+ \rightleftharpoons (2,1,2)^{2-} + nH_2O$, where L is the carbohydrate. All formation constants have been obtained under conditions where activity coefficients (not equal to 1) have been incorporated into the formation constant. Accuracy of log K_{212}, ± 0.10 (63). Not determined = nd. Values determined by potentiometry (0.1 *M* KCl; *t*, 25°C).
[b]Reference (236).
[c]Reference (238).
[d]Reference (229).
[e]Reference (239). Slow: equilibrium is not reached within 1 h. Fast: equilibrium is reached within 10 min.

(243, 244). For each dinuclear W^{VI}–carbohydrate complex, ^{183}W NMR spectroscopy shows two sharp signals generally coupled to protons via measurable vicinal $^3J_{W,H}$ coupling constants. These couplings can effectively be used for 2D heteronuclear ^1H ^{183}W experiments (245–250), which reveal the site of chelation of the ligand.

i. ***Mo^{VI} Complexes of Alditols.*** Two types of dinuclear Mo^{VI} complexes are known in which alditols act as tetradentate donors (211, 226). In both types of complexes, the alditol adopts a conformation in which four adjacent hydroxyl groups point toward the direction of the dinuclear metal core. Complexes of type E are formed with alditols that contain a central erythro diol in a sickle arrangement (erythritol, arabinitol, or galactitol) (Fig. 11). Complexes of type Q are formed with alditols that contain a central threo diol in zigzag arrangement (threitol or xylitol) (Fig. 12). Complexes of type E are stronger than those of type Q (Table VI).

Symmetrical erythro alditols such as galactitol and erythritol afford a single Mo^{VI} complex of type E, in which the CIS pattern of the site of chelation is not symmetrical, since one central carbon is more strongly deshielded ($\Delta\delta$ mean value = 19.5 ppm) and the three other carbons ($\Delta\delta$ mean value = 9 ppm). This CIS pattern was attributed to the twisting of the site of chelation, in which one central carbon atom must move toward the oxo ion core (Fig. 11).

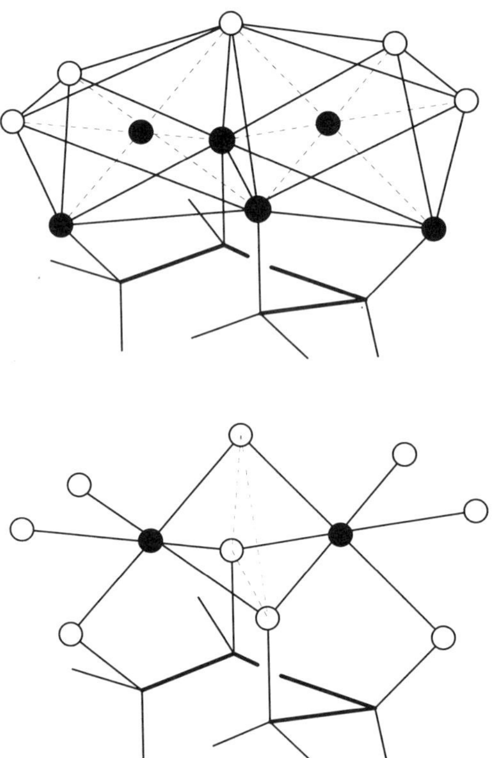

Figure 11. Structure of the Mo^{VI} and W^{VI} complexes of tetradentate alditols of erythro configuration (type E). The alditol is in the sickle conformation. Two MO_6 octahedra share a face. For the $[M_2O_5]^{2+}$ group the metal atoms are filled circles, the oxygen atoms are open circles, and the oxygen donors of the ligand are gray circles.

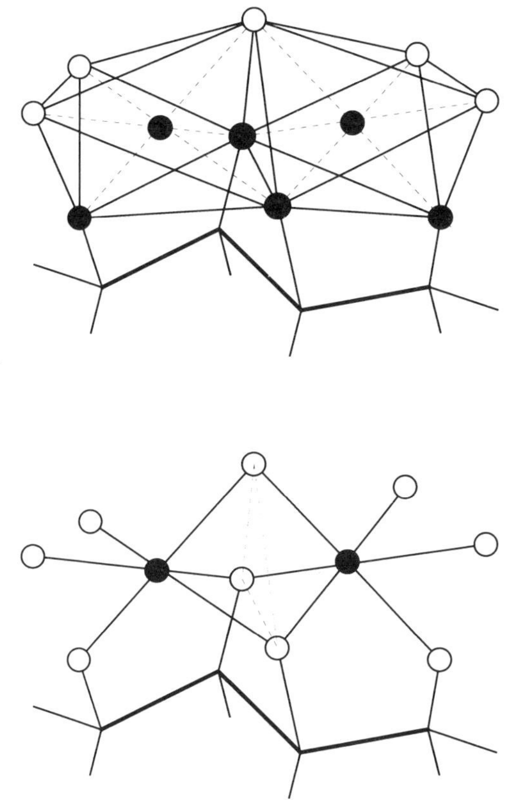

Figure 12. Structure of the Mo^{VI} and W^{VI} complexes of tetradentate alditols of threo configuration (type Q). The alditol is in zigzag conformation. Two MO_6 octahedra share a face. For the $[M_2O_5]^{2+}$ group the metal atoms are filled circles, the oxygen atoms are open circles, and the oxygen donors of the ligand are gray circles.

Alditols in which the lateral carbons of the chelating site possess different substituents R and R′, such as D-mannitol, D-arabinitol, perseitol, and volemitol, form a pair of isomeric Mo^{VI} complexes of type E (211, 224, 251), whereas only one species could be crystallized (214, 215). For example, D-mannitol affords a pair of complexes involving the arabino system coordinating through O1,2,3,4 and O4,3,2,1, respectively. The second isomer was initially erroneously described as a threo complex involving the manno site O2,3,4,5 in zigzag conformation (226).

In the complexes of threo alditols, the tetradentate ligand in zigzag conformation chelates Mo^{VI} in a symmetrical fashion (211), that is, the two central carbon atoms are equally deshielded ($\Delta\delta$ mean value = 12.5 ppm) as well as the lateral ones ($\Delta\delta$ mean value = 10.5 ppm). These NMR data suggest that the solution structure (Fig. 12) is similar to the crystal structure of the Mo^{VI} complex of dithiothreitol (212). The Mo atoms are bridged by two triply bonded oxygen atoms of the central diol system of the alditol.

The alditols that possess both erythro and threo diol systems form mixtures of Mo^{VI} complexes of types E and Q. In particular, D-glucitol afforded four complexes of which two are of type E and two are of type Q (211).

The high affinity of alditols for Mo^{VI} has been used for the design of a postcolumn detection method after HPLC separation of alditols on a cation-exchange column in Ca^{2+} form at t, 85 °C (252). The eluted alditols are detected by indirect photometry at λ = 347 nm, as the formation of their nonabsorbing Mo^{VI} complexes dissociates a weak, colored complex of Mo^{VI} present in the eluent, inducing negative peaks in the constant baseline due to the absorbing eluent. The method is especially suitable for the determination of alditols in food since D-glucose, which forms a weak Mo^{VI} complex, does not interfere.

ii. W^{VI} Complexes of Alditols. The analogy between Mo^{VI}– and W^{VI}–alditol complexes has only been confirmed with erythro alditols, which afford similar Mo^{VI} and W^{VI} complexes of type E (224). On the contrary, threo alditols, which give Mo^{VI} complexes of type Q only, can form three different types (Q, T, and P) of W^{VI} complexes.

The W^{VI} complexes of type E were characterized by ^{183}W NMR spectroscopy (234, 242, 251, 253). The ^{1}H-coupled ^{183}W spectrum of each dinuclear complex contains two signals of equal intensities (Table VIII). The values of chemical shifts (referenced to Na_2WO_4 in alkaline D_2O) are in agreement with tungsten being in the +VI oxidation state (244). Each tungsten atom is mainly coupled to one alditol proton ($^{3}J_{WH}$ = 8–10 Hz). The assignment of the protons coupling patterns were made by 2D heteronuclear ^{1}H–^{183}W experiments via long-range coupling constants. In the W^{VI}–erythritol complex, W1 is chelated by the adjacent O2,3,4 atoms and W2 by O1,2,4, in which only two oxygen atoms are adjacent. Thus, O2,4 are bridging oxygen atoms whereas O1,3 are bound to only one tungsten atom, in a solution structure similar with that reported for the crystalline Mo^{VI}–erythritol complex (213) (Fig. 11).

Erythro alditols with a tetradentate site of chelation in the ribo or altro configurations, which contain three hydroxyl groups in cis,cis orientation, afford single W^{VI} or Mo^{VI} complexes (type E′) instead of pairs of isomeric complexes of type E (Fig. 13). Specifically, ribitol forms Mo^{VI} and W^{VI} complexes at the ribo O1,2,3,4 site (251), D-altritol forms a Mo^{VI} complex at the altro site O2,3,4,5 (226), and volemitol forms minor Mo^{VI} and W^{VI} complexes at the altro site O3,4,5,6 (251). Complexes of type E′ are weaker than complexes of type E (Tables VI and VII). The origin of this difference in complex stability lies in the orientation of the R^1 and R^2 substituents attached to the lateral carbon

TABLE VIII
The 16.67-MHz ^{183}W NMR Chemical Shifts (δ in ppm) of W^{VI} Complexes of Alditols with erythro Sites of Chelation[a]

Alditol	Complex	Type	W1	W2
Erythritol[b]		E	−76.1	−79.5
D-Arabinitol[c]	A_1	E	−75.3	−82.4
	A_2	E	−78.1	−82.1
Ribitol[c]		E′	−72.1	−83.9
Galactitol[c]		E	−79.3	−85.8
D-Mannitol[c]	M_1	E	−74.4	−81.6
	M_2	E	−73.1	−83.1
D-Glucitol[c]	G_2	E	−73.7	−79.2
	G_3	E	−76.2	−78.7
Volemitol[d]	V_1	E	−76.9	−85.8
	V_2	E	−74.6	−79.1
	V_3	E′	−73.6	−83.4
Perseitol[c]	P_1	E	−77.0	−83.7
	P_2	E	−72.8	−85.6

[a]Chemical shifts are referenced to Na_2WO_4 in alkaline D_2O.
[b]Reference (242).
[c]Reference (234).
[d]Reference (25).

atoms of the tetradentate site of chelation in sickle conformation, as shown in Fig. 14. In complexes of type E, R^1 and R^2 are oriented away from the chelating site. On the contrary, in complexes of type E′, the R^1 substituent creates a steric constraint in the alditol conformation favorable for complex formation that destabilizes the complex and prevents the formation of one of both isomers.

The threo alditols generally form three different types of dinuclear W^{VI} complexes depending on the pH: a type Q at neutral pH with a tetradentate ligand, a type T at pH 8–9 with a tridentate ligand, and a type P at pH 9–12 with a pentadentate ligand (253). In contrast, the corresponding Mo^{VI} complexes are of type Q over the entire pH range.

The W^{VI} complexes of type Q (Fig. 12) give broad ^{183}W NMR signals at $\delta \approx -93$ (Table IX), contrary to all other types that give sharp signals (253).

In W^{VI} complexes of type T, chelation always occurs at a tridentate xylo site in a zigzag conformation (Fig. 15). The ^{13}C CIS patterns are symmetrical, with a highly deshielded central carbon atom ($\Delta\delta$ 13–14 ppm) and two equivalent lateral carbon atoms ($\Delta\delta$ 9–10 ppm). These complexes display two specific, sharp ^{183}W NMR signals near −60 and −120 ppm (Table IX). Examination of molecular models showed that one tungsten atom (W1, δ −60) is bound to three adjacent oxygen atoms, whereas the second (W2, δ −120) is only chelated by two oxygen atoms (242, 253).

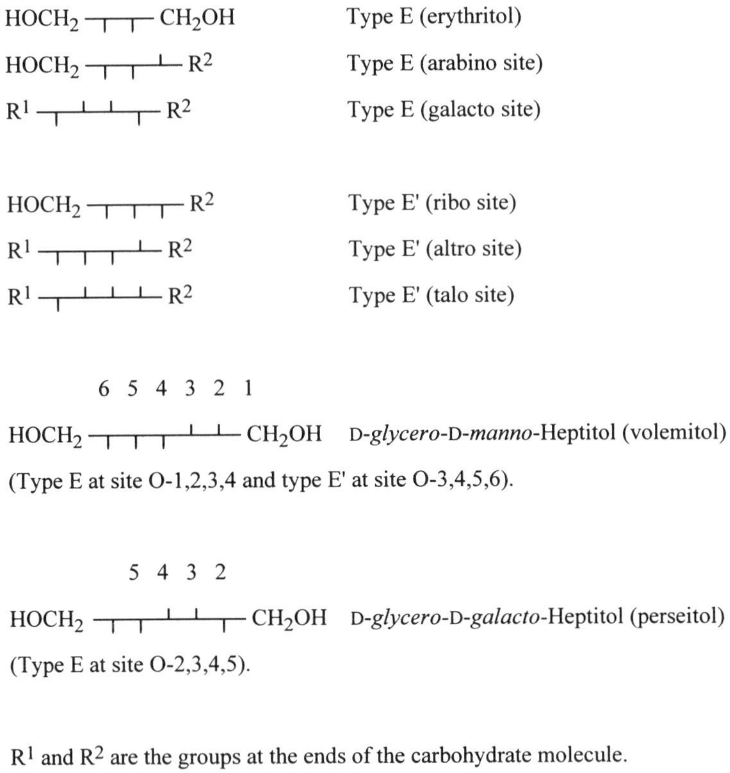

$HOCH_2$ ┬┬ CH_2OH — Type E (erythritol)

$HOCH_2$ ┬┬┴ R^2 — Type E (arabino site)

R^1 ┬┴┴┬ R^2 — Type E (galacto site)

$HOCH_2$ ┬┬┬ R^2 — Type E' (ribo site)

R^1 ┬┬┬┴ R^2 — Type E' (altro site)

R^1 ┬┴┴┴ R^2 — Type E' (talo site)

6 5 4 3 2 1

$HOCH_2$ ┬┬┬┴┴ CH_2OH D-*glycero*-D-*manno*-Heptitol (volemitol)

(Type E at site O-1,2,3,4 and type E' at site O-3,4,5,6).

5 4 3 2

$HOCH_2$ ┬┬┴┴┬ CH_2OH D-*glycero*-D-*galacto*-Heptitol (perseitol)

(Type E at site O-2,3,4,5).

R^1 and R^2 are the groups at the ends of the carbohydrate molecule.

Figure 13. The configurations of tetradentate alditols in relation with the formation of Mo^{VI} and W^{VI} complexes of types E and E′. Example of two heptitols.

The W^{VI}–polyol complexes of type P are obtained in alkaline medium (pH 9–12) with alditols that contain an all-threo site of chelation in a zigzag conformation (Fig. 16), such as xylitol, L-iditol and a heptitol (*meso-glycero-ido*-heptitol) (253). The ^{13}C NMR CIS pattern ($\Delta\delta$ 15-13-10-13-15) indicates that the alditols are pentadentate with a symmetrical structure. The ^{183}W NMR spectra are characterized by signals with positive chemical shifts, whereas all other types of complexes give signals with negative chemical shifts. The L-iditol complex (Table IX) shows two sharp signals of equal intensities at δ 82.3 and 93.4. For the symmetrical xylitol complex, a single signal is observed at δ 93.3 ppm for both equivalent W atoms. A likely structure for W^{VI}–alditol complexes of type P (Fig. 17), is related to that of tungstate polyanions (27), since the WO_6 octahedra share a common edge, contrary to Mo^{VI} and W^{VI} complexes of types E, Q, and T in which the MO_6 octahedra share a face.

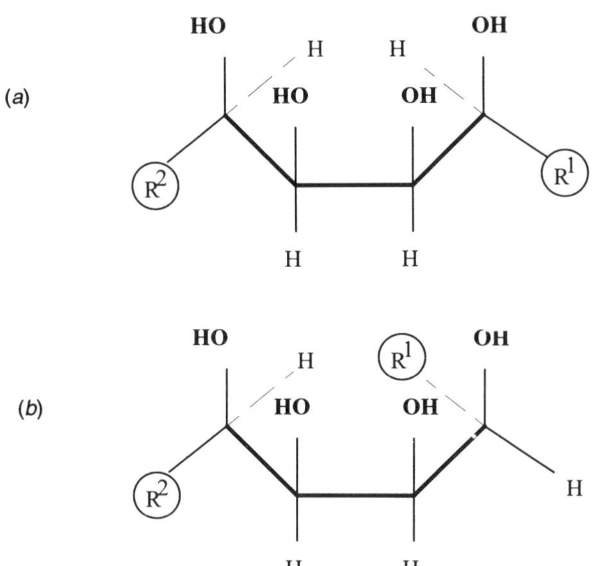

Figure 14. Structural differences between the ligands in Mo^{VI} and W^{VI} complexes of tetradentate erythro alditols in sickle conformation. (*a*) Type E and (*b*) type E′. Steric hindrance between substituent R^1 an H is indicated (251).

TABLE IX
The 16.67-MHz ^{183}W NMR Chemical Shifts δ (ppm) of W^{VI} Complexes of Alditols with threo Sites of Chelation[a]

Alditol	Type	W1	W2
DL-Threitol[b]	T	−59.3	−118.3
Xylitol[c]	T	−61.7	−120.8
L-Iditol[c]	T	−61.5	−118.6
D-Arabinitol[d]	T	−55.5	−120.0
D-Glucitol[d]	T	−57.7	−121.6
Xylitol[c]	Q	−92.0[e]	−92.0[e]
L-Iditol[c]	Q	−93.8[e]	−93.8[e]
Xylitol[c]	P	93.3	93.3
L-Iditol[c]	P	93.4	82.3
Heptonic acid[f]	P″	108.1	102.3

[a]Chemical shifts are referenced to Na_2WO_4 in alkaline D_2O. Accuracy: $\delta \pm 0.1$ ppm.
[b]Reference (242).
[c]Reference (253).
[d]Reference (234).
[e]Broad signals, line widths $\Delta\nu$ = 45 Hz for xylitol and 140 Hz for L-iditol (253).
[f]D-*Glycero*-D-*gulo*-heptonic acid (254).

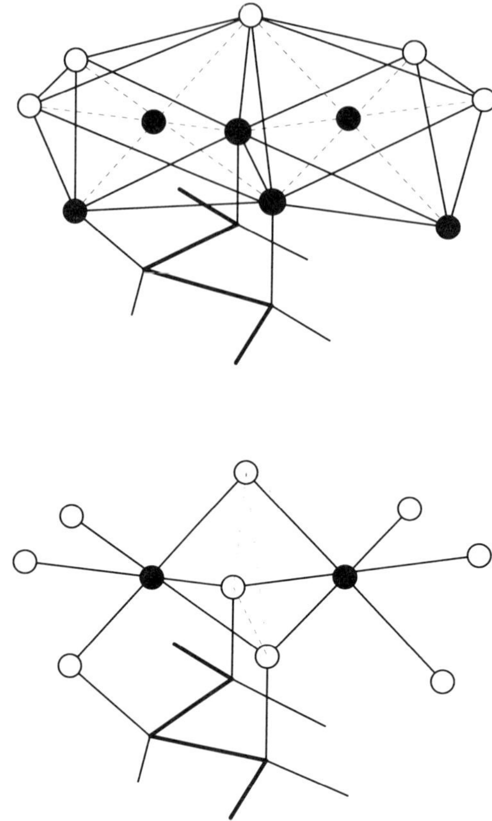

Figure 15. Structure of the W^{VI} complexes of tridentate alditols of threo configuration (type T). The alditol is in zigzag conformation. Two WO_6 octahedra share a face. For the $[W_2O_5]^{2+}$ group, the metal atoms are filled circles, the oxygen atoms are open circles, and the oxygen donors of the ligand are gray circles.

Related W^{VI} complexes of pentadentate ligands (heptitols and a heptonic acid) formed at pH 12.2, were characterized as modified complexes of type P and types P′ and P″ (254, 255). A multinuclear NMR study indicated that the carboxylate group of the heptonate is not coordinated, since the site of chelation is O2,3,4,5,6 (254). The differences between types P, P′, and P″ (Fig. 16) were attributed to the orientations of the lateral C2,6 atoms with respect to the central xylo system C3,4,5. In complexes of type P, the lateral substituents are pointing away from the site of chelation, whereas in types P′ and P″, one or two lateral substituents create a steric strain by pointing toward C4.

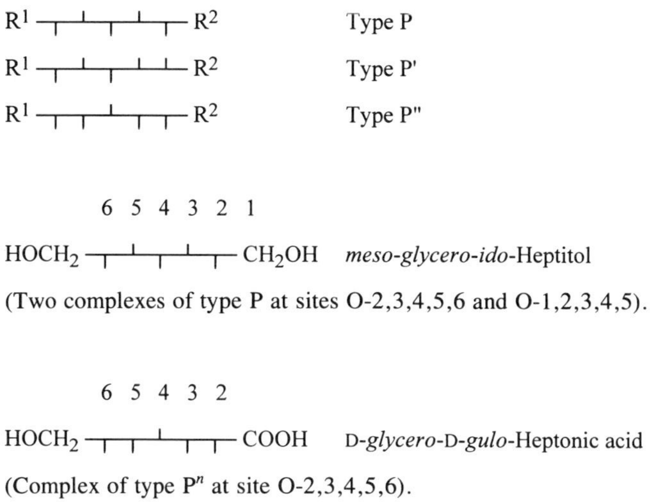

R^1 ┬┴┬┴┬ R^2 Type P

R^1 ┬┴┬┴┴ R^2 Type P'

R^1 ┬┬┴┬┬ R^2 Type P"

6 5 4 3 2 1

$HOCH_2$ ┬┴┬┴┬ CH_2OH *meso-glycero-ido*-Heptitol

(Two complexes of type P at sites O-2,3,4,5,6 and O-1,2,3,4,5).

6 5 4 3 2

$HOCH_2$ ┬┬┴┬┬ COOH D-*glycero*-D-*gulo*-Heptonic acid

(Complex of type P″ at site O-2,3,4,5,6).

R^1 and R^2 are the groups at the ends of the carbohydrate molecule.

Figure 16. The configurations of pentadentate polyols in relation with the formation of W^{VI} complexes of types P, P′, and P″.

iii. Mo^{VI} and W^{VI} Complexes of Cyclitols. Both Mo^{VI} and W^{VI} complexes of cyclitols are generally weaker than those of acyclic alditols, presumably because of the rigidity of cyclitols. The Mo^{VI} and W^{VI} complexes of 44 cyclitols were studied by electrophoresis, at pH 5.0 and pH 5.7, respectively (256). The dinuclear species formed with inositols were formulated as $Na_2\{M_2O_5[C_6H_6(OH)_2O_4]\}$, where M is Mo or W. For polyhydroxycyclohexanes, the more stable complexes are formed when five or six OH groups are cis, in *cis*-inositol (O1,2,3,4,5,6) and *epi*-inositol (O1,2,3,4,5), for example. Less stable complexes are obtained with *allo*-inositol (O1,2,3,4) and *myo*-inositol (O1,2,3,5). The same trend is apparent with polyhydroxycyclopentanes: (O1,2,3,4,5) > (O1,2,3,4) > (O1,2,3,4). Thus, the minimum structural requirement for complex formation is the presence of four cis-hydroxyl groups. Ligands that form the most stable complexes are those containing cis-HO1,2,3,5 (1*a*, 2*e*, 3*a*, 5*a*), and those with cis-HO1,2,3,4 (1*a*, 2*e*, 3*a*, 4*e*). This stability pattern favors complex formation with *epi*-inositol, which can be isolated as its Mo^{VI} complex and subsequent treatment with an ion-exchange resin. This process allows separation from all isomers that do not react with molybdate (256).

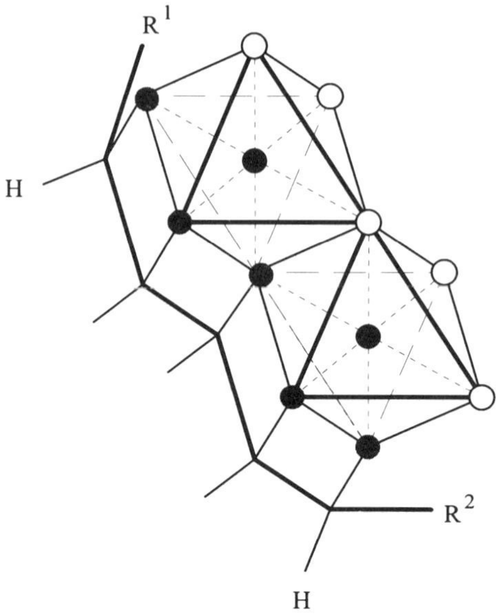

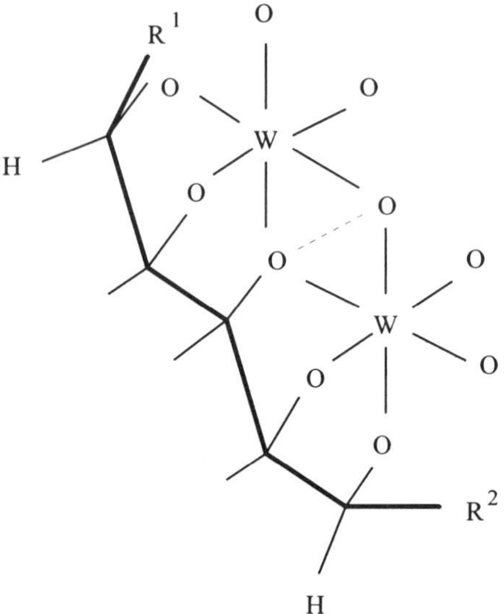

Figure 17. Structure of the W^{VI} complexes of pentadentate alditols of xylo–ido configuration (type P). The alditol is in zigzag conformation. Two WO_6 octahedra share an edge. For the $[W_2O_5]^{2+}$ group, the metal atoms are filled circles, the oxygen atoms are open circles, and the oxygen donors of the ligand are gray circles (253).

d. Complexes of Mo^{VI} and W^{VI} with Sugars in Solution. Sugars can form a greater number of Mo^{VI} and W^{VI} complexes than the acyclic alditols, since aldoses and ketoses, which normally exist in cyclic forms, may also open to give the aldehyde, ketone, and hydrated forms in aqueous solution. A key parameter for complex formation is the orientation of the hydroxyl groups of the sugar in the cyclic forms. The stability of the complexes decreases in the following order of configurations: lyxo > ribo > xylo > arabino (Table VI). The major sugar complexes of the lyxo and ribo series maintain the ligand in cyclic form. On the contrary, the weaker sugar complexes of the xylo and arabino series involve only the acyclic ligands (257, 258).

The solution structures of the Mo^{VI} complexes of aldoses shed light on the nature of the types of intermediate species that are involved in the C2 epimerization reaction of aldoses, which is catalyzed by Mo^{VI} in acidic aqueous solution. In this respect, it is important to recognize that although W^{VI} and Mo^{VI} generally react similarly (27), tungstate does not catalyze the epimerization reaction, despite its ability to form carbohydrate complexes that are often isostructural with their Mo^{VI} congeners. Another application of the formation of Mo^{VI}–sugar complexes is the separation of mixtures of sugars by paper electrophoresis in molybdate solution (223, 228–230) based on the large dependence of the complex stabilities on the configuration of the carbohydrate. Thin-layer chromatography is also a common technique for carbohydrate separation, either on plates coated with a cation-exchange resin loaded with tungstate (91, 259) or on cellulose powder impregnated with tungsten (260–262). A scale of pseudo stability constants was determined for the tungstate–carbohydrate complexes. These constants are defined as the ratio of the retention factors of the sugar, referenced to glucose, in the absence and in the presence of tungstate (261).

Early polarimetric studies had suggested the existence of 1 : 1 Mo^{VI}–sugar species (263–265), but it is now generally recognized that Mo^{VI} and W^{VI} form dinuclear 2 : 1 complexes (230, 266, 267) with sugars. The structural identification of the Mo^{VI} and W^{VI} complexes of aldoses and ketoses is commonly carried out by multinuclear NMR spectroscopy (225, 226, 239, 253, 268–278).

i. Mo^{VI} Complexes of Acyclic Aldoses. Sugars in solution are found essentially in cyclic *p* and *f* forms. Thus, the existence of Mo^{VI} complexes of acyclic sugars demonstrates that complexation may stabilize a form of the sugar that exist in low amount at equilibrium. The prevalence of complexes of cyclic or acyclic forms depends on the specific geometry of all conformers and the configuration of HO2,3 in the aldose. When this system is erythro, HO2,3 are cis in the cyclic forms and complexes of the cyclic forms exist (lyxo and ribo series). On the contrary, when this system is threo, HO2,3 are trans in the cyclic forms and only complexes of the acyclic form exist in observable concentration (arabino and xylo series). The Mo^{VI} complexes formed by aldoses in acyclic form are much weaker than those formed by aldoses in cyclic forms, because the free cyclic sugars are converted to a less stable form before chelating the dimetallic core. The energetic cost of this unfavorable opening step decreases the overall formation constant. The results of thermodynamical calculations were consistent with the presence of less than 0.5% of open form in the uncomplexed aldoses (39).

Aldotetroses (erythrose and threose) give Mo^{VI} complexes only in their acyclic hydrated form (275). The absence of complexes involving D-erythrose in α-f form, which contains three cis-HO groups, documents the low complexing ability of the tridentate f forms of sugars. A ^{13}C NMR study revealed that the erythrose and threose complexes have similar chelation modes as the erythritol and threitol complexes. The tetroses are tetradentate ligands, that is, one of the HO1 groups is part of the complexing site. Since the chelating sites are unsymmetrical, erythrose and threose form pairs of complexes, E_1–E_2 and T_1–T_2, respectively. The erythrose complexes involve the ligand in sickle conformation and are related to type E, whereas those of threose contain the ligand in zigzag conformation and are related to type Q.

The NMR data revealed that aldoses of the xylo series (xylose and glucose) give small amounts of complexes analogous to the T_1 complex of threose, in which the sites of chelation are the HO1,2,3,4 systems (258). In the case of xylose, two tetradentate sites in zigzag form are available: the O1,2,3,4 site (threose type) or O2,3,4,5 (threitol type). However, the absence of the complex with the O2,3,4,5 site indicates that an HO group of the hydrated carbonyl has a higher chelating affinity for Mo^{VI} than a CH_2OH group.

Aldoses of the arabino series (arabinose and galactose) form pairs of Mo^{VI} complexes in small amounts ($\approx 100\%$) in which the ligands are in acyclic hydrated form and in sickle conformation. In these complexes, the chelating site (O2,3,4,5), which does not involve an HO1 group, differs from that in the erythrose complexes, but is similar to that in the Mo^{VI}–alditol (arabinitol and galactitol) counterparts of type E (39, 257).

Aldoses of the lyxo series (lyxose, mannose, and gulose) (274) and of the ribo series (ribose, allose, and talose) (257, 272) mainly afford Mo^{VI} complexes in their cyclic forms. However, they also yield minor species that involve the acyclic, hydrated ligand in sickle form binding at the O1,2,3,4 site (258, 278). Such complexes, in which the central diol group HO2,3 is erythro, are referred to as type A and are analogous to the molybdate E_1 complex of erythrose (275). A multinuclear NMR study (^1H, ^{13}C, and ^{95}Mo) revealed that the Mo^{VI} complexes of D-ribose and 5-deoxy-L-ribose were isostructural with that of 5-deoxyribitol and thus involved the acyclic ligands with the tetradentate O1,2,3,4 site (275).

ii. Mo^{VI} and W^{VI} Complexes of Acyclic Ketoses. The main factor that influences the formation of acyclic complexes of ketoses is, as for aldoses, the

TABLE X
The 15.005-MHz ^{183}W NMR Chemical Shifts (δ, ppm) for the W^{VI} Complexes of Types L and M of Sugars of the lyxo Series in f Form (277) and for the W^{VI} Complexes of D-Fructose and L-Sorbose in Acyclic Hydrated (AH) Form (239).[a]

Ligand	Type	W1	W2
D-Lyxose	L	−79.6	−71.1
D-Mannose	L	−73.75	−90.9
L-Rhamnose	L	−73.4	−92.3
D-Tagatose	L	−68.1	−78.4
D-*manno*-Heptulose	L	−70.85	−89.75
D-Mannose	M	15.0	0.3
D-*manno*-Heptulose	M	13.7	−3.6
D-Fructose	AH	−95.2	−106.5
L-Sorbose	AH	−96.5	−114.5

[a]Chemical shifts are referenced to Na_2WO_4 in alkaline D_2O. Accuracy: $\delta \pm 0.1$ ppm.

configuration of the ring hydroxyl groups, here HO3,4. First, we will describe the major complexes and then provide information on some minor complexes. In the arabino and xylo series, the HO3,4 system is threo and only complexes of the acylic form may exist. A multinuclear NMR study (239) showed that D-fructose and L-sorbose, which possess a threo HO3,4 diol system, form the same type of major Mo^{VI} and W^{VI} complexes containing the ketoses in acyclic hydrated form with the tetradentate O1,2,2′,4 site of chelation. The ^{183}W chemical shifts for the two W^{VI}–ketose complexes are given in Table X. Molecular models show that each metal atom is bound to three oxygen atoms, which are of two types. Both O1 and O4 are bound to a single metal atom, whereas both O2,2′ of the hydrated carbonyl group bridge the metal atoms as triply bound donors (Fig. 18). Moreover, the formation of minor Mo^{VI} complexes in which

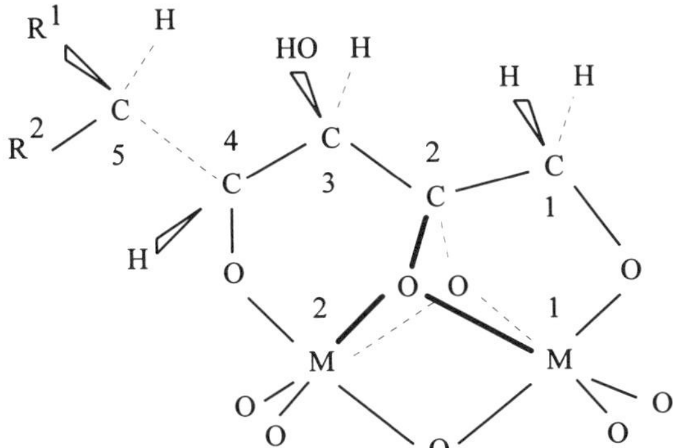

Figure 18. Proposed structures for the Mo^{VI} and W^{VI} complexes of two hexoketoses in acyclic hydrated form. D-Fructose, R^1 = HO and R^2 = CH_2OH. L-Sorbose, R^1 = CH_2OH and R^2 = HO (239).

D-fructose and L-sorbose are in acyclic keto form was revealed by the observation of C2 signals for carbonyl groups near δ 215 ppm. These minor complexes, in which the sites of chelation of the ketoses involve the tetradentate O3,4,5,6 systems, are similar to the complexes of D-mannitol (type E) and D-glucitol (type Q), respectively.

iii. Complexes of Mo^{VI} and W^{VI} with Cyclic Aldoses. The Mo^{VI} and W^{VI} complexes of aldoses of the lyxo series in cyclic form are remarkably stable (225). Although crystallographic data were available for a solid dinuclear Mo^{VI}–lyxofuranose complex (216), the major complex of lyxo aldoses in solution (referred to as type L) was repeatedly suggested to contain the ligand in the *p* form (225, 241, 257, 258, 268). However, subsequent ^{13}C NMR studies demonstrated that D-lyxose, D-mannose, and L-rhamnose are indeed in *f* form with a tetradentate O1,2,3,5 site of chelation. For these dinuclear complexes, the ^{95}Mo NMR spectra show two broad signals (210), whereas the ^{183}W NMR spectra contain two sharp signals with chemical shifts in the range −68.0 to −90.9 ppm (Table X). Furthermore, 2D ^{1}H–^{183}W NMR data demonstrated that, besides the ring O1,2,3 atoms, the O5 atom of the side chain is involved in chelation (277). Thus, the crystal structure (216) of the Mo^{VI}–lyxose complex (Fig. 19) is retained in solution and the suggestion that the Mo^{VI} complex of D-gulofuranose would be unique in utilizing a tridentate O1,2,3 site (276) should be questioned.

Each anomer of the lyxo furanose sugars, acting as a tetradentate ligand, forms a distinct type of W^{VI} and Mo^{VI} complex (277, 278). Complexes of type L are obtained when the anomeric HO group is cis to both ring HO groups. In the second type (type M) characterized with mannose and mannoheptose (D-*glycero*-L-*manno*-heptose), the ligand is the anomer (Fig. 20) in which HO1 is trans to the ring HO2,3 and thus is nonchelating. The site of chelation

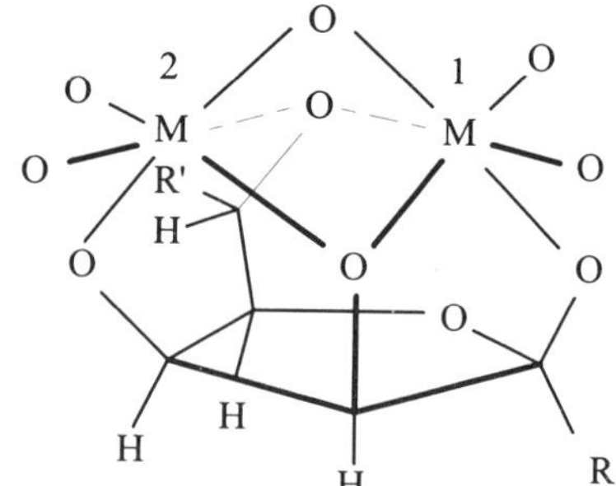

Figure 19. Structure of the Mo^{VI} and W^{VI} complexes of type L formed with tetradentate sugars of lyxo configuration (R = H, aldoses; R = CH_2OH, ketoses) (216, 274).

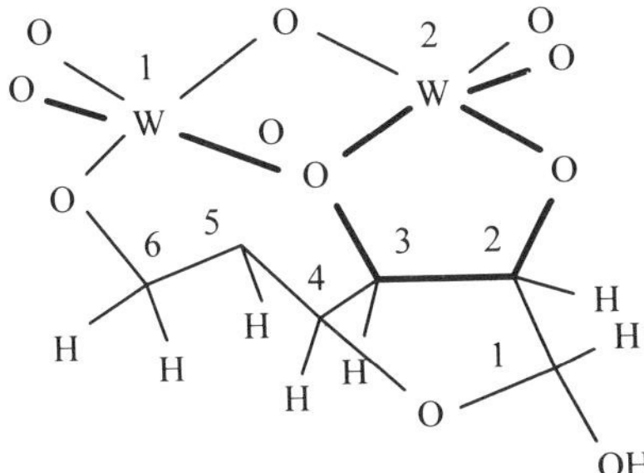

Figure 20. Structure of the W^{VI} complex of type M formed with tetradentate (O2,3 : 5,6) α-D-mannofuranose (278).

O2,3 : 5,6 involves two adjacent hydroxyl groups of the side chain. The corresponding ^{183}W NMR data are given in Table X. All complexes of types L and M contain the ligand in furanose form. It documents the stabilization of forms of the sugar that are not normally prevailing in solution.

Sugars of the ribo series afford two types of Mo^{VI} complexes. The major complexes involve the cyclic sugar (257, 272) and minor complexes contain the acyclic form (275). For the major Mo^{VI} complexes of ribose (272), allose, talose, and L-*glycero*-L-*talo*-heptose (257), NMR data rule out the sugars being present in acyclic or in *f* forms. Therefore, the sugars must be in *p* form and act as tridentate (O2,3,4) donors. Examination of molecular models suggests that the sugars adopt a boat conformation described as $B_{1,4}$ for D-ribose and $^{1,4}B$ for D-talose (257).

iv. ***Mo^VI and W^VI Complexes of Cyclic Ketoses.*** Complexes of the cyclic forms only exist with ketoses of the lyxo and ribo series, in which the HO3,4 system is erythro. Thus, the report that ribo, lyxo, xylo, and arabino ketohexoses form a common type of Mo^{VI} complexes in which the ligands are in *p* form and chelate Mo^{VI} through the O1,2,3 system (241) is erroneous. The formation of Mo^{VI} complexes involving the ketoses in *f* form with the O1,3,4 site of chelation (type K) was demonstrated by a NMR study of the complex of L-*erythro*-pentulose, which cannot adopt the *p* form (274). The Mo^{VI} complexes of ketoses that contain an erythro HO3,4 diol system were shown to contain the α anomer of D-tagatose and the β anomer of D-psicose (274). The ketoses of the lyxo series (D-tagatose and D-*manno*-heptulose) do not afford complexes of type K, but yield Mo^{VI} and W^{VI} complexes of the tetradentate *f* forms analogous to types L and M (277) observed with lyxo aldoses. The formation constants of the Mo^{VI} and W^{VI} complexes of ketoses decrease in the following order: type L > type M > type K > complexes of acyclic species (239, 274, 277).

e. Mo^{VI} and W^{VI} Complexes of Sugars Acids. Multinuclear NMR studies have established that sugar acids readily bind Mo^{VI} and W^{VI} at the bidentate CHOH—COOH system, forming bis-chelated complexes analogous to those of general formula $MO_2[R—CH(O)—COO]_2$ obtained with simple α-hydroxy-acids (279). In acidic medium, these 1 : 2 complexes are generally stronger than those of alditols (280–282). However, a polarographic study of the Mo^{VI}-gluconate complexes (283) indicated other stoichiometries, since at pH 2, the Mo^{VI} and Mo^{V} complexes are Mo_2L species, whereas at pH 4, a 1 : 1 Mo^{VI}-gluconate complex is formed. This 1 : 1 complex is reduced in three monoelectronic steps to Mo^{V} and Mo^{IV} complexes prior to formation of Mo^{III}. The surprising formation of Mo^{IV} may be explained by its stabilization as a gluconate complex.

At pH ≥ 7, aldaric acids $HOOC—(CHOH)_4—COOH$ form Mo^{VI} and W^{VI} complexes analogous to those of alditols, in which the two carboxylate groups are not coordinated to the metal ion. The nature of these M_2L complexes depends on the configuration of the tetradentate O2,3,4,5 site of chelation. In the D-glucarate complexes of type Q (280), the site of chelation is a gluco system in zigzag conformation. The galactarate complexes of Mo^{VI} (281) and W^{VI} (282) involve a galacto system in sickle conformation and belong to type E. The mannarate complexes of Mo^{VI} (281) and W^{VI} (282) involve a manno system in zigzag conformation, and thus belong to type Q. This manno site of chelation is symmetrical, as indicated by the ^{13}C CIS patterns: Δδ 6-13-13-6 for Mo^{VI} and 6-12-12-6 for W^{VI}. Accordingly, single NMR signals are found for the equivalent Mo (δ 23) and W (δ −84.9) atoms. This observation is particularly interesting because chelation of Mo^{VI} and W^{VI} by the manno site of D-mannitol had not been previously reported (211). Thus, for the mannarate ligand, complexation at the manno site is forced by the absence of more suitable sites.

In the case of D-*xylo*-5-hexulosonic acid, $HOOC—(CHOH)_3—CO—CH_2OH$, a ketoacid, which only exists in *f* forms, four Mo^{VI}, and six W^{VI} complexes were identified by multinuclear NMR spectroscopy (284). In acidic medium of pH 3, the Mo^{VI} and W^{VI} complexes are pairs of 1 : 2 species in which the ligand, in acyclic keto form, acts as an α-hydroxy acid chelating by O1,2. At pH > 5, Mo^{VI} and W^{VI} form mixtures of complexes involving the *f* forms and the site of chelation is O1,3,5,6. At pH > 5, two W^{VI} complexes of the *f* forms were detected, in which the COOH group is not chelating. At pH > 7, another W^{VI} complex involves the tetradentate ligand (O1,2,3,4) in acyclic keto form (284). All W^{VI} complexes formed at pH > 3 need structural characterization.

With alduronic acids, $HOOC—(CHOH)_4—CHO$, such as D-galacturonic acid, W^{VI} and Mo^{VI} are bis-chelated by the CHOH—COOH system (O5,6). Complex formation alters the stereochemistry of the ligand, since the chelating OH group adjacent to the carboxylate group is no longer free for hemiacetal formation, and the sugar acid is shifted from the normal *p* forms into the *f* forms (50, 285). This result documents the fact that the furanoses or acyclic forms of carbohydrates form stable complexes upon coordination with metal ions.

3. *The Epimerization of Aldoses Catalyzed by Mo^{VI}, Ni^{II}, and Other Metal Ions*

Bilik and co-workers (286–291) discovered in 1971 that molybdic acid catalyzed the epimerization (i.e., inversion of configuration) of aldoses at C2 in aqueous solution. This slow reaction (2–13 h at 95°C) represents a simple method for the preparation of rare sugars such as D-talose (292), L-fucose (293), and L-talose (294) from natural aldoses. An equilibrium mixture is obtained in which the threo HO2,3 epimer prevails over the erythro epimer, corresponding to the relative stabilities of the uncomplexed aldoses. This remarkable epimerization in acidic medium is completely different from the well-known isomerization of aldoses and ketoses, known as the "Lobry de Bruyn-Alberda van Ekenstein rearrangement," which proceeds in alkaline medium and involves a 1,2-enediol intermediate (295). The epimerization reaction is also catalyzed by dioxobis(2,4-pentanedionate-O,O′)molybdenum(VI), "molybdenyl acetylacetonate" in DMF at 50°C (296).

The mechanism was initially believed to involve C1,2 hydride shift, because isotopic exchange was demonstrated using ^{3}H-labeled glucose (297), until a ^{13}C NMR study revealed that 1-[^{13}C]-aldoses yield their 2-[^{13}C]-epimer (298). The generally accepted mechanism (298, 299) postulates an intermediate Mo^{VI} complex in which the sugar is bound by O1,2,3,4. Once C1,2,3 are in suitable tricyclic arrangement, a bond shifts between C1,3 and C2,3 with C1/C2 transposition (Fig. 21). Epimerization is induced because the configurations of the initial and final C2 of the ligand are opposite.

The kinetics of the glucose–mannose interconversion were studied. The maximum rate was observed at pH 2.5 (300). The rate of reaction of mannose was slower than that of glucose, which is consistent with the Mo^{VI}–mannose complex being less reactive than the Mo^{VI}–glucose complex. It suggests that the stable Mo^{VI} complexes of acyclic or cyclic aldoses discussed above are inert with respect to epimerization, and that the catalytic molybdate (or polymolybdate) complexes of aldoses must be of a different complex type.

The reaction initially was believed to be specific for Mo^{VI}, since W^{VI} is not a catalyst, which represents an interesting difference between these metal ions. However, similar, faster epimerizations (5 min at 60°C, 30 min at 25°C) were later found to occur with Ni^{II} complexes of substituted ethylenediamines in methanolic solution (301, 302). In the absence of diamine, Ni^{II} does not interact with the aldoses and no epimerization occurs. A ternary complex of Ni^{II} is formed, containing a *N*-glycoside bond between the erythro isomer of the sugar

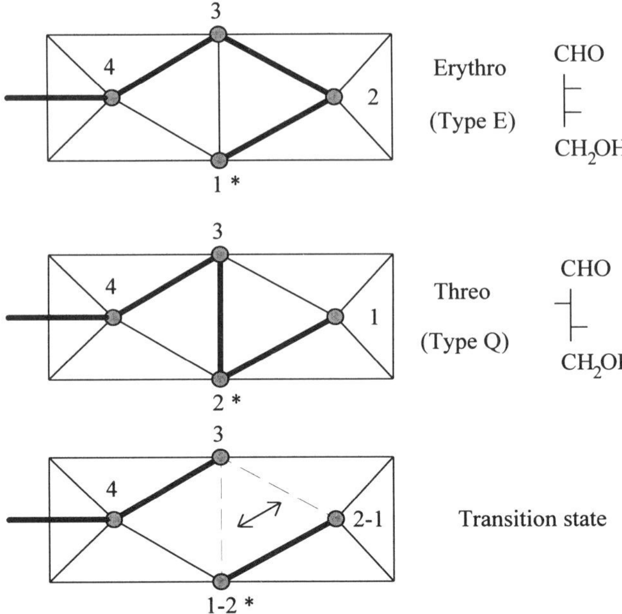

Figure 21. Epimerization of aldoses catalyzed by molybdic acid. The intermediate dinuclear Mo^{VI} complexes of the hydrated aldoses are represented from above (see Figs. 11 and 12). Bond transfer takes place within the trinuclear C1,2,3 system. An asterisk (*) shows the carbon atom marked for demonstration of C1/C2 exchange. [Adapted from (298).]

and the diamine (303). However, the finding that *N*,*N*,*N*′,*N*′-tetramethylethylenediamine, which cannot form *N*-glycosides with aldoses, is an efficient catalyst demonstrates that the epimerization reaction does not involve intermediate *N*-glycosides (302). The site of chelation of the aldose was defined by testing various deoxy-aldoses, showing that HO1,2, and 4 were essential, but not HO3. As in the molybdate reaction, evidence for C1/C2 transposition was obtained in the Ni reaction by using 1-[^{13}C]-aldoses (302, 304). From EXAFS data, the intermediate complex is believed to have a mononuclear octahedral structure in which the Ni atom is bound to two solvent molecules, one molecule of diamine and the chelating O1,2,4 site of the aldose in the aldehyde form. The mechanism of this reaction presents similarities with a stereospecific pinacol-type 1,2-carbon shift reaction (302). Electron withdrawing at C1, which is anchored to Ni and the terminal amino group, would produce an electron-poor carbon atom, thus allowing bond transfer within the C1,2,3 system (Scheme 5a). A likely structure of the transition state is represented in Fig. 22 (302). Using long-chain *N*-alkylated en ligands for the chelation of Ni^{II} resulted in the formation of "metallomicelles" with increased catalytic activity (305).

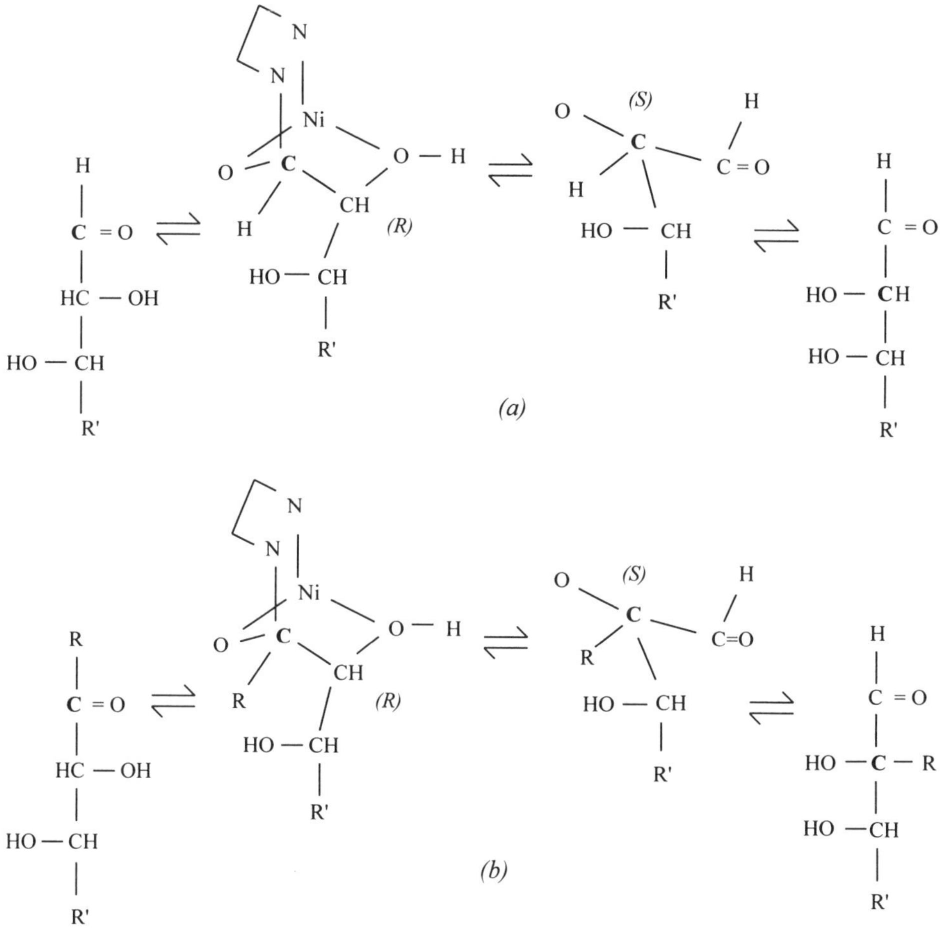

Scheme 5. Reactions catalyzed by Ni^{II}–en complexes. Simplified schemes showing only the isomerization of the sugars. Here R′ is the rest of the sugar molecule. The carbon of the carbonyl group is marked in boldface. (*a*) Epimerization of aldoses. (*b*) Branching reaction of ketoses (R = CH_2OH). [Adapted from (302).]

Similarly, 2-ketoses, in the presence of amino complexes of Ni^{II}, undergo a very stereospecific rearrangement (Eq. 3) yielding branched sugars (306, 307). The mechanism shown in Scheme 5b resembles that for the epimerization reaction.

$$HOCH_2-(CHOH)_{n-4}-CHOH-CO-CH_2OH \longrightarrow$$

$$HOCH_2-(CHOH)_{n-4}-C(OH)(CH_2OH)-CHO \quad (3)$$

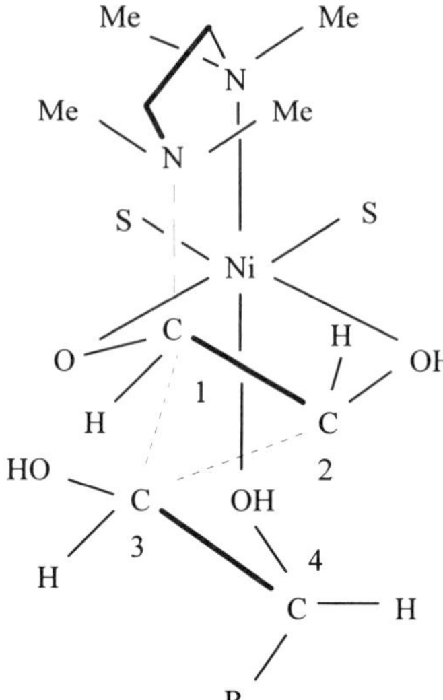

Figure 22. The transition state for the epimerization of aldoses by the Ni^{II}–tetramethylethylenediamine complex. [Adapted from (302).] [R is the rest of the aldose molecule and S is the solvent (methanol).]

This reaction was applied to fructose, psicose, tagatose, and sorbose for the preparation of the four corresponding 2-*C*-(hydroxymethyl)aldoses (307). It suggests that the reactions of ketoses with Mo^{VI} would need further studies in order to detect a similar rearrangement.

The epimerization reaction is now known to occur with other metal ions including Ca^{II}, Co^{II}, Sr^{II}, Pr^{III}, and Ce^{III} in the presence of *N*-alkylated ethylenediamines (308–311). Studies of the complexes of transition metal ions with *N*-glycosylamines prepared by the reaction of a diamine and a monosaccharide have possible relevance to this reaction (16, 19).

Other systems containing the calcium cation and *N*-alkylated monoamines (e.g., triethylamine) in methanol were found to be synthetically useful for the epimerization of reducing (1 → 4) and (1 → 6) disaccharides (312). Moreover, aldoses such as xylose and glucose are also epimerized by $Ca(OH)_2$ in aqueous solution of pH 12.5, in the absence of amines. The reaction of glucose affords small amounts of fructose and occurs with C1,2 rearrangement. In contrast, with alkali metal ions, Sr^{II} and Ba^{II}, the epimerization of glucose to mannose produced large amounts of fructose, indicating that the enediol rearrangement was the prevailing reaction (313). Experiments with other metal hydroxides such as Al, Co, Ni, Cu, and Mg in aqueous alkaline medium were not successful, suggesting that the reaction may be specific to Ca^{II} ions (313).

Several points can be made from the above studies. It is clear that several mechanisms or catalytic systems with very different geometries are now known.

The possibility that one type of metal–sugar geometry explains all observations can be safely ruled out. In all systems, the function of the metal ion is probably to organize the C1,2,3 chelating center in which the bond transfers occur in a specific geometry (Figs. 21, 22, and Scheme 5).

However, Sunjic and co-workers (314) pointed to catalysis by the heptamolybdate ion, by suggesting that the rate-limiting transition state for epimerization may not only be dictated by the steric strain imposed on the aldose by complexation, but also by the d–π interaction between the central Mo atom of the $[Mo_7O_{24}]^{6-}$ ion and the carbonyl group of the acyclic aldose, leading to a stabilized transition state. Several kinetic studies of the reversible epimerization reactions have been performed, showing typical Michaelis–Menten behaviors, which reveal the formation of intermediate complexes (231, 315, 316). The epimerization reaction is favored when acetonitrile (80% v/v) is added to the aqueous medium (317). This result suggests that the rate-determining transition state is less polar than the reagents and/or the initial metal–carbohydrate complex.

4. *The Mutarotation of Aldoses Catalyzed by* Mo^{VI}, W^{VI}, *and Other Metal Ions*

The mutarotation of α-D-glucose (anomerization at C1) is catalyzed by metal ions, which assist the electrophilic attack by association with the oxygen atoms on the anomeric carbon and supply a correctly oriented water molecule (318–320). The reaction with Cu^{II} ions was studied in detail. The measurements of the volumes of activation supported the expectation that the aqua metal ion was a better proton donor than water alone (319). The dimeric species $[Cu_2(OH)_2]^{2+}$ is by far a more efficient catalyst than Cu^{2+} ions, in agreement with the finding that copper complexes of carbohydrates are dimers (321). The mutarotation of α-D-glucose is also subject to general base catalysis by oxo ions (319). The efficiency of catalysts decreases in the following order: tetraborate (pH 8.4) > heptatungstate (pH 6.3) > heptamolybdate (pH 5.2) > tungstate (pH 8.0) > molybdate (pH 8.0). Plots of the rate constants versus the catalyst concentrations were almost linear in the presence of tetraborate, tungstate, and molybdate. In contrast, both polyanions gave curved plots revealing an initial complex-forming step and saturation kinetics. The kinetic data fit Lineweaver–Burke plots. Formation constants K were determined for the 1 : 1 complex between glucose and heptatungstate (K = 530) or heptamolybdate (K = 255) (320). In view of the known structure of the Mo^{VI}–glucose complex, which contains the acyclic hydrated sugar (39, 258), the catalysis of mutarotation may be due to the facilitation of the ring-opening reaction of glucose by the oxometalate.

5. *Uranium*

Although the analogy of U with Mo and W is limited (27), the complexes of these elements are often compared. An important difference is that the dioxo groups of $[M^{VI}O_2]^{2+}$ are cis for Mo and W, but trans in the linear uranyl ion $[UO_2]^{2+}$. A stable 1:1 uranyl–gluconate complex was characterized by photometry and polarimetry (322). The uranyl complexes of D-mannose and D-ribose were studied by ^1H and ^{13}C NMR spectroscopies in D_2O at pH > 10 (272). It was concluded that the bidentate sugars were in *β-p* (mannose, 1:1 complex) or *p* and *f* forms (ribose, 1:1 and 1:2 complexes). Various sugar derivatives form complexes in strongly alkaline solution with uranyl ions immobilized on a cation-exchange resin (323). Such columns were used for ligand-exchange chromatography of carbohydrates.

F. Group 7 (VIIB) Metal Ions

Transition metals of Group 7 (VIIB) include Mn, Tc, and Re. Most studies were devoted to manganese–carbohydrate complexes, which are of biological interest. The interest of technetium complexes lies in the use of ^{99m}Tc isotope in nuclear medicine as imaging agent. Among many compounds marketed for the imaging of kidneys, the single carbohydrate complex is ^{99m}Tc–gluceptate, which is believed to be a Tc^V complex of D-glucoheptonate (324). In the proposed structure, a Tc^V monooxocore is chelated by two glucoheptonate ions acting as bidentate ligands (O1,2). This weak complex is often used in ligand-exchange experiments for the syntheses of more stable Tc^V complexes (324).

Complexes of Re with carbohydrates do not seem to have been characterized. However, this element can be chelated by 1,2-diols since stable Re^V complexes containing ethyleneglycol and a cyclic triamino ligand (1,4,7-triazacyclononane, L) were prepared by reaction of $[ReOCl_3(PPh_3)_2]$ with both ligands in dry THF. These complexes contain the blue $[LReO(O_2C_2H_4)]^+$ cation and were isolated as halide salts. A crystal structure has been reported for the bromide compound, confirming the chelation of the ReO group by both L and ethyleneglycol (325). Another complex of Re^V with 1,3,5-trideoxy-1,3,5-tris((2-hydroxybenzyl)amino)-*cis*-inositol, a derivative of taci, has been prepared from the same Re precursor. An X-ray structure determination demonstrated that the ReO group is chelated by N1,3 and O2 of the inositol ring. Two oxygen atoms of the phenol groups attached to N1,3 complete the octahedral (N_2, O_4) coordination shell of Re^V (326).

1. *Electrochemical Studies of Manganese–Carbohydrate Complexes*

Many studies have investigated the electrochemical behavior of carbohydrate–manganese complexes, since the significant role of manganese in several biological systems is probably due to the involvement of the II, III, and IV oxidation states in electron-transfer reactions. Dolezal and Gürtler (327) performed extensive polarographic studies of the complexes of alditols such as D-mannitol (327), D-glucitol and galactitol, with Mn^{II} and other cations (328). In an alkaline (1 *M* KOH) solution of D-glucitol, many metal ions such as Mn^{II}, Tl^{I}, Cd^{II}, Pb^{II}, Zn^{II}, and $[UO_2]^{2+}$ give reproducible waves (328). Generally, two anodic waves are observed for the Mn^{II}–alditol complexes, which are thus oxidized to the Mn^{III} and Mn^{IV} complexes. The stoichiometry (MnL_2) and stability constants (β_2) of the Mn^{II}–alditol complexes were determined (Table XI) by displacing the Mn^{II}–edta complex with increasing amounts of alditols, showing that the complex of edta is weaker than those of alditols. By considering the stability sequence, it is assumed that threo diol systems are favorable sites of chelation. The proposed sites are given in Table XI. The results were compared with that obtained for *myo*-inositol by the same method (329).

2. *Solid-State Structures*

The crystal structures of Mn^{III} and Mn^{IV} complexes of simple α-hydroxyacids such as lactic acid (H_2Lac) were determined (330). These structure are likely models for the polyol complexes. 2-Hydroxyisobutyric acid, H_2HIB, forms a tris-chelated anion $[Mn^{IV}(HIB)_3]^{2-}$ in which Mn^{IV} is octahedral (330). This species is similar to two tris(catecholato) manganese(IV) anions (331, 332). A lactate complex of Mn^{III} containing the $[Mn_2^{III}(Lac)_4(HLac)]^{3-}$ ion was also characterized (330). The two Mn atoms were doubly bridged by the alcoholic oxygens of two lactate ions, and by the carboxylate group of the hydrogen lactate ion. Each Mn^{III} ion has two bidentate lactate dianions bound in a *cis*-propeller configuration.

Solid complexes of Mn^{II} with D-galactose, D-fructose, D-glucose, D-xylose, D-ribose, and maltose were synthesized (333) by reacting Mn^{II} salts with sugar

TABLE XI
Stability Constants of the MnL_2 Complexes of Alditols and the Proposed Sites of Chelation[a]

Alditol	Log β_2	Proposed Site
Glucitol	16.29[b]	O2,3,4
Mannitol	16.02[b]	O2,3,4
Galactitol	15.94[b]	O1,2,3
myo-Inositol	12.06[c]	nd

[a]The stability constant of the MnL_2 complex is β_2 ($L^2\ mol^{-2}$), according to the reaction: $Mn^{II} + 2\ L \rightleftharpoons MnL_2$, where L is the alditol. Not determined = nd.

[b]Reference (328).

[c]Reference (329).

anions formed by a strong base in methanol and were studied by cyclic voltammetry. Compounds with a 2:2 stoichiometry are obtained in nonaqueous solutions of $[TEA]_2[MnCl_2Br_2]$, but 3:3 compounds are produced when the reagent is $MnCl_2 \cdot 4H_2O$. Further structural characterization was not performed.

3. *Solution Studies*

A complete study in solution of the Mn^{II}, Mn^{III}, and Mn^{IV} complexes of many polyols and aldaric acids has been made by Sawyer and co-workers (334) who also reviewed earlier literature. The Mn^{III} complexes exist as two types. Oxidation of the 1:2 complex of Mn^{II} yields the Mn^{III_A} complex, which is believed to be a 2:4 species on the basis of its low magnetic moment, attributed to electron pairing. Then the Mn^{III_A} complex rearranges to the Mn^{III_B} complex, which is probably a 1:3 species with a structure analogous to that of the Mn^{IV} complex. Results with polyols and aldaric acids parallel those obtained with D-gluconic acid (335). The D-glucitol complexes were studied in more detail with similar results (336). Since a dimerization equilibrium was demonstrated for the Mn^{III_A}–glucitol complex, three Mn^{III} species indeed exist. UV–vis spectral data were reported for the $Mn^{III}L_2$, $Mn_2^{III}L_4$, $Mn^{III}L_3$, and $Mn^{IV}L_3$ complexes. A dinuclear Mn^{III} complex with β-cyclodextrin (β-CD) was isolated from a DMF–ethanol solution and characterized by various spectroscopic methods (337). The bis(μ-hydroxo)-bridged cation $[Mn_2(OH)_2]^{4+}$ cation is believed to be bis-chelated within the β-CD internal cavity.

A kinetic study of the oxidation of monosaccharides by Mn^{VI} has been reported (338). On the other hand, D-fructose reduces Mn^{III} to Mn^{II} in a light-induced reaction. Since Mn^{II} is reoxidized by atmospheric oxygen, the overall process is the oxidation of fructose by oxygen in the presence of catalytic amounts of manganese ions (339). This reaction is also catalyzed by iron ions and its mechanism is discussed in the iron section. Complex formation may also occur with oligosaccharides.

G. Groups 8, 9, 10 (VIII) Metal Ions

Groups 8, 9, 10 (VIII) transition metals include Group 8 (VIII) containing Fe, Ru, and Os; Group 9 (VIII) containing Co, Rh, and Ir; and Group 10 (VIII) containing Ni, Pd, and Pt. As in the cases of other transition metals, most work has been carried out with the first row transition metals, Fe, Co, and Ni. The Ni^{II} complexes are of particular interest, because they are involved in useful epimerization reactions and are discussed in this section. Limited information is available on metal ions of the platinum group, however, these metal ions are soft Lewis acids and afford organometallic species described at the end of the chapter (Section II.M).

1. *Iron*

The crystal structure of a Fe^{III} complex of 1,3,5-triamino-2,4,6-trihydroxy-cyclohexane, taci, containing the $[Fe(taci)_2]^{3+}$ ion, reveals that one of the ligands binds the ferric cation by three oxygen donors, whereas the second is coordinated via three nitrogen donors (340). This documents the existence of complexes of Fe^{III} with cyclic ligands containing a 1,3,5-triol system. However, studies of Fe^{III} complexes with selected diols and glycerol demonstrate that the most stable complexes are formed between Fe^{3+} and adjacent diols. Both glycerol-1-methylether and 1,3-propanediol form only very weak complexes if any, whereas different 1,2-diols form stable complexes (341). In alkaline solution ($[OH^-] > 0.1\ M$) the mononuclear Fe^{3+}–1,2-diol complexes have a 1 : 1 stoichiometry as determined by UV–vis spectroscopy and magnetic measurements, while Fe^{3+} reacts with D-glucitol (sorbitol) to give both mononuclear and dinuclear complexes (342, 343). Upon acidifying an alkaline solution of Fe^{III}, precipitates of hydrolyzed ferric complexes form. However, when glycerol or sorbitol are used as ligands, the polynuclear complex remains in solution. The polynuclear material of Fe^{III}–sorbitol has a Fe/sorbitol ratio of 1 : 1 and consists of an iron oxide–hydroxide core, which is surrounded by a shell of polyols coordinated on the surface of the core. The carbohydrate OH groups presumably enhance the solubility (341). The EXAFS, Mössbauer, and EPR measurements have been used to study the fructose, mannitol, or gluconate complexes of iron(III). A dinuclear model structure was proposed (Fig. 23), which is not only based on EXAFS and XANES results (48), but is also supported by Mössbauer and EPR measurements (49).

Iron(III) complexes of glucose, mannose, galactose, fructose, maltose, lactose, sucrose, and glycine esters of D-glucopyranosyl esters have been isolated and characterized in the solid state by elemental analysis, IR, and Mössbauer spectroscopy, and magnetic susceptibility methods (344, 345). These com-

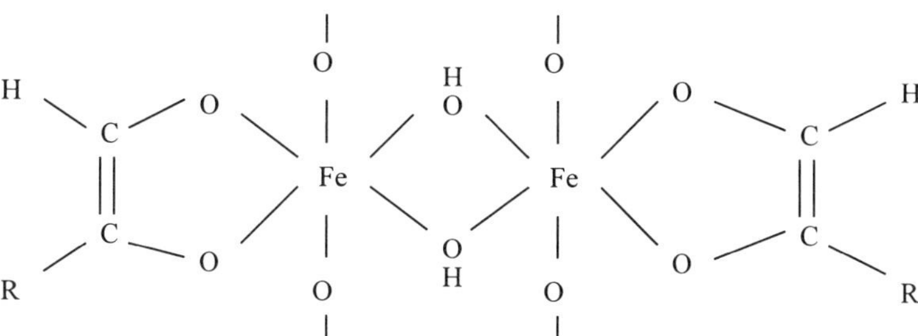

Figure 23. Proposed structure for the Fe^{III}–alditol complexes, showing the enediol complexing site, where R is the rest of the alditol molecule. [Adapted from (48).]

plexes were reported to have varying stoichiometries. In solution, most studies indicate that the complexes are mononuclear (1 : 1 or 1 : 2) or dinuclear (2 : 4) at pH > 13, whereas polymeric complexes exist at lower pH (341).

The chromatography of mixtures of carbohydrates has been studied on cation-exchange resins loaded with Fe^{III} ions (346). The mobile phase was 0.1 *M* NaOH. The results were compared with those obtained using columns of immobilized Ln^{III} ions, namely, Eu (346) and Tb (347).

The photochemically induced oxidative degradation of monosaccharides by iron(III) has been investigated in aerobic conditions (339, 348, 349). The reaction occurs in alkaline medium. Iron(III) is reduced to iron(II), which is eventually reoxidized by dissolved oxygen. Consequently, the process can be used for the oxidation of sugars by atmospheric oxygen, catalyzed by the Fe^{III}/Fe^{II} couple (Scheme 6).

The degradation processes are selective depending on the isomer and conformation of the sugar. In the presence of Fe^{III}, D-fructose (Fru), and D-arabinose were degraded to D-erythrose. D-Glucose and D-mannose were first degraded to D-arabinose, which was subsequently degraded to D-erythrose, while D-ribose was directly degraded to both D-erythrose and D-glyceraldehyde. These reactions were proposed to proceed via the formation of Fe^{III}–monosaccharide complexes. For fructose, the reaction is very specific and only the C2–C3 bond is cleaved, which presumably reflects the coordination of Fe^{III} with HO2,3 in the β-D-fructofuranose form, when the two hydroxyl groups are cis to one another. This is, however, inconsistent with a recent report using ^{13}C NMR spectroscopy to examine the Fe^{III}–fructose complex, which concluded that the ligand was in β-*p* form (46). The reactions of all aldoses and the ketose in the presence of Fe^{III} involve cleavage near the anomeric carbon. Besides, for ribose, cleavage also affects the C2–C3 bond, not involving the anomeric carbon. The cleavage pattern for ribose is attributed to the fact that HO2,3 are cis in ribofuranoses, favoring metal coordination, while the corresponding hydroxyl groups are trans in arabinose. The nature of the fructose–Fe^{III} complex in acidic (pH 1–3) solution was examined (350) and by assuming the formation of a 1 : 1 complex (Eq. 4), the equilibrium constant at *t*, 20°C is log $K = -2.8$.

$$Fe^{3+} + \text{Fru} \rightleftharpoons \text{Complex} + H^{+} \tag{4}$$

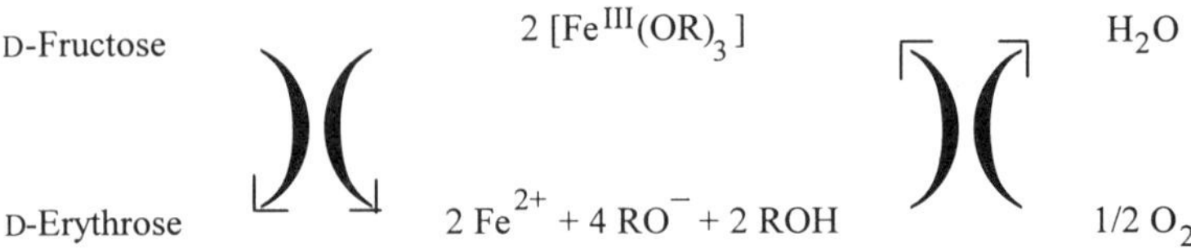

Scheme 6. Photochemically induced oxidation of fructose to erythrose by Fe^{III} with regeneration of Fe^{III} by reduction of oxygen. [Adapted from (339).]

A similar reaction occurs with Mn^{III} and was discussed in the manganese section (Section II.F) (339).

2. *Ruthenium*

A limited number of complexes of ruthenium with carbohydrate derivatives have been examined (351–353). The Ru^{III} ions have been shown to be strongly complexed by sodium gluconate (351). The complex is assigned a 1 : 1 stoichiometry based on spectrophotometric and polarographic studies. No structural information is available, yet the coordination likely involves the carboxylate group (351).

A complex was isolated from the reaction between the cluster $Ru_3(CO)_{12}$ and 1,2-*O*-isopropylidene-α-D-glucofuranose (H_2L) (352). Its crystal structure revealed the formula $Ru_3(CO)_8(L)$. The formation of this complex involved the HO3 group since no corresponding complex could be obtained from 1,2 : 5,6-di-*O*-isopropylidene-α-D-allofuranose in which the C3 configuration is reversed (352). Mixed-ligand complexes of Ru^{III}–edta with nucleosides were examined (353). One of the nucleoside coordination sites is assigned to the N atom of the base as determined by ^1H NMR spectroscopy. When the nucleoside is cytidine, Ru is coordinated in a bidentate manner through N3 and HO2 of the base (353).

3. *Osmium*

Two kinds of Os^{VI} glycolates have been reported both in solid state and in solution: One complex has a 2 : 2 stoichiometry [Fig. 24(*a*)], while the other has a 1 : 2 stoichiometry [Fig. 24(*b*)] (354–356). The 2 : 2 complexes are prepared by reacting osmium tetraoxide with alkenes (tetramethylethylene, cyclopentene, 2,3-dimethylpent-2-ene, camphene, and cholesterol) in nonaqueous solution, and can generate cis-1,2-diols upon hydrolysis (357, 358). The synthetic utility of this reaction for the synthesis of 1,2-diols has been reviewed (359). The X-ray crystal structure of the pinacol complex $\{OsO_2[Me_2(O)C{-}C(O)Me_2]\}_2$ shows that both Os atoms are in a square pyramidal geometry. The Os atoms are linked by two bridging oxygen atoms, and coordinated in a bidentate manner with one 1,2-diol to form a five-membered ring (360). The ^1H NMR spectroscopy, IR in solid state and solution, and molecular weight measurements of these complexes suggest that the dimeric structure is retained in solution. The 1 : 2 complexes are prepared by the slow reaction of OsO_4 with ethyleneglycol or pinacol in carbon tetrachloride (360), or alternatively by reaction of $K_2[OsO_2(OMe)_4]$ with pinacol followed by acid hydrolysis (358). As proposed by IR studies, the solution structure of these complexes remains that observed by X-ray crystallography (361). Thus, Os^{VI} is in a five-coordinated square pyramidal geometry coordinated by two bidentate 1,2-diol ligands. The possible structures of intermediate diol complexes generated during the os-

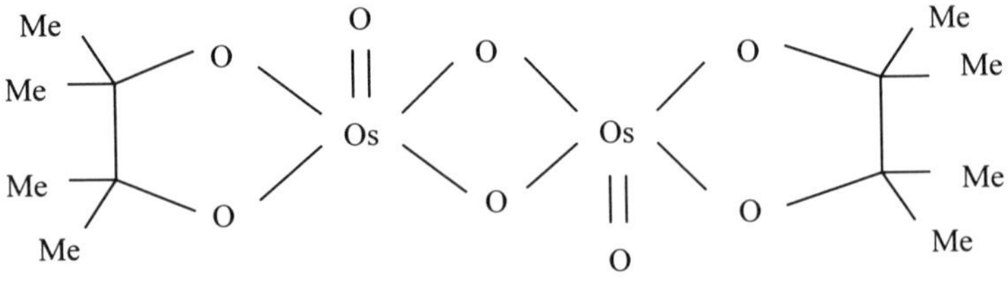

(*a*)

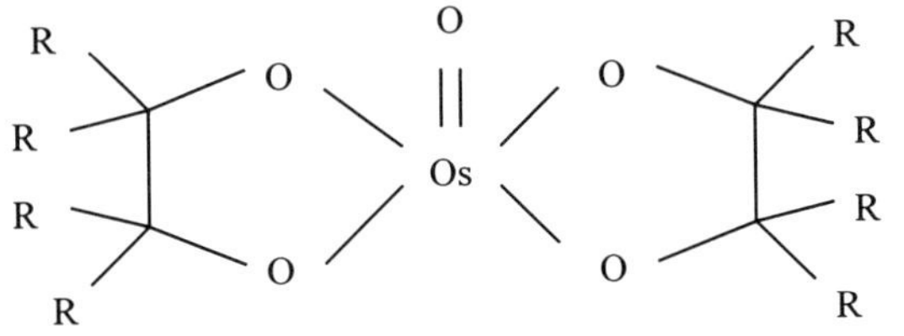

(*b*)

Figure 24. Schematic drawing of (*a*) the structure for $[Os_2O_4(O_2C_2Me_4)_2]$ and (*b*) the structure for $[OsO(O_2C_2R_2)_2]$. [Adapted from (355).]

mium-catalyzed dihydroxylation reaction of alkenes have been discussed elsewhere (359).

The Os^{VI}, Os^{IV}, and Os^{III} ions form complexes with sodium gluconate in alkaline solution (362). The redox chemistry of these species is essentially reversible. All these complexes have been assigned a 1 : 2 osmium/gluconate stoichiometry by UV studies. Although structural determinations have not been carried out, presumably the Os^{VI} species have similar structures as the Os^{VI}–cis-diol complexes.

4. *Cobalt*

The Co^{II} complexes of carbohydrates do not differ from those of other divalent cations. On the other hand, Co^{III} has a strong affinity for nitrogen donors and generally forms mixed-ligand complexes containing amines and the carbohydrate surrounding the octahedral Co atom. The crystal structures of Co^{III} complexes containing two en ligands and a carbohydrate molecule have been reported. In the erythritol-, methyl α-D-mannopyranoside- and methyl β-D-galactopyranoside-Co^{III} complexes, the sugar molecule is deprotonated twice and chelates the Co atom at a trans-1,2-diol site (363).

The complexes of Co^{II} and Co^{III} with D-mannitol were studied by spectrophotometry in alkaline aqueous solution (KOH, 0.1–4 *M*) (364). The 1 : 1 mannitol–Co^{II} complex is oxidized by air in 20 min to the 1 : 1 olive-green mannitol–Co^{III} complex. In the blue Co^{II} species, the Co atom is tetrahedral. The stability constant of CoL^{2+} is log $\beta_1 = 5.3$ in 4 *M* KOH (365). In alkaline (3 *M* KOH) solutions containing 0.5 *M* D-mannitol, Co^{III} and Mn^{IV} are reduced by Fe^{II}. Thus, the mannitol–Fe^{II} complex can be used as a strongly reducing analytical reagent ($E_{1/2} = -1.15$ V vs. SCE) (365). The determinations included the reductions of Co^{III} to Co^{II} (and also Mn^{IV} to Mn^{III} and Cu^{II} to Cu^{I}), all these metal ions being present as mannitol complexes.

Cage-type complexes of Co^{II} with *N*-glycosides derived from tris(2-aminoethyl)amine and a monosaccharide (mannose or rhamnose) have been synthesized and shown to undergo a very peculiar inversion of configuration around the seven-coordinated cobalt center (366). This reaction was preceded by the reversible addition or removal of a sulfate ion, "just like flowers open and close." This reaction may be the first example of sophisticated stereoselective chemistry based on ligands with a carbohydrate backbone. Many other examples of relevant complexes of amino sugars can be found in reviews by Yano and co-workers (16, 19).

The inorganic complex *cis*-$[Co(NH_3)_4(H_2O)_2]_2(SO_4)_3 \cdot 3H_2O$ reacts with D-ribose and L-sorbose (sugar = L) to give pink solids characterized as $[Co(NH_3)_4(L)_2]_2(SO_4)_2$ (367). Data for optical activity and UV spectroscopy were obtained. The sugars were considered to be bidentate ligands through a cis-diol group. With D-arabinose in aqueous solution of pH 6, two complexes were isolated, but their structures were not determined (368). One is a dimeric 2:2 complex described as a mixed-valence Co^{II}–Co^{III} paramagnetic species. The other is a mononuclear 1 : 1 Co^{III}–arabinose complex of low stability in aqueous medium.

Aldonic acids such as D-gluconic and L-mannonic acids (H_2L) form pairs of diastereoisomeric Co^{III} complexes formulated as $[Co(en)_2(L)]ClO_4H_2O$ in which the chelating sites are the CHOH—COOH systems (369).

5. *Platinum*

Platinum(II) presents a higher affinity for nitrogen than for oxygen donors. Thus, in the complexes formed with diamino-dideoxy-carbohydrates (370, 371), all the ligands chelate Pt^{II} by two adjacent nitrogen atoms only, even when free HO groups were available. The reported examples include glucose, galactose, and mannose derivatives in which HO2,3 were replaced by amino groups (370) and alditols derivatives in which HO2,3 or HO3,4 were substituted (371). Crystalline complexes formulated as $PtCl_2$(sugar–*N*,*N*) were isolated and the crystal structure of *cis*-dichloro(3,4-diamino-3,4-dideoxy-D-iditol-*N*,*N*′)-

TABLE XII
Regioselectivities in the Complex-Forming Equilibria of (dppp)Pt(alditolate) Complexes (373)[a]

Alditol Site of Chelation	Proportion (%) O1,2	Proportion (%) O2,3	Proportion (%) O3,4
Erythritol	89	11 (e)	
Threitol	40	60 (t)	
Ribitol	83	17 (e)	
Xylitol	14	86 (t)	
Galactitol	11	71 (t)	18 (e)
Mannitol	<1	17 (e)	82 (t)

[a]1,3-Bis(diphenylphosphino)propane = dppp; e = erythro; t = threo.

platinum was determined (371). These compounds present antitumor activity and are more soluble in water than cisplatin, the well-known antitumor drug.

However, the existence of platinum(II) alkoxides suggested the possibility of preparing optically active platinum complexes of bidentate (O1,2) glycerol. Thus, the reaction of the bisphosphine racemic complex (R^*,R^*)-{$Pt(OMe)_2[1,2\text{-}C_6H_4(PMePh)_2]$} with glycerol in benzene–methanol solution produced an equilibrium mixture of diastereoisomers, epimeric at O2, that exist in dichloromethane predominantly as internally hydrogen-bonded monomers (372). Crystallization yielded a solid compound characterized as a centrosymmetric dimer composed of asymmetric monomers of opposite helicity. Another study was devoted to the reactions in CH_2Cl_2 of a series of unprotected diols, triols, and alditols with bis(phosphine)platinum(II) carbonate complexes (dppp)Pt(CO_3), where dppp is 1,3-bis(diphenylphosphino)propane (373). An exchange reaction occurred in which carbonate was replaced by the alditol acting as a bidentate ligand through a 1,2-diol group, in agreement with the finding that the complexes of 1,2-diols are much stronger than those of 1,3-diols. Multinuclear ^{13}C and ^{31}P NMR data have been obtained. Alditols with threo diol groups form stronger complexes than those with erythro diol groups (Table XII). The X-ray structure of the mannitolate complex shows that platinum is bound at the O3,4 site (Fig. 25).

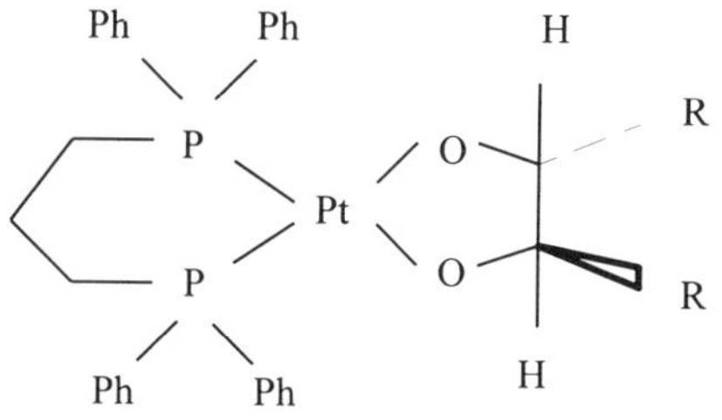

Figure 25. Schematic structure of the (dppp)-Pt(D-mannitolate) complex, where R = $CHOH-CH_2OH$, dppp = 1,3-bis(diphenylphosphino)propane. [Adapted from (373).]

H. Group 11 (IB) Metal Ions

Elements of Group 11 (IB) are Cu, Ag, and Au. Stable Cu^{II} or Ag^{I} complexes cannot be expected to form with reducing sugars, which are readily oxidized by Cu^{II} and Ag^{I} ions in alkaline media. Such reactions have received well-known analytical applications. Moreover, Cu^{II} ions are known to present a higher affinity for nitrogen than for oxygen donors, as shown by the reaction with taci, which yields a solid compound containing $[Cu(taci)_2]^{2+}$ ions in which Cu^{II} is bound to six nitrogen atoms (83). Such stable complexes of Cu^{II} with amino sugars are important in biology and this topic has been recently reviewed (374).

The solid-state structures of the Cu^{II} complexes of nonreducing carbohydrates show analogies with those of Ca^{II} and Pb^{II} complexes, since polymeric species are obtained in which Cu atoms and polyol molecules are mutually bridging, forming polyolatometalates. The crystal structures of the Cu^{II} complexes of β-cyclodextrin (β-CD) (375), erythritol, and galactitol (376) have been reported. They consist of coordination polymers containing the fourfold deprotonated ligand as the building block and protonated polyol molecules that ensure cross-linking of the polymer chain by the formation of hydrogen bonds with the tetraoxocuprate groups (376). The inclusion of Cu^{2+} within the internal cavity of β-CD was studied by ^{1}H NMR spectroscopy. From spin-lattice relaxation time measurements, the average distance between Cu^{II} and the glucose H1,2,3,4,6 was found to be in the range 5–6 Å (377).

Some experimental data indicate that sugar derivatives (alditols, cyclitols, and nonreducing disaccharides) are strongly complexed in Cu^{II}–ammonia solution (378) and more generally in neutral or slightly alkaline media. Two Cu^{II}–mannitol complexes, $[CuL]^{2+}$ and $[Cu_2L]^{4+}$, were studied in alkaline aqueous solution (KOH 0.1–4 *M*) by spectrophotometry (379). The stability constant of CuL^{2+} is $\beta_1 = 6.3 \times 10^7$ in 0.5 *M* KOH. However, the TLC analysis of carbohydrates on plates coated with resins in Cu^{II} form gave contradictory results, depending on the use of alkaline ammonia solutions or neutral copper salts (91, 380). A reexamination of a previous work (91) under comparable conditions (93) confirmed that sugars are poorly complexed in copper sulfate solutions. The discrepancies were explained in a careful work by Angyal (321). The Cu^{2+} ion itself forms only weak complexes, like those of Mg^{2+}, for example. When complexation occurs at pH $\geq$ 5, the active species present in an aqueous solution of copper acetate is believed to be a dimeric $[Cu_2(OH)_2]^{2+}$ cation of yet unknown structure. A likely possibility is that both copper ions are linked by two μ-hydroxo groups. A possible structure for a dinuclear Cu^{II}–inositol complex is shown in Fig. 26. The influence of the stereochemistry of the polyol on the stability of the complexes (deduced from the R_f values in TLC) indicates that a threo arrangement of the central diol group is particularly

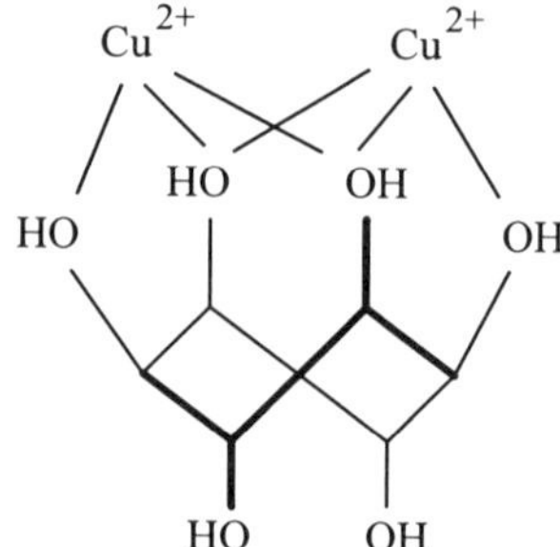

Figure 26. Schematic view of the mode of binding in the stable dinuclear Cu^{II} complex of *muco*-inositol. [Adapted from (321).]

favorable. The hypothesis that the alditol ligands are tetradentate was supported by ^{13}C NMR line-broadening experiments carried out with D-glucitol. For the dinuclear Cu^{II} complex, broadening occurred for the C1,2,3,4 signals of glucitol, with a large effect at C1 (321), whereas lanthanide cations are bound at O2,3,4 (103).

The weak mononuclear complexes of Cu^{II} are analogous to those of other divalent cations, whereas the dinuclear complexes have more specific structural requirements and present noteworthy structural analogies with the Mo^{VI} and W^{VI} complexes (321). However, Mo^{VI} and W^{VI} form stronger complexes with tetradentate alditols of the erythro configuration, whereas Cu^{II} seems to favor threo alditols.

I. Group 12 (IIB) Metal Ions

The corresponding elements are Zn, Cd, and Hg. Their M^{II} ions do not present a particular affinity for oxygen donors. For example, the versatile ligand taci forms Zn^{II} [83] and Cd^{II} (82) complexes containing $[M(taci)_2]^{2+}$ cations in which the metal atoms are bound only to nitrogen atoms. In solution, the Zn^{II}–carbohydrate complexes are essentially similar with those of Ca^{II}. Recently, solid complexes of Zn^{II} with D-galactose, D-fructose, D-glucose, D-xylose, D-ribose, and maltose were synthesised from various inorganic precursors (381). The Zn/saccharide ratio is 1 : 1 for monosaccharides and 2 : 1 for maltose. These complexes were studied by cyclic voltammetry in the pH range 3.7–10.3.

J. Group 13 (IIIA) Metal Ions

The metals of this group are Al, Ga, In, and Tl. The affinities of their M^{III} ions for carbohydrate-type ligands were compared by using taci, a ligand that may bind metal cations either at a N1,3,5 or a O2,4,6 tridentate site. Solid compounds containing the $[M(taci)_2]^{3+}$ cations were isolated by reacting Al^{III},

Ga^{III}, and Tl^{III} with taci (382). The X-ray determination of their crystal structures revealed AlO_6, GaN_3O_3, and TlN_6 coordination spheres, in agreement with the increasing soft character of the metal cation. These structures are retained in aqueous solution (382). Thus, Al^{III} may be expected to form the stronger complexes with carbohydrates.

The interactions of hydrolyzed Al^{III} with fructose and sucrose were studied by ^{13}C NMR spectroscopy, showing that sucrose is bound through its fructose moiety (383). With glucose and glucuronic acid, the sites of chelation were assigned to O4,6 of the α- and β-*p* forms (384). The Al^{III} complex of a protected glucofuranose has been prepared by ligand-exchange between $AlEt_3$ and DAG (HL) (132). This AlL_3 complex forms adducts with pyridine, AlL_3(py), and 4,4′-bipyridine, AlL_3(bpy). The crystal structure of the pyridine adduct was determined, showing that Al is bound to the monodentate sugars at O3 and forms a trigonal pyramid. Similar complexes of DAG with M^{IV} metals (M = Ti, Zr, and Hf) (132) and V^{III} (190) have been described.

K. Group 14 (IVA) Metal Ions

Both Sn and Pb belong to this column. The complexes of Pb^{II} were considered with those of calcium and other divalent cations. The main interest of the chemistry of tin–carbohydrate complexes lies in their applications in synthetic organic chemistry.

1. Solid-State Structures of Tin–Carbohydrate Complexes

The two main types of tin-substituted carbohydrates are the trialkylstannyl ethers L—O—SnR_3 and the dialkylstannylene acetals L—O,O—SnR_2, where L—O or L—O,O, respectively, represent a monodentate or a bidentate carbohydrate and R are alkyl groups, generally *n*-Bu.

The trialkyl ethers are obtained by reaction of the carbohydrates with tributyltin oxide (385) or tributyltin methoxide (386) in refluxing toluene. The X-ray structure of α-1-tributyltin-*O*-2,3-bisacetyl-4,6-ethylidene-D-glucose was determined, showing that this compound is monomeric and contains a tetrahedral tin atom (386).

Other tin–carbohydrate complexes are known in which tin is bound to a carbon atom of the ligand rather than to an oxygen atom. Two tetrasubstituted organotin derivatives $R_2(R'—CH_2)$SnI and $R_3(R'—CH_2)$Sn in which R are alkyl or phenyl groups and R′ is a C-bound carbohydrate derivative were prepared and characterized in solution and in the solid state (387, 388). For the iodo complex, 1,2:5,6-di-*O*-isopropylidene-3-*C*-dibutyliodostannylmethyl-α-D-allofuranose, the crystal structure revealed that the tin atom is not tetrahedral, but has a trigonal bipyramidal geometry, since Sn is bound to O3 with a bond length of 2.68 Å (397). The other complex, 1,2:5,6-di-*O*-isopropylidene-3-*C*-triphenylstannylmethyl-α-D-allofuranose was prepared by reaction of Ph_3Sn-CH_2-Li with the protected 3-ketose precursor. The crystal structure and NMR spectra of this triphenyl compound have been reported (388) and reveal a slightly distorted tetrahedral tim atom, with a bond distance of 3.01 Å due to a weak Sn· · ·O3 interaction.

The dialkyl acetals are commonly prepared by reaction of the unprotected ligand with dialkyltin oxide or dialkyltin diethoxide (389) in benzene. A series of dialkylstannylene acetals of carbohydrates (aldoses and ketoses) were synthesized in anhydrous methanol (390). X-Ray studies have demonstrated a dramatic influence of the configuration of the sugar on the structure of carbohydrate dibutylstannylenes. In the 2:2 adduct of the gluco compound, methyl 4,6-*O*-benzylidene-2,3-*O*-dibutylstannylene-α-D-glucopyranoside, both tin atoms are positioned as a trigonal bipyramid forming part of a dimer by association through a common edge (Fig. 27) (391, 392). Both Sn—O—Sn bridges are unsymmetrical with a short (2.09 Å) and a long (2.23–2.24 Å) bond. On the other hand, the crystal structure of the adduct of the manno derivative, methyl 4,6-*O*-benzylidene-α-D-mannopyranoside, revealed the formation of a 5:5 polymer (393) in which the central Sn atoms are octahedral, whereas the terminal Sn atoms are pentacoordinate.

2. *Solution Studies of Tin–Carbohydrate Complexes*

The electrophoretic migration of polyols in aqueous sodium stannate has been demonstrated a long time ago (394). The nature of the tin complexes in solution can be investigated by NMR spectroscopy, since ^{119}Sn NMR chemical shifts are diagnostic for the geometry of the tin atom in the adducts. Characteristic

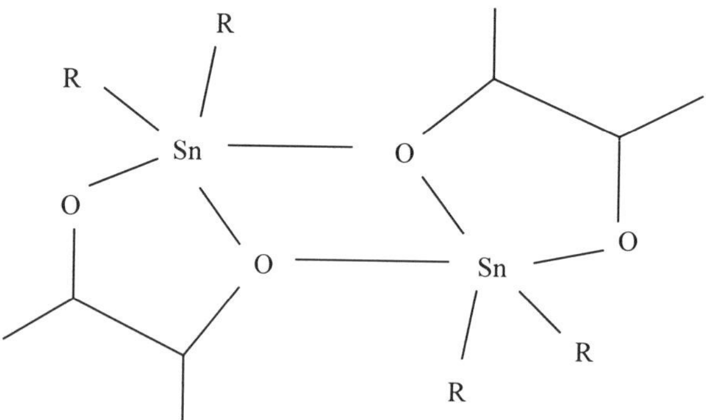

Figure 27. The dimeric structure of dialkylstannylene acetals formed by vicinal diols. R is an alkyl group, generally Bu. [Adapted from (391 and 392).]

values relative to tetramethyltin are $\delta > -40$ for tetracoordinate Sn atoms in dialkoxydialkyltin compounds (monomers); $-80 > \delta > -190$ for pentacoordinate Sn atoms (dimers); and $-200 > \delta > -300$ for hexacoordinate Sn atoms (higher oligomers, up to hexamers) (395, 396).

The tributylstannylene ethers exist in solution as monomers containing a tetrahedral tin atom. Typical ^{119}Sn NMR chemical shifts in C_6D_6 are $75 < \delta < 120$ ppm (397). On the other hand, the spectra of most dialkylstannylene acetals display a single sharp signal with a chemical shift near -120 ppm, typical of pentacoordinate Sn atoms. It is consistent with the presence of a single symmetric dimer, sometimes accompanied by small amounts of higher oligomers. The ^{13}C NMR spectra are in agreement with the presence of only one dimer. The carbon atoms of the complexing site are deshielded by less than 5 ppm.

Further studies by ^{119}Sn NMR spectroscopy in chloroform demonstrated that the tin adducts of diols are dynamic species subject to aggregation equilibria. The prevalent species are dimers, along with higher oligomers (395). Moreover, an intramolecular exchange process was observed within the dimer, in which a scrambling of tin atoms occurs between the two diol moieties of the dimer. With simple diols, the corresponding rate is faster than that of most organic reactions, but the equilibrium is assumed to be slower in the carbohydrate complexes, thus explaining the observed selectivities of reactions of organotin derivatives (396).

Mössbauer spectroscopy has been used for the study of dibutyltin(IV) (398) and dibenzyltin(IV) (399) complexes of carbohydrates. A comparison of the experimental quadrupole splitting values with those calculated revealed that the dibutyltin species are of four types: with the two central Sn^{IV} atoms surrounded by donor atoms in a trigonal–bipyramidal, octahedral, both a trigonal–bipyramidal and octahedral, or a tetrahedral arrangement. The results were compared with those for a similar study of diethyltin(IV) complexes (398). The reaction between dibenzyltin(IV)–dichloride and sugar-type ligands led to reaction products containing tin(IV) oxide besides the dibenzyltin(IV)–sugar complex (399).

3. *Tin–Carbohydrate Complexes in Organic Synthesis*

Tin(IV)–sugar complexes are particularly useful intermediates for the regioselective substitution by electrophiles including esterification, alkylation, arylation, oxidation (400), sulfation (401), and phosphorylation (402). It offers a convenient route for performing reactions at only one or two hydroxyl groups of the molecule, thereby avoiding the numerous protection–deprotection steps that make carbohydrate chemistry so tedious. One-pot, fast reactions may be achieved by preparing the tin adducts in benzene or toluene with commercially available dibutyltin dimethoxide (389).

The presence of an adjacent diol system in the carbohydrate is sufficient for the chelation of tin. Therefore, carbohydrates could potentially afford mixtures of complexes in which several sites would be activated for reaction. In such a case, this problem is avoided by using partially protected sugars that possess only two free hydroxyl groups. The activation of carbohydrates by the formation of a stannyl ether is due to an enhancement of the nucleophilicity of one oxygen atom compared to the original hydroxyl group. Thus, subsequent attack on the organotin intermediate by a suitable electrophilic reagent occurs selectively at the position of stannylation, rather than at the less reactive other free hydroxyl groups. Moreover, whereas two oxygen atoms are bound to Sn, one is generally much more reactive than the other, leading mainly to the formation of a monosubstituted product. This is the case for 4,6-*O*-benzylidene-α-D-glucopyranoside, which binds tin at the O2,3 diol group and reacts exclusively at O2 (392).

Since their first use in 1974 (403, 404) dibutylstannylene acetals obtained from various carbohydrates have been found to give mainly monosubstitution products with several types of electrophiles (405, 406). The selectivity patterns (as products ratio) in the benzylation of the dibutylstannylene acetals of *myo*-inositol protected at various positions were determined (407). The early suggestion (408) that the nature of the product is determined by the structure of the intermediate tin adduct was experimentally verified. It was postulated that the lability of the apical Sn—O bonds was likely to reflect the enhanced reactivity toward electrophilic reagents of one of the two chelating oxygen atoms (409). The factors that influence regioselectivity in reactions of dibutylstannylene acetals have been determined by ^{1}H, ^{13}C, and ^{119}Sn NMR studies (410, 411). The sugar derivatives that react regioselectively are those that form a single 2:2 stannylene complex in which the site of chelation is an *a–e* diol system. On the contrary, sugar derivatives that give mixtures of products are those that form multiple stannylene complexes. Sugars with an *e–e* diol system form a noninterconverting mixture of dimers, whereas those with an *a–a* diol system form rapidly interconverting mixtures of 1:1 dimers, trimers, and tetramers. For example, the structure of a tetramer is illustrated in Fig. 28.

A mechanistic scheme has been developed that explains the regioselectivity observed in reactions of stannylene acetals (412, 413). The key factor is the equilibrium between the symmetrical and the asymmetrical dimers. For slow reactions like esterification, regioselectivity is determined by competition between the position of the equilibrium and the relative reactivities of the oxygen atoms. For fast reactions, such as oxidation by bromine or *N*-bromosuccinimide (NBS), regioselectivity is determined by the position of the equilibrium.

Regioselectivity was also observed in acylations (385, 414) and alkylations (397) of carbohydrate derivatives mediated by tributyltin ethers. These ethers interact with *N*-methylimidazole, a well-known catalyst of tin-catalyzed acylations. The results of ^{119}Sn and ^{13}C NMR studies of the effect of *N*-methylim-

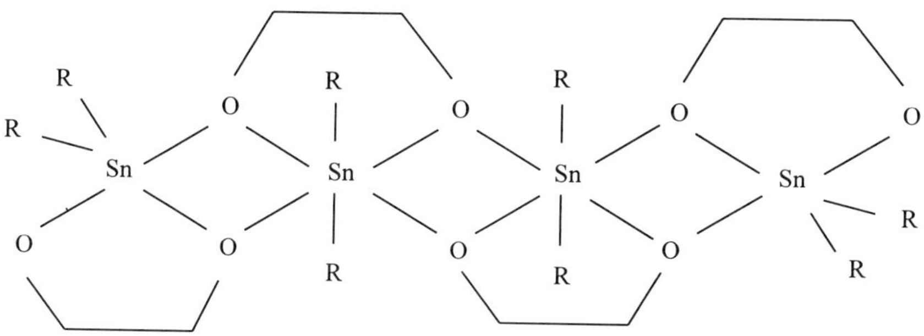

Figure 28. Structure of a tetrameric dialkylstannylene acetal formed by a vicinal diol. R is an alkyl group, generally Bu. The related trimers and pentamers have shorter or longer polymeric chains of the same type. [Adapted from (393).]

idazole have been interpreted as indicating the selective formation of pentacoordinate tin species that would react more easily with the alkylating agent (397). Tributyltin ethers, which involve the anomeric carbon atom of the sugar, allow simple preparations of glycosides or thioglycosides. α-1-Tributyltin-*O*-2,3-bisacetyl-4,6-ethylidene-D-glucose has been proposed as a convenient glycosidation reagent (386).

The dibutyltin oxide mediated process for the regioselective monoacetylation of sucrose in solution is a reaction of industrial importance (415, 416). The single product is 6-*O*-acetyl sucrose, a precursor of the chlorinated sweetener sucralose. This method has recently been improved by substituting dibutyltin oxide with polymer-supported butyltin(IV) reagents, which are nontoxic and practical catalysts (417).

L. Group 16 (VIA) Metal Ions

Only Te is considered in this section. The chemistry of this element presents analogies with that of Sb in Group 15 (VA). The reaction between the tellurate ion and triols was studied as a model for the first step of the oxidation of sugars by periodate, which is assumed to involve a transient cyclic triester (418). The concept was that, since tellurate and periodate ions are of similar size, the compounds would be of comparable stabilities, but the tellurate complexes would not undergo subsequent oxidation steps. Complex formation is revealed by the fact that the pH of a solution of telluric acid, initially adjusted to pH 8.00 with sodium hydroxide, slowly decreases after the addition of several carbohydrates. Some sugars give rise to noticeable pH variations, which establish the relative stabilities of the complexes. The corresponding values of pH after 60 and 120 min are D-lyxose (6.88; 6.41), D-ribose (7.28; 6.46), Me β-D-ribopyranoside (7.11; 6.54), and D-arabinose (7.58; 7.35). For three sugars, the pH drop was small and took place in only 30 min: D-xylose (7.67), D-glucose (7.80), Me α-D-glucopyranoside (7.81) (418).

Conductimetric and potentiometric studies indicated the formation of 1 : 1 complexes between telluric acid and polyols (419, 420). With the acyclic hexitols, the Te/polyol stoichiometries are 1 : 1 in acidic media and 3 : 1 in strongly alkaline media. Cyclic alditols (inositols) only yield weak 1 : 1 complexes at any pH. Antimonate(V) also gives 1 : 1 complexes with alditols. The order of stability was related to the configuration of the alditol : galactitol > D-mannitol and allitol > altritol, D-glucitol > inositols, showing that alditols with an erythro site of chelation form stronger complexes than those with threo sites (419).

M. Organometallic Species

All metal–carbohydrate complexes heretofore examined in this chapter involve metal–oxygen bonding. However, with "soft" metal ions such as platinum, metal–carbon bonding has been reported to compete with metal–oxygen bonding. This finding is the case for bis-ascorbate complexes of platinum that have demonstrated good antitumor activity (421). The solution chemistry of platinum–en complexes was investigated by multinuclear (^{195}Pt, ^{15}N, and ^{13}C) NMR spectroscopy (422). The reaction of $[Pt(^{15}en)(H_2O)_2]^{2+}$ with aqueous sodium ascorbate yields three kinetic products with oxygen bonding in the initial stage ($t < 1$ h), $[Pt(^{15}en)(H_2O)(O^3\text{-Hasc})]^+$, $[Pt(^{15}en)(O^3\text{-Hasc})_2]$, and $[Pt(^{15}en)(O^2,O^3\text{-asc})]$ of which the latter is a chelate of the O2,3 enediol group. The reaction ultimately produces two carbon-bound ascorbate chelates, $[Pt(^{15}en)(C^2,O^5\text{-asc})]$ and $[Pt(^{15}en)(C^2\text{-Hasc})(O^3\text{-Hasc})]$. A similar chemistry was reported for the ammonia analogues (423) in which the following oxygen adducts were initially characterized: *cis*-$[Pt(^{15}NH_3)_2(H_2O)(O^3\text{-Hasc})]^+$, *cis*-$[Pt(^{15}NH_3)_2(O^3\text{-Hasc})_2]$, and chelate *cis*-$[Pt(^{15}NH_3)_2(O^2,O^3\text{-asc})]$. The slower formation of the more stable carbon adducts involves exchange of ammonia by a water molecule, affording the chelates $[Pt(^{15}NH_3)(H_2O)(C^2,O^5\text{-asc})]$ and $[Pt(^{15}NH_3)(O^3\text{-Hasc})(C^2,O^5\text{-asc})]^-$ and the diammines *cis*-$[Pt(^{15}NH_3)_2(C^2,O^5\text{-asc})]$ and *cis*-$[(Pt(^{15}NH_3)_2(C^2\text{-Hasc})(O^3\text{-Hasc})]$.

Few organometallic complexes of protected carbohydrates are known (424, 425). Glycosyl complexes were prepared with good stereoselectivity by reacting protected glycopyranosyl bromides with suitable organometallic compounds of cobalt (426), iron (427), and manganese (428, 429). The preparations and reactions of glycosylmanganese pentacarbonyl complexes $RMn(CO)_5$ have been reviewed (430, 431). Pyranosyl and furanosyl complexes have been obtained and shown to undergo various insertion reactions resulting in the formation of *C*-glycosyl derivatives (430). The reaction rates depend on the configuration of the anomeric center bound to the metal (431). The capacity of sugars to serve

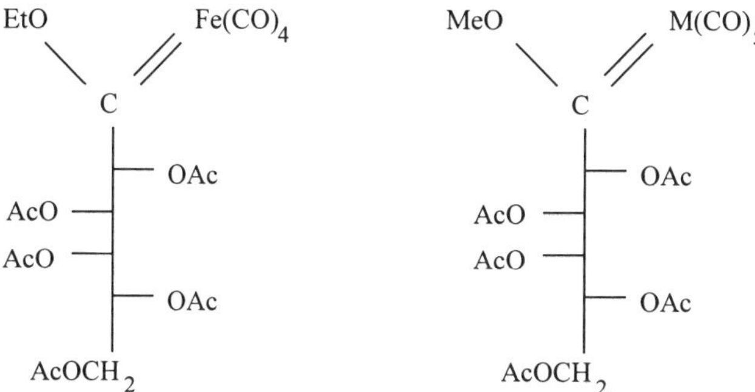

Figure 29. Sugar–carbene complexes derived from 2,3,4,5,6-penta-*O*-acetyl-D-galactonic alkyl esters. [Adapted from (434).] M = Cr, Mo, or W.

as optically active ligands for chiral induction is convincingly documented for reactions mediated by titanium (432).

Carbene complexes of carbohydrates have been synthesized (433, 434). The first examples were vinylcarbene complexes $(CO)_5M{=}C(OEt){-}CH{=}C(OR){-}Ph$ in which M = Cr or W, and ROH is a partially protected sugar such as DAG (433). Carbene complexes were also prepared in which the metal atom is directly attached to the carbohydrate carbon backbone. For example, the fully acetylated galactonic acid chloride $AcOCH_2{-}(CHOAc)_4{-}COCl$ reacted under mild conditions (THF, −60°C) with $Na_2Fe(CO)_4$ or $K_2M(CO)_5$ (in which M = Cr, Mo, or W) to give acylmetalates $AcOCH_2{-}(CHOAc)_4{-}C(O^-){=}M(CO)_n$, which afforded the carbenes (Fig. 29) by reaction with alkylating agents in CH_2Cl_2 (434). The organometallic compounds were isolated in the solid state and fully characterized (Table XIII).

Whereas homogeneous organotransition metal catalysts are known to possess high reactivity and selectivity, the recent statement that the potential applications of organometallic complexes of carbohydrates are virtually unexplored

TABLE XIII
100-MHz ^{13}C NMR data (δ, ppm) in $CDCl_3$ for Sugar–Carbenes Derived form 2,3,4,5,6-Penta-*O*-acetyl-D-Galactonic Alkyl Esters $AcOCH_2{-}(CHOAc)_4{-}C({=}Z)OR$ (434)

Metal Group Z	R	C1	C2	Color	mp (°C)[a]
$Fe(CO)_4$	Ethyl	327.6	84.9	Red	91–93
$Cr(CO)_5$	Methyl	355.0	84.5	Orange	113–115
$Mo(CO)_5$	Methyl	346.2	84.4	Yellow	115–117
$W(CO)_5$	Methyl	328.2	86.0	Red	130–132

[a]melting point = mp.

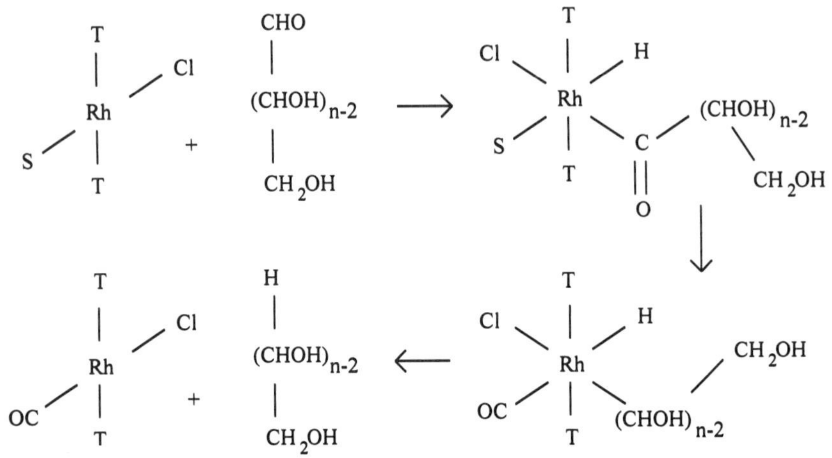

Scheme 7. Decarbonylation of aldoses using the rhodium catalyst $RhCl(Ph_3P)_3$. Solvent = S and triphenylphosphine = T. [Adapted from (441).]

remains valid (435). There is a serious need for fundamental studies of the interactions of organometallic species with carbohydrates.

Some reactions involving organometallics have been extended in the carbohydrate field. One of those is the decarbonylation of aldehydes by tris(triphenylphosphine)rhodium(I) chloride, $Rh(PPh_3)_3Cl$ (Wilkinson's catalyst) (436–439). The mechanism is known to involve transient sugar–rhodium complexes (Scheme 7). This overall reaction was successfully used for the stoichiometric shortening of unprotected C_n aldoses to the corresponding C_{n-1} alditols (440, 441). For example, glucose affords arabinitol in 88% yield (Eq. 5) (442).

$$HOCH_2{-}(CHOH)_4{-}CHO \longrightarrow HOCH_2{-}(CHOH)_3{-}CH_2OH \qquad (5)$$

Since a very low concentration of aldose is present in aldehyde form, the reaction must be performed at 130°C versus 70°C for simple aldehydes. The preferred solvent is NMP, which can dissolve both the hydrophilic sugar and the hydrophobic catalyst.

III. OCCURRENCE OF METAL COMPLEXES IN SYNTHESIS REACTIONS

Mixtures of products are generally obtained in substitution reactions of carbohydrates, such as formation of ethers or esters, because the polyol chain possesses several secondary hydroxyl groups of comparable reactivities (the terminal primary HO group is generally more reactive as it is less hindered). Consequently, regioselective functionalization of carbohydrates is currently performed by lengthy processes. The initial protection of the hydroxyl groups that are to remain unchanged is followed by the activation of hydroxyl groups that are to be converted by introduction of the reagent and the final deprotection step to yield the products. Therefore, the design of procedures for the specific substitution of unprotected carbohydrates remains an attractive synthetic challenge. Common strategies are based on the formation of metal complexes that are expected to have a dual effect by protecting and/or activating specific hydroxyl groups.

The regioselective protection of the carbohydrate backbone is illustrated by reports that the pattern of periodate oxidation of polyols is influenced by the addition of tungstate (230) or molybdate (226) ions. The complexes were formed at pH 5 for 1 h, then oxidation by periodate was performed at pH 8. Since the oxidation pathway begins by the chelation of periodate at two adjacent hydroxyl groups, the carbon–carbon bonds located within the site of chelation are not cleaved relative to the others. For example, galactitol and altritol, which are complexed at the O2,3,4,5 site, are almost not oxidized. By contrast, mannitol and allitol, which are complexed at the O1,2,3,4 site, respectively, afford arabinose and ribose, corresponding to cleavage of the C5,6 bond. Such patterns of periodate oxidation of heptitols in the presence of molybdate were used to confirm the identification of the prevailing site of chelation (224, 251).

In the activation theory, the approach is based on the expectation that metal bonding to the carbohydrate ligand should modify the reactivity of the hydroxyl groups, either by electronic effects or by steric effects. Many reports have used well-known synthetic reactions on preformed carbohydrate complexes or simply on carbohydrates in the presence of metal salts. Generally, the metal-catalyzed reactions afford the same products in a modified ratio, or even products that are not obtained in the original reaction, showing that the metal complexation is altering kinetically controlled reactions. A frequent approach is to search for trends in the variations of the reactivity of the carbohydrate substrates within a series of related metal ions. Until now, these results were rarely discussed in relation with the structures or the stabilities of the postulated intermediate complexes. The Ca^{II}- and Sr^{II}-mediated synthesis of selected methyl glycosides by Angyal et al. (443) were targets defined by selection of the chelation-stabilized forms of each sugar. This method favors the formation of furanosides and was used to synthesize the previously unknown methyl α- and β-D-talofuranosides.

Few metal salts other than tin(IV) compounds are now established reagents for carbohydrate synthesis. Nevertheless, promising results have been reported for iron(III) (444–447). The reactions of sugars with fatty alcohols yield *O*-glycosides, which are of interest as neutral surfactants, and the stereoselectivity of the glycosylation reactions of aldoses is strongly affected by $FeCl_3$. It was initially reported that $FeCl_3$ orients the reaction of glucose and galactose with methanol to the formation of methyl furanosides (444). It was also observed that fully acetylated methyl β-glycopyranosides were converted into their α anomers by $FeCl_3$ in dichloromethane (445). The proposed mechanism involves a complex in which Fe^{III} is bound to the ring oxygen atom, inducing mostly the endocyclic C—O bond cleavage.

Several unprotected aldoses (D-glucose, D-galactose, and D-mannose) and a ketose (D-fructose) were reacted with various fatty alcohols at room temperature in THF (446). The use of $BF_3 \cdot OEt_2$ as the catalyst yielded the thermodynamically more stable alkyl α-pyranosides. In the presence of anhydrous $FeCl_3$, the aldoses afforded a mixture of the kinetically favored alkyl α- and β-furanosides in good yield, whereas fructose reacted much faster, but exclusively yielded the alkyl β-pyranosides (25%). Anhydrous $ZnCl_2$ did not catalyze the reactions under the same conditions, showing that Fe^{III} does not act only as a Lewis acid, but specifically complexes the aldoses. Moreover, the addition of $BaCl_2$ or $CaCl_2$ to the medium increased the proportion of the β-furanosides in the case of aldoses. Similar trends were reported for glycuronic acids (447). The reaction of unprotected D-galacturonic acid with fatty alcohols in THF (heterogeneous medium and room temperature) does not afford the ester when a Lewis acid such as $BF_3 \cdot OEt_2$ or $FeCl_3$ (4 equiv) is used as catalyst, but results in glycosidation yielding the alkyl α-pyranoside as the major product. In contrast, the major product is the alkyl β-furanoside when the reaction is performed with $FeCl_3$ (3 equiv) in the presence of $CaCl_2$ (2 equiv). Although structural studies are still missing, these results are undoubtedly due to the formation of intermediate complexes of sugars with Fe^{III} and Ca^{II} ions.

The influence of the nature of the metal ion on the regioselective alkylation and acylation of carbohydrates engaged in metal complexes was examined (448). The anions of carbohydrates were prepared by reaction with sodium hydride in THF. Then a metal salt was added, followed by addition of an alkyl or allyl iodide or an acylating agent. Alkylation is catalyzed by Cu^{II} ions, with modification of the regioselectivity. Alkylation of hydroxyls in positions 3 and 4 is favored with respect to positions 2 and 6, whereas HO2,6 are the more reactive in the absence of metal ion. Acylation is promoted by Hg^{II} ions, however, the selectivity is reversed, since hydroxyl groups in positions 2 and 6 are preferred over hydroxyl groups in positions 3 and 4.

Zinc(II) chloride was used for the selective acetylation of methyl α-D-hexopyranosides (gluco and manno derivatives) (449). Selectivity was also demonstrated in the benzoylation of metal chelates of sucrose (450). Ionic complexes with various metal ions were prepared in DMF from the dianion of sucrose obtained by the action of sodium hydride. The resulting complexes reacted at low temperature (0°C) with benzoic anhydride. In the absence of metal ion, O2 (glucose moiety) is the more reactive position, but addition of calcium(II) chloride gave a mixture of various benzoates, and little selectivity was observed with Ni^{II}, Cu^{II}, Zn^{II}, and Hg^{II} salts. In contrast, acetylation at O′3 (fructose moiety) was observed with Mn^{II} chloride in 80% yield and Co^{II} chloride in 88% yield (450).

The formation of isopropylidene derivatives of sugars and alditols by condensation with dry acetone is acid–catalyzed (451). Several metal ions exhibited interesting catalytic activity, which is likely due to the formation of intermediate carbohydrate complexes. Like other Lewis acid catalysts, anhydrous Fe^{III} chloride (452) and Al^{III} chloride (453) were reported to form mainly the thermodynamically more stable product. However, when anhydrous Cu^{II} sulfate is used, some kinetic control is introduced (454–456). For example, D-galactose yields 1,2:5,6-di-*O*-isopropylidene-α-D-galactofuranose instead of 1,2:3,4-di-*O*-isopropylidene-α-D-galactopyranose, which is the thermodynamically favored product (454). Recently, zeolite HY, obtained by calcination of ammonium exchanged NaY zeolite at 500°C over dioxygen, proved to be a suitable heterogeneous catalyst (457). Moreover, products obtained under kinetic control were formed in higher yield with zeolite HY than with a Cu^{II} catalyst, as was demonstrated in the acetonation of D-galactose, D-arabinose, and L-sorbose. Acetonation is also carried out by reaction of sugars with 2,2-dimethoxypropane in 1,2-dimethoxyethane. In the presence of catalytic amounts of Sn^{II} chloride, D-fructose affords 1,2-*O*-isopropylidene-β-D-fructofuranose instead of the mixture of di-*O*-isopropylidene-fructopyranoses formed in the absence of $SnCl_2$ (458). With D-mannitol, the same catalytic system oriented the reaction toward the formation of 1,2:5,6-di-*O*-isopropylidene-D-mannitol instead of the 1,2:3,4:5,6-tri-*O*-isopropylidene derivative (459).

Metal–carbohydrate interactions are also involved in the dissolution of cellulose in aqueous solutions of metal salts or complexes. Cu^{II}, Zn^{II}, or the Cu and Ni ammine complexes have been used in processes for the preparations of "artificial silk." Complexes formed from aqueous $ZnCl_2$ and cellulose-related D-glucopyranosides were studied because of their industrial relevance to the mechanism of swelling of cellulose (460).

IV. CONCLUSION

This chapter makes clear that, although the chelating ability of simple carbohydrates has been underestimated for a long time, real progress is now being made, in particular with the help of newer spectroscopic methods and the increasing number of structures characterized by X-ray crystallography. An obvious reason for the previous reluctance of researchers to invest in this field is that carbohydrates contain many functionalities with various possibilities of isomerism and conformational changes. However, acyclic polyols and alditols, hydroxyacids, glycosides, and partially protected sugars provide valuable models, which allows a much simpler approach to the sugar systems.

Studies oriented toward the biological applications of carbohydrate–metal interactions possess several interesting targets such as the design of nontoxic chelators for the removal of overloads of metal ions and the preparation of carbohydrate complexes of lanthanide ions such as Gd^{III} as contrast agents for magnetic resonance imaging. Enhancement of the relaxation of water has been obtained using paramagnetic metal complexes of polysaccharides (461). The applications of the complexing properties of polysaccharides to gel formation and extraction of metal ions have been mentioned in a previous review (18). The understanding of the role of metal cations such as Ca^{II} in the interaction of glycans with lectins is certainly another challenge in biochemistry because of its relevance to the mechanism of cell adhesion (56). The recent characterization by X-ray of a lectin–Ca^{2+}–carbohydrate complex (119) documents the first structural characterization of a metal ion–carbohydrate complex in biology and will undoubtedly be followed by more examples. Moreover, the recent discovery that many natural carbohydrates are substituted, for example, with sulfate groups, demonstrates that identification of new classes of ligands with unsuspected properties will still occur in the years to come.

Considerable progress was recently achieved in understanding the modes of bonding in solution of carbohydrates with metal ions such as vanadium, molybdenum, tungsten, and tin. These metal ions were previously neglected, perhaps because their biological roles were less widely recognized. Potential applications of such metal–carbohydrate complexes as catalysts for various organic reactions have, however, stimulated this interest. Although the uses of metal ions in synthetic organic reactions of carbohydrates are presently still in their infancy, the use of tin derivatives will probably be the first example of such uses and the beginning of a field focusing on specific activation modes of saccharides. A permanent factor to continue to develop the chemistry of carbohydrate transformation is their availability in large quantities and their renewable sources from the biomass (62).

NOTE ADDED IN PROOF

A Cu^{II} complex of D-glucitol (462) obtained in alkaline medium was shown to consist of 16 Cu atoms linked by 8 fully or partly deprotonated polyol ligands, forming a toroidal cluster. The structural determination of M_2L complexes of fivefold deprotonated D-mannose (L) with various M^{III} ions (M = Fe, V, Cr, Al and Ga) revealed that the ligand was in the *β-f* form (463).

Since the completion of this work, additional information related to the subject of this chapter was reported. The electrochemical properties and in vitro RNase inhibition activity of oxovanadium(IV) complexes of saccharides were described (464). Co^{II} (465) and Ni^{II} (466) complexes of aldoses were characterized in solid form. NMR spectroscopic studies of the reactions of D-galactonic and L-mannonic acid with W^{VI} and Mo^{VI} showed the existence of complexes formed at the CHOH—COOH group in acidic medium and at $(CHOH)_n$ sites in alkaline medium (467, 468). The epimerization of aldoses by Ca^{II} was shown to involve an intermediate complex of the tetradentate ligand (469). In the Cu^{II} complexes of α-CD, characterized by X-ray crystallography, the ligand adopts a cylindrical shape instead of the normal toric geometry (470).

ACKNOWLEDGMENTS

We thank the National Institute of Health for funding (to DCC) and the staff of the library of Institut National des Sciences Appliquées, Mont-Saint-Aignan, for their assistance.

ABBREVIATIONS

Å	Angstrom = 10^{-10} m
a	Axial
acac	Acetylacetonate
Ad	Adenosine
ADP	Adenosine diphosphate
AMP	Adenosine monophosphate
asc	Ascorbate dianion
ATP	Adenosine triphosphate
bpy	2,2′-Bipyridine = 2,2′-dipyridyl
Bu	Butyl
cat	Catecholate (dianion of 1,2-dihydroxybenzene)
CD	Circular dichroism
α-CD, β-CD, γ-CD	Cyclodextrins, natural cyclic (1 → 4) oligomers containing six, seven, or eight α-D-glucopyranose units, respectively
CIS	Coordination induced shift
COSY	Correlation spectroscopy
CRD	Carbohydrate recognition domain
2D	Two dimensional
DAG	1,2 : 5,6-Di-*O*-isopropylidene-α-D-glucofuranose
dbc	Dianion of 3,5-di-*tert*-butylcatechol
dhn	Dianion of 2,3-dihydroxynaphthalene

DMF	*N*,*N*-Dimethylformamide
dppp	1,3-Bis(diphenylphosphino)propane
dtt	1,4-Dithiothreitolate ion $(C_4H_6O_2S_2)^{4-}$
e	Equatorial
$E_{1/2}$	Half-wave potential
edta	Ethylenediaminetetraacetic acid
en	Ethylenediamine
ENDOR	Electron nuclear double resonance
EPR	Electron paramagnetic resonance
equiv	Equivalent
Ery	Erythritolate ion $(C_4H_6O_4)^{4-}$
Et	Ethyl
EXAFS	Extended X-ray absorption fine structure
f	Furanose
FAB	Fast atom bombardment
Fru	Fructose
Glc	Glucose
HEPES	4-(2-Hydroxyethyl)-1-piperazineethanesulfonic acid
H_2HIB	2-Hydroxyisobutyric acid (2-hydroxy-2-methylpropanoic acid)
H_2Lac	Lactic acid (2-hydroxypropanoic acid)
HPLC	High-performance liquid chromatography
IR	Infrared
Ln	Lanthanide
Lyx	Lyxose
Man	Mannitolate ion $(C_6H_{10}O_6)^{4-}$
MBP	Mannose-binding protein
Me	Methyl
MS	Mass spectrometry
MS/MS	Tandem mass spectrometry
NAD	Nicotinamide adenine dinucleotide
NADP	NAD phosphate
NBS	*N*-Bromosuccinimide
nd	Not determined
nf	Not found
NMP	*N*-Methyl-2-pyrrolidone
NMR	Nuclear magnetic resonance
NOE	Nuclear Overhauser effect
ORD	Optical rotatory dispersion
ox	Oxalate dianion
p	Pyranose
Ph	Phenyl

phen	1,10-Phenanthroline
ppm	Parts per million
pq	9,10-Phenanthrenequinone
py	Pyridine
R_f	Retardation factor
SCE	Standard calomel electrode
taci	1,3,5-Triamino-1,3,5-trideoxy-*cis*-inositol
TBA	Tetra-*n*-butylammonium ion
TEA	Tetraethylammonium ion
THF	Tetrahydrofuran
TLC	Thin-layer chromatography
UV	Ultraviolet
Vis	Visible
XANES	X-ray absorption near-edge structure

REFERENCES

1. R. A. Kwek, *Chem. Rev.*, *96*, 683 (1996).
2. J. A. Rendleman, Jr, *Adv. Carbohydr. Chem. Biochem.*, *21*, 209 (1966).
3. *Carbohydrates in Solution*, R. F. Gould, Ed., *Adv. Chem. Ser.*, *117* (1973).
4. J. A. Rendleman, Jr., *Adv. Chem. Ser.*, *117*, 51 (1973).
5. S. J. Angyal, *Adv. Chem. Ser.*, *117*, 106 (1973).
6. R. Montgomery, *Adv. Chem. Ser.*, *117*, 197 (1973).
7. A. H. Haines, *Adv. Carbohydr. Chem. Biochem.*, *33*, 11 (1976).
8. K. Bock and C. Pedersen, *Adv. Carbohydr. Chem. Biochem.*, *41*, 27 (1983).
9. K. Bock, C. Pedersen, and H. Pedersen, *Adv. Carbohydr. Chem. Biochem.*, *42*, 193 (1984).
10. K. Dill and R. D. Carter, *Adv. Carbohydr. Chem. Biochem.*, *47*, 125 (1989).
11. S. J. Angyal, *Tetrahedron*, *30*, 1695 (1974).
12. S. J. Angyal, *Chem. Soc. Rev.*, *9*, 415 (1980).
13. S. J. Angyal, *Pure Appl. Chem.*, *35*, 131 (1973).
14. S. J. Angyal, *Adv. Carbohydr. Chem. Biochem.*, *47*, 1 (1989).
15. K. B. Hicks, *Adv. Carbohydr. Chem. Biochem.*, *46*, 17 (1988).
16. S. Yano, *Coord. Chem. Rev.*, *92*, 113 (1988).
17. K. Burger and L. Nagy, in *Biocoordination Chemistry: Coordination Equilibria in Biologically Active Systems*, K. Burger, Ed., Ellis Horwood, New York, 1990, pp. 236–283.
18. D. M. Whitfield, S. Stojkovski, and B. Sarkar, *Coord. Chem. Rev.*, *122*, 171 (1993).
19. S. Yano and T. Otsuka, *Metal Ions Biol. Systems*, *32*, 27 (1996).

20. H. Kozlowski and M. Jesowska-Bojczuk, *Handb. Met-Ligand Interact. Biol. Fluids: Bioinorg. Chem.*, *1*, 679 (1995).
21. U. Piarulli and C. Floriani, *Progress in Inorganic Chemistry*, Wiley-Interscience, New York, 1997, Vol. 45, p. 393.
22. S. J. Angyal, *Angew. Chem. Int. Ed. Engl.*, *8*, 157 (1969).
23. S. J. Angyal, *Adv. Carbohydr. Chem. Biochem.*, *42*, 15 (1984).
24. S. J. Angyal, *Adv. Carbohydr. Chem. Biochem.*, *49*, 19 (1991).
25. D. T. Sawyer, *Chem. Rev.*, *64*, 633 (1964).
26. Rules for carbohydrate nomenclature, *J. Org. Chem.*, *28*, 281 (1963). Definitive Rules for Carbohydrate Nomenclature, *Pure Appl. Chem.*, *68*, 1989 (1996); *Carbohydr. Res.*, *297*, 1, (1997).
27. F. A. Cotton and G. Wilkinson, *Advanced Inorganic Chemistry*, 5th ed., Wiley-Interscience, New York, 1988.
28. J. Böseken, *Adv. Carbohydr. Chem.*, *4*, 189 (1949).
29. D. C. Crans, H. Chen, O. P. Anderson, and M. M. Miller, *J. Am. Chem. Soc.*, *115*, 6769 (1993).
30. A. Beltrán, F. Caturla, A. Cervilla, and J. Beltrán, *J. Inorg. Nucl. Chem.*, *43*, 3277 (1981).
31. J. J. Cruywagen, J. B. Heyns, and R. F. van de Water, *J. Chem. Soc. Dalton Trans.*, 1857 (1986).
32. M. T. Pope and A. Muller, Eds., *Polyoxometalates: From Platonic Solids to Anti Retroviral Activity*, Kluwer Academic, Dordrecht, The Netherlands, 1994.
33. D. Leggett, *Computational Methods for the Calculation of Stability Constants*, Plenum, New York, 1985.
34. M. T. Beck and I. Nagypál, *Chemistry of Complex Equilibria*, Ellis Horwood, Chichester, UK, 1990, pp. 263–286.
35. A. E. Martell and R. J. Moketaikis, *Determination and Use of Stability Constants*, VCH, New York, 1992, pp. 1–5.
36. N. K. Richtmyer and C. S. Hudson, *J. Am. Chem. Soc.*, *73*, 2249 (1951).
37. M. Hamon, C. Morin, and R. Bourdon, *Anal. Chim. Acta*, *46*, 255 (1969).
38. A. P. G. Kieboom, H. M. A. Bourmans, L. K. van Leeuwen, and H. J. van Benschop, *Recl. Trav. Chim. Pays-Bas*, *98*, 393 (1979).
39. J.-P. Sauvage, S. Chapelle, A. M. Dona, and J.-F. Verchère, *Carbohydr. Res.*, *243*, 293 (1993).
40. W. P. Aue, E. Bartholdi, and R. R. J. Ernst, *J. Chem. Phys.*, *64*, 2229 (1976).
41. R. Freeman and H. D. W. Hill, in *Dynamic NMR Spectroscopy*, L. M. Jackman and F. A. Cotton, Eds., Academic, London, 1975, p. 131.
42. A. Bax and G. Morris, *J. Magn. Reson.*, *42*, 501 (1981).
43. I. Bertini, P. Turano, and A. J. Vila, *Chem. Rev.*, *93*, 2833 (1993).
44. L. Banci, I. Bertini, and C. Luchinat, *Methods Enzymol.*, *239*, 485 (1994).
45. L. J. Berliner and J. Reuben, Eds., *Biological Magnetic Resonance*, Vol. 12, *NMR of Paramagnetic Molecules*, Plenum, New York, 1993.

46. M. Tonkovic, *Carbohydr. Res.*, *254*, 277 (1994).
47. D. C. Koningsberger and R. Prins, Eds., *X-Ray Absorption: Principles, Applications, Techniques of EXAFS, SEXAFS and XANES*, Wiley, New York, 1988.
48. L. Nagy, H. Ohtaki, T. Yamaguchi, and M. Nomura, *Inorg. Chim. Acta*, *159*, 201 (1989).
49. L. Nagy, K. Burger, J. Kuerti, M. A. Mostafa, L. Korecz, and I. Kiricsi, *Inorg. Chim. Acta*, *124*, 55 (1986).
50. M. Hlaïbi, M. Benaïssa, S. Chapelle, and J.-F. Verchère, *Carbohydr. Lett.*, *2*, 9 (1996).
51. L. Pettersson, *Acta Chem. Scand.*, *26*, 4067 (1972).
52. T. A. Beyer, J. E. Sadler, J. I. Rearick, J. C. Paulson, and R. L. Hill, *Adv. Enzymol. Relat. Areas Mol. Biol.*, *52*, 23 (1981).
53. Z. Derewenda, J. Yariv, J. R. Helliwell, A. J. Kalb, E. J. Dodson, M. Z. Papiz, T. Wan, and J. Campbell, *EMBO J.*, *8*, 2189 (1989).
54. J. Bouckaert, F. Poortmans, L. Wyns, and R. J. Loris, *J. Biol. Chem.*, *271*, 16144 (1996).
55. K. Drickamer, *J. Biol. Chem.*, *263*, 9557 (1988).
56. L. A. Lasky, *Science*, *258*, 964 (1992).
57. K. Drickamer, *Biochem. Soc. Trans.*, *21*, 456 (1993).
58. K. Drickamer, *Biochem. Soc. Trans.*, *24*, 146 (1996).
59. D. O'Connell, A. Koenig, S. Jennings, B. Hicke, H.-L. Han, T. Fitzwater, Y.-F. Chang, N. Varki, D. Parma, and A. Varki, *Proc. Natl. Acad. Sci. USA*, *93*, 5883 (1996).
60. W. E. Harte, Jr., and J. Bajorath, *J. Am. Chem. Soc.*, *116*, 10394 (1994).
61. G. Liang, R. K. Schmidt, H.-A. Yu, D. A. Cumming, and J. W. Brady, *J. Phys. Chem.*, *100*, 2528 (1996).
62. F. W. Lichtenthaler Ed., *Carbohydrates as Organic Raw Materials*, VCH Publishers, Weinheim, Vol. 1, 1991; G. Descotes, Ed., Vol. 2, 1993; H. van Bekkum, Ed., Vol. 3, 1996.
63. J.-F. Verchère, J.-P. Sauvage, and G. R. Rapaumbya, *Analyst*, *115*, 637 (1990).
64. K. B. Wiberg, Ed., *Oxidation in Organic Chemistry, Part A*, Academic, New York, 1965.
65. D. Benson, *Mechanisms of Oxidation by Metal Ions*, Elsevier, Amsterdam, The Netherlands, 1976.
66. W. G. Nigh, in *Oxidation in Organic Chemistry, Part B*, W. S. Trahanovsky, Ed., Academic, New York, Chapter I, 1973, p. 1.
67. D. G. Lee and M. van den Engh, in *Oxidation in Organic Chemistry, Part B*, W. S. Trahanovsky Ed., Academic, New York, Chapter IV, 1973, p. 177.
68. C. A. Bunton, in *Oxidation in Organic Chemistry, Part A*, K. B. Wiberg, Ed., Academic, New York, Chapter VI, 1965, p. 367.
69. J. M. Harrowfield, M. Mocerino, B. W. Skelton, W. Wei, and A. H. White, *J. Chem. Soc. Dalton Trans.*, 783 (1995).

70. J. A. Rendleman, Jr., *J. Org. Chem.*, *31*, 1839 (1966).
71. J. A. Rendleman, Jr., *J. Org. Chem.*, *31*, 1845 (1966).
72. J. A. Leary and S. F. Pedersen, *J. Org. Chem.*, *54*, 5650 (1989).
73. Z. Zhou, S. Ogden, and J. A. Leary, *J. Org. Chem.*, *55*, 5444 (1990).
74. G. E. Hofmeister, Z. Zhou, and J. A. Leary, *J. Am. Chem. Soc.*, *113*, 5964 (1991).
75. A. Staempfli, Z. Zhou, and J. A. Leary, *J. Org. Chem.*, *57*, 3590 (1992).
76. A. Fura and J. A. Leary, *Anal. Chem.*, *65*, 2805 (1993).
77. M. Kohler and J. A. Leary, *Anal. Chem.*, *67*, 3501 (1995).
78. A. P. G. Kieboom, T. Spoormaker, A. Sinnema, J. M. van den Toorn, and H. van Bekkum, *Recl. Trav. Chim. Pays-Bas*, *94*, 53 (1975).
79. R. D. Hancock and K. Hegetschweiler, *J. Chem. Soc. Dalton Trans.*, 2137 (1993).
80. M. L. Dheu-Andries and S. Pérez, *Carbohydr. Res.*, *124*, 324 (1983).
81. P. Klüfers and J. Schuhmacher, *Angew. Chem.*, *106*, 1925 (1994); *Angew. Chem., Int. Ed. Engl.*, *33*, 1863 (1994).
82. K. Hegetschweiler, R. D. Hancock, M. Ghisletta, T. Kradolfer, V. Gramlich, and H. W. Schmalle, *Inorg. Chem.*, *32*, 5273 (1993).
83. K. Hegetschweiler, V. Gramlich, M. Ghisletta, and H. Samaras, *Inorg. Chem.*, *31*, 2341 (1992).
84. J. A. Mills, *Biochem. Biophys. Res. Commun.*, *6*, 418 (1962).
85. R. W. Goulding, *J. Chromatogr.*, *103*, 229 (1975).
86. S. J. Angyal, G. S. Bethell, and R. J. Beveridge, *Carbohydr. Res.*, *73*, 9 (1979).
87. L. Petruš, V. Bílik, L. Kuniak, and L. Stankovic, *Chem. Zvesti*, *34*, 530 (1980).
88. J. K. N. Jones and R. A. Wall, *Can. J. Chem.*, *38*, 2290 (1960).
89. V. Bílik, L. Petruš, and J. Alföldi, *Chem. Zvesti*, *30*, 698 (1976).
90. H. D. Scobell and K. M. Brobst, *J. Chromatogr.*, *212*, 51 (1981).
91. J. Briggs, P. Finch, M. C. Matulewicz, and H. Weigel, *Carbohydr. Res.*, *97*, 181 (1981).
92. S. J. Angyal and J. A. Mills, *Aust. J. Chem.*, *38*, 1279 (1985).
93. M. M. Hämäläinen and H. Lönnberg, *Carbohydr. Res.*, *215*, 357 (1991).
94. H. Caruel, L. Rigal, and A. Gaset, *J. Chromatogr.*, *558*, 89 (1991).
95. H. Lönnberg and A. Vesala, *Carbohydr. Res.*, *78*, 53 (1980).
96. S. J. Angyal and D. C. Craig, *Carbohydr. Res.*, *241*, 1 (1993).
97. S. J. Angyal and K. P. Davies, *J. Chem. Soc. Chem. Commun.*, 500 (1971).
98. S. J. Angyal, *Aust. J. Chem.*, *25*, 1957 (1972).
99. R. E. Lenkinski and J. Reuben, *J. Am. Chem. Soc.*, *98*, 3089 (1976).
100. M. C. R. Symons, J. A. Benbow, and H. Pelmore, *J. Chem. Soc. Faraday Trans. I*, *78*, 3671 (1982).
101. M. C. R. Symons, J. A. Benbow, and H. Pelmore, *J. Chem. Soc. Faraday Trans. I*, *80*, 1999 (1984).

102. J. K. Beattie and M. T. Kelso, *Aust. J. Chem.*, *34*, 2563 (1981).
103. A. P. G. Kieboom, A. Sinnema, J. M. van der Toorn, and H. van Bekkum, *Recl. Trav. Chim. Pays-Bas*, *96*, 35 (1977).
104. S. J. Angyal, D. C. Craig, J. Defaye, and A. Gadelle, *Can. J. Chem.*, *68*, 1140 (1990).
105. Z. Chen, N. Morel-Desrosiers, J.-P. Morel, and C. Detellier, *Can. J. Chem.*, *72*, 1753 (1994).
106. W. J. Evans and V. L. Frampton, *Carbohydr. Res.*, *59*, 571 (1977).
107. J.-P. Morel and C. Lhermet, *Can. J. Chem.*, *63*, 2639 (1985).
108. J.-P. Morel, C. Lhermet, and N. Morel-Desrosiers, *Can. J. Chem.*, *64*, 996 (1986).
109. A. M. Alvarez, N. Morel-Desrosiers, and J.-P. Morel, *Can. J. Chem.*, *65*, 2656 (1987).
110. J.-P. Morel, C. Lhermet, and N. Morel-Desrosiers, *J. Chem. Soc. Faraday Trans.*, *84*, 2567 (1988).
111. N. Morel-Desrosiers and J.-P. Morel, *J. Chem. Soc. Faraday Trans.*, *85*, 3461 (1989).
112. N. Morel-Desrosiers, C. Lhermet, and J.-P. Morel, *J. Chem. Soc. Faraday Trans.*, *87*, 2173 (1991).
113. N. Morel-Desrosiers, C. Lhermet, and J.-P. Morel, *J. Chem. Soc. Faraday Trans.*, *89*, 1223 (1993).
114. P. Rongère, N. Morel-Desrosiers, and J.-P. Morel, *J. Chem. Soc. Faraday Trans.*, *91*, 2771 (1995).
115. P. Rongère, N. Morel-Desrosiers, and J.-P. Morel, *J. Sol. Chem.*, *23*, 351 (1994).
116. H.-A. Tajmir-Riahi, *J. Inorg. Biochem.*, *27*, 65 (1986).
117. Y. Israëli, J.-P. Morel, and N. Morel-Desrosiers, *Carbohydr. Res.*, *263*, 25 (1994).
118. Y. Israëli, C. Lhermet, J.-P. Morel, and N. Morel-Desrosiers, *Carbohydr. Res.*, *289*, 1 (1996).
119. W. I. Weis, K. Drickamer, and W. A. Hendrickson, *Nature(London)*, *360*, 127 (1992).
120. K. Drickamer, *Nature(London)* , *360*, 183 (1992).
121. J. M. de Bruijn, A. P. G. Kieboom, and H. van Bekkum, *Recl. Trav. Chim. Pays-Bas*, *106*, 35 (1987).
122. A. Vesala and H. Lönnberg, *Acta Chem. Scand. Ser. A*, *35A*, 123 (1981).
123. A. Vesala, H. Lönnberg, R. Käppi, and J. Arpalahti, *Carbohydr. Res.*, *102*, 312 (1982).
124. H. Lönnberg, A. Vesala, and R. Käppi, *Carbohydr. Res.*, *86*, 137 (1980).
125. A. Vesala, R. Käppi, and P. Lehikoinen, *Finn. Chem. Lett.*, 45 (1983).
126. A. Vesala and R. Käppi, *Polyhedron*, *4*, 1047 (1985).
127. A. Vesala and R. Käppi, *Finn. Chem. Lett.*, *13*, 27 (1986).

128. P. A. Kooreman and J. B. F. N. Engberts, *Recl. Trav. Chim. Pays-Bas*, *114*, 421 (1995).
129. D. J. O'Leary and Y. Kishi, *Tetrahedron Lett.*, *35*, 5591 (1994).
130. D. J. O'Leary and Y. Kishi, *J. Org. Chem.*, *58*, 304 (1993).
131. M. Riediker, A. Hafner, U. Piantini, G. Rihs, and A. Togni, *Angew. Chem.*, *101*, 493 (1989); *Angew. Chem. Int. Ed. Engl.*, *28*, 499 (1989).
132. D. N. Williams, U. Piarulli, C. Floriani, A. Chiesi-Villa, and C. Rizzoli, *J. Chem. Soc. Dalton Trans.*, 1243 (1994).
133. U. Piarulli, D. N. Williams, C. Floriani, G. Gervasio, and D. Viterbo, *J. Organomet. Chem.*, *503*, 185 (1995).
134. M. Riediker and R. O. Duthaler, *Angew. Chem.*, *101*, 488 (1989); *Angew. Chem. Int. Ed. Engl.*, *28*, 494 (1989).
135. G. Bold, R. O. Duthaler, and M. Riediker, *Angew. Chem.*, *101*, 491 (1989); *Angew. Chem. Int. Ed. Engl.*, *28*, 497 (1989).
136. R. O. Duthaler, P. Herold, W. Lottenbach, K. Oertle, and M. Riediker, *Angew. Chem.*, *101*, 490 (1989); *Angew. Chem. Int. Ed. Engl.*, *28*, 495 (1989).
137. C. F. G. C. Geraldes and M. M. C. A. Castro, *J. Inorg. Biochem.*, *35*, 79 (1989).
138. M. J. Gresser and A. S. Tracey, *J. Am. Chem. Soc.*, *108*, 1935 (1986).
139. A. S. Tracey and C. H. Leon-Lai, *Inorg. Chem.*, *30*, 3200 (1991).
140. J. Richter and D. Rehder, *Z. Naturforsch.*, *46 b*, 1613 (1991).
141. D. C. Crans, S. E. Harnung, E. Larsen, P. K. Shin, L. A. Theisen, and I. Trabjerg, *Acta Chem. Scand.*, *45*, 456 (1991).
142. S. E. Harnung, E. Larsen, and E. J. Pedersen, *Acta Chem. Scand.*, *47*, 674 (1993).
143. B. Zhang, S. Zhang, and K. Wang, *J. Chem. Soc. Dalton Trans.*, 3257 (1996).
144. S. J. Anugs-Dunne, R. J. Batchelor, A. S. Tracey, and F. W. B. Einstein, *J. Am. Chem. Soc.*, *117*, 5292 (1995).
145. D. C. Crans, R. A. Felty, and M. M. Miller, *J. Am. Chem. Soc.*, *113*, 265 (1991).
146. G. Barr-David, T. W. Hambley, J. A. Irwin, R. J. Judd, P. A. Lay, B. D. Martin, R. Bramley, N. E. Dixon, P. Hendry, J.-Y. Ji, R. S. U. Baker, and A. M. Bonin, *Inorg. Chem.*, *31*, 4906 (1992).
147. T. W. Hambley, R. J. Judd, and P. A. Lay, *Inorg. Chem.*, *31*, 343 (1992).
148. D. C. Crans, R. W. Marshman, M. S. Gottlieb, O. P. Anderson, and M. M. Miller, *Inorg. Chem.*, *31*, 4939 (1992).
149. D. C. Crans, R. A. Felty, H. Eckert, and N. Das, *Inorg. Chem.*, *33*, 2427 (1994).
150. M. I. Khan, Q. Chen, D. P. Goshorn, and J. Zubieta, *Inorg. Chem.*, *32*, 672 (1993).
151. M. I. Khan, Q. Chen, H. Höpe, S. Parkin, C. J. O'Connor, and J. Zubieta, *Inorg. Chem.*, *32*, 2929 (1993).
152. R. Codd, T. W. Hambley, and P. A. Lay, *Inorg. Chem.*, *34*, 877 (1995).
153. A. S. Tracey, J. S. Jaswal, M. J. Gresser, and D. Rehder, *Inorg. Chem.*, *29*, 4283 (1990).

154. O. W. Howarth and J. R. Trainor, *Inorg. Chim. Acta*, *127*, L27 (1987).
155. W. J. Ray, Jr, D. C. Crans, J. Zheng, J. W. Burgner, II, H. Deng, and M. Mahroof-Tahir, *J. Am. Chem. Soc.*, *117*, 6015 (1995).
156. D. C. Crans, H. Chen, and R. A. Felty ,*J. Am. Chem. Soc.*, *114*, 4543 (1992).
157. M. J. Gresser and A. S. Tracey, *J. Am. Chem. Soc.*, *107*, 4215 (1985).
158. A. S. Tracey, M. J. Gresser, and B. Galeffi, *Inorg. Chem.*, *27*, 157 (1988).
159. A. S. Tracey and M. J. Gresser, *Can. J. Chem.*, *66*, 2570 (1988).
160. A. S. Tracey, B. Galeffi, and S. Mahjour, *Can. J. Chem.*, *66*, 2294 (1988).
161. D. C. Crans, in *Metal Ions in Biological Systems*, H. Sigel and A. Sigel, Eds., Marcel-Dekker, New York, 1995, pp. 147–209.
162. A. S. Tracey and M. J. Gresser, *Inorg. Chem.*, *27*, 2695 (1988).
163. C. F. G. C. Geraldes and M. M. C. A. Castro, *J. Inorg. Biochem.*, *35*, 79 (1989).
164. D. Rehder, H. Holst, R. Quaas, W. Hinrichs, U. Hahn, and W. Saenger, *J. Inorg. Biochem.*, *37*, 141 (1989).
165. C. F. G. C. Geraldes and M. M. C. A. Castro, *J. Inorg. Biochem.*, *37*, 213 (1989).
166. A. S. Tracey and M. J. Gresser, *Inorg. Chem.*, *27*, 1269 (1988).
167. (a) F. Hillerns and D. Rehder, *Chem. Ber.*, *124*, 2249 (1991). (b) F. Hillerns, F. Olbrich, U. Behrens, and D. Rehder, *Angew. Chem. Int. Ed. Engl.*, *31*, 3628 (1992).
168. D. Crans, unpublished results.
169. K. Elvingston, P. M. Ehde, D. C. Crans, and L. Pettersson, unpublished work.
170. R. N. Lindquist, J. L. Lynn, Jr, and G. E. Lienhard, *J. Am. Chem. Soc.*, *95*, 8762 (1973).
171. B. Borah, C. W. Chen, W. Egan, M. Miller, A. Wlodawer, and J. S. Cohen, *Biochemistry*, *24*, 2058 (1985).
172. A. Wlodawer, M. Miller, and L. Sjölin, *Proc. Natl. Acad. Sci.*, *80*, 3628 (1983).
173. A. S. Tracey, M. J. Gresser, and S. Liu, *J. Am. Chem. Soc.*, *110*, 5869 (1988).
174. E. Alberico, D. Dewaele, T. Kiss, and G. Micera, *J. Chem. Soc. Dalton Trans.*, 425 (1995).
175. D. C. Crans, C. M. Simone, and J. S. Blanchard, *J. Am. Chem. Soc.*, *114*, 4926 (1992).
176. D. C. Crans, R. W. Marshman, R. Nielsen, and I. Felty, *J. Org. Chem.*, *58*, 2244 (1993).
177. K. K. S. Gupta and S. N. Basu, *Carbohydr. Res.*, *80*, 223 (1980).
178. P. O. I. Virtanen, S. Kurkisuo, H. Nevala, and S. Pohjola, *Acta Chem. Scand. Ser. A*, *40A*, 200 (1986).
179. P. O. I. Virtanen and R. Lindroos-Heinänen, *Acta Chem. Scand. Ser. B*, *42B*, 411 (1988).
180. P. O. I. Virtanen, T. Kuokkanen, and T. Rauma, *Carbohydr. Res.*, *180*, 29 (1988).
181. P. O. I. Virtanen and S. Kurkisuo, *Carbohydr. Res.*, *138*, 215 (1985).

182. G. Micera, *Carbohydr. Res.*, *188*, 25 (1989).
183. M. Branca, G. Micera, D. Sanna, A. Dessi, and H. Kozlowski, *J. Chem. Soc. Dalton Trans.*, 1997 (1990).
184. H. Kozlowski, S. Bouhsina, P. Decock, G. Micera, and J. Swiatek, *J. Coord. Chem.*, *24*, 319 (1991).
185. M. W. Makinen and D. Mustafi, in *Metal Ions in Biological Systems*, H. Sigel and A. Sigel, Eds., Marcel-Dekker, New York, 1995, pp. 89–128.
186. E. J. Baran, in *Metal Ions in Biological Systems*, H. Sigel and A. Sigel, Eds., Marcel-Dekker, New York, 1995, pp. 129–146.
187. M. Branca, G. Micera, A. Dessi, and D. Sanna, *J. Inorg. Biochem.*, *45*, 169 (1992).
188. S. P. Kaiwar and C. P. Rao, *Carbohydr. Res.*, *237*, 203 (1992).
189. A. Sreedhara, M. S. S. Raghavan, and C. P. Rao, *Carbohydr. Res.*, *264*, 227 (1994).
190. J. Ruiz, C. Floriani, A. Chiesi-Villa, and C. Guastini, *J. Chem. Soc. Dalton Trans.*, 2467 (1991).
191. D. A. Brown, M. G. H. Wallbridge, and N. W. Alcock, *J. Chem. Soc. Dalton Trans.*, 2037 (1993).
192. D. A. Brown, W. Errington, and M. G. H. Wallbridge, *J. Chem. Soc. Dalton Trans.*, 1163 (1993).
193. G. J. Moore, T. J. Boyle, and D. B. Dimos, *Novel 'sol-gel' precursor solutions for generation of PMN thin films*, in 13th ACS Rocky Mountain Regional Meeting, 1996, Lakewood, CO.
194. S. P. Kaiwar, M. S. S. Raghavan, and C. P. Rao, *Carbohydr. Res.*, *256*, 29 (1994).
195. H. W. Schmalle, K. Hegetschweiler, and M. Ghisletta, *Acta Crystallogr. Sect. C*, *47*, 2047 (1991).
196. G. Kovacs, J. Gyarmati, L. Somsak, and K. Micskei, *Tetrahedron Lett.*, *37*, 1293 (1996).
197. R. Bramley, J.-Y. Ji, and P. A. Lay, *Inorg. Chem.*, *30*, 1557 (1991).
198. M. Branca, G. Micera, U. Segre, and A. Dessi, *Inorg. Chem.*, *31*, 2404 (1992).
199. M. Krumpholc, B. G. DeBoer, and J. Rocek, *J. Am. Chem. Soc.*, *100*, 145 (1978).
200. M. Krumpholc and J. Rocek, *J. Am. Chem. Soc.*, *101*, 3206 (1979).
201. R. J. Judd, T. W. Hambley, and P. A. Lay, *J. Chem. Soc. Dalton Trans.*, 2205 (1989).
202. R. Bramley, J.-Y. Ji, R. J. Judd, and P. A. Lay, *Inorg. Chem.*, *29*, 3089 (1990).
203. M. Branca, A. Dessi, G. Micera, and D. Sanna, *Inorg. Chem.*, *32*, 578 (1993).
204. G. Barr-David, M. Charara, R. Codd, R. P. Farrell, J. A. Irwin, P. A. Lay, R. Bramley, S. Brumby, J.-Y. Ji, and G. R. Hanson, *J. Chem. Soc. Faraday Trans.*, *91*, 1207 (1995).
205. M. Charara, B.Sc. Thesis, University of Sydney, Australia, 1988.

206. D. H. Brown and J. MacPherson, *J. Inorg. Nucl. Chem.*, *32*, 3309 (1970).
207. J. T. Spence and M. Heydanek, *Inorg. Chem.*, *6*, 1489 (1967).
208. D. H. Brown and J. MacPherson, *J. Inorg. Nucl. Chem.*, *33*, 4203 (1971).
209. D. H. Brown and J. MacPherson, *J. Inorg. Nucl. Chem.*, *34*, 1705 (1972).
210. D. H. Brown and D. Neumann, *J. Inorg. Nucl. Chem.*, *37*, 330 (1975).
211. J.-F. Verchère, S. Chapelle, and J.-P. Sauvage, *Polyhedron*, *9*, 1225 (1990).
212. S. J. N. Burgmaier and E. I. Stiefel, *Inorg. Chem.*, *27*, 2518 (1988).
213. L. Ma, S. Liu, and J. Zubieta, *Polyhedron*, *8*, 1571 (1989).
214. J. E. Godfrey and J. M. Waters, *Cryst. Struct. Commun.*, *4*, 5 (1975).
215. B. Hedman, *Acta Crystallogr. Sect. B*, *33*, 3077 (1977).
216. G. E. Taylor and J. M. Waters, *Tetrahedron Lett.*, *22*, 1277 (1981).
217. L. O. Atovmyan, V. V. Tkachev, and T. G. Shishova, *Dokl. Phys. Chem. Sect.*, *205*, 622 (1972); *Dokl. Akad. Nauk. SSR*, *205*, 609 (1972).
218. C. G. Pierpont and R. M. Buchanan, *Inorg. Chem.*, *21*, 652 (1982).
219. A. M. El-Hendawy, W. P. Griffith, C. A. O'Mahoney, and D. J. Williams, *Polyhedron*, *8*, 519 (1989).
220. C. G. Pierpont and R. M. Buchanan, *J. Am. Chem. Soc.*, *97*, 6450 (1975).
221. D. D. Dexter and J. V. Silverton, *J. Am. Chem. Soc.*, *90*, 3589 (1968).
222. F. A. Cotton, S. M. Morehouse, and J. S. Wood, *Inorg. Chem.*, *3*, 1603 (1964).
223. H. Weigel, *Adv. Carbohydr. Chem.*, *18*, 61 (1963).
224. S. Chapelle and J.-F. Verchère, *Carbohydr. Res.*, *211*, 279 (1991).
225. J.-F. Verchère and S. Chapelle, *Polyhedron*, *8*, 333 (1989).
226. M. Matulova, V. Bílik, and J. Alföldi, *Chem. Papers*, *43*, 403 (1989).
227. W. Voelter, E. Bayer, R. Records, E. Bunnenberg, and C. Djerassi, *Chem. Ber.*, *102*, 1005 (1969).
228. E. J. Bourne, D. H. Hutson, and H. Weigel, *J. Chem. Soc.*, 4252 (1960).
229. E. J. Bourne, D. H. Hutson, and H. Weigel, *J. Chem. Soc.*, *35* (1961).
230. H. J. F. Angus, E. J. Bourne, and H. Weigel, *J. Chem. Soc.*, 21 (1965).
231. G. Snatzke, J. Guo, Z. Raza, and V. Šunjic, *Croat. Chim. Acta*, *64*, 501 (1991).
232. A. Cervilla, J. A. Ramirez, and A. Beltran-Porter, *Transition Met. Chem.*, *8*, 21 (1983).
233. E. Llopis, J. A. Ramirez, and A. Cervilla, *Polyhedron*, *5*, 2069 (1986).
234. S. Chapelle, J.-P. Sauvage, and J.-F. Verchère, *Inorg. Chem.*, *33*, 1966 (1994).
235. H. Günther, *NMR Spectroscopy*, 2nd ed., Wiley, Chichester, UK, 1995.
236. M. Mikesova and M. Bartusek, *Coll. Czech. Chem. Commun.*, *43*, 1867 (1978).
237. E. Mikanova and M. Bartusek, *Scr. Fac. Sci. Natl. Univ. Purk. Brun.*, *11*, 451 (1981).
238. W. H. Lee, *Kumsok Hakhoe Chi.*, *10*, 207 (1972).
239. J.-P. Sauvage, J.-F. Verchère, and S. Chapelle, *Carbohydr. Res.*, *286*, 67 (1996).

240. J.-F. Verchère and J.-P. Sauvage, *Bull. Soc. Chim. Fr.*, *115*, 263 (1988).
241. M. Matulova and V. Bílik, *Chem. Papers*, *44*, 97 (1990).
242. S. Chapelle and J.-F. Verchère, *Inorg. Chem.*, *31*, 648 (1992).
243. R. I. Maksimovskaya and K. G. Burtseva, *Polyhedron*, *4*, 1559 (1985).
244. M. Minelli, J. H. Enemark, R. T. C. Brownlee, M. J. O'Connor, and A. G. Wedd, *Coord. Chem. Rev.*, *68*, 169 (1985).
245. L. Mueller, *J. Am. Chem. Soc.*, *101*, 4481 (1979).
246. V. Sklenar and A. Bax, *J. Magn. Reson.*, *71*, 379 (1987).
247. R. Benn, C. Brevard, A. Rufinska, and G. Schroth, *Organometallics*, *6*, 938 (1987).
248. R. Benn and A. Rufinska, *Magn. Reson. Chem.*, *26*, 895 (1988).
249. R. Benn, A. Rufinska, M. A. King, C. E. Osterberg, and T. G. Richmond, *J. Organomet. Chem.*, *376*, 359 (1989).
250. A. Bax, R. H. Griffey, and B. L. Hawkins, *J. Magn. Reson.*, *55*, 301 (1983).
251. S. Chapelle and J.-F. Verchère, *Carbohydr. Res.*, *266*, 161 (1995).
252. A. M. Dona and J.-F. Verchère, *J. Chromatogr. A*, *689*, 13 (1995).
253. S. Chapelle, J.-P. Sauvage, P. Köll, and J.-F. Verchère, *Inorg. Chem.*, *34*, 918 (1995).
254. M. Hlaïbi, M. Benaïssa, C. Busatto, J.-F. Verchère, and S. Chapelle, *Carbohydr. Res.*, *278*, 227 (1995).
255. S. Chapelle, P. Köll, and J.-F. Verchère, unpublished results.
256. T. Posternak, D. Janjic, E. A. C. Lucken, and A. Szente, *Helv. Chim. Acta*, *50*, 1027 (1967).
257. M. Matulova and V. Bílik, *Chem. Papers*, *44*, 77 (1990).
258. M. Matulova and V. Bílik, *Chem. Papers*, *44*, 257 (1990).
259. E. J. Bourne, F. Searle, and H. Weigel, *Carbohydr. Res.*, *16*, 185 (1971).
260. T. Mezzetti, M. Lato, S. Rufini, and G. Suiffini, *J. Chromatogr.*, *63*, 329 (1971).
261. H. J. F. Angus, J. Briggs, N. A. Sufi, and H. Weigel, *Carbohydr. Res.*, *66*, 25 (1978).
262. J. Briggs, I. R. Chambers, P. Finch, I. R. Slaiding, and H. Weigel, *Carbohydr. Res.*, *78*, 365 (1980).
263. J. T. Spence and S. Kiang, *J. Org. Chem.*, *28*, 244 (1963).
264. L. Velluz and M. Legrand, *C. R. Acad. Sci. Paris*, *263*, 1429 (1966).
265. W. Voelter, E. Bayer, G. Barth, E. Bunnenberg, and C. Djerassi, *Chem. Ber.*, *102*, 2003 (1969).
266. H. J. F. Angus, E. J. Bourne, F. Searle, and H. Weigel, *Tetrahedron Lett.*, 55 (1964).
267. H. J. F. Angus and H. Weigel, *J. Chem. Soc.*, 3994 (1964).
268. J. Alföldi, L. Petrus, and V. Bílik, *Collect. Czech. Chem. Commun.*, *43*, 1476 (1978).

269. J. Alföldi, L. Petrus, and V. Bílik, *Collect. Czech. Chem. Commun.*, *45*, 123 (1980).
270. V. Bílik, J. Alföldi, and E. Sookyova, *Chem. Zvesti*, *38*, 499 (1984).
271. V. Bílik, J. Alföldi, and M. Matulova, *Chem. Papers*, *40*, 763 (1986).
272. C. F. G. C. Geraldes, M. M. C. A. Castro, M. E. Saraiva, M. Aureliano, and B. A. Dias, *J. Coord. Chem.*, *17*, 205 (1988).
273. M. Matulova and V. Bílik, *Chem. Papers*, *44*, 703 (1990).
274. J.-P. Sauvage, S. Chapelle, and J.-F. Verchère, *Carbohydr. Res.*, *237*, 23 (1992).
275. M. Matulova and V. Bílik, *Chem. Papers*, *46*, 253 (1992).
276. M. Matulova and V. Bílik, *Carbohydr. Res.*, *250*, 203 (1993).
277. S. Chapelle and J.-F. Verchère, *Carbohydr. Res.*, *277*, 39 (1995).
278. M. Matulova, J.-F. Verchère, and S. Chapelle, *Carbohydr. Res.*, *287*, 37 (1996).
279. M. Hlaïbi, S. Chapelle, M. Benaïssa, and J.-F. Verchère, *Inorg. Chem.*, *34*, 4434 (1995).
280. M. L. Ramos, M. M. Caldeira, and V. M. S. Gil, *Inorg. Chim. Acta*, *180*, 219 (1991).
281. M. L. Ramos, M. M. Caldeira, V. M. S. Gil, H. van Bekkum, and J. A. Peters, *Polyhedron*, *13*, 1825 (1994).
282. M. L. Ramos, M. M. Caldeira, V. M. S. Gil, H. van Bekkum, and J. A. Peters, *J. Coord. Chem.*, *33*, 319 (1994).
283. J. T. Spence and G. Kallos, *Inorg. Chem.*, *2*, 710 (1963).
284. M. M. Caldeira, M. L. Ramos, V. M. S. Gil, H. van Bekkum, and J. A. Peters, *Inorg. Chim. Acta*, *221*, 69 (1994).
285. M. L. D. Ramos, M. M. M. Caldeira, and V. M. S. Gil, *Carbohydr. Res.*, *286*, 1 (1996).
286. V. Bílik, W. Voelter, and E. Bayer, *Angew. Chem.*, *83*, 967 (1971); *Angew. Chem. Int. Ed. Engl.*, *10*, 909 (1971).
287. V. Bílik, *Chem. Zvesti*, *26*, 76 (1972).
288. V. Bílik, *Chem. Zvesti*, *26*, 183 (1972).
289. V. Bílik, *Chem. Zvesti*, *26*, 187 (1972).
290. V. Bílik, *Chem. Zvesti*, *26*, 372 (1972).
291. V. Bílik, W. Voelter, and E. Bayer, *Justus Liebigs Ann. Chem.*, *759*, 189 (1972).
292. V. Bílik, W. Voelter, and E. Bayer, *Justus Liebigs Ann. Chem.*, 1162 (1974).
293. J. Defaye, A. Gadelle, and S. J. Angyal, *Carbohydr. Res.*, *126*, 165 (1984).
294. K. B. Hicks, E. V. Symanski, and P. E. Pfeffer, *Carbohydr. Res.*, *112*, 37 (1983).
295. C. A. Lobry de Bruyn and W. Alberda van Ekenstein, *Recl. Trav. Chim. Pays-Bas*, *14*, 203 (1895).
296. Y. Abe, T. Takizawa, and T. Kunieda, *Chem. Pharm. Bull.*, *28*, 1324 (1980).
297. V. Bílik, V. Farkas, and L. Petrus, *Chem. Zvesti*, *29*, 690 (1975).

298. M. L. Hayes, N. J. Pennings, A. S. Serianni, and R. Barker, *J. Am. Chem. Soc.*, *104*, 6764 (1982).
299. E. L. Clark, Jr., M. L. Hayes, and R. Barker, *Carbohydr. Res.*, *153*, 263 (1986).
300. A. Cybulski, B. F. M. Kuster, and G. B. Marin, *J. Mol. Catal.*, *68*, 87 (1991).
301. T. Tanase, F. Shimizu, S. Yano, and S. Yoshikawa, *J. Chem. Soc. Chem. Commun.*, 1001 (1986).
302. T. Tanase, F. Shimizu, M. Kuse, S. Yano, M. Hidai, and S. Yoshikawa, *Inorg. Chem.*, *27*, 4085 (1988).
303. T. Tanase, K. Kurihara, S. Yano, K. Kobayashi, T. Sakurai, and S. Yoshikawa, *J. Chem. Soc. Chem. Commun.*, 1562 (1985).
304. R. E. London, *J. Chem. Soc. Chem. Commun.*, 661 (1987).
305. S. Osanai, R. Yanagihara, K. Uematsu, A. Okumura, and S. Yoshikawa, *J. Chem. Soc. Perkin Trans.*, 1937 (1993).
306. R. Yanagihara, S. Osanai, and S. Yoshikawa, *Chem. Lett.*, *1*, 89 (1992).
307. R. Yanagihara, J. Egashira, S. Yoshikawa, and S. Osanai, *Bull. Chem. Soc. Jpn.*, *68*, 237 (1995).
308. T. Tanase, F. Shimizu, M. Kuse, S. Yano, S. Yoshikawa, and M. Hidai, *J. Chem. Soc. Chem. Commun.*, 659 (1987).
309. T. Tanase, T. Murata, S. Yano, M. Hidai, and S. Yoshikawa, *Chem. Lett.*, 1409 (1987).
310. T. Tanase, K. Ishida, T. Watanabe, M. Komiyama, K. Koumoto, S. Yano, M. Hidai, and S. Yoshikawa, *Chem. Lett.*, 327 (1988).
311. T. Yamauchi, K. Fukushima, R. Yanagihara, S. Osanai, and S. Yoshikawa, *Carbohydr. Res.*, *204*, 233 (1990).
312. T. Takei, T. Tanase, S. Yano, and M. Hidai, *Chem. Lett.*, 1629 (1991).
313. R. Yanagihara, K. Soeda, S. Shiina, S. Osanai, and S. Yoshikawa, *Bull. Chem. Soc. Jpn.*, *66*, 2268 (1993).
314. B. Klaic, Z. Raza, M. Sankovic, and V. Šunjic, *Helv. Chim. Acta*, *70*, 59 (1987).
315. M. Sankovic, S. Emini, S. Rusman, and V. Šunjic, *J. Mol. Catal.*, *61*, 247 (1990).
316. S. Kolaric, M. Gelo, M. Sankovic, and V. Šunjic, *J. Mol. Catal.*, *79*, 365 (1993).
317. S. Kolaric, M. Gelo, M. Sankovic, and V. Šunjic, *J. Mol. Catal.*, *89*, 247 (1994).
318. C. J. O'Connor, A. L. Odell, and A. A. T. Barley, *Aust. J. Chem.*, *35*, 951 (1982).
319. C. J. O'Connor, A. L. Odell, and A. A. T. Barley, *Aust. J. Chem.*, *36*, 279 (1983).
320. C. J. O'Connor and A. A. T. Barley, *Aust. J. Chem.*, *37*, 1411 (1984).
321. S. J. Angyal, *Carbohydr. Res.*, *200*, 181 (1990).
322. D. T. Sawyer and R. J. Kula, *Inorg. Chem.*, *1*, 303 (1962).
323. M. Stefansson and D. Westerlund, *Chromatographia*, *35*, 199 (1993).
324. S. Jurisson, D. Berning, W. Jia, and D. Ma, *Chem. Rev.*, *93*, 1137 (1993).
325. G. Böhm, K. Wieghardt, B. Nuber, and J. Weiss, *Inorg. Chem.*, *30*, 3464 (1991).

326. K. Hegetschweiler, A. Egli, R. Alberto, and H. W. Schmalle, *Inorg. Chem.*, *31*, 4027 (1992).
327. J. Dolezal and O. Gürtler, *Talanta*, *15*, 299 (1968).
328. B. L. Velikov and J. Dolezal, *J. Electroanal. Chem.*, *71*, 91 (1976).
329. J. Dolezal and H. Kekulova, *J. Electroanal. Chem.*, *69*, 239 (1976).
330. S. M. Saadeh, M. S. Lah, and V. L. Pecoraro, *Inorg. Chem.*, *30*, 8 (1991).
331. D. H. Chin, D. T. Sawyer, W. P. Schaefer, and C. J. Simmons, *Inorg. Chem.*, *22*, 752 (1983).
332. J. A. Hartman, B. M. Foxman, and S. R. Cooper, *Inorg. Chem.*, *23*, 1381 (1984).
333. R. P. Bandwar and C. P. Rao, *Carbohydr. Res.*, *287*, 157 (1996).
334. K. D. Magers, C. G. Smith, and D. T. Sawyer, *Inorg. Chem.*, *17*, 515 (1978).
335. M. E. Bodini, L. A. Willis, T. L. Riechel, and D. T. Sawyer, *Inorg. Chem.*, *15*, 1538 (1976).
336. D. T. Richens, C. G. Smith, and D. T. Sawyer, *Inorg. Chem.*, *18*, 706 (1979).
337. B. U. Nair and G. C. Dismukes, *Carbohydr. Res.*, *105*, 124 (1983).
338. K. V. Rao, M. T. Rao, and M. Adinarayana, *Int. J. Chem. Kinet.*, *27*, 555 (1995).
339. K. Araki and S. Shiraishi, *Bull. Chem. Soc. Jpn.*, *59*, 229 (1986).
340. K. Hegetschweiler, M. Ghisletta, L. Hausherr-Primo, T. Kradolfer, H. W. Schmalle, and V. Gramlich, *Inorg. Chem.*, *34*, 1950 (1995).
341. H. W. Rich, K. Hegetschweiler, H. M. Streit, I. Erni, and W. Schneider, *Inorg. Chim. Acta*, *187*, 9 (1991).
342. W. Schneider, I. Erni, D. Hametner, B. Magyar, B. Schwayn, and F. van Steenwijk, in *The Biochemistry and Physiology of Iron*, P. Saltman and J. Hegenauer, Eds., Elsevier Biomedical, New York, 1982.
343. W. Schneider, *Chimia*, *42*, 9 (1988).
344. K. Geetha, M. S. S. Raghavan, S. K. Kulshreshtha, R. Sasikala, and C. P. Rao, *Carbohydr. Res.*, *271*, 163 (1995).
345. M. Tonkovic, S. Horvat, J. Horvat, S. Music, and O. Hadzija, *Polyhedron*, *9*, 2895 (1990).
346. M. Stefansson, *J. Chromatogr.*, *630*, 123 (1993).
347. M. Stefansson and D. Westerlund, *J. Chromatogr. Sci.*, *32*, 46 (1994).
348. K. Araki, M. Sakuma, and S. Shiraishi, *Chem. Lett.*, 665 (1983).
349. K. Araki, M. Sakuma, and S. Shiraishi, *Bull. Chem. Soc. Jpn.*, *57*, 997 (1984).
350. K. Araki and S. Shiraishi, *Bull. Chem. Soc. Jpn.*, *59*, 3661 (1986).
351. D. T. Sawyer, R. S. George, and J. B. Bagger, *J. Am. Chem. Soc.*, *81*, 5893 (1959).
352. S. Bhadrui, N. Sapre, H. Khwaja, and P. G. Jones, *J. Organometallic Chem.*, *426*, C12 (1992).
353. B. T. Khan and K. Annapoorna, *Polyhedron*, *10*, 2465 (1991).
354. W. P. Griffith and R. Rossetti, *J. Chem. Soc. Dalton Trans.*, 1449 (1972).

355. R. J. Collins, J. Jones, and W. P. Griffith, *J. Chem. Soc. Dalton Trans.*, 1094 (1974).

356. D. V. McGrath, G. D. Brabson, K. B. Sharpless, and L. Andrews, *Inorg. Chem.*, *32*, 4164 (1993).

357. J. Böseken, *Recl. Trav. Chim. Pays-Bas*, *41*, 199 (1922).

358. R. Criegee, *Annalen*, *522*, 75 (1936).

359. H. C. Kolb, M. S. VanNieuwenhze, and K. B. Sharpless, *Chem. Rev.*, *94*, 2483 (1994).

360. R. J. Collins, W. P. Griffith, F. Phillips, and A. C. Skapski, *Biochim. Biophys. Acta*, *320*, 745 (1973).

361. F. L. Phillips and A. C. Skapski, *Acta Crystallogr. Sect. B*, *B31*, 1814 (1975).

362. D. T. Sawyer and D. S. Tinti, *Inorg. Chem.*, *2*, 796 (1963).

363. J. Burger and P. Klüfers, *Chem. Ber.*, *128*, 75 (1995).

364. J. Dolezal, K. S. Klausen, and F. J. Langmyhr, *Anal. Chim. Acta*, *63*, 71 (1973).

365. J. Dolezal and F. J. Langmyhr, *Anal. Chim. Acta*, *61*, 73 (1972).

366. T. Tanase, M. Nakagoshi, A. Teratani, M. Kato, Y. Yamamoto, and S. Yano, *Inorg. Chem.*, *33*, 5 (1994).

367. S. Bunel and C. Ibarra, *Polyhedron*, *4*, 1537 (1985).

368. S. Bunel, C. Ibarra, V. Calvo, and E. Moraga, *Polyhedron*, *8*, 2023 (1989).

369. T. Tsubomura, S. Yano, and S. Yoshikawa, *Bull. Chem. Soc. Jpn.*, *61*, 3497 (1988).

370. T. Tsubomura, M. Ogawa, S. Yano, K. Kobayashi, T. Sakurai, and S. Yoshikawa, *Inorg. Chem.*, *29*, 2622 (1990).

371. S. Hanessian, J.-Y. Gauthier, K. Okamoto, A. L. Beauchamp, and T. Theophanides, *Can. J. Chem.*, *71*, 880 (1993).

372. A. Appelt, A. C. Willis, and S. B. Wild, *J. Chem. Soc. Chem. Commun.*, 938 (1988).

373. M. A. Andrews, E. J. Voss, G. L. Gould, W. T. Klooster, and T. F. Koetzle, *J. Am. Chem. Soc.*, *116*, 5730 (1994).

374. G. Micera and H. Kozlowski, *Handb. Met-Ligand Interact. Biol. Fluids: Bioinorg. Chem.*, *1*, 707 (1995).

375. R. Fuchs, N. Habermann, and P. Klüfers, *Angew. Chem.*, *105*, 895 (1993); *Angew. Chem. Int. Ed. Engl.*, *32*, 852 (1993).

376. P. Klüfers and J. Schuhmacher, *Angew. Chem.*, *106*, 1839 (1994); *Angew. Chem. Int. Ed. Engl.*, *33*, 1742 (1994).

377. S. Divikar, *J. Inclusion Phenom. Mol. Recognit. Chem.*, *17*, 119 (1994).

378. R. E. Reeves, *Adv. Carbohydr. Chem.*, *6*, 107 (1951).

379. J. Dolezal, K. S. Klausen, and F. J. Langmyhr, *Anal.Chim. Acta*, *63*, 71 (1973).

380. E. J. Bourne, F. Searle, and H. Weigel, *Carbohydr. Res.*, *16*, 185 (1971).

381. R. P. Bandwar, M. Giralt, J. Hidalgo, and C. P. Rao, *Carbohydr. Res.*, *284*, 73 (1996).

382. K. Hegetschweiler, M. Ghisletta, T. F. Fässler, R. Nesper, H. W. Schmalle, and G. Rihs, *Inorg. Chem.*, *32*, 2032 (1993).
383. M. Tonkovic, H. Bilinski, and M. E. Smith, *Inorg. Chim. Acta*, *197*, 59 (1992).
384. M. Tonkovic, H. Bilinski, and M. E. Smith, *Polyhedron*, *14*, 1025 (1995).
385. T. Ogawa and M. Matsui, *Tetrahedron*, *37*, 2363 (1981).
386. K. Vogel, J. Sterling, Y. Herzig, and A. Nudelman, *Tetrahedron*, *52*, 3049 (1996).
387. P. J. Cox, S. M. S. V. Doidge-Harrison, R. A. Howie, I. W. Nowell, O. J. Taylor, and J. L. Wardell, *J. Chem. Soc. Perkin Trans. 1*, 2017 (1989).
388. L. A. Burnett, S. M. S. V. Doidge-Harrison, S. J. Garden, R. A. Howie, O. J. Taylor, and J. L. Wardell, *J. Chem. Soc. Perkin Trans. 1*, 1621 (1993).
389. G.-J. Boons, G. H. Castle, J. A. Clase, P. Grice, S. V. Ley, and C. Pinel, *Synlett*, 913 (1993).
390. J. D. Donaldson, S. M. Grimes, L. Pellerito, M. A. Girasolo, P. J. Smith, A. Cambria, and M. Famá, *Polyhedron*, *6*, 383 (1987).
391. S. David, C. Pascard, and M. Cesario, *Nouv. J. Chim.*, *3*, 63 (1979).
392. T. S. Cameron, P. K. Bakshi, R. Thangasara, and T. B. Grindley, *Can. J. Chem.*, *70*, 1623 (1992).
393. C. W. Holzapfel, J. M. Koekemoer, C. F. Marais, G. J. Kruger, and J. A. Pretorius, *S. Afr. J. Chem.*, *35*, 80 (1982).
394. E. M. Lees and H. Weigel, *J. Chromatogr.*, *16*, 360 (1964).
395. S. Roelens and M. Taddei, *J. Chem. Soc. Perkin Trans. 2*, 799 (1985).
396. C. Luchinat and S. Roelens, *J. Org. Chem.*, *52*, 4444 (1987).
397. C. Cruzado, M. Barnabé, and M. Martin-Lomas, *J. Org. Chem.*, *54*, 465 (1989).
398. K. Burger, L. Nagy, N. Buzás, A. Vértes, and H. Mehner, *J. Chem. Soc. Dalton Trans.*, 2499 (1993).
399. N. Buzas, M. A. Pujar, L. Nagy, A. Vértes, E. Kuzmann, and H. Mehner, *J. Radioanal. Nucl. Chem.*, *189*, 237 (1995).
400. S. David and S. Hanessian, *Tetrahedron*, *41*, 643 (1985).
401. A. Lubineau and R. Lemoine, *Tetrahedron Lett.*, *35*, 8795 (1994).
402. D. D. Manning, C. R. Bertozzi, S. D. Rosen, and L. L. Kiessling, *Tetrahedron Lett.*, *37*, 1953 (1996).
403. D. Wagner, J. Verheyden, and J. G. Moffatt, *J. Org. Chem.*, *39*, 24 (1974).
404. S. David, *C. R. Acad. Sci. Paris, Ser. C*, *278*, 1051 (1974).
405. S. J. Blunden, P. A. Cusack, and P. J. Smith, *J. Organomet. Chem.*, *325*, 141 (1987).
406. M. Pereyre, J.-P. Quintard, and A. Rahm, *Tin in Organic Synthesis*, Butterworths, London, 1987, Chapter 11, pp. 261–323.
407. P. J. Garegg, B. Lindberg, I. Kvarnstrom, and S. C. T. Svensson, *Carbohydr. Res.*, *173*, 205 (1988).
408. S. David, A. Thieffry, and A. Veyrieres, *J. Chem. Soc. Perkin Trans. 1*, 1796 (1981).

409. S. David, A. Thieffry, and A. Forchioni, *Tetrahedron Lett.*, *22*, 2647 (1981).
410. T. B. Grindley and R. Thangarasa, *Can. J. Chem.*, *68*, 1007 (1990).
411. T. B. Grindley and R. Thangarasa, *J. Am. Chem. Soc.*, *112*, 1364 (1990).
412. X. Kong and T. B. Grindley, *Can. J. Chem.*, *72*, 2396 (1994).
413. X. Kong and T. B. Grindley, *Can. J. Chem.*, *72*, 2405 (1994).
414. C. Cruzado and M. Martin-Lomas, *Carbohydr. Res.*, *175*, 193 (1988).
415. R. Khan and K. Mufti, *Br. Patent* 2 079 749 A, 1981.
416. J. L. Navia, *U.S. Patent* 4 950 746 (1990).
417. W. M. Macindoe, A. Williams, and R. Khan, *Carbohydr. Res.*, *283*, 17 (1996).
418. G. R. Barker and D. F. Shaw, *J. Chem. Soc.*, 584 (1959).
419. J. Mbabazi, *Polyhedron*, *4*, 75 (1985).
420. J. Mbabazi and W. J. Popiel, *J. Inorg. Nucl. Chem.*, *41*, 1491 (1979).
421. L. S. Hollis, A. R. Amundsen, and E. W. Stern, *J. Am. Chem. Soc.*, *107*, 274 (1985).
422. L. S. Hollis, E. W. Stern, A. R. Amundsen, A. V. Miller, and S. L. Doran, *J. Am. Chem. Soc.*, *109*, 3596 (1987).
423. H. Basch, M. Krauss, and W. J. Stevens, *Inorg. Chem.*, *25*, 4777 (1986).
424. G. Descotes, D. Sinou, and J.-P. Praly, *Carbohydr. Res.*, *78*, 25 (1980).
425. G. D. Daves, *Acc. Chem. Res.*, *23*, 201 (1990).
426. A. Rosenthal and H. J. Koch, *Tetrahedron Lett.*, 871 (1967).
427. G. L. Trainor and B. E. Smart, *J. Org. Chem.*, *48*, 2447 (1983).
428. P. DeShong, G. A. Slough, V. Elango, and G. L. Trainor, *J. Am. Chem. Soc.*, *107*, 7788 (1985).
429. P. DeShong, G. A. Slough, and V. Elango, *Carbohydr. Res.*, *171*, 342 (1987).
430. P. DeShong, G. A. Slough, D. R. Sidler, V. Elango, P. J. Rybczynski, L. J. Smith, T. A. Lessen, T. X. Le, and G. B. Anderson, *ACS Symp. Ser.*, *494*, 112 (1992).
431. P. DeShong, T. A. Lessen, T. X. Le, G. B. Anderson, D. R. Sidler, G. A. Slough, W. von Philipsborn, M. Voehler, and O. Zerbe, *ACS Symp. Ser.*, *539*, 227 (1993).
432. R. O. Duthaler and A. Hafner, *Chem. Rev.*, *92*, 807 (1992).
433. R. Aumann, *Chem. Ber.*, *125*, 2773 (1992).
434. K. H. Dötz, W. Straub, R. Ehlenz, K. Peseke, and R. Meisel, *Angew. Chem. Int. Ed. Engl.*, *34*, 1856 (1995).
435. M. A. Andrews, S. A. Klaeren, and G. L. Gould, in *Carbohydrates as Organic Raw Materials*, G. Descotes, Ed., VCH, Weinheim, Germany, 1993, Vol. 2, pp. 3-25.
436. J. Tsuji and K. Ohno, *Tetrahedron Lett.*, 3969 (1965).
437. K. Ohno and J. Tsuji, *J. Am. Chem. Soc.*, *90*, 99 (1968).
438. M. C. Baird, C. J. Nyman, and G. Wilkinson, *J. Chem. Soc. (A)*, 348 (1968).
439. J. Tsuji and K. Ohno, *Synthesis*, 157 (1969).
440. M. A. Andrews and S. A. Klaeren, *J. Chem. Soc. Chem. Commun.*, 1266 (1988).

441. M. A. Andrews, G. L. Gould, and S. A. Klaeren, *J. Org. Chem.*, *54*, 5257 (1989).
442. M. A. Andrews and S. A. Klaeren, *Statut. Invent. Regist.* US 918; *Chem. Abstr.*, *115*, 115008u, 1991.
443. S. J. Angyal, C. L. Bodkin, and F. W. Parrish, *Aust. J. Chem.*, *28*, 1541 (1975).
444. A. Lubineau and J. C. Fischer, *Synth. Commun.*, *21*, 815 (1991).
445. N. Ikemoto, O. K. Kim, L. C. Lo, V. Satyanarayana, M. Chang, and K. Nakanishi, *Tetrahedron Lett.*, *33*, 4295 (1992).
446. V. Ferrières, J.-N. Bertho, and D. Plusquellec, *Tetrahedron Lett.*, *36*, 2749 (1995).
447. J.-N. Bertho, V. Ferrières, and D. Plusquellec, *J. Chem. Soc. Chem. Commun.*, 1391 (1995).
448. R. Eby, K. T. Webster, and C. Schuerch, *Carbohydr. Res.*, *129*, 111 (1984).
449. S. Hanessian and M. Kagotani, *Carbohydr. Res.*, *202*, 67 (1990).
450. J. L. Navia, R. A. Roberts, and R. E. Wingard, Jr., *J. Carbohydr. Chem.*, *14*, 465 (1995).
451. O. T. Schmidt, in *Methods in Carbohydrate Chemistry*, R. L. Whistler and M. L. Wolfrom, Eds., Academic, New York, 1963, Vol. 2, p. 318.
452. P. P. Singh, M. M. Gharia, F. Dasgupta, and H. C. Srivastava, *Tetrahedron Lett.*, 439 (1977).
453. B. Lal, R. M. Gidwany, and R. H. Rupp, *Synthesis*, 711 (1989).
454. S. Morgenlie, *Acta Chem. Scand.*, *27*, 3609 (1973).
455. S. Morgenlie, *Acta Chem. Scand.*, *29*, 367 (1975).
456. S. Morgenlie, *Carbohydr. Res.*, *41*, 77 (1975).
457. A. P. Rauter, F. Ramôa-Ribeiro, A. C. Fernandez, and J. A. Figueiredo, *Tetrahedron*, *51*, 6529 (1995).
458. G. F. J. Chittenden, *J. Chem. Soc. Chem. Commun.*, 882 (1980).
459. G. F. J. Chittenden, *Carbohydr. Res.*, *84*, 350 (1980).
460. N. J. Richards and D. G. Williams, *Carbohydr. Res.*, *12*, 409 (1970).
461. J. H. Braybrook and L. D. Hall, *Carbohydr. Res.*, *187*, c6 (1989).
462. P. Klüfers and J. Schuhmacher, *Angew. Chem.*, *107*, 2290 (1995); *Angew. Chem. Int. Ed. Engl.*, *34*, 2119 (1995).
463. J. Burger, C. Gack, and P. Klüfers, *Angew. Chem.*, *107*, 2950 (1995); *Angew. Chem. Int. Ed. Engl.*, *34*, 2647 (1995).
464. A. Sreedhara, C. P. Rao, and B. J. Rao, *Carbohydr. Res.*, *289*, 39 (1996).
465. R. P. Bandwar, M. D. Sastry, R. M. Kadam, and C. P. Rao, *Carbohydr. Res.*, *297*, 333 (1997).
466. R. P. Bandwar and C. P. Rao, *Carbohydr. Res.*, *297*, 341 (1997).
467. M. L. Ramos, M. M. Caldeira and V. M. S. Gil, *Carbohydr. Res.*, *297*, 191 (1997).
468. M. L. Ramos, M. M. Caldeira, and V. M. S. Gil, *Carbohydr. Res.*, *299*, 209 (1997).
469. S. J. Angyal, *Carbohydr. Res.*, *300*, 279 (1997).
470. P. Klüfers, H. Piotrowski, and J. Uhlendorf, *Chem. Eur. J.*, *3*, 601 (1997).

Subject Index

Cumulative Index, Volumes 1–47